# Esse Medicine
in In agement

Annu Navani • Sairam Atluri
Mahendra Sanapati
Editors

# Essentials of Regenerative Medicine in Interventional Pain Management

Second Edition

Editor-in-Chief Laxmaiah Manchikanti

Springer

*Editors*
Annu Navani
Comprehensive Spine & Sports Center
Campbell, CA, USA

Sairam Atluri
Tri-State Spine Care Institute
Cincinnati, OH, USA

Mahendra Sanapati
Pain Management Centers Of America
Evansville, IN, USA

*Editor-in-Chief*
Laxmaiah Manchikanti
Pain Management Centers of America
Paducah, KY, USA

ISBN 978-3-031-50359-7 ISBN 978-3-031-50357-3 (eBook)
https://doi.org/10.1007/978-3-031-50357-3



This Springer imprint is published by the registered company Springer Nature Switzerland AG
The registered company address is: Gewerbestrasse 11, 6330 Cham, Switzerland

Paper in this product is recyclable.

***Manchikanti's Essentials of Regenerative Medicine in Interventional Pain Management, 2nd edition,*** *is dedicated to my grandmother, the late Gopamma Manchikanti, who was responsible for my entering into a medical career and who taught me the value of education and giving back; to my parents, the late Yadagiry and Laxmamma Manchikanti for their unconditional love and support; to my wife, Chandrakala, for her dedication and unceasing support; to our children, Anu and her husband, Alvaro, Sunil and his wife, Neena, Kavita (an interventional pain physician); and our grandchildren, Nikhil and Mehr, who have provided us the true meaning of life with pleasure.*

*I also dedicate this book to all interventional pain physicians and regenerative pain medicine physicians across the globe and patients who suffer from chronic pain seeking relief, and, finally, the great staff of the American Society of Interventional Pain Physicians (ASIPP), Pain Management Centers, and Ambulatory Surgery Center.*

*Laxmaiah Manchikanti, MD*

*Editor-in-Chief*

# Foreword by Steven E. Sampson

Following my PM&R residency in 2006 and venturing into the unknown realm of clinical practice, I was introduced to the concept of Platelet-Rich Plasma. My journey began at an orthopedic practice in Santa Monica, CA, working with the United States men's national soccer team. We had word of a new treatment emerging from Spain using one's own blood to heal soft tissue, tendon, and ligament injuries. Since that time, the field of orthobiologics or more popularly termed regenerative medicine has rapidly evolved beyond my expectations.

After being disenchanted with the limitations of medications and corticosteroids, the promise and fundamental basic science behind cell therapy as a natural remedy for pain and musculoskeletal conditions were encouraging. At the same time, technology was rapidly advancing with musculoskeletal ultrasound and fluoroscopic image-guided delivery.

Eager to explore this realm and seek resources to build a solid foundation, I quickly found there were limited options. There were no orthopedic medical conferences or societies focusing on Platelet-Rich Plasma (PRP) and cell therapies. Therefore, I created TOBI (The Orthobiologic Institute) to gather multiple specialties to collaborate and promote best practices. The conference echoed the field's growth from 20 physicians at the initial meeting to hundreds of physicians attending from over 40 countries.

In 2006, there was a single PRP dental and craniofacial textbook available by Robert Marx. This resource described the premise behind platelets and growth factors affecting tissue and wound healing. This foundational textbook was a helpful resource providing the confidence to further explore orthobiologics. There was no regenerative medicine textbook or comprehensive literature reviews for common musculoskeletal conditions available for support.

Nowadays following the efforts of many pioneering colleagues and friends, editor-in-chief Laxmaiah Manchikanti, MD, and editors Annu Navani, MD, Sairam Atluri, and Mahendra Sanapati, MD, have provided a complete resource for practitioners at all skill levels. As a visionary, Laxmaiah Manchikanti, MD, founded ASIPP (American Society of Interventional Pain Physicians) and cultivated interventional pain management as a specialty. Dr. Manchikanti recognizes the vast potential and future trajectory in orthobiologics and, with a herculean effort, has produced this critical resource.

This essentials textbook of regenerative medicine in interventional pain management contains 34 chapters from leading authorities in the field. Its topics vastly cover critical subjects from autologous PRP, bone marrow concentrate, and adipose, to allogenic products, regulatory guidelines, ethical considerations, and more. It is a must-have for anyone practicing orthobiologics and applicable to multiple specialties confronting patients suffering from pain and musculoskeletal disease.

Orthopedic regenerative medicine is an exciting and rapidly changing field with potential to reshape the treatment of musculoskeletal injuries and pain management. There remain many challenges ahead from standardization, protocols, regulatory compliance, data collection, and vetting new technology. Having this bedside textbook resource available for clinicians is a major step forward as the field continues to move closer to mainstream.

The achievement of producing this comprehensive textbook will inspire the next generation of practitioners and hopefully heal the next generation of patients.

Pomona, CA, USA
Los Angeles, CA, USA

Steven E. Sampson

# Foreword by Gabor B. Racz

In our daily life as physicians, one hears about successes, hopes, and failures, but also an undeniable generally accepted willingness to learn and find out more. Upon reading the updated version of *Manchikanti's Essentials of Regenerative Medicine in Interventional Pain Management*, 2nd edition, by Laxmaiah Manchikanti, MD, Annu Navani, MD, Sairam Atluri, MD, and Mahendra Sanapati, MD, I find an amazing knowledge base presented in a manner that finds the reader wanting to learn even more. The presentation is clear, the composition of sentences is anything but tiring, and it provides answers even to topics that the reader failed to consider. The body of knowledge comes from way over 2725 references, 34 chapters over 333 figures and 76 tables, through the diligent works of more than 57 authors.

To overcome the resistance to write another book, Laxmaiah Manchikanti, Annu Navani, Sairam Atluri, and Mahendra Sanapati deserve real credit because they recognized the need for broad-based multidisciplinary physician involvement. The field is large including other science-based disciplines such as engineering, molecular sciences, researchers, and healthcare-related national and international organizations getting into providing guidance in the last decade. It may have started as prolotherapy that organized medicine could not completely accept; however, determination of prolotherapy continues to take what was being gradually accepted not as prolotherapy but evolving into regenerative medicine. The separation and the regenerative powers of platelet-rich plasma, stem cells, and *mesenchymal stem cells* focusing on the ability to change function and repair, reconstruct even regrow, not just groups of cells but structures, are finding a way to broader and broader acceptance, basic science, and clinical information.

No question, the 34-chapters book effectively gathers the pertinent information without omitting any details for the reader, fulfilling its purpose as the Essentials of Regenerative Medicine for Interventional Pain Management. Truth has a way of winning and regenerative medicine is not necessarily the precursor or first step before surgery. Interventional pain with neuroplasty and lysis of adhesions is also growing and producing long-lasting results with inhibited fibroblast and reduction of inflammation through the use of hypertonic saline and hyaluronidase that can address multi-level pain and complement regenerative procedures. The notion of repairing, re-growing, and reforming body parts is very attractive, and the evolution of the knowledge, when one looks at the description of responding and nonresponding cell types, ensures that the outcome is not assured with any single patient. But as we know more, we recognize which therapies must have roles, when to consider neuromodulation, functional restoration, or other procedures based on evidence, and when surgery is the solution.

The big attraction is also through the figures and tables that lend themselves to a quick way of presenting lectures to others that have not had the opportunity to get to the essentials that really should be part of the armamentarium of decision-making on how to treat patients. The medical, legal, and ethical considerations are beautifully presented, and I feel that this book should be part of the personal library of all physicians in interventional pain management and other related fields.

The authors deserve our gratitude for the beautiful work that *Manchikanti's Essentials of Regenerative Medicine in Interventional Pain Management*, 2nd edition, has become.

Austin, TX, USA
Winston-Salem, NC, USA
Paducah, KY, USA
Frisco, TX, USA
Carrollton, TX, USA

Gabor B. Racz

# Prologue

Having been an early adopter and innovator in Musculoskeletal Regenerative Medicine since my first PRP injection in 2008, I have witnessed rapid advancements in understanding and the complexity of Orthobiologics. However, there are still significant barriers to adoption and controversies in the field. These barriers exist in scientific, political, and clinical aspects. We are entering an era where an incredible amount of literature is published, for example more than 43 Randomized Controlled Trials for PRP, yet surprisingly slow adoption by institutions and insurers. The reasons are multifactorial and complex. In addition, we are seeing a dilution of research by studies that on paper have excellent methodology, yet utilize poor quality equipment and suffer from poor characterization of final orthobiologic product. These subtleties are not discerned by the beginner in this field, nor insurance and regulatory decision makers. Likewise, a trend towards "Assembly Line" regenerative medicine does not meet the highest standard of care, and threatens the whole field, with potential for poor outcomes at scale. We are in a time of desperate need to bring clarity and ethics into our field. This textbook is an attempt to lay a foundation in this regard. The authors collectively make up a passionate, disciplined, and ethical group, many of whom are driving the field forward with seminal research, development of new procedures, and equipment, as well as teaching on the national and international stage, and in active discussions with regulatory bodies. With additional time, research, adoption, and regulatory clarity, it is my hope that these therapies may be much more available to the masses, as an accepted way to meet the challenge of treating those in the treatment gap between conservative care and major surgical intervention. Until then, please enjoy learning with this comprehensive textbook, as a first step to competency in this field.

Rowan Paul
Geisel Dartmouth School of Medicine,
San Francisco, CA, USA

# Preface

*Manchikanti's Essentials of Regenerative Medicine in Interventional Pain Management*, 2nd edition, is an updated version of the book published in 2019 to bring concise, collective, and comprehensive information to all physicians practicing regenerative medicine, including interventional pain physicians and others with its applications in managing pain. Regenerative medicine is an integral part of interventional pain management within the definitions of interventional pain management and interventional techniques. Interventional techniques are defined as minimally invasive procedures including percutaneous precision needle placement, with placement of drugs in targeted areas or ablation of targeted nerves, and some surgical techniques such as laser or endoscopic diskectomy, intrathecal infusion pumps, and spinal cord stimulators for the diagnosis and management of chronic, persistent, or intractable pain. Interventional pain management is defined as the discipline of medicine devoted to the diagnosis and treatment of pain-related disorders principally with the application of interventional techniques in managing subacute, chronic, persistent, and intractable pain, independently or in conjunction with other modalities of treatment. While interventional pain management has been recognized as a specialty of its own since 2001, its origins date back to 1901 when the first caudal epidural injection was administered to manage low back and lower extremity pain. Interventional pain management continues to be an evolving and dynamic specialty. More recently entering the scene, regenerative medicine was defined in 1992 by Leland Kaiser as, "a new branch of medicine that attempts to change the course of chronic disease and in many instances will regenerate tired and failing organ systems." Subsequently, numerous authors with widespread literature have established the unique ability of stem cells to differentiate into all the cell types of the human body (pluripotency) and envisioned the immense potential to develop a new kind of regenerative therapy; at the same time, inevitable growth of regenerative medicine, at times dynamic, has also led to chaos and uncertainty creating multiple ethical and legal issues, and leading to an uncertain regulatory atmosphere.

Regenerative medicine by definition is a multidisciplinary specialty and a branch of translational research in tissue engineering and molecular biology, which encompasses the process of replacing, engineering, or regenerating human cells, tissues, or organs to restore or establish normal function. Consequently, regenerative medicine offers a transformative approach to health care, with the promise of engineering damaged tissues and organs by stimulating the body's own repair mechanisms to functionally heal previously irreparable tissues or organs. In the modern era, two major components of regenerative medicine in managing chronic pain include platelet-rich plasma and stem cell therapy. There is increasing literature in support of these modalities in managing spinal and musculoskeletal disorders. Enthusiastic support for the evidence-based practice of regenerative medicine continues to emerge. Even then, along with optimism and future directions, substantial controversy, misuse, abuse, and regulatory impact have taken center stage in recent years.

The goal of this comprehensive text is to provide an evidence-based approach to the application of principles of regenerative medicine in managing chronic pain of spinal, neurological, and musculoskeletal origins. While this is the first comprehensive text on the application of principles of regenerative medicine with appropriate descriptions of technical aspects for interventional pain physicians, we were committed to ensuring that this text was neither too long nor too cumbersome, and, yet, we have striven to make this work encyclopedic in its coverage. With this book covering the entire field of regenerative medicine applicable to interventional pain physicians it provides not only didactic materials but also focuses on technical aspects. From across the globe, leading experts in their respective fields have contributed chapters on specific topics following a single format: to present a cogent and integrative understanding of the field of regenerative medicine as applicable for interventional pain physicians.

We have maintained the overall unique structure of this text to our previous publications with an introduction of the subject, historical context, pathophysiology, applicability of regenerative medicine with its evidence base, indications, anatomy, technical aspects, complications, and precautions for each topic when available and applicable. This comprehensive book consists of 34 chapters, more than 333 figures, and 76 tables under the main sections of:

Part I: Basic Considerations

Part II: Regenerative Medicine in Musculoskeletal Disorders

Part III: Regenerative Medicine in the Spine

Part IV: Miscellaneous Applications

The development of this monumental task in the form of a book publication is attributed to the administrative and logistical expert eyes of the editors and staff of the American Society of Interventional Pain Physicians (ASIPP). This has placed a substantial burden on all involved and their families. The editors wish to thank all the contributors for their time, efforts, and development. In addition to the significant work by editors and contributors, an insurmountable effort has been forwarded by Tonie Hatton, Diane Neihoff, Savannah Gold, and Vidyasagar Pampati, MSc. We are also indebted to many of the world leaders in regenerative medicine and ethics and their families, without whose guidance and patience, this work would have been impossible. We are grateful to a world leader in interventional pain management, Gabor Racz, MD, and a world leader in regenerative medicine, Steven E. Sampson, DO, for writing the foreword.

This second edition will serve as a unique text for regenerative medicine physicians, including interventional pain physicians—a work in progress, which will see continued developments and changes. The overall focus continues to be patient safety in all applications of regenerative medicine to reduce pain and suffering. It is our goal to expand the evidence base and curb misuse and abuse, while at the same time reducing controversy by the implementation of appropriate regulations and guidelines. In this way, we can be optimistic about the future.

| | |
|---|---|
| Paducah, KY, USA | Laxmaiah Manchikanti |
| Campbell, CA, USA | Annu Navani |
| Cincinnati, OH, USA | Sairam Atluri |
| Evansville, IN, USA | Mahendra Sanapati |

# Important Notice

The information provided in this book is based on the best available evidence and does not constitute inflexible treatment recommendations. Due to the changing body of evidence, this document is not intended to be a "standard of care." The material and guidance in this book are meant to provide a basis for the understanding behind the role of biologics in healing, to provide a source of appropriate indications for the use of biologics, to facilitate and help standardize biologic therapy, and to encourage the performance of high-quality studies in an effort to document outcomes and advance this field.

This book has been prepared with utmost care and diligence. Appropriate care has been taken to confirm the accuracy of the information presented and to describe generally accepted practices. However, the authors, editors, and publishers are not responsible for errors or omissions or for any consequences from application of the information in this book and make no warranty, expressed or implied, with respect to the contents of this book.

Readers must apply standard safety precautions and perform the techniques described only with appropriate experience. As new research and clinical experience broaden our knowledge of interventional pain management, changes in interventional modalities in regenerative medicine arena and drug therapy become necessary and appropriate. Readers are advised to keep abreast of the latest developments by staying informed about Food and Drug Administration (FDA) regulations, by reading journals, and checking the product information provided by the manufacturer of each drug to be administered to verify the recommended dosage, the method and duration of administration, and the contraindications. Some biologics presented in this publication may have FDA approval only for limited use. It is the responsibility of the healthcare provider to ascertain the FDA's approval status for each biologic, drug, or device planned for use in his/her clinical practice. This book will only provide guidance; however, it does not provide any standards. It is the responsibility of the treating physician, relying on individual experience and technical expertise, along with appropriate history, physical examination, and laboratory investigations of the patient to determine the appropriateness of regenerative medicine and drugs administered for each individual patient.

Neither the publisher, the editors, nor the contributors assume any responsibility for any injury and/or damage to persons or property arising from this publication.

# Contents

# About the Editors

**Laxmaiah Manchikanti** is a physician specializing in interventional pain management, a professor, philanthropist, and author.

He is the founder of American Society of Interventional Pain Physicians, Society of Interventional Pain Management Surgery Centers, multiple state societies, American Board of Interventional Pain Physicians, as well as *Pain Physician* and *Pain Medicine Case Reports* medical journals.

He has been credited with advancing the evolution and development of interventional pain management as a specialty. Interventional pain management was recognized as a specialty in 2002 by the Centers for Medicare and Medicaid Services. In March 2005, interventional pain management was provided with mandatory representation on the Medicare Carrier Advisory Committee.

He founded the MGM scholarship fund, the Manchikanti Charitable Trust, Chandrakala and Laxmaiah Manchikanti Foundation, and sponsored the Manchikanti Library at McCracken County High School. In addition, he also founded and operates the Manchikanti Gopamma Primary School, Manchikanti Gopamma High School, and Manchikanti Yadagiry Junior College, donated fully equipped facilities for Manchikanti Yadagiry Memorial welfare residential school, and sponsored multiple other activities in Bodangparthy, Nalgonda District, Telangana state, and other areas. He is cofounder of the Gandhi Medical College Global Alliance and sponsored Manchikanti Auditorium at Gandhi Medical College, Hyderabad, Telangana state.

Dr. Manchikanti has been serving as Clinical Professor of Anesthesiology and Perioperative Medicine at the University of Louisville, Louisville, KY, since 2012, and Professor of Anesthesiology-Research, Department of Anesthesiology, School of Medicine, at the LSU Health Sciences Center, New Orleans and Shreveport, LA, since 2017.

He is serving as Chairman of the Board and Chief Executive Officer of American Society of Interventional Pain Physicians (ASIPP) and Society of Interventional Pain Management Surgery Centers (SIPMS). He has also served as President of ASIPP and SIPMS before being named Chairman of the Board of both ASIPP and SIPMS.

He is subspecialty certified in pain medicine by the American Board of Anesthesiology, Diplomate of American Board of

Interventional Pain Physicians and American Board of Pain Medicine, and competency certified in regenerative medicine.

He has been in practice in Paducah, KY, since completion of a fellowship in anesthesiology and critical care medicine in 1980. He graduated from Gandhi Medical College, Osmania University, Hyderabad, India. He completed his internship and residency in anesthesiology at Gandhi Hospital, Youngstown Hospital Association (Northeastern Ohio School of Medicine), Allegheny General Hospital, and his fellowship in anesthesiology and critical care medicine at the University of Pittsburgh School of Medicine.

Dr. Manchikanti has published over 600 publications and serves on several editorial boards. He is also the editor of 14 books designed for interventionalists.

Dr. Manchikanti has been recognized for his dedicated efforts by several prestigious organizations, including:

**AAPI Presidential Recognition Award,** 2023

**Trailblazer Award,** World Institute of Pain, 2019

**Giants in Pain Management Award,** American Society of Interventional Pain Physicians, 2017

**Distinguished Alumni Award,** Gandhi Medical College Global Association, 2016

**Becker's Healthcare Leadership Award,** Becker's Hospital Review

**Interventional Pain management Visionary Award,** National Spine & Pain Centers, 2015

**Distinguished Service Award,** Osmania, Gandhi and Kakatiya Medical Alumni Association of America, 2006

**House of Representatives Presented in honor of Dr. Laxmaiah Manchikanti, MD, by Congressman Ed Whitfield, 109th Congress First Session,** 2005–2006

**Annu Navani** is the Founder and CEO of Comprehensive Spine and Sports Center, the leader in interventional and interdisciplinary spine and orthopedic care. She founded Comprehensive Spine and Sports Center in 2008 with the vision of bringing evidence-based, cutting-edge therapies to the community. She also serves as the Medical Director of Le Reve Regenerative Wellness where she offers conventional and regenerative medicine therapies along with her team of interdisciplinary medical providers.

Dr. Navani completed her Anesthesiology residency from Medical College of Wisconsin, Milwaukee, and fellowship in Pain Medicine from the University of California, Davis. She is board certified by the American Board of Anesthesiology in Anesthesiology along with subspecialty certification in Pain Medicine (ABA) and through American Board of Interventional Pain Physicians (ABIPP). She is board certified in Regenerative Medicine by American Board of Regenerative Medicine and competency certified by American Board of Interventional Pain Physicians.

She is widely published in the field of Interventional Spine and Regenerative Medicine and has contributed via several peer-reviewed journal articles, visionary editorials, and book chapters including the first-ever clinical practice guidelines for Spine Regenerative Medicine. Her clinical and research activities involve spine and orthopedic biologics, with a special interest in the application of biologic therapies for the restoration of structure and function of disrupted discs.

Her vision is to forward the field of regenerative medicine by rigorous research in biomedical engineering and cellular regeneration in an effort to offer it as a standard treatment to create a functional and fulfilling life for people.

**Sairam Atluri** is the founder of StemCures and medical director of Tristate Pain Management Institute in Cincinnati, Ohio. He has been practicing interventional pain medicine for more than 20 years after completing anesthesia residency and pain management fellowship at the University of Cincinnati. Dr. Atluri is board certified in anesthesia and pain medicine. He is certified in regenerative medicine by the American Society of Interventional Pain Physicians (ASIPP) along with certification in Pain Medicine by the American Board of Interventional Pain Physicians (ABIPP). Additionally, he is also a Fellow in Interventional Pain Practice (FIPP).

Dr. Atluri has published many articles in the field of Regenerative Medicine including the first and the largest controlled trial using bone marrow stem cells to treat severe chronic low back pain. He has written numerous chapters in four regenerative medicine textbooks. Dr. Atluri is recognized as one of the top 70 pain management physicians in the US by Becker Review.

Dr. Atluri's interest is in using stem cell therapy to treat chronic spine and joint pain. He is an instructor for the cadaver courses conducted by ASIPP and a speaker at the annual meetings of ASIPP and Ohio Society of Interventional Pain Physicians regarding stem cells. He is also a speaker at the top regenerative medicine organizations like TOBI (The Orthobiologic Institute) and IOF (Interventional Orthobiologics Foundation).

**Mahendra Sanapati** is Director of the Pain Management Centers of America and Medical Director of Pain Management Centers of America, Evansville, IN, and the Advanced Ambulatory Surgery Center, Evansville, IN. He is the founder of Advanced Pain Care Clinic in Evansville, IN, and the Global Scientific Innovations, Evansville, IN.

He is the Executive Vice President of the American Society of Interventional Pain Physicians, President of the Indiana Society of Interventional Pain Physicians, as well as Vice President of the Society of Interventional Pain Management Surgery Centers. Dr. Sanapati completed his residency at New York Medical College Affiliated Hospitals in Valhalla, NY, and a Fellow in Pain Management.

He has board certifications as a Diplomate with the American Board of Anesthesiology and American Board of Interventional

Pain Physicians, and a subspecialty certification in Pain Medicine from the American Board of Anesthesiology. He is also certified in regenerative medicine.

Dr. Sanapati has authored multiple manuscripts and has been a presenter and panelist for multiple meetings, covering a variety of topics in pain management. He is a member of the American Society of Interventional Pain Physicians, Kentucky Society of Interventional Pain Physicians, Indiana Society of Interventional Pain Physicians, the International Neuromodulation Society, the American Medical Association, and the American Society of Regional Anesthesia and Pain Management.

# Part I

# Basic Considerations

# Evolution of Regenerative Medicine in Managing Musculoskeletal and Spinal Disorders

**1**

Laxmaiah Manchikanti, Annu Navani, and Mahendra Sanapati

## Introduction

Regenerative medicine is the science of repair, restoration, and regeneration of diseased or injured cells, tissues, or organs to restore homeostasis. This multidisciplinary specialty includes advanced research leading up to clinical applications from the art of tissue engineering all the way to the science of cellular biology producing a variety of injectates, implants, scaffolds to simulate normal morphology and function in tissues [1–3]. In addition, regenerative medicine offers a transformative approach to health care, with a promise of engineering damaged tissues and organs by stimulating the body's own repair mechanisms to functionally heal previously irreparable tissues or organs [4]. Cell biology is *the branch of biology that studies the structure, function, and behavior of cells,* the structural, functional, and biological unit of all organisms. Tissue engineering, on the other hand, is the use of a combination of cells, engineering and materials, methods, and suitable biochemical and physical chemical factors to improve or replace biological tissues. The term regenerative medicine is often used synonymously with tissue engineering, even though scientists and practitioners involved in regenerative medicine place more emphasis on the use of multiple techniques beginning with prolotherapy, platelet-rich plasma (PRP), and stem cell therapy.

In the modern era, two major components of regenerative medicine in managing chronic pain include PRP and stem cell therapy. An increasing number of physicians believe that the injection of PRP or stem cells are an effective treatment for various spinal and musculoskeletal disorders [5–10]. Even though these treatments have not been well established, and the literature continues to emerge, there is enthusiastic support for the evidence-based practice of regenerative medicine [11]. Even then, along with optimism and future directions, substantial controversy, misuse, abuse, and regulatory impact has taken center stage in recent years [12–26].

## Historical Context

The concept of regeneration dates to 2000 years ago when Aristotle recorded Initial observations on the regeneration of the tail in lizards, as reported in his book, *History of Animals*, cited by Dinsmore in his publication [27].

In 1992, Leland Kaiser [28] in an article describing future technologies utilized the term regenerative medicine stating that, "a new branch of medicine will develop that attempts to change the course of chronic disease and in many instances will regenerate tired and failing organ systems" [29]. Subsequently, the wide-spread use of the term regenerative medicine has been attributed to the founder of human genome sciences, William A. Haseltine [30]. Speaking on the isolation of human embryonic stem cells, Haseltine recognized that unique ability of these cells to differentiate into all the cell types of the human body (pluripotency) and envisioned that there was a potential to develop a new kind of regenerative therapy [31–34]. Consequently, he used the term "regenerative medicine" in the way that it is used today describing an approach to therapy that employs human genes, proteins, and cells to re-grow, restore or provide mechanical replacements for tissues that have been injured by trauma, damaged by diseases or worn by time and offering the prospect of curing diseases that cannot be treated effectively including those related to aging [30].

In a breakthrough in 1908, Russian histologist Alexander Maximow [35, 36] proposed the term "stem cell" for scien-

L. Manchikanti (✉)
Pain Management Centers of America, Paducah, KY, USA
e-mail: drlm@thepainmd.com

A. Navani
LeReve Regenerative Wellness, Campbell, CA, USA

Boomerang Healthcare, Walnut Creek, CA, USA

Stanford University School of Medicine, Stanford, CA, USA
e-mail: anavani@cssctr.com

M. Sanapati
Pain Management Centers of America, Evansville, IN, USA



A. Navani et al. (eds.), *Essentials of Regenerative Medicine in Interventional Pain Management*,
https://doi.org/10.1007/978-3-031-50357-3_1

tific usage. After a hiatus of a few decades, a series of discoveries led to renewed interest and acceleration in research focused on clinical applications. The scientific evidence of adult neurogenesis in ongoing stem cell activity in the brain was presented by Joseph Altman et al. in 1960 [37]. In 1963, McCulloch and Till illustrated the presence of self-renewing cells in mouse bone marrow [38] and in 1968, bone marrow transplant was first used between 2 siblings to treat severe combined immunodeficiency. In 1981, mouse embryonic cells were derived from inner cell mass [39] and in 1992, neural stem cells were cultured in vitro as neurospheres [36]. Then in 1981, embryonic stem cells were first identified in mice and soon the first artificial animal clone, Dolly the sheep, took birth in 1997 [40]. Soon researchers claimed that similar hybrids could be made by fusing human embryonic stem cells with adult cells from a particular person to create genetically matched tissue and organs. From 1995 to 1998, Michael D. West, PhD, organized and managed the research between Geron Corporation and its academic collaborators, James Thomson at the University of Wisconsin, Madison, and John Gearhart of Johns Hopkins University, that led to the first isolation of human embryonic stem and human embryonic germ cells, respectively [41].

In 1998, Thomson et al. isolated human embryonic stem cells [42]. In 2006, Shinya Yamanaka of Kyoto University announced the arrival of "induced pluripotent stem cells;" adult cells that were genetically induced to revert to embryonic-like cells [43]. In 2007, Martin Evans shared the Nobel Prize for medicine with Mario Capecchi and Oliver Smithies for work on genetics and embryonic stem cells [44]. In 2012, Yamanaka won the Nobel prize for creating induced pluripotent stem cells, which he shared with John Gurdon of the University of Cambridge [45, 46].

A new type of stem cell was discovered by scientists of Wake Forest University and Harvard University led by Dr. Anthony Atala in early 2007 [47]. In the same year (2007), research reported by three different groups showed that normal skin cells can be reprogrammed to an embryonic state in mice [48]. On September 12, 2014, Masayo Takahashi at the Riken Centre at the Institute of Biomedical Research and Innovation Hospital in Kobe, Japan, transplanted a 1.3/3.0-millimeter sheet of retinal pigment epithelium cells, which were differentiated from induced pluripotenti stem cells (iPS) cells through directed differentiation, into an eye of an elderly woman, who suffered from age-related macular degeneration [48]. The scientific work in this field is progressing rapidly, opening new frontiers to solving some of the greatest medical challenges in history.

Regenerative medicine is also described interchangeably as tissue engineering aiming to regenerate damaged tissues, instead of replacing them, by developing biological substitutes that restore, maintain, or improve tissue function [49]. The term "tissue engineering" was officially coined at the National Science Foundation workshop in 1988 to mean, "the application of principles and methods of engineering and life sciences toward the fundamental understanding of structure-function relationships in normal and pathological mammalian tissues and the development of biological substitutes to restore, maintain or improve tissue function." The field of tissue engineering is multidisciplinary, drawing expertise from engineering and life sciences including clinical medicine, materials science, genetics, mechanical engineering, and other related branches. Tissue engineering relies extensively on the use of scaffolds—artificial three-dimensional frame structures that serve as a mimic extracellular matrix for cellular adhesion, migration, proliferation, and tissue regeneration in three dimensions. The scaffolds essentially act as a template for tissue formation and are typically seeded with cells and occasionally growth factors or subjected to biophysical stimuli on the form of a bioreactor; a device or system which applies different types of mechanical or chemical stimuli to cells [50]. Thus, cell-seeded scaffolds are either cultured in vitro to synthesize tissues which can then be implanted into an injured site or are implanted directly into the injured site using the body's own systems, where regeneration of tissues or organs is induced in vitro.

Interventional pain management is defined as, the discipline of medicine devoted to the diagnosis and treatment of pain related disorders principally with the application of interventional techniques in managing subacute, chronic, persistent, and intractable pain, independently or in conjunction with other modalities of treatment [51]. In addition, interventional techniques are defined as, minimally invasive procedures including, percutaneous precision needle placement, with placement of drugs in targeted areas or ablation of targeted nerves; and some surgical techniques such as laser or endoscopic diskectomy, intrathecal infusion pumps and spinal cord stimulators, for the diagnosis and management of chronic, persistent, or intractable pain [52]. Regenerative medicine has significant potential for interventional pain physicians to incorporate into daily practices of chronic pain management. The past 20 years have seen a tremendous growth of interventional pain management to a level of explosive increases until 2009, followed be deceleration through 2019 and decreases due to COVID-19 pandemic providing minimally invasive, interventional techniques in managing chronic pain [53–61]. However, recent data and analysis have shown a reversal of growth patterns from 2009 to 2016 [53–61]. Further, interventional pain management has suffered with reimbursement issues [62–65], coverage policies, and finally debate on the effectiveness of interventional techniques [64, 65]. Consequently, regenerative medicine has offered a potential for growth with research and applications in interventional pain management.

Consequently, the therapeutic application of cells or cell-based therapy, forms a fundamental part of the landscape of regenerative medicine or tissue engineering and the attendant exploding applications, development, revenues, and associated regulations, fraud, and abuse [12, 66]. While regenerative medicine has extensive applications, its applications in managing chronic pain and musculoskeletal disorders are of paramount interest for interventional pain physicians.

## Evolution

Once the discovery started, the adaptation and growth of Regenerative Medicine was inevitable. The existing short-lived surgical solutions for musculoskeletal and spine conditions were now open to a minimally invasive solution that would have the potential for lasting a lifetime [66]. Also, efforts went into conceptualizing solutions that would be healing while offering efficacy and sustainability. There was also a consumer driven interest towards natural and holistic solutions that could preserve and heal native tissues to restore homeostasis as opposed to the traditional concept of excision and replacement of the damaged tissues. In vitro and animal studies have appealed to researchers and clinicians alike who have exponentially moved the field forward over the last decade. The last few decades have seen a tremendous growth in clinical trials around the world as shown in Fig. 1.1 [67]. These can be further traced to identify the variations of publications between different countries of the world.

Figure 1.2a, b depict the global trends of worldwide clinical trials for mesenchymal stem cells (MSCs) (Fig. 1.2a) and stem cells in general (Fig. 1.2b) [68]. It is interesting to note that most MSC trials are being conducted in China (35%) followed by the U.S. (21%) and Europe (23%). In contrast, the U.S. (61%) is by far the world leader in stem cell clinical trials, followed by Europe (23%), China (7%), followed

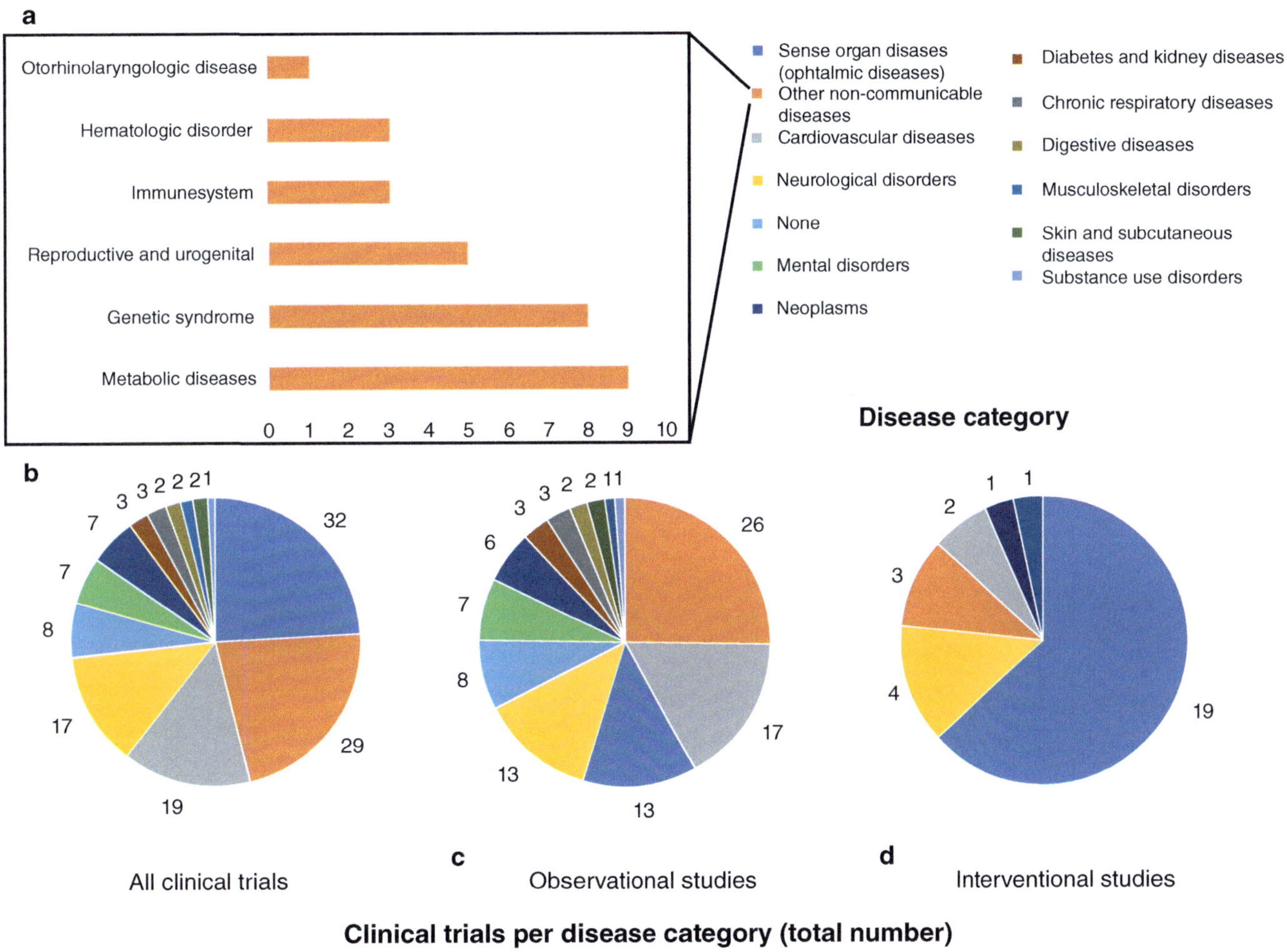


**Fig. 1.1** Disease categories targeted in clinical trials involving human pluripotent stem cells. The legend shown on the top right is color coordinated with all three pie charts. (**a**) Sub-classification of the category "non-communicable diseases". (**b**) Distribution of clinical trials involving pluripotent stem cells dependent on the targeted disease. (**c**) Distribution of observational clinical trials dependent on the targeted disease. (**d**) Distribution of interventional clinical trials dependent on the targeted disease. Source: Deinsberger J, Reisinger D, Weber B. Global trends in clinical trials involving pluripotent stem cells: A systematic multi-database analysis. *NPJ Regen Med* 2020; 5:15 [67]

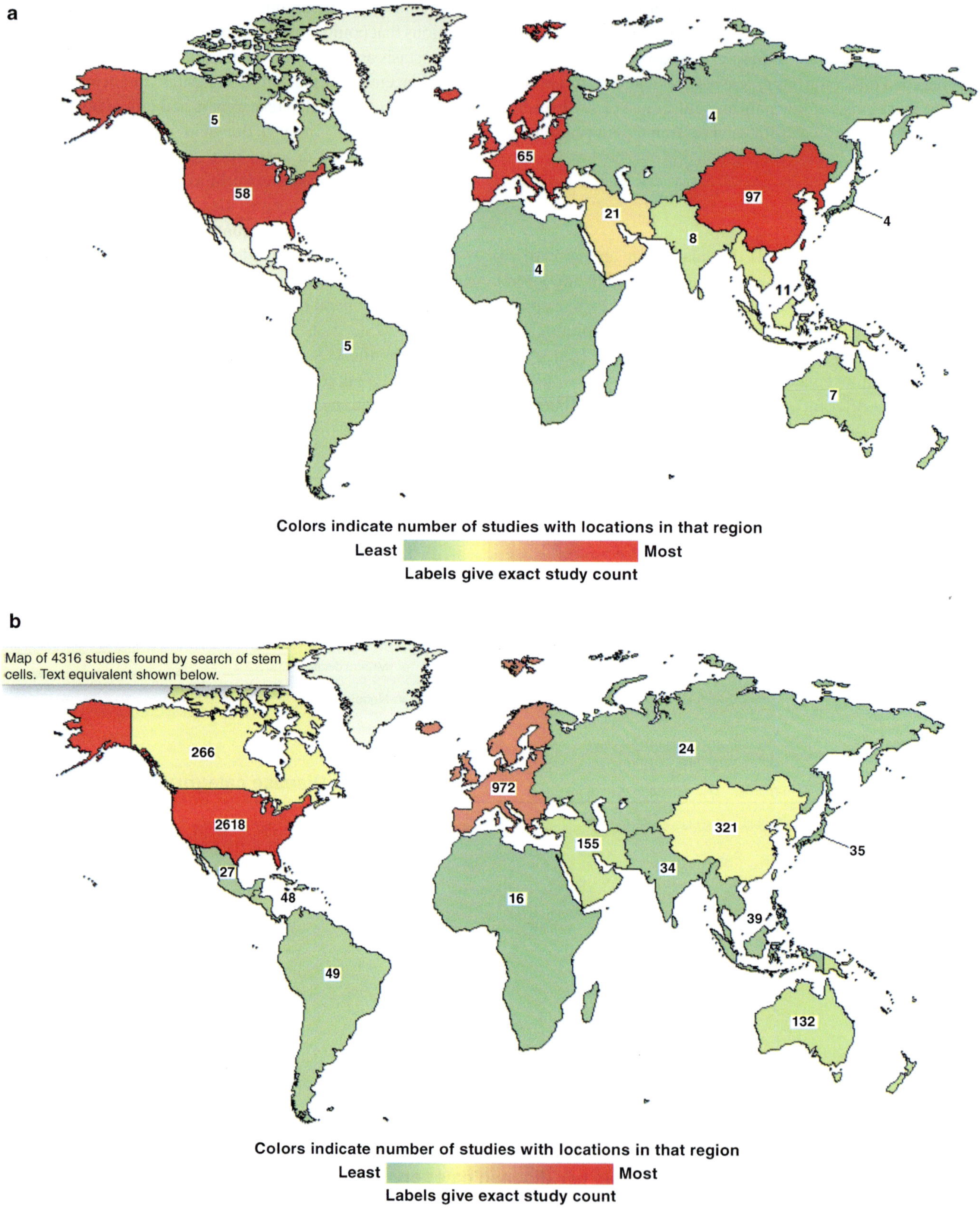


**Fig. 1.2** Global trends of worldwide clinical trials for mesenchymal stem cells (**a**) and stem cells in general (**b**). Mapping global trends in MSC & stem cell clinical trials: Unexpected findings by Professor Paul Knoepfler, Ph.D/2 Comments/December 10, 2012/Blog, stem cells

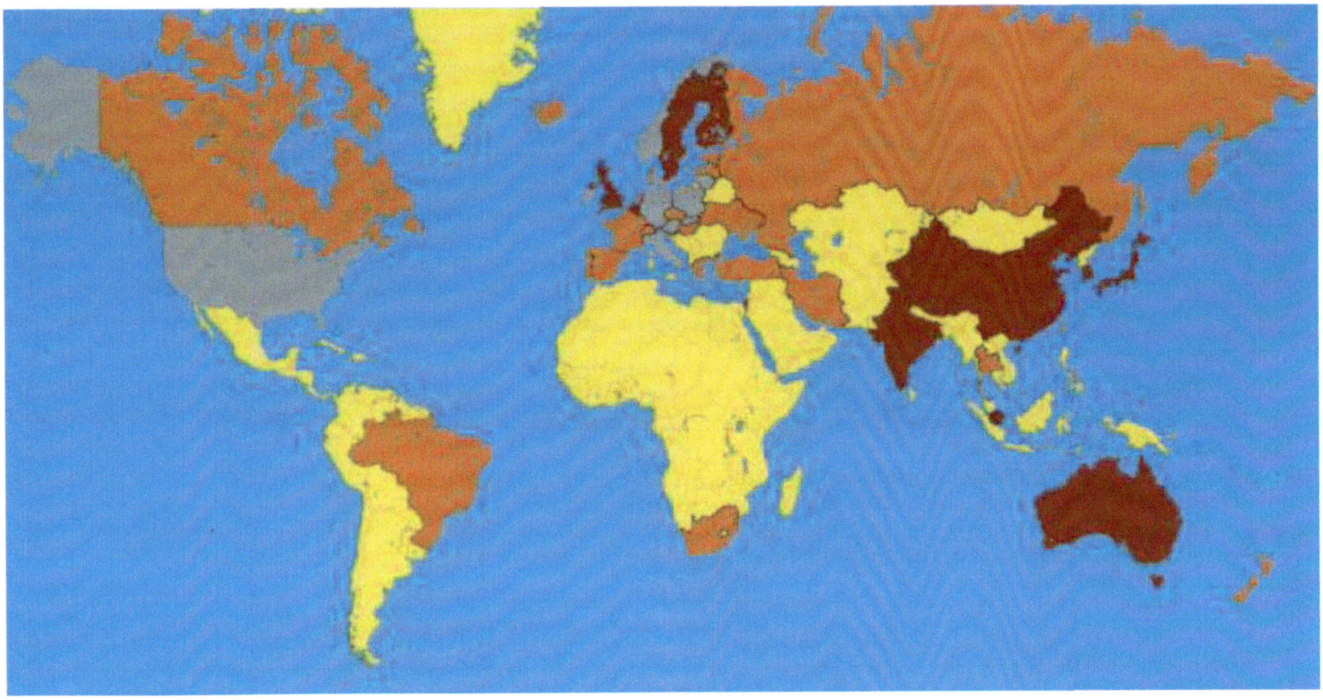

**Fig. 1.3** World stem cell policy map showing countries with permissive (dark brown), flexible (light brown), or restrictive or no policy on ES cell research. Permissive policy allows various laboratory techniques to create embryonic stem cell lines, including nuclear transfer/research cloning and the extraction of stem cells from embryos that remain unused after IVF treatments. Flexible policy allows the creation of stem cell lines from donated embryos unused after IVF treatments. Countries colored in brown represent about 3.5 billion people, more than half the world's population. Reproduced from: scmap.html (reprinted with permission of William Hoffman, MBBNet, University of Minnesota)

closed by Canada, with India and Australia after that [68]. These trends in growth have been dictated by politics and policies around the world. Figure 1.3 outlines the world stem cell policy map showing countries with permissive (dark brown), flexible (light brown), or restrictive or no policy on embryonic stem cell research [69].

In 2016, the search of PubMed database for scientific publications containing the term "mesenchymal stem cell" or "mesenchymal stem cells" results in nearly 30,000 results as reported by Bioinformant Inc. [70]. Of those, 62% i.e., 17,804 of those MSC articles were released between 2011–2016 noted in Fig. 1.4 [70] and more than half of the total publications, i.e., 15,808 of all MSC publications pertain to "Humans," rather than "Other Animals" as noted in Fig. 1.5 [70]. A current analysis through June 2023 showed rapid increase in publications in the last 7 years (Fig. 1.6).

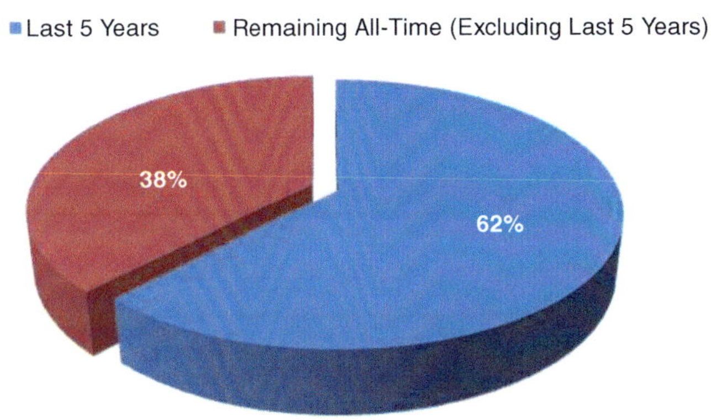


**Fig. 1.4** Percentage of MSC scientific publications released in the last 5 years. Source: Hildreth C. Two-thirds of all mesenchymal stem cell (MSC) publications were released in the past 5 years. *BioInformat.* 2016. https://bioinformant.com/two-thirds-of-all-mesenchymal-stem-cell-msc-publications-were-released-in-the-past-5-years/ [70]

The trends in the total number of newly registered trials that increased linearly in each year from 2007 to 2012, and more than quadrupled during this period is depicted in Fig. 1.7a thereafter slowing in 2018. When newly registered trials are divided by phase, the number of new phase 2 trials increased through 2011, after which they appear to have plateaued Fig. 1.7b. The number of phase 3 trials increased transiently in 2012–2014 to a peak of ~12% of all trials reported but only accounted for ~6% of all trials since 2015 [71].

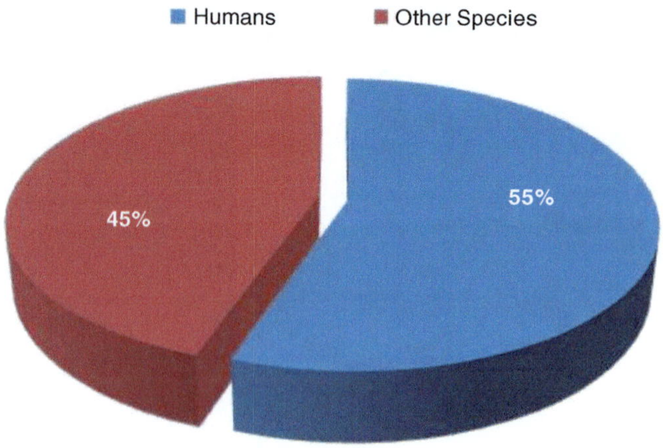


**Fig. 1.5** Breakdown of MSC publications by species. Source: Hildreth C. Two-thirds of all mesenchymal stem cell (MSC) publications were released in the past 5 years. *BioInformat*, July 15, 2016. https://bioinformant.com/two-thirds-of-all-mesenchymal-stem-cell-msc-publications-were-released-in-the-past-5-years/ [70]

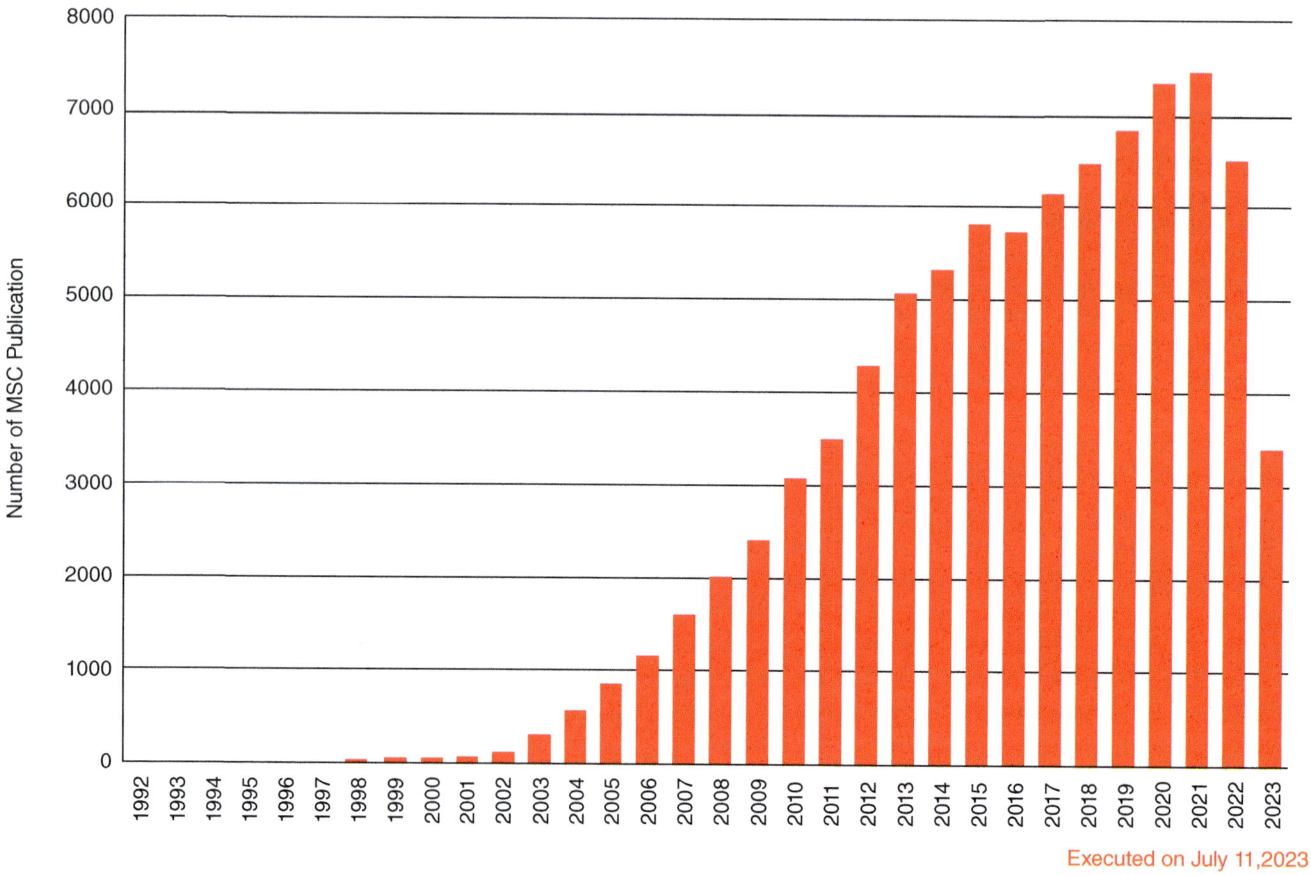


**Fig. 1.6** Half of all mesenchymal stem cell (MSC) publications on PubMed were published in the past 7 years

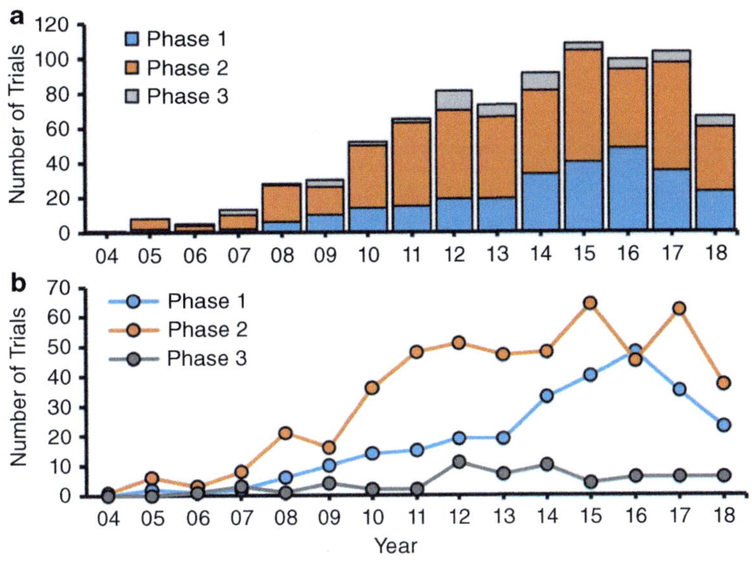


**Fig. 1.7** Trends in the total number of newly registered trials increased linearly in each year from 2007 to 2012, and more than quadrupled during this period. (**a**) The rate of increase in new trial listings slowed thereafter and dropped dramatically in 2018. When newly registered trials are divided by phase, the number of new phase 2 trials increased through 2011, after which they appear to have plateaued (**b**). Source: Bergel S. Uber wirkungen des fibrins. *Deutsche Medizinische Wochenschrift* 1909; 35:663 [71]

## Origin and Types of Regenerative Medicine

The present focus on regenerative medicine is based on therapy with (PRP) and adult stem cell therapy in chronic pain, even though, a multitude of other biologicals have been described and utilized.

### Platelet-Rich Plasma (PRP)

After the initial descriptions of prolotherapy, the use of PRP has been popularized [72–76]. PRP is defined as, "a sample of autologous blood with concentrations of platelets above baseline values [72]." PRP has been used since 1909 for its potential for clinical use as autologous fibrin glue [66, 72]. Since 1950, PRP has been used to manage dermatologic and oral maxillofacial conditions [72]. Over the years, PRP gradually has been studied and used in several branches of medicine including musculoskeletal medicine, orthopedic surgery, and specifically spine [5–8, 72, 75].

The nomenclature of PRP products is confusing with multiple terms used including platelet concentrates, autologous growth factors, plasma rich growth factors, platelet gel, and platelet-rich fibrin matrix [76, 77]. In addition, classification categorizing PRP formulations based on the presence of leukocytes and the fibrin architecture of the formulation have been described [78]:

- Pure platelet-rich plasma (PPRP) referring to leukocyte-poor PRP.
- Leukocyte- and platelet-rich plasma (LPRP) referring to PRP that contains leukocytes and has a low-density fibrin network after activation.
- Pure platelet rich plasma fibrin (PPRPF) which designates preparations without leukocytes and with a high-density fibrin network, which are also called leukocyte-poor platelet rich fibrin.
- LPRF designated to describe the preparations that contain leukocytes in a high-density fibrin network.

Another classification system described by DeLong et al. [79] also known as the PAW Classification System of PRP is based on the absolute number of platelets (P), the way platelet activation (A) occurs, and the presence of absence of leukocytes (W).

Platelets are a nuclear 2–3 μm diameter fragments of megakaryocytes from the bone marrow and contain an abundance of growth, chemotactic, and clotting factors (Fig. 1.8) [80–82]. Platelets are capable of adhering to and pulling together torn tissue using their tentacle-like filopodia with an internal network of actin and myosin (Fig. 1.9) [82, 83]. PRP is simply a platelet concentrate, platelets in plasma concentrated 3–5× or higher than in whole blood [73, 84].

Platelets are activated to adhere to collagen and release their growth factors. Platelet activation depends on external stimuli such as thrombin, adenosine diphosphate (ADP), calcium [85–87], or the presence of certain structural proteins that are not present in the endothelium such as collagen [88]. Platelet degranulation involves the fusion of alpha and dense (delta) granules to their membranes with release of growth and anabolic factors [89].

Platelet degranulation results in inflammation [90, 91], recruitment and proliferation of several cell types including leukocytes (days 0–2) [92, 93], promotion of neovascularization (days 2–3), migration and division of fibroblasts, synthesis of collagens type I and III, and new tissue formation (days 3–5) [91].

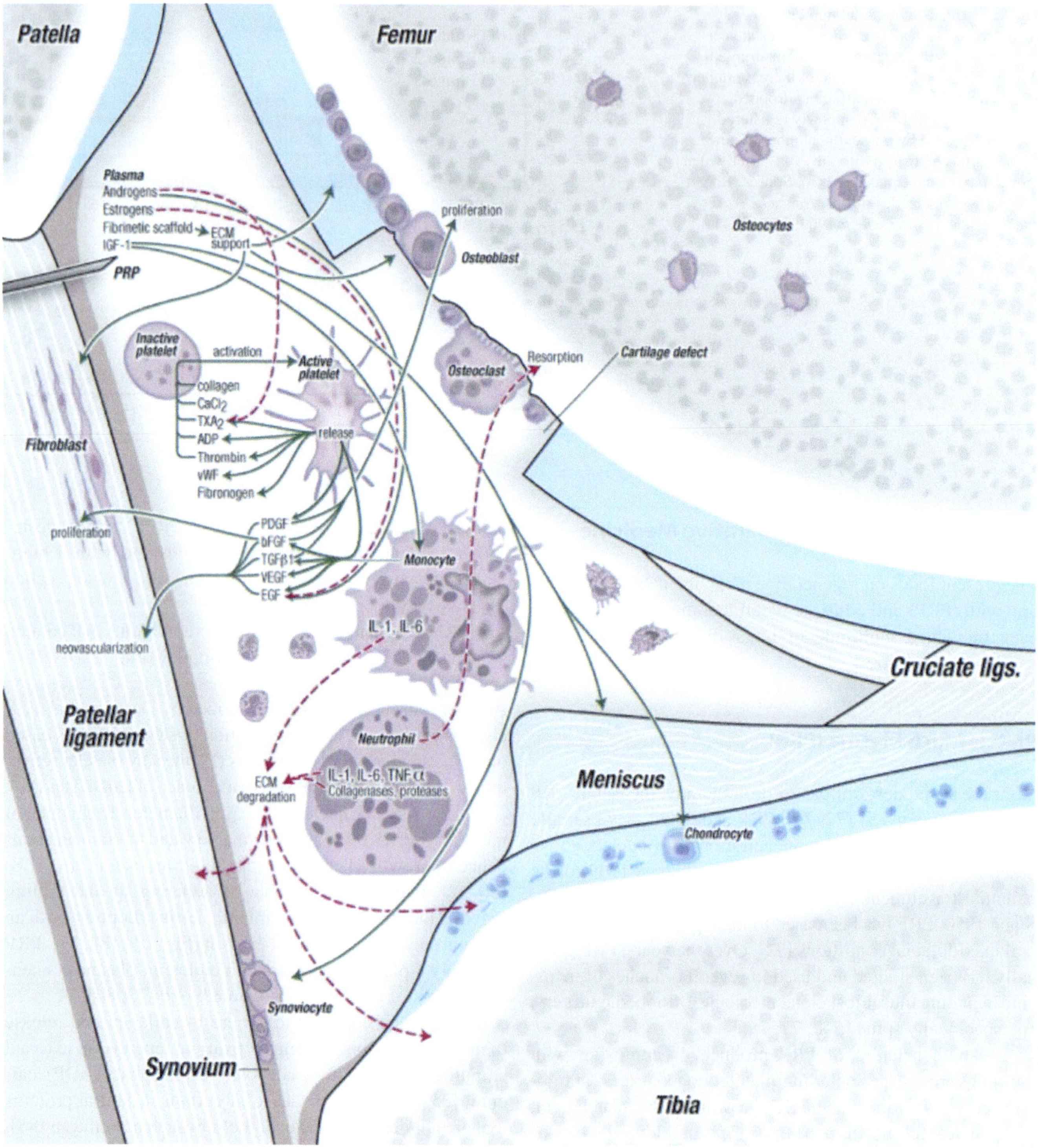


**Fig. 1.8** Platelets and their growth factors upon PRP injection into the knee. PRP has leukocytes to a variable degree, including neutrophils and monocytes. Soluble hormones such as IGF-1 and target cells such as synovial, meniscal, and ligamentous fibroblasts, chondrocytes, and osteocytes are shown. Arrows indicate positive (green) and negative (red) effects. *ADP* adenosine diphosphate; *$CaCl_2$* calcium chloride, *ECM* extracellular matrix, *EGF* epidermal growth factor, *IL-1* interleukin 1, *IL-6* interleukin 6, *ligs* ligaments, *TNF* tumor necrosis factor, *VEGF* vascular endothelial growth factor, *vWF* von Willebrand factor. Source: Boswell SG, Cole BJ, Sundman EA, et al. Platelet-rich plasma: A milieu of bioactive factors. Arthroscopy 2012; 28:429–439 [80]

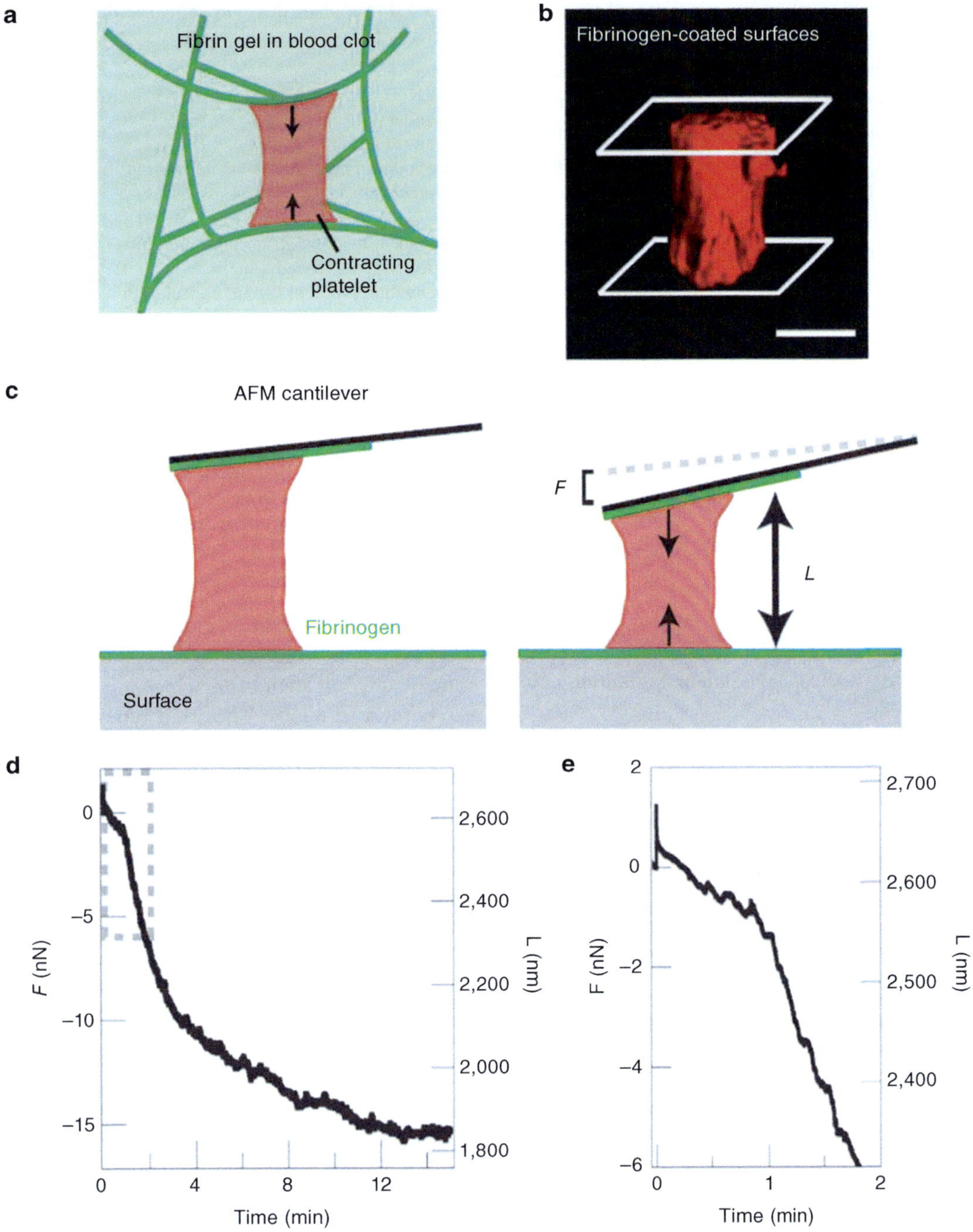


**Fig. 1.9** Measuring contraction of single platelets using an atomic force micrometer (AFM). Diagrams of (**a**) platelet contraction in a blood clot, (**b**) platelet within experimental set-up, (**c**) AFM cantilever measuring platelet contraction force. Graph (**d**) of activated platelet force (nN) and displacement (nm) versus time (min), and (**e**) close-up of left upper corner of (**d**). Reproduced from: Lam WA, Chaudhuri O, Crow A, et al. Mechanics and contraction dynamics of single platelets and implications for clot stiffening. Nat Mater 2011; 10:61–66 [82]

Platelets release platelet derived growth factor (PDGF), vascular endothelial growth factor (VEGF), transforming growth factor beta-1 (TGF-B1), interleukin 1B (IL-1B), ADP, and histamine [94, 95]. These factors stimulate leukocytes to release inflammatory cytokines (IL-1B, TNF-á, IL-6) [96], enhance the expression of degradative enzymes of the matrix metalloproteinase (MMP) family [97], and prepare tissue for repair and regeneration [97].

Platelet-related growth factors and neovascularization stimulate fibroblast homing and proliferation in tendons and ligaments [81]. Fibroblasts orient at the site of injury marked by platelet aggregates and neovessels and synthesize collagen [98].

Lam et al. [82] have described the mechanical properties of platelets. Platelets are capable of adhering to and pulling together torn collagen using tentacle-like filopodia and an internal network of actin and myosin, the same contractile proteins in skeletal muscle. Platelets can generate a dynamic force of 29 nN and a static force of 70 nN, approximately one-third that of slow twitch muscle fibers [82, 83].

In preparation of PRP the manipulation of blood to obtain the final product is minimal, facilitating the process of regulatory approval [99]. The American Academy of Orthopaedic Surgeons (AAOS) [77] in a symposium identified several obstacles to the advancement of PRP therapies and made recommendations on how future research efforts should be directed to overcome the challenges. The consensus statements of future research and multiple variables of PRP of AAOS are shown in Tables 1.1 and 1.2.

**Table 1.2** Variables that may influence the growth factor profile of platelet-rich plasma

| |
|---|
| Donor |
| • Age |
| • Gender |
| • Comorbidities |
| • Concurrent medications (including anti-inflammatories |
| • Nutritional status |
| Processing |
| • Blood collection and storage conditions |
| • Spin protocol (speed, time) |
| • Activation protocol (agent, concentration, timing) |
| • Storage |
| Delivery |
| • Form of delivery (gel, solution) |
| • Timing of delivery in relation to isolation |
| • Timing of delivery in relation to activation |
| • Host factors (similar to donor factors) |
| • Injury chronicity |

Adapted and modified from: LaPrade RF, Dragoo JL, Koh JL, et al. AAOS research symposium updates and consensus: Biologic treatment of orthopaedic injuries. J Am Acad Orthop Surg 2016; 24:e62–e78 [77]

## Mesenchymal Stem cell or Medicinal Signaling Cell Therapy

The second therapy of regenerative medicine is with adult MSCs, more recently called "medicinal signaling cells" [100–106].

A stem cell has two main characteristic features that qualifies it to be the stem cell. The ability to proliferate and the ability to differentiate into cells of any lineage. Stem cells can be classified as embryonic stem cells, which are found in the inner cell mass of the blastocyst (embryo), or as somatic or adult stem cells, which are found in tissues in the newborn or in adults. The adult type can be obtained from bone marrow, peripheral blood, umbilical cord, fetal liver, neural tissue, synovium, periosteum, muscle, or adipose tissue [107].

**Table 1.1** Consensus statements on platelet-rich plasma

| |
|---|
| 1. Nomenclature and classification system that encompasses autologous blood/plasma products and categorizes preparations in sufficient detail to facilitate comparison across studies is not available<br>• A widespread system must be developed with involvement of academics, clinicians, and industry representatives |
| 2. Quality assessment with influence of donor variance and processing and delivery factors on the composition of PRP must be established |
| 3. A validated assay of the efficacy of PRP should be established for each clinical application<br>• Specific formulations of PRP should be matched with specific pathologic indications<br>• Methods for establishing proof of safety and efficacy of PRP should be determined. This process may require evidence of phenotype stability or viability for each indication |
| 4. The relationship between PRP composition and efficacy must be established<br>• Minimum standards of reporting for all studies (preclinical and clinical) evaluating PRP must be established to facilitate communication and the interpretation and synthesis of scientific investigations. These standards must include measured characteristics of the PRP, factors relating to the donor, processing, and delivery of the PRP and outcome parameters |

*PRP* platelet-rich plasma

Adapted and modified from: LaPrade RF, Dragoo JL, Koh JL, et al. AAOS research symposium updates and consensus: Biologic treatment of orthopaedic injuries. J Am Acad Orthop Surg 2016; 24:e62–e78 [77]

In the bone marrow, there are at least two populations of progenitor stem cells: hematopoietic progenitor cells and mesenchymal progenitor cells. Hematopoietic progenitor cells express a membrane glycoprotein that makes it possible to identify and quantify such cells: the CD34 antigen. The MSCs differentiate into bone, cartilage, and other soft tissues associated with musculoskeletal system. For spine and musculoskeletal medicine, these mesenchymal cells are of prime interest. For most clinical applications pertaining to spine and musculoskeletal system, the MSCs are obtained from bone marrow and adipose tissue due to ease of harvest and accessibility. Recently, the Food and Drug Administration (FDA) has issued guidance for the clinical use of cells stressing on minimal manipulation and homologous use [24].

Pluripotent stem cells are the primitive cells in the lineage capable of differentiating into virtually any cell type in the body and therefore of great research interest. This property opens the door to clinical applications such as cell and organ replacement, disease modeling such as cancer and transplantation and thereby, drug development [68]. There are different kinds of pluripotent stem cells: embryonic stem cells are sourced from a blastocyst (an early embryo), and when sourced from human blastocysts are called human embryonic stem cells. Recently, the induced pluripotent stem cells have enjoyed a lot of attention with regards to drug development. These, as noted earlier, were discovered in mid-2000 by genetic reprogramming of body cells to become pluripotent [67].

Stem cell therapy is the most revolutionized and controversial application in managing chronic painful conditions.

Multiple attempts have been made to standardize the nomenclature used in MSC research; however, such standardization efforts have been unsuccessful based on variations in methods of isolation, culture, and assay resulting in complicated and at times misleading nomenclature. Even then, in 2006, the International Society for Cellular Therapy (ISCT) produced a position statement suggesting the minimum criteria required to define MSCs [107]. The statement was inclusive of the facts that MSCs must be plastic adherent, must express the cell surface antigens CD105, CD73, and CD90, must not express the cell surface antigens (CD45, CD34, CD14, CD11b, CD79a, CD19, or HLA-DR) and they must differentiate into osteoblasts, adipocytes, and chondroblasts in vitro. However, these criteria have several limitations as they are based on attributes observed in the laboratory environment and may not reflect characteristics of MSCs in their native environment [107] and multiple distinct populations of cells have been found to fulfill the same criteria [108–110]. Thus, future definitions of MSCs must reflect their identity in vivo and account for the heterogeneity of populations identified [77].

The contributions of MSCs to regeneration in native tissues is not well understood, despite rather extensive use of MSCs in clinical trials [77]. Even though clinical studies clearly demonstrate that MSCs have the capacity to perform multiple functions with release of growth factors with differentiation and immunomodulation, the timing of and reasons for the adoption of each phenotype continues to be an enigma.

As described earlier, MSCs are obtained from multiple sources. They were first isolated from bone marrow, and most studies to date have evaluated cells from this source, based on FDA recommendations. However, MSCs also have been isolated from multiple organs and tissues with multiple advantages and disadvantages of each source still emerging. Apart from intersource variation, MSCs may differ between anatomic regions of the same tissue type in terms of yield and characteristics [77]. The equivalency of MSC populations derived from distinct anatomic origins continues to be contentious. MSC populations from multiple sources have shown differences with respect to the immunophenotype, secreted cytokine profile, and results of proteome analysis [111–113]. However, it also has been shown that cloned human MSCs isolated from fat default to adipogenic potential, whereas those isolated from bone marrow default to osteogenic potential, indicating the importance of origin, which ultimately influences the characteristics of MSCs [111]. Consequently, four key pathways that have been identified in tissue regeneration, namely osteogenesis, adipogenesis, chondrogenesis [114, 115], and fibrogenesis [116–118] seem to be offered with differing potentials from MSCs from bone marrow compared to adipose tissue.

However, MSCs represent only a small proportion of cells present within any given organ and concentrates prepared for administration. Thus, presently available MSC preparations vary in the degree to which MSCs are separated from other types of cells and in the method by which this task is accomplished [77]. Bone marrow concentrate and the stromal vascular fraction (SVF) of adipose tissue are commonly used preparations, even though SVF is a highly controversial issue [23–25]. Both preparations represent highly heterogeneous cell populations that include inflammatory cells, hematopoietic cells, endothelial cells, nonviable cells, and other cell types. Studies also have shown that poor and unreliable tissue formation or lower regenerative efficacy of heterogeneous populations compared cultured MSCs [119, 120]. Thus, laboratory culture has historically been used to enrich and expand the MSC population [121], taking advantage of properties of rapid expansion of MSCs in laboratory conditions, and growing of other cell types including hematopoietic cells and endothelial cells that are less suited to these conditions. However, laboratory cultures are not approved by the FDA in the United States [23–25]. Within the confines of the FDA regulation with minimal manipulation, SVF is not recommended, and laboratory cultures are outside the realm of FDA regulations [23]. Table 1.3 shows consensus statement on stem cell therapy from AAOS.

Apart from PRP and stem cells, scaffolds also have been utilized which are categorized based on the presence or

**Table 1.3** Consensus statements on stem cell therapy

1. It is essential to identify the factors contributing to tissue development, regeneration, and healing in each specific tissue
   - The mechanisms regulating these contributions must be characterized
2. The optimum preparation of stem cells for each indication must be established in a systematic fashion
   - Considerations should include cell number, concomitant use of growth factors, predifferentization, and vehicle
   - Mesenchymal stem cells isolated from different tissues must be compared to identify the most appropriate cell source for each specific indication
3. The mechanism responsible for therapeutic effects observed in applications and appropriate outcome parameters must be established
   - A standardized assay of stem cell efficacy is needed
   - Methods for establishing proof of safety of stem cell therapy should be determined in collaboration with industry and regulatory agencies. This process may require evidence of phenotype stability or viability
   - The most appropriate control for clinical studies evaluating stem cell therapy in each indication must be identified
   - The most appropriate, replicable outcomes must be established

Adapted and modified from: LaPrade RF, Dragoo JL, Koh JL, et al. AAOS research symposium updates and consensus: Biologic treatment of orthopaedic injuries. J Am Acad Orthop Surg 2016; 24:e62–e78 [77] absence of cells [118]. The synthetic or biologic origin of the scaffold, mechanical properties of the scaffold, and the target location and tissue type replicated [77]. The use of scaffolds in the treatment of articular cartilage and osteochondral defects, meniscal pathology, and rotator cuff tears has been evaluated in preclinical studies, in vitro studies, and some early clinical reports [77].

## Clinical Effectiveness

Regenerative medicine with biologic therapies is applied in musculoskeletal and spine conditions in acute, subacute, and chronic disorders involving ligament, meniscus, tendon tear, annular tear, disc protrusion, osteoarthritis, and degenerative disc disease and other spinal conditions. Multiple publications include numerous observational reports, but also include basic science and clinical evidence with randomized controlled trials (RCTs) and systematic reviews [5–11, 66–79, 81, 82, 122–156]. Recently, guidelines were developed [12] for responsible, safe, and effective use of biologics in the lumbar spine.

### Platelet Rich Plasma

PRP injections have been shown to be effective in disorders of bone, tendons, muscles, ligaments, joints, and discs. Among the major systematic reviews assessing the effectiveness of platelet rich therapies for musculoskeletal soft tissue injuries, the Cochrane database systematic review performed in 2014 [11] assessed the effects of platelet rich therapies for musculoskeletal soft tissue injuries. They included only randomized and quasi RCTs that compared platelet rich therapy with either placebo, autologous whole blood, dry needling, or no platelet rich therapy for patients with acute or chronic musculoskeletal soft tissue injuries. In this review, 9 small single center trials with 17 randomized and 2 quasi randomized trials with 1088 participants that compared platelet rich therapy with placebo, autologous whole blood, dry needling, or no platelet rich therapy were included. In this analysis, multiple conditions were assessed:

- Rotator cuff tears, arthroscopic repair, 6 trials
- Shoulder impingement syndrome surgery, 1 trial
- Elbow epicondylitis, 3 trials
- Anterior cruciate ligament (ACL) reconstruction, 4 trials
- ACL reconstruction (donor graph site application), 2 trials
- Patellar tendinopathy, 1 trial
- Achilles tendinopathy, 1 trial
- Acute Achilles rupture, surgical repair, 1 trial

Cochrane review methodology also grouped trials into tendinopathies "where platelet rich therapy injections were the main treatment involving 5 trials, and surgical augmentation procedures where PRP was applied during surgery in 14 trials." Essentially, PRP therapy in chronic conditions involved 5 trials. There were wide variations in methods of preparing PRP without standardization and quantification of the PRP applied to the patients. Overall, the evidence was of low quality yielding the conclusion that overall and for the individual clinical conditions, there was insufficient evidence to support the use of PRP for treating musculoskeletal soft tissue injuries [11].

In contrast, Navani et al. [5] in a narrative review showed limited effectiveness of PRP in musculoskeletal pathology. However, they cautioned that a variety of factors such as method of preparation, composition, medical condition of the patient, anatomic location of the lesion, and tissue type can alter outcome.

One of the extensively studied conditions is knee osteoarthritis, including multiple RCTs [77, 130, 131, 135, 136, 138–142, 144–146]. In fact, the majority of all systematic reviews have shown positive results demonstrating the effectiveness of PRP therapy in knee osteoarthritis [130–134, 138–142]. Campbell et al. [141] in a 2015 publication concluded that intraarticular PRP was a viable treatment for knee osteoarthritis and has the potential to lead to symptomatic relief for up to 12 months. They also reported that there was an increased risk of local adverse reactions after multiple PRP injections. However, Milants et al. [130] have emphasized that there was lack of standardization in PRP preparation techniques for knee osteoarthritis. Khoshbin et al. [136] published the systematic review with quantitative synthesis in 2013 utilizing 6 controlled trials with 4 randomized and 2 prospective nonrandomized including 577 patients. They concluded that compared with hyaluronic acid or sodium chloride injection, multiple sequential intraarticular PRP injections may have beneficial effects in the treatment of adult patients with mild to moderate knee osteoarthritis at approximately 6 months. Lai et al. [140] conducted a systematic review and published the results in 2015 with inclusion of 8 articles with only 2 RCTs and 4 comparative studies. They reached the conclusion that intraarticular injections of the PRP of the knee may be an effective alternate treatment for knee osteoarthritis. Meheux et al. [135] in 2016 with the inclusion of 6 articles concluded that in patients with symptomatic knee osteoarthritis, PRP injection resulted in significant clinical improvement up to 12 months post injection, with a stronger conclusion than others. In 2017, 3 systematic reviews were published [138, 139, 142]. Shen et al. [139] included 14 RCTs comprising 1423 participants, Laudy et al. [138] incorporated 10 trials, and Dai et al. [142] incorporated 10 RCTs with 1069 patients. All 3 trials concluded that current evidence indicates intraarticular PRP injection, is probably more efficacious in the treatment of knee osteoarthritis in terms of pain relief and self-reported function improvement at 3, 6 and 12-month follow-up compared to other injections, including saline placebo, hyaluronic acid, ozone, and corticosteroids.

PRP also extensively has been utilized in various other conditions including shoulder, elbow, and wrist disorders with multiple publications [5, 12, 70, 149, 157–174]. Bodor et al. [72] described the clinical applications of PRP in managing ligament or fibrocartilage injuries [175–185], tears of the tendons, and tendonitis [176, 177, 184–191]. Further, Sadoghi et al. [192] in a review of 150 studies analyzed 14 high quality studies in animals and humans and concluded that PRP for Achilles tendinopathy has a medium to large positive effect for tendon tears; however, they concluded that it was not effective for tendinosis. Fitzpatrick et al. [193] in a metanalysis of 8 high quality studies included that there was good evidence to support a single ultrasound-guided injection of leukocyte-poor PRP for tendinopathy. Evidence also has been presented for hip and ankle [185, 194, 195] osteoarthritis along with other disorders [189–192]. Of importance to interventional pain physicians is the management of spinal disorder [149, 163–174]. Even though there is a lack of systematic reviews, there have been multiple observational studies and a few controlled trials in managing spinal pain with PRP [149, 163–174]. Based on these assessments, results of PRP are encouraging. Significant evidence has been presented with preclinical studies in animal models showing the beneficial effects of autologous PRP on intervertebral disc degeneration [172–174]. Obata et al. [172] in assessing autologous PRP on intervertebral disc degeneration in the rabbit model showed that the administration of active PRP-releasate induced a reparative effect on rabbit degeneration of intervertebral discs, concluding that the use of autologous PRP is safe and can lead to a clinical application for intervertebral disc degeneration.

Among the studies assessing the role of PRP in discogenic pain, Basso et al. [154] in a systematic review identified 3 studies [149, 164, 166] utilizing PRP in managing intervertebral disc degeneration. They concluded that use of PRP in intradiscal degeneration has shown preclinical and clinical positive results. They recommended additional, more powered, high-quality studies to appreciate the long-term safety of PRP in intervertebral disc degeneration. The results showed at 18 months greater than 15% relief was seen in 94% of the patients compared to 93% at 6 months.

Sanapati et al. [196] in a systematic review with qualitative and quantitative analysis performed a single arm meta-analysis in assessing the evidence of PRP injections with lumbar disc injections, lumbar epidural injections, lumbar facet joint injections, and sacroiliac joint injections. Based on the available evidence, including one high quality RCT for injection of PRP, with multiple moderate quality observational studies and evidence from a systematic review [154], showed the qualitative evidence for disc injections of PRP of Level III on a scale of Level I to V using a qualitative modified approach to grading of evidence based on best evidence synthesis, implicating Level I as strong and Level V evidence based on consensus. The evidence also showed for lumbar epidural injections as Level IV, or limited, based on one high quality RCT, multiple relevant moderate quality observational studies, and a single arm meta-analysis. Further, evidence was also Level IV based on high quality RCT and 2 moderate quality observational studies for lumbar facet joint injections and Level IV for sacroiliac joint injections based on one high quality RCT, one moderate quality observational study, and one low quality case report.

PRP also has been utilized in managing neuropathic pain [147, 153, 156, 157].

## Mesenchymal Stem Cells or Medicinal Signaling Cells

Even though the literature is replete with numerous case reports, observational studies, and reviews, there is a paucity of literature with controlled trials in reference to MSCs in managing chronic painful spinal and musculoskeletal conditions. Few RCTs have been published in showing the effectiveness in knee osteoarthritis as well as discogenic pain, even though these studies are very small, and the results have not been replicated.

In the treatment of chronic low back pain, stem cell therapy has been used more frequently. Stem cell research is focused on healing of degenerated intervertebral discs with cell replacement and an increase in the synthesis of proteoglycans and type 2 collagen [127]. The high osmolarity and low PH of disc may be deleterious factors to the viability of stem cells [196]; however, low glucose conditions may facilitate with a favorable factor for survival of stem cells and their regenerative effect [196]. Stem cell therapy has been studied in multiple animal models with improvement demonstrated in disc cellular degeneration. In an animal study of rabbits [197], at 16 weeks, the MSCs with the hydrogel carriers demonstrated recovery of disc height compared with the transplanted carrier alone and the controls. Histological scoring of disc repair also improved in the MSCs group compared with both the carrier alone and the control group. Cao et al. [198] showed the bone marrow MSCs slow intervertebral disc degeneration through the nuclear factor kappa B (NF-êB) pathway. In a recent article by Jezierska-Wozniak et al. [199] ability of CD271+ MSCs to differentiate into chondrocyte-like cells of the nucleus pulposus of intervertebral disc was investigated. They concluded that low affinity MSCs were superior to plastic adherence-MSCs for nucleus pulposus restorative treatment.

Basso et al. [154] in a systematic review identified three articles exploring MSCs use in intervertebral disc disease [149, 164, 166]. They concluded that use of MSCs in intradiscal degeneration has shown preclinical and clinical positive results. They recommended however, additional, high-quality studies to determine the long-term safety and effectiveness of MScs in intervertebral disc degeneration.

In a systematic review and single-arm metanalysis of cell-based therapies for lumbar discogenic low back pain [200], the authors identified 6 studies eligible for the review [201–209]. However, there were no randomized control trials. Of these, one study was published on 3 occasions [202–204]. The authors concluded that MSC and chondrocyte therapy for discogenic low back pain are associated with improved pain relief and functional status. They also commented that even though it seems to be a safe and effective treatment of discogenic low back pain, there are still many questions that future studies will hopefully answer based on the literature to date. They cautioned that optimal cell therapy protocol remains unclear and cell therapy is not a standard treatment of discogenic pain.

In assessing effectiveness of regenerative therapy in low back pain, Sanapati et al. [196] showed Level III evidence based on a single arm metanalysis, one RCT, and observational studies for disc injections of MSCs. There were no studies available qualifying for evidence synthesis with MSCs with lumbar facet joint injections and sacroiliac joint injections.

While there is not significant literature in reference to various painful conditions, there have been some reports of effectiveness of stem cell therapy in neuropathic pain [6, 209], chronic spinal cord injury [210–215], chronic intractable migraine [216], and chronic pancreatitis [151]. There are numerous anecdotal reports of effectiveness of stem cell therapy in various types of disorders; however, also in conjunction with numerous complications related to stem cell injection therapy [217, 218].

Stem cell therapy has been widely utilized in musculoskeletal conditions; however, most of the publications are related to knee osteoarthritis with limited evidence from controlled trials. Zhao et al. [137] in a comprehensive review of stem cells for treatment of knee osteoarthritis included 18 articles which met the inclusion criteria after reviewing 268 articles. The review of these controlled trials with large population [219–236] demonstrated that stem cell therapy provided good evidence in the treatment of osteoarthritis of the knee; however, they also cautioned that greater in-depth study to explore a more ideal way to overcome present difficulties, including standardization of sources of cells, is warranted. Vega et al. [234] published a RCT of treatment of knee osteoarthritis with bone marrow MSCs. They randomized 30 patients with chronic knee pain unresponsive to conservative treatments and showing radiological evidence of osteoarthritis into 2 groups of 15 patients. The test group was treated with bone marrow MSCs by intraarticular injection, whereas control group received intraarticular hyaluronic acid. They also assessed articular cartilage quality by quantitative magnetic resonance imaging. The results showed that the MSC-treated patients displayed significant improvement in algofunctional indices versus the active controls treated with hyaluronic acid. They also showed improvement in quantification of cartilage quality. Soler et al. [237], in a report of results of a phase 1–2 trial using expanded autologous mesenchymal stromal cells for the treatment of osteoarthritis of the knee, confirmed that single intraarticular infusion of MSCs is a safe and well tolerated treatment with a long-lasting amelioration of pain, improvement of quality of life up to 4 years and signs of cartilage repair; however, this study included only 15 patients. Assessment of clinical and MRI outcomes after MSC implantation in patients with knee osteoarthritis also showed encouraging clinical and MRI outcomes with MSC implantation in 20 patients with 24 knees at 2-year follow-up [238]. In another small study of 18 patients, Koh et al. [239] provided encouraging results showing effectiveness of MSCs for reducing pain and improving knee function in patients with knee osteoarthritis. Mesenchymal stromal cells also have been reported for cartilage repair in osteoarthritis in a review [240].

Stem cells also have been utilized in almost all types of musculoskeletal and articular pathologies [77, 143–146, 152, 155, 241, 242].

## Ethical and Regulatory Challenges

A multitude of regulatory and ethical challenges exist in the application of regenerative medicine principles in managing various chronic painful conditions [243–253]. Of the 3 commonly used modalities of regenerative medicine, PRP and bone marrow aspirate applications appear to be least controversial, whereas adipose-derived stem cells appear to be the most controversial.

Taylor-Weiner and Graff Zivin [18] described unlicensed stem cell clinics in the United States as medicine's Wild West. They described that, even though stem-cell therapy may hold great potential, the field is less advanced than the public has been led to believe. They also described that most therapies offered by stem cell clinics use adipose-derived stem cells packaged as a product called stromal vascular fraction (SVF) and procedures using SVF have become increasingly popular because of the relative ease of acquiring the cells. This article related that to produce SVF, clinics collect liposuction aspirate from a patient, separate the cells from the surrounding fat tissue, and administer the isolated cells back to the patient intravenously or by injection into the tissue to be treated. They also state that it is easy to obtain SVF, however, the treatments are unproven and stem cell clinics have been, in effect, operating without regulation due to ambiguous FDA guidelines. In multiple articles, the Food and Drug Administration (FDA) [22–24] also warned that the current excitement over the potential for stem cell therapy to improve patient outcomes or even cure diseases must be accompanied by understanding its risks and benefits with the development of therapeutic approaches based on sound science. They continue to express concerns in reference to stem cell therapy due to lack of evidence. Turner described multiple issues related to adipose-derived autologous stem cell interventions [16].

The center of contentious arguments is based on the same surgical procedure exception; however, to date, the FDA has not approved any adipose-derived stem cell medical products for the U.S. marketplace. The providers performing adipose-derived stem cell interventions claim that they are engaged in the practice of medicine, and the FDA does not directly regulate the practice of medicine [13–16, 22–25, 246, 247]. Turner [16] believes that these procedures do not fall under the same surgical procedure exception identified in 21 CFR 1271; however, others disagree [247]. Guidance for the industry issued by the FDA addresses what actions clinicians and manufacturers can perform while remaining within the scope of the exception [14, 23, 24].

## Human Cells, Tissues, and Cellular and Tissue-Based Products

Marks and Gottlieb [15] described regulatory context for regenerative medicine. To implement a comprehensive policy framework, the FDA's statutory authority in this area is based on, in part, the Public Health Service (PHS) Act. Section 351 of this provides the FDA with authority surrounding the licensure of biologic products, and Section 361 mandates that the agency will issue and enforce regulations necessary to prevent the introduction, transmission, or spread of infectious disease. That regulatory framework is risk-based and divides human cells, tissues, and cellular and tissue-based products (HCT/Ps) [15].

The products that are regulated under both Sections 351 and 361 of the Act are biologic products and must be studied under the provisions for investigational new drugs [14]. Further, the manufacturers of such products are required to submit a biologics license application to the FDA for approval before marketing. However, the products that are regulated solely under Section 361 and under the implementing regulations do not need premarket approval. Instead of premarket approval, these products require registration and listing with the FDA before marketing. However, they must be produced in compliance with the appropriate provisions to prevent the transmission of infectious diseases [14].

The decision in reference to the regulatory pathway to be followed for a given product rests on in part whether the product meets or does not meet the criteria of the regulations promulgated under the Code of Federal Regulations part 1271 of the Title 21 of 2005. Under this section, products that are regulated solely under Section 361 generally are those that do not undergo substantial processing (minimal manipulation), are used in a manner in the recipient that is similar to that in the donor (homologous use), are not combined with another drug or biologic product, and do not have a systematic effect, unless they are designed for autologous transplantation, first- or second-degree-related allogeneic transplantation, or reproductive use as shown in Table 1.4 [15]. Thus, any HCT/P products other than tissue such as corneas and heart valves, are regulated as drugs, biologics, or devices and require appropriate regulatory submissions for the conduct of clinical trials and marketing [15]. Figure 1.10 illustrates in a flow chart how manufacturers and health care providers should apply the criteria outlined in 21 CFR 1271 to determine if the product meets the definition of an HCT/P [23, 24].

In summary, an HCT/P regulated solely under Section 361 of the PHS Act and 21 CFR Part 1271 if it meets all the following criteria [14, 23, 24]:

**Table 1.4** Categories of human cells, tissues, or cellular and tissue-based products (HCT/Ps)

| |
|---|
| HCT/P |
| Cardiovascular tissue |
| Cell-derived therapeutic products (e.g., pancreatic islets, mesenchymal stem or stromal cells, fibroblasts |
| Dura mater |
| Hematopoietic progenitor cells derived from peripheral or cord blood (including hematopoietic stem cells) |
| Musculoskeletal tissue (include adipose-derived stem cells) |
| Ocular tissue |
| Placenta or amnion |
| Reproductive cells and tissues |
| Skin |
| Not HCT/P |
| Blood vessels that are a part of an organ intended for transplantation |
| Human collagen |
| Human milk |
| In vitro diagnostic products |
| Minimally manipulated bone marrow for homologous use and not combined with another article (with a few exceptions described in the tissue regulations) |
| Nonhuman cells, tissues, or organs |
| Vascular composite allografts |
| Vascularized human organs for transplantation |
| Whole blood or blood components, including platelet-rich plasma |

Source: Marks P, Gottlieb S. Balancing safety and innovation for cell-based regenerative medicine. N Engl J Med 2018 378:954–959 [15]

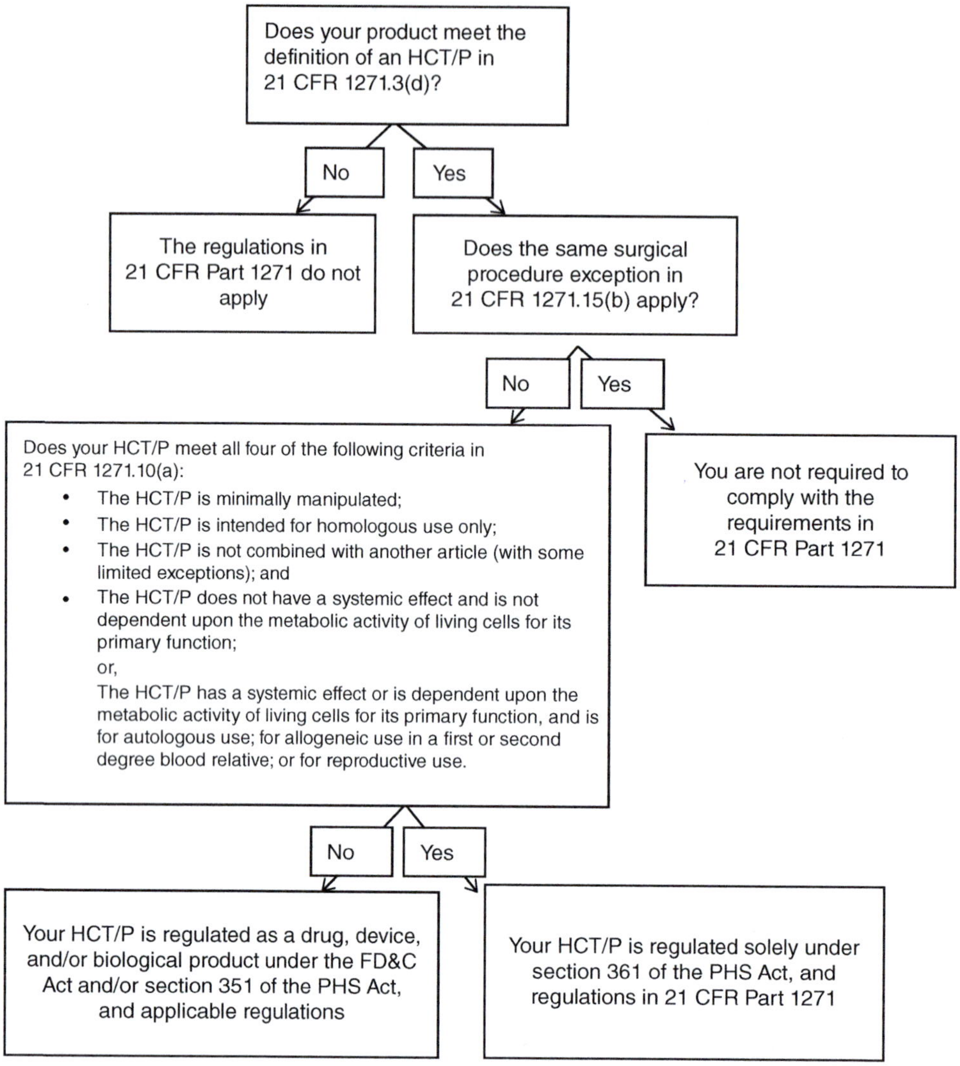


**Fig. 1.10** Flowchart to illustrate the definition of an HCT/P criteria in 21 CFR 1271.15(b) and 1271.10(a). Source: Regulatory Considerations for Human Cell, Tissues, and Cellular and Tissue-Based Products: Minimal Manipulation and Homologous Use. Guidance for Industry and Food and Drug Administration Staff. 2017. https://www.fda.gov/downloads/BiologicsBloodVaccines/GuidanceComplianceRegulatoryInformation/Guidances/CellularandGeneTherapy/UCM585403.pdf [24]

1. The HCT/P is minimally manipulated;
2. The HCT/P is intended for homologous use only, as reflected by the labeling, advertising, or other indications of the manufacturer's objective intent;
3. The manufacture of the HCT/P does not involve the combination of the cells or tissues with another article, except for water, crystalloids, or a sterilizing, preserving, or storage agent, provided that the addition of water, crystalloids, or the sterilizing, preserving, or storage agent does not raise new clinical safety concerns with respect to the HCT/P; and
4. Either:
   (a) The HCT/P does not have a systemic effect and is not dependent upon the metabolic activity of living cells for its primary function; or
   (b) The HCT/P has a systemic effect or is dependent upon the metabolic activity of living cells for its primary function, and:
      Is for autologous use;
      Is for allogeneic use in a first-degree or second-degree blood relative; or
      Is for reproductive use.
5. If an HCT/P does not meet the criteria set out in 21 CFR1271.10(a), and the establishment that manufactures the HCT/P does not qualify for any of the exceptions in 21 CFR 1271.15, the HCT/P will be regulated as a drug, device, and/or biological product under the FD&C Act, and/or section 351 of the PHS Act (42 U.S.C. 262), and applicable regulations, including 21 CFR Part 1271, and premarket review will be required.

## Minimal Manipulation

Section 1271.10(a)(1) (21 CFR 1271.10(a)(1)) provides that one of the criteria for an HCT/P to be regulated solely under section 361 of the PHS Act and the regulations in Part 1271 is that the HCT/P is minimally manipulated [14, 23, 24]. As defined in 21 CFR 1271.3(f), minimal manipulation means:

1. For structural tissue, processing does not alter the original relevant characteristics of the tissue relating to the tissue's utility for reconstruction, repair, or replacement.
2. For cells or nonstructural tissues, processing does not alter the relevant biological characteristics of cells or tissues.

The FDA noted that if information does not exist to show that the processing meets the definition of minimal manipulation, the FDA considers the processing of an HCT/P to be "more than minimal manipulation" that cannot qualify for regulation solely under section 361 of the PHS Act and 21 CFR Part 1271 [14, 23, 24].

Section 1271.3(f) provides two definitions of minimal manipulation, one that applies to structural tissue and one that applies to cells or nonstructural tissues. For structural tissue, minimal manipulation means that the processing of the HCT/P does not alter the original relevant characteristics of the tissue relating to the tissue's utility for reconstruction, repair, or replacement (21 CFR 1271.3(f)(1)). For cells or nonstructural tissues, minimal manipulation means that the processing of the HCT/P does not alter the relevant biological characteristics of cells or tissues (21 CFR 1271.3(f)(2)).

Original relevant characteristics of structural tissues generally include the properties of that tissue in the donor that contribute to the tissue's function or functions. Similarly, relevant biological characteristics of cells or nonstructural tissues generally include the properties of the cells or nonstructural tissues in the donor that contribute to the cells or tissue's function(s). Processing that alters the original characteristics of the HCT/P raises increased safety and effectiveness concerns for the HCT/P because there would be less basis on which to predict the product's function after transplantation [23]. Thus, the determination of whether an HCT/P is minimally manipulated is based on the effect of manufacturing on the original relevant characteristics of the HCT/P as the HCT/P exists in the donor, and not based on the intended use of the HCT/P in the recipient.

Processing is defined as any activity performed on an HCT/P, other than recovery, donor screening, donor testing, storage, labeling, packaging, or distribution, such as testing for microorganisms, preparation, sterilization, steps to inactivate or remove adventitious agents, preservation for storage, and removal from storage (21 CFR 1271.3(ff)). Processing also includes cutting, grinding, shaping, culturing, enzymatic digestion, and decellularization [23].

Tissues that physically support or serve as a barrier or conduit, or connect, cover, or cushion in the donor are generally considered structural tissues for the purposes of determining the applicable regulatory definition (i.e., bone, skin).

Adipose tissue is considered a structural tissue for the purpose of applying the HCT/P regulatory framework. Adipose tissue is typically defined as a connective tissue composed of clusters of cells (adipocytes) surrounded by a reticular fiber network and interspersed small blood vessels, divided into lobes and lobules by connective tissue septa [249]. Additionally, adipose tissue contains other cells, including preadipocytes, fibroblasts, vascular endothelial cells, and macrophages [254, 255]. Adipose tissue provides cushioning and support for other tissues, including the skin and internal organs, stores energy in the form of lipids, and insulates the body, among other functions. While adipose tissue has multiple functions, because it is predominantly composed of adipocytes and surrounding connective tissues that provide cushioning and support to the body, the FDA considers adipose tissue to be a structural tissue for the purpose of applying the HCT/P regulatory framework.

To evaluate whether processing of adipose tissue would meet the regulatory definition of minimal manipulation, one should consider whether the processing alters the original relevant characteristics of the adipose tissue relating to its utility to provide cushioning and support (Fig. 1.10).

## Homologous Use

Section 1271.10(a)(2) (21 CFR 1271.10(a)(2)) provides that one of the criteria for an HCT/P to be regulated solely under section 361 of the PHS Act and the regulations in Part 1271 is that the "HCT/P is intended for homologous use only, as reflected by the labeling, advertising, or other indications of the manufacturer's objective intent."

As defined in 21 CFR 1271.3(c), homologous use means the repair, reconstruction, replacement, or supplementation of a recipient's cells or tissues with an HCT/P that performs the same basic function or functions in the recipient as in the donor. This criterion reflects the FDA's conclusion that there would be increased safety and effectiveness concerns for HCT/Ps that are intended for a non-homologous use, because there is less basis on which to predict the product's behavior, whereas HCT/Ps for homologous use can reasonably be expected to function appropriately (assuming all the other criteria are also met) [23, 24].

In applying the homologous use criterion, the FDA will determine what the intended use of the HCT/P is, as reflected by the labeling, advertising, and other indications of a manufacturer's objective intent, and will then apply the homologous use definition.

Homologous use means the repair, reconstruction, replacement, or supplementation of a recipient's cells or tissues with an HCT/P that performs the same basic function or functions in the recipient as in the donor (21 CFR 1271.3(c)), including when such cells or tissues are for autologous use. We generally consider an HCT/P to be for homologous use when it is used to repair, reconstruct, replace, or supplement:

- Recipient cells or tissues that are identical (e.g., skin for skin) to the donor cells or tissues, and perform one or more of the same basic functions in the recipient as the cells or tissues performed in the donor; or,
- Recipient cells or tissues that may not be identical to the donor's cells or tissues, but that perform one or more of the same basic functions in the recipient as the cells or tissues performed in the donor [256, 257].

For applying the HCT/P regulatory framework, the same basic function or functions of HCT/Ps are those basic functions the HCT/P performs in the body of the donor, which, when transplanted, implanted, infused, or transferred, the HCT/P would be expected to perform in the recipient. It is not necessary for the HCT/P in the recipient to perform all the basic functions it performed in the donor to meet the definition of homologous use. However, to meet the definition of homologous use, any of the basic functions that the HCT/P is expected to perform in the recipient must be a basic function that the HCT/P performed in the donor.

The use of an HCT/P from adipose tissue for the repair, reconstruction, replacement, or supplementation of adipose tissue would be considered a homologous use. In these situations, FDA would consider the HCT/P from adipose tissue to be performing the same basic function in the recipient as in the donor. In contrast, the use of an HCT/P from adipose tissue for the treatment of a degenerative, inflammatory, or demyelinating disorder would generally be considered a non-homologous use. Figure 1.10 illustrates how manufacturers and healthcare providers should apply the criteria outlined in 21 CFR 1271.15(b) and 1271.10(a) for HCT/Ps [23].

## Regulatory Scope and Compliance Policy

This guidance only applies to products and establishments that are subject to FDA's regulations in 21 CFR Part 1271. Establishments that meet the same surgical procedure exception in 21 CFR 1271.15(b) are not subject to FDA's regulations in 21 CFR Part 1271.

This guidance also does not apply to products that fall outside the definition of HCT/P in 21 CFR 1271.3(d). For example, PRP (blood taken from an individual and given back to the same individual as platelet rich plasma) is not an HCT/P under Part 1271 because it is a blood product. Accordingly, FDA does not apply the criteria in 21 CFR 1271.10(a) to PRP, and PRP is outside the scope of this guidance.

## FDA's Comprehensive Framework for Regenerative Medicine

As per Marks and Gottlieb [15], the FDA recognizes the time and effort that go into the creation of regulatory submission and the effect that working through the regulatory process can have on timelines for the development of innovative products. Consequently, in November 2017, building on this policy and scientific opportunities, the FDA released a comprehensive framework for the oversight of regenerative medicine to help the field continue to advance. This regulatory framework is articulated in 2 final and 2 draft guidance documents as shown in Table 1.5 [15]. The FDA's new policy framework more clearly describes the developers of regenerative medicine to provide distinction between therapies that require premarket authorization and those that do not. Consequently, the FDA provided appropriate criteria for minimal manipulation and homologous use. As part of the regulatory framework, the FDA also articulated a risk-based compliance and enforcement policy.

**Table 1.5** Four guidance documents describing the regenerative medicine framework[a]

| Document | Summary | Example |
|---|---|---|
| Same Surgical Procedure Exception under 21 CFR 12.71.15(b): Questions and Answers Regarding the Scope of the Exception—final | Addresses the criteria required for the exception, the types of procedures generally considered to be the same surgical procedure, and what processing steps can be undertaken to still meet the exception. In essence, this guidance clarifies how the regulations apply in order to facilitate the optimal care of patients undergoing surgical procedures | A situation in which this guidance would apply is when a piece of the skull is removed for decompression after traumatic head injury. The bone may be minimally processed, stored, and then returned to the patient a few weeks later when the acute event is over, without the need for regulatory interaction with FDA |

**Table 1.5** (continued)

| Document | Summary | Example |
|---|---|---|
| Regulatory Considerations for Human Cells, Tissues, and Cellular and Tissue-Based Products: Minimal Manipulation and Homologous use—final | Provides FDA's interpretation of the existing regulatory definitions of minimal manipulation and homologous use. The guidance clarifies that these are distinct concepts and notes how to determine whether an HCT/P has been minimally manipulated or is intended for homologous use The guidance also describes the compliance and enforcement policy that the FDA will use for HCT/Ps. For the first 36 months after issuance of the final guidance in November 2017, the FDA intends to exercise enforcement discretion for certain products that pose a low risk to public health so that sponsors will be able to have a dialogue with the agency and file the appropriate regulatory documentation | Adipose tissue is considered to be a structural tissue for the purpose of the regulatory framework. This is relevant to determining the appropriate regulatory pathway for stem cells derived from adipose tissue, which in many applications will be regulated under both Sections 351 and 361 of the Public Health Service Act |
| Evaluation of Devices Used with Regenerative Medicine Advanced Therapies—draft | Provides a comprehensive resource to developers of devices used with RMATs. Topics covered include how the FDA will simplify and streamline its application of regulatory requirements for devices and cell-tissue combination products | Under certain circumstances, a device that is used with an RMAT might be classified as a class III device or be limited to a specific intended use with only one type of cell |
| Expedited Programs for Regenerative Medicine Therapies for Serious Conditions—draft | Provides information about the expedited programs available to RMATs, including fast-track and breakthrough-therapy designations, and describes the FDA's considerations in implementing the new expedited program for RMATs. The guidance also describes an innovative program using cooperative development open to regenerative medicine products | Multiple sites that manufacture a product using a common process may collaborate on clinical trials as part of a development program, which ultimately results in biologics licenses for each of the individual sites |

RMAT denotes regenerative medicine advanced therapy

[a] The listed guidance documents can be accessed at: www.fda.gov

## Expediting the Development of New Therapies

The FDA has developed a process of expediting the development of new therapies. The FDA traditionally has focused on ensuring the quality, safety, and efficacy of medical products. However, its mandate has expanded to encompass a role in expediting the developments of new therapies, particularly those aimed at serious or life-threatening conditions. Consequently, the expedited programs including:

- Fast track designation
- Priority review
- Accelerated approval
- Designation as a breakthrough therapy

The FDA described that these programs have been successful in accomplishing the goal of expediting the development of new therapies [258, 259]. Further, the FDA also described its role in facilitating innovation while upholding the agency's approval standards, especially when it comes to areas of unmet medical need and new technologies as expressed in the legislative initiative contained in the 21st Century Cures Act, which was enacted on December 13, 2016 [260]. To facilitate therapeutic advances from stem cell therapies, along with other HCT/Ps, the 21st Century Cures Act introduced an additional expedited program in which a product is designated as a regenerative advanced therapy, or RMAT. This designation provides sponsors of a qualified regenerative medicine product that is intended for the treatment of serious or life-threatening conditions with advantages like those of breakthrough therapy designation, provided that preliminary clinical evidence indicates that the therapy addresses unmet medical needs. The requirement for preliminary clinical evidence of efficacy distinguishes RMAT from the breakthrough designation, which requires preliminary clinical evidence of a substantial improvement over existing therapies. Further, RMAT-designated products that receive accelerated approval may be eligible to use an expanded range of options to fulfill their post approval commitments. Such options include the use of traditional studies, as well as the submission of patient registries or other sources of real-world evidence [259–262]. Multiple requests have been submitted for RMAT designation with FDA approving at least one-third of them.

#### Implementation of Comprehensive Framework

The FDA described that it strove to take a modern approach to existing regulations and statutes, balancing the objective of fostering expedient development of innovation products for patients who have medical needs with the need to ensure that such therapies are both safe and effective. In addition, as part of the regulatory framework, the FDA also articulated a risk-based compliance and enforcement policy. This policy essentially allows developers of lower risk products up to 36 months from November 16, 2017, to determine whether they need to apply for investigational new drug or a marketing application considering the recently published guidance documents and, if such an application is needed, to prepare the new-drug or marketing application. However, the FDA intends to take additional enforcement actions in cases in which it believes unproven products may put patients at risk.

Finally, working within the existing regulatory framework, the FDA will make use of all available regulatory pathways and will adopt the use of some new principles that will make the appropriate premarket evaluation of stem cell-based therapies more efficient. On a large scale, the FDA will be incorporating some new concepts for how small investigators and firms can see and meet approval standards for products through efficient and expedited pathways. To achieve this goal, the FDA will provide tools to encourage individual or small groups of physicians to collaborate in support of the development of a stem cell or other regenerative medicine product, which will ultimately lead to the receipt of a biologics license by each of the physicians or groups (Fig. 1.11).

### State Medical Board Regulations/Guidelines

On April 21, 2017, Senator Lamar Alexander requested the Federation of State Medical Boards (FSMB) [263] to respond to multiple questions including:

1. How do state medical boards investigate complaints against stem cell clinics?
2. How are the existing false claims best practices enforced or used by state medical boards?
3. Are there standards or best practices regarding the use and communication of novel technology, such as adult stem cells?
4. Are there standards for education necessary before implementing novel technology, such as adult stem cell procedures?

Consequently, the FSMB formed a working group to develop a report on stem cells and regenerative medicine including the problem of direct-to-consumer clinics [264]. The working group provided concerning information and expressed concern over the landscape of stem cell clinics in the United States. They expressed that stem cell clinics market the presence of multiple elements including affiliations with a professional society or network, partnerships with academic institutions, FDA approval or explicit mention of exemption from FDA oversight, mention of official endorsement from a local or other authority or professional accreditation, listing of patents granted, and the statement that clinical trials of investigational stem cells based on interventions are being conducted [265]. Table 1.6 shows co-opted tokens of scientific legitimacy as described by Sipp et al. [266].

The working group expressed concern that the lack of a formal mechanism for reporting outcomes of unproven stem cells interventions, both positive and negative, adds to the difficulty involved in generating data on the effectiveness of such interventions, as does the fact that there is neither a requirement, nor a mechanism, for reporting adverse events related to interventions administered outside of the clinical trials and investigations. The working group emphasized on existing regulations and made 10 recommendations in their draft report as shown in Table 1.7.

These recommendations as shown in Table 1.7 are as follows [264]:

1. State Medical Boards raise awareness among licensee of applicable federal and state legislation and guidelines regarding regenerative and stem cell therapies, including "right to try" legislation existing or pending at the state and federal levels. State medical boards should also keep their licensees and the public apprised of new develop-

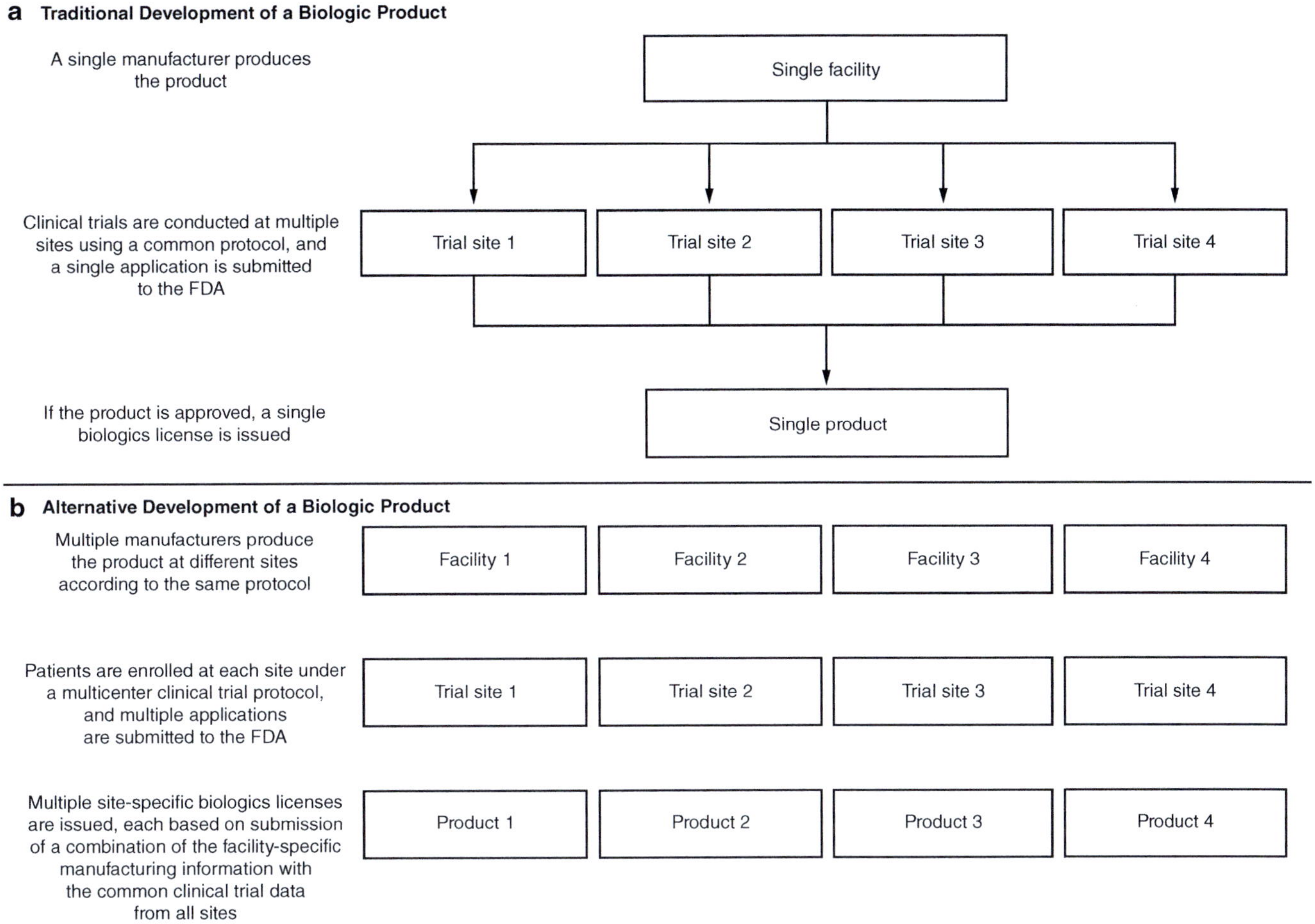


**Fig. 1.11** Traditional versus alternative development of a biologic product. In the traditional development pipeline (**a**), a single manufacturer produces the product at a single manufacturing facility and sponsors the clinical trials, which are conducted at multiple clinical sites. The manufacturer ensures that the product is manufactured consistently with appropriate quality control for use at each site and that it is administered pursuant to the protocol. The manufacturer then collects and analyzes the data from the clinical trials and submits a biologics licensing application to the Food and Drug Administration (FDA). If the product is approved, the manufacturer then receives a biologics license to produce and distribute the product. As an alternative to this process (**b**), multiple manufacturers, which may be individual physicians or groups of physicians, enter into a cooperative development agreement. These manufacturers then produce the product at different sites according the same protocol, which includes appropriate quality-control procedures to help ensure consistency between different lots produced at different sites. Patients are enrolled at each of the sites that are manufacturing the product in a multicenter clinical trial protocol. Once the data from the multicenter trial are analyzed to evaluate the safety and efficacy of the product, the individual physicians or groups of physicians submit a biologics licensing application that includes the manufacturing protocol used, the clinical data obtained at the individual site, and the results of the multicenter clinical trial showing safety and efficacy. This ultimately results in the issuance of a site-specific biologics license for the product made by each physician or group of physicians. Source: Marks P, Gottlieb S. Balancing safety and innovation for cell-based regenerative medicine. N Engl J Med 2018 378:954–959 [15]

**Table 1.6** Co-oped tokens of scientific legitimacy

| | |
|---|---|
| Accreditations and awards | Asserting certification of products or practices by international standards organizations or claiming training certification |
| Boards and advisers | Convening scientific or medical advisory boards featuring prominent business leaders and academic faculty members |
| Clinical study registration | Registering trials whose apparent purpose is solely to attract patients willing to pay to participate in them |
| Ethics review | Using the imprimatur of "ethics review" to convey a sense of legitimacy to their products or procedures |
| Location | Renting of laboratory or business space within a legitimate scientific or government institution |
| Membership | Joining established academic or professional societies to suggest legitimacy by association |

(continued)

**Table 1.6** (continued)

| | |
|---|---|
| Outcome registries | Publication of open ended voluntary monitoring data sets rather than undertaking controlled clinical trials |
| Patenting | Suggesting that patent applications or grants indicate clinical utility rather than initiation of an application process or recognition of novelty and inventiveness |
| Publication | Publishing research and commentary in journals with limited anonymous peer review |
| Rationales | Citing preclinical and other research findings to justify clinical application without sufficient efficacy testing in humans |
| Self regulation | Forming organizations to self regulate in ways that support premature commercialization |
| Technical language | Using scientific sounding words that imply academic rigor<br>Testimonials and endorsements |
| Testimonials and endorsements | Providing expert opinions or celebrity comments on unsupported clinical uses or standing of the provider |

Source: Sipp D, Caulfield T, Kaye J, et al. Marketing of unproven stem cell based interventions: A call to action. Sci Transl Med 2017; 9(397) [266]

**Table 1.7** Draft FSMB stem cell recommendations

| |
|---|
| 1. Med Boards should raise awareness amongst licensees |
| 2. Examine rules on consent, potentially update |
| 3. Review clinic claims and marketing materials |
| 4. Follow-up on FDA warnings |
| 5. Physicians should know patients well, stay in area of training |
| 6. Use caution in areas w/o evidence, exhaust standards of care 1st |
| 7. Avoid deceptive claims, raising expectations |
| 8. Don't use gag orders, disclaimers to avoid liability |
| 9. Support claims w/evidence including peer reviewed papers |
| 10. Avoid excessive fees |

Source: Draft report of the FSMB Workgroup to study regenerative and stem cell therapy practices, December 2017 [264]

ments and regulations in the field of regenerative and stem cell therapies including educational sources, guidance documents, and appropriate industry and stakeholder information on a state medical board's website. State medical boards should further provide information as to reporting procedures of adverse actions related to stem cell interventions.

2. State medical boards should examine their policies and rules addressing informed consent and consider expanding these to include a shared decision-making framework that includes the following general elements at a minimum:
   (a) An explanation, discussion, and comparison of treatment options with the patient
   (b) An assessment of the patient's values and preferences
   (c) Arrival at a decision in partnership with the patient
   (d) An evaluation of the patient's decision in partnership with the patient
3. State medical boards should review professional marketing materials and claims, including any office/clinic and/or doctor websites, and information publicly available about an office/clinic or licensee on online blogs or social media, as information sources in the investigation of complaints made against physicians.
4. State medical boards should proactively monitor warning letters sent to licensees that are made publicly available on the FDA website to ascertain information, and consider opening an investigation, about licensees who may be engaged in other unscrupulous or unprofessional practices related to the provision of regenerative and stem cell therapy. State medical boards should investigate such practices, when appropriate, in conjunction with applicable state laws, policies, and procedures [264].
5. Physicians must only offer treatments to patients for which they have a bona fide physician patient relationship. Physicians must have received adequate and appropriate training and be able to perform any proposed interventions safely and competently [264].
6. Where evidence is unavailable for a particular treatment in the form of clinical trials or case studies, physicians must only proceed with an appropriate rationale for the proposed treatment, and justification of its use, in relation to the patient's symptoms or condition. Novel, experimental, and unproven interventions should only be proposed when traditional or accepted proven treatment modalities have been exhausted. In such instances, there must still be a basis in theory or peer reviewed acknowledged practice [264].
7. Physicians should employ a "shared decision making" process when discussing treatment options with patients. Physicians must avoid any claims that may be deceptive or are intentionally or knowingly false or misleading, especially in terms of making promises about uncertain or unrealistic outcomes.
8. Physicians should not use gag orders (rulings that a case must not be discussed publicly) or disclaimers to circumvent liability.
9. Physicians should be prepared to support any claims made about benefits of treatments or devices with documented evidence, for example with studies published in peer reviewed publications.
10. Physicians should refrain from charging excessive fees for treatments provided. Further, physicians should not

recommend, provide, or charge for unnecessary medical services, nor should they make intentional misrepresentations to increase the level of payment they receive [265–267].

## Current Applications and Future Directions

Multiple chapters in this book describe clinical applications of PRP, bone marrow aspirate, as well as adipose-derived SVF. Overall, there is emerging evidence to support clinical applications of regenerative medicine with PRP and bone marrow concentrate. There is strong evidence in favor of PRP in knee osteoarthritis with moderate evidence for stem cells in managing osteoarthritis of the knee with limited evidence for all other conditions with both PRP and stem cells. The path of regenerative medicine seems to be moving faster outside the US, even those countries with appropriate regulations [10] compared to the United States [9, 14–26]. Corbett et al. [10] described innovative regenerative medicine in the European Union hoping for a better future in evidence. They described that several factors have contributed to the scarcity of cutting-edge regenerative medicines in clinical practice. Some of the factors included the great expense and difficulties involved in planning how individual therapies will be developed, manufactured to commercial levels, and ultimately successfully delivered to patients. They also described specific challenges without a license from the European Medicine Agency, under "hospital exemption" from the European Commission (EC) legislation. For products which are licensed, alternate financing approaches by health care providers may be needed, since many therapies will have significant up-front costs, but certain benefits and harms in the long-term. However, authors described that increasing political interest and more flexible mechanisms for licensing and financing of therapies are now evidence. These could be key to the future growth and development of regenerative medicine in clinical practice in the European Union.

In the United States, Mao and Mooney [9] report proposed directions for current and future regenerative medicine therapies. In the United States, even though significant stem cell therapies are provided, none of the therapies are approved by payers, consequently, it creates a substantial burden on patients. In addition, in the United States, FDA regulations are also difficult to meet, specifically with adipose cell derivative injections.

In 2015, the AAOS held a symposium to identify knowledge gaps related to biologic treatment of orthopedic injury and to establish at least some consensus on how future research might be best directed. They developed a consensus statement on working with regulatory bodies and industry in partnership with academia to modernize regulations of biologics. They also described that an internal FDA review showed critical challenges facing the FDA and its mission including funding, staffing, and necessary scientific expertise. The AAOS developed consensus statements for therapy with PRP and stem cells. They recommended dialog with international partners and collaboration among investigators, industry, and regulators to advocate for the field while ensuring the safety of the patient. They also described that it is essential to develop accepted nomenclature, validated assays of efficacy, minimum standards for reporting of all studies, and finally, methods for establishing proof of safety and efficacy. Thus, it is essential to provide nomenclature, classification, and standardization of preparations.

Recently, the FDA has published multiple articles [13–15] clarifying stem cell therapy's benefit and risks, rejuvenating regenerative medicine regulation, and finally balancing safety and innovation for cell-based regenerative medicine. The FDA recognizes that regenerative medicine is a field that involves replacing, engineering, or regenerative human cells, tissues, or organs to establish, restore, or enhance normal function. Further, the FDA has accepted that regenerative medicine is an area with great promise that goes directly to the role of the FDA in helping to facilitate the availability of safe and effective treatments. Consequently, the FDA has described the expanding use of stem cell-based products, regulatory context for regenerative medicine, and comprehensive framework for regenerative medicine. These developments are exciting in that the FDA also has provided a pathway for development of a biologic product as shown in Fig. 1.11. This will essentially facilitate investigators who manufacture the product to agree on and follow a common manufacturing protocol and develop a common clinical trial protocol. Each site will then produce the product to treat the patients who are enrolled in the clinical trial at its own site. Subsequently, the pooled safety and efficacy data from the various sites that are participating in the trial will be submitted as part of biologics license application for each. Further, if the clinical data that are submitted in conjunction with the manufacturing information show a favorable benefit-risk profile, the FDA could rely on that pooled data in determining whether the product is safe and effective. The FDA also has facilitated therapeutic advances based on 21st Century CURES Act with changing requirements for preliminary clinical evidence for regenerative medicine advanced therapy. These actions by the FDA in the United States are exciting for the future of regenerative medicine with application of interventional techniques, with application of real-world evidence [257, 258, 268–273].

In spite of regenerative medicine being one of the most discussed and controversial topics in medicine and biotech-

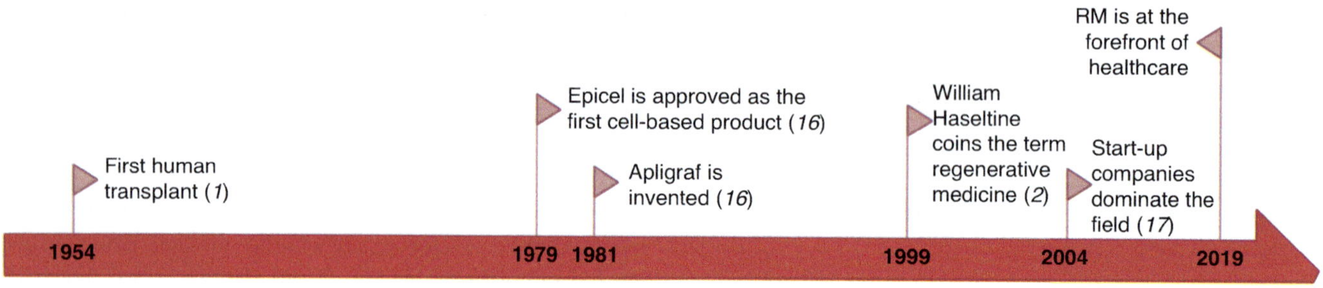


**Fig. 1.12** A summary timeline of the recent history of regenerative medicine (RM). Selected milestones in the development of RM are presented starting from the 1950s all the way up to the present day. Source: Jacques E, Suuronen EJ. The progression of regenerative medicine and its impact on therapy translation. *Clin Transl Sci* 2020; 13:440–450 [268]

nology for the past 3 decades, it is acknowledged that the fields performance with clinical applications has been disappointing. While this may be linked to the novelty of these technologies and their disruptive nature, which brought an increasing level of complexity, multiple issues related to the research, clinical applications and business practices have caused significant damage to the field. Jacques and Suuronen [268] in a publication described a summary timeline of the recent history of regenerative medicine starting from the first human transplant in 1954 to a large number of startup companies dominating the field and, finally, regenerative medicine taking shape at the forefront of healthcare (Fig. 1.12). Thus, not only research and literature continue to grow, ethical and appropriate applications of regenerative medicine in clinical settings is also growing. As of now, the literature is significantly limited except in cases of treatment of the lateral epicondylitis and knee with PRP or stem cells and to some extent discogenic pain and with very limited evidence for all other musculoskeletal and spinal conditions producing chronic pain. In conclusion, stem cell therapy holds great promise because of its potential of modulating cellular biology, structure, and function and therefore, addressing a variety of medical conditions [268, 274]. In order for us to realize this potential to its full extent, we would need to advance this field carefully and systematically in an ethically and clinically appropriate fashion while avoiding misuse, overuse, and abuse.

**Acknowledgments** This book chapter is modified and updated from a previous book chapter, "Evolution of regenerative medicine in managing chronic pain" by Laxmaiah Manchikanti, MD, and Annu Navani, MD, in *Essentials of Regenerative Medicine in Interventional Pain Management* published by ASIPP Publishing.

## Key Points

1. Regenerative medicine is a multidisciplinary field focused on repairing, restoring, and regenerating diseased or injured cells, tissues, or organs to restore homeostasis. It encompasses a wide range of scientific disciplines and includes advanced research that translates into clinical applications.
2. Tissue engineering, which involves the use of cells, engineering techniques, materials, and biochemical and physical factors, plays a crucial role in regenerative medicine.
3. Regenerative medicine is often used interchangeably with tissue engineering. Tissue engineering aims to regenerate damaged tissues by developing biological substitutes that restore, maintain, or improve tissue function. It involves multidisciplinary expertise and relies on the use of scaffolds as templates for tissue formation.
4. Once the potential of regenerative medicine was discovered, its adaptation and growth became inevitable. Traditional surgical solutions for musculoskeletal and spine conditions often provided short-term relief. Regenerative medicine offered a minimally invasive alternative that had the potential for long-lasting results, which appealed to patients and practitioners alike.
5. Stem cells can be classified into two main types based on their origin—embryonic stem cells and somatic or adult stem cells. Embryonic stem cells are found in the inner cell mass of the blastocyst, while somatic or adult stem cells are found in various tissues of newborns or adults such as bone marrow, peripheral blood, umbilical cord, etc.
6. In the bone marrow, there are two populations of progenitor stem cells—hematopoietic progenitor cells and mesenchymal progenitor cells. Mesenchymal progenitor cells (MSCs) have the ability to differentiate into bone, cartilage, and other soft tissues associated with the musculoskeletal system, making them of particular interest in spine and musculoskeletal medicine.
7. Pluripotent stem cells are primitive cells that have the capacity to differentiate into virtually any cell type in the body. They include embryonic stem cells sourced from blastocysts and induced pluripotent stem cells created through genetic reprogramming of body cells. Pluripotent stem cells have potential applications in cell and organ replacement, disease modeling, transplantation, and drug development.
8. The nomenclature and standardization of stem cell research have faced challenges due to variations in isola-

tion, culture, and assay methods. Efforts have been made to define minimum criteria for MSCs, including plastic adherence, specific antigen expression, and in vitro differentiation capabilities.

9. Processing refers to any activity performed on a human cell, tissue, or cellular and tissue-based product (HCT/P) that alters its original characteristics. It includes various steps such as testing, sterilization, preservation, and enzymatic digestion, among others.
10. Adipose tissue, which consists of adipocytes and surrounding connective tissues, is considered a structural tissue. It provides cushioning and support to other tissues, stores energy, and insulates the body.
11. Homologous use refers to the repair, reconstruction, replacement, or supplementation of a recipient's cells or tissues with an HCT/P that performs the same basic function in the recipient as it did in the donor.

## References


1. Manchikanti L, Navani A. Evolution of regenerative medicine in managing chronic pain. In: Manchikanti L, Navani A, Atluri S, editors. Essentials of regenerative medicine in interventional pain management. Paducah: ASIPP Publishing; 2019. p. 1–38.
2. Kokubu C, Takeda J. When half is better than the whole: advances in haploid embryonic stem cell technology. Cell Stem Cell. 2014;14:265–7.
3. Lytal C. New health facility aims to translate stem cell science into therapies. USC News, May 9, 2014. http://news.usc.edu/62466/new-health-facility-aims-to-translate-stem-cell-science-into-treatment-therapies/.
4. Mason C, Dunnill P. A brief definition of regenerative medicine. Regen Med. 2008;3:1–5.
5. Navani A, Li G, Chrystal J. Platelet rich plasma in musculoskeletal pathology: a necessary rescue or a lost cause? Pain Physician. 2017;20:E345–56.
6. Chakravarthy K, Chen Y, He C, et al. Stem cell therapy for chronic pain management: review of uses, advances, and adverse effects. Pain Physician. 2017;20:293–305.
7. Hsu WK, Mishra A, Rodeo SR, et al. Platelet-rich plasma in orthopaedic applications: evidence-based recommendations for treatment. J Am Acad Orthop Surg. 2013;21:739–48.
8. Grambart ST. Sports medicine and platelet-rich plasma: nonsurgical therapy. Clin Podiatr Med Surg. 2015;32:99–107.
9. Mao AS, Mooney DJ. Regenerative medicine: current therapies and future directions. Proc Natl Acad Sci U S A. 2015; 112:14452–9.
10. Corbett MS, Webster A, Hawkins R, et al. Innovative regenerative medicines in the EU: a better future in evidence? BMC Med. 2017;15:49.
11. Moraes VY, Lenza M, Tamaoki MJ, et al. Platelet-rich therapies for musculoskeletal soft tissue injuries. Cochrane Database Syst Rev. 2014;4:CD010071.
12. Navani A, Manchikanti L, Albers SL, et al. Responsible, safe, and effective use of biologics in the management of low back pain: American Society of Interventional Pain Physicians (ASIPP) guidelines. Pain Physician. 2019;22:S1–S74.
13. Charo A, Sipp D. Rejuvenating regenerative medicine regulation. N Engl J Med. 2018;378:504–5.
14. Marks PW, Witten CM, Califf RM. Clarifying stem-cell therapy's benefits and risks. N Engl J Med. 2017;376:1007–9.
15. Marks P, Gottlieb S. Balancing safety and innovation for cell-based regenerative medicine. N Engl J Med. 2018;378:954–9.
16. Turner L. US stem cell clinics, patient safety, and the FDA. Trends Mol Med. 2015;21:271–3.
17. Trounson A. Potential pitfall of pluripotent stem cells. N Engl J Med. 2017;377:490–1.
18. Taylor-Weiner H, Graff ZJ. Medicine's Wild West–unlicensed stem-cell clinics in the United States. N Engl J Med. 2015;373:985–7.
19. Halme DG, Kessler DA. FDA regulation of stem-cell-based therapies. N Engl J Med. 2006;355:1730–5.
20. Knoepfler PS. From bench to FDA to bedside: US regulatory trends for new stem cell therapies. Adv Drug Deliv Rev. 2015;82-83:192–6.
21. Okie S. Stem-cell politics. N Engl J Med. 2006;355:1633–7.
22. Daley GQ. Polar extremes in the clinical use of stem cells. N Engl J Med. 2017;376:1075–7.
23. FDA News Release. FDA acts to remove unproven, potentially harmful treatment used in 'stem cell' centers targeting vulnerable patients. Vaccinia Virus Vaccine (Live) seized after being used inappropriately in vulnerable cancer patients. 2017. www.fda.gov/NewsEvents/Newsroom/PressAnnouncements/ucm573427.htm. Accessed 11 January 2023.
24. Regulatory Considerations for Human Cell, Tissues, and Cellular and Tissue-Based Products: Minimal Manipulation and Homologous Use. Guidance for Industry and Food and Drug Administration Staff, 2017. https://www.fda.gov/downloads/BiologicsBloodVaccines/GuidanceComplianceRegulatoryInformation/Guidances/CellularandGeneTherapy/UCM585403.pdf. Accessed 11 January 2023.
25. U.S. Department of Health and Human Services. Same Surgical Procedure Exception under 21 CFR 1271.15(b): Questions and Answers Regarding the Scope of the Exception. Guidance for Industry. 2017. www.fda.gov/downloads/BiologicsBloodVaccines/GuidanceComplianceRegulatoryInformation/Guidances/Tissue/UCM419926.pdf. Accessed 11 January 2023.
26. Rubin R. Study: business booms for unlicensed and unproven stem cell treatments. JAMA. 2022;327:414–5.
27. Dinsmore CH. Urodele limb and tail regeneration in early biological thought: an essay on scientific controversy and social change. Int J Dev Biol. 1996;40:621–7.
28. Kaiser LR. The future of multihospital systems. Top Health Care Financ. 1992;18:32–45.
29. Lysaght MJ, Crager J. Origins. Tissue Eng Part A. 2009;15:1449–50.
30. Haseltine WA. The emergence of regenerative medicine: a new field and a new society. J Reg Med. 2004;2:17–23.
31. Viola J, Lal B, Grad O. The emergence of tissue engineering as a research field. Arlington: National Science Foundation; 2003. www.nsf.gov/pubs/2004/nsf0450/. Accessed 11 January 2023.
32. Bailey R. Liberation biology: the scientific and moral case for the biotech revolution. New York: Prometheus Books; 2005.
33. Alexander B. Don't die, stay pretty: the exploding science of superlongevity. Wired. 2000. www.wired.com/2000/01/forever/. Accessed 11 January 2023.
34. U.S. Food and Drug Administration. Regenerative medicine advanced therapy designation. 2017. www.fda.gov. Accessed 11 January 2023.
35. Konstantinov IE. In search of Alexander A. Maximow: the man behind the unitarian theory of hematopoiesis. Perspect Biol Med. 2000;43:269–76.
36. Kumar R, Sharma A, Pattnaik AK, et al. Stem cells: an overview with respect to cardiovascular and renal disease. J Nat Sci Biol Med. 2010;1:43–52.
37. Friedenstein AJ, Deriglasova UF, Kulagina NN, et al. Precursors for fibroblasts in different populations of hematopoietic cells

as detected by the in vitro colony assay method. Exp Hematol. 1973;2:83–92.
38. Becker AJ, McCulloch EA, Till JE. Cytological demonstration of the clonal nature of spleen colonies derived from transplanted mouse marrow cells. Nature. 1963;197:452–4.
39. Evans MJ, Kaufman MH. Establishment in culture of pluripotential cells from mouse embryos. Nature. 1981;292:154–6.
40. Wilmut I. A tribute to Keith Campbell: the birth of the first clone of an adult vertebrate, "Dolly" the sheep. Cell Reprogram. 2013;15:339–43.
41. Bloomberg Longevity Economy Conference 2013 Panelist Bio. Archived from the original on 2013-08-03.
42. Thomson JA, Itskovitz-Eldor J, Shapiro SS, et al. Embryonic stem cell lines derived from human blastocysts. Science. 1998;282:1145–7.
43. Takahashi K, Yamanaka S. Induction of pluripotent stem cells from mouse embryonic and adult fibroblast cultures by defined factors. Cell. 2006;126:663–76.
44. The Nobel Prize. http://www.Nobelprize.org. Accessed 12 January 2023.
45. The Nobel Prize in Physiology or Medicine 2012. Press Release: The Nobel Prize in Physiology or Medicine 2012 to Sir John B. Gurdon, Shinya Yamanaka. 2012. https://www.nobelprize.org/nobel_prizes/medicine/laureates/2012/press.html. Accessed 12 January 2023.
46. De-Coppi P, Bartsch G, Siddiqui MM, et al. Isolation of amniotic stem cell lines with potential for therapy. Nat Biotechnol. 2007;25:100–6.
47. Mitalipov SM, Zhou Q, Byrne JA, et al. Reprogramming following somatic cell nuclear transfer in primates is dependent upon nuclear remodeling. Hum Reprod. 2007;22:2232–42.
48. Cyranoski D. Next-generation stem cells transplanted in human for the first time. Nature. 2014. https://www.scientificamerican.com/article/next-generation-stem-cells-transplanted-in-human-for-the-first-time/. Accessed 12 January 2023.
49. O'Brien FJ. Biomaterials and scaffolds for tissue engineering. Mater Today. 2011;14:88–95.
50. Martin I, Wendt D, Heberer M. The role of bioreactors in tissue engineering. Trends Biotechnol. 2004;22:80–6.
51. The National Uniform Claims Committee. Specialty Designation for Interventional Pain Management-09. www.cms.hhs.gov/transmittals/Downloads/r1779b3.pdf. Accessed 11 September 2023.
52. Medicare Payment Advisory Commission. Report to the congress: paying for interventional pain services in ambulatory settings. Washington: MedPAC; 2001. https://permanent.fdlp.gov/lps21261/dec2001PainManagement.pdf. Accessed 27 July 2021.
53. Manchikanti L, Pampati V, Sanapati MR, et al. COVID-19 pandemic reduced utilization of interventional techniques 18.7% in managing chronic pain in the Medicare population in 2020: analysis of utilization data from 2000 to 2020. Pain Physician. 2022(25):223–38.
54. Manchikanti L, Pampati V, Knezevic NN, et al. The influence of COVID-19 on utilization of epidural procedures in managing chronic spinal pain in the Medicare population. Spine. 2023;48:950–61.
55. Manchikanti L, Kaye AD, Latchaw RE, et al. Impact of COVID-19 pandemic on utilization patterns of facet joint interventions in managing spinal pain in Medicare population. Pain Ther. 2023;12:505–27.
56. Manchikanti L, Senapathi SHV, Milburn JM, et al. Utilization and expenditures of vertebral augmentation continue to decline: an analysis in fee-for-service (FFS) Recipients from 2009 to 2018. Pain Physician. 2021;24:401–15.
57. Manchikanti L, Kosanovic R, Pampati V, Sanapati MR, Hirsch JA. Outcomes of cervical therapeutic medial branch blocks and radiofrequency neurotomy: clinical outcomes and cost utility are equivalent. Pain Physician. 2022;25:35–47.
58. Manchikanti L, Pampati V, Vangala BP, et al. Spinal cord stimulation trends of utilization and expenditures in fee-for-service (FFS) Medicare population from 2009 to 2018. Pain Physician. 2021;24:293–308.
59. Manchikanti L, Kosanovic R, Pampati V, Kaye AD. Declining utilization patterns of percutaneous adhesiolysis procedures in the fee-for-service (FFS) Medicare population. Pain Physician. 2021;24:17–29.
60. Manchikanti L, Pampati V, Sanapati SP, Sanapati MR, Kaye AD, Hirsch JA. Evaluation of cost-utility of thoracic interlaminar epidural injections. Curr Pain Headache Rep. 2020;24:5.
61. Manchikanti L, Sanapati MR, Pampati V, Boswell MV, Kaye AD, Hirsch JA. Update on reversal and decline of growth of utilization of interventional techniques in managing chronic pain in the Medicare population from 2000 to 2018. Pain Physician. 2019;22:521–36.
62. Manchikanti L, Pampati V, Kaye AD, Hirsch JA. Therapeutic lumbar facet joint nerve blocks in the treatment of chronic low back pain: cost utility analysis based on a randomized controlled trial. Korean J Pain. 2018;31:27–38.
63. Department of Health and Human Services, Centers for Medicare & Medicaid Services. 42 CFR Parts 405, 410, 411, 414, 415, 423, 424, 425, and 455 [CMS–1770–F, CMS–1751–F2, CMS–1744– F2, CMS–5531–IFC]. Medicare and Medicaid Programs; CY 2023 Payment Policies Under the Physician Fee Schedule and Other Changes to Part B Payment and Coverage Policies; Medicare Shared Savings Program Requirements; Implementing Requirements for Manufacturers of Certain Single-dose Container or Single-use Package Drugs to Provide Refunds with Respect to Discarded Amounts; and COVID–19. Final rule and interim final rules. November 18, 2022.
64. Manchikanti L, Knezevic NN, Navani A, et al. Epidural interventions in the management of chronic spinal pain: American Society of Interventional Pain Physicians (ASIPP) comprehensive evidence-based guidelines. Pain Physician. 2021;24:S27–S208.
65. Manchikanti L, Kaye AD, Soin A, et al. Comprehensive evidence-based guidelines for facet joint interventions in the management of chronic spinal pain: American Society of Interventional Pain Physicians (ASIPP) guidelines. Pain Physician. 2020;23:S1–S127.
66. Manchikanti L, Centeno CJ, Atluri S, et al. Bone marrow concentrate (BMC) therapy in musculoskeletal disorders: evidence-based policy position statement of American Society of Interventional Pain Physicians (ASIPP). Pain Physician. 2020;23:E85–E131.
67. Deinsberger J, Reisinger D, Weber B. Global trends in clinical trials involving pluripotent stem cells: a systematic multi-database analysis. NPJ Regen Med. 2020;5:15.
68. MacGuire P, Tucker J. Evaluation of the regenerative medicine & nanomedicine initiative. Canadian Institutes of Health Research, 2013. http://www.cihr-irsc.gc.ca/e/47302.html.
69. van Servellen A, Oba I. Stem cell research: trends in and perspectives on the evolving international landscape. Res Trends. 2014;36(1):11. https://www.researchtrends.com/researchtrends/vol1/iss36/11. Accessed 12 January 2023.
70. Hildreth C. Two-thirds of all mesenchymal stem cell (MSC) publications were released in the past 5 years. BioInformat. 2016. https://bioinformant.com/two-thirds-of-all-mesenchymal-stem-cell-msc-publications-were-released-in-the-past-5-years/. Accessed 14 February 2023.
71. Bergel S. Uber wirkungen des fibrins. Deut Med Wochensch. 1909;35:663.
72. Bodor M, Dregalla R, Uribe Y. Platelet rich plasma. In: Manchikanti L, Navani A, Atluri S, editors. Essentials of regenerative medicine in interventional pain management. Paducah: ASIPP Publishing; 2018. p. 67–82.

73. Marx RE. Platelet-rich plasma: evidence to support its use. J Oral Maxillofac Surg. 2004;62:489–96.
74. Tayapongsak P, O'Brien DA, Monteiro CB, et al. Autologous fibrin adhesive in mandibular reconstruction with particulate cancellous bone and marrow. J Oral Maxillofac Surg. 1994;52:161–5; discussion 166.
75. Lana JF, Weglein A, Vicente E, et al. Platelet rich plasma and its growth factors: the state of the art. In: Lana JF, Santana MHA, Belangero WD, et al., editors. Platelet-rich plasma: regenerative medicine: sports medicine, orthopedic, and recovery of musculoskeletal injuries. New York: Springer; 2014. p. 1–60.
76. Hall MP, Band PA, Meislin RJ, et al. Platelet-rich plasma: current concepts and application in sports medicine. J Am Acad Orthop Surg. 2009;17:602–8.
77. LaPrade RF, Dragoo JL, Koh JL, et al. AAOS research symposium updates and consensus: biologic treatment of orthopaedic injuries. J Am Acad Orthop Surg. 2016;24:e62–78.
78. Dohan Ehrenfest DM, Rasmusson L, Albrektsson T. Classification of platelet concentrates: from pure platelet-rich plasma (P-PRP) to leucocyte- and platelet-rich fibrin (L-PRF). Trends Biotechnol. 2009;27:158–67.
79. DeLong JM, Russell RP, Mazzocca AD. Platelet-rich plasma: the PAW classification system. Arthroscopy. 2012;28:998–1009.
80. Boswell SG, Cole BJ, Sundman EA, et al. Platelet-rich plasma: a milieu of bioactive factors. Arthroscopy. 2012;28:429–39.
81. Blair P, Flaumenhaft R. Platelet alpha-granules: basic biology and clinical correlates. Blood Rev. 2009;23:177–89.
82. Lam WA, Chaudhuri O, Crow A, et al. Mechanics and contraction dynamics of single platelets and implications for clot stiffening. Nat Mater. 2011;10:61–6.
83. Bodor M, Toy A, Aufiero D. Disc regeneration with platelets and growth factors. In: Lana JF, et al., editors. Platelet-rich plasma. regenerative medicine: sports medicine, orthopedic, and recovery of musculoskeletal injuries. Berlin: Springer; 2014. p. 265–79.
84. Woodell-May JE, Ridderman DN, Swift MJ, et al. Producing accurate platelet counts for platelet rich plasma: validation of a hematology analyzer and preparation techniques for counting. J Craniofac Surg. 2005;16:749–59.
85. Brass LF. Thrombin and platelet activation. Chest. 2003;124:18S–25S.
86. Amable PR, Carias RB, Teixeira MV, et al. Platelet-rich plasma preparation for regenerative medicine: optimization and quantification of cytokines and growth factors. Stem Cell Res Ther. 2013;4:67.
87. Puri RN, Colman RW. ADP-induced platelet activation. Crit Rev Biochem Mol Biol. 1997;32:437–502.
88. Roberts DE, McNicol A, Bose R. Mechanism of collagen activation in human platelets. J Biol Chem. 2004;279:19421–30.
89. Flaumenhaft R. Molecular basis of platelet granule secretion. Arterioscler Thromb Vasc Biol. 2003;23:1152–60.
90. Dimauro I, Grasso L, Fittipaldi S, et al. Platelet-rich plasma and skeletal muscle healing: a molecular analysis of the early phases of the regeneration process in an experimental animal model. PLoS One. 2014;9:e102993.
91. Mishra A, Woodall J Jr, Vieira A. Treatment of tendon and muscle using platelet-rich plasma. Clin Sports Med. 2009;28:113–25.
92. El-Sharkawy H, Kantarci A, Deady J, et al. Platelet-rich plasma: Growth factors and pro- and anti-inflammatory properties. J Periodontol. 2007;78:661–9.
93. Pierce GF, Mustoe TA, Deuel TF. Transforming growth factor beta induces increased directed cellular migration and tissue repair in rats. Prog Clin Biol Res. 1988;266:93–102.
94. Morrell CN, Aggrey AA, Chapman LM, et al. Emerging roles for platelets as immune and inflammatory cells. Blood. 2014;123:2759–67.
95. Manne BK, Xiang SC, Rondina MT. Platelet secretion in inflammatory and infectious diseases. Platelets. 2017;28:155–64.
96. Sokolove J, Lepus CM. Role of inflammation in the pathogenesis of osteoarthritis: latest findings and interpretations. Ther Adv Musculoskelet Dis. 2013;5:77–94.
97. Santos-Martinez MJ, Medina C, Jurasz P, et al. Role of metalloproteinases in platelet function. Thromb Res. 2008;121:535–42.
98. Gill SE, Parks WC. Metalloproteinases and their inhibitors: regulators of wound healing. Int J Biochem Cell Biol. 2008;40:1334–47.
99. Cole BJ, Seroyer ST, Filardo G, et al. Platelet-rich plasma: where are we now and where are we going? Sports Health. 2010;2:203–10.
100. Caplan AI. Adult mesenchymal stem cells for tissue engineering versus regenerative medicine. J Cell Physiol. 2007;213:341–7.
101. Caplan AI. Cell delivery and tissue regeneration. J Control Release. 1989;11:157–65.
102. Caplan AI. Mesenchymal stem cells. J Orthop Res. 1991;9: 641–50.
103. Caplan AI. Review: mesenchymal stem cells: cell-based reconstructive therapy in orthopedics. Tissue Eng. 2005;11:1198–211.
104. Caplan AI. Mesenchymal stem cells: time to change the name! Stem Cells Transl Med. 2017;6:1445–51.
105. Sell S. Stem cells handbook. New York: Springer; 2013.
106. Kobolak J, Dinnyes A, Memic A, et al. Mesenchymal stem cells: identification, phenotypic characterization, biological properties and potential for regenerative medicine through biomaterial micro-engineering of their niche. Methods. 2016;99:62–8.
107. Murray IR, West CC, Hardy WR, et al. Natural history of mesenchymal stem cells, from vessel walls to culture vessels. Cell Mol Life Sci. 2014;71:1353–74.
108. Jiang Y, Jahagirdar BN, Reinhardt RL, et al. Pluripotency of mesenchymal stem cells derived from adult marrow. Nature. 2002;418:41–9.
109. D'Ippolito G, Diabira S, Howard GA, et al. Marrow isolated adult multilineage inducible (MIAMI) cells, a unique population of postnatal young and old human cells with extensive expansion and differentiation potential. J Cell Sci. 2004;117:2971–81.
110. Beltrami AP, Cesselli D, Bergamin N, et al. Multipotent cells can be generated in vitro from several adult human organs (heart, liver, and bone marrow). Blood. 2007;110:3438–46.
111. Katz AJ, Tholpady A, Tholpady SS, et al. Cell surface and transcriptional characterization of human adipose-derived adherent stromal (hADAS) cells. Stem Cells. 2005;23:412–23.
112. Mitchell JB, McIntosh K, Zvonic S, et al. Immunophenotype of human adipose derived cells: temporal changes in stromal-associated and stem cell associated markers. Stem Cells. 2006;24:376–85.
113. Kilroy GE, Foster SJ, Wu X, et al. Cytokine profile of human adipose derived stem cells: expression of angiogenic, hematopoietic, and proinflammatory factors. J Cell Physiol. 2007;212:702–9.
114. Day TF, Guo X, Garrett-Beal L, et al. Wnt/beta-catenin signaling in mesenchymal progenitors controls osteoblast and chondrocyte differentiation during vertebrate skeletogenesis. Dev Cell. 2005;8:739–50.
115. Cawthorn WP, Bree AJ, Yao Y, et al. Wnt6, Wnt10a and Wnt10b inhibit adipogenesis and stimulate osteoblastogenesis through a b-catenindependent mechanism. Bone. 2012;50:477–89.
116. Li Y, Foster W, Deasy BM, et al. Transforming growth factor-beta1 induces the differentiation of myogenic cells into fibrotic cells in injured skeletal muscle: a key event in muscle fibrogenesis. Am J Pathol. 2004;164:1007–19.
117. Serra R, Johnson M, Filvaroff EH, et al. Expression of a truncated, kinase defective TGF-beta type II receptor in mouse skeletal tissue promotes terminal chondrocyte differentiation and osteoarthritis. J Cell Biol. 1997;139:541–52.

118. Paredes B, Santana A, Arribas MI, et al. Phenotypic differences during the osteogenic differentiation of single cell derived clones isolated from human lipo aspirates. J Tissue Eng Regen Med. 2011;5:589–99.
119. Müller AM, Mehrkens A, Schäfer DJ, et al. Towards an intraoperative engineering of osteogenic and vasculogenic grafts from the stromal vascular fraction of human adipose tissue. Eur Cell Mater. 2010;19:127–35.
120. Cheung WK, Working DM, Galuppo LD, et al. Osteogenic comparison of expanded and uncultured adipose stromal cells. Cytotherapy. 2010;12:554–62.
121. Murphy MB, Moncivais K, Caplan AI. Mesenchymal stem cells: environmentally responsive therapeutics for regenerative medicine. Exp Mol Med. 2013;45:e54.
122. Kon E, Roffi A, Filardo G, et al. Scaffold-based cartilage treatments: with or without cells? A systematic review of preclinical and clinical evidence. Arthroscopy. 2015;31:767–75.
123. Manchikanti L, Boswell MV, Hirsch JA. Innovations in interventional pain management of chronic spinal pain. Expert Rev Neurother. 2016;16:1033–42.
124. Lana JF, Purita J, Rava CA, et al. Bone marrow concentrate. In: Manchikanti L, Navani A, Atluri S, editors. Essentials of regenerative medicine in interventional pain management. Paducah: ASIPP Publishing; 2018. p. 83–98.
125. Knezevic NN, Mandalia S, Raasch J, et al. Treatment of chronic low back pain - new approaches on the horizon. J Pain Res. 2017;10:1111–23.
126. Wei C, Lin AB, Hung S. Mesenchymal stem cells in regenerative medicine for musculoskeletal diseases: bench, bedside, and industry. Cell Transplant. 2014;23:505–12.
127. Liu H, Pan H, Yang H, et al. LIM mineralization protein-1 suppresses TNF-α induced intervertebral disc degeneration by maintaining nucleus pulposus extracellular matrix production and inhibiting matrix metalloproteinases expression. J Orthop Res. 2015;33:294–303.
128. Navani A, Ambach MA, Wei JJ, et al. Biologic therapies for intervertebral degenerative disc disease: a review of novel applications. J Stem Cells Res Rev Rep. 2017;4:1023–31.
129. Ochi M, Nakasa T, Kamei G, et al. Regenerative medicine in orthopedics using cells, scaffold, and microRNA. J Orthop Sci. 2014;19:521–8.
130. Milants C, Bruyère O, Kaux JF. Responders to platelet-rich plasma in osteoarthritis: a technical analysis. Biomed Res Int. 2017;2017:7538604.
131. Muchedzi TA, Roberts SB. A systematic review of the effects of platelet rich plasma on outcomes for patients with knee osteoarthritis and following total knee arthroplasty. Surgeon. 2018;16:250–8.
132. Saltzman BM, Jain A, Campbell KA, et al. Does the use of platelet-rich plasma at the time of surgery improve clinical outcomes in arthroscopic rotator cuff repair when compared with control cohorts? A systematic review of meta-analyses. Arthroscopy. 2016;32:906–18.
133. Anitua E, Andia I, Ardanza B, et al. Autologous platelets as a source of proteins for healing and tissue regeneration. Thromb Haemost. 2004;91:4–15.
134. Anitua E, Andia I, Sanchez M, et al. Autologous preparations rich in growth factors promote proliferation and induce VEGF and HGF production by human tendon cells in culture. J Orthop Res. 2005;23:281–6.
135. Meheux CJ, McCulloch PC, Lintner DM, et al. Efficacy of intra-articular platelet-rich plasma injections in knee osteoarthritis: a systematic review. Arthroscopy. 2016;32:495–505.
136. Khoshbin A, Leroux T, Wasserstein D, et al. The efficacy of platelet-rich plasma in the treatment of symptomatic knee osteoarthritis: a systematic review with quantitative synthesis. Arthroscopy. 2013;29:2037–48.
137. Zhao L, Kaye A, Abd-Elsayed A. Stem cells for the treatment of knee osteoarthritis: a comprehensive review. Pain Physician. 2018;21:229–42.
138. Laudy AB, Bakker EW, Rekers M, et al. Efficacy of platelet-rich plasma injections in osteoarthritis of the knee: a systematic review and meta-analysis. Br J Sports Med. 2015;49:657–72.
139. Shen L, Yuan T, Chen S, et al. The temporal effect of platelet-rich plasma on pain and physical function in the treatment of knee osteoarthritis: systematic review and meta-analysis of randomized controlled trials. J Orthop Surg Res. 2017;12:16.
140. Lai LP, Stitik TP, Foye PM, et al. Use of platelet-rich plasma in intra-articular knee injections for osteoarthritis: a systematic review. PM R. 2015;7:637–48.
141. Campbell KA, Saltzman BM, Mascarenhas R, et al. Does intra-articular platelet-rich plasma injection provide clinically superior outcomes compared with other therapies in the treatment of knee osteoarthritis? A systematic review of overlapping meta-analyses. Arthroscopy. 2015;31:2213–21.
142. Dai WL, Zhou AG, Zhang H, et al. Efficacy of platelet-rich plasma in the treatment of knee osteoarthritis: a meta-analysis of randomized controlled trials. Arthroscopy. 2017;33:659–70.
143. Mascarinas A, Harrison J, Boachie-Adjei K, et al. Regenerative treatments for spinal conditions. Phys Med Rehabil Clin N Am. 2016;27:1003–17.
144. Viganò M, Sansone V, d'Agostino MC, et al. Mesenchymal stem cells as therapeutic target of biophysical stimulation for the treatment of musculoskeletal disorders. J Orthop Surg Res. 2016;11:163.
145. Labusca L, Zugun-Eloae F, Mashayekhi K. Stem cells for the treatment of musculoskeletal pain. World J Stem Cells. 2015;7:96–105.
146. Cianca JC, Jayaram P. Musculoskeletal injuries and regenerative medicine in the elderly patient. Phys Med Rehabil Clin N Am. 2017;28:777–94.
147. Malahias M-A, Chytas D, Babis GC, et al. Platelet-rich plasma guided injections: clinical application in peripheral neuropathies. Front Surg. 2014;1:41.
148. Wehling P, Evans C, Wehling J, et al. Effectiveness of intra-articular therapies in osteoarthritis: a literature review. Ther Adv Musculoskelet Dis. 2017;9:183–96.
149. Akeda K, Ohishi K, Masuda K, et al. Intradiscal injection of autologous platelet-rich plasma releasate to treat discogenic low back pain: a preliminary clinical trial. Asian Spine J. 2017;11:380–9.
150. Kirchner F, Anitua E. Intradiscal and intra-articular facet infiltrations with plasma rich in growth factors reduce pain in patients with chronic low back pain. J Craniovertebr Junction Spine. 2016;7:250–6.
151. El-Tahawy FN, Rifaai RA, Saber EA, et al. Effect of platelet rich plasma (PRP) injection on the endocrine pancreas of the experimentally induced diabetes in male albino rats: a histological and immunohistochemical study. J Diabetes Metab. 2017;8:3.
152. Pintat J, Silvestre A, Magalon G, et al. Intra-articular injection of mesenchymal stem cells and platelet-rich plasma to treat patellofemoral osteoarthritis: preliminary results of a long-term pilot study. J Vasc Interv Radiol. 2017;28:1708–13.
153. Santiago-Figueroa J, Sosa IJ, Reyes O, et al. A novel technique for reducing and eliminating peripheral neuropathic pain: a clinical study. J Pain Manag. 2011;4:387–94.
154. Basso M, Cavagnaro L, Zanirato A, et al. What is the clinical evidence on regenerative medicine in intervertebral disc degeneration? Musculoskelet Surg. 2017;101:93–104.
155. Bashir J, Sherman A, Lee H, et al. Mesenchymal stem cell therapies in the treatment of musculoskeletal diseases. PM R. 2014;6:61–9.
156. Cunha S. The role of platelet rich plasma in pain management and decrease in opioid use. Nurse Anesth Capstones. 2017;18:27.

157. Kuffler DP. Platelet-rich plasma and the elimination of neuropathic pain. Mol Neurobiol. 2013;48:315–32.
158. Tahririan MA, Moezi M, Motififard M, et al. Ultrasound guided platelet-rich plasma injection for the treatment of rotator cuff tendinopathy. Adv Biomed Res. 2016;5:200.
159. Varshney A, Maheshwari R, Juyal A, et al. Autologous platelet-rich plasma versus corticosteroid in the management of elbow epicondylitis: a randomized study. Int J Appl Basic Med Res. 2017;7:125–8.
160. Andia I, Maffulli N. Platelet-rich plasma for muscle injury and tendinopathy. Sports Med Arthrosc. 2013;21:191–8.
161. Wang SZ, Jin JY, Guo YD, et al. Intervertebral disc regeneration using platelet-rich plasma-containing bone marrow-derived mesenchymal stem cells: a preliminary investigation. Mol Med Rep. 2016;13:3475–81.
162. Malahias MA, Nikolaou VS, Johnson EO, et al. Platelet-rich plasma ultrasound-guided injection in the treatment of carpal tunnel syndrome: a placebo-controlled clinical study. J Tissue Eng Regen Med. 2018;12:e1480–8.
163. Navani A, Ambach MA, Navani R, et al. Biologics and lumbar discogenic pain: 18-month follow-up for safety and efficacy. IPM Rep. 2018;2:111–8.
164. Levi D, Horn S, Tyszko S, et al. Intradiscal platelet-rich plasma injection for chronic discogenic low back pain: preliminary results from a prospective trial. Pain Med. 2016;17:1010–22.
165. Monfett M, Harrison J, Boachie-Adjei K, et al. Intradiscal platelet-rich plasma (PRP) injections for discogenic low back pain: an update. Int Orthop. 2016;40:1321–8.
166. Tuakli-Wosornu YA, Terry A, Boachie-Adjei K, et al. Lumbar intradiskal platelet-rich plasma (PRP) injections: a prospective, double-blind, randomized controlled study. PM R. 2016;8:1–10.
167. Wu J, Du Z, Lv Y, et al. A new technique for the treatment of lumbar facet joint syndrome using intra-articular injection with autologous platelet rich plasma. Pain Physician. 2016;19:617–25.
168. Wu J, Zhou J, Liu C, et al. A prospective study comparing platelet-rich plasma and local anesthetic (LA)/corticosteroid in intra-articular injection for the treatment of lumbar facet joint syndrome. Pain Pract. 2017;17:914–24.
169. Navani A, Gupta D. Role of intra-articular platelet-rich plasma in sacroiliac joint pain. Reg Anesth Pain Med. 2015;19:54–9.
170. Ko GD, Mindra S, Lawson GE, et al. Case series of ultrasound-guided platelet-rich plasma injections for sacroiliac joint dysfunction. J Back Musculoskelet Rehabil. 2017;30:363–70.
171. Singla V, Batra YK, Bharti N, et al. Steroid vs. platelet-rich plasma in ultrasound-guided sacroiliac joint injection for chronic low back pain. Pain Pract. 2017;17:782–91.
172. Obata S, Akeda K, Imanishi T, et al. Effect of autologous platelet-rich plasma-releasate on intervertebral disc degeneration in the rabbit anular puncture model: a preclinical study. Arthritis Res Ther. 2012;14:R241.
173. Akeda K, An HS, Okuma M, et al. Platelet-rich plasma stimulates porcine articular chondrocyte proliferation and matrix biosynthesis. Osteoarthr Cartil. 2006;14:1272–80.
174. Akeda K, An HS, Pichika R, et al. Platelet-rich plasma (PRP) stimulates the extracellular matrix metabolism of porcine nucleus pulposus and annulus fibrosus cells cultured in alginate beads. Spine. 2006;31:959–66.
175. Colio S, McAuliffe M, Uribe Y, et al. Regenerative medicine for muscle and ligament problems: technical aspects and evidence. Tech Reg Anesth Pain Manag. 2016;19:80–4.
176. Bodor M, Lesher JM, Colio S. Ultrasound-guided hand, wrist and elbow injections. In: Narouze S, editor. Atlas of ultrasound guided procedures for pain management. 1st ed. New York: Springer; 2011. p. 307–23.
177. Bodor M, Colio S, Bonzon C. Hand and wrist injections: ultrasound. In: Narouze S, editor. Multi-modality imaging guidance in interventional pain management. Oxford: Oxford University Press; 2017. p. 554–74.
178. Urzen JM, Fullerton BD. Nonsurgical resolution of a bucket handle meniscal tear: a case report. PM R. 2016;8:1115–8.
179. Pujol N, Salle De Chou E, Boisrenoult P, et al. Platelet-rich plasma for open meniscal repair in young patients: any benefit? Knee Surg Sports Traumatol Arthrosc. 2015;23:51–8.
180. Podesta L, Crow SA, Volkmer D, et al. Treatment of partial ulnar collateral ligament tears in the elbow with platelet-rich plasma. Am J Sports Med. 2013;41:1689–94.
181. Seijas R, Ares O, Cuscó X, et al. Partial anterior cruciate ligament tears treated with intraligamentary plasma rich in growth factors. World J Orthop. 2014;5:373–8.
182. Griffin JW, Hadeed MM, Werner BC, et al. Platelet-rich plasma in meniscal repair: does augmentation improve surgical outcomes? Clin Orthop Relat Res. 2015;473:1665–72.
183. Redmond JM, Gupta A, Stake CE, et al. Clinical results of hip arthroscopy for labral tears: a comparison between intraoperative platelet-rich plasma and bupivacaine injection. Arthroscopy. 2015;31:445–53.
184. Filardo G, Di Matteo B, Kon E, et al. Platelet-rich plasma in tendon-related disorders: results and indications. Knee Surg Sports Traumatol Arthrosc. 2018;26:1984–99.
185. Battaglia M, Guaraldi F, Vannini F, et al. Efficacy of ultrasound-guided intra-articular injections of platelet-rich plasma versus hyaluronic acid for hip osteoarthritis. Orthopedics. 2013;36:E1501–8.
186. Edwards SG, Calandruccio JH. Autologous blood injections for refractory lateral epicondylitis. J Hand Surg Am. 2003;28:272–8.
187. Martin JI, Merino J, Atilano L, et al. Platelet-rich plasma (PRP) in chronic epicondylitis: study protocol for a randomized controlled trial. Trials. 2013;14:410.
188. Gustas CN, Lee KS. Multimodality imaging of the painful elbow: current imaging concepts and image-guided treatments for the injured thrower's elbow. Radiol Clin N Am. 2016;54:817–39.
189. Mautner K, Colberg RE, Malanga G, et al. Outcomes after ultrasound-guided platelet-rich plasma injections for chronic tendinopathy: a multicenter, retrospective review. PM R. 2013;5(3):169–75.
190. Mishra AK, Skrepnik NV, Edwards SG, et al. Efficacy of platelet-rich plasma for chronic tennis elbow: a double-blind, prospective, multicenter, randomized controlled trial of 230 patients. Am J Sports Med. 2014;42:463–71.
191. Gosens T, Den Oudsten BL, Fievez E, et al. Pain and activity levels before and after platelet-rich plasma injection treatment of patellar tendinopathy: a prospective cohort study and the influence of previous treatments. Int Orthop. 2012;36:1941–6.
192. Sadoghi P, Rosso C, Valderrabano V, et al. The role of platelets in the treatment of Achilles tendon injuries. J Orthop Res. 2013;31:111–8.
193. Fitzpatrick J, Bulsara M, Zheng MH. The effectiveness of platelet-rich plasma in the treatment of tendinopathy. Am J Sports Med. 2017;45:226–33.
194. Dallari D, Stagni C, Rani N, et al. Ultrasound-guided injection of platelet-rich plasma and hyaluronic acid, separately and in combination, for hip osteoarthritis: a randomized controlled study. Am J Sports Med. 2016;44:664–71.
195. Di Sante L, Villani C, Santilli V, et al. Intra-articular hyaluronic acid vs platelet-rich plasma in the treatment of hip osteoarthritis. Med Ultrason. 2016;18:463–8.
196. Sanapati J, Manchikanti L, Atluri S, et al. Do regenerative medicine therapies provide long-term relief in chronic low back pain: a systematic review and metaanalysis. Pain Physician. 2018;21:515–40.
197. Subhan RA, Puvanan K, Murali MR, et al. Fluoroscopy assisted minimally invasive transplantation of allogenic mesenchymal

stromal cells embedded in HyStem reduces the progression of nucleus pulposus degeneration in the damaged intervertebral [corrected] disc: a preliminary study in rabbits. Sci World J. 2014;2014:818502.
198. Cao C, Zou J, Liu X, et al. Bone marrow mesenchymal stem cells slow intervertebral disc degeneration through the NF-κB pathway. Spine J. 2015;15:530–8.
199. Jezierska-Wozniak K, Barczewska M, Habich A, et al. The feasibility of the CD271+ and CD271- mesenchymal stromal cell enrichment toward nucleus pulposus-like cells. Folia Histochem Cytobiol. 2017;55:114–23.
200. Wu T, Song HX, Dong Y, et al. Cell-based therapies for lumbar discogenic low back pain: systematic review and single-arm meta-analysis. Spine. 2018;43:49–57.
201. Mochida J, Sakai D, Nakamura Y, et al. Intervertebral disc repair with activated nucleus pulposus cell transplantation: a three-year, prospective clinical study of its safety. Eur Cell Mater. 2015;29:202–12.
202. Pettine KA, Murphy MB, Suzuki RK, Sand TT. Percutaneous injection of autologous bone marrow concentrate cells significantly reduces lumbar discogenic pain through 12 months. Stem Cells. 2015;33:146–56.
203. Pettine K, Suzuki R, Sand T, et al. Treatment of discogenic back pain with autologous bone marrow concentrate injection with minimum two year follow-up. Int Orthop. 2016;40:135–40.
204. Pettine KA, Suzuki RK, Sand TT, et al. Autologous bone marrow concentrate intradiscal injection for the treatment of degenerative disc disease with three-year follow-up. Int Orthop. 2017;41:2097–103.
205. Pang X, Yang H, Peng B. Human umbilical cord mesenchymal stem cell transplantation for the treatment of chronic discogenic low back pain. Pain Physician. 2014;17:E525–30.
206. Coric D, Pettine K, Sumich A, et al. Prospective study of disc repair with allogeneic chondrocytes presented at the 2012 Joint Spine Section Meeting. J Neurosurg Spine. 2013;18:85–95.
207. Orozco L, Soler R, Morera C, et al. Intervertebral disc repair by autologous mesenchymal bone marrow cells: a pilot study. Transplantation. 2011;92:822–8.
208. Meisel HJ, Ganey T, Hutton WC, et al. Clinical experience in cell based therapeutics: intervention and outcome. Eur Spine J. 2006;15:S397–405.
209. Sivakamasundari V, Lufkin T. Stemming the degeneration: IVD stem cells and stem cell regenerative therapy for degenerative disc disease. Adv Stem Cells. 2013;2013:724547.
210. Vickers ER, Karsten E, Flood J, et al. A preliminary report on stem cell therapy for neuropathic pain in humans. J Pain Res. 2014;7:255–63.
211. Mendonça MV, Larocca TF, de Freitas Souza BS, et al. Safety and neurological assessments after autologous transplantation of bone marrow mesenchymal stem cells in subjects with chronic spinal cord injury. Stem Cell Res Ther. 2014;5:126.
212. Penha EM, Aguiar PH, Barrouin-Melo SM, et al. Clinical neurofunctional rehabilitation of a cat with spinal cord injury after hemilaminectomy and autologous stem cell transplantation. Int J Stem Cells. 2012;5:146–50.
213. Chen L, Huang H, Sharma HS, et al. Cell transplantation as a pain therapy targets both analgesia and neural repair. Cell Transplant. 2013;22:S11–9.
214. Escalhão CCM, Ramos IP, Hochman-Mendez C, et al. Safety of allogeneic canine adipose tissue-derived mesenchymal stem cell intraspinal transplantation in dogs with chronic spinal cord injury. Stem Cells Int. 2017;2017:3053759.
215. Forouzanfar F, Amin B, Ghorbani A, et al. New approach for the treatment of neuropathic pain: fibroblast growth factor 1 gene-transfected adipose-derived mesenchymal stem cells. Eur J Pain. 2018;22:295–310.
216. Mauskop A, Rothaus KO. Stem cells in the treatment of refractory chronic migraines. Case Rep Neurol. 2017;9:149–55.
217. Song L, Sun Z, Kim D, et al. Adipose stem cells from chronic pancreatitis patients improve mouse and human islet survival and function. Stem Cell Res Ther. 2017;8:192.
218. Wu Z, Zhang S, Zhou L, et al. Thromboembolism induced by umbilical cord mesenchymal stem cell infusion: a report of two cases and literature review. Transplant Proc. 2017;49:1656–8.
219. Randsborg PH, Brinchmann J, Løken S, et al. Focal cartilage defects in the knee - a randomized controlled trial comparing autologous chondrocyte implantation with arthroscopic debridement. BMC Musculoskelet Disord. 2016;17:117.
220. Freitag J, Ford J, Bates D, et al. Adipose derived mesenchymal stem cell therapy in the treatment of isolated knee chondral lesions: design of a randomised controlled pilot study comparing arthroscopic microfracture versus arthroscopic microfracture combined with postoperative mesenchymal stem cell injections. BMJ Open. 2015;5:e009332.
221. Koh YG, Kwon OR, Kim YS, et al. Adipose-derived mesenchymal stem cells with microfracture versus microfracture alone: 2-year follow-up of a prospective randomized trial. Arthroscopy. 2016;32:97–109.
222. Akgun I, Unlu MC, Erdal OA. Matrix-induced autologous mesenchymal stem cell implantation versus matrix-induced autologous chondrocyte implantation in the treatment of chondral defects of the knee: a 2-year randomized study. Arch Orthop Trauma Surg. 2015;135:251–63.
223. Wong KL, Lee KB, Tai BC, et al. Injectable cultured bone marrow-derived mesenchymal stem cells in varus knees with cartilage defects undergoing high tibial osteotomy: a prospective, randomized controlled clinical trial with 2 years' follow-up. Arthroscopy. 2013;29:2020–8.
224. Aghdami N, Ghorbani Liastani M, et al. Repeated intra articular injection of bone marrow derived mesenchymal stem cell in knee osteoarthritis: double blind randomized clinical trial. Cytotherapy. 2014;16:S14.
225. Pham PV. A controlled clinical trial of adipose derived stem cell transplantation for osteoarthritis. Cytotherapy. 2016;18:S12.
226. Gupta PK, Chullikana A, Rengasamy M, et al. Efficacy and safety of adult human bone marrow-derived, cultured, pooled, allogeneic mesenchymal stromal cells (Stempeucel): preclinical and clinical trial in osteoarthritis of the knee joint. Arthritis Res Ther. 2016;18:301.
227. Bhattacharya N. Clinical use of amniotic fluid in osteoarthritis: a source of cell therapy. Transplantation. 2010;90:466.
228. Vangsness CT Jr, Farr J, Boyd J, et al. Adult human mesenchymal stem cells delivered via intra-articular injection to the knee following partial medial meniscectomy: a randomized, double-blind, controlled study. J Bone Joint Surg Am. 2014;96:90–8.
229. March L, Hunter D, Ward C, et al. A randomised placebo controlled pilot study of autologous non-expanded adipose-derived mesenchymal stem cells in the treatment of knee osteoarthritis. Intern Med J. 2013;43:4–5.
230. Lamo-Espinosa JM, Mora G, Blanco JF, et al. Intra-articular injection of two different doses of autologous bone marrow mesenchymal stem cells versus hyaluronic acid in the treatment of knee osteoarthritis: multicenter randomized controlled clinical trial (phase I/II). J Transl Med. 2016;14:246.
231. Lu LJ, Song Y, Hui D, et al. Treatment with human adipose-derived mesenchymal stem cells for knee OA: evidences from a randomized and double-blinded phase I/II a clinical trial. Int J Rheum Dis. 2016;19:134.
232. Koh YG, Kwon OR, Kim YS, et al. Comparative outcomes of open-wedge high tibial osteotomy with platelet-rich plasma alone or in combination with mesenchymal stem cell treatment: a prospective study. Arthroscopy. 2014;30:1453–60.

233. Koh YG, Choi YJ. Infrapatellar fat pad-derived mesenchymal stem cell therapy for knee osteoarthritis. Knee. 2012;19:902–7.
234. Vega A, Martín-Ferrero MA, Del Canto F, et al. Treatment of knee osteoarthritis with allogeneic bone marrow mesenchymal stem cells: a randomized controlled trial. Transplantation. 2015;99:1681–90.
235. Varma HS, Dadarya B, Vidyarthi A. The new avenues in the management of osteo-arthritis of knee-stem cells. J Indian Med Assoc. 2010;108:583–5.
236. Saw KY, Anz A, Siew-Yoke Jee C, et al. Articular cartilage regeneration with autologous peripheral blood stem cells versus hyaluronic acid: a randomized controlled trial. Arthroscopy. 2013;29:684–94.
237. Soler R, Orozco L, Munar A, et al. Final results of a phase I-II trial using ex vivo expanded autologous Mesenchymal Stromal Cells for the treatment of osteoarthritis of the knee confirming safety and suggesting cartilage regeneration. Knee. 2016;23:647–54.
238. Kim YS, Choi YJ, Lee SW, et al. Assessment of clinical and MRI outcomes after mesenchymal stem cell implantation in patients with knee osteoarthritis: a prospective study. Osteoarthr Cartil. 2016;24:237–45.
239. Koh YG, Jo SB, Kwon OR, et al. Mesenchymal stem cell injections improve symptoms of knee osteoarthritis. Arthroscopy. 2013;29:748–55.
240. Mamidi MK, Das AK, Zakaria Z, et al. Mesenchymal stromal cells for cartilage repair in osteoarthritis. Osteoarthr Cartil. 2016;24:1307–16.
241. Lee SY, Kim W, Lim C, et al. Treatment of lateral epicondylosis by using allogeneic adipose-derived mesenchymal stem cells: a pilot study. Stem Cells. 2015;33:2995–3005.
242. Dall'Oca C, Cengarle M, Costanzo A, et al. Current concepts in treatment of early knee osteoarthritis and osteochondral lesions; the role of biological augmentations. Acta Biomed. 2017;88:5–10.
243. Shook JR, Giordano J. Ethical and policy issues in practices of regenerative medicine for the treatment of chronic pain. In: Manchikanti L, Navani A, Atluri S, editors. Essentials of regenerative medicine in interventional pain management. Paducah: ASIPP Publishing; 2018. p. 641–50.
244. Rosenthal RM, Murphy M. Regulatory aspects of regenerative therapy stem cell therapy: current perspectives. In: Manchikanti L, Navani A, Atluri S, editors. Essentials of regenerative medicine in interventional pain management. Paducah: ASIPP Publishing; 2018. p. 621–8.
245. Draft Guidance for Industry: Same Surgical Procedure Exception Under 21 CFR 1271.15(b): Questions and Answers Regarding the Scope of the Exception. www.fda.gov/BiologicsBloodVaccines/GuidanceComplianceRegulatoryInformation/Guidances/Tissue/ucm419911.htm.
246. Aicher RH. Is the FDA regulating the practice of medicine? Aesthet Surg J. 2013;33:452–5.
247. Berman M, Lander E. Regulatory issues with human cells: an alternate view. In: Manchikanti L, Navani A, Atluri S, editors. Essentials of regenerative medicine in interventional pain management. Paducah: ASIPP Publishing; 2018. p. 629–40.
248. Malarkey MA. Warning letter to Dr. Steven Victor, Chief Executive Officer. Rockville: IntelliCell Biosciences Inc.; 2012.
249. Mescher AL. Adipose tissue. In: Junqueira's basic histology: text & atlas. 13th ed. New York: McGraw-Hill; 2013. p. 124–9.
250. Malarkey MA. Warning letter to Thomas E. Young, MD, owner and medical director, young medical spa. Rockville: Public Health Service, Food and Drug Administration; 2012.
251. Warning Letter to John S. Arnone, American CryoStem Corporation from Elizabeth Waltrip, Acting Program Division Director, Office of Biological Products Operations – Division 1, U.S. Food and Drug Administration.
252. Associated Press. North Dakota stem cell clinic investigated after complaints. The Seattle Times. 2017. https://www.seattletimes.com/nation-world/bismarck-stem-cell-clinic-investigated-after-complaints/. Accessed 12 January 2023.
253. Kuriyan AE, Albini TA, Townsend JH, et al. Vision loss after intravitreal injection of autologous "stem cells" for AMD. N Engl J Med. 2017;376:1047–53.
254. Malarkey MA. Warning letter to David G. Eller, Chief Executive Officer and President, Celltex Therapeutics Corporation. Rockville: Public Health Service, Food and Drug Administration; 2012.
255. Brown SA, Levi B, Lequeux C, et al. Basic science review on adipose tissue for clinicians. Plast Reconstr Surg. 2010;126:1936–46.
256. Department of Health and Human Services, Food and Drug Administration. 21 CFR Parts 207, 807, and 1271. Establishment Registration and Listing for Manufacturers of Human Cellular and Tissue-Based Products. Proposed Rule, May 14, 1998.
257. Department of Health and Human Services, U.S. Food & Drug Administration. 21 CFR Part 1271. Human Cells, Tissues, and Cellular and Tissue-Based Products From Adipose Tissue: Regulatory Considerations; Draft Guidance for Industry; Availability; Proposed Rule. Federal Register, Wednesday, December 24, 2014.
258. Department of Health and Human Services, Food and Drug Administration. 21 CFR Parts 207, 807, and 1271. Human Cells, Tissues, and Cellular and Tissue-Based Products; Establishment Registration and Listing. Final Rule, January 19, 2001.
259. Sherman RE, Li J, Shapley S, et al. Expediting drug development — the FDA's new "breakthrough therapy" designation. N Engl J Med. 2013;369:1877–80.
260. H.R. 34 – 21st Century Cures Act. P.L. 114-255, December 13, 2016.
261. Jarow JP, LaVange L, Woodcock J. Multidimensional evidence generation and FDA regulatory decision making: defining and using "real-world" data. JAMA. 2017;318:703–4.
262. Sherman RE, Anderson SA, Dal Pan GJ, et al. Real-world evidence – what is it and what can it tell us? N Engl J Med. 2016;375:2293–7.
263. Letter to Gregory B. Snyder, MD, Federation of State Medical Boards from Senator Lamar Alexander RE: Stem cell therapy, April 21, 2017.
264. Federation of State Medical Boards. Regenerative and Stem Cell Therapy Practices: Report and Recommendations of the Workgroup to Study Regenerative and Stem Cell Therapy Practices. 2018. https://www.fsmb.org/siteassets/advocacy/policies/fsmb-stem-cell-workgroup-report.pdf. Accessed 14 February 2023.
265. Berger I, Ahmad A, Bansal A, et al. Global distribution of businesses marketing stem cell based interventions. Cell Stem Cell. 2016;19:158–62.
266. Sipp D, Caulfield T, Kaye J, et al. Marketing of unproven stem cell based interventions: a call to action. Sci Transl Med. 2017;9:397.
267. FDA's warning letters. www.fda.gov/ICECI/EnforcementActions/WarningLetters/default.htm. Accessed 12 January 2023.
268. Jacques E, Suuronen EJ. The progression of regenerative medicine and its impact on therapy translation. Clin Transl Sci. 2020;13:440–50.
269. White BD, Gelinas LC. Balancing the surgeon's responsibility to individuals and society. In: Stain SC, et al., editors. The SAGES manual ethics of surgical innovation. Cham: Springer; 2016. p. 191–211.
270. American Medical Association, Code of Medical Ethics, Opinion 11.3.1.
271. Sheth U, Nguyen NA, Gaines S, et al. New orthopedic devices and the FDA. J Long-Term Eff Med Implants. 2009;19:173–84.

272. FDA Science Board: FDA science and mission at risk: report of the Subcommittee on Science and Technology. 2007. https://www.supremecourt.gov/opinions/URLs_Cited/OT2008/06-1249/06_1249_1.pdf. Accessed 12 January 2023.
273. ASIPP Quality Registry and national interventional pain management – qualified clinical data registry (NIPM-QCDR). https://nipmqcdr.org/about-nipm-qcdr/. Accessed 11 January 2023.
274. Khorraminejad-Shirazi M, Farahmandnia M, Kardeh B, et al. Aging and stem cell therapy: AMPK as an applicable pharmacological target for rejuvenation of aged stem cells and achieving higher efficacy in stem cell therapy. Hematol Oncol Stem Cell Ther. 2018;11:189–94.

# Principles of Evidence-Based Medicine

# 2

Nebojsa Nick Knezevic, Laxmaiah Manchikanti, and Joshua A. Hirsch

## Introduction

The trend to develop and implement research in support of evidence-based practice could be considered the convention in medicine for the past decade. This emphasis has been fostered, at least partly, by a perceived (if not defined and called for) need to improve patient care through applied clinical decision-making in diagnosis and treatment. Evidence-based practice or medicine (EBM) evolves through a methodical, rational accumulation, analysis, and understanding of the evidentiary knowledge that can be applied in the clinical setting(s) [1–3]. Over the years, due to criticism of EBM, it also has been called by various names including "eminence-based" medicine and "confidence-based" medicine as shown in Table 2.1 [4].

Evidence can be defined as any ground or reason for knowledge or certitude in knowledge; proof, whether immediate or derived by inference; a fact or body of facts on which a proof, belief, or judgment is based [1, 2]. However, it is the nature of belief and the foundation upon which it rests that provides the utility of evidence [5]. For medical purposes, evidence can be any data or information, whether it be strong or weak, obtained through experience, observational or experimental research [1, 2, 6]. Thus, EBM is about solving clinical problems [1, 7]. In contrast to the traditional paradigm of medical practice, EBM acknowledges that intuition, unsystematic clinical experience, and pathophysiologic rationale are insufficient grounds for clinical decision-making, while it stresses the examination of evidence from clinical research [1, 2, 8]. EBM suggests that a formal set of rules must complement medical training and common sense for clinicians to interpret the results of clinical research effectively. Finally, EBM places a lower value on authority than does the traditional medical paradigm [1, 9–14].

**Table 2.1** Seven alternative terms to evidence-based medicine

| |
|---|
| Eminence-based medicine |
| Confidence-based medicine |
| Vehemence-based medicine |
| Eloquence-based medicine |
| Providence-based medicine |
| Diffidence-based medicine |
| Nervousness-based medicine |

Comparative effectiveness research (CER) is another term utilized to describe EBM. The driving purpose of high-quality CER studies is to generate real-world evidence demonstrating the effectiveness and cost-effectiveness of interventions using real-world data [15]. The International Society for Pharmacoeconomics and Outcomes Research (ISPOR) has convened several task forces to publish good practice guidelines for reporting the findings of different types of CER studies [16–18]. Similar to EBM, the reviews of comparative effectiveness are considered systematic, with investigation of variations, randomized controlled trials (RCTs), and cost-benefit analysis. CER, previously identified as the scientific interest of clinical and health services researchers who compared medical treatments, has now become one of the most important methodologies of health care assessment [13, 14].

## History

Before implementing EBM in practice, all decisions in patients' diagnostics and treatments were based on experience [19]. Even though the historical basis of EBM could be linked to Thomas Beddoes (1760–1808), Pierre Charles

N. N. Knezevic
Department of Anesthesiology, Advocate Illinois Masonic Medical Center, Chicago, IL, USA

College of Medicine, University of Illinois, Chicago, IL, USA

L. Manchikanti (✉)
Pain Management Centers of America, Paducah, KY, USA
e-mail: drlm@thepainmd.com

J. A. Hirsch
Massachusetts General Hospital and Harvard Medical School, Boston, MA, USA
e-mail: jahirsch@mgh.harvard.edu


A. Navani et al. (eds.), *Essentials of Regenerative Medicine in Interventional Pain Management*, https://doi.org/10.1007/978-3-031-50357-3_2

Alexander Louis (1787–1872) and Sir William Osler (1849–1919), EBM was developed more in the second half of the twentieth century by Archie Cochran (1909–1988), a Scottish physician who assisted other physicians in organizing a massive pool of contradictory data and information [19]. He expressed strong criticism of physicians for not organizing critical information from RCTs and other studies by specialties and sub-specialties [20]. The term EBM was published in 1991 for the first time [7].

Limited and inconsistent data regarding the effectiveness of different medical treatments, increased health care cost, and the capability to measure and analyze the outcomes with better tools become three major factors that drive the EBM today [10, 19]. All of these concepts require clinicians and researchers to look for high-quality evidence as support of treatment effectiveness, safety, and cost-saving [10]. Increased health care cost pushes insurance companies to mandate EBM data in order to prove the efficacy for covered services [21]. Insurance companies will often choose to not pay for therapeutic options that lack EBM data, claiming them as "investigational therapies" [21]. At the same time, physicians have the responsibility to provide effective therapeutic options to their patients and reduce the potential harm to them by avoiding treatments with negative risk/benefit ratios [21].

The Institute of Medicine (IOM) introduced EBM as "the conscientious, explicit, and judicious use of current best evidence in making decisions about the care of individual patients" [22]. They categorized EBM into 3 levels [22, 23]:

Level A1 (likely reliable) including RCTs and systematic reviews of RCTs

Level B2 (mid-level) evidence from less quality RCTs and non-randomized comparison studies

Level C and D3 (lacking direct evidence) including evidence from case reports, case series, and expert opinions.

With multiple advances and evolution in synthesis and reporting of EBM recommendations, a qualitative modified approach to grading of evidence has been recommended by the American Society of Interventional Pain Physicians (ASIPP) (Table 2.2) [24]. Similarly, the Agency for Healthcare Research and Quality (AHRQ) has also recommended similar strength of recommendations as shown in Table 2.3 [25, 26].

**Table 2.2** Qualitative modified approach to grading of evidence

| | |
|---|---|
| Level I | Evidence obtained from multiple relevant high quality randomized controlled trials<br>or<br>Evidence obtained from multiple high quality diagnostic accuracy studies |
| Level II | Evidence obtained from at least one relevant high quality randomized controlled trial or multiple relevant moderate or low quality randomized controlled trials<br>or<br>Evidence obtained from at least one high quality diagnostic accuracy study or multiple moderate or low quality diagnostic accuracy studies |
| Level III | Evidence obtained from at least one relevant moderate or low quality randomized controlled trial study<br>or<br>Evidence obtained from at least one relevant high quality non-randomized trial or observational study with multiple moderate or low quality observational studies<br>or<br>Evidence obtained from at least one moderate quality diagnostic accuracy study in addition to low quality studies |
| Level IV | Evidence obtained from multiple moderate or low quality relevant observational studies<br>or<br>Evidence obtained from multiple relevant low quality diagnostic accuracy studies |
| Level V | Opinion or consensus of large group of clinicians and/or scientists |

At least 60% of studies in the direction of the objective being assessed
Source: Manchikanti L, Falco FJE, Benyamin RM, et al. A modified approach to grading of evidence. *Pain Physician* 2014; 17:E319–E325 [24]

**Table 2.3** Guide for strength of recommendations

| Rating for strength of recommendation | |
|---|---|
| Strong | There is high confidence that the recommendation reflects best practice. This is based on: (a) strong evidence for a true net effect (e.g., benefits exceed harms); (b) consistent results, with no or minor exceptions; (c) minor or no concerns about study quality; and/or (d) the extent the panelists' agreement. Other compelling considerations (discussed in the guideline's literature review and analyses) may also warrant a strong recommendation |
| Moderate | There is moderate confidence that the recommendation reflects best practice. This is based on: (a) good evidence for a true net effect (e.g. benefits exceed harms); (b) consistent results, with minor and/or few exceptions; (c) minor and/or few concerns about study quality; and/or (d) the extent of panelists' agreement. Other compelling considerations (discussed in the guideline's literature review and analyses) may also warrant a moderate recommendation |
| Weak | There is some confidence that the recommendation offers the best current guidance for practice. This is based on: (a) limited evidence for a true net effect (e.g., benefits exceed harms); (b) consistent results, but with important exceptions; (c) concerns about study quality; and/or (d) the extent of panelists' agreement. Other considerations (discussed in the guideline's literature review and analyses) may also warrant a weak recommendation |

Source: National Guideline Clearinghouse Extent Adherence to Trustworthy Standards (NEATS) instrument [25]

## EBM in Regenerative Medicine in Chronic Pain

The EBM practice sometimes compels "extraordinary obligations from ordinary clinicians," especially in "the subjective world of pain." Pain, as a multi-dimensional subjective biopsychosocial entity is unique and requires special considerations [14]. Similar problems occur in other disciplines, such as occupational health, psychotherapy, physiotherapy, etc. when it is very difficult to use the classical EBM approach [19]. Patients suffering from severe pain are desperate and sometimes begging physicians to try any treatment that offers even a small chance for improvement [27].

Interventional regenerative medicine techniques achieved an important role in the management of different chronic pain conditions, and they underwent a quick evolution over the last decade. However, the evidence in regenerative medicine is only evolving [11, 28]. The main goal of interventional pain research in regenerative medicine is to standardize techniques, improve effectiveness, and reduce the risk of adverse events [11, 28].

Though research and evidence are expanding at fast rates, the roles of biologicals in interventional techniques remain to be defined. The methodological approaches of EBM can be relied upon to evaluate the effectiveness of regenerative medicine therapies and their potential applications in the management of chronic pain. To provide high quality evidence per EBM, a systematic review and meta-analysis was conducted to determine the effectiveness of regenerative medicine therapies in the treatment of low back and lower extremity pain [29]. The methodology involved a literature search of several databases, followed by data extraction and methodological quality assessment using Cochrane review methodologic quality assessment, Interventional Pain Management Techniques—Quality Appraisal of Reliability and Risk of Bias Assessment (IPM-QRB) and Interventional Pain Management Techniques—Quality Appraisal of Reliability and Risk of Bias Assessment for Nonrandomized Studies (IPM-QRBNR). This resulted in the inclusion of 21 studies, and the evidence synthesis yielded Level 3 evidence for disc injections of platelet rich plasma (PRP) and mesenchymal stem cells (MSCs), Level 4 evidence for epidural injections, and Level 4 evidence for lumbar facet joint injections and sacroiliac joint injections [29].

Once the evidence synthesis is computed, evidence-based recommendations and guidelines can be produced. Summaries of evidence on regenerative medicine have been used to create guidelines and policy position statements by the ASIPP on the use of biologics in the management of low back pain [28], bone marrow concentrate therapy in musculoskeletal disorders [11], and epidural interventions in the management of chronic spinal pain [9].

## Randomized Controlled Trials

RCTs are considered the gold standard of EBM. However, RCTs have some limitations, especially when investigating the treatment options for pain management because of too strict inclusion and exclusion criteria and confounding factors (age, gender, past therapy, etc.) that do not allow generalization [19].

In RCTs, patients receive either an investigational (targeted) or placebo/control treatment, and the order of treatments is randomized, and patients and clinicians are blinded to allocation [30]. However, there is extensive debate regarding placebo controls and study design. Hill writing on the issues related to the debate about using placebos described that the essential medical question at issue is how the new treatment compares with the old one, not whether the new treatment is better than nothing [30, 31]. There are extensive discussions also in reference to the role of placebo in RCTs along with nocebo effects in determining the results of placebo-controlled trials. Consequently, appropriate placebo design is crucial [32, 33].

The null hypothesis in randomized trials assumes that there is no difference between a new and a placebo/control treatment. Properly conducted RCTs minimize bias by acknowledging known and unknown factors, which may affect treatment effect.

Consolidated Standards of Reporting Trials (CONSORT) acknowledges that properly conducted RCTs provide the best evidence on the efficacy of a health care intervention. The biased interpretation of the results, in favor or against a particular treatment, and incorrect understanding of randomized trials leads to poor quality data interpretation or drawn conclusions. The CONSORT statement for reporting randomized trials [34] provides a checklist of items to include: study question, study population, randomization, blinding, interventions, outcomes, statistical analysis, results, discussion, and funding. Thus, reports of RCTs should contain the CONSORT flow diagram available at: www.consort-statement.org/consort-statement/flow-diagram.

Randomization is the process of dividing participants to experimental (treatment) or control groups, assuming that each participant has an equal chance to be assigned to one of these groups [35, 36]. However, randomized design may not be most applicable in interventional pain management, as demonstrated by Manchikanti and Pampati [37].

Randomization procedure refers to generating a random sequence of allocations, which may be a simple random assignment of patients to any of the equally probable trial groups [36]. However, randomization may have some advantages of eliminating bias in assignment of treatments, and by facilitating blinding, and permits the use of probability theory to express the likelihood that any outcome difference between intervention groups merely reflects chance [38]. Thus, preventing selection and confounding biases is the most important advantage of randomization [39]. Simple randomization is based on a single sequence of random assignments [36, 40]. For simple participant randomization, a random number table or computer-generated random numbers can be used. Restricted randomization describes any procedure that helps achieve balance between groups in size or characteristics by blocking, stratification, or covariate adaptation. Blocking ensures a close balance of the numbers in each group during any time of the trial. However, certain covariates among participants, such as comorbid medical conditions, could undermine proper comparison of the groups and negatively influence the results. Stratification ensures that participants receiving each intervention are well matched per baseline characteristics, such as age and stage of disease. Stratified randomization may be difficult to implement if numerous covariates must be controlled [41–45].

In a crossover RCT, 2 or more treatments are provided to subjects in a randomized sequence. The simplest crossover design is an AB/BA sequence, where subjects receive treatment A and, after the effects of treatment A have subsided, they receive treatment B. The greatest advantage of a crossover design is that each subject serves as his or her own control, removing inter-subject variability. It also allows for high power and statistical efficiency. Challenges of crossover study designs are in managing possible carryover effect, period effects, sequence effect, and period-by-treatment interaction. Other disadvantages include the required stability of the subjects' condition, washout periods affected by treatment duration, and processing of missing data [46].

Other aspects of randomized-controlled trials include blinding. In an unblinded study, the information about assigned interventions is available to all the groups involved. Blinding is the process of withholding information that has the potential to influence study results, from one or more of the parties involved. This process may prevent several forms of bias which cannot be corrected through analytical techniques. Unblinded study subjects may introduce bias into their expectations, adherence to protocol, treatment-seeking behavior outside the trial, and assessment of the effectiveness of an intervention. On the other hand, unblinded research staff may influence outcomes through differential treatment, attention, or attitudes towards subjects [47].

Allocation concealment is a technique where the allocation sequence is hidden from those assigning participants to the intervention groups, until the moment of assignment. This prevents researchers inadvertently, or otherwise, from influencing the assigning process [48, 49]. Allocation concealment includes 4 standard methods: central randomization; pharmacy controlled; sequentially numbered, opaque, sealed envelopes (SNOSE); and sequentially numbered containers [39].

Since most studies have very strict eligibility criteria upon which they choose treatment groups, conclusions may only be applicable to those who fit such criteria [50]. Placebo-controlled RCTs may fail to apply to the general patient population. Due to the disadvantages of placebo-controlled RCTs, some physicians focus on practical clinical trials, which ensure that participants are diverse, that they come from a heterogeneous group of practice settings and geographic locations, and that they reflect the underlying affected population, at the same time focusing that the trial's endpoints reflect a broad range of meaningful clinical outcomes [31, 39, 51].

The choice of proper control group is critical when designing a clinical trial because it affects conclusions drawn from the generated results, the degree of minimizing bias in the study, types of participants that could be recruited, credibility of the results, and interpretation of the study. The control group provides us with the conclusion of what would have happened if patients had not received the test treatment or a comparison if they had received a different treatment [40].

In placebo-controlled trials, patients are randomly allocated to receive either a treatment or identical-appearing drug-free product. These trials are typically double-blind, and groups like control or placebo serve to investigate potential influences on the course of the disease, other than those derived from the pharmacologic action of the test drug. The implementation of randomization and blinding in a placebo-controlled trial minimizes participant and investigator bias but remains vulnerable to breaking blind through perception of the pharmacologic effects of one treatment. There are usually no ethical issues comparing the new treatment to placebo when a treatment is tested for a condition for which no effect is yet known. Advantages of a randomized double-blind placebo-controlled trial include exploring the effectiveness of a treatment, freedom from assumptions, and the ability to distinguish adverse effects caused by a drug or procedure from those resulting from underlying disease. Disadvantages include ethical concerns, lack of generalizability, and lack of effectiveness comparativeness [39, 40, 52].

Active (positive) control is used when a new treatment is compared to an active control treatment group. Such trials are usually, but not always, double-blind, and demonstrate the efficacy of the tested treatment by showing that it is as good as a known effective treatment or superior to the active control. They may also be used to compare the efficacy and/or safety of the 2 newly developed treatments. A randomized, blinded, active control trial generally minimizes participant and investigator bias. Active controls pose fewer ethical and practical problems than placebo-controlled trials. Disadvantages relate to problems in quantitating safety outcomes, lack of direct evaluation of effect size, and need for large sample sizes tested. Furthermore, some arguments can be made that omitting a true placebo arm does not eliminate the placebo response, but instead renders it invisible and uncontrollable [53].

In no-treatment control groups, participants are randomly assigned to a test treatment or to a no treatment study group. In this type of controlled design, neither participants nor the investigators are blind to the treatment assignment and makes it applicable only when it is too difficult or impossible to double-blind. This type of control is used both in interventional pain management and surgery [39].

In a dose-response control design, participants are randomized to one out of several fixed-dose groups, initially assigning them on either fixed dose or titrating them to a certain dose gradually. These trials are usually double-blind and may include a placebo and/or an active group. If the study is blinded, bias is minimized, and this design may be acceptable in interventional pain management or surgery trials. Advantages include efficiency and possible ethical advantage and enable the determination of efficacy and safety in situations where a placebo-controlled trial has similar credibility. Disadvantages include the necessity to recognize that a positive dose-response without significant pair-wise differences may cause confusion in determination of which dose, other than the highest, is genuinely effective.

Placebo-controlled trials, active-controlled trials, and dose-response controlled trials elicit and provide various aspects of understanding of effectiveness of a modality as shown in Table 2.4 [31]. Consequently, a placebo-controlled trial provides information on the existence of effect and measurable absolute effect size; however, it does not provide any dose response relationship or compare therapies. In contrast, an active-control trial shows the existence of effect and also comparative effectiveness but does not measure absolute effect size nor does it show dose response relationship. To obtain the results on more than 2 aspects of existence of effect, effect size, dose relationship, and comparativeness of therapies multiple combinations of designs may be utilized as shown in Table 2.4. Multiple manuscripts have described these variations in the results; however, even experts at AHRQ often ignore these crucial aspects of design [54–60].

Table 2.4 demonstrates the most crucial aspect in designing and conducting of the trials or studies, systematic reviews and meta-analysis, and finally evidence synthesis and its clinical applications. Figure 2.1 shows study design.

The details for conducting a study are defined by the study protocol. Prior to the beginning of a trial, the Institutional Review Board (IRB) evaluates the investigation to assess study design quality, ethical questions, safeguards for patients, and a review of the informed consent statement. The CONSORT statement FOR reporting randomized trials alleviates the problem of inadequate reporting of RCTs [34], and further addresses equivalence and non-inferiority trials. Equivalence trials serve to determine if one intervention is therapeutically similar to another, whereas a non-inferiority trial seeks to determine if a treatment is no worse than the reference treatment [54]. From a statistical perspective, comparative effectiveness studies cannot hypothesize superiority of a compound over another and can only claim the null hypothesis of non-inferiority [53]. Critical appraisal of a clinical trial's quality is often incomplete and confounded by poor methodology [34]. All trials must be registered with the U.S. National Institutes of Health (NIH) Clinical Trial Registry of the United States at www.clinicaltrials.gov.

## Assessment of the Bias and Quality of the Manuscripts

Multiple instruments have been developed over the years to assess the methodological quality, along with bias, in RCTs

**Table 2.4** Usefulness of specific control types in various situations

| | Type of control | | | | | | |
|---|---|---|---|---|---|---|---|
| Trial objective | Placebo control | Active control | Dose response (D/R) | Placebo + active | Placebo + D/R | Active + D/R | Placebo + active + D/R |
| Measure absolute effect size | Y | N | N | Y | Y | N | Y |
| Show existence of effect | Y | Y | Y | Y | Y | Y | Y |
| Show dose-response relationship | N | N | Y | N | Y | Y | Y |
| Compare therapies | N | Y | N | Y | N | P | Y |

Y = Yes, N = No, P = Possible, depending on whether there is historical evidence of sensitivity to drug effects

Source: International Conference on Harmonisation of Technical Requirements for Registration of Pharmaceuticals for Human Use. ICH Harmonised Tripartite Guideline. Choice of Control Group and Related Issues in Clinical Trials E10. July 20, 2000 [31]

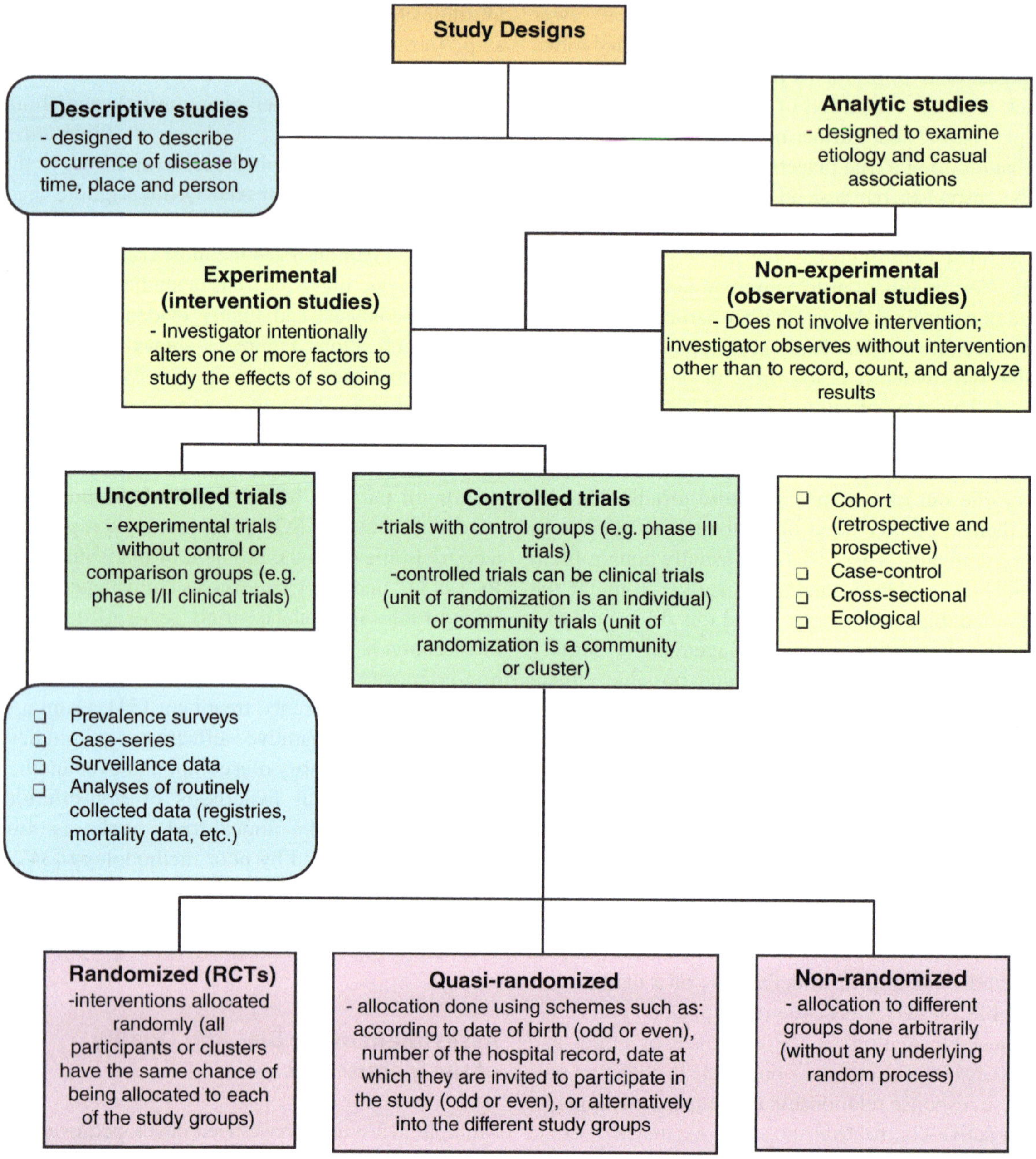


**Fig. 2.1** Classification of study designs

[61, 62]. The Cochrane review editorial board has developed its own criteria, which has been utilized extensively and has been modified over the years. Table 2.5 shows Cochrane review criteria [62] and Table 2.6 shows criteria developed by interventional pain physicians with a specific item checklist for assessment of RCTs of interventional pain management techniques [61]. The criteria developed by Manchikanti et al. [61] is specific to interventional techniques, and assesses the design, patient inclusion criteria, conduct, outcomes, randomization, allocation concealment, blinding, and conflicts of interest. Consequently, the instrument developed by Manchikanti et al will be the most appropriate instrument for utilization in studies relating to regenerative medicine to assess their quality. Methodology for systematic reviews and meta-analysis of randomized trials and observational studies has been described [63–65].

**Table 2.5** Sources of risk of bias

| Bias domain | Source of bias | Possible answers |
|---|---|---|
| Selection | (1) Was the method of randomization adequate? | Yes/No/ unsure |
| Selection | (2) Was the treatment allocation concealed? | Yes/No/ unsure |
| Performance | (3) WAs the patient blinded to the intervention? | Yes/No/ unsure |
| Performance | (4) Was the care provider blinded to the intervention? | Yes/No/ unsure |
| Detection | (5) Was the outcome assessor blinded to the intervention? | Yes/No/ unsure |
| Attrition | (6) Was the drop-out rate described and acceptable? | Yes/No/ unsure |
| Attrition | (7) Were all randomized participants analyzed in the group to which they were allocated? | Yes/No/ unsure |
| Reporting | (8) Are reports of the study free of suggestion of selective outcome reporting? | Yes/No/ unsure |
| Selection | (9) Were the groups similar at baseline regarding the most important prognosis indications? | Yes/No/ unsure |
| Performance | (10) Were cointerventions avoided or similar? | Yes/No/ unsure |
| Performance | (11) Was the compliance acceptable in all groups? | Yes/No/ unsure |
| Detection | (12) Was the timing of the outcome assessment similar in all groups? | Yes/No/ unsure |
| Other | (13) Are the other sources of potential bias unlikely? | Yes/No/ unsure |

Source: Furlan AD, Malmivaara A, Chou R, et al; Editorial Board of the Cochrane Back, Neck Group. 2015 Updated Method Guideline for Systematic Reviews in the Cochrane Back and Neck Group. *Spine (Phila Pa 1976)* 2015; 40:1660–1673 [62]

**Table 2.6** Item checklist for assessment of randomized controlled trials of IPM techniques utilizing IPM–QRB

| | | Scoring |
|---|---|---|
| 1. | Trial design and guidance reporting | 0–3 |
| 2. | Type and design of trial | 0–3 |
| 3. | Setting/physician | 0–2 |
| 4. | Imaging | 0–3 |
| 5. | Sample size | 0–3 |
| 6. | Statistical methodology | 0–1 |
| 7. | Inclusiveness of population | 0–2 |
| 8. | Duration of pain | 0–2 |
| 9. | Previous treatments | 0–2 |
| 10. | Duration of follow-up with appropriate interventions | 0–3 |
| 11. | Outcomes assessment criteria for significant improvement | 0–4 |
| 12. | Analysis of all randomized participants in the groups | 0–2 |
| 13. | Description of drop out rate | 0–2 |
| 14. | Similarity of groups at baseline for important prognostic indicators | 0–2 |
| 15. | Role of co-interventions | 0–1 |
| 16. | Method of randomization | 0–2 |
| 17. | Concealed treatment allocation | 0–2 |
| 18. | Patient blinding | 0–1 |
| 19. | Care provider blinding | 0–1 |
| 20. | Outcome assessor blinding | 0–1 |
| 21. | Funding and sponsorship | −3 to +3 |
| 22. | Conflicts of Interest | −3 to +3 |
| Total maximum | | 48 |

Source: Manchikanti L, et al. Assessment of methodologic quality of randomized trials of interventional techniques: Development of an interventional pain management specific instrument. *Pain Physician* 2014; 17:E263–E290 [61]

## Observational Studies

Most of the research in clinical practice comes from observational studies [65–69]. Proponents of observational studies describe them being just as effective as RCTs, however, from a methodological perspective, these 2 studies are considered complimentary rather than opposing [69]. Reporting of observational research is often insufficient to make statements about the strengths and weaknesses of the investigations [65–70]. Incorporation of high-quality evidence into clinical practice, to improve the effectiveness and safety of patient care, is not always limited to randomized, double-blind, placebo-controlled trials.

There are three main types of observational studies: cohort, case-control, and cross-sectional designs, and represent different approaches in an attempt to investigate the occurrence of health-related events in a given population and time period [66]. The STROBE (Strengthening the Reporting of Observational studies in Epidemiology) statement [67] with explanation and elaboration [66] for reporting observational studies addresses the 3 main types of observational studies and ensures clear presentation of what was done in any observational study [71]. The STROBE checklist is available at: www.strobe-statement.org/fileadmin/Strobe/uploads/checklists/STROBE_checklist_v4_combined.pdf.

Cohort studies are the best methods for determining the incidence and natural history of the condition [70, 72]. In a prospective cohort study, individuals without the disease, but who may or may not have been exposed, are selected, and followed over a period of time. In a retrospective cohort study, information on exposure and disease is already collected and is used to evaluate the relationship between exposure and disease over a period of time. Cohort studies describe incidence or natural history, analyze risk factors, and thereby calculate relative risk. A major strength of cohort studies is the ability to study multiple outcomes that can be associated with a single or multiple exposures in a single study. Compared to RCTs, cohort studies have broader inclusion and fewer exclusion criteria, and investigators may obtain large samples and reach greater power in statistical analysis. They also allow for the study of rare exposures and may be more generalizable in clinical practice [73].

In case-control studies, two groups are selected, either diseased or non-diseased, exposure is measured in both groups, and the association of exposure to the disease is calculated. Even though they are described as only retrospective by some [70], they also can be prospective, and are particularly useful for studying infrequent events. They are simple to organize, useful for hypothesis generation, and allow assessment of the influence of predictors on outcome via calculation of an odds ratio. The nested case-control is a variation of case-control design performed within a cohort study. Cohort subject exposures and characteristics are assessed at the start of the study, and case and control subjects are identified at a later time. The nested study design facilitates the selection of control subjects, as it is selected from cohort members who do not meet the requirements of a case [74].

Cross-sectional studies are used to measure the prevalence of health outcomes, understand determinants of health, and describe features of a population. They are relatively quick and inexpensive, do not differentiate between cause and effect, and can study multiple outcomes. Unlike other observational study designs, cross-sectional studies do not follow subjects over time. Cross-sectional studies are also highly susceptible to sampling bias as the study sample needs to be selected from a large and heterogeneous population [75].

Similar to the checklist for RCTs, Manchikanti et al. developed a comprehensive instrument that is helpful in assessing the methodological quality of non-randomized trials and is specific to interventional techniques (Table 2.7) [76].

**Table 2.7** IPM checklist for assessment of nonrandomized or observational studies of IPM techniques utilizing IPM-QRBNR

| | | Scoring |
|---|---|---|
| 1. | Study design guidance and reporting | 0–4 |
| 2. | Study design and type | 0–4 |
| 3. | Setting/physician | 0–2 |
| 4. | Imaging | 0–3 |
| 5. | Sample size | 0–4 |
| 6. | Statistical methodology | 0–2 |
| 7. | Inclusiveness of population | 1–4 |
| 8. | Duration of pain | 0–2 |
| 9. | Previous treatments | 0–2 |
| 10. | Duration of follow-up with appropriate interventions | 1–4 |
| 11. | Outcomes assessment criteria for significant improvement | 0–4 |
| 12. | Description of drop out rate | 0–2 |
| 13. | Similarity of groups at baseline for important prognostic indicators | 0–2 |
| 14. | Role of co-interventions | 1–2 |
| 15. | Method of assignment of participants | 1–4 |
| 16. | Funding and sponsorship | −3 to +3 |
| Total maximum | | 48 |

Source: Manchikanti L, Hirsch JA, Heavner JE, et al. Development of an interventional pain management specific instrument for methodologic quality assessment of nonrandomized studies of interventional techniques. Pain Physician 2014; 17: E291–E317 [76]

## Outcomes in Regenerative Medicine Studies in Chronic Pain

To properly address the possible association between exposures and outcomes in the conducted study, clearly defined primary and secondary outcome measures are essential. The majority of the current available musculoskeletal and spine Regenerative Medicine literature measures pain as one of the end points. Because of its subjective nature, interpreting pain is difficult. Chronic pain includes emotional, physical, and functional aspects besides the sensory phenomenon [77, 78]. Two widely used scales in RCTs include numeric rating scale (NRS) and visual analog scale (VAS) [38]. VAS scales are described on a scale 0–100 mm or 0–10 cm and VAS scores are more suitable for parametric analyses. NRS uses 0–10, 0–20 or even 0–100-point scales and is used more widely than VAS [38]. Statistical and clinical significance are not necessarily equivalent; treatment may result in a statistically significant change and yet be clinically meaningless [79].

Quality of life could be evaluated by a number of instruments: Short-Form-36 (SF-36) [80], Oswestry Disability Index (ODI) [81], Roland-Morris Disability Questionnaire (RDQ) [82], and Neck Disability Index (NDI) [83]. SF-36 is a health survey with 36 questions and measures physical and mental health summary scores [80]. SF-36 is sensitive to change, may serve as a validation tool in developing new disease-specific instruments, and is able to differentiate between treatment responders and non-responders.

ODI is the most common condition-specific outcome measure for low back pain assessment [84] demonstrating face and content validity and test to re-test stability. RDQ, similar to ODI, has been used as a condition-specific health status measure for low back pain [82, 85] and is designed to assess physical disability in patients with this condition. RDQ focuses on several physical functions such as walking, bending over, sitting, lying down, sleeping, self-care etc. RDQ scores range from 0 (no disability) to 24 (maximum disability) and although designed on paper, it can also be administered on computer and by telephone [85]. RDQ fails to address psychological or social problems, but its restricted nature of the domains covered makes the score easy to understand and interpret [85].

NDI has been validated in several study populations [86]. Its strength lies in its validity against multiple measures of function, pain, and clinical signs/symptoms [87]. A significant correlation between physical and mental health components of SF-36 and NDI was identified [87].

## Systematic Review and Meta-analysis

Treatment decisions are based on the hierarchy of strength of the evidence. The best sources of evidence include systematic reviews and meta-analyses. Systematic review is defined as, "the application of scientific strategies that limit bias by the systematic assembly, critical appraisal and synthesis of all relevant studies on a specific topic" [88, 89]. Systematic reviews, unlike narrative reviews, minimize bias through a comprehensive search as well as selection of articles for review and provides assessment of the methodological quality of the studies (Table 2.8) [90, 91]. Oxman et al. [92] provided guidance for critical appraisal of the evidence since reviews are prone to bias: The best approach would be to include the maximum possible number of relevant trials, with a detailed description of their strengths and weaknesses.

A systematic review in interventional pain medicine should include a comprehensive review using at least three sources/databases (Medline, EMBASE, Cochrane library), hand search of references for eligible trials, and direct contact with corresponding authors of eligible trials asking for additional published or unpublished trials [93]. Empiric research has shown that not all reviews are truly systematic, and that the quality of the systematic reviews is highly inconsistent [90].

A properly designed systematic review with meta-analysis of the literature without conflict or confluence of interest is currently the best, and least biased way to organize, evaluate, and integrate the research evidence [91]. Systematic review serves various purposes [94]: extracts relevant information, assesses consistency of the results among studies and their generalizability, combines information from individual studies, limits bias, and improves reliability and accuracy of recommendations. In contrast to a systematic review, meta-analysis is a statistical procedure, which integrates results of several compatible independent studies. Whereas a conventional meta-analysis makes direct comparisons between trials in a pairwise manner, a network meta-analysis can evaluate multiple treatments in a single analysis, by using logical inference to incorporate evidence from indirect comparisons (i.e., comparing A to C by using studies which compare A to B and B to C) [95]. Meta-analysis must be carefully planned, the graphical display from individual studies allows for the examination of degree of heterogeneity between studies, different methods can be used to combine data, and the definition of eligibility criteria for studies to be included in comprehensive search is imperative. Heterogeneity is a common finding among meta-analysis [96]. Bias in meta-analysis may be detected by a simple, graphical test [97]. In order to reduce bias in meta-analysis, effort must be put to find all published studies, available but unpublished work, and careful quality assessment component studies.

**Table 2.8** Differences between narrative and systematic reviews

| Core feature | Narrative review | Systematic review |
|---|---|---|
| Study question | Often broad in scope | Often a focused clinical question |
| Data sources and search strategy | Specifications of database searched and search strategy are not typically provided | Comprehensive search of many databases as well as the so-called gray literature. Explicit search strategy provided |
| Selection of articles for study | Not usually specified. If specified, potentially biased | Criterion-based selection, uniformly applied |
| Article review or approval | Variable, depending on who is conducting the review | Rigorous critical appraisal, typically using a data extraction form |
| Study quality | Usually not assessed. If assessed, may not use formal quality assessment | Some assessment of quality is almost always included as part of the data extraction process |
| Synthesis | Often a qualitative summary | Quantitative or qualitative summary |
| Inferences | Occasionally evidence based | Usually evidence based |

Adapted from: Manchikanti L. Evidence-based medicine, systematic reviews, and guidelines in interventional pain management, Part I: Introduction and general considerations. *Pain Physician* 2008; 11:161–186 [90] and Atkins D, et al. Systems for grading the quality of evidence and the strength of recommendations I: critical appraisal of existing approaches The GRADE Working Group. *BMC Health Serv Res* 2004; 4:38 [89]

Meta-analyses are performed using data collected from independent primary studies which answer the same question, to yield a quantitative estimate of the studied phenomenon [98]. Different methodology can be used depending on the question being asked. To study the effect of a single intervention, a single-arm meta-analysis can be used. For example, significant decreases in pain scores following lumbar disc injections of PRP and MSCs were detected by Sanapati et al. [29] using a single-arm meta-analysis of randomized trials, observational studies, and case reports. Alternatively, a dual-arm meta-analysis is used to compare two treatments, with the same outcome measure. For example, Knezevic et al. [99] used a series of dual-arm meta-analyses of randomized trials to qualify superiority or equivalency of epidural injections of local anesthetic alone or combined with steroids, on pain levels, functionality, and disability.

Shrier et al. [100] demonstrated the advantages of using both observational studies and RCTs in meta-analysis in many situations to reduce false conclusions based solely on RCTs. They also justified reasons to include observational studies in systematic reviews, based on the findings that RCTs sometimes contradict the findings of highly publicized observational studies, and sometimes also highly publicized RCTs [101]. It is important to evaluate the discrepancies that occur between studies as they provide information that can be utilized for appropriate clinical rea-

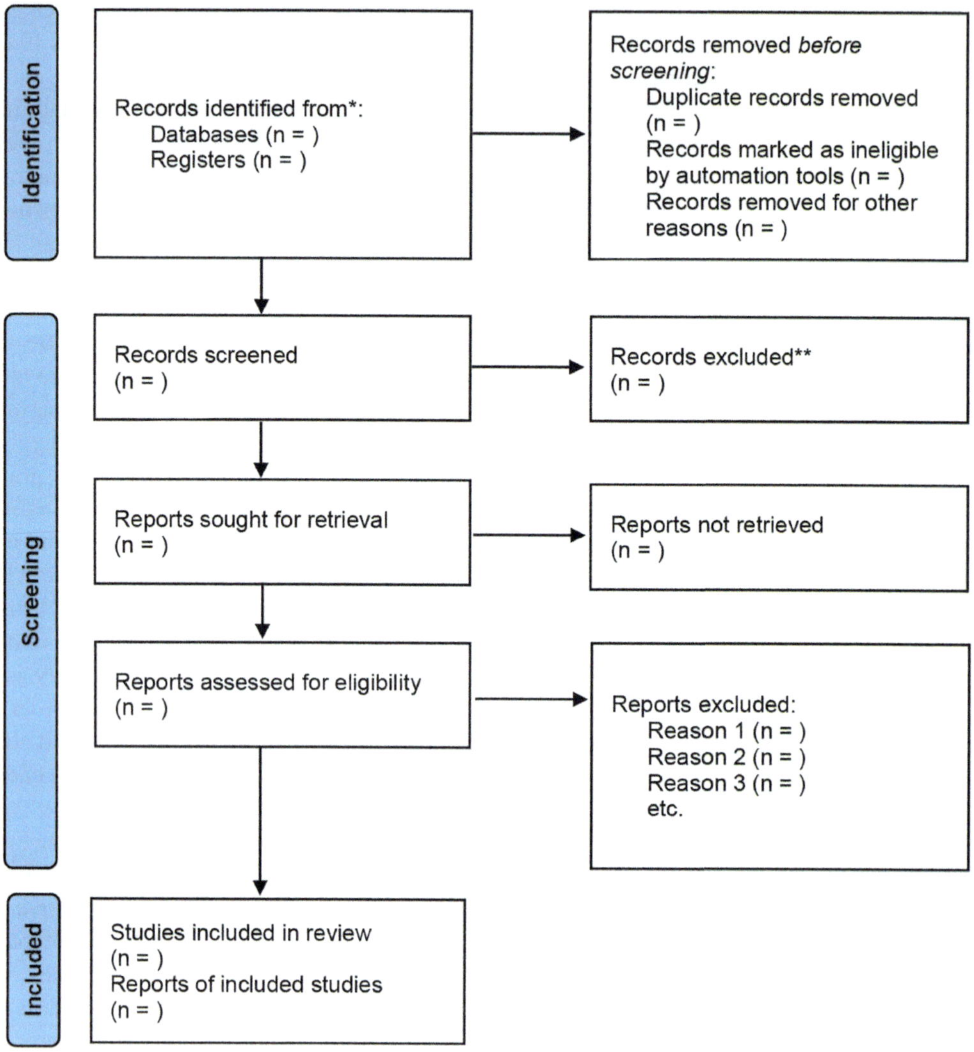


**Fig. 2.2** Preferred reporting items for systematic reviews and meta-analyses (PRISMA) 2020 flow diagram. Adapted and modified from: Page MJ, McKenzie JE, Bossuyt PM et al. The PRISMA 2020 statement: An updated guideline for reporting systematic reviews. BMJ 2021; 372:n71 [102]

soning and causal inferences [100]. It was described that even though the randomized trials were identified as "best evidence" on the hierarchy of evidence, they still were imperfect. In addition, non-randomized studies are useful because they can provide information about a larger and more diverse population of patients, along with their tendency to follow patients over a long period of time. If particular attention was not paid to whether potential confounding factors, such as age or gender, relatively proportionally occurred in the treatment and control groups, then a hierarchy of evidence that favored randomized or non-randomized was not necessary.

The reporting evidence was updated in the form of PRISMA (Preferred Reporting Items for Systematic reviews and Meta-Analyses) statement, and it is usually followed by the PRISMA flow diagram available at: http://prisma-statement.org/PRISMAStatement/FlowDiagram.aspx (Fig. 2.2) [63, 64, 102].

## Conclusion

Evidence provides insight to facts, and we need to recognize that insight's reflective meanings are differentially relevant to distinct groups of moral agents (e.g., patients, physicians, etc.). Evidence is a tool, and different types of evidence (as different types of tools) serve particular purposes for those who use them [1, 2]. Subsequently, the tools used must be well constructed and durable – while remaining revisable to accommodate the purpose of the task at hand and the knowledge that defines both the tasks and the utility of the tool(s) [1, 2].

If the task at hand is the ongoing improvement of patient-centered pain medicine, and if we are to succeed at this task, then we must be empowered to (1) advance well-conducted research, (2) expand education, and (3) enthuse good practice by recognizing what is effective and what is not and develop new techniques and approaches to therapeutics.

Evidence synthesis is complex and difficult, similar to the practice of interventional pain management. Difficult tasks require diligence, commitment, and knowledge (i.e., the "right" knowledge to inform and compel right and good decisions and actions). Such knowledge is powerful. It is hoped that by working together, as a community of clinicians, patients, researchers, administrators, and policymakers, we can become empowered to maintain a global effort to enhance the standards and practices of regenerative medicine.

**Key Points**

1. Evidence-based medicine (EBM) has gained prominence in medicine over the past decade, driven by a desire to enhance patient care through informed clinical decision-making. EBM involves the systematic accumulation, analysis, and application of evidentiary knowledge in clinical settings.
2. Comparative effectiveness research (CER) is a term often used interchangeably with EBM. CER aims to generate real-world evidence on the effectiveness and cost-effectiveness of interventions using real-world data. It involves systematic reviews, investigation of treatment variations, randomized controlled trials, and cost-benefit analysis, and has become a crucial methodology for healthcare assessment.
3. The development of evidence-based medicine (EBM) can be traced back to the second half of the twentieth century, driven by the need to organize contradictory data and improve clinical decision-making.
4. Factors such as limited and inconsistent data on treatment effectiveness, rising healthcare costs, and the demand for measurable outcomes have contributed to the widespread adoption of EBM.
5. The Institute of Medicine (IOM) defined EBM as the conscientious use of current best evidence in patient care and categorized evidence into different levels.
6. Various organizations, such as the American Society of Interventional Pain Physicians (ASIPP) and the Agency for Healthcare Research and Quality (AHRQ), have recommended modified approaches to grading evidence and providing strength of recommendations.
7. The subjective nature of pain poses challenges for evidence-based medicine (EBM) in the field of chronic pain management. Patients with severe pain may seek any treatment with even a small chance of improvement, requiring clinicians to balance evidence-based practices with the desperate needs of their patients.
8. Interventional regenerative medicine techniques have gained importance in managing chronic pain conditions, but the evidence in this field is still evolving. The goal of research in this area is to standardize techniques, improve effectiveness, and minimize the risk of adverse events. EBM methodology, including systematic reviews and meta-analyses, can be utilized to evaluate the effectiveness of regenerative medicine therapies and inform evidence-based recommendations and guidelines for their use in chronic pain management.
9. Randomized controlled trials (RCTs) are considered the gold standard in evidence-based medicine (EBM), but they have limitations when studying pain management due to strict inclusion criteria and confounding factors. Debate exists regarding placebo controls and study design.
10. Observational studies, including cohort, case-control, and cross-sectional designs, play a crucial role in clinical research and complement randomized controlled trials (RCTs). Cohort studies provide insights into incidence and natural history, while case-control studies assess the association between exposure and disease. Cross-sectional studies measure prevalence and describe population features. The STROBE statement provides guidelines for reporting observational studies, and the checklist developed by Manchikanti et al helps assess the methodological quality of non-randomized trials in interventional techniques.

**Questions and Answers**

1. What is the Null Hypothesis in a randomized controlled trial?
   (A) The intervention is superior to the control
   (B) There is no difference between the intervention and the control
   (C) The control is superior to the intervention
   (D) The intervention and control both significantly influence outcome(s)

   **Answer: B**

   **Explanation:** The null hypothesis in randomized trials assumes that there is no difference between a new and a placebo/control treatment.
2. Which of the following is the evidence-based, minimum set of recommendations for reporting randomized trials?
   (A) Consolidated Standards of Reporting Trials (CONSORT)
   (B) Strengthening of Reporting of Observational Studies in Epidemiology (STROBE)
   (C) Institutional Review Board (IRB)
   (D) Preferred Reporting Items for Systematic Reviews and Meta-Analysis (PRISMA)

   **Answer: A**

   **Explanation:** CONSORT (Consolidated Standards of Reporting Trials) encompasses various initiatives developed by the CONSORT Group to alleviate the problems arising from inadequate reporting of randomized controlled trials. The CONSORT statement for reporting randomized trials provides a checklist of items to include: study question, study population, randomization, blinding, interventions, outcomes, statistical analysis, results, discussion, and funding.

3. Which of the following sources of evidence allows for the examination of degree of heterogeneity between studies?
   (A) Narrative Review
   (B) Meta-Analysis
   (C) Systematic Review
   (D) Randomized Controlled Trial

   **Answer: B**

   **Explanation**: Meta-analysis must be carefully planned, the graphical display from individual studies allows for the examination of degree of heterogeneity between studies.

4. Which of the following study designs involves a randomization element?
   (A) Cohort
   (B) Cross-sectional
   (C) Case-control
   (D) Crossover

   **Answer: D**

   **Explanation:** In a crossover randomized controlled trial, two or more treatments are provided to subjects in a randomized sequence.

5. Which of the following types of control is appropriate for measuring absolute effect size?
   (A) Dose-Response Control
   (B) Placebo Control
   (C) Active Control
   (D) No Control Group

   **Answer: B**

   **Explanation:** A placebo-controlled trial provides information on the existence of effect and measurable absolute effect size.

**Acknowledgments** This book chapter is modified and updated from a previous book chapter, "Principles of Evidence-Based Medicine" by Nebojsa Nick Knezevic, MD, PhD, MD, Laxmaiah Manchikanti, MD, Annu Navani, MD, and Joshua A. Hirsch, MD, in *Essentials of Regenerative Medicine in Interventional Pain Management* published by ASIPP Publishing.

## References

1. Knezevic NN, Manchikanti L, Navani A, Hirsch JA. Principles of evidence-based medicine. In: Manchikanti L, Navani A, Atluri S, editors. Essentials of regenerative medicine in interventional pain management. Paducah: ASIPP Publishing; 2019. p. 39–54.
2. Manchikanti L, Atluri S, Boswell MV, et al. Methodology for evidence synthesis and development of comprehensive evidence-based guidelines for interventional techniques in chronic spinal pain. Pain Physician. 2021;24:S1–S26.
3. Jenicek M. How do we see medicine, health and disease? A basic set of rules and fundamental paradigms. In: Foundations of evidence-based medicine. New York: The Parthenon Publishing Group, Inc.; 2005. p. 3–13.
4. Isaacs D, Fitzgerald D. Seven alternatives to evidence based medicine. BMJ. 1999;319:1618.
5. Jonas WB. The evidence house: how to build an inclusive base for complementary medicine. West J Med. 2001;175:79–80.
6. Jenicek M. The work of physicians with individuals and communities. Epidemiology and other partners in evidence-based medicine. In: Foundations of evidence-based medicine. New York: The Parthenon Publishing Group, Inc.; 2005. p. 3–13.
7. Haynes RB, Sackett RB, Gray JMA, et al. Transferring evidence from research into practice, 1: the role of clinical care research evidence in clinical decisions. ACP J Club. 1996;125:A14–6.
8. Napodano RJ. Values in medical practice. New York: Human Sciences Press; 1986.
9. Manchikanti L, Knezevic NN, Navani A, et al. Epidural interventions in the management of chronic spinal pain: American Society of Interventional Pain Physicians (ASIPP) comprehensive evidence-based guidelines. Pain Physician. 2021;24:S27–S208.
10. Manchikanti L, Kaye AD, Soin A, et al. Comprehensive evidence-based guidelines for facet joint interventions in the management of chronic spinal pain: American Society of Interventional Pain Physicians (ASIPP) guidelines. Pain Physician. 2020;23:S1–S127.
11. Manchikanti L, Centeno CJ, Atluri S, et al. Bone marrow concentrate (BMC) therapy in musculoskeletal disorders: evidence-based policy position statement of American Society of Interventional Pain Physicians (ASIPP). Pain Physician. 2020;23:E85–E131.
12. Manchikanti L, Abdi S, Atluri S, et al. An update of comprehensive evidence-based guidelines for interventional techniques of chronic spinal pain: part II: guidance and recommendations. Pain Physician. 2013;16:S49–S283.
13. Manchikanti L, Falco FJE, Boswell MV, et al. Facts, fallacies, and politics of comparative effectiveness research: part 1. Basic considerations. Pain Physician. 2010;13:E23–54.
14. Manchikanti L, Falco FJE, Boswell MV, et al. Facts, fallacies, and politics of comparative effectiveness research: part 2. Implications for interventional pain management. Pain Physician. 2010;13:E55–79.
15. Massarweh NN, Haukoos JS, Ghaferi AA. ISPOR reporting guidelines for comparative effectiveness research. JAMA Surg. 2021;156:673–4.
16. Berger ML, Mamdani M, Atkins D, Johnson ML. Good research practices for comparative effectiveness research: defining, reporting and interpreting nonrandomized studies of treatment effects using secondary data sources: The ISPOR Good Research Practices for Retrospective Database Analysis Task Force Report—part I. Value Health. 2009;12:1044–52.
17. Cox E, Martin BC, Van Staa T, et al. Good research practices for comparative effectiveness research: approaches to mitigate bias and confounding in the design of non-randomized studies of treatment effects using secondary data sources: the ISPOR Good Research Practices for Retrospective Database Analysis Task Force–part II. Value Health. 2009;12:1053–61.
18. Johnson ML, Crown W, Martin BC, et al. Good research practices for comparative effectiveness research: analytic methods to improve causal inference from nonrandomized studies of treatment effects using secondary data sources: the ISPOR Good Research Practices for Retrospective Database Analysis Task Force Report—part III. Value Health. 2009;12:1062–73.
19. Cahana A. Ethical and epistemological problems when applying evidence-based medicine to pain management. Pain Pract. 2005;5:298–302.
20. Cochrane A. 1931-1971: a critical review, with particular reference to the medical profession. Medicines for the Year 2000. Office of Health Economics, London, 1979.
21. Hartrick CT. Evidence-based medicine: our responsibility to our patients. Pain Pract. 2007;7:219–20.

22. Olsen LA. The learning healthcare system: workshop summary. Washington: Institute of Medicine; 2007.
23. Clark LL. The value of the case report in the age of evidence-based medicine. Pain Med. 2011;12:692–4.
24. Manchikanti L, Falco FJE, Benyamin RM, et al. A modified approach to grading of evidence. Pain Physician. 2014;17:E319–25.
25. National guideline clearinghouse extent adherence to trustworthy standards (NEATS) instrument.
26. Manchikanti L, Kaye AM, Knezevic NN, et al. Responsible, safe, and effective prescription of opioids for chronic non-cancer pain: American Society of Interventional Pain Physicians (ASIPP) guidelines. Pain Physician. 2017;20:S3–S92.
27. Rathmell JP, Carr DB. The scientific method, evidence-based medicine, and rational use of interventional pain treatments. Reg Anesth Pain Med. 2003;28:498–501.
28. Navani A, Manchikanti L, Albers SL, et al. Responsible, safe, and effective use of biologics in the management of low back pain: American Society of Interventional Pain Physicians (ASIPP) guidelines. Pain Physician. 2019;22:S1–S74.
29. Sanapati J, Manchikanti L, Atluri S, et al. Do regenerative medicine therapies provide long-term relief in chronic low back pain: a systematic review and metaanalysis. Pain Physician. 2018;21:515–40.
30. Hill AB. Medical ethics and controlled trials. BMJ. 1963;1:1043–9.
31. International Conference on Harmonisation of Technical Requirements for Registration of Pharmaceuticals for Human Use. ICH Harmonised Tripartite Guideline. Choice of Control Group and Related Issues in Clinical Trials E10. July 20, 2000.
32. Manchikanti L, Giordano J, Fellows B, Hirsch JA. Placebo and nocebo in interventional pain management: a friend or a foe – or simply foes? Pain Physician. 2011;14:E157–75.
33. Manchikanti L, Boswell MV, Kaye AD, Helm S II, Hirsch JA. Therapeutic role of placebo: evolution of a new paradigm in understanding research and clinical practice. Pain Physician. 2017;20:363–86.
34. Turner L, Shamseer L, Altman DG, et al. Consolidated standards of reporting trials (CONSORT) and the completeness of reporting of randomized controlled trials (RCTs) published in medical journals. Cochrane Database Syst Rev. 2012;11:MR000030.
35. Kao LS, Tyson JE, Blakely ML, et al. Clinical research methodology I: introduction to randomized trials. J Am Coll Surg. 2008;206:361–9.
36. Kang M, Ragan BG, Park JH. Issues in outcomes research: an overview of randomization techniques for clinical trials. J Athl Train. 2008;43:215–21.
37. Manchikanti L, Pampati V. Research designs in interventional pain management: is randomization superior, desirable or essential? Pain Physician. 2002;5:275–84.
38. Greenland S. Randomization, statistics, and causal inference. Epidemiology. 1990;1:421–9.
39. Manchikanti L, Hirsch JA, Smith HS. Evidence-based medicine, systematic reviews, and guidelines in interventional pain management: part 2: randomized controlled trials. Pain Physician. 2008;11:717–73.
40. Altman DG, Bland JM. How to inimizat. BMJ. 1999;319:703–4.
41. Treasure T, MacRae KD. Minimisation: the platinum standard for trials? Randomisation doesn't guarantee similarity of groups; minimization does. BMJ. 1998;317:362–3.
42. Donner A, Klar N. Pitfalls of and controversies in cluster randomization trials. Am J Public Health. 2004;94:416–22.
43. Cornfield J. Randomization by group: a formal analysis. Am J Epidemiol. 1978;108:100–2.
44. Halloran ME, Longini IM Jr, Struchiner CJ. Design and interpretation of vaccine field studies. Epidemiol Rev. 1999;21:73–88.
45. Weir CJ, Lees KR. Comparison of stratification and adaptive methods for treatment allocation in an acute stroke clinical trial. Stat Med. 2003;22:705–26.
46. Lim CY, In J. Considerations for crossover design in clinical study. Korean J Anesthesiol. 2021;74:293–9.
47. Monaghan TF, Agudelo CW, Rahman SN, et al. Blinding in clinical trials: seeing the big picture. Medicina. 2021;57(7):647.
48. Pildal J, Chan AW, Hrobjartsson A, et al. Comparison of descriptions of allocation concealment in trial protocols and the published reports: cohort study. BMJ. 2005;330:1049.
49. Forder PM, Gebski VJ, Keech AC. Allocation concealment and blinding: when ignorance is bliss. Med J Aust. 2005;182:87–9.
50. Feinstein AR, Horwitz RI. Problems in the "evidence" of "evidence-based medicine". Am J Med. 1997;103:529–35.
51. Siderowf AD. Evidence from clinical trials: can we do better? NeuroRx. 2004;1:363–71.
52. Stoessl AJ. Deception and the ethics of placebo. Int Rev Neurobiol. 2020;153:147–63.
53. Enck P, Klosterhalfen S. Placebos and the placebo effect in drug trials. Handb Exp Pharmacol. 2019;260:399–431.
54. Piaggio G, Elbourne DR, Altman DG, et al. Reporting of noninferiority and equivalence randomized trials: an extension of the CONSORT statement. JAMA. 2006;295:1152–60.
55. Chou R, Hashimoto R, Friedly J, et al. Pain management injection therapies for low back pain. In: Technology assessment report ESIB0813. (Prepared by the Pacific Northwest Evidence-based Practice Center under Contract No. HHSA 290-2012-00014-I.). Rockville: Agency for Healthcare Research and Quality; 2015.
56. Manchikanti L, Knezevic NN, Boswell MV, et al. Epidural injections for lumbar radiculopathy and spinal stenosis: a comparative systematic review and meta-analysis. Pain Physician. 2016;19:E365–410.
57. Boswell MV, Manchikanti L. Appropriate design and methodologic quality assessment, clinically relevant outcomes are essential to determine the role of epidural corticosteroid injections. Evid Based Med. 2016;21:89.
58. Chou R, Atlas SJ, Loeser JD, et al. Guideline warfare over interventional therapies for low back pain: can we raise the level of discourse? J Pain. 2011;12:833–9.
59. Manchikanti L, Benyamin RM, Falco FJE, et al. Guidelines warfare over interventional techniques: is there a lack of discourse or straw man? Pain Physician. 2012;15:E1–E26.
60. Manchikanti L, Kaye AD, Hirsch JA. Comment RE: Chou R, Hashimoto R, Friedly J, et al. Epidural corticosteroid injections for radiculopathy and spinal stenosis: a systematic review and meta-analysis. Ann Intern Med. 2015;163:373–81.
61. Manchikanti L, Hirsch JA, Cohen SP, et al. Assessment of methodologic quality of randomized trials of interventional techniques: development of an interventional pain management specific instrument. Pain Physician. 2014;17:E263–90.
62. Furlan AD, Malmivaara A, Chou R, et al. 2015 Updated method guideline for systematic reviews in the Cochrane back and neck group. Spine. 2015;40:1660–73.
63. Liberati A, Altman DG, Tetzlaff J, et al. The PRISMA statement for reporting systematic reviews and meta-analyses of studies that evaluate health care interventions: explanation and elaboration. Ann Intern Med. 2009;151:W65–94.
64. Moher D, Liberati A, Tetzlaff J, et al. Preferred reporting items for systematic reviews and meta-analyses: The PRISMA statement. BMJ. 2009;339:b2535.
65. Manchikanti L, Datta S, Smith HS, et al. Evidence-based medicine, systematic reviews, and guidelines in interventional pain management: part 6. Systematic reviews and meta-analyses of observational studies. Pain Physician. 2009;12:819–50.

66. Vandenbroucke JP, von Elm E, Altman DG, et al. Strengthening the reporting of observational studies in epidemiology (STROBE): explanation and elaboration. Ann Intern Med. 2007;147:W163–94.
67. von Elm E, Altman DG, Egger M, et al. The strengthening the reporting of observational studies in epidemiology (STROBE) Statement: guidelines for reporting observational studies. Int J Surg. 2014;12:1495–9.
68. Greene T. Are observational studies 'just as effective' as randomized clinical trials? Blood Purif. 2000;18:317–22.
69. Manchikanti L, Singh V, Smith HS, et al. Evidence-based medicine, systematic reviews, and guidelines in interventional pain management: part 4: observational studies. Pain Physician. 2009;12:73–108.
70. Mann CJ. Observational research methods. Research design II: cohort, cross sectional, and case-control studies. Emerg Med J. 2003;20:54–60.
71. Ghaferi AA, Schwartz TA, Pawlik TM. STROBE reporting guidelines for observational studies. JAMA Surg. 2021;156:577–8.
72. Setia MS. Observational studies: how to go about them? Indian J Dermatol Venereol Leprol. 2008;74:288–91.
73. Wang X, Kattan MW. Cohort studies: design, analysis, and reporting. Chest. 2020;158:S72–8.
74. Dey T, Mukherjee A, Chakraborty S. A practical overview of case-control studies in clinical practice. Chest. 2020;158:S57–64.
75. Wang X, Cheng Z. Cross-sectional studies: strengths, weaknesses, and recommendations. Chest. 2020;158:S65–71.
76. Manchikanti L, Hirsch JA, Heavner JE, et al. Development of an interventional pain management specific instrument for methodologic quality assessment of nonrandomized studies of interventional techniques. Pain Physician. 2014;17:E291–317.
77. Gracely RH, Kwilosz DM. The descriptor differential scale: applying psychophysical principles to clinical pain assessment. Pain. 1988;35:279–88.
78. Leavitt F, Garron DC, Whisler WW, et al. Affective and sensory dimensions of back pain. Pain. 1978;4:273–81.
79. Turk DC. Statistical significance and clinical significance are not synonyms! Clin J Pain. 2000;16:185–7.
80. Ware JE Jr. SF-36 health survey update. Spine. 2000;25: 3130–9.
81. Fairbank JC, Pynsent PB. The Oswestry disability index. Spine. 2000;25:2940–52. discussion 2952
82. Roland M, Morris R. A study of the natural history of back pain. Part I: development of a reliable and sensitive measure of disability in low-back pain. Spine. 1983;8:141–4.
83. Vernon H, Mior S. The neck disability index: a study of reliability and validity. J Manip Physiol Ther. 1991;14:409–15.
84. Fairbank JC, Couper J, Davies JB, et al. The Oswestry low back pain disability questionnaire. Physiotherapy. 1980;66:271–3.
85. Roland M, Fairbank J. The Roland-Morris disability questionnaire and the Oswestry disability questionnaire. Spine. 2000;25:3115–24.
86. Cleland JA, Childs JD, Whitman JM. Psychometric properties of the Neck Disability Index and Numeric Pain Rating Scale in patients with mechanical neck pain. Arch Phys Med Rehabil. 2008;89:69–74.
87. Pietrobon R, Coeytaux RR, Carey TS, et al. Standard scales for measurement of functional outcome for cervical pain or dysfunction: a systematic review. Spine. 2002;27:515–22.
88. Wright RW, Brand RA, Dunn W, et al. How to write a systematic review. Clin Orthop Relat Res. 2007;455:23–9.
89. Atkins D, Eccles M, Flottorp S, et al. Systems for grading the quality of evidence and the strength of recommendations I: critical appraisal of existing approaches The GRADE Working Group. BMC Health Serv Res. 2004;4:38.
90. Manchikanti L. Evidence-based medicine, systematic reviews, and guidelines in interventional pain management, Part I: introduction and general considerations. Pain Physician. 2008;11:161–86.
91. Manchikanti L, Benyamin RM, Helm S, et al. Evidence-based medicine, systematic reviews, and guidelines in interventional pain management: part 3: systematic reviews and meta-analyses of randomized trials. Pain Physician. 2009;12:35–72.
92. Oxman AD, Schunemann HJ, Fretheim A. Improving the use of research evidence in guideline development: 8. Synthesis and presentation of evidence. Health Res Policy Syst. 2006;4:20.
93. Cook DJ, Guyatt GH, Ryan G, et al. Should unpublished data be included in meta-analyses? Current convictions and controversies. JAMA. 1993;269:2749–53.
94. Petrie A, Bulman JS, Osborn JF. Further statistics in dentistry Part 8: systematic reviews and meta-analyses. Br Dent J. 2003;194:73–8.
95. Rouse B, Chaimani A, Li T. Network meta-analysis: an introduction for clinicians. Intern Emerg Med. 2017;12:103–11.
96. Engels EA, Schmid CH, Terrin N, et al. Heterogeneity and statistical significance in meta-analysis: an empirical study of 125 meta-analyses. Stat Med. 2000;19:1707–28.
97. Egger M, Davey Smith G, Schneider M, et al. Bias in meta-analysis detected by a simple, graphical test. BMJ. 1997;315:629–34.
98. Mikolajewicz N, Komarova SV. Meta-analytic methodology for basic research: a practical guide. Front Physiol. 2019;10:203.
99. Knezevic NN, Manchikanti L, Urits I, et al. Lack of superiority of epidural injections with lidocaine with steroids compared to without steroids in spinal pain: a systematic review and meta-analysis. Pain Physician. 2020;23:S239–70.
100. Shrier I, Boivin JF, Steele RJ, et al. Should meta-analyses of interventions include observational studies in addition to randomized controlled trials? A critical examination of underlying principles. Am J Epidemiol. 2007;166:1203–9.
101. Ioannidis JP. Contradicted and initially stronger effects in highly cited clinical research. JAMA. 2005;294:218–28.
102. Page MJ, McKenzie JE, Bossuyt PM, et al. The PRISMA 2020 statement: an updated guideline for reporting systematic reviews. BMJ. 2021;372:n71.

# 3 Evidence-Based Regenerative Medicine

Swarnima Vardhan, Ankur A. Patel, and Sudhir Diwan

## Introduction

The field of regenerative medicine has expanded the horizon for management of musculoskeletal and spinal pain. Regenerative medicine is focused on human cells and incorporates biomedical, biochemical, and biomechanical technologies to improve cellular migration, replication, and modeling. Biomedical approaches include the injection of mesenchymal stem cells (MSCs) or progenitor cells called cell therapy, the induction of regeneration by biologically active molecules administered alone or as a complex of infused cells referred to as immunomodulation therapy, and transplantation of in vitro grown organs and tissues that fall under the umbrella of tissue engineering [1]. Biological therapies reduce the inflammation and provide symptomatic relief in short term, alongside, address the root cause and delay the degenerative process to aid in long-term regeneration of the damaged tissue. In this chapter, we will discuss about two main pillars of regenerative medicine used for management of spinal and musculoskeletal conditions, namely platelet-rich plasma (PRP) and MSCs. The American Society of Interventional Pain Physicians (ASIPP) has developed guidelines on performing regenerative medicine procedures along with a position statement on bone marrow concentrate (BMC) injections for musculoskeletal and spinal use [1, 2].

S. Vardhan
Advanced Spine on Park Avenue, New York, NY, USA

Yale New Haven Health, Bridgeport, CT, USA

A. A. Patel
New York-Presbyterian Hospital, Weill Cornell Medical Center, New York, NY, USA

S. Diwan (✉)
Advanced Spine on Park Avenue, New York, NY, USA

## Regenerative Therapies

### Platelet-Rich Plasma

PRP is a concentrate of whole blood that is centrifuged to obtain plasma rich in platelets and hence growth factors. Growth factors in PRP which are essential for the healing process include endothelial growth factor (EGF), platelet-derived growth factor (PDGF), vascular endothelial growth factor (VEGF), and basic fibroblast growth factor (b-FGF) [3]. These growth factors promote angiogenesis, increase fibroblast or osteoblast metabolic activity, which contributes to new tissue formation and aid in the healing process [1].

Platelets are obtained from whole blood through a process of differential centrifugation. There are two basic methods used for preparation of platelet concentrate, one is PRP-based and other buffy-coat based. The buffy coat is a layer rich in platelets and leukocytes visible within a sample of fractionate blood after centrifuging. The inclusion of the buffy coat in the injectate has been a question of debate. The presence of white blood cells is associated with a reduced incidence of infection and accentuated inflammatory response due to the increased concentrations of inflammatory cytokines, such as interleukin-1β (IL-1β), TNF-α, IL-6, and IL-8, which augments the healing process. However, many studies have demonstrated that addition of leukocytes to PRP causes increased pain and discomfort after the procedure. The platelet poor fraction of blood is also known to have substances that may be of use for regenerative approaches, including fibrinogen, alpha-2-macroglobulin (A2M) and exosomes.

### Mesenchymal Stem Cells or Medicinal Signaling Cells (MSCs)

Stem cells have long been believed to be unspecialized cells capable of self-renewal [4]. Caplan proposed new nomenclature for these cells to be identified as "medicinal signaling




A. Navani et al. (eds.), *Essentials of Regenerative Medicine in Interventional Pain Management*,
https://doi.org/10.1007/978-3-031-50357-3_3

cells" as growing literature treatise against their believed multipotent effect, rather demonstrating that these cells play a part in the regeneration process by releasing cytokines and growth factors that promote healing via paracrine signals. This reflects that directly injected MSCs did not differentiate in the injured tissue, but home to the injury site to secrete bioactive factors, thereby resulting in therapeutic effects [5].

Multiple tissue sources have been described for MSCs, both autologous and allogeneic. These include bone marrow (BM), adipose tissue, amniotic-tissue derived like Wharton's jelly and umbilical cord. BM and adipose derived (AD) MSCs are the most frequently utilized sources for spinal pain [6]. The MSCs and marrow elements obtained from BM aspiration is termed bone marrow aspirate concentrate (BMAC). The maximum concentration of MSCs has been shown to be from posterior iliac crest, the most common site used for BM aspiration [7].

Both autologous and allogeneic MSCs lack the histocompatibility complex class II (MHC II) surface proteins which renders them to have low immunogenic potential, thereby, decreasing the risk of rejection during allogeneic transfer [1].

## Other Biologic Preparations

A2M is a plasma glycoprotein which has the ability to inhibit metalloprotease and inflammatory cytokines. Cartilage degradation is associated with increased activity of proteases and cytokines, as a result, A2M has been explored for cartilage-based pathology in disc degeneration and inflammatory arthritis [8].

Exosomes or extracellular vesicles are secretory products, which play an important role in intracellular communication. They are released by nearly all cells, tissues, and body fluids, containing microRNA, messenger RNA and other proteins [9]. Acellular MSC derived exosomes appear as a promising treatment option, promoting tissue regeneration and homeostasis, with no associated risks of immunogenicity. The data from various in-vitro and in-vivo studies on exosomes show pronounced therapeutic competence for tissue recovery in intervertebral disc degeneration (IVDD) through the maintenance of their endogenous stem cells, promoting proliferation of nucleus pulposus (NP) cells, inhibition of apoptosis with immune modulation, downregulation of inflammation, and stimulation of angiogenesis. However, studies demonstrating the efficacy of utilization of exosomes in treatment of IVDD in clinical settings remains to be elucidated [6].

## Regenerative Medicine for Spinal Pain

Chronic spinal pain is a global pandemic associated with increased healthcare costs and socioeconomic burden. It is believed that the prevalence and incidence of low back pain increases with ageing. According to the Global Burden of Disease study done across 195 countries assessing the incidence, prevalence, and years lived with disability for 354 medical conditions found low back pain to be the leading cause of worldwide productivity loss, and the top cause of years lived with disability in 126 countries [10]. A systematic review of 13 studies from North America, northern Europe, and Israel published in 2019 reported the prevalence of low back pain to range between 1.4% and 20%, and the annual incidence ranging between 0.024% and 7%, being highest in the USA [11].

Chronic spinal pain is a complex and multifactorial disease process, of which low back pain is the most common. One study noted that the estimates of regional spine pain vary, with the highest prevalence of 43% noted in the low back, followed by the neck at 32%, and the lowest in the thoracic spine [12]. Lumbar intervertebral discs, facet joints, sacroiliac joints, ligaments, fascia, muscles, and nerve root dura are all pain generators of low back and lower extremity pain [13]. The management of spinal pain and other musculoskeletal disorders requires the appropriate diagnoses and the use of effective therapeutic modalities, under the auspices of evidence-based medicine, in a cost-effective manner.

## Pain Generators in Spine

### Intervertebral Disc

The intervertebral discs are composed of a central gelatinous layer called nucleus pulposus (NP), the outer fibrous region termed annulus fibrosus (AF), and the cartilaginous endplate (CEP). The NP cells produce components of extracellular matrix (ECM) such as the type I and type II collagen networks responsible for tensile strength, and water-binding proteoglycans, such as aggrecan, which are vital for disc integrity. The disc is an avascular structure and depends on the diffusion of nutrients and oxygen across the subchondral plate. It is postulated that fracture and degeneration of endplates may trigger the degenerative disc process.

The disc is a dynamic structure with the microenvironment milieu dependent on a harmonious balance between anabolic and catabolic factors, which are important for normal disc cell turnover [14]. Anabolic factors include growth factors like TGF-α, BMP, GDF5 and IL-GF, etc. Catabolic metabolism is accomplished by several factors like catabolic enzymes, inflammatory cytokines, proteinases, interleukin-1 (IL-1), tumor necrosis factor-alpha (TNF-α).

IVD degeneration (IVDD) is a chronic and irreversible process marked with elevated matrix degradation, NP proteoglycan loss and hydration, disruption of the disc structure and reduced disc height. This stimulates peripheral inflammatory cell infiltration, followed by upregulation of the local cytokines like IL-1β, IL-1α, TNF-α, vascular and nerve growth factors, and other catabolic factors. Simultaneously, matrix degradation can increase absorption of IVD tissue and activation of peripheral nerve endings, which raises peripheral nociceptors sensitization, and enhances pain sensation [15].

Degenerative disc disease (DDD) is characterized by nuclear dehydration and fibrosis with resultant disc space narrowing. This redistributes the axial mechanical forces on the disc, vertebral body and facet joints. Disc degeneration is a sequential process possibly starting with a circumferential tear in the AF that progresses to a radial tear, herniation, loss of disc height and resorption. Progressive degeneration may eventually result in discogenic low back pain.

## Evidence for PRP in Intervertebral Disc Degeneration (IVDD)

Clinical evidence on the use of PRP treatment of IVDD in humans has been reported since 2011 [16]. A summary of nine studies [17–25] focusing on the use of PRP to treat IVDD is shown in Table 3.1. There have been several prospective and retrospective trials and case reports studying PRP, however, only a single randomized controlled trial (RCT) and one meta-analysis have been published to this date. The number of patients in these studies ranged from 11 to 65 with varied follow-up periods, the shortest being 3 months. All the studies reported decreased pain scores and functional improvements, affirming the efficacy of intradiscal PRP in the treatment of IVDD.

**Table 3.1** Evidence for PRP in Intervertebral disc degeneration

| Author | Study details | Level of evidence |
|---|---|---|
| Tuakli et al. [17] | Prospective, double-blind randomized controlled trial<br>**N = 47**<br>Follow up—12 months<br>Single, Intradiscal injections of PRP showed improvement in pain scores (NRS), function (FRI) and satisfaction scores (outcomes with respect to control not estimated after 8 weeks) | I |
| Levi et al. [18] | Prospective trial<br>**N = 22**<br>Follow-up—6 months<br>Single injections of PRP at one or more levels (up to 5)—categorical success rates were 14% after 1 month, 32% after 2 months and 47% at the end of 6 months | III |
| Akeda et al. [19] | Preliminary clinical trial<br>**N = 14**<br>Follow-up—12 months<br>PRP releasate (PRPr), isolated from clotted PRP, was injected into center of nucleus pulposus. Significant improvement in VAS at 1 month which was sustained throughout the follow-up period. No adverse events were observed. | III |
| Akeda et al. [20] | Long-term follow-up study for the previous clinical trial [4]<br>**N = 11**<br>Results showed that the VAS scores for low back pain and disability scores remained decreased (more than 30% decrease from baseline) for an average of 5.9 years after the intradiscal injection of PRPr in 91% patients | III |
| Navani et al. [21] | Prospective case series<br>**N = 20**<br>Follow-up—18 months<br>18 patients were followed up for 6 months, 94% reported improved pain scores (VAS) and 100% had improved SF-36 scores. At 18 months, 15 patients were followed up, 14 patients (93%) experienced more than 50% relief in pain | IV |
| Jain et al. [22] | Prospective single-arm interventional study<br>**N = 20**<br>Follow-up—3 and 6 months<br>PRP injections at single or multiple levels given to patients with discogenic pain<br>The improvement in NRS, ODI scores was positively correlated with platelet concentrations in the PRP sample. Authors concluded that higher platelet counts elicit a more favorable response | III |
| Chang et al. [23] | **Meta-analysis**<br>Included 3 studies, results showed Intradiscal PRP injections are effective in relieving discogenic lower back pain and improving disabilities. These effects manifest two or six months after treatment | II |
| Lutz et al. [24] | Retrospective study<br>**N = 37**<br>Follow up—3 to 43 months (mean 18.3 ± 13.3 months)<br>Intradiscal injections of higher concentration (>10×) PRP resulted in significant improvements in pain scores. Also, the satisfaction rates were higher in patients receiving >10× PRP (81%) compared to <5× PRP (55%) | IV |
| Kirchner et al. [25] | Retrospective study<br>**N = 65** (18 cervical pain, 47 lumbar back pain)<br>Patients were given 2 or more infiltrations of plasma rich in growth factors (PRGF), which is leukocyte-free PRP, one month apart. Significant improvements in NRS were reported | IV |

*PRP* platelet rich plasma, *VAS* visual analogue scale, *NRS* numerical rating scale, *ODI* Oswestry Disability Index, *FRI* functional rating index

Tuakli-Wosornu et al. [17] performed the first double-blind RCT of intradiscal PRP therapy in 2016 on 47 participants with chronic lumbar discogenic pain. The study reported statistically significant reduction in pain scores, functional improvement and patient satisfaction in the treatment group. Notably, the effects of PRP were sustained for a period of at least 1 year with significant improvement in Functional Rating Index (FRI). No complications were reported.

The first meta-analysis to analyze the effect of intradiscal PRP injection on pain reduction in patients with discogenic lower back pain was conducted by Chang et al. [23]. The final analysis included one RCT and two prospective observational studies. The pain scores (VAS scores) were significantly reduced two and six months after intradiscal PRP injection, but not after one month. As PRP promotes the endogenous healing process by regulating the inflammatory process and enhancing angiogenesis, the clinical effects are evident after a few weeks.

Authors concluded that intradiscal PRP injections have large positive pain-reducing and functional improving effects in patients with discogenic lower back pain.

The safety concerns with PRP are limited as autologous blood is used. PRP provides the degenerated disc with certain factors that may aid repair. However, if the disc cells are already severely damaged, some studies suggest that no amount of PRP may make a difference [26]. Thus, PRP may be more efficient when applied at an earlier stage of degeneration, in a patient that has relatively a healthy number of functioning cells. The stem cells have a theoretical advantage over PRP as they can restore severely degenerated cells.

Evaluation of the available literature and utilizing best evidence synthesis principles based on systematic reviews [23] and individual studies, the evidence for PRP in management of lumbar discogenic pain has been assessed as Level III of V.

## Role of Alpha-2-Macroglobulin in Degenerative Disc Disease

In 2017, Pasquale et al. [8] published a prospective study to determine the role of autologous A2M in relieving back pain in DDD. 24 patients with MRI-concordant DDD were enrolled and given a single intradiscal injection of platelet poor plasma rich in A2M under fluoroscopic guidance. Alongside, they analyzed the presence or absence of a cartilage degradation product, the Fibronectin-aggrecan complex (FAC) in patients with DDD. Oswestry Disability Index (ODI) and Visual Analog Scale (VAS) were noted at baseline, 3-month and 6-month follow-up. Statistically significant improvement in mean VAS scores and ODI was noted in FAC positive patients in comparison to FAC negative. When a 20-point ODI improvement cut-off was applied to evaluate clinical improvement, 77% FAC-positive patients and 27% FAC-negative patients met this criterion. A2M can prevent the formation of G3 domain of aggrecan, thereby reducing the FAC. Authors suggest that autologous A2M may be an efficacious biologic treatment in discogenic pain and FAC may be an important biomarker in patient selection. Even though the study was limited by small sample size and short duration of follow-up, it did provide an impetus for role and plausible beneficial effect of A2M in degenerative disc disease.

## Evidence of Mesenchymal Stem Cells (MSCs) in Intervertebral Disc Degeneration

MSCs are unspecialized cells that have extensive proliferative capacity and multi-lineage potential in vitro and in vivo [6]. Zeckser et al. [14] stated that all interventional therapies should optimally strive to fulfill a trio of treatment objectives inclusive of resolution of primary nociceptive disc pain with functional improvement, slowing and/or reversal of the catabolic metabolism and associated degradation within the disc environment, and partial or complete restoration of disc tissue [1]. The transplanted cells replenish the cell population in degenerated IVDs, alongside enhance the production of ECM and aggrecan, resulting in an increase in IVD height [1].

Table 3.2 outlines the relevant studies [27–32] to date which focus on stem cell therapy to treat degenerative disc disease. A meta-analysis by Wu et al. [29] suggests that MSC and chondrocyte therapy for discogenic low back pain correlates with improved pain relief and function metrics.

A clinical study by Mochida et al. [33] investigated the safety of transplanting activated NP cells via MSC co-culture in humans. Nine patients with Grade III disc degeneration were followed for 3 years, during which time no adverse events were reported. All patients showed no further degeneration of their discs, and one patient showed notable improvement of the disc over the 3 years of observation. Furthermore, patients reported improvement in low back pain scores. This study demonstrated that NP transplantation may serve a role in the treatment of disc degeneration and that it can be performed safely in human subjects.

In 2020, Beall et al. [34] evaluated a novel technology that provides the benefits of allograft supplementation combined with a matrix replete with viable cells called VAST trial—Viable Allograft Supplemented Disc Regeneration in the Treatment of Patients with Low Back Pain. They conducted a randomized, prospective multicenter study using allogeneic grafts combining MSCs mixed with micronized NP as the vehicle ("viable structural allograft," or VIA Disc Matrix™) for treatment of DDD. This study was an extension of the VAST trial (NCT03709901) which meant to establish preliminary safety and efficacy of this particular allogeneic graft.

Twenty-four patients with lumbar DDD at one or more vertebral level were included. Patients were randomized at 3.5:1:1 to receive either a supplemental allograft at up to 2 levels, saline as placebo or continue medical management. The supplemental viable allograft was prepared from human NP allograft containing a minimum of $6 \times 10^6$ allogeneic

**Table 3.2** Evidence for MSCs in intervertebral disc degeneration

| Author | Injectate | Study details | Level of evidence |
|---|---|---|---|
| Noriega et al. [27] | Allogeneic bone marrow MSCs | Randomized controlled trial<br>**N = 24**<br>Follow-up—12 months<br>Intradiscal injection of allogeneic bone marrow MSCs (25 × $10^6$ cells per segment) compared with sham infiltration of paravertebral musculature with anesthetic<br>Both lumbar pain and disability were significantly reduced at 3 months, maintained at 6 and 12 months after MSC treatment | II |
| Pettine et al. [28] | Autologous bone marrow concentrate | Prospective, non-randomized study<br>**N = 26**<br>Follow-up—36 months<br>2 ml autologous bone marrow concentrate injected into the nucleus pulposus of the treated discs. After 36 months, six patients progressed to surgery. The remaining 20 patients (77%) reported significant improvement in average ODI and VAS scores | III |
| Wu et al. [29] | Autologous and allogeneic stem cells | Systematic review and single-arm meta-analysis<br>Included 6 studies<br>Authors identified cell-based therapy for discogenic pain is associated with decreased pain scores and ODI. No adverse effects were reported by the included studies | II |
| Wolff et al. [30] | Bone marrow aspirate | Retrospective analysis<br>**N = 33**<br>Intradiscal injection of concentrated bone marrow aspirate (cBMA) at one or more levels—autologous 3 mL from posterior superior iliac spine (PSIS)<br>Participants reported improvement in NRS, SF-36 and ODI at follow-up, but these results were not statistically significant | III |
| Amirdelfan et al. [31] | Allogeneic mesenchymal precursor cells (Mesoblast phase 3 trial) | Multicenter, randomized controlled study<br>**N = 100**<br>Follow-up—36 months<br>Single intradiscal injection of adult allogeneic mesenchymal precursor cells (MPCs) combined with Hyaluronic acid (HA). Patients were randomized in a 3:3:2:2 ratio to receive 6 million MPCs with HA, 18 million MPCs with HA, HA vehicle control, or saline control (placebo) treatment<br>Significant differences between the control and MPC group for improvement in VAS and ODI scores<br>Intradiscal injection of MPC with HA is safe, effective and durable for chronic LBP with moderate DDD | II |
| Her et al. [32] | Autologous or allogeneic BMAC and Bone Marrow MSCs | Systematic review<br>16 studies, including 607 participants included<br>Intradiscal autologous or allogeneic BMAC and culture-expanded BM-MSCs improved discogenic pain compared to baseline along with improvements in physical functioning and positive anatomical changes on spine MRI (improved disc height, disc water content, Pfirrmann grading) although anatomical findings were inconsistent across studies. However, the overall evidence level for this study was low due to heterogeneity, small sample sizes and poor generalizability | II |

*VAS* visual analogue scale, *NRS* numerical rating scale, *ODI* Oswestry Disability Index, *DDD* degenerative disc disease, *LBP* low back pain, *MRI* magnetic resonance imaging

viable cells. At 3 months, subjects in conservative care crossed over to allograft group as they had no relief and 1 subject in saline group crossed over to allograft at 6 months. In total, 21 of 24 subjects received the VIA Disc Tissue Matrix either randomized or open label. There were clinically meaningful improvements in VAS and ODI in both the allograft and saline groups at 12 months. At 12-months, VAS back pain improved from 54.81 to 12.27 in the allograft, 55.25 to 19.67 in the saline group, and 62.25 to 6.0 in the nonsurgical management group with crossover, respectively. A responder analysis showed a clinically meaningful reduction in ODI of ≥15 points at 12 months, which was statistically significant; 76.5% of the allograft subjects were responders ($P = 0.03$) compared to 56.7% in the saline group. Subjects who received the allograft had larger reduction in pain and greater functional improvement than other cohorts.

Recently, Atluri et al. [35] published a prospective nonrandomized study to evaluate the effectiveness of autologous BM-MSCs in the treatment of chronic low back pain due to lumbar spinal degeneration with the involvement of multiple structures. 40 patients in the treatment group were injected with autologous BM into the discs, facet joints, sacroiliac joints and around the spinal nerves depending on the primary determinant of pain. Patients in the control group received conventional treatment with opioids, spinal injections or exercise as indicated. At the 12-month follow-up, 67% of patients in the study group showed significant improvement in pain and function. Pain relief was also seen with a 2-point difference in 56% of patients in the study group at 12 months compared to only 8% in the control group. The study group also showed reduced opioid usage. This is the first of its kind study illustrating the benefit of injecting BM-MSC in multiple structures in one setting in chronic spinal degeneration. The technique used in this study resulted in consistently high concentrations of total nucleated cells (TNC) and colony forming unit-fibroblasts (CFU-F), which are synonymous with MSCs. This emphasizes the need for proper aspiration and concentration techniques which would significantly impact clinical efficacy.

Stem cell therapy has the potential to be a standard treatment in patients that have not adequately responded to nonoperative management [29]. However, definitive studies with an optimal cell therapy protocol are lacking in this regard. Based on evaluation of the available literature and utilizing best evidence synthesis principles based on RCTs, metanalysis and observational studies, the evidence for MSCs in management of lumbar discogenic pain has been assessed as Level II of V.

Overall, stem cell therapy has shown promising results in different aspects. Meanwhile, the indications and sources for stem cell therapy for internal disc disruption (IDD) have gradually expanded, laying a cornerstone for the experimental and clinical application of stem cell therapy for IDD. Although cell therapy appears to have great potential for IVD regeneration, it requires more evidence regarding safety, long-term complications, effectiveness in different patient populations, and surgical cost-effectiveness.

## Bone Marrow Concentrate for Cervical Disc

Pettine [36] conducted a prospective, nonrandomized study in 2017 comprising 182 patients with symptomatic discogenic cervical pain. BMC was injected in one or more discs (0.5 ml in each disc). Follow-up at 6, 12 and 24 months showed statistically significant decreased VAS and Neck Disability Index (NDI) scores. No patient underwent surgery, and no adverse events were reported. These preliminary results may offer cervical pain patients with a nonsurgical option for management [36].

## Exosomes

Recently, an increasing number of in vitro and animal studies have shown that exosomes derived from stem cells play an important role in restoring tissue homeostasis and promoting tissue regeneration. However, no published human trials are available on management of discogenic pain with MSC-exosomes.

### Radicular Pain

Low back pain is multifactorial. Compression or inflammation of the exiting spinal nerve roots may lead to radicular pain symptoms, commonly in the setting of a disc herniation or spondylotic changes of the spine [37]. The use of epidural injections to treat radicular pain has been studied extensively using steroids but we have limited data on the use of PRP for the same. Theoretical advantages of using epidural PRP include the ability to treat bilateral pain and use of a lower volume.

## Evidence for Lumbar Epidural PRP Injections

A summary of five studies focusing on epidural PRP injections is detailed in Table 3.3. This includes 2 observational studies and 3 RCTs that have evaluated epidurally administered biologics for the treatment of radicular pain [38–42]. The only systematic review assessing epidural biologic injections was performed by Sanapati et al. [43].

In 2018, Sanapati et al. [43] published a systematic review assessing the effectiveness of PRP and MSCs in treatment of back pain. This metanalysis included 21 studies comprising RCTs, observational studies and case reports. There were 12 lumbar disc injections, 5 epidural, 3 lumbar facet joint, and 3 sacroiliac joint studies. This study demonstrated Level 3 evidence for disc injections of PRP and MSCs, while the evidence for epidural, facet joints and sacroiliac joints was Level IV (on a scale through evidence Levels I–V).

According to the authors, based on systematic reviews and individual studies, the evidence for management of radicular pain using lumbar epidural PRP injection has been assessed at Level III.

**Table 3.3** Lumbar epidural PRP injections for radicular pain

| Author | Study details | Level of evidence |
|---|---|---|
| Centeno et al. [38] | Retrospective study<br>**N = 470**<br>Follow up—24 months<br>Fluoroscopy-guided transforaminal or interlaminar injection of PL (platelet lysate)<br>Patients treated with platelet lysate epidurals reported significant improvements in pain and functional rating index (FRI) through 2-year follow-up | III |
| Jose et al. [39] | Prospective, observational non-randomized study<br>N = 250<br>Follow-up—12 months<br>Epidural injection of PRGF in cervical spine (30%) or lumbar spine (70%)<br>Improvement in mean VAS scores in 85% patients and decreased use of opioid rescue from 96% to 0% at 12 months | III |
| Xu et al. [40] | Randomized controlled trial<br>*N* = 124 (61 in PRP, 63 in ESI group)<br>Follow-up—12 months<br>Ultrasound-guided transforaminal injection of PRP compared with steroid in lumbar disc herniation. Statistically significant improvement in VAS, ODI, PPT and SF-36 observed at 1 month and maintained at 12 months | II |
| Ruiz-Lopez et al. [41] | Randomized controlled trial, double blind<br>N = 50 (25 in triamcinolone group, 25 in PRP group)<br>Follow-up—6 months<br>Fluoroscopy-guided caudal epidural injection of Leukocyte-rich PRP compared with steroid injection<br>Improved VAS score in steroid group at 1 month, the PRP group had improved VAS and SF-36 scores at 3 and 6 months | I |
| Núñez et al. [42] | Randomized controlled trial<br>N = 93 (46 in PRP, 47 in ESI group)<br>Follow-up—12 months<br>Two injections were given at weekly intervals of lumbar interlaminar epidural under fluoroscopic guidance<br>Clinical improvement was measured by Lattinen index. PRP group showed sustained improvement for 12 months. ESI group had improved scores at 1 month which showed worsening at 3 and 12 months. Also, the ESI group had five times more adverse events | II |

*PL* platelet lysate, *FRI* functional rating index, *PRGF* platelet-rich plasma, *ESI* epidural steroid injection, *VAS* visual analogue scale, *ODI* Oswestry Disability Index, *PPT* physical performance test, *SF-36* 36-item short form health survey

## Evidence for Intra-articular MSCs in Radicular Pain

No studies to date have been published on the use of MSCs administered via epidural placement.

## Facet Joint

Facet joints are synovial joints innervated by the medial branches of the dorsal rami. Facet joint as a source of low back pain was first described by Goldthwaite in 1911, and Ghormhley who used the term "facet syndrome" to describe a symptom originating from the facet joint [44, 45]. Facet joint osteoarthritis is the most common cause of facet joint pain. Lumbar facet joints constitute a common source of pain with a prevalence of 27–40% in patients with chronic low back pain [46]. The origin of low back pain in facet joint degeneration is supported by presence of substance P nerve fibers in subchondral bone. Inflammatory mediators such as prostaglandins and cytokines (IL6, TNF-α) have also been found in degenerative facet joints [47].

## Evidence for PRP in Facet Joint Pain

The use of PRP in the treatment of facet pathology is very promising. The platelet-derived growth factors are considered essential for the healing process. The facet joint is a smaller joint which poses the challenge of evaluating the adequate injectate volume of PRP so as to be effective. PRP has an advantage over stem cells as it is not known to promote osteophyte formation or pathological bone growth within the facet joints.

Table 3.4 summarizes study results demonstrating the outcomes of injecting PRP for facet joint pain [48–50]. The studies insofar have been limited with patient size ranging from 19 to 86 and follow-up periods extending up to 6 months. In the RCT comparing PRP with intra-articular local anesthetic/corticosteroid, Wu et al. [49] concluded that patients receiving PRP had higher subjective satisfaction and objective success rates with sustained increase in efficacy over 6 months. Longer follow up studies with larger number of patients remains an unmet area of research in this subgroup.

According to the authors, based on systematic reviews and individual studies, the evidence for management of lumbar facet disorders using PRP has been assessed at Level III.

## Evidence for Intra-articular MSCs in Facet Pain

No studies to date have been published on the use of MSCs administered to the facet joints.

## Sacroiliac Joint

The sacroiliac joint is the largest axial joint and acts to transmit forces from the upper body to the lower body. These joints are a common cause of disabling backpain, accounting for 10–27% of chronic low backpain [51]. Sacroiliac joint-mediated pain can stem from the osteoarthritic degeneration or disruption of the ligaments. Many patients with sacroiliac

**Table 3.4** Evidence for PRP in facet joint pain

| Author | Study details | Level of evidence |
|---|---|---|
| Wu et al. [48] | Prospective clinical trial<br>N = 19<br>Follow-up—3 months<br>Autologous PRP injections into lumbar facet joints—79% of the patients (15/19) reported significant pain relief along with improvement in lumbar functional capacity<br>Limitations—Short follow-up period, small sample size | III |
| Wu et al. [49] | Prospective randomized trial<br>**N = 46**<br>Follow-up—6 months<br>Patients were randomized to receive either intra-articular PRP or local anesthetic/corticosteroid<br>Both the groups showed improvement in VAS, ODI and RMQ at one month follow-up. Only the PRP group had sustained improvement at 6 months | II |
| Kirchner et al. [50] | Observational retrospective study<br>**N = 86**<br>Follow-up—6 months<br>One intradiscal, one intra-articular facet, and one transforaminal epidural injection of PRGF was carried out in 86 patients with chronic LBP | IV |

*VAS* visual analogue scale, *NRS* numerical rating scale, *ODI* Oswestry Disability Index, *RMQ* Roland-Morris Disability Questionnaire, *PRGF* platelet rich growth factor

joint pain do not respond to conservative care or steroid injections and may require fusion surgery. The evidence continues to evolve to support the use of regenerative medicine therapies like prolotherapy or PRP injections in sacroiliac joint-related disorders, which aim at healing the damaged tissue.

## Evidence for PRP in Sacroiliac Joint Pain

Table 3.5 outlines the relevant studies to date which focus on PRP therapy to treat sacroiliac joint pain [52–55]. Few studies have been published showing clinical evidence on the use of PRP treatment sacroiliac joint pain. All the studies are either prospective cohort study, non-randomized or RCTs with sample size ranging from 40–186 patients. Follow-up period ranged from 3 to 12 months. Saunders et al. [52] injected PRP into and around the dorsal interosseous ligament, whereas, in other studies PRP was injected into the synovial portion of the joint. All the studies reported significantly improved pain scores in comparison to steroid injections.

In 2021, Ling et al. [56] conducted the first systematic review comparing patients treated with PRP or corticosteroids for lumbar spondylosis and sacroiliac arthropathy. They included 5 studies comprising 242 patients with chronic low back pain. Three studies were randomized and two were nonrandomized trials. Three studies reported a significant improvement in pain scores among the PRP groups. Authors concluded that both PRP and corticosteroid injections are safe and effective options for the treatment of lumbar spondylosis and sacroiliac arthropathy. However, there is some evidence that PRP injection is a more effective option at long-term follow-up as compared with corticosteroid injection. This review was limited by inclusion of studies analyzing a variety of spinal pathologies.

**Table 3.5** Evidence for PRP in sacroiliac joint pain

| Author | Study details | Level of evidence |
|---|---|---|
| Saunders et al. [52] | Prospective cohort study<br>**N = 45**<br>Follow-up—12 months<br>Patients with mechanical dysfunction of SIJ received ultrasound-guided PRP injection into and around the DIOL rather than the synovial portion of the SIJ under ultrasound guidance and compared to 25 patients who historically underwent prolotherapy<br>The improvement in pain scores and functional status was superior in the PRP group after 12 months | III |
| Singla et al. [53] | Prospective randomized open blinded end point (PROBE) study<br>**N = 40**<br>Follow-up—3 months<br>Ultrasound-guided intra-articular SIJ injection with leukocyte-free PRP in 20 patients compared with methylprednisolone injection in the steroid group<br>90% patients in the PRP group reported decreased VAS scores compared to 25% in the steroid group | II |
| Eldin et al. [54] | Prospective, non-randomized controlled trial<br>**N = 186** (PRF patients—124; PRP patients—62)<br>Follow-up—6 months<br>Fluoroscopy-guided intra-articular SIJ injection with two different forms of platelet concentrates<br>At 6-month follow-up, significant improvement in VAS score was noted among patients who were treated with intra-articular PRF compared with PRP, but both groups improved with statistical significance 6 months post injection | III |
| Wallace et al. [55] | Prospective, non-randomized interventional study<br>**N = 50**<br>Follow-up—6 months<br>Ultrasound-guided PRP injections in the SIJ are effective at reducing disability and pain with most improvement (NRS and ODI) seen within 4 weeks after injection and with sustained reduction at 6 months | III |

*SIJ* sacroiliac joint, *PRP* platelet rich plasma, *DIOL* dorsal interosseous ligament, *PROBE* prospective randomized open blinded end point, *PRF* platelet rich fibrin, *NRS* numerical rating scale, *ODI* Oswestry Disability Index

According to the authors, based on systematic reviews and individual studies, the evidence for management of sacroiliac joint pain using PRP has been assessed at Level III of V.

### Evidence for Intra-articular MSCs in Sacroiliac Joint Pain

No studies have been published on the use of MSCs in the sacroiliac joints and remains an area for future research.

## Regenerative Medicine for Musculoskeletal Pain

### Tendinopathies

Tendons are connective tissues connecting muscle to bone. Playing a role in movement and joint stability, tendons are susceptible to overuse and degeneration which may result in tendinopathy [57]. Common sites of tendinopathies include the rotator cuff, medial and lateral epicondyles, patellar, and Achilles tendons [58]. The hypovascularity of tendons poses a challenge when considering treatment options [59].

Rehabilitation protocols specifically eccentric based strengthening exercises have shown to be effective in the treatment of tendinopathies [60]. Corticosteroid injections have been a common treatment option given the short-term pain relief; however, it is important to note that steroids may negatively affect tendon remodeling and may result in further progression of pathology and/or spontaneous rupture [61, 62].

PRP produces an anabolic and anti-inflammatory response through the secretion of growth factors which have been shown to provide beneficial effects on tendon remodeling and healing [59]. The focus of this section is to review the literature evaluating the use of regenerative therapies including PRP in the treatment of chronic tendinopathies. When evaluating the current literature, it is important to consider that there are significant variations in dosing protocol and preparations in various studies which makes it challenging to evaluate the efficacy.

### Patellar Tendinopathy

Patellar tendinopathy is a common condition that presents with anterior knee pain that is related to mechanical loading [63]. Commonly, it has been related to sports activities involving jumping such as long-distance running, volleyball and track and field involving long and high jumps. Notably, it is a degenerative disorder in the setting of overuse rather than an inflammatory response. There are various treatment options ranging from eccentric based strengthening program to surgical correction. The focus of this section will be to review the literature on the use of PRP and other regenerative therapies in the treatment of chronic patellar tendinopathy compared to other conservative treatment options.

### PRP Vs. Saline

In 2019, Scott et al. [64] published an RCT comparing PRP, either leukocyte-rich or leukocyte-poor, versus saline in the treatment of patellar tendinopathy. Fifty-seven subjects with greater than 6 months of patellar tendinopathy were recruited and randomized into 3 arms: leukocyte-rich PRP, leukocyte-poor PRP, and saline. All patients received a single injection and underwent a 6-week rehabilitation protocol consisting of concentric, muscle shortening, eccentric, muscle lengthening, and heavy slow resistance training. Three outcome measures, Victorian Institute of Sport Assessment Patellar Tendon (VISA-P), an eight-question assessment evaluating symptom severity of patellar tendinopathy, pain during activity, and global rating of change, were assessed at 6 weeks, and 3, 6 and 12 months. There was no significant difference among the groups, leukocyte-poor, leukocyte-rich, and saline at the primary end point or at 12 months, suggesting a PRP injection was no more effective than a saline injection in those with chronic patellar tendinopathy.

### PRP Vs Dry Needling

In addition to PRP, dry needling is another technique utilized to promote healing in tendinopathies. In dry needling, a fine needle is inserted into the target tissue to mechanically disrupt the irritable local tissue. There are multiple proposed mechanisms of action of dry needling. The mechanical disruption is thought to increase local blood flow and encourage fibroblastic proliferation [65]. Secondly, it is thought an involuntary spinal reflex is generated causing a local muscle contraction altering muscle fiber length and tension [66]. Lastly, dry needling affects the local and central nervous system responses reducing peripheral and central sensitization to pain [67].

The 3 months results in the study by Scott et al. [64] differed from a study published by Dragoo et al. [68] in 2014. In this RCT, 23 patients underwent either ultrasound-guided PRP with dry needling versus ultrasound-guided dry needling alone. Both groups had a standardized eccentric exercise protocol post-intervention. At 12 weeks, the group treated with ultrasound-guided PRP and dry needling had improved significantly more than the dry needling alone group; however, these effects were not sustained at greater than 26 weeks. The authors concluded that eccentric-based exercises with PRP and dry needling may help accelerate recovery compared to eccentric-based exercises with dry needling alone, but the effects of PRP dissipate over time.

In both the studies by Dragoo et al. [68] and Scott et al. [64], the dosing protocol included a single PRP injection. This poses the question if there is a need for multiple consecutive PRP injections. A randomized prospective study conducted by Zayni et al. [69] compared the clinical outcomes of a single versus two consecutive PRP injections in the management of chronic patellar tendinopathy. The study concluded that those that received two PRP injections 2 weeks apart had significantly better pain and functional outcomes at 2 years post-injection compared to those that received a single injection. Compared to the study conducted by Scott et al. the authors reported a significant limitation to this study was the lack of a control group. In addition, the preparation of PRP in this study did not measure the cellular content or growth factors, though it is thought the composition is classified as LP-PRP as the study states "platelet concentration is known to be two times higher that of whole blood with no leukocytes." Although this study suggests that patients may do better with 2 consecutive injections compared to a single injection, there is a no standardized protocol or dosing guidelines when using PRP for chronic patellar tendinopathy.

## PRP Vs Extracorporeal Shock Wave Therapy (ESWT)

Another treatment option utilized in tendinopathies is extracorporeal shock wave therapy (ESWT), a non-invasive therapeutic modality. In ESWT, emitted shockwaves propagate through tissues generating mechanical and cellular effects, such as vascularization and protein biosynthesis [70]. This is thought to lead to tissue regeneration with subsequent improvement in pain and function. ESWT has shown positive effects in the treatment of various soft tissue pathologies and tendinopathies. The non-invasive modality has gained popularity amongst physicians and patients, leading to more research comparing the efficacy of ESWT and PRP for tendinous pathology.

Vetrano et al. [71] published a RCT with 46 patients with patellar tendinopathy comparing the effectiveness of 2 ultrasound-guided PRP injections over 2 weeks versus 3 sessions of focused ESWT. At 2 months, there was no significant difference between the two groups in terms of VISA-P, VAS scores and modified Blazina scale scores, a 5-stage functional outcome score. However, there was significant improvement in the PRP group compared to ESWT group in VISA-A and VAS scores at 6 and 12 months, and modified Blazina scale score at 12 months. It is important to note that although the PRP preparations had mean platelet concentration that was 3 to 5 times greater than the baseline concentration, the authors did not specify if the PRP preparation was LR or LP-PRP. The single-spin preparation of PRP would lead one to assume that it was a LP-PRP preparation; however, there was no information regarding leukocyte count to support this. Additionally, the authors hypothesized that the PRP group achieved better results compared to the ESWT group because the PRP group had the optimal number of platelets and growth factors thus stimulating more effectively the regenerative mechanism of tendon repair compared to ESWT. This poses the question: what is the optimal ESWT settings and dosing protocol for the treatment of patellar tendinopathy? In this study, subjects received 3 treatments consisting of 2400 impulses at 0.17–0.25 mJ/mm/session which was separated by 48–72 h. Currently, there is no consensus on the optimal setting and dosing for ESWT in the management of patellar tendinopathy. Further studies evaluating the regenerative effects of ESWT, and the effects of various dosing protocols are necessary. As the regenerative literature regarding chronic patellar tendinopathy continues to develop, we hope there will be more studies attempting to answer these critical questions as well as assessing the long-term (>12 months) pain and functional outcomes of PRP and ESWT.

## Utilization of Cell-Based Therapy

Recently, Rodas et al. [72] published a prospective, double-blinded, RCT comparing the effectiveness of MSC (mesenchymal stem cells or medicinal signaling cells) versus PRP in subjects with proximal patellar tendinopathy. A total of 20 subjects were recruited and randomized into two arms: injection of bone marrow derived MSCs (BM-MSC) and LP-PRP. Both arms received post-intervention rehabilitation protocol. At the conclusion of the study (6 months), there was significant reduction in pain measured by VAS scores and increased VISA-P scores compared to baseline in both groups, but no significant difference between the two groups. The results from the PRP group are similar to other studies at 6 months [71, 73]. Interestingly, the MSC group had statistically significant greater improvement in tendon structure on ultrasound and restoration of tendon architecture on MRI at 6 months compared to PRP group. The authors report that the regenerative effects of tendons injected with LP-PRP may lag behind those receiving BM-MSC. Thus, more long-term data is necessary to assess the efficiency of these two treatment arms more definitively.

In 2021, Khoury et al. [74] published a pilot study evaluating the effects of ultrasound-guided adipose derived MSCs (AD-MSC) injection for chronic patellar tendinopathy. A total of 16 knees were recruited and subjects underwent three bi-weekly injections of AD-MSC followed by physical therapy. There was statistically significant improvement in VISA-P and VAS scores at 3, 6, and 12 months follow up. Notably, there were significant changes in tendon thickness, tear length, tear width and tear thickness based on MRI imaging at 6 months.

Although these limited studies report initial functional and pain improvement as well as soft tissue architecture with the use of MSCs, additional larger, long-term studies are needed to further assess the use of this therapy in patients with chronic patellar tendinopathy.

## Rotator Cuff Tendinopathy

The rotator cuff muscles include the supraspinatus, infraspinatus, teres minor, and subscapularis muscles, which assist in stabilizing the glenohumeral joint. Rotator cuff tendinopathy commonly presents with pain and weakness in the shoulder region that can lead to overall reduced shoulder function and biomechanical changes within shoulder motion. Pain is worse when lying on the affected shoulder or with overhead activities [75]. Weakness in shoulder external rotators, shoulder abductors and signs of impingement may also be present. There are various proposed mechanisms that can cause the development of rotator cuff tendinopathy including extrinsic compression and intrinsic factors [76, 77]. Extrinsic compression includes mechanical impingement between the rotator cuff and the undersurface of the acromion. The rotator cuff may be predisposed to greater compression in the setting of morphological changes of the acromion such as hooked-type acromion. Intrinsic factors include vascular changes that may be due to age-related changes or due to external compression. The attachment of the rotator cuff is hypovascular, predisposing it to pathology and may lead to challenges when treating [78].

## PRP Vs Active Controlled Group

In 2013, Rha et al. [79] conducted a prospective double-blind RCT comparing the efficacy of PRP injection and dry needling. Thirty-nine subjects were included in the study and underwent either two dry needling procedures or two PRP injections. The PRP group was found to have a significant reduction in shoulder pain and disability index and improvement in range of motion when compared to dry needling at 6 months in the management of supraspinatus tendon pathology.

However, Kesikburun et al. [80] conducted a double-blinded RCT comparing the effects of a single PRP injection versus saline injection in 40 patients. The authors found no difference in Western Ontario Rotator Cuff Index (WORC), a patient reported score for rotator cuff disorders, Shoulder Pain and Disability Index (SPADI), a 13-item score measuring shoulder pain and disability, and VAS scores compared to the saline arm at 12 month follow up [81, 82]. It can be argued that the authors had used a single PRP injection and that multiple injections may be necessary. Importantly, Rha et al. [79] comment that further research is necessary to understand the most effective PRP preparation, dosing, interval between injections, and post PRP injection rehabilitation protocol for rotator cuff tendinopathy.

More recently, Prodromos et al. [83] published a prospective study in 2021 evaluating the efficacy and safety of dual PRP shoulder injections in patients who have underlying rotator cuff pathology and have failed conservative treatment. Sixty-eight patients underwent PRP injection (4 mL) into the supraspinatus tendon and glenohumeral intra-articular space. Thirteen patients received a second dose of dual injections at 10 days to 20 months post-initial injection. An additional two patients received two additional dual injections between 1 and 20 months post initial dual injection. There was a statistically significant reduction in mean VAS scores and improvement in Quick Disabilities of the Arm, Shoulder, and Hand (QuickDASH) at 6, 12, and 24 months. QuickDASH is a shortened scoring system used to evaluate the function and symptoms of upper extremity musculoskeletal disorders [84]. Importantly, no patients with a partial tear had evidence of a progression to full thickness tear. Although the subjects continued conservative treatment throughout the duration of the study, the authors were quite confident the improvement seen after PRP was due to the injection as all the subjects had failed conservative therapy prior to being enrolled in the study. This was the first study to evaluate outcomes at 24 months and to report positive results with the use of PRP in the management of partial rotator cuff tears.

## PRP Vs Corticosteroid Injections

Corticosteroid injections are widely used in the management of tendinopathies. Current literature supports that corticosteroid injection is beneficial in providing short term relief in pain and improvement in function [85]; however, there are significant long-term effects on tendon repair and remodeling that need to be considered. Studies have shown reduced cell viability, collagen organization and synthesis, and cell proliferation. Additionally, local administration of corticosteroid alters tendon biomechanical properties including reducing tendon weight, inhibiting formation of adhesions, and decreasing amount of load that can be taken before failure [86].

Kwong et al. [87] conducted a double-blinded RCT with 99 enrolled subjects comparing a single injection of LP-PRP

(3–5 mL) and corticosteroid (40 mg of triamcinolone) in patients with rotator cuff pathology, specifically rotator cuff tendinopathy and partial-thickness rotator cuff tears. They found PRP group had significant improvement in VAS and WORC outcomes at 3 months compared to corticosteroid injection group; however, this effect was not sustained at 12 months. It is important to note that despite randomization, there was significant difference in the baseline pain and outcome scores, with the PRP group having a higher baseline pain score and worse patient-reported outcome scores compared to the corticosteroid injection group. Similarly, Shams et al. [88] found similar short-term effects of PRP in the treatment of partial rotator cuff tears. However, this study did not specify if the PRP preparation was LR or LP and ultrasound guidance was not used for needle placement. Both of these studies reported similar findings when comparing the use of a single PRP versus corticosteroid injection for rotator cuff tendinopathy. As further research is conducted, it would be beneficial to determine if there is an added benefit for additional PRP injections in regards to more long-term relief.

Chen et al. [89] published a systematic review and meta-analysis in 2020 based on level 1 randomized control trial studies assessing the efficacy of PRP for rotator cuff pathology. A total of 17 studies encompassing 1116 patients were included. The PRP group had significant lower long-term re-tear rates, and better VAS and short- and long-term Constant-Murley Shoulder Outcome Scores, a 100-point scale assessing functionality after shoulder injury treatment. However, it is important to note that the PRP group did not reach the 5% minimal clinical importance difference threshold suggesting it may not provide clinically meaningful improvement in pain or function. The authors emphasize a common theme that makes it challenging to compare various regenerative medicine studies. Based on the studies included, there was high variability in the preparation of PRP including temperature, LR versus LP-PRP, volume of injectate, platelet concentration and yield, and dosing protocol which limits the interstudy comparability.

As there continues to be more literature on the use of regenerative therapies for rotator cuff pathology, there is a strong need for more high-quality, larger scale studies with an emphasis on PRP preparation and dosing protocol to help guide clinical practice.

## Utilization of Cell-Based Therapy

There is a paucity of literature on the use of MSC in the treatment of rotator cuff pathology. In 2015, Centeno et al. [90] published a multi-center prospective study evaluating a total of 115 shoulders that were treated with autologous BMC for the management of rotator cuff pathology and/or glenohumeral joint arthritis. At the 2-year follow up, there was a significant reduction in DASH, Numeric Pain Scale (NPS) and an average subjective improvement of 48.8%. It is important to note that the injectate in this study not only included BMC, but also PRP and platelet lysate, a growth factor rich solution which plays a role in healing. The authors mentioned that PRP and platelet lysate products were used due to previous studies demonstrating proliferative effects they have on MSC.

In 2020, Centeno et al. [91] conducted a randomized, crossover trial comparing BMC + PRP and platelet lysate versus exercise therapy in the management of partial and full-thickness supraspinatus tears. A total of 25 subjects were enrolled and were randomized to either exercise therapy group or ultrasound-guided BMC with PRP and platelet lysate injection. Patients were eligible to cross over to BMC + PRP and platelet lysate group after 3 months of therapy. The BMC group demonstrated statistically significant improvement in DASH scores at 3 months compared to exercise group. Secondary outcomes including Single Assessment Numeric Evaluation (SANE), a single question patient reported outcome measure, revealed significant improvement at 3 months, but not at 6 months. However, NPS was significantly better at 3 and 6 months for the BMC + PRP and platelet lysate group compared to the exercise group. Overall, the BMC + PRP and platelet lysate group had statistically significant improvement in pain and function compared to baseline starting at 3 months through a 24-month period.

Similarly, Kim et al. [92] evaluated 12 subjects treated with a combination of ultrasound-guided BMAC and PRP versus 3-month home exercise program. There was significant improvement in VAS and American Shoulder and Elbow Surgeons (ASES) scores at 3 months, but no difference between manual muscle testing strength between the two groups.

Although these studies report positive findings associated with the use of orthobiologics in the management of non-surgical rotator cuff pathology, the heterogeneity of current protocols and outcomes measures makes it challenging to make definitive recommendations in the treatment of non-surgical rotator cuff pathology. Further high-quality studies with the adoption of standardized protocols are imperative when evaluating long term outcomes of the use of these products.

## Common Extensor Tendinopathy

Common extensor tendinopathy (CET) commonly presents with lateral elbow pain secondary to overuse and eccentric overload of the proximal origin of the wrist common extensors. The most commonly affected extensor tendon is the extensor carpi radialis brevis. The cascade is initiated with an early inflammatory reaction. This further develops into microtears and tendon structure disruption with an influx of local angiogenesis with fibroblast proliferation. As time progresses, structural failure can be observed and ultimately lead to calcification or fibrosis. There is growing evidence evaluating the use of regenerative therapies for the treatment of CET. The purpose of this section is to review the literature on the use of PRP and other regenerative therapies in the treatment of CET compared to other conservative treatment options.

## PRP Vs Active Controlled Group

In 2006, Mishra and Pavelko [93] published a pilot study with 20 subjects evaluating the use to PRP in the treatment of CET in those who had failed non-operative treatment. Fifteen subjects received unactivated LR-PRP and 5 subjects received bupivacaine with epinephrine. At 8 weeks, the LR-PRP group reported 60% improvement in VAS compared to 16% improvement in the control group (bupivacaine with epinephrine). Additionally, the LR-PRP group reported a 93% reduction in pain scores compared to pre-procedure at an average of 25.6 months follow up. Although this small study showed favorable results for the PRP group, it was significantly underpowered.

In 2014, Mishra et al. [94] published a multicenter, double-blinded, prospective RCT evaluating the efficacy of LR-PRP versus dry needling in patients with CET who failed conventional therapy. A total of 230 subjects were enrolled and subjects underwent palpation guided injection of either LR-PRP or dry needling. There were no significant differences found at 12 weeks between both groups; however, at 24 weeks, clinically meaningful improvements were observed in the PRP group.

Behera et al. [95] conducted a similar study evaluating the effectiveness of LP-PRP versus control group (local anesthetic only). At 3 and 6 months, the PRP group had significant improvement in VAS, modified Mayo Clinic Performance Index for Elbow Function (MMCPIE) and Nirschl scores for activity-related pain compared to the control group.

## PRP Vs Corticosteroid Injection

Gautam et al. [96] published a RCT in 2015 comparing PRP versus corticosteroid injection in the treatment of CET. Thirty subjects were recruited and underwent either PRP or corticosteroid injection. There was improvement compared to baseline in both groups at 6-month follow up. However, diagnostic ultrasonographic evaluation revealed that the PRP group had biological healing of the lesion compared to the corticosteroid injection group that resulted increased tendon degeneration.

Yadav et al. [97] recruited 65 subjects and randomized into either PRP or corticosteroid injection group. The study reported there was better improvement in VAS scores and functional outcomes measured in a QuickDASH score in the corticosteroid injection group at 15 and 30 days. However, at the end of the 3-month period, the PRP group had significantly better improvement compared to the corticosteroid injection group, suggesting corticosteroid injection may be better in the short-term period (<3 months) compared to PRP.

In effort to evaluate long term efficacy, Gosen et al. [98] conducted a double-blind RCT evaluating the effectiveness of LR-PRP versus corticosteroid injection in the treatment of chronic CET with a follow up of 2 years. Initially, the corticosteroid injection group had improved VAS and DASH scores, however these effects started to dissipate over time. On the contrary, the PRP group continued to have sustained effects even at 2 years follow up, suggesting PRP has a more sustainable effect compared to corticosteroid injection. It is important to note that this study in combination with the pilot study by Mishra and Pavelko [93] and follow up study by Mishra et al. [94] supports the use of LR-PRP preparation in the treatment of CET in those who have failed other conservative therapies.

More recently, the results of these studies were supported by a meta-analysis published by Li et al. [99]. The authors concluded that PRP had better outcomes in the long term (>24 weeks) follow up even though corticosteroid injection had better results in the short term (4–8 weeks). Specifically, there is growing evidence supporting the use of LR-PRP in the treatment of CET; however, there is a paucity of literature discussing specific dosing protocol including the number of injections and time interval between sessions. Future research evaluating dosing and post injection rehabilitation protocol would be beneficial to help further guide the treatment of CET.

## PRP Vs ESWT

Şahbaz et al. [100] conducted a prospective RCT evaluating the effectiveness of PRP versus ESWT for CET. A total of 74 patients were randomized into either home exercise only, ESWT (once a week for 3 weeks) + exercise, or PRP (single injection) + exercise groups. At 6 months, there was a significant improvement in VAS and DASH scores in the PRP + exercise group compared to exercise alone group. However, there was no statistically significant difference between PRP + exercise and ESWT + exercise in VAS and DASH scores. Regarding Patient-Rated Tennis Elbow Evaluation (PRTEE) scores, a questionnaire measuring disability and pain in patients with lateral epicondylitis, the PRP + exercise group had statistically significant improvement compared to exercise alone and ESWT + exercise groups.

Alessio-Mazzola et al. [101] published a study with 63 patients who underwent either ultrasound-guided ESWT (3-4 sessions per week) or ultrasound-guided injection of PRP (single injection) for chronic CET. At 2 years, there was a statistically significantly higher improvement in VAS, qDASH and PRTEE scores in both groups compared to baseline. However, there was no statistically significant difference between the two groups. Overall, both PRP and ESWT provide significant benefit in the treatment of chronic CET; however, there is conflicting result regarding if one therapy is better than the other.

## Utilization of Cell-Based Therapy

Khoury et al. [74] conducted a pilot study evaluating the effects of AD-MSC in the treatment of chronic CET. A total of 18 subjects were recruited and underwent a single injection of AD-MSC at the origin of the common extensor tendon with ultrasound guidance. At 6 and 12 month follow up, there was a statistically significant improvement in VAS and QuickDASH scores. Singh et al. [102] published a prospective study assessing the efficacy of a single injection of BMAC. At 2, 6, and 12 weeks there was statistically significant improvement in PRTEE scores. Although there is growing literature on the use of cell therapy for chronic CET, further well-designed trials are necessary to determine the role of this therapy.

## Achilles Tendinopathy

Achilles tendinopathy is common overuse condition that can result in localized ankle swelling and pain that is worse with tendon-loading activities such as running and impact exercises [103]. It is classified as a degenerative process due to cumulative tissue microtrauma [103]. A recent systematic review identified risk factors that may predispose an individual to develop Achilles tendinopathy including prior lower extremity tendinopathy, use of fluroquinolones, moderate alcohol use, training during cold weather, and biomechanical factors including decreased forward progression of propulsion during gait and isokinetic plantar flexor strength [104]. Achilles tendinopathy can be divided into two broad categories based on the anatomical location of pathology: insertional tendinopathy (at the calcaneus-Achilles tendon junction) and midportion tendinopathy (2–6 cm proximal to the insertion of the Achilles tendon onto the calcaneus) [105]. Currently, there are various conservative treatment options for chronic Achilles tendinopathy. The focus of this section is to review the literature on the use of PRP and other regenerative therapies in the treatment of Achilles tendinopathy compared to other conservative treatment options.

## PRP Vs Saline

In 2010, De Vos et al. [106] published one of the earlier double-blinded, randomized, placebo-controlled trials comparing eccentric exercises + single ultrasound-guided PRP injection versus eccentric exercises + single ultrasound-guided saline injection for chronic midportion Achilles tendinopathy. A total of 54 patients were recruited and primary outcome measure was VISA-A questionnaire. Although there was significant improvement at 24 weeks in VISA-A scores in both groups, this improvement was not statistically significant between the two groups. At 1-year follow up, the results were similar to the findings at 6 months that there was no statistically significant difference between the PRP and saline groups [107]. Although both studies report no significant difference between the PRP and saline groups, Borg-Stein and Borgstorm comment on major issues regarding the study designs [108]. First, patients were excluded from the study protocol if they had previously participated in an eccentric based rehabilitation protocol. The authors argue that this exclusion criteria negates those with chronic refractory tendinosis, which is one of the main indications for the use of PRP. Additionally, imaging modalities were not used to grade the severity of tendinopathy which the authors argued could have been a cofounding factor.

In 2016, Krogh et al. [109] conducted a RCT comparing single ultrasound guided PRP injection versus saline injection for Achilles tendinopathy. The authors of this study used an ultrasonographic criteria when selecting patients to control for severity of pathology. At 3 months, PRP injection group did not result in significant difference in improvement of VISA-A scores compared to the saline group. Long-term (>3 months) evaluation was limited due to a high drop off rate as the study design allowed for patients to discontinue therapy starting at 3 months if they did not achieve satisfy treatment response. It can be argued that 3 months may not be enough time to appreciate functional benefits and long-term evaluations are necessary. Notably, the study did demonstrate a statistically significant difference in tendon thickness in the PRP injection group compared to saline injection group at 3 months.

## PRP Vs Sham Injection

More recently, Kearney et al. [110] published a multicenter RCT that included 240 participants comparing the effects of single intra-tendinous PRP injection or a sham injection, which consisted of insertion of a subcutaneous dry needle.

At 6 months, there was no significant difference in VISA-A scores between the PRP group and the sham group. Additionally, there was no significant difference in pain or quality of life at 2 weeks, 3 months or 6 months. Although this study does not support the use of PRP for chronic Achilles tendinopathy, there are a few flaws to the study design that were discussed by the authors. First, the injection was not done with ultrasound guidance which questions where the PRP product was injected. Secondly, as Malanga et al. [111] argued, the authors failed to describe the PRP product that was injected. Lastly, there was no standardized rehabilitation protocol post injection. As the literature regarding the use of PRP for Achilles tendinopathy continues to grow, it is important to carefully evaluate study designs and the type of PRP product used.

### PRP Vs ESWT

To the authors' knowledge, there are no direct studies evaluating the effectiveness of PRP versus ESWT in the treatment of chronic Achilles tendinopathy at the time this chapter was written.

### Utilization of Cell-Based Therapy

Currently, there are a paucity of studies evaluating the effectiveness and safety of MSCs in the treatment of chronic Achilles tendinopathy. Usuelli et al. [112] conducted one of the first RCT studies comparing the efficacy of PRP versus stromal vascular fraction (SVF) within adipose tissue in the treatment of midportion Achilles tendinopathy. In this study, 44 subjects were recruited and underwent a single injection of either PRP or SVF with ultrasound guidance. Post-intervention, there was no specific rehabilitation protocol and subjects were allowed to progressively resume normal activities. At 15 and 30 days follow up, the SVF group had significantly better VAS, American Orthopaedic Foot and Ankle Society (AOFAS) Ankle-Hindfoot Score and VISA-A compared to the PRP group. However, there was no statistically significant difference between the two groups at 6 months. This early study is one of the first evaluating the use of AD-MSCs in the treatment of chronic midportion Achilles tendinopathy. To date, there are no prospective studies investigating the use of BM-MSC for the treatment of chronic Achilles tendinopathy. Thueakthong et al. [113] published a retrospective chart review evaluating the outcomes of BMAC injection for the treatment of refractory Achilles tendinopathy. A total of 15 cases (5 insertional, 8 non-insertional, and 2 combined) that underwent a single BMAC injection demonstrated significant sustained improvement in pain scores based on NRS at 48-week follow up with no complications. Although this study reported positive findings, it is important to note that this study was a retrospective analysis with a relatively small sample size and no control group. As the literature continues to grow, future prospective studies will hopefully provide more evidence in the utilization of MSCs in the long-term management of chronic Achilles tendinopathy.

## Intra-articular Pathology

Osteoarthritis (OA) is a progressive, non-inflammatory, degenerative musculoskeletal joint disease that leads structural changes and loss of the articular cartilage, subchondral bone sclerosis, osteophyte formation, and changes in synovium composition [114]. OA can be further classified into primary OA and secondary OA. Primary OA is more common and refers to articular degeneration without an underlying cause, whereas secondary OA occurs in the setting of underlying pathology such as previous trauma [115]. The knees and hips are particularly weight-bearing joints which makes them more susceptible to OA. Commonly, OA presents with joint pain and stiffness, and can lead to loss of range of motion and joint instability [116]. Current treatment options for mild to moderate OA include oral and topical pharmacological agents, exercise-based therapy, interventions including corticosteroid injections, hyaluronic acid injections, and regenerative therapies including PRP and MSC. In severe cases of OA, surgical intervention may be necessary as a definitive treatment.

### Osteoarthritis of the Knee

#### PRP Vs Corticosteroid Injection

Currently, there are a plethora of studies evaluating the use of PRP in the management of knee osteoarthritis (OA). A recent systematic review by McLarnon et al. [117] analyzed a total of 8 studies including 648 patients evaluating the effects of intra-articular PRP injection versus intra-articular corticosteroid injection for the symptomatic management of osteoarthritis of the knee. PRP was found to be significantly better in reducing OA symptoms, specifically pain, stiffness, and functionality, at 3, 6, and 9 months. At 6 months, PRP allowed greater return to sporting activities than corticosteroid injection, which was measured by the Knee Injury and Osteoarthritis Outcome Score (KOOS), scoring system evaluating pain, symptoms, activities of daily living, function in sports, and quality of life in patients with knee injury and osteoarthritis [118]. Of note, the study also concluded that administering three PRP injections (separated by a week) was more effective than one injection.

## PRP Vs Hyaluronic Acid Products

A systematic review and meta-analysis by Belk et al. [119] compared the efficacy of PRP and hyaluronic acid. A total of 18 level 1 studies were included with 811 patients. The PRP group had significantly higher improvement for Western Ontario and McMaster Universities Arthritis Index (WOMAC) scores compared to the hyaluronic acid (HA) group. Additionally, those that received LP-PRP compared to LR-PRP had significantly better International Knee Documentation Committee (IKDC) scores, a subjective functional scale that focuses on symptoms, knee function, and sports activity. This sub analysis suggests that LP-PRP may be a superior treatment option compared to LR-PRP in the treatment of knee OA, although there are no further studies directly comparing the PRP leukocyte concentration in the treatment of knee OA. Another meta-analysis that included 20 RCTs reported similar results that the PRP group had significantly better WOMAC total scores at 6 and 12-month follow ups compared to HA group [120].

Migliorini et al. [121] published the first Bayesian network meta-analysis in 2021 comparing the efficacy of various intra-articular injections including PRP, corticosteroid injection, HA and placebo in patients with knee OA. A total of 30 RCTs with 3463 patients were included with up to 12 months follow-up. The authors had concluded that PRP injections resulted in the greatest reduction of pain based on VAS scores and greatest overall outcome based on WOMAC compared to corticosteroid injection, HA and placebo at 3, 6, and 12-months follow-up. Additionally, there was no significant difference found between corticosteroid injection, HA and placebo in terms of VAS and WOMAC scores. A significant limitation of this study was the heterogenicity amongst the studies in regard to number of injections and latency between doses, which appears to be a common theme within the musculoskeletal regenerative literature.

Although these meta-analyses and others have supported that PRP injections might be more effective than HA in the management of osteoarthritis of the knee, there is a need for more research to determine the optimal preparation (LR versus LP) and concentration of PRP, frequency of injections as well as dosage [119, 120, 122–124].

## Utilization of Cell-Based Therapies

Over the past several years, there has been growing evidence evaluating the efficacy of AD-MSC and BM-MSC in the treatment of OA of the knee. Migliorini et al. [125] published a systematic review in 2020 evaluating the effects of MSC injections in the treatment of knee OA. A total of 18 studies were included comprising of 1069 treated knees. 72% of the studies used BM-MSC and 28% used AD-MSC. Compared to baseline, there was a significant improvement in VAS, WOMAC, mean walking distancing, mean Lesquesne scale, and KOOS score at 6- and 12-month follow-up. No differences were found between the AD-MSC and BM-MSC groups. Additionally, the study noted that those patients who were treated at earlier stages of OA reported statistically significant greater VAS, WOMAC and walking distance. Although this study reports promising results, the authors mention that the data must be interpreted with caution. Similar to many other studies within the field of regenerative medicine, MSCs harvesting, preparation, and administrating protocols vary significantly between studies. This supports the need for more high-quality research evaluating the best effective methods of harvesting, preparing, administrating and dosing MSCs in order to better understand its use in the treatment of knee OA.

Zhao et al. [126] conducted a systematic review and meta-analysis evaluating the efficacy of HA, LP-PRP, LR-PRP, BM-MSC, AD-MSC and saline in the treatment of knee OA. Forty-three studies were included with a total of 5575 patients. At 6-months, AD-MSC was found to be the most effective in relieving pain while LP-PRP was the most effective for functional improvement. At 12-month follow up, both AD-MSC and LP-PRP showed improvement in knee pain, while LP-PRP showed functional improvement based on WOMAC scores. Compared to placebo, both BM-MSC and AD-MSC have significant improvement in pain and function at 6-months. However, it was found that AD-MSC had better clinical efficacy compared to BM-MSC. It is important to note that there were more grade 4 (based on Kellgren-Lawrence grade) knee OA patients included in the BM-MSC arm indicating further progression in disease which could explain the results.

Previously mentioned study by Migliorini et al. [125] did not find a difference between the AD-MSC and BM-MSC groups in the treatment of OA. Bolia et al. [127] published a systematic review in 2022 evaluating single injection of BMAC and SVF in the treatment of knee OA. Ten studies with 427 patients were included. A short-term follow up (6–22 months), single injection of SVF resulted in greater reduction in knee pain based on VAS scores. When considering functional improvement (based on WOMAC scores), both BMAC and SVF were equivalent.

These studies reveal positive findings with the use of cell-based therapies including BM-MSC and AD-MSC in the treatment of knee OA. A common limitation found within these initial studies includes variability between preparation protocols and study methodology which poses a challenge when comparing studies. As research continues to develop within the field of regenerative medicine, future studies should focus on comparing aggregate data from studies with similar baseline patient characteristics, preparation protocol, and degree of pathology based on radiological imaging.

## Osteoarthritis of the Hip

### PRP Vs HA

Compared to the treatment of knee osteoarthritis, there are limited studies RCT evaluating the efficacy of PRP in the treatment of hip osteoarthritis. Nouri et al. [128] conducted a RCT comparing the effects of ultrasound-guided intra-articular injection of PRP, HA and PRP + HA in the management of osteoarthritis of the hip. A total of 105 subjects with either grade 2 or 3 hip OA based on Kellgren and Lawrence classification were enrolled and randomized into one of three groups: PRP, HA, and PRP + HA. At 6 months, there was a significant difference in performance, ADLs and disability in the PRP and PRP + HA group compared to HA group. However, the combination of PRP + HA did not lead to significant improvement compared to PRP alone, which was also supported by Dallari et al. [129]. A prospective, double-blinded RCT by Doria et al. reported both PRP and HA revealed significant improvement in WOMAC, VAS and Harris Hip Score at 6 and 12 months compared to baseline [130]. However, the PRP group failed to show better results compared to HA group. Similarly, Di Sante et al. concluded that compared to HA, PRP showed a significant reduction in VAS scores at early follow up (1 month) which dissipated at long term follow up (4 months) [131]. Overall, these studies report varying results, and it is unclear at this time if PRP is superior compared to HA in the treatment of hip osteoarthritis.

Additionally, Kraeutler et al. [132] conducted the first double-blind, RCT evaluating effects of intra-articular injection of either LP-PRP or HA on conversion to surgical interventions including total hip arthroplasty or hip resurfacing procedure. A total of 36 hip were randomized to receive a series of 3 weekly intra-articular injection of either LP-PRP or HA. This study revealed that a significantly greater number of patients had converted to surgical interventions in the HA group compared to the PRP group ($P = 0.035$), suggesting PRP may delay the need for total hip replacement or hip resurfacing procedure compared to HA. Although these initial results are positive for PRP, further long-term studies are needed to determine how effective PRP is in hip preservation.

## Utilization of Cell-Based Therapy

The studies evaluating the use of cell-based therapy in the treatment of hip OA are largely limited to prospective or retrospective case series without a comparison arm. Whitney et al. [133] published a case series in 2020 with 24 subjects who underwent single BMC injection. There was significant improvement in Numeric Rating Scale (NRS) score with and without activity at 6 weeks, 3 months and 6 months. Additionally, the authors noted improvement in modified Harris Hip Score, WOMAC, and Hip Outcome Score-Activities of Daily Living. Notably, there were no adverse events or complications during this study. Tsitsilianos et al. [134] conducted a retrospective analysis of 31 subjects who underwent intra-articular injection of bone marrow aspirate (BMA). At 6 and 12 months, there was a statistically significant improvement in NRS and Hip Disability and Osteoarthritis Outcome Score Jr scores. Both pilot studies show positive results with the use of cell-based therapy in the treatment of hip osteoarthritis. This warrants the need for prospective RCT trials to further assess the effectiveness of cell-based therapy in the treatment of hip osteoarthritis.

## Conclusion

The use of biologics for spine and orthopedic conditions seems to be an active area for ongoing research. These therapies have ushered a new era for nonsurgical management of chronic pain. This field has, therefore, witnessed exponential growth since its inception primarily to drive precise and tailored patient management. It is worth noting that while regenerative therapies to treat spinal and musculoskeletal pain have shown remarkable progress in the past decade, however, outcomes in clinical studies have remained short from stellar. The absence of technical and ethical oversight for the use of biologics has led to nonuniformity in clinical practice. Additional research is therefore, needed to better standardize the utilization of these biological substances. Recent guidelines put forth by ASIPP are a significant step toward this addressing this grand challenge. Such guidelines also help lay the necessary groundwork required for increasing the likelihood of biological regenerative medicine injection therapy to become the standard of care.

**Key Points**

1. The main generators of pain in the lumbar spine include intervertebral discs, facet joints, sacroiliac joints and neural connections.
2. Both PRP and MSCs are used to aid the natural healing process and target the underlying root cause, rather than providing mere symptomatic relief.
3. There is plethora of literature with level of evidence ranging from Level-I to Level-IV, supporting the role of PRP in management of chronic spinal pain
4. The evidence regarding the use of MSCs for the treatment of discogenic low backpain centers on the use of intradiscal injections.
5. The published literature on facet joints and sacroiliac joints primarily focuses on the use of PRP injections,

with no available studies investigating the potential effects of MSCs in these contexts.

6. Recent systematic reviews and meta-analyses have demonstrated significant improvements in both pain scores and quality of life, indicating the potential efficacy of biologics for the spine.
7. Most studies are limited by small study populations and short follow-up duration, narrowing the scope of clinical application.
8. Majority of studies focus on the lumbar spine, more controlled studies for the thoracic and cervical spine are needed to demonstrate efficacy.
9. The literature on the use of regenerative therapies, including PRP and BMC, for the management of rotator cuff pathologies continues to evolve. Although recent studies have reported positive findings, there is high variability in preparation methods and outcome measures which poses a challenge to make definitive recommendations at this time.
10. Common extensor tendinopathy has been traditionally managed with corticosteroid injections. However, there is growing literature supporting the use of PRP, specifically LR-PRP preparation, in the treatment of common extensor tendinopathy as it provides better outcomes in long term follow up compared to corticosteroid injections. There remains a paucity of literature discussing specific dosing protocol including dosing, interval between injections and post injection rehabilitation protocol.
11. The current studies evaluating the use of PRP compared to sham or saline injection for Achilles tendinopathy do not support the use of PRP. However, as noted in this chapter, some researchers have argued these studies have methodological flaws, which may confound the current data. Further high-quality studies are needed to evaluate the effectiveness of PRP in the management of Achilles tendinopathy.
12. PRP has been shown to be more effective compared to hyaluronic acid in the treatment of knee osteoarthritis. Furthermore, there is growing evidence suggesting that LP-PRP preparation may be a superior treatment option. Further studies directly comparing PRP leukocyte concentration in the treatment of knee osteoarthritis are necessary.
13. Based on the reported studies evaluating the use of PRP compared to hyaluronic acid for hip osteoarthritis, there are varying results. At this time, it is not clear if PRP is superior to hyaluronic acid in the treatment of hip OA.
14. There has been a vast increase in the literature evaluating the use of regenerative therapies for the management of common sites of tendinopathies and intra-articular pathologies. However, the heterogeneity between the studies with varying preparation and dosing protocols makes it challenging to determine if regenerative therapies are superior. There is a strong need for more high-quality, larger scale studies with an emphasis on PRP and cell-based preparations and dosing protocols to help guide clinical practice.

## References


1. Navani A, Manchikanti L, Albers SL, et al. Responsible, safe, and effective use of biologics in the management of low back pain: American Society of Interventional Pain Physicians (ASIPP) guidelines. Pain Physician. 2019;22:S1–S74.
2. Manchikanti L, Centeno CJ, Atluri S, et al. Bone marrow concentrate (BMC) therapy in musculoskeletal disorders: evidence-based policy position statement of American Society of Interventional Pain Physicians (ASIPP). Pain Physician. 2020;23:E85–E131.
3. Beitzel K, Allen D, Apostolakos J, et al. US Definitions, current use, and FDA stance on use of platelet rich plasma in sports medicine. J Knee Surg. 2015;28:29–34.
4. Berebichez-Fridman R, Montero-Olvera PR. Sources and clinical applications of mesenchymal stem cells: state-of-the-art review. Sultan Qaboos Univ Med J. 2018;18:e264–77.
5. Caplan AI. Mesenchymal stem cells: time to change the name! Stem Cells Transl Med. 2017;6:1445–51.
6. Krut Z, Pelled G, Gazit D, et al. Stem cells and exosomes: new therapies for intervertebral disc degeneration. Cell. 2021;10:2241.
7. Desai MJ, Mansfield JT, Robinson DM, et al. Regenerative medicine for axial and radicular spine-related pain: a narrative review. Pain Pract. 2020;20:437–53.
8. Pasquale XM, Jason MC, Gaetano JS. Intradiscal injection of an autologous alpha-2-macroglobulin (A2M) concentrate alleviates back pain in FAC-positive patients. Ortho Rheum. 2017;4:555634.
9. Spiel D. The role of exosomes in regenerative medicine. In: Manchikanti L, Navani A, Atluri S, editors. Essentials of regenerative medicine in interventional pain management. Paducah: ASIPP Publishing; 2019. p. 115–28.
10. GBD 2017 Disease and Injury Incidence and Prevalence Collaborators. Global, regional, and national incidence, prevalence, and years lived with disability for 354 diseases and injuries for 195 countries and territories, 1990-2017: a systematic analysis for the Global Burden of Disease Study 2017. Lancet. 2018;392:1789–858.
11. Fatoye F, Gebrye T, Odeyemi I. Real-world incidence and prevalence of low back pain using routinely collected data. Rheumatol Int. 2019;39:619–26.
12. Leboeuf-Yde C, Nielsen J, Kyvik KO, et al. Pain in the lumbar, thoracic or cervical regions: do age or gender matter? A population-based study of 34,902 Danish twins 20–71 years of age. BMC Musculoskelet Disord. 2009;10:39.
13. Manchikanti L, Abdi S, Atluri S, et al. An update of comprehensive evidence-based guidelines for interventional techniques of chronic spinal pain: part II: guidance and recommendations. Pain Physician. 2013;16:S49–S283.
14. Zeckser J, Wolff M, Tucker J, et al. Multipotent mesenchymal stem cell treatment for discogenic low back pain and disc degeneration. Stem Cells Int. 2016;2016:3908389.
15. Li W, Gong Y, Liu J, et al. Peripheral and central pathological mechanisms of chronic low back pain: a narrative review. J Pain Res. 2021;14:1483–94.

16. Akeda K, Imanishi T, Ohishi K, et al. Intradiscal injection of autologous serum isolated from platelet-rich-plasma for the treatment of discogenic low back pain: preliminary prospective clinical trial: GP141. Spine Affiliated Society Meeting Abstracts; 2011.
17. Tuakli-Wosornu YA, Terry A, Boachie-Adjei K, et al. Lumbar intradiskal platelet-rich plasma (PRP) injections: a prospective, double-blind, randomized controlled study. PM R. 2016;8:1–10.
18. Levi D, Horn S, Tyszko S, et al. Intradiscal platelet-rich plasma injection for chronic discogenic low back pain: preliminary results from a prospective trial. Pain Med. 2016;17:1010–22.
19. Akeda K, Ohishi K, Masuda K, et al. Intradiscal injection of autologous platelet-rich plasma releasate to treat discogenic low back pain: a preliminary clinical trial. Asian Spine J. 2017;11:380–9.
20. Akeda K, Takegami N, Yamada J, et al. Platelet-rich plasma-releasate (PRPr) for the treatment of discogenic low back pain patients: long-term follow-up survey. Medicina. 2022;58:428.
21. Navani A, Ambach M, Navani R, et al. Biologics for lumbar discogenic pain: 18-month follow up for safety and efficacy. Interv Pain Manag Rep. 2018;2:111–8.
22. Jain D, Goyal T, Verma N, et al. Intradiscal platelet-rich plasma injection for discogenic low back pain and correlation with platelet concentration: a prospective clinical trial. Pain Med. 2020;11:2719–25.
23. Chang MC, Park D. The effect of intradiscal platelet-rich plasma injection for management of discogenic lower back pain: a meta-analysis. J Pain Res. 2021;14:505–12.
24. Lutz C, Cheng J, Prysak M, et al. Clinical outcomes following intradiscal injections of higher-concentration platelet-rich plasma in patients with chronic lumbar discogenic pain. Int Orthop 2022; 46: 1381-1385.
25. Kirchner F, Milani I, Martinez A, et al. Plasma rich in growth factors (PRGF) in the treatment of cervical and lumbar back pain: a retrospective observational clinical study. Pain Physician. 2021;24:E649–60.
26. Mohammed S, Yu J. Platelet-rich plasma injections: an emerging therapy for chronic discogenic low back pain. J Spine Surg. 2018;4:115–22.
27. Noriega DC, Ardura F, Hernández-Ramajo R, et al. Intervertebral disc repair by allogeneic mesenchymal bone marrow cells: a randomized controlled trial. Transplantation. 2017;101:1945–51.
28. Pettine KA, Suzuki RK, Sand TT, et al. Autologous bone marrow concentrate intradiscal injection for the treatment of degenerative disc disease with three-year follow-up. Int Orthop. 2017;41:2097–103.
29. Wu T, Song HX, Dong Y, et al. Cell-based therapies for lumbar discogenic low back pain: systematic review and single-arm meta-analysis. Spine. 2018;43:49–57.
30. Wolff M, Shillington JM, Rathbone C, et al. Injections of concentrated bone marrow aspirate as treatment for discogenic pain: a retrospective analysis. BMC Musculoskelet Disord. 2020;21:135.
31. Amirdelfan K, Bae H, McJunkin T, et al. Allogeneic mesenchymal precursor cells treatment for chronic low back pain associated with degenerative disc disease: a prospective randomized, placebo-controlled 36-month study of safety and efficacy. Spine J. 2021;21:212–30.
32. Her YF, Kubrova E, Martinez Alvarez GA, et al. The analgesic efficacy of intradiscal injection of bone marrow aspirate concentrate and culture-expanded bone marrow mesenchymal stromal cells in discogenic pain: a systematic review. J Pain Res. 2022;15:3299–318.
33. Mochida J, Sakai D, Nakamura Y, et al. Intervertebral disc repair with activated nucleus pulposus cell transplantation: a three-year, prospective clinical study of its safety. Eur Cell Mater. 2015;29:202–12.
34. Beall DP, Wilson GL, Bishop R, et al. VAST clinical trial: Safely supplementing tissue lost to degenerative disc disease. Int J Spine Surg. 2020;14:239–53.
35. Atluri S, Murphy MB, Dragella R, et al. Evaluation of the effectiveness of autologous bone marrow mesenchymal stem cells in the treatment of chronic low back pain due to severe lumbar spinal degeneration: a 12-month, open-label, prospective controlled trial. Pain Physician. 2022;25:193–207.
36. Pettine KA. Two-year follow-up results of treating cervical degenerated discs with bone marrow concentrate to avoid surgery. Stem Cell Res Ther. 2017;2:80–7.
37. Manchikanti L, Knezevic NN, Boswell MV, et al. Epidural injections for lumbar radiculopathy and spinal stenosis: a comparative systematic review and meta-analysis. Pain Physician. 2016;19:E365–410.
38. Centeno C, Markle J, Dodson E, et al. The use of lumbar epidural injection of platelet lysate for treatment of radicular pain. J Exp Orthop. 2017;4:38.
39. Jose C, Henry C, Patricia A, et al. Epidural plasma rich in growth factors for degenerative disc disease: a valuable alternative to conventional "palliative medicine". Int J Anesth Clin Med. 2019;7:1–6.
40. Xu Z, Wu S, Li X, et al. Ultrasound-guided transforaminal injections of platelet-rich plasma compared with steroid in lumbar disc herniation: a prospective, randomized, controlled study. Neural Plast. 2021;2021:5558138.
41. Ruiz-Lopez R, Tsai YC. A randomized double-blind controlled pilot study comparing leucocyte-rich platelet-rich plasma and corticosteroid in caudal epidural injection for complex chronic degenerative spinal pain. Pain Pract. 2020;20:639–46.
42. Benitez N. Efficacy of parasargital translaminar epidural application of growth factors derived from platelet rich plasma as a treatment for unilateral root pain caused by multisegmental disc disease. Investig Med. 2021;13:1.
43. Sanapati J, Manchikanti L, Atluri S, et al. Do regenerative medicine therapies provide long-term relief in chronic low back pain: a systematic review and metaanalysis. Pain Physician. 2018;21:515–40.
44. Goldthwait JE. The lumbosacral articulation: an explanation of many cases of lumbago, sciatica, and paraplegia. Boston Med Surg J. 1911;164:365–72.
45. Ghormley RK. Low back pain with special reference to the articular facets, with presentation of an operative procedure. JAMA. 1933;101:773.
46. Datta S, Lee M, Falco FJ, et al. Systematic assessment of diagnostic accuracy and therapeutic utility of lumbar facet joint interventions. Pain Physician. 2009;12:437–60.
47. Perolat R, Kastler A, Nicot B, et al. Facet joint syndrome: from diagnosis to interventional management. Insights Imaging. 2018;9:773–89.
48. Wu J, Du Z, Lv Y, et al. A new technique for the treatment of lumbar facet joint syndrome using intraarticular injection with autologous platelet rich plasma. Pain Physician. 2016;19:617–25.
49. Wu J, Zhou J, Liu C, et al. A prospective study comparing platelet-rich plasma and local anesthetic (LA)/corticosteroid in intra-articular injection for the treatment of lumbar facet joint syndrome. Pain Pract. 2017;17:914–24.
50. Kirchner F, Anitua E. Intradiscal and intra-articular facet infiltrations with plasma rich in growth factors reduce pain in patients with chronic low back pain. J Craniovertebr Junction Spine. 2016;7:250–6.
51. Simopoulos TT, Manchikanti L, Gupta S, et al. Systematic review of the diagnostic accuracy and therapeutic effectiveness of sacroiliac joint interventions. Pain Physician. 2015;18:E713–56.

52. Saunders J. A comparison of ultrasound guided PRP injection and prolotherapy for mechanical dysfunction of the sacroiliac joint. J Prolother. 2018;10:e992–9.
53. Singla V, Batra YK, Bharti N, et al. Steroid vs. platelet-rich plasma in ultrasound-guided sacroiliac joint injection for chronic low back pain. Pain Pract. 2017;17:782–91.
54. Mohi Eldin M, Sorour OO, Hassan ASA, et al. Percutaneous injection of autologous platelet-rich fibrin versus platelet-rich plasma in sacroiliac joint dysfunction: an applied comparative study. J Back Musculoskelet Rehabil. 2019;32:511–8.
55. Wallace P, Bezjian Wallace L, Tamura S, et al. Effectiveness of ultrasound-guided platelet-rich plasma injections in relieving sacroiliac joint dysfunction. Am J Phys Med Rehabil. 2020;99:689–93.
56. Ling JF, Wininger AE, Hirase T. Platelet-rich plasma versus corticosteroid injection for lumbar spondylosis and sacroiliac arthropathy: a systematic review of comparative studies. Cureus. 2021;13:e14062.
57. Girdwood M, Docking S, Rio E, et al. Pathophysiology of tendinopathy. In: Canata GL, d'Hooghe P, Hunt KJ, editors. Muscle and tendon injuries: evaluation and management. Berlin: Springer; 2017. p. 23–44.
58. Millar NL, Silbernagel KG, Thorborg K, et al. Tendinopathy. Nat Rev Dis Primers. 2021;7:1.
59. Zhou Y, Wang JHC. PRP treatment efficacy for tendinopathy: a review of basic science studies. Biomed Res Int. 2016;2016:9103792.
60. Murtaugh B, Ihm JM. Eccentric training for the treatment of tendinopathies. Curr Sports Med Rep. 2013;12:175–82.
61. Abate M, Salini V, Schiavone C, et al. Clinical benefits and drawbacks of local corticosteroids injections in tendinopathies. Expert Opin Drug Saf. 2017;16:341–9.
62. Dean BJF, Lostis E, Oakley T, et al. The risks and benefits of glucocorticoid treatment for tendinopathy: a systematic review of the effects of local glucocorticoid on tendon. Semin Arthritis Rheum. 2014;43:570–6.
63. Figueroa D, Figueroa F, Calvo R. Patellar tendinopathy: diagnosis and treatment. J Am Acad Orthop Surg. 2016;24:e184–92.
64. Scott A, LaPrade RF, Harmon KG, et al. Platelet-rich plasma for patellar tendinopathy: a randomized controlled trial of leukocyte-rich PRP or leukocyte-poor PRP versus saline. Am J Sports Med. 2019;47:1654–61.
65. Stoychev V, Finestone AS, Kalichman L. Dry needling as a treatment modality for tendinopathy: a narrative review. Curr Rev Musculoskelet Med. 2020;13:133–40.
66. Cagnie B, Dewitte V, Barbe T, et al. Physiologic effects of dry needling topical collection on myofascial pain. Curr Pain Headache Rep. 2013;17:348.
67. Gattie E, Cleland JA, Snodgrass S. The effectiveness of trigger point dry needling for musculoskeletal conditions by physical therapists: a systematic review and meta-analysis. J Orthop Sports Phys Ther. 2017;47:133–49.
68. Dragoo JL, Wasterlain AS, Braun HJ, et al. Platelet-rich plasma as a treatment for patellar tendinopathy: a double-blind, randomized controlled trial. Am J Sports Med. 2014;42:610–8.
69. Zayni R, Thaunat M, Fayard JM, et al. Platelet-rich plasma as a treatment for chronic patellar tendinopathy: comparison of a single versus two consecutive injections. Muscles Ligaments Tendons J. 2015;5:92–8.
70. Simplicio CL, Purita J, Murrell W, et al. Extracorporeal shock wave therapy mechanisms in musculoskeletal regenerative medicine. J Clin Orthop Trauma. 2020;11:S309–18.
71. Vetrano M, Castorina A, Vulpiani MC, et al. Platelet-rich plasma versus focused shock waves in the treatment of Jumper's knee in athletes. Am J Sports Med. 2013;41:795–803.
72. Rodas G, Soler-Rich R, Rius-Tarruella J, et al. Effect of autologous expanded bone marrow mesenchymal stem cells or leukocyte-poor platelet-rich plasma in chronic patellar tendinopathy (with gap >3 mm): preliminary outcomes after 6 months of a double-blind, randomized, prospective study. Am J Sports Med. 2021;49:1492–504.
73. Smith J, Sellon JL. Comparing PRP injections with ESWT for athletes with chronic patellar tendinopathy. Clin J Sport Med. 2014;24:88–9.
74. Khoury M, Tabben M, Rolón AU, et al. Promising improvement of chronic lateral elbow tendinopathy by using adipose derived mesenchymal stromal cells: a pilot study. J Exp Orthop. 2021;8:6.
75. Factor D, Dale B. Current concepts of rotator cuff tendinopathy. Int J Sports Phys Ther. 2014;9:274–88.
76. Seitz AL, McClure PW, Finucane S, et al. Mechanisms of rotator cuff tendinopathy: Intrinsic, extrinsic, or both? Clin Biomech. 2011;26:1–12.
77. Maffulli N, Longo UG, Berton A, et al. Biological factors in the pathogenesis of rotator cuff tears. Sports Med Arthrosc Rev. 2011;19:194–201.
78. Sambandam SN, Khanna V, Gul A, et al. Rotator cuff tears: an evidence based approach. World J Orthop. 2015;6:902–18.
79. Rha DW, Park GY, Kim YK, et al. Comparison of the therapeutic effects of ultrasound-guided platelet-rich plasma injection and dry needling in rotator cuff disease: a randomized controlled trial. Clin Rehabil. 2013;27:113–22.
80. Kesikburun S, Tan AK, Yilmaz B, et al. Platelet-rich plasma injections in the treatment of chronic rotator cuff tendinopathy: a randomized controlled trial with 1-year follow-up. Am J Sports Med. 2013;41:2609–16.
81. Wessel RN, Wolterbeek N, Fermont AJM, et al. Responsiveness and disease specificity of the Western Ontario Rotator Cuff index. J Orthop. 2018;15:337–42.
82. Breckenridge JD, McAuley JH. Shoulder pain and disability index (SPADI). J Physiother. 2011;57:197.
83. Prodromos CC, Finkle S, Prodromos A, et al. Treatment of rotator cuff tears with platelet rich plasma: a prospective study with 2 year follow-up. BMC Musculoskelet Disord. 2021;22:499.
84. Beaton DE, Wright JG, Katz JN, et al. Development of the QuickDASH: comparison of three item-reduction approaches. J Bone Jt Surg Am. 2005;87:1038–46.
85. Hart L. Corticosteroid and other injections in the management of tendinopathies: a review. Clin J Sport Med. 2011;21:540–1.
86. Kapetanos G. The effect of the local corticosteroids on the healing and biomechanical properties of the partially injured tendon. Clin Orthop Relat Res. 1982;163:170–9.
87. Kwong CA, Woodmass JM, Gusnowski EM, et al. Platelet-rich plasma in patients with partial-thickness rotator cuff tears or tendinopathy leads to significantly improved short-term pain relief and function compared with corticosteroid injection: a double-blind randomized controlled trial. Arthroscopy. 2021;37:510–7.
88. Shams A, El-Sayed M, Gamal O, et al. Subacromial injection of autologous platelet-rich plasma versus corticosteroid for the treatment of symptomatic partial rotator cuff tears. Eur J Orthop Surg Traumatol. 2016;26:837–42.
89. Chen X, Jones IA, Togashi R, et al. Use of platelet-rich plasma for the improvement of pain and function in rotator cuff tears: a systematic review and meta-analysis with bias assessment. Am J Sports Med. 2020;48:2028–41.
90. Centeno CJ, Al-Sayegh H, Bashir J, et al. A prospective multi-Site registry study of a specific protocol of autologous bone marrow concentrate for the treatment of shoulder rotator cuff tears and osteoarthritis. J Pain Res. 2015;8:269–76.
91. Centeno C, Fausel Z, Stemper I, et al. A randomized controlled trial of the treatment of rotator cuff tears with bone marrow con-

centrate and platelet products compared to exercise therapy: a midterm analysis. Stem Cells Int. 2020;2020:5962354.
92. Kim SJ, Kim EK, Kim SJ, et al. Effects of bone marrow aspirate concentrate and platelet-rich plasma on patients with partial tear of the rotator cuff tendon. J Orthop Surg Res. 2018;13:1.
93. Mishra A, Pavelko T. Treatment of chronic elbow tendinosis with buffered platelet-rich plasma. Am J Sports Med. 2006;34:1774–8.
94. Mishra AK, Skrepnik NV, Edwards SG, et al. Efficacy of platelet-rich plasma for chronic tennis elbow: a double-blind, prospective, multicenter, randomized controlled trial of 230 patients. Am J Sports Med. 2014;42:463–71.
95. Behera P, Dhillon M, Aggarwal S, et al. Leukocyte-poor platelet-rich plasma versus bupivacaine for recalcitrant lateral epicondylar tendinopathy. J Orthop Surg. 2015;23:6–10.
96. Gautam VK, Verma S, Batra S, et al. Platelet-rich plasma versus corticosteroid injection for recalcitrant lateral epicondylitis: clinical and ultrasonographic evaluation. J Orthop Surg. 2015;23:1–5.
97. Yadav R, Kothari SY, Borah D. Comparison of local injection of platelet rich plasma and corticosteroids in the treatment of lateral epicondylitis of humerus. J Clin Diagn Res. 2015;9:RC05–7.
98. Gosens T, Peerbooms JC, van Laar W, et al. Ongoing positive effect of platelet-rich plasma versus corticosteroid injection in lateral epicondylitis: a double-blind randomized controlled trial with 2-year follow-up. Am J Sports Med. 2011;39:1200–8.
99. Li A, Wang H, Yu Z, et al. Platelet-rich plasma vs corticosteroids for elbow epicondylitis: a systematic review and meta-analysis. Medicine. 2019;98:e18358.
100. Şahbaz T, Ceylan CM, Karacay BÇ, et al. Comparison of platelet-rich plasma and extracorporeal shock wave therapy in patients with chronic lateral epicondylitis: a prospective, randomized-controlled study. Turk J Phys Med Rehabil. 2021;67:490–501.
101. Alessio-Mazzola M, Repetto I, Biti B, et al. Autologous US-guided PRP injection versus us-guided focal extracorporeal shock wave therapy for chronic lateral epicondylitis: a minimum of 2-year follow-up retrospective comparative study. J Orthop Surg. 2018;26:2309499017749986.
102. Singh A, Gangwar DS, Singh S. Bone marrow injection: a novel treatment for tennis elbow. J Nat Sci Biol Med. 2014;5:389–91.
103. Von Rickenbach KJ, Borgstrom H, Tenforde A, et al. Achilles tendinopathy: evaluation, rehabilitation, and prevention. Curr Sports Med Rep. 2021;20:327–34.
104. Van Der Vlist AC, Breda SJ, et al. Clinical risk factors for Achilles tendinopathy: a systematic review. Br J Sports Med. 2019;53:1352–61.
105. Li HY, Hua YH. Achilles tendinopathy: current concepts about the basic science and clinical treatments. Biomed Res Int. 2017;2017:9534125.
106. De Vos RJ, Weir A, Van Schie HTM, et al. Platelet-rich plasma injection for chronic achilles tendinopathy: a randomized controlled trial. JAMA. 2010;303:144–9.
107. De Jonge S, De Vos RJ, Weir A, et al. One-year follow-up of platelet-rich plasma treatment in chronic achilles tendinopathy: a double-blind randomized placebo-controlled trial. Am J Sports Med. 2011;39:1623–9.
108. Tenforde A, Robinson D, Borg-Stein J, et al. Extracorporeal shockwave therapy versus platelet-rich plasma for achilles tendinopathy. PM R. 2020;12:1169–76.
109. Krogh TP, Ellingsen T, Christensen R, Jensen P, Fredberg U. Ultrasound-guided injection therapy of achilles tendinopathy with platelet-rich plasma or saline: a randomized, blinded, placebo-controlled trial. Am J Sports Med. 2016;44:1990–7.
110. Kearney RS, Ji C, Warwick J, et al. Effect of Platelet-rich plasma injection vs sham injection on tendon dysfunction in patients with chronic midportion achilles tendinopathy: a randomized clinical trial. JAMA. 2021;326:137–44.
111. Malanga GA, Mautner K, Buford D. Platelet-rich plasma injection vs sham injection and tendon dysfunction in patients with chronic midportion achilles tendinopathy. JAMA. 2021;326:1974–5.
112. Usuelli FG, Grassi M, Maccario C, et al. Intratendinous adipose-derived stromal vascular fraction (SVF) injection provides a safe, efficacious treatment for Achilles tendinopathy: results of a randomized controlled clinical trial at a 6-month follow-up. Knee Surg Sports Traumatol Arthrosc. 2018;26:2000–10.
113. Thueakthong W, de Cesar NC, Garnjanagoonchorn A, et al. Outcomes of iliac crest bone marrow aspirate injection for the treatment of recalcitrant Achilles tendinopathy. Int Orthop. 2021;45:2423–8.
114. Chen D, Shen J, Zhao W, et al. Osteoarthritis: toward a comprehensive understanding of pathological mechanism. Bone Res. 2017;5:16044.
115. Cucchiarini M, de Girolamo L, Filardo G, et al. Basic science of osteoarthritis. J Exp Orthop. 2016;3:22.
116. Kean WF, Kean R, Buchanan WW. Osteoarthritis: symptoms, signs and source of pain. Inflammopharmacology. 2004;12:3–31.
117. McLarnon M, Heron N. Intra-articular platelet-rich plasma injections versus intra-articular corticosteroid injections for symptomatic management of knee osteoarthritis: systematic review and meta-analysis. BMC Musculoskelet Disord. 2021;22:550.
118. Roos EM, Lohmander LS. The knee injury and osteoarthritis outcome score (KOOS): from joint injury to osteoarthritis. Health Qual Life Outcomes. 2003;1:64.
119. Belk JW, Kraeutler MJ, Houck DA, et al. Platelet-rich plasma versus hyaluronic acid for knee osteoarthritis: a systematic review and meta-analysis of randomized controlled trials. Am J Sports Med. 2021;49:249–60.
120. Tang JZ, Nie MJ, Zhao JZ, et al. Platelet-rich plasma versus hyaluronic acid in the treatment of knee osteoarthritis: a meta-analysis. J Orthop Surg Res. 2020;15:403.
121. Migliorini F, Driessen A, Quack V, et al. Comparison between intra-articular infiltrations of placebo, steroids, hyaluronic and PRP for knee osteoarthritis: a Bayesian network meta-analysis. Arch Orthop Trauma Surg. 2021;141:1473–90.
122. Han Y, Huang H, Pan J, et al. Meta-analysis comparing platelet-rich plasma vs hyaluronic acid injection in patients with knee osteoarthritis. Pain Med. 2019;20:1418–29.
123. Wu Q, Luo X, Xiong Y, et al. Platelet-rich plasma versus hyaluronic acid in knee osteoarthritis: a meta-analysis with the consistent ratio of injection. J Orthop Surg. 2020;28:2309499019887660.
124. Tan J, Chen H, Zhao L, et al. Platelet-rich plasma versus hyaluronic acid in the treatment of knee osteoarthritis: a meta-analysis of 26 randomized controlled trials. Arthroscopy. 2021;37:309–25.
125. Migliorini F, Rath B, Colarossi G, et al. Improved outcomes after mesenchymal stem cells injections for knee osteoarthritis: results at 12-months follow-up: a systematic review of the literature. Arch Orthop Trauma Surg. 2020;140:853–68.
126. Zhao D, Pan JK, Yang WY, et al. Intra-articular injections of platelet-rich plasma, adipose mesenchymal stem cells, and bone marrow mesenchymal stem cells associated with better outcomes than hyaluronic acid and saline in knee osteoarthritis: a systematic review and network meta-analysis. Arthroscopy. 2021;37:2298–314.
127. Bolia IK, Bougioukli S, Hill WJ, et al. Clinical efficacy of bone marrow aspirate concentrate versus stromal vascular fraction injection in patients with knee osteoarthritis: a systematic review and meta-analysis. Am J Sports Med. 2022;50:1451–61.
128. Nouri F, Babaee M, Peydayesh P, et al. Comparison between the effects of ultrasound guided intra-articular injections of platelet-rich plasma (PRP), high molecular weight hyaluronic acid, and their combination in hip osteoarthritis: a randomized clinical trial. BMC Musculoskelet Disord. 2022;23:1–12.

129. Dallari D, Stagni C, Rani N, et al. Ultrasound-guided injection of platelet-rich plasma and hyaluronic acid, separately and in combination, for hip osteoarthritis. Am J Sports Med. 2016;44:664–71.
130. Doria C, Mosele GR, Caggiari G, et al. Treatment of early hip osteoarthritis: ultrasound-guided platelet rich plasma versus hyaluronic acid injections in a randomized clinical trial. Joints. 2017;5:152–5.
131. Di Sante L, Villani C, Santilli V, et al. Intra-articular hyaluronic acid vs platelet-rich plasma in the treatment of hip osteoarthritis. Med Ultrason. 2016;18:463–8.
132. Kraeutler MJ, Houck DA, Garabekyan T, et al. Comparing intra-articular injections of leukocyte-poor platelet-rich plasma versus low–molecular weight hyaluronic acid for the treatment of symptomatic osteoarthritis of the hip: a double-blind, randomized pilot study. Orthop J Sports Med. 2021;9:2325967120969210.
133. Whitney KE, Briggs KK, Chamness C, et al. Bone marrow concentrate injection treatment improves short-term outcomes in symptomatic hip osteoarthritis patients: a pilot study. Orthop J Sports Med. 2020;8:2325967120966162.
134. Tsitsilianos N, Shirazi Z, Lu J, et al. Interventional pain medicine bone marrow aspirate injection for osteoarthritis of the hip: a pilot study. Interv Pain Med. 2022;4:100163.

# Biologic and Cellular Therapies in the Clinical Setting: Standardization and Collaboration

4

David C. Karli and Theodore T. Sand

## Introduction

Biologic and cellular therapy development continues to grow and diversify [1]. Over the past 20 years, the field of orthobiologics has established a foundation as a safe and efficacious treatment option for a variety of musculoskeletal (MSK) injuries, including osteoarthritis (OA) and chronic tendinopathies. In more recent years, clinical applications in pain management and interventional spine techniques have started to receive more significant research attention, driven by an aging patient population, as well as a high incidence of spinal related pain and disability in adults have expanded [2]. The increased patient demand for treatment has progressively led to physicians attempting to integrate regenerative options into their practices. However, as the growth of orthobiologics continues to expand, coordinated research efforts and clarity regarding government guidelines for compliant use have not been able to match the same trajectory, resulting in an inadequacy of high-level studies, and confusion regarding product efficacy and regulatory compliance. As the number of physicians utilizing orthobiologics continues to rise, the need for cohesiveness and standardization in an attempt to establish the validity, consistency, safety and efficacy of best practices, has never been more critical to sustain a future for the field. In order to satisfy this responsibility and advance the field of orthobiologics, it is important to establish common definitions of current orthobiologic options, quality standards for producing cell-based products, access to physician education, best practices for clinical application and promotion of research collaboration throughout the global medical community.

Orthobiologic treatments, as they pertain to the MSK field, are defined as any treatment modality that utilizes cellular components or derivatives of cellular components and directs their use towards damaged or diseased tissues [3, 4]. They are often concentrated versions of the body's naturally occurring cell depots, such as blood, bone marrow, or adipose tissue. Most commonly, they are utilized as an injectable treatment for joints, tendons, or ligaments. Most orthobiologic injections are performed under image guidance, with either ultrasound or fluoroscopy [5, 6]. Arthroscopy may also be utilized to provide real-time, high-definition color visualization for accurate cellular delivery to target pathology as a surgical adjunct. While clinical application of cellular therapeutics is under development in multiple medical disciplines, it is not yet a standard element of residency or fellowship training programs. As a result, normal variability exists among practitioners across the field pertaining to general knowledge, procedural accuracy and skill.

The core of orthobiologic treatments can be classified by primary product categorizations: hyaluronic acid (HA), platelet-rich plasma (PRP), bone marrow concentrate (BMC), and adipose tissue-derived mesenchymal stromal cells (AD-MSC) or lipoaspirate. Cells or cell derivatives isolated to produce biologic products can originate from within the same patient (autologous) or from another person (allogeneic). This has important implications on the regulatory classification of numerous products being marketed to clinicians, and the regulatory compliance of their use.

## Hyaluronic Acid

Hyaluronic Acid (HA) is considered as the first generation of orthobiologics, which has been used as an intraarticular injectable in the treatment of OA since the late 1990s. HA is a naturally occurring protein in the body with viscoelastic properties that helps to decrease frictional forces within synovial joints [7]. With aging and the onset of OA changes, the natural concentration of HA within synovial fluid decreases and HA distribution shifts toward lower molecular weight variants [8], which has been associated with increased

D. C. Karli (✉)
Karli Center, Miami, FL, USA
e-mail: davidkarli@greyledgebiotech.com; david@karlicenter.com

T. T. Sand
Sand Consulting, Poway, CA, USA
e-mail: tsand@biotechexpert.com


A. Navani et al. (eds.), *Essentials of Regenerative Medicine in Interventional Pain Management*,
https://doi.org/10.1007/978-3-031-50357-3_4

pain in OA [9]. Intraarticular HA administration has been used to restore the deficient HA concentration to its non-pathological concentration. However, many other theorized therapeutic mechanisms of HA have been postulated, including shock absorption, joint lubrication, anti-inflammatory effects, chrondroprotection, proteoglycan synthesis, and cartilage matrix alterations [8, 10]. Although, several intrinsic mechanisms have been postulated, much of the chrondroprotective and anti-inflammatory mechanisms are correlated with HA binding to the cluster of differentiation 44 (CD44) [11] cell surface marker, which inhibits the pro-inflammatory effects of Interleukin-1beta (IL-1beta), resulting in down regulation of many matrix metalloproteinases (MMPs) associated with cartilage degradation [12].

Currently, all commercially available injectable versions of HA are derived from xenogenic (animal tissues) or biosynthetic sources (recombinant expression via bacterial fermentation) and differ in formulation depending on manufacturer and production technique. Some evidence suggests larger molecular weight HA provides greater anti-inflammatory effects, proteoglycan synthesis, joint lubrication, and viscoelastic maintenance compared to lower molecular weight HA counterparts [8]. In addition, avian-derived HA has shown a less favorable safety profile with increased risk for localized intraarticular pseudo-septic reactions when compared to HA derived from biological fermentation [13].

The efficacy of intraarticular HA for the treatment of painful symptoms associated with OA has been demonstrated in several clinical trials [10], while also demonstrating a superior safety profile when compared to continuous non-steroidal anti-inflammatory drug (NSAID) use for pain control [14–16]. It also has been shown to lengthen the time from diagnosis of OA to time of knee arthroplasty in Medicare patients [17]. Recent Osteoarthritis Research Society International (OARSI) guidelines for the treatment of OA suggest a "good" level of evidence for the treatment of OA with intraarticular HA [18]. However, previous meta-analyses have illustrated between-study heterogeneity in the efficacy of HA for OA, with lower quality studies revealing more efficacious results [19, 20], which provides further support for the necessity of high level of evidence studies on the topic.

Although the volume of evidence for HA continues to expand for joints such as the hip, knee, or shoulder, its application and efficacy for spinal pathology is largely under-researched. Most studies utilizing HA for low back pain and spinal disorders have examined intraarticular facet injections for the treatment of facet arthropathy and facet-mediated pain. Some preliminary results illustrate pain relief and improvements in quality of life, comparable to traditional steroid injections with non-radicular low back pain [21], demonstrating efficacy lasting up to 6 months [22]. Other research has reported no therapeutic benefit of intraarticular HA for facet-mediated low back pain [23]. However, all trials, both efficacious and not, are nonrandomized and underpowered by low treatment numbers. A much newer and experimental area of HA application in spine pathology involves the use of injectable HA-based hydrogels for the treatment of degenerative disc disease [24, 25]. The goal of the HA-based hydrogel implantation is to restore native, non-degenerate nucleus pulposus mechanics to support overall disc function. In theory, the injectable hydrogels provide support to existing nucleus pulposus cells through their ability to swell and maintain hydrostatic pressure [24]. HA modulation of the disc environment was investigated in a bovine disc organ culture model [26]. The authors placed bovine discs into culture, which were not treated, underwent a punch biopsy to create a defect, or were cultured in the presence of an inflammatory cytokine (IFN-alpha(2)-beta). HA injection was shown to shift the microenvironment of the disc from pro-inflammatory to reduced inflammation and a more anabolic condition, with increased synthesis of aggrecan and collagen I [26]. Similar mitigation of a pro-inflammatory disc microenvironment was demonstrated in a rabbit disc puncture model, in which HA-treated discs maintained disc height in the injured disc compared to the untreated control discs [27]. This application of HA is highly experimental at this time and is based primarily on animal and *in vitro* research.

## Platelet Rich Plasma

The second generation of orthobiologics, platelet-rich plasma (PRP), was the first autologous point of care orthobiologic. Early reports on "PRP" involved processing whole blood to generate a fibrin gel from activated platelets for use in wound care [28], or platelet-enriched autologous plasma units were produced pre-operatively for use during cardiac surgery [29]. However, what is generally considered to be PRP was described by Marx in 2001 [30], who distinguished platelet enrichment over the whole blood baseline platelet level (PRP) from preparations that are platelet concentrates (very low plasma volume). Since then, PRP has been used in a myriad of different medical fields including ear, nose, and throat (ENT); maxillofacial surgery; ophthalmology; urology; dentistry; cosmetic and neurosurgery orthopedics; vascular disease and wound healing. Recent publications have shown an expanding role of platelets beyond their critical involvement in hemostasis. For example, platelets have been found to express toll-like receptors (TLR), which indicates that platelets are involved in innate and adaptive immune responses to both viral and bacterial agents [31]. Platelets release alpha and dense granules upon activation, which are a rich source of growth factors and other biological modifiers [32]. Platelets also release lysosomes that are a source of degradative enzymes that act on proteins and carbohydrates, which are common components in the extracellular matrix [31], and extracellular vesicles (EVs) are released by platelets [33]. Exosomes, a subclass of EVs, also are released from platelets, and contain a variety of biological cargo, including mRNAs, microRNAs, proteins, and lipids [33, 34]. Clearly, platelets are a substantial source of growth factors and other biologic modifiers to support their critical role in the complex and time-dependent healing cascade [35]. The alpha granules within platelets act as the primary storage center for an array of biologically active components and growth factors including transforming growth factor beta (TGFbeta), vascular endothelial growth factor (VEGF), platelet-derived growth factor (PDGF), and epithelial growth factor (EGF) which are thought to contribute to the platelet's therapeutic potential [36, 37]. Platelets also contain dense granules, which release a different profile of bio-active components (e.g., serotonin, histamine, ATP, ADP) [31]. An emerging area of research and clinical interest is focused on the cargo of exosomes released from platelets upon activation, which have been shown to contain numerous cytokines, proteins and other biological agents that contribute to the therapeutic impact of PRP treatments [35]. More recent hypotheses on the mechanism of PRP suggest that intraarticular application potentially may alter the entire joint environment through its effects on the signaling cascade, creating a more advantageous stimulatory environment for healing [38]. Changes in the cellular milieu may exert therapeutic benefits by acting on localized joint tissue cells such as synoviocytes or meniscal cells, promoting chemoattraction of the body's native healing cells to damaged tissue [39, 40], or through direct analgesic effect [41]. Although research continues to investigate such hypotheses, the specific mechanism(s) of action for PRP's clinical benefits is ill-defined. It is likely the etiology of its therapeutic effects is multifactorial and potentially variable across different tissue types.

PRP is derived from autologous venous blood. Blood is withdrawn from a patient's vein, and centrifuged, inducing cell and platelet separation in suspension based on their densities. Fractionally, red blood cells (RBCs) are the heaviest and layer at the bottom, followed by white blood cells (WBCs), platelets and plasma in decreasing order. More specifically, WBCs and platelets form a layered "buffy coat", which can be selectively harvested to create a PRP with highly concentrated platelets and varying levels of WBCs and RBCs [42]. Historically, inclusion or exclusion of WBCs in PRP broadly categorizes two general sub forms: LR-PRP (Leukocyte rich PRP) and LP-PRP (Leukocyte poor PRP) [37] (Fig. 4.1). Many clinicians have started to emphasize that not all PRP preparations are created equal, especially when there are numerous cellular processing techniques and kit technologies for producing PRP. Most practitioners utilize designated medical devices with disposable sterile devices, which vary widely by manufacturer in regard to the output of cellular composition with no built-in mechanism for measuring cell or platelet levels within the sample. Other practitioners perform more individualized techniques, utilizing single and double spin centrifugation cycles with more precise laboratory procedures [43]. A recent report [44] has summarized a variety of technical details (input blood volume, PRP volume, and levels of platelets, WBCs and RBCs, etc.) on the use of 14 commercial PRP kits from multiple published studies. The compositional analysis listed in the report clearly illustrates the significant differences in technical performance of the commercial devices. For example, % platelet recovery varied from a low of 17.2% to a high of 94.0% [44].

Hemoanalysis of PRP preparations typically requires utilization of point of care medical instruments designed to measure components in blood samples as are often found in most clinical diagnostic laboratories (Fig. 4.2). In an attempt to facilitate uniform PRP classification, researchers have established the PLRA categorization (platelet, leukocyte, red blood cell, and PRP activation), which classifies PRP based on platelet concentration, number of WBCs, presence or absence of red blood cells, and activation status [45]. PRP is "activated" when anticoagulation is reversed prior to implantation in the body stimulating platelet degranulation and the coagulation cascade. Additional approaches to classifying types of PRP have been summarized [44]. Routine classification based on hemoanalysis and study of cellular concentration-response relationships will lead to more customized PRP formulations, working towards maximizing

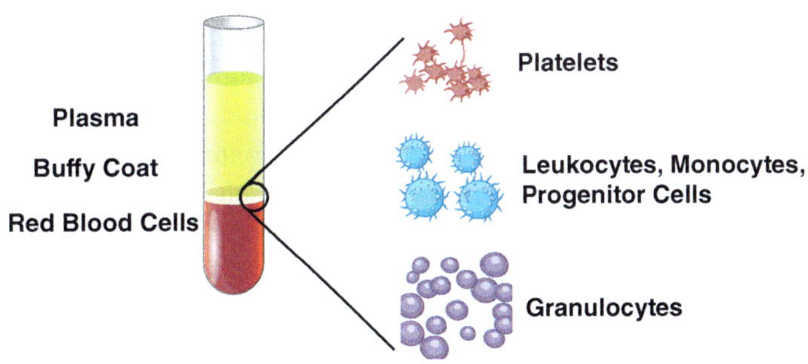


**Fig. 4.1** After venous blood is processed in a centrifuge, it is separated into 3 layers: plasma, buffy coat, and red blood cells. The buffy coat layer contains the platelets, which store a variety of potent growth factors, thought to be a primary mechanism of PRP. Image courtesy of Ted Sand, PhD

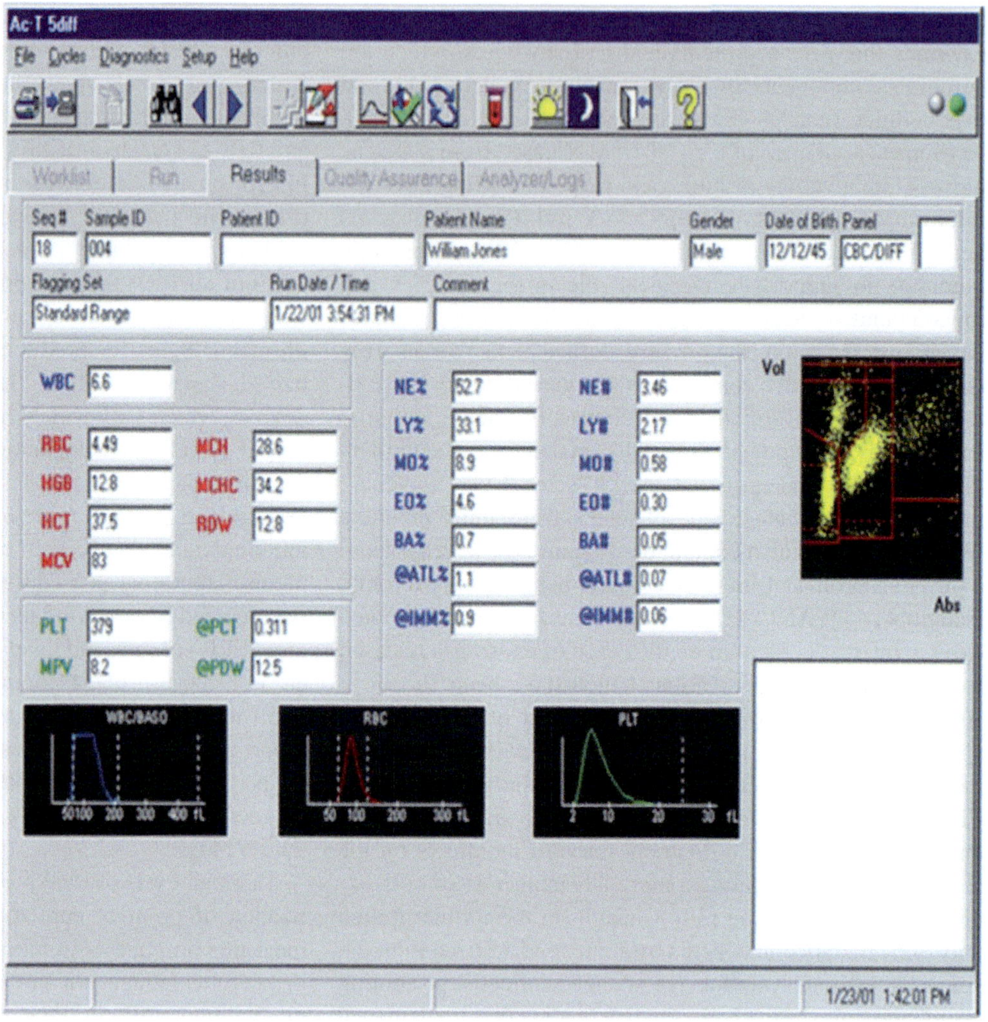


**Fig. 4.2** Beckman Coulter ACT 5 DIFF CTL PLUS: cell cytometry analysis of venous blood. This sample is an example of pre-centrifugation data, noted by the normal concentration of platelets. The cytometry data is used to classify PRP based on the PLRA PRP classification system

therapeutic efficacy for specific MSK disorders, as well as addressing inter-patient variability and aid with interpretation of clinical trials. For example, initial research suggested that Leukocyte poor-PRP may have stronger efficacy with intraarticular application, although this has recently been called into question and remains under investigation [46, 47].

Most of the clinical literature on PRP consists of small case series with mixed results without comparative evaluation of the system used to produce the PRP, while very few studies consider cellular counts in the PRP preparation as it pertains to study outcome. Larger randomized controlled trials have demonstrated general trends of efficacy in areas such as chronic tendinopathies [48, 49], and early to moderate severity knee OA [50]. A recent meta-study [51] combined published clinical data on 30 randomized clinical studies (3463 knees) comparing PRP with control (HA, steroids, placebo), and concluded that PRP demonstrated superior Western Ontario and McMaster Universities Arthritis Index (WOMAC) and VAS scores at 1-year follow-up compared to control. Research also has been published suggesting therapeutic benefits of combining PRP with other orthobiologic treatments such as HA [52] or mesenchymal stromal cells (MSCs) [53, 54], as well as utilizing multiple orthobiologics in a specific sequence as a treatment protocol for OA [55]. Furthermore, protocols have been established for post-PRP recovery and rehabilitative exercises, establishing a preliminary framework for doctors and therapists to provide optimal treatment for return to activity or sport following treatment [43].

While most studies and treatment protocols have been developed for peripheral joints, the use of PRP for spinal pathology is a growing area of orthobiologics. Although many of the existing trials are largely based on animal, preclinical, or in vitro designs [56, 57], preliminary human trials have illustrated improvements in pain and function with intradiscal injections for the treatment of degenerative disc disease and chronic discogenic low back pain [57–60]. A small preliminary study examined the efficacy of intradiscal PRP injections on 22 patients with chronic discogenic low back pain and reported that 47% of patients had >50% reduction in Visual Analog Scale (VAS) scores and >30% improvement in Oswestry Disability Index (ODI) at 6 months follow-up [58]. Another study involving 29 intradiscal injections for chronic discogenic low back pain revealed statistically significant improvements in pain and function at up to 2-year follow-up [59]. A study that used PRP releasate (serum-only was collected from a clotted PRP) reported durable benefits as assessed by VAS and Roland-Morris Disability Questionnaire (RDQ) out to an average of 5.9 years follow-up [61]. In addition, studies also have shown potential therapeutic benefit from PRP injections directed towards facet mediated pain [62, 63], although less clinical support exists for this application. One retrospective pilot study examined the effectiveness of combining intradiscal, epidural, and intraarticular application of PRGF (a dilute PRP preparation) under fluoroscopic guidance for chronic low back pain in 86 patients [64]. The results revealed significant improvements in VAS score at 6 months follow-up, with 91% of patients having a sustained VAS score between 0–3. No major adverse events were documented from the study [64].

## Bone Marrow Concentrate

Bone marrow concentrate (BMC) is considered as a third generation orthobiologic treatment. It is comprised of a heterogeneous mixture of MSCs, hematopoietic stem cells (HSCs), endothelial progenitor cells (EPCs), WBCs and platelets, at various stages of maturation. General scientific and clinical observation has noted BMC's capacity to stimulate anti-inflammatory, analgesic, immunomodulatory, osteogenic and chondrogenic properties, which act as foundations for its regenerative potential [65, 66]. The physiologic activity of BMC as it applies to therapeutic benefit is likely complex and paracrine/secretory in nature. While exact mechanisms remain unknown, it is evident that BMC plays an immunomodulatory role in reducing inflammation, while also inducing differentiation and proliferation of resident tissue cells [65, 66].

Bone marrow is most commonly aspirated from the posterior iliac crest, utilizing ultrasound or fluoroscopic guidance. The bone marrow aspirate undergoes processing via similar centrifugation protocols as PRP, producing a buffy coat separation and subsequent selective harvest. Physicians currently have multiple options for marrow concentration, either via designated medical devices or individualized laboratory processes, very similar to PRP production. Individual practitioners have started to utilize cell cytometry with BMC procedures, similar to the PLRA PRP classification; however, no standardized system of classification exists to date. At least one peer reviewed publication has described the validity and reproducibility of an automated system for hemoanalysis in quantifying and phenotyping nucleated cells, RBCs and platelets in bone marrow derived samples [67]. Flow cytometry also has been used to identify cell populations in BMC based on cell surface markers, which have been correlated with therapeutic success and failure in multiple MSK applications [68]. Continued research and classification of BMC is a much-needed step, and crucial for further development in the field. As one of the newer generations of orthobiologics, BMC clinical research evidence is mounting to further validate safety, efficacy and utility. Much of the background evidence for BMC utility and application is credited to French surgeon, Dr. Philippe Hernigou (Henri Modor University Hospital, Crétiel France), who began using autologous BMC to treat MSK pathologies in 1989 [69]. Subsequently, he and his group have published landmark studies on the safety (no increased risk of cancer) of BMC [70], and the therapeutic benefits of BMC treatment in: long bone non-union [71], rotator cuff repair [72], osteonecrosis of the femoral head [73], diabetic ankle non-union [74] and a 15-yr follow-up of a randomized controlled trial in which patients with bilateral knee OA were treated with equal volumes of BMC either via the IA or IO routes [75]. Another recent report also has demonstrated patient safety and efficacy of intraarticular BMC to treat joint OA (Fig. 4.3) [55].

Targeted use of BMC in OA and spinal applications have expanded in recent years. In a continuation of research conducted in animal studies, early human trials have illustrated potential efficacy for mitigating disc mediated low back pain, as well as facet arthropathy. Currently, a majority of the research has focused on degenerative disc disease treatments targeting intradiscal applications of BMC and its potent milieu of MSCs (Figs. 4.4 and 4.5). A small pilot study

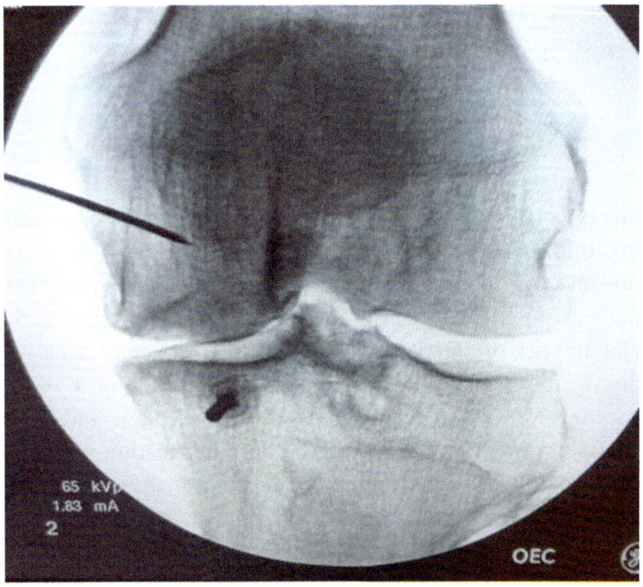


**Fig. 4.3** Example of needle placement for intraosseous (IO) BMC implantation in the knee

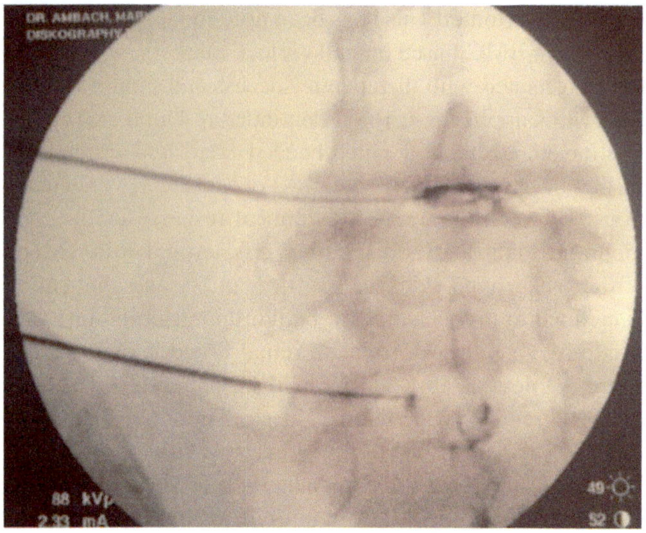

**Fig. 4.4** AP view of fluoroscopic guided intradiscal injection

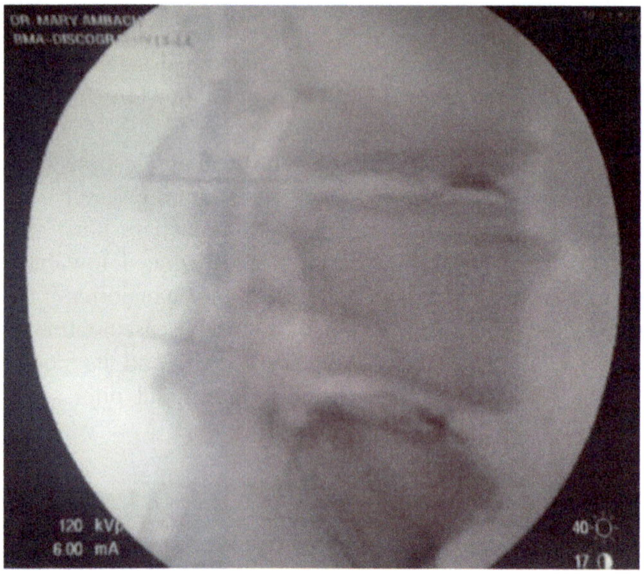

**Fig. 4.5** Lateral view of fluoroscopic guided intradiscal injection

involving 10 patients showed improvements in pain and function, in 9 out of 10 participants, up to 12 months after injection of cells cultured from autologous BMC into the nucleus pulposus [76]. Another study examined the beneficial impact of intradiscal autologous BMC injections on discogenic back pain in 26 patients, including CFU-F (Colony Forming Unit-Fibroblast) quantification and flow cytometry analysis of the BMC with confirmation of BMC sample cellularity and treatment success at the 1-year follow-up [68]. At the 3-yr follow-up [77], the 20 patients (77%) remaining in the study reported a 69% decrease in mean ODI and an 83% decrease in mean VAS. The authors reported that there was a trend in improved outcomes for those patients receiving higher levels of MSCs (as measured in the CFU-F assay) at the time of treatment [77]. More recently, a clinical study was performed in which BMC was injected into multiple pain generators (facets, disc, SI joint and/or epidural) as appropriate based on clinical assessment of each study participant to treat lower back pain [78]. The authors reported statistically meaningful improvements in ODI at all post-treatment milestones for the treatment group compared to the ODI changes in the control group, with 67% of the treatment group compared to just 7.7% of the control group achieving at least a 10-point decrease in ODI (i.e., the minimal clinically important difference for lower back pain) at the 12-month milestone [78].

As with many of the iterations of orthobiologics, researchers are examining the combination of multiple types of biologics to examine potential additive benefits, although most of the studies have evaluated applications in peripheral joints. However, there are reports on the use of orthobiologic combination treatments in the spine. Results from a Phase 1b/2a intradiscal treatment with allogeneic, cultured bone marrow derived stem cells (MPCs) with and without HA were recently reported [79]. The authors reported that approximately 30% of the study participants receiving the cell therapy had a mean VAS reduction of slightly less than 80% at the 3-year follow-up.

In a recent study [55], the effects of a BMC treatment involving a single intraarticular BMC injection followed by PRP injection at 8 weeks post-BMC treatment in 125 cases were reported. The study investigated the effects on various joints including two cases for cervical facet arthropathy. The results for the cervical spine injections showed a median absolute change in VAS score of −7.0, which equated to an 89% improvement in pre- and post-injection pain level [55].

## Adipose Derived Mesenchymal Stem Cells

As the field of orthobiologics continues to develop, research efforts and clinical observation continue to refine our scientific understanding, opening possibilities for future generations of orthobiologics. Recent literature has suggested a perivascular origin of MSCs and other cells with therapeutic potential [68, 80]. This has led to exploration of other autologous sources of MSCs, including the most recent fourth generation of orthobiologics: lipoaspirate/adipose-derived cell therapeutics. Lipoaspirate is harvested under local anesthesia and vacuum-assisted lipectomy to the abdomen (most frequent), posterior superior buttock or lateral thigh. There are two types of adipose tissue-derived therapeutics. Stromal vascular fraction (SVF) is isolated from a patient's adipose tissue with a protocol involving enzymatic digestion and extensive washing. The result is a single cell suspension [81, 82]. The other type of adipose tissue-derived therapeutic preparation involves treating the lipoaspirated adipose tissue with mechanical force, which results in the collapse of the lipid vacuoles in the adipocytes present in the adipose tissue and yields microfragmented adipose tissue (MFAT) [81]. Another approach involves gravity or low speed centrifugation induced separation of natural oils from fat stroma to isolate nucleated cells from the suctioned sample [83]. Like BMC, processed lipoaspirate contains AD-MSCs (adipose derived mesenchymal stromal cells) that have exhibited differentiation into chondrogenic, osteogenic, adipogenic, myogenic, and neurogenic lineages in the presence of lineage-specific induction factors [84, 85]. AD-MSCs cultured from adipose tissue demonstrated a higher potential for inhibiting PBMC proliferation compared to bone marrow-derived cultured MSCs [86], although the relevance to clinical efficacy has yet to be demonstrated.

Preliminary research suggests that AD-MSCs exhibit an anti-inflammatory effect on chondrocytes and synoviocytes in patients with OA [87], illustrating obvious potential applications for patients with joint OA, including low back pain caused by facet arthropathy. In addition, *in vitro* studies also have examined the potential benefits of AD-MSCs for degenerative disc disease. One study found that AD-MSCs, placed in a culture medium of Type II collagen, promoted differentiation of AD-MSCs into nucleus pulposus-type cells [88]. Other studies have shown AD-MSCs potentially to have a protective quality for nucleus pulposus cells, by altering gene expression for proteins involved with maintenance of the extracellular matrix integrity of the nucleus pulposus [89], or even stimulating cell proliferation of degenerated nucleus pulposus cells [90]. To date, there are no published randomized controlled clinical studies that have evaluated the efficacy of adipose tissue-derived autologous SVF or MFAT for treating low back pain or spine pathology. Furthermore, the therapeutic assessment of SVF or MFAT for treating spinal pathologies should be limited to Institutional Review Board (IRB)-approved clinical studies.

## Future Directions

Allogeneic models for cellular therapies represent an attractive path for exploration, considering no autologous harvest is needed to acquire and manufacture a final product. To date, all non-autologous viable cell therapeutics have been categorized as biological drugs per the FDA's Code of Federal Regulations (21CFR1271), requiring premarket approval (PMA) as a drug or biological product. Thus, products derived from allogeneic tissues in whatever form (fluid, cultured cells, etc.) are not compliant for routine clinical use outside of an IRB-approved clinical trial, or until approved for marketing via an IND or IDE pathway. Emerging allogeneic orthobiologic sources include banked adult, culture expanded cells and human perinatal tissues (amniotic, placental and Wharton's Jelly tissues). These tissues also have been shown to contain a variety of progenitor cells, including MSCs [91–93]. While sharing a number of phenotypic markers, MSCs from the tissues demonstrate *in vitro* differences, such as proliferation rate in culture [93]. Few human trials exist for human amniotic membrane applications, but small case studies have shown efficacy for elbow tendinopathy [94] and plantar fasciitis [95], while preliminary animal studies have suggested potentially positive applications for tendon injuries [96] and OA [97]. Peri-natal tissue as a source of MSCs is an area of intense commercialization and thus is an emerging future generation of orthobiologics. However, there are no published randomized controlled clinical studies to date involving the use of cells or other therapeutic materials derived from peri-natal tissues for spine pathology or low back pain. Another area of intense commercialization and research involves the assessment of the therapeutic potential of EVs, including a subset known as exosomes. All cells release EVs, are thought to participate in cell-to-cell communication and are found naturally in PRP, BMC, MFAT and cultured cell-based therapies. EVs are a cell-free therapeutic option and have been studied in animal models for treating OA [98]. For example, a recent publication showed that EVs obtained from human adipose tissue-derived cultured MSCs were able to diminish disease progression in rat and mouse OA models [99].

Genetic modification of cells represents yet another appealing progression as technological capabilities with CRISPR and gene splicing continue to advance under stringent research investigation and regulatory oversight. Genetic engineering holds great potential for multiple applications to enhance creation of autologous and allogeneic cell lines for addressing important native target tissue pathologies and systemic applications. The concept of "programming" a cell to perform a specific physiologic and therapeutic function has obvious relevance to truly customizable cell therapy applications. Although still early in terms of broad accessibility, induced pluripotent stem cells (iPSC) are widely seen as the ultimate in personalized regenerative therapy [100]. iPS cells are obtained from patients or donor-derived adult cells that are manipulated *in vitro* to obtain a more primitive stem cell that can be cultured indefinitely and differentiated into all adult tissues. It is important to note that iPSC preparations always must be differentiated prior to use as a therapeutic treatment to avoid formation of teratomas [100]. While having great promise, the scale-up of iPSC production and efficient differentiation protocols to generate cells safe for therapy in humans present significant challenges [100].

The field of orthobiologics is faced with the burden of balancing immense growth and diversification with a firm scientific foundation. As this separation widens, the lack of standardization and uniformity amongst practitioners is becoming significantly more apparent. The field as a whole continues to expand at an impressive rate, risking dilution over validation of the core principles of orthobiologics. If not matched by coordinated research efforts, product standardization and quality control, hemoanalysis driven outcomes, data collection and standardization of physician education and training, the field will struggle to scale with adequate scientific validity.

Future research efforts for enhancing the therapeutic effectiveness of orthobiologics will require performing clinical studies with higher quality data, including analysis of what is being injected in terms of the preparation's cellularity and phenotyping, in order to establish dose-benefit guidelines. Preliminary attempts to coordinate research have relied on the use of international digital data registry software to monitor patient outcomes, track patient safety, and record cellular compositions for biologic formulations. Larger amounts of quantified PRP, BMC, and lipoaspirate data will allow for improved efficacy, optimization, and standardization of treatments. It also is possible that predictive analytics and integrating precision medicine, a medical model that proposes customization of health care to each individual, may help guide research efforts and formulate treatment protocols in the future.

Although practitioners are rapidly investigating a myriad of methods for extracting, processing, and injecting orthobiologics, these efforts are largely uncoordinated, resulting in a lack of clearly defined, compliant best practices. It will be imperative to further establish protocols for each phase of orthobiologic processing from aspiration techniques, cellular processing, cell quantification, as well as specific injection protocols, driven by coordinated multidisciplinary collaboration.

## Conclusion

Overall, the field of orthobiologics has expanded from the early days of limited options, while continuing to explore more effective treatment protocols and combination therapies. Standardization of orthobiologics remains a critical unmet need. Current orthobiologics lack a standardized and validated definition as to what they are, their mechanisms of action and how they should be used. To date, we can confidently state based on strong and growing safety and efficacy data, that HA, PRP, BMC, and adipose tissue have established a place for themselves in the future of orthopedic and MSK care. However, investigation into their therapeutic applications, optimal cellular compositions, and treatment protocols trail their widespread use. With more commercial biologic options rapidly surfacing, it behooves the industry to establish a proper framework for pre-existing biologic treatment options, with head-to-head focused comparative analyses of different therapeutics in matched cohorts. Although it is not clear where the field will be in another 10 years, by establishing and/or maturing the pathways described in this chapter, we can help increase orthobiologics' utility to mitigate pain, and potentially reduce progression of common MSK diseases.

**Key Points**

1. Orthobiologic treatments consisting of hyaluronic acid, and the autologous preparations of PRP, BMC and SVF are being used at greater frequencies to treat a wide variety of musculoskeletal pathologies.
2. Hyaluronic acid originally was obtained from animal sources, but more recently has been obtained by microbial fermentation; with the latter type of HA having an enhanced safety profile.
3. PRP began to be used more frequently after kits and centrifuges were commercialized for use in non-hospital medical facilities for providing same-day treatments.
4. The number of Platelets in PRP generally has been shown to correlate with improved outcomes, but benefit might depend on the number of RBCs and WBCs present in the PRP.
5. There is wide variation in the performance of commercially available kits and technologies for producing PRP and BMC.
6. BMC treatments began with Dr. Philippe Hernigou in 1989, and its use has expanded dramatically as an adjunctive treatment for both bony (e.g., long-bone non-union) and soft tissue (e.g., rotator cuff repair) pathologies, while also being used as a percutaneous therapy for conditions like knee OA.
7. PRP is most frequently used in treating knee OA via intraarticular delivery.
8. A recent study showed that BMC when delivered via the intra-osseous route was able to delay the need for a total knee replacement for an average of 14 years, suggesting that this type of orthobiologic preparation not only can reduce pain, but slow disease progression.
9. Adipose tissue-derived therapies originally were associated with enzymatic digestion, yielding SVF; an alternative approach has recently been commercialized that results in a microfragmented fat tissue preparation (MFAT).

**Questions and Answers**

1. Which of the following best describes the main therapeutic benefit of Hyaluronic Acid when injected into a joint?
   (A) It binds to receptors on free-floating cells.
   (B) It binds to cells in the synovial lining.

(C) It acts as a lubricant.
(D) It acts as a nutrient.

**Answer: C**

**Explanation:** It acts a lubricant. Hyaluronic acid is a polymer that can bind to free-floating cells or cells in the synovial lining, which might provide signaling. However, there is evidence that higher molecular weight HAs provide better pain relief, suggesting a mechanical roll for its therapeutic benefit.

2. Of the four treatment modalities listed, which was the earliest form of a biologic used as a percutaneous treatment in musculoskeletal pathologies?
(A) Platelet-rich plasma (PRP)
(B) microfragmented adipose tissue (MFAT)
(C) stromal vascular fraction (SVF)
(D) Hyaluronic acid (HA)

**Answer: D**

**Explanation:** Hyaluronic acid (HA) was the first biologic of the four to be used to treat musculoskeletal pathologies.

3. What granule is not found in Platelets?
(A) Primary
(B) Alpha
(C) Dense
(D) Exosome

**Answer: A**

**Explanation:** The Primary granule is found in granulocytes, while the other three granules are found in platelets.

4. What is the least dense component found in blood?
(A) Red blood cells (RBCs)
(B) Platelets
(C) Neutrophils
(D) Granulocytes

**Answer: B**

**Explanation:** Platelets are the least dense of the components listed, with red blood cells (RBCs) being the most dense and granulocytes/neutrophils being slightly less dense then RBCs.

5. Which of the following types of cells is not found in bone marrow aspirate?
(A) Hematopoietic stem cell
(B) Endothelial progenitor cell
(C) Mesenchymal stem cell
(D) Embryonic stem cell

**Answer: D**

**Explanation:** Embryonic stem cell, which is found only in an embryo.

6. Which of the following is not a type of platelet-rich plasma (PRP)?
(A) Leukocyte Rich
(B) Leukocyte Poor
(C) Homogenized
(D) Releasate

**Answer: C**

**Explanation:** Homogenized is not a form of PRP, while the others describe different compositions of PRP published in the literature.

7. What bioassay is used to enumerate the number of Mesenchymal Stromal/Stem Cells (MSCs) in bone marrow concentrate?
(A) Colony Forming Unit-Fibroblast (CFU-F)
(B) Colony Forming Unit-Chondrocyte (CFU-C)
(C) Colony Forming Unit-Osteoblast (CFU-O)
(D) Colony Forming Unit-granulocyte–macrophage (CFU-GM)

**Answer: A**

**Explanation:** Colony Forming Unit-Fibroblast (CFU-F) is the bioassay, since Mesenchymal Stromal/Stem Cells (MSCs) will grow attached in a fibroblastic shape in culture.

8. In the current regulatory environment in the USA, what treatment should not be used outside of an Institutional Review Board (IRB)-approved clinical trial for treating MSK pathologies?
(A) Platelet-rich plasma (PRP)
(B) Stromal vascular fraction (SVF)
(C) Bone marrow concentrate (BMC)
(D) Hyaluronic acid (HA)

**Answer: B**

**Explanation:** Stromal vascular fraction (SVF) use is considered by the United States Food and Drug Administration (USFDA) as non-compliant if used to treat musculoskeletal pathologies, so it should be used only under an Institutional Review Board (IRB)-approved clinical study to avoid the risk of being considered to be treating patients with an experimental drug.

9. Which of the following statements best describes the function of exosomes in the body?
(A) Exosomes are a waste product released by cells
(B) Exosomes are only released when there is an infection
(C) Exosomes are only released by platelets when they are activated
(D) Exosomes are found in all cells and participate in cell-to-cell communication

**Answer: D**

**Explanation:** Exosomes are found in all cells in the body and are considered to participate in cell-to-cell communication.

10. Which of the following statements doesn't apply to induced pluripotent stem cell (iPSC) treatments?
(A) Can be a virtually unlimited source of differentiated cells

(B) Can be used to obtain autologous progenitor cells
(C) Can be used without differentiation
(D) Can be used to obtain allogeneic progenitor cells

**Answer: C**

**Explanation:** Induced pluripotent stem cell (iPSC) preparations must be differentiated into a more adult form of cell, since pluripotent stem cells when injected into humans will form teratomas.

**Acknowledgements** This book chapter is modified and updated from a previous book chapter, "Biologics in Clinical Setting: Standardization and Collaboration" by Steven E. Sampson, DO, Hunter Vincent, DO, and Mary Ambach, MD, in *Essentials of Regenerative Medicine in Interventional Pain Management* published by ASIPP Publishing.

## References

1. Sampson SE, Vincent H, Ambach M. Biologics in clinical setting: standardization and collaboration. In: Manchikanti L, Navani A, Atluri S, editors. Essentials of regenerative medicine in interventional pain management. Paducah: ASIPP Publishing; 2019. p. 55–66.
2. Dieleman JL, Baral R, Birger M, et al. US spending on personal health care and public health, 1996-2013. JAMA. 2016;316:2627–46.
3. Malanga G, Abdelshahed D, Jayaram P. Orthobiologic interventions using ultrasound guidance. Phys Med Rehabil Clin N Am. 2016;27:717–31.
4. Dhillon MS, Behera P, Patel S, et al. Orthobiologics and platelet rich plasma. Ind J Orthop. 2014;48:1–9.
5. Trescot A, Watanabe A. Principles of ultrasound imaging in spine and musculoskeletal medicine. In: Manchikanti L, Navani A, Atluri S, editors. Essentials of regenerative medicine interventional pain management. Paducah: ASIPP Publishing; 2018. p. 155–72.
6. Desai R, Manning T, Olson A, et al. Principles of fluoroscopy imaging in spine and musculoskeletal medicine. In: Manchikanti L, Navani A, Atluri S, editors. Essentials of regenerative medicine interventional pain management. Paducah: ASIPP Publishing; 2018. p. 145–54.
7. Miller LE, Block JE. US-approved intra-articular hyaluronic acid injections are safe and effective in patients with knee osteoarthritis: systematic review and meta-analysis of randomized, saline-controlled trials. Clin Med Insights Arthritis Musculoskelet Disord. 2013;6:57–63.
8. Altman RD, Manjoo A, Fierlinger A, et al. The mechanism of action for hyaluronic acid treatment in the osteoarthritic knee: a systematic review. BMC Musculoskelet Disord. 2015;16:321.
9. Band PA, Heeter J, Wisniewski HG, et al. Hyaluronan molecular weight distribution is associated with the risk of knee osteoarthritis progression. Osteoarthr Cartil. 2015;23:70–6.
10. Moreland LW. Intra-articular hyaluronan (hyaluronic acid) and Hylan for the treatment of osteoarthritis: mechanisms of action. Arthritis Res Ther. 2003;5:54–67.
11. Karna E, Miltyk W, Surażyński A, et al. Protective effect of hyaluronic acid on interleukin-1-induced deregulation of β1-integrin and insulin-like growth factor-I receptor signaling and collagen biosynthesis in cultured human chondrocytes. Mol Cell Biochem. 2008;308:57–64.
12. Waddell DD, Kolomytkin OV, Dunn S, et al. Hyaluronan suppresses IL-1beta-induced metalloproteinase activity from synovial tissue. Clin Orthop Relat Res. 2007;465:241–8.
13. Goldberg VM, Coutts RD. Pseudoseptic reactions to Hylan viscosupplementation: diagnosis and treatment. Clin Orthop Relat Res. 2004;419:130–7.
14. Bannuru RR, Vaysbrot EE, Sullivan MC, et al. Relative efficacy of hyaluronic acid in comparison with NSAIDs for knee osteoarthritis: a systematic review and meta-analysis. Semin Arthritis Rheum. 2014;43:593–9.
15. Bannuru RR, Natov NS, Dasi UR, et al. Therapeutic trajectory following intra-articular hyaluronic acid injection in knee osteoarthritis–meta-analysis. Osteoarthr Cartil. 2011;19:611–9.
16. Colen S, van den Bekerom MP, Mulier M, et al. Hyaluronic acid in the treatment of knee osteoarthritis: a systematic review and meta-analysis with emphasis on the efficacy of different products. BioDrugs. 2012;26:257–68.
17. Ong KL, Anderson AF, Niazi F, et al. Hyaluronic acid injections in Medicare knee osteoarthritis patients are associated with longer time to knee arthroplasty. J Arthroplast. 2016;31:1667–73.
18. McAlindon TE, Bannuru RR, Sullivan MC, et al. OARSI guidelines for the non-surgical management of knee osteoarthritis. Osteoarthr Cartil. 2014;22:363–88.
19. Wang CT, Lin J, Chang CJ, et al. Therapeutic effects of hyaluronic acid on osteoarthritis of the knee. A meta-analysis of randomized controlled trials. J Bone Joint Surg Am. 2004;86:538–45.
20. Lo GH, LaValley M, McAlindon T, et al. Intra-articular hyaluronic acid in treatment of knee osteoarthritis. JAMA. 2003;290:3115–21.
21. Fuchs S, Erbe T, Fischer HL, et al. Intraarticular hyaluronic acid versus glucocorticoid injections for nonradicular pain in the lumbar spine. J Vasc Interv Radiol. 2005;16:1493–8.
22. DePalma MJ, Ketchum JM, Queler ED, et al. Prospective pilot study of painful lumbar facet joint arthropathy after intra-articular injection of Hylan G-F 20. PM R. 2009;1:908–15.
23. Cleary M, Keating C, Poynton AR. Viscosupplementation in lumbar facet joint arthropathy. J Spinal Disord Tech. 2008;21:29–32.
24. Cloyd JM, Malhotra NR, Weng L, et al. Material properties in unconfined compression of human nucleus pulposus, injectable hyaluronic acid-based hydrogels and tissue engineering scaffolds. Eur Spine J. 2007;16:1892–8.
25. Tsaryk R, Gloria A, Russo T, et al. Collagen-low molecular weight hyaluronic acid semi-interpenetrating network loaded with gelatin microspheres for cell and growth factor delivery for nucleus pulposus regeneration. Acta Biomater. 2015;20:10–21.
26. Kazezian Z, Li Z, Alini M, et al. Injectable hyaluronic acid down-regulates interferon signaling molecules, IGFBP3 and IFIT3 in the bovine intervertebral disc. Acta Biomater. 2017;52:118–29.
27. Inoue M, Isa ILM, Orita S, et al. An injectable hyaluronic acid hydrogel promotes intervertebral disc repair in a rabbit model. Spine. 2021;46:E810–6.
28. Knighton DR, Fiegel VD, Austin LL, et al. Classification and treatment of chronic nonhealing wounds successful treatment with autologous platelet-derived wound healing factors (PD WHF). Ann Surg. 1986;204:322–9.
29. Ferrari M, Zia S, Valbonesi M, et al. A new technique for hemodilution, preparation of autologous platelet-rich plasma and intraoperative blood salvage in cardiac surgery. Int J Artif Organs. 1987;10:47–50.
30. Marx RE. Platelet-rich plasma (PRP): what is PRP and what is not PRP? Implant Denistry. 2001;10:225–8.
31. Ostermeier B, Soriano-Sarabia N, et al. Platelet-released factors: their role in viral disease and applications for extracellular vesicle (EV) therapy. Int J Mol Sci. 2022;23:2321.

32. Manne BK, Xiang SC, Rondina MT. Platelet secretion in inflammatory and infectious diseases. Platelets. 2017;28:155–64.
33. Eustes AS, Dayal S. The role of platelet-derived extracellular vesicles in immune-mediated thrombosis. Int J Mol Sci. 2022;23:7837.
34. Rui SL, Yuan Y, Du CZ, et al. Comparison and investigation of exosomes derived from platelet-rich plasma activated by different agonists. Cell Transplant. 2021;30:1–13.
35. Saumell-Esnaola M, Delgado D, García del Caño G, et al. Isolation of platelet-derived exosomes from human platelet-rich plasma: biochemical and morphological characterization. Int J Mol Sci. 2022;23:2861.
36. Alsousou J, Thompson M, Hulley P, et al. The biology of platelet-rich plasma and its application in trauma and orthopaedic surgery: a review of the literature. J Bone Joint Surg. 2009;91:987–96.
37. Sampson S, Gerhardt M, Mandelbaum B. Platelet rich plasma injection grafts for musculoskeletal injuries: a review. Curr Rev Musculoskelet Med. 2008;1:165–74.
38. Filardo G, Kon E, Roffi A, et al. Platelet-rich plasma: why intra-articular? A systematic review of preclinical studies and clinical evidence on PRP for joint degeneration. Knee Surg Sports Traumatol Arthrosc. 2015;23:2459–74.
39. Krüger JP, Hondke S, Endres M, et al. Human platelet-rich plasma stimulates migration and chondrogenic differentiation of human subchondral progenitor cells. J Orthop Res. 2012;30:845–52.
40. Krüger JP, Ketzmar AK, Endres M, et al. Human platelet-rich plasma induces chondrogenic differentiation of subchondral progenitor cells in polyglycolic acid-hyaluronan scaffolds. J Biomed Mater Res Part B Appl Biomater. 2014;102:681–92.
41. Lee HR, Park KM, Joung YK, et al. Platelet-rich plasma loaded hydrogel scaffold enhances chondrogenic differentiation and maturation with up-regulation of CB1 and CB2. J Control Release. 2012;159:332–7.
42. Alves R, Grimalt R. A review of platelet-rich plasma: history, biology, mechanism of action, and classification. Skin Appendage Disord. 2018;4:18–24.
43. Maffuli N. Platelet rich plasma in musculoskeletal practice. London: Springer; 2016.
44. Magalon J, Brandin T, Francois P, et al. Technical and biological review of authorized medical devices for platelets-rich plasma preparation in the field of regenerative medicine. Platelets. 2021;32:201–9.
45. Mautner K, Malanga GA, Smith J, et al. A call for a standard classification system for future biologic research: the rationale for new PRP nomenclature. PM R. 2015;7:S53–9.
46. Braun HJ, Kim HJ, Cr C, et al. The effect of platelet-rich plasma formulations and blood products on human synoviocytes: implications for intra-articular injury and therapy. Am J Sports Med. 2014;42:1204–10.
47. Riboh JC, Saltzman BM, Yanke AB, et al. Effect of leukocyte concentration on the efficacy of platelet-rich plasma in the treatment of knee osteoarthritis. Am J Sports Med. 2016;44:792–800.
48. Mishra AK, Skrepnik NV, Edwards SG, et al. Efficacy of platelet-rich plasma for chronic tennis elbow: a double-blind, prospective, multicenter, randomized controlled trial of 230 patients. Am J Sports Med. 2014;42:463–71.
49. Rha DW, Park GY, Kim YK, et al. Comparison of the therapeutic effects of ultrasound-guided platelet-rich plasma injection and dry needling in rotator cuff disease: a randomized controlled trial. Clin Rehabil. 2013;27:113–22.
50. Patel S, Dhillon MS, Aggarwal S, et al. Treatment with platelet-rich plasma is more effective than placebo for knee osteoarthritis: a prospective, double-blind, randomized trial. Am J Sports Med. 2013;41:356–64.
51. Migliorini F, Driessen A, Quack V, et al. Comparison between intra-articular infiltrations of placebo, steroids, hyaluronic and PRP for knee osteoarthritis: a Bayesian network meta-analysis. Arch Orthop Trauma Surg. 2021;141:1473–90.
52. Lana JF, Weglein A, Sampson SE, et al. Randomized controlled trial comparing hyaluronic acid, platelet-rich plasma and the combination of both in the treatment of mild and moderate osteoarthritis of the knee. J Stem Cells Regen Med. 2016;12:69–78.
53. Koh YG, Kwon OR, Kim YS, et al. Comparative outcomes of open-wedge high tibial osteotomy with platelet-rich plasma alone or in combination with mesenchymal stem cell treatment: a prospective study. Arthrosc J Arthrosc Relat Surg. 2014;30:1453–60.
54. Pak J, Lee JH, Kartolo WA, et al. Cartilage regeneration in human with adipose tissue-derived stem cells: current status in clinical implications. Biomed Res Int. 2016;2016:4702674.
55. Sampson S, Smith J, Vincent H, et al. Intra-articular bone marrow concentrate injection protocol: short-term efficacy in osteoarthritis. Regen Med. 2016;11:511–20.
56. Charchian B, Tribuzio B, Zappaterra M, et al. Regenerative spinal therapies for low back pain. Curr Phys Med Rehabil Rep. 2014;2:41–7.
57. Navani A, Ambach MA, Wei JJ, et al. Biologic therapies for intervertebral degenerative disc disease: a review of novel applications. J Stem Cells Res Rev Rep. 2017;4:1023.
58. Levi D, Horn S, Tyszko S, et al. Intradiscal platelet-rich plasma injection for chronic discogenic low back pain: preliminary results from a prospective trial. Pain Med. 2016;17:1010–22.
59. Monfett M, Harrison J, Boachie-Adjei K, et al. Intradiscal platelet-rich plasma (PRP) injections for discogenic low back pain: an update. Int Orthop. 2016;40:1321–8.
60. Navani A, Hames A. Platelet-rich plasma injections for lumbar discogenic pain: a preliminary assessment of structural and functional changes. Tech Reg Anesth Pain Manag. 2015;9:38–44.
61. Akeda K, Takegami N, Yamada J, et al. Platelet-rich plasma-releasate (PRPr) for the treatment of discogenic low back pain patients: long-term follow-up survey. Medicina. 2022;58:428.
62. Aufiero D, Vincent H, Sampson S, et al. Regenerative injection treatment in the spine: review and case series with platelet rich plasma. J Stem Cell Res Rev Rep. 2015;2:1019.
63. Tolbert G, Roy D, Walker V. Ultrasound guided dextrose prolotherapy and platelet rich plasma therapy in chronic low back pain: three case reports. Int J Phys Med Rehabil. 2013;1:149.
64. Kirchner F, Anitua E. Intradiscal and intra-articular facet infiltrations with plasma rich in growth factors reduce pain in patients with chronic low back pain. J Craniovertebr Junction Spine. 2016;7:250–6.
65. Sampson S, Botto-van Bemden A, Aufiero D. Autologous bone marrow concentrate: review and application of a novel intra-articular orthobiologic for cartilage disease. Phys Sportsmed. 2013;41:7–18.
66. Murphy MB, Moncivais K, Caplan AI. Mesenchymal stem cells: environmentally responsive therapeutics for regenerative medicine. Exp Mol Med. 2013;45:e54.
67. Mantripragada VJ, Piuzzi NS, Bova GJ, et al. Reliable assessment of bone marrow and bone marrow concentrates using automated hematology analyzer. Regen Med. 2019;14:639–46.
68. Pettine KA, Murphy MB, Suzuki RK, et al. Percutaneous injection of autologous bone marrow concentrate cells significantly reduces lumbar discogenic pain through 12 months. Stem Cells. 2015;33:146–56.
69. Hernigou P, Trousselier M, Roubineau F, et al. Stem cell therapy for the treatment of hip osteonecrosis: a 30-year review of progress. Clin Orthop Surg. 2016;8:1–8.

70. Hernigou P, Homma Y, Flouzat-Lachaniette C-H, et al. Cancer risk is not increased in patients treated for orthopaedic diseases with autologous bone marrow cell concentrate. J Bone Joint Surg Am. 2013;95:2215–21.
71. Hernigou P, Poignard A, Beaujean F, et al. Percutaneous autologous bone-marrow grafting for nonunions. Influence of the number and concentration of progenitor cells. J Bone Joint Surg Am. 2005;87:1430–7.
72. Hernigou P, Flouzat-Lanchaniette CH, Delambre J, et al. Biologic augmentation of rotator cuff repair with mesenchymal stem cells during arthroscopy improves healing and prevents further tears: a case-controlled study. Int Orthop. 2014;38:1811–8.
73. Hernigou P, Poignard A, Zilber S, et al. Cell therapy of hip osteonecrosis with autologous bone marrow grafting. Indian J Orthop. 2009;43:40–5.
74. Hernigou P, Guissou I, Homma Y, et al. Percutaneous injection of bone marrow mesenchymal stem cells for ankle non-unions decreases complications in patients with diabetes. Int Orthop. 2015;39:1639–43.
75. Hernigou P, Bouthors C, Bastard C, et al. Subchondral bone or intra-articular injection of bone marrow concentrate mesenchymal stem cells in bilateral knee osteoarthritis: what better postpone knee arthroplasty at fifteen years? A randomized study. Int Orthop. 2021;45:391–9.
76. Orozco L, Soler R, Morera C, et al. Intervertebral disc repair by autologous mesenchymal bone marrow cells: a pilot study. Transplantation. 2011;92:822–8.
77. Pettine KA, Suzuki RK, Sand TT, et al. Autologous bone marrow concentrate intradiscal injection for the treatment of degenerative disc disease with three-year follow-up. Int Orthop. 2017;41:2097–103.
78. Atluri S, Murphy MB, Dregalla R, et al. Evaluation of the effectiveness of autologous bone marrow mesenchymal stem cells in the treatment of chronic low back pain due to severe lumbar spinal degeneration: a 12-month, open-label, prospective controlled trial. Pain Physician. 2022;25:193–207.
79. Amirdelfan K, Bae H, McJunkin T, et al. Allogeneic mesenchymal precursor cells treatment for chronic low back pain associated with degenerative disc disease: a prospective randomized, placebo-controlled 36-month study of safety and efficacy. Spine J. 2021;21:212–30.
80. Da Silva Meirelles L, Sand TT, Harman RJ, et al. MSC frequency correlates with blood vessel density in equine adipose tissue. Tissue Eng A. 2009;15:221–9.
81. Polancec D, Zenic L, Hudetz D, et al. Immunophenotyping of a stromal vascular fraction from microfragmented lipoaspirate used in osteoarthritis cartilage treatment and its lipoaspirate counterpart. Gene. 2019;10:474.
82. Ragni E, Viganò M, Torretta E, et al. Characterization of microfragmented adipose tissue architecture, mesenchymal stromal cell content and release of paracrine mediators. J Clin Med. 2022;11:2231.
83. Condé-Green A, Kotamarti VS, Sherman LS, et al. Shift toward mechanical isolation of adipose-derived stromal vascular fraction: review of upcoming techniques. Plast Reconstr Surg Glob Open. 2016;4:e1017.
84. Zuk PA, Zhu M, Mizuno H, et al. Multilineage cells from human adipose tissue: implications for cell-based therapies. Tissue Eng. 2001;7:211–28.
85. Rodriquez AM, Elabd C, Amri EZ, et al. The human adipose tissue is a source of multipotent stem cells. Biochimie. 2005;87:125–8.
86. Melief SM, Zwaginga JJ, Fibbe WE, Roelofs H. Adipose tissue-derived multipotent stromal cells have a higher immunomodulatory capacity than their bone marrow-derived counterparts. Stem Cells Transl Med. 2013;2:455–63.
87. Manferdini C, Maumus M, Gabusi E, et al. Adipose-derived mesenchymal stem cells exert antiinflammatory effects on chondrocytes and synoviocytes from osteoarthritis patients through prostaglandin E2. Arthritis Rheum. 2013;65:1271–81.
88. Tao Y, Zhou X, Liu D, et al. Proportion of collagen type II in the extracellular matrix promotes the differentiation of human adipose-derived mesenchymal stem cells into nucleus pulposus cells. Biofactors. 2016;42:212–23.
89. Sun Z, Luo B, Liu ZH, et al. Adipose-derived stromal cells protect intervertebral disc cells in compression: implications for stem cell regenerative disc therapy. Int J Biol Sci. 2015;11:133–43.
90. Song K, Gu T, Shuang F, Tang J, et al. Adipose-derived stem cells improve the viability of nucleus pulposus cells in degenerated intervertebral discs. Mol Med Rep. 2015;12:4664–8.
91. Si YL, Zhao YL, Hao HJ, et al. MSCs: biological characteristics, clinical applications and their outstanding concerns. Ageing Res Rev. 2011;10:93–103.
92. Díaz-Prado S, Muiños-López E, Hermida-Gómez T, et al. Human amniotic membrane as an alternative source of stem cells for regenerative medicine. Differentiation. 2011;81:162–71.
93. Heo JS, Choi YJ, Kim HS, et al. Comparison of molecular profiles of human mesenchymal stem cells derived from bone marrow, umbilical cord blood, placenta and adipose tissue. Int J Mol Med. 2016;37:115–25.
94. Aufiero D, Sampson S, Onishi K, et al. Treatment of medial and lateral elbow tendinosis with an injectable amniotic membrane allograft – a retrospective case series. J Pain Reli. 2016;5:242.
95. Zelen CM, Poka A, Andrews J. Prospective, randomized, blinded, comparative study of injectable micronized dehydrated amniotic/chorionic membrane allograft for plantar fasciitis – a feasibility study. Foot Ankle Int. 2013;34:1332–9.
96. He Q, Li Q, Chen B, Wang Z. Repair of flexor tendon defects of rabbit with tissue engineering method. Chin J Traumatol. 2002;5:200–8.
97. Willett NJ, Thote T, Lin ASP, et al. Intra-articular injection of micronized dehydrated human amnion/chorion membrane attenuates osteoarthritis development. Arthritis Res Ther. 2014;16:R47.
98. Zhou T, Yuan ZN, Weng JY, et al. Challenges and advances in clinical applications of mesenchymal stromal cells. J Hematol Oncol. 2021;14:24.
99. Woo CH, Kim HK, Jung GY, et al. Small extracellular vesicles from human adipose-derived stem cells attenuate cartilage degeneration. J Extracellular Vesicles. 2020;9:1735249.
100. Liu P, Li KE, Xu S. The future of iPS cells in advancing regenerative medicine. Genet Res Camb. 2016;98:e4.

# 5 Fluoroscopy for Interventional Techniques in Regenerative Medicine

David M. Schultz and Scott Stayner

## History

On November 8, 1895, physics professor Wilhelm Roentgen (Fig. 5.1) was working alone at night in his laboratory conducting an experiment using a cathode ray tube in which he created a potential difference between the cathode and an anode placed several inches away. He knew from previous experiments that a stream of charged particles would be created in the cathode and move toward the anode. Since he was working at night, his laboratory was dark. The cathode-anode apparatus was encased inside a cardboard housing and, by chance, a screen of fluorescent material lay on a table not far away. As the experiment progressed, Professor Roentgen noticed a mysterious light appearing on the tabletop fluorescent screen. He repeated the experiment and each time he generated the potential difference, he observed the glowing light. He realized this light could not be caused by the cathode rays themselves since they were unable to penetrate the cathode ray tube. Furthermore, visible light could not be the source of the mysterious illumination since the tube apparatus was covered by an opaque material. He postulated that he must have generated some previously unknown type of radiation, which he called "X-rays."

Roentgen discovered that objects placed between the tube and the fluorescent screen produced shadows on the screen that could be captured on photographic film. As the story goes, Roentgen's wife placed her hand on a photographic plate while he projected the mysterious rays through it for 15 min, creating the first X-ray image (Fig. 5.2) [1–4]. Roentgen ushered in a new era of X-ray imaging and received the 1901 Nobel Prize in physics for his discovery.

**Fig. 5.1** Professor Wilhelm Roentgen. (Adapted from: Schultz DM. Fluoroscopy in interventional pain management. In: Manchikanti L, Kaye AD, Falco FJE, Hirsch JA (eds). *Essentials of Interventional Techniques in Managing Chronic Spinal Pain*. Springer, New York, NY, 2018, pp. 109–124 [4])

Early, unprocessed fluoroscopic images were quite dim and with early techniques, the fluoroscopist needed to sit in a darkened room for about 20 min until vision was dark-adapted simply to see the images. With typical X-ray imaging at the beginning of the twentieth century, the radiologist would sit or stand in front of the patient, the X-ray tube was placed behind the patient and the fluorescent screen was interposed between the patient and the doctor. Images were routinely dim and of poor quality until the advent of the image intensifier in 1953.

D. M. Schultz (✉) · S. Stayner
Nura Precision Pain Management, Minneapolis, MN, USA
e-mail: dschultz@nuraclinics.com; sstayner@nuraclinics.com


A. Navani et al. (eds.), *Essentials of Regenerative Medicine in Interventional Pain Management*,
https://doi.org/10.1007/978-3-031-50357-3_5

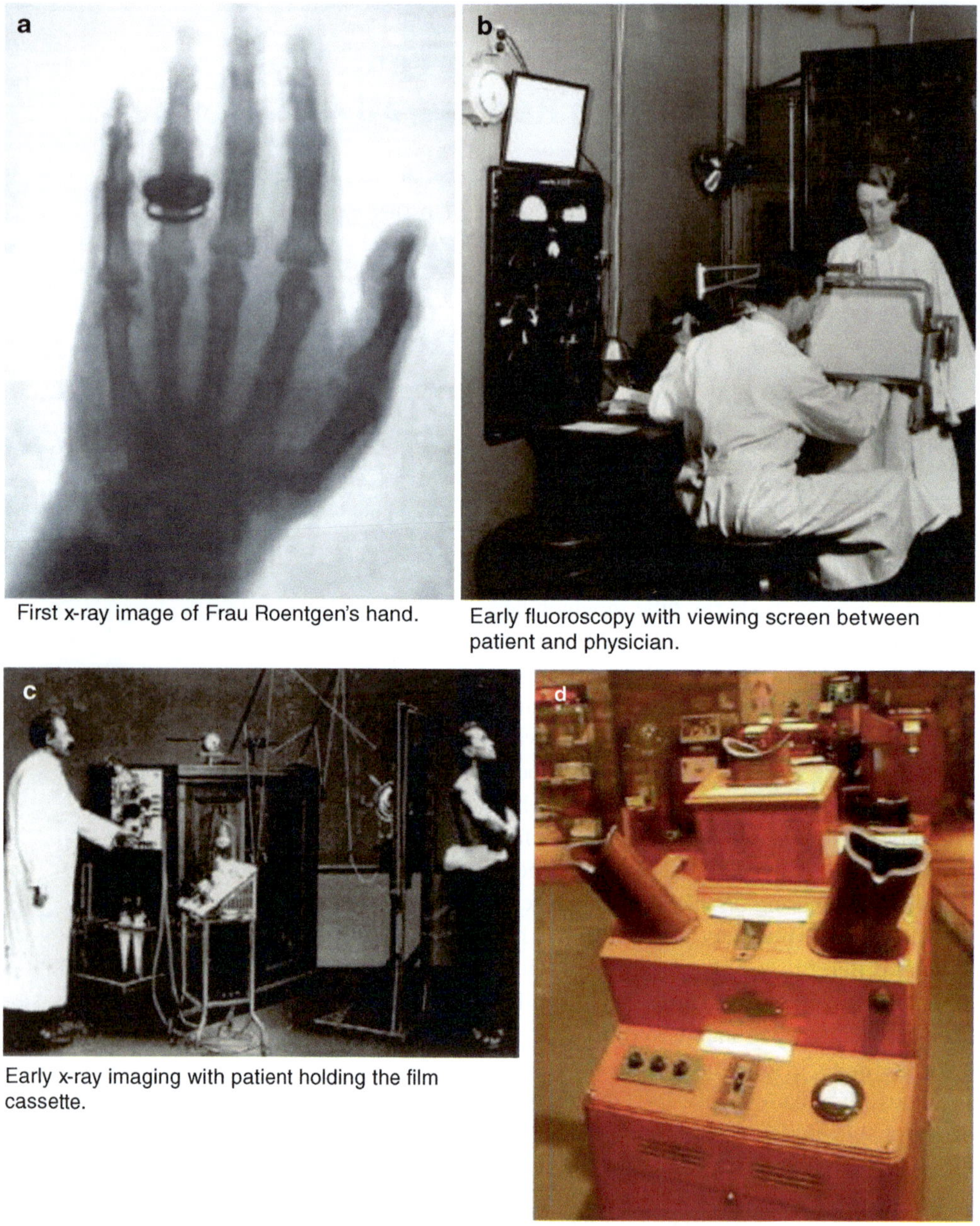


First x-ray image of Frau Roentgen's hand.

Early fluoroscopy with viewing screen between patient and physician.

Early x-ray imaging with patient holding the film cassette.

Fluoroscopic shoe-fitting device.

**Fig. 5.2** Early fluoroscopic and X-ray imaging equipment. (Adapted from: Schultz DM. Fluoroscopy in interventional pain management. In: Manchikanti L, Kaye AD, Falco FJE, Hirsch JA (eds). *Essentials of Interventional Techniques in Managing Chronic Spinal Pain*. Springer, New York, NY, 2018, pp. 109–124 [4])

By the 1940s, the use of fluoroscopy had become widespread in medical practice and also became popular for various nonmedical uses. The fluoroscope has evolved steadily over the past century to become a powerful and flexible tool for viewing both static and real-time X-ray images of the body. Continuing improvements in fluoroscopic imaging, combined with increasing knowledge of the mechanisms and anatomical sources for chronic pain, make it possible for the interventional pain specialist to provide increasingly safe and effective interventional pain management. Similarly, the developments in regenerative medicine have led to image-guided procedures, either with fluoroscopy or ultrasound. Fluoroscopy is most commonly utilized by interventional pain physicians performing interventional techniques with administration of biologics. Fluoroscopy is predominantly utilized for spinal interventional techniques, whereas ultrasound is commonly utilized for joint injections [4–6].

## Technical Aspects

When high-velocity electrons collide with metal, the kinetic energy contained within the electrons is converted to electromagnetic energy released in the form of X-rays. Kinetic energy is the energy of motion and is a property of all moving projectiles. The amount of kinetic energy contained within a projectile is a function of the projectile's mass and velocity as determined by the following equation: $KE = 1/2\ mv^2$ where KE is the kinetic energy in joules, m is the mass, and v is the velocity. It follows from the above relationship that, with regard to kinetic energy, velocity is exponentially more important than mass. Therefore, low-mass particles, like electrons traveling at high speed, can have high kinetic energy.

The function of the fluoroscope, like any X-ray machine, is to provide a steady stream of high-velocity electrons (projectiles) to bombard a metal target in order to produce a continuous, controllable stream of X-rays. The X-ray tube of the fluoroscope serves to focus high-energy electrons contained within an electrical current onto a small point within a metal target. As the kinetic energy of the electrons within the electric current is increased, the number and energy of X-rays produced from the collision with the metal target are increased. The fluoroscope allows for precise control of the number of X-rays generated (measured in milliamps) and the energy of the generated X-rays (measured in kilovolts). The X-ray tube housing is lined with lead and directs the X-rays out of an aperture, through tissue and onto a fluorescent screen within the fluoroscope. The imaging chain within the fluoroscope then amplifies and processes the fluoroscopic image to create visible X-ray images on the monitor.

## Fluorescence

The term fluorescence describes a property of certain materials that emit visible light when exposed to stimulation by chemicals, electricity, or ionizing radiation. X-rays are a form of ionizing radiation powerful enough to cause the phosphorous material within a fluorescent screen to glow. Early fluoroscopic systems used barium platinocyanoide as the fluorescent material. Barium was subsequently replaced by cadmium tungstate and later by zinc-cadmium sulfide, as these more effective and efficient phosphors became available.

The fluoroscopic picture that is produced is composed of shadows created as X-rays are absorbed preferentially by body tissues of various densities. Air allows most of the X-rays to penetrate through to the imaging medium, thus exposing the medium to unblocked X-rays. Bone and metal are denser and absorb or deflect X-rays, allowing fewer of them to penetrate through to the underlying plate. With plain X-rays, a photographic plate is used as the imaging media. The plate starts out as a dark background and becomes whiter in places as X-rays penetrate lower-density tissues and reach the plate to expose it. The plate, therefore, shows "shadows" of dense tissue structures as white areas on the picture.

With fluoroscopy, the density of the viewed image is reversed, with the higher-density tissues casting shadows that appear to be darker on the displayed image. Higher-density tissue appears as a darker area on the image screen in fluoroscopy because as penetrating X-rays hit the phosphor screen, they stimulate a fluorescent reaction within the screen, causing that portion of the screen to become bright (Fig. 5.3). The area that is not stimulated by penetrating X-rays is darker since X-rays that are absorbed by various tissues do not penetrate to the screen to cause phosphorescence. The denser the tissues, the fewer X-rays penetrate through to cause fluorescence on the underlying screen. The fluoroscopic image is analogous to the photographic negative; the developed X-ray film is analogous to the photographic picture. The principle of increasing absorption of X-rays by increasingly dense tissue gives rise to the five basic observed radiological densities: air, fat, water (soft tissue), bone, and metal.

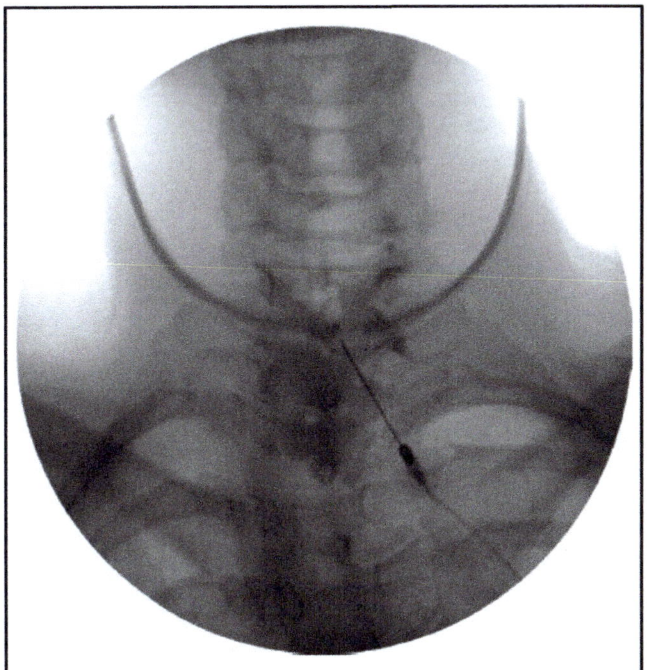


**Fig. 5.3** Fluoroscopic image illustrating density. (Adapted from: Schultz DM. Fluoroscopy in interventional pain management. In: Manchikanti L, Kaye AD, Falco FJE, Hirsch JA (eds). *Essentials of Interventional Techniques in Managing Chronic Spinal Pain*. Springer, New York, NY, 2018, pp. 109–124 [4])

## The Fluoroscopic Imaging Chain

The fluoroscopic imaging chain consists of various components including an X-ray tube with a cathode and anode, an image intensifier or flat panel detector, and a fluorescent phosphor screen to capture the image created (Fig. 5.4).

## The X-Ray Generator

As with any X-ray machine, the fluoroscope converts electrical energy to X-rays. Alternating electrical current from the power source is transformed into high-voltage direct current by the electrical transformer and rectifiers within the fluoro-

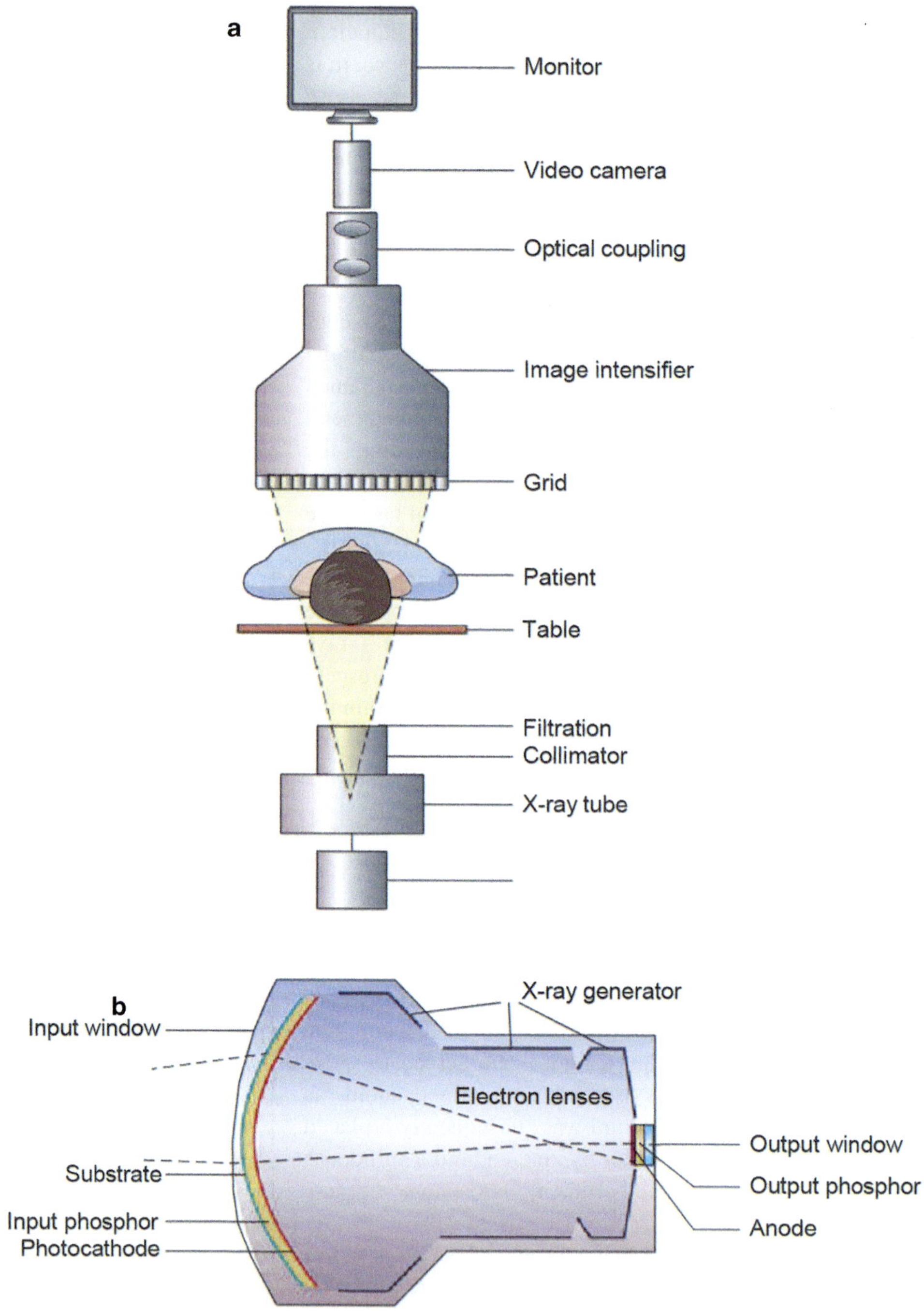


**Fig. 5.4** Fluoroscopic equipment. (**a**) Fluoroscopic imaging chain. (**b**) Image intensifier with electron paths from input window to output phosphor. (Adapted from: Schultz DM. Fluoroscopy in interventional pain management. In: Manchikanti L, Kaye AD, Falco FJE, Hirsch JA (eds). *Essentials of Interventional Techniques in Managing Chronic Spinal Pain*. Springer, New York, NY, 2018, pp. 109–124 [4])

scopic X-ray generator. The X-ray generator allows for selection of the amount of electrical current (measured in terms of amperage) and voltage of the current (measured in terms of kilovolt peak) that is delivered to the X-ray tube for subsequent X-ray production.

The milliampere (mA) of the current determines the number of electrons released to the X-ray tube and the intensity of the generated X-rays and is proportional to the density of X-rays produced. The voltage (kVp) of the current applied across the anode and cathode of the X-ray tube determines the wavelength, and thus the energy and penetrating ability of the X-rays. The milliamperes and kilovoltage of the current determine the quality of the image produced on the monitor by determining the quantity and energy respectively of the X-rays directed into the patient.

The X-ray generator assembly includes an automatic brightness control (ABC) system that allows automatic adjustment of kVp and mA to optimize the image brightness on the monitor, as the fluoroscope moves across areas of varying tissue density. Optimal mA and kVp settings can be set automatically by the modern fluoroscope or can be manually set by the operator.

## The X-Ray Tube

During an X-ray exposure, direct current from the X-ray generator is transmitted into the X-ray tube. The high voltage current heats the filament within the X-ray tube, causing electrons to be accelerated from the cathode of the tube to the positively charged tungsten anode. These accelerated electrons strike the anode with enough kinetic energy to cause the electrons within the heavy atoms of tungsten to be displaced to lower energy levels, thus releasing energy from the tungsten atom in the form of high-energy (X-ray) photons. The moving electrical current creates thermal energy (heat) and radiant energy in the form of X-rays. Radiant energy generated in this fashion is called "characteristic radiation."

Additional X-ray photons are produced from "bremsstrahlung radiation," which creates energy from the collisions of the projectile electrons with the tungsten nucleus independent of any orbital drops of the tungsten electrons. Bremsstrahlung is the German word for braking, and this type of radiation produces X-ray photons from the energy released by the projectile electrons as they collide with, and are slowed down by, the tungsten nucleus. The X-ray tube is inherently inefficient with 99% of energy converting to heat and only about 1% to X-rays.

The entire X-ray tube apparatus is housed in a shielded vacuum tube within the fluoroscope (Fig. 5.5). The electrons released from the X-ray tube are high-energy X-ray photons capable of penetrating tissue. The X-rays that are directed through the patient may be absorbed or scattered by the tissues of the patient or may pass through the patient to be captured by the fluorescent imaging screen.

## Image Intensifier

Early fluoroscopic images were dim and difficult to view. The image intensifier was developed to brighten the viewable image. Early image intensifiers consisted of a system of lenses and mirrors. Later versions added a video camera and monitor for viewing intensified images on the monitor screen.

The intensifier system takes the incoming X-rays and converts them to electrons in a series of steps that take place within the input layer. The X-rays enter the input window of the image intensifier, which is made of a thin layer of glass or metal. The X-rays are then directed to an input phosphor layer made of cesium iodide, where they are converted to light photons. There is an initial weak multiplying effect at

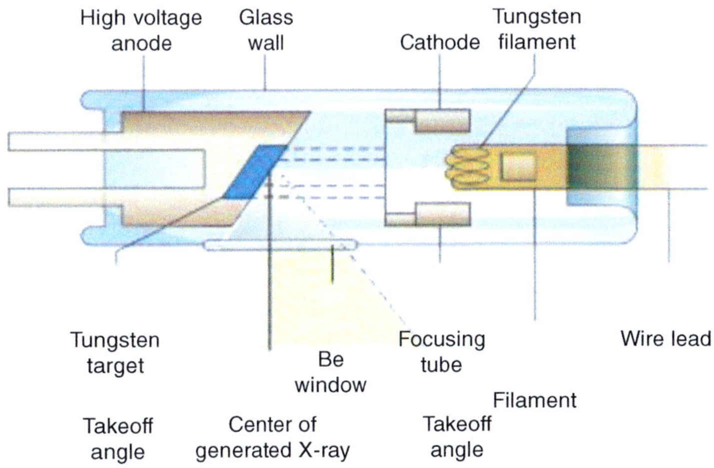


**Fig. 5.5** Components of the X-ray tube. (Adapted from: Schultz DM. Fluoroscopy in interventional pain management. In: Manchikanti L, Kaye AD, Falco FJE, Hirsch JA (eds). *Essentials of Interventional Techniques in Managing Chronic Spinal Pain*. Springer, New York, NY, 2018, pp. 109–124 [4])

this level, as a single X-ray photon can yield several hundred visible light photons. The input phosphor produces the instantaneous dim fluoroscopic image (first viewed by Roentgen) and these light photons are absorbed by the photocathode and converted to electrons. The energy from this fluorescence is magnified (intensified) several 1000-fold as those electrons converted from the light photons are accelerated from the photocathode through a vacuum to an anode within the output layer where they are available for viewing. Intensification comes from the fact that these energized, accelerated electrons now have enough energy to yield several thousand visible photons when they interact with the atoms of the output phosphor layer. The image is further intensified via an output phosphor window that is much smaller than the input surface of the intensifier, in a process known as minification. The intensification process produces a much brighter, more discernible image for viewing. Brightness is amplified by the image intensifier on the order of 5000- to 20,000-fold.

Modern image intensifiers are of varying sizes depending on the field of view necessary for clinical application. Small, portable C-arm systems used for visualizing extremities may only require a 9-in. diameter input window for adequate imaging, while systems used for visualizing the abdomen or trunk may use input windows as large as 20 in. in diameter. Typically, fluoroscopes used for interventional pain management have image intensifiers with 12-in. diameter input windows (Fig. 5.6).

## Flat-Panel Detectors

In 2009, flat-panel detectors (FPD) were introduced in the United States as a better alternative to the decades-old image intensifier (Fig. 5.7). Whereas image intensifiers convert X-rays to light photons, FPDs use solid-state technology, similar to image sensors in digital photography and video, to intensify the image viewed on the monitor. Although FPDs cost more than image intensifiers, advantages include better image quality, a smaller device profile with longer component life, less heat creation, and dramatically reduced radiation dose to patients and staff.

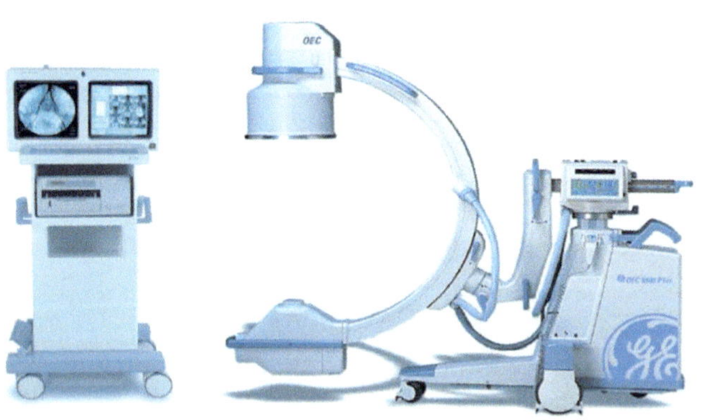

**Fig. 5.6** Typical fluoroscopic unit with image intensifier. (Adapted from: Schultz DM. Fluoroscopy in interventional pain management. In: Manchikanti L, Kaye AD, Falco FJE, Hirsch JA (eds). *Essentials of Interventional Techniques in Managing Chronic Spinal Pain*. Springer, New York, NY, 2018, pp. 109–124 [4])

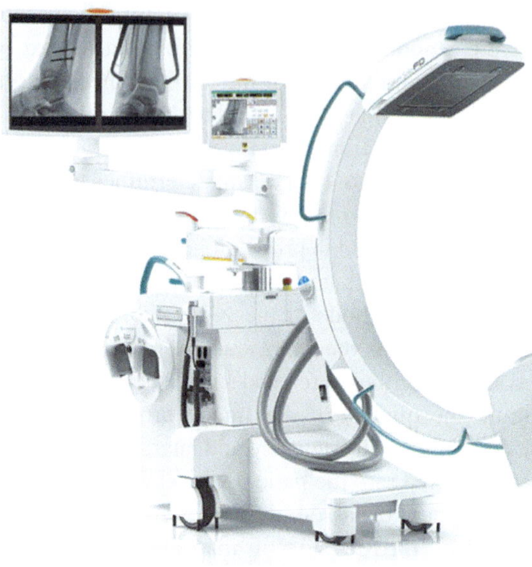

**Fig. 5.7** Typical fluoroscopic unit with flat panel detector. (Courtesy of Ziehm Imaging)

FPD arrays contain thousands of pixels, each tens or hundreds of micrometers long depending on the spatial resolution. The pixels consist of a photodiode and a switch. During X-ray imaging, the pixels receive and store short bursts of radiation. Pixels are then converted by readout integrated circuits (ROIC) and gate drivers into digital outputs. A computer then compiles the digital output of the flat panel detector into an image. FPD systems tend to cost more than traditional digital image intensifiers and can be more costly to repair. However, they tend to break down less frequently than digital image intensifiers when properly maintained [7].

## Collimation

The modern fluoroscope provides for internal collimation to optimally define the size and shape of the X-ray beam to better conform to the field of view. Unattenuated X-rays may cause glare on the image screen at the edges of the patient's body, resulting in poor image quality. Collimation reduces the amount of unattenuated radiation at the edges of the viewing area, thus reducing glare as well as overall patient radiation exposure by minimizing both direct and scatter radiation. Reduced scatter also means less radiation exposure to the operator and staff. Optimizing the size and shape of the X-ray beam through collimation reduces the volume of tissue exposed to X-rays, thereby reducing scatter, and improving image quality by reducing image degradation.

The collimator consists of radiopaque shutters that move into the X-ray path, attenuating areas of the X-ray beam. Collimator blades are typically circular (iris) or rectangular in configuration and many fluoroscopes allow for both circular and rectangular collimation. The collimator automatically adjusts the X-ray beam to conform to the field of view as the fluoroscope moves across areas of varying body tissue density. The operator can manually adjust the collimation window to conform to a particular region of clinical interest (Fig. 5.8a).

## Optical Coupling

Modern fluoroscopic systems incorporate optical coupling chains to process the images to optimize viewing capabilities. Once the light leaves the image intensifier or flat panel detector, it is available for viewing as an X-ray image. In C-arm systems, the image signal is typically routed to a video camera, which converts the X-ray image to a voltage signal that is viewable by closed-circuit television. This television monitor displays a real-time video image during continuous fluoroscopy.

A typical mobile C-arm fluoroscope can simultaneously display a static image by using a second television monitor for viewing the last image in the video sequence (Fig. 5.8b). This "last image hold" capability allows the interventionalist to minimize radiation exposure by performing procedures utilizing a series of static images to follow needle placement. The typical modern fluoroscope incorporates two monitors mounted on a mobile video display unit to allow several people to view real-time and static images simultaneously, side-by-side.

## Analog to Digital Converter

Newer fluoroscopic imaging systems incorporate digital image conversion technology to utilize digital image enhancement techniques for improved image quality and to facilitate com-

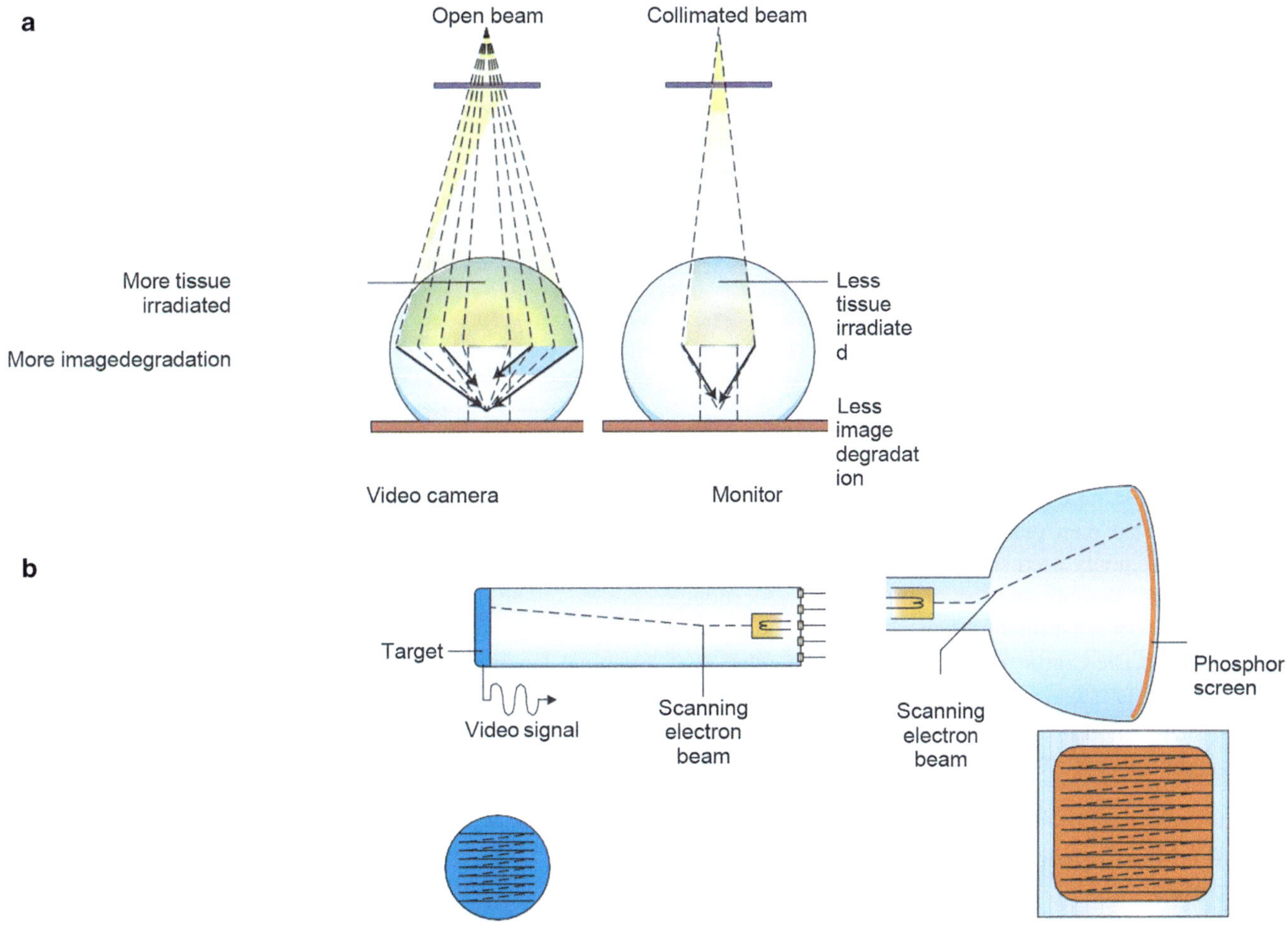


**Fig. 5.8** Illustration of optical coupling and collimation. (**a**) Effects of collimation. (**b**) Optical coupling. (Adapted from: Schultz DM. Fluoroscopy in interventional pain management. In: Manchikanti L, Kaye AD, Falco FJE, Hirsch JA (eds). *Essentials of Interventional Techniques in Managing Chronic Spinal Pain*. Springer, New York, NY, 2018, pp. 109–124 [4])

puter processing, storage, and distribution of images. With digital conversion, the analog video signal is digitized and stored in computer memory. Subsequent computer enhancement of the fluoroscopic image may achieve image clarity approaching that of X-ray film but with the use of less radiation. Digital images can also be quickly and conveniently distributed via computer networks and stored on computer workstations or archived into various digital storage media for later retrieval.

## Fluoroscopy Table and Pad

Materials placed between the X-ray tube and the image intensifier will tend to absorb X-rays. Differential absorption of X-rays by differing densities of patient tissues creates meaningful shadows that convey clinically useful information. Table and pad materials that have inconsistent density or produce high attenuation of X-rays will create a loss of image contrast, may cast artifactual shadows on the imaging screen, and can ultimately result in increased patient radiation dose. The X-ray tables and pads used with the fluoroscope are designed to optimize fluoroscopic imaging. Newer composite materials, such as carbon fiber, used in fluoroscopy tables provide adequate strength to support large patients while minimizing X-ray attenuation and distortion. Thin foam pads overlying the table have negligible effect on X-rays, whereas large gel supports or irregularly folded pillows may create significant X-ray attenuation and/or distortion and artifact. Diving board configuration of the table will allow for easier imaging of upper body structures during interventional pain procedures.

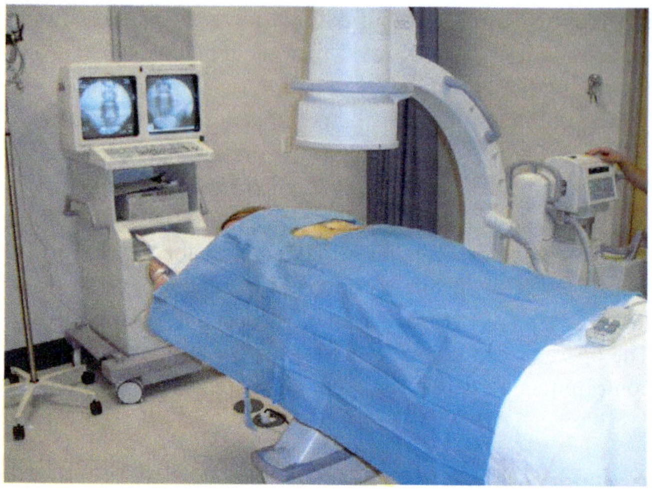

**Fig. 5.9** Carbon fiber diving board style table with pad and C-arm focused. (Adapted from: Schultz DM. Fluoroscopy in interventional pain management. In: Manchikanti L, Kaye AD, Falco FJE, Hirsch JA (eds). *Essentials of Interventional Techniques in Managing Chronic Spinal Pain*. Springer, New York, NY, 2018, pp. 109–124 [4])

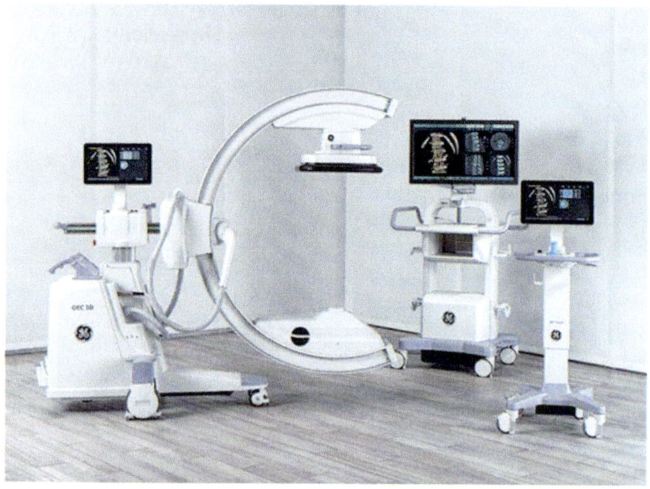

**Fig. 5.10** GE 3D C-arm. (Courtesy GE Imaging)

## Configuration of Fluoroscopic Systems

Fluoroscopic systems can be configured to meet the specific needs of the clinical application. The fluoroscopic components described above can be configured into fixed or mobile units. For interventional pain management, mobile C-arm fluoroscopes are commonly used since the compact design is both economical and flexible and the configuration can produce high-quality images, especially when used with a carbon fiber table (Fig. 5.9). The C-arm configuration allows angulation of the imaging chain into various orientations about the patient's body axis, improving visualization and access to anatomic structures during interventional pain procedures.

## Advanced Radiographic Imaging for Complex Procedures

More recently, computer-assisted navigation C-arms (Fig. 5.10) and O-arms (Fig. 5.11) have been developed to allow physicians to track instruments while performing minimally invasive spine and orthopedic surgeries. Initially, a reference point that can be optically tracked is placed non-invasively on the body or invasively attached to an anatomical landmark such as an iliac crest or spinous process through a small incision. Then a 3-D scan, with resolution similar to a computed tomography (CT) scan is generated. The instruments and hardware to be used in the procedure or surgery are then "registered" or paired optically to cameras outside the surgical field and to the initial reference marker. A computer-assisted 3-D reconstruction of the area of interest on the body is then generated. The location of the surgical instruments can then be tracked on the computer-generated 3-D image while the procedure is performed without radiation.

For minimally invasive spinal fusion surgeries, pedicle screws and other hardware may be placed with no further

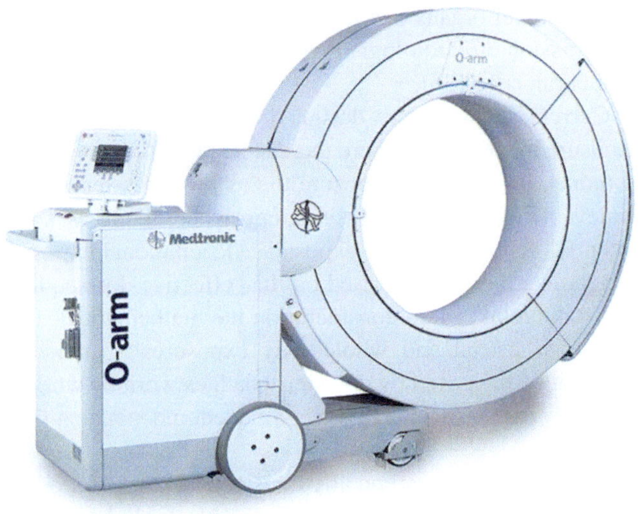


**Fig. 5.11** Medtronic O-arm. (Courtesy Medtronic)

radiation once the initial CT "spin" is completed, as long as the patient does not move. To assure that the patient is completely stationary throughout the case, navigation typically requires general anesthesia with intubation. If the patient's body does move during the case, the C-arm CT scanner must be brought back into the room to perform a second spin with recalibration of the navigation reference marker and external cameras. Failure to recalibrate after patient movement could provide inaccurate information as to the location of instruments and hardware within the body, leading to patient injury.

Similarly, for kyphoplasty and vertebroplasty using navigation, no further X-ray images are necessary after the initial CT spin for placement of the trocar through the pedicle and into the vertebral body so long as the patient does not move during the procedure. Live C-arm imaging is necessary, however, to confirm the location of non-registered devices such as the contrast dye-filled balloon or infusion of polymethyl methacrylate into the vertebral body.

Several comparative studies have shown that radiation exposure to the patient for kyphoplasty or vertebroplasty procedures is significantly reduced when computer-assisted navigation is used compared to traditional A-P and lateral C-arm imaging [8, 9].

## Radiation Safety

Soon after the discoveries of X-rays in 1895 and radium in 1897, it became apparent that radiation had the capacity to cause injury [1–4, 10, 11]. Reports of skin irritation from radiation exposure soon began to appear in the scientific and lay literature. By 1900, it was well-accepted in the scientific community that prolonged exposure to X-rays could produce skin burns and by 1905, medical malpractice suits for X-ray injuries were being settled in favor of patients.

There were early recommendations to limit the time and frequency of X-ray exposure, to collimate the X-ray beam to protect patients and to increase the distance from the X-ray apparatus for operator protection. Although the focus during the first quarter of the twentieth century was primarily on the discovery of new X-ray applications, there was increasing concern regarding the potential danger of X-ray exposure.

Radiation safety became a major concern by the mid-twentieth century and modern standards of X-ray protection have evolved steadily since about 1950. The National Council on Radiation Protection & Measurements (NCRP) was established in 1929 and continues to be the primary source of radiation safety recommendations in the United States today [11].

Health hazards from X-rays (and all forms of ionizing radiation) stem from the fact that energy imparted to the body from radiation has the capacity to cause cellular injury and cell death. When ionizing radiation contacts atoms of matter, an orbital electron of the matter molecule is converted to an ion. When ions are created within living cells they behave as free radicals and cause other atoms within the cell molecules to ionize, resulting in damage to the cell structure. Sometimes the damage is slight and can be repaired by the cell, but higher levels of radiation may cause irreversible cell injury.

Radiation may cause direct cell injury with alterations to the structure of DNA (somatic mutations) and other chemical changes within cellular matter. Somatic mutation may cause pathology that is initially silent but is manifested in subsequent generations. Cell injury may also be indirect as radiation triggers chemical chain reactions that result in cell membrane permeability changes that ultimately translate into cellular dysfunction.

Fluoroscopic guidance is frequently utilized in performing many interventional techniques in regenerative medicine with injection of biologics. It has been estimated that approximately four to ten million interventional procedures are performed annually in the United States, with at least 50% using fluoroscopic guidance [12–25]. While data is not available for application of fluoroscopy in interventional techniques for injection of biologics, it continues to increase.

The major purpose of fluoroscopy in interventional pain management is to verify correct needle placement and to ensure target specificity and accurate delivery of the injectate [12–23, 26–33]. Incorrect needle placement has been described for multiple spinal procedures when fluoroscopy is not used. The most commonly used fluoroscopy system for interventional pain management procedures consists of a mobile C-arm fluoroscope with image intensification or flat panel detection.

Radiation exposure may be associated with risks to the physician, patient, and personnel [1–4, 10, 11, 14, 34–36]. Most interventional procedures in the management of chronic pain require fluoroscopic exposure only for short periods of time. In 1994, the Center for Devices and Radiological Health (CDRH) of the Food and Drug Administration (FDA) issued an advisory, warning healthcare facilities of the potential for radiation-induced burns to patients from prolonged fluoroscopic procedures [11]. The same warning also applies to physicians and other staff members of the team. Physicians are more likely to have side effects and significant radiation exposure due to the cumulative effect of radiation exposure.

A number of interventional procedures, including radiofrequency cardiac catheter ablation, percutaneous transluminal angioplasty, vascular embolization, stent and filter placement, thrombolytic and fibrinolytic procedures, percutaneous transhepatic cholangiography, endoscopic retrograde cholangiopancreatography, transjugular intrahepatic portosystemic shunt placement, percutaneous nephrostomy, and biliary drainage or urinary or biliary stone removal are high-risk procedures from a radiation exposure perspective since they require prolonged use of continuous fluoroscopy [34]. However, none of the procedures listed by the CHRH included interventional pain techniques which typically require the brief use of intermittent fluoroscopy.

Biologic effects of radiation can be broadly grouped as stochastic or non-stochastic effects [34]. A stochastic effect is one in which the probability of the effect, rather than its severity, increases with dose. The probability of radiation-induced leukemia is substantially greater after exposure to 1 Gy (100 rad) than after exposure to 1 cGy (1 rad), but there will be no difference in the severity of the disease if it occurs. Stochastic effects are believed to lack a threshold dose because injury to a few cells or even a single cell could theoretically result in production of the effect. Non-stochastic effects are other effects for which the probability of causing certain types of harm will be zero at small radiation doses. Above a threshold level, damage will become apparent, with severity increasing as the dose rises above the threshold. Cataracts, erythema, epilation, and even death are examples of the non-stochastic, deterministic effects that can result from high radiation exposures.

The goals of radiation protection standards are to limit the incidence of radiation-induced illness and disease in patients and workers and to reduce the risk of genetic injuries in their progeny. Radiation risks to the physician and assisting personnel are evaluated using the maximum safe allowable exposure limits, which have been established by the NCRP [35]. The current estimation of risk from radiographic exposure to a specific body part is based on the biologic effects of whole-body exposure converted by weight factors, specific for individual organs and tissues. In 1991, the International Commission on Radiologic Protections (ICRP) adapted specific organ risks [36].

Current standards are designed to prevent any radiation exposure in situations where there is no potential benefit and to reduce unavoidable exposure to "as low as is reasonably achievable" (ALARA) in situations where there is some potential benefit from the exposure. Occupational exposure standards are set limits based on the "effective dose equivalent" which takes into consideration the mathematical summation of partial and whole-body exposures. Cumulative dose can be measured by dosimetry badges worn by workers and analyzed periodically by a contracted, independent third party.

Dosimetry badges typically consist of a solid-state radiation detection device encased inside a plastic holder outfitted with various filters. The X-rays entering the badge alter the solid-state radiation detector. Subsequent light output of the detection device is proportional to the radiation absorbed, and therefore periodic readings of the detector can determine the amount of X-ray exposure for the individual. For optimal monitoring, two badges are worn, one outside and one inside the lead apron to determine the total external exposure and the efficacy of the lead apron protection. Badges are analyzed monthly and allow for monitoring of cumulative radiation exposure over time. Current occupational exposure recommendations set the upper effective dose equivalent of 50 mSv/year. (5 roentgen-equivalent-man [REM]/year) and a cumulative dose not to exceed 10 mSv (1 REM) times the age of the worker. Thus, lifetime exposure for a 50-year-old radiation worker would be 500 mSv (50 rem).

In the early days of radiology, an X-ray image of a body part typically required 10 to 15 min of direct X-ray exposure and radiation injuries were commonplace. Modern X-rays are now completed within milliseconds and typical radiation exposure is a small fraction of what it was 100 years ago. However, with modern fluoroscopes, there is again the possibility of delivering high-energy X-rays to the patient for prolonged periods of time as increasingly complex procedures are performed using continuous fluoroscopy with high-powered fluoroscopes.

In 1994 alone, there were more than 50 reports of patient injury from prolonged exposure to X-rays during fluoroscopic procedures, leading the FDA to issue a health advisory in that year [11]. Since 1994, several case reports of serious patient injuries from radiation during fluoroscopy have been published [37–44]. These injuries are often delayed since the effects of excessive radiation exposure are usually not immediately apparent. Injuries have included skin burns serious enough to require skin grafting. Since mandatory reporting of fluoroscopy injury is not required, the actual extent of this problem is unknown.

Occupational radiation exposure can be reduced to "as low as reasonably achievable" (ALARA) through adherence to three basic principles:

1. Reduce time of exposure.
2. Increase distance from the radiation source (X-ray tube, scatter from patient).
3. Shield yourself and your patient from direct and scatter radiation.

Additional safety enhancements include:

- Pulsed fluoroscopy
- Removal of grid
- Image magnification
- Dose level settings
- Dose spreading
- Electronic collimation
- Adjustment of beam quality
- Appropriate training of operators

The longer the time one is exposed to a radiation field, the greater the total exposure. Limiting time of exposure is a simple, common-sense method for reducing risk. With modern fluoroscopic systems, a short burst of radiation used with "last image hold" capability allows the operator to identify needle position with minimal X-ray time. With last image hold technique, the interventionalist takes a fluoroscopic "snapshot" of the field, holds the static image on the monitor, moves the needle a small distance toward the target, and then obtains another brief fluoroscopic image to determine the new needle position. This type of technique will reduce total fluoroscopy time, as compared to continuous fluoroscopy, to track needle progress.

The inverse square law states that the amount of radiation exposure is proportional to the inverse square of the distance from the source. Therefore, the amount of radiation exposure declines exponentially as the operator moves away from the source. With fluoroscopy, distances of 6 ft or more from the X-ray tube and the patient result in minimal radiation exposure if the operator is not in the direct path of the X-ray beam. However, direct exposure to the X-ray beam, such as when the operator's hand is visible on the fluoroscopic screen, or close exposure to scatter from the patient, results in significantly higher risk.

Lead aprons are mandatory, and thyroid shields are recommended as standard garb within the procedure room when fluoroscopy is used. Lead shielding can be tailored to body contours and rendered reasonably comfortable while providing an effective barrier to radiation exposure, especially to the thyroid and pelvis. To reduce direct and scatter radiation exposure respectively, a wide variety of lead glass screens are available for placement between the operator and the X-ray source and patient. Lead glass screens can be configured to fit the particular procedure room space and can be attached to ceiling mounts or used with rolling frames on the floor. Lead glass is an effective barrier to eye exposure and can be configured with optical correction, although lead glass lenses may be heavy and uncomfortable. Regular glass lenses also afford some protection, and other alternative materials for eye protection are becoming available. Eyeglasses that do not provide wrap-around capability may not provide adequate protection.

The use of leaded gloves has been advocated by some to reduce exposure of the interventionalist's hands during procedures. With lead protection worn on the hands, there is less chance of direct exposure to the hands from both direct and scatter X-rays. However, the increased density of lead gloves within the X-ray field will cause the fluoroscope to increase output as it tries to penetrate the increased density. Therefore, leaded gloves within the field of view (i.e., visible on the monitor) will cause an automatic increase in mA, kVp, and scatter radiation. Furthermore, lead gloves are expensive and may decrease tactile sensation, which is so important to needle placement. Keeping hands completely out of the beam (and therefore never visible on the monitor) is advisable with or without lead gloves.

It has been described that risks can be minimized with a healthy respect for electromagnetic radiation, continued radiation safety education, radiation monitoring, and safe, common-sense practices, along with keeping in mind basic principles of ALARA (as low as reasonably achievable), time, distances, and shielding. The top ten measures for minimizing risks from fluoroscopic imaging in interventional pain management are listed in Table 5.1 [45].

**Table 5.1** Ten measures for minimizing risks from fluoroscopic imaging in interventional pain management

| | |
|---|---|
| 1 | Recognize that dose rates are greater and dose accumulates faster in larger patients |
| 2 | Keep the tube current as low as possible |
| 3 | Keep the kVp as high as possible and mA as low as possible, to achieve the appropriate compromise between image quality and low patient dose |
| 4 | Keep the patient at maximum distance from the X-ray tube |
| 5 | Keep the image intensifier as close to the patient as possible |
| 6 | Do not overuse geometric or electronic magnification |
| 7 | If the image quality is not compromised, remove the grid during procedures on small patients or when the image intensifier cannot be placed close to the patients |
| 8 | Always collimate down to the area of interest |
| 9 | Personnel must wear protective aprons, use shielding, monitor their doses, and know how to position themselves and the machines for minimum exposure |
| 10 | Keep beam-on time to an absolute minimum |

Adapted and modified from: Wagner LK, Archer BR. *Minimizing Risks from Fluoroscopic X-rays*, 3rd ed. RM Partnership, The Woodlands, TX, 2000 [45]

## Indications and Clinical Applications

The major purpose of fluoroscopy is to ensure correct needle placement for accurate delivery of injectate and solutions to increase clinical efficacy, decrease possible side effects, and enhance patient safety. The importance of fluoroscopy in correct needle placement has been demonstrated repeatedly in multiple publications.

Botwin et al. [16, 17, 29, 30] and Manchikanti et al. [12, 19, 46] have prospectively evaluated the radiation exposure to physicians performing fluoroscopically-guided interventional procedures in private practice. These studies found low radiation exposure, leading the authors to conclude that interventional procedures could be performed safely under optimal conditions with appropriate safety precautions.

Manchikanti et al. [19] evaluated many patients in a private practice setting. In the first study in 2002, 1000 consecutive patients with chronic pain, undergoing interventional procedures by a single physician, were evaluated. Table 5.2 illustrates radiation exposure in 1000 patients. Manchikanti et al. [19] showed that mean radiation exposure time was higher per patient and per procedure in obese patients.

Manchikanti et al. [12] evaluated radiation exposure to various regions with physicians at various levels of training. Groups were based on experience of the physician with Group III undergoing procedures with a physician having the highest level of experience; Group II with an intermediately-experienced physician; and Group I with the least experienced physician (Fig. 5.12). Table 5.3 shows the procedural characteristics and radiation exposure in seconds.

Manchikanti et al. [12] also evaluated scatter radiation inside and outside the apron. Table 5.4 shows scatter radiation exposure outside and inside the apron to various physicians. Manchikanti et al. [46] have shown the effectiveness of protective measures in interventional pain management as illustrated in Tables 5.5 and 5.6.

Botwin et al. [16, 17, 29, 30] evaluated radiation exposure to a physician performing fluoroscopically-guided caudal epidural steroid injections, lumbar transforaminal epidural steroid injections, and lumbar discography. The results showed that total fluoroscopic time was 15.6 s on average for transforaminal epidural steroid injections [16], 12.55 s for caudal epidural steroids injections [17], and 57.24 s for lumbar discography [30].

Zhou et al. [47] evaluated fluoroscopic radiation safety for spine interventional pain procedures in university teaching hospitals. The results of their study showed that fluoroscopic exposure time for various interventional procedures performed in university settings was significantly higher than the radiation exposures in private practice settings as shown in Fig. 5.9. This study also showed significant differences among physicians in the same university setting.

**Table 5.2** Illustration of radiation exposure in 1000 patients

| | Number | Radiation exposure in seconds | Range |
|---|---|---|---|
| Per procedure | – | 7.7 ± 0.21 | 1–69 |
| Per patient | – | 13.2 + 0.33 | 1–97 |
| Cervical facet joint nerve blocks | 547 | 5.9 ± 0.07 | 1–19 |
| Caudal/interlaminar epidurals | 353 | 3.75 ± 0.13 | 1–27 |
| Lumbar facet joint nerve blocks | 352 | 5.7 ± 0.09 | 1–14 |
| Intercostal nerve blocks | 96 | 4.0 ± 0.18 | 1–8 |
| Percutaneous adhesiolysis | 85 | 18.9 ± 1.32 | 4–45 |
| Lumbar transforaminal epidurals | 72 | 10.9 ± 0.72 | 3–32 |
| Thoracic facet joint nerve blocks | 68 | 5.1 ± 0.33 | 1–13 |
| Joint injections | 67 | 7.5 ± 1.28 | 1–52 |
| Cervical transforaminal epidurals | 28 | 8.8 ± 0.78 | 3–17 |
| Stellate ganglion blocks | 20 | 6.5 ± 0.68 | 2–11 |
| Lumbar sympathetic blocks | 17 | 12.0 ± 1.49 | 6–27 |
| Medial branch neurotomy | 16 | 12.7 ± 1.49 | 6–23 |
| Spinal endoscopy | 8 | 39.6 ± 2.31 | 25–46 |

Reproduced from Manchikanti L, Cash KA, Moss TL, et al. Risk of whole body radiation exposure and protective measures in fluoroscopically guided interventional techniques: A prospective evaluation. *BMC Anesthesiol* 2003; 3:2 [12]

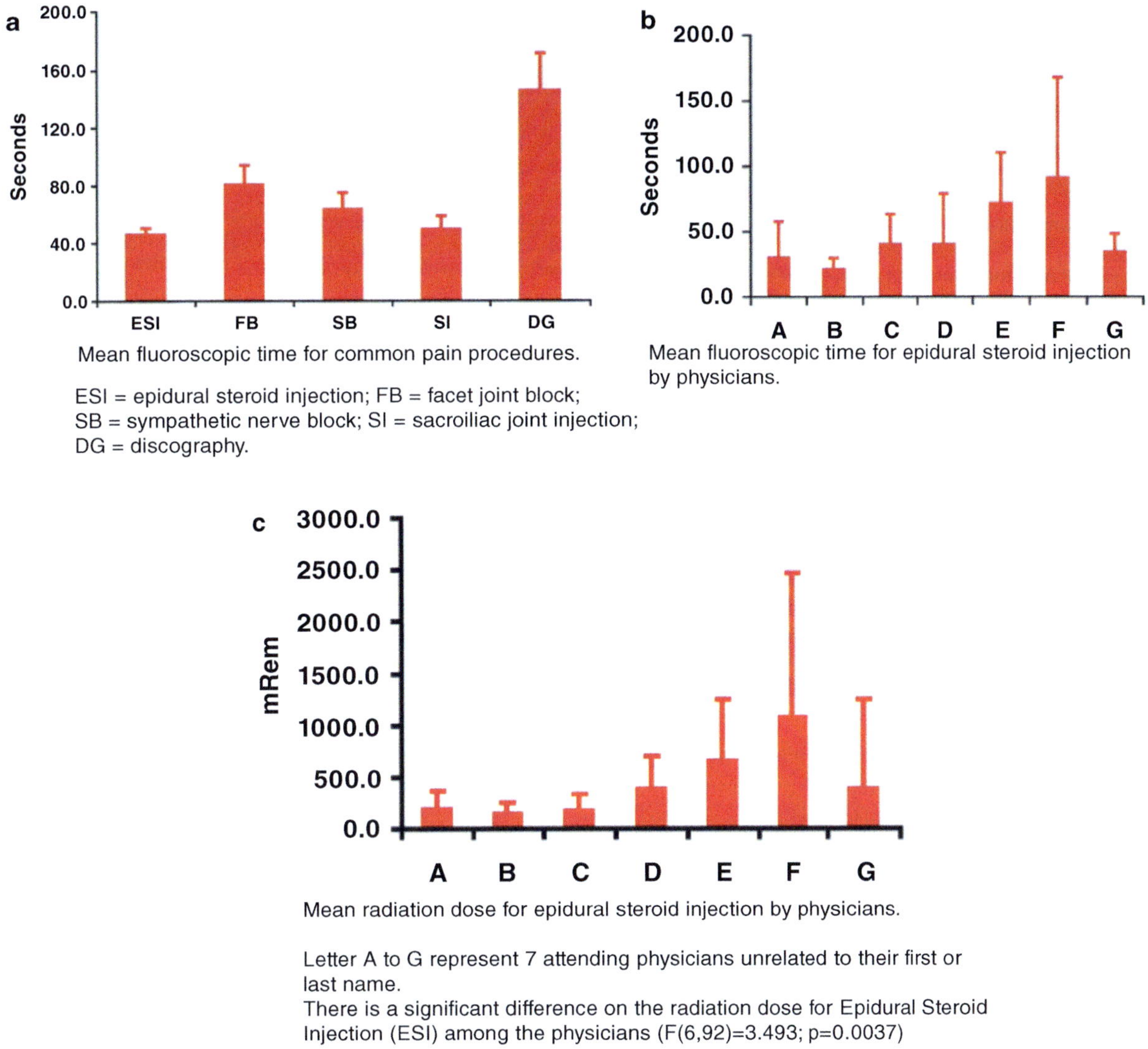


**Fig. 5.12** Illustration of fluoroscopic exposure for interventional procedures in an academic setting. (Adapted from: Schultz DM. Fluoroscopy in interventional pain management. In: Manchikanti L, Kaye AD, Falco FJE, Hirsch JA (eds). *Essentials of Interventional Techniques in Managing Chronic Spinal Pain*. Springer, New York, NY, 2018, pp. 109–124 [4])

**Table 5.3** Illustration of procedural characteristics and radiation exposure in seconds

| | Group I | Group II | Group III |
|---|---|---|---|
| Per procedure | 12.0[a] ± 0.49 (330) | 9.0[b] ± 0.37 (662) | 7.5 ± 0.27 (827) |
| Lumbar facet joint nerve blocks | 11.7[a] ± 0.56 (143) | 7.0[b] ± 0.32 (208) | 5.7 ± 0.17 (180) |
| Cervical facet joint nerve blocks | 10.4[a] ± 0.65 (73) | 5.5 ± 0.23 (131) | 5.9 ± 0.14 (265) |
| Caudal/interlaminar epidurals | 11.7[a] ± 1.41 (56) | 8.7[b] ± 0.96 (95) | 3.7 ± 0.29 (141) |
| Lumbar/cervical transforaminal epidurals | 14.0 ± 1.77 (26) | 15.0[b] ± 1.23 (80) | 10.6 ± 0.60 (92) |
| Percutaneous adhesiolysis | 20.8 ± 5.65 (5) | 14.5 ± 1.69 (28) | 18.9 ± 1.72 (50) |
| Thoracic facet joint nerve blocks | 11.5[a] ± 2.17 (13) | 7.9 ± 1.14 (29) | 5.6 ± 0.65 (29) |
| Sacroiliac joint injection | 15.0[a] ± 4.89 (5) | 3.7 ± 0.28 (50) | 7.5 ± 3.20 (4) |
| Intercostal/paravertebral/lumbar sympathetic nerve blocks | 10.0 (1) | 12.5 ± 3.40 (13) | 7.4 ± 1.09 (27) |

() Indicates number of procedures
Procedures performed less than 25 times were not listed in this table; however, they were utilized in calculating the exposure and per procedure
Reproduced from Manchikanti L, Cash KA, Moss TL, et al. Risk of whole body radiation exposure and protective measures in fluoroscopically guided interventional techniques: A prospective evaluation. *BMC Anesthesiol* 2003; 3:2 [12]
[a]Indicates significant difference: Group III vs. Groups I and II
[b]Indicates significant difference: Group II vs. Group I

**Table 5.4** Illustration of scatter radiation exposure in mREM outside and inside lead apron

| Location of dosimetry badge | Group I (330) | Group II (662) | Group III (827) |
|---|---|---|---|
| Chest [outside] | 510 | 535 | 690 |
| Neck [inside] | 68 | 25 | 0 |
| Groin [inside] | 0 | 0 | 15 |
| Groin [outside] | 1260 | 400 | 1152 |

() Indicates number of procedures
Reproduced from Manchikanti L, Cash KA, Moss TL, et al. Risk of whole body radiation exposure and protective measures in fluoroscopically guided interventional techniques: A prospective evaluation. *BMC Anesthesiol* 2003; 3:2 [12]

**Table 5.5** Illustration of procedural characteristics and radiation exposure in seconds

| | Group I (509) | Group II (500) |
|---|---|---|
| Per patient | 12.5 ± 0.40 | 8.9[a] ± 0.40 |
| Per procedure | 7.5 ± 0.27 (827) | 4.9[a] ± 0.11 (865) |
| Facet joint nerve blocks (cervical/lumbar/thoracic) | 5.8 ± 0.11 (474) | 4.5[a] ± 0.07 (481) |
| Epidurals (caudal/interlaminar) | 3.7 ± 0.29 (141) | 2.7[a] ± 0.27 (160) |
| Transforaminal (cervical/lumbar) | 10.6 ± 0.60 (92) | 8.4[a] ± 0.50 (102) |
| Percutaneous adhesiolysis | 18.9 ± 1.72 (50) | 11.8[a] ± 2.39 (35) |
| Intercostal/sympathetic blocks (cervical or lumbar) | 7.4 ± 1.09 (27) | 4.0[a] ± 0.35 (35) |

() Indicates number of procedures
Procedures performed less than 25 times were not listed in this table; however, they were utilized in calculating the exposure and per procedure
Reproduced from Manchikanti L, Cash KA, Moss TL, et al. Effectiveness of protective measures in reducing risk of radiation exposure in interventional pain management: A prospective evaluation. *Pain Physician* 2003; 6:301–305 [46]
[a]Indicates significant difference

**Table 5.6** Illustration of scatter radiation exposure in mREM outside and inside lead apron

| Location of Dosimetry Badge | Group I | Group II |
|---|---|---|
| Number of patients | 509 | 500 |
| Number of procedures | 827 | 865 |
| Neck [inside] | 0 | 0 |
| Chest [outside] | 690 | 313 |
| Chest [inside] | NA | 0 |
| Groin [outside] | 1152 | 176 |
| Groin [inside] | 15 | 13 |

*NA* not available

Reproduced from Manchikanti L, Cash KA, Moss TL, et al. Effectiveness of protective measures in reducing risk of radiation exposure in interventional pain management: A prospective evaluation. *Pain Physician* 2003; 6:301–305 [46]

## Precautions

Radiation exposure must be reduced substantially to as low as reasonably achievable (ALARA) through adherence to the three basic principles. The top ten measures for minimizing risks from fluoroscopic imaging in interventional pain management as described in Table 5.1 include principles of ALARA and understanding of radiation and application of safety measures. In summary, fluoroscopy is an essential part of interventional pain management. Multiple risks are associated with fluoroscopy. The risks can be reduced with appropriate clinical preventive measures.

### Key Points

1. Over the past 30 years, fluoroscopy has revolutionized the medical treatment of pain with interventional techniques now extending to interventional techniques with administration of biologics.
2. The history of fluoroscopy dates back to November 8, 1895, when Professor Wilhelm Roentgen discovered X-rays and invented the first crude fluoroscope.
3. When high-velocity electrons collide with metal, the kinetic energy contained within the electrons is converted to electromagnetic energy released in the form of X-rays.
4. The function of the fluoroscope is to provide a steady stream of high-velocity electrons to produce a continuous, controllable stream of X-ray radiation. The term fluorescence describes a property of certain materials that emit visible light when exposed to stimulation by chemicals, electricity, or ionizing radiation.
5. The fluoroscopic picture that is produced is composed of shadows created as X-rays are absorbed preferentially by body tissues of various densities. The fluoroscopic imaging chain consists of various components including an X-ray tube with a cathode and anode, an image intensifier, and a fluorescent phosphor screen to capture the image created.
6. Direct current from the X-ray generator is transmitted into the X-ray tube during an X-ray exposure.
7. The image intensifier takes the incoming X-rays and converts them to electrons in a series of steps that take place within the input layer and brightens the viewable image.
8. The modern fluoroscope provides the ability to collimate and incorporates optical coupling.
9. Radiation exposure may be associated with risks to the physician, patient, and personnel.
10. Most interventional procedures in the management of chronic pain require fluoroscopic exposure only for short periods of time. Biologic effects of radiation are grouped as stochastic and nonstochastic and follow the principles of ALARA, "as low as is reasonably achievable," all the time.

**Acknowledgments** This book chapter is modified and updated from a previous book chapter, "Fluoroscopy in interventional pain management" by David M. Schultz, MD, in *Essentials of Interventional Techniques in Managing Chronic Spinal Pain* published by Springer.

## References

1. Wagner LK. Foundations of safety in fluoroscopy. In: Manchikanti L, Singh V, editors. Interventional techniques in chronic spinal pain. Paducah, KY: ASIPP Publishing; 2007. p. 89–100.
2. Wagner LK. Production, nature, and effects in fluoroscopy. In: Manchikanti L, Singh V, editors. Interventional techniques in chronic spinal pain. Paducah, KY: ASIPP Publishing; 2007. p. 101–12.
3. Wagner LK. Radiation management for patients and protection of staff. In: Manchikanti L, Singh V, editors. Interventional techniques in chronic spinal pain. Paducah, KY: ASIPP Publishing; 2007. p. 113–24.
4. Schultz DM. Fluoroscopy in interventional pain management. In: Manchikanti L, Kaye AD, Falco FJE, Hirsch JA, editors. Essentials of interventional techniques in managing chronic spinal pain. New York, NY: Springer; 2018. p. 109–24.
5. Manchikanti L, Navani A. Evolution of regenerative medicine in managing chronic pain. In: Manchikanti L, Navani A, Atluri S, editors. Essentials of regenerative medicine in interventional pain management. Paducah, KY: ASIPP Publishing; 2019. p. 1–38.
6. Manchikanti L, Atluri S, Albers SL, Latchaw RE. Epidural administration of PRP and stem cells. In: Manchikanti L, Navani A, Atluri S, editors. Essentials of regenerative medicine in interventional pain management. Paducah, KY: ASIPP Publishing; 2019. p. 421–58.
7. Lee JJ, Venna AM, McCarthy I, et al. Flat panel detector C-arms are associated with dramatically reduced radiation exposure during uteroscopy and produce superior images. J Endourol. 2021;35:780–94.
8. Chang CC, Chang HK, Wu JC, et al. Comparison of radiation exposure between O-arm navigated and C-arm guided screw placement in minimally invasive transforaminal lumbar interbody fusion. World Neurosurg. 2020;139:e489–95.
9. Chang CJ, Lin GL, Tse A, Chu HY, Tseng CS. Registration of 2D C-arm and DD CT images for a C-arm image-assisted navigation system for spinal surgery. Appl Bionics Biomech. 2015;2015:478062.
10. Mayo Clinic. www.mayoclinic.com. Accessed 16 Dec 2022.

11. FDA public health advisory: avoidance of serious x-ray-induced skin injuries to patients during fluoroscopically guided procedures. Rockville, MD: Food and Drug Administration; 1994.
12. Manchikanti L, Cash KA, Moss TL, et al. Risk of whole-body radiation exposure and protective measures in fluoroscopically guided interventional techniques: a prospective evaluation. BMC Anesthesiol. 2003;3:2.
13. Manchikanti L, Abdi S, Atluri S, et al. An update of comprehensive evidence-based guidelines for interventional techniques of chronic spinal pain: part II: guidance and recommendations. Pain Physician. 2013;16:S49–S283.
14. Manchikanti L, Bakhit CE, Pakanati RR, et al. Fluoroscopy is medically necessary for the performance of epidural steroids. Anesth Analg. 1999;89:1330–1.
15. Paulson EK, Sheafor DH, Enterline DS, et al. CT fluoroscopy-guided interventional procedures: techniques and radiation dose to radiologists. Radiology. 2001;220:161–7.
16. Botwin KP, Thomas S, Gruber RD, et al. Radiation exposure of the spinal interventionalist performing fluoroscopically guided lumbar transforaminal epidural steroid injections. Arch Phys Med Rehabil. 2002;83:697–701.
17. Botwin KP, Freeman ED, Gruber RD, et al. Radiation exposure to the physician performing fluoroscopically guided caudal epidural steroid injections. Pain Physician. 2001;4:343–8.
18. Schade CM. Fluoroscopic safety. In: Proceedings of the Worldwide Pain Conference, San Francisco, CA; 2000.
19. Manchikanti L, Cash KA, Moss TL, et al. Radiation exposure to the physician in interventional pain management. Pain Physician. 2002;5:385–93.
20. Fredman B, Nun MB, Zohar E, et al. Epidural steroids for treating "failed back surgery syndrome": is fluoroscopy really necessary? Anesth Analg. 1999;88:367–72.
21. Stitz MY, Sommer HM. Accuracy of blind versus fluoroscopically guided caudal epidural injection. Spine. 1999;24:1371–6.
22. Stojanovic MP, Vu TN, Caneris O, et al. The role of fluoroscopy in cervical epidural steroid injections: an analysis of contrast dispersal patterns. Spine (Phila Pa 1976). 2002;27:509–14.
23. Cluff R, Mehio AK, Cohen SP, et al. The technical aspects of epidural steroid injections: a national survey. Anesth Analg. 2002;95:403–8.
24. Manchikanti L, Knezevic NN, Navani A, et al. Epidural interventions in the management of chronic spinal pain: American Society of Interventional Pain Physicians (ASIPP) comprehensive evidence-based guidelines. Pain Physician. 2021;24:S27–S208.
25. Manchikanti L, Kaye AD, Soin A, et al. Comprehensive evidence-based guidelines for facet joint interventions in the management of chronic spinal pain: American Society of Interventional Pain Physicians (ASIPP) guidelines. Pain Physician. 2020;23:S1–S127.
26. Manchikanti L, Kaye AD, Latchaw RE, et al. Impact of COVID-19 pandemic on utilization patterns of facet joint interventions in managing spinal pain in Medicare population. Pain Ther. 2023;12:505–27.
27. Manchikanti L, Pampati V, Knezevic NN, et al. The influence of COVID-19 on utilization of epidural procedures in managing chronic spinal pain in the Medicare population. Spine (Phila Pa 1976). 2023;48:950–61.
28. Manchikanti L, Simopoulos TT, Pampati V, et al. Impact of COVID-19 pandemic and updated utilization patterns of sacroiliac joint injections from 2000 to 2020 in the fee-for-service (FFS) Medicare population. Pain Physician. 2022;25:239–50.
29. Manchikanti L, Pampati V, Sanapati MR, et al. COVID-19 pandemic reduced utilization of interventional techniques 18.7% in managing chronic pain in the Medicare population in 2020: Analysis of utilization data from 2000 to 2020. Pain Physician. 2022;25:223–38.
30. Manchikanti L, Pampati V, Soin A, Sanapati MR, Kaye AD, Hirsch JA. Declining utilization and inflation-adjusted expenditures for epidural procedures in chronic spinal pain in the Medicare population. Pain Physician. 2021;24:1–15.
31. Manchikanti L, Pampati V, Soin A, et al. Trends of expenditures and utilization of facet joint interventions in fee-for-service (FFS) Medicare population from 2009-2018. Pain Physician. 2020;23:S129–47.
32. Goel A, Pollan JJ. Contrast flow characteristics in the cervical epidural space: an analysis of cervical epidurograms. Spine (Phila Pa 1976). 2006;31:1576–9.
33. Furman MB, Giovanniello MT, O'Brien EM. Incidence of intravascular penetration in transforaminal cervical epidural steroid injections. Spine. 2003;28:21–5.
34. Mahesh M. Fluoroscopy: patient radiation exposure issues. Radiographics. 2001;21:1033–45.
35. National Council on radiation protection and measurements. Ionizing radiation exposures of the population of the United States. Report No. 116. Washington, DC: NCRP; 1993.
36. International Commission on Radiological Protection. Recommendation of the international commission on radiation protection 26. Ann Int Commission Radiat Prot. 1977;1:1–53.
37. Lichtenstein DA, Klapholz L, Vardy DA, et al. Chronic radiodermatitis following cardiac catheterization. Arch Dermatol. 1996;132:663–7.
38. Rosenthal LS, Beck TJ, Williams J, et al. Acute radiation dermatitis following radiofrequency catheter ablation of atrioventricular nodal reentrant tachycardia. Pacing Clin Electrophysiol. 1997;20:1834–9.
39. Sovik E, Klow NE, Hellesnes J, et al. Radiation-induced skin injury after percutaneous transluminal coronary angioplasty: case report. Acta Radiol. 1996;37:305–6.
40. Knautz MA, Abele DC, Reynolds TL. Radiodermatitis after transjugular intrahepatic portosystemic shunt. South Med J. 1997;90:352–6.
41. Vano E, Gonzalez L, Beneytez F, et al. Lens injuries induced by occupational exposure in non-optimized interventional radiology laboratories. Br J Radiol. 1998;71:728–33.
42. Marx VM, Niklason L, Mauge EA. Occupational radiation exposure to interventional radiologists: a prospective study. J Vasc Interv Radiol. 1992;3:597–606.
43. Adelstein SJ. Uncertainty and relative risks of radiation exposure. JAMA. 1987;258:655–7.
44. Hernandez RJ, Goodsitt MM. Reduction of radiation dose in pediatric patients using pulsed fluoroscopy. Am J Roentgerol. 1996;167:1247–53.
45. Wagner LK, Archer BR. Minimizing risks from fluoroscopic X-rays. 3rd ed. The Woodlands, TX: RM Partnership; 2000.
46. Manchikanti L, Cash KA, Moss TL, et al. Effectiveness of protective measures in reducing risk of radiation exposure in interventional pain management: a prospective evaluation. Pain Physician. 2003;6:301–5.
47. Zhou Y, Singh N, Abdi S, et al. Fluoroscopy radiation safety for spine interventional pain procedures in university teaching hospitals. Pain Physician. 2005;8:49–53.

# Principles of Ultrasound Imaging in Musculoskeletal Medicine

6

Tolu Alugo

## Introduction

Ultrasound is an imaging modality based on the information acquired from the propagation of a high frequency (millions of cycles per second) sound waves (mechanical pressure waves) emitted by a transducer to characterize differences in tissue [1, 2]. The transducer converts electrical energy into transmitted sound waves and receives returned scattered ultrasound waves, converting them back into electrical signals, and then into bright scale images. This enables real-time imaging of moving and stationary structures. Ultrasound provides information complementary to, and in many cases supplemental to, other imaging modalities, e.g., radiography, computed tomography (CT), magnetic resonance imaging (MRI). Acoustic physics properties of sound waves include compression, rarefaction, reflection, and impedance, which provide localization and characterization of different tissues in the body. Medical ultrasound sound wave frequencies are in the megahertz (MHz) range, which can be contrasted to human hearing frequency range of 20 kHz.

Ultrasound has expanded the field of pain management by providing visualization of structures such as tendons, ligaments, muscle fibers and nerves that have become targets of therapy. Ultrasound has promoted dynamic diagnostic and therapeutic techniques and may be unique in its ability to diagnose and treat enthesopathies, for which regenerative medicine is particularly effective.

## Historical Background

Ultrasonographic instruments have been used to image the human body for at least 50 years [3]. The first documented use of sound waves for spatial orientation was by Spallanazini in 1794 who analyzed how bats obtained spatial orientation. In 1880, Galton created the first device capable of producing sound waves at a frequency of 40,000 Hz. The Curie brothers discovered that the quartz crystal under mechanical vibration produced electricity, a phenomenon known as the piezoelectric effect, as well as the inverse piezoelectric effect, which was the ability of a liquid crystal to produce electricity under the vibrations produced by an ultrasound wave. In 1912, Richardson created an echo locator based on sonography used in navigation in water. In 1929, Sokolov proposed the sound transmission theory and in the early 1930s diagnosed metal defects in construction.

In 1937, the Dussig brothers attempted to use ultrasound to image the brain but were unsuccessful due to the skull blocking ultrasound waves. Ludwig and Stuthers Dussig used pulse ultrasound to image gallstones. Ultrasound use in diagnostic medicine was introduced by Ian Donald in 1956 using amplitude mode (A-Mode) one dimensional imaging to evaluate the bi-parietal diameter of the fetal head. Subsequently, Brown invented a two-dimensional scanner which enabled imaging of tissue density and was considered a significant milestone in medical ultrasound. In 1963, brightness mode (B-mode) scanners became available, enabling two-dimensional imaging. In the mid-1970s, the grey scale was introduced by Kossoff and Garret, which allowed real-time imaging. A decade later, Doppler imaging was developed, enabling blood circulation imaging [2–4].

## Ultrasound Equipment Overview

Ultrasound equipment is composed of a transducer that serves as a sender and receiver of ultrasound waves (Fig. 6.1). The reflected (received) sound wave is then processed to provide spatial and contrast resolution, producing the anatomical information in the form of a two-dimensional gray scale image [4]. Other features include Doppler, which detects frequency shifts in sound waves, which indicates whether the tissue is

T. Alugo (✉)
Department of Anesthesiology and Pain Medicine, Faculty of Medicine, Saint John Regional Hospital, Dalhousie University, Halifax, NS, Canada


A. Navani et al. (eds.), *Essentials of Regenerative Medicine in Interventional Pain Management*, https://doi.org/10.1007/978-3-031-50357-3_6

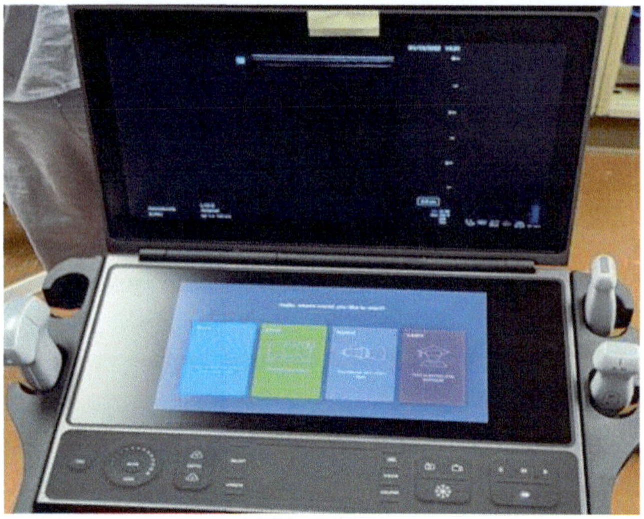

**Fig. 6.1** Modern ultrasound machine. (Image courtesy of Tolu Alugo, MD)

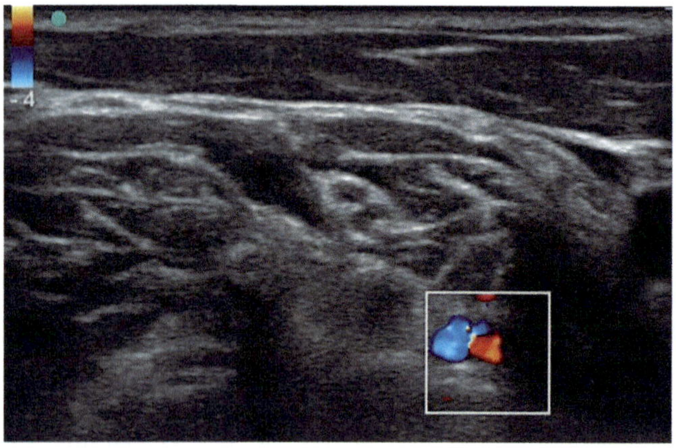

**Fig. 6.2** Ultrasound image showing color Doppler identifying vascular flow

moving towards or away from the transducer. In addition, color imaging shows similar information visually with colors to demonstrate flowing fluids, e.g., blood (Fig. 6.2). New developments beyond the scope of this introductory discussion include three-dimensional imaging, elastography, and contrast-enhanced ultrasound with micro-bubbles [5].

The advantages of ultrasound are multiple:

- Sound waves are not known to be associated with deleterious tissue/DNA injury seen with ionizing radiation, i.e., radiography, CT, and does not require the use of contrast adjuvants with their attendant risks of allergic reactions, such as nephrotoxicity.
- Ultrasound is readily available and less expensive than CT or MRI.
- Newer ultrasound machines approach the size of a large laptop computer (and more recently the size of a table), which accordingly makes the equipment highly portable. These laptop ultrasound machines are considerably more affordable than the large ultrasound units frequently used in hospitals and imaging centers.
- Ultrasound also provides dynamic imaging with the ability to image structures in range of motion and flowing fluids.
- No radiation exposure to the patient or provider, and therefore no need for shielding.

The disadvantages of ultrasound are:

- Highly dependent on the skills and experience of the operator.
- Ultrasound is not useful for imaging tissues with high acoustic impedance (see discussion below) such as bone.
- There is a minor risk that ultrasound may produce thermal heating, especially with implantable stimulators.
- Ultrasound has unique, inherent artifacts which may reduce image quality, accuracy, and as a result, may result in a misdiagnosis/misinterpretation.
- Injection requires "3 hands" and "wiggling the needle."
- The probe is directly in the field, requiring probe covers versus "no touch" technique.
- There is currently no gold standard for technique.
- Because of poor image capture and the difficulty of interpreting isolated images, there is a potential for unscrupulous use and misuse.
- Ultrasound utility may be limited in patients with large body habitus.
- There is also a concern that the structures like nerves and tendons are best viewed in cross-section, but that view is perpendicular to the most *effective* angle for injection. This view is parallel to the structure (for instance, trying to enter the biceps tendon sheath or the use of hydrodissection). However, it is important to note that there is no gold standard on the where the most effective angle is.

## Basic Ultrasound Physics

Radiographic imaging involves X-rays that travel in a straight line between the X-ray generator and the image intensifier, while ultrasound imaging uses sound wave that bounce off the structures, providing mechanical longitudinal waves of energy [6].

Ultrasound probes (also known as transducers) contain piezoelectric crystals. When electric current is applied to these crystals, which are usually quartz, the resulting vibration produces mechanical sound waves. This is known as the piezoelectric effect. The resulting sound waves are characterized by alternating components of compression and rarefaction as they propagate through the tissues of the body. The ultrasound frequency used in medical imaging is in the MHz range, which is far above 20 kHz (the upper limit of human hearing). Ultrasound waves are a mechanical compression and classified as either longitudinal (parallel to the direction of travel) or transverse waves. Ultrasound waves have inherent acoustical properties that include frequency, pressure, wavelength, velocity, power, and intensity.

Frequency is the number of complete oscillations or cycles per second. Diagnostic ultrasound uses frequencies in the range of 5–30 MHz. Wavelength and frequency are inversely proportional, meaning high frequency pulses have a short wavelength and low frequency sound waves have a long wavelength. The higher frequencies are associated with narrower beam width and smaller beam divergence. Small beam widths are important for good lateral resolution.

Selecting probes with the correct frequency range for the anatomy of interest is important. High frequency transducers (e.g., 10–15 MHz) are indicated for imaging superficial anatomy versus low frequency transducers (e.g., 2–5 MHz), which are indicated for imaging deeper structures. There is always a tradeoff between using different frequency probes. High frequency sound waves generate better resolution but are more easily attenuated, so they are limited to imaging only superficial structures. Low frequency waves, although able to penetrate deeper, generally produce images of lower resolution.

Sound velocity is the rate at which the ultrasound wave is transmitted in its propagation direction. Velocity is a vector quantity, which implies direction as well as magnitude. Speed is a descriptor used when a direction is not specified. Ultrasound wave propagation speeds are 5× greater in water and soft tissues than air, and the speed in bone and metal is 10× greater than air. Most ultrasound machines use an average soft tissue speed of 1540 m/s to measure internal distance measurements.

Depth information with ultrasound is a function of the time between the emitted and received ultrasound wave. It is given by the formula:

$$depth = \left(sound\ speed \times time\ of\ flight\right) / 2$$

Power and intensity are ultrasound strength measures. Power is expressed in watts and is the total energy passing through surface unit per time, but power is subject to attenuation as it passes through tissue. Intensity is the concentration of energy per unit time (energy/unit area/unit time), which changes as a function of ultrasound beam width and at the narrowest portion of a focused beam.

Attenuation is a result of energy loss to the tissue, with the resulting loss of acoustic energy per unit distance as an ultrasound wave travels through tissue. The ultrasound beam may be (1) partially transmitted to deeper tissues, (2) partially reflected as an echo to the transducer, (3) partially scattered, and (4) lastly, partially converted to heat. This is a result of the three causes of attenuation: divergence, deflection, and absorption. Ultrasound beam divergence spreads the acoustic energy over a larger beam area, reducing beam intensity. Deflection also reduces intensity due to reflection, refraction, and scattering. The greatest loss of ultrasound beam energy is by far through absorption, and the amount of absorption is a function of ultrasound beam frequency. The relationship is considered linear for most tissues of the human body.

Acoustic impedance (Z) is an inherent tissue physical property, describing ultrasound beam resistance passing through various body tissues. It is a function of the tissue's density (d measured in $kg/m^3$) and sound wave velocity (c measured in m/s). It is given by the formula: $Z = d \times c$. Therefore, impedance will increase as the density of a tissue increases, and, as the speed of sound increases in a tissue, impedance will increase. Acoustic impedance is most evident at the interface between two tissues of different density. If the difference in the acoustic impedance is sufficient, the ultrasound beam will be reflected to the probe, creating tissue contrast resolution in the image. The amount of reflection that occurs is expressed as the reflection fraction [7].

## Biological Effects of Ultrasound

Ultrasound is non-ionizing, so the risk of damage to DNA is extremely small. However, at sufficient intensities, ultrasound can produce temperature elevation, mechanical effects, certain cavitation and chemical effects of tissues. In its diagnostic use in medicine, these effects are not clinically significant, and, in over 40 years of diagnostic imaging, there have been no published, confirmed reports of any adverse effects. Regardless, clinicians using ultrasound should be aware of the small, but real potential adverse effects.

## Optimizing the Ultrasound Image

Before you start, review the user manual with attention to entering patient data, capturing, storing, and reviewing the ultrasound images. Be sure that the power cord is plugged in and/or the battery is fully charged.

Five function keys are basic to optimize an ultrasound image [8].

1. **Probe frequency:** Use the ultrasound transducer with a frequency range appropriate to visualize the target tissue. Since ultrasound energy is attenuated by transmission through tissue, the higher the frequency, the more prominent the attenuation and the less distance/depth is visualized. Use a low-frequency range transducer to scan deeper structures but expect somewhat reduced image resolution.

2. **Depth:** Depth controls the location of the target tissue in 1 cm gradations. The target tissue must always be in the center of the image, which is the key for proper focusing. This ensures optimal resolution of the target tissue as well as adjacent anatomic structures. However, increasing depth comes with a loss of resolution and frame rate.
3. **Focus:** Lateral resolution is improved by using higher frequency transducers and by focusing the ultrasound beam. This is designated the "focal zone." Transmission focus can be used to optimize image intensity in the near and far fields to improve spatial resolution.
4. **Gain:** Screen image brightness is controlled by two buttons: (a) gain and (b) time-gain compensation sliders (TGC, aka depth compensation). Too much or insufficient gain setting will result in "unsharpness" of tissue boundaries and visualization, i.e., loss of spatial resolution due to noise. The goal is to adjust gain to obtain the best contrast between the target tissues versus adjacent structures. The TGC sliders enable adjusting the image such that all signals are of similar intensity independent of the depth (distance traveled) and can be manipulated to best image structures deep and/or superficial to the target tissue. Adjust gain to the minimum level to obtain an adequate image that shows a complete range from low (dark gray) to high (white) signals.
5. **Doppler:** Color Doppler and power Doppler are useful adjuncts to identify vascular flow. If, when performing an ultrasound -guided intervention with the needle not optimally seen, applying power Doppler function allows the location of the injectate flow to be visualized.

For more advanced controls, consult the manual of the ultrasound machine and refer to some of the references cited in the recommended reading section.

## Common Ultrasound Artifacts

Ultrasound artifacts produce an image aberration that is not representative of anatomy or its appearance [9–12]. The most frequent artifacts are:

1. **Anisotropy:** This is a loss of image brightness due to the ultrasound beam being not perfectly perpendicular to the object. Toggling and/or rocking the ultrasound probe slowly and observing the changes in the target tissue can diagnose it. This artifact can be misdiagnosed as a lesion.
2. **Shadowing:** This is a prominent attenuation of echo signals, resulting in loss of image (hypoechoic or hypoechogenic) due to calcium deposits and osseous structures.
3. **Through transmission/enhancement:** Ultrasound waves traversing through low attenuating areas such as fluid, cyst, or larger vascular structure result in enhanced echo signals due to the fluid, cyst, etc. Changing the ultrasound beam angle or orientation may be helpful to prevent this artifact from obscuring the target tissue.
4. **Reverberation:** Manifested by equidistant linear echoes due to reflectors in the near field of the image. Changing the ultrasound beam direction and/or orientation or changing the ultrasound probe frequency may help.
5. **Mirror Image:** Results from an object positioned on a side of a highly reflective interface appearing on the opposite side as well. Both the virtual and artificial images are equidistant to the reflector structure. Changing the ultrasound beam direction and/or orientation may minimize the artifact.

The choice of transducers depends on the target structure. Most of the structures treated in pain management are relatively superficial (within the reach of a 3.5-in. needle), and so the high frequency linear transducer is generally all that is needed. Some of the deeper structures, such as lumbar facets or the psoas muscle, are best seen using the low frequency curvilinear transducer. The transducer usually has an indicator to identify the direction of orientation. The type of needle used depends also on the depth of the target and the operator preference. Some physicians prefer a 25-gauge 2-in. MonojetÒ needle with an added curve at the tip for superficial structures. Other interventionists use Quincke (subarachnoid), B-bevel stimulating, or specific ultrasound-designed needles.

There are several ways to describe the planes of ultrasound scanning. The sagittal plane divides the body into right and left sections, which are parallel to the long axis of the body; by convention, the transducer position indicator should be pointing to the head. The transverse plane divides the body into superior and inferior sections that are perpendicular to the long axis of the body; the transducer indicator should be pointing to the right side of the body. The coronal plane divides the body into an anterior and posterior section, which is perpendicular to the sagittal plane and parallel to the long axis; in this plane, the indicator should point to the patient's head.

## Scanning Technique Basics

Use careful scanning technique to avoid anisotropy (and possible misdiagnosis).

- Before scanning, know the anatomy, including bony landmarks, origins, and insertions of tendons, ligaments.
- Position the patient. Most operators place the patient between themselves and the ultrasound monitor. It is often useful to show your patient the anatomy and pathology initially as an educational tool (Fig. 6.3).

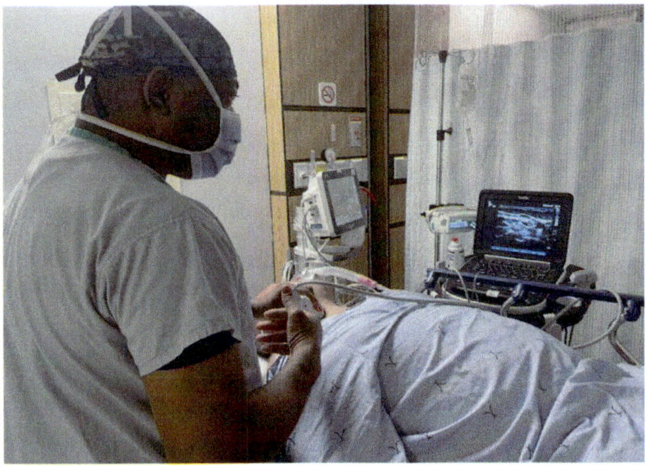

**Fig. 6.3** Ergonomics: Patient positioned between the physician and the ultrasound machine. (Image courtesy of Tolu Alugo, MD)

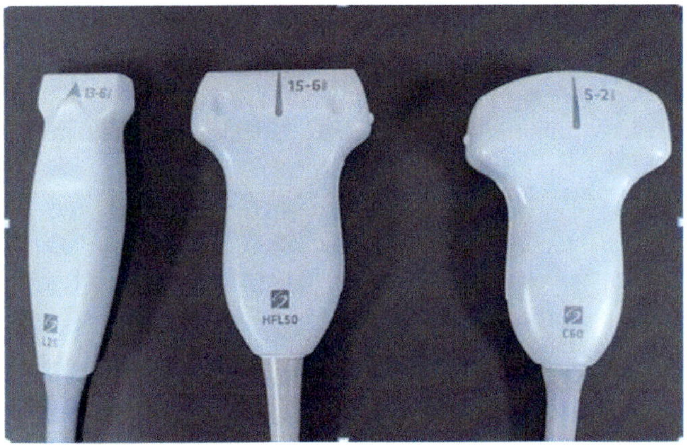

**Fig. 6.4** Ultrasound probes—linear and curvilinear. (Image courtesy of Tolu Alugo, MD)

- Before placing the ultrasound probe on the patient, examine the anatomy by palpation to identify pain generators and bony landmarks. It is often helpful to mark bony landmarks with a surgical marker to help with orientation.
- Depending upon the target tissue of interest, select the appropriate frequency transducer (a linear high frequency probe for superficial structures and a curvilinear low frequency probe for deeper anatomy) (Fig. 6.4).
- Hold the probe with the thumb and index/middle fingers while using the ring and pinkie finger as a "bridge" or support on the patient, similar to holding a pool cue. This type of technique stabilizes the probe, which can frequently "skate" away due to the ultrasound gel. The ultrasound gel is an acoustic coupler that ensures good contact between the probe and the patient by displacing air, which will compromise the image. When using a sterile probe cover, be sure to "massage" away air bubbles that may be trapped in the gel between the probe and the probe cover to avoid artifacts.
  – Start scanning by identifying bony landmarks as your base, then evaluate the target anatomy in both the longitudinal (long axis) and transverse (short axis) planes (see below).
  – Identify normal versus abnormal anatomy. The key is to train your eyes to recognize normal anatomy. If your eye has the "sight picture" of normal anatomy, pathology almost becomes self-evident.
  – If you are uncertain about anatomy or pathology, compare sides. If you get lost, go back to the known bony landmarks, and start over.
  – Ultrasound is a dynamic imaging modality, so moving appropriate anatomy can be a useful diagnostic tool.

There are two main techniques used to view the interventional target. When using the long axis view, the transducer is oriented along the path of the structure, so that the structure appears as a line. The short axis view orients the transducer perpendicular to the structure, so that it appears as a circle (Fig. 6.5). The needle can be parallel (in plane) or perpendicular (out of plane) to the transducer (Fig. 6.6). There are then four potential relationships between the ultrasound image orientation and the needle placement:

- Needle parallel to the ultrasound beam, which is parallel to the structure
- Long axis, in plane (LAX IP) (Fig. 6.7a)
- Needle parallel to the ultrasound beam, which is perpendicular to the structure
- Short axis, in plane (SAX IP) (Fig. 6.7b)
- Needle perpendicular to the ultrasound beam, which is parallel to the structure
- Long axis, out of plane (LAX OP) (Fig. 6.7c)
- Needle perpendicular to the ultrasound beam, which is perpendicular to the structure
- Short axis, out of plane (SAX OP) (Fig. 6.7d)

## Tips to Keep in Mind

- Beam steering or compounding, if available on the machine you are using, can help to minimize the dreaded anisotropy artifact in linear structures such as tendons. Use a careful scanning technique to avoid anisotropy (and possible misdiagnosis).
- Use color / power Doppler capabilities when assessing vessels or vascularity of a structure. Power Doppler can be used to see flow of the injectate from the tip of the needle and is a useful tool when the needle is not well seen. Power Doppler can also image increased blood flow, which can be an index for inflammation.

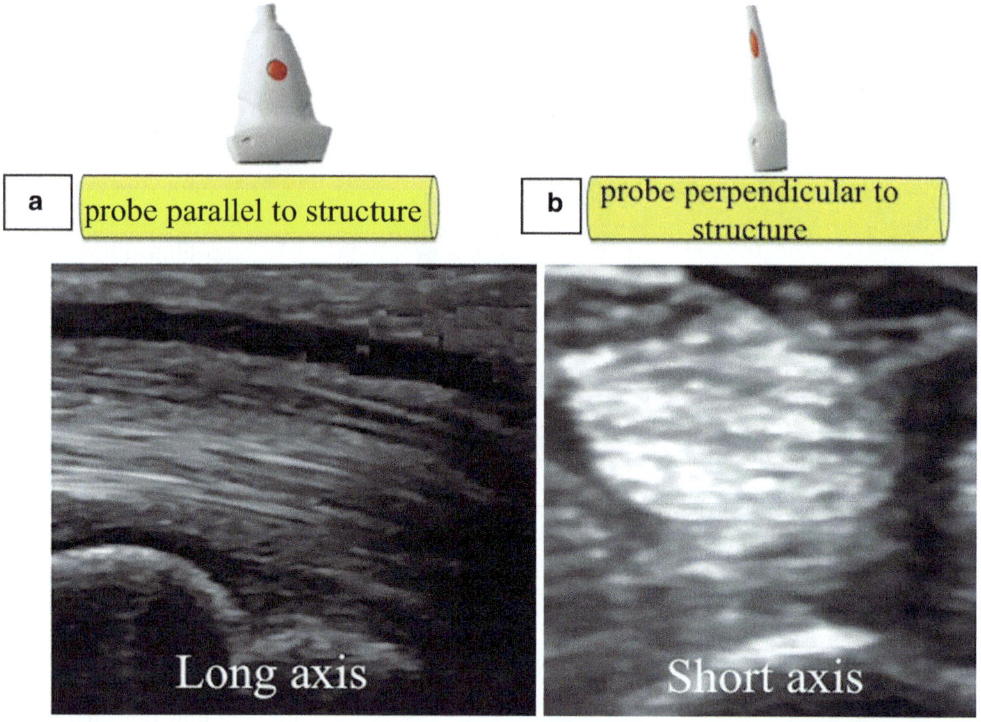


**Fig. 6.5** Ultrasound images showing long axis (**a**) and short access (**b**) views

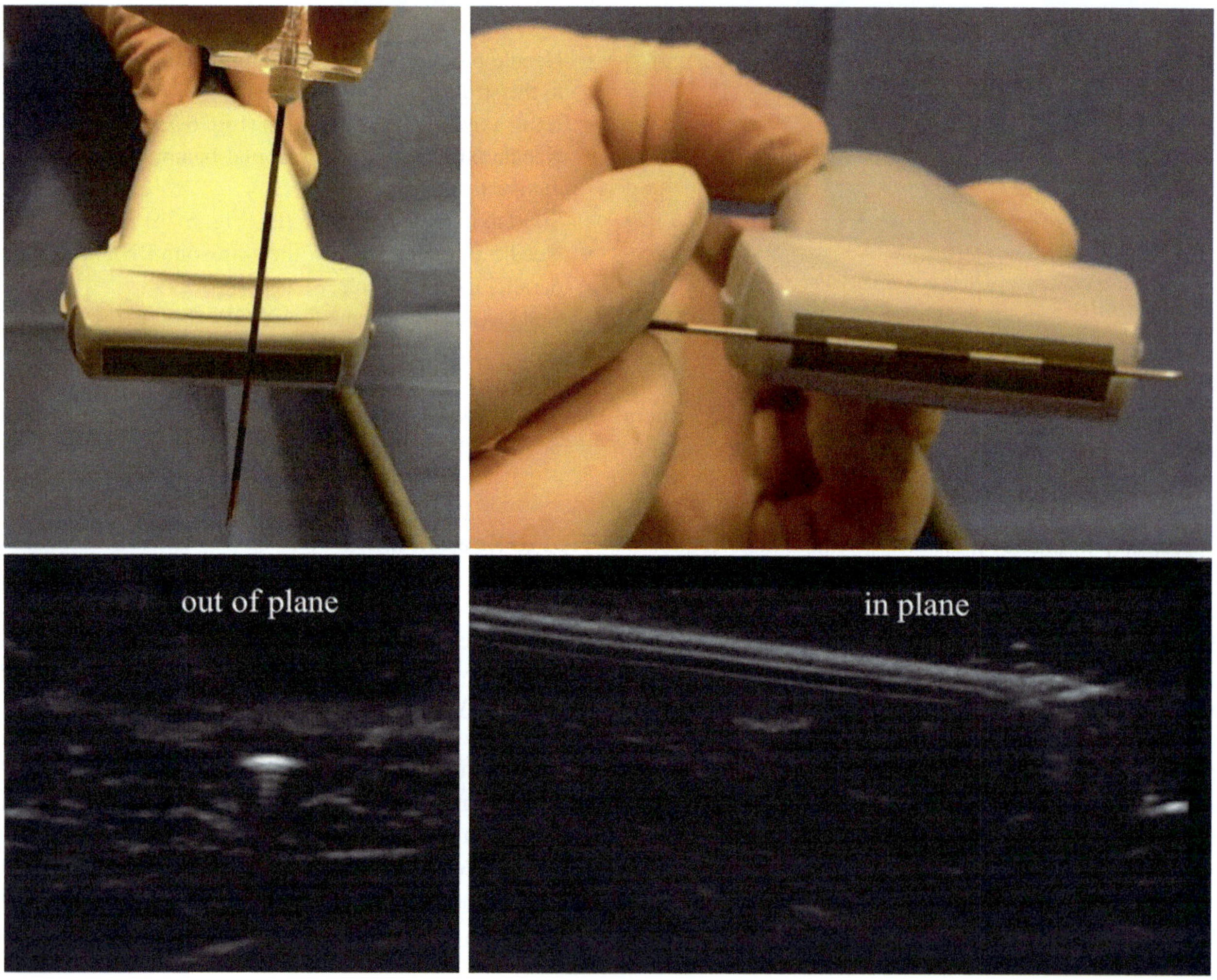


**Fig. 6.6** Relationship of a needle to the ultrasound probe. (Image courtesy of Thiago Nouer, MD, with permission)

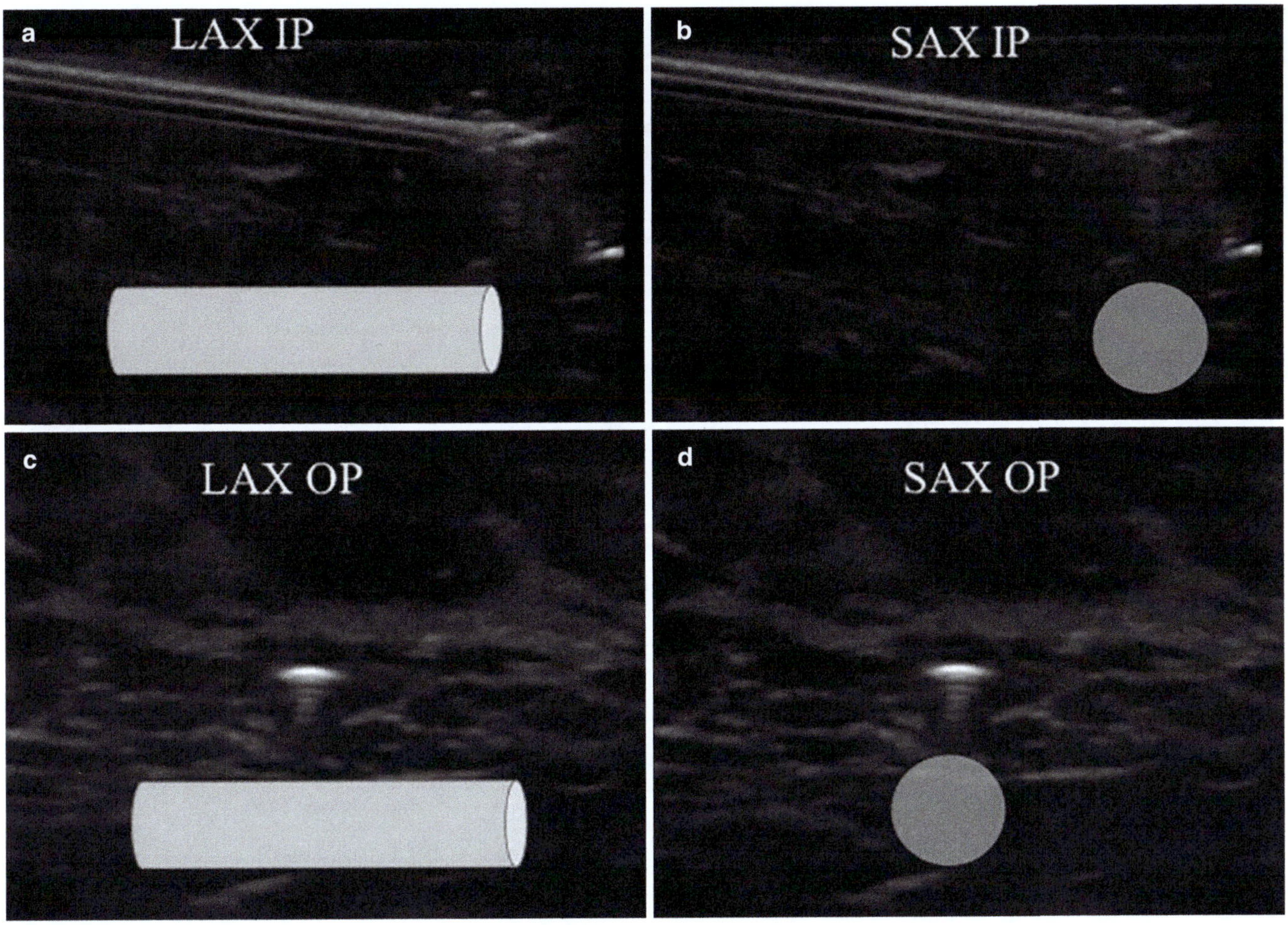


**Fig. 6.7** Relationship between needle and target. (**a**) Needle parallel to ultrasound beam, parallel to target (LAX IP). (**b**) Needle parallel to ultrasound beam, perpendicular to target (SAX IP). (**c**) Needle perpendicular to ultrasound beam, parallel to target (LAX OP). (**d**) Needle perpendicular to ultrasound beam, perpendicular to target (SAX OP). (Image courtesy of Thiago Nouer, MD, with permission)

- Be prepared to change the frequency of the probe you are using if you are not able to image the structure adequately and to assess both superficial and deeper structures.
- The size of the patient can limit the visualization of the normal anatomical landmarks. Also, be aware that some patients seem to be more "ultrasound photogenic" than others.
- Be sure to document the normal anatomy and any pathology in both planes, including measurements and any vascularity.

Ultrasound in regional anesthesiology has become virtually part of standard of care. However, ultrasound in chronic pain treatment is still in its infancy (or "early teens"), with still relatively poor image quality, so that fluoroscopy is still the gold standard in pain treatment. There continues to be advances in the use of ultrasound, and there are several specific areas of chronic pain treatment where ultrasound is the imaging modality of choice.

## Ultrasound Imaging

### Ultrasound of Stellate Ganglion

The stellate ganglion is located on each side of the neck and is formed by the fusion of the inferior cervical ganglion and first thoracic ganglion. Blockade of the stellate ganglion is the most common sympathetic ganglion injection utilized by pain physicians for pain arising from the head, neck and upper limbs.

Ultrasound of the stellate was first described in 1995 [13], which allowed visualization of all the relevant anatomic structures of the middle cervical ganglion region, with the probe placed between the carotid and trachea at C6 or C7. However, the more recent technique involves a lateral in plane approach (Fig. 6.8). Ghai et al. [14] studied 20 patients; all the patients had Horner's syndromes after injections.

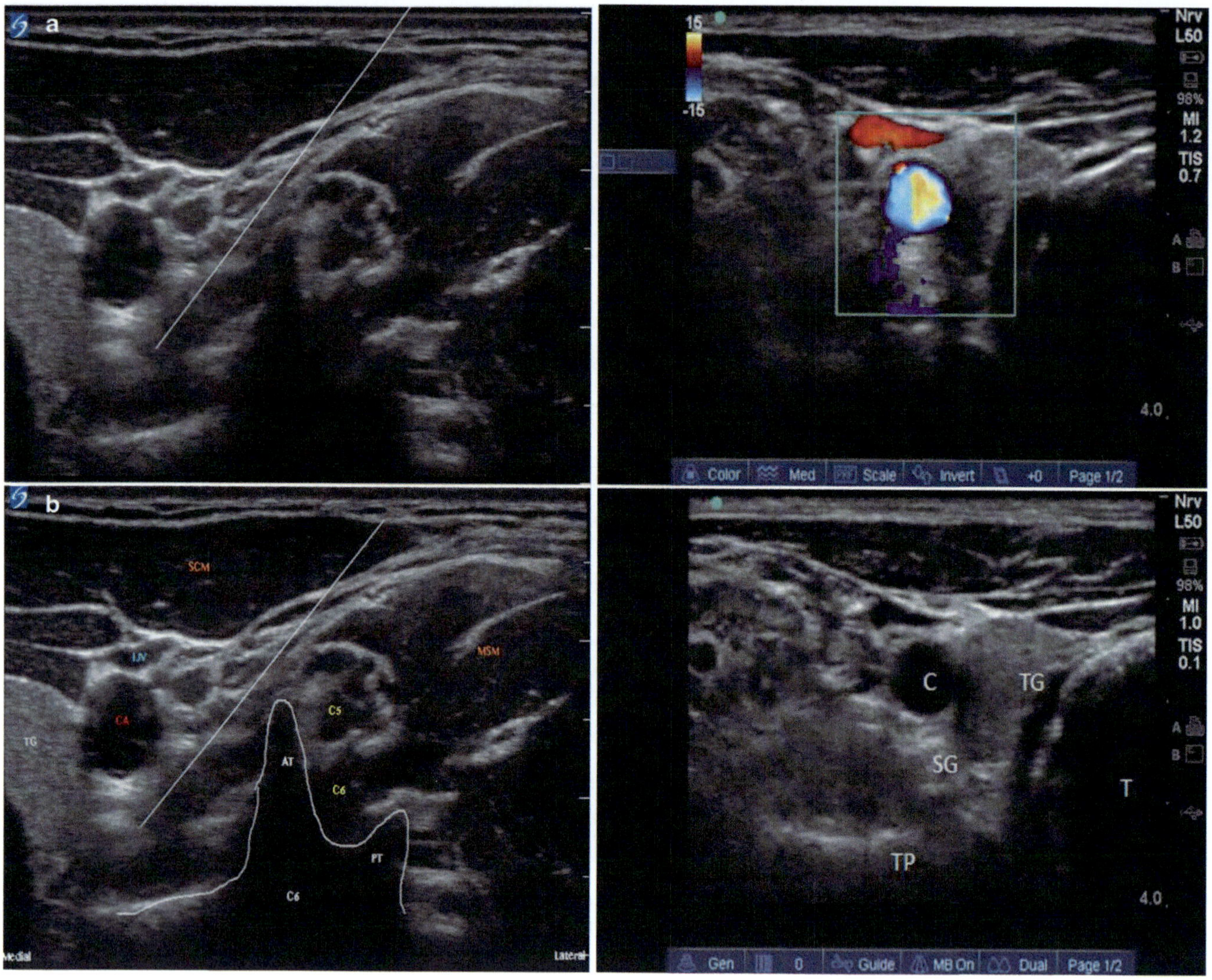


**Fig. 6.8** Stellate ultrasound simulated injection. (**a**) unlabeled image; (**b**) labeled image. *AT* anterior tubercle of C6, *PT* posterior tubercle of C6, *MSM* middle scalene muscle, *SCM* sternocleidomastoid muscle, *CA* carotid artery, *IJV* internal jugular vein, *TG* thyroid gland. **Note:** {C} carotid artery, {TG} thyroid gland, {SG}, stellate ganglion, {TP} transverse process, {T} trachea. (Image courtesy of Agnes Stogicza & Tolu Alugo, MD, with permission)

## Ultrasound of Sacroiliac Joint

The sacroiliac joint is a diarthrodial joint formed by the articular surfaces of the ilium and sacrum. The joint is innervated by the lateral branches of the first through third sacral roots, with varying contribution from the fourth sacral root and fifth lumbar dorsal ramus.

The sacroiliac joint is also easily visualized. The patient is placed prone, and the curvilinear probe is placed transversely at the level of the sacrum to intensify the sacral cornua. The probe is then moved laterally until the lateral edge of the sacrum is identified, and the probe is then moved cephalad to identify the ilium. The cleft between the two structures represents the sacroiliac joint; the needle is then advanced from medially to laterally into the joint. Pekkafahil et al. [15] looked at 60 sacroiliac joint injections under ultrasound guidance; although the early intentions were only 60% successful, the last 30 patients were 93% successful when confirmed by fluoroscopy. The mean procedure time was 9 min.

## Ultrasound for Musculoskeletal Imaging

Ultrasound has revolutionized the approach to joint and muscle evaluation. By visualizing the structure and real-time movement of muscles, joints, and tendons, a specific diagnosis can be made [16], and ultrasound can direct the injection of steroids and regenerative medicine in and around the specific joint [17].

## Shoulder

The shoulder is a synovial joint of the ball and socket type. It's important to note that a diseased shoulder could have multiple potential pain generators making treatment complicated.

The complicated shoulder joint is well suited for evaluation by ultrasound; these superficial structures are oriented in multiple directions, and the real-time movement evaluation has led to functional evaluations not visualized by other modalities (Fig. 6.9). Friedman et al. [18] retrospectively reviewed almost 1000 charts with musculoskeletal shoulder evaluations; 63.2% had a post-ultrasound treatment plan that was different than the pre-ultrasound treatment plan.

## Elbow

Although the elbow structure is superficial, the irregular surface makes good contact of the probe with the skin difficult, requiring smaller probes or extra gel to maintain contact (Fig. 6.10). However, dynamic evaluation of the stability of the elbow collateral ligaments provides unique information. Kim et al. [19] retrospectively evaluated 41 baseball players who had already undergone at least 6 weeks of rehabilitation without relief. They compared the ultrasound evaluation of instability of the ulnar collateral ligament to the MRI evaluation to simple palpation and looked at the subsequent need for surgery. They concluded that the addition of ultrasound was helpful in predicting the rehabilitation outcome.

## Hip

The hip joint can be evaluated and treated by ultrasound guided injection, especially to evaluate dynamic conditions such as snapping hip syndrome [17]. The structures around the hip joint can also cause pain. For example, Chen et al. [20] described ultrasound injection of the obturator internus tendon to treat a persistent gluteal pain, and Kim et al. [21] described the use of ultrasound to direct a cooled RF probe to lesion the articular branch of the femoral nerve to treat chronic post-arthroplasty hip pain.

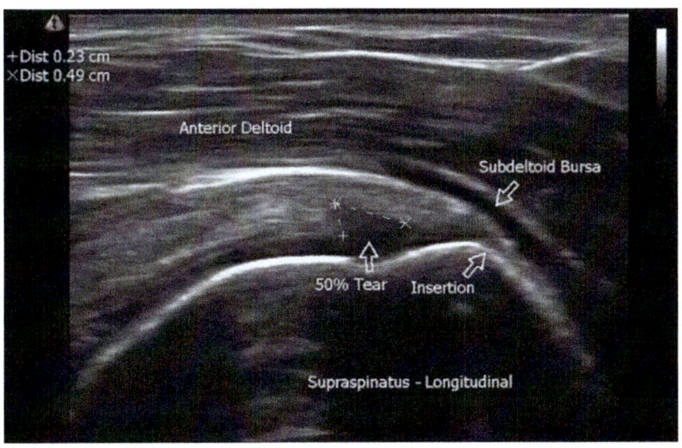


**Fig. 6.9** Shoulder ultrasound evaluation showing supraspinatus partial tear. (Image courtesy of Mark Koester, Terason with permission)

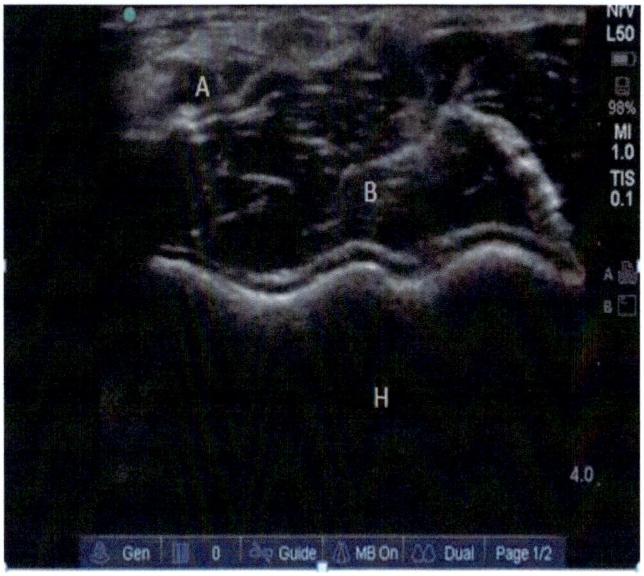


**Fig. 6.10** The elbow joint crease. *A* Brachial artery, *B* Brachialis muscle, *H* Humerus. (Image courtesy of Tolu Alugo, MD)

## Knee

Ultrasound of the knee is useful for evaluating the quadriceps and patellar tendons, the medial and lateral collateral ligaments, joint effusions, the distal hamstrings tendons, the iliotibial tract, the common peroneal nerve, the popliteal vessels, and Baker cysts (Fig. 6.11) [22].

## Ankle

The ankle is prone to injury, and ultrasound can be used to evaluate ligamentous injuries, small fractures, and vascular as well as nerve pathologies (Fig. 6.12). Hsu et al. [23] evaluated 23 patients with ankle trauma in whom ultrasound showed a fracture; all were subsequently confirmed by bone scan, even though several had negative X-rays.

## Peripheral Nerves

Peripheral nerves rarely have clear bony landmarks. They are often difficult to identify by fluoroscopy or CT. Blind injections are often performed with large volumes to overcome the lack of precision. Ultrasound, on the other hand, allows more precise injections, which should improve outcomes. In addition, pathology of the nerves, such as scarring and compression, can be visualized (Fig. 6.13). Cryoneuroablation (freezing the nerve) can be specifically useful for some of these peripheral nerves [24].

## Intercostal Nerves

Intercostal nerves are rarely directly seen with ultrasound because they sit close to or are covered by the caudal edge of

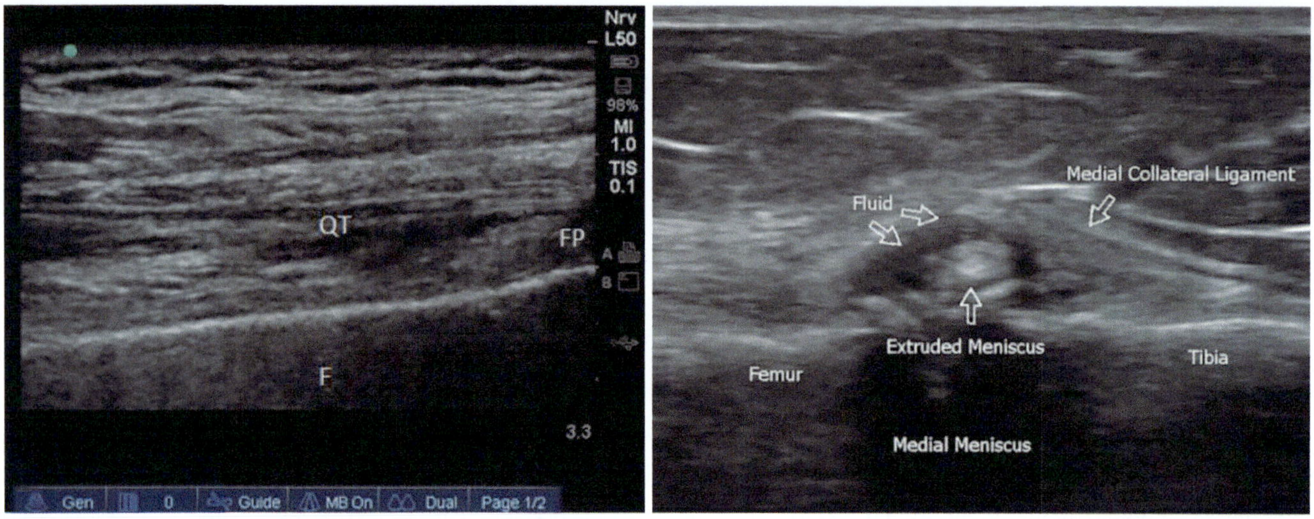


**Fig. 6.11** Knee ultrasound evaluation showing an extruded meniscus. {QT} = quadriceps tendons; {FP} = fat pad; {F} = femur. (Image courtesy of Mark Koeste & Tolu Alugo, MD, Terason with permission)

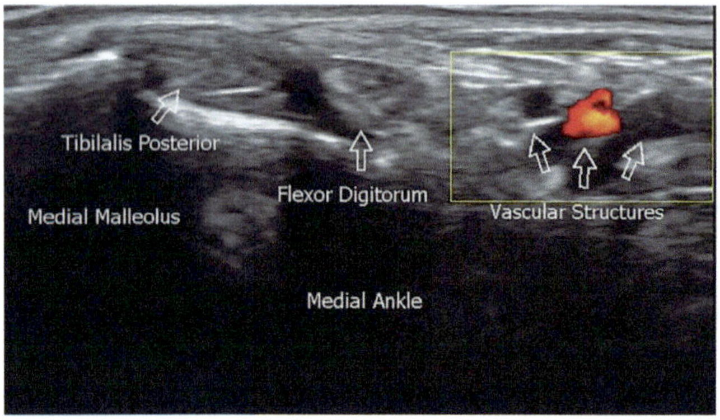


**Fig. 6.12** Ankle ultrasound evaluation of the tarsal tunnel. (Image courtesy of Mark Koester, Terason with permission)

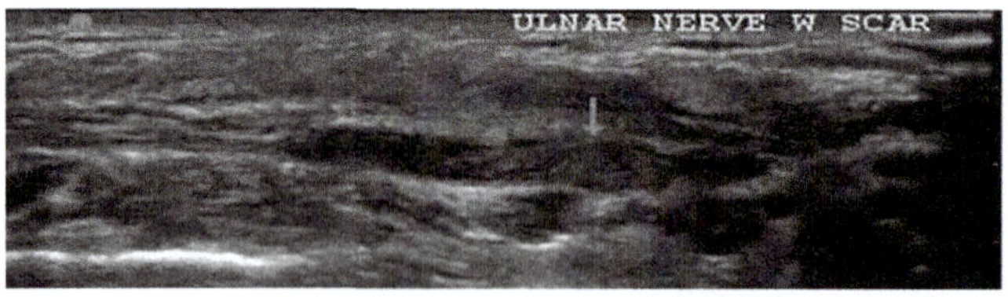


**Fig. 6.13** Image of the ulnar nerve showing a narrowing of the nerve, consistent with scarring. (Image courtesy of Michael N. Brown, MD with permission)

the rib. The needles are thin and relatively difficult to visualize, though the spread of the injectate is easily seen (Fig. 6.14). Cryoneuroablation, because of the size of the probe, can be visualized very easily using the linear probe (Fig. 6.15).

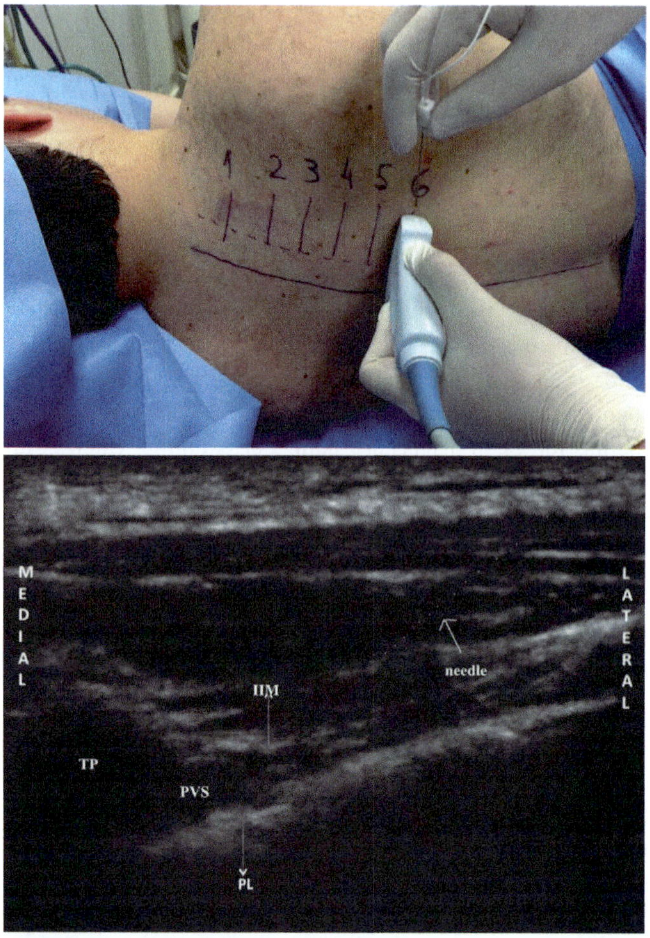


**Fig. 6.14** Ultrasound injection of intercostal space, from an in-plane approach. *PL* pleural space, *TP* transverse process, *PVS* paravertebral space, *IIM* internal intercostal membrane. (Images courtesy of Thialgo Nouer Frederico, MD, with permission)

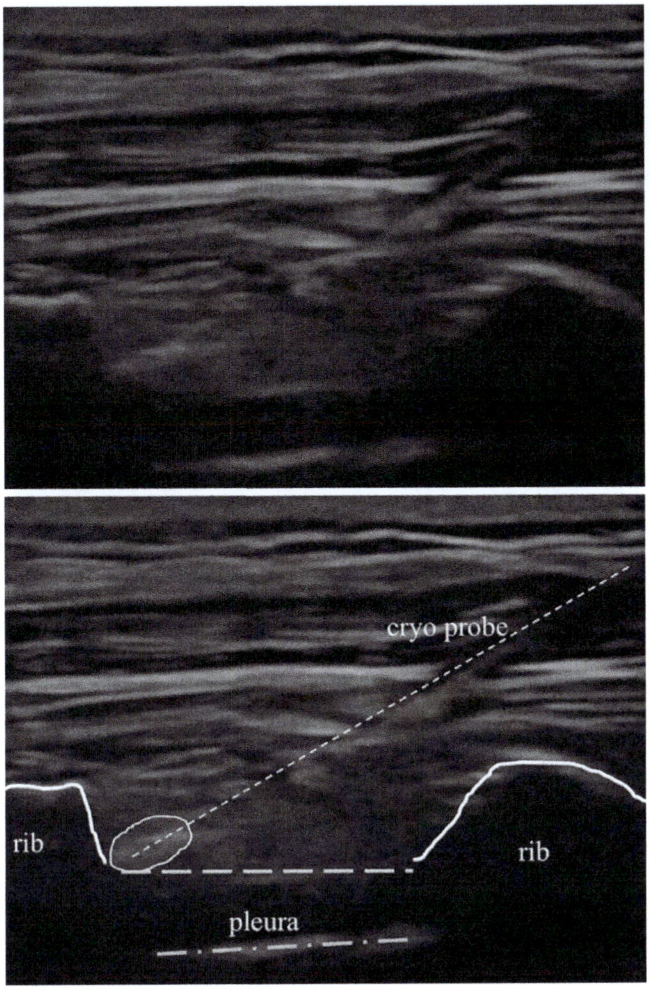


**Fig. 6.15** Cryoneuroablation probe positioned on the intercostal nerve under ultrasound direction. (Image courtesy of Christ Declerck, MD, with permission)

## Iliohypogastric/Ilioinguinal Nerves

Entrapment of the ilioinguinal and iliohypogastric nerves cause groin and abdominal pain, called the Border-nerve syndrome. The nerves are most often entrapped at the rectus border, though the easiest place to find the nerves is at the anterior superior iliac spine (ASIS) (Fig. 6.16). One technique of injection involves positioning the patient supine, with the lateral end of the linear probe placed on the ASIS, perpendicular to the inguinal line. The probe is tilted until all three layers of the abdominal wall are visualized (transverse abdominus, internal oblique, external oblique). The nerves lie between the transverse abdominus and the internal oblique, and the nerve can be approached from an in plane or out of plane approach, and cryoneuroablation can be effective. Eichenberger et al. [25] evaluated 37 ilioinguinal/iliohypogastric injections of dye performed in cadavers under ultrasound guidance; anatomic dissection after those injections showed correct placement in 95%.

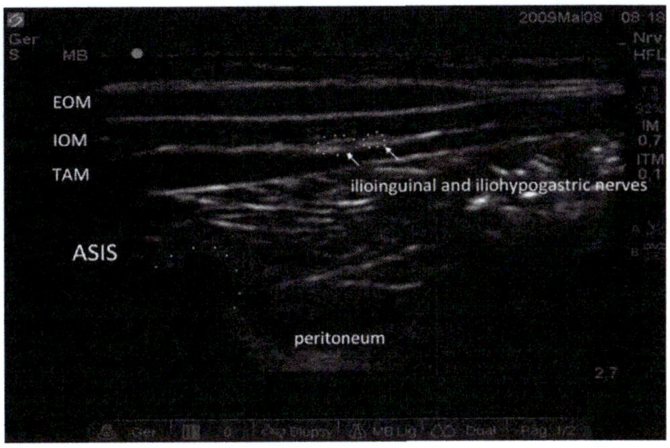


**Fig. 6.16** Ultrasound image of the ilioinguinal and iliohypogastric nerves at the iliac crest. *ASIS* anterior superior iliac spine, *EOM* external oblique muscle, *IOM* internal oblique muscle, *TAM* transverse abdominus muscle. (Image courtesy of Thiago Nouer Frederico, MD, modified by Charles de Oliveira, MD with permission)

## Occipital Nerves

The occipital nerves are implicated in the pathophysiology of different types of chronic headaches. Patients may present with this includes primary headaches and secondary headache disorders. Ultrasound guided targeting of the aforementioned nerves gives some symptomatic good pain relief and amelioration of the pain experienced.

The greater occipital nerve (GON) is a branch of the dorsal ramus of the C2 spinal nerve, and the lesser occipital nerve (LON) arises from the superficial cervical plexus, ventral rami of C2 and C3.

There are several approaches to the occipital nerves (greater, lesser, and third occipital nerves), a distal and proximal approach has been described. Crisp et al. [26] described a distal injection, placing the linear transducer on the occipital tubercle at the nuchal ridge in the sagittal plane. A 25-gauge needle is advanced from a medial in plane approach to the nerve; all the patients had dense anesthesia after 1–2 mL of local anesthetic injection.

There is also a more proximal entrapment at the level of C2 and the inferior oblique muscle, which, because of the proximity of the vertebral artery, needs to be performed under ultrasound [27]. If there is only temporary relief, cryoneuroablation can be given relief (Fig. 6.17) [28].

## Pudendal Nerve

The pudendal nerve has an extremely complicated path, from its origin at S1, S2, and S3, over the sacrospinous ligament, under the sacrotuberous ligament, over the obturator internus muscle, and then over the medial aspect of the pubic ramus. It can be addressed with ultrasound at the level of the sacrospinous ligament (Fig. 6.18), as well as Alcock's canal [29].

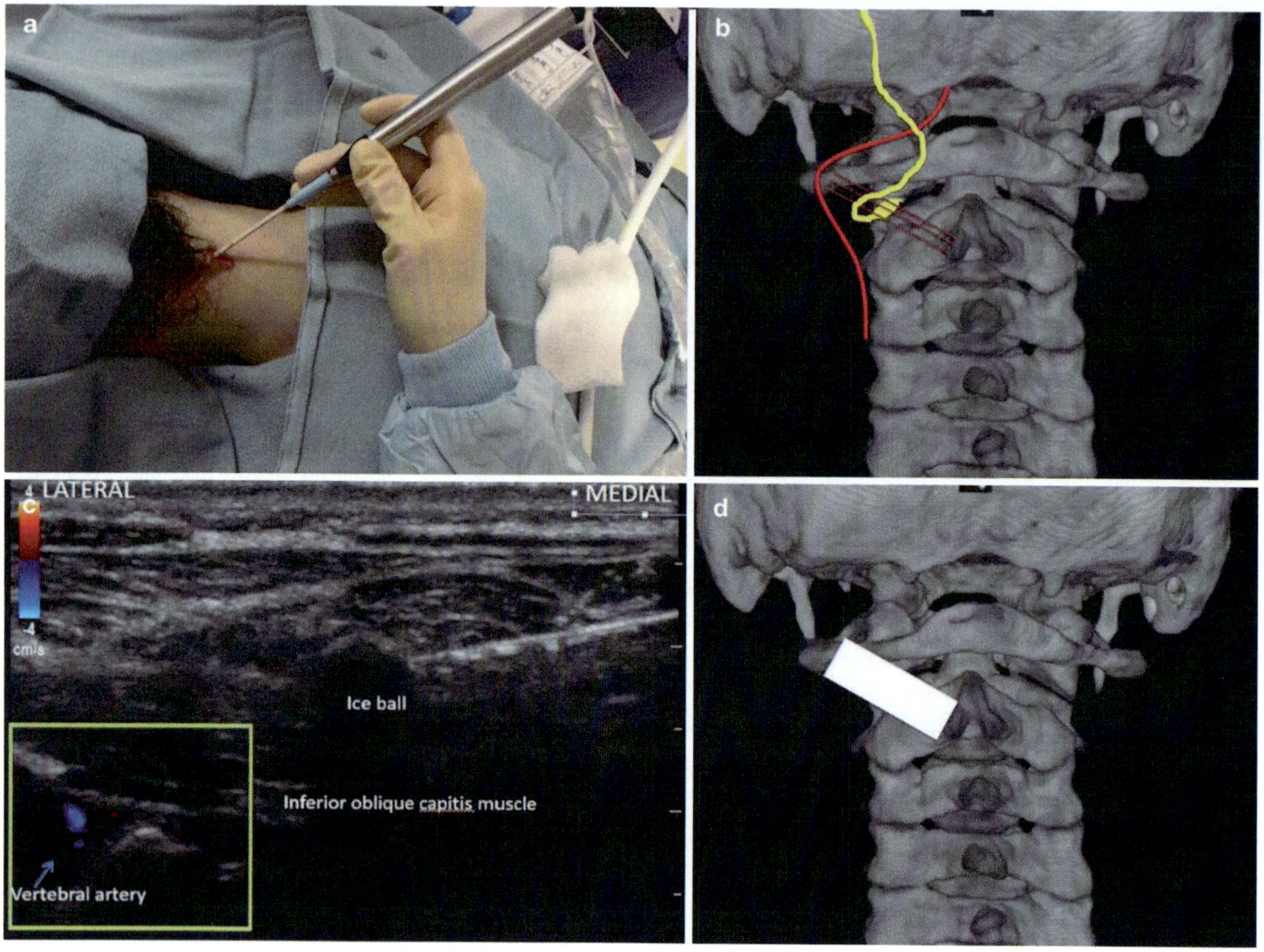


**Fig. 6.17** Cryoneuroablation of the proximal occipital nerve under ultrasound. (**a**) Cryoneuroablation probe placement; (**b**) Anatomy of the greater occipital nerve; (**c**) ultrasound image of the cryoneuroablation probe on the greater occipital nerve; (**d**) location of the ultrasound probe. (Image courtesy of Agnes Stogicza, MD, and Andrea Trescot, MD with permission)

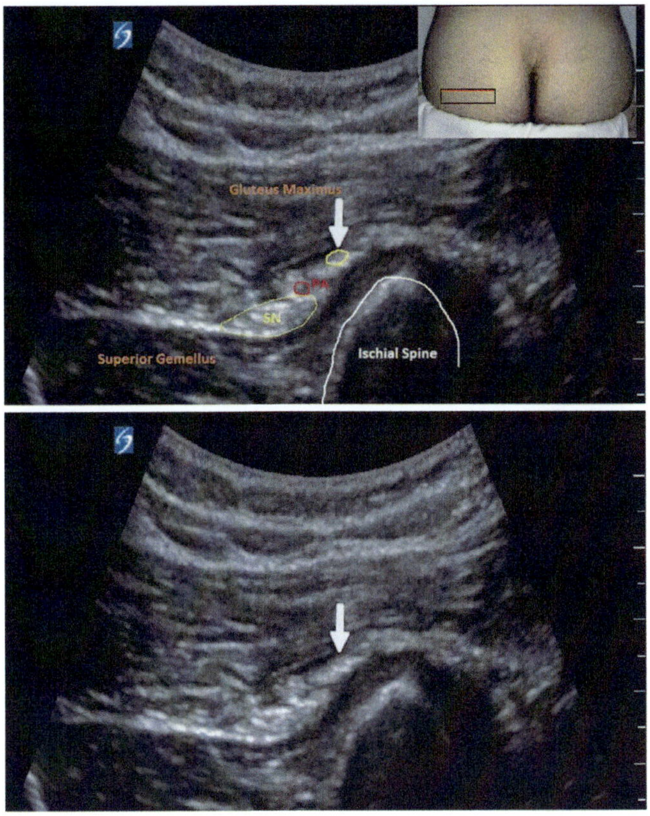


**Fig. 6.18** Ultrasound image of the pudendal nerve. *SN* sciatic nerve, *PA* pudendal artery, *white arrow* pudendal nerve. (Image courtesy of Agnes Stogicza, MD, with permission)

## Conclusion

Imaging with ultrasound utilizes high frequency sound waves transmitted by a transducer, sending sound waves through the body. As a transmitted ultrasound pulse encounters a structure in the human body, a portion of its energy is reflected and collected by the transducer. The ultrasound echo's time of flight enables calculation by computer algorithms that determine the structure's depth. Image formation uses the pulse echo's parameters, including the amplitude of the echo and the structure's spatial position, which are displayed as gray scale images, with brighter echoes having higher amplitudes. In addition, the sound wave frequency shifts identify differences between the transmitted and received echo, i.e., Doppler effect. The Doppler spectrum information is used to encode flowing structures such as blood, and, by convention, with color flow ultrasound imaging, blue represents motion towards the transducer and red represents flow away from the transducer. Ultrasound use in regenerative medicine is based on the ability to identify soft tissue anatomical structures, real time images enabling dynamic evaluation and needle placement, relative patient safety compared to X-rays, portability with the newer smaller laptop-sized machines, and cost effectiveness.

Ultrasound use by pain physicians is expanding, but it continues to evolve, with new techniques and new indications. Portability, reduced radiation, real time imaging, and identification of vascular structures are some of the many benefits from ultrasonography. The use of ultrasonography is important to adapt for patient pain management by physicians across the globe.

**Key Points**

1. Ultrasound is an imaging modality based on the information acquired from the propagation of a high frequency sound waves this in turn helps to characterize differences in tissues.
2. Accurate visualization of structures including ligaments, tendons, muscle fibers and the neurovascular bundle have become target of therapy.
3. Ultrasound is readily available at patients bedside and less expensive than CT or MRI.
4. Ultrasound provides dynamic imaging with the ability to image structures in range of motion and flowing fluids.
5. The use of ultrasound in regenerative medicine is expanding with the added advantage of identification of soft tissue abnormalities, real time imaging and precise needle placement for therapeutic purposes.

**Questions**

1. Diagnostic ultrasound uses frequencies in the range of?
   A. 5–10 MHz
   B. 3–5 MHz
   C. 5–30 MHz
   D. 2–5 MHz
   **Answer: C**
   **Explanation**: 5–30 MHz. Frequency is the number of complete oscillations of cycles per second. Diagnostic ultrasound uses frequencies in the range of 5–30 MHz. The ultrasound frequency that is used in medical imaging is in the megahertz (MHz) range, which is far above 20 KHz (the upper limit of human hearing).
2. A 36-year-old woman presents with symptoms and signs of complex regional pain syndrome of her left upper limb following an open reduction and internal fixation of a distal radial bone fracture. The hand and forearm were swollen, and computerized tomography scan confirms oedema and osteogenic changes in the bony architecture. For pain control you planned a stellate ganglion block. What position do you place the ultrasound probe?
   A. At the level of C2-C3
   B. C4-C5 Cervical spine level
   C. Placed between the carotid and trachea at C6-C7
   D. Supraclavicular region

**Answer: C**

**Explanation**: The stellate ganglion block is performed to interrupt the sympathetic fibers supplying the head, neck, and upper limbs. Blockade of the stellate ganglion help differentiate sympathetically maintained pain (SMP) from sympathetically independent pain (SIP). The ganglion is located at the level of the Sixth Cervical Vertebra (C6). The ultrasound probe is placed between the carotid and trachea at C6 or C7 where the inferior Cervico-thoracic part of the ganglion is located.

3. A 72-year-old man presented with left sided lower abdominal pain. He had a sigmoidectomy performed for an infiltrating intentional polyp. The pain he described as stabbing over the surgical scar. He has failed trial of topiceuticals and oral pharmacotherapy. To perform an ultrasound diagnostic transverses abdominus plane block. What level does the nerve lie?
   A. Between the External oblique and internal oblique
   B. Between the layers of the Rectus abdominus muscle
   C. Between the transverse abdominus and internal oblique
   D. Between the lower intercostal muscles

   **Answer: C**

   **Explanation**: Anterior Cutaneous nerve entrapment syndrome (ACNES) is one of the causes of abdominal pain. This syndrome occurs when there is entrapment of cutaneous nerve within the abdominal wall muscle. The patient presents with sharp, stabbing and burning pain. The nerve lies between the transverse abdominus and the internal oblique, and the nerve can be approached from an in plane or out of plane approach.

4. The pudendal nerve originates from:
   A. L5–S2
   B. S1, S2, S3
   C. L4–L5
   D. L1–L3

   **Answer: B**

   **Explanation**: The pudendal nerve originates at S1, S2 and S3 over the sacrospinous ligament under the Sacro tuberous ligament over the obturator internus muscle and then over the medial aspect of the pubic ramus.

5. The structures located within the Guyon's canal includes:
   A. Redial artery flexor digitorum tendons and median nerve
   B. Extensor wrist tendons and radial nerve
   C. Ulnar artery, Vein, and nerve
   D. Brachial artery and Median nerve

   **Answer: C**

   **Explanation**: The ulnar artery, vein, and nerve are located within Guyon's canal. The Guyon's canal or ulnar canal, is an anatomical fibro-osseus canal found on the medial border of the hand. The boundaries are the pisiform bone and the hook of the hamate.

6. A 65-year of woman with Rheumatoid arthritis diagnosed 6 years previously presents with ongoing pain involving her right hand. She describes paresthesia and numbness overlying the Thenar eminence. The pain is worse during sleeping hours and she has been woken up several times with the pain. A trial of median nerve block with local anesthesia and triamcinolone gave her a temporary relief. During one of her ultrasound examination the sonologist identified a bifid appearance of the median Nerve. What percentage of people have the bifid median nerve variation?
   A. 4%–5%
   B. 10%–15%
   C. 40%–55%
   D. 50%–60%

   **Answer: B**

   **Explanation**: 1015% of the population are found to have a bifid Median nerve. This is usually associated with a persistent median artery between the two nerve trunks. Most patients are asymptomatic with this finding. It is also essential to rule out other abnormalities such as hamartomas, neuromas, schwannomas, lipofibromas and ganglion cysts among others.

7. Appropriate statements regarding the use of ultrasound-guided injections include the following except:
   A. Ultrasound is readily available and more expensive than CT and MRI
   B. No radiation exposure to the patient or provider, and therefore no need for shielding
   C. Sound waves may be deleterious to tissue DNA
   D. Imaging requires no skill or experience of the operator

   **Answer: B**

   **Explanation**: No radiation exposure to patient or provider an essential principle of ultrasonography is a unique imaging method in that sound waves are used rather than ionizing radiation for image production. The piezoelectric effect of the ultrasound crystals allows electrical signal to be changed to ultrasonic energy and hence image production.

8. A 54-year-old left hand dominant male presented at your pain clinic with progressively worsening pain over his left elbow. The pain is a dull continuous ache with limits his activities and he says the pain is worse on the outside of his elbow. He works at an engineering firm and his jobs involves extensive manual labour using his hands. Examination findings were essentially normal aside from pain on supination and wrist extension. Neurovascular examination was normal. What is the most likely diagnosis in this patient?
   A. Cervical radiculopathy
   B. Radial Nerve entrapment syndrome
   C. Lateral Epicondylitis
   D. Medial epicondylitis

**Answer: C**

**Explanation:** Lateral epicondylitis known as "tennis elbow is a common condition affecting the elbow. It results from excessive overuse and frequent wrist extension. The pain experienced by the patient results from motor impairment and alteration in both local tendon pathology and pain perception. Microscopic findings of lateral epicondylitis include collagen disarray and degeneration. Its responds to NSAID, rest and intralesional injection with corticosteroid or platelet rich plasma.

9. A 28-year-old male presented to the emergency department with pain over the posterior scalp more on the right side for over 3 years. He describes the pain as sharp and radiates to the left mastoid region. He has failed trial of pharmacotherapies. To perform a proximal injection of the greater occipital nerve at the level of the second cervical vertebrae, what artery should the interventionist be careful of?
   A. Carotid artery
   B. Inferior thyroid artery
   C. Basilar artery
   D. Vertebral artery

   **Answer: D**

   **Explanation:** There are several approaches to the occipital nerves (greater, lesser and third occipital nerves). The entrapment of the greater occipital nerve could be distal or proximal. The more proximal entrapment is at the level of the C2 and inferior oblique muscle, which, because of the proximity of the vertebral artery, needs to be performed under ultrasound guidance.

10. The Sacroiliac joint is best visualized with which ultrasound probe?
    A. Linear probe
    B. Curvilinear probe
    C. Cardiac probe
    D. Focused array probe

    **Answer: B**

    **Explanation**: The sacroiliac joint can be easily visualized, the patient is positioned prone and a curvilinear probe is placed transversely at the level of the sacrum. The probe is moved laterally until the lateral edge of the sacrum is identified. The cleft between the ilium and sacrum represents the sacroiliac joint; the needle is then advanced from medial to lateral into the joint.

**Acknowledgments** This book chapter is modified and updated from a previous book chapter, "Principles of Ultrasound Imaging in Musculoskeletal Medicine" by Andrea Trescot, MD, and Art S. Watanabe, MD, in *Essentials of Regenerative Medicine in Interventional Pain Management* published by ASIPP Publishing.

## References

1. Trescot A, Watanabe AS. Principles of ultrasound imaging in musculoskeletal medicine. In: Manchikanti L, Navani A, Atluri S, editors. Essentials of regenerative medicine in interventional pain management. Paducah, KY: ASIPP Publishing; 2019. p. 155–72.
2. Hangiandreou NJ. AAPM/RSNA physics tutorial for residents. Topics in US: B-mode US: basic concepts and new technology. Radiographics. 2003;23:1019–33.
3. Kaproth-Joslin KA, Nicola R, Dogra VS. The history of US: from bats and boats to the bedside and beyond: RSNA centennial article. Radiographics. 2015;35:960–70.
4. Spawls P. The physical principles of medical imaging. 2nd ed. Madison, WI: Medical Physics Publishing; 1995.
5. Kremkau FW. Sonography principles and instruments. 8th ed. St. Louis, MO: Saunders; 2011.
6. Ziskin MC. Fundamental physics of ultrasound and its propagation in tissue. Radiographics. 1993;13:705–9.
7. Goldstein A. Overview of the physics of US. Radiographics. 1993;13:701–4.
8. Heller LB. Knobology for image optimization. Seattle, WA: University of Washington Medical Center.
9. Sites BD, Brull R, Chan VW, et al. Artifacts and pitfall errors associated with ultrasound-guided regional anesthesia. Part I: understanding the basic principles of ultrasound physics and machine operations. Reg Anesth Pain Med. 2007;32:412–8.
10. Sites BD, Brull R, Chan VW, et al. Artifacts and pitfall errors associated with ultrasound-guided regional anesthesia. Part II: a pictorial approach to understanding and avoidance. Reg Anesth Pain Med. 2007;32:419–33.
11. Feldman MK, Katyal S, Blackwood MS. US artifacts. Radiographics. 2009;29:1179–89.
12. Xu D, Meirsman SD, Schreurs R. Optimizing an ultrasound image: *NYSORA*. https://www.nysora.com/topics/equipment/optimizing-ultrasound-image/. Accessed 16 Feb 2023.
13. Kapral S, Krafft P, Gosch M, et al. Ultrasound imaging for stellate ganglion block: direct visualization of puncture site and local anesthetic spread. A pilot study. Reg Anesth. 1995;20:323–8.
14. Ghai A, Kaushik T, Kundu ZS, et al. Evaluation of new approach to ultrasound guided stellate ganglion block. Saudi J Anaesth. 2016;10:161–7.
15. Pekkafahli MZ, Kiralp MZ, Basekim CC, et al. Sacroiliac joint injections performed with sonographic guidance. J Ultrasound Med. 2003;22:553–9.
16. Chen CP, Tang SF, Hsu CC, et al. A novel approach to sonographic examination in a patient with a calf muscle tear: a case report. J Med Case Reports. 2009;3:7291.
17. Zbojniewicz AM. US for diagnosis of musculoskeletal conditions in the young athlete: emphasis on dynamic assessment. Radiographics. 2014;34:1145–62.
18. Friedman MV, Hillen TJ, Holland DV, et al. Impact of shoulder sonography on clinical decision making. J Ultrasound Med. 2017;36:1365–71.
19. Kim NR, Moon SG, Park JY, et al. Stress ultrasound in baseball players with ulnar collateral ligament injuries: additional value for predicting rehabilitation outcome. J Shoulder Elbow Surg. 2017;26:815–23.
20. Chen B, Rispoli L, Stitik T, et al. Successful treatment of gluteal pain from obturator internus tendinitis and bursitis with ultrasound-guided injection. Am J Phys Med Rehabil. 2017;96:e181–4.
21. Kim DJ, Shen S, Hanna GM. Ultrasound-guided radiofrequency lesioning of the articular branches of the femoral nerve for the treatment of chronic post-arthroplasty hip pain. Pain Physician. 2017;20:E323–7.

22. Alves TI, Girish G, Kalume Brigido M, et al. US of the knee: scanning techniques, pitfalls, and pathologic conditions. Radiographics. 2016;36:1759–75.
23. Hsu CC, Tsai WC, Chen CP, et al. Ultrasonographic examination for inversion ankle sprains associated with osseous injuries. Am J Phys Med Rehabil. 2006;85:785–92.
24. Trescot AM. Cryoanalgesia in interventional pain management. Pain Physician. 2003;6:345–60.
25. Eichenberger U, Greher M, Kirchmair L, et al. Ultrasound-guided blocks of the ilioinguinal and iliohypogastric nerve: accuracy of a selective new technique confirmed by anatomical dissection. Br J Anaesth. 2006;97:238–43.
26. Crisp VW, Weingarten TN, Smith J. Ultrasound guided greater occipital nerve block: description of a new technique. In: AAPM poster presentation, Orlando, FL; 2006.
27. Greher M, Moriggl B, Peng PW, et al. Ultrasound-guided approach for L5 dorsal ramus block and fluoroscopic evaluation in unpreselected cadavers. Reg Anesth Pain Med. 2015;40:713–7.
28. Stiogicza A, Trescot AM. New technique for cryoneuroablation of the proximal occipital nerve. New York, NY: World Institute of Pain World Congress Abstract; 2016.
29. Bendtsen TF, Parras T, Moriggl B, et al. Ultrasound-guided pudendal nerve block at the entrance of the pudendal (Alcock) canal: description of anatomy and clinical technique. Reg Anesth Pain Med. 2016;41:140–5.

# 7 Platelet-Rich Plasma

Glenn M. Flanagan, John Knab, Joshua Rothenberg, and Peter A. Everts

## Introduction

### Platelet Functions

Platelet production, or thrombocytogenesis, most commonly occurs in the bone marrow. There, hematopoietic stem cells undergo multiple rounds of endomitosis to form megakaryocytes. These megakaryocytes then produce membrane extensions termed "proplatelets," which undergo shear stress and release fragments termed "preplatelets." These preplatelets are finally processed in the circulation to become mature platelets [1]. Circulating platelets are small discoid anuclear blood cells with diameters around 2–4 μm circulating in a resting state where they require a trigger before activation in processes like hemostasis, wound healing, angiogenesis, and immune system responses. The average platelet count ranges from 1.5 to 3.0 × $10^5$/μL in circulating blood, and the in vivo half-life is about 7 days [2, 3].

### Hemostasis

Hemostasis is the most well-known role of platelets. It is commonly initiated when glycoprotein receptors on "resting" platelets contact damaged tissue, leading to a cascade of changes, including secretions in the form of adhesive proteins, growth factors, and cytokines from platelet-rich plasma (PRP) alpha, dense, lysosome and T-granules [4]. The glycoprotein receptors also form bridges between platelets and the surrounding protein network to form a platelet plug. Actin-myosin interactions then pull the clot into a more compact structure, aided by ligand sites found in the extracellular matrix (ECM) and the proteins released by platelets.

Numerous small and complex interactions between the platelets, surrounding proteins, ECM, and blood flow through the vessel coordinate platelet shape changes, clot formation, and then clot retraction [5]. The clotting process, if not accounted for during PRP processing, can alter the final total deliverable platelet counts and give inaccurate readings on hematology analyzers. Therefore, physicians need to carefully follow protocol to prevent this clotting process while preparing a PRP injectate for interventional orthopedic procedures.

### Tissue Healing

Tissue healing is a complex biophysiological process often described simply as having four distinct phases: hemostasis, inflammation, proliferation, and remodeling. These phases are highly integrated and overlapping [6]. The first 4–6 days after injury is described as an inflammatory response categorized by vasoconstriction and clot formation. A few days after an injury, growth factors released by platelets help direct a transition to local vasodilation, allowing neutrophils, then monocytes, and finally macrophages to the area.

Macrophages play key roles in transitioning to the proliferation phase, as they shift from a classically activated M1 inflammatory phenotype to the alternatively activated M2 anti-inflammatory phenotype [7]. For around 2 weeks, angiogenesis and fibroplasia lead proliferation. Fibroblasts are recruited to initiate collagen formation which leads to the maturation and remodeling phase. Tissue remodeling can

G. M. Flanagan (✉)
Naples Regenerative Institute, Naples, FL, USA

J. Knab
Center for Pain Management, Wilmington, NC, USA
e-mail: jknab@portcitypain.com

J. Rothenberg
Baptist Health Orthopedic Care, Sports Medicine, Boca Raton, FL, USA
e-mail: jrothenberg@baptisthealth.net

P. A. Everts
Gulf Coast Biologics, Research and Science Department, Fort Myers, FL, USA

OrthoRegen-Max Planck University, Indaiatuba, Sao Paulo, Brazil
e-mail: peter@gulfcoastbiologics.com


A. Navani et al. (eds.), *Essentials of Regenerative Medicine in Interventional Pain Management*, https://doi.org/10.1007/978-3-031-50357-3_7

continue out to a year and beyond while the ECM and tissue network is organized [8].

The precise timing of injections after the onset of tissue damage may have important implications on outcomes. Additionally, recovery times and postprocedure protocols may also play an equally significant role in allowing the tissue to appropriately move through the phases.

## The Evolution of PRP for Clinical Use

In attempts to mimic the initiation of these innate tissue healing processes previously described, orthobiologic research has focused on concentrated PRP (C-PRP) for decades. The idea that a supraphysiological dose of growth factors released from a concentration of platelets in plasma was the guiding principle. The understanding of the complex communication between growth factors, chemokines, and cytokines has grown considerably during the last 20 years. We will review the cellular biological processes in more detail later in this chapter.

## Describing Platelet Content in PRP

### Concentration Factor (How Concentrated Is the PRP Product?)

Marx et al. [9] published clinical data in 1998 showing that PRP mixed with autogenous bone grafts helped improve mandibular continuity defects. This initial definition of PRP was simply a preparation of concentrated platelets above the baseline in plasma. If 10 mL of whole autologous blood were drawn from the patient, and concentrated by centrifuge to 5 mL, that would be a "2×" product. Assuming all the platelets in the blood draw ended up in the final product, the platelet concentration could be estimated by doubling the platelet number from a pretreatment complete blood count (CBC).

In clinics, we do not see 100% platelet capture during the process of making PRP, so platelet capture should always be accounted for when estimating platelet content in PRP. Fundamental problems with describing PRP by this early definition is that it does not account for varying patient platelet counts in whole blood, differences in PRP product design, or variations in methods of centrifugation and preparation. Two separate PRP formulations described as "2×" could have tremendous differences in the number of platelets per milliliter due to differences in whole blood platelet counts between patients, even using the same PRP products, centrifugation parameters, and methods of preparation.

*Patient A: whole blood platelet count 150,000/μL, 2× = 300,000/μL (assume 100% platelet capture)*

*Patient B: whole blood platelet count 250,000/μL, 2× = 500,000/μL (assume 100% platelet capture)*

*As seen, these 2 PRP formulations have very different concentrations of platelets, but the concentration factor description of them both is 2×.*

### Absolute Platelet Count (How Many Platelets Are in the PRP Product?)

In 2009, Giusti et al. [10] showed that the optimal concentration of platelets in vitro for the stimulation of angiogenesis is $1.5 \times 10^6$ platelets/μL [10]. This highlights a transition from describing PRP in changes of concentration by volume reduction to measuring total deliverable platelet numbers—a "dose" of platelets in the final product based on the platelet concentration and volume of injectate. This platelet concentration level was adopted by some researchers and clinicians around the world as a target concentration for platelets in PRP, but unfortunately many publications continue to advise treating patients with PRP that has platelet contents below this number, and use concentration factor times baseline as the guiding calculation.

As an example, if 20 mL of blood is drawn and concentrated to 5 mL for the treatment, that is described as a "4×" treatment formulation, which makes no direct reference to the number of platelets and ignores the loss of platelets during preparation. If a physician wants to make sure the concentration described by Giusti et al. [10] is reached in order to optimally stimulate angiogenesis, the concentration factor alone will not provide enough information. Therefore, precisely calculated platelet concentrations and total deliverable platelet numbers serve as a more meaningful parameter when describing PRP used in treatments.

A simple mathematic estimation of platelets based on volumetric reduction is insufficient to accurately calculate platelet dose, which requires analysis of the final PRP preparation via a bedside hematology analyzer. Ideally, using a hematology analyzer will help provide a concentration of platelets which can be multiplied by the volume of injectate used for treatment, thereby precisely furnishing the physician with the total number of platelets delivered to the treated area.

## Treatment Volume

Another key factor to consider is the volume needed for the pathoanatomic area to be treated. If a physician wishes to inject 2 mL of PRP into a carpometacarpal joint and surrounding ligaments, 30 mL of blood might be drawn and concentrated to a final PRP volume of 2 mL, or a 15× platelet concentration compared to whole blood. If the physician plans to treat a knee joint with 7 mL volume of PRP, and draws the same 30 mL to start, the final PRP formulation is only about 4×.

Deciding on the treatment volume is a crucial step to be done before preparing PRP for clinical use, so enough whole blood can be drawn from the patient. In 2021, Bansal et al. [11] concluded that injecting knee joints with ten billion platelets in a volume of 8 mL provides significant potential chondro-protection and alleviates symptoms compared to a control in knee osteoarthritis. Preparing PRP in the office to meet this goal must consider absolute platelet numbers, i.e., platelet dose, instead of just platelet concentration. The most accurate way to calculate this is to have a hematology analyzer in the office, which allows physicians to rapidly test the concentration of platelets in the PRP. A less precise method involves calculating expected platelet counts in the final product based on recent lab work, using the concentration factor estimated by the amount of volume reduction, and accounting for platelet capture during the PRP preparation. Platelet capture rates are highly variable between products and preparation methods.

*If a treatment is designed to deliver at least ten billion platelets in 8 mL of PRP for a knee treatment, that means there should be about 1.25 billion (10 billion/8 mL) platelets in each mL of the PRP.*

*If the patient's platelet count on their lab work is 250,000 platelets/μL, that number can be used to estimate the volume of whole blood needed to arrive at the desired platelet dose for treatment in milliliters. A baseline value of 250,000/μL is converted to mL by multiplying × 1000 = 250,000,000/mL. This is the number of platelets in one mL of whole blood, prior to centrifugation.*

*This would need to be concentrated fivefold (5×) if there is 100% platelet capture to reach 1.25 billion/mL (250,000,000 × 5 = 1.25 billion). This would require 40 mL of whole blood (40 ÷ 5 = 8).*

*If platelet capture is 80%, the whole blood needed would be 48 mL (40 × 1.2 = 48).*

Notice the change from reporting whole blood platelet numbers in microliters to using milliliters when "dosing" the volume of PRP that will be used for treatment. This concept is important to understand when reviewing the literature and hematology analyzer reports in the office. In orthobiologic medicine, when calculating platelet dose, it is important to remember that platelet numbers on hematology reports are traditionally reported in number of platelets—in thousands—*per microliter (μL)*. In contrast, in orthobiologics, the platelet dose of a final volume of injectate is described in terms of platelets *per milliliter (mL)*.

*If a patient has a baseline platelet count of "150" this means that there are:*

*(150,000 platelets/μL) × (1000 μL/mL) = 150,000,000 platelets/mL.*

*If the final PRP product has a platelet count of "1250," this means that there are:*

*(1250 platelets/μL) × (1000 μL/mL) = 1250,000,000 platelets/mL.*

*If the final PRP volume delivered is 5 mL, then the total delivered platelet dose will be:*

*(1250,000,000 platelet/mL) × 5 mL = 6,250,000,000 platelets.*

## Platelet Capture

As described previously, it cannot be assumed that 100% of the platelets in the starting blood draw are captured and present in the final PRP preparation. One of the authors has used a protocol that typically captures about 80% of the starting platelets, which must be factored in when calculating the final dose of platelets.

*If the platelet value is 150,000 platelets/μL on recent lab work, and it is concentrated 6× from 60 mL whole blood to 10 mL PRP, the estimated platelet concentration would be 900,000/μL (150,000 × 6), and 85% capture would leave 765,000 platelets/μL (900,000 × .85) in the final PRP. As referenced above in Giusti's 2009 study* [10], *a goal to stimulate angiogenesis is a platelet count more than 1,500,000/μL, so this example would fall below that ideal range.*

*The concentration factor needs to be doubled to prepare the PRP with a platelet concentration of around 1,500,000/μL (2 × 765,000 = 1,530,000/μL). To do this with a 60 mL draw, you would prepare 5 mL of PRP instead of 10 mL. Alternatively, you could start with 120 mL to prepare 10 mL for treatment. Either the volume of blood drawn or the final PRP volume must be adjusted.*

*The concentration factor estimate needed in this scenario is about 12×, accounting for 85% platelet capture. If 3 mL of PRP is required for treatment, the method of preparation would require 36 mL whole blood concentrated to 3 mL. If 6 mL PRP is required, 72 mL of whole blood would be needed.*

## PRP Safety Profile

The use of PRP in clinical applications continues to have a good safety profile after decades of research. Since PRP is prepared from autologous blood, there are theoretically minimal risks for disease transmission and immune reactions [3]. Numerous reviews during the last two decades have concluded that PRP is safe for use and this trend has continued in recent years. A registry review of 470 patients who received platelet lysate epidurals showed no additional risk beyond traditional steroid epidurals [12]; these results were confirmed in a review of more than 1000 patients who received platelet-related treatments in the epidural space [13]. A review of more than 1900 patients treated with PRP for ligament and tendon pathology concluded that it is a safe treatment for these tissues [14]. A review comparing different PRP treatments used for knee osteoarthritis showed PRP was safe in more than 1500 patients [15].

Like other interventional procedures, PRP injections carry inherent risks related to the technical components of injections, like bleeding, infection, and tissue damage. After PRP injections, there is usually a painful "flare" for a few days of varying degrees based on the type of PRP used and the area treated. The American Society of Interventional Pain Physicians (ASIPP) guidelines published in 2019 for safe and effective use of biologic treatments for low back pain did not recognize any additional safety concerns related to adverse events caused by using PRP [16].

## Adding PRP to Clinical Practice

Musculoskeletal pain is a major public problem, especially with a rapidly increasing aging population [17]. Current spending on musculoskeletal disorders in the United States likely exceeds $183 billion dollars [16, 17]. PRP treatments can be safe alternatives that decrease pain and improve function more favorably than many traditional treatments [18–20].

A bibliometric analysis published in 2022 [21] shed light on the wide range of PRP applications and the numerous sources where important publications can be found. The top 100 most cited articles on PRP use in regenerative medicine spanned 19 years, had first authors from 17 different countries, pooled articles from 53 different journals, and focused on 7 disciplines. The *American Journal of Sports Medicine* had the most cited articles, with 12 of the top 100; most of the articles were from Western Europe (n = 46) and North America (n = 38); and 54 of the articles were focused on applications related to the musculoskeletal system/orthopedics [21]. Treatment of musculoskeletal pain is an integral part of pain management, so the growing evidence for PRP in this field translates well for adoption into pain management, musculoskeletal medicine, and interventional sports and spine practices.

## Preprocedure Planning

### Lab Work

Prior to initiating PRP treatment, it is imperative that updated lab work is obtained to verify that it is safe to move forward with injection-based treatments. Platelet count, hemoglobin and hematocrit values, and leukocyte values should all be reviewed. The whole blood platelet number will also help to determine the volume of blood draw required to achieve the desired platelet dose for treatment.

### Medications

Many medications and supplements may negatively affect the function of platelets as well as tissue healing, which are crucial for platelet-based therapies with PRP. Although this area has not been studied in detail, to optimize PRP treatments a variety of medications should be considered.

Platelet function is targeted by many commonly used prescription medications. PRP treatments are not often recommended for patients on anti-platelet medications but if they can be safely stopped before treatment, this would improve platelet functions for treatment. Many nonselective nonsteroidal anti-inflammatory drugs (NSAIDs) have been shown to decrease platelet function, so it is recommended they be stopped prior to treatment [22]. COX-2 selective NSAIDs can be continued if necessary. Acetaminophen has been shown to decrease platelet aggregation in a dose dependent manner and should be stopped if possible [23]. Since the inflammatory processes involved in tissue healing can be altered by medications like NSAIDs and steroids, cessation of and abstinence from these medications for a pretreatment period of 2–4 weeks and 2–4 weeks posttreatment are often recommended.

In 2022, a study by Lampron et al. [24] on cannabis showed that in vitro exposure of whole blood to a cannabinoid mixture impairs the quality of red blood cells and platelets, so it may be wise to have patients stop cannabis before and after PRP treatments.

### Comorbid Disease

If a patient has an active process like a blood disorder or cancer, it is generally recommended not to perform a PRP treatment that involves transferring the patient's blood products from one area to another. For such cases, consulting with the patient's care team prior to treatment is recommended. In addition, contraindications to these procedures would also include fever, cutaneous infections in the area to be injected, and severe psychiatric impairments.

## Preparing PRP in the Office

Recalling the concentration of platelets for optimal stimulation of angiogenesis from the study by Giusti et al. [10] is a good starting point when preparing PRP. The concentration is 1.5 × $10^6$ platelets/μL, or 1.5 million platelets/μL, or 1.5 billion platelets/mL (1.5 × $10^9$ platelets/mL). An accepted goal for a therapeutic platelet concentration in PRP is often defined as at least one million platelets/μL [25–27]. The total dose of platelets delivered to a treated area would vary based on the volume of injectate—a tendon treated with 1 mL would deliver half as many platelets as a tendon treated with 2 mL, despite the same concentration of platelets per volume.

If starting with a platelet count of 250,000/μL, it can be estimated that a sixfold increase, or 6× concentration factor, would yield approximately the desired number of platelets in many situations. Considering the patient's baseline platelet count, the final volume of PRP to be injected and the platelet capture percentage in the methods used will lead to preparing a superior PRP product for treatment. If a patient has a platelet count of 150,000/μL the protocol would call for around a tenfold increase. If the protocol used has 85% platelet capture, even higher concentrations would be called for. If evidence suggests a certain type of PRP is more effective in specific areas to be treated, this must also be factored in to prepare the ideal formulation.

## Blood Draw

PRP preparation in the office begins with a blood draw performed under aseptic guidelines. The goals of the blood draw are to get an adequate amount of blood for the planned treatment, with as little damage as possible to the cells being drawn from the patient. A proper draw can prevent early platelet activation and red blood cell (RBC) lysis. Most protocols recommend large bore needles (>21 G){should this be <21 G?} to draw blood [28]. The needle and syringe used for the blood draw should be prefilled with anticoagulant to prevent clotting. Commonly used anticoagulants are sodium citrate and anticoagulant citrate dextrose solution, solution A (ACD-A). Ethylenediaminetetraacetic acid (EDTA) suppresses platelet degranulation and is not recommended [29].

## Centrifugation Process

After the blood is drawn, it is transferred to undergo the centrifuge processing. Exposure to the atmosphere can increase pH by $CO_2$ diffusion out of the plasma so small diameter tubes with caps are recommended to limit this process [29]. Some kits are available with closed tubes for processing that avoid this problem. Although there are many PRP preparation protocols and over 50 commercial systems have been identified, double spin centrifuge techniques have been shown to consistently capture higher platelet numbers [30].

After the first centrifuge spin, the blood separates into layers based on cellular size and density (Fig. 7.1). The plasma layer is on top, the RBC layer is on the bottom, and in between lies the cellular layer consisting of platelets and white blood cells (WBCs). The cellular layer between the RBCs and plasma is often referred to as the buffy coat layer, but the formal buffy coat layer is formed after the second spin in two-spin protocols.

In Fig. 7.1a, one can see how density separation principles lead to layering of the cells and plasma after the first spin—notice the overlap in values. In Fig. 7.1b, one can see the PRP prepared in the clinic, after the first spin of a two-spin protocol. In Fig. 7.1c, one can see a depiction of how the

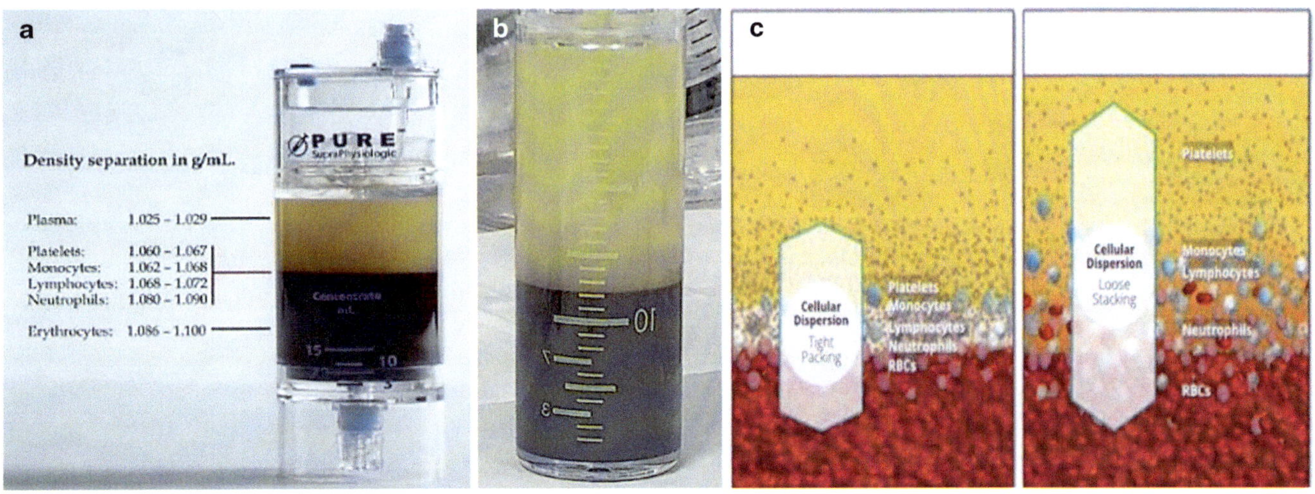


**Fig. 7.1** PRP layers during preparation after the first spin of a two-spin protocol. (Source(s): (**a**) First spin images from EmCyte Corporation, Fort Myers, FL website; (**b**) From author's clinic; (**c**) Image adapted from Cervos Medical, Marshfield, MA website)

components in the formulation might vary after undergoing different centrifugation protocols.

Depending on the centrifugation specifications, regarding aspects like duration, acceleration, processed volume, and force, the gradient of cells within the formulation can vary at each step of the process [31]. On the left, there is tighter packing of the cellular layer more clearly visible between the plasma and RBC layers. On the right the cells are more loosely packed resulting in more platelets suspended in the plasma, which can be referred to as a platelet plasma suspension (PPS). This is not to be confused with platelet poor plasma (PPP), which is the remaining plasma after final PRP preparation—ideally platelet capture is maximized in preparing the PRP, leaving few platelets in the discarded plasma.

The plasma and RBC poor cellular mixture is then recentrifuged in the second spin (Fig. 7.2). At this crucial step, and depending on the device and methods, various formulations can be made that include different combinations of leukocytes to be delivered back to the patient during treatment. What is commonly referred to as a leukocyte poor PRP (LP-PRP) preparation would be processed by obtaining the plasma/buffy coat with minimal content from the RBC layer below. A leukocyte rich PRP (LR-PRP) preparation would be processed with the full buffy coat layer and a small part of the upper RBC layer to maximize leukocyte content without getting too many RBCs. The LR-PRP will be redder in color, because even a small fraction of RBCs in the final PRP will change the color significantly from amber to red.

Keep in mind that these general principles can vary from protocol to protocol, and it is important for physicians preparing PRP to understand the intricacies of the entire process in order to produce the specific type of PRP best suited for each specific treatment [25, 26, 32, 33]. It is possible to alter the preparation methods to specifically change the number of monocytes, neutrophils, RBCs, and platelets in the final PRP formulation.

Figure 7.2 shows the steps involved in a 2-spin protocol. After the blood is drawn from the patient it is transferred into a closed tube and placed in the centrifuge for the first spin. The result is a distinct separation of the RBC layer from the plasma, with a layer of platelets and WBCs in between. The plasma and cellular layers are separated from the RBC layer and placed back in the centrifuge for the second spin. In this example, the cells collect at the bottom of the tube. The PPP not required for treatment is removed, and then the cellular layer is mixed with the remaining plasma for the injection. The final volume should vary based on the area to be treated and is decided before preparation.

Having the flexibility to decide on treatment volumes ahead of time is important to prepare high quality products for orthobiologic procedures. Ideally, the preparation process should allow physicians to capture as many platelets as clinically indicated in a final volume of PRP that will be completely delivered to the patient. A suboptimal treatment protocol would be one that produces 10 mL of PRP when only 2 mL is required for treatment. When a physician delivers only 2 mL of this product to the patient, the wasted 8 mL leaves 80% of the platelets in the PRP undelivered during the treatment.

Additionally, if a physician prefers to prepare PRP with predictable and different combinations of leukocytes, the physician should become familiar with different products and choose one that allows adjustments to be made as needed from patient to patient, or even for different target tissues on the same patient. A case presented by Everts et al. [34] showed how four different PRP formulations can be pre-

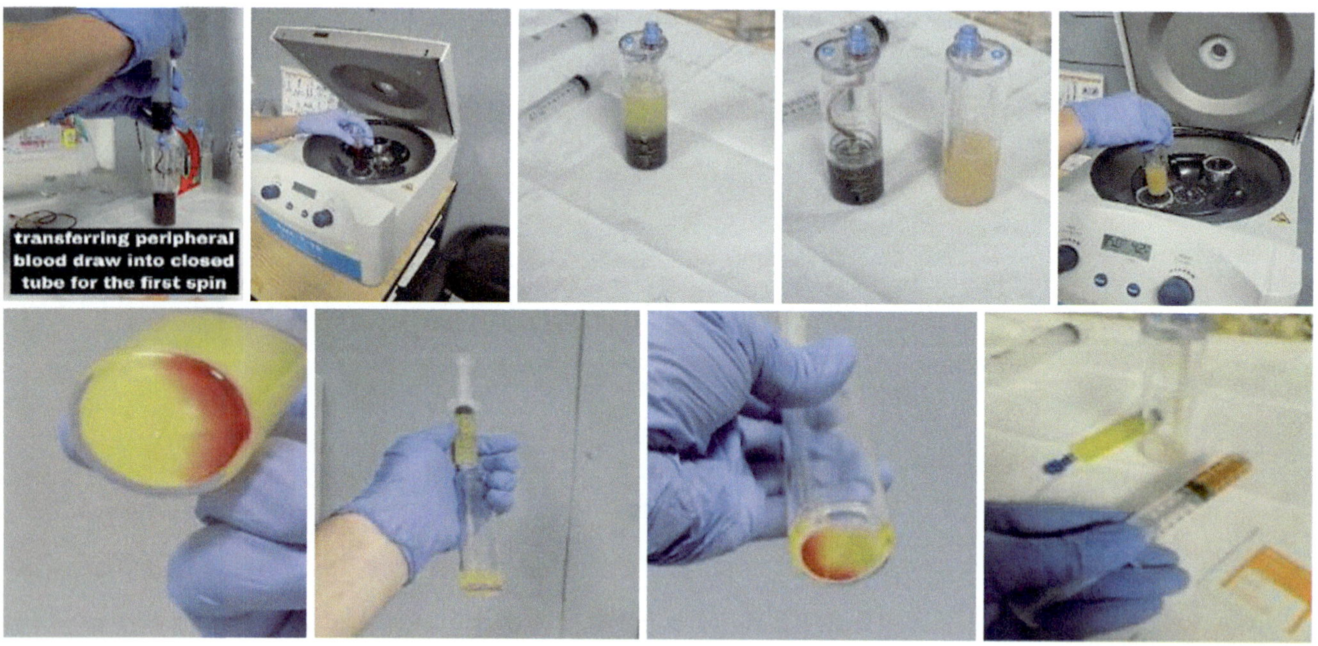


**Fig. 7.2** Two spin centrifuge process with closed tubes

pared to treat seven different targets during a single procedure. During that case, a bedside cell analyzer was used by the physician to prepare the final formulations for treatment. We will explore the roles of different cells available in PRP in more detail later in this chapter.

## Activation of PRP Prior to Treatment

When PRP is injected, there is a natural endogenous activation by local biological tissues, via platelet tissue factor pathways, or collagen Type I. Exogenous activation prior to injection by combination with thrombin, calcium chloride, or a combination of the two is also possible and can affect the amount and duration of growth factor release from PRP [35]. Growth factor release in response to different activation protocols can be achieved in a burst fashion over minutes or in a sustained fashion over days [33, 36]. If specific time-release protocols are shown to be beneficial for treatment, activation could play a key role in optimizing outcomes. Additionally, activation protocols must be closely examined and considered when comparing results across clinical studies.

## Scaffold and Gel Preparation

There has been interest in producing a more stable structure than liquid PRP for a number of years, indicated by the study of different fibrin products and platelet gels [3]. While preparing PRP in the office, different protocols allow production of various fibrin or platelet gel products that can mimic a natural ECM and release bioactive factors in different patterns unachievable by more common liquid PRP products [37].

Methods continue to evolve and recently there has been a method of gel formation described by heating along with the addition of hyaluronic acid [38]. An analysis of level one studies showed that patients receiving PRP gel did better on shoulder function Constant scores than patients getting treatments other than gel [39]. If outcomes like this continue to be published, the use of scaffolds, gels, and matrixes during treatments may increase in popularity. It is possible that these products may be helpful to support and fill gaps during injections into tissues like ligaments, tendons, and bones.

## Confusing Terminology Regarding PRP Formulations

PRP nomenclature has expanded rapidly over the last 20 years. It is important to describe PRP products accurately for comparison across studies and for translation into clinical practice, but the growing list of product descriptions (Table 7.1) is unfortunately a common source of confusion [40, 41].

**Table 7.1** List of PRP and fibrin preparations

| | | Fibrin-rich formulations |
|---|---|---|
| A-PRF | Advanced Platelet-Rich Fibrin | ✓ |
| ACP | Autologous conditioned plasma | |
| AGF | Autologous growth factors | |
| APG | Autologous platelet gel | |
| C-PRP | Clinical platelet-rich plasma | |
| c-PRF | Concentrated PRF | ✓ |
| FPRGF | Fibrin-plasma rich in growth factors | ✓ |
| i-PRF | Injectable platelet-rich fibrin | ✓ |
| iL-PRF | Injectable leukocyte platelet-rich fibrin | ✓ |
| LP-PRP | Leukocyte-poor platelet-rich plasma | |
| LR-PRP | Leukocyte-rich platelet-rich plasma | |
| L-PRF | Leukocyte platelet-rich fibrin | ✓ |
| PFC | Platelet-derived factor concentrate | |
| P-PRP | Pure platelet rich plasma | |
| PFS | Platelet fibrin sealant | ✓ |
| PLG | Platelet-leukocyte gel | |
| PL | Platelet lysate | |
| PRF | Platelet-rich fibrin | ✓ |
| PRFM | Platelet-rich fibrin matrix | ✓ |
| PRGF | Platelets rich in growth factors | |

Adapted and modified from:
Everts P, Onishi K, Jayaram P, Lana JF, Mautner K. Platelet-rich plasma: New performance understandings and therapeutic considerations in 2020. *Int J Mol Sci* 2020; 21):7794 [40]
Everts PA, Sadeghi P, Smith DR. Basic science of autologous orthobiologics: Part 1. platelet-rich plasma. *Phys Med Rehabil Clin N Am* 2023; 34:1–23 [41]

New products were first classified individually as they emerged, but the growing number of products made it evident that more structured classification was needed [42] leading to numerous proposals over the last 15 years. The evolution of these classification proposals tells a story of its own, about the increasingly sophisticated products available, and the important variables to consider for accurate comparison of the many different PRP formulations used across studies.

The Dohan Ehrenfest classification published in 2009 was based on the leukocyte count and fibrin architecture in addition to concentrated platelet formulations. The MISHRA classification published in June 2012 called for measuring the concentration of platelets above baseline, the leukocyte concentration, and whether the product was activated. Shortly afterwards, in July 2012, the PAW classification was published, which was different in that it called for documenting the absolute number of platelets instead of platelet concentration, as well as a description of the method used for platelet activation. In 2015, the PLRA system suggested quantifying the leukocyte and RBC contents in more detail, in addition to platelet count and activation. The DEPA classification proposal published in 2016 recommended adding device efficiency data regarding the recovery of platelets as well as the purity of PRP by comparing the relative composition of platelets, leukocytes, and RBCs in the PRP product. The MARSPILL classification published in 2017 saw the additional recommendations of classifying based on one-spin or 2-spin protocols, whether image guidance was used for treatment, and if light activation of the PRP was used. In 2018, the Scientific and Standardization Committee (SSC) proposal saw the additional recommendation of including the differential leukocyte counts regarding neutrophils, lymphocytes, and monocytes [40, 42, 43]. Table 7.2 shows multiple considerations for PRP classification and outcome tracking.

**Table 7.2** Considerations for PRP classification and outcome tracking

| | |
|---|---|
| Preprocedure plan | Medications—continued meds, timing of discontinued meds<br>Pathology—history and physical findings, imaging findings<br>Comorbidities—chronic medical issues<br>Failed treatments—therapy, bracing, laser, shockwave, injections, medications, surgeries |
| Product description | Autologous vs. Allogenic<br>Cellular composition—absolute platelet number, specify WBC content, RBC content by hematocrit<br>Additives (anything added directly to the final product)—other biologics (fibrin matrix, BMAC, ATC, PL), antibiotics, steroids, local anesthetics, hyaluronic acid, dextrose, ozone<br>Liquid vs. Gel—describe gel formation process<br>Volume—for each prepared product |
| Preparation method | Technology—centrifuge, kit, manual processing, number of spins, G-forces, duration of spins<br>Anticoagulation—product and amount<br>Activation—product and description of protocol<br>Blood draw volume<br>Starting CBC with differential |
| Treatment technique | Application process—surgical adjunct, injection, topical<br>Guidance for injection—X-ray, ultrasound, CT, arthroscopic<br>Supplies—injection needles, syringes, trocars<br>Anesthesia—local, oral sedation, iv sedation, general, nerve blocks |
| Targeted area(s) | Each treated area listed—scar, wound, nerve, muscle, tendon, ligament, intra-articular, epidural space, intra-thecal space, intradiscal, intraosseous<br>Description of product used for each target—volume, detailed product description, application/injection method |
| Post procedure plan | Rehabilitation protocol—immobilization, weight bearing, physical therapy, chiropractic<br>Follow up imaging<br>Medications to be restarted<br>Additional therapies—laser, shockwave |

*ATC* adipose tissue concentrate, *BMAC* bone marrow aspirate concentrate, *CBC* complete blood count, *CT* computed tomography, *PL* platelet lysate

## PRP Mechanism of Action

Since the clotting cascade, tissue healing mechanisms, and immune system defenses are tightly interwoven shortly after platelet aggregation, it is necessary to understand the cross-talk mechanisms at a cellular level. With this knowledge, physicians can apply different platelet-based products appropriately to meet tissue repair and pain management treatment goals for each individual patient. A deep dive into the cellular makeup of office-based PRP formulations shows both the complexity and diverse potential of these treatments.

### Platelets

Platelets have been the most extensively studied component of PRP products in regenerative musculoskeletal medicine. After PRP is injected, the platelets interact with exposed endothelial tissue initiating hemostasis followed by immune and repair processes [44]. The activated platelets release contents from alpha, dense, lysosomal, and T granules into the local environment through exocytosis [44, 45]. The numerous proteins released by platelets quickly initiate a cascade that leads to fibrin formation, producing a gel-like scaffold involved in tissue repair while simultaneously acting as an air-liquid biofilm to prevent against bacterial proliferation and dissemination. Depending on the local environment, the response of platelets can direct a strong immune response to combat infection and a strong regenerative response to repair tissue through interactions with local tissues and cells responding to platelet-derived growth factors, chemokines, and cytokines [44, 46, 47]. Table 7.3 shows a partial listing of notable growth factors, chemokines, and cytokines.

### Leukocytes

While the dose of platelets has been the primary focus of PRP injectates for clinical use, there's abundant evidence that leukocyte content is important in defense against infections as well as tissue healing [48]. Recent classification proposals have begun to include variations on reporting these cells in treatment products [32, 40, 42]. Although LR-PRP containing WBCs has been avoided, particularly to keep neutrophils out of the injectate, more recent research has shown interactions between platelets and WBCs to have important effects toward regenerative processes [49]. Additionally, a review of six randomized controlled trials (Level I evidence) and three prospective studies (Level II evidence) with 1055 patients concluded that LP-PRP and LR-PRP have similar safety profiles and that adverse reactions to PRP may not be related directly to leukocytes in the products used.

**Table 7.3** Notable growth factors, chemokines, and cytokines (partial list)

| | |
|---|---|
| *Pro angiogenesis* | |
| ANG-1 | Stimulates endothelial cell proliferation, stabilizes blood vessel development via pericytes |
| BFGF | Promotes stem cell differentiation and cell proliferation; collagen production and tissue repair |
| CTGF | Stimulates leukocyte migration, activates myofibroblasts for ECM deposition and remodeling |
| EGF | Accelerates re-epithelialization, promotes M2 differentiation, fibroblast/epithelial/endothelial proliferation, increases wound strength and organizes granulation tissue |
| HGF | Regulates cell growth and motility in epithelial/endothelial cells, supports epithelial repair and neovascularization during wound healing |
| IL-8 | Pro angiogenetic |
| PDGF | Attracts neutrophils, macrophages, fibroblasts, endothelial cells, mesenchymal cells; regulates fibroblasts and collagen production; promotes M2 differentiation; helps wound contraction |
| TGF-alpha-beta | Epithelial development, fibroblast proliferation, fibroblast and monocyte migration, collagen synthesis |
| VEGF | Stimulates migration and proliferation of endothelial cells; collagen synthesis; attracts macrophages and granulocytes |
| *Anti angiogenesis* | |
| Angiostatin | Angiogenesis inhibitor; results from cleavage of plasminogen |
| Endostatin | Angiogenesis inhibitor via actions on VEGF subtypes and bFGF |
| PF4 | Attracts leukocytes and regulates their activation; microbiocidal activities |
| PAI | Promotes fibrinolysis, regulates ECM processes |
| TIMPS | Inhibit MMPs; inhibit VEGF and FGF activities |
| TSP | Antagonize VEGF; directly inhibit endothelial cell processes |
| *Others* | |
| FGF | Attracts macrophages, fibroblasts, endothelial cells and regulates their migration and proliferation; collagen synthesis; wound contraction |
| IGF-1 | Attracts fibroblasts |
| KGF | Regulates epithelial migration and proliferation |
| SDF1alpha | Attracts CD34+ cells and induces homing, proliferation, and differentiation into endothelial progenitor cells; attracts mesenchymal stem cells and leucocytes |
| TNF-alpha | Regulates monocyte migration, fibroblast proliferation and macrophage activation |
| IL-1 | Promotes systemic inflammation; crucial in innate immune responses |
| IL-6 | Pro inflammation and anti inflammation; links innate to acquired immune responses |
| IL-8 | Promotes endothelial and vascular smooth muscle cell migration |
| IL-10 | Regulates extracellular matrix and fibroblast activity |
| MIP-1alpha | Critical macrophage and T cell chemoattractant |

*GF* growth factor, *IL* interleukin, *ANG-1* angiopoietin-1, *BFGF* basic fibroblast, *CTGF* connective tissue, *EGF* epidermal, *FGF* fibroblast, *HGF* hepatocyte, *IGF-1* insulin-like, *KGF* keratinocyte, *Mip-1alpha* macrophage inflammatory protein, *MMPs* matrix metalloproteinases, *PAI* plasminogen activator inhibitor, *PDGF* platelet derived, *PF4* platelet factor, *SDF1* alpha stromal cell derived factor, *TGF* transforming, *TIMPS* tissue inhibitor of metalloproteinases, *TNF* tumor necrosis factor, *TSP* thrombospondins, *VEGF* vascular endothelial, *IL-1* interleukin 1, *IL-4* interleukin 4, *IL-6* interleukin 6

Describing the leukocyte fractions in PRP continues to be an area of confusion since individual leukocyte contents are not commonly referenced. Preparation protocols vary remarkably, as does the meaning of LR-PRP. It is possible to prepare bioformulations that keep many WBCs for treatment except neutrophils, which can be described more specifically as neutrophil-poor PRP. This type of product has been described as "leukocyte poor" at times in reference to lacking neutrophils, which is not accurate since other leukocytes are present in the final product. Therefore, more correct nomenclature would include a description of a neutrophil-rich or poor product. However, the number of ideal neutrophils for this definition has not yet been consistently established.

## Monocytes and Macrophages

Monocytes and macrophages appear to exist in a continuum of two distinct spectrums. Inflammatory monocytes (IMs) and M1 macrophages are proinflammatory and cytotoxic while anti-inflammatory monocytes (AMs) and M2 macrophages participate in tissue remodeling, wound healing, and angiogenesis. Platelets have been shown to shift populations of these cells toward the AM subset which may be of importance in regenerative treatments [50, 51].

Macrophages have been shown to adopt the default resident M2 state in tissues and are even capable of low levels of proliferation and self-renewal at steady state conditions. This has led to a hypothesis that they may have stem cell-like properties, and it's noted that proliferation can increase greatly in response to macrophage depletion during inflammatory challenges [52]. Interestingly, recent research on Behcet disease—a systemic vascular inflammatory condition characterized by recurrent ulcers—has uncovered a genetically driven component that leads to lower M2 macrophage activity as a likely part of the pathogenesis [53]. The evolving research on monocytes and macrophages shows they have important roles in tissue healing, and PRP injectates prepared in a way to be rich in platelets and monocytes may prove more beneficial than monocyte-poor products in certain situations [20, 54].

## Neutrophils

Neutrophil content in PRP products is an area of interest because these leukocytes can exhibit antimicrobial [55] and tissue healing functions [56]. Neutrophils are the most abundant WBC in circulation and the first immune cells recruited to infiltrate wounds. They are recruited by damage-associated molecular patterns (DAMPs) via mechanisms not completely understood but neutrophil-mediated tissue repair is possible through clearing necrotic debris, promoting angiogenesis and regeneration, and resolution of inflammation by release of anti-inflammatory and reparative cytokines when apoptotic neutrophils undergo phagocytosis [56, 57].

Clinically, neutrophils are available in LR-PRP when a full buffy coat product is prepared for treatment. Neutrophils are the leukocytes in the buffy coat layer closest to the RBCs (Fig. 7.1). These neutrophil-rich PRP bioformulations have been shown to inhibit the growth of staphylococcus aureus and other bacteria in vitro [58] and have been mentioned as treatment options for chronic tendinopathy [27, 29]. Despite these promising roles in regenerative treatments, there is continued concern because of the traditional idea that neutrophils are predominantly involved in catabolic and tissue destroying cellular processes [59].

## Lymphocytes

T and B cells are more concentrated than any other leukocyte in C-PRP. In response to inflammation or infection, research suggests that platelets may facilitate recruitment of lymphocytes where they are mediated by factors released from platelets like platelet factor 4 or serotonin [60]. T-cells have been shown to play important roles in macrophage polarization, and vice versa [52]. A systematic review by D'asta et al. [48] showed that despite platelet concentration in PRP products ranging from 2× to 10× above baseline, leukocyte content in those same formulations ranged from a 3.9× decrease up to a 4× increase, indicating wider adoption of more specific protocols are needed to prepare PRP with or without lymphocytes as desired for clinical use.

## Fibrinogen and Fibrin

Formation of a stable fibrin clot is vital to the initiation of tissue healing, providing a matrix for various cellular interactions, including migration of cells and tissue growth. Fibrinogen is a complex structure linked by disulfide bridges that are cleaved to form fibrin when signaled by thrombin to initiate hemostasis and tissue healing. Fibrin polymerization and cross-linking supports the interactions among platelets, leukocytes, fibroblasts, progenitor cells, and endothelial cells. Varying the centrifugation process can lead to leukocyte- and platelet-rich fibrin products, or a liquid fibrinogen product with potentially different clinical applications [61, 62].

Like PRP products, platelet fibrin products have been prepared with and without leukocytes resulting in differences in growth factor release signatures that should be considered and studied in more detail [63]. Different fibrin-based products can also be characterized by their mechanical strength and many types of matrixes have been investigated with potentially various clinical applications [46] since certain clinical situations may call for treatments with a more stable matrix or scaffold instead of a liquid product. Autologous fibrin matrix products have been used in specialties including surgery, sports medicine, dermatology, ophthalmology, nerve regeneration, and wound healing [47] and continue to be an area of interest.

## Erythrocytes (RBCs)

RBCs are not established as having key functions in tissue regeneration. Transporting oxygen to tissues and removing carbon dioxide by way of iron and heme components are the primary roles of RBCs. Current recommendations are to decrease RBC counts because of the potentially negative effects secondary to disintegration of RBCs during orthobiologic procedures [64]. A study by Harrison et al. [65] showed that five different PRP preparation methods led to significant differences in WBCs and RBCs in the final products, so care must be taken to address final RBC content as well as WBCs in PRP prepared for clinical use. RBC release of macrophage inhibitory factor (MIH) [66] and plasma free hemoglobin (PFH) may be key components driving cellular dysfunction due to inflammation and oxidative stress [67]. Erythrophagocytosis, accomplished by macrophages, is described as the central defense against these stresses on tissue homeostasis due to premature dysfunction and RBC breakdown [68].

## Innate and Adaptive Immunity

The platelets and leukocytes found in PRP formulations for orthobiologic treatments initiate complex biological responses that may play pivotal roles in addressing orthopedic conditions. Research on osteoarthritis suggests a paradigm shift is likely needed, viewing the joint as an organ responding biologically to biomechanics, inflammation, and the immune system, similar to a chronic wound [69].

The innate immune system is biologically programmed to respond to threats like tissue damage and infection, requiring inflammation, wound healing, and immune processes to coordinate quick responses. Monocytes, macrophages, neutrophils [57], and natural killer cells [70] have important roles in the innate response. Lymphocytes are more important in the adaptive mechanisms [69], categorized by antigen-specific responses when a host has subsequent encounters with a pathogen.

Platelets can be activated through their cell surface receptors causing release of numerous molecules influencing both the innate and adaptive immune responses. The diverse platelet signaling receptors known originally to be critical in hemostasis also have inflammatory and immunomodulatory functions, acting by secretion and translocation of more than 300 proteins and lipids [71]. As first responders of the innate system, formation of platelet-leukocyte aggregates with monocytes and neutrophils in circulation seems to be crucial for acute and chronic inflammatory processes while adaptive immunity is modulated by specific ligand expression on platelets recognized by T-cells [72].

## Opposing Inflammatory Effects After PRP Treatment

Inflammation in vitro is an ongoing balancing act of pro- and anti-inflammatory cellular processes resulting in anabolic and catabolic environments. PRP treatments alter the local environments where they are applied, having significant effects on orthobiologic processes in response to changes in growth factors, cytokines, protein expression, and angiogenesis. A concern has been that more pro-inflammatory cells like neutrophils could lead to increasingly painful posttreatment flares and possibly excessive tissue damage, which is why many recommendations over the years have recommended using LP-PRP [15, 73].

In comparison, there has also been evidence for the use of LR-PRP in certain situations. Tendons were responsive to LR-PRP in a study on subacromial injections for rotator cuff pathology [27] and a review by Fitzpatrick et al. [74] found that LR-PRP had a greater positive effect on tendon healing than LP-PRP. LR-PRP has also been shown to be efficacious in treating ligaments after years of follow-up data [75, 76]. Unlike in vivo studies, it is suggested that clinical PRP rich in leukocytes may be checked and balanced by the local tissue environment in vitro as well as other components in the PRP injection in a way that prevents deleterious effects. The result could be that these in vitro mechanisms are important to the resolution of inflammatory phases and transition to proliferative phases of healing, and are not deleterious as previously thought [56, 77]. In particular, this is believed because neutrophils are an essential part of the classical healing cascade.

## PRP Analgesic Effects

Many musculoskeletal studies on PRP rely on pain and functional outcome scores as signs of efficacy. Although the exact mechanisms of pain reduction are not well understood, the positive findings regarding analgesia after PRP treatments are promising. A systematic review and meta-analysis from 2019 [78] realized the limited methods of many prior reviews on the use of PRP in orthopedics and looked at patient-reported pain as the outcome of interest. The analysis of 78 randomized controlled trials showed that PRP leads to a reduction in pain across indications, even though the study heterogeneity remained high, and the trials covered many indications with various protocols [78].

Pain reduction after PRP treatments has been documented in many musculoskeletal indications. These indications include, but are not limited to: low back pain after injections into lumbar multifidi muscles [79], peripheral nerve pain after perineural and intraneural injections [80], radiculopathy and discogenic pain after epidurals and intradiscal treatments [12, 13, 16, 54, 81–83], sacroiliac and facet joint pain [16, 83], vertebrogenic and discogenic pain with treatments both intradiscally and vertebral intraosseous injections [84], tendon pathology [14, 19, 20, 85, 86], ligament pathology [14, 75], intraarticular knee and hip injections [11, 73, 87, 88], intraarticular injections in patients with rheumatoid arthritis [89], intraosseous injections for osteonecrosis [90], use in addition to shoulder decompression surgery [91], use in addition to meniscus repair [92], and combined with hyaluronic acid at the completion of knee arthroscopic debridement [93]. This is not a comprehensive list by any means, but it shows various indications where physicians can consider using PRP to reduce pain.

The mechanisms of analgesia after PRP treatments may be due to a variety of biological and structural changes. Months after a PRP treatment, when tissue healing, angiogenesis, and remodeling has taken place in the treated area, more functional tissue may be less painful with activities. This could explain pain relief seen months to years out in the numerous studies indicating PRP for use as an interventional pain management treatment.

PRP's anti-inflammatory effects have been well documented, for instance the synovial fluid in knee joints were studied after intraarticular treatments with PRP showing shifts from an environment of painful, inflammatory, catabolic substances to less inflammatory anabolic substances. Some of the documented changes were highlighted in a review of molecular mechanisms [88] and included decreased matrix metalloproteinase activity, increased hyaluronic acid and cartilage synthetic activity, reduced synovial thickness, reduced interleukin (IL)-1, decreased apoptosis, and release of VEGF, IGF, FGF, TNF, EGF and TGF-beta with inhibition of nuclear factor-κB.

Serotonin (5-HT) is released in exceedingly high supraphysiological amounts after PRP treatments and may also play a role in reducing pain after PRP treatments. In 2008 a study [91] documented that injection with autologous platelet-leukocyte gel (PLG) after shoulder surgery resulted in less postoperative pain and decreased pain medication requirements, as well as quicker recovery of shoulder function compared to controls. In addition to PRP's anti-inflammatory effects and initiation of tissue healing mechanisms, it was suggested that 5-HT release from dense granules may play a role in pain control after treatment [94]. Considering the role of managing 5-HT in pain management clinics already, this is an area that warrants further clinical investigation.

## Conclusion

Concentrating platelets for PRP treatments was the initial goal for clinical use. The earliest definition of PRP was simply a platelet concentration above baseline [9]. This means a 15 mL blood draw concentrated to 5 mL by centrifugation for treatment is 3× over a baseline PRP formulation. Despite many publications calling for a shift to referencing platelet concentrations and doses in PRP treatments, the concentration factor is still the most common way PRP is referenced.

Research has indicated specific platelet numbers are needed to achieve effects like bone and soft tissue healing, angiogenesis, immunomodulatory activities, and stimulation of other cellular processes, which requires more detailed descriptions of platelet dosing. Moving forward, characterizing PRP by absolute platelet numbers based on concentration and treatment volumes is recommended in clinical and research settings [95]. Studies that have shown poor results were often completed without a description of the PRP used or with protocols resulting in low-dose PRP used in treatments.

Drawing conclusions from studies without knowledge of the PRP used is not recommended, and when well-defined *low-dose PRP* is studied, be careful not to generalize those outcomes to all PRP treatments. Despite the lack of standardization in classification, preparation, and protocols, a recent review of level I and II studies showed 66% of the 90 musculoskeletal PRP studies found PRP favorable over control treatments in subjective outcome reporting [96]. There is also growing evidence that different PRP preparations can optimize healing when used appropriately in the setting of specific injuries and tissue types. Understanding concepts related to PRP preparation, dosing, and clinical applications will have important implications in optimizing musculoskeletal regenerative medicine outcomes in interventional pain management moving forward [97].

**Key Points**

1. Interventional pain management of musculoskeletal disorders is rapidly evolving with evidence-based medicine utilizing autologous orthobiologics.
2. Interventional pain physicians have the skillset to precisely deliver platelet-rich plasma (PRP), bone marrow concentrate (BMC) and adipose tissue concentrate (ATC) with image-guided injections.
3. Regenerative medicine treatments are potential catalysts in plans designed to aid tissue repair, decrease pain, and improve function. However, slight variations in preparation can lead to vastly different PRP formulations, but a lack of standardized preparation protocols continues to be an area of concern.

   Case in point, an article by Paget et al. published in *The Journal of the American Medical Association* did not support PRP injections for ankle osteoarthritis; however, the PRP was prepared from low volumes, which correlates with low platelet doses. Obviously, the conclusions derived that PRP was ineffective are inaccurate.
4. The understanding of the complex communication among growth factors, chemokines, and cytokines has grown considerably over the last 20 years.
5. In adding PRP to clinical practice, certain important preprocedure planning must be performed, along with procedural performance and postprocedural follow-up.
6. Table 7.2 describes considerations for PRP classification and outcome tracking describing a preprocedure plan, product description, preparation method, treatment technique, treatment technique and postprocedure plan.
7. Table 7.3 shows a partial list of notable growth factors, chemokines and cytokines, including proangiogenesis and anti-angiogenesis.
8. Pain reduction after PRP treatments has been documented in many musculoskeletal indications. There is also growing evidence that different PRP preparations can optimize healing when used appropriately in the setting of specific injuries and tissue types.
9. Understanding concepts related to PRP preparation, dosing, and clinical applications will have important implications in optimizing musculoskeletal regenerative medicine outcomes in interventional pain management moving forward.

**Questions**

1. The phases of tissue healing in their correct order are:
   A. hemostasis, inflammation, proliferation, and remodeling
   B. Inflammation, hemostasis, proliferation, and remodeling
   C. Proliferation, inflammation, hemostasis, and remodeling
   D. Hemostasis, proliferation, inflammation, and remodeling

   **Answer: A**

   **Explanation**: The correct order of tissue healing begins with hemostasis, and then progresses to inflammation, proliferation, and remodeling. This is the correct sequence of tissue healing. Answers that suggest inflammation or proliferation as the first phase, or proliferation preceding inflammation are incorrect.

2. In 2009, research conducted demonstrated that the optimal concentration of platelets to stimulate angiogenesis was:
   A. Giusti et al. showed that the optimal concentration of platelets in vitro for the stimulation of angiogenesis is $1.5 \times 10^5$ platelets/mL
   B. Giusti et al. showed that the optimal concentration of platelets in vitro for the stimulation of angiogenesis is $1.5 \times 10^6$ platelets/μL
   C. Giusti et al. showed that the optimal concentration of platelets in vitro for the stimulation of angiogenesis is $1 \times 10^6$ platelets/μL
   D. Giusti et al. showed that the optimal concentration of platelets in vitro for the stimulation of angiogenesis is $1.5 \times 10^6$ platelets/mL

   **Answer: B**

   **Explanation**: The publication by Giusti et al. that demonstrated angiogenesis was stimulated in vitro with an optimal concentration of 1.5 × 106 platelets/μL. Answers in (A) and (D) demonstrate platelets/ mL, which would be far less than the number needed to stimulate angiogenesis. Answer (C) demonstrates 1 × 106 platelets/μL, which is close, but less than the optimal concentration shown to optimally stimulate angiogenesis.
3. A patient presented to a clinic for a platelet-rich plasma injection. He was told he was going to be getting a 5 mL injection after a total of 10 mL of blood was taken for processing. If you had a blood draw of 10 mL and had a final volume of injectable PRP of 5 mL, the actual final concentration of platelets/mL in the product compared to whole blood would be?
   A. 2× concentration of platelets due to 100% platelet capture during processing
   B. Less than 2× concentration due to the inability to capture 100% of platelets during processing
   C. 3× concentration of platelets due to 150% platelet capture during processing
   D. 4× concentration of platelets due to 200% platelet capture during processing

   **Answer: B**

   **Explanation**: Due to processing, inherently some platelets will be lost, and therefore, 100% platelet capture does not occur. When concentrating from 10 mL to a 5 mL injection, this would indicate a 2× concentration of platelets per milliliter. However, factoring in a loss of platelets during processing, the final concentration will be less than 2× concentration. Therefore, answer (B) is correct, and answer (A) is incorrect as it does not factor in platelet loss during processing. Answers (C) and (D) are incorrect because the final volume would have to be more concentrated than 5 mL to have a 3 or 4× concentration, and you can't capture a greater % of platelets than are available.
4. A patient presented with right knee osteoarthritis that was moderate in severity both radiographically and symptomatically. Patient had failed physical therapy, activity modification, low impact aerobic exercise programs, cortisone and viscosupplementation injections. In 2021, Bansal et al. concluded that injecting a volume of 8 mL of PRP at a specific dose provided significant potential chondro-protection and alleviates symptoms compared to a control in knee OA. What was the total deliverable number of platelets that would lead to chondro-protection and alleviate symptoms based on this study?
   A. 5 billion platelets in 8 mL
   B. 1 billion platelets in 8 mL
   C. 1 million platelets in 8 mL
   D. 10 billion platelets in 8 mL

   **Answer: D**

   **Explanation**: Bansal et al.'s research suggested that ten billion platelets in 8 mL led to chondro-protection and alleviation of symptoms. This would equate to a dose of 1.25 billion platelets/mL of PRP (10 billion platelets/8 mL). Answers (A), (B), and (C) fail to meet the suggested platelet dose that reaches chondro-protection with symptom alleviation per Bansal et al.'s research.
5. Patient A came into the clinic for a PRP injection to the left knee due to knee osteoarthritis. He brought a list of medications to include metoprolol, meloxicam, and atorvastatin. Patient B came into the clinic the same day for a PRP injection to his right knee and brought a list of medications to include an albuterol inhaler and rosuvastatin. The provider decided to reschedule Patient A but proceeded with Patient B's PRP treatment. What was their likely reasoning for cancelling Patient A's treatment?
   A. Patient A likely had blood pressure of 145/85 and couldn't proceed
   B. Patient B was likely healthier and had a better chance of success
   C. Patient A was on an NSAID, and this could alter the inflammatory process
   D. Patient B likely had milder arthritis and was a better candidate

   **Answer: C**

   **Explanation**: Since the inflammatory processes involved in tissue healing can be altered by medications like NSAIDs and steroids, cessation of and abstinence from these medications for a period of 2–4 weeks before and 2–4 weeks after treatment is often recommended. Based on current knowledge, (A) high blood pressure does not have an impact on the treatment. Although answer (B) and (D) may lead to more successful treatments, it would not cause a physician to reschedule an appointment as suggested in this question.

6. When processing PRP, the terminology leukocyte rich and leukocyte poor are often used. In order to prepare a leukocyte rich product, what would the person processing the PRP need to know in a buffy coat based PRP system?
   A. A leukocyte rich PRP (LR-PRP) product would be processed with the full buffy coat layer and a small portion of the upper RBC layer to maximize leukocyte content
   B. A leukocyte rich PRP (LR-PRP) product would be processed with only a portion of the buffy coat layer to maximize which leukocytes are captured
   C. A leukocyte rich PRP (LR-PRP) product would be processed with only the plasma layer and no part of the buffy coat layer
   D. A leukocyte rich PRP (LR-PRP) is not possible with buffy coat based PRP systems

   **Answer: A**

   **Explanation**: A reference to the terminology of a leukocyte poor PRP (LP-PRP) product in a buffy coat based PRP system would be processed by obtaining the plasma/buffy coat with drawing minimal content from the RBC layer below. On the contrary, a leukocyte rich PRP (LR-PRP) product would be processed with the full buffy coat layer and a small portion of the upper RBC layer to maximize leukocyte content (and fully capture leukocytes based on density) without getting too many RBCs in the final product. Answer (B) is partially correct but to fully maximize a leukocyte rich product, you would draw a full buffy coat layer and a small portion of the RBC layer. Answers (C) and (D) are incorrect as they are untrue.

7. What is the mechanism leading to platelet rich plasma (PRP) activation and hemostasis after injection into tissue?
   A. PRP goes straight to the inflammatory stage prior to entering hemostasis
   B. PRP has already started hemostasis prior to the injection
   C. PRP interacts with exposed endothelial tissue which initiates hemostasis
   D. PRP does not enter hemostasis in orthobiologic treatments

   **Answer: C**

   **Explanation**: After platelet rich plasma (PRP) is injected, the platelets interact with exposed endothelial tissue initiating hemostasis followed by immune and repair processes. The activated platelets release contents from alpha, dense, lysosomal and T granules into the local environment through exocytosis. The numerous proteins released by platelets quickly initiate a cascade that leads to fibrin formation, producing a gel-like scaffold involved in tissue repair while simultaneously acting as an air-liquid biofilm to prevent against bacterial proliferation and dissemination. Although PRP does enter the inflammatory phase, answer (A) is untrue as it enters hemostasis prior to the inflammatory phase. Answer (B) a (D) are incorrect, as PRP initially enters hemostasis once exposed to endothelial tissue.

8. Which type of monocytes are involved in tissue remodeling, wound healing, and angiogenesis?
   A. M1 macrophages
   B. M2 macrophages
   C. M3 macrophages
   D. M4 macrophages

   **Answer: B**

   **Explanation**: Monocytes and macrophages appear to exist in a continuum of two distinct spectrums. Inflammatory monocytes (IMs) and M1 macrophages are proinflammatory and cytotoxic while anti-inflammatory monocytes (Ams) and M2 macrophages participate in tissue remodeling, wound healing, and angiogenesis. Platelets have been shown to shift populations of these cells towards the AM subset which may be of importance in regenerative treatments. The evolving research on monocytes and macrophages shows that they have important roles in tissue healing, and PRP injectates prepared in a way to be rich in platelets and monocytes may prove more beneficial than monocyte poor products in certain situations. Although answer (A) is a type of macrophage, it is proinflammatory and cytotoxic. Answers (C) and (D) are not used as current classifications for subtypes of macrophages.

9. Which are the most abundant white blood cells (WBCs) in circulation and the first immune cells recruited to infiltrate wounds that also promote angiogenesis and regeneration in platelet rich plasma (PRP)?
   A. Neutrophils
   B. Basophils
   C. Monocytes
   D. Red blood cells

   **Answer: A**

   **Explanation**: Neutrophils are the most abundant white blood cells (WBCs) in circulation and the first immune cells recruited to infiltrate wounds. Neutrophil-mediated tissue repair is possible through clearing necrotic debris, promoting angiogenesis and regeneration, and resolution of inflammation by release of anti-inflammatory and reparative cytokines when apoptotic neutrophils undergo phagocytosis. Clinically, more neutrophils are available in LR-PRP when a full buffy coat product is prepared for treatment. Neutrophils are the leukocytes in the buffy coat layer closest to the red blood cells (RBCs). Answer (C) monocytes do play an important role in clearing necrotic debris, promoting angiogenesis, and regeneration but they are not the most abundant WBC in circulation nor the first immune cells to infiltrate the wound. Answer (B) and (D) are not relevant answers.

10. According to the available literature, the blood products most important to measure and characterize at the current time before and after a platelet rich plasma (PRP) injection that could aid the regenerative field in the future would include:
    A. Platelets and red blood cells (RBCs) only
    B. Neutrophils and monocytes
    C. Platelets, neutrophils, and RBCs
    D. Platelets, neutrophils, monocytes, and RBCs

    **Answer: D**

    **Explanation**: When delivering a platelet rich plasma (PRP) injection, ideally values for platelets, neutrophils, monocytes, and red blood cells (RBCs) are measured pre- and post procedure while also measuring standardized clinical outcomes. This will lead to more information to deliver the highest quality treatments and contribute to the research being conducted. Answer (A), (B), and (C) are all true, however, ideally all blood product values that can be measured should be measured.

## References


1. Cunin P, Nigrovic PA. Megakaryocytes as immune cells. J Leukoc Biol. 2019;105:1111–21.
2. Blair P, Flaumenhaft R. Platelet alpha-granules: basic biology and clinical correlates. Blood Rev. 2009;23:177–89.
3. Everts PA, Knape JT, Weibrich G, et al. Platelet-rich plasma and platelet gel: a review. J Extra Corpor Technol. 2006;38:174–87.
4. Heijnen H, van der Sluijs P. Platelet secretory behaviour: as diverse as the granules … or not? J Thromb Haemost. 2015;13:2141–51.
5. Feghhi S, Sniadecki NJ. Mechanobiology of platelets: techniques to study the role of fluid flow and platelet retraction forces at the micro- and nano-scale. Int J Mol Sci. 2011;12:9009–30.
6. Guo S, Dipietro LA. Factors affecting wound healing. J Dent Res. 2010;89:219–29.
7. Martin KE, García AJ. Macrophage phenotypes in tissue repair and the foreign body response: implications for biomaterial-based regenerative medicine strategies. Acta Biomater. 2021;133:4–16.
8. Centeno CJ, Pastoriza SM. Past, current and future interventional orthobiologics techniques and how they relate to regenerative rehabilitation: a clinical commentary. Int J Sports Phys Ther. 2020;15:301–25.
9. Marx RE. Platelet-rich plasma (PRP): what is PRP and what is not PRP? Implant Dent. 2001;10:225–8.
10. Giusti I, Rughetti A, D'Ascenzo S, et al. Identification of an optimal concentration of platelet gel for promoting angiogenesis in human endothelial cells. Transfusion. 2009;49:771–8.
11. Bansal H, Leon J, Pont JL, et al. Platelet-rich plasma (PRP) in osteoarthritis (OA) knee: correct dose critical for long term clinical efficacy. Sci Rep. 2021;11:3971. Erratum in: Sci Rep 2021; 11:18612.
12. Centeno C, Markle J, Dodson E, et al. The use of lumbar epidural injection of platelet lysate for treatment of radicular pain. J Exp Orthop. 2017;4:38.
13. Kubrova E, Martinez Alvarez GA, Her YF, et al. Platelet rich plasma and platelet-related products in the treatment of radiculopathy—a systematic review of the literature. Biomedicine. 2022;10:2813.
14. Chen X, Jones IA, Park C, et al. The efficacy of platelet-rich plasma on tendon and ligament healing: a systematic review and meta-analysis with bias assessment. Am J Sports Med. 2018;46:2020–32.
15. Kim JH, Park YB, Ha CW, et al. Adverse reactions and clinical outcomes for leukocyte-poor versus leukocyte-rich platelet-rich plasma in knee osteoarthritis: a systematic review and meta-analysis. Orthop J Sports Med. 2021;9:23259671211011948.
16. Navani A, Manchikanti L, Albers SL, et al. Responsible, safe, and effective use of biologics in the management of low back pain: American Society of Interventional Pain Physicians (ASIPP) guidelines. Pain Physician. 2019;22:S1–S74.
17. Noroozian M, Raeesi S, Hashemi R, et al. Pain: the neglect issue in old people's life. Open Access Maced J Med Sci. 2018;6:1773–8.
18. Tramś E, Malesa K, Pomianowski S, et al. Role of platelets in osteoarthritis-updated systematic review and meta-analysis on the role of platelet-rich plasma in osteoarthritis. Cell. 2022;11:1080.
19. Arirachakaran A, Sukthuayat A, Sisayanarane T, et al. Platelet-rich plasma versus autologous blood versus steroid injection in lateral epicondylitis: systematic review and network meta-analysis. J Orthop Traumatol. 2016;17:101–12.
20. Ladurner A, Fitzpatrick J, O'Donnell JM. Treatment of gluteal tendinopathy: a systematic review and stage-adjusted treatment recommendation. Orthop J Sports Med. 2021;9:23259671211016850.
21. Coulange Zavarro A, De Girolamo L, Laver L, et al. The top 100 most cited articles on platelet-rich plasma use in regenerative medicine-a bibliometric analysis-from the ESSKA Orthobiologic initiative. Bioengineering (Basel). 2022;9:580.
22. Schippinger G, Prüller F, Divjak M, et al. Autologous platelet-rich plasma preparations: influence of nonsteroidal anti-inflammatory drugs on platelet function. Orthop J Sports Med. 2015;3:2325967115588896.
23. Kao DS, Zhang SW, Vap AR. A systematic review on the effect of common medications on platelet count and function: which medications should be stopped before getting a platelet-rich plasma injection? Orthop J Sports Med. 2022;10:23259671221088820.
24. Lampron MC, Desbiens-Tremblay C, Loubaki L. In vitro exposure of whole blood to a cannabinoid mixture impairs the quality of red blood cells and platelets. Blood Transfus. 2023;21:240–50.
25. Dhurat R, Sukesh M. Principles and methods of preparation of platelet-rich plasma: a review and author's perspective. J Cutan Aesthet Surg. 2014;7:189–97.
26. Fadadu PP, Mazzola AJ, Hunter CW, et al. Review of concentration yields in commercially available platelet-rich plasma (PRP) systems: a call for PRP standardization. Reg Anesth Pain Med. 2019:rapm-2018-100356.
27. Rossi LA, Piuzzi N, Giunta D, et al. Subacromial platelet-rich plasma injections decrease pain and improve functional outcomes in patients with refractory rotator cuff tendinopathy. Arthroscopy. 2021;37:2745–53.
28. Dashore S, Chouhan K, Nanda S, et al. Preparation of platelet-rich plasma: national IADVL PRP taskforce recommendations. Indian Dermatol Online J. 2021;12:S12–23.

29. Collins T, Alexander D, Barkatali B. Platelet-rich plasma: a narrative review. EFORT Open Rev. 2021;6:225–35.
30. Muthu S, Krishnan A, Ramanathan KR. Standardization and validation of a conventional high yield platelet-rich plasma preparation protocol. Ann Med Surg (Lond). 2022;82:104593.
31. Perez AG, Lana JF, Rodrigues AA, Luzo AC, Belangero WD, Santana MH. Relevant aspects of centrifugation step in the preparation of platelet-rich plasma. ISRN Hematol. 2014;2014:176060.
32. Magalon J, Brandin T, Francois P, et al. Technical and biological review of authorized medical devices for platelets-rich plasma preparation in the field of regenerative medicine. Platelets. 2021;32:200–8.
33. Roh YH, Kim W, Park KU, Oh JH. Cytokine-release kinetics of platelet-rich plasma according to various activation protocols. Bone Joint Res. 2016;5:37–45.
34. Everts PA, Mazzola T, Mautner K, Randelli PS, Podesta L. Modifying orthobiological PRP therapies are imperative for the advancement of treatment outcomes in musculoskeletal pathologies. Biomedicine. 2022;10:2933.
35. Cavallo C, Roffi A, Grigolo B, et al. Platelet-rich plasma: the choice of activation method affects the release of bioactive molecules. Biomed Res Int. 2016;2016:6591717.
36. Peterson LJ. Dentistry and oral and maxillofacial surgery. Oral Surg Oral Med Oral Pathol Oral Radiol Endod. 1998;86:1.
37. Giusti I, D'Ascenzo S, Macchiarelli G, Dolo V. In vitro evidence supporting applications of platelet derivatives in regenerative medicine. Blood Transfus. 2020;18:117–29.
38. Godoi TTF, Rodrigues BL, Huber SC, et al. Platelet-rich plasma gel matrix (PRP-GM): description of a new technique. Bioengineering (Basel). 2022;9:817.
39. Chen X, Jones IA, Togashi R, Park C, Vangsness CT Jr. Use of platelet-rich plasma for the improvement of pain and function in rotator cuff tears: a systematic review and meta-analysis with bias assessment. Am J Sports Med. 2020;48:2028–41.
40. Everts P, Onishi K, Jayaram P, Lana JF, Mautner K. Platelet-rich plasma: new performance understandings and therapeutic considerations in 2020. Int J Mol Sci. 2020;21:7794.
41. Everts PA, Sadeghi P, Smith DR. Basic science of autologous orthobiologics. Part 1. Platelet-rich plasma. Phys Med Rehabil Clin N Am. 2023;34:1–23.
42. Lana JFSD, Purita J, Paulus C, et al. Contributions for classification of platelet rich plasma—proposal of a new classification: MARSPILL. Regen Med. 2017;12:565–74.
43. Magalon J, Chateau AL, Bertrand B, et al. DEPA classification: a proposal for standardising PRP use and a retrospective application of available devices. BMJ Open Sport Exerc Med. 2016;2:e000060.
44. Cognasse F, Laradi S, Berthelot P, et al. Platelet inflammatory response to stress. Front Immunol. 2019;10:1478.
45. Morrell CN, Aggrey AA, Chapman LM, Modjeski KL. Emerging roles for platelets as immune and inflammatory cells. Blood. 2014;123:2759–67.
46. Santos SCNDS, Sigurjonsson ÓE, Custódio CA, Mano JFCDL. Blood plasma derivatives for tissue engineering and regenerative medicine therapies. Tissue Eng Part B Rev. 2018;24:454–62.
47. Anitua E, Prado R, Padilla S. Evolutionary insight into immunothrombosis as a healing mechanism. Int J Mol Sci. 2022;23:8346.
48. D'asta F, Halstead F, Harrison P, Zecchi Orlandini S, Moiemen N, Lord J. The contribution of leucocytes to the antimicrobial activity of platelet-rich plasma preparations: a systematic review. Platelets. 2018;29:9–20.
49. Lana JF, Macedo A, Ingrao ILG, Huber SC, Santos GS, Santana MHA. Leukocyte-rich PRP for knee osteoarthritis: current concepts. J Clin Orthop Trauma. 2019;10:S179–82.
50. Ogle ME, Segar CE, Sridhar S, Botchwey EA. Monocytes and macrophages in tissue repair: implications for immunoregenerative biomaterial design. Exp Biol Med (Maywood). 2016;241:1084–97.
51. Vasina EM, Cauwenberghs S, Feijge MA, Heemskerk JW, Weber C, Koenen RR. Microparticles from apoptotic platelets promote resident macrophage differentiation. Cell Death Dis. 2011;2:e211.
52. Italiani P, Boraschi D. From monocytes to M1/M2 macrophages: phenotypical vs. functional differentiation. Front Immunol. 2014;5:514.
53. Hirahara L, Takase-Minegishi K, Kirino Y, et al. The roles of monocytes and macrophages in Behçet's disease with focus on M1 and M2 polarization. Front Immunol. 2022;13:852297.
54. Cheng J, Santiago KA, Nguyen JT, Solomon JL, Lutz GE. Treatment of symptomatic degenerative intervertebral discs with autologous platelet-rich plasma: follow-up at 5-9 years. Regen Med. 2019;14:831–40.
55. Moojen DJ, Everts PA, Schure RM, et al. Antimicrobial activity of platelet-leukocyte gel against *Staphylococcus aureus*. J Orthop Res. 2008;26:404–10.
56. Wang J. Neutrophils in tissue injury and repair. Cell Tissue Res. 2018;371:531–9.
57. Brazil JC, Quiros M, Nusrat A, Parkos CA. Innate immune cell-epithelial crosstalk during wound repair. J Clin Invest. 2019;129:2983–93.
58. Cieślik-Bielecka A, Bold T, Ziółkowski G, Pierchała M, Królikowska A, Reichert P. Antibacterial activity of leukocyte- and platelet-rich plasma: an *in vitro* study. Biomed Res Int. 2018;2018:9471723–8.
59. Zhou Y, Wang JH. PRP treatment efficacy for tendinopathy: a review of basic science studies. Biomed Res Int. 2016;2016:9103792–8.
60. Koupenova M, Clancy L, Corkrey HA, Freedman JE. Circulating platelets as mediators of immunity, inflammation, and thrombosis. Circ Res. 2018;122:337–51.
61. Bayer IS. Advances in fibrin-based materials in wound repair: a review. Molecules. 2022;27:4504.
62. Serafini G, Lopreiato M, Lollobrigida M, et al. Platelet rich fibrin (PRF) and its related products: biomolecular characterization of the liquid fibrinogen. J Clin Med. 2020;9:1099.
63. Dohan Ehrenfest DM, Andia I, Zumstein MA, Zhang CQ, Pinto NR, Bielecki T. Classification of platelet concentrates (platelet-rich plasma-prp, platelet-rich fibrin-PRF) for topical and infiltrative use in orthopedic and sports medicine: current consensus, clinical implications and perspectives. Muscles Ligaments Tendons J. 2014;4:3–9.
64. Everts PA, Malanga GA, Paul RV, Rothenberg JB, Stephens N, Mautner KR. Assessing clinical implications and perspectives of the pathophysiological effects of erythrocytes and plasma free hemoglobin in autologous biologics for use in musculoskeletal regenerative medicine therapies. A review. Regen Ther. 2019;11:56–64.
65. Harrison TE, Bowler J, Levins TN, Reeves KD, Cheng AL. Platelet-rich plasma centrifugation changes leukocyte ratios. Cureus. 2021;13:e14470.
66. Hofmann E, Soppert J, Ruhl T, et al. The role of macrophage migration inhibitory factor in adipose-derived stem cells under hypoxia. Front Physiol. 2021;12:638448.
67. Schaer DJ, Buehler PW, Alayash AI, Belcher JD, Vercellotti GM. Hemolysis and free hemoglobin revisited: exploring hemoglobin and hemin scavengers as a novel class of therapeutic proteins. Blood. 2013;121:1276–84.
68. Humar R, Schaer DJ, Vallelian F. Erythrophagocytes in hemolytic anemia, wound healing, and cancer. Trends Mol Med. 2022;28:906–15.
69. Woodell-May JE, Sommerfeld SD. Role of inflammation and the immune system in the progression of osteoarthritis. J Orthop Res. 2020;38:253–7.

70. Vivier E, Tomasello E, Baratin M, Walzer T, Ugolini S. Functions of natural killer cells. Nat Immunol. 2008;9:503–10.
71. Vieira-de-Abreu A, Campbell RA, Weyrich AS, Zimmerman GA. Platelets: versatile effector cells in hemostasis, inflammation, and the immune continuum. Semin Immunopathol. 2012;34:5–30.
72. Nicolai L, Gaertner F, Massberg S. Platelets in host defense: experimental and clinical insights. Trends Immunol. 2019;40:922–38.
73. Lim A, Zhu JB, Khanduja V. The use of intra-articular platelet-rich plasma as a therapeutic intervention for hip osteoarthritis: a systematic review and meta-analysis. Am J Sports Med. 2022;51:3635465221095563–2497.
74. Fitzpatrick J, Bulsara M, Zheng MH. The effectiveness of platelet-rich plasma in the treatment of tendinopathy: a meta-analysis of randomized controlled clinical trials. Am J Sports Med. 2017;45:226–33.
75. Podesta L, Crow SA, Volkmer D, Bert T, Yocum LA. Treatment of partial ulnar collateral ligament tears in the elbow with platelet-rich plasma. Am J Sports Med. 2013;41:1689–94.
76. Deal JB, Smith E, Heard W, O'Brien MJ, Savoie FH 3rd. Platelet-rich plasma for primary treatment of partial ulnar collateral ligament tears: MRI correlation with results. Orthop J Sports Med. 2017;5:2325967117738238.
77. Parrish WR, Roides B. Physiology of blood components in wound healing: an appreciation of cellular co-operativity in platelet rich plasma action. J Exerc Sports Orthop. 2017;4:1–14.
78. Johal H, Khan M, Yung SP, et al. Impact of platelet-rich plasma use on pain in orthopaedic surgery: a systematic review and meta-analysis. Sports Health. 2019;11:355–66.
79. Hussein M, Hussein T. Effect of autologous platelet leukocyte rich plasma injections on atrophied lumbar multifidus muscle in low back pain patients with monosegmental degenerative disc disease. SICOT J. 2016;2:12.
80. Sánchez M, Garate A, Bilbao AM, et al. Platelet-rich plasma for injured peripheral nerves: biological repair process and clinical application guidelines. In: Bozzetto Ambrosi P, editor. Demystifying polyneuropathy—recent advances and new directions. London: IntechOpen; 2019.
81. Le VT, Nguyen Dao LT, Nguyen AM. Transforaminal injection of autologous platelet-rich plasma for lumbar disc herniation: a single-center prospective study in Vietnam. Asian J Surg. 2023;46:438–43.
82. Akeda K, Ohishi K, Takegami N, et al. Platelet-rich plasma releasate versus corticosteroid for the treatment of discogenic low back pain: a double-blind randomized controlled trial. J Clin Med. 2022;11:304.
83. Desai MJ, Mansfield JT, Robinson DM, Miller BC, Borg-Stein J. Regenerative medicine for axial and radicular spine-related pain: a narrative review. Pain Pract. 2020;20:437–53.
84. Kirchner F, Pinar A, Milani I, Prado R, Padilla S, Anitua E. Vertebral intraosseous plasma rich in growth factor (PRGF-Endoret) infiltrations as a novel strategy for the treatment of degenerative lesions of endplate in lumbar pathology: description of technique and case presentation. J Orthop Surg Res. 2020;15:72.
85. Sampson S, Gerhardt M, Mandelbaum B. Platelet rich plasma injection grafts for musculoskeletal injuries: a review. Curr Rev Musculoskelet Med. 2008;1:165–74.
86. Prodromos CC, Finkle S, Prodromos A, Chen JL, Schwartz A, Wathen L. Treatment of rotator cuff tears with platelet rich plasma: a prospective study with 2 year follow-up. BMC Musculoskelet Disord. 2021;22:499.
87. McLarnon M, Heron N. Intra-articular platelet-rich plasma injections versus intra-articular corticosteroid injections for symptomatic management of knee osteoarthritis: systematic review and meta-analysis. BMC Musculoskelet Disord. 2021;22:550.
88. Rodríguez-Merchán EC. Intra-articular platelet-rich plasma injections in knee osteoarthritis: a review of their current molecular mechanisms of action and their degree of efficacy. Int J Mol Sci. 2022;23:1301.
89. Moeda F, Melo X, Hatia M, et al. The effects of intra-articular platelet-rich plasma injections in rheumatoid arthritis: a narrative review. Cureus. 2022;14:e28182.
90. Street A, Jerome MA, Williams CJ. Atraumatic medullary osteonecrosis of the tibia and femur treated with intraosseous orthobiologics. Cureus. 2021;13:e16677.
91. Everts PA, Devilee RJ, Brown Mahoney C, et al. Exogenous application of platelet-leukocyte gel during open subacromial decompression contributes to improved patient outcome. A prospective randomized double-blind study. Eur Surg Res. 2008;40:203–10.
92. Li Z, Weng X. Platelet-rich plasma use in meniscus repair treatment: a systematic review and meta-analysis of clinical studies. J Orthop Surg Res. 2022;17:446.
93. Lu M, Jin Y. Efficacy evaluation of the combined platelet-rich plasma and hyaluronic acid after arthroscopic joint debridement in treating knee osteoarthritis. Scanning. 2022;2022:6994017–5.
94. Sprott H, Franke S, Kluge H, Hein G. Pain treatment of fibromyalgia by acupuncture. Rheumatol Int. 1998;18:35–6.
95. Everts PA, Flanagan G, Podesta L. Autologous orthobiologics. In: Mostoufi SA, George TK, Tria AJ, editors. Clinical guide to musculoskeletal medicine. Cham: Springer International; 2022. p. 651–79.
96. Nazaroff J, Oyadomari S, Brown N, Wang D. Reporting in clinical studies on platelet-rich plasma therapy among all medical specialties: a systematic review of level I and II studies. PLoS One. 2021;16:e0250007.
97. Bec C, Rousset A, Brandin T, et al. A retrospective analysis of characteristic features of responders and impaired patients to a single injection of pure platelet-rich plasma in knee osteoarthritis. J Clin Med. 2021;10:1748.

# Platelet Lysate

# 8

John Knab, Ben Rawson, and David Harris

## Introduction

The field of regenerative medicine is rapidly expanding with the growing recognition that healing of injured and degenerated connective tissues using naturally derived growth factors is both possible and proving to be safe and effective. Growth factors are becoming better categorized as they are discovered and have been found to stimulate healing via immunomodulatory effects, both pro-inflammatory and anti-inflammatory. When activated, platelets in the circulation release the contents of their alpha granules, which facilitate several physiologic processes including initiation of blood clotting, recruitment of white blood cells to both fight infection and debride damaged tissue, and stimulation of fibroblasts and mesenchymal stem cells (MSCs) to facilitate tissue healing.

Standard preparations of platelet-rich plasma (PRP) are known to contain elevated levels of growth factors (GFs). However, with PRP, GF delivery to pathologic tissue is highly dependent on a number of factors including local tissue environment, activation status of PRP, and the degree to which platelets are able to degranulate and release GFs contained in their alpha granules. For these, and other reasons, clinicians began to look at platelet lysate as a potential substitute and/or replacement for PRP in a variety of clinical conditions (Figs. 8.1 and 8.2).

Platelet lysate is not a new discovery. It was first used in the 1980s as a substitute for fetal bovine serum (FBS) in fibroblast and endothelial cell cultures. In 2005, Doucet found that platelet lysate outperformed FBS, resulting in larger colony-forming unit (CFU) counts and improved stem cell differentiation [1]. The use of FBS as a cell culture supplement is discouraged by regulatory authorities to limit the risk of zoonosis and xenogeneic immune reactions in the transplanted host. Additionally, FBS production came under scrutiny due to animal welfare concerns. Platelet derivatives have been proposed as FBS substitutes for the ex-vivo expansion of mesenchymal stem/stromal cells since platelet-derived growth factors can promote MSC ex-vivo expansion [2]. Platelet lysate has been proven to be helpful in cell culture media to accelerate the proliferation of cell types including MSCs, tenocytes, fibroblasts, chondrocytes, keratinocytes, and osteoblasts [3].

Jiang and colleagues appear to have been the first report on the clinical use of platelet lysate in the treatment of tibia non-union in 2016. Since that initial publication, other publications have followed, suggesting that platelet lysate may be a useful tool in the management of a variety of musculoskeletal, spinal, and neuropathic conditions [4–7].

To create platelet lysate, autologous blood is first minimally processed using centrifugation to produce platelet-rich plasma (PRP). This can be further processed using techniques described in this chapter to produce an acellular supernatant called platelet lysate and/or platelet releasate (PR), which is are rich in cytokines and growth factors and which may clinically appear to have a generally more anti-inflammatory profile compared to than PRP.

There are currently no published studies comparing the clinical effectiveness of Platelet Lysate and PRP to date. In theory, there should be extensive overlap between the growth factor profiles of each. The primary differences that separate platelet lysate from PRP include the elimination of cellular membrane material, additional uncategorized growth factor activation mechanisms, and the release of enzymes during the alpha granule activation process which occurs during the lysing process. Platelet lysate is acellular and non-particulate; these may be desirable attributes in certain clinical conditions. GF levels in platelet lysate are also highly quantifiable,

J. Knab (✉)
Center for Pain Management, Wilmington, NC, USA
e-mail: jknab@portcitypain.com

B. Rawson · D. Harris
CHARM Austin, Austin, TX, USA
e-mail: dharris@charmaustin.com


A. Navani et al. (eds.), *Essentials of Regenerative Medicine in Interventional Pain Management*,
https://doi.org/10.1007/978-3-031-50357-3_8

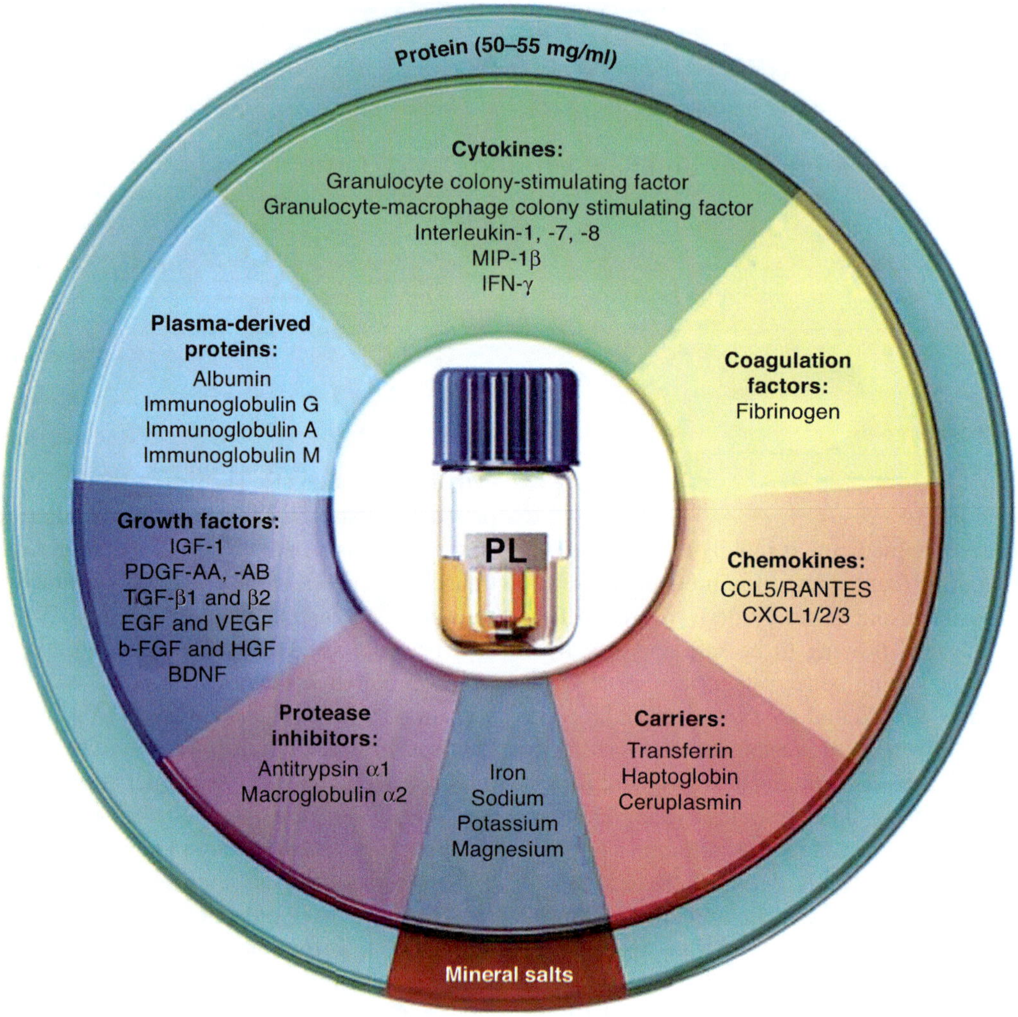


**Fig. 8.1** Platelet lysate composition. *PRP* platelet-rich plasma, *GF* growth factor, *BDNF* bone-derived growth factor, *EGF* epidermal growth factor, *HB-EGF* heparin-binding EGF-like growth factor, *IL-1 RA* interleukin-1 receptor antagonist, *PDGF-BB* platelet-derived growth factor with 2 β subunits, *VEGF* vascular endothelial growth factor. (Source: Meftahpour V, et al. Platelet lysate: A promising candidate in regenerative medicine. *Regen Med* 2021; 16:71–85)

with *in vitro* models suggesting that platelet lysate from aged donors yielded a dose-response relationship in tenocyte behavior, with higher platelet lysate concentrations resulting in increased tenocyte proliferation and migration [8]. Being able to deliver condition-specific levels of GFs may enhance the clinical utility of platelet lysate therapy.

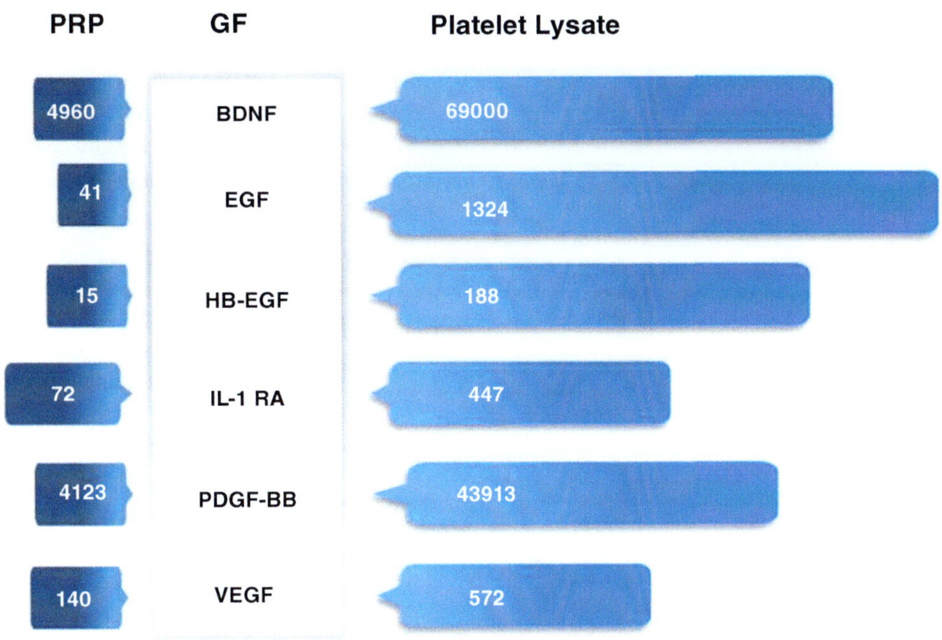


**Fig. 8.2** Average GF levels in PRP vs. platelet lysate prepared via ultrasonic probe method in author's (BR, DH) study. (Source: Rawson B, Torres H, Harris D. Comparative methods for producing platelet lysate. Submitted for publication)

## Methods of Producing Platelet Lysate

The process of creating a solution rich in proteins and growth factors from platelet alpha granules starts with a platelet-rich plasma (PRP) solution. PRP is commonly prepared from whole blood using a double centrifugation method. The concentrated platelets are suspended in plasma to create platelet-rich plasma solution. The PRP can then be processed by additional steps to create platelet lysate or releasate solution (Fig. 8.3).

Of the several methods of producing platelet lysate, there is no consensus on which results in the most substantial concentration of proteins and growth factors [9–12].

Platelets can either be stimulated to release their alpha granules for a PR (using CaCL or CaCl + thrombin) versus mechanically lysed to create a platelet lysate solution (freeze thaw or ultrasound). Methods that produce platelet lysis are described as beneficial over those of platelet degranulation, in which proteins may become trapped in the fibrin meshwork and are consequently unavailable in the resulting solution [9].

Common methods of producing platelet lysate (Tables 8.1, 8.2 and 8.3) include use of calcium chloride ($CaCl_2$), freeze-thaw techniques, or ultrasonic probe homogenization and Ultrasonic bath (USb) homogenization [10, 13–18]. The addition of $CaCl_2$ to PRP is a well-studied method of activation. $CaCl_2$ stimulates the formation of a clot and platelet degranulation with the resulting release of the intracorpuscular products into the plasma [19].

The sample can then be centrifuged to remove cellular products and the clot, leaving growth factors and other cytokines concentrated in the plasma. Protocols contain variability in the concentration of $CaCl_2$ used, the time allowed for clot formation, and the concomitant administration of other activators, such as thrombin [20–23].

A potential complication of the $CaCl_2$ method is the loss of growth factors during centrifugation because the proteins can become trapped within the cellular matrix formed during the clot [17, 19].

### Freeze-Thaw Method

Freeze-thaw cycles are one of the common methods described for creating effective platelet lysate. These also have considerable variability in methodology including the number and length of time of cycles, and the range of temperatures used [9, 10, 24, 25] (Figs. 8.4 and 8.5).

During each freeze cycle, water crystals form and lyse the cellular membranes of platelets, allowing their intracellular products to be released [10].

There are concerns for the stability of proteins during the freeze-thaw process, though this effect during the activation of PRP has yet to be described [10, 26].

Strandberg and colleagues took PRP from 10 healthy donors and made platelet lysate by freezing the PRP at −70°C for 10 min and then thawing it in a 37°C water bath for 7.5 min. They analyzed growth factor levels after each of 30 sequential freeze thaw cycles. Optimal levels of epidermal growth factor (EGF) and platelet-derived growth factor (PDGF) were achieved at cycles 3–5 (Fig. 8.6), and the clinical significance of using a greater number (more than 3–5) of cycles is likely limited [27].

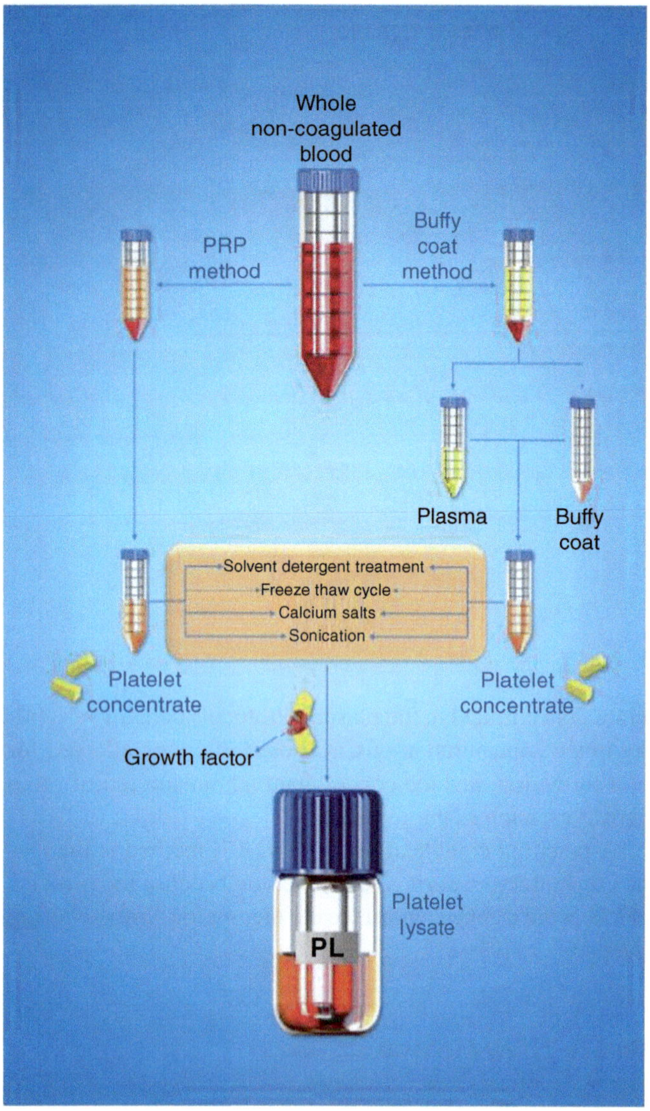


**Fig. 8.3** Various methods of producing platelet lysate. (Source: Meftahpour V, et al. Platelet lysate: A promising candidate in regenerative medicine. *Regen Med* 2021; 16:71–85)

## Ultrasonic Homogenization

Ultrasonic homogenization is another common laboratory procedure used to lyse cells. Sound waves, produced at frequencies of >20 kHz, rapidly compress and decompress dissolved gasses causing violent collisions resulting that in the rupture of cell membranes and subsequent release of intracellular proteins. Ultrasonic homogenization can transfer heat, possibly denaturing proteins. Although this phenomenon is suggested to be an inefficient transfer, submersion of the sample into an ice bath is commonly often used to decrease this transfer of heat [14].

Although there is limited evidence that ultrasonic techniques are effective in producing platelet lysate, two methods of ultrasonic homogenization have been studied: USb (Fig. 8.7) and ultrasonic probe (Fig. 8.8). USb methods rely on use of closed containers in which the PRP sample is placed in a water bath and ultrasonic waves are then transmitted through the bath. Bernardi et al. have described the successful use of USb in producing factor-concentrated platelet lysate used for stimulating mesenchymal stromal cell differentiation [13, 14, 28].

The ultrasonic probe (Fig. 8.8) method relies on a probe directly inserted into the PRP solution for direct transfer of ultrasonic waves. This has the potential to provide a more efficient transfer of energy to the PRP. In order to minimize the risk of contamination, the sterilized probe is placed into a sterile ultrasound sleeve before placement into the PRP. Fortunato et al. [13] demonstrated the successful use of an ultrasonic probe to efficiently and effectively create platelet lysate, which contained high growth factor concentrations and stimulated endothelial colony forming cells to form and expand microvascular networks.

## Final Centrifugation and Filtering

Following processing of PRP into platelet lysate using one of the methods described above, the solution will still contain

**Table 8.1** PDGF levels (pg/mL) in positive control (Bovine Thrombin + $CaCl_2$) plasma vs. 4 different releasate and lysate methods used during validation process in author's (JK) clinic

| Donor ID | Negative control | Positive control | Osmotic lysis | Freeze-thaw lysis | Autologous thrombin | Autologous thrombin + $CaCl_2$ |
|---|---|---|---|---|---|---|
| 2301 | 11,630.41 | 42,053.11 | 15,506.35 | 18,151.76 | 18,249.17 | 18,921.70 |
| 2302 | ND | 35,712.47 | ND | 18,494.61 | 15,549.20 | 20,488.11 |
| 2303 | 10,587.56 | 42,013.82 | ND | 21,094.58 | 16,406.34 | 21,301.67 |
| 2304 | ND | 34,958.19 | ND | 19,823.17 | 17,134.90 | 20,330.26 |
| 2305 | 10,458.99 | 39,884.56 | 15,088.50 | 24,930.25 | 16,031.34 | 17,416.01 |
| Mean | 10,892.32 | 38,924.43 | 15,297.42 | 20,498.88 | 16,674.19 | 19,691.55 |
| St dev | 524.54 | 3043.11 | 208.93 | 2448.45 | 942.70 | 1371.55 |

*ND* not detected

**Table 8.2** Serum Markers produced by each lysate method in the author's (BR, DH) study (Rawson B, Torres H, Harris D. Comparative methods for producing platelet lysate. Submitted for publication)

| | BDNF (pg/mL) | EGF (pg/mL) | HB-EGF (pg/mL) | IL-1 RA (pg/mL) | PDGF-BB (pg/mL) | VEGF (pg/mL) |
|---|---|---|---|---|---|---|
| Control | 4960 | 41 | 15 | 72 | 4123 | 140 |
| Ozone | 17,138 | 357 | 84 | 78 | 10,013 | 245 |
| USp | $54500^{a\,b\,c\,d}$ | $1324^{a\,b\,c\,d}$ | $188^{a}$ | $447^{a\,b\,d}$ | $43913^{a\,b\,c\,d}$ | $572^{a\,b\,d}$ |
| USb | $69000^{a\,b\,c\,d}$ | $1390^{a\,d}$ | $234^{a\,b\,d}$ | 259 | $46813^{a\,b\,c\,d}$ | $713^{a\,d}$ |
| cFT | 22,625 | 600 | 123 | $471^{a\,b\,d}$ | 20,625 | 310 |
| wFT | 24,838 | 269 | 98 | 107 | 21,024 | 279 |

*USp* ultrasound probe, *USb* ultrasound bath, *cFT* cold freeze-thaw, *wFT* warm freeze-thaw, *BDNF* bone-derived growth factor, *EGF* epidermal growth factor, *HB-EGF* heparin-binding EGF-like growth factor, *IL-1 RA* interleukin-1 receptor antagonist, *PDGF-BB* platelet-derived growth factor with 2β subunits, *VEGF* vascular endothelial growth factor

This table shows the average marker concentrations produced by each method from the individual samples. The superscripts represent marker levels that were produced at significantly ($p<.05$) higher concentrations than the corresponding marker levels for Control ($^{a}$), Ozone ($^{b}$), cFT ($^{c}$), or wFT ($^{d}$). (*pg/mL* picogram per milliliter)

**Table 8.3** Serum markers produced by each lysate method (ozone activation, ultrasound probe (USp), ultrasound bath (USb), cold freeze-thaw (cFT), warm freeze-thaw (wFT)) in the author's (BR, DH) study (Rawson B, Torres H, Harris D. Comparative methods for producing platelet lysate. Submitted for publication)

| | BDNF (pg/mL) | EGF (pg/mL) | HB-EGF (pg/mL) | IL-1 RA (pg/mL) | PDGF-BB (pg/mL) | VEGF (pg/mL) |
|---|---|---|---|---|---|---|
| Control | 4960 | 41 | 15 | 72 | 4123 | 140 |
| Ozone | 17,138 | 357 | 84 | 78 | 10,013 | 245 |
| USp | 54,500 | 1324 | 188 | 447 | 43,913 | 572 |
| USb | 69,000 | 1390 | 234 | 259 | 46,813 | 713 |
| cFT | 22,625 | 600 | 123 | 471 | 20,625 | 310 |
| wFT | 24,838 | 269 | 98 | 107 | 21,024 | 279 |

*pg/mL* picogram per milliliter

*BDNF* bone-derived growth factor, *EGF* epidermal growth factor, *HB-EGF* heparin-binding EGF-like growth factor, *IL-1 RA* interleukin-1 receptor antagonist, *PDGF-BB* platelet-derived growth factor with 2β subunits, *VEGF* vascular endothelial growth factor

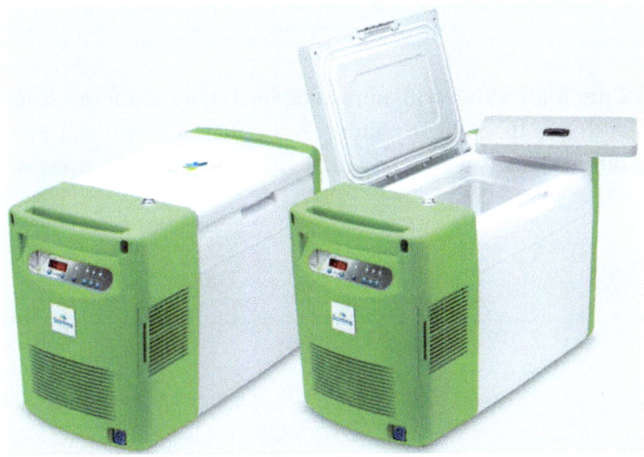

**Fig. 8.4** Benchtop ultracold freezer for preparation of platelet lysate via freeze-thaw method. (Source: Biolife Solutions. https://www.biolifesolutions.com/storage/stirling-ultra-low/portable-ult-freezer/)

**Fig. 8.5** Digital warming oven for preparation of platelet lysate via freeze-thaw method. (Source: Benchtop Oven. Thomas Scientific. https://www.thomassci.com/scientific-supplies/Benchtop-Oven)

platelet fragments and debris along with the desired growth factors. Growth factors can be isolated in solution by filtering out the platelet debris. This is accomplished by a final centrifugation at high speed to create a platelet fragment pellet (PFP). The supernatant above the PFP is siphoned off, and then filtered through a terminal filtration step down to 0.2 μm. The final filtration step may also be accomplished by passing the supernatant through a series of filters, e.g., 15 μm → 5 μm → 2 μm → 0.2 μm (Fig. 8.9); doing so may help to prevent clogging of the smaller filters. The final 0.2 μm filter will likely also filter out potential bacterial and fungal pathogens.

The ultrasound devices described above have traditionally been used for sterility [29]. The ultrasound waves effectively kill any bacteria or viruses by destroying the cell wall or

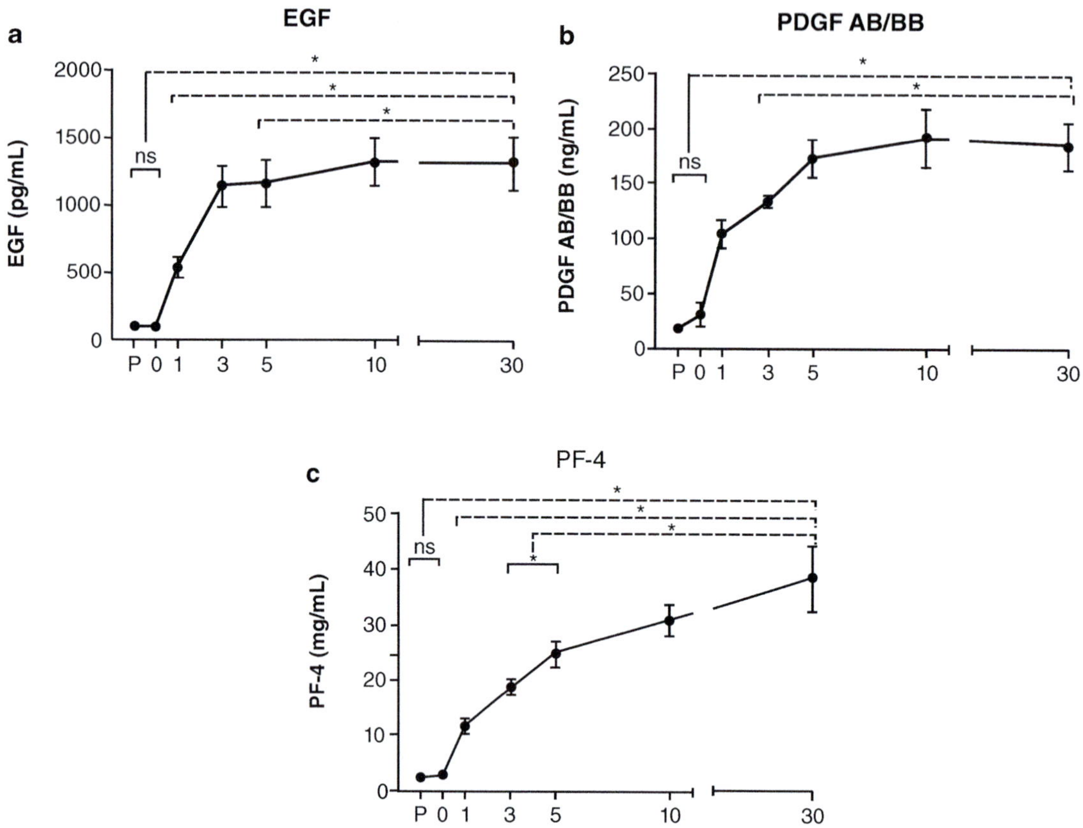


**Fig. 8.6** Graphs showing optimization of GF levels in platelet lysate after 3–5 freeze-thaw cycles. Graphs **a**, **b**, and **c** demonstrating EGF, PDGF AB/BB, and PF-4 levels with sequential freeze-thaw cycles in pg/mL, ng/mL, and mg/mL, respectively. (Source: Strandberg G, Sellberg F, Sommar P, et al. Standardizing the freeze-thaw preparation of growth factors from platelet lysate. *Transfusion* 2017; 57:1058–1065 [27])

membrane. This may prove to be beneficial for creation of platelet lysate by creating a more sterile solution compared to the other methods described. The combined use of a 0.2 μm filter, along with ultrasonication, may result in a solution that is theoretically more sterile than a traditional PRP solution. This is then ready to be used for clinical purposes.

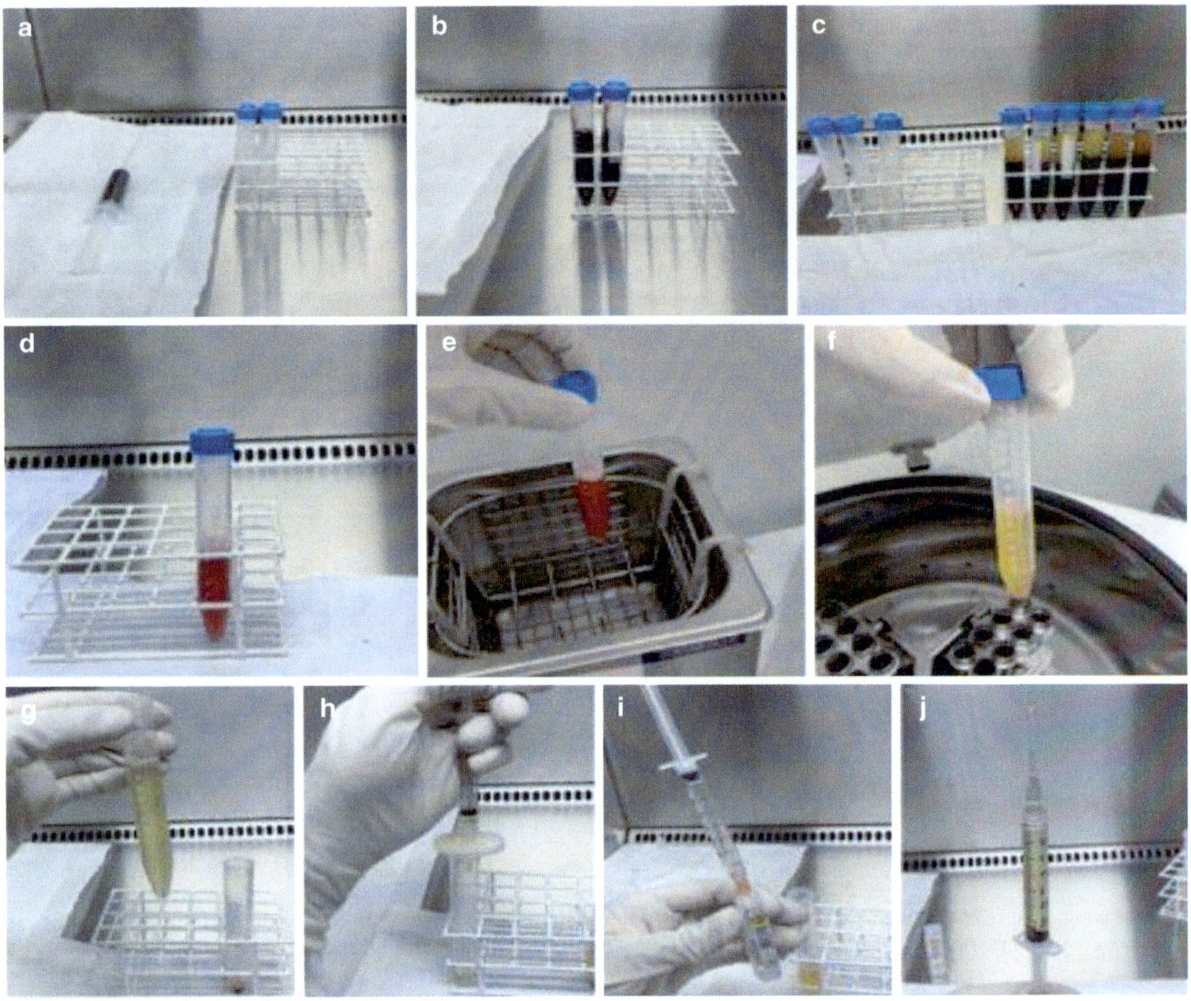


**Fig. 8.7** Platelet lysate preparation via ultrasonic bath method. (Source: da Fonseca L, Santos GS, Huber SC, et al. Human platelet lysate—A potent (and overlooked) orthobiologic. *J Clin Orthop Trauma* 2021; 21:101534 [28])

**Fig. 8.8** Ultrasonic probe setup for preparation of platelet lysate. (Source: QSonica. https://www.sonicator.com/products/q700-sonicator?_pos=6&_sid=ce4a2fa23&_ss=r&variant=36028875344)

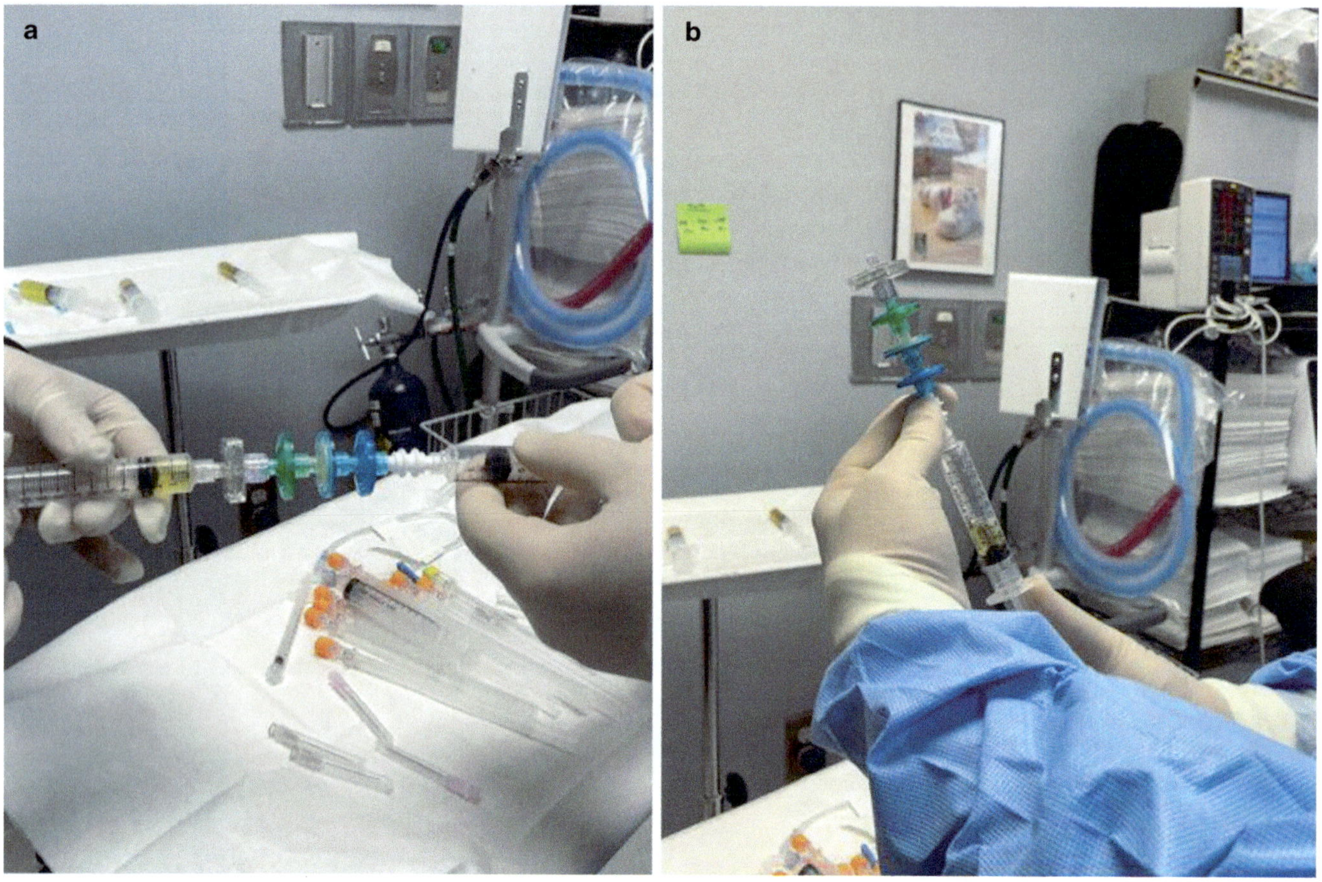

**Fig. 8.9** (**a**) Filtration of platelet lysate after final centrifugation from platelet lysate syringe (left syringe) through a series of filters (smallest = 0.2 μm) into a collection syringe (right syringe). (**b**) Platelet lysate in collection syringe ready for injection. (Source: JK personal photo)

## Platelet Lysate and Releasate Current and Future Clinical Use: Spine and Musculoskeletal Conditions

### Spinal Pathology

Low back pain is a significant cause of disability resulting in major social and economic burdens. From 1992–2015, the number of people with disabilities due to low back pain increased by 54% [30]. Inflammation and activation of signaling pathways play an important role in onset and progression of disc degeneration [31]. In the degenerated intervertebral disc, there is an upregulation of pro-inflammatory molecules such as tumor necrosis factor (TNF) alpha, interleukin (IL)-1 data, IL-4, IL-6, IL-8, IL-12, prostaglandin E2 (PGE2), interferon gamma, and nitric oxide (NO), which are found at elevated levels in painful, herniated intervertebral discs [32–34]. These molecules may also play a role in the ingrowth of nociceptive fibers into the annulus fibrosus [35, 36].

In a review by Lyu et al. [31] in 2021, the authors point out that disc degeneration is a significant contributor to low back pain. The article focused on advances in the treatment of inflammation-related pain in intervertebral disc disease (IDD). Factors contained in PRP, such as platelet derived growth factor, TGF beta, EGF and insulin-like growth factor 1 and vascular endothelial growth factor were suggested as playing an important role in dampening inflammation and promoting tissue regeneration. They refer to a study by Akeda et al. [6] in which "6 low back pain patients were treated with 2 mL of PRP" [31]; however, in the Akeda study, the substance injected into the intervertebral disks was a platelet releasate (PR), and not PRP [6, 31].

In 2017, Akeda et al. [6] reported on 14 patients who were receiving diagnostic discography for suspected discogenic low back pain from April 2009 to March 2012 (Fig. 8.10). Inclusion criteria included age over 18, one or more painful lumbar intervertebral discs (L3–L4 to L5–S1) on discography, with evidence of disc degeneration on magnetic resonance imaging (MRI). A minimum of 50% or more of normal disc height was required. Exclusion criteria included radiculopathy, stenosis, spondylolisthesis, and inflammatory arthritis. PRP releasate was prepared by collecting 200 mL of whole blood with ACD which was centrifuged to create PRP (3.7× baseline). Autologous serum (containing autologous thrombin) was combined with 2% $CaCl_2$ and PRP to form a PRP clot. The clot was incubated for 60 min at room tem-

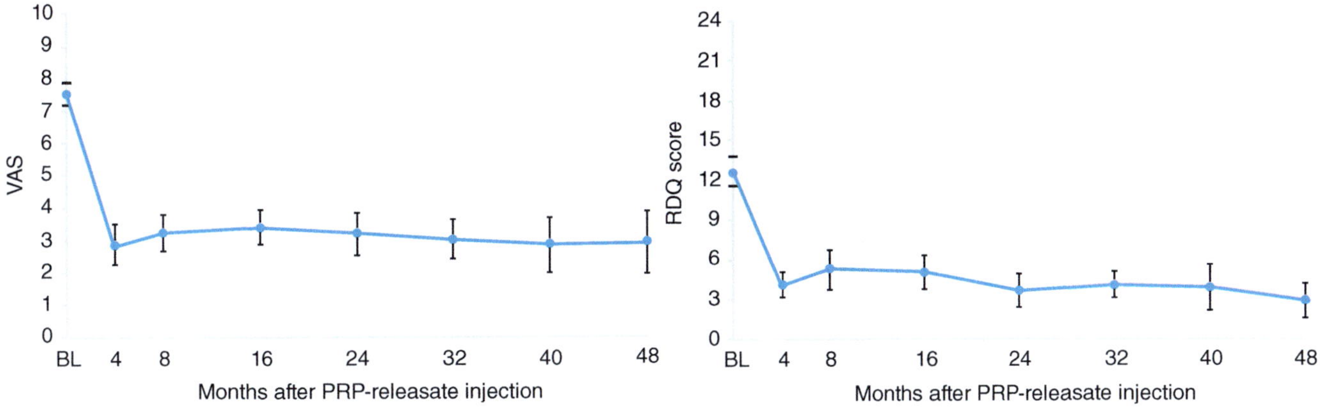


**Fig. 8.10** VAS and RMDQ scores over 48 months in 14 patients after intradiscal platelet releasate injection. (Source: Akeda K, Ohishi K, Masuda K, et al. Intradiscal injection of autologous platelet-rich plasma releasate to treat discogenic low back pain: A preliminary clinical trial. *Asian Spine J* 2017; 11:380–389 [6])

perature and supernatant (PRP releasate) was isolated from the PRP clot by centrifugation. 2 mL of platelet lysate was injected into each symptomatic disc. Patients were evaluated at baseline, 4, 8, 16, 24, 32, 40, and 48 weeks after treatment by means of Visual Analog Scale (VAS) and Roland Morris Disability Questionnaire (RMDQ). Results included more than 50% reduction of low back pain in 71% of patients within 4 weeks after injection, with maintenance of pain reduction throughout the observation period. RMDQ scores were significantly reduced in 79% of patients 4 weeks after PRP releasate injection, with maintenance of improvements for 48 weeks after treatment. The authors reported that no patient exhibited neurologic deterioration or symptoms of discitis, and no adverse events other than transient numbness lasting less than 7 days in two patients [6].

Centeno et al. [7] reported on the use of lumbar epidural injection of platelet lysate for treatment of radicular pain in 2017. Registry data was collected on 470 patients suffering from lumbar radicular pain who had MRI findings consistent with symptoms. Leukocyte poor PRP was created by withdrawing 60 mL of heparinized venous blood which was centrifuged at 200 g for 10 min. PRP samples were placed in a −80°C freezer for 5–10 min followed by thawing. Platelet lysate was re-centrifuged in order to pellet any remaining platelet bodies, and the supernatant was extracted and sent to bedside for use. Prior to injection, the injectate was filtered through a 0.22 μ filter to remove all remaining lysed platelet membrane debris. Platelet lysate was delivered using C-arm fluoroscopy via either transforaminal or interlaminar epidural route, with 3–5 mL of a mixture of platelet lysate, lidocaine, and nanogram dose hydrocortisone being injected per symptomatic level. No serious adverse events such as infection, paralysis, or neurologic deficit were reported. Transient adverse events including dural puncture headache in 3 patients were reported in a total of 29 patients.

Results included a 42.4% improvement at an average of 16.6 months in modified Single Assessment Numeric Evaluation (SANE) rating response. Patients reported significantly less pain after treatment, which was maintained through 24 months (Fig. 8.11). The 9-point MCID for Functional Rating Index (FRI) was met or exceeded at all time points beyond 1 month. The authors state that platelet lysate may be preferable to PRP in spinal indications as the former does not carry the potential for platelet adhesion and aggregation, which increases the risk of vascular occlusion. Given the fact that several tragic cases of ischemic spinal cord lesions have been reported via particulate steroid injection of radicular arteries, the authors suggest that scrutiny concerning the embolic nature of injections is warranted [7].

In 2019, Rawson [5] reported on the use of platelet-rich plasma (PRP) and epidural platelet lysate for treatment of lumbar disc herniation. He described two patients with symptomatic herniated discs who were suffering from low back pain and lumbosacral/sacral radiculopathy. PRP was prepared from 180 mL of whole blood using a double centrifugation method in an in-house laboratory. PRP was classified as HA-/RBC-P/Sp2/Plt 4-6/G+/Lc-R/A.

A portion of PRP was further processed to produce platelet lysate by mechanical rupture of platelets using an ultrasonic homogenization method. Platelet lysate solution was filtered to remove cellular debris and platelet growth factor concentrations were measured. PRP was used to inject the posterior structures of the spine (spinal ligaments, facet joints), and 3 mL of platelet lysate was injected into the epidural space using transforaminal, interlaminar, or caudal approach, based on patient pathology.

The first patient presented with a 1-year history of progressively worsening pain including radiculopathy which was felt to be due to a large herniated/extruded disc at L4–L5 with displacement of the L5 and S1 nerve roots (Fig. 8.12). She received an injection of PRP/platelet lysate,

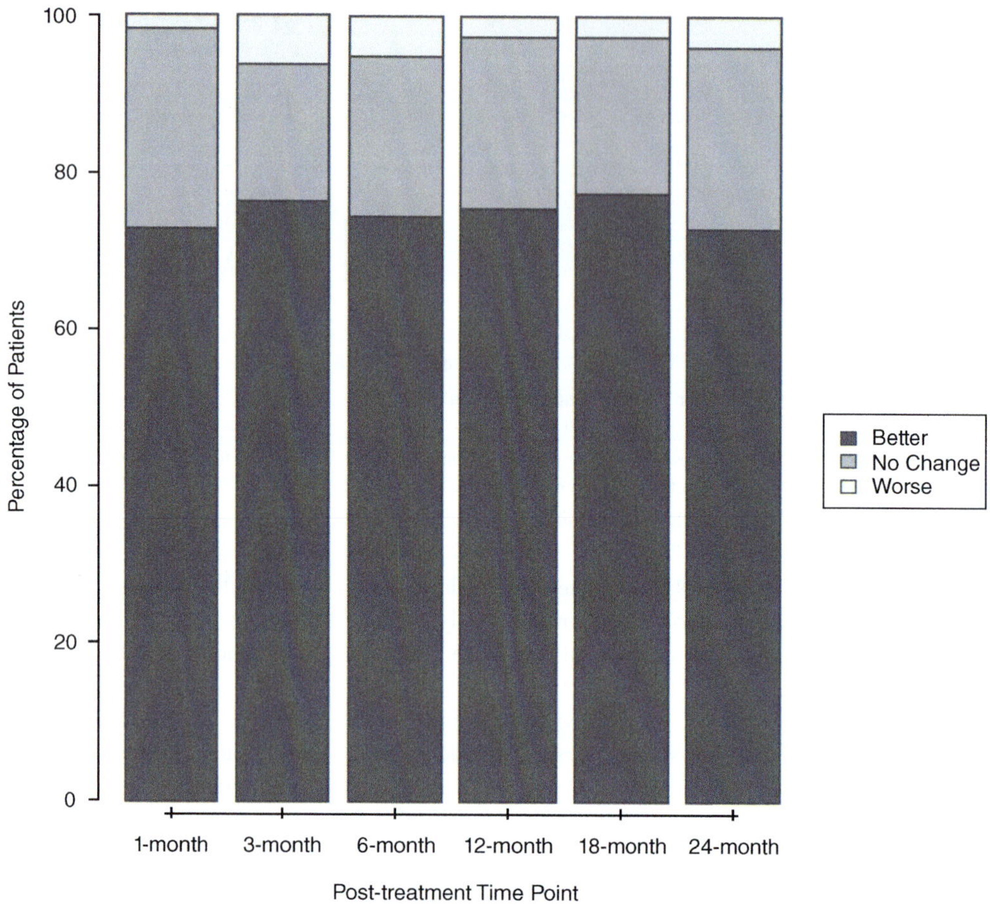


**Fig. 8.11** Percentage of patients reporting changes of "better" (dark gray) "no change" (med gray) or "worse" (light gray) following epidural injection of platelet lysate for treatment of radiculopathy. (Source: Centeno C, Markle J, Dodson E, et al. The use of lumbar epidural injection of platelet lysate for treatment of radicular pain. *J Exp Orthop* 2017; 4:38 [7])

including a fluoroscopically guided, left-sided L4–L5 paramedian interlaminar injection of platelet lysate. At 4 weeks following treatment, she was reporting 50% improvement in pain and underwent a second PRP/platelet lysate treatment. At 4 weeks following the second procedure, the patient reported almost complete resolution of pain and function with no adverse effects. Repeat lumbar MRI revealed significant resorption of disc herniation with only mild residual disc material which was not compressing the spinal nerves. At six-month follow-up significant overall improvement in pain and function continued.

The second patient (Fig. 8.13), also suffering from radiculopathy due to a large left paracentral and caudally extruded L4-L5 disc herniation. The patient received 2 separate injections of PRP/platelet lysate at 1 and 3 months after the initial visit. At 3 months following the second treatment the patient was describing complete resolution of pain, numbness, tingling, and weakness along with improved function. No adverse effects were reported by the patient. MRI showed almost complete resorption of disc material with no evidence of ongoing neural impingement [5].

## Refractory Lateral Epicondylitis

Pain and dysfunction of the lateral epicondyle, commonly referred to as "lateral epicondylitis" or "tennis elbow" is a common medical condition affecting up to 3% of the general population, peaking in the fourth to fifth decade of life [37–39].

Current clinical and surgical data suggests that lateral epicondylitis is not an inflammatory process but instead a chronic degenerative process involving tendinopathic change of the extensor carpi radialis brevis muscle at its origin on the lateral epicondyle of the humerus. Pathology of tendon fibers is often the result of repetitive stress injury, specifically activities involving active and resisted wrist extension and/or repetitive rotation, extension, and flexion of the forearm [40].

Recognizing that autologous platelet lysates (APL) contain a myriad of biologic growth factors such as vascular endothelial growth factor, insulin-like growth factor, and transforming growth factor, which might potentially influence tissue repair through angiogenesis, cellular chemotaxis, removal of tissue debris, and reconstruction of extracellular

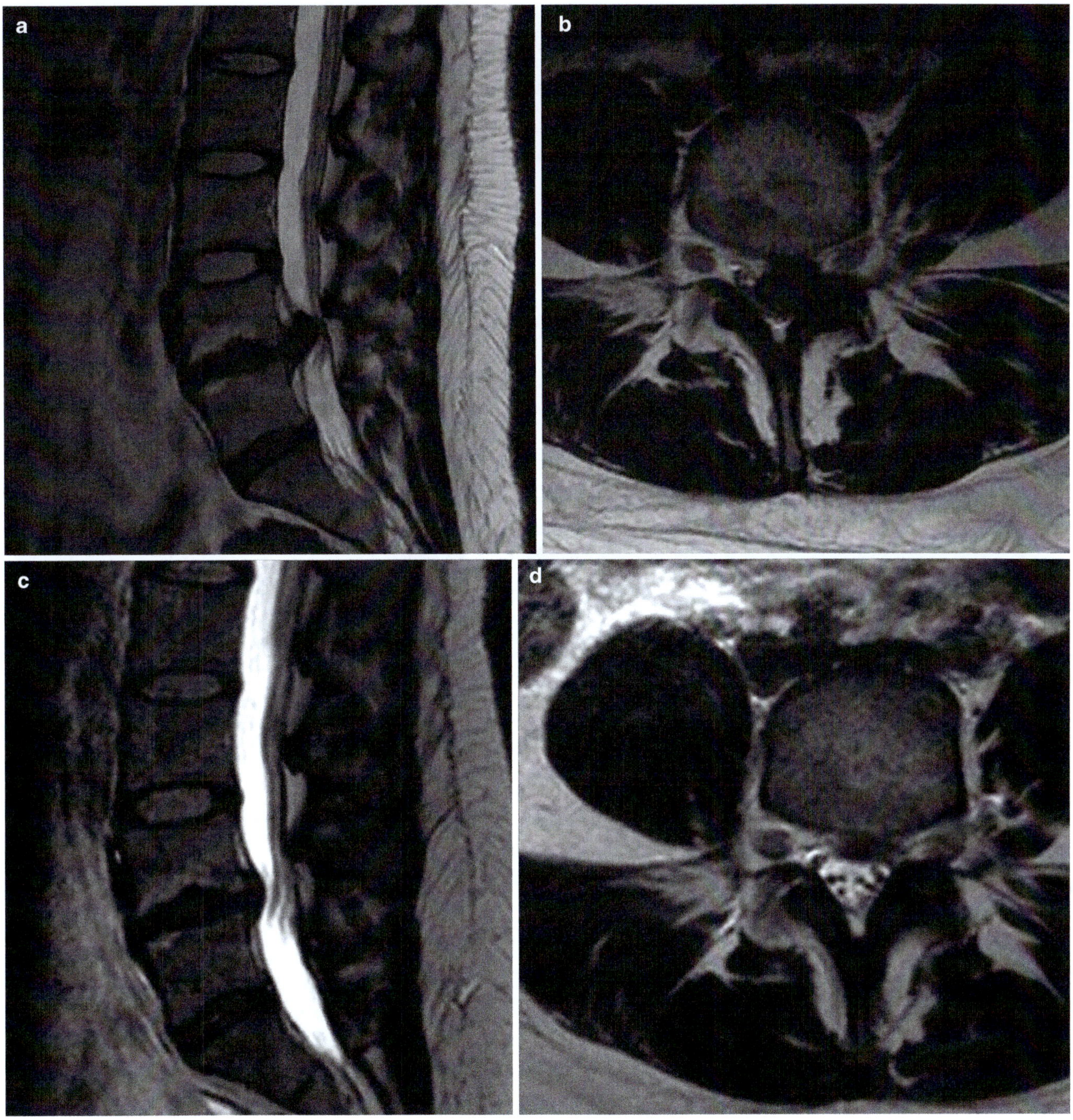


**Fig. 8.12** MRI images from patient 1 in Rawson study before (**a**, **b**) and at 8 weeks after (**c**, **d**) two consecutive platelet lysate/PRP procedures showing significant resorption of L4/5 disc extrusion. (Source: Rawson B. Platelet-rich plasma and epidural platelet lysate: Novel treatment for lumbar disk herniation. *J Am Osteopath Assoc* 2020; 120:201–207 [5])

matrix, Tan et al. [40] performed in retrospect of study to evaluate the short-term efficacy of local APL injections in reducing pain and improving function in patients with refractory lateral epicondylitis [4, 41–43].

They reported results in 56 patients presenting to their clinic between 2012 and 2014 ranging in ages from 36 to 62 with average duration of clinical symptoms of 9.7 months. All patients had failed conservative treatments, and symptoms were present for at least 3 months. Diagnosis of tendon pathology was confirmed via ultrasound. Platelet lysate was prepared by collecting 60 mL of peripheral blood mixed with 6 cc ACD. The PRP preparation containing platelets at a concentration of 3× over baseline was prepared in a non-activated form. Platelet lysate was next prepared by cryopreservation

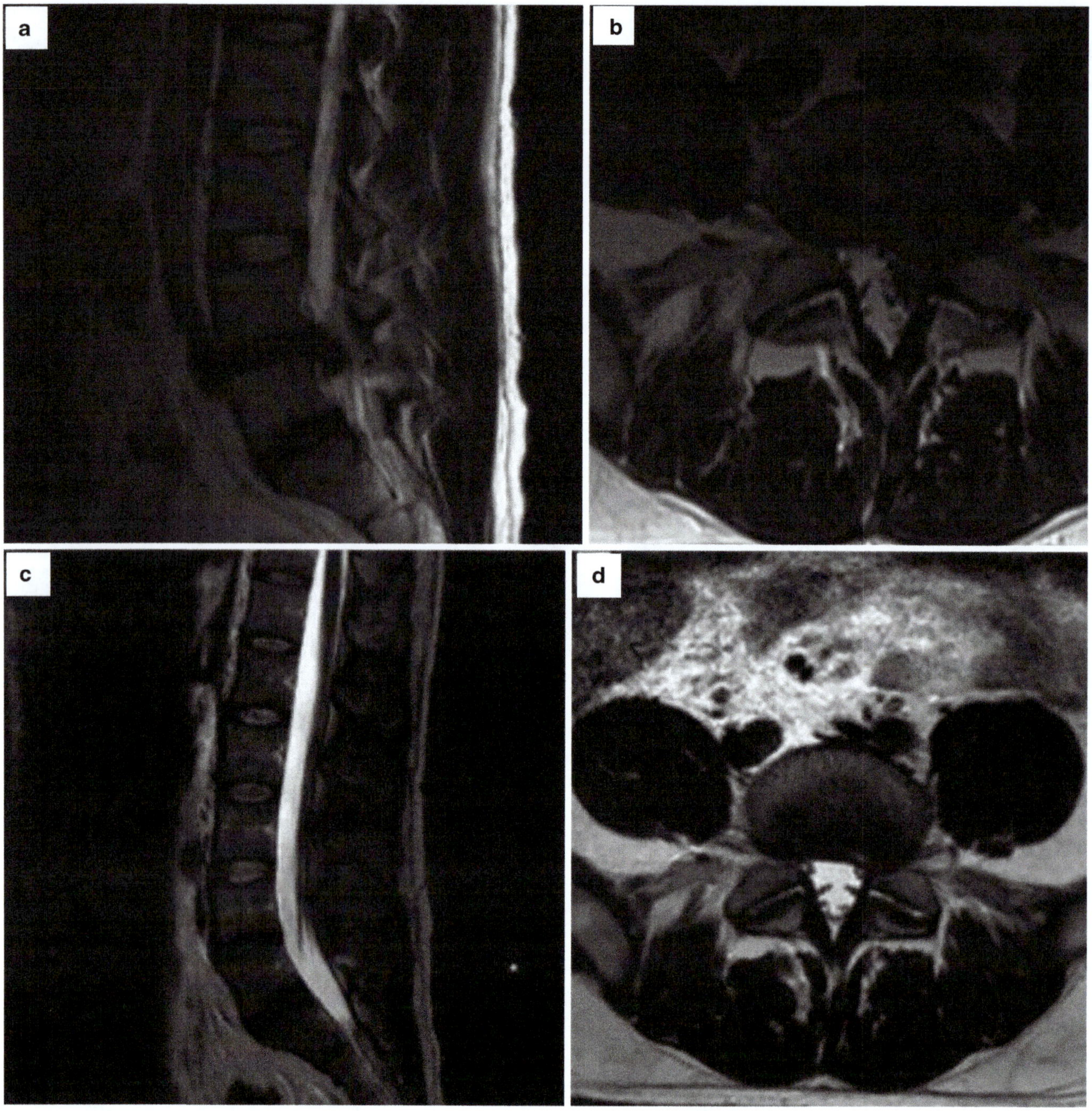


**Fig. 8.13** MRI images from patient 2 in Rawson study prior to (**a**, **b**) and at 7 weeks following (**c**, **d**) 2 consecutive PRP/platelet lysate treatments. (Source: Rawson B. Platelet-rich plasma and epidural platelet lysate: Novel treatment for lumbar disk herniation. *J Am Osteopath Assoc* 2020; 120:201–207 [5])

at −80° overnight and then thawed and a 37°C water bath for 5 min. The resulting platelet lysate volume was 3 mL.

Injections were performed with ultrasonic guidance, via 25-gauge needle, with targeted injections into areas of ultrasonic abnormalities. A single skin portal with multiple penetrations into areas of tendon pathology was made. Post treatment, patients were instructed to refrain from strong gripping or lifting activities for 1 week and to not use nonsteroidal anti-inflammatory medications for 4 weeks. After 2 weeks, gradual strengthening and return to activity were initiated.

Patients were evaluated using a VAS to assess pain, and Mayo score to assess elbow function and patient's satisfaction. Longitudinal ultrasonography was performed on all patients prior to injection and 1 month after injection. All patients were evaluated at 1, 6, and 12 months after injection.

No adverse events were reported at any point during the study.

VAS improved from baseline of 72.8 to 19.8 and 8.2 at 1 month and 12 months, respectively. Mayo scores improved from a baseline of 61.9 to 89.6, 92.5, and 97.6 at 1, 6, and 12 months respectively after injection (Fig. 8.14).

Improvements in color Doppler activity on ultrasound examination was seen as early as 1 month in three patients following treatment (Fig. 8.15).

## Fracture Nonunion

Jiang et al. [4] described a 64-year-old woman with nonunion of the tibia at 4 months following open reduction and internal fixation (ORIF), despite appropriate healing of the simultaneously fractured fibula. Platelet lysate was prepared from 40 mL of peripheral blood, and the patient received three weekly fluoroscopically guided injections of 5 mL of platelet lysate into the areas of abnormality. Weightbearing was prohibited for 24 h with progression to full weightbearing at 8 weeks. Radiographs at 2, 4, and 6 months (Fig. 8.16) following platelet lysate treatment showed gradual and abundant callus formation with near complete healing at 8 months following injection.

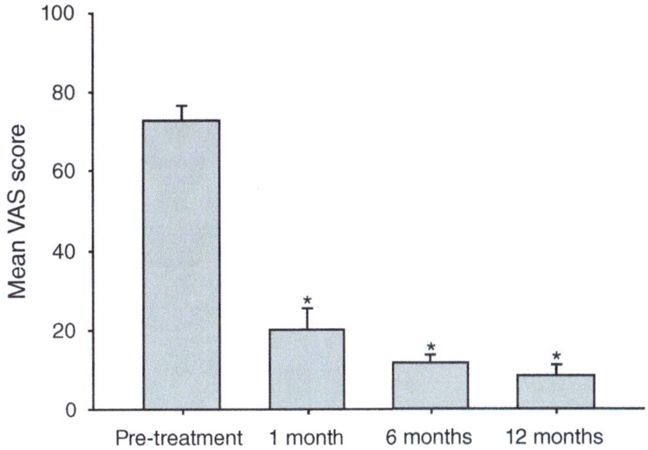


**Fig. 8.14** VAS Scores in refractory epicondylitis after platelet lysate injection. (Source: Tan XX, Ju HY, Yan W, et al. Autologous platelet lysate local injections for the treatment of refractory lateral epicondylitis. *J Orthop Surg Res* 2016; 11:17 [40])

**Fig. 8.15** 3 representative cases before (**a**, **b**, **c**) and 1 month after (**d**, **e**, **f**) platelet lysate injection showing reduced doppler activity. (Source: Tan XX, Ju HY, Yan W, et al. Autologous platelet lysate local injections for the treatment of refractory lateral epicondylitis. *J Orthop Surg Res* 2016; 11:17 [40])

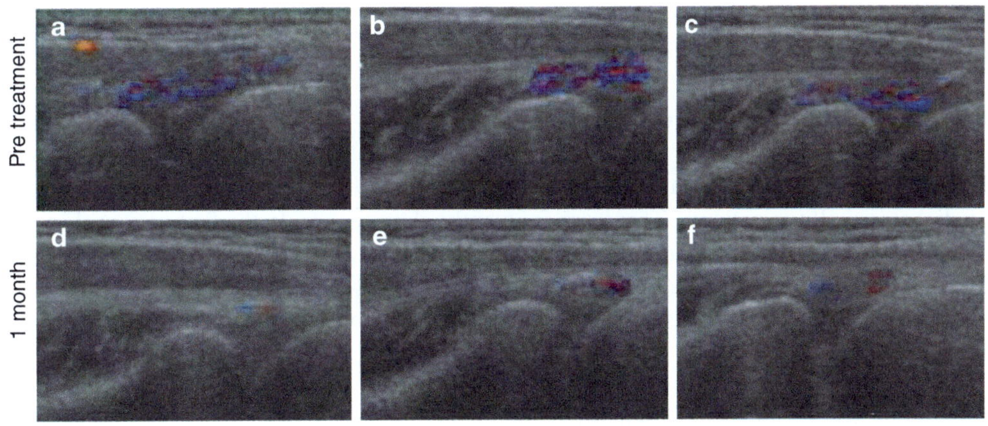

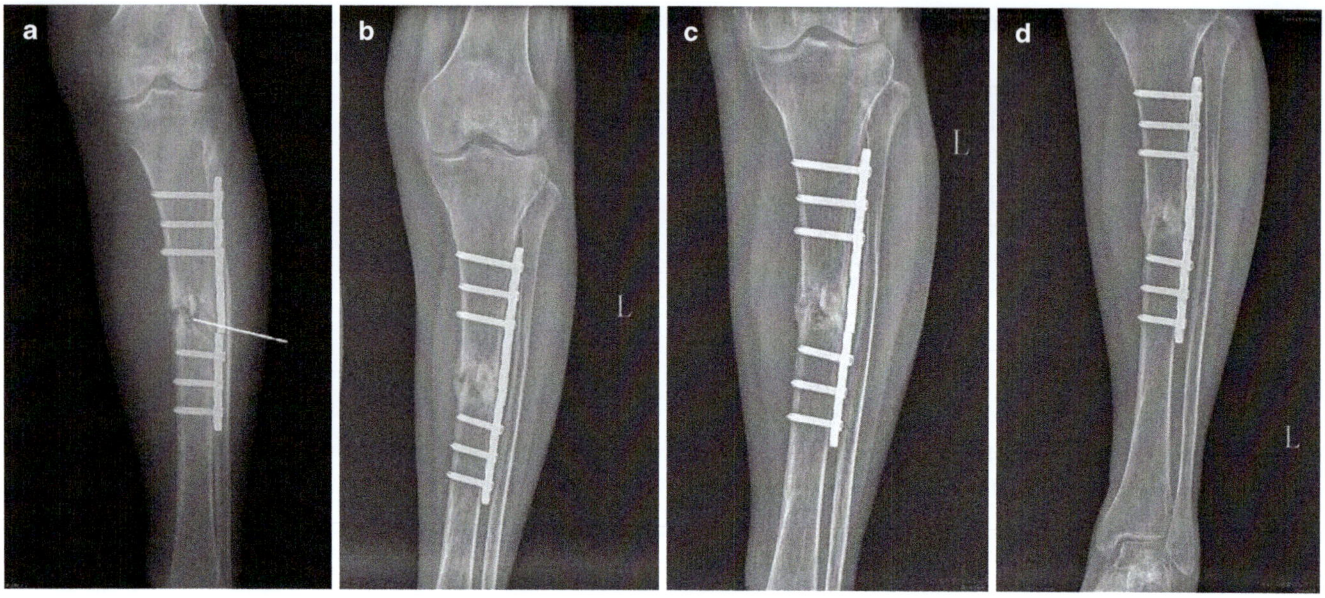


**Fig. 8.16** Images **a**, **b**, **c**, and **d** showing tibia nonunion at baseline, 2, 4, and 6 mos following platelet lysate injection. (Source: Jiang HJ, Tan XX, Ju HY, et al. Autologous platelet lysates local injections for treatment of tibia non-union with breakage of the nickelclad: A case report. *Springerplus* 2016; 5:2013 [4])

## Platelet Lysate and Releasate: Current and Future Clinical Use: Non-Musculoskeletal Conditions

### Longstanding Diabetic Foot Ulcers

Diabetic foot ulcers (DFUs) are a chronic problem for 15% of diabetes mellitus patients. These are frequently slow to heal and may become infected leading to complications. Defects in re-epithelialization, angiogenesis, as well as cytokine and growth factor deficits are the main impediments to healing [44–48].

Cell migration and proliferation are initiated and regulated by growth factors and cytokines that are released from the wounded epithelium. Fibroblasts and keratinocytes are particularly important in this process [49]. Numerous recombinant growth factors have been used in clinical practice for wound healing for more than two decades, however widespread use is limited by lack of reliability, limited shelf-life, and excessive cost [50].

Jafar et al. [48] presented their findings in two patients participating in an ongoing clinical trial examining both autologous and allogeneic platelet lysate in the treatment of diabetic foot ulcers. Participants in this study were those with type 2 diabetes mellitus between the ages of 18 and 70 with an ulcer of at least 4 weeks duration ulcers were located on the plantar, medial, or lateral aspect of the foot, measuring between 0.5 and 20 $cm^2$.

Platelet lysate was prepared from PRP obtained from a 20 mL autologous whole blood collection, resulting in 5 mL of platelet lysate. A single freeze/thaw cycle to −80° C was used to prepare platelet lysate. Platelet lysate was injected perilesional at ulcer margins by multiple punctures per each session of therapy, with a total of 5–7 mL/session. A total of four injection sessions were used (day 0, 14, 28, and 42), with a total of 80 mL of WB taken from patients during the entire course of treatment.

In the first patient (Fig. 8.17a, c) with an elliptical shaped ulcer located on the lateral aspect of the right foot, the ulcer area was reduced by 56% by week 2. Two weeks after the second injection there was a further decrease in the ulcer size by 71% with granulation tissue noted at the ulcer floor. Two weeks after the fourth injection, the ulcer was completely healed.

The second patient (Fig. 8.17b, d) had a non-infected, triangular shaped ulcer on the plantar aspect of the left heel. Two weeks after the first perilesional platelet lysate injection, ulcer size was reduced by 55%, with granulation tissue covering the base of the ulcer. Two weeks after the second injection, the ulcer surface area decreased by 82%. By day 56, 2 weeks after the fourth injection, the ulcer was fully healed [48].

### Ocular Surface Disease (OSD) and Refractory Primary Sjogren's Syndrome

Both serum eye drops (SEDs) and platelet lysate eye drops (PLEDs) have been shown to be beneficial in the management of ocular surface diseases (OSDs) such as persistent

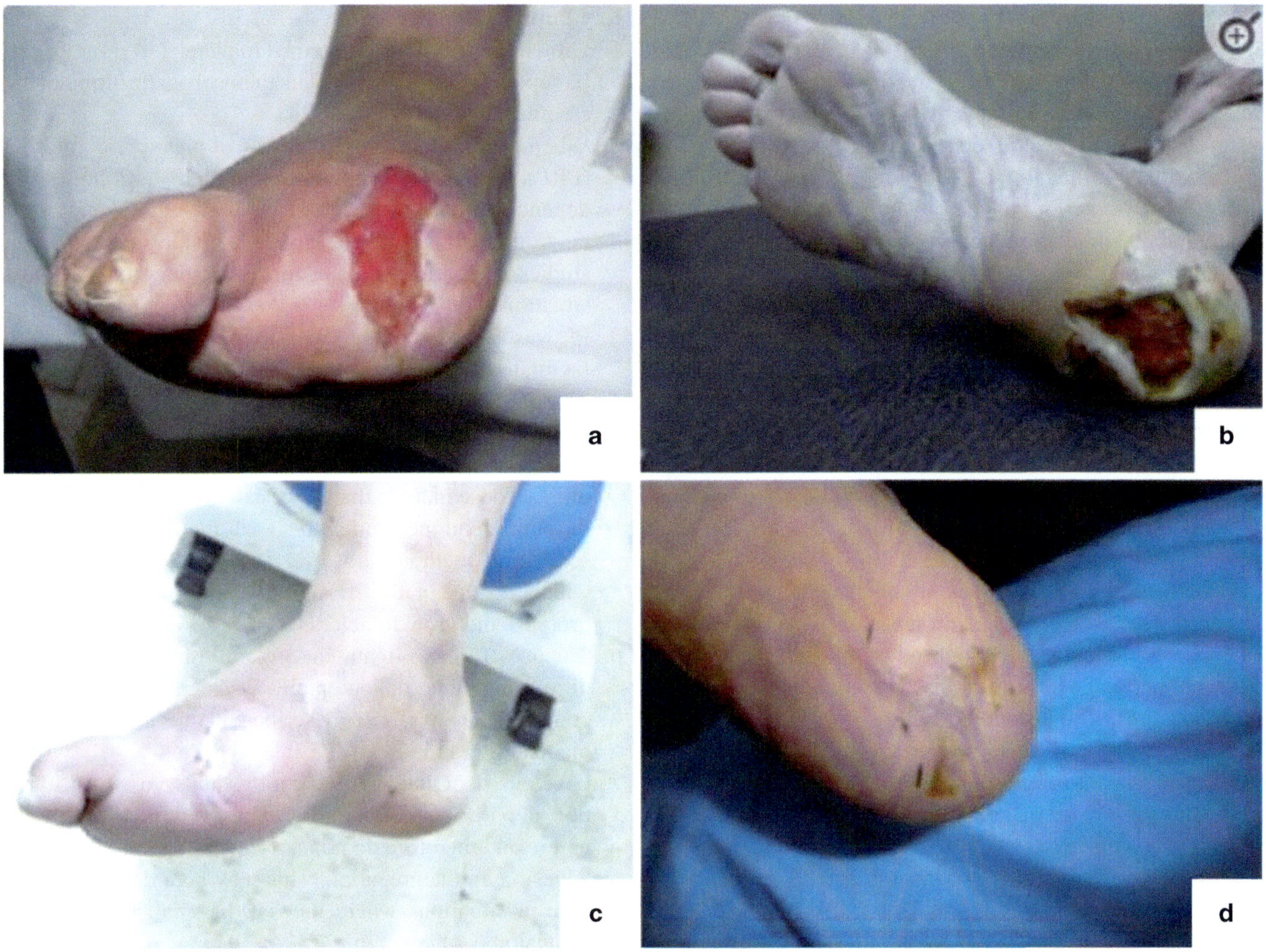

**Fig. 8.17** Representative images of nonhealing ulcers in 2 patients at baseline (**a** and **b**) and day 56 (**c** and **d**) following perilesional platelet lysate injection. (Source: Jafar H, Hasan M, Al-Hattab D, et al. Platelet lysate promotes the healing of long-standing diabetic foot ulcers: A report of two cases and *in vitro* study. *Heliyon* 2020; 6:e03929 [48])

epithelial defects, severe dry eye, Stevens-Johnson syndrome, neurotrophic keratopathy, And other diseases. Treatment with SCDs and PLEDs accelerate corneal epithelial wound healing due to the greater concentration of growth factor and serum compared with normal tear composition [51, 52].

Fea et al. [53] performed a prospective case-control study in 30 patients (60 eyes) suffering from Sjogren syndrome (SS) refractory for more than 2 months with conventional therapy. Patients were randomized to treatment with either a 90-day course of either artificial tears (0.2% hyaluronic acid and TS–polysaccharide 0.2%) or autologous platelet lysate eye drops (APLEDs). APLEDs were prepared by first withdrawing 250–300 mL of venous peripheral blood which was processed into PRP. PRP was then subjected to three consecutive thermal shocks followed by a final centrifugation. Patients were instructed to keep APLEDs frozen and to then thaw the single daily dose the night before administration. Dose of APLEDs was two drops q.i.d. for 90 days. Degree of disease was measured before and after 90 days of therapy via Standard ophthalmologic assessment tests, as well as *in vivo* corneal confocal microscopy assessment of basal epithelial cell density, number of subbasal nerve fibers, nerve fiber density, and other parameters.

Ocular Surface Disease Index (OSDI) scores decreased in patients receiving PLEDs compared to artificial tear therapy, and differences were statistically significant between the two groups after 90 days. Statistically significant increases in basal epithelium cells density, subbasal nerve fiber number and density of innervation were seen at 90 days in patients receiving APLEDs. The authors hypothesized that APLEDs, containing higher levels of growth factors compared to SEDs, may be a more effective form of therapy in a select group of patients suffering from OSD.

Pezzotta et al. [54] reported on the long-term safety and efficacy of autologous platelet lysate drops for treatment of ocular graft versus host disease (oGvHD). They described the long-term continuous treatment of ocular GVHD with autologous platelet lysate drops in 31 patients refractory to topical conventional therapy. At 36 months, all patients were classified as responders, with the authors reporting statistically significant decreases in glaucoma symptom scale (GSS). No serious adverse events occurred, and it was concluded that pain and function can be improved via long-term administration of APLEDs. Furthermore, the authors concluded that the early use of APLEDs in oGvHD may prevent permanent ocular impairment.

## Other Potential Future Considerations

Nebie et al. [55] tested the therapeutic potential of human platelet lysate using *in vitro* and *in vivo* models of traumatic brain injury (TBI). Two mouse models were used to investigate the neuroprotective potential of platelet lysate. They demonstrated that topical and intranasal application of platelet lysate consistently improved mouse motor function and mitigated cortical neuroinflammation and oxidative stress in the injury area. They proposed that brain administration of platelet lysate is a therapeutic strategy that deserves serious and urgent consideration for universal brain trauma treatment. In a follow-up paper, Nebie et al. [56] proposes that platelet lysate may also be of therapeutic benefit in neurodegenerative diseases such as Parkinson's disease, Alzheimer's disease, and stroke.

In a 2021 paper, Jeyaraman et al. [57] propose that because of its acellular nature, and its regenerative and immunomodulatory potential, that nebulized platelet lysate may emerge as a pivotal therapeutic agent in the treatment of COVID-19 pneumonia.

## Conclusion

Platelet lysate is emerging as a potentially useful tool in the armamentarium of treatments for the orthobiologic physician. It is easily prepared by means of a variety of methods which mechanically lyse platelets, thereby allowing for immediate delivery of growth factors to the site of tissue pathology. Published case series both outside as well as in the field of orthopedics, musculoskeletal and spinal medicine show promising results. Additional well-designed studies in these fields are needed to realize the full potential utility of this autologous point-of-care orthobiologic agent. Information is still lacking in terms of which specific growth factors (GFs) in platelet lysate are important, and at this point optimal levels of those GFs are still unknown.

**Key Points**

1. Growth factors (GFs) contained in platelet α-granules can be separated from the platelets themselves resulting in the creation of platelet lysate (PL).
2. When using standard preparations of platelet rich plasma (PRP), GF delivery to tissue is highly variable, and dependent on a multitude of factors including local tissue environment. For this reason, platelet lysate has been studies as a replacement for PRP in certain clinical situations.
3. PL was first used in the 1980s in the laboratory setting to facilitate fibroblast and endothelial cell cultures.
4. The first reported clinical use of PL was in the treatment of a tibial nonunion in 2013.
5. Various methods have been described for the preparation of PL. These include the use of calcium chloride ($CaCl_2$), thrombin, and mechanical lysis via freeze-thaw, or ultrasonication methods, however there is no consensus on which results in the most substantial concentration of proteins and growth factors.
6. The use of PL for treatment of painful lumbar spine conditions including radiculopathy, disc herniations, and discogenic pain has been described in the pure reviewed medical literature.
7. Given the fact that PL preparations undergo a final filtration step through a 0.2 μ filter they are non-particulate in nature and therefore do not carry the potential for platelet adhesion and aggregation. This may be a desirable attribute in situations where the risk of small blood vessel occlusion is of concern.
8. The use of PL in the treatment of a variety of orthopedic conditions including refractory lateral epicondylitis and fracture nonunion has been described in the medical literature.
9. PL has been proposed as a viable treatment option for non-orthopedic conditions including ocular surface disease, Sjogren's syndrome, long-standing diabetic foot ulcers, traumatic brain injuries, and COVID–19 pneumonia.

**Questions**

1. A healthy 53-year-old female presents to your practice with a 9 month history of severe axial low back and right lower extremity pain. Symptoms have not improved with conservative care, OTC medication, and 2 rounds of physical therapy. On examination she has sensory deficits in the L5 and S1 dermatomes and MRI reveals a right paracentral disc herniation with annular tear compressing the exiting L5 and traversing S1 nerve roots. Her history is significant for type 2 diabetes mellitus, and she has experienced significant escalations in blood sugar and hemoglobin A1c with prior steroid injections. She is ask-

ing if there are any nonsurgical treatments not involving the use of steroids that might be able to alleviate her symptoms. You discussed the idea of using platelet lysate and platelet releasate with her, and she would like to proceed. When considering treatment methods for preparation of platelet releasate, which of the following methods may be considered?

A. Calcium chloride ($CaCl_2$) and thrombin may be added to PRP, stimulating the formation of a clot, signaling platelets to degranulate and release their intracellular contents into the plasma.
B. PRP can be subjected to multiple freeze-thaw cycles, resulting in water crystals forming within the platelets, causing cellular membrane rupture and release of intracellular products into the plasma.
C. PRP can be placed into an ultrasound bath with frequencies over 20 kHz for 30 min, resulting in 74% release of PDGF contained in platelets into plasma.
D. A sterilized, ultrasonic homogenization probe, covered by a sterile liner can be placed into the PRP solution, causing release of growth factors and other biologic agents into the plasma.
E. PRP can be left in a refrigerator and injected within 48 h, by which time platelets have released 80% of their growth factors into the plasma.

**Answer: A**

**Explanation**: The addition of $CaCl_2$ and thrombin PRP will stimulate platelets to aggregate and release the contents of their alpha granules into the plasma. This process results in the creation of platelet releasate as opposed to platelet lysate. The methods described in answers B, C, and D have all been studied as a means of creating platelet lysate. Answer E is not appropriate since this is not a method known to produce platelet lysate or platelet releasate.

2. A colleague contacts you because they have just purchased a new ultra-low temperature freezer so that they can prepare platelet lysate for research purposes. You have been using platelet lysate as a replacement for fetal bovine serum in your studies on fibroblast proliferation and stem cell differentiation. When making recommendations regarding the optimal number of freeze thaw cycles to maximize growth factor levels in platelet lysate, what is your recommendation to your colleague?

A. A single freeze thaw cycle to −20°C and subsequent thawing at room temperature should be sufficient. You tell your colleague that the ultra-low temperature freezer can be returned and a standard freezer can be used.
B. 3–5 freeze thaw cycles, freezing at −70°C for 10 min and thawing in a 37°C water bath for 7.5 min provides optimal levels of growth factor release.
C. A single freeze thaw cycle at −70°C overnight followed by following at room temperature provides optimal levels of growth factor release.
D. 30 freeze thaw cycles are required to achieve optimal levels of growth factor release.
E. 10–20 freeze thaw cycles are sufficient for optimization of growth factor release.

**Answer: B**

**Explanation**: Strandberg and colleagues took PRP from 10 healthy donors and made platelet lysate by freezing the PRP and −70°C for 10 min and then thawing it in a 37°C water bath for 7-1/2 min. They analyzed growth factor levels after one freeze thaw cycle, and then 2, then 3 cycles, going all the way out to 30 cycles. They demonstrated that optimal levels of PDGF were achieved at cycles 3–5 and that the clinical significance of using a greater number of cycles is likely limited.

None of the other methods described have been shown to provide optimal levels of growth factor release when preparing platelet lysate. Freezing PRP at temperatures above −70°C have not been validated as methods of providing adequate lysis of platelets.

Strandberg G, Sellberg F, Sommar P, Ronaghi M, Lubenow N, Knutson F, Berglund D. Standardizing the freeze-thaw preparation of growth factors from platelet lysate. Transfusion. 2017 Apr;57 (4):1058–1065. doi: 10.1111/trf.13998. Epub 2017 Feb 9. PMID: 28182293.

3. A 68-year-old female presents with a 5-year history of progressively worsening lumbar radiculopathy due to a combination of multilevel facet hypertrophy and foraminal disc bulging resulting in severe neural foraminal stenosis. She has previously undergone lumbar discectomy and fusion at L3–S1 and is suffering from what appears to be adjacent segment disease at the level above the fusion. Signs and symptoms, as well as physical exam are consistent with bilateral L2 and L3 radiculopathy. Transforaminal epidural steroid injection using triamcinolone has provided relief lasting only 6 weeks. She is hesitant to pursue additional surgery since her surgeon is recommending an additional three level fusion due to multiple areas of degeneration in the spine. You recommend a trial of injection of platelet lysate via transforaminal epidural route at L2–L3. Which of the following statements is true about epidural platelet lysate injections?

A. Platelet lysate injections exert their effect via growth factors and anti-inflammatory cytokines. Therefore they have a longer duration of effect compared to steroid injections.
B. Epidural platelet lysate injections carry a lower risk of infection compared to epidural steroid injections.
C. The preparation of platelet lysate results in an acellular, non-particulate injectate which is filtered through a 0.22 μ filter immediately prior to injection, which

may mitigate the risk of embolic stroke due to retrograde injection of particulate substances.

D. A maximum of 2 transforaminal epidural platelet lysate injections may be performed at one time; either bilaterally at a single level, or vertically at 2 levels simultaneously.

E. Transforaminal epidural platelet lysate injections may not be performed in a patient who has been taking nonsteroidal anti-inflammatory drugs (NSAIDs).

**Answer: C**

**Explanation**: Given that several tragic cases of ischemic spinal cord lesions have been reported via particulate steroid occlusion of radicular arteries, scrutiny concerning the embolic nature of injectates is warranted (Pountos et al. 2016).

There are no known comparative studies showing superiority of epidural platelet lysate injections over epidural steroid injections with regard to efficacy, duration of effect or infection risk.

Since epidural platelet lysate injection should not be billed to any insurance carrier, insurance restrictions regarding the number of levels that may be treated simultaneously are not applicable.

The effect of NSAIDs on platelet aggregation, and subsequent deformation may not play as important role when using platelet lysate, since growth factors are mechanically released from platelets during the process of freezing and thawing.

Obeid TH, Gornas MA, El-Mesallamy AA. Spinal cord ischemia following cervical transforaminal epidural steroid injection. Neurosciences (Riyadh). 2014 Oct;19 (4):326–7. PMID: 25274595; PMCID: PMC4727674.

Epstein NE. The risks of epidural and transforaminal steroid injections in the Spine: Commentary and a comprehensive review of the literature. Surg Neurol Int. 2013 Mar 22;4(Suppl 2):S74–93. doi: 10.4103/2152-7806.109446. PMID: 23646278; PMCID: PMC3642757.

Centeno C, Markle J, Dodson E, Stemper I, Hyzy M, Williams C, Freeman M. The use of lumbar epidural injection of platelet lysate for treatment of radicular pain. J Exp Orthop. 2017 Nov 25;4 (1):38. doi: 10.1186/s40634-017-0113-5. PMID: 29177632; PMCID: PMC5701904.

4. You are seeing a 65-year-old male with painful, recalcitrant tendinosis of the common extensor tendon of the elbow. He has undergone prior treatment at an outside clinic with two separate treatments of platelet rich plasma injections. He has had only modest improvements in pain and function and is asking for your advice regarding additional treatments. You obtain records from his prior treatments and determine that he has undergone injections with both leukocyte-rich and leukocyte-poor preparations. Each of the blood draws for these procedures involved approximately 8–10 cc of whole blood. In-depth characterization of the PRP injected was not performed. He is asking about possible injection of platelet lysate before considering other treatment options. How do you advise the patient?

A. Abandon further attempts at using platelet–derived procedures for healing of his condition since he has failed two prior attempts using both leukocyte rich and leukocyte poor preparations.

B. Bone marrow concentrate and micronized fat injections are the treatment of choice for recalcitrant lateral epicondylitis and should be pursued in favor of additional platelet–based injections.

C. Tell him that his PRP injections may have been understood. A higher volume blood draw (100 cc+) will result in a higher platelet count withing the PRP, and therefore a higher concentration of growth factors in any platelet–based injectate, including platelet lysate.

D. Since he has failed to cell-based therapy procedures he should proceed with surgery.

E. Platelet lysate has never been studied for treatment of recalcitrant lateral epicondylitis and therefore you cannot advise him either way.

**Answer: C**

**Explanation**: A low-volume blood draw will limit the number of platelets and growth factors that can be obtained in any PRP or platelet lysate preparation. This may have an effect on healing: "...higher PL concentrations promote increased tenocyte proliferation and migration when derived from ageing, but not young, donors. These results suggest that increasing the platelet concentration within PRP may be a useful biological strategy for helping tendon regeneration in an ageing patient population."

There are no studies showing superiority of surgery, platelet, bone marrow concentrate, or adipose derived injectates in the setting of recalcitrant lateral epicondylitis. Platelet lysate has been described in the treatment of recalcitrant lateral epicondylitis.

Berger DR, Centeno CJ, Steinmetz NJ. Platelet lysates from aged donors promote human tenocyte proliferation and migration in a concentration-dependent manner. Bone Joint Res. 2019 Feb 2;8 (1):32–40. doi: 10.1302/2046-3758.81.BJR-2018-0164.R1. PMID: 30800297; PMCID: PMC6359887.

Tan, Xx., Ju, Hy., Yan, W. et al. Autologous platelet lysate local injections for the treatment of refractory lateral epicondylitis. *J Orthop Surg Res* 11, 17 (2016). https://doi.org/10.1186/s13018-016-0349-2

5. You are asked to speak at grand rounds to a group of your colleagues regarding regenerative medicine. You plan to include discussions regarding the history and rationale, operation methods, regulatory issues, ethical issues and current opinions regarding compliant clinical use in the

United States and elsewhere. Which of the following statements is true regarding peer-reviewed data supporting the clinical use of platelet lysate?

A. There are currently no peer-reviewed studies regarding the clinical use of platelet lysate.
B. Platelet lysate has similar efficacy when compared in head-to-head clinical trials against platelet rich plasma for certain indications including osteoarthritis of the knee.
C. Platelet lysate prepared from leukocyte-rich PRP has a higher degree of efficacy compared to platelet lysate made from leukocyte-poor PRP.
D. The use of platelet lysate has been described in peer-reviewed medical journals for treatment of Sjogren syndrome, lateral epicondylitis, non-healing diabetic foot ulcers, lumbar disc herniation, and ocular surface diseases.
E. Platelet lysate is considered more than minimally manipulated and therefore is regulated as a section 351 HCT/P.

**Answer: D**

**Explanation**: There are no comparator trials showing superiority of platelet lysate over PRP, nor are there any descriptions comparing platelet lysate made from LR-PRP compared to LP-PRP. Regarding the regulatory status of platelet lysate, "...certain products including blood products such as PRP are exempt and therefore do not follow the FDA's traditional regulatory pathway that includes animal studies and clinical trials."

Beitzel K, Allen D, Apostolakos J, Russell RP, McCarthy MB, Gallo GJ, Cote MP, Mazzocca AD. US definitions, current use, and FDA stance on use of platelet-rich plasma in sports medicine. J Knee Surg. 2015 Feb;28 (1):29–34. doi: 10.1055/s-0034-1390030. Epub 2014 Sep 30. PMID: 25268794.

Rawson B. Platelet-Rich Plasma and Epidural Platelet Lysate: Novel Treatment for Lumbar Disk Herniation. J Am Osteopath Assoc. 2020 Mar 1;120 (3):201–207. doi: 10.7556/jaoa.2020.032. PMID: 32091562.

Jafar H, Hasan M, Al-Hattab D, Saleh M, Ameereh LA, Khraisha S, Younes N, Awidi A. Platelet lysate promotes the healing of long-standing diabetic foot ulcers: A report of two cases and in vitro study. Heliyon. 2020 May 4;6 (5):e03929. doi: 10.1016/j.heliyon.2020.e03929. PMID: 32420494; PMCID: PMC7218073.

Antonio M. Fea, Vittoria Aragno, Valeria Testa, Federica Machetta, Simone Parisi, Sergio D'Antico, Roberta Spinetta, Enrico Fusaro, Federico M. Grignolo, "The Effect of Autologous Platelet Lysate Eye Drops: An In Vivo Confocal Microscopy Study", BioMed Research International, vol. 2016, Article ID 8406832, 10 pages, 2016. https://doi.org/10.1155/2016/8406832

Tan, Xx., Ju, Hy., Yan, W. et al. Autologous platelet lysate local injections for the treatment of refractory lateral epicondylitis. *J Orthop Surg Res* 11, 17 (2016). https://doi.org/10.1186/s13018-016-0349-2

## References


1. Doucet C, Ernou I, Zhang Y, et al. Platelet lysates promote mesenchymal stem cell expansion: a safety substitute for animal serum in cell-based therapy applications. J Cell Physiol. 2005;205:228–36.
2. Astori G, Amati E, Bambi F, et al. Platelet lysate as a substitute for animal serum for the ex-vivo expansion of mesenchymal stem/stromal cells: present and future. Stem Cell Res Ther. 2016;7:93.
3. Hara Y, Steiner M, Baldini MG. Platelets as a source of growth-promoting factor(s) for tumor cells. Cancer Res. 1980;40:1212–6.
4. Jiang HJ, Tan XX, Ju HY, et al. Autologous platelet lysates local injections for treatment of tibia non-union with breakage of the nickelclad: a case report. Springerplus. 2016;5:2013.
5. Rawson B. Platelet-rich plasma and epidural platelet lysate: novel treatment for lumbar disk herniation. J Am Osteopath Assoc. 2020;120:201–7.
6. Akeda K, Ohishi K, Masuda K, et al. Intradiscal injection of autologous platelet-rich plasma releasate to treat discogenic low back pain: a preliminary clinical trial. Asian Spine J. 2017;11:380–9.
7. Centeno C, Markle J, Dodson E, et al. The use of lumbar epidural injection of platelet lysate for treatment of radicular pain. J Exp Orthop. 2017;4:38.
8. Berger DR, Centeno CJ, Steinmetz NJ. Platelet lysates from aged donors promote human tenocyte proliferation and migration in a concentration-dependent manner. Bone Joint Res. 2019;8:32–40.
9. Altaie A, Owston H, Jones E. Use of platelet lysate for bone regeneration—are we ready for clinical translation? World J Stem Cells. 2016;8:47–55.
10. Roffi A, Filardo G, Assirelli E, et al. Does platelet-rich plasma freeze-thawing influence growth factor release and their effects on chondrocytes and synoviocytes? Biomed Res Int. 2014;2014:692913.
11. Arnoczky SP, Sheibani-Rad S. The basic science of platelet-rich plasma (PRP): what clinicians need to know. Sports Med Arthrosc Rev. 2013;21:180–5.
12. Mautner K, Malanga GA, Smith J, et al. A call for a standard classification system for future biologic research: the rationale for new PRP nomenclature. PMR. 2015;7:S53–9.
13. Fortunato TM, Beltrami C, Emanueli C, et al. Platelet lysate gel and endothelial progenitors stimulate microvascular network formation in vitro: tissue engineering implications. Sci Rep. 2016;6:25326.
14. Bernardi M, Albiero E, Alghisi A, et al. Production of human platelet lysate by use of ultrasound for ex vivo expansion of human bone marrow-derived mesenchymal stromal cells. Cytotherapy. 2013;15:920–9.
15. Franklin SP, Birdwhistell KE, Strelchik A, et al. Influence of cellular composition and exogenous activation on growth factor and cytokine concentrations in canine platelet-rich plasmas. Front Vet Sci. 2017;4:40.
16. Zimmermann R, Arnold D, Strasser E, et al. Sample preparation technique and white cell content influence the detectable levels of growth factors in platelet concentrates. Vox Sang. 2003;85:283–9.

17. Renn T-Y, Kao YH, Wang CC, et al. Anti-inflammatory effects of platelet biomaterials in a macrophage cellular model. Vox Sang. 2015;109:138–47.
18. Sonker A, Dubey A. Determining the effect of preparation and storage: an effort to streamline platelet components as a source of growth factors for clinical application. Transfus Med Hemother. 2015;42:174–80.
19. Anitua E, Andia I, Ardanza B, et al. Autologous platelets as a source of proteins for healing and tissue regeneration. Thromb Haemost. 2004;91:4–15.
20. Eppley BL, Woodell JE, Higgins J. Platelet quantification and growth factor analysis from platelet-rich plasma: implications for wound healing. Plast Reconstr Surg. 2004;114:1502–8.
21. Amable PR, Carias RB, Teixeira MV, et al. Platelet-rich plasma preparation for regenerative medicine: optimization and quantification of cytokines and growth factors. Stem Cell Res Ther. 2013;4:67.
22. Borghese C, Agostini F, Durante C, et al. Clinical-grade quality platelet-rich plasma releasate (PRP-R/SRGF) from CaCl2-activated platelet concentrates promoted expansion of mesenchymal stromal cells. Vox Sang. 2016;111:197–205.
23. do Amaral RJ, Matsiko A, Tomazette MR, et al. Platelet-rich plasma releasate differently stimulates cellular commitment toward the chondrogenic lineage according to concentration. J Tissue Eng. 2015;6:2041731415594127.
24. Iudicone P, Fioravanti D, Bonanno G, et al. Pathogen-free, plasma-poor platelet lysate and expansion of human mesenchymal stem cells. J Transl Med. 2014;12:28.
25. Krüger JP, Freymannx U, Vetterlein S, et al. Bioactive factors in platelet-rich plasma obtained by apheresis. Transfus Med Hemother. 2013;40:432–40.
26. Arakawa T, Prestrelski SJ, Kenney WC, et al. Factors affecting short-term and long-term stabilities of proteins. Adv Drug Deliv Rev. 2001;46:307–26.
27. Strandberg G, Sellberg F, Sommar P, et al. Standardizing the freeze-thaw preparation of growth factors from platelet lysate. Transfusion. 2017;57:1058–65.
28. da Fonseca L, Santos GS, Huber SC, et al. Human platelet lysate—a potent (and overlooked) orthobiologic. J Clin Orthop Trauma. 2021;21:101534.
29. Majid I, Nayik GA, Nanda V. Ultrasonication and food technology: a review. Cogent Food Agric. 2015;1:1071022.
30. Jackson T, Thomas S, Stabile V, et al. Prevalence of chronic pain in low-income and middle-income countries: a systematic review and meta-analysis. Lancet. 2015;385:S10.
31. Lyu FJ, Cui H, Pan H, et al. Painful intervertebral disc degeneration and inflammation: from laboratory evidence to clinical interventions. Bone Res. 2021;9:7.
32. Risbud MV, Shapiro IM. Role of cytokines in intervertebral disc degeneration: pain and disc content. Nat Rev Rheumatol. 2014;10:44–56.
33. Shamji MF, Setton LA, Jarvis W, et al. Proinflammatory cytokine expression profile in degenerated and herniated human intervertebral disc tissues. Arthritis Rheum. 2010;62:1974–82.
34. Yang G, Liao W, Shen M, et al. Insight into neural mechanisms underlying discogenic back pain. J Int Med Res. 2018;46:4427–36.
35. Hayashi S, Taira A, Inoue G, et al. TNF-alpha in nucleus pulposus induces sensory nerve growth: a study of the mechanism of discogenic low back pain using TNF-alpha-deficient mice. Spine (Phila Pa 1976). 2008;33:1542–6.
36. Evashwick-Rogler TW, Lai A, Watanabe H, et al. Inhibiting tumor necrosis factor-alpha at time of induced intervertebral disc injury limits long-term pain and degeneration in a rat model. JOR Spine. 2018;1:e1014.
37. Gliedt JA, Daniels CJ. Chiropractic treatment of lateral epicondylitis: a case report utilizing active release techniques. J Chiropr Med. 2014;13:104–9.
38. Serafini G. Treatment of lateral epicondylitis. In: Sconfienza LM, Serafini G, Silvestri E, editors. Ultrasound-guided musculoskeletal procedures. Milano: Springer; 2012. p. 67–71.
39. Ruch DS, Orr SB, Richard MJ, et al. A comparison of debridement with and without anconeus muscle flap for treatment of refractory lateral epicondylitis. J Shoulder Elbow Surg. 2015;24:236–41.
40. Tan XX, Ju HY, Yan W, et al. Autologous platelet lysate local injections for the treatment of refractory lateral epicondylitis. J Orthop Surg Res. 2016;11:17.
41. Kumai T, Muneta T, Tsuchiya A, et al. The short-term effect after a single injection of high-molecular-weight hyaluronic acid in patients with enthesopathies (lateral epicondylitis, patellar tendinopathy, insertional Achilles tendinopathy, and plantar fasciitis): a preliminary study. J Orthop Sci. 2014;19:603–11.
42. Ruggiu A, Ulivi V, Sanguineti F, et al. The effect of platelet lysate on osteoblast proliferation is associated with a transient increase of the inflammatory response in bone regeneration. Biomaterials. 2013;34:9318–30.
43. Leotot J, Coquelin L, Bodivit G, et al. Platelet lysate coating on scaffolds directly and indirectly enhances cell migration, improving bone and blood vessel formation. Acta Biomater. 2013;9:6630–40.
44. Reiber GE, Boyko EJ, Smith DG. Lower extremity foot ulcers and complications in diabetes. In: Diabetes in America. 2nd ed. Bethesda, MD: National Institutes of Health; 1995. p. 409–28.
45. Reiber GE, Vileikyte L, Boyko EJ, et al. Causal pathways for incident lower-extremity ulcers in patients with diabetes from two settings. Diabetes Care. 1999;22:157–62.
46. Falanga V. Wound healing and its impairment in the diabetic foot. Lancet. 2005;366:1736–43.
47. Brem H, Tomic-Canic M. Cellular and molecular basis of wound healing in diabetes. J Clin Invest. 2007;117:1219–22.
48. Jafar H, Hasan M, Al-Hattab D, et al. Platelet lysate promotes the healing of long-standing diabetic foot ulcers: a report of two cases and *in vitro* study. Heliyon. 2020;6:e03929.
49. Lubin FD, Segal M, Mcgee DW. Regulation of epithelial cell cytokine responses by the a3b1 integrin. Immunology. 2003;108:204–10.
50. Barsotti MC, Losi P, Briganti E, et al. Effect of platelet lysate on human cells involved in different phases of wound healing. PLoS One. 2013;8:1–11.
51. Zhang J, Crimmins D, Faed JM, et al. Characteristics of platelet lysate compared to autologous and allogeneic serum eye drops. Transl Vis Sci Technol. 2020;9:24.
52. Hartwig D, Harloff S, Liu L, et al. Epitheliotrophic capacity of a growth factor preparation produced from platelet concentrates on corneal epithelial cells: a potential agent for the treatment of ocular surface defects? Transfusion. 2004;44:1724–31.
53. Fea AM, Aragno V, Testa V, et al. The effect of autologous platelet lysate eye drops: an in vivo confocal microscopy study. Biomed Res Int. 2016;2016:8406832–10.
54. Pezzotta S, Del Fante C, Scudeller L, et al. Long-term safety and efficacy of autologous platelet lysate drops for treatment of ocular GvHD. Bone Marrow Transplant. 2017;52:101–6.
55. Nebie O, Carvalho K, Barro L, et al. Human platelet lysate biotherapy for traumatic brain injury: preclinical assessment. Brain. 2021;144:3142–58.
56. Nebie O, Buée L, Blum D, et al. Can the administration of platelet lysates to the brain help treat neurological disorders? Cell Mol Life Sci. 2022;79:379.
57. Jeyaraman M, Muthu S, Khanna M, et al. Platelet lysate for COVID-19 pneumonia-a newer adjunctive therapeutic avenue. Stem Cell Investig. 2021;8:11.

# Bone Marrow Concentrate

9

Matthew B. Murphy, Sairam Atluri, and Navneet Boddu

## Introduction

While bone marrow has been known as a source of healing and regenerative factors from time immemorial, its usage in human transplants began in the 1960s for the known presence of adult stem and progenitor cells [1–5]. Throughout the twentieth century, physicians utilized bone marrow aspirate (BMA) to promote healing in fractures, non-unions, and other forms of trauma [6–10]. It was not until the 1970s that scientists identified many of the progenitor cells within BMA responsible for this healing capacity. This chapter will review the biology of components in human BMA and bone marrow concentrate (BMC, also referred to as bone marrow aspirate concentrate or BMAC), methods of preparation, and clinical data of its safety and efficacy in the treatment of several orthopedic conditions.

## Biologic Composition of Bone Marrow Aspirate

### Cellular Components

BMA is primarily comprised of blood that has been enriched with components of the bone marrow stroma. Because blood is the central component of BMA, it includes platelets, erythrocytes, leukocytes, and plasma containing cytokines and other proteins. Within the heterogeneous leukocyte population, there are two main types of adult stem or progenitor cells in BMA. The presence of these stem cells is the primary distinction between BMC and platelet-rich plasma (PRP) prepared from whole blood and the reason for BMC's greater regenerative potential.

Mesenchymal stem cells (MSCs), also referred to as mesenchymal progenitor cells or "medicinal signaling cells," exist as pericytes on the outer surface of capillaries, veins, and arteries throughout the skeletal connective tissues. After an injury, activated platelets release chemotactic cytokines that, along with inflammatory signals such as interleukin-1 and TNF-alpha, recruit MSCs to leave the blood vessel and into the injury site. MSCs are potent in the healing of skeletal tissues due to their innate immunomodulatory, stimulatory, anti-microbial, and anti-apoptotic properties.

Hematopoietic stem cells (HSCs) are adult progenitors responsible for generating Erythrocytes, Platelets and leukocytes that comprise the blood and immune system. This was originally believed to be their sole purpose, however, they are able to provide their own immunomodulatory function, cell signaling, and work synergistically with MSCs.

### Protein Components

The protein composition of BMA, from a regenerative perspective, can be grouped between cytokines and inflammation modulators. Platelets and plasma contain cytokines and growth factors derived from platelets including platelet derived growth factors (PDGF-AA, AB, and BB), vascular endothelial growth factor (VEGF), fibroblast growth factors (e.g. FGF-1, 2), insulin-like growth factor (IGF-1), transforming growth factor beta 1 (TGF-b1), and others. PDGF stimulates cell recruitment, extracellular matrix protein production, and angiogenesis. VEGF stimulates angiogenesis and the recruitment of endothelial cells into an area of remodeling and regenerating tissues. FGFs support cell MSC recruitment and early differentiation, while also stimulating migration, proliferation, and differentiation of many fibroblast type cells. IGF-1 is also derived from platelets and stimulates collagen production critical in orthopedic extra-

M. B. Murphy
Murphy Technology Consulting, Houston, TX, USA

S. Atluri
Tristate Spine Care Institute, Cincinnati, OH, USA

N. Boddu (✉)
Advanced Pain and Regenerative Specialists, Oceanside, CA, USA


A. Navani et al. (eds.), *Essentials of Regenerative Medicine in Interventional Pain Management*,
https://doi.org/10.1007/978-3-031-50357-3_9

cellular matrices. TGF-b1 is critical in osteogenic and chondrogenic differentiation of MSCs and fibroblasts.

Many patients who received BMA or BMC injections anecdotally reported immediate pain relief that was often ascribed to a placebo effect. In the past decade though, two specific proteins have emerged as responsible for immediate modulation of inflammatory and mitigation of degenerative enzymes. Interleukin-1 receptor antagonist protein (IRAP or IL-1Ra) is a natural antagonist of interleukin-1 alpha and beta (IL-1a, IL-1b) by binding or blocking their receptors. This nullifies the signaling pathway for IL-1a and IL-1b, and the corresponding downstream symptoms including inflammation-induced pain. IRAP is synthesized by leukocytes and is present in (platelet-poor) plasma. Plasma from BMA contains greater than 20 times more IRAP (micrograms per mL) than plasma derived from whole blood. Thus, BMC has 20× more IRAP than an equivalent volume of PRP prepared for injection or implantation and a greater anti-inflammatory potential in the near term.

Alpha-2-macroglobulin (A2M) is a broad-spectrum protease inhibitor [11]. It is secreted in the liver and is present in (platelet-poor) plasma. A2M binds and inactivates degradative enzymes such as matrix-metalloproteinases (MMPs) present in inflamed and degenerative joints. There is no reported difference in A2M levels between BMA and whole blood derived plasma.

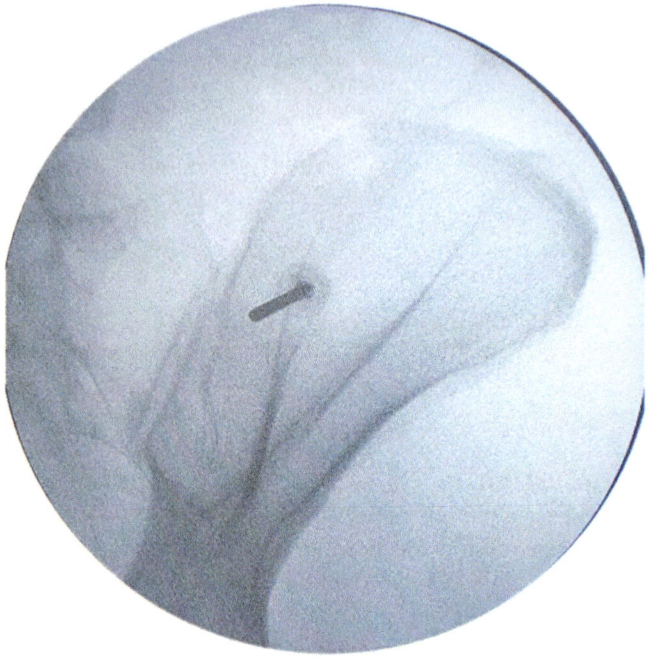

**Fig. 9.1** Bone marrow aspiration of posterior iliac crest. (Image courtesy of authors)

## Harvesting of Bone Marrow Aspirate

BMA is harvested through a needle or trocar that is inserted through the cortex or cortical bone and into the trabecular bone (Fig. 9.1). This porous bone is filled with fatty stromal tissue and blood vessels. As MSCs are also known to originate as pericytes on the outside of blood vessels, this niche is the true source of adult progenitor cells. Hernigou et al. [12] reported that a threefold increase in MSCs was attainable by using small volume syringes (10 mL) with small, rapid pulls. Biologically, this technique delivers rapid suction (vacuum) at the distal holes of the needle, rupturing capillaries, liberating pericytes (MSCs), and permitting peripheral blood to flow through the needle into the syringe, carrying progenitor cells with it [13]. Bone marrow aspiration performed at multiple sites with small volume aspirations can maximize nucleated cells including progenitor cells resulting in better clinical outcomes. Aging decreased the number of MSCs in the long bones and knees. However, this decrease is less dramatic in the iliac crest which makes it an ideal source.

## Preparation of Bone Marrow Concentrate

In the 1990s, physicians and scientists attempted to increase the healing properties of peripheral blood and BMA by concentrating the cells known to elicit chemotactic, mitogenic, and anti-inflammatory properties. This concentration was achieved via centrifugation of the fluid, causing the erythrocytes (red blood cells, the densest cells in blood and BMA at 1.10 g/mL) to sink downward and form to the bottom portion of separated blood/BMA. Acellular plasma (density 1.020 g/mL) rises to form the upper portion [14]. In the middle was a thin white layer referred to as the "buffy coat" that spans from 1.07 to 1.085 g/mL, containing various leukocyte populations (white blood cells, density 1.06–1.08 g/mL) [15]. In the case of BMA, among the leukocytes are MSCs and HSCs. Platelets, the least dense cells in blood and BMA, range in density from 1.06 to 1.09 g/mL and accumulate mostly on the upper portion of the buffy coat [16]. In the United States and Europe, it is common to centrifuge 60–120 mL BMA and collect a volume of 3–10 mL that includes the buffy coat, resulting in a concentration of progenitors and platelets 5–10 times above baseline. This is generally performed by centrifuging BMA for 10–20 min at relative centrifugal force of 3000 to 4000 revolutions per minute. For clinical applications that require limited volume (e.g., small joints or intervertebral discs).

## Characterization of Bone Marrow Aspirate and Concentrate

There are several methods of characterizing the cell and protein composition of BMA and BMC. The most accessible approach for physicians is a complete blood cell (CBC) count to measure the number and phenotype of leukocytes and the number of platelets and erythrocytes. An even simpler measurement is for total nucleated cells (TNCs). These methods do not quantify stem or progenitor cells, but do confirm how enriched BMA is above baseline levels of leukocytes in peripheral blood. For example, peripheral blood commonly contains 5–10 million white blood cells per milliliter, but high-quality BMA may have 20 or even 40 million nucleated cells per milliliter. After concentration via centrifugation, TNCs may be enriched five to tenfold in BMC over BMA.

More sophisticated approaches to characterizing BMA and BMC include flow cytometry and assays for colony forming units-fibroblast (CFU-F). Flow cytometry identifies cell phenotypes by proteins and receptors on their surface call called clusters of differentiation (CD) markers. MSCs are often defined by the presence of fibroblastic markers such as CD90, CD105, and CD73, while being negative for the hematopoietic/blood cell marker CD45. Because MSCs are pericytes, they also express CD146 (endothelial cell adhesion molecule). The cellular composition of BMA is described in Table 9.1 [17–23].

## Clinical Efficacy and Safety of Bone Marrow Concentrate

### Osteoarthritis of the Knee

The most common musculoskeletal condition treated with BMC is osteoarthritis. There are increasing numbers of clinical studies published annually that demonstrate the efficacy of BMC in reduction of pain and/or return of function for these conditions. Kim et al. [24] reported a 50% reduction in knee pain from a single intra-articular injection of BMC after 12 months in a Level 2, 75 patient study with average subject age of 60 years. Oliver et al. [25] demonstrated similar results (50% pain reduction, 30% functional improvement) after 2 years in a larger Level 3 study (254 patients, single intra-articular knee injection).

Due to differences in aspiration technique and the efficiency of concentration processes or devices, it is critical to assess the cellular quality of the BMA and BMC before drawing comparative conclusions between studies. One example of this phenomena is a study by Anz et al. [26] that concluded BMC was not statistically different than PRP for treating knee arthritis. However, the substandard aspiration method yielded low cell counts, as demonstrated by the average TNCs in BMC being similar to that of the PRP group. Another study found no statistical difference between BMC and a saline placebo control, but utilized poor aspiration methods and a centrifuge that recovers less than 50% of progenitor cells to prepare the BMC [27, 28]. The average nucleated cell, progenitor cell, and platelet counts reported in this study were approximately equivalent to blood, not BMC or even BMA.

Hernigou et al. [29, 30] sought to determine the efficacy of intra-osseous injections of autologous BMC compared to intra-articular injections as well as to total knee arthroplasty (TKA) replacement surgery. In the former study, 60 patients with bilateral knee osteoarthritis received 20 mL BMC intra-articularly in one knee, while receiving 20 mL BMC intra-osseously (10 mL in tibia, 10 mL in femur) in the contralateral knee. Magnetic resonance imaging (MRI) was superior for intraosseous injected knees after 2 years, while the rate of

**Table 9.1** Cellular composition of Bone Marrow Aspirate/Concentrate (1, 17–23)

| Cell type | Description | Phenotype | Frequency in BMA |
|---|---|---|---|
| Mesenchymal stem cells (MSCs) | Heterogenous population of connective tissue progenitors of pericyte origin; known for regenerative and immunomodulatory properties | CD146+, CD90+, CD73+, CD90+, CD45−, CD31−. (may be CD34+ in fresh BMA/BMC, CD34− after in vitro expansion) Also quantified by CFU-F in vitro assay | 0.001–0.01% of TNCs. Higher frequencies associated small volume aspiration, harvest from iliac crest vs. limb bones, younger donor age |
| Hematopoeitic stem cells (HSCs) | Progenitors of immune cells; synergetic with MSCs for immunomodulation | CD34+, CD45+, CD31− | 0.05–1% of TNCs |
| Endothelial progenitor cells (EPCs) | Progenitors associated with formation or regeneration of endothelial vascular tissues | CD34+, CD31+, CD45− | >0.9% |
| Other leukocytes or total nucleated cells (TNCs) | T-cells, B-cells, NK cells, monocytes, macrophages, etc. | CD34−, CD45+ | $10–40 \times 10^6$/mL |
| Platelets | Anuclear cells from blood containing high levels of cytokines | CD41+, CD62p+, CD63+ | $60–100 \times 10^6$/mL |
| Erythrocytes (RBCs) | Anuclear cells from blood containing hemoglobin for oxygen transport | CD235a+ | $6 \times 10^9$/mL |

advancement to TKA was approximately three time greater for intra-articular injections. The second study of similar design compared intra-osseous BMC injection to contralateral TKA in 30 patients (average age 28 years). After 12-year average follow-up, pain reduction and functional score improvement was equivalent between the groups. Besides the greater risks associated with surgery, 6 of 30 TKA knees required revision surgery, while only one of 30 intra-osseous BMC injection knees required TKA. The study was repeated with an older patient population (median age 75 years, range 65–90) and 140 patients [31]. The bone marrow lesions reduced on average 1.5 $cm^3$ after 2 years following intra-osseous injection of BMC. Progression to TKA was statistically equivalent to original TKA group knees requiring revision (approximately 1.19% per year). After 10 year follow-up, only 18% of knees treated with intra-osseous BMC advanced to TKA.

### Avascular Necrosis of the Hip

Hip avascular necrosis (AVN) is a condition with few non-surgical treatment options. Hernigou et al. [32] demonstrated that intra-osseous injection of BMC into the lesion(s) within the femoral head could significantly delay progression of collapse in the bone and advancement to total hip arthroplasty (THA). With 371 patients and average follow-up of 12 years, MRI showed an average reduction of lesion volume from 26 to 12 $cm^3$. Only 9 of 145 hips treated in pre-collapse state required THA by 10 years after a single injection of BMC.

### Non-Union Fractures

Treatment of long-term non-unions in the tibia with percutaneous BMC injection showed healing in 88% of patients. Interestingly, 100% of non-unions healed when the MSC concentrations in the BMC were greater than 1500 CFU-F/mL [33]. A later study treated infected tibial non-unions with osteomyelitis by percutaneous injection of BMC and found that 38 of 40 non-unions achieved union without recurrence of infection compared to only 70% treated with antibiotics [34]. Another trial comparing diabetic and non-diabetic patients, it has shown that tibial non-unions fractures in diabetic patients were healed by percutaneous BMC in 76% of patients compared to 91% of non-diabetic patients [35].

### Rotator Cuff Tears

Shoulder surgeons have sought to supplement surgical repair of torn rotator cuff tendons by augmenting with biologics including BMC and PRP. The study with most extensive characterization of BMC and corresponding outcomes was performed by Hernigou et al. [36], where 45 patients with full thickness supraspinatus tendon tears were surgically repaired followed by injection of 8 mL BMC into the humerus and 4 mL BMC into the bone-tendon junction. Retrospectively, 45 patients matched by injury treated with surgery alone (no biologic augmentation) were considered as a control. After 6 months, 100% of BMC-injected tendons were completely healed according to MRI, compared to 67% of tendons repaired without BMC. After 10-year follow-up, 87% of tendons treated with BMC remained radiographically intact, compared to only 44% of tendons in the control group. The average reported dose of MSCs in this study's BMC was 4300 CFU-F/mL.

### Degenerative Disc Disease in Lumbar Spine

Early non-surgical, biologic interventions to treat lumbar degenerative disc disease (DDD) utilized therapies considered drugs by the United States Food and Drug Administration (FDA) under regulated clinical trials. Pettine et al. [37] explored autologous BMC as a non-drug alterative in similarly designed study. The study (26 patients, 13 single level, 13 two level, median age 40 years) discovered a correlation between MSC content in BMC and clinical improvement (pain reduction, MRI Pfirrmann score) with overwhelming improvement in patients whose BMC contained greater than 2000 CFU-F/mL [38, 39]. The average Oswestry Disability Index (ODI) score reduced by 67% and Visual Analogue Score (VAS) reduced by 73% through 3 years following intradiscal injection of 2 to 3 mL BMC.

A recent study by Atluri et al. [40] treated a more inclusive demographic of patients suffering from lumbar DDD (40 subjects, average age 61 years) via intradiscal BMC and included a control group of conservative treatments (40 subjects, average age 59 years). The average ODI and VAS scores of the group treated with BMC were reduced by 15 points (100-point scale) and 2.9 points (10 point scale) after 12 months, while ODI and VAS scores in the control group worsened during the same period. The authors reported average cellularity of 45 million nucleated cells and 1220 MSC (CFU-F) per mL in BMA, while the average post-centrifuged BMC used for injection contained 239 million nucleated cells and 4987 MSCs per mL. This study showed similar benefits of BMC to Pettine et al. [37–39] with older patients but greater progenitor cell counts in the BMC. A summary of mesenchymal progenitor cell counts correlating with improved clinical outcomes is provided in Table 9.2 [9, 32, 35–39, 41]. Metanalysis performed revealed that there is level 3 evidence for treating discogenic pain with biologics and level 4 evidence for facet, sacroiliac and epidural administration [42]. The American Society of Interventional Pain Physicians (ASIPP) published practice guidelines for safe use of biologics for the treatment of low back pain [43].

### Safety of Autologous Bone Marrow Concentrate Injections

Due to the association with "stem cells," the safety of autologous BMC is sometimes questioned relative to

**Table 9.2** Mesenchymal stem cell (MSC) concentrations associated with positive clinical outcomes for MSK indications

| Clinical application | Critical MSC (or CFU-F) concentration | Clinical outcomes |
|---|---|---|
| Avascular necrosis of the hip (injection) (116 patients, 10-year follow-up) [32] | 31,000 total | • 50% reduction osteonecrosis volume (avg.)<br>• 15/116 total resolution of lesion by MRI<br>• Only 12.5% collapse rate @ 5 years |
| Non-union tibial fractures (percutaneous injection) (60 patients, >3-year follow-up) [9, 35] | 1500 per cc<br>54,000 total | • Bone union achieved in all 53 patients with >1500 MSC/cc<br>• 7 continued non-unions received avg. 630 MSC/cc or 19,000 total |
| Posterolateral lumbar spinal fusion [41] | 1134 per cc<br>18,000 per level | • Equivalent fusion rates to iliac crest bone grafting (considered gold standard)<br>• 14/25 patients were smokers |
| Intervertebral disc degeneration (injection) (26 patients, 3-year follow-up) [37–39] | 2000 per cc<br>4500 total | • >70% improvement in pain and disability scores through 3 years (all patients)<br>• >86% improvement in patients w/ >2000 MSC/cc<br>• MRI improvement in 40% of patients |
| Rotator cuff repair (surgical) (45 patients, 10-year follow-up) [36] | 4200 per cc<br>54,000 total | • 100% Healing at 12 months<br>• 87% intact cuffs after 10 years |

embryonic or allogeneic cell therapies. One safety study reported that with short term follow-up (average 14 months) after orthopedic surgical use of BMC, no adverse events were observed in all 101 patients treated [44]. A larger study (2372 patients whose treatments included BMC at 18 different clinical facilities) with average follow-up of 2.2 years reported 325 adverse events after 3012 procedures, mostly short-term post-procedure pain or pain associated with degenerative joint progression [45]. It was noted that the lowest adverse event rate was in patients treated by BMC injection alone.

## Regulatory Considerations for Use of Bone Marrow Concentrate

Biologics are generally regulated in the United States by the FDA. Authorization of FDA is granted for human cell, tissue, and cell-tissue products (HCT/Ps) under 21 CFR 1271. However, BMA is not considered an HCT/P so long as it is minimally manipulated and used in a homologous fashion to its original biological function. The process of centrifugation of blood and BMA the separate or concentrate cellular components is not considered more than minimal manipulation. Thus, autologous BMC used for homologous clinical applications is not regulated as an HCT/P and does not require drug or device approval from FDA. Manchikanti along with other leading regenerative specialists published a position statement that BMC is homologous and minimally manipulated and thus is FDA compliant to treat spine and joint arthritis [46]. Hernigou et al. [47] observed that rates of new or recurring cancers in 1873 patients treated with autologous BMC for orthopedic conditions were not increased or statistically different from the general population with a mean follow-up of 12.5 years.

## Conclusion

Preliminary clinical data shows that use of autologous BMC for treatment of musculoskeletal conditions is safe and efficacious. Several studies have shown that treatment with autologous BMC to be equally effective and is a safe alternative to surgery and other traditional treatments.

Properly aspirated and concentrate bone marrow is rich in mesenchymal and hematopoietic progenitor cells, platelets, IRAP, and A2M. These factors offer a multimodal treatment to address chronic inflammation, degeneration, avascularity, apoptosis, necrosis, and/or infection. As with most regenerative medicine options, a key benefits of BMC is its capability in promoting healing, with control of pain and with functional improvement without adversely affecting the anatomy or integrity of the part treated.

**Key Points**

1. Bone marrow has two kind of progenitor cells, hematopoietic and mesenchymal.
2. Mesenchymal stem cells are responsible for repair and regeneration of musculoskeletal tissue.
3. BM-MSC are found attached to capillaries and small arterioles.
4. Multiple sites aspiration of bone marrow has shown to yield more BM-MSC.
5. Use of small size syringes and rapid aspiration of small volumes of bone marrow is shown to yield higher BM-MSC.
6. There is no increased risk of carcinogenicity using BMC for treatment purposes.
7. Autologous BMC used for homologous clinical applications is not regulated as an HCT/P and does not require drug or device approval from FDA.

**Questions**

1. Which statement is true regarding outcomes of injection of bone marrow concentrate for treatment of knee arthritis
   A. Intraosseous injection is superior than intraarticular injection
   B. Intraarticular injection is superior than intraosseous injection
   C. Both intraarticular and intraosseous injections are equally good.
   D. Neither intraosseous nor intraarticular injection has shown to improve knee arthritis

   Answer: **A**. Intraosseous injection is superior to intraarticular injection for knee arthritis. P Hernigou showed in randomized control trial that Subchondral injection of bone marrow concentrates had better clinical and imaging (MRI) improvement than intraarticular injection.
2. Bone marrow mesenchymal stem cells aspirates can be optimized by
   A. Using large volume syringes with slow aspiration
   B. Using single site aspiration
   C. Using small volume syringes and rapid aspiration
   D. Bone marrow aspiration at the shaft of long bones.

   Answer: **C**. Hernigou demonstrated that using 10 mL syringes and aspirating small volume of BMA rapidly acquired more BM-MSCs than using 60 mL syringes.
3. After Centrifugation of BMA, colony forming unit-fibroblast are present in
   A. Free floating in plasma
   B. In the sedimented erythrocytes
   C. Adipose tissue layer
   D. Buffy coat layer

   Answer: **D**. After the centrifugation of BMA the RBCs will sediments first, above which the nucleated cells and platelets sediment in the buffy coat layer. Plasma does not have any CFU-F, the adipose tissue is the top layer.
4. Use of biologics in United States of America is regulated by
   A. State medical board
   B. Center for Disease Control
   C. Food and Drug Administration
   D. American Medical Association

   Answer: **E**. The Food and Drug Administration regulates use of biologics. Authorization of FDA is granted for human cell, tissue, and cell-tissue products (HCT/Ps) under 21 CFR 1271. However, BMA is not considered an HCT/P so long as it is minimally manipulated and used in a homologous fashion to its original biological function. The process of centrifugation of blood and BMA the separate or concentrate cellular components is not considered more than minimal manipulation.
5. Studies have shown that minimally manipulated autologous bone marrow concentrates are used in treatment of
   A. Treatment of osteosarcoma
   B. Pain after total knee replacement
   C. reducing bone collapse after avascular necrosis of femoral head
   D. Vertebral hemangiomas

   Answer: **C**. Hernigou et al. [32] demonstrated that intraosseous injection of BMC into the lesion(s) within the femoral head could significantly delay progression of collapse in the bone and advancement to total hip arthroplasty (THA). With 371 patients and average follow-up of 12 years, MRI showed an average reduction of lesion volume from 26 to 12 cm$^3$. Only 9 of 145 hips treated in pre-collapse state required THA by 10 years after a single injection of BMC.

**Acknowledgments** This book chapter is modified and updated from a previous book chapter, "Bone Marrow Concentrate" by José Fábio Lana, MD, Joseph Purita, MD, Carlos Augusto Silva Rava, MD, Mathew Atkerson Nicholls, MD, PhD, Carlos Amilcar Parada, PhD, André Paráiso Forti, MD, Inácio Facó Ventura Vierira, MD, João Lopo Madureira Junior, MD, Paulo David Fortis Gusmão, MD, Lauro Schledorn de Camargo, MD, and Carolina Caliari-Oliveira, PhD, in *Essentials of Regenerative Medicine in Interventional Pain Management* published by ASIPP Publishing.

## References

1. Lana JF, Purita J, Rava CA, et al. Bone marrow concentrate. In: Manchikanti L, Navani A, Atluri S, editors. Essentials of regenerative medicine in interventional pain management. Paducah, KY: ASIPP Publishing; 2019. p. 83–98.
2. Murphy MB, Moncivais K, Caplan AI. Mesenchymal stem cells: environmentally responsive therapeutics for regenerative medicine. Exp Mol Med. 2013;45:e54.
3. Doulatov S, Notta F, Laurenti E, Dick JE. Hematopoiesis: a human perspective. Cell Stem Cell. 2012;10:120–36.
4. Friedenstein AJ, Piatetzky-Shapiro II, Petrakova KV. Osteogenesis in transplants of bone marrow cells. J Embryol Exp Morphol. 1966;16:381–90.
5. Caplan AI. Mesenchymal stem cells. J Orthop Res. 1991;9:641–50.
6. Undale A, Fraser D, Hefferan T, et al. Induction of fracture repair by mesenchymal cells derived from human embryonic stem cells or bone marrow. J Orthop Res. 2011;29:1804–11.
7. Shekkeris AS, Jaiswal PK, Khan WS. Clinical applications of mesenchymal stem cells in the treatment of fracture non-union and bone defects. Curr Stem Cell Res Ther. 2012;7:127–33.
8. Centeno CJ, Schultz JR, Cheever M. A case series of percutaneous treatment of non-union fractures with autologous, culture expanded, bone marrow derived, mesenchymal stem cells and platelet lysate. J Bioeng Biomed Sci. 2011;1:2–7.
9. Goel A, Sangwan SS, Siwach RC, Ali AM. Percutaneous bone marrow grafting for the treatment of tibial non-union. Injury. 2005;36:203–6.
10. Galois L, Bensoussan D, Diligent J, et al. Autologous bone marrow graft and treatment of delayed and non-unions of long bones: technical aspects. Biomed Mater Eng. 2009;19:277–81.
11. Vandooren J, Itoh Y. Alpha-2-macroglobulin in inflammation immunity and infections. Front Immunol. 2021;12:5411.
12. Hernigou P, Homma Y, Flouzat Lachaniette CH, et al. Benefits of small volume and small syringe for bone marrow aspirations of mesenchymal stem cells. Int Orthop. 2013;37:2279–87.
13. Murphy MB, Terrazas JA, Buford DA. Bone marrow concentrate and platelet-rich plasma acquisition and preparation: why technique matters. Tech Reg Anesth Pain Manag. 2015;19:19–25.

14. Trudnowski RJ, Rico RC. Specific gravity of blood and plasma at 4 and 37°C. Clin Chem. 1974;20:615–6.
15. Zipursky A, Bow E, Seshadri RS, Brown EJ. Leukocyte density and volume in normal subjects and in patients with acute lymphoblastic leukemia. Blood. 1976;48:361–71.
16. Van Oost BA, Timmermans APM, Sixma JJ. Evidence that platelet density depends on the alpha-granule content in platelets. Blood. 1984;63:482–5.
17. Rebolj K, Veber M, Drobnič M, Maličev E. Hematopoietic stem cell and mesenchymal stem cell population size in bone marrow samples depends on patient's age and harvesting technique. Cytotechnology. 2018;70:1575–83.
18. Ghem C, Dias LD, Sant'Anna RT, Kalil RAK, Markoski M, Nardi NB. Combined analysis of endothelial, hematopoietic, and mesenchymal stem cell compartments shows simultaneous but independent effects of age and heart disease. Stem Cells Int. 2017;2017:5237634–13.
19. Hristov M, Weber C. Endothelial progenitor cells: characterization, pathophysiology, and possible clinical relevance. J Cell Mol Med. 2004;8:498–508.
20. Schäfer R, Debaun MR, Fleck E, et al. Quantitation of progenitor cell populations and growth factors after bone marrow aspirate concentration. J Transl Med. 2019;17:1–9.
21. Sugaya H, Yoshioka T, Kato T, et al. Comparative analysis of cellular and growth factor composition in bone marrow aspirate concentrate and platelet-rich plasma. Bone Marrow Res. 2018;2018:1549826–9.
22. Lykov AP, Poveshchenko OV, Cherniavsky AM, et al. Phenotype of bone-marrow mononuclear cells before and after short-time precondition with erythropoietin from patients with ischemic heart failure. Russian Open Med J. 2018;7:1–7.
23. Herrmann M, Binder A, Menzel U, Zeiter S, Alini M, Verrier S. CD34/CD133 enriched bone marrow progenitor cells promote neovascularization of tissue engineered constructs in vivo. Stem Cell Res. 2014;13:465–77.
24. Do KJ, Lee GW, Jung GH, et al. Clinical outcome of autologous bone marrow aspirates concentrate (BMAC) injection in degenerative arthritis of the knee. Eur J Orthop Surg Traumatol. 2014;24:15051511.
25. Oliver KS, Bayes M, Crane D, Briggs E. Autologous bone marrow concentrate-clinical efficacy in the treatment of knee osteoarthritis. 2016.
26. Anz AW, Hubbard R, Rendos NK, Everts PA, Andrews JR, Hackel JG. Bone marrow aspirate concentrate is equivalent to platelet-rich plasma for the treatment of knee osteoarthritis at 1 year: a prospective, randomized trial. Orthop J Sports Med. 2020;8:2325967119900958.
27. Shapiro SA, Kazmerchak SE, Heckman MG, Zubair AC, O'Connor MI. A prospective, single-blind, placebo-controlled trial of bone marrow aspirate concentrate for knee osteoarthritis. Am J Sports Med. 2017;45:82–90.
28. Hegde V, Shonuga O, Ellis S, et al. A prospective comparison of 3 approved systems for autologous bone marrow concentration demonstrated nonequivalency in progenitor cell number and concentration. J Orthop Trauma. 2014;28:591–8.
29. Hernigou P, Bouthors C, Bastard C, et al. Subchondral bone or intra-articular injection of bone marrow concentrate mesenchymal stem cells in bilateral knee osteoarthritis: what better postpone knee arthroplasty at fifteen years? A randomized study. Int Orthop. 2021;45:391–9.
30. Hernigou P, Auregan JC, Dubory A, Flouzat-Lachaniette CH, Chevallier N, Rouard H. Subchondral stem cell therapy versus contralateral total knee arthroplasty for osteoarthritis following secondary osteonecrosis of the knee. Int Orthop. 2018;42:2563–71.
31. Hernigou P, Delambre J, Quiennec S, Poignard A. Human bone marrow mesenchymal stem cell injection in subchondral lesions of knee osteoarthritis: a prospective randomized study versus contralateral arthroplasty at a mean fifteen year follow-up. Int Orthop. 2021;45:365–73.
32. Hernigou P, Daltro G, Filippini P, Mukasa MM, Manicom O. Percutaneous implantation of autologous bone marrow osteoprogenitor cells as treatment of bone avascular necrosis related to sickle cell disease. Open Orthop J. 2008;2:62–5.
33. Hernigou P, Poignard A, Beaujean F, Rouard H. Percutaneous autologous bone-marrow grafting for nonunions. J Bone Joint Surg. 2005;87:1430–7.
34. Hernigou P, Dubory A, Homma Y, Flouzat Lachaniette CH, Chevallier N, Rouard H. Single-stage treatment of infected tibial non-unions and osteomyelitis with bone marrow granulocytes precursors protecting bone graft. Int Orthop. 2018;42:2443–50.
35. Flouzat-Lachaniette CH, Heyberger C, Bouthors C, et al. Osteogenic progenitors in bone marrow aspirates have clinical potential for tibial non-unions healing in diabetic patients. Int Orthop. 2016;40:1375–9.
36. Hernigou P, Flouzat Lachaniette CH, Delambre J, et al. Biologic augmentation of rotator cuff repair with mesenchymal stem cells during arthroscopy improves healing and prevents further tears: a case-controlled study. Int Orthop. 2014;38:1811–8.
37. Pettine KA, Murphy MB, Suzuki RK, Sand TT. Percutaneous injection of autologous bone marrow concentrate cells significantly reduces lumbar discogenic pain through 12 months. Stem Cells. 2015;33:146–56.
38. Pettine K, Suzuki R, Sand T, Murphy M. Treatment of discogenic back pain with autologous bone marrow concentrate injection with minimum two year follow-up. Int Orthop. 2016;40:135–40.
39. Pettine KA, Suzuki RK, Sand TT, Murphy MB. Autologous bone marrow concentrate intradiscal injection for the treatment of degenerative disc disease with three-year follow-up. Int Orthop. 2017;41:2097–103.
40. Atluri S, Murphy MB, Dragella R, et al. Evaluation of the effectiveness of autologous bone marrow mesenchymal stem cells in the treatment of chronic low back pain due to severe lumbar spinal degeneration: a 12-month, open-label, prospective controlled trial. Pain Physician. 2022;25:193–207.
41. Johnson RG. Bone marrow concentrate with allograft equivalent to autograft in lumbar fusions. Spine (Phila Pa 1976). 2014;39:695–700.
42. Sanapati J, Manchikanti L, Atluri S, et al. Do regenerative medicine therapies provide long-term relief in chronic low back pain: a systematic review and metaanalysis. Pain Physician. 2018;21:515–40.
43. Navani A, Manchikanti L, Albers SL, et al. Responsible, safe, and effective use of biologics in the management of low back pain: American Society of Interventional Pain Physicians (ASIPP) guidelines. Pain Physician. 2019;22:S1–S74.
44. Hendrich C, Franz E, Waertel G, Krebs R, Jäger M. Safety of autologous bone marrow aspiration concentrate transplantation: initial experiences in 101 patients. Orthop Rev (Pavia). 2009;1:e32.
45. Centeno CJ, Al-Sayegh H, Freeman MD, Smith J, Murrell WD, Bubnov R. A multi-center analysis of adverse events among two thousand, three hundred and seventy two adult patients undergoing adult autologous stem cell therapy for orthopaedic conditions. Int Orthop. 2016;40:1755–65.
46. Manchikanti L, Centeno CJ, Atluri S, et al. Bone marrow concentrate (BMC) therapy in musculoskeletal disorders: evidence-based policy position statement of American Society of Interventional Pain Physicians (ASIPP). Pain Physician. 2020;23:E85–E131.
47. Hernigou P, Homma Y, Flouzat-Lachaniette CH, Poignard A, Chevallier N, Rouard H. Cancer risk is not increased in patients treated for orthopaedic diseases with autologous bone marrow cell concentrate. J Bone Joint Surg Am. 2013;95:2215–21.

# 10 Adipose Tissue-Derived Stromal Vascular Fraction

Theodore T. Sand

## Introduction

Enzymatic digestion of adipose tissue produces a suspension of single cells that is comprised of a combination of progenitor and adult cells [1]. An early report [2] used collagenase digestion of adipose tissue that resulted in the isolation of adipocytes and stromal vascular cells, with the released adipocytes the focus of the report. Subsequent studies [3, 4] relied on the use of collagenase-based enzymatic digestion of adipose tissue in order to evaluate adipocyte precursor cells found among the stromal vascular cells. The discovery and description of mesenchymal stromal cells (MSCs) present in adipose tissue was first reported in a publication in 2001 [5], and further described in 2002 [6]. Adipose tissue-derived MSCs (AD-MSC) were associated with the single cell suspension obtained by centrifugation of the enzymatic digestate of adipose tissue, also referred to as the "stromal vascular fraction" (SVF) of the released cells [5]. The discovery of MSCs in human adipose tissue was the starting point for the exploration of the potential of using a patient's own adipose tissue as a source of therapeutic treatments.

MSCs are widely distributed in mammalian tissues [7], which is due in part to the presence of MSCs residing in the perivascular space (pericytes) [8, 9]. MSCs are characterized by functional properties, including adherence to tissue culture containers and a capability of being differentiated under tissue culture conditions into multiple adult cell lineages [5]:

- Chondrogenic, osteogenic, adipogenic and myogenic cell lineages [5, 6]
- Neurogenic cell lineage [10]

As a result of the AD-MSC's capacity for being differentiated into multiple cell lineages, it is considered to be a multipotent progenitor cell [11]. AD-MSCs also grow in colonies, with a fibroblastic (i.e., crescent shaped) morphology [5]. The therapeutic potential of MSCs was first associated with bone marrow-derived MSCs in the 1990s [12]. Subsequent investigation into the MSC's potential therapeutic mechanisms of action have revealed that MSCs might differentiate into a multiplicity of adult tissue cell lineages, as well as secrete a wide variety of cytokines, growth factors and proteins that act as agents with immunomodulatory, mitogenic, rescue of dying cells (anti-apoptotic), wound-healing and anti-scarring effects [13, 14].

Use of lipoaspirate as a source of adipose tissue was recognized early-on as a convenient, safe and well-tolerated procedure, and when digested, resulted in the release of substantial numbers of MSCs and other cells [5, 6, 15]. Clinical translation of SVF preparations started in 2004 in the USA, which allowed veterinarians to treat musculoskeletal pathologies in equine patients with the preparation of autologous SVF through a centralized processing laboratory [16]. Subsequently, this commercial service was expanded to dogs and cats, and included treating osteoarthritic joints based on clinical studies in the canine that reported a therapeutic benefit of intraarticular injection of the canine patient's SVF into osteoarthritic joints [17, 18]. A case study (calvarial defects; [19]) and a clinical study (lower extremity ulcers; [20]) involving human-derived SVF preparations provided support for the use of adipose tissue-derived SVF in treating a variety of human pathologies. Commercialization of a Point-of-Care (POC) SVF therapy to treat human pathologies was initiated in the USA in 2011 [21]. It should be pointed out that the U.S. Food and Drug Administration (USFDA) had indicated in 2012 that autologous SVF therapy was not compliant with the provisions of 21 CFR 1271.10 [22, 23], which means that SVF therapy was then and remains an experimental drug or biologic product and as such should be used for human treatment only within an Institutional Review Board (IRB) approved clinical study [24].

T. T. Sand (✉)
Sand Consulting, Poway, CA, USA
e-mail: tsand@biotechexpert.com


A. Navani et al. (eds.), *Essentials of Regenerative Medicine in Interventional Pain Management*,
https://doi.org/10.1007/978-3-031-50357-3_10

## Preparation and Biology

### Preparation of SVF

As noted above, lipoaspirate is a ready source of adipose tissue that can be processed to produce an autologous SVF preparation [15]. Lipoaspiration typically is performed with a tumescent fluid and gentle aspiration of adipose tissue [25]. The lipoaspirate tissue is condensed at the top of the container following standing or centrifugation. In a typical manual-based digestion protocol, the recovered adipose tissue is washed with phosphate buffered saline (PBS) and incubated with an enzyme digestion fluid containing collagenase typically at 37 °C for 30 min or more with mechanical agitation [25, 26]. Other enzymes have been used along with collagenase, including dispase (or other neutral proteases) and hyaluronidase [25] to enhance release of single cells from the adipose matrix. After the digestion step is completed, the digestate is centrifuged, the supernatant is removed, the pellet resuspended in PBS or other physiologically compatible buffer and recentrifuged to wash out the enzyme(s) used for the digestion [26, 27]. If the SVF preparation is intended for therapy, a cell count and viability might be performed prior to treatment [17]. An early evaluation of enzymatic digestion of lipoaspirates compared to cells recovered from the fluid portion of the lipoaspirate showed that while there were an equivalent number of viable cells present in the SVF from both the enzymatic digest and the centrifuged lipoaspirate fluid, the SVF obtained with enzymatic digestion resulted in a substantially higher number of AD-MSCs [15].

Although a manual method of mincing adipose tissue, followed by enzymatic digestion and washing with PBS was commercialized for producing autologous SVF therapeutic preparations for use in veterinary patients as early as 2004 [16], commercialization of SVF for human therapy generally has been associated with the development of automated or semi-automated technologies [28], which have been used in IRB-approved clinical studies [29, 30]. However, some manufacturers' kits and technology intended for enzymatic digestion of a patient's adipose tissue have been used without an IRB-approved prospective clinical study [21], since currently there are no SVF production methods or technologies available to use to treat humans that have been approved by the USFDA. Thus, it is important for physicians considering the use of automated or semi-automated technologies for providing SVF therapy at POC to use the technology only within an IRB-approved clinical trial. Registration of a "trial" with ClinicalTrials.gov doesn't mean that the clinical trial has been approved by an IRB.

### Biology of SVF

A wide variety of cells are found in the stromal vascular fraction, including lymphocytes, red blood cells, endothelial cells, endothelial precursor cells (EPCs), hematopoietic stem cells (HSCs), AD-MSCs, neutrophils, macrophages, pericyte precursor cells, pericytes, fibroblasts, smooth muscle cells and supra-adventitial adipose stem cells [11, 15, 31]. Also, a recent report identified the presence in SVF of a variety of adipocyte progenitor cells (APCs) [32]. This diverse cellularity reflects the composition of adipose tissue, which contains extracellular matrix (ECM) and adipocytic components that form the vascular and connective tissues of adipose. In fact, adipose tissue is considered to be an organ, which regulates a wide variety of physiological and metabolic processes via autocrine, paracrine and endocrine pathways [33].

While AD-MSCs present in SVF are thought to provide the main therapeutic benefit, clearly there are a variety of cells present when implanted in the site of pathology. Thus, both SVF and bone marrow-derived preparations (bone marrow concentrate [BMC]) are composed of a diverse set of cell types, although a key distinction is that BMC contains far more HSCs per unit volume compared to the number of HSCs previously reported for SVF [15]. The cellular diversity and variable numbers of individual cell types in SVF depend on the source of the adipose tissue (abdomen, flank, back, etc.), the harvesting technique, the digestion protocol and patient-dependent factors [34]. The inherent variability in cell number and composition in SVF preparations presents a challenge in attempting to establish a relationship between a "dose" of SVF and treatment outcomes. Naturally, much effort has been expended on identifying the cells found in SVF. Clusters of differentiation markers (CD markers) have been used extensively to try to fashion a unique identity for AD-MSCs. For example, a consensus summary of CD markers has been published that depends on a combination of positive and negative markers of freshly isolated cells in SVF. AD-MSCs are negative for CD45, CD235a, and CD31, while positive for CD34, CD13, CD73, CD90, and CD105. CD45 is a marker associated with adult leukocytes, while CD31 is associated with adult and progenitor endothelial cells [11]. CD146 is used to identify pericytes, which are a type of multipotent progenitor cell found in the perivascular space in tissues including adipose tissue [31]. There is some disagreement about the CD marker profile of "AD-MSCs", since in a recent publication the CD marker profile of "supra-adventitial adipose stem cells" was described as negative for CD45, CD31, CD146 and CD105, while being positive for CD34, CD73 and CD90 [31], which differs slightly from the consensus profile in which freshly isolated AD-MSCs are variably CD105 positive [11]. Since there isn't a single CD marker that is unique to AD-MSCs, phenotyping of SVF cells requires a panel of both positive and negative CD markers [11], and benefits from the use of a three-laser flow cytometer [31].

AD-MSCs attach and proliferate in tissue culture containers, which can be exploited in two ways. When cells from an SVF preparation are diluted, counted, distributed at different seeding levels (cells/culture well) and cultured for more than 9–10 days, it is possible to obtain an estimate of the number of colonies that grow out under those conditions. This approach for quantifying the AD-MSC content of a SVF preparation is called a Colony Forming Unit-Fibroblast (CFU-F) assay, since the AD-MSCs will grow in a fibroblastic shape (i.e., crescent shaped morphology) [11]. AD-MSCs also can be cultured in the presence of specific induction agents in order to demonstrate that the attached AD-MSCs can be differentiated into a variety of adult cell lineages, with the most common being osteogenic, chondrogenic and adipogenic cell lineages [11].

Genetic probe-based analyses have aided in better identification strategies in conjunction with more traditional CD marker analysis [11]. Furthermore, the use of single-cell RNA analysis has been exploited to provide a more nuanced profile of adipocytic progenitor cells [32], which points to a complex dynamic of cell regulation in a patient's adipose tissue. Recent evidence of functional heterogeneity in bone marrow-derived MSCs assessed by performing single-cell transcriptomics revealed the existence of several subpopulations of MSCs that differed in their mechanosensitivity and differentiability [35]. Similar heterogeneity of AD-MSCs was found in three unrelated donor SVF preparations, which was influenced primarily by the adipose tissue ECM [36].

## Mechanism of Action

Numerous experiments have been performed in an attempt to sort out how progenitor-cell containing SVF preparations might provide a therapeutic benefit in treating musculoskeletal pathologies. The original thinking about how MSCs might be therapeutic focused on their ability to differentiate into adult tissue cells (e.g., adipocytes, chondrocytes, etc.) *in vitro*, which was assumed to be the situation *in vivo* [37]. Currently, it is thought that pericytes (the MSC that resides in the perivascular space) respond to soluble signaling factors, become activated into MSCs, which migrate to sites of pathology, and initiate a secretory response that is appropriate to the pathologic milieu [38]. This mechanism of action is referred to as the paracrine effect [14], and encompasses the release of cytokines, growth factors, proteins and other biologically active agents that have broad activities for immunomodulation, neo-angiogenesis, enhanced tissue cell proliferation, wound repair, anti-apoptotic, and anti-microbial activity [14, 38]. Thus, pericytes found in SVF diminish the frequently pro-inflammatory environment at sites of pathology (e.g., osteoarthritic joints) by regulating the activity of macrophages, monocytes and neutrophils [38]. Reduced inflammation allows resident tissue cells to return to important activities like secreting ECM proteins and maintaining ECM. The recovering tissue cells become more responsive to other molecules secreted by the MSCs, thereby aiding in the transition to a remodeling phase that leads ultimately to tissue restoration with the potential for delaying disease progression with pathologies like osteoarthritis (OA). There is recent clinical evidence that MSC-containing autologous BMC preparations when injected either via the intraarticular (IA) or intraosseous (IO) routes to treat knee OA were able to delay progression to total knee arthroplasty by an average of 7 (IA) or 14 (IO) years [39], which suggests that autologous, MSC-containing therapies like SVF and BMC have the potential to reduce pain and delay disease progression. One advantage of the use of adipose tissue to obtain access to progenitor stromal/stem cells is that adipose tissue has been reported to contain 10–100x the number of progenitor cells on a volume basis compared to bone marrow, and that approximately 3% of the cells present in SVF will adhere to tissue culture containers, suggesting that these cells might be AD-MSCs [25].

The paracrine effect of AD-MSCs described above reflects the release of a wide variety of biochemical agents by the MSC itself [14]. However, all cells in the body release extracellular vesicles (EVs), with a recent report reviewing release of a variety of EVs, including exosomes (also known as small EVs), from adipose tissue cells [40]. Of particular clinical interest are the EVs released from AD-MSCs [41]. It is important to appreciate that once the viable cells in an SVF preparation are implanted in a pathologic site, the cells will release exosomes and other types of EVs, which has two benefits. First, the "cargo" of exosomes contains lipids, proteins (including cytokines), and a variety of nucleic acids (e.g., microRNA, RNA, DNA), all of which have been shown to interact with and modify target cells [40, 41]. Thus, the cargo of exosomes supports the paracrine activity of the AD-MSCs themselves. Second, release of EVs directly depends on AD-MSCs, which means that each AD-MSC is able to amplify the impact of its paracrine effect beyond the number of AD-MSCs actually present in the pathologic site.

While there is no doubt that AD-MSCs will release EVs *in vivo*, the number of EVs released per AD-MSC *in vivo* has yet to be reported. In a recent publication a human plasma sample was found to have approximately $6 \times 10^{11}$ particles per mL of plasma [42]. Another report considered the impact of a variety of culture conditions on the release of "particles" (i.e., EVs) from a cultured MSC cell line. The authors reported that under some conditions up to $1 \times 10^{6}$ EVs/cultured cell were recovered in the spent culture medium after a 24-h culture interval [43]. Since the numbers of EVs present in plasma and spent tissue culture media routinely are described to be in the billions, it is reasonable to speculate that AD-MSCs probably release dozens of exosomes (if not many more) while residing in a pathologic site. Thus, the release of exosomes can significantly amplify the therapeutic benefit of individual AD-MSCs present in the SVF. However, the EV-based amplification factor becomes another of the unknowns in providing an autologous SVF therapy. For example, a patient's AD-MSCs might not release a sufficient number of EVs to aid in achieving a therapeutic benefit, or their EVs' cargo might not contain components that are sufficiently or appropriately therapeutic, since functional heterogeneity of AD-MSCs has been reported that was dependent on the ECM of the adipose tissue from which the AD-MSCs were isolated [36]. Furthermore, AD-MSCs will continue to release EVs over time, adjusting to the microenvironment as remodeling of the pathology evolves. Since AD-MSCs are responsive to the signals from the microenvironment [38], most likely the contents of the cargo in the exosomes and other EVs could change to reflect changes in the microenvironment [41].

The versatility of EVs for treating a wide spectrum of disease pathology, from wound healing to Covid 19, was reviewed recently [41, 44]. It is important to note that EVs are unable to replicate, which means that EVs, if used as a standalone therapy, might need to be tailored to address the evolving stages of the time-dependent remodeling of a pathologic site. In that regard, they are like the platelets in platelet-rich plasma (PRP) preparations—once the platelets are activated, they degranulate and then are recycled, which has led to the use of multiple PRP treatments at timed intervals to maximize their therapeutic benefit (e.g., in treating knee OA [45]). A similar approach to treatment with EV-containing preparations might be required if an EV-based, cell-free therapeutic paradigm demonstrates superior therapeutic benefits. On the other hand, AD-MSCs responding to the signals in an evolving pathologic site is one of the important advantages to the use of autologous SVF therapeutic preparations over cell-free preparations, like exosomes obtained from cell culture, that are unable to respond to the feedback loops governing a pathologic site or condition.

## Clinical Applications and Safety

### Clinical Applications in Cosmetic Dermatology and Reconstructive Plastic Surgery

The grafting of autologous adipose tissue (lipotransfer, fat transfer) has been used for decades for a variety of indications. For example, lipotransfer has been used for treating age-related cosmetic changes on the face or used for reconstructive surgery and breast augmentation [46]. One of the issues with fat transfer is the survival of adipocytes in the grafted volume of adipose tissue, since graft resorption that reduces the volume of the graft over time is a common occurrence [46, 47]. Use of an autologous SVF preparation mixed with the fat transfer has been evaluated in an effort to enhance volume retention of fat grafts. In an early report on the clinical use of fat transfer for breast augmentation, autologous SVF prepared from a portion of the lipoaspirate with an automated instrument, was combined with the lipoaspirate prior to treatment. A total of 40 study participants underwent the "Cell-assisted Lipotransfer (CAL)" procedure. Breast implant volume averaged 270 mL, which underwent resorption but reached a stable volume of between 100 and 200 mL in the study participants 2 months after treatment. The authors indicated that the increase in breast circumference achieved with the CAL-treated study participants ranged from 4–8 cm, while in a cohort of patients treated with conventional lipotransfer the average increase in breast circumference was 2–3 cm 2 months after treatment [46].

A recent report on the use of SVF prepared by a semi-automated method for augmenting lipotransfers used for breast augmentation failed to find an advantage in adding SVF to the lipoaspirated adipose tissue [48]. However, the method used to collect the lipoaspirate started with an in-situ emulsification of the adipose tissue with ultrasonic energy, which subsequently was harvested and processed to produce the SVF. It isn't clear that treating adipose tissue with ultrasound that is able to emulsify the tissue would yield healthy cells. The authors checked the viability of the SVF preparation and sent an aliquot to a lab for growing attached cells for flow cytometric analysis. However, it is possible that some of the ultrasound-treated cells in the SVF preparations might have entered a programmed cell-death sequence known as apoptosis, which would diminish or eliminate any benefit of using SVF to augment the lipotransfers. Standard lipoaspirate procedures and SVF handling were performed in the study that showed a benefit to adding SVF to the lipotransfers [46].

Aging-induced alterations in the area of the temple are treated with lipotransfer [49]. In a randomized controlled, split-face design, 15 study participants underwent a lipotransfer with or without SVF (prepared with a manual method) for bilateral temple augmentation. The skin layer thickness of the implanted area was assessed by ultrasound biomicroscopy, the degree of concavity at the temple was assessed (Temple Hollow Score [THS]), along with assessing participant visual analog satisfaction scores, and the frequency of adverse events. The authors reported that at 6-month follow-up there was no difference in the patient visual analog satisfaction score between the two temples. However, there was a substantial and durable increase in the thickness of the temples receiving SVF-lipotransfer compared to the temples receiving standard treatment as measured by ultrasound biomicroscopy, and the THS also was meaningfully reduced (improved) for the SVF-augmented lipotransfer compared to the standard treatment. There was no difference reported in the frequencies of adverse events (e.g., lumping, edema, ecchymosis) for the two treatments. The authors concluded that the use of SVF to augment lipotransfer might be considered for improving the contour in the temple area [49].

SVF prepared in an automated system was used to treat androgenic alopecia (AGA) in a small, split-scalp clinical study of four males and five females [50]. Autologous SVF with between 7–9 × $10^6$ viable cells was injected into 48 sites in the treated scalp. Both untreated and treated scalp sections were evaluated prior to treatment for several parameters, including keratin score, hair density and hair thickness. There were durable, meaningful differences in the keratin score and hair density in the treated scalp compared to the untreated scalp when assessed at the 6-month milestone. Although there was a trend of higher hair thickness in the SVF-treated scalp compared to the untreated scalp it wasn't meaningfully different at 6 months, but the authors indicated that a longer follow-up might show a larger difference in hair thickness [50].

Burn injuries have been treated with SVF, but most of the published papers demonstrating improved healing with SVF treatment are in animal pre-clinical models [51]. A recent review of the use of SVF for treating scar tissue and wounds pointed to the frequent combination of SVF with a bioscaffold (e.g., nano-fat, fat-grafting). The authors concluded that there was substantial evidence to support the use of SVF in augmenting treatments for dealing with scar tissue remodeling and enhancing wound healing [52].

## Clinical Applications in Musculoskeletal Pathologies

An early indication that autologous SVF preparations prepared with a manual digestion method had therapeutic potential for treating musculoskeletal pathologies was reported in a Level 1 double blinded, placebo-controlled clinical evaluation of the use of SVF for treating canine OA of the hip. Enrolled test subjects received autologous SVF in an intraarticular injection into the hip, while the control group received sterile normal saline. An assessment of treatment effect was made at 3-months. The mean effect size for a composite pain score (including all of the individual metrics, like lameness at a trot, range of pain-free motion, etc.) was 1.34 for the SVF-treated animals, which the authors considered to be "large", especially when compared to an aggregate effect size of 0.44 reported for glucosamine treatment in human knee OA [17]. A follow-up open label study of SVF treatment of canine elbow OA showed statistically meaningful improvement in the composite pain score and all individual metrics except pain on manipulation at the 6-month milestone compared to pre-treatment values [18].

Knee OA is one of the most frequently reported human musculoskeletal pathologies in assessing SVF treatment efficacy. As was recently noted [53], 22 clinical studies on the treatment of knee OA with SVF therapy (or in combination with other treatments, e.g., PRP) were published between 2016 and 2020. Half of the studies were conducted as open-label, uncontrolled studies. Overall, from the results of the studies in the publications reviewed, the authors concluded that there was evidence of safety and therapeutic benefit in treating knee OA with SVF [53]. Dose-escalation in SVF clinical studies is rarely reported, with the default treatment being to provide all of the cells obtained after processing [53]. However, a recent Level 1, prospective double-blinded, placebo-controlled clinical study compared outcomes of a high dose (3 × $10^7$ cells) versus a low dose (1.5 × $10^7$ cells) of autologous SVF cells prepared by an automated method that were injected via the intraarticular route for treating knee OA [54]. The primary outcome measure was the Western Ontario and McMaster Universities Osteoarthritis Index (WOMAC) score. The following outcomes were reported [54]:

- Both the low dose and high dose SVF treated groups showed a reduction in median WOMAC scores at 1-year post-treatment of 89.5%, 68.2% and 0% for the high dose, low dose and placebo (unspecified) groups, respectively.
- The effect sizes for both SVF doses were reported as >0.7, which was considered to be a "large" treatment effect.
- No differences were observed in the metrics used to assess articular cartilage health for the three groups based on magnetic resonance imaging (MRI) evaluation pre-treatment and at 1-year post-treatment.

Another study looked at the benefit of using 2.5 × $10^7$ SVF cells compared to 5.0 × $10^7$ SVF cells prepared with an automated method that were injected via the intraarticular route to treat knee OA [55]. The report showed there were higher Knee Injury and Osteoarthritis Outcome Scores (KOOS) for the high SVF dose group compared to the low SVF dose group at 1-year follow-up. However, there was no dose-related difference in the imaging evaluations of the two groups at 1-year follow-up [55]. In a randomized controlled clinical study [56] in which enrolled study participants suffering from bilateral knee OA acted as their own controls, SVF cells prepared by a manual method were used to treat knees and control knees received hyaluronic acid injections. The outcomes of the study were as follows [56]:

- SVF knees showed substantial improvement in VAS (Visual Analog Scale) pain scores and WOMAC scores and improved range of motion at 12-months.
- Control knees showed poorer outcomes for VAS and WOMAC.
- Whole-organ magnetic resonance imaging score of the knee (WORMS) was found to have improved for SVF knees, but not for control knees.
- Magnetic resonance assessment of cartilage tissue (MOCART) of SVF-treated knees showed clear improvement in filling cartilage defects, while control knees showed poorer outcomes.

SVF treatment has been studied as an adjunctive therapy following arthroscopic removal of inflamed tissue in the treatment of knee OA [57]. In the non-randomized study, all participants underwent an arthroscopic procedure followed by microfracture, with the control group receiving no further treatment, while the other group received an intraarticular injection of autologous SVF cells prepared with a manual method. In addition to using VAS and WOMAC metrics, MRI was used to assess the cartilage status and bone marrow edema (BME). The study reported the following outcomes [57]:

- VAS and WOMAC scores decreased for treatment and control groups from pre-treatment to 12-month milestone.
- Control group metrics increased, while SVF treated group metrics decreased further at 24 months.
- The Outerbridge Score (OS) for the control and treatment groups showed no meaningful difference pre-treatment, while the SVF treatment group OS value was substantially lower (improved) compared to the control OS value which showed virtually no change at 24 months.
- Control group BME lesions increased in length, but decreased substantially for the SVF treatment group at the 24-month study endpoint.
- Greater improvement in VAS and WOMAC was associated with individuals that had a pre-treatment Kellgren-Lawrence (KL) Grade 3 score compared to a KL Grade 2 score.

The authors suggested that the control arthroscopic procedure had a therapeutic benefit, but that the SVF adjunctive therapy extended the benefit of the arthroscopic procedure at least out to 2 years [57].

Overall, there is a general consensus that SVF injections delivered via the intraarticular route provide pain relief for knee OA, but repair of cartilage damage is not consistently observed. The possibility of using SVF preparations as an adjunctive therapy following arthroscopic treatment of osteoarthritic knees might be an additional option for obtaining not only pain relief but repair of damaged cartilage. Another factor that might influence the approach for using SVF therapy for knee OA was illustrated by the greater improvement observed with KL3 knees [57] that had undergone an arthroscopic procedure prior to delivery of the SVF preparation. As was recently noted [58], the most popular biologic for treating tendinopathies in preclinical animal models was PRP, with relatively few studies reported with SVF or BMC. Thus, it should be no surprise that most of the high-quality human clinical studies with SVF therapy published in the literature have been limited to knee OA. Regardless of the musculoskeletal pathology, extending study endpoints beyond 1 or 2 years is needed to refine best practices in achieving more durable pain relief, enhanced tissue remodeling and regeneration, and to slow disease progression in orthopedic pathologies being treated with autologous SVF.

## Safety of SVF

As reviewed in this chapter, SVF refers to the single cells isolated from adipose tissue, usually by enzymatic means. However, appending SVF to other types of preparations is becoming increasingly popular. For example, micronized/microfragmented adipose tissue (MAT) has been referred to as "MAT-SVF" [58]. It is important to acknowledge that single cell suspensions, like SVF, can be injected IV, while tissue-containing preparations, like MAT, can't be injected IV. Referring to tissue-containing biologics as "SVF" might lead to confusion and the inappropriate delivery of the MAT preparation.

An important safety concern with SVF preparations relates to the use of collagenase during the processing of the condensed adipose tissue to produce SVF. Following digestion, extensive washing of the cell pellet is required to reduce any residual collagenase that might be present. Four commercial systems for processing adipose tissue and producing an injectable SVF preparation were assessed for residual collagenase levels in the SVF [59]. Three of the systems had a residual collagenase activity in the SVF that was less than 0.01 U/mL, with one of those systems having approximately 0.001 U/mL collagenase activity. The fourth system in the study had approximately 0.045 U/mL collagenase activity. While it isn't apparent how damaging these levels of residual collagenase might be if injected into a knee, for example, it is important to understand how a commercial system operates and how a semi-automated or automated system ensures that the SVF preparation is washed to remove the enzymatic digestion fluid. A failure to reduce the collagenase level would be quite destructive [59].

There were no serious adverse events reported in the individual studies reviewed above [54–57] or in a recent review of SVF treatments for knee OA [53]. Some discomfort at the site of lipoaspiration has been reported [56], while minor swelling and pain were occasionally reported for injected knees, which resolved within 72 h [55] or 2 weeks [56]. However, there have been very serious adverse events reported with the use of SVF to treat age-related macular degeneration [60]. Three patients who had undergone bilateral intravitreal injections of autologous SVF produced at the same facility suffered severe bilateral vision loss, with some patients found to have no light perception at 1-year following the treatment [60]. Thus, when considering the safety of SVF therapy, the indications for use covered in this chapter have been limited to treatments of musculoskeletal pathologies or in the areas of lipotransfers, reconstructive surgery, wound healing and scar remodeling.

An early concern with the use of SVF therapy in breast augmentation was the issue of cyst formation or microcalcification; conditions that appeared in a limited number of patients in an early study on combining SVF with a patient's lipotransfer. However, these conditions were assessed as non-interfering with a breast cancer evaluation [46]. More recently, a report looked at the use of SVF augmentation of lipotransfers used in reconstructive surgery following mastectomy. No increased risk was found for a recurrence of breast cancer in patients receiving the SVF therapy compared to patients who received lipotransfers, but without SVF augmentation [61].

## Future Directions

An obvious need exists to increase the breadth of the clinical studies performed on assessing the therapeutic potential of SVF in treating musculoskeletal pathologies beyond human knee OA. This is especially true in the areas of treating tendinopathies, scar remodeling and wound healing. There is some evidence that the number of SVF cells plays a role in enhancing therapeutic outcomes [54, 55], which should be confirmed in clinical studies with higher numbers of study participants. Unlike clinical studies of the treatment of musculoskeletal pathologies with BMC where patients with higher CFU-F levels (i.e., BM-MSCs) were shown to have more efficacious outcomes [39], SVF analysis was limited to total viable cells [54, 55]. In view of the recent discovery of different types of AD-MSCs found in adipose tissue [36], the use of genetic probe analysis (e.g., single cell transcriptomics) might enable the identification of a critical sub-population of AD-MSCs that is more highly correlated to better therapeutic outcomes. The finding that there are different sub-types of adipose progenitor cells that play a role in modulating the extracellular milieu [32] suggests that it might be possible to develop a method to shift a patient's adipose toward a more therapeutically beneficial ECM, which in turn might boost the levels of potentially critical sub-populations of AD-MSCs prior to harvest. In the meantime, it might be prudent to continue to urge patients to stop smoking, consume less alcohol and lose weight prior to being treated with SVF, or other biologics.

## Conclusion

SVF treatments have been shown to reduce pain and support the remodeling of cartilage tissues under some conditions in treating knee OA. The use of SVF treatment also has demonstrated some efficacy in enhancing wound healing, scar remodeling and in reducing pain and improving functionality of tendinopathies. Adding SVF to lipotransfers in the areas of reconstructive surgery and breast augmentation has shown a benefit for graft volume retention. Most importantly, no serious adverse events have been reported when SVF was used in the areas of orthopedics, dermatology or plastic surgery. Thus, SVF will continue to play an important role in the expanding uses of autologous biological therapies in the future.

**Key Points**

1. Adipose tissue-derived MSCs (AD-MSC) are present in SVF, which is the single cell suspension obtained by centrifugation of the enzymatic digestate of adipose tissue.
2. The AD-MSC is considered to be a multipotent progenitor cell.
3. MSCs can differentiate into a multiplicity of adult tissue cell lineages.
4. AD-MSCs secrete a wide variety of biomolecules including cytokines, growth factors and proteins that constitute the paracrine effect of AD-MSCs.
5. AD-MSCs diminish pro-inflammatory environments (osteoarthritic joints) and promote tissue remodeling through the paracrine effect by secreting biomolecules that have immunomodulatory, mitogenic, anti-apoptotic, wound-healing and anti-scarring effects.
6. Extracellular vesicles are released by AD-MSCs, amplifying the paracrine effect of implanted AD-MSCs.
7. SVF has been used to treat a wide range of musculoskeletal pathologies from tendinopathies to osteoarthritic joints.
8. Most clinical studies of SVF have focused on treating knee OA, with patients reporting pain mitigation and improved functionality.
9. SVF has been combined with lipotransfers to improve retained graft volume for use in reconstructive surgeries and dermatology conditions.
10. Automated methods are being developed to provide consistent and safe SVF preparations at point of care.

**Questions**

1. Stromal vascular fraction (SVF) is composed of which of the following?
   A. Adipose tissue pieces
   B. Single cells isolated from adipose tissue
   C. Cells from the blood
   D. Lipids from adipose tissue
   **Answer: B**
   **Explanation**: Stromal vascular fraction (SVF) is composed of single cells isolated from adipose tissue.
2. What single enzyme preparation produces the most cells when used to digest adipose tissue?
   A. Hyaluronidase
   B. Neutral protease
   C. Trypsin
   D. Collagenase
   **Answer: D**
   **Explanation**: The typical enzyme cocktail for digesting adipose tissue has as the primary component Collagenase, which usually is combined with Neutral Protease. Hyaluronidase has been reported to enhance cell release from adipose tissue in a digestion with Collagenase, while Trypsin is rarely used due to lower yields and poor viability of the released cells.
3. Registration of an SVF clinical study on ClinicalTrials.gov means?
   A. The study is a prospective clinical study
   B. The study still needs to be approved by an institutional review board or ethics committee
   C. The U.S. Food and Drug Administration (USFDA) has approved the clinical study protocol
   D. The study participants don't need to provide informed consent to participate in the study beyond a standard treatment consent
   **Answer: B**
   **Explanation**: The SVF study needs to be approved by an institutional review board or an ethics committee of the medical facility where the study is to be performed. Treating patients with just a registration of the clinical study on ClinicalTrials.gov means that the physician might be using experimental drugs on their patients in violation of state medical boards' restrictions on experimenting on patients.
4. One advantage of using stromal vascular fraction (SVF) compared to bone marrow concentrate (BMC) is?
   A. Mesenchymal stromal cells (MSCs) are present at higher frequencies in SVF
   B. There is a greater diversity of cell types
   C. The number of lymphocytes is higher
   D. There are more cytokines present
   **Answer: A**
   **Explanation**: Mesenchymal stromal cells (MSCs) are present at higher frequencies per unit volume of tissue compared to the MSCs in bone marrow concentrate (BMC).
5. The most reliable way to identify adipose-derived mesenchymal stromal cells (AD-MSC) is?
   A. Use an enzyme-linked immunosorbent assay (ELISA)
   B. Use a panel of clusters of differentiation (CD) markers
   C. Use a hemoanalyzer
   D. Use a viability stain
   **Answer: B**
   **Explanation**: Use of a panel of clusters of differentiation (CD) markers enables the identification of a wide variety of cell types during a single experiment.
6. Adipose-derived mesenchymal stromal cells (AD-MSCs) are considered to be multipotent because of what characteristic?
   A. They are quite numerous in adipose tissue
   B. They can grow as colonies in tissue culture containers in normal tissue culture media
   C. They can be differentiated in induction medium into more than one adult cell lineage
   D. They are able to release exosomes
   **Answer: C**
   **Explanation**: They can be differentiated in induction medium into more than one adult cell lineage. Some progenitor cells, like EPCs (endothelial progenitor cells) are unipotent, since they can only differentiate into adult epithelial cells.
7. Therapeutic efficacy of stromal vascular fraction (SVF) preparations has been shown to depend on what characteristic?
   A. The use of tumescent fluid to obtain the lipoaspirate tissue to digest
   B. The anatomical site from which the lipoaspirate is obtained
   C. The number of cells present in the SVF
   D. The diluent used to resuspend the cells in to produce the SVF preparation
   **Answer: C**
   **Explanation**: The number of cells present in the stromal vascular fraction (SVF) has been shown to influence the therapeutic potential of SVF treatments, at least in treating knee osteoarthritis.
8. Stromal vascular fraction (SVF) contains cells diluted with what type of fluid?
   A. Plasma
   B. Blood
   C. PRP
   D. Physiologically compatible fluid (e.g., Saline, phosphate buffered saline (PBS))
   **Answer: D**
   **Explanation**: A physiologically compatible fluid like saline or phosphate buffered saline (PBS) is commonly used to dilute the stromal vascular fraction (SVF) cells. Platelet rich plasma (PRP) has been mixed with SVF preparations in some clinical studies, but the usual diluent is saline or PBS.

9. Which of the following doesn't describe a characteristic of exosomes released from adipose-derived mesenchymal stromal cells (AD-MSCs)?
   A. AD-MSC exosomes contain "cargo"
   B. AD-MSC exosomes can replicate
   C. AD-MSC exosomes are released *in vivo*
   D. AD-MSC exosomes bind to target cells
   **Answer: B**
   **Explanation**: Exosomes from any type of cell are unable to replicate. Only cells can generate exosomes.
10. Serious adverse events have been reported with the injection of stromal vascular fraction (SVF) to treat what pathology?
   A. Alopecia
   B. Tissue deficit in the temple area
   C. Knee osteoarthritis
   D. Macular Degeneration
   **Answer: D**
   **Explanation**: So far only stromal vascular fraction (SVF) preparations delivered intravitreally to treat macular degeneration have resulted in serious adverse events.

**Acknowledgements** This book chapter is modified and updated from a previous book chapter, "Adipose Cells or Stromal Vascular Fraction" by Elliot B. Lander, MD, in *Essentials of Regenerative Medicine in Interventional Pain Management* published by ASIPP Publishing.

## References

1. Lander EB. Adipose cells or stromal vascular fraction. In: Manchikanti L, Navani A, Atluri S, editors. Essentials of regenerative medicine in interventional pain management. Paducah, KY: ASIPP Publishing; 2019. p. 99–106.
2. Rodbell M. Metabolism of isolated fat cells I. Effects of hormones on glucose metabolism and lipolysis. J Biol Chem. 1964;239:375–80.
3. Björntorp P, Karlsson M, Pertoft H, et al. Isolation and characterization of cells from rat adipose tissue developing into adipocytes. J Lipid Res. 1978;19:316–24.
4. Hauner H, Entenmann G, Wabitsch M, et al. Promoting effect of glucocorticoids on the differentiation of human adipocyte precursor cells cultured in a chemically defined medium. J Clin Invest. 1989;84:1663–70.
5. Zuk PA, Zhu M, Mizuno H, et al. Multilineage cells from human adipose tissue: implications for cell-based therapies. Tissue Eng. 2001;7:211–28.
6. Zuk PA, Zhu M, Ashjian P, et al. Human adipose tissue is a source of multipotent stem cells. Mol Biol Cell. 2002;13:4279–95.
7. Crisan M, Yap S, Casteilla L, et al. A perivascular origin for mesenchymal stem cells in multiple human organs. Cell Stem Cell. 2008;3:301–13.
8. Caplan A. All MSCs are pericytes? Cell Stem Cell. 2008;3:229–30.
9. da Silva ML, Sand TT, Harman RJ, et al. MSC frequency correlates with blood vessel density in equine adipose tissue. Tiss Eng Part A. 2009;15:221–9.
10. Safford KM, Hicok KC, Safford SD, et al. Neurogenic differentiation of murine and human adipose-derived stromal cells. Biochem Biophys Res Commun. 2002;294:371–9.
11. Bourin P, Bunnell B, Casteilla L, et al. Stromal cells from the adipose tissue-derived stromal vascular fraction and culture expanded adipose tissue-derived stromal/stem cells: a joint statement of the International Federation of Adipose Therapeutics (IFATS) and science and the International Society for Cellular Therapy (ISCT). Cytotherapy. 2013;15:641–8.
12. Caplan AI. Mesenchymal stem cells. J Orthop Res. 1991;9:641–50.
13. DiMarino AM, Caplan AI, Bonfield TL. Mesenchymal stem cells in tissue repair. Front Immunol. 2013;4:201.
14. Murphy MB, Moncivais K, Caplan AI. Mesenchymal stem cells: environmentally responsive therapeutics for regenerative medicine. Exp Mol Med. 2013;45:e54.
15. Yoshimura K, Shigeura T, Matsumoto D, et al. Characterization of freshly isolated and cultured cells derived from the fatty and fluid portions of liposuction aspirates. J Cell Physiol. 2006;208:64–76.
16. Vet-Stem website. https://vetstem.com/. Accessed 8 Mar 2023.
17. Black LL, Gaynor J, Gahring D, et al. Effect of adipose-derived mesenchymal stem and regenerative cells on lameness in dogs with chronic osteoarthritis of the coxofemoral joints: a randomized, double-blinded, multicenter, controlled trial. Vet Ther. 2007;8:272–84.
18. Black LL, Gaynor J, Adams C, et al. Effect of intraarticular injection of autologous adipose-derived mesenchymal stem and regenerative cells on clinical signs of chronic osteoarthritis of the elbow joint in dogs. Vet Ther. 2008;9:192–200.
19. Lendeckel S, Jödicke A, Christophis P, et al. Autologous stem cells (adipose) and fibrin glue used to treat widespread traumatic calvarial defects: case report. J Craniomaxillofac Surg. 2004;32:370–3.
20. Cervelli V, Gentileb P, De Angelis B, et al. Application of enhanced stromal vascular fraction and fat grafting mixed with PRP in post-traumatic lower extremity ulcers. Stem Cell Res. 2011;6:103–11.
21. Berman M, Lander E. A prospective safety study of autologous adipose-derived stromal vascular fraction using a specialized surgical processing system. Am J Cosmet Surg. 2017:1–14.
22. U.S. Food and Drug Administration. Warning letter to Dr. Thomas E Young (Young Medical Spa). Accessed 20 Apr 2012.
23. U.S. Food and Drug Administration. Warning Letter to Celltex. Accessed 24 Sept 2012.
24. U.S. Department of Health and Human Services Guidance for Industry and Food and Drug Administration Staff. Regulatory considerations for human cells, tissues, and cellular and tissue-based products: minimal manipulation and homologous use. 2020. https://www.fda.gov/regulatory-information/search-fda-guidance-documents/regulatory-considerations-human-cells-tissues-and-cellular-and-tissue-based-products-minimal. Accessed 8 Mar 2023.
25. Gimble JM, Grayson W, Guilak F, et al. Adipose tissue as a stem cell source for musculo-skeletal regeneration. Front Biosci. 2011;3:69–81.
26. Harman R, Sand T. Methods of processing and using novel stem cell compositions and kits comprising same. United States Patent Application Publication. Pub. No.: US2017/0196912 A1. Accessed 13 Jul 2017.
27. Seaman SA, Tannan ST, Cao Y, et al. Differential effects of processing time and duration of collagenase digestion on human and murine fat grafts. Plast Reconstr Surg J. 2015;136:189e–99e.
28. Gentile P, Calabrese C, De Angelis B, et al. Impact of the different preparation methods to obtain human adipose-derived stromal vascular fraction cells (AD-SVFs) and human adipose-derived mesenchymal stem cells (AD-MSCs): enzymatic digestion versus mechanical centrifugation. Int J Mol Sci. 2019;20:5471.
29. Domenis R, Lazzaro L, Calabrese S, et al. Adipose tissue derived stem cells: in vitro and in vivo analysis of a standard and three commercially available cell-assisted lipotransfer techniques. Stem Cell Res Ther. 2015;6:2.
30. Doi K, Tanaka S, Iida H, et al. Stromal vascular fraction isolated from lipo-aspirates using an automated processing system: bench and bed analysis. J Tissue Eng Regen Med. 2013;7:864–70.
31. Polancec D, Zenic L, Hudetz D, et al. Immunophenotyping of a stromal vascular fraction from microfragmented lipoaspirate used

in osteoarthritis cartilage treatment and its lipoaspirate counterpart. Genes. 2019;10:474.
32. Deutsch A, Feng D, Pessin JE, et al. The impact of single-cell genomics on adipose tissue research. Int J Mol Sci. 2020;21:4773.
33. Estrada-Gutierrez G, Bravo-Flores E, Ortega-Castillo V, et al. Isolation of viable adipocytes and stromal vascular fraction from human visceral adipose tissue suitable for RNA analysis and macrophage phenotyping. J Vis Exp. 2020;(164).
34. Ramakrishnan VM, Boyd NL. The adipose stromal vascular fraction as a complex cellular source for tissue engineering applications. Tiss Eng Pt B Rev. 2018;24:289–99.
35. Briellea S, Bavlid D, Motzikd A, et al. Delineating the heterogeneity of matrix-directed differentiation toward soft and stiff tissue lineages via single-cell profiling. Proc Natl Acad Sci. 2021;118:e2016322118.
36. Wang Z, Chai CY, Wang R, et al. Single-cell transcriptome atlas of human mesenchymal stem cells exploring cellular heterogeneity. Clin Transl Med. 2021;11:e650.
37. Caplan AI. New era of cell-based orthopedic therapies. Tiss Eng Pt B Rev. 2009;15:195–200.
38. Caplan AI. New MSC: MSCs as pericytes are sentinels and gatekeepers. J Orthop Res. 2017;35:1151–9.
39. Hernigou P, Bouthors C, Bastard C, et al. Subchondral bone or intra-articular injection of bone marrow concentrate mesenchymal stem cells in bilateral knee osteoarthritis: what better postpone knee arthroplasty at fifteen years? A randomized study. Int Orthop. 2021;45:391–9.
40. Li CJ, Fang QH, Liu ML, et al. Current understanding of the role of adipose-derived extracellular vesicles in metabolic homeostasis and diseases: communication from the distance between cells/tissues. Theranostics. 2020;10:7422–35.
41. McLaughlin C, Datta P, Singh YP, et al. Mesenchymal stem cell-derived extracellular vesicles for therapeutic use and in bioengineering applications. Cell. 2022;11:3366.
42. Li WJ, Chen H, Tong ML, et al. Comparison of the yield and purity of plasma exosomes extracted by ultracentrifugation, precipitation, and membrane-based approaches. Open Chem. 2022;20:182–91.
43. Patel DP, Gray KM, Santharam Y, et al. Impact of cell culture parameters on production and vascularization bioactivity of mesenchymal stem cell-derived extracellular vesicles. Bioeng Transl Med. 2017;2:170–9.
44. Tsiapalis D, O'Driscoll L. Mesenchymal stem cell derived extracellular vesicles for tissue engineering and regenerative medicine applications. Cell. 2020;9:991.
45. Yurtbay A, Say F, Çinka H, et al. Multiple platelet-rich plasma injections are superior to single PRP injections or saline in osteoarthritis of the knee: the 2-year results of a randomized, double-blind, placebo-controlled clinical trial. Arch Orthop Trauma Surg. 2022;142:2755–68.
46. Yoshimura K, Sato K, Aoi N, et al. Cell-assisted lipotransfer for cosmetic breast augmentation: supportive use of adipose-derived stem/stromal cells. Aesthetic Plast Surg. 2008;32:48–55.
47. Piccinno MS, Petrachi T, Pignatti M, et al. Human adipose mesenchymal stromal/stem cells improve fat transplantation performance. Cell. 2022;11:2799.
48. Chiu CH. Does stromal vascular fraction ensure a higher survival in autologous fat grafting for breast augmentation? A volumetric study using 3-dimensional laser scanning. Aesthet Surg J. 2019;39:41–52.
49. Roshdy OH, Abdallah WI, Farid CI, et al. Stromal vascular fraction improves the durability of autologous fat temple augmentation-a split-face randomized study using ultrasound biomicroscopy. J Plast Reconstr Aesthet Surg. 2022;75:1870–7.
50. Kim JY, Rhim WK, Yoo YI, et al. Defined MSC exosome with high yield and purity to improve regenerative activity. J Tiss Eng. 2021;12:1–15.
51. Fakiha K. Adipose stromal vascular fraction: a promising treatment for severe burn injury. Hum Cell. 2022;35:1323–37.
52. Gentile P, Sterodimas A, Calabrese C, et al. Systematic review: advances of fat tissue engineering as bioactive scaffold, bioactive material, and source for adipose-derived mesenchymal stem cells in wound and scar treatment. Stem Cell Res Ther. 2021;12:318.
53. Tang Q, Zhao XS, Guo A, et al. Therapeutic applications of adipose-derived stromal vascular fractions in osteoarthritis. World J Stem Cells. 2022;14:744–55.
54. Garza JR, Campbell RE, Tjoumakaris FP, et al. Clinical efficacy of intra-articular mesenchymal stromal cells for the treatment of knee osteoarthritis: a double-blinded prospective randomized controlled clinical trial. Am J Sports Med. 2020;48:588–98.
55. Tsubosaka M, Matsumoto T, Sobajima S. Comparison of clinical and imaging outcomes of different doses of adipose-derived stromal vascular fraction cell treatment for knee osteoarthritis. Cell Transplant. 2021;30:1–12.
56. Hong ZP, Chen JH, Zhang SJ, et al. Intra-articular injection of autologous adipose-derived stromal vascular fractions for knee osteoarthritis: a double-blind randomized self-controlled trial. Int Orthop. 2019;43:1123–34.
57. Tran TDX, Wu CM, Dubey NK, et al. Time- and Kellgren–Lawrence grade-dependent changes in intra-articularly transplanted stromal vascular fraction in osteoarthritic patients. Cell. 2019;8:308.
58. Ude CC, Shah S, Ogueri KS, et al. Stromal vascular fraction for osteoarthritis of the knee regenerative engineering. Regen Eng Transl Med. 2022;8:210–24.
59. Aronowitz JA, Ellenhorn JDI. Adipose stromal vascular fraction isolation: a head-to-head comparison of four commercial cell separation systems. Plast Reconstr Surg. 2013;136:932e–9e.
60. Kuriyan AE, Albini TA, Townsend JH, et al. Vision loss after intravitreal injection of autologous "stem cells" for AMD. N Engl J Med. 2017;376:1047–53.
61. Mazur S, Zołocińska A, Siennicka K, et al. Safety of adipose-derived cell (stromal vascular fraction–SVF) augmentation for surgical breast reconstruction in cancer patients. Adv Clin Exp Med. 2018;27:1085–90.

# Wharton Jelly MSC for Treating Local and Systemic Illnesses

# 11

Sairam Atluri, Jaya Sanapati, Navneet Boddu, and Guido Moll

## Introduction

Among the most recent technological developments in the medical field, multipotent mesenchymal stromal/stem cell (MSC) therapeutics currently rank as one of the most promising, which can potentially change millions of lives [1–4]. The new prospect of adopting the tissue regenerative attributes of nature to repair organ dysfunction is very alluring. Furthermore, easy accessibility and low immunogenicity of adult MSCs make them promising candidates for allogeneic cell therapy. Amidst the various available stem cells, adult tissue derived stem cells (e.g., perinatal, bone marrow and adipose) all seem to be well positioned for clinical translation due to their established safety and efficacy in numerous preclinical trials and preliminary human clinical trials, although differences exist in the properties of MSCs from different sources that need to be taken into account for their safe and efficient clinical use [5, 6]. Here, allogeneic adult MSCs can be manufactured as "off the shelf products" for emergency use. On the contrary, embryonic (derived from the inner cell mass of embryonal blastocyst), fetal (from fetal abortus) and induced pluripotent (converted from adult cells via reprogramming with pluripotent transcription factors OCT3/4, c-Myc, KLF4, SOX2) stem cells albeit demonstrating excellent differentiation and proliferative potential can be tumorigenic (two of the Yamanaka-factors, c-Myc and Klf-4 are potent oncogenes). In addition, embryonic and fetal stem cells pose considerable ethical challenges. Currently, induced pluripotent cells are being explored in preclinical and clinical studies for differentiating them into specific organ cells and administering them to repair tissue degeneration. However, differentiated cells develop immunogenicity mandating the use of immunosuppressants. Inefficient reprogramming, karyotype abnormalities, and point mutations may generate other clinical risks. These above characteristics of the embryonic, fetal and induced pluripotent stem cells currently present barriers to patient care.

Since bone marrow-derived MSCs (BM-MSCs) were originally the most frequently researched and tested cell type, they were considered to be the gold standard [5, 6]. Subsequently, adipose derived MSCs (AD-MSCs) have also gained popularity by the virtue of possessing higher concentrations of MSCs in the starting material. However, invasive harvesting, donor morbidity, limited stem cell concentrations (0.01–0.001% of all cells extracted from bone marrow and 2–10% of adipose cells are MSCs), as well as epigenetic changes (e.g., due to external environmental and internal metabolic insults) and a potential detrimental negative impact of donor aging *per se* considering the amount and quality of the extractable cell numbers (and their respective proliferation/differentiation/stemness capability) has compelled the exploration for superior MSC alternatives [5, 7]. The above concerns exist in particular for culture expanded autologous BM-MSCs and AD-MSCs, but also their use as allogeneic therapies. Perinatal MSCs like Wharton jelly MSCs (WJ-MSCs) and placental stem cells which are relatively absolved from the above limitations are now entering mainstream as reflected by the increase in the number of clinical trials during the last decade [5]. Although WJ-MSCs were first isolated only in 1991, among the perinatal MSCs,

S. Atluri (✉)
Tri-State Spine Care Institute, Cincinnati, OH, USA

J. Sanapati
Pain Management Centers of America, Evansville, IN, USA
e-mail: jaya@apccweb.com

N. Boddu
Advanced Pain and Regenerative Specialists, Oceanside, CA, USA

G. Moll
BIH Center for Regenerative Therapies (BCRT) and Berlin-Brandenburg School for Regenerative Therapies (BSRT), and Department of Nephrology and Internal Intensive Care Medicine, all Part of Charité Universitätsmedizin Berlin, Freie Universität Berlin, Humboldt-Universität Zu Berlin, and Berlin Institute of Health(BIH), Berlin, Germany
e-mail: Guido.moll@charite.de



A. Navani et al. (eds.), *Essentials of Regenerative Medicine in Interventional Pain Management*,
https://doi.org/10.1007/978-3-031-50357-3_11

some judiciously posit that WJ-MSCs are the "gold standard" [8], for the following reasons:

1. No ethical issues as the umbilical cord is generally considered as "medical waste"
2. Plentiful source due to millions of births worldwide.
3. Wharton jelly contains the highest concentration of MSCs per milliliter than any other tissue (up to 4,700,000 MSCs/cm) and has significantly higher numbers than bone marrow (317,400 MSCs/ml) and adipose tissue (1,550,000 MSCs/ml) [9].
4. Efficiency of primary isolation from WJ exceeds other available tissues like cord blood, amniotic fluid, endothelial and subendothelial layers of umbilical vein of the cord, adipose tissue and bone marrow.
5. WJ-MSCs are regarded as day "0" cells which have not yet undergone any cellular divisions unlike BM-MSCs and AD-MSCs that are inevitably "aged" and numerous studies have shown that in vivo aging is detrimental to the stem cell "fitness". Aging not only attenuates the function of the MSCs but the numbers too. The CFU-f dramatically decreases with aging from 1:10000 in a newborn to 1:200000 in an 80-year-old [10].
6. Among all adult MSCs, WJ-MSCs seem to be the most proliferative [9] avowed with the lowest population doubling (PD) times (21 h) among all adult tissues like BM or adipose tissue and some perinatal tissues.
7. WJ-MSCs are less prone to senescence [11] and maintain karyotype stability during culturing and passaging when compared to BM and AD-MSCs due to their primitive status and also because they are endowed with longer telomeres.
8. Although WJ-MSCs express more pluripotent markers Tra 1–60, Tra 1–81, SSEA-4, SSEA-3, and pluripotent genes OCT4A, SOX2, OCT2, OCT4B, OCT1 and NANOG when compared to MSCs from other sources, implying superior "stemness" [12], they do not develop teratomas [13]. Thus far, no clinical cases of malignant transformation have been reported for WJ-MSC or any other adult MSCs [8].
9. Although their differentiation potential is less robust than that of ES cells and iPSCs, numerous studies have shown the proficient differentiation capacity of WJ-MSCs [12] as they express higher amounts of CD-146, a marker that is linked to multipotency and they are capable of broad differentiation into all the 3 germ layers: Ectoderm, Mesoderm and Endoderm.
10. Recently, for the first time, WJ-MSCs could be successfully reprogrammed into iPSCs [9]. This feat has not yet been attempted for any other adult MSCs.
11. Amongst all the adult MSCs, WJ-MSCs seem to possess the least immunogenicity as they are known to have the lowest MHC class II (escaping CD $4^+$ T cell detection) expression compared to BM-MSCs/ADSCs. Moreover, low levels of MHC Class 1 molecules protect them from natural killer cell lysis. Furthermore, MSCs do not express co-stimulatory molecules (CD 40,80,86) required for effector T-cell induction.
12. Although WJ-MSCs akin to other MSCs, can influence almost all cells of the immune system, WJ-MSCs can suppress allogenic-stimulated immune cells to a greater extent than either BM-MSCs or adipose-derived MSCs.
13. In the last decade it has been increasingly recognized that the exosomes generated by MSCs have significant therapeutic effects. Exosome yield from WJ-MSCs exceeds that of BM-MSCs and AD-MSCs and are known to be more regenerative [14].

Overall, UC-MSCs' plentiful supply, rich source, ease of extraction, primitive nature yet non-tumorigenic, exemption from environmental/metabolic insults, greater proliferative capacity with less senescence along with maintenance of stemness/genetic/karyotype stability, broad differentiation potential, higher immunosuppressive effects, higher exosome yield and hypo-immunogenic features make them suitable for large-scale, universal production [8, 12].

## Mechanism of Action

Although the complete mechanism of action of MSCs still has to be fully elucidated, MSCs therapeutic effect appears to stem from the secretion of soluble molecules, extracellular vesicles and cell-to-cell contact, which induces changes in the immunobiology of both innate and adaptive immune cells (e.g. the suppression of immune cell proliferation/maturation and transformation into anti-inflammatory phenotypes) (Fig. 11.1) [2–4, 15, 16], thereby establishing immune homeostasis that results in a local and systemic regenerative environment.

Although MSCs are rapidly eliminated (within 24–48 h) after infusion due to the instant triggering of innate immune cascade systems (IBMIR) upon blood contact [5, 6, 17–19], their long-term restorative effects despite their "hit and run" mechanism seems to be from immunoregulatory modulation of the immune cells some of which (macrophages) last for months and years.

However, MSC are not constitutively immunosuppressive, as they require an initial 'licensing' step provided by molecules of acute phase inflammation for the transformation from pro-inflammatory phenotype (MSC1) to the

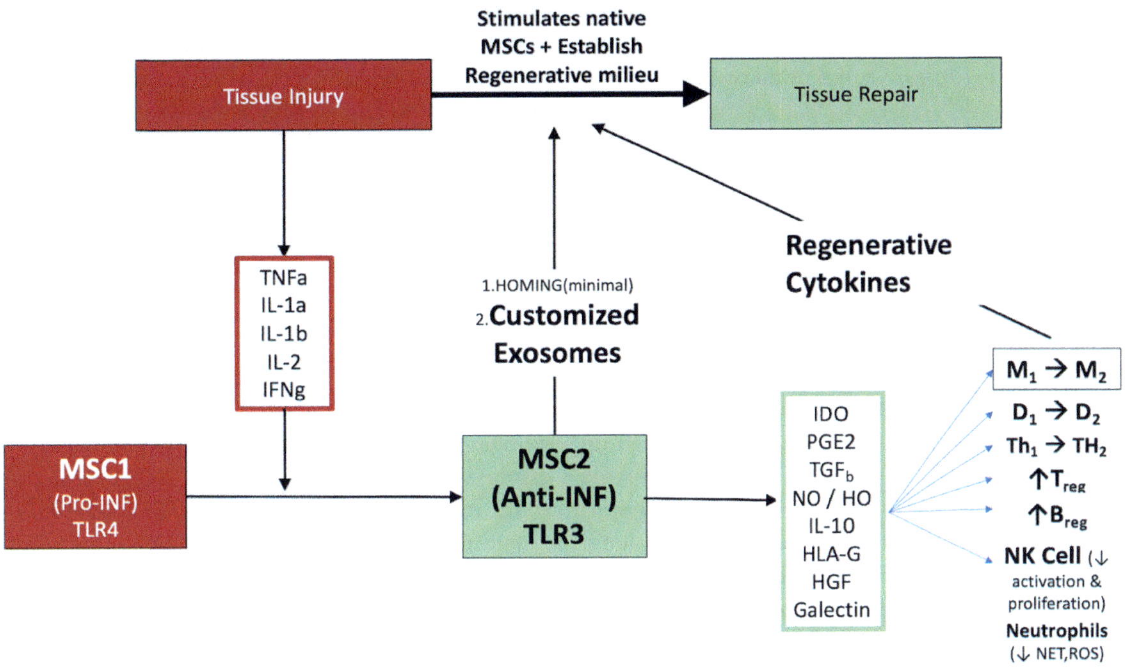


**Fig. 11.1** *Mechanism of action of MSCs. TNFa* tumor necrosis factor-alpha, *IL-1a* interleukin-1 alpha, *IL-1-b* interleukin-1 beta, *IL-2* interleukin-2, *IFNg* interferon gamma, *TLR4* toll like receptor 4, *TLR3* toll like receptor 3, *IDO* indoleamine 2,3-dioxygenase, *PGE2* prostaglandin E2, *TFGb* transforming growth factor beta, *NO* nitrous oxide, *HO* hemoxygenase, *IL-10* interleukin-10, *HLA-G* human leukocyte antigen-G, *HGF* hepatocyte growth factor, *M1* M1 macrophage, *M2* M2 macrophage, *D1* D1 dendritic cell, *D2* D2 dendritic cell, *Th1* T helper cell 1, *Th2* T helper cell 2, *Treg* T regulatory cell, *Breg* B regulatory cell, *NK cell* natural killer cell

immunomodulatory anti-inflammatory phenotype (MSC2) [15]. Since the activation of MSC is required for the pharmacological effects, MSCs could be considered as a "prodrug".

Switching the pro-inflammatory M1 phenotype to anti-inflammatory M2 phenotype appears to be a crucial part of MSC immunomodulation as despite loss of secretome and lack of any effects on T and B cells by heat inactivated (non-functional) MSCs, beneficial effects were still seen through activated Macrophages [20].

MSCs stimulate the differentiation of naive CD 4+ T cells into immunomodulatory $CD4^+CD25^+Foxp3^+$ regulatory T cells (Tregs) and this upregulation of Tregs does not result from an expansion of pre-existing natural Tregs but via induction of Tregs from conventional T cells. Furthermore, MSCs suppress T cell proliferation of CD4+ and CD8+ T cell subsets, shift pro-inflammatory Th1 to anti-inflammatory Th2 cells and convert immunogenic Th17 cells to Treg cells thereby decreasing the levels of the inflammatory IL-17 [21].

MSCs decrease B cell proliferation/differentiation, abrogate plasmablast formation and enhance the induction of regulatory Bregs culminating in anti-inflammatory actions including converting T cells to Treg cells [21].

MSCs prevent the differentiation of monocytes into DCs (dendritic cells) and also impair their antigen-presenting ability. Additionally, they reduce DC apoptosis and convert the pro-inflammatory conventional DCs (cDCs) to anti-inflammatory plasmacytoid DCs (pDCs) that possess strong immunoregulatory properties [21]. Hyperactivity of NK (natural killer) cells are linked to many diseases. MSC interaction decreases their activity and proliferation [21]. MSCs rescue neutrophils from apoptosis and decrease neutrophil-derived ROS, the release of matrix metalloproteinases (MMPs) and NET production [21].

## Clinical Trials

MSCs are under consideration as "broad spectrum" drugs capable of treating various diseases and the route of clinical administration largely depends on the type of the condition [4–6, 22]. Recently, favorable outcomes seen in numerous COVID-19 trials piqued the interest of regulatory agencies [14, 15, 23].

In the musculoskeletal arena, direct injection of MSCs into the joints, tendons, ligaments, subchondral bone and discs seems to provide the best outcomes for spine [24–28] and joint pains [29]. When intradiscal injection of expanded autologous BM-MSCs was performed in low back pain patients, they not only survived the harsh hypoxic, hypoglycemic and acidic environment, they engrafted, proliferated and also differentiated into chondrocytes and secreted extracellular matrix resulting in pain relief in many patients [30, 31].

On the contrary, in order to treat systemic diseases, administration of MSCs is most commonly performed via intravenous infusion (IV) [5]. After an IV infusion, within the first hours, most of the cells are entrapped in the microvasculature (average size of cultured MSC is 30 μm with a range of 16–53 μm and size of the capillary is approximately 6 μm and it is often unappreciated that relatively small increases in cell diameter translate into significant increases in cell volume, because this value increases as the cube of the radius) of the lungs and by 6 h the numbers decrease from the lung and subsequently increase predominantly in the spleen with some gain in the liver [16, 32]. This trend continues for 10 days when only about 3% are seen in the lung, whereas the accumulation in the spleen and liver is approximately 35% and 15% respectively [33]. Most of the infused MSCs are cleared by macrophages which are attracted by the MSCs, and long-term persistence of MSCs post-transplantation (>2 weeks) is less than 2% and most MSCs are not viable after 24 h when given IV [16].

Therefore, the long term immunomodulatory and regenerative effects of infused MSC must be mediated most likely via repolarizing other immune cells. Two human trials show that despite intra-arterial (IA) administration there is minimal presence of MSCs in the target tissues. Thus, minimal homing seen with IV administration is not enhanced significantly by IA route despite the invasiveness [34, 35]. Risk of thrombosis is another concern with IA infusions [6, 17]. Intriguingly, IA application results in MSCs bypassing the lung (unlike IV) with their accumulation seen mostly in the capillary beds of the liver, spleen and kidney.

Numerous animal trials have been performed purporting the safety and efficacy of WJ-MSCs in the various disease conditions, which are unfortunately beyond the scope of this chapter. The few published clinical trials of WJ-MSCs in humans have substantiated primarily their safety with some displaying first signs of efficacy. An general overview on some of these human studies in COVID-19 [25, 36–42] (Table 11.1), liver failure [43–48] (Table 11.2), diabetes [49–54] and heart failure [55–58] is shown in Table 11.1.

Additional studies have been done in the musculoskeletal arena [59–61], GVHD [62], Crohn's disease [63, 64], neurological diseases [65–68], autoimmune diseases [69, 70], etc., making the MSCs pharmaceutically versatile. Saliently, no major complications were noted in any of these trials and beneficial outcomes were reported in the majority of these studies.

**Table 11.1** UC-MSC controlled trials in COVID-19

| Trial, Year | No. of patients. Active vs control | Severity of Covid-19 | MSC type and dose | Outcome | Complications |
|---|---|---|---|---|---|
| Lanzoni et al. 2021 [36] | 12 vs 12 RCT | ARDS, ventilated | Subepithelial lining of the umbilical cord. 100 million cells given 3d apart. Passage 5. Rapid thaw | 91% vs 42% survival. Faster recovery. Decrease in inflammatory markers esp. GM-CSF. No change in viral load | None |
| Shi et al. 2021 [37] | 65 vs 35 (RCT) | Severe, non ventilated | WJ MSC. 40 million cells × 3 doses, 3 days apart. Passage 5. Non cryopreserved | Improvement in CT scan scores | None |
| Dilogo et al. 2021 [38] | 20 vs 20 (RCT) | Severe | WJ MSC, 100 million cells. 1 dose. Passage 5–6 | 2.5 times increase in survival w/o comorbidities and 4.5 times w morbidities. Decrease in IL-6 and LIF | None |
| Shu et al. 2020 [39] | 12 vs 29 (RCT) | Severe | WJ-MSC. Two million/kg. Passage 3–5. One dose | Faster recovery. Decreased mortality. Decreased IL-6, CRP. Increased lymphocytes. Quicker CT changes resolution | None |
| Meng et al. 2020 [40] | 9 VS 9 (RCT) | Severe | WJ MSC. 30 million cells × 3 doses, 3 days apart. Passage 7 | Decreased need for ventilation, 1 vs 4 patients | None |
| Wei et al. 2021 [41] | 12 vs 13 (CT) | Severe | WJ MSC. One million cells/kg. One dose | Improved oxygenation index and CT changes. Decreased IGM levels | None |
| Adas et al. 2021 [42] | 10 vs 10 (CT) | Critical | WJ MSC, P4, 3 mil/kg | 30% vs 60% survival, decreased ferritin, fibrinogen and CRP | None |

**Table 11.2** WJ-MSCs trials in liver failure

| Author, Year, Type of Study | # of Patients in treatment group | BMA (cc) | Number of SC | Type of stem cell | Delivery method | Result | follow-up (months) | Condition | Findings |
|---|---|---|---|---|---|---|---|---|---|
| Zhang et al. 2012 [43] PCT | 30 | NA | 0.5 million/kg | WJ MSC (pas 3–4) | IV | YES | 12 | Cirrhosis | ALB increased for 48 weeks and CHE increased for 4 weeks. TBIL decreased for 48 weeks. MELD decreased, Plts increased and ascites decreased for 48 weeks. Serum laminin, hyaluronic acid, PIIINP, and type IV collagen which are markers for disease severity of liver fibrosis were significantly decreased at weeks 24 and 48 |
| Wang et al. 2013 [44] Case series | 7 | NA | 0.5 million/kg × 3 doses | WJ MSC (pas 4) | IV | YES | 48 weeks | PBC | Fatigue, pruritus and ascites improved but no major change in other parameters. Serum ALP levels and MRS improved (2 key parameters for the definition of response to treatment in patients with PBC) during a 48-week follow-up period |
| Shi et al. 2012 [45] PCT | 43 | NA | 0.5 million/kg × 3 doses | WJ MSC (pas 3–4) | IV | YES | 12 or 18 | Acute on chronic liver failure | MELD improved. Albumin, CHE and PTA increased showing improved liver functioning. ALT and TBIL levels decreased. AFP levels increased showing liver proliferation. Improvement in survival seen |
| Xue et al. 2015 [47] Case series | 50 | NA | 30 million | WJ MSC (pas 3) | HAI | YES | 6 | ESLD | The frequency of hospitalization decreased, because of improved ascites, hypoproteinemia, and quality of life. Bilirubin and ALT interestingly were increased for 24 weeks. PTT was decreased up to 24w but no change in PT. MELD decreased at 24 weeks |
| Li et al. 2016 [48] PCT | 11 | NA | 100 million (pas 3–4) | WJ MSC (pas 3–4) | HAI | YES | 24 | Acute on chronic liver failure | Increased survival. Improved MELD at 4 weeks. At 2 years., there was improvement in AST, albumin, PT, INR. However, there was no change in ALT, TBIL, DBIL, Cr, WBC, HGB, PLT and ascites |
| Liang et al. 2017 [46] Case series | 26 | NA | 1 million/kg | Mostly UC MSC (pas 2–5) | IV | YES | 24 | Cirrhosis | Bilirubin decreased up to 1 year. ALB increased for 2 years. PT lowered for 6 Months. PLT ct increased for 4 years. no change in MELD |

## Safety

To date more than 1000 clinical trials with adult MSCs involving tens of thousands of patients are being conducted or completed with an impressive safety profile. MSC infusion is safe, well tolerated, and only gives rise to mild transient systemic reactions like fever, rash, headaches in a limited number of patients, but no major complications have been reported in multiple studies and metanalysis [5, 6, 71–75].

However, administration of improper stem cells (fetal, embryonic and induced Pluripotent) and adult MSCs (not subjected to proper testing, accepted manufacturing standards and clinical supervision) can pose serious risks [5, 6]. Additionally, complications can arise from inappropriate dosing and route of administration (intraarterial, intrathecal, intraorgan etc). Unethical and cavalier medical providers also contribute to maladies seen with this therapy. At least 4 complications (cardiac arrest × 2, pulmonary embolism and neurological) were reported after IV MSC infusions [76, 77]. The 3 cases of PE that were mentioned above were attributed to the high expression of TF/CD142 on MSCs, which triggers instant blood-mediated inflammatory reaction (IBMIR) causing potential life-threatening thromboembolism primarily from MSC incompatibility with the innate immune cascade systems of the blood and was first reported by Moll et al. [17]. The IBMIR risk increases with IV infusion, extensively passaged cells, fresh thawed cells and with doses higher than three million cells/kg. The lowest to highest expression of highly procoagulant TF/CD142 is for BM-MSCs, WJ-MSCs, PT-MSCs, and AT-MSCs, respectively [4]. To enhance safety, calls to include TF/CD142 as a phenotypic safety marker as one of the release criteria have been made [6].

Although tumors have been reported, they occurred from pluripotent and HSCs and no tumors from adult MSCs were reported in the numerous clinical trials.

## Optimizing Clinical Outcomes

There are more than 800 ongoing clinical trials with MSCs across the world with more than 55,000 existing publications [24]. Although preliminary small clinical trials in university settings offered hope that MSCs can treat challenging medical conditions, this promise could not be translated with large-scale commercial manufacturing resulting in rare MSC drug approvals across the globe [24]. Moreover, despite obtaining preliminary regulatory approval, many MSC-based therapies have failed to show effectiveness in Phase III clinical trials.

Manufacturing MSCs of high quality is the first step to ensure the safety and efficacy in clinical trials [16]. Donor variability, excessive passaging, post-thaw poor cell viability, suboptimal route of delivery, and immunogenicity have been speculated as reasons for the poor performance of MSCs manufactured for commercial purposes [16, 19, 24, 78–81]. Furthermore, the International Society for Cell & Gene Therapy (ISCT) definition of MSCs that is currently universally applied is inadequate as many other cells including fibroblasts meet the same criteria. Calls for a more pragmatic characterization of MSCs have been made [6].

As MSCs age in vitro, they undergo replicative senescence compromising their regenerative and immunomodulatory properties [7]. When comparing the efficacy of MSCs in treating acute graft-versus-host disease (GvHD) patients with early-passage MSCs (passages 1–2) had significantly better outcomes than those treated with later passage MSCs (passages 3–4), with increased patient survival and treatment response rates [82]. This was even more evident when taking into account the use of fresh vs. thawed cells, with superior performance of minimal-expanded fresh cells [19].

To provide sufficient MSCs for commercial use, MSCs need to be amplified in a large scale from limited starting material, which invariably leads to MSCs senescence and decreased proliferation, modifications of gene expressions with mutations, karyotype instability, telomere shortening, poor paracrine secretion of cytokines [83] and exosomes [84], premature unwanted differentiation or loss of differentiation, increased immunogenicity, increased propensity for IBMIR and in extreme cases, tumorigenicity [6, 85].

Although seldom reported in clinical trials and by manufacturers, PD is a superior measure than passage number to assess MSC health and function and minimizing PDs should be mandated by regulatory authorities. Ideally for clinical use, only MSCs around a PD 10 should be employed to avoid replicative concerns mentioned above. Comprehensive MSC characterization and release criteria is indispensable prior to clinical use especially for the manufactured cells.

The majority of the commercial MSCs are cryopreserved and are rapidly thawed immediately preceding administration. This "heat shock" from rapidly thawed cells upregulates heat shock proteins and impairs the immunomodulatory function of the MSCs and also augments the IBMIR response [19, 86]. Cryopreserved MSCs are particularly susceptible to the triggering of IBMIR and complement lysis, which is subsequently amplified by host immune cells lysis [12, 86]. Besides, their viability is inferior to fresh MSCs [19].

Moreover, in an elegant trial involving patients with GVHD, freshly harvested cells in low passage had a 100% response rate, twice the response rate of 50% observed in a comparable group of patients treated with freeze-thawed cells at higher passage [19]. Similarly, another group postulated that the poor results seen in their study were probably from poor viability of the immediately thawed MSCs [87]. Furthermore, an animal study suggested the superiority of fresh MSCs compared to thawed [86]. Fortunately, rescue culture for 24–48 h [88] seems to reverse the insalubrious effects of cryopreservation.

## Conclusion

MSC therapy potentially can treat many conditions that are deemed incurable and adult MSCs currently seem to be well positioned for clinical induction. Amongst adult MSCs, WJ-MSCs seem to be apt for clinical use due to plentiful supply, easy isolation, karyotype stability with passaging and robust proliferative properties with least immunogenicity and effective immunomodulation. Preliminary clinical data suggests that WJ-MSCs are safe, efficacious as well as promising to treat multiple conditions, however, rigorous and large/well designed trials must be conducted. Importantly cost-effective mass production under GMP conditions must be developed before this treatment can become mainstream. Limiting PDs and cryopreservation "recovery" are critical for clinical success. Additionally, MSC preconditioning, injection timing, route of administration, dosage, number of doses and interval between injections also need to be optimized to improve the safety and effectiveness of WJ-MSC therapy in order to treat a myriad of diseases which currently have poor treatment options.

**Key Points**

1. Regenerative medicine holds a lot of promise in the treatment of various medical conditions that have poor options in conventional medicine.
2. Among the various biologic products available, MSCs have been the most studied in the non-cancer arena. They are considered as a "broad spectrum" therapy.
3. Preliminary clinical trials have displayed the efficacy of MSCs in many diseases
4. When properly administered, based on medical literature, their safety has been assured
5. Adult MSCs are not known to be tumorigenic
6. Among all adult tissues MSC sources, perinatal MSCs seem to be well positioned for allogeneic, "off the shelf" clinical application with some considering them as "Gold Standard".
7. Infused MSCs (both autologous and allogenic) are not known to survive, home, engraft nor differentiate inVivo. They exert beneficial effects through exosomes and paracrine secretion resulting in immunomodulation.
8. MSCs exert their salubrious effects on virtually every immune cell. Taking advantage of the plasticity of immune cells and thereby positively influencing their paracrine secretory profile, MSCs transform them from a pro-inflammatory to an anti-inflammatory status thus enabling tissue regeneration.
9. Although various routes of infusion for MSCs are available to treat systemic diseases, IV route seems preferable due to ease of administration. Currently, the superiority of intra-arterial or intra-organ injection in clinical outcomes has not been established. Additional morbidity and costs are also deterrents.
10. However, to treat spine and joint arthritis, local administration seems superior to systemic infusions
11. Efficacy of MSC therapy can be improved upon by proper manufacturing which comprises of limiting population doublings, cryopreservation recovery of frozen cells, decreasing IBMIR and xenofree processing.

**Questions**

1. Wharton jelly mesenchymal stem cells (WJ-MSC) have the following attributes
   A. Since they have pluripotent markers like embryonic stem cells, tumorigenicity is a concern with clinical use
   B. Among all adult MSC sources, WJ are the most "stem like" because they have pluripotent cell surface markers similar to embryonic stem cells
   C. Bone marrow stem cells as of date are the most studied stem cells and are preferred source because of higher proliferative rates, ease of isolation and maintenance of karyotype stability
   D. Cord blood which is very commonly banked has more MSCs than Wharton jelly.

   **Answer: B.**

   **Explanation:** Although Wharton jelly mesenchymal stem cells (WJ-MSCs) have pluripotent markers like embryonic stem cells, they are not known to be tumorigenic.
2. The true statement regarding mechanism of action of mesenchymal stem cells (MSCs) is
   A. MSCs orchestrate tissue regeneration by engrafting, proliferating and differentiating in the injured/damaged organs.
   B. Transformation of anti-inflammatory M2 macrophages to the pro-inflammatory M1 macrophages is one the beneficial effects of MSCs on organ regeneration.
   C. The therapeutic effects of MSCs are noted despite their cytotoxic engulfment by the host's immune cells.
   D. MSCs impact all the immune cells except neutrophils.

   **Answer: C.**

   **Explanation:** Efferocytosis (different from phagocytosis) by macrophages and their transformation to anti-inflammatory/anabolic M2 phenotype that are subsequently released into circulation is the main mechanism of action by which tissue regeneration is noted.
3. True statement regarding Culture expanded allogeneic Wharton jelly mesenchymal stem cells (WJ-MSCs) and autologous/allogeneic bone marrow (BM) or adipose derived mesenchymal stem cells (AD-MSCs) is
   A. AD-MSCs have the most MSCs per cm of tissue
   B. BM-MSCs are the safest because they have been most researched among all adult MSCs
   C. Since WJ-MSCs have fetal origin MSCs, there are ethical concerns.
   D. WJ-MSCs are the most proliferative MSCs among adult MSCs.

   **Answer: D.**

   **Explanation:** Wharton jelly (WJ) is the known richest source of mesenchymal stem cells (MSCs). Safety of WJ-MSCs has been established and are equally safe as bone marrow derived mesenchymal stem cells (BM-MSCs). Since umbilical cord is considered as medical waste, there are no ethical concerns.
4. Induced pluripotent MSCs (iPSC) have the following advantages over adult mesenchymal stem cells (MSCs)
   A. They do not exhibit genetic mutations
   B. They are not tumorigencic
   C. They have high proliferation rates
   D. They have excellent differentiation capability

   **Answer: D.**

   **Explanation:** Induced pluripotent MSCs (iPSCs) have genetic instability and are known to be tumorigenic.

5. Clinical trials in university settings have shown promising results. However, these positive outcomes have not been reproduced commercially because.
   A. Large scale manufacturing significantly alters stem cell surface markers
   B. Large scale manufacturing usually relies on extensive passaging
   C. Cryopreservation significantly damages the stem cells.
   D. Passage number is important to report to the regulatory agencies than the population doubling number

   **Answer: B.**

   **Explanation:** Excessive passaging decreases the potency of mesenchymal stem cells (MSCs).

## References


1. Pittenger MF, Discher DE, Péault BM, Phinney DG, Hare JM, Caplan AI. Mesenchymal stem cell perspective: cell biology to clinical progress. NPJ Regen Med. 2019;4:22.
2. Moll G, Hoogduijn MJ, Ankrum JA. Editorial: safety, efficacy and mechanisms of action of mesenchymal stem cell therapies. Front Immunol. 2020;11:243.
3. Capilla-González V, Herranz-Pérez V, Sarabia-Estrada R, Kadri N, Moll G. Editorial: mesenchymal stromal cell therapy for regenerative medicine. Front Cell Neurosci. 2022;16:932281.
4. Ringdén O, Moll G, Gustafsson B, Sadeghi B. Mesenchymal stromal cells for enhancing hematopoietic engraftment and treatment of graft-versus-host disease, hemorrhages and acute respiratory distress syndrome. Front Immunol. 2022;13:839844.
5. Moll G, Ankrum JA, Kamhieh-Milz J, et al. Intravascular mesenchymal stromal/stem cell therapy product diversification: time for new clinical guidelines. Trends Mol Med. 2019;25:149–63.
6. Moll G, Ankrum JA, Olson SD, et al. Improved MSC minimal criteria to maximize patient safety: a call to embrace tissue factor and hemocompatibility assessment of MSC products. Stem Cells Transl Med. 2022;11:2–13.
7. Andrzejewska A, Catar R, Schoon J, et al. Multi-parameter analysis of biobanked human bone marrow stromal cells shows little influence for donor age and mild comorbidities on phenotypic and functional properties. Front Immunol. 2019;10:2474.
8. El Omar R, Beroud J, Stoltz JF, et al. Umbilical cord mesenchymal stem cells: the new gold standard for mesenchymal stem cell-based therapies? Tissue Eng Part B Rev. 2014;20:523–44.
9. Wang Q, Wang Y, Chang C, et al. Comparative analysis of mesenchymal stem/stromal cells derived from human induced pluripotent stem cells and the cognate umbilical cord mesenchymal stem/stromal cells. Heliyon. 2023;9:e12683.
10. Caplan AI. Adult mesenchymal stem cells for tissue engineering versus regenerative medicine. J Cell Physiol. 2007;213:341–7.
11. Karahuseyinoglu S, Cinar O, Kilic E, et al. Biology of stem cells in human umbilical cord stroma: in situ and in vitro surveys. Stem Cells. 2007;25:319–31.
12. Subramanian A, Fong CY, Biswas A, et al. Comparative characterization of cells from the various compartments of the human umbilical cord shows that the Wharton's jelly compartment provides the best source of clinically utilizable mesenchymal stem cells. PLoS One. 2015;10:e0127992.
13. Gauthaman K, Fong CY, Suganya CA, et al. Extra-embryonic human Wharton's jelly stem cells do not induce tumorigenesis, unlike human embryonic stem cells. Reprod Biomed Online. 2012;24:235–46.
14. Gazdic M, Volarevic V, Arsenijevic N, et al. Mesenchymal stem cells: a friend or foe in immune-mediated diseases. Stem Cell Rev Rep. 2015;11:280–7.
15. Doorn J, Moll G, Le Blanc K, van Blitterswijk C, de Boer J. Therapeutic applications of mesenchymal stromal cells: paracrine effects and potential improvements. Tissue Eng Part B Rev. 2012;18:101–15.
16. Moll G, Le Blanc K. Engineering more efficient multipotent mesenchymal stromal (stem) cells for systemic delivery as cellular therapy. ISBT Sci Ser. 2015;10:357–65.
17. Moll G, Rasmusson-Duprez I, von Bahr L, et al. Are therapeutic human mesenchymal stromal cells compatible with human blood? Stem Cells. 2012;30:1565–74.
18. von Bahr L, Batsis I, Moll G, et al. Analysis of tissues following mesenchymal stromal cell therapy in humans indicates limited long-term engraftment and no ectopic tissue formation. Stem Cells. 2012;30:1575–8.
19. Moll G, Alm JJ, Davies LC, et al. Do cryopreserved mesenchymal stromal cells display impaired immunomodulatory and therapeutic properties? Stem Cells. 2014;32:2430–42.
20. Luk F, de Witte SF, Korevaar SS, et al. Inactivated mesenchymal stem cells maintain immunomodulatory capacity. Stem Cells Dev. 2016;25:1342–54.
21. Weiss ML, Anderson C, Medicetty S, et al. Immune properties of human umbilical cord Wharton's jelly-derived cells. Stem Cells. 2008;26:2865–74.
22. Caplan H, Olson SD, Kumar A, et al. Mesenchymal stromal cell therapeutic delivery: translational challenges to clinical application. Front Immunol. 2019;10:1645.
23. Atluri S, Manchikanti L, Hirsch JA. Expanded umbilical cord mesenchymal stem cells (UC-MSCs) as a therapeutic strategy in managing critically ill COVID-19 patients: the case for compassionate use. Pain Physician. 2020;23:E71–83.
24. Moll G, Drzeniek N, Kamhieh-Milz J, Geissler S, Volk HD, Reinke P. MSC therapies for COVID-19: importance of patient coagulopathy, thromboprophylaxis, cell product quality and mode of delivery for treatment safety and efficacy. Front Immunol. 2020;11:1091.
25. Couto PS, Al-Arawe N, Filgueiras IS, et al. Systematic review and meta-analysis of cell therapy for COVID-19: global clinical trial landscape, published safety/efficacy outcomes, cell product manufacturing and clinical delivery. Front Immunol. 2023;14:1200180.
26. Manchikanti L, Centeno CJ, Atluri S, et al. Bone marrow concentrate (BMC) therapy in musculoskeletal disorders: evidence-based policy position statement of American Society of Interventional Pain Physicians (ASIPP). Pain Physician. 2020;23:E85–E131.
27. Sanapati J, Manchikanti L, Atluri S, et al. Do regenerative medicine therapies provide long-term relief in chronic low back pain: a systematic review and meta-analysis. Pain Physician. 2018;21:515–40.
28. Navani A, Manchikanti L, Albers SL, et al. Responsible, safe, and effective use of biologics in the management of low back pain: American Society of Interventional Pain Physicians (ASIPP) guidelines. Pain Physician. 2019;22:S1–S74.
29. Awad ME, Hussein KA, Helwa I, et al. Meta-analysis and evidence base for the efficacy of autologous bone marrow mesenchymal stem cells in knee cartilage repair: methodological guidelines and quality assessment. Stem Cells Int. 2019;2019:3826054–15.
30. Henriksson HB, Papadimitriou N, Hingert D, et al. The traceability of mesenchymal stromal cells after injection into degenerated discs in patients with low back pain. Stem Cells Dev. 2019;28:1203–11.
31. Atluri S, Murphy MB, Dragella R, et al. Evaluation of the effectiveness of autologous bone marrow mesenchymal stem cells in the treatment of chronic low back pain due to severe lumbar spinal degeneration: a 12-month, open-label, prospective controlled trial. Pain Physician. 2022;25:193–207.

32. Moll G, Ignatowicz L, Catar R, Sadeghi B. Different procoagulant activity of therapeutic mesenchymal stromal cells derived from bone marrow and placental decidua. Transpl Int. 2015;28:50–1.
33. Gholamrezanezhad A, Mirpour S, Bagheri M, et al. In vivo tracking of 111 in-oxine labeled mesenchymal stem cells following infusion in patients with advanced cirrhosis. Nucl Med Biol. 2011;38:961–7.
34. Rosado-de-Castro PH, Schmidt Fda R, Battistella V, et al. Biodistribution of bone marrow mononuclear cells after intra-arterial or intravenous transplantation in subacute stroke patients. Regen Med. 2013;8:145–55.
35. Lezaic L, Socan A, Peitl PK, et al. Imaging and 1-day kinetics of intracoronary stem cell transplantation in patients with idiopathic dilated cardiomyopathy. Nucl Med Biol. 2016;43:410–4.
36. Lanzoni G, Linetsky E, Correa D, et al. Umbilical cord mesenchymal stem cells for COVID-19 acute respiratory distress syndrome: a double-blind, phase 1/2 a, randomized controlled trial. Stem Cells Transl Med. 2021;10:660–73.
37. Shi L, Huang H, Lu X, et al. Effect of human umbilical cord-derived mesenchymal stem cells on lung damage in severe COVID-19 patients: a randomized, double-blind, placebo-controlled phase 2 trial. Signal Transduct Target Ther. 2021;6:58.
38. Dilogo IH, Aditianingsih D, Sugiarto A, et al. Umbilical cord mesenchymal stromal cells as critical COVID-19 adjuvant therapy: a randomized controlled trial. Stem Cells Transl Med. 2021;10:1279–87.
39. Shu L, Niu C, Li R, et al. Treatment of severe COVID-19 with human umbilical cord mesenchymal stem cells. Stem Cell Res Ther. 2020;11:361.
40. Meng F, Xu R, Wang S, et al. Human umbilical cord-derived mesenchymal stem cell therapy in patients with COVID-19: a phase 1 clinical trial. Signal Transduct Target Ther. 2020;5:172.
41. Wei F, Kong D, Li T, et al. Efficacy and safety of umbilical cord mesenchymal stem cells for the treatment of patients with COVID-19. Clinics (Sao Paulo). 2021;76:e2604.
42. Adas G, Cukurova Z, Yasar KK, et al. The systematic effect of mesenchymal stem cell therapy in critical COVID-19 patients: a prospective double controlled trial. Cell Transplant. 2021;30:9636897211024942.
43. Zhang Z, Lin H, Shi M, et al. 2 human umbilical cord mesenchymal stem cells improve liver function and ascites in decompensated liver cirrhosis patients. J Gastroenterol Hepatol. 2012;27:112–20.
44. Wang L, Li J, Liu H, et al. Pilot study of umbilical cord-derived mesenchymal stem cell transfusion in patients with primary biliary cirrhosis. J Gastroenterol Hepatol. 2013;28:85–92.
45. Shi M, Zhang Z, Xu R, et al. Human mesenchymal stem cell transfusion is safe and improves liver function in acute-on-chronic liver failure patients. Stem Cells Transl Med. 2012;1:725–31.
46. Liang J, Zhang H, Zhao C, et al. Effects of allogeneic mesenchymal stem cell transplantation in the treatment of liver cirrhosis caused by autoimmune diseases. Int J Rheum Dis. 2017;20:1219–26.
47. Xue HL, Zeng WZ, Wu XL, et al. Clinical therapeutic effects of human umbilical cord-derived mesenchymal stem cells transplantation in the treatment of end-stage liver disease. Transplant Proc. 2015;47:412–8.
48. Li YH, Xu Y, Wu HM, et al. Umbilical cord-derived mesenchymal stem cell transplantation in hepatitis b virus related acute-on-chronic liver failure treated with plasma exchange and entecavir: a 24-month prospective study. Stem Cell Rev Rep. 2016;12:645–53.
49. Hu J, Yu X, Wang Z, et al. Long-term effects of the implantation of Wharton's jelly-derived mesenchymal stem cells from the umbilical cord for newly-onset type 1 diabetes mellitus. Endocr J. 2013;60:347–57.
50. Cai J, Wu Z, Xu X, et al. Umbilical cord mesenchymal stromal cell with autologous bone marrow cell transplantation in established type 1 diabetes: a pilot randomized controlled open-label clinical study to assess safety and impact on insulin secretion. Diabetes Care. 2016;39:149–57.
51. Liu X, Zheng P, Wang X, et al. A preliminary evaluation of efficacy and safety of Wharton's jelly mesenchymal stem cell transplantation in patients with type 2 diabetes mellitus. Stem Cell Res Ther. 2014;5:57.
52. Lian XF, Lu DH, Liu HL, et al. Effectiveness and safety of human umbilical cord-mesenchymal stem cells for treating type 2 diabetes mellitus. World J Diabetes. 2022;13:877–87.
53. Hu J, Wang Y, Gong H, et al. Long term effect and safety of Wharton's jelly-derived mesenchymal stem cells on type 2 diabetes. Exp Ther Med. 2016;12:1857–66.
54. Zang L, Li Y, Hao H, et al. Efficacy and safety of umbilical cord-derived mesenchymal stem cells in Chinese adults with type 2 diabetes: a single-center, double-blinded, randomized, placebo-controlled phase II trial. Stem Cell Res Ther. 2022;13:180.
55. Bartolucci J, Verdugo FJ, González PL, et al. Safety and efficacy of the intravenous infusion of umbilical cord mesenchymal stem cells in patients with heart failure: a phase 1/2 randomized controlled trial (RIMECARD trial (randomized clinical trial of intravenous infusion umbilical cord mesenchymal stem cells on cardiopathy)). Circ Res. 2017;121:1192–204.
56. Zhao XF, Xu Y, Zhu ZY, Gao CY, Shi YN. Clinical observation of umbilical cord mesenchymal stem cell treatment of severe systolic heart failure. Genet Mol Res. 2015;14:3010–7.
57. Gao LR, Chen Y, Zhang NK, et al. Intracoronary infusion of Wharton's jelly-derived mesenchymal stem cells in acute myocardial infarction: double-blind, randomized controlled trial. BMC Med. 2015;13:162.
58. Li X, Hu YD, Guo Y, et al. Safety and efficacy of intracoronary human umbilical cord-derived mesenchymal stem cell treatment for very old patients with coronary chronic total occlusion. Curr Pharm Des. 2015;21:1426–32.
59. Matas J, Orrego M, Amenabar D, et al. Umbilical cord-derived mesenchymal stromal cells (MSCs) for knee osteoarthritis: repeated MSC dosing is superior to a single MSC dose and to hyaluronic acid in a controlled randomized phase I/II trial. Stem Cells Transl Med. 2019;8:215–24.
60. Dilogo IH, Canintika AF, Hanitya AL, et al. Umbilical cord-derived mesenchymal stem cells for treating osteoarthritis of the knee: a single-arm, open-label study. Eur J Orthop Surg Traumatol. 2020;30:799–807.
61. Wang Y, Jin W, Liu H, et al. Curative effect of human umbilical cord mesenchymal stem cells by intra-articular injection for degenerative knee osteoarthritis. Zhongguo Xiu Fu Chong Jian Wai Ke Za Zhi. 2016;30:1472–7.
62. Gao L, Zhang Y, Hu B, et al. Phase II multicenter, randomized, double-blind controlled study of efficacy and safety of umbilical cord-derived mesenchymal stromal cells in the prophylaxis of chronic graft-versus-host disease after HLA-haploidentical stem-cell transplantation. J Clin Oncol. 2016;34:2843–50.
63. Zhang J, Lv S, Liu X, et al. Umbilical cord mesenchymal stem cell treatment for Crohn's disease: a randomized controlled clinical trial. Gut Liver. 2018;12:73–8.
64. Forbes GM, Sturm MJ, Leong RW, et al. A phase 2 study of allogeneic mesenchymal stromal cells for luminal Crohn's disease refractory to biologic therapy. Clin Gastroenterol Hepatol. 2014;12:64–71.
65. Wang S, Cheng H, Dai G, et al. Umbilical cord mesenchymal stem cell transplantation significantly improves neurological function in patients with sequelae of traumatic brain injury. Brain Res. 2013;1532:76–84.
66. Barczewska M, Maksymowicz S, Zdolińska-Malinowska I, et al. Umbilical cord mesenchymal stem cells in amyotrophic lateral sclerosis: an original study. Stem Cell Rev Rep. 2020;16:922–32.

67. Riordan NH, Morales I, Fernández G, et al. Clinical feasibility of umbilical cord tissue-derived mesenchymal stem cells in the treatment of multiple sclerosis. J Transl Med. 2018;16:57.
68. Li JF, Zhang DJ, Geng T, et al. The potential of human umbilical cord-derived mesenchymal stem cells as a novel cellular therapy for multiple sclerosis. Cell Transplant. 2014;23:S113–22.
69. Wang L, Huang S, Li S, et al. Efficacy and safety of umbilical cord mesenchymal stem cell therapy for rheumatoid arthritis patients: a prospective phase I/II study. Drug Des Devel Ther. 2019;13:4331–40.
70. Wang L, Wang L, Cong X, et al. Human umbilical cord mesenchymal stem cell therapy for patients with active rheumatoid arthritis: safety and efficacy. Stem Cells Dev. 2013;22:3192–202.
71. Lalu MM, McIntyre L, Pugliese C, et al. Safety of cell therapy with mesenchymal stromal cells (SafeCell): a systematic review and meta-analysis of clinical trials. PLoS One. 2012;7:e47559.
72. Lalu MM, Sullivan KJ, Mei SH, et al. Evaluating mesenchymal stem cell therapy for sepsis with preclinical meta-analyses prior to initiating a first-in-human trial. elife. 2016;5:e17850.
73. Liang J, Zhang H, Kong W, et al. Safety analysis in patients with autoimmune disease receiving allogeneic mesenchymal stem cells infusion: a long-term retrospective study. Stem Cell Res Ther. 2018;9:312.
74. Thompson M, Mei SHJ, Wolfe D, et al. Cell therapy with intravascular administration of mesenchymal stromal cells continues to appear safe: an updated systematic review and meta-analysis. EclinicalMedicine. 2020;19:100249.
75. Atluri S, Manocha V, Boddu N, et al. Safety and effectiveness of intravascular mesenchymal stem cells to treat organ failure and possible application in COVID-19 complications. Pain Physician. 2020;23:S391–420.
76. Bauer G, Elsallab M, Abou-El-Enein M. Concise review: a comprehensive analysis of reported adverse events in patients receiving unproven stem cell-based interventions. Stem Cells Transl Med. 2018;7:676–85.
77. Issue brief: harms linked to unapproved stem cell interventions highlight need for greater FDA enforcement. Pew. 2021. Accessed 7 June 2023. https://www.pewtrusts.org/-/media/assets/2021/05/harms-linked-to-unapproved-stem-cell-interventions-highlight-need-for-greater-fda-enforcement.pdf.
78. Moll G, Geißler S, Catar R, et al. Cryopreserved or fresh mesenchymal stromal cells: only a matter of taste or key to unleash the full clinical potential of msc therapy? Adv Exp Med Biol. 2016;951:77–98.
79. Cottle C, Porter AP, Lipat A, et al. Impact of cryopreservation and freeze-thawing on therapeutic properties of mesenchymal stromal/stem cells and other common cellular therapeutics. Curr Stem Cell Rep. 2022;8:72–92.
80. Moll G, Hult A, von Bahr L, et al. Do ABO blood group antigens hamper the therapeutic efficacy of mesenchymal stromal cells? PLoS One. 2014;9:e85040.
81. Galipeau J. The mesenchymal stromal cells dilemma—does a negative phase III trial of random donor mesenchymal stromal cells in steroid-resistant graft-versus-host disease represent a death knell or a bump in the road? Cytotherapy. 2013;15:2–8.
82. von Bahr L, Sundberg B, Lönnies L, et al. Long-term complications, immunologic effects, and role of passage for outcome in mesenchymal stromal cell therapy. Biol Blood Marrow Transplant. 2012;18:557–64.
83. Jin HJ, Bae YK, Kim M, et al. Comparative analysis of human mesenchymal stem cells from bone marrow, adipose tissue, and umbilical cord blood as sources of cell therapy. Int J Mol Sci. 2013;14:17986–8001.
84. Patel DB, Gray KM, Santharam Y, et al. Impact of cell culture parameters on production and vascularization bioactivity of mesenchymal stem cell-derived extracellular vesicles. Bioeng Transl Med. 2017;2:170–9.
85. Røsland GV, Svendsen A, Torsvik A, et al. Long-term cultures of bone marrow-derived human mesenchymal stem cells frequently undergo spontaneous malignant transformation. Cancer Res. 2009;69:5331–9.
86. Chinnadurai R, Copland IB, Garcia MA, et al. Cryopreserved mesenchymal stromal cells are susceptible to T-cell mediated apoptosis which is partly rescued by IFNγ licensing. Stem Cells. 2016;34:2429–42.
87. Matthay MA, Calfee CS, Zhuo H, et al. Treatment with allogeneic mesenchymal stromal cells for moderate to severe acute respiratory distress syndrome (START study): a randomized phase 2a safety trial. Lancet Respir Med. 2019;7:154–62.
88. Chinnadurai R, Garcia MA, Sakurai Y, et al. Actin cytoskeletal disruption following cryopreservation alters the biodistribution of human mesenchymal stromal cells in vivo. Stem Cell Rep. 2014;3:60–72.

# Non-autologous Biologics

# 12

Theodore T. Sand

## Introduction

While it is possible to eliminate a painful musculoskeletal condition by treating just the painful symptoms, an advantage of using regenerative materials is that pain can be reduced along with repair/remodeling at the pathologic site, which enables in some conditions to modify disease progression [1]. Consequently, the focus of this chapter is on non-autologous biologic products with a regenerative potential, which includes the following types of materials:

- Placental tissue associated products (e.g., amniotic fluid, micronized and sheet placental tissue)
- Cadaveric tissue-derived products
- Cord blood and cord tissue-derived products
- Cultured cell products (cultured cells, conditioned medium and induced pluripotent stem cells [iPSCs])
- Exosome containing products

The topics will be restricted to those products that are derived from donor sources, meaning that they fall into the allogeneic category of therapeutic treatment. Since the source materials are donor-derived, the companies offering these products must register their facilities with the United States Food and Drug Administration (USFDA) and produce the products under stringent manufacturing conditions known as "Current Good Manufacturing Practices" (cGMP). As with any donor-derived material, there always is a risk of an allogeneic product being contaminated with viral or microbial agents.

While a wide variety of donor-derived materials are used as part of a surgical procedure, the focus will be on allogeneic materials that have been used therapeutically for nonsurgical, point-of-care treatments. Thus, the primary mode of delivery is either via percutaneous injection or as a wound covering.

T. T. Sand (✉)
Biotechnology and Regenerative Medicine, Sand Consulting, Poway, CA, USA
e-mail: tsand@biotechexpert.com

## History

One of the first examples of the use of donor-derived materials in a controlled, medical setting involved donor-derived fetal membranes as reported in a publication in 1910 by Davis [2], which documented the therapeutic benefit of amniotic membrane in skin grafting. Corneas and soft tissues like tendons and ligaments recovered intact from cadavers also have had a long history of use.

The rise of the use of injectable, donor-derived tissues has occurred over roughly the past 15 years. These materials can be obtained from cadaveric sources or from living donors (e.g., placental tissue or amniotic fluid), which subsequently are processed to create injectable products.

## Regulatory Considerations

The main regulatory rule that the USFDA relies on to cover donor-derived materials is known as 21 CFR 1271 Human Cells, Tissues, Cellular and Tissue-Based Products (HCT/Ps) [3]. Multiple parts of 1271 have a bearing on how manufacturers produce and market their products. However, the one element of 1271 that manufacturers of donor-derived tissue strictly adhere to is the requirement to register their manufacturing facility with the USFDA. Occasionally, a manufacturer will claim on their website that they are fully compliant with USFDA requirements, which is based on just the registration of their company and the formal listing of any HCT/P-derived products they plan on marketing with the USFDA. However, a company manufacturing and/or marketing HCT/P-derived products is required by 1271 Subpart B to register and list their products, which has nothing to do with the regulatory status of those products. Furthermore, the USFDA doesn't endorse companies or products appearing in the registration and listing entries, nor does the Agency endorse products that are cleared as a part of the pre-market notification process (i.e., 510(k)) or approved by the pre-


A. Navani et al. (eds.), *Essentials of Regenerative Medicine in Interventional Pain Management*,
https://doi.org/10.1007/978-3-031-50357-3_12

market approval (PMA) process (i.e., Investigational Device Exemption [IDE] or Investigational New Drug [IND]).

There are four subsections to 1271.10(a) that cover minimal manipulation, homologous use, no adulteration and 1271.10(a)(4) (with edits):

(4) Either:

1. The HCT/P does not have a systemic effect and is not dependent upon the metabolic activity of living cells for its primary function; or
2. The HCT/P has a systemic effect or is dependent upon the metabolic activity of living cells for its primary function, and:
   (a) Is for autologous use;
   (b) Is for allogeneic use in a first-degree or second-degree blood relative; or
   (c) Is for reproductive use.

The first issue for a manufacturer is to determine if the product to be sold has a systemic effect or not. Unfortunately, it would seem that some manufacturers stop their regulatory assessment at that point. But the wording concerning the presence of cells in HCT/P products suggests that any tissue, stromal, stem and/or progenitor cells present should be dead ("…is not dependent on the metabolic activity of living cells…").

How can a physician be protected from using donor-derived products that are USFDA noncompliant? Consider the following:

- Don't rely on sales representatives or distributors for regulatory advice
- Ask for a letter on corporate letterhead signed by a corporate officer of the manufacturer in which the corporate officer states that the USFDA has reviewed the product being offered for sale and agrees with the company's assessment that the product doesn't require pre-market studies prior to sale
- Send the sales literature/company letter for the product to the Tissue Reference Group at the USFDA (email to: TissueReferenceGroup@fda.hhs.gov) and ask for an assessment of the product's regulatory status for treating pathologies of interest.

As long as manufacturers are registered with the USFDA and follow the required procedures for ensuring the safety of a product, they might choose to sell the product regardless of whether it is USFDA-compliant or not, until the USFDA audits them and issues a determination that their product or products are being sold without the appropriate pre-market studies. Therefore, it is important to be certain about the regulatory status of an allogeneic-derived regenerative therapeutic product, in order to avoid injecting patients with an experimental treatment of an unapproved drug or biological product. The regulatory status of potential sources of or already commercialized allogeneic tissue-derived materials is shown in Table 12.1. It also is important to understand that registering a clinical study on Clinicaltrials.gov isn't the same as having the clinical study approved by an Institutional Review Board (IRB). If a non-autologous biologic is considered to be a drug and biological product (due to its failure to meet the four elements listed in 1271.10(a)), the manufacturer will need to follow a PMA process that includes conducting clinical studies for an IND or an IDE to obtain clinical evidence of the product's safety and efficacy to submit to the USFDA.

## Placental Tissue Associated Materials

### Placental Tissue-Derived Materials

An early documented use of placental tissue-derived materials in a medical setting was published by Davis in 1910 [2]. The author provided evidence that the use of amniotic membranes for skin grafting was superior to the use of a xenograft or cadaveric skin graft. There are two main, distinctive membranes of the amniotic sac that have been used to create therapeutic products. The inner layer is the amniotic membrane, and the outer layer is the chorionic membrane. There is an intermediate layer that links the two together. Each of these layers is divided into sub-layers [4]. Placental tissue-derived products typically include both the amnion and chorion, or may include one layer or the other. The membrane form of dehydrated Human Amniotic Chorionic Membrane (dHACM) has been used in several clinical trials of diabetic ulcers [5, 6], and burns [7], while other placental tissue-derived membranes have been used in treating a wide variety of topical wounds, ophthalmological wounds and to prevent post-surgical adhesions [8].

The foregoing indications for using placental tissue-derived membranes depend on the physical ability of the products to cover wounds, so these products usually take the form of sheets. However, more recently, products based on the micronization of placental tissue have become available. These products are created by subjecting pieces of the placenta (some manufacturers use either the amnion or chorion membrane) to a high-shear process. The resulting disruption of the tissue produces much smaller particles that are injectable.

In addition to the appropriate evaluation of micronized placental tissue-derived products through the use of clinical

**Table 12.1** Donor tissue-derived materials and their regulatory status

| Tissue source | Product forms | Clinical application | HCT/P[a] (Yes/No); 351/361 | Notes |
|---|---|---|---|---|
| Amniotic fluid (no cells) | Amniotic fluid | Depends on the instructions for use of the IND | No; still considered as a 351 | Amniotic fluid without cells is a secretion and excluded from 1271[b] by definition; it is considered to be a biological drug regulated by section 351 of the PHSA[c] |
| Amniotic fluid (with viable cells) | Cryopreserved vials of fluid | Depends on the indications for use of the IND | Yes; 351 | Materials containing donor-derived viable cells do not meet the criterion in 1271.10(a)(4) |
| Intradiscal nucleus pulposus (cadaveric) | Injectable tissue | Restoration of the intervertebral disc | Yes; 361 | Nucleus pulposus cells must not be viable |
| Placental tissue (dried/lyophilized) | Sheets | Wound healing, covering or barrier | Yes; 361 | Placental-derived tissues must not contain viable cells |
| Placental tissue (dried/lyophilized) | Micronized particles that are flowable or injectable | Will depend on the IDE indications for use | Yes; 351 | Micronization of placental tissue alters the physical state of the tissue: Above minimal manipulation |
| Cord Blood | Whole, cell enriched | Recapitulation of a patient's ablated bone marrow | Yes; 351 | Whole cord blood and its components are 351 biological drugs |
| Cord Blood | Plasma only | Depends on the IND indications for use | Yes; 351 | Any component of cord blood is regulated as a 351 biological drug |
| Umbilical Cord tissue | Micronized | Depends on the IND indications for use | Yes; 351 | Umbilical Cord is considered a structural tissue and micronization alters the structural nature of the tissue |
| Exosomes (obtained from cultured cells) | Fluid | No IND-cleared product | Yes; 351 | Exosomes are collected from culture fluid of various cell types like MSCs—which makes them a 351 material |
| Exosomes obtained from platelet packs or Cord Blood | Fluid | Various | Yes; 351 | Any exosome containing product used to treat diseases or conditions is a drug and biologic product; regulated by Section 351 of the PHSA |
| Cultured Cells (donor) | Cryopreserved fluid | Depends on the IND indications for use | Yes; 351 | Cultured cells rise above minimal manipulation—351 material |
| Allogeneic products containing viable cells with metabolic activity | Any | Any | Yes; 351 | HCT/Ps with viable cells that have a systemic or metabolic effect must be used on the donor or First/second degree blood relative for 361 status; otherwise it is a 351 |

*IND* Investigational new drug, *IDE* Investigational Device Exemption (IDE)
[a] HCT/P Human cells, tissues, cellular and tissue-based products
[b] 1271 21 CFR Part 1271 Human cells, tissues, cellular and tissue-based products
[c] PHSA-Public Health Service Act

studies and INDs, the USFDA has warned a number of companies for selling micronized placental tissue products without the benefit of a PMA process. The USFDA early on sent an untitled letter to Surgical Biologics (MiMedx Group; AmnioFix) [9] pointing out that injectable products made with micronized amnion/chorion membranes didn't meet the requirement for minimal manipulation (1271.10(a)(1)). The USFDA stated in the company's letter that manipulation of placental tissue resulting in an injectable form of the tissue alters "... the original relevant characteristics of the structural tissue relating to the tissue's utility for reconstruction, repair or replacement". Consequently, the company's product was considered by the USFDA to be a drug and a biological product and was required to follow the PMA process prior to being marketed. MiMedx subsequently placed their product on the PMA pathway to obtain formal approval.

## Clinical Studies of dHACM

A recent Level 1, multi-center study of a commercially available dHACM product to treat lower extremity ulcers (e.g., diabetic foot ulcers) of 4-week duration that failed initial treatment showed that 70% of the study participants receiving the dHACM product (with off-loading, as appropriate) had complete wound closure at 12 weeks, compared to 50% of the participants in the control group (continued alginate wound dressing and off-loading). At the final follow-up of 16 weeks, 95% of the dHACM group's wounds had remained closed, while 86% of the control group had closed wounds [10].

In a recently published report, the clinical results of several dHACM products evaluated in multiple, randomly controlled studies were reviewed, including a comparison of clinical outcomes of the products to each other [6]. In a clinically challenging application, dHACM has been used to treat pediatric burns as recently summarized in a case series report [7]. The dHACM graft was used on a range of burns from superficial partial to full thickness burns for which split thickness skin grafts (STSGs) are the standard of care. The authors reviewed outcomes of treating 30 patients with a dHACM product. An important finding was the authors' observation that healing of burns over bone and tendon was facilitated, which are areas that normally present challenges with STSG treatment. Pain was well-managed during the treatment period (0–1 on the Wong-Baker FACES score scale), and time to heal was faster with the dHACM compared to published reports of STSG healing rates [7]. Clearly, dHACM products have demonstrated therapeutic efficacy to treat ulcers and burns, with reduced scarring and more rapid wound closure in many cases.

### Clinical Studies of Micronized Placental Tissue-Derived Products

Most of the published clinical studies for placental tissue-derived products utilize products in sheet form. However, there have been several recent reports published on clinical studies of micronized forms of placental tissue-derived products.

The use of a micronized dehydrated human amniotic/chorionic membrane (mdHACM) product to treat chronic plantar fasciitis has been reported [8]. The Level 1 study enrolled 45 patients in three treatment groups: control (1.25 mL sterile saline), lower dose (0.5 mL mdHACM) and higher dose (1.5 mL mdHACM). Each patient received an injection of 2.0 mL of 0.5% Marcaine, followed by a control or mdHACM injection. After 8 weeks (the endpoint of the study), control patients reported an average improvement of 12.9 points on the American Orthopedic Foot & Ankle Society (AOFAS) Hindfoot metric, which includes an assessment of pain. Both of the groups receiving mdHACM treatment reported an improvement of greater than 50 points. The authors indicated that while there was no statistical difference between the improvement shown by the patients receiving a mdHACM treatment, both test groups were statistically meaningfully different from control [8]. However, in another Level 1 study with a "cryopreserved" micronized human amniotic membrane product in treating plantar fasciitis [11], no meaningful differences were observed at the study end point of 12 or 18 weeks. Although two patients who had received two injections showed better outcomes compared to the single-injection group [11].

A retrospective study was performed on 82 patients who received injections in 100 knees of a mdHACM commercial product to treat knee osteoarthritis (OA). The Knee Injury and Osteoarthritis Outcome Score (KOOS) was the outcome metric that was assessed pre-treatment and at 6-months [12]. The authors reported that the cohort mean KOOS was 40 at pre-treatment, increasing to 65 at the 6-month milestone, with Quality of Life and Sports/Recreation domains increasing by 111 and 118% over the study period, respectively. While no serious adverse events were noted, 68% of the patients reported pain at the injection site that resolved within 2–7 days [12].

## Amniotic Fluid

Amniotic fluid serves to protect the developing fetus from physical trauma, is a source of nutrients for the amniotic sac, and acts as a protective barrier against infectious agents. While the composition of amniotic fluid changes with gestational age, the main component is water containing electrolytes, biochemicals (proteins and lipids, hormones, etc.), lanugo hair, meconium (fetal fecal matter) and cells [13]. Amniotic fluid is obtained during amniocentesis, in which 3–5 mL of fluid is obtained. As recently reported [14], 1 mL of amniotic fluid contains an average of 100,000 nucleated cells with a viability that ranges from 60–90%. The amniotic fluid cells are quite heterogenous, including adult cells like epithelial cells and fibroblasts, as well as two types of stem cells (amniotic fluid stem cells and amniotic fluid mesenchymal stem cells [MSCs]). Approximately 1% of the nucleated cells found in amniotic fluid are stem cells [13]. Extensive analysis of these stem cells suggest that they are at least multipotent, which is similar to MSCs found in adipose or bone marrow, but have proliferative advantages when placed in culture [13, 14].

Recently, a study was published in which commercially available amniotic fluid products were analyzed for the number and type of viable cells contained in the products. Several vendors declined to provide samples for the study, but of the two products that were analyzed, neither company's vials of product from multiple production lots contained MSCs (as assessed in a bioassay for colony-forming units-fibroblast, while the positive control of human bone marrow were positive for colonies). Viability of the cells showed substantial variation, with most vials having no viable cells present. The authors concluded that clinicians should be wary of commercial products that claim the presence of amniotic fluid-derived MSCs, since most of the cells could be dead, including any MSCs that initially might have been present in the amniotic fluid [15]. This direct assessment of MSC content in commercially available products confirms that it is unlikely that amniotic fluid contains stem cells at the levels physicians frequently are led to believe (i.e., millions). There also is an issue with the recovery of viable cells in this type of product, since for the cells to remain viable they must be cryopreserved, which means that the products with viable cells must be stored, thawed and used under tightly controlled conditions [15].

Although amniotic fluid-derived products containing donor-derived viable cells are not compliant with 1271.10(a), and therefore are considered to be drugs and biological products, an acellular product containing amniotic fluid is not subject to the requirements of 1271.10, since secretions like amniotic fluid are excluded by definition [1271.3(d)(3)] from being regulated as HCT/Ps [3]. Nonetheless, the USFDA has determined that products made from or containing amniotic fluid that are used to treat human diseases or conditions are drugs under the Federal Food, Drug, & Cosmetics (FD&C) Act, and are biological products regulated under Section 351 of the Public Health Service Act (PHSA). The USFDA issued a Warning Letter to BioLab Sciences stating that the company's amniotic fluid-containing products were considered to be drugs and biological products and were being sold without following the appropriate regulatory pathway [16].

While there are products on the market that contain just amniotic fluid, there also are products that contain both micronized placental tissue and amniotic fluid, including viable cells. The process of micronization generally results in dead cells in the tissue, especially if the resulting tissue product is injectable through a 22–25 gauge needle, so tissue is micronized separately and subsequently combined with the cell fraction isolated from the same donor's amniotic fluid. For example, a commercially available product (ReNu, Organogenesis, Canton, MA) recently was described as containing micronized amniotic tissue, which is combined with separately packaged viable cells right before injection [17].

### Clinical Studies of Amniotic Fluid

Much of the clinical assessment of amniotic fluid cells conducted to-date involves cultured cell preparations and pre-clinical studies in animal models. However, there is a recent report of the use of a commercial product (ReNu, Organogenesis, Canton, MA) with micronized amniotic membrane and cells that was used in a limited study for treating hip OA [17]. Nine study participants were treated with the commercial product and reported statistically meaningful improvement in the modified Harris Hip Score (Mhhs) results at 12-month follow-up compared to pre-treatment. The maximal visual analog scale (VAS) pain score for the preceding three-days was statistically meaningfully improved from pre-treatment to the 12-month milestone, but the average VAS pain score for the preceding three-days was not statistically meaningfully different at 12-months versus pre-treatment. Overall, the authors concluded that the preponderance of patient reported outcomes showed statistically meaningfully improved clinical outcomes without evidence of joint space narrowing at 12-months [17].

## Cadaveric Tissue Derived Products

Most tissues recovered from cadaveric sources are implanted in surgical procedures. For example, cadaveric tendons are used to replace torn tendons. However, a product composed of nucleus pulposus (NP) tissue obtained from cadaveric intervertebral discs has been introduced in the market (https://gotviadisc.com/). This product doesn't appear to contain viable cells, which means that it most likely meets the four criteria of 1271.10(a) and would be regulated solely by 1271. The NP product is delivered into the disc through a 22-g needle. In a Level 1 study (Clinicaltrials.gov Identifier: NCT03709901) cells were added to the NP tissue product just prior to injection into the degenerated disc, which is a drug treatment. The study subjects receiving the NP/Cells injection showed a substantial decrease in both VAS and Oswestry Disability Index (ODI) over the first year, while no adverse or serious adverse events were reported [18].

## Cord Blood and Cord Tissue Derived Products

The scope of the chapter is limited to those donor-derived products currently being used as a regenerative therapy in the treatment of pain or as a therapy for wounds. Thus, some types of donor-derived biological materials have been excluded. For example, cord blood and bone marrow units have been used for decades as a source of donor-derived cell therapy. However, this type of cell therapy requires that the donor material should be matched by human leukocyte antigen (HLA)-typing to the recipient [19]. In order to eliminate the need for HLA-matching, adherent cells from a wide variety of donor tissues have been cultured during the past 20 years, since MSCs (which grow as an adherent cell in plastic culture containers) don't express a full set of HLA markers [19]. The use of a cultured cell product derived from a cord blood unit or from cord tissue might be a potential therapy, but it would require approval by the USFDA through the PMA process.

As shown in a clinical review of human umbilical cord MSCs (UC-MSCs), there is substantial evidence from in vitro studies, pre-clinical studies and IRB-approved clinical studies for the therapeutic potential of UC-MSCs to treat a wide variety of pathologies including OA, hepatic diseases, spinal cord injury, and autoimmune disorders [20]. While rheumatoid arthritis (RA) is an autoimmune disease that results frequently in severe stiffness, pain and swelling of joints, a recent clinical study (Clinicaltrials.gov Identifier: NCT01547091) examined the impact of combining an infusion of UC-MSCs in study participants who had been diagnosed with RA for a minimum of 6-months and who were receiving standard drug therapy. RA-associated serological markers and the Disease Activity Score 28 (DAS28) were used to assess clinical status of the study participants for 3 years. A statistically meaningful decrease in the mean DAS28 score was observed between the 1-year milestone and pre-treatment, and also between the 1-year and 3-year milestones. Study participants remained on the low-dose drug regimen used prior to UC-MSC treatment for the study period, but with a substantial improvement in their physical condition. The authors indicated that RA-drug treatment helped to manage the autoimmune aspects of the study participants' disease, but suggested that the UC-MSC therapy aided in the mitigation of the typical manifestations of joint swelling, pain and stiffness [21]. While no placebo group was included in this RA treatment study, the results point to the potential for using MSC therapy in combination with established drug treatments for other autoimmune conditions in order to mitigate the musculoskeletal sequalae of an autoimmune disorder. The UC-MSC treatment was performed via IV infusion, which was well tolerated [21], suggesting that multiple joints benefited from the infusion, since the DAS28 score is comprised of an evaluation of 28 joints.

A number of companies have been sent Warning Letters by the USFDA for offering noncompliant products derived from umbilical cord blood [22–24] or umbilical cord tissue [23, 25, 26]. Thus, umbilical cord blood and cord tissue probably are the most popular donor tissues for marketing noncompliant therapies.

## Cultured Cell Products

### Cultured Cells

Stem/stromal cells have two features that distinguish them from other cells in the adult: they can self-renew and differentiate into all of the adult cells in the body. Self-renewal means that a stem/stromal cell can proliferate in culture, without changing its developmental status, and can be expanded in culture to create millions of cultured stem/stromal cells. Thus, large numbers of cultured cells can be placed in a vial and cryopreserved, for subsequent use as a primary off-the-shelf therapy. In particular, the MSC is a multipotent stromal cell that can differentiate into adult cells, including osteoblasts, chondrocytes and adipocytes [27]. The use of cultured, allogeneic stromal cells, like the MSC, greatly diminishes but doesn't eliminate the chance for an immune response in the recipient [28].

The MSC traditionally has been isolated from donor bone marrow, but MSCs are found in all tissues of the body [27]. A number of commercial allogeneic cultured MSC-based products are being studied in clinical trials, most of which have been isolated from donor bone marrow. However, some of the products have been derived from placental tissue via enzymatic digestion [29], yielding stromal cells that share many biochemical markers with bone marrow-derived MSCs, but have additional biomarkers that indicate the isolated cells share features with embryonic stem cells, which are very primitive [30].

The use of placental tissue-associated cells is emerging as a favorite source of more primitive progenitor cells for several reasons. The types of stromal cells from placental tissue (including umbilical cord) and amniotic fluid are thought to be in a "...middle ground between ES [embryonic stem] cells and adult stem cells..." [14]. The isolation of cells from placental tissue/fluid is straightforward [13, 14, 29, 31]. These cells are reported to proliferate rapidly in culture [14], while maintaining their multipotent status, and have favorable immunomodulatory properties [29]. In contrast, while cells from cord blood can be cultured easily for therapeutic use, the expanded cells mostly supported hematopoietic functions [14]. Another popular source of allogeneic MSC-type stromal cells for culture expansion is adipose tissue [28]. Finally, MSC-type stromal cells from placental tissue aren't teratogenic, meaning that they don't form tumors when injected, which is a major concern with embryonic stem cells [13, 14, 31].

### Clinical Studies

Due to the allogeneic source of stem cells and the culture expansion used to create these products, they are biologic products according to the USFDA [3] and need to undergo extensive pre-market study, usually as an IND. As might be expected, most of the pending clinical studies involving cultured, allogeneic cells have been directed at conditions like Graft-vs-Host Disease, Crohn's Disease, idiopathic pulmonary fibrosis and other substantial and life-threatening pathologies [31].

There are dozens of studies listed on ClinicalTrials.gov that use some form of allogeneic tissue derived cultured MSC as the treatment for a wide range of pathologies. To cite one example, an IND study (Clinicaltrials.gov Identifier: NCT01290367) that used a bone marrow-derived, cultured MSC preparation to treat painful discs of study participants has been published [30]. In a Phase Ib/2a clinical trial, enrolled participants diagnosed with discogenic pain and degeneration at a single-level disc diagnosed with modified Pfirrmann II-VII scores (approximately 80% of participants receiving cell therapy had scores of II-IV) were injected with a proprietary cultured MSC product at two doses, and compared to control injections of saline or hyaluronic acid in saline. Approximately 43% of the study participants receiving the lower cell dose had a durable 50% decrease in VAS at 36 months compared to baseline, with 20% of participants who received saline reporting a 50% decrease in VAS at 36 months [30]. In another report of a randomized, controlled study, 12 study participants diagnosed with discogenic pain were injected with a preparation of 25 × $10^6$ cultured, allogeneic MSCs. Five of the 12 patients showed a rapid, but durable improvement in pain out to 1-year [32].

### USFDA Approved Product

A cultured cell containing product recently was approved by the USFDA for use in treating deep partial-thickness burns. Stratagraft® (Stratagraft Corporation, Madison, WI USA) is an allogeneic cellularized scaffold comprising a rat collagen scaffold that has been conditioned by human dermal fibroblasts and seeded with a cultured human-derived keratinocyte cell line (although the original formulation relied on a mouse-derived cell line), which means it is an allogeneic composite product [33]. The product is placed on the surgically debrided burn, which allows the patient's cells to migrate into and remodel the allograft over time. In a recent summary of a Phase 3 clinical study, StrataGraft® was able to accelerate wound closure in 83% of the study participants by month 3, without the need for autograft supplemental treatment. Reducing the need for using autograft, with its donor site cosmetic complications and potential wound healing issues, is an important objective in treating deep partial-thickness burns [33].

## Induced Pluripotent Stem Cells

Since publication of the first paper that described the creation of iPSCs in 2007, the field has exploded [34]. iPSCs were created by reprogramming adult, fully differentiated cells by a process that resulted in a cell that resembled an embryonic stem cell in many respects. In particular, iPSCs are able to proliferate in culture indefinitely and still retain the ability to differentiate into all of the cell types of the body. Originally, the attraction of iPSCs was that anyone's cells could be converted to a pluripotent stem cell, but the use of donor-derived material to create banks of iPSCs also is attracting attention [28].

It is important to note that iPSCs in their most primitive form can't be used directly in the body, due to the risk of generating teratomas. Consequently, before there can be broad clinical translation of iPSC-derived cell preparations, a number of technical details need to be sorted out. Furthermore, not all adult cells can be reprogrammed with the same efficiency [35]. It also is very important to understand what the differentiation target is when treating a pathology. For example, should the iPSCs be differentiated to a unipotent stem cell stage (i.e., endothelial progenitor cell) or just to the multipotent stem cell stage? The outcome will depend on an explicit knowledge of the intended use of the therapeutic cell preparation and a putative mechanism of action. The potential for the differentiated cell to raise an immune response in the recipient also needs to be considered [35]. There is a clinical study underway in which autologous iPSCs differentiated into an adult cell have been used in treating macular degeneration [34]. There also was a recent report that characterized a teratoma that was discovered in a study participant who had been injected with a therapeutic cell therapy derived from autologous iPSCs that had been differentiated to produce islet beta-cells. The immature teratoma was found at the site of the injection and was assessed as aggressive, and while the frequency of such reports is rare, as noted by the authors, nonetheless the use of cells differentiated from iPSCs always has a potential for teratoma formation [36].

## Cell Culture Fluid-Derived Products

Donor-derived materials like amniotic membranes have demonstrated therapeutic benefit for more than a century [37], which has led to an exploration of the role played by cells in the tissue. It was observed that cells isolated from the amniotic membrane of donated placentas when grown in culture secreted a wide variety of potent growth factors and proteins, and this type of cell culture-derived product currently is being studied for its potential therapeutic benefit in promoting wound healing [38]. Extracellular vesicles (EVs) released into the cell culture fluid also have shown a potential to play an important role in providing therapeutic benefits, as shown in *in vitro* and animal model studies [39]. The approach of using cultured cells to generate an enriched culture fluid is enticing given the opportunity to create a nearly infinite supply of iPSCs to grow in culture in order to harvest their supernatants. The real therapeutic potential of iPSCs may lie in the ability to differentiate iPSCs to a particular stage, culture the cells, and harvest the culture fluid containing a specific profile of exosomes, factors and/or proteins.

A product obtained from the spent culture fluid of multipotent cells obtained in a proprietary manner from donor amniotic tissue has been used to treat persistent corneal epithelial defects in an IND Phase 2 clinical study (Clinicaltrials.gov Identifier: NCT03687632). Results from a recent report on the Phase 2 clinical study indicated that after following a treatment protocol of daily eye drops of the test preparation for 28 days, all 12 study participants showed a statistically meaningful reduction in the size of their corneal epithelial defects when assessed after 1 week of no treatment, with five patients showing complete re-epithelialization. No drug-related adverse events or serious adverse events were reported [40].

## Exosome Containing Products

One of the most important advances in the area of non-autologous biologics is the on-going development and clinical evaluation of therapeutic preparations based on exosomes. Exosomes are an important part of cell-to-cell communication *in vivo*, and thus can be found in most fluids and tissues in the body [41]. Exosomes carry "cargo" that has been shown to influence target cells and plays a role in angiogenesis, cell proliferation and immune modulation [41]. Exosome containing products offer a cell-free therapy that can be customized to meet specific therapeutic objectives but shouldn't have issues with immune compatibility commonly associated with HLA-mismatched MSC therapies [42]. Donor tissue-derived exosome containing products have been evaluated, with a few clinical trials registered on Clinicaltrials.gov [43]. Exosomes isolated from 100 mL of human cord blood plasma (for each dose) have been used in a study on biomarkers of aging assessed before and after a series of intramuscular injections of the exosome preparation [44]. The study demonstrated that after 10 injections at weekly intervals, the study participants showed decreased levels of Immunoglobulin G (IgG), Immunoglobulin M (IgM) and Immunoglobulin A (IgA), with no change in Immunoglobulin E (IgE), compared to pre-treatment levels. Eosinophils and basophils were statistically meaningfully elevated, but there were no adverse or serious adverse events reported following treatment. Some "age-related" biomarkers were improved after treatment, but durability of the treatment wasn't stated [44]. In a proprietary process, exosomes were released from donor platelets placed in suspension and further purified to produce a therapeutic preparation containing $5 \times 10^{12}$ exosomes [45]. This therapeutic preparation currently is under investigation to treat skin graft donor site wounds (ClinicalTrials.gov Identifier: NCT04664738) and also as an adjunctive therapy following percutaneous coronary interventions (e.g., stent placement and dilation) to treat heart attacks (ClinicalTrials.gov Identifier: NCT04327635) [45]. The use of tissue culture to grow a variety of cells from which released exosomes can be isolated from spent culture media recently was reviewed [41].

In contrast to the clinical studies cited above, exosome containing products are being sold in the USA that are not approved for treating any pathology or disease. Invitrx, Inc. received a Warning Letter from the USFDA concerning their sale of exosome containing products, as well as for their umbilical cord-derived cell containing products [26]. The USFDA also noted in the Warning Letter sent to Stemell, Inc. [23] that they had been manufacturing exosome containing products and reminded them that exosomes are regulated as drugs and biological products, since they are materials that are sold for treating medical pathologies and conditions. Thus, the position of the USFDA is clear that exosomes from donor tissue-derived sources are drugs and biological products and subject to the IND/Biologics License Application procedure of the PMA process prior to being marketing.

## Mechanisms of Therapeutic Action

Despite the variety of forms that allogeneic products take, from cells to micronized tissue to culture fluid, the therapeutic mechanisms can be broadly broken down into those with a potential for providing an evolving therapeutic benefit and those that have a pre-formed therapeutic benefit. Borrowing from the autologous world, preparations like bone marrow concentrate can be considered to have an evolving therapeutic benefit, since they contain viable multipotent stromal/progenitor cells that are responsive to the microenvironment [46], whereas preparations like platelet-rich plasma (PRP) have a pre-formed therapeutic benefit due to the presence of alpha, dense and lysosomal granules in platelets that are released upon activation [47, 48]. EVs are not able to replicate [49], unlike progenitor cells, so EVs act as a pre-formed therapeutic agent—similar to the granules found in platelets. However, cargoes of EVs can vary depending on the signals in the microenvironment in which progenitor cells reside [49]. Thus, the role of EVs in providing a therapeutic benefit parallels that of the cells or platelets that release the EVs.

### Evolving Therapeutic Benefit

As with autologous bone marrow concentrate, preparations that contain viable stem/stromal cells are able to exert a therapeutic impact on a pathology as a result of the cells' sensing the microenvironment at the site of the pathology [49, 50]. For example, cultured cell products and products containing viable cells from amniotic fluid provide a therapeutic benefit as a direct result of the metabolic activity of the viable cells. One of the very important properties of cultured MSCs is their ability to modulate the immune response of the recipient. Thus, cultured allogeneic MSCs have shown a therapeutic benefit when used in patients suffering from Graft-versus-Host disease by diminishing the recipient's immune response, which has allowed the patients to recapitulate their bone marrow following an implant of HLA-matched cord blood or bone marrow [28].

In addition, there are a variety of growth factors and proteins released by MSCs that are able to influence adult cells present at the site of pathology. An extensive review of the various paracrine functions of MSCs with respect to anti-microbial, anti-inflammatory, and pro-angiogenic outcomes (among others) has been published [50]. In fact, there is growing evidence that the main therapeutic benefit of MSCs is related to their paracrine activities [27, 49, 50]. Although MSCs are thought to act primarily via the paracrine mechanism [49], MSCs also are able to differentiate into adult tissue cells and could aid in repair and recovery of damaged tissue by direct differentiation. For example, results obtained in animal studies involving placental-derived stem cells demonstrated that the implanted stem cells differentiated into cardiomyocytes in a rat myocardial infarct model [31]. Finally, the release of EVs by MSCs are able to amplify the therapeutic benefit of implanted MSCs, due to delivery of the EVs' cargo into target cells, since the contents of the EVs cargo is thought to act via the paracrine mechanism [51]. Importantly, implanted MSCs have been shown to respond to microenvironments and respond to changes in their microenvironment, which in turn could change the contents of the MSCs' EV cargo [49].

## Pre-formed Therapeutic Benefit

A variety of allogeneic products are thought to provide a therapeutic benefit primarily by providing growth factors in the absence of viable progenitor cells. For example, micronized placental tissue-derived products won't contain viable cells, but do retain a wide variety of structural proteins, enzyme inhibitors and growth factors [4]. Some micronized placental tissue-derived preparations were found to contain an excess of inhibitors of matrix metalloproteinases, which are enzymes that are responsible for degrading collagen-containing tissues. Thus, not only can allogeneic tissue implants directly act as a source of beneficial growth factors, which will influence local tissue cells, but agents like enzyme inhibitors are present that can reduce the rate of local tissue degradation, which in turn might reduce pro-inflammatory activity [4].

The presence of micronized intact tissue, albeit in very small particles, also might serve to stimulate cells that migrate into the tissue particles, since the architecture of the amniotic membrane is retained. This type of mechanism is thought to contribute to the wound healing properties of larger sheets of amniotic membranes [5]. There also is the possibility that EVs, especially smaller EVs, might survive the manufacturing processes to obtain sheets of placental tissue-derived products or micronized tissues, which if true, suggests that intact EVs present in the implanted allogeneic tissue could interact with and influence cells that migrate into the tissue grafts.

In the case of the recently USFDA-approved cellularized allograft, Stratagraft®, no implanted genetic material was observed at 3-months post-treatment, which was interpreted to mean that the study participant's cells had migrated into and completely remodeled the allograft. Thus, the biophysical properties of the cellularized allograft were able to attract and promote the proliferation of host cells, which subsequently remodeled the allograft into a durable closed wound [33].

Finally, culture fluid obtained from the *in vitro* culture of progenitor cells has been shown to be a rich source of growth factors and other proteins at an appropriate physiological level and with a therapeutically beneficial mixture of growth factors, proteins and exosomes associated with extracelluar matrix. Providing these factors in a localized site of pathology is thought to stimulate local tissue cells to respond [38].

## Conclusion

Multiple non-autologous biologic products have been shown to have regenerative potential, which include placental tissue associated products (e.g., amniotic membranes—sheets and micronized, and amniotic fluid) and a multitude of products under development, including cultured cells, cord blood/tissue-derived products, exosomes and iPSCs.

These materials are donor-derived, so facilities providing these products are registered with the USFDA and are produced under stringent manufacturing conditions defined under "cGMP." Regardless of the type of allogeneic product (cells, exosomes, tissues), there always is a risk that any donor-derived material could be contaminated with viral or microbial agents. Based in part on *in vitro* and pre-clinical studies, products containing exosomes hold great potential for providing a cell-free therapeutic treatment, and are the focus of intense commercialization efforts. Overall, there has been a dramatic rise in the use of injectable, donor tissue-derived products over the past 15 years.

The USFDA regulation dictating the handling of donor-derived materials is described under 21 CFR 1271 [3]. It is essential for physicians to inquire if the manufacturer is compliant with relevant USFDA regulations, either by contacting the Tissue Reference Group at the USFDA (email to: TissueReferenceGroup@fda.hhs.gov) or by reviewing regulatory-related documentation of a product's status provided by the manufacturer.

In summary, there is significant *in vitro* and pre-clinical evidence demonstrating the potential therapeutic value of placental tissue-derived materials, including amniotic fluid, exosomes and micronized tissues. There is a focus on addressing challenges in ensuring the safety of these products, along with establishing their clinical efficacy. Demonstration of the safety and efficacy of these products will enable manufacturers to meet the requirements of the PMA application process for drugs and biologic products to gain product approval. The marketing of these types of allogenic-derived preparations will provide an expanding array of therapeutic options for treating patients in order to mitigate pain and to possibly slow or reverse disease progression.

**Key Points**

1. Non-autologous biologics take many forms, including cells, exosomes, culture fluid, sheets of tissue, micronized tissue and composite products.
2. Although any donor tissue might be exploited for developing a non-autologous product, manufacturers generally have focused on using peri-natal tissues (placental, umbilical cord) and fluids (cord blood, amniotic fluid) for developing products for clinical studies.
3. Regulation of the marketing of non-autologous biologics depends on what the manufacturer does with the donor tissue, which will influence how the product ranks with the four components of 21 CFR 1271 (the primary document used by the USFDA for controlling all human-derived biological materials considered to be HCT/Ps).
4. Generally speaking, if a product is obtained from a sheet of tissue (e.g., amniotic membrane) but is injectable, that product will be considered to be a drug and biological product.
5. If a product contains viable cells, that product will be considered to be a drug and biological product.
6. Sheets of placental tissue for covering wounds and burns, and cadaveric nucleus pulposus matrix without viable cells are two types of non-autologous biologics that currently can be marketed without following the PMA process in the USA.
7. Dehydrated human amnion/chorion membrane products generally have shown therapeutic benefit in aiding wound closure and treating burns.
8. Micronized placental tissue products have shown therapeutic benefit in treating musculoskeletal pathologies like OA in joints and wound healing.
9. A potentially new application of allogeneic cell therapy is to combine the cell treatment with drug treatments that manage immune aspects of autoimmune diseases, in order to remodel the musculoskeletal sequalae and damage caused by the dysfunctional immune system.
10. Development of non-autologous biologics that incorporate exosomes presents an attractive opportunity for exploiting cell-free therapies.

**Questions**

1. Which of the following isn't associated with exosomes?
   A. Self-replication
   B. Binding to target cells
   C. Released by mesenchymal stem cells (MSCs)
   D. Carry "cargo"

   **Answer: A.**

   **Explanation:** Exosomes are unable to replicate, which means that parent cells, like mesenchymal stromal cells (MSCs), actively and continually produce exosomes internally, which subsequently are released to bind to target cells, and release their cargo into those cells.
2. Donor-derived nucleus pulposus matrix that has no viable cells is considered by the United States Food and Drug Administration (USFDA) to be what type of product?
   A. Excluded by definition from being considered as an Human Cells, Tissues, Cellular and Tissue-Based Products (HCT/P)
   B. An HCT/P that is regulated as a biological product
   C. An HCT/P regulated by 21 CFR 1271 only
   D. An HCT/P that is regulated as a drug

   **Answer: C.**

   **Explanation:** Nucleus pulposus matrix is not excluded by definition from 1271, but without viable cells it is solely regulated by the provisions of 1271, meaning the manufacturer is exempt from having to follow a Pre-market approval process prior to placing the product on the market.
3. A non-autologous biologic made from living cells is regulated in the USA according to what regulations?
   A. Public Health Services Act (PHSA) and the Federal Food, Drug, & Cosmetics (FD&C) Act
   B. 21 CFR 1271, the PHSA and the FD&C Act
   C. FD&C Act only
   D. 21 CFR 1271 only

   **Answer: B.**

   **Explanation:** A non-autologous biologic with viable cells is considered to be a drug (FD&C Act) and a biological product (PHSA) and needs to conform to most of the provisions in 21 CFR 1271.
4. Umbilical Cord mesenchymal stromal cells (UC-MSCs) have been found to have a therapeutic benefit in treating what sequalae for patients suffering from rheumatoid arthritis (RA)?
   A. Adverse drug reactions
   B. Immune dysfunction
   C. Elevated T-cells
   D. The musculoskeletal damage of rheumatoid arthritis (RA)

   **Answer: D.**

   **Explanation:** Umbilical Cord mesenchymal stromal cells (UC-MSCs) in a preliminary study were shown to reduce the physical manifestations of rheumatoid arthritis (RA), including swollen joints.
5. Fluid from cultured cells is a rich source of what potentially therapeutic agents?
   A. Cells

B. Proteins and cytokines
C. Membrane fragments
D. Glucose

**Answer: B.**

**Explanation:** Fluid from tissue culture has been processed to remove all cells or fragmented membranes, leaving secreted proteins and cytokines for further concentration to produce a therapeutic preparation.

6. Exosomes are most similar to what other therapeutic component?
A. Proteins
B. Mitochondria
C. Granules in platelets
D. Nuclei in MSCs

**Answer: C.**

**Explanation:** Exosomes most resemble granules in platelets, since upon platelet activation the granules release their contents, while exosomes are released and upon binding to a target cell release the contents of their cargo into the target cell.

7. Dehydrated Human Amnion/Chorion Membrane (dHACM) can be used in what form without USFDA approval?
A. Sheets that contain viable cells added back in
B. Sheets that are intended to cover wounds
C. Micronized tissue
D. Sheets that are intended to reduce pain

**Answer: B.**

**Explanation:** Dehydrated Human Amnion/Chorion Membrane (dHACM) in the form of sheets that are advertised as wound coverings are allowed to be marketed without a pre-market approval (PMA) process. Sheets that are advertised for pain reduction and/or scar mitigation will be regulated as a drug and biological product, since these outcomes aren't a homologous use of placental membranes, which act as a barrier or covering in utero. The process of micronization of this type of tissue exceeds the minimal manipulation standard for biological tissues, since the teeny tiny tissue pieces no longer can act as a barrier in a wound covering.

8. Which of the following are able to sense the microenvironment where they are implanted?
A. Exosomes
B. Proteins
C. Cytokines
D. Viable cells

**Answer: D.**

**Explanation:** Only viable cells are able to engage with and respond to the local environment into which they have been implanted. Exosomes will bind to cells, but aren't known to change their physiology in response to pro-inflammatory factors present in, for example, an arthritic knee, while mesenchymal stromal cells (MSCs) are known to actively mitigate a pro-inflammatory milieu.

9. Which of the following properties of a mesenchymal stromal cell (MSC) doesn't directly contribute to its paracrine function?
A. Release of exosomes
B. Release of cytokines and proteins
C. Viscosity of its cytoplasm
D. Receptors on its plasma membrane

**Answer: C.**

**Explanation:** A mesenchymal stromal cell's (MSC's) paracrine function is amplified by release of exosomes, can involve the direct release of cytokines and proteins and senses the microenvironment through the receptors on its outer plasma membrane as it continually updates the cell's paracrine profile. However, the cytoplasm viscosity of an MSC has yet to be directly linked to the paracrine function.

10. Which of the following isn't a beneficial property of induced pluripotent stem cells (iPSCs) when used for therapeutic treatments?
A. Nearly limitless supply of iPSCs can be made/maintained
B. Potential to form teratomas in vivo
C. Can be made by dedifferentiation of a patient's adult cells
D. Can be differentiated into less primitive cells for therapeutic purposes

**Answer: B.**

**Explanation:** Formation of teratomas is the primary danger of using induced pluripotent stem cells (iPSCs) as a source of starting cell material to be differentiated into targeted adult cells (i.e., less primitive cells) for therapeutic treatments. Sometimes the differentiation process isn't complete and some primitive cells remain that can form a teratoma.

**Acknowledgments** This book chapter is modified and updated from a previous book chapter, "Non-Autologous Biologics" by Theodore T. Sand, PhD, in *Essentials of Regenerative Medicine in Interventional Pain Management* published by ASIPP Publishing.

## References

1. Sand TT. Non-autologous biologics. In: Manchikanti L, Navani A, Atluri S, editors. Essentials of regenerative medicine in interventional pain management. Paducah, KY: ASIPP Publishing; 2019. p. 107–14.
2. Davis J. Skin transplantation with a review of 550 cases at the Johns Hopkins Hospital. Johns Hopkins Med J. 1910;15:307–96.
3. U.S. Government Publishing Office. Electronic code of federal regulations. Title 21, Chapter I, Subchapter L, Part 1271. In: Human cells, tissues, cellular and tissue-based products. Washington, DC: Government Publishing Office. Accessed 13 Mar 2023. https://

www.ecfr.gov/current/title-21/chapter-I/subchapter-L/part-1271.; 2023.

4. Lei J, Priddy LB, Lim JJ, et al. Identification of extracellular matrix components and biological factors in micronized dehydrated human amnion/chorion membrane. Adv Wound Care (New Rochelle). 2017;6:43–53.
5. Silini AR, Cargoni A, Magatti M, et al. The long path of human placenta, and its derivatives, in regenerative medicine. Front Bioeng Biotechnol. 2015;3:162.
6. Au AS, Leung WY, Stavosky JW. Efficacy of dehydrated human amnion chorion membrane in the treatment of diabetic foot ulcers: a review. J Am Podiatr Med Assoc. 2021;111:Article_13.
7. Ahuja N, Jin R, Powers C, et al. Dehydrated human amnion chorion membrane as treatment for pediatric burns. Adv Wound Care. 2019;9:602–11.
8. Zelen CM, Poka A, Andrews J. Prospective, randomized, blinded, comparative study of injectable micronized dehydrated amniotic/chorionic membrane allograft for plantar fasciitis—a feasibility study. Foot Ankle Int. 2013;34:1332–9.
9. USFDA. Untitled Letter issued to Surgical Biologics, (MiMedx Corporation), August 28, 2013; as referenced in The Niche, Knoepfler lab stem cell blog, MiMedx (MDXG) placental "stem cell magnet" drugs attract FDA attention and lawsuit, September 22, 2013. Accessed 13 Mar 2023. https://wayback.archive-it.org/6850/20161105113627/http://www.ipscell.com/tag/amniofixtm-injectable/.
10. Tettelbach W, Cazzell S, Reyzelman AM, et al. A confirmatory study on the efficacy of dehydrated human amnion/chorion membrane dHACM allograft in the management of diabetic foot ulcers: a prospective, multicentre, randomised, controlled study of 110 patients from 14 wound clinics. Int Wound J. 2019;16:19–29.
11. Hanselman AE, Tidwell JE, Santrock RD. Cryopreserved human amniotic membrane injection for plantar fasciitis: a randomized, controlled, double-blind pilot study. Foot Ankle Int. 2015;36:151–8.
12. Alden KJ, Harris S, Hubbs B, et al. Micronized dehydrated human amnion chorion membrane injection in the treatment of knee osteoarthritis—a large retrospective case series. J Knee Surg. 2021;34:841–5.
13. Loukogeorgakis SP, De Coppi P. Stem cells from amniotic fluid—potential for regenerative medicine. Best Prac Res Clin Obstet Gynaecol. 2016;31:45–57.
14. Dziadosz M, Basch RS, Young BK. Human amniotic fluid: a source of stem cells for possible therapeutic use. Am J Obst Gynecol. 2016;214:321–7.
15. Panero AJ, Hirahara AM, Anderson WJ, et al. Are amniotic fluid products stem cell therapies? A study of amniotic fluid preparations for mesenchymal stem cells with bone marrow comparison. Am J Sports Med. 2019;47:1230–5.
16. USFDA. Warning letter issued to BioLab sciences, Inc. 2022. Accessed 13 Mar 2023. https://www.fda.gov/inspections--compliance-enforcement-and-criminal-investigations/warning-letters/biolab-sciences-inc-621465-08232022.
17. Meadows MC, Elisman K, Nho SJ, et al. A single injection of amniotic suspension allograft is safe and effective for treatment of mild to moderate hip osteoarthritis: a prospective study. Arthrosc J Arthrosc Related Surg. 2022;38:325–31.
18. Beall DP, Wilson GL, Bishop R, et al. VAST clinical trial: safely supplementing tissue lost to degenerative disc disease. Int J Spine Surg. 2020;14:239–53.
19. Ng AP, Alexander WS. Haematopoietic stem cells: past, present and future. Cell Death Discov. 2017;3:17002.
20. Xie QX, Liu R, Jiang J, et al. What is the impact of human umbilical cord mesenchymal stem cell transplantation on clinical treatment? Stem Cell Res Ther. 2020;11:519.
21. Wang LM, Huang SG, Li SM, et al. Efficacy and safety of umbilical cord mesenchymal stem cell therapy for rheumatoid arthritis patients: a prospective phase I/II study. Drug Design Dev Ther. 2019;13:4331–40.
22. USFDA. Warning letter issued to smart surgical, Inc dba Burst Biologics. 2022. Accessed 13 Mar 2023. https://www.fda.gov/inspections-compliance-enforcement-and-criminal-investigations/warning-letters/smart-surgical-inc-dba-burst-biologics-614361-02022022.
23. USFDA. Warning letter issued to Stemell, Inc. 2019. Accessed 13 Mar 2023. https://www.fda.gov/inspections-compliance-enforcement-and-criminal-investigations/warning-letters/stemell-inc-579013-08282019.
24. USFDA. Warning letter issued to Cord for life, Inc. 2019. Accessed 13 Mar 2023. https://www.fda.gov/inspections-compliance-enforcement-and-criminal-investigations/warning-letters/cord-life-inc-572770-03292019.
25. USFDA. Warning letter issued to Surgenex, LLC. 2021. Accessed 13 Mar 2023. https://www.fda.gov/inspections-compliance-enforcement-and-criminal-investigations/warning-letters/surgenex-llc-615254-11152021.
26. USFDA. Warning letter issued to Invitrx therapeutics, Inc; November 9, 2022. Accessed 13 Mar 2023. https://www.fda.gov/inspections-compliance-enforcement-and-criminal-investigations/warning-letters/invitrx-therapeutics-inc-630712-11092022.
27. Freitag J, Bates D, Boyd R, et al. Mesenchymal stem cell therapy in the treatment of osteoarthritis: reparative pathways, safety and efficacy—a review. BMC Musculoskelet Disord. 2016;17:230.
28. Girdlestone J. Mesenchymal stromal cells with enhanced therapeutic properties. Immunotherapy. 2016;8:1405–16.
29. Murphy SV, Atala A. Amniotic fluid and placental membranes: unexpected sources of highly multipotent cells. Semin Reprod Med. 2013;31:62–8.
30. Amirdelfan K, Bae H, McJunkin T, et al. Allogeneic mesenchymal precursor cells treatment for chronic low back pain associated with degenerative disc disease: a prospective randomized, placebo-controlled 36-month study of safety and efficacy. Spine J. 2021;21:212–30.
31. Antoniadou E, David AL. Placental stem cells. Best Pract Res Clin Obstet Gynaecol. 2016;31:13–29.
32. Noriega DC, Ardura F, Hernández-Ramajo R, et al. Intervertebral disc repair by allogeneic mesenchymal bone marrow cells: a randomized controlled trial. Transplantation. 2017;101:1945–51.
33. Gibson ALF, Holmes JH IV, Shupp JW, et al. A phase 3, open-label, controlled, randomized, multicenter trial evaluating the efficacy and safety of StrataGraft® construct in patients with deep partial-thickness thermal burns. Burns. 2021;47:1024–37.
34. Kumar D, Anand T, Kues WA. Clinical potential of human-induced pluripotent stem cells: perspectives of induced pluripotent stem cells. Cell Biol Toxicol. 2017;33:99–112.
35. Khazaei M, Ahuja CS, Fehlings MG. Induced pluripotent stem cells for traumatic spinal cord injury. Front Cell Dev Biol. 2017;4:152.
36. Han L, He H, Yang YH, et al. Distinctive clinical and pathologic features of immature teratomas arising from induced pluripotent stem cell-derived beta cell injection in a diabetes patient. Stem Cells Dev. 2022;31:97–101.
37. Gupta A, Kedige SD, Jain K. Amnion and chorion membranes: potential stem cell reservoir with wide applications in periodontics. Int J Biomater. 2015;2015:1–9.
38. Steed DL, Trumpower C, Duffy D, et al. Amnion-derived cellular cytokine solution: a physiological combination of cytokines for wound healing. Eplasty. 2008;8:e18.
39. Kolios G, Paspaliaris V. Mesenchyme stem cell-derived conditioned medium as a potential therapeutic tool in idiopathic pulmonary fibrosis. Biomedicine. 2022;10:2298.
40. Jeng BH, Hamrah P, Kirshner ZZ, et al. Exploratory phase II multicenter, open-label, clinical trial of ST266, a novel secretome for

treatment of persistent corneal epithelial defects. Transl Vis Sci Technol. 2022;11:8.
41. Costa A, Quarto R, Bollini S. Small extracellular vesicles from human amniotic fluid samples as promising theranostics. Int J Mol Sci. 2022;23:590.
42. Liu ZC, Zhuang Y, Fang LF, et al. Breakthrough of extracellular vesicles in pathogenesis, diagnosis and treatment of osteoarthritis. Bioactive Mater. 2023;22:423–52.
43. Rezaie J, Feghhi M, Etemadi T. A review on exosomes application in clinical trials: perspective, questions, and challenges. Cell Comm Signaling. 2022;20:145.
44. Clement J, Yan Q, Agrawal M, et al. Umbilical cord plasma concentrate has beneficial effects on DNA methylation GrimAge and human clinical biomarkers. Aging Cell. 2022;21:e13696.
45. Shi A, Li JL, Qiu XY, et al. TGF-β loaded exosome enhances ischemic wound healing in vitro and in vivo. Theranostics. 2021;11:6616–31.
46. Caplan AI. New MSC: MSCs as pericytes are sentinels and gatekeepers. J Orthop Res. 2017;35:1151–9.
47. Ostermeier B, Soriano-Sarabia N, Maggirwar SB. Platelet-released factors: their role in viral disease and applications for extracellular vesicle (EV) therapy. Int J Mol Sci. 2022;23:2321.
48. Manne BK, Xiang SC, Rondina MT. Platelet secretion in inflammatory and infectious diseases. Platelets. 2017;28:155–64.
49. McLaughlin C, Datta P, Singh YP, et al. Mesenchymal stem cell-derived extracellular vesicles for therapeutic use and in bioengineering applications. Cell. 2022;11:3366.
50. Murphy MB, Moncivais K, Caplan AI. Mesenchymal stem cells: environmentally responsive therapeutics for regenerative medicine. Exp Mol Med. 2013;45:e54.
51. Tsiapalis D, O'Driscoll L. Mesenchymal stem cell derived extracellular vesicles for tissue engineering and regenerative medicine applications. Cell. 2020;9:991.

# The Role of Exosomes in Regenerative Medicine

# 13

Sheldon Jordan and Jonathan Haroon

## Introduction

"Regenerative therapy is a microcosm of medicine focused on recapitulating the conditions of our youth in order to restore, rather than repair, morphological and physiological aberrancies as they appear. As a medical society, it is far more attractive to offer our patients the hope of regeneration in order to afford them the ability to restore a younger version of themselves rather than just repairing their injuries. Whether rehabilitating a disc, a joint, a nerve or the entire individual the dream has always been to live longer and better" [1].

One of the main aims of regenerative medicine is to find ways to stimulate the body's own regenerative processes, such as the natural ability of some tissues to repair themselves after injury. Researchers in this field are exploring a wide range of potential treatments, including the use of stem cells to regenerate damaged tissues, the development of biomaterials that can support the growth of new cells, and the use of gene editing technologies to correct genetic defects that cause diseases.

The use of exosomes in regenerative medicine is a rapidly growing area of research, with the potential to improve the treatment of a range of diseases and conditions. In this chapter, we will discuss what exosomes are, how they are produced and released by cells, the advantages and disadvantages and their potential applications in regenerative medicine.

S. Jordan (✉) · J. Haroon
Neurological Associates of West Los Angeles, Santa Monica, CA, USA
e-mail: sjordan@theNeuroAssociates.com; jharoon@theNeuroAssociates.com

## What Are MSCS and Exosomes?

Medicinal signaling cells (MSCs) and exosomes are two different elements that need to be defined. MSCs are involved in the body's response to illness or injury [2]. These cells help to coordinate the body's immune and repair response partly by releasing exosomes in a targeted zone.

Exosomes, on the other hand, are small vesicles that are released by cells, including MSCs, and play a role in intercellular communication [3]. In 1998, it was shown in a study that dendritic cells also secrete exosomes. These findings supported the idea that exosomes can actively contribute to intercellular communication [4]. Exosomes are also involved in the transfer of molecules such as proteins and ribonucleic acid (RNA) from one cell to another, helping to regulate various cellular processes, such as the immune response [5], and cell repair [6].

MSCs and exosomes differ in size. MSCs are much larger than exosomes. The latter are typically only 30–100 nanometers in diameter [7]. A review published in 2021 classifies extracellular vesicles (EVs) into three groups: exosomes (30–200 nm), microvesicles (MVs) (100–1000 nm) and apoptotic bodies (>1000 nm) [8]. Additionally, while MSCs are involved in the body's response to illness or injury, exosomes released from other cells may have other ways of affecting intercellular communication [9].

## Benefits of Extracellular Vesicles

Exosomes are created by a double invagination of the extracellular membrane and are therefore composed of a phospholipid bilayer embedded with lipids and proteins [10].

One of the key functions of exosomes is to transfer biomolecules between cells. This ability makes them attractive as potential therapeutic agents, as they can be used to deliver therapeutic molecules to specific cells in the body [11, 12]. For example, exosomes have been used to deliver small


A. Navani et al. (eds.), *Essentials of Regenerative Medicine in Interventional Pain Management*,
https://doi.org/10.1007/978-3-031-50357-3_13

interfering RNA molecules to target cells, which can silence specific genes and modulate cell function. In a 2012 study, exogenous short interfering RNA (siRNA) was delivered to human blood cells using plasma exosomes (exosomes from peripheral blood), which were used as a gene delivery vector (GDV). Mitogen-activated protein kinase 1's gene was selectively silenced because of the vesicles' efficient delivery of the injected siRNA into monocytes and lymphocytes. According to these findings, human exosomes may be employed as a GDV to deliver heterologous nucleic acids to cells, such as therapeutic siRNAs [13].

Prior to release, exosomes are packaged with cargo that includes a variety of nucleic acid and protein. Non-coding RNA is an important source of epigenetic control that is provided by a target cell's absorption of exosome material. For example, non-coding RNA has been shown to have a significant effect in tumor metastasis, in part through their direct interaction with proteins [14].

Regenerative medicine aims to take advantage of this epigenetic effect while promoting repair and regeneration in targeted tissues [15, 16].

## Difference Between Stem Cells and Exosomes in Regenerative Medicine Application

Stem cells and exosomes are both types of biological agents that have potential uses in regenerative medicine. However, there are some key differences between the two that may have a potential differential effect in outcome.

One major disadvantage of stem cells is that they may not be able to survive the freeze and thaw cycle involved in deployment [17]. A recent review has estimated that 60–70% of viability and function can be lost in MSC post-thaw recovery; additionally, changes in cryopreservation can result in chromosomal damage and epigenetic changes [18]. By contrast, exosomes are more likely to maintain their therapeutic effectiveness while going through a freeze/thaw process [19].

For stem cells, another theoretical disadvantage of stem cells is that depending on the source there is a potential for transformation for neoplastic growth. On the other hand, exosomes are devoid of a capacity for potential generation of tumor growth.

Stem cells do have the capacity for motility, and therefore they may have a better ability to target sites within tissue [20]. While exosomes can target, they do so through passive diffusion, which may ultimately make a difference in effective targeting [21, 22]. In addition, stem cells typically will reside in a target zone for a time frame of up to several weeks, in which case they are releasing exosomes over a long-time frame, which may have certain advantages [23]. Preclinical studies utilizing MSCs in rat ischemia models have demonstrated that MSCs release a variety of growth factors, including brain-derived neurotrophic factor (BDNF), vascular endothelial growth factor (VEGF), glial cell-derived neurotrophic factor (GDNF), and transforming growth factor (TGF) [24, 25].

Exosomes, on the other hand, also release a variety of growth factors, as well as the previously mentioned RNAs and iRNAs [26]. However, exosomes are a one-time effect and for prolonged and repetitive effects, infusion may need to be repeated over the course of a certain time frame.

## Extracellular Vesicles in Musculoskeletal Regeneration

Exosomes have been shown to have potential therapeutic benefits for musculoskeletal injuries. For example, exosomes can be isolated from MSCs and used to deliver therapeutic molecules to injured tissues, as these molecules can help to promote the growth of new blood vessels, reduce inflammation, and stimulate the production of new cells, which can help to improve tissue repair and accelerate the healing process.

Recent studies have suggested that exosomes may have the capacity to improve symptoms of Sarcopenia, or the age-related decline of muscle function and mass. Sarcopenia affects more than 50 million people today and will affect more than 200 million in the next 40 years, even with a modest estimate of prevalence. Sarcopenia has a significant negative impact on older persons, and its effects are assessed in terms of death, morbidity, disability, and excessive healthcare expenditures [27].

Despite the plethora of clinical trials have been registered to assess the potential of sEV-based diagnostic and therapeutic potential in a variety of fields, including cancer diagnosis and therapy, infection, SARS-CoV-2, psychiatry, dermatologic applications, and neurodegenerative conditions, including Parkinson's and Alzheimer's disease, none of the trials have been registered to examine their role in orthopedic applications.

### Arthritis

Osteoarthritis (OA), a prevalent painful condition marked by chronic articular cartilage deterioration, is the leading contributor to elderly incapacity. The incidence rate of OA rises yearly due to the acceleration of global aging. The three main approaches to treating OA now are surgery, pharmaceutical treatments, and nonpharmacological treatments. As the main form of treatment for early OA, non-pharmacological approaches such as patient education, physical therapy, and exercise are advised. Despite providing momentary symptom alleviation, these treatments do not address the pathophysiology of OA or stop its development [28].

Exosomes have shown significant promise in applications such as knee OA and cartilage restoration, and several preclinical studies have been published using exosomes in animal models of OA [29, 30].

TGF, interleukin-1 (IL-1), interleukin-6 (IL-6), and tumor necrosis factor-a (TNF-a) are among the signaling pathways connected to OA that are controlled by micro ribonucleic acid (miRNA) genes, and these miRNA genes have been shown to be able to be modulated by MSC derived exosomes, leading to a reduction in inflammation and pain symptoms from knee OA [31].

Additionally, previous research has implicated chondrogenesis and chondrocyte metabolism as a major contributing factor to OA [32]. A recent preclinical study demonstrated that exosomes could block the release of TNF-a, enhance OA chondrocyte proliferation, and drastically lower the OA chondrocyte apoptosis rate in OA chondrocytes [33]. Another similar study demonstrated that MSC-exosomes were able to inhibit the pathogenesis of OA by inhibiting syndecan-1, a transmembrane proteoglycan that can bind a variety of cytokines and modulate their activity [34].

## Tendonitis

Tendon regeneration is another crucial area which may be served by exosome research, as 50% of musculoskeletal ailments are diagnosed as ligament and tendon injuries [35]. Many tendon and ligament injuries, like simple sprains and strains, recover on their own without the need for surgery, though surgery is frequently needed to address more severe injuries including whole tendon or ligament tears. The effectiveness of these surgical procedures varies. Although reconstructive techniques have been quite effective, these treatments frequently lead to the development of inferior scar tissue, which is more vulnerable to re-rupture, among other issues [36].

A 2018 preclinical study examined tendon stem cells (TSCs) with a chitosan/b-glycerophosphate/collagen hydrogel in rat Achilles tendinopathy models. The study demonstrated that in comparison to the control group, hydrogel, and TSCs groups, the combination of TSCs and C/GP/Co hydrogel improves tendon repair. The improvement in histology and immunohistochemistry results as well as the rise in the biomechanical qualities of the regenerated tissue at both 4 and 6 weeks after the injury were indicators of the enhanced healing [37]. Another study found that the injection of TSCs *and* exosomes individually greatly reduced the production of matrix metalloproteinases (MMPs) and raised the expression of tissue inhibitor of metalloproteinases (TIMPs) in rat Achilles tendinopathy models [38].

## Bone Fractures

Exosomes also have been examined for their potential in the regeneration of bone defects and injuries. Specifically, exosomes have shown promise in the healing osteoporosis (OP), as well as nonunion, a common complication after bone fracture [39]. Because they primarily differentiate into osteoblasts and chondrocytes, bone marrow mesenchymal stem cells (BMSCs) have significant influence on the production of bone and cartilage and have been examined for their orthopedic applications.

Numerous studies have demonstrated that exosomes produced by bone BMSCs could improve OP by inducing osteoblast activation, differentiation, and proliferation as well as preventing cell death [40, 41]. Additionally, some research suggested that exosomes derived from BMSCs with the goal of improving osteoclast activity and maturation could be beneficial for OP [42]. Regarding non-union healing of fractures, it may be possible to enhance BMSCs propensity to proliferate and differentiate into osteoblasts and chondrocytes, which also play crucial roles in fracture healing [43].

Regarding bone fracture applications, the most pressing hurdle to overcome is the clarification of the functions of exosomes in the occurrence and healing of osteoporotic fractures.

## Muscle Conditions

One of the human body's most important tissues is skeletal muscle. Skeletal muscle in humans makes up about 40% of the total weight of the body and includes 50–75% of all the proteins [44]. As such, musculoskeletal anomalies, traumatic traumas, or muscle loss may negatively affect many aspects of health. Multiple pharmacologic medicines have been used in skeletal muscle regeneration treatments; however, the lack of functional improvement and unanticipated complications have prevented the practical application of these drugs in muscle regeneration [45].

Cell-based muscle regeneration therapies are beginning to be extensively explored, as exosomes are thought to play a key role in myogenesis and muscle formation as carriers of proteins, mRNA, miRNA, and other non-coding RNAs. Exosomes generated from human skeletal myoblasts (HSkM) that carry a variety of signal molecules can help muscle regeneration, as demonstrated in a mouse model of muscle injury. Exosomes may also improve muscle protein synthesis and hypertrophy [46].

A preclinical study using exosomes derived from adipose-tissue derived MSCs (AD-MSCs) demonstrated a muscle-regenerating effect in muscle defect models in rodents [47]. This study utilized AD-MSCs for their myogenic lineage, and greater capacity for rapid expansion when compared to BMSC-Exosomes, as demonstrated in a previous study [48]. The study by Byun et al. concluded that systemic infusion of AD-MSCs or AD-MSC-derived exosomes may be an alternative to the pharmaceutical strategy, which did not show functional improvement, as a treatment for this muscle loss.

However, BMSCs may hold promise in prevention of muscle loss, as a recent study found that bone marrow mesenchymal stem cell derived exosomes (BMSC-Exos) protected against the atrophy of skeletal muscle by attenuating the shrinkage of myotube diameter [49].

In terms of therapeutic applications, creating methods to collect enough exosomes is a significant difficulty. Only a small number of exosomes are produced by current exosome isolation techniques like ultracentrifugation and ultrafiltration [50].

## Exosomes in Neurological Regeneration

In neurological applications, the method for the delivery of small molecules which has gained interest is intranasal administration. One recent preclinical study examined the use of intranasal exosomes for regeneration of spinal cord injury. After immunological staining, it was confirmed that MSC-exosomes were able to pass through the blood brain barrier (BBB) when administered intranasally in rats. Additionally, this resulted in considerable improvements in motor function, sensory recovery, and speedier urine reflex restoration. Increased axonal regeneration and angiogenesis, decreased neuroinflammation and gliosis, improved structural and electrophysiological conditions, were all shown to have contributed to the functional recovery [51]. Another preclinical study examining intranasal exosomes in rat models of Autism showed an improvement in social interactions, as well as a reduction of repetitive behaviors. These two behavior patterns are hallmark features of autism-spectrum disorders [52].

Focused ultrasound has also become an area of focus for non-invasive delivery of molecules. It has been demonstrated that focused ultrasound, at a variety of differing intensities, has the capacity to transiently open the BBB. In animal models, it has been demonstrated that using microbubbles intravenously delivered in combination with magnetic resonance-guided focused ultrasound temporarily opens the blood-brain barrier [53, 54]. Additionally, a study examining BBB permeability after focused ultrasound both with and without microbubbles demonstrated that the BBB was also transiently opened without microbubbles, at a frequency of 1.47 mHz [55]. This finding is important, as focused ultrasound paired with microbubbles increases the chance for adverse events such as hemorrhage and brain bleeding and being able to open the BBB without the need for microbubbles decreases the overall risk of the procedure.

In a study by Lipsman et al. [56], five human individuals with early to moderate Alzheimer's disease had the blood-brain barrier opened using focused ultrasound. The blood-brain barrier was repeatedly, safely, and reversibly opened in the target volume in all patients. There were no notable clinical or radiographic side effects from breaking down the blood-brain barrier, and there was no clinically significant decline in cognitive scores at 3 months compared to baseline [56].

As methods of delivery through the BBB become more available, research into the positive effects of targeted delivery of exosomes to specific brain regions have sparked new methods of regeneration. A study published in Nature discussed the role of the hypothalamus in aging; specifically, Zhang et al. [57] implicate hypothalamic stem cells in aging, and identified that this process is partly controlled by exosome miRNAs. Additionally, this study showed that hypothalamic stem cell derived exosomes, when injected intracranially into the hypothalamus of aged rats, slowed the aging process [57]. This study suggests that, if appropriately targeted, some of the negative effects of aging may be ameliorated via exosome influence on the hypothalamus in humans.

**Key Points**

1. The main aim of regenerative medicine is to find ways to stimulate the body's own regenerative processes, such as the natural ability of some tissues to repair themselves after injury.
2. Exosomes are small vesicles released by cells, including MSCs, that play a role in intercellular communication and transfer of molecules such as proteins and RNA from one cell to another.
3. Exosomes have the potential to be used as therapeutic agents due to their ability to transfer biomolecules between cells, making them useful for delivering therapeutic molecules to specific cells in the body.
4. Exosomes have been used to deliver small interfering RNA molecules which can silence specific genes and modulate cell function.
5. Non-coding RNA provided by a target cell's absorption of exosome material has a significant effect in tumor metastasis and regenerative medicine aims to take advantage of this epigenetic effect to promote repair and regeneration in targeted tissues.
6. Exosomes have potential therapeutic benefits for musculoskeletal injuries, including promoting growth of new blood vessels, reducing inflammation, and stimulating

cell production to improve tissue repair and accelerate healing.
7. Studies have suggested exosomes may have the capacity to improve symptoms of Sarcopenia, an age-related decline of muscle function and mass that affects millions of people.
8. OA is a prevalent condition marked by chronic cartilage deterioration, and is a leading contributor to elderly incapacity.
9. Exosomes have shown promise in applications such as knee OA and cartilage restoration in preclinical studies.
10. Tendon regeneration is another area where exosome research may be beneficial, as many tendon and ligament injuries can lead to the development of inferior scar tissue which can be more vulnerable to re-rupture.

**Questions**

1. What are exosomes?
   A. Small vesicles that are released by cells and play a role in intercellular communication
   B. Medicinal signaling cells that are involved in the body's response to illness or injury
   C. Biomaterials that can support the growth of new cells
   D. Gene editing technologies to correct genetic defects that cause diseases

   **Answer: A**

   **Explanation:** Small vesicles that are released by cells and play a role in intercellular communication Exosomes are small extracellular vesicles that are released by cells and can be used for therapeutic purposes.
2. Medicinal signaling cells (MSCs) are similar to exosomes in all of the following ways, except for their:
   A. Ability to home to an area of inflammation when given intravenously
   B. Ability to be frozen and thawed for remote deployment
   C. Motility
   D. Ability to persist locally when injected into an inflamed joint

   **Answer: C**

   **Explanation:** Stem cells demonstrate an ability to move from place to place, but exosomes merely travel by way of diffusion passively.
3. Exosomes generally demonstrate the following characteristics except for:
   A. Can transfer antiviral elements
   B. Contain large quantities of DNA
   C. Tend to get trapped in the lung when given intravenously
   D. Contain mitochondria

   **Answer: B**

   **Explanation**: Exosomes typically do not contain notable quantities of DNA, if any at all. They contain RNA (either messenger RNA or noncoding RNA).
4. In the preparation of platelet rich plasma, the post-centrifugation supernatant contains enlarged quantities of all of the following except for:
   A. Exosomes
   B. Alpha-2-Macroglobulin
   C. Fibrinogen
   D. Platelets

   **Answer: D**

   **Explanation**: Platelets are the supernatent in the precipitating fraction, as the purpose is the separation of platelet for injection.
5. Exosomes are found in which of the following fluids:
   A. Cow's milk
   B. Normal saline
   C. Distilled water
   D. Human serum albumin solution

   **Answer: A**

   **Explanation**: Exosomes are indeed found in cow's milk, as part of certain secretions.
6. Exosomes are injected into a knee join for the treatment of an instance of osteoarthritis; a maximum benefit would be expected in which of the following timeframes:
   A. Immediately (Up to 1-hour post-injection)
   B. 1 day post-injection
   C. 1 month post-injection
   D. 6-months post-injection

   **Answer: C**

   **Explanation:** There is a delay in the onset of maximum benefit for exosomes, as exosomes work in part by way of epigenetic reprogramming and via growth factors.
7. After intra-articular injection of exosomes, which additional interventions is likely to keep the exosomes within the target region?
   A. Pre-treatment with shockwave therapy 20-minutes prior to injection
   B. Laser light therapy 2-hours post injection
   C. Pre-treatment with steroids
   D. Continued treatment with non-steroidal anti-inflammatory agents

   **Answer: A**

   **Explanation:** Induction of an inflammatory reaction to a given site prior to administration has been shown to increase the amount of time the exosomes will persist in a specific area.
8. How do MSCs and exosomes differ in size?
   A. MSCs are larger than exosomes
   B. Exosomes are larger than MSCs
   C. They are the same size
   D. There is no difference in size

**Answer: A**

**Explanation:** Exosomes are smaller in diameter than MSCs, which is important to consider when determining the location in which to administer treatment.

9. What are the three groups of extracellular vesicles?
   A. Exosomes, microvesicles, and apoptotic bodies
   B. Microvesicles, exosomes, and therapeutic siRNA
   C. Apoptotic bodies, exosomes, and microvesicles
   D. Exosomes, therapeutic siRNA, and apoptotic bodies

**Answer: A**

**Explanation**: It is important to distinguish that while exosomes are technically extracellular vesicles, microvesicles and apoptotic bodies also fall under this category. It is therefore necessary to be specific when discussing 'extracellular vesicles'.

10. What is currently **not** one of the main forms of treatment for Osteoarthritis (OA)?
   A. Surgery
   B. Pharmaceutical treatments
   C. Non-pharmacological treatments such as patient education, physical therapy, and exercise
   D. Exosome therapy

**Answer: D**

**Explanation:** While research has demonstrated that exosome therapy shows promise for the treatment of OA, it is not currently one of the main forms of treatment for the condition.

**Acknowledgments** This book chapter is modified and updated from a previous book chapter, "The Role of Exosomes in Regenerative Medicine" by Doug Spiel, MD, in *Essentials of Regenerative Medicine in Interventional Pain Management* published by ASIPP Publishing.

## References

1. Spiel D. The role of exosomes in regenerative medicine. In: Manchikanti L, Navani A, Atluri S, editors. Essentials of regenerative medicine in interventional pain management. Paducah, KY: ASIPP Publishing; 2019. p. 115–28.
2. Caplan AI. Mesenchymal stem cells: time to change the name! Stem Cells Transl Med. 2017;6:1445–51.
3. Théry C. Exosomes: secreted vesicles and intercellular communications. F1000 Biol Rep. 2011;3:15.
4. Zitvogel L, Regnault A, Lozier A, et al. Eradication of established murine tumors using a novel cell-free vaccine: dendritic cell-derived exosomes. Nat Med. 1998;4:594–600.
5. Schwarzenbach H, Gahan PB. Exosomes in immune regulation. Noncoding RNA. 2021;7:4.
6. Basu J, Ludlow JW. Exosomes for repair, regeneration and rejuvenation. Expert Opin Biol Ther. 2016;16:489–506.
7. Gowen A, Shahjin F, Chand S, et al. Mesenchymal stem cell-derived extracellular vesicles: challenges in clinical applications. Front Cell Dev Biol. 2020;8:149.
8. Gurung S, Perocheau D, Touramanidou L, et al. The exosome journey: from biogenesis to uptake and intracellular signalling. Cell Commun Signal. 2021;19:47.
9. Rustad KC, Gurtner GC. Mesenchymal stem cells home to sites of injury and inflammation. Adv Wound Care (New Rochelle). 2012;1:147–52.
10. Skotland T, Sandvig K, Llorente A. Lipids in exosomes: current knowledge and the way forward. Prog Lipid Res. 2017;66:30–41.
11. Vader P, Mol EA, Pasterkamp G, Schiffelers RM. Extracellular vesicles for drug delivery. Adv Drug Deliv Rev. 2016;106:148–56.
12. Fuhrmann G, Serio A, Mazo M, et al. Active loading into extracellular vesicles significantly improves the cellular uptake and photodynamic effect of porphyrins. J Control Release. 2015;205:35–44.
13. Wahlgren J, Karlson DLT, Brisslert M, et al. Plasma exosomes can deliver exogenous short interfering RNA to monocytes and lymphocytes. Nucleic Acids Res. 2012;40:e130.
14. Wang D, Zhang W, Zhang C, et al. Exosomal non-coding RNAs have a significant effect on tumor metastasis. Mol Ther Nucleic Acids. 2022;29:16–35.
15. Behbahani GD, Khani S, Hosseini HM, et al. The role of exosomes contents on genetic and epigenetic alterations of recipient cancer cells. Iran J Basic Med Sci. 2016;19:1031–9.
16. Mansouri V, Beheshtizadeh N, Gharibshahian M, et al. Recent advances in regenerative medicine strategies for cancer treatment. Biomed Pharmacother. 2021;141:111875.
17. Hunt CJ. Cryopreservation of human stem cells for clinical application: a review. Transfus Med Hemother. 2011;38:107–23.
18. Hunt CJ. Technical considerations in the freezing, low-temperature storage and thawing of stem cells for cellular therapies. Transfus Med Hemother. 2019;46:134–50.
19. Cheng Y, Zeng Q, Han Q, et al. Effect of pH, temperature and freezing-thawing on quantity changes and cellular uptake of exosomes. Protein Cell. 2019;10:295–9.
20. Danielyan L, Schwab M, Siegel G, et al. Cell motility and migration as determinants of stem cell efficacy. EBioMedicine. 2020;60:102989.
21. Wang J, Yeung BZ, Cui M, et al. Exosome is a mechanism of intercellular drug transfer: application of quantitative pharmacology. J Control Release. 2017;268:147–58.
22. Sancho-Albero M, Navascués N, Mendoza G, et al. Exosome origin determines cell targeting and the transfer of therapeutic nanoparticles towards target cells. J Nanobiotechnol. 2019;17:16.
23. Zhou G, Wang Y, Gao S, et al. Potential mechanisms and perspectives in ischemic stroke treatment using stem cell therapies. Front Cell Dev Biol. 2021;9:646927.
24. Wakabayashi K, Nagai A, Sheikh AM, et al. Transplantation of human mesenchymal stem cells promotes functional improvement and increased expression of neurotrophic factors in a rat focal cerebral ischemia model. J Neurosci Res. 2010;88:1017–25.
25. Wagenaar N, de Theije CGM, de Vries LS, et al. Promoting neuroregeneration after perinatal arterial ischemic stroke: neurotrophic factors and mesenchymal stem cells. Pediatr Res. 2018;83:372–84.
26. Yin K, Wang S, Zhao RC. Exosomes from mesenchymal stem/stromal cells: a new therapeutic paradigm. Biomark Res. 2019;7:8.
27. Santilli V, Bernetti A, Mangone M, et al. Clinical definition of sarcopenia. Clin Cases Miner Bone Metab. 2014;11:177–80.
28. Murray CJ, Vos T, Lozano R, et al. Disability-adjusted life years (DALYs) for 291 diseases and injuries in 21 regions, 1990-2010: a systematic analysis for the global burden of disease study 2010. Lancet. 2012;380:2197–223.
29. Miao C, Zhou W, Wang X, et al. The research progress of exosomes in osteoarthritis, with particular emphasis on the mediating roles of miRNAs and lncRNAs. Front Pharmacol. 2021;12:685623.
30. Tao SC, Yuan T, Zhang YL, et al. Exosomes derived from miR-140-5p-overexpressing human synovial mesenchymal stem cells enhance cartilage tissue regeneration and prevent osteoarthritis of the knee in a rat model. Theranostics. 2017;7:180–95.

31. Jeyaraman M, Muthu S, Gulati A, et al. Mesenchymal stem cell-derived exosomes: a potential therapeutic avenue in knee osteoarthritis. Cartilage. 2021;13:1572S–85S.
32. Meng F, Li Z, Zhang Z, et al. MicroRNA-193b-3p regulates chondrogenesis and chondrocyte metabolism by targeting HDAC3. Theranostics. 2018;8:2862–83.
33. He L, He T, Xing J, et al. Bone marrow mesenchymal stem cell-derived exosomes protect cartilage damage and relieve knee osteoarthritis pain in a rat model of osteoarthritis. Stem Cell Res Ther. 2020;11:276.
34. Jin Z, Ren J, Qi S. Exosomal miR-9-5p secreted by bone marrow-derived mesenchymal stem cells alleviates osteoarthritis by inhibiting syndecan-1. Cell Tissue Res. 2020;381:99–114.
35. Wu F, Nerlich M, Docheva D. Tendon injuries: basic science and new repair proposals. EFORT Open Rev. 2017;2:332–42.
36. Rashid MS, Cooper C, Cook J, et al. Increasing age and tear size reduce rotator cuff repair healing rate at 1 year. Acta Orthop. 2017;88:606–11.
37. Yang Z, Cao H, Gao S, et al. Effect of tendon stem cells in chitosan/β-glycerophosphate/collagen hydrogel on achilles tendon healing in a rat model. Med Sci Monit. 2017;23:4633–43.
38. Wang Y, He G, Guo Y, et al. Exosomes from tendon stem cells promote injury tendon healing through balancing synthesis and degradation of the tendon extracellular matrix. J Cell Mol Med. 2019;23:5475–85.
39. Meng F, Xue X, Yin Z, et al. Research progress of exosomes in bone diseases: mechanism, diagnosis and therapy. Front Bioeng Biotechnol. 2022;10:866627.
40. Xie Y, Hu JH, Wu H, et al. Bone marrow stem cells derived exosomes improve osteoporosis by promoting osteoblast proliferation and inhibiting cell apoptosis. Eur Rev Med Pharmacol Sci. 2019;23:1214–20.
41. Zhao P, Xiao L, Peng J, et al. Exosomes derived from bone marrow mesenchymal stem cells improve osteoporosis through promoting osteoblast proliferation via MAPK pathway. Eur Rev Med Pharmacol Sci. 2018;22:3962–70.
42. Cheng P, Chen C, He HB, et al. miR-148a regulates osteoclastogenesis by targeting V-maf musculoaponeurotic fibrosarcoma oncogene homolog B. J Bone Miner Res. 2013;28:1180–90.
43. Claes L, Recknagel S, Ignatius A. Fracture healing under healthy and inflammatory conditions. Nat Rev Rheumatol. 2012;8:133–43.
44. Frontera WR, Ochala J. Skeletal muscle: a brief review of structure and function. Calcif Tissue Int. 2015;96:183–95.
45. Walpurgis K, Thomas A, Thevis M. Detection of the myostatin-neutralizing antibody Domagrozumab in serum by means of Western blotting and LC-HRMS. Drug Test Anal. 2019;11:1714–23.
46. Rong S, Wang L, Peng Z, et al. The mechanisms and treatments for sarcopenia: could exosomes be a perspective research strategy in the future? J Cachexia Sarcopenia Muscle. 2020;11:348–65.
47. Byun SE, Sim C, Chung Y, et al. Skeletal muscle regeneration by the exosomes of adipose tissue-derived mesenchymal stem cells. Curr Issues Mol Biol. 2021;43:1473–88.
48. Bayati V, Hashemitabar M, Gazor R, et al. Expression of surface markers and myogenic potential of rat bone marrow- and adipose-derived stem cells: a comparative study. Anat Cell Biol. 2013;46:113–21.
49. Li Z, Liu C, Li S, et al. BMSC-derived exosomes inhibit dexamethasone-induced muscle atrophy *via* the miR-486-5p/FoxO1 Axis. Front Endocrinol (Lausanne). 2021;12:681267.
50. Peinado H, Alečković M, Lavotshkin S, et al. Melanoma exosomes educate bone marrow progenitor cells toward a pro-metastatic phenotype through MET. Nat Med. 2012;18:883–91.
51. Guo S, Perets N, Betzer O, et al. Intranasal delivery of mesenchymal stem cell derived exosomes loaded with phosphatase and tensin homolog siRNA repairs complete spinal cord injury. ACS Nano. 2019;13:10015–28.
52. Perets N, Hertz S, London M, et al. Intranasal administration of exosomes derived from mesenchymal stem cells ameliorates autistic-like behaviors of BTBR mice. Mol Autism. 2018;9:57.
53. Hynynen K, McDannold N, Vykhodtseva N, et al. Noninvasive MR imaging-guided focal opening of the blood-brain barrier in rabbits. Radiology. 2001;220:640–6.
54. Raymond SB, Treat LH, Dewey JD, et al. Ultrasound enhanced delivery of molecular imaging and therapeutic agents in Alzheimer's disease mouse models. PLoS One. 2008;3:e2175.
55. Alptekin A, Khan MB, Ara R, et al. Pulsed focal ultrasound as a non-invasive method to deliver exosomes in the brain/stroke. J Biomed Nanotechnol. 2021;17:1170–83.
56. Lipsman N, Meng Y, Bethune AJ, et al. Blood-brain barrier opening in Alzheimer's disease using MR-guided focused ultrasound. Nat Commun. 2018;9:2336.
57. Zhang Y, Kim MS, Jia B, et al. Hypothalamic stem cells control ageing speed partly through exosomal miRNAs. Nature. 2017;548:52–7.

# 14 Platelet Rich Plasma, Bone Marrow Concentrate and Adipose Derived Cells: Quantitative Analysis and Quality Control—What Does It All Mean?

David C. Karli and Theodore T. Sand

## Introduction

Point of care (POC) autologous biologic therapies continue to play a vital role in the progression of the field of regenerative medicine. To date, clinical utilization and development has been dominated by Class II automated medical devices, which separate and isolate blood-based components through density gradient centrifugation and a variety of mechanical sorting mechanisms to produce platelet-rich plasma (PRP) and bone marrow concentrate (BMC). Processing fat tissue to produce a single cell suspension yields stromal vascular fraction (SVF) or a microfragmented nucleated cellular fraction, which usually involves enzymatic digestion or mechanical agitation, respectively, prior to isolation and collection [1].

Limited independent research has routinely explored the compositions of PRP, BMC, microfragmented adipose tissue (MFAT) and SVF. Data that does exist identifies significant variability in concentrating cellular elements of PRP and BMC [2]. SVF study has also shown significant variability in cellular composition [3]. MFAT analysis requires an additional digestion step to release its cells for analysis. Not surprisingly, recent analysis also shows considerable variability in cellular composition of MFAT [4]. Most clinical publications lack a detailed cellular analysis of the POC treatment product provided, since none of the existing on-market medical device options available to clinicians provide any real time quantitative analytics. As a result, the growing body of clinical literature lacks clear scientific quantitative validation of efficacy and optimization of use concerning these promising therapeutics.

D. C. Karli (✉)
Karli Center, Miami, FL, USA
e-mail: davidkarli@greyledgebiotech.com; david@karlicenter.com

T. T. Sand
Biotechnology and Regenerative Medicine, Sand Consulting, Poway, CA, USA

## Why Should We Measure?

One of the primary limitations to clinical validation of POC biologics (e.g., PRP, BMC, SVF, MFAT) and subsequent optimization of their use in treating patients relates to the fact that most clinical utilization and research to-date has been limited by a lack of detailed sample cellular analysis, quantification, and quality control when using available commercial device-based processing systems. Patient-derived biologics are a complex mixture of cells, proteins, and biologically active material with an ill-defined understanding of an optimized composition or mechanism of therapeutic action. The lack of analytical data and reliance on single-use devices to produce POC biologics limits customization of the biologic and doesn't support evaluation of a real-time dose-response relationship for the therapeutic preparation being provided to the patient. Consequently, the lack of compositional analysis limits the potential to optimize the practice of regenerative medicine relating to these therapeutic preparations.

The platform of quantitative analysis relies on a strategy involving accurate identification and measurement of cell populations, control and consistency of processing, confirmation of final target composition, and adaptation to optimize therapeutic potential. This strategy can be broken down further:

- **Measurement**—Modern technology focused on cellular analytics creates the capability to rapidly, accurately, and electronically quantify cellular components within blood-based products. Micronized adipose tissue and SVF require enzymatic digestions to isolate cells from native tissue and require a combination of a cell counter and flow cytometry to quantify and perform a detailed phenotyping of the reported nucleated cell populations [4–6].
- **Control/Confirmation**—Using modern analytical technology to evaluate pre-processed samples (whole blood and/or bone marrow aspirate, adipose tissue), processing


A. Navani et al. (eds.), *Essentials of Regenerative Medicine in Interventional Pain Management*,
https://doi.org/10.1007/978-3-031-50357-3_14

protocols can be developed to target end-product parameters to achieve specific cellular and platelet concentrations. Subsequently these targets can be confirmed through repeat analysis of final products, prior to implantation at a variety of cellular and sub-cellular levels:

- **Macrocellular**—currently available commercially—white blood cells (WBCs), red blood cells (RBCs), platelet counts, WBC differential, non-specific nucleated cells
- **Microcellular**—technology available to measure biologically active proteins, secretomes/exosomes, trace elements, but not readily utilized outside of academia
- **Molecular**—technology available but not readily utilized outside of academia (e.g., deoxyribonucleic acid [DNA], ribonucleic acid [RNA], genes, gene expression)
- **Consistency**—Cellular analysis and compositional standardization to end-target parameters allow for more valid comparison of outcomes, accounting for expected patient-patient variability of pre-processed samples.
- **Adaptation**—Once final sample preparation standardization is achieved, clinical outcomes can be compared to different sample compositions to identify what concentrations and combinations of cellular and non-cellular components yield the greatest potential for a positive therapeutic effect specific to patient and clinical indication.

In addition, formal documentation of quantified biologic preparations ensures the greatest potential for patient safety, product purity, and consistency. This documentation generates a historical record of precisely what was implanted into the patient, creating the ability to manage physician risk and liability. Quantification ensures the greatest potential for therapeutic success and creates a platform from which the field and practitioner can refine the practice of regenerative medicine.

## What Should We Measure?

PRP and BMC in particular are autologous biotherapeutic preparations obtained from peripheral or marrow derived blood, which have been processed in order to concentrate specific blood components. MFAT and SVF usually require additional processing steps, like enzymatic or mechanical disruption, as well as washing steps and concentration of the single cell suspension [7]. As previously mentioned, both cellular and non-cellular elements are potentially measurable.

## Platelet Rich Plasma

For PRP WBCs, RBCs and platelets are variably concentrated via most commercial systems. Depending on the device used, WBCs can be enriched or depleted. An early suggestion for classification was published by Mishra et al. [8] (Table 14.1). This system suggested differentiating PRP based on the presence or absence of WBCs, platelet concentrations either above or below 5× baseline concentration, and if the sample was activated by reversal of anticoagulation to induce platelet degranulation. While specific quantification parameters are absent, the classification scheme identified several critical elements useful in a binary categorization of different PRP preps.

In an effort to enhance clinical understanding of PRP treatments, Mautner et al. [9] published an alternative classification system, in which cellular components of PRP are specified by their respective concentrations (components/μL), focusing on platelet count, leukocyte presence, RBC presence, and activation status. The authors indicate that the threshold cutoff for RBCs and WBCs should be set at 1%. They also stated that the neutrophil concentration could be substituted for the WBC component element. A recent report has summarized the elements of a variety of PRP classification methods [10].

As a comprehensive system of classification, all specific cellular components of PRP can be defined and reported as actual concentrations (components/μL) or total dose (total components in final sample implanted), creating a new categorization that would take into account each cell type and platelet concentration:

WBC/μL
Lymphocytes/μL
Neutrophils/μL
Monocytes/μL
Basophils/μL
Eosinophils/μL
Immature Granulocytes/μL
RBCs/μL
Platelets/μL

**Table 14.1** PRP classification from Mishra et al. [8]

| | White blood cells | Activated? |
|---|---|---|
| Type 1 | Increased over baseline | No |
| Type 2 | Increased over baseline | Yes |
| Type 3 | Minimal or no WBCs | No |
| Type 4 | Minimal or no WBCs | Yes |
| | A: > 5× platelets | |
| | B: < 5× platelets | |

## Bone Marrow Concentrate

For BMC, similar parameters can be obtained from automated hemoanalyzers, but due to the heterogeneity of cells at various stages of maturity, validation of the results presents a significant challenge. To overcome this heterogeneity, it is suggested that broader measures be quantified, namely total nucleated cell count (TNC/μL), total mononucleated cell count (TMNC/μL), Immature Granulocytes (IG/μL) and platelet count (PLT/μL). As a complication, BMC contains two identified progenitor stromal cell populations (endothelial progenitor cells [EPCs] and mesenchymal stromal cells [MSCs], along with hematopoietic stem cells (HSCs). However, in order to quantify EPCs, MSCs and HSCs it is necessary to perform more sophisticated assays, including the colony forming unit assay (CFU-F) for MSCs and flow cytometry for MSCs and HSCs. Both are discussed in more detail later in the chapter.

Using hemoanalyzed cell concentrations for PRP and/or BMC, total number of the individual components can be calculated in the following formula:

$$\text{Total Units} = \left(\text{Units} / \mu L\right) \times 1000 \times \left(\text{total volume}\left(\text{mL}\right)\text{of the implant}\right)$$

As a result of the calculation, data for both concentration and total unit quantity ("dose") can ultimately be determined to aid in identifying optimized levels, yielding a biologic preparation with high therapeutic potential.

In view of emerging theories of the paracrine effects of POC biologics [10, 11], there is a need for quantification of biologically active material at the sub-cellular level. A number of secreted bio-molecular factors have been linked with physiologic response, including growth factors, proteins related to the immune response, enzymatic inhibitors (e.g., α-2-Macroglobulin) [12], anti-inflammatory proteins (e.g., Interleukin-1 receptor antagonist protein: IL-1RAP) [13], and others as yet to be identified. These factors can be measured with lab-based protein assays (i.e., enzyme-linked immunosorbent assays or ELISAs), which although technically demanding, yield quantifiable data.

## Microfragmented Adipose Tissue (MFAT)

The analysis of MFAT preparations is more complicated, since the preparation contains remnants of adipose tissue, and clearly is not a single cellular preparation.

MFAT is obtained through the use of mechanical methods based on physically disrupting the adipose tissue through agitation and filtration, producing fragmented tissue. The fact that the material produced by such technology can be characterized as "fragmented" adipose tissue highlights the distinction between a single cell suspension and the more heterogeneous MFAT. At the present time, it is difficult to envision a way to provide a rapid validated quantitative analysis or to set quality standards when working with a fragmented adipose product. Thus, for quantitative analysis on MFAT preparations it is necessary to digest the fragmented tissue to release the cells [4, 14].

Once cells are released from the fragmented tissue, the cells can be centrifuged, washed free of particulate debris and counted using cell counters, which provides a nucleated cell number and viability. However, enzymatic digestion protocols used for this purpose generally take 30–90 min [15], which means that the cell count and viability are not available prior to use of the MFAT preparation for treating the patient. On the other hand, the quantitative data can be entered into the patient's electronic medical record for retrospective review. The cells also could be processed for analysis on a flow cytometer to provide a more detailed phenotypic profile, but this level of analysis is highly technical and not routinely found in a clinic setting.

## Stromal Vascular Fraction

The quantitative assessment of cellular components in SVF works only because the components are single cells in suspension. Enzymatic digestion of adipose tissue results in a single cell suspension for use as a therapeutic preparation. Enzymatic digestions of lipoaspirate generally are performed at 37 °C with an incubation time of at least 30 min [15], which means that the released cells aren't available for analysis (e.g., cell counting) until after the preparation has been provided to the physician for treatment. An alternative route to obtaining a single cell suspension involves the recovery of cells released during the lipoaspiration procedure. For example, use of physical manipulation (e.g., vortexing or shaking) and centrifugation of the lipoaspirate, or just centrifugation of the lipoaspirate, yields a cell pellet that can be recovered, as reviewed in a recent publication [7]. The authors noted, however, that the yield of viable cells from limited, mechanically processed adipose tissue was lower and, in some cases, substantially lower compared to enzymatically digested adipose tissue cellular yields.

While nucleated cells found in SVF can be counted in a hemoanalyzer [5], the validity of such a count remains to be established. In this regard, the approach adopted for BMC (TNC, TMNC, etc.) probably is the most relevant but remains under investigation.

## How Should we Measure?

Quantifying specific components in and/or associated with POC biologics is a function of available and accessible technology. Factors that should be considered include:

1. Acquisition cost of the equipment and on-going operational costs
2. Licensure/clearance of the equipment for use
3. Space needed to house equipment and resources to operate
4. Efficiency of processing
5. Time cost to produce useable and valid results
6. Level of technical expertise needed to operate the technology
7. Ready access to samples for analysis—minimizing delays in analysis

### Counting Cells: Manual Method

Options for quantification vary in terms of automation. The simplest approach involves standard light microscopy and manual counting using hemocytometers with standard grids. This method usually involves extrapolation of cell concentrations following serial dilutions of the original sample. The dilution steps create a significant potential for measurement error considering the very high concentrations of platelets, WBCs, and RBCs present in PRP and BMC preparations (Fig. 14.1).

### Counting Cells: Automated Method

A number of different automated hemoanalyzers exist, which have the capability to rapidly differentiate cell types and platelets using light scatter and optical scanning technology. The hemoanalyzer class of medical instrumentation is frequently cleared by the Food and Drug Administration (FDA) for diagnostic analysis of peripheral blood cell counts and blood chemistries. Although most are capable of running a variety of diagnostic blood tests, for the purposes of blood and marrow concentrate analyses, a standard complete blood count (CBC) and 6-part differential can be used to measure concentrations of various types of cells present in the preparation, as well as platelet counts. Unfortunately, clinical grade hemoanalyzers are expensive to acquire and require technical expertise to operate and maintain.

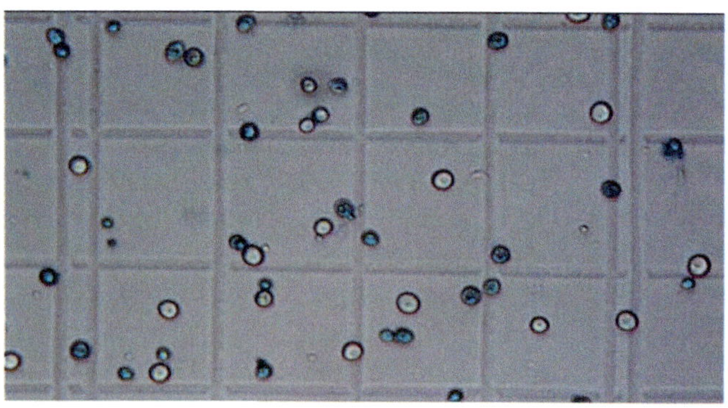

**Fig. 14.1** Standard hemocytometer grid under light microscopy

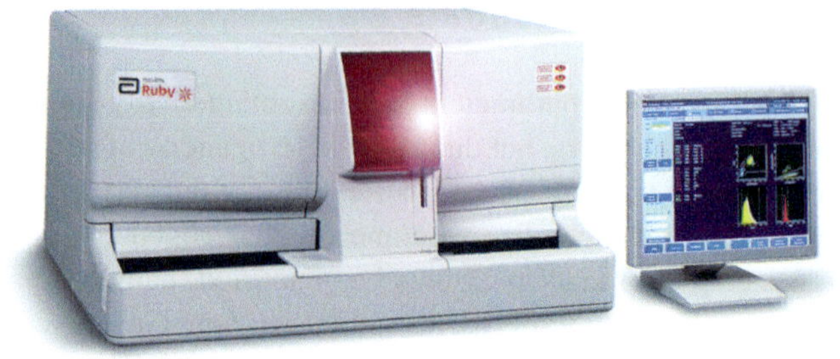

**Fig. 14.2** Automated hemoanalyzer—Abbott Ruby System

Standard hemoanalyzers (Fig. 14.2) have variable capabilities to measure cell and platelet counts in the highly concentrated ranges found in POC biologics. In addition, these devices are not cleared by the FDA for validated analysis of marrow derived blood samples. This poses challenges for standardizing quantifiable metrics for BMC. At least one publication exists validating the use of one automated hemoanalyzer in quantifying marrow derived cells as reliable and accurate [16].

Figure 14.3 shows a typical scatter plot generated for a standard blood-based analysis.

### Counting MSCs: Colony Forming Unit Assays (CFU-F)

The use of a tissue culture-based assay has been available since the MSC was first identified in culture by Freidenstein et al. [17]. A sample is placed in culture media and diluted under controlled conditions and incubated at 37 °C in a $CO_2$ incubator for approximately 9 days. During the incubation, MSCs will attach to the plastic culture container and begin to proliferate as a colony. The longer the plates are incubated, the higher the number of cells in a colony. At the end of a fixed period of incubation time, the medium is removed, and the adherent cells are stained to aid in being counted using an inverted microscope. While the medium used to culture the cells can play a role in how robust the adherent cells proliferate, the number of cells in a colony for which a colony is counted also can vary. Thus, it is important to exercise caution when viewing CFU-F data, since some researchers

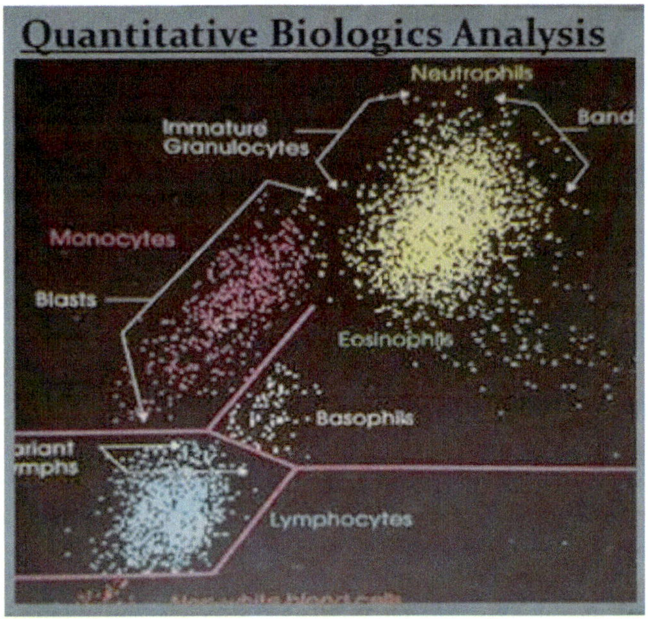


**Fig. 14.3** Example of a scatter plot demonstrating how different cell types can be measured with an automated hemoanalyzer

might use a colony cell count of 20, while others might require 50 cells in a colony. CFU-F, the unit of measure of MSCs, is determined by the number of colonies in the assay. Thus, it is important to keep a consistent time in culture when comparing samples from different patients within a clinical study.

## Counting Cells: Flow Cytometry

Flow cytometers are sophisticated instruments used for cell counting, sorting, and biomarker detection. The technology involves use of laser fluorescence and light scattering to count cells present in a single file flow chamber. Frequently, the cell suspension is exposed to fluorescing agents/dyes in order to increase the specificity of the analysis. Modern analyzers can analyze thousands of cells rapidly in real time. Cells are exposed to fluorescent labels or fluorophores attached to purified antibodies that recognize specified cell surface antigens on the cell membrane, although internal antigens also can be characterized with a permeabilization and fixation protocol. As the fluorescing agent is excited by light or laser activation, the resulting pattern can be measured to identify "clusters of differentiation" or CD markers. CD markers are antigens associated with specific cell types. For example, an antibody to CD 4 will detect a type of helper T-cell. Flow cytometers are technically complex, and accurate interpretation usually requires extensive (and frequently PhD-level) expertise.

In addition to the technical complexity for performing CD marker-based assays, another limiting factor is a lack of consensus regarding what panel of CD markers represents freshly isolated MSCs [18]. Both positive and negative CD marker outcomes are needed to identify a specific type of cell. For example, a freshly isolated MSC might be CD31-, CD34+, CD45-, or CD146+ [19]. Given the complexity of antigens expressed on cell surfaces, overlap of available markers has made acceptance of a set of 3–6 antigenic identifiers challenging for researchers to validate (Table 14.2) [20–22]. Recent flow cytometric analysis of single cell preparations obtained from MFAT preparations have revealed the presence of endothelial cells (adult and progenitor), several types of stromal cells ("adipose" stromal cells, MSCs and pericytes—an MSC found in the perivascular space) and adult CD45+ cells (e.g., leukocytes) [4–6]. The use of three-laser flow cytometry analysis has been described that identified a type of adipose tissue-associated stromal cell— supra-adventitial adipose stromal cells (SA-ASC) (Table 14.2) present in SVF and matched MFAT-derived SVF [4]. Furthermore, the use of sophisticated analytical techniques like flow cytometry has revealed some

**Table 14.2** Published cell surface antigens (CD Markers) for adult progenitor/stromal cells

- MSC (cultured) [18, 20]
  - Positive:
    CD: 105, 90, 44, 73, 29, 13, 34(?), 146, 106, 54, 166
    STRO-1, SH2, SH3, SH4, HLA-A, HLA-B, HLA-C, HLA-DR, HLA-I, DP, EMA, CDIO5, Oct 4, Oct 4A, Nanog, Sox-2, TERT, Stat-3, fibroblast surface antigen, smooth muscle alpha-actin, vimentin, integrin subunits alpha4, alpha5, beta1, integrins alphavbeta3, alphavbeta5, ICAM-1
  - Negative:
    CD: HLA-DR, CD31, CD45, CD106, CD271, CD34, CD11b/CD14 and CD19
- Cells in freshly isolated bone marrow [21, 22]
  - Lymphocytes, monocytes, neutrophils, basophils, eosinophils, immature granulocytes (CD45+)
  - Endothelial lineage cells (CD31+)
  - Hematopoietic stem cells (CD34+, CD45-, CD31-)
  - MSCs (CD271+, CD90+, CD105+, CD146+, CD73+, among others and CD31-, CD45-)
- Freshly Isolated Cells in SVF/MFAT [4]
  - Supra-adventitial adipose stromal cells (SA-ASC): CD31–, CD34+, CD73$^{high}$, CD90+, CD105–, CD146–, (CD45–)
  - Pericytes: CD31–, CD34–, CD73±, CD90+, CD105–, CD146+, (CD45–)
  - Transitional Pericytes (TP): CD31–, CD34+, CD73±, CD90+, CD105–, CD146+, (CD45–)
  - Endothelial mature cells: CD31+, CD34–, CD73±, CD90±, CD105–, CD146±, (CD45–)
  - Endothelial progenitor cells: CD31+, CD34+, CD73±, CD90±, CD105±, CD146±, (CD45–)
  - Leukocytes: CD45+, CD31–, CD34–, CD146–, CD73–, CD90–, CD105–

trends in patient derived MFAT preparations. For example, a recent report [5] demonstrated that the number of "stromal cells" found in enzymatic digests of lipoaspirate and matched MFAT preparations were not meaningfully different. However, it also has been reported that there were sex-linked differences in the frequencies of SA-ASCs and endothelial stromal cells in both the lipoaspirate and the matched MFAT preparations that were used for therapy [6].

In order to avoid artifacts associated with cell aggregates, debris and non-viable cells, viability markers like Propidium Iodide are used during flow cytometric characterization. Characterizing cells in enzymatic digests of MFAT preparations is complicated by the physical disruption of adipocytes and other cells that results in tissue remnants and cellular debris ("naked nuclei"). A recent report [4] relied on a nuclear staining dye to limit the analysis to intact and viable cells. The authors indicated that this approach was necessary to reduce the background and limit the inclusion of non-nuclear counts.

## Correlation with Published Data

Peer reviewed clinical evidence supports a relationship between threshold concentration and/or finite quantities of certain cell types and more favorable therapeutic outcomes. Four examples of such data sets are summarized below. In addition, quantitative analysis of bone marrow aspirates can be used to assess the quality of the technique employed.

### Intradiscal BMC Data

Pettine et al. [19] studied 26 patients with lumbar degenerative disc disease, who had been treated with percutaneous implantation of autologous BMC into one or two discs. As summarized in a report with 3-year follow-up [23], 20 study participants (77% of the original 26) reported significant subjective and functional improvement. The average percent decrease in the Oswestry Disability Index (ODI) for the study participants was 69%, with a 83% subjective decrease in visual analog scores (VAS) at the 3-year milestone [23].

Beyond the positive and sustained outcomes, what was equally important was the correlation of MSC and HSC concentrations with both subjective and functional improvements. Patients with BMC CFU-F counts of less than 2000 MSCs/mL of injectate averaged 41% functional (ODI) and 48% subjective (VAS) improvements at 3 years. Patients with BMC CFU-F counts greater than 2000 MSCs/mL of injectate averaged 86% functional and 90% subjective improvements within the surviving cohort of 20 patients at 3 years [23].

HSCs were identified by CD34 positivity on flow cytometry. Patients at 3-year follow-up with BMC CD34+ counts less than one million cells/mL of injectate averaged a 38% functional (ODI) and 42% subjective (VAS) average improvement. Those patients with BMC CD34+ counts of greater than two million cells/mL of injectate averaged 79% functional and 83% subjective improvements by comparison [23].

### BMC in Rotator Cuff Repair

In a study on the use of BMC to augment surgical rotator cuff repair, 45 study participants had BMC implanted into the tendon after surgical repair, as well as into the subchondral space below the tendon repair within the proximal humeral head. A control group ($n$ = 45) received the surgical repair without BMC augmentation [24]. All of the BMC-treated patients showed complete healing of their repair by 6 months, whereas just 67% of controls in the non-BMC group had similar healing rates as defined by ultrasonographic evaluation. Rotator cuffs were intact at 10 years in 87% of BMC-treated group, whereas cuff repairs were intact in 44% of controls for the same interval as defined by blinded magnetic resonance imaging (MRI) evaluation.

Within the BMC group, 39 patients with intact cuff repair at 10 years were injected with an average of 4200 MSCs/mL of injectate and greater than 54,000 total MSCs (as defined by CFU-F assay). Six BMC-treated patients with failed cuff repairs at 10 years averaged a significantly reduced MSC concentration (1500 MSCs/mL of injectate) and had an average total MSC count less than 14,000. Lastly, patients with greater than 30,000 total MSCs/BMC implant took 4 months for their tendon tear to heal, while patients with fewer than 30,000 total MSCs/BMC implant required 6 months on average to heal as defined by ultrasound evaluation [24].

### BMC in Knee Osteoarthritis

Long-term outcomes for osteoarthritic knees treated with BMC delivered via the intraosseous route (IO) or by intraarticular injection (IA) in patients presenting with bilateral knee osteoarthritis (OA) were assessed [25]. The treatment each knee received was determined by the patient and both treatments were performed during the same surgical procedure. An average of 155 mL of bone marrow aspirate was obtained from the iliac crests, processed to create BMC, which was divided into two equal volumes of 20 mL each. The CFU-F level was found to be on average 5752 MSC/mL of the BMC preparations. The IO BMC volume was divided, and 10 mL was injected into the subchondral compartments

**Table 14.3** Outcomes at 15-year follow-up for intraarticular (IA) and intraosseous (IO) BMC treatments for knee OA [25]

| Treatment route | Number of TKAs at 15-year follow-up | Mean time to TKA (year) | Annual rate of TKA (% knee-year) | Mean knee score |
|---|---|---|---|---|
| **IA** | **42 (70%)** | **7** | **4.6%** | **58.4** |
| **IO** | **12 (20%)** | **14** | **1.3%** | **76.2** |

of the condyle and the tibial plateau of the IO knee, while 20 mL was injected into the drained knee joint. Ultrasound and fluoroscopy guidance were used for accurate delivery. IO treated knees showed a statistically meaningfully higher Knee Society knee score at 2-year follow-up compared to the IA treated knees, with both IO and IA Knee Society scores statistically meaningfully higher compared to pre-treatment. VAS scores followed a similar pattern at the 2-year follow-up, with IO knees showing significantly meaningfully lower scores, while the IA knees were not statistically meaningfully lower compared to pretreatment. Also, a total of 11 IA knees showed an increased Kellgren-Lawrence knee score at 2-year follow-up, while just two IO knees were increased. Significantly, bone marrow lesion volume was statistically meaningfully decreased in IO knees at the 2-year follow-up, but had not changed for the IA knees. Table 14.3 shows the results for total knee arthroplasty (TKA) for an average of 15-year follow-up. As shown in the Table 14.3, 70% of the IA knees had progressed to TKA with an average time of 7 year. In contrast, just 20% of the IO knees had progressed to TKA with an average time of 14 year. The authors also noted that there was a reduced risk of progressing to a TKA as the number of progenitors cells injected increased for IO knees, while the time to TKA lengthened for knees receiving higher progenitor cell numbers in the subchondral compartments [25].

## PRP Dosing: "Low Dose" Vs "High Dose"

The term "PRP" refers to a preparation that has an enriched level of platelets compared to the source whole blood. However, there are a wide variety of methods and commercially available kits that have been used to produce PRP preparations for use in clinical studies, and they don't all share the same efficiency in recovery of platelets [10]. Methods of producing PRP were reviewed recently, and clinical study outcomes for treating lesions of bone, cartilage, muscle, and tendon with PRP also were summarized [26]. The authors concluded by pointing to the poor disclosure of PRP methods, and the uncertainty related to the quality and composition of the PRP used to treat musculoskeletal pathologies in publications. This issue has become more acute as more studies are published on the clinical outcomes of PRP made from a limited volume (e.g., 20 mL or less) of whole blood. In an *in vitro* assessment of four commercial PRP devices, Fitzpatrick et al. [27] reported that the Arthrex Autologous Conditioned Plasma (ACP) kit yielded a PRP with a 1.53X increase in the platelet level compared to whole blood for a processed volume of 15 mL of whole blood, while other PRP kits that were able to accommodate >50 mL of whole blood each, produced an average of 4.13X increase in the PRP platelet level compared to whole blood. Clearly, the volume of whole blood that is processed with a specific kit technology (device design and use) will dictate the total number of platelets available for injection, but the platelet recovery of the kit and specific protocol (e.g., reducing RBC and/or WBC levels) also could play a role in the therapeutic benefit of the PRP [10, 27].

Several recent publications highlight differences in treating knee OA based on the use of low volume or higher volume PRP device technology. A recent clinical study was published in which PRP made from 20 mL of whole blood with a commercial device in a Level 1 study was compared to saline as the placebo treatment. They reported no difference in a number of clinical outcomes with a 1-year follow-up between the PRP they used and saline control injections. The authors identified their type of PRP as "leucocyte poor" PRP made with the Arthrex ACP device technology [28]. In another Level 1 clinical study recently published [29], a PRP kit technology was used to process 20 mL of whole blood, which resulted in an average of approximately $1.6 \times 10^9$ platelets injected via the intraarticular route, and was compared with saline as a placebo. The authors reported no statistically meaningful differences between the two treatments at 1-year follow-up for pain score or cartilage volume. In contrast to these two low platelet dose PRP studies, a clinical study was performed in which 32 mL of whole blood was processed in a manual method performed in a Biological Safety Cabinet (to provide a sterile environment—ISO 5) that resulted in a Leukocyte Rich (LR)-PRP preparation containing on average $12.8 \times 10^9$ platelets per injected treatment preparation [30]. The authors' Level 1 study evaluated a three-injection protocol of freshly prepared LR-PRP (at study initiation, at 1 month and 2 months). Saline was used as the placebo and was injected on the same schedule. Two other arms of the study involved just a single injection of LR-PRP and a corresponding saline control injection. Three LR-PRP injections provided better outcomes through the 1-year follow-up compared to the single LR-PRP injection. More importantly, both LR-PRP treatment groups demonstrated statistically meaningful improvement over the saline control groups through the 1-year milestone. However, all groups had overlapping scores by the 2-year follow-up. The authors concluded that the use of a LR-PRP with an average of more than 12 billion platelets per treatment provided greater therapeutic benefit compared to a saline control through 6-months, but returned to pre-treatment scores by the 2-year follow-up [30].

Clearly, there should be an obligation on authors of articles that provide clinical study outcomes involving PRP to include sufficient information in the article to at least assess the mean platelet dose used in the treatment. However, reporting of the number of platelets injected in a clinical study is not universally embraced. For example, when a critique [21] was leveled against the Level 1 study using Arthrex's ACP device publication [28], one of the authors (on behalf of all the authors) stated that an explicit PRP characterization of the PRP used in their clinical study wasn't necessary, since the performance of the commercial kit was well known [31]. The fact that the kit used to produce the PRP has been identified as providing only an average of a 2X increase in the platelets in the PRP over whole blood [32], suggests that the platelet dose is low and might have played a role in the equivocal results reported [28].

As more clinical results for the use of PRP treatments are published, it will be vital to have direct access to the number of platelets injected in the publications, since there is no universal set of characteristics that describe what is in a PRP preparation. Hence the need, at a minimum, for reporting the number of platelets present in the injectate.

## Validation of Bone Marrow Aspiration Technique

Two primary techniques for aspirating bone marrow samples for BMC production from the iliac crest have gained favor in recent years. Hernigou, et al. [33] developed and have used a single channel approach in parallel to the cortical shelves of the iliac crest (depending on the target goal for bone marrow to be aspirated) for more than 30 years with documented MSC levels on large quantitative BMC data sets. A proposed alternative is to use a lateral approach, perpendicular to the iliac crest cortical shelves, which requires careful positioning between the tables of the iliac crest, and presumably confirmation of the trocar's position by ultrasound and/or fluoroscopy.

Data for the lateral aspiration approach is scarce. Hernigou and co-workers have published three papers [33–35] that describe in detail their aspiration technique, including the use of a 10 mL syringe and aspirating between 2–4 mL per needle position within the medullary cavity. The Hernigou technique utilizes harvest at varying depths within the medullary channel in an effort to maximize the sample area, starting at the greatest depth. At each depth, the bone marrow aspiration (BMA) needle is rotated 90 or 120 degrees, aspirated again, then rotated a second time. The needle depth is then changed by withdrawing the tip toward the cortical surface, where the aspiration sequence at the new level is repeated. Depending on how deep the initial needle penetration is (which depends on the location of the channel—posterior usually offers a deeper option compared to anterior iliac crest) [35], it is possible to harvest 60 mLs of high-quality bone marrow from a single channel. If additional bone marrow is required, Dr. Hernigou recommends moving the starting point for an additional channel 2 cm away from the starting channel to repeat the same sequence and process. An angling technique from the same insertion point to create multiple channels from a single cortical puncture has also been utilized in an attempt to maximize cellular harvest. Additional details covering the anatomy of the iliac crest, including the use of sectors to characterize the safest zones to access, have been published [34, 35].

In contrast to the well documented technique for traditional parallel access between the iliac tables, the lateral aspiration approach requires multiple punctures of the iliac crest, which means that the local volume of bone marrow is very limited—essentially whatever is present between the width of the tables of the iliac crest at the point of entry. One issue with the lateral approach is that a number of neural and vascular features are potentially put at risk, in addition to a risk of breaching the opposite table [34]. Area and volume of marrow to harvest from is limited to the width of the cortical shelves at a single point.

Adopting a quantitative, hemoanalytic approach for assessing the cellularity of BMA and BMC preparations allows physicians to track key cellular components on each patient. While stem and progenitor cell analysis require labor-intensive, multi-day protocols, especially for MSC assessment, it is possible to track the mononuclear cell levels with a hemoanalyzer, which provides an indirect indication of the MSC content, since MSCs will be characterized as mononuclear cells based on their size, shape and cell density.

## Quality Control Considerations

In view of the complexity of quantifying the components of POC biologics, when doing so it is important to work within a quality framework. Operating within a quality management system helps to ensure that what is being measured is accurate and validated. The following describes the elements for operating a quality-controlled program:

### Proficiency Testing

Several levels of proficiency testing exist and can be applied to verify accuracy of automated technology used in quantifying elements in POC biologics. Each level is independently critical and when combined can make up a sound platform to validate hemoanalytic data.

- **Level 1:** Involves a daily or weekly internal calibration, typically contained within the software programming of most modern hemoanalyzers. This requires no external inputs and follows manufacturer-established protocols.
- **Level 2:** Involves testing of commercially available standardized control samples obtained from the device manufacturer or an external vendor.
- **Level 3:** Testing requires access to multiple hemoanalyzers, from which a single human blood sample is analyzed and compared internally with other instruments. This practice is known as split testing.
- **Level 4:** Is the most comprehensive form of proficiency testing, and involves participation in a formal multicenter comparison of analyses using a control sample with known cellular and acellular parameters compared against matched hemoanalyzer devices. One such program is administered by the College of American Pathologists (CAP), which involves standardized control sample analysis 3 times per year.

All four layers of quality testing are suggested to ensure accuracy and validation with use of modern automated hemoanalyzers to quantify POC biologics and represent important elements of a quality management system.

## Benefits of Measurement

Biologic therapies are not without contraindications or potential complications, and maximal safety cannot be assured if product composition is not known. In addition, if validated measures are used, a record of the implant can be created for inclusion into a patient's medical record to document compliance, safety, efficacy potential, and risk management.

- From a clinical research perspective, understanding composition will allow for comparison to clinical outcomes to establish concentration or total "dose." These comparisons may lead to optimization of product composition for patient populations and/or tissue specific applications. As research continues to mature, final product preparation can be adjusted to meet these emerging targets.
- Recent field trends have suggested that leukodepletion of PRP preparations may be beneficial to prevent target tissue exposure to catabolic secretions from granulocytes. Hemoanalysis provides a direct count of the leukocytes (including granulocytes) within a PRP preparation. This allows for correlation of an actual granulocyte concentration to clinical outcomes and potential refinement of protocols to optimize therapeutic potential. Hemoanalysis creates the potential to define specified thresholds for "WBC(-) or leukodepleted" preparations beyond simple binary presence or absence of these cells [36].
- Similarly, literature to date has suggested that an optimal target for platelets in most musculoskeletal applications lies at a concentration between 1–1.5 million platelets/μL [8]. However, in view of the equivocal outcomes when using less than five billion platelets per injection to treat knee OA [28, 29], treatment of joints probably will require substantially more platelets in the PRP injectate to obtain a durable therapeutic benefit [30]. Most of the published clinical literature lacks explicit characterization of the injected PRP sample preparations, including platelet concentration. A recent publication [2] illustrates the wide average variability of residual RBCs, WBCs, neutrophils and platelets found in PRP produced in a variety of commercially available device technologies. This variability further supports the need for quantification of POC biologics.
- For BMC products, more specific challenges exist to establish validated quantitative analytics as described previously. As discussed, modern automated hemoanalyzers are not designed to analyze blood/tissue obtained from a medullary cavity. While a standard CBC and 5 or 6-point differential can be obtained, due to the significant level of cellular heterogeneity and maturity, the likelihood of measurement error must be accounted for with testing across medical device options.
- TMNCs (Monocytes/μL) can be measured with reasonable accuracy in a hemoanalyzer. From the TMNC, an estimation of the number of MSCs can be mathematically generated, based of previously reported studies, suggesting that MSCs represent roughly 0.001% of the TMNC [17].
- More specific and accurate MSC and HSC counts can be obtained through CFU-F assays and flow cytometry, respectively. From these metrics, minimum and optimal therapeutic ranges can be established by looking at concentrations and total MSC and HSC counts present in the injected BMC compared with clinical outcomes.

Little to no attention has been given to measuring the platelet concentrations in BMC preparations to date. Via similar mechanisms to the proposed paracrine effect of platelets in PRP preparations [37], it is logical to consider that platelets present in marrow-derived blood/tissue may have similar capability and subsequently may contribute to a therapeutic effect in BMC. Automated hemoanalyzers can obtain a rapid platelet count that can be considered as a potentially important contributor to therapeutic outcomes in BMC treatments (Table 14.4).

**Table 14.4** Bone marrow concentrate (BMC) composition

| |
|---|
| • Non-homogenous, multi-lineage concentrate with even more variability than PRP |
| – Multipotent cell fraction |
| **MSCs (0.01–0.001% of total nucleated cells)** |
| HSCs |
| EPCs (endothelial progenitor cells) |
| Megakaryocytes/platelets |
| RBCs |
| Macrophage/granulocyte adult and precursor cells |
| • Cells present at various stages of maturity and delineation |

## Conclusion

Building on the platform of quantitative hemoanalysis of POC biologics creates a wealth of future opportunity to further understand the science behind biologic therapies and how to improve consistency and efficacy in treatment outcomes. The use of hemoanalysis to characterize POC biologics faces several technical challenges.

CFU-F determination poses challenges for routine use given the technical nature of the assay as well as the time to reach a result. Compressing the interval to achieve a faster result would make the testing more readily accessible and practical to apply routinely to better characterize the therapeutic potential of BMC.

Similarly, with flow cytometry, establishing a validated set of CD marker assessments to accurately identify MSCs of freshly isolated, uncultured BMC presents a technical challenge to using the method as an alternative to CFU-F assays.

Finally, establishing patient screening methods to aid in predicting probability of therapeutic success would have an impact on patient counseling, treatment decision making, and customization of biologic preparations. Opportunity to develop predictive screening methods depends on successful implementation of hemoanalytics protocols and relationship to outcomes to identify variables with the potential for predictive value.

### Key Points

1. Cell counting of PRP, BMC and SVF are relatively straightforward since these preparations are single cell suspensions.
2. MFAT preparations contain cells, which are embedded in tissue, and need to be released by enzymatic digestion in order to obtain a cell count.
3. Cell counts are the starting point for analyzing the complex compositions of the autologous PRP, BMC, SVF and MFAT preparations, while the use of flow cytometry and CD marker reagents provide much more detailed profiles of the types and frequencies of cells present in these preparations.
4. Beyond the technical aspects of doing detailed quantitative analysis, it also is necessary to include proficiency testing of the operators performing the analysis to ensure consistency in the results obtained.
5. Commercial hemoanalyers can be exploited for analyzing the single cell preparations PRP, BMC and to a lesser extent, SVF, which will provide rapid results for a five- or six-part differential of the WBCs present, RBCs, Platelets and hematocrit.
6. Estimates of the number of MSCs present in BMC have been reported in some clinical studies, which have demonstrated that patients having higher levels of MSCs generally achieve a better therapeutic outcome, including faster healing, or providing a more durable therapeutic outcome.
7. Platelet number for PRP preparations used in clinical studies highlight differences between less therapeutically beneficial results with "low-dose" PRP preparations and PRP with much higher numbers of Platelets that demonstrate meaningfully improved outcomes compared to a saline control in treating knee OA.
8. Quantitative analysis of the Platelet level and residual levels of WBCs and RBCs in PRP preparations would provide more definitive characterization of these preparations, compared to relying on the terms "Leukocyte-poor" and "Leukocyte-rich" to describe the PRP being used in a clinical study.
9. It is important to be aware of the technical limits of any instrumentation that is used for quantitative analysis of a sample like concentrated bone marrow, since not all hemoanalyzers have a wide enough linear range to avoid the need to dilute the BMC sample.
10. Having access to cell counts and other rapid quantitative assessments of the therapeutic preparations used for treatment could aid in helping to define best practices for using PRP, BMC or SVF for treating MSK pathologies.

### Questions

1. Which of the following best describes the analysis of PRP, BMC, SVF or MFAT at Point-of-Care?
   A. All are routinely analyzed for cellularity
   B. All are routinely analyzed for cellularity and cytokines
   C. All can be analyzed for cellularity with varying degrees of challenge and effort
   D. All can be analyzed for cellularity without further processing

   **Answer: C.**

   **Explanation:** All of the preparations can be analyzed at POC, but it is easier and more straightforward to analyze PRP, BMC and SVF since these preparations are all single cell suspensions. MFAT contains adipose tissue and would need to be digested to release the cells prior to analysis.

2. When summarizing PRP, which of the following should be provided?
   A. Fold increase of platelets
   B. Fold increase of platelets and hematocrit
   C. Concentration of platelets
   D. Concentrations of WBC, RBC and platelets

   **Answer: D.**

   **Explanation:** Concentrations of WBC, RBC and platelets provide an easily interpretable profile of what actually is in the PRP being injected, and can be translated into "dose" of platelets by multiplying the platelet concentration and the PRP volume.
3. Which of the following represents the best justification to adopt quantitative analysis of preparations like PRP and BMC?
   A. Can be used to meet state or federal requirements
   B. Can be used to meet data quality requirements for publication
   C. Can be performed to assess patient outcomes versus component levels in the preparations
   D. Can be performed to meet in-house performance goals

   **Answer: C.**

   **Explanation:** Tracking levels of components in the therapeutic preparations delivered to patients with patient outcomes can provide useful insight in refining best practices to optimize therapeutic benefit for the various MSK pathologies being treated.
4. When publishing results from a clinical study in which PRP is used, which of the following will provide the most useful information?
   A. Describe the Platelet fold increase in the PRP over baseline
   B. Describe the platelet dose as total platelets per injected volume
   C. Describe the type of PRP produced (i.e., Leukocyte Rich/Poor)
   D. Describe the device technology used to produce the PRP

   **Answer: B.**

   **Explanation:** Describing the total number of platelets provided in the injectate along with the injectate volume provides the most critical information about the PRP used in the study. It is hard to know if dose matters if not enough information is provided to establish the dose.
5. Which of the following approaches for bone marrow aspiration yields the best quality of the aspirate with the least discomfort?
   A. Direct the trocar between the tables of the iliac crest through a single channel, angle the trocar to collect additional aspirate, and use a 10 mL syringe to aspirate
   B. Direct the trocar through the table of the iliac crest taking a lateral approach in a single channel, and use a 10 mL syringe to aspirate
   C. Direct the trocar through the table of the iliac crest through multiple channels, and use a 10 mL syringe to aspirate
   D. Direct the trocar between the tables of the iliac crest through multiple channels, while aspirating at a single level in each channel, and use a 10 mL syringe to aspirate

   **Answer: A.**

   **Explanation:** Using a single channel between the tables of the iliac crest, with multiple aspirations per channel (changing the depth for each aspiration) and using angling to increase the aspirate collected reduces the risk of multiple penetrations of the iliac crest. Lateral bone marrow aspiration presents an increased risk due to the presence of blood vessels and nerves. A publication demonstrated that a 10 mL syringe with a "quick" draw-back of the plunger provided the highest level of CFU-F (MSCs) compared to slower draws and larger volume syringes.
6. Which of the following is most likely to provide a strong level of quality control when producing PRP and BMC?
   A. Written protocols
   B. Written protocols with periodic discussions with the operator
   C. Written protocols with periodic proficiency testing of personnel
   D. Written protocols with documented training and periodic proficiency testing of personnel

   **Answer: D.**

   **Explanation:** Having written protocols with the training of personnel documented is a minimum for demonstrating a commitment to quality, but periodic proficiency testing is a standardized approach to demonstrate the impact of the quality system with real world samples.
7. Which of the following are the key components to analyze in a BMC preparation?
   A. Total Mononuclear Cells (TMNC)
   B. TMNC and Platelets
   C. Total Nucleated Cells (TNCs) and RBCs
   D. TNCs, platelets and RBCs

   **Answer: B.**

   **Explanation:** TMNC can be used as a rough estimator of the number of MSCs present in a BMC preparation (by applying a percentage of MSCs reported in the literature to be present in bone marrow aspirates), since MSCs will be counted in the monocyte channel of a hemoanalyzer. Platelets are the therapeutic agent of interest in PRP, so platelets in BMC also might contribute to a therapeutic benefit. RBC levels haven't been reported to correlate to a therapeutic benefit in published clinical studies in which a hemoanalysis has been performed.

8. Which of the following are the key component(s) to analyze in a PRP preparation?
   A. Platelets
   B. Platelets and hematocrit
   C. Platelets, RBCs, and neutrophils
   D. Platelets and Total Nucleated Cells
   **Answer: C.**
   **Explanation:** Platelets are the most important component to quantify in PRP, but there also is interest in characterizing the RBC and neutrophil levels. So called "Pure PRP" will be depleted of RBCs and WBCs, but since having neutrophils in the PRP is thought to be counterproductive, quantifying the neutrophil level is more frequently reported compared to TNC.
9. Which of the following statements best describes what a "cluster of differentiation" (CD) is?
   A. CDs are an indication of a cell's auto fluorescence
   B. CDs are proteins/receptors on the cell's surface that are unique to that type of cell
   C. CDs are proteins on the surface of the cell that aggregate together
   D. CDs indicate the type of differentiation a cell can undergo
   **Answer: B.**
   **Explanation:** CDs are proteins/receptors that typically are expressed on the surface of the cell that are unique to that type of cell. For example, a CD4 positive cell is uniquely identified as a T-helper cell.
10. Which of the following attributes applies equally to PRP, SVF, BMC and MFAT?
   A. Ease of analysis
   B. Ease of preparation
   C. Ease of delivery
   D. Ease of raw material acquisition
   **Answer: C.**
   **Explanation:** Each of the preparations are placed in syringes after processing for return to the patient. MFAT is requires enzymatic digestion to assess its cellularity, SVF requires enzymatic digestion, and invasive protocols are used to obtain the raw materials for producing BMC, SVF and MFAT.

**Acknowledgments** This book chapter is modified and updated from a previous book chapter, "Platelet Rich Plasma, Bone Marrow Concentrate and Adipose Mesenchymal Stem Cells: Hemoanalysis and Quality Control—What Does it All Mean?" by David C. Karli, MD, MBA, in *Essentials of Regenerative Medicine in Interventional Pain Management* published by ASIPP Publishing.

## References

1. Karli DC. Platelet rich plasma, bone marrow concentrate and adipose mesenchymal stem cells: hemoanalysis and quality control—what does it all mean? In: Manchikanti L, Navani A, Atluri S, editors. Essentials of regenerative medicine in interventional pain management. Paducah, KY: ASIPP Publishing; 2019. p. 173–82.
2. Degen RM, Bernard JA, Oliver KS, et al. Commercial separation systems designed for preparation of platelet-rich plasma yield differences in cellular composition. HSS J. 2017;13:75–80.
3. Yoshimura K, Shigeura T, Matsumoto D, et al. Characterization of freshly isolated and cultured cells derived from the fatty and fluid portions of liposuction aspirates. J Cellular Physiol. 2006;208:64–76.
4. Polancec D, Zenic L, Hudetz D, et al. Immunophenotyping of a stromal vascular fraction from microfragmented lipoaspirate used in osteoarthritis cartilage treatment and its lipoaspirate counterpart. Genes (Basel). 2019;10:474.
5. Ragni E, Viganò M, Torretta E, et al. Characterization of microfragmented adipose tissue architecture, mesenchymal stromal cell content and release of paracrine mediators. J Clin Med. 2022;11:2231.
6. Zenic L, Polancec D, Hudetz D, et al. Polychromatic flow cytometric analysis of stromal vascular fraction from lipoaspirate and microfragmented counterparts reveals sex-related immunophenotype differences. Genes (Basel). 2021;12:1999.
7. Condé-Green A, Kotamarti VS, Sherman LS, et al. Shift toward mechanical isolation of adipose-derived stromal vascular fraction: review of upcoming techniques. Plast Reconstr Surg Glob Open. 2016;4:e1017.
8. Mishra A, Harmon K, Woodall J, et al. Sports medicine applications of platelet rich plasma. Curr Pharm Biotechnol. 2012;13:1185–95.
9. Mautner K, Malanga GA, Smith J, et al. A call for a standard classification system for future biologic research: the rationale for new PRP nomenclature. PM R. 2015;7:S53–9.
10. Magalon J, Brandin T, Francois P, et al. Technical and biological review of authorized medical devices for platelets-rich plasma preparation in the field of regenerative medicine. Platelets. 2021;32:201–9.
11. Murphy MB, Moncivais K, Caplan AI. Mesenchymal stem cells: environmentally responsive therapeutics for regenerative medicine. Exp Mol Med. 2013;45:e54.
12. Wang SW, Wei XC, Zhou JM, et al. Identification of α 2-macroglobulin as a master inhibitor of cartilage-degrading factors that attenuates the progression of posttraumatic osteoarthritis. Arth Rheum. 2014;66:1843–53.
13. Cassano JM, Kennedy JG, Ross KA, et al. Bone marrow concentrate and platelet-rich plasma differ in cell distribution and interleukin 1 receptor antagonist protein concentration. Knee Surg Sports Traumatol Arthrosc. 2018;26:333–42.
14. van Dongen JA, Stevens HP, Parvizi M, et al. The fractionation of adipose tissue procedure to obtain stromal vascular fractions for regenerative purposes. Wound Rep Reg. 2016;24:994–1003.
15. Seaman SA, Tannan ST, Cao Y, et al. Differential effects of processing time and duration of collagenase digestion on human and murine fat grafts. Plast Reconstr Surg. 2015;136:189e–99e.

16. Mantripragada VJ, Piuzzi NS, Bova GJ, et al. Reliable assessment of bone marrow and bone marrow concentrates using automated heamatology analyzer. Regen Med. 2019;14:639–46.
17. Freidenstein AJ, Chailakhyan RK, Latsinik NV, et al. Stromal cells responsible for transferring the microenvironment of the hemopoietic tissues. Transplantation. 1974;17:331–40.
18. Mafi P, Hindocha S, Mafi R, et al. Adult mesenchymal stem cells and cell surface characterization—a systematic review of the literature. Open Orthop J. 2011;5:253–60.
19. Pettine KA, Murphy MB, Suzuki RK, et al. Percutaneous injection of autologous bone marrow concentrate cells significantly reduces lumbar discogenic pain through 12 months. Stem Cells. 2015;33:146–56.
20. Mckinnirey F, Herbert B, Vesey G, et al. Immune modulation via adipose derived mesenchymal stem cells is driven by donor sex in vitro. Sci Rep. 2021;11:12454.
21. Fransson M, Benson M, Erjefält JS, et al. Expression of toll-like receptor 9 in nose, peripheral blood and bone marrow during symptomatic allergic rhinitis. Resp Res. 2007;8:17.
22. Li HZ, Ghazanfari R, Zacharaki D, et al. Isolation and characterization of primary bone marrow mesenchymal stromal cells. Ann N Y Acad Sci. 2016;1370:109–18.
23. Pettine KA, Suzuki RK, Sand TT, et al. Autologous bone marrow concentrate intradiscal injection for the treatment of degenerative disc disease with three-year follow-up. Int Orthop. 2017;41:2097–103.
24. Hernigou P, Flouzat-Lanchaniette CH, Delambre J, et al. Biologic augmentation of rotator cuff repair with mesenchymal stem cells during arthroscopy improves healing and prevents further tears: a case-controlled study. Int Orthop. 2014;38:1811–8.
25. Hernigou P, Bouthors C, Bastard C, et al. Subchondral bone or intra-articular injection of bone marrow concentrate mesenchymal stem cells in bilateral knee osteoarthritis: what better postpone knee arthroplasty at fifteen years? A randomized study. Int Orthop. 2021;45:391–9.
26. Marques LF, Stessuk T, Camargo IC, et al. Platelet-rich plasma (PRP): methodological aspects and clinical applications. Platelets. 2015;26:101–13.
27. Fitzpatrick J, Bulsara MK, McCrory PR, et al. Analysis of platelet-rich plasma extraction: variations in platelet and blood components between 4 common commercial kits. Ortho J Sports Med. 2017;5:2325967116675272.
28. Lewis E, Merghani K, Robertson I, et al. The effectiveness of leucocyte-poor platelet-rich plasma injections on symptomatic early osteoarthritis of the knee: the PEAK randomized controlled trial. Bone Joint J. 2022;104-B:663–71.
29. Bennell KL, Paterson KL, Metcalf BR, et al. Effect of intra-articular platelet-rich plasma vs placebo injection on pain and medial tibial cartilage volume in patients with knee osteoarthritis. The RESTORE randomized clinical trial. JAMA. 2021;326:2021–30.
30. Yurtbay A, Say F, Çinka H, et al. Multiple platelet-rich plasma injections are superior to single PRP injections or saline in osteoarthritis of the knee: the 2-year results of a randomized, double-blind, placebo-controlled clinical trial. Arch Orthop Trauma Surg. 2022;142:2755–68.
31. Mulford J. Authors' reply to Sussman, et al letter to the editor. Bone Joint J. 2022;104:663.
32. Sussman WI, Murrell WD, Jayaram P, et al. Letter to the editor in response to Lewis, et al. [The effectiveness of leucocyte-poor platelet-rich plasma injections on symptomatic early osteoarthritis of the knee: the PEAK randomized controlled trial. Bone Joint J 2022; 104-B(6):663–671] Bone Joint J. 2022.
33. Hernigou P, Homma Y, Flouzat-Lachaniette CH, et al. Benefits of small volume and small syringe for bone marrow aspirations of mesenchymal stem cells. Int Ortho. 2013;37:2279–87.
34. Hernigou J, Alves A, Homma Y, et al. Anatomy of the ilium for bone marrow aspiration: map of sectors and implications for safe trocar placement. Int Ortho. 2014;38:2585–90.
35. Hernigou J, Picard L, Alves A, et al. Understanding bone safety zones during bone marrow aspiration from the iliac crest: the sector rule. Int Ortho. 2014;38:2377–84.
36. Riboh JC, Saltzman BM, Yanke AB, et al. Effect of leukocyte concentration on the efficacy of platelet-rich plasma in the treatment of knee osteoarthritis. Am J Sports Med. 2016;44:793–800.
37. Alves R, Grimalt R. A review of platelet-rich plasma: history, biology, mechanism of action, and classification. Skin Appendage Disord. 2018;4:18–24.

# Antiplatelet and Anticoagulant Therapy in Regenerative Medicine

# 15

Laxmaiah Manchikanti, Mahendra Sanapati, Alan D. Kaye, and Adam M. Kaye

Implications of antithrombotic and anticoagulant therapy during interventional procedures have been updated and multiple guidelines have been developed by various organizations [1–6]. The use of prescription medications to manage thrombosis risk and the ease of availability of over-the-counter (OTC) and herbal products that mediate or modulate the coagulation cascade is increasing [3–20]. Interventional pain physicians frequently encounter the challenge of the potential risk of bleeding or thrombosis in perioperative management of the patients receiving antithrombotic and anticoagulant therapy [15, 19, 21–43]. Modulation of anticoagulant and antithrombotic therapy during performance of interventional techniques in general, and injection of biologics in particular, is one of the major clinical decisions often made without precise evidence-based literature to support existing opinions [3, 15, 19, 22–44].

Regenerative medicine is an emerging specialty with ever-increasing scientific developments, evidence-based indications, and utilization patterns for various musculoskeletal and spinal disorders. While interventional techniques are frequently performed for injections of biologics, the major impetus of regenerative therapy is related to intraarticular injections and advanced procedures with entering bone marrow. Effects of antithrombotic therapy on platelet-rich plasma (PRP) and stem cell therapy, as well as risks of various interventions in regenerative medicine have not been specifically described [44]. In addition to the bleeding, which is based on biochemical pathways of coagulation cascade, by disrupting platelet surfaces or prematurely activating platelets, may influence the outcomes.

Leading causes of morbidity and mortality worldwide include cardiovascular and cerebrovascular disease [6, 7, 11, 14, 45–48]. Additionally, one of the leading causes of disability and functional impairment across the globe is chronic persistent pain [49–59]. Therefore, the overlap between chronic persistent pain and cardiovascular disease has a synergistic impact on physical and psychological health, affecting the performance of social responsibilities, including work and family life. Antithrombotic therapy has a clear evidence-based foundation with a favorable risk-benefit profile for prevention and management of cardiovascular disease, including limiting the present and future burden of cardiac or cerebrovascular infarcts [3, 5–7, 20, 60–73]. A significant portion of patients with established cerebrovascular, cardiovascular, or peripheral vascular disease who are receiving antithrombotic therapy, are commonly in need of interventional pain management.

Based on published guidelines, derived by clinical case reports and consensus, a large subset of clinicians report stopping antiplatelet therapy and consider this concept as a standard of care [3, 19, 24, 27, 29, 74–80]. Although the overall incidence of bleeding complications and epidural hematoma in the nonobstetric epidural literature has been reduced, the incidence has been higher with procedures involving cervical and thoracic spine in interventional procedures and is growing, with or without anticoagulant therapy [3, 22–32, 42–44, 81–100]. Of note, the clinical literature reports of incidences of epidural hematoma cases accompa-

L. Manchikanti (✉)
Pain Management Centers of America, Paducah, KY, USA
e-mail: drlm@thepainmd.com

M. Sanapati
Pain Management Centers of America, Evansville, IN, USA

A. D. Kaye
Pharmacology, Toxicology, and Neurosciences, LSU School of Medicine, Shreveport, LA, USA

Anesthesiology and Pharmacology, LSU School of Medicine, New Orleans, LA, USA

Anesthesiology and Pharmacology, Tulane School of Medicine, New Orleans, LA, USA
e-mail: alan.kaye@lsuhs.edu

A. M. Kaye
Department of Pharmacy Practice, Thomas J. Long School of Pharmacy and Health Sciences, University of the Pacific, Stockton, CA, USA
e-mail: akaye@pacific.edu



A. Navani et al. (eds.), *Essentials of Regenerative Medicine in Interventional Pain Management*,
https://doi.org/10.1007/978-3-031-50357-3_15

nying interventional techniques and neuraxial techniques are increasing rapidly [3, 19, 22, 32, 44, 101–117]. While some reports indicate a decreasing incidence of bleeding complications related to neuraxial techniques [114], multiple studies have been performed assessing the prevalence and risks related to bleeding complications and epidural hematoma with neuraxial procedures, specifically epidural injections [3, 19, 22–32, 40, 41, 44, 109–117]. In fact, anticoagulants have been described to cause the most serious adverse events in the US. According to the Centers for Disease Control and Prevention (CDC), in the US in 2011, 48% of the US population were taking a prescription medicine in any given month, and 11% were taking 5 or more prescribed medicines. The US Food and Drug Administration (FDA) estimated that the number of reports it receives represents only the tip of the "iceberg" [11]. Among the healthcare providers and patients, anticoagulant drugs Warfarin and dabigatran were on the top [11].

Significant risks of withdrawing antiplatelet therapy include cardiovascular, cerebrovascular, and peripheral vascular thrombosis. In addition, the risks are higher in chronic pain patients as chronic psychosocial stress causes a hypercoagulable state, as reflected by increased procoagulant molecular fibrinogen (or coagulation factor 7), reduced fibrinolytic capacity and increased platelet activity [3, 4, 117–119]. Stress has been shown to affect coagulation activity via an influence on the regulation of genes coding for coagulation and fibrinolysis molecules [119] with increase in hormonal levels [120–122] and an underlying surge of catecholamine and cortisol induced hypercoagulability [122]. A prior systematic review and meta-analysis explored the hazards of discontinuing or altering aspirin regimens among patients at risk for coronary artery disease [62]. Importantly, in patients at moderate to high risk for coronary artery disease, withdrawal or noncompliance with aspirin therapy was associated with three-fold higher risk of major adverse cardiac events and the risk was magnified in patients with coronary stents. The findings support the recommendation that aspirin discontinuation in this patient population should be advocated only under circumstances where the risk of adverse outcomes caused by bleeding risk clearly outweighs that of catastrophic atherothrombotic events. These findings have been confirmed in later studies [63–70, 117]. Conversely, recently published large-scale evidence [16–18] shows lack of benefit of aspirin for primary prevention of cardiovascular events, but aspirin therapy is associated with increased bleeding episodes. Thus, current evidence suggests that the risks of coronary events related to patients abstaining from their antiplatelet medications during the perioperative period are more serious when compared to the risks of continuing antiplatelet therapy through the perioperative period.

Multiple publications [3, 23, 24, 32, 36, 109, 112] have also supported the concept of continuing antiplatelet/anticoagulant agents in patients undergoing various interventional pain procedures in light of identical complication rates as compared to the patients who stop taking these for a particular recommended period. It should be noted that there are only a few clinical reports of epidural hematoma available in patients undergoing interventional therapies for chronic pain, which included patients both continuing and discontinuing antithrombotic therapy.

However, there are no reports available of hematoma formation related to injection of biologics in regenerative medicine practices.

## Prevalence and Risk Assessment of Studies

Multiple studies have been published assessing the prevalence, as well as the risk of bleeding with interventional techniques [22–25, 28, 31, 32, 40–43]. In a survey of practice patterns among interventional pain physicians in 2012, Manchikanti et al. [22] showed that the majority of physicians discontinued antithrombotic agents; however, this study also showed that there were a significantly higher number of complications related to thromboembolic events of a total of 162 compared to hemorrhagic complications of a total of 55 in this population.

In a prospective evaluation of bleeding risks for interventional techniques in chronic pain, Manchikanti et al. [23] assessed the rates of adverse events in patients undergoing interventional techniques on antithrombotic therapy with cessation or without cessation and compared them to a group of patients without antithrombotic therapy. While the results showed differences in milder complications, there were no reports of hemorrhagic complications requiring any type of treatment. In this assessment, the authors studied all types of procedures with 1227 of 1831 continuing aspirin compared to 604 of 1831 discontinuing them. Similarly, they also studied 100 patients on clopidogrel with continuation, whereas 226 patients were discontinued. Further, there were 128 patients with aspirin and other agents with continuation and 151 were discontinued. The procedures performed included cervical epidural injections with continued aspirin in 249 patients, thoracic epidural in 30 patients, lumbar interlaminar epidural in 128 patients, lumbar transforaminal in 144 patients, whereas 528 patients for caudal epidural injections, and 148 for percutaneous adhesiolysis. In reference to clopidogrel, it was continued in 10 patients undergoing cervical epidural, one patient with thoracic epidural, 14 patients with lumbar epidural, 44 patients with caudal epidurals, 10 with lumbar transforaminal epidural, and 21 with percutaneous adhesiolysis. There were a large number of facet joint interventions and other treatments.

Warner et al. [43] in a publication describing bleeding and neurological complications in 58,000 interventional pain procedures showed that preprocedural aspirin or nonsteroidal anti-inflammatory drug (NSAID) therapy was prevalent in 17,825 procedures or 30.7% of the procedures without significant bleeding complications. Out of total of 58,066 procedures performed in the study, 22.4% of the procedures were performed with perioperative administration of aspirin within 7 days, 12.1% of the patients with administration of NSAIDs within 7 days, 1.6% of clopidogrel within 7 days. The study also included 3880 lumbar epidural injections, 304 thoracic interlaminar injections with a large number of epidural injections with over 50% of the epidural injections not assigned to a region. They concluded that bleeding complications were rare in patients undergoing low or intermediate risk pain procedures even in the presence of antiplatelet medication.

The comprehensive review of the studies assessing the risk of thrombosis and bleeding with interventional pain management techniques. All studies except 2 studies in this assessment are related to performing interventional techniques without cessation of antithrombotic therapy [22–25, 28, 30–32, 39, 41]. Only one study [22] was related to an online physician survey and study of spinal hematoma with neuraxial blocks. Among all the studies, only 2 studies included epidural injections [23, 32]. All others have performed a large portion of procedures with low risk or intermediate risk including transforaminal epidural injections.

A systematic review of risks and benefits of seizing or continuing anticoagulant medication for image-guided procedures for spine pain by Smith et al. [24], including 14 articles assessing the role of antithrombotics in interventional pain management. They showed that procedures involving interlaminar access carry a nonzero risk of hemorrhagic complications, regardless of whether anticoagulants are seized or continued. For other procedures, hemorrhagic complications have not been reported, and case series indicate that they are safe when performed in patients who continue anticoagulants. Among the reports they reviewed, 3 of them reported the adverse effects of seizing anticoagulants, with serious consequences, including death. They concluded that other than for interlaminar procedures, the evidence does not support the view that anticoagulant and antiplatelet medication must be seized before image-guided spine pain procedures.

Thus, based on the evidence presented from reviews and primary studies, it appears that there is no significant difference whether antithrombotic therapy is discontinued or continued in reference to the bleeding. A majority of the authors have studied intermediate and low risk procedures without the inclusion of epidural injections. The only one study reviewing spinal cord stimulation also showed lack of increased risk with continuation of aspirin and other NSAIDs. Consequently, cessation of anticoagulant medication is recommended only for interlaminar epidural injections in consultation with the patient and other healthcare providers managing antithrombotic medications, instead of seizing by providing them with mandatory instructions to stop for 7–10 days, or even longer.

## Reports of Thromboembolic Events

There have been multiple reports of thromboembolic events with the discontinuation of antithrombotics and anticoagulants prior to performance of interventional procedures [1]. Manchikanti et al. [22] in an assessment of practice patterns of perioperative management of antiplatelet and anticoagulant therapy in interventional pain management reported 162 thromboembolic events compared to 55 serious bleeding complications from epidural hematomas. This study showed thromboembolic events were 3 times more frequent than bleeding complications. Further, they also showed bleeding complications from epidural hematomas were similar whether antiplatelet therapy was continued or discontinued with an occurrence of 26 versus 29 respectively; in this survey the sample sizes were not provided. Consequently, it is difficult to assess the exact risk of bleeding complications and similarly thromboembolic events. Endres et al. [32] reported 9 patients with thromboembolic events after cessation of anticoagulant therapy out of 1626 procedures. These complications included 2 deaths 5 strokes, one pulmonary embolism, and one myocardial infarction in patients when anticoagulants were stopped; however, they have not reported any bleeding complications in patients where anticoagulants were continued. Kumar et al. [123] reported a case of pulmonary embolism after discontinuation of warfarin during a spinal cord stimulation trial. Linn et al. [124] also reported right middle cerebral artery infarction with persistent left hemiparesis, neglect and dysarthria with L5-S1 epidural steroid injection after discontinuation of warfarin for 9 days pre-procedure. Manchikanti et al. [109] in providing 2 case reports and a literature review described 2 cases of thromboembolic complications with cessation of antithrombotic therapy. Table 15.1 shows reported thromboembolic and cardiovascular complications related to discontinuation of antiplatelet or anticoagulation therapy.

## Reports of Bleeding Complications

There were multiple case reports discovered citing epidural hematoma in patients with or without continued antithrombotic therapy during an interventional technique [101–110, 125–146].

Multiple reports of bleeding and epidural hematoma associated with interventional techniques in patients without antithrombotic therapy have been published [11, 81–83, 85, 86, 88–99, 107, 109, 112–126, 129, 131, 135, 137, 138, 142, 144, 146–153], whereas several reports of bleeding in patients with discontinued antithrombotic therapy also have been published [11, 101, 104–106, 109, 110, 115–127, 132–134, 136]. Reports of bleeding complications and epidural hematoma in patients with continuation of antiplatelet therapy with interventional techniques also have been published. Multiple reports of bleeding complications with continuation of anticoagulant therapy during interventional techniques were identified [102–104, 107, 108, 128, 143, 145].

**Table 15.1** Thromboembolic and cardiovascular complications related to discontinuation of antiplatelet or anticoagulation therapy

| Study | Type of study | Complications |
|---|---|---|
| Endres et al. [32] | Observational report of interventional techniques | • 2 patients died<br>• 5 suffered strokes<br>• 1 suffered pulmonary embolism<br>• 1 suffered myocardial infarction |
| Kumar et al. [123] | Case report of dorsal column stimulator trial | • Pulmonary embolism without lasting complications |
| Linn et al. [124] | Case report of L5/S1 epidural steroid injection | • Right middle cerebral artery infarction with persistent left hemiparesis, neglect, and dysarthria |
| Manchikanti et al. [22] | Online survey | • Reports of epidural hematoma: 55<br>• Reports of thromboembolic complications: 162 |
| Manchikanti et al. [109] | Case report and literature review of interventional techniques | • 2 cases of thromboembolic complications with cessation of antithrombotic therapy. |

Manchikanti L, et al. Epidural interventions in the management of chronic spinal pain: American Society of Interventional Pain Physicians (ASIPP) comprehensive evidence-based guidelines. *Pain Physician* 2021; 24:S27-S208 [1] and Kaye AD, et al. Responsible, safe, and effective use of antithrombotics and anticoagulants in patients undergoing interventional techniques: American Society of Interventional Pain Physicians (ASIPP) guidelines. *Pain Physician* 2019; 22:S75-S128 [3]

Figures 15.1, 15.2 and 15.3 show summary reports of several epidural hematomas with epidural injections, acupuncture and dry needling, and spinal cord stimulation lead placement in patients without antithrombotic therapy, with antithrombotic therapy withheld for an appropriate duration, and with antithrombotic therapy continued. A total of 46 epidural hematomas were described in 43 case reports. There were 23 in the cervical spine, 8 in the thoracic spine, and 15 in the lumbar spine. Of these, 21 patients were not on antiplatelet therapy, 9 patients had their antithrombotic therapy discontinued, 14 continued antithrombotic therapy, and 2 cases were due to fish oil and one case of ketorolac and paroxetine. Further, as shown in Fig. 15.3, epidural injections were responsible for 33 cases of hematomas, 7 cases were secondary to acupuncture or dry needling and 6 were related to spinal cord stimulation. There was one case report of caudal epidural injection with cilostazol [145], and in one case report, we were unable to obtain the full article [150]. Cases of chronic subdural hematoma and cases of abdominal hematomas were not included.

## Guidelines and Recommendations

Guidelines and recommendations were based on risk stratification, pharmacology of antithrombotics and anticoagulants, and the application of available evidence.

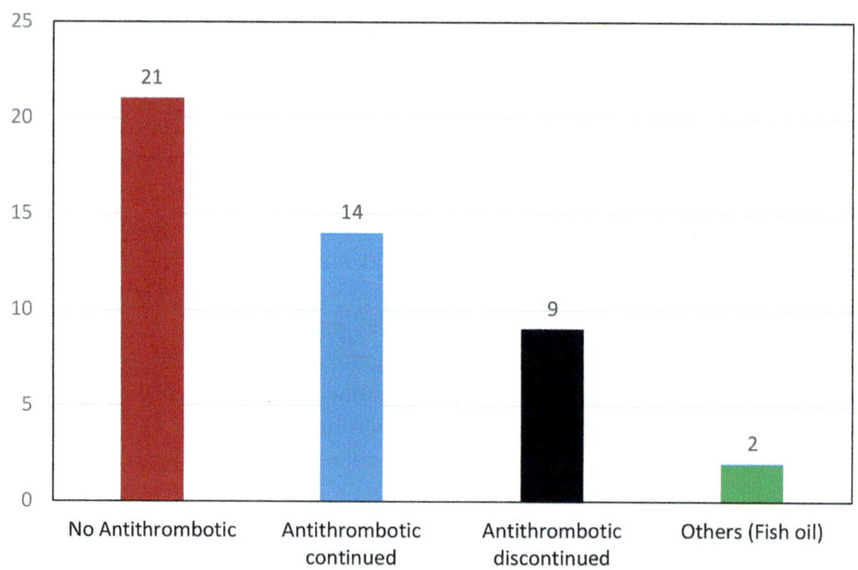


**Fig. 15.1** *Epidural hematoma incidence based on presence or absence of anticoagulant/antiplatelet therapy following administration of epidural injections.* (Source: Manchikanti L, et al. Epidural interventions in the management of chronic spinal pain: American Society of Interventional Pain Physicians (ASIPP) comprehensive evidence-based guidelines. *Pain Physician* 2021; 24:S27-S208 [1])

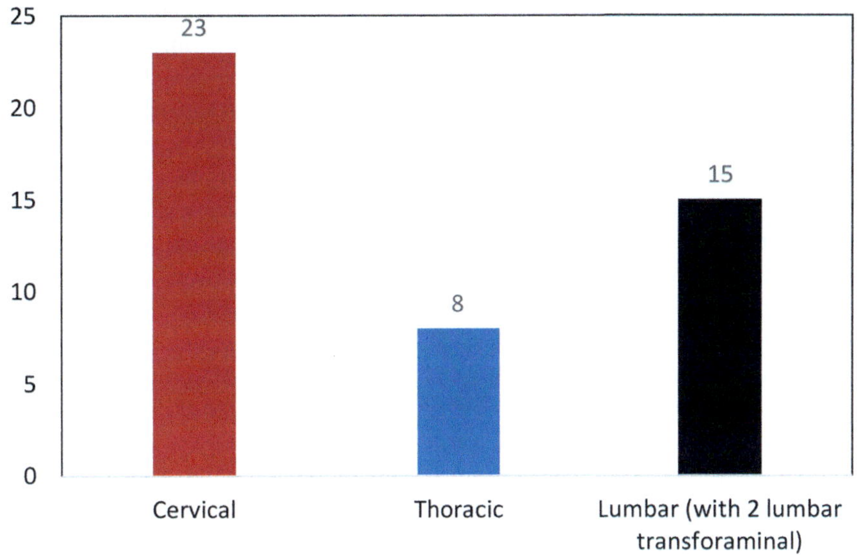


**Fig. 15.2** *Epidural hematoma incidence based on spinal regions.* (Source: Manchikanti L, et al. Epidural interventions in the management of chronic spinal pain: American Society of Interventional Pain Physicians (ASIPP) comprehensive evidence-based guidelines. *Pain Physician* 2021; 24:S27-S208 [1])

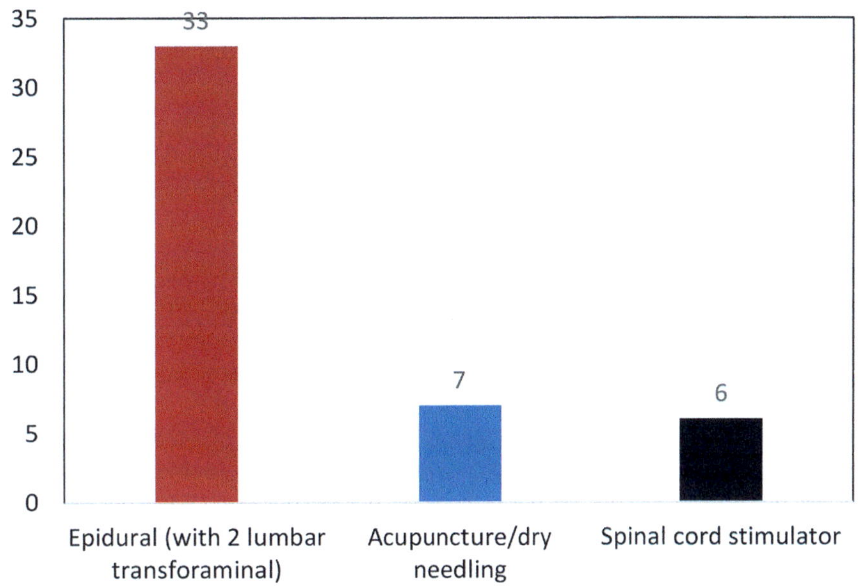


**Fig. 15.3** *Epidural hematoma cases reported based type of procedure.* (Source: Manchikanti L, et al. Epidural interventions in the management of chronic spinal pain: American Society of Interventional Pain Physicians (ASIPP) comprehensive evidence-based guidelines. *Pain Physician* 2021; 24:S27-S208 [1])

## Risk Stratification

Interventional techniques performed in the spine and other regions for chronic cancer and noncancer pain patients face variable risks depending on anticoagulant or antithrombotic therapy, age, anatomy, the specific region of interest, and obesity. Consequently, various authors have described procedural classifications according to the potential risk for serious bleeding.

Raj et al. [154] stratified risk scores based on technique related bleeding risk and patient-related bleeding risk factors. This risk classification took various factors into consideration including a sharp or blunt needle, as well as the use of fluoroscopy and lack of fluoroscopy.

Breivik et al. [27], in a comprehensive topical review of reducing risk of spinal hematoma from spinal epidural and pain procedures, based their recommendations on an extensive review of 166 case reports published from 1994 through 2015 [42, 43], pharmacology of drugs, and available clinical evidence relating to complications whether the antithrombotics were continued or discontinued.

Narouze et al. [4] provided guidance for interventional spine and pain procedures in patients on antiplatelet and anticoagulant medications. They provided an extensive review of the literature, considered pharmacology, and current literature with development of risk stratification.

Deer et al. [5] provided recommendations on bleeding and coagulation management in neurostimulation devices. They also provided bleeding risk stratifications for neuromodulation procedures. Their classification showed spinal cord stimulation trial and implant, dorsal root ganglion stimulation, and intrathecal catheter and pump implant as high to intermediate risk neuromodulation procedures. Further, they classified deep brain stimulation and motor cortex stimulation as high-risk neuromodulation procedures. This is in contrast to other guidance where spinal cord stimulation trial and implant, dorsal root ganglion stimulation, and intrathecal catheter and pump implant are considered as high risk.

Oprea et al. [19] published risk stratification, perioperative, and periprocedural management of patients receiving anticoagulant therapy based on bleeding risk for each procedure, pharmacology, and evidence of the risk of the development of bleeding complications as well as thromboembolic risks.

Bogduk [155] published an editorial in reference to ceasing anticoagulants to interventional pain procedures. Reviewing multiple studies, he opined that for lumbar transforaminal injections and lumbar facet joint injections with 2 anticoagulants (warfarin and clopidogrel) with large sample sizes, showed statistical significance that continuing anticoagulants was safer than ceasing anticoagulants.

Buchner et al. [156] published evaluation of a clinical decision support system for the most evidence-based approach to managing perioperative coagulation. In doing so, authors utilized European guidelines; however, they discussed various other guidelines. These guidelines were developed by anesthesiologists for surgical procedures. They started with the premise that management of anticoagulant therapy is complex, particularly the decision to pause and restart or to bridge anticoagulation is crucial, because any incorrect treatment increases the risk for perioperative bleeding, thromboembolism, and mortality. In this clinical decision support system, they utilized similar philosophy as Raj et al. [154] described years ago, identifying the anticoagulant medication associating with intervention and anatomy leading to risk evaluation and finally bleeding risk to summarize the requirements and provide recommendations.

Multiple factors and the available literature in reference to the adverse consequences of anticoagulant and antithrombotic therapy with interventional techniques were utilized in developing risk stratification. The following classification for interventional techniques has been developed (Table 15.2) [3, 19, 22–30, 42, 43, 143, 154]. The classification describes low risk, intermediate risk, and high-risk procedures. However, based on comorbid medical conditions and other risk factors of coagulopathies and concurrent use of other anticoagulants and antiplatelets, the procedural risk classification may be changed from low risk to intermediate, and intermediate to high risk.

Of all the low-risk procedures as shown in Table 15.2, the caudal epidural injection was reported with an epidural hematoma with continuation of cilostazol [145]. Justifiably, we have included that caudal percutaneous adhesiolysis has not been reported with epidural hematomas. Lumbar transforaminal epidural injections at L4, L5, S1 have been provided with 2 case reports resulting in hematoma without antithrombotic therapy. This is extremely unusual and probably coincidental. Consequently, with caudal epidural injections and other low risk procedures, antithrombotic and anticoagulant therapy may be continued with the appropriate guidelines.

Intermediate risk procedures include multiple procedures performed, constituting a great proportion of the procedural prevalence. This risk stratification is somewhat different from the one proposed by American Society of Regional Anesthesia and Pain Medicine (ASRA) [4]. Procedures performed with interlaminar epidural access are included in the high-risk procedures. A majority of the lumbar epidural hematomas developed despite the discontinuation of antithrombotic therapy, and these were performed above the L5-S1 interspace. If epidural hematoma develops at L5-S1 it will have significantly higher space availability to be asymptomatic and to be managed conservatively.

The high-risk procedures include the majority of the procedures performed in the cervical and thoracic spine. These guidelines are in contrast to ASRA guidelines with the inclusion of cervical, thoracic, and lumbar due to the available literature and potential issues related to epidural hematoma requiring surgical exploration and spinal cord damage.

## Pharmacologic Aspects and Hemostasis Monitoring

The main categories of antithrombotics and anticoagulants are described as; platelet inhibitors, interfering with platelet aggregation (clumping) and thrombus formation; anticoagulants interfering with formation of the clotting, thereby reducing fibrin formation and preventing clots from forming and expanding; and fibrinolytics interfering with the final clot.

Monitoring of hemostasis is performed with multiple standard tests including platelet count, activated partial thromboplastin time (APTTa), and international normalized ratio (INR) [27]. A normal INR is considered as 0.9–1.2. Consequently, for epidural injections, an INR of less than 1.5 is ideal for high risk and moderate risk procedures, and less than 1.8–2.00 is ideal for low-risk procedures.

Additional advanced hemostatic tests for monitoring of hemostasis include multiple viscoelastic tests like thrombo-

**Table 15.2** ASIPP's classification of interventional techniques based on the potential risk for bleeding compared to ASRA guidelines [4]

| Low-risk procedures | | Intermediate-risk procedures [a] | | High-risk procedures [a] | |
|---|---|---|---|---|---|
| ASIPP | ASRA | ASIPP | ASRA[a] | ASIPP | ASRA[a] |
| 1. Trigger point and muscular injections (including piriformis injection)<br>2. Peripheral nerve blocks<br>3. Sacroiliac joint and ligament injections and nerve blocks<br>4. Facet joint interventions (intra-articular injections, nerve blocks and radiofrequency neurotomy)<br>5. Intraarticular injections of extremity joints<br>6. Pocket revision and implantable pulse generator/ intrathecal pump replacement<br>7. Peripheral nerve stimulation trial and implant<br>8. Caudal epidural injections<br>9. Lumbar transforaminal epidural injections at L4, L5, S1<br>10. Caudal percutaneous adhesiolysis<br>11. Ganglion impar blocks | 1. Trigger point injections including piriformis injection<br>2. Peripheral nerve blocks<br>3. Sacroiliac joint injection and sacral lateral branch blocks<br>4. Thoracic and lumbar facet medial branch nerve block and radiofrequency ablation<br>5. Peripheral joints and musculoskeletal injections<br>6. Pocket revision and implantable pulse generator/ intrathecal pump replacement<br>7. Peripheral nerve stimulation trial and implant [b] | 1. Cervical, thoracic and lumbar transforaminal epidural injections above L3<br>2. Lumbar intradiscal procedures<br>3. Hypogastric plexus blocks<br>4. Lumbar sympathetic blocks<br>5. Lumbar interlaminar below L4 | 1. Transforaminal epidural steroid injections (cervical, thoracic, lumbar, sacral)<br>2. Intradiscal procedures (cervical, thoracic, lumbar)<br>3. Sympathetic blocks (stellate, thoracic, splanchnic, celiac, lumbar, hypogastric)<br>4. Interlaminar epidural steroid injections (cervical, thoracic, lumbar, sacral)<br>5. Cervical [c] facet medial branch nerve block and radiofrequency ablation<br>6. Trigeminal and sphenopalatine ganglia blocks | 1. Spinal cord stimulator trial and implant<br>2. Intrathecal catheter and pump implant<br>3. Vertebral augmentation, lumbar, thoracic and cervical<br>4. Percutaneous disc decompression<br>5. Minimally invasive lumbar decompression (MILD)<br>6. Cervical, thoracic, and lumbar interlaminar epidurals above L4<br>7. Trigeminal and cranial nerve stimulation<br>8. Percutaneous adhesiolysis with interlaminar or transforaminal approach<br>9. Discography<br>10. Intradiscal procedures<br>11. Endoscopic disc decompression<br>12. Sympathetic blocks (stellate ganglion; thoracic splanchnic, celiac plexus)<br>13. Trigeminal ganglion and sphenopalatine ganglion blocks<br>14. Intervertebral spinous prosthesis<br>15. Lumbar lateral spine fusion with arthrodesis of vertebral space<br>16. SI joint fusion | 1. Spinal cord stimulation trial and implant<br>2. Intrathecal catheter and pump implant<br>3. Vertebral augmentation (vertebroplasty and kyphoplasty)<br>4. Percutaneous decompression laminotomy<br>5. Epiduroscopy and epidural decompression |

**Source(s):** Manchikanti L, et al. Epidural interventions in the management of chronic spinal pain: American Society of Interventional Pain Physicians (ASIPP) comprehensive evidence-based guidelines. *Pain Physician* 2021; 24:S27-S208 [1]; Kaye AD, et al. Responsible, safe, and effective use of antithrombotics and anticoagulants in patients undergoing interventional techniques: American Society of Interventional Pain Physicians (ASIPP) guidelines. *Pain Physician* 2019; 22:S75-S128 [3]; and Narouze S, Benzon HT, Provenzano D, et al. Interventional spine and pain procedures in patients on antiplatelet and anticoagulant medications (Second Edition): Guidelines from the American Society of Regional Anesthesia and Pain Medicine, the European Society of Regional Anaesthesia and Pain Therapy, the American Academy of Pain Medicine, the International Neuromodulation Society, the North American Neuromodulation Society, and the World Institute of Pain. *Reg Anesth Pain Med* 2018; 43:225–262 [4]

[a] Patients with high risk of bleeding (e.g., old age, history of bleeding tendency, concurrent uses of other anticoagulants/antiplatelets, liver cirrhosis or advanced liver disease, and advanced renal disease) undergoing low or intermediate-risk procedures should be treated as intermediate or high risk, respectively

[b] Peripheral neuromodulation is low to intermediate risk, depending on the location of the targeted nerve in relation to critical vessels and the invasiveness of the procedure

[c] There is rich neck vascularity in the vicinity of the target structure(s) (refer to the section entitled Anatomical Considerations for Hematoma Development in Spinal and Nonspinal Areas)

elastography or thromboelastrometry extensively used to evaluate liver disease [157]. However, this parameter has been studied for safe epidural catheter removal with the conclusion that the tests were not well validated in this context and there were frequent false negative test results [158, 159]. Even then, a clearly abnormal curve indicates deranged hemostasis and must be taken seriously [159, 160].

Platelet function in patients on aspirin prior to interventional pain management procedures has been measured by the closure time using collagen-epinephrine cartridge in a commercial platelet-functional analyzer [161]. Seo et al. [161] in a cross- sectional study showed the high prevalence of normal platelet function (56.4%) in patients taking aspirin suggesting no necessity of discontinuation before procedures in such patients. Abnormal platelet function can also occur in patients who are not taking aspirin.

Pharmacological aspects of NSAIDs and aspirin (Table 15.3), Adenosine Diphosphate Receptor Inhibitors (Table 15.4), Phosphodiesterase (PDE) Inhibitors, Glycoprotein GPIIb/IIIa Inhibitors, Warfarin, direct thrombin inhibitors (Table 15.5), direct factor Xa Inhibitors (Table 15.6), and thrombolytic agents have been described.

Cannabis (THC, CBD), omega-3 fatty acids, and vitamin E, selective serotonin inhibitors (SSRIs), and multiple herbal alternative therapies also have been described.

**Table 15.3** *Characteristics of aspirin and NSAIDs*

| | Aspirin or acetylsalicylic acid (Oral Low Dose) | NSAIDs |
|---|---|---|
| Target | COX-1 irreversible | COX-1 reversible, COX-2 |
| Time to peak effect | 0.5 h | Varies |
| Plasma Half Life | 0.5 h | Variable from 1 to 72 h |
| Renal elimination | + | + |
| Time to 50% recovery of platelet function | 3 days | 1 day |
| Hours to C-Max | 0.5 h | ~0.5 h |
| Metabolism | Hepatic | Hepatic |
| Bioavailability | 60% | 50–95% |
| Antihemostatic effect | ++ | + |

*NSAID* non-steroidal anti-inflammatory drug, *COX* cyclo-oxygenase
Rating of antihemostatic effect and renal elimination: (+) insignificant, + low, ++ moderate, +++ pronounced, ++++ high
Manchikanti L, et al. Epidural interventions in the management of chronic spinal pain: American Society of Interventional Pain Physicians (ASIPP) comprehensive evidence-based guidelines. *Pain Physician* 2021; 24: S27-S208 [1] and Kaye AD, et al. Responsible, safe, and effective use of antithrombotics and anticoagulants in patients undergoing interventional techniques: American Society of Interventional Pain Physicians (ASIPP) guidelines. *Pain Physician* 2019; 22: S75-S128 [3]

**Table 15.4** Comparative pharmacokinetics/pharmacodynamics of ADP-receptor inhibitors

| | Clopidogrel (Plavix®) | Prasugrel (Effient®) | Ticlopidine (Ticlid®) | Ticagrelor (Brilinta®) |
|---|---|---|---|---|
| Target | P2Y12ADP | P2Y12ADP | P2Y12ADP, also inhibits liver CYP2C19 and CYP2B6 | P2Y12ADP |
| Antithrombotic activity | ++ | +++ | ++ | ++++ |
| Time to Cmax | 3–7 days | 3–5 days | 8–11 days | 2–4 h |
| Time to peak effect | 4 h to 4 days | 1 h | 3–5 days | 2.5 h |
| CYP metabolism | CYP1A2, CYP2B6, CYP2C9, CYP2C19 and CYP3A4/5 | CYP450-mediated (primarily CYP3A4 and CYP2B6) | Cytochromes P450 | CYP3A4 |
| Bioavailability | > 50% | ≥ 79% | > 80% | 36% |
| Protein binding | 94–98% | Active metabolite: ~98% | 98% | >99.7% |
| Plasma half-life | 7–8 h (inactive metabolite) | ~7 h (range 2–15 h) | 12 h (single dose) 4–5 days (repeated dose) | 7 h (ticagrelor), 8.5 h (active metabolite AR-C124910XX) |
| Renal elimination | 50% kidney, 46% biliary | Urine (~68% inactive metabolites); feces (27% inactive metabolites) | Renal and fecal | Biliary |
| Time to 50% recovery of platelet function | 3 days | 3 days | 6 days | 1.5 days |

*CYP* cytochrome P450
Manchikanti L, et al. Epidural interventions in the management of chronic spinal pain: American Society of Interventional Pain Physicians (ASIPP) comprehensive evidence-based guidelines. *Pain Physician* 2021; 24:S27-S208 [1] and Kaye AD, et al. Responsible, safe, and effective use of antithrombotics and anticoagulants in patients undergoing interventional techniques: American Society of Interventional Pain Physicians (ASIPP) guidelines. *Pain Physician* 2019; 22:S75-S128 [3]

**Table 15.5** Comparative pharmacokinetics/pharmacodynamics of direct thrombin inhibitor

| | Dabigatran (Pradaxa) | Argatroban (Acova) | Bivalirudin (Angiomax) | Lepirudin (Refludan) | Desirudin (IPRIVASK) | Hirudin |
|---|---|---|---|---|---|---|
| Target | Direct thrombin inhibitor | Direct thrombin inhibitor | Reversible direct thrombin inhibitor | Direct thrombin inhibitor | Direct thrombin inhibitor | Naturally occurring peptide anticoagulant |
| Time to Cmax | 2–4 h | 1–3 h | 2 min | 4 h | 1–3 h | 3 h |
| Time to peak effect | 0.5–2 h | 2 h | 15 min | 0.5–2 h | 2 h | 3–4 h |
| Metabolism | Metabolized via conjugation into 4 acyl glucuronides, not mediated by CYP450 | CYP3A4 | Proteolytic cleavage | Lepirudin is thought to be metabolized by release of amino acids via catabolic hydrolysis of the parent drug | Metabolized by stepwise degradation from the C-terminus possibly catalyzed by carboxypeptidase(s) such as carboxypeptidase A | Proteolytic cleavage |
| Bioavailability | 3–7%(Oral) | 100% IV | 100% IV application only | 100% (injection or infusion) | 100% | 100% IV |
| Protein binding | 35% | 54% | no | n/a | n/a | n/a |
| Plasma half-life | 13 h | 50 min | ~25 min in patients with normal renal function | 1.3 h | 2–3 h | 80 min |
| Renal elimination | 80% urine | Liver | Yes | Yes | Yes | Renal, about 48% (35% unchanged) |
| Linear PK | Yes | Yes | Yes | Yes | Yes | Yes |
| Time to 50% recovery of thrombin function | 12 h | 2 h | 0.5 h | 1.5 h | 2 h | 2 h |
| Reversal agents | Praxbind | NA | NA | NA | NA | NA |

*CYP* cytochrome P450, *IV* intravenous

Manchikanti L, Knezevic NN, Navani A, et al. Epidural interventions in the management of chronic spinal pain: American Society of Interventional Pain Physicians (ASIPP) comprehensive evidence-based guidelines. *Pain Physician* 2021; 24:S27-S208 [1] and Kaye AD, et al. Responsible, safe, and effective use of antithrombotics and anticoagulants in patients undergoing interventional techniques: American Society of Interventional Pain Physicians (ASIPP) guidelines. *Pain Physician* 2019; 22:S75-S128 [3]

**Table 15.6** Comparative pharmacokinetics/pharmacodynamics of direct factor Xa inhibitor

| | Apixaban (Eliquis) | Rivaroxaban (Xarelto) | Edoxaban (Savaysa, Lixiana) | Betrixaban (Bevyxxa) | Fondaparinux (Arixtra) |
|---|---|---|---|---|---|
| Target | Xa | Xa | Xa | Xa | Xa |
| Time to Cmax | 1–3 h | 2–4 h | 1–2 h | 3–4 h | 2 h |
| Time to peak effect | 3–4 h | 2–4 h | 1–2 h | 3–4 h | 2–3 h |
| CYP Metabolism | 15% | 32% | NR | NR | n/a |
| Bioavailability | 66% | 80% | >45% | 34% | 100% |
| Transporter | P-gp | P-gp/BCRP | P-gp | P-gp | P-gp |
| Protein binding | 87% | >90% | 55% | 60% | 94% |
| Plasma Half-life | 8–15 h | 9–13 h | 8–10 h | 37 h | 17–21 h |
| Renal elimination | 25% | 33% | 35% | <1% | 100% |
| Linear PK | Yes | No | Yes | n/a | Yes |
| Time to 50% recovery of Xa | 12 h | 12 h | 12 h | 19–27 h | 12 h |
| Reversal AGENTS | Andexxa | Andexxa | NA | NA | NA |

*BCRP* breast cancer resistance protein, *CYP* cytochrome P450, *NR* not reported, *P-gp* P-glycoprotein

Manchikanti L, et al. Epidural interventions in the management of chronic spinal pain: American Society of Interventional Pain Physicians (ASIPP) comprehensive evidence-based guidelines. *Pain Physician* 2021; 24:S27-S208 [1] and Kaye AD, et al. Responsible, safe, and effective use of antithrombotics and anticoagulants in patients undergoing interventional techniques: American Society of Interventional Pain Physicians (ASIPP) guidelines. *Pain Physician* 2019; 22:S75-S128 [3]

## Effects of Antithrombotic Therapy on Injection of Biologics

The significance of an integral platelet surface membrane cannot be understated, given that it allows for the appropriate dissemination of healing biological proteins and growth factors essential in regenerative medicine. The biochemical pathways of the coagulation cascade are key in that disruption of platelet surfaces or prematurely activated platelets may cause the limited efficacy of the clinical procedure [162]. Furthermore, with the recent advent of mesenchymal stem cell (MSC) technology in regenerative medicine, it is important to consider the viability of these cells in the context of antithrombotic therapy. To that end, the effects of heparin on ex vivo MSCs recovered from bone marrow showed that supplementation of even low doses of heparin could adversely impact the growth and differentiation potential of this pleiotropic cell type [163]. Therefore, it is important to evaluate bone marrow tissue in patients receiving heparin treatment particularly if ex vivo expansion hMSCs is to occur.

The use of PRP is increasing due to its ligament and tendon healing potential. It is also considered to be a natural alternative to surgery. However, anticoagulants and antiplatelet drugs are commonly used in patients who are candidates for receipt of PRP. Alternatively, the use of an anticoagulant, while not taken systemically, is required to process PRP to prevent automatic activation. Since antithrombotic agents influence the stability of platelets, they are, therefore, likely to exert an effect on PRP efficacy as well and should be ceased at a suitable timeframe preceding injection therapy [162]. PRP has been shown to significantly improve the proliferation of differentiated cells, enhance synthesis of collagen, and prompt angiogenesis and revascularization, all of which aid regeneration. The seminal study by Sutherland and colleagues [164] in 2005 eloquently demonstrated the utility of autologous MSCs in regenerative medicine using sheep through reconstruction of stem cell tissue-engineered heart valves.

The healing influence of the dispensation of PRP depends on the bioactive amalgams such as cytokines and growth factors that are being released (via activation and aggregation) precisely to the location of the injury. The characteristic of the platelets confined in autologous PRP (mainly exhibited by the potential of activation and aggregation of platelets) may be essential [165].

NSAIDs are depicted to have negative influences on platelets such as inhibition of activation, a decrease in storage of alpha granules, and aggregation of platelets. The negative impact of these medications on platelet activation and aggregation may be significant since a substandard quality of autologous PRP is emitted post-NSAID therapy. The coherent substantial inhibition of platelet function such as aggregation acquired post stimulation using arachidonic acid within the NSAID designated study group was found to be regardless of duration and type of NSAID ingestion and the blood compendium technique used for PRP formulation [165]. Without the NSAIDs intake, such effects could not be observed in the healthy control group. NSAIDs impede cyclo-oxygenase–mediated oxygen utilization and therefore prevent platelet activation and aggregation [166]. Moreover, bioactive compounds such as transforming growth factor–β, growth factors, and platelet factor 4 stored in the alpha granules, cannot be released adequately if NSAIDs bar this pathway, and platelet function is drastically compromised. Thus, such stipulations validate the hypothesis that autologous PRP generated after NSAID intake is of inferior attribute, and therefore, may negatively influence the healing outcomes.

NSAIDs are essential to control pain in the post-traumatic setting. While NSAIDs can interfere with bone healing, some contradict these findings [167]. Although NSAID analgesic potency is well documented, clinicians continue to question the associated safety issues. A 2012 study of the effects of NSAIDs on bone healing in animals reveal very mixed results [167]. However, the conclusion of the authors correctly notes that the "lack of evidence does not constitute the absence of an effect" and further cautions that the clinician should (continue to) treat NSAIDs as a risk factor for bone healing impairment and should be avoided in high-risk patients" [167].

Ramsook and Danesh [162] described issues related to PRP injections and antithrombotic therapy. They discussed the importance of understanding the intrinsic and extensive pathways of the coagulation cascade and that any disruption of this mechanism may result in prematurely activated platelets and therefore limited efficacy. They also made clear that antithrombotic agents affect platelet stability and will have an effect on PRP efficacy and must be discontinued at an appropriate time frame prior to injection therapy. Overall, the general rules appear to be the same for antithrombotic and anticoagulant agents similar to utilization in interventional techniques as described above.

However, there is a significant paucity of literature in regard to the safety, efficacy, and timing of PRP injections in patients with concomitant antithrombotic therapy. The importance of an intact platelet surface membrane allows for the appropriate release of the healing bioproteins and growth factors granting PRP therapy its efficacy. Antithrombotic agents that affect the stability of platelets will have an effect on PRP efficacy and must be discontinued at an appropriate time frame prior to injection therapy. With future research, appropriate guidelines may be established not only for PRP, but also for stem cell therapy.

## Recommendations

Table 15.7 shows guidelines for antithrombotic medication management during interventional spine procedures. We were unable to provide comparisons of the American Society of Interventional Pain Physicians (ASIPP) and ASRA guidelines due to differences in risk levels. These recommendations differ from our previously published guidelines, as well as from ASRA guidelines based on the present analysis of the evidence.

General recommendations based on ASRA guidelines [4] are to discontinue the drugs for 5 half-lives; however, the exception to the 5 half-life recommendation should occur in individuals with hepatic dysfunction, and renal dysfunction including nephrotic syndrome.

For safe and efficient administration of biologics, the same principles as interventional techniques are followed.

Guidelines on anticoagulants and antithrombotics are described [168] and also on the ASIPP app, which is regularly updated.

**Key Points**

1. Interventional pain management is a specialty that utilizes minimally invasive procedures to diagnose and treat chronic pain in patients taking exogenous anticoagulants and antithrombotics.
2. Regenerative medicine is an emerging specialty with injection of biologics with application of similar principles as interventional techniques in relation to the technical aspects and risks associated with performance of these procedures.
3. The risk of major bleeding is very small, but the consequences can be catastrophic not only with interventional techniques, but also with injection of biologics.
4. Based on multiple reviews and guidelines for interventional techniques, there is good evidence for the risk of a thromboembolic phenomenon in patients who stop antithrombotic therapy.
5. There is good evidence for the lack of necessity for stopping NSAIDs, prior to performing interventional techniques.
6. There is fair evidence that excessive bleeding, including epidural hematoma formation, may occur with interventional techniques when antithrombotic therapy is continued.
7. There is fair evidence that the risk of a thromboembolic phenomenon is higher than the risk of epidural hematomas when antiplatelet therapy is stopped prior to interventional techniques.
8. As shown in Table 15.1, risk is stratified as low risk, medium risk, and high risk.
9. There is fair evidence for continuing phosphodiesterase inhibitors (dipyridamole [Persantine], cilostazol [Pletal], and Aggrenox [aspirin and dipyridamole]).
10. There is limited evidence for stopping antiplatelet therapy including aspirin, in high risk procedures, platelet aggregation inhibitors to avoid bleeding and epidural hematomas and/or to continue antiplatelet therapy with clopidogrel (Plavix), ticlopidine (Ticlid), or prasugrel (Effient) during interventional techniques to avoid cerebrovascular and cardiovascular thromboembolic fatalities.
11. There is limited evidence in reference to antithrombotic agents dabigatran (Pradaxa®) and rivaroxaban (Xarelto®) to discontinue usage to avoid bleeding and epidural hematomas during interventional techniques and to continue to avoid cerebrovascular and cardiovascular thromboembolic events.
12. There is no specific evidence available with application of injections of biologics.

**Table 15.7** Guidelines for antithrombotic medication management and epidural interventions (risk stratification described in Table 15.2)

| Medication | Time to wait after last dose of medication before interventional techniques are performed | | | | | | Timing of therapy restoration or restarting | |
|---|---|---|---|---|---|---|---|---|
| | Low risk procedures [a] | | Intermediate risk procedures [a] | | High-risk procedures[a] | | | |
| | ASIPP | ASRA | ASIPP | ASRA | ASIPP | ASRA | ASIPP | ASRA |
| | (peripheral nerve blocks, sacroiliac joint and ligament injections and nerve blocks, caudal epidural injections, facet joint interventions, lumbar transforaminal epidural injections at L4, L5, S1, peripheral nerve stimulation trial and implant, etc.) | (peripheral nerve blocks, sacroiliac joint injection and sacral lateral branch blocks, thoracic and lumbar facet medial branch nerve block and radiofrequency ablation, pocket revision, peripheral nerve stimulation trial and implant) | (cervical, thoracic and lumbar transforaminal epidural injections above L3, lumbar intradiscal procedures, hypogastric plexus blocks, lumbar sympathetic blocks, lumbar interlaminar below L4) | Transforaminal epidural injections, Intradiscal procedures, sympathetic blocks, Interlaminar epidural injections, cervical† facet medial branch nerve block and radiofrequency ablation, trigeminal and sphenopalatine ganglia blocks | (Interlaminar epidurals above L4, spinal cord stimulator trial and implant, percutaneous disc decompression, sympathetic blocks, vertebral augmentation, intrathecal catheter and pump implant, etc.) | Spinal cord stimulation trial and implant, intrathecal catheter and pump implant, vertebral augmentation (vertebroplasty and kyphoplasty), percutaneous decompression laminotomy, Epiduroscopy and epidural decompression | | |
| NSAIDS (COX 1) (COX 2) | May continue or stop 1–10 days due to lack of protective effect | Stop 1–10 days due to lack of protective effect | May continue or stop 1–10 days due to lack of protective effect | Stop 1–10 days due to lack of protective effect | May continue or stop 1–10 days due to lack of protective effect | Stop 1–10 days due to lack of protective effect | 24 h | 24 h |
| THC/CBD | May continue or stop 1–10 days | N/A | May continue or stop 1–10 days | N/A | Stop for 5 days | N/A | 24 h | N/A |
| **Aspirin** | | | | | | | | |
| Low-dose aspirin | Continue or may stop for 3 days | Stop for 4 days | Continue or may stop for 3 days | Stop for 4 days | Stop for 5 days | Stop for 6 days | 24 h | 24 h |
| High dose aspirin | Continue or may stop for 3 days | Stop for 4 days | Continue or may stop for 3 days | Stop for 4 days | Stop for 5 days | Stop for 6 days | 24 h | 24 h |
| **Antiplatelet agents** (phosphodiesterase inhibitors) | | | | | | | | |
| Dipyridamole (Persantine) | May continue | May continue | May continue | May continue | May continue or stop for 2 days | Stop for 2 days | 12 h | 12 h |
| Cilostazol (Pletal) | May continue | May continue | May continue | May continue | May continue or stop for 2 days | Stop for 2 days | 12 h | 12 h |
| Aggrenox (dipyridamole plus aspirin) | May continue | Stop for 4 days | May continue | Stop for 4 days | Stop for 5 days | Stop for 6 days | 24 h | 24 h |
| **Platelet aggregation inhibitors** | | | | | | | | |
| Clopidogrel (Plavix) | May continue | May continue | May continue or stop for 5 days | Stop for 7 days | Stop for 5 days | Stop for 7 days | 12 h | 12 h |
| Prasugrel (Effient) | May continue | May continue | May continue or stop for 6 days | Stop for 7–10 days | Stop for 6 days | Stop for 7–10 days | 24 h | 24 h |
| Ticlopidine (Ticlid) | May continue | NA | May continue or stop for 7 days | NA | Stop for 7–10 days | NA | 24 h | 24 h |
| Ticagrelor (Brilinta) | May continue | Continue | May continue or stop for 3 days | NA | Stop for 3–5 days | Stop for 5–10 days | 24 h | 24 h |

| | | | | | | | | |
|---|---|---|---|---|---|---|---|---|
| **Vitamin K antagonists** | | | | | | | | |
| Warfarin | May stop for 2 days INR ≤ 3.0 | INR < 3.0 | Stop for 2–5 days INR ≤ 2 | Stop for 5 days INR normalize | Stop for 2–5 days INR ≤ 1.5 | Stop for 5 days INR normalize | 24 h | 24 h |
| **Thrombin inhibitors** | | | | | | | | |
| Dabigatran (Pradaxa) | May continue or stop for 2 days | May continue or stop for 2 days | Stop for 4–5 days 6 days - renal | Stop for 4–5 days 6 days - renal | Stop for 4–5 days 6 days - renal | Stop for 4–5 days 6 days - renal | 24 h | 24 h |
| **Anti-Xa agents** | | | | | | | | |
| Apixaban (Eliquis) | May continue or stop for 2 days | May continue or stop for 2 days | Stop for 3–5 days | Stop for 3–5 days | Stop for 3–5 days | Stop for 3–5 days | 24 h | 24 h |
| Rivaroxaban (Xarelto) | May continue or stop for 1 day | May continue or stop for 1 day | Stop for 2 days | Stop for 3 days | Stop for 2 days | Stop for 3 days | 24 h | 24 h |
| Edoxaban (Savaysa, Lixiana) | May continue or stop for 1 day | NA | Stop for 3 days | NA | Stop for 3 days | NA | 24 h | 24 h |
| **Heparins** | | | | | | | | |
| Heparin (treatment) - IV | Discontinue for 4 h | Discontinue for 4 h | Discontinue for 4 h | Discontinue for 4 h | Discontinue for 4 h | Discontinue for 4 h | 24 h | 24 h |
| Heparin (treatment) - SC | Discontinue for 8–10 h | Discontinue for 8–10 h | Discontinue for 8–10 h | Discontinue for 8–10 h | Discontinue for 8–10 h | Discontinue for 8–10 h | 24 h | 24 h |
| Low molecular weight heparin | Discontinue for 24 h | Discontinue for 24 h | Discontinue for 24 h | Discontinue for 24 h | Discontinue for 24 h | Discontinue for 24 h | 24 h | 24 h |
| **Thrombolytic agents** | | | | | | | | |
| TPA, streptokinase, Alteplase, Reteplase | May continue | May continue | Stop for 2 days | Stop for 2 days | Stop for 2 days | Stop for 2 days | 24 h | 24 h |
| **GPIIb/IIIa inhibitors** | | | | | | | | |
| Abciximab (ReoPro) | May continue | May continue | Stop for 1–2 days | Stop for 2–5 days | Stop for 1–2 days | Stop for 2–5 days | 8–12 h | 8–12 h |
| Eptifibatide (Integrilin) | May continue | May continue | Stop for 8 h | Stop for 8–24 h | Stop for 8 h | Stop for 8–24 h | 8–12 h | 8–12 h |
| Tirofiban (Aggrastat) | May continue | May continue | Stop for 8 h | Stop for 8–24 h | Stop for 8 h | Stop for 8–24 h | 8–12 h | 8–12 h |
| **Miscellaneous** | | | | | | | | |
| Fondaparinux (Arixtra) | May continue | May continue | Stop for 4 days | Stop for 4 days | Stop for 4 days | Stop for 4 days | 8–12 h | 8–12 h |

Manchikanti L, et al. Epidural interventions in the management of chronic spinal pain: American Society of Interventional Pain Physicians (ASIPP) comprehensive evidence-based guidelines. *Pain Physician* 2021; 24:S27-S208 [1] and Kaye AD, et al. Responsible, safe, and effective use of antithrombotics and anticoagulants in patients undergoing interventional techniques: American Society of Interventional Pain Physicians (ASIPP) guidelines. *Pain Physician* 2019; 22:S75-S128 [3] and Narouze S, et al. Interventional spine and pain procedures in patients on antiplatelet and anticoagulant medications (Second Edition): Guidelines from the American Society of Regional Anesthesia and Pain Medicine, the European Society of Regional Anaesthesia and Pain Therapy, the American Academy of Pain Medicine, the International Neuromodulation Society, the North American Neuromodulation Society, and the World Institute of Pain. *Reg Anesth Pain Med* 2018; 43:225–262 [4]

[a] For full list of procedures please see Table 15.2. These differ from ASIPP classification to ASRA classification

## References


1. Manchikanti L, Knezevic NN, Navani A, et al. Epidural interventions in the management of chronic spinal pain: American Society of Interventional Pain Physicians (ASIPP) comprehensive evidence-based guidelines. Pain Physician. 2021;24:S27–S208.
2. Manchikanti L, Kaye AD, Soin A, et al. Comprehensive evidence-based guidelines for facet joint interventions in the management of chronic spinal pain: American Society of Interventional Pain Physicians (ASIPP) guidelines. Pain Physician. 2020;23:S1–S127.
3. Kaye AD, Manchikanti L, Novitch MB, et al. Responsible, safe, and effective use of antithrombotics and anticoagulants in patients undergoing interventional techniques: American Society of Interventional Pain Physicians (ASIPP) guidelines. Pain Physician. 2019;22:S75–S128.
4. Narouze S, Benzon HT, Provenzano D, et al. Interventional spine and pain procedures in patients on antiplatelet and anticoagulant medications (second edition): guidelines from the American Society of Regional Anesthesia and Pain Medicine, the European Society of Regional Anaesthesia and Pain Therapy, the American Academy of pain medicine, the international neuromodulation society, the north American neuromodulation society, and the world Institute of Pain. Reg Anesth Pain Med. 2018;43:225–62.
5. Deer TR, Narouze S, Provenzano DA, et al. The Neurostimulation appropriateness consensus committee (NACC): recommendations on bleeding and coagulation management in neurostimulation devices. Neuromodulation. 2017;20:51–62.
6. Jenkins AT, Kantorovich A, Burman L. Contemporary use of oral antithrombotic agents: focus on dual and triple therapeutic approaches. Pharmacotherapy. 2017;37:1545–64.
7. Colonna P, Andreotti F, Ageno W, et al. Clinical conundrums in antithrombotic therapy management: a Delphi consensus panel. Int J Cardiol. 2017;249:249–56.
8. Levine GN, Bates ER, Bittl JA, et al. 2016 ACC/AHA guideline focused update on duration of dual antiplatelet therapy in patients with coronary artery disease. J Am Coll Cardiol. 2016;68:1082–115.
9. Zhao Q, Zhu Y, Xu Z, et al. Effect of ticagrelor plus aspirin, ticagrelor alone, or aspirin alone on saphenous vein graft patency 1 year after coronary artery bypass grafting: a randomized clinical trial. JAMA. 2018;319:1677–86.
10. Johnston SC, Easton JD, Farrant M, et al. Clopidogrel and aspirin in acute ischemic stroke and high-risk TIA. N Engl J Med. 2018;379:215–25.
11. Lenzer J. Anticoagulants cause the most serious adverse events, finds US analysis. BMJ. 2012;344:e3989.
12. Roule V, Blanchart K, Humbert X, et al. Antithrombotic therapy for ACS in elderly patients. Cardiovasc Drugs Ther. 2017;31:609–18.
13. Proietti M, Romanazzi I, Romiti GF, et al. Real-world use of Apixaban for stroke prevention in atrial fibrillation: a systematic review and meta-analysis. Stroke. 2018;49:98–106.
14. Gerstein NS, Albrechtsen CL, Mercado N, et al. A comprehensive update on aspirin management during noncardiac surgery. Anesth Analg. 2020;131:1111–23.
15. Goes R, Muskens IS, Smith TR, et al. Risk of aspirin continuation in spinal surgery: a systematic review and meta-analysis. Spine J. 2017;17:1939–46.
16. McNeil JJ, Wolfe R, Woods RL, et al. Effect of aspirin on cardiovascular events and bleeding in the healthy elderly. N Engl J Med. 2018;379:1509–18.
17. McNeil JJ, Woods RL, Nelson MR, et al. Effect of aspirin on disability-free survival in the healthy elderly. N Engl J Med. 2018;379:1499–508.
18. Ridker PM. Should aspirin be used for primary prevention in the post-statin era? N Engl J Med. 2018;379:1572–4.
19. Oprea AD, Noto CJ, Halaszynski TM. Risk stratification, perioperative and periprocedural management of the patient receiving anticoagulant therapy. J Clin Anesth. 2016;34:586–99.
20. Hart RG, Sharma M, Mundl H, et al. Rivaroxaban for stroke prevention after embolic stroke of undetermined source. N Engl J Med. 2018;378:2191–201.
21. Kusher FH, Olson JC. Retinal hemorrhage as a consequence of epidural steroid injection. Arch Opthalmol. 1995;113:309–13.
22. Manchikanti L, Benyamin RM, Swicegood JR, et al. Assessment of practice patterns of perioperative management of antiplatelet and anticoagulant therapy in interventional pain management. Pain Physician. 2012;15:E955–68.
23. Manchikanti L, Malla Y, Wargo BW, et al. A prospective evaluation of bleeding risk of interventional techniques in chronic pain. Pain Physician. 2011;14:317–29.
24. Smith CC, Schneider B, McCormick ZL, et al. Risks and benefits of ceasing or continuing anticoagulant medication for image-guided procedures for spine pain: a systematic review. Pain Med. 2018;19:438–48.
25. Moeschler SM, Warner NS, Lamer TJ, et al. Bleeding complications in patients undergoing percutaneous spinal cord stimulator trials and implantations. Pain Med. 2016;17:2076–81.
26. LaVallee J, Royer R, Smith G. Prevalence of bleeding complications following ultrasound-guided botulinum toxin injections in patients on anticoagulation or antiplatelet therapy. PM R. 2017;9:1217–24.
27. Breivik H, Norum H, Fenger-Eriksen C, et al. Reducing risk of spinal haematoma from spinal and epidural pain procedures. Scand J Pain. 2018;18:129–50.
28. Horlocker TT, Bajwa ZH, Ashraf Z, et al. Risk assessment of hemorrhagic complications associated with nonsteroidal antiinflammatory medications in ambulatory pain clinic patients undergoing epidural steroid injection. Anesth Analg. 2002;95:1691–7.
29. Ahmed SU, Tonidandel W, Trella J, et al. Peri-procedural protocols for interventional pain management techniques: a survey of US pain centers. Pain Physician. 2005;8:181–5.
30. McCormick ZL, Popescu A, Smith C. Spine intervention Society's patient safety committee. Fact finders for patient safety: risk of bleeding with nonaspirin nonsteroidal anti-inflammatory drugs before spine procedures. Pain Med. 2018;19:2322–3.
31. Petraglia FW 3rd, Farber SH, Gramer R, et al. The incidence of spinal cord injury in implantation of percutaneous and paddle electrodes for spinal cord stimulation. Neuromodulation. 2016;19:85–90.
32. Endres S, Shufelt A, Bogduk N. The risks of continuing or discontinuing anticoagulants for patients undergoing common interventional pain procedures. Pain Med. 2017;18:403–9.
33. Selak V, Kerr A, Poppe K, et al. Annual risk of major bleeding among persons without cardiovascular disease not receiving antiplatelet therapy. JAMA. 2018;319:2507–20.
34. Chen LW, Yin HL. A literature review of antithrombotic and anticoagulating agents on sexual function. Andrologia. 2017;49:10.
35. Sáez-Alcaide LM, Sola-Martín C, Molinero-Mourelle P, et al. Dental management in patients with antiplatelet therapy: a systematic review. J Clin Exp Dent. 2017;9:e1044–50.
36. Kent TL, Custer PL. Bleeding complications in both anticoagulated and nonanticoagulated surgical patients. Ophthalmic Plast Reconstr Surg. 2013;29:113–7.
37. Veitch AM. Endoscopy in patients on antiplatelet agents and anticoagulants. Curr Treat Options Gastroenterol. 2017;15:256–67.
38. Baron TH, Kamath PS, McBane RD. Management of antithrombotic therapy in patients undergoing invasive procedures. N Engl J Med. 2013;368:2113–24.
39. Lucijanic M, Ziga S, Grgurevic I. Comment to: "management and outcome of gastrointestinal bleeding in patients taking oral anticoagulants or antiplatelet drugs". J Gastroenterol. 2017;52:1075–6.

40. Goodman BS, House LM, Vallabhaneni S, et al. Anticoagulant and antiplatelet management for spinal procedures: a prospective, descriptive study and interpretation of guidelines. Pain Med. 2017;18:1218–24.
41. van Helmond N, Day W, Chapman KB. Continuing anti-thrombotic medication during low-to-intermediate risk spinal procedures: a retrospective evaluation. Pain Physician. 2017;20:437–43.
42. Lagerkranser M, Lindquist C. Neuraxial blocks and spinal haematoma: Review of 166 cases published 1994–2015. Part 2: diagnosis, treatment, and outcome. Scand J Pain. 2017;15:130–6.
43. Warner NS, Hooten WM, Warner MA, et al. Bleeding and neurologic complications in 58,000 interventional pain procedures. Reg Anesth Pain Med. 2017;42:782–7.
44. Daly-Camacho R, Cabaret J, Novitch M, Cornett E, Kaye AD. Implications of antithrombotic therapy in regenerative medicine. In: Manchikanti L, Navani A, Atluri S, editors. Essentials of regenerative medicine in interventional pain management. Paducah, KY: ASIPP Publishing; 2019. p. 183–98.
45. Andrade JG, Macle L, Nattel S, et al. Contemporary atrial fibrillation management: a comparison of the current AHA/ACC/HRS, CCS, and ESC guidelines. Can J Cardiol. 2017;33:965–76.
46. Steinberg BA, Washam JB. Appropriate dosing of nonvitamin K antagonist oral anticoagulants for stroke prevention in atrial fibrillation. Trends Cardiovasc Med. 2017;27:567–72.
47. Erath JW, Hohnloser SH. Anticoagulation in atrial fibrillation: Current evidence and guideline recommendations. Herz. 2018;43:2–10.
48. CDC, NCHS. Underlying Cause of Death 1999–2013 on CDC WONDER Online Database, released 2015. Data are from the Multiple Cause of Death Files, 1999–2013, as compiled from data provided by the 57 vital statistics jurisdictions through the Vital Statistics Cooperative Program.
49. Hoy DG, Bain C, Williams G, et al. A systematic review of the global prevalence of low back pain. Arthritis Rheum. 2012;64:2028–37.
50. Manchikanti L, Abdi S, Atluri S, et al. An update of comprehensive evidence-based guidelines for interventional techniques of chronic spinal pain: part II: guidance and recommendations. Pain Physician. 2013;16:S49–S283.
51. Dieleman JL, Baral R, Birger M, et al. US spending on personal health care and public health, 1996-2013. JAMA. 2016;316:2627–46.
52. Dieleman JL, Cao J, Chapin A, et al. US health care spending by payer and health condition, 1996-2016. JAMA. 2020;323:863–84.
53. Manchikanti L, Atluri S, Candido KD, et al. Zohydro™ approval by Food and Drug Administration: controversial or frightening? Pain Physician. 2014;17:E437–50.
54. Manchikanti L, Kosanovic R, Pampati V, et al. Declining utilization patterns of percutaneous adhesiolysis procedures in the fee-for-service (FFS) medicare population. Pain Physician. 2021;24:17–29.
55. Manchikanti L, Manchikanti MV, Vanaparthy R, et al. Utilization patterns of sacroiliac joint injections from 2000 to 2018 in fee-for-service medicare population. Pain Physician. 2020;23:439–50.
56. Manchikanti L, Soin A, Mann DP, et al. Utilization patterns of facet joint interventions in managing spinal pain: a retrospective cohort study in the US fee-for-service Medicare population. Curr Pain Headache Rep. 2019;23:73.
57. Manchikanti L, Sanapati MR, Pampati V, et al. Update of utilization patterns of facet joint interventions in managing spinal pain from 2000 to 2018 in the US fee-for-service medicare population. Pain Physician. 2020;23:E133–49.
58. Manchikanti L, Soin A, Mann DP, et al. Comparative analysis of utilization of epidural procedures in managing chronic pain in the Medicare population: pre and post affordable care act. Spine (Phila Pa 1976). 2019;44:220–32.
59. Manchikanti L, Sanapati J, Benyamin RM, et al. Reframing the prevention strategies of the opioid crisis: focusing on prescription opioids, fentanyl, and heroin epidemic. Pain Physician. 2018;21:309–26.
60. Becattini C, Franco L, Agnelli G. Risk of death in patients with major bleedings while on treatment with oral anticoagulants. Int J Cardiol. 2017;235:200.
61. Melmed KR, Lyden P, Gellada N, et al. Intracerebral hemorrhagic expansion occurs in patients using non-vitamin K antagonist oral anticoagulants comparable with patients using warfarin. J Stroke Cerebrovasc Dis. 2017;26:1874–82.
62. Biondi-Zoccai GG, Lotrionte M, Agostoni P, et al. A systematic review and meta-analysis on the hazards of discontinuing or not adhering to aspirin among 50,279 patients at risk for coronary artery disease. Eur Heart J. 2006;27:2667–74.
63. Gurbel PA, Bliden KP, Hiatt BL, et al. Clopidogrel for coronary stenting: response variability, drug resistance, and the effect of pretreatment platelet reactivity. Circulation. 2003;107:2908–13.
64. Cerrato E, D'Ascenzo F, Biondi-Zoccai GG, et al. Dual antiplatelet therapy after drug-eluting stent implantation: when is "enough" enough? J Cardiovasc Pharmacol. 2014;64:38–40.
65. Zullo A, Hassan C, Radaelli F. Gastrointestinal endoscopy in patients on anticoagulant therapy and antiplatelet agents. Ann Gastroenterol. 2017;30:7–14.
66. Saia F. Surgery after drug-eluting stent implantation: it's not all doom and gloom! J Thorac Dis. 2017;9:E373–7.
67. Huynh K. Pharmacotherapy: aspirin discontinuation increases risk of cardiovascular events. Nat Rev Cardiol. 2017;14:696–7.
68. Luni FK, Riaz H, Khan AR, et al. Clinical outcomes associated with per-operative discontinuation of aspirin in patients with coronary artery disease: a systematic review and meta-analysis. Catheter Cardiovasc Interv. 2017;89:1168–75.
69. Hastings S, Myles PS, McIlroy DR. Aspirin and coronary artery surgery: an updated meta-analysis. Br J Anaesth. 2016;116:716–7.
70. Ho PM, Peterson ED, Wang L, et al. Incidence of death and acute myocardial infarction associated with stopping clopidogrel after acute coronary syndrome. JAMA. 2008;299:532–9. Erratum in: JAMA 2008; 299:2390.
71. Marso SP, Amin AP, House JA, et al. Association between use of bleeding avoidance strategies and risk of periprocedural bleeding among patients undergoing percutaneous coronary intervention. JAMA. 2010;303:2156–64.
72. Sprigg N, Gray LJ, England T, et al. A randomised controlled trial of triple antiplatelet therapy (aspirin, clopidogrel and dipyridamole) in the secondary prevention of stroke: safety, tolerability and feasibility. PLoS One. 2008;3:e2852.
73. Walker CW, Dawley CA, Fletcher SF. Aspirin combined with clopidogrel (Plavix) decreases cardiovascular events in patients with acute coronary syndrome. Am Fam Physician. 2007;76:1643–5.
74. Bellini M, Barbieri M. Coagulation management in epidural steroid injection. Anaesthesiol Intensive Ther. 2014;46:195–9.
75. Gogarten W, Van Aken H, Buttner J, et al. Neuraxial blockade and thromboembolism prophylaxis/antithrombotic therapy: revised recommendations of the German society of anaesthesiology and intensive care. Anaesth Intensivmed. 2003;44:218–30.
76. Llau JV, de Andrés J, Gomar C, et al. Drugs that alter hemostasis and regional anesthetic techniques: safety guidelines. Consensus conference. Rev Esp Anestesiol Reanim. 2001;48:270–8.
77. Bugada D, Massimo A, Nicola Z, et al. Regional anesthesia and anticoagulant drugs: a survey of current Italian practice. Eur J Pain Suppl. 2011;5:335–43.
78. Gogarten W, Buerkle H, Van Aken H. The use of concomitant antiplatelet drugs during neuraxial anesthesia is contraindicated in Germany. Reg Anesth Pain Med. 2003;28:585–6. author reply 586.

79. Llau Pitarch JV, De Andrés IJ, Gomar Sancho C, et al. Hemostasis-altering drugs and techniques for regional anesthesia and analgesia: safety recommendations. Rev Esp Anestesiol Reanim. 2005;52:248–50.
80. Kozek-Langenecker SA, Fries D, Gütl M, et al. Locoregional anesthesia and coagulation inhibitors. Recommendations of the task force on perioperative coagulation of the Austrian society for anesthesiology and intensive care medicine. Anaesthesist. 2005;54:476–84.
81. Beasley D, Goree JH. Cervical epidural hematoma following interlaminar epidural steroid injection via the contralateral oblique view in patient taking omega-3 fatty acids. Reg Anesth Pain Med. 2019;44:253–5.
82. Karri J, Chien G, Polson G, et al. Case of an epidural hematoma following a lumbar epidural steroid injection in a patient taking aspirin and duloxetine. PMCR. 2020;4:121–6.
83. Petro J, Asgerally A, Simopoulos T, et al. Diagnosis and spontaneous resolution of an epidural hematoma in a patient presenting after cervical epidural steroid injection. IPM Rep. 2018;2:213–9.
84. Yland MJ. Letter to editor. Risk of spinal cord injury with interlaminar cervical epidural steroid injections, related to the dorsal muscular artery, and retrograde flow. IPM Rep. 2019;3:193–4.
85. Noori S, Laufer I, Gulati A. Acute epidural hematoma occurring after removal of percutaneous spinal cord stimulator trial leads in a cancer patient with chronic thrombocytopenia: a case report. IPM Rep. 2018;2:27–33.
86. Lam CM, Monroe BR, Novosat T. Spontaneous resolution of presumed acute epidural hematoma formation after lumbar epidural steroid injection. IPM Rep. 2017;1:157–61.
87. Candido KD. Epidural hematoma formation following neuraxial interventional pain management procedures. Have we even begun to comprehend the mechanisms and risk factors? IPM Rep. 2017;1:7–10.
88. Sanders RA, Bendel MA, Moeschler SM, et al. Epidural hematoma following interlaminar epidural injection in patient taking aspirin. Reg Anesth Pain Med. 2018;43:310–2.
89. Sawaya C, Sawaya R. Central nervous system bleeding after a lumbar puncture: still an ongoing complication. Am J Case Rep. 2018;19:1103–7.
90. De Cassai A, Correale C, Sandei L. Neuraxial and perineural bleeding after neuraxial techniques: an overview of the last year. Eurasian J Med. 2020;52:211–6.
91. Khan TW, Yacoub A. Pitfalls in interventional pain medicine: hyponatremia after DDAVP for a patient with Von Willebrand disease undergoing an epidural steroid injection. Case Rep Anesthesiol. 2017;2017:6467090.
92. Banik RK, Chen Chen CC. Spinal epidural hematoma after interlaminar cervical epidural steroid injection. Anesthesiology. 2019;131:1342–3.
93. Mehta N. Intracranial hemorrhage and pneumocephaly after cervical epidural injection. Clin Pract Cases Emerg Med. 2019;3:369–71.
94. Kim SI, Lee DH, Kim SH, et al. Spinal epidural hematoma occurring at a distance from the transforaminal epidural injection site: a case report. Medicine (Baltimore). 2019;98:e16654.
95. Gungor S, Aiyer R. Epidural hematoma development contralateral to dura after lumbar transforaminal epidural steroid injection. Pain Manag. 2017;7:367–75.
96. Zhang B, Chen J, Zou N, et al. Spontaneous resolution and complete recovery of spontaneous cervical epidural hematoma: report of two cases and literature review. Neurochirurgie. 2019;65:27–31.
97. Teles P, Correia JP, Pappamikail L, et al. A spontaneous cervical epidural hematoma mimicking a stroke—a case report. Surg Neurol Int. 2020;11:157.
98. Emamhadi M, Ghadarjani S, Alijani B, et al. Spontaneous cervical epidural hematoma with stroke manifestations. Asian J Neurosurg. 2019;14:286–8.
99. Mohamed EH, Dsouza LB, Elnabawy WA, et al. Acute spinal extradural hematoma and cord compression: case report and a literature review. Cureus. 2020;12:e11603.
100. Gallice M, Rouberol F, Albaledejo P, et al. Managing antithrombotic therapy in vitreoretinal surgery. J Fr Ophtalmol. 2015;38:61–73.
101. Benyamin RM, Vallejo R, Wang V, et al. Acute epidural hematoma formation in cervical spine after interlaminar epidural steroid injection despite discontinuation of clopidogrel. Reg Anesth Pain Med. 2016;41:398–401.
102. Buvanendran A, Young AC. Spinal epidural hematoma after spinal cord stimulator trial lead placement in a patient taking aspirin. Reg Anesth Pain Med. 2014;39:70–2.
103. Chien GC, McCormick Z, Araujo M, et al. The potential contributing effect of ketorolac and fluoxetine to a spinal epidural hematoma following a cervical interlaminar epidural steroid injection: a case report and narrative review. Pain Physician. 2014;17:E385–95.
104. Giberson CE, Barbosa J, Brooks ES, et al. Epidural hematomas after removal of percutaneous spinal cord stimulator trial leads: two case reports. Reg Anesth Pain Med. 2014;39:73–7.
105. Page J, Moisi M, Oskouian RJ. Lumbar epidural hematoma following interlaminar fluoroscopically guided epidural steroid injection. Reg Anesth Pain Med. 2016;41:402–4.
106. Swicegood J, Manchikanti L, Benyamin R, et al. A report of acute thoracic epidural hematoma after interlaminar epidural injection. IPM Rep. 2017;1:33–8.
107. Swicegood J, Manchikanti L, Benyamin R, et al. Two cases of acute epidural hematoma formation after cervical interlaminar epidural steroid injections. IPM Rep. 2017;1:27–32.
108. Jenkie E, Benyamin R, Manchikanti L. Fish oil as a potential contributor to epidural hematoma following cervical epidural steroid injection: a case report and focused literature review. IPM Rep. 2017;1:19–26.
109. Manchikanti L, Malla Y, Benyamin R, et al. Prevalence of epidural hematoma following cervical epidural injections in interventional pain management settings: literature review with two case reports. IPM Rep. 2017;1:11–7.
110. Kim SH, Han YJ, Kim YH, et al. Spontaneous absorption of a lumbar epidural hematoma after interlaminar epidural steroid injection in a patient with spinal stenosis: close observation as a treatment strategy. Chin Med J. 2018;131:117–8.
111. Castro VM, Gallagher PJ, Clements CC, et al. Incident user cohort study of risk for gastrointestinal bleed and stroke in individuals with major depressive disorder treated with antidepressants. BMJ Open. 2012;2:e000544.
112. Manchikanti L, Gruber TJ, Prabhakar H, et al. Acute epidural hematoma following a cervical epidural injection in a patient without bleeding risk. IPM Rep. 2018;2:119–26.
113. Berrigan WA, Whitehair C, Zorowitz R. Acute spinal epidural hematoma as a complication of dry needling: a case report. PMR. 2019;11:313–6.
114. Cameron CM, Scott DA, McDonald WM, et al. A review of neuraxial epidural morbidity: experience of more than 8,000 cases at a single teaching hospital. Anesthesiology. 2007;106:997–1002.
115. Caputo AM, Gottfried ON, Nimjee SM, et al. Spinal epidural hematoma following epidural steroid injection in a patient treated with dabigatran: a case report. JBJS Case Connect. 2013;3:e64.
116. Warner NS, Moeschler SM, Warner MA, et al. Bleeding complications in patients undergoing celiac plexus block. Reg Anesth Pain Med. 2016;41:488–93.

117. Lagerkranser M, Johnsson H, Ljungström KG. Management of thrombocyte inhibitors prior to surgery. Too early withdrawal can result in severe risks for the patient. Lakartidningen. 2008;105:2188–9.
118. Gould MK, Garcia DA, Wren SM, et al. Prevention of VTE in nonorthopedic surgical patients: antithrombotic therapy and prevention of thrombosis, 9th ed: American College of Chest Physicians evidence-based clinical practice guidelines. Chest. 2012;141:e227S–77S.
119. Falck-Ytter Y, Francis CW, Johanson NA, et al. Prevention of VTE in orthopedic surgery patients: antithrombotic therapy and prevention of thrombosis, 9th ed: American College of Chest Physicians evidence-based clinical practice guidelines. Chest. 2012;141:e278S–325S.
120. ACCP-NHLBI National Conference on Antithrombotic Therapy. American College of Chest Physicians and the National Heart, lung and blood institute. Chest. 1986;89:1S–106S.
121. Guyatt GH, Akl EA, Crowther M, et al. Introduction to the ninth edition: antithrombotic therapy and prevention of thrombosis, 9th ed: American College of Chest Physicians evidence-based clinical practice guidelines. Chest. 2012;141:48S–52S.
122. Eikelboom JW, Hirsh J, Spencer FA, et al. Antiplatelet drugs: antithrombotic therapy and prevention of thrombosis, 9th ed: American College of Chest Physicians evidence-based clinical practice guidelines. Chest. 2012;141:e89S–e119S.
123. Kumar V, Turakhia PB, Wunnava MS. Pulmonary embolism during dorsal column stimulator (DCS) trial. Reg Anesth Pain Med. 2013;38:1.
124. Linn AJ, Desilva C, Peeters-Asdourian C. Thromboembolic stroke: a rare complication associated with peri-procedural management of an epidural steroid injection. Pain Physician. 2009;12:159–62.
125. Yoo HS, Park SW, Han JH, et al. Paraplegia caused by an epidural hematoma in a patient with unrecognized chronic idiopathic thrombocytopenic purpura following an epidural steroid injection. Spine (Phila Pa 1976). 2009;34:E376–9.
126. Desai MJ, Dua S. Perineural hematoma following lumbar transforaminal steroid injection causing acute-on-chronic lumbar radiculopathy: a case report. Pain Pract. 2014;14:271–7.
127. Shanthanna H, Park J. Acute epidural haematoma following epidural steroid injection in a patient with spinal stenosis. Anaesthesia. 2011;66:837–9.
128. Williams KN, Jackowski A, Evans PJ. Epidural haematoma requiring surgical decompression following repeated cervical epidural steroid injections for chronic pain. Pain. 1990;42:197–9.
129. Ghaly RF. Recovery after high-dose methylprednisolone and delayed evacuation: a case of spinal epidural hematoma. J Neurosurg Anesthesiol. 2001;13:323–8.
130. Weller RS, Gerancher JC, Crews JC, et al. Extensive retroperitoneal hematoma without neurologic deficit in two patients who underwent lumbar plexus block and were later anticoagulated. Anesthesiology. 2003;98:581–5.
131. Stoll A, Sanchez M. Epidural hematoma after epidural block: implications for its use in pain management. Surg Neurol. 2002;57:235–40.
132. Chiravuri S, Wasserman R, Chawla A, et al. Subdural hematoma following spinal cord stimulator implant. Pain Physician. 2008;11:97–101.
133. Xu R, Bydon M, Gokaslan ZL, Wolinsky JP, et al. Epidural steroid injection resulting in epidural hematoma in a patient despite strict adherence to anticoagulation guidelines. J Neurosurg Spine. 2009;11:358–64.
134. Ain RJ, Vance MB. Epidural hematoma after epidural steroid injection in a patient withholding enoxaparin per guidelines. Anesthesiology. 2005;102:701–3.
135. Lee JH, Lee H, Jo DJ. An acute cervical epidural hematoma as a complication of dry needling. Spine (Phila Pa 1976). 2011;36:E891–3.
136. Reitman CA, Watters W 3rd. Subdural hematoma after cervical epidural steroid injection. Spine (Phila Pa 1976). 2002;27:E174–6.
137. Eftekhar B, Ketabchi E, Ghodsi M, et al. Lumbar epidural hematoma due to lumbar acupunctures. Neurol India. 2005;53:245–6.
138. Ozdemir O, Calisaneller T, Yildirim E, et al. Acute intracranial subdural hematoma after epidural steroid injection: a case report. J Manip Physiol Ther. 2007;30:536–8.
139. Chen JC, Chen Y, Lin SM, et al. Acute spinal epidural hematoma after acupuncture. J Trauma. 2006;60:414–6.
140. Lee JY, Nassr A, Ponnappan RK. Epidural hematoma causing paraplegia after a fluoroscopically guided cervical nerve-root injection. A case report. J Bone Joint Surg Am. 2007;89:2037–9.
141. Bose B. Quadriparesis following cervical epidural steroid injections: case report and review of the literature. Spine J. 2005;5:558–63.
142. Domenicucci M, Marruzzo D, Pesce A, et al. Acute spinal epidural hematoma after acupuncture: personal case and literature review. World Neurosurg. 2017;102:695.e11–4.
143. Maier C, Gleim M, Weiss T, et al. Severe bleeding following lumbar sympathetic blockade in two patients under medication with irreversible platelet aggregation inhibitors. Anesthesiology. 2002;97:740–3.
144. Keane JR, Ahmadi J, Gruen P. Spinal epidural hematoma with subarachnoid hemorrhage caused by acupuncture. AJNR Am J Neuroradiol. 1993;14:365–6.
145. Choi JJ, Chang YJ, Jung WS, et al. Discordant lumbar epidural hematoma after caudal steroid injection: a case report (CARE-compliant). Medicine (Baltimore). 2017;96:e7127.
146. Kim M, Park KS. Intracranial chronic subdural hematoma presenting with intractable headache after cervical epidural steroid injection. J Korean Neurosurg Soc. 2015;58:144–6.
147. Takawira N, Han RJ, Nguyen TQ, et al. Spinal cord stimulator and epidural haematoma. Br J Anaesth. 2012;109:649–50.
148. Kloss BT, Sullivan AM, Rodriguez E. Epidural hematoma following spinal cord stimulator implant. Int J Emerg Med. 2010;3:483–4.
149. Smith CC, Lin JL, Shokat M, et al. A report of paraparesis following spinal cord stimulator trial, implantation and revision. Pain Physician. 2010;13:357–63.
150. Santiago FM, Santiago J, Prieto M, et al. Dorsal epidural hematoma after implantation of a dorsal nerve stimulator. Rev Esp Anestesiol Reanim. 2005;52:440–1.
151. Nam KH, Hwa Choi CH, Yang MS, et al. Spinal epidural hematoma after pain control procedure. J Korean Neurosurg Soc. 2010;48:281–4.
152. Chen CY, Liu GC, Sheu RS, et al. Bacterial meningitis and lumbar epidural hematoma due to lumbar acupunctures: a case report. Kaohsiung J Med Sci. 1997;13:328–31.
153. Dehaene S, Biesemans J, Van Boxem K, et al. Post-dural puncture headache evolving to a subdural hematoma: a case report. Pain Pract. 2021;21:83–7.
154. Raj PP, Shah RV, Kaye AD, et al. Bleeding risk in interventional pain practice: Assessment, management, and review of the literature. Pain Physician. 2004;7:3–51.
155. Bogduk N. Ceasing anticoagulants for interventional pain procedures. Pain Med. 2020;21:881–2.
156. Buchner LM, Park EJ, Bendz P, et al. Evaluation of a clinical decision support system for the most evidence-based approach to managing perioperative anticoagulation. J Clin Anesth. 2022;80:110877.
157. Mallett SV. Clinical utility of viscoelastic tests of coagulation (TEG/ROTEM) in patients with liver disease and during liver transplantation. Semin Thromb Hemost. 2015;41:527–37.

158. Thomas O, Rein H, Strandberg K, et al. Coagulative safety of epidural catheters after major upper gastrointestinal surgery: advanced and routine coagulation analysis in 38 patients. Perioper Med (Lond). 2016;5:28.
159. Thomas OD. Haemostatic safety in epidural analgesia. Lund: Lund University, Faculty of Medicine, Doctoral Dissertation Series 2016:117. Accessed 12 Dec 2022. http://portal.research.lu.se/portal/files/15557976/Haemostatic_safety_in_epidural_analgesia_minus_manus.pdf.
160. Forkin KT, Colquhoun DA, Nemergut EC, et al. The coagulation profile of end-stage liver disease and considerations for intraoperative management. Anesth Analg. 2018;126:46–61.
161. Seo KH, Nahm FS, Han WK, et al. Aspirin cessation before interventional procedures: not blindly following guidelines but making test-based decisions. Pain Physician. 2022;25:501–7.
162. Ramsook RR, Danesh H. Timing of platelet rich plasma injections during antithrombotic therapy. Pain Physician. 2016;19:E1055–61.
163. Ling L, Camilleri ET, Helledie T, et al. Effect of heparin on the biological properties and molecular signature of human mesenchymal stem cells. Gene. 2016;576:292–303.
164. Sutherland FW, Perry TE, Yu Y, et al. From stem cells to viable autologous semilunar heart valve. Circulation. 2005;111:2783–91.
165. Schippinger G, Prüller F, Divjak M, et al. Autologous platelet-rich plasma preparations. Orthop J Sport Med. 2015;3:2325967115558889.
166. Benzon HT, Avram MJ, Green D, et al. New oral anticoagulants and regional anaesthesia. Br J Anaesth. 2013;111:i96–i113.
167. Pountos I, Georgouli T, Calori GM, et al. Do nonsteroidal anti-inflammatory drugs affect bone healing? A critical analysis. Sci World J. 2012;2012:1–14.
168. Manchikanti L, Sanapati M, Kaye AD, Kaye AM. Antiplatelet and anticoagulant therapy in interventional pain management. In: Manchikanti L, Singh V, Falco FJE, Kaye AD, Soin A, Hirsch JA, editors. Manchikanti's essentials of interventional techniques in managing chronic pain. 2nd ed. New York, NY: Springer; 2023.

# Part II

# Regenerative Medicine in Musculoskeletal Disorders

# 16 Pathophysiology and Mechanisms of Chronic Musculoskeletal Pain

Mary A. Ambach

## Background

Chronic Musculoskeletal Pain affects approximately 37% of the United States population with an economic burden of up to $635 billion per year [1]. The annual U.S. expenditures related to pain, including direct medical costs and lost wages, are higher than those for cancer, heart disease, and diabetes combined [2]. According to the Global Burden of Disease 2019 data, approximately 1.71 billion people live with musculoskeletal conditions worldwide [3].

Musculoskeletal conditions are also the biggest contributor to years lived with disability (YLDs) worldwide with approximately 149 million YLDs, accounting for 17% of all YLDs worldwide [4].

These are the top contributors to the overall burden of musculoskeletal conditions:

1. Low back pain: is the main contributor to the overall burden of musculoskeletal conditions with 570 million prevalent cases worldwide, responsible for 7.4% of global YLDs.
2. Osteoarthritis—528 million people (19 million YLDs)
3. Fractures–440 million people globally (26 million YLDs)
4. Neck pain—222 million people (22 million YLDs)
5. Amputations—180 million people (5.5 million YLDs)
6. Gout—54 million people (1.7 million YLDs)
7. Rheumatoid arthritis—18 million people (2.4 million YLDs)
8. Other musculoskeletal conditions—453 million people (38 million YLDs)

## Definitions

The current definition of pain by the International Association for the Study of Pain (IASP) revised in 2020 is "An unpleasant sensory and emotional experience associated with, or resembling that associated with, actual or potential tissue damage" [5].

**Acute pain** is defined as a biologic symptom of an apparent nociceptive stimulus. Example would be tissue damage due to disease or trauma [6]. Acute pain can be self-limiting as the pain stimulus decreases, or it can persist to chronic pain. **Chronic pain** is defined as pain that lasts or recurs for longer than 3 months [7]. It is a complex disease process that has neurologic, psychological and physiologic components and that lasts more than the anticipated usual course of an acute disease or injury.

The Pain Task Force of the (IASP), defines **Chronic Primary Musculoskeletal Pain** as "chronic pain in the muscles, bones, joints, or tendons that is characterized by significant emotional distress or functional disability" [8].

Musculoskeletal pain can be caused by repetitive strain, overuse, work-related musculoskeletal disorders, or disease conditions. Any condition that affect the muscles (myositis and neuromuscular disorders), bone (osteoporosis and fractures,) joints (osteoarthritis, rheumatoid arthritis and other spondyloarthritis) and the contiguous connective tissue can cause musculoskeletal pain [4]. Systemic diseases like infection, connective tissue disease, rheumatologic disease, inflammatory conditions and autoimmune diseases like vasculitis can have musculoskeletal manifestations [9].

The IASP Task force for the classification of chronic pain developed a practical classification system of chronic pain for the 11th revision of International Classification of Diseases (ICD-11) [8]. In this classification, Chronic musculoskeletal pain was categorized into seven categories, based primarily on pain etiology: (1) Chronic musculoskeletal pain from persistent inflammation, (2) Chronic musculoskeletal pain from structural osteoarticular changes, (3) Chronic

M. A. Ambach (✉)
San Diego Orthobiologics Medical Group, Carlsbad, CA, USA

University of California San Diego School of Medicine and Department of Orthopedic Surgery, San Diego, CA, USA
e-mail: ambach@sdomg.com


A. Navani et al. (eds.), *Essentials of Regenerative Medicine in Interventional Pain Management*,
https://doi.org/10.1007/978-3-031-50357-3_16

musculoskeletal pain due to disease of the nervous system, (4) Chronic nonspecific musculoskeletal pain, (5) Other chronic musculoskeletal pain syndromes, and (6) Chronic musculoskeletal pain not otherwise specified.

There is ongoing debate and confusion on how to properly classify chronic pain and chronic musculoskeletal pain conditions. To illustrate, the ICD-11 categories above on chronic musculoskeletal pain include *nociceptive type* pain only and does not include pain that may be perceived in musculoskeletal tissues but does not arise therefrom, such as neuropathic pain or referred pain. There are well-described musculoskeletal conditions for which the causes are incompletely understood, such as nonspecific back pain or chronic widespread pain. These are classified as chronic primary pain which is a separate section on its own. In the same manner, there are certain conditions where there is still uncertainty and confusion on its etiology, pathology and diagnostic criteria. An example of this is myofascial pan syndromes where categorizing may be complicated.

This highlights the recognized complexity of chronic pain, where perspectives can be different, and no universal consensus on many of the chronic pain conditions are realized.

## Pathophysiology of Pain

Pain can be classified based on pathophysiology into nociceptive, neuropathic or mixed type categories [10, 11]. **Nociceptive pain** is the most common cause of musculoskeletal pain and the most common type following tissue injury. It results from actual or threatened damage to non-neural tissue and is due to excitation of peripheral pain receptors which activates specific sensory nuclei and spinal cord pathways [12, 13]. Nociceptive pain can be either somatic or visceral type. **Somatic pain** arises from inflammation or injury to the skin and subcutaneous tissue (superficial somatic pain); muscle, tendon, joint and bones (deep somatic pain). Superficial somatic pain is typically sharp, throbbing, well localized, intermittent to constant pain. Deep somatic pain is less well localized and has a deep and aching quality. **Visceral pain** is due to an abnormal function or pathology of the deep visceral organs. This is characterized by deep, dull, diffuse, aching pain that is poorly localized.

The other major category of pain is **Neuropathic pain** which is caused by an injury or disease of the central and peripheral nervous system. The pain is typically along a nerve distribution and is commonly associated with sensory changes like hypo/hyperesthesia (decreased/increased sensitivity to stimulation), paresthesia (abnormal sensation), hypo/hyperalgesia (decrease/increase pain in response to a normally painful stimulus), and allodynia (pain from a stimulus that does not normally cause pain) [13]. Neuropathic pain is commonly described as burning in quality, electric-like, pins and needles sensation, numbness, or shooting. **Mixed pain** occurs with the presence of both nociceptive and neuropathic pain components. Patients who have chronic low back pain with radiculopathy for example, will have both nociceptive and neuropathic pain symptoms.

IASP has a new classification of pain termed **Nociplastic Pain** which is defined as "Pain that arises from altered nociception despite no clear evidence of actual or threatened tissue damage causing the activation of peripheral nociceptors or evidence for disease or lesion of the somatosensory system causing the pain" [13]. These are typically chronic pain conditions with a non-discrete pain distribution and clinical signs of pain hypersensitivity.

## Pain Mechanisms

### Nociceptors

Pain is conducted through neuronal pathways that transmit noxious stimuli from the periphery to the cerebral cortex. A nociceptor is defined as a high-threshold sensory receptor of the peripheral somatosensory nervous system that is capable of transducing and encoding noxious stimuli [13]. Nociceptive information is transmitted by two main pain nerve fibers. The myelinated fast A-delta fibers convey sharp, easily localized pain signals to the cortical regions of the brain; while the unmyelinated slow C fibers conducts a more generalized, aching type of pain to the subcortical regions [14]. The skin is supplied by A-delta mechanoreceptors, A-delta mechanothermal nociceptors and the C polymodal nociceptors. Muscles, joints, fascia and other deep somatic structures are supplied by both C and A delta fibers [15]. The A beta fibers are reserved primarily for vibration transmission [14].

### Algesic Substances

Tissue damage caused by injury, disease or inflammation releases chemicals into the extracellular fluid that cause pain by direct excitatory action on the membrane of nociceptors. These pain-producing substances include histamine, found in platelets, basophils, and mast cell granules; serotonin, found in mast cells and platelets; prostaglandins and other members of the arachidonic acid cascade located in the tissues; kinins (Bradykinin), located in the plasma; Substance P which are in the nerve terminals; $K^+$, and $H^+$ [15].

### Arachidonic Acid Pathway

Cell injury leads to a cascade of events that starts with the action of phospholipase A2 on the phospholipids released from cell membranes to form arachidonic acid. The arachidonic acid is transformed into prostacyclin, prostaglandins, and thromboxanes through the cyclooxygenase (COX) pathway; and into leukotrienes through the lipoxygenase pathway [16]. These metabolites are involved in vasonstriction/dilatation, cell chemotaxis, neurotransmitter release, regulation of platelet aggregation and a whole host of processes involved in the proper function of the immune system, promotion of allergies, inflammation modulation, mood, and appetite.

### Pain Pathway

The primary afferent neurons, called the first-order neurons, carry signals from the peripheral tissues to the dorsal horn of the spinal cord. The painful impulses from the nociceptors are projected to the second-order neurons in the spinal cord and brainstem [17]. With enough excitation of the second-order neurons, the nociceptive input gets transmitted to the different areas of the central nervous system through different ascending tracts of the spinal cord. These signals are projected to the third-order neurons in the thalamus which then carry the signals to the cerebral cortex where perception, localization and emotional components of the pain are realized. Structures in the brain involved in pain processing and perception include the primary and secondary somatosensory cortex, anterior and medial cingulate cortex, insula, and prefrontal cortex [18]. A complicated interaction of sensory, motivational and cognitive processes determines the complex expression of pain behavior.

## Modulation of Pain

### Augmentation

Nociceptive input to the nociceptors can be augmented by several factors. With repeated stimulation, **sensitization** occurs in the nerve endings which results to enhanced sensitivity, lowered threshold to stimulation and enhanced response to stimulation. This is called **primary hyperalgesia** and can occur after tissue damage caused by injury or disease [11]. This can expand to a bigger area of hyperalgesia and allodynia around the site of injury called **secondary hyperalgesia or neurogenic inflammation**. This phenomenon of **Peripheral Sensitization** involves the action of several cells such as macrophages, mast cells, platelets, endothelial cells and other immune cells. These cells release a host of inflammatory factors including cytokines, histamine, bradykinin, prostaglandin $E_2$ ($PGE_2$), adenosine triphosphate (ATP), and others that lead to excitability of the peripheral sensory nerve endings, which then release substance P and calcitonin gene-related peptide (CGRP) contributing more to the process [11]. Biochemical agents, including $H^+$, nitrogen oxide (NO), $K^+$, ATP, nerve growth factor (NGF), tumor necrosis factor alpha (TNF-$\alpha$), interleukin 6 (IL-6),and glutamate have been involved in this process of peripheral sensitization [19].

Sensitization can also happen centrally in the spinal cord. **Central Sensitization or Facilitation** mechanisms involve the glutamate and N- methyl D-aspartate (NMDA) receptor—mediated sensitization, wind-up (perceived increase in intensity of pain over time when a painful stimulus is delivered repeatedly above a critical rate), disinhibition of spinal cord interneurons, receptive field expansion and hyperexcitability of reflexes and microglial activation [11, 20]. Neurochemical mediators of central sensitization include substance P, CGRP, vasoactive intestinal peptide (VIP), cholecystokinin (CCK), angiotensin, and galanin, as well as the excitatory amino acids L-glutamate and L-aspartate which activate the NMDA receptors [11].

### Inhibition

On the other hand, inhibition of pain transmission can occur in the spinal cord by segmental activity in the cord itself and/or by descending pathways from supraspinal centers. Glycine and $\gamma$ Aminobutyric acid (GABA) are the inhibitory neutotransmitters that likely play a role in the segmental inhibitions of pain in the spinal cord. While Norepinephrine, Serotonin and the endogenous opioid system which act to inhibit the release of Substance P, are all involved in supraspinal inhibition.

Neurotransmitters modulating pain [11]:

1. Excitatory Neurotrasmitters: Substance P, Calcitonin gene related peptide CGRP, glutamate, aspartate, ATP
2. Inhibitory Neurotransmitters: Acetylcholine, Enkephalins, Beta Endorphin, Norepineprhine, Adenosine, Serotonin, GABA, Glycine, Somatostatin

## Chronic Pain

The pathophysiology of chronic pain involves a combination of peripheral, central and psychologic mechanisms. Peripheral mechanisms include the processes of sensitization, neuropathic pain and neural inflammation. Central mechanisms include sensitization, disinhibition, wind-up and re-organization of neural connections.

The Sympathetic nervous system can play a role in patients with chronic pain although the exact pathophysiology remains unclear. Animal studies have shown sensitization of primary afferent receptors by catecholaminergic sympathetic outflow while human studies show sympathetic-induced catecholamine sensitivity following nerve lesions and upregulation of skin adrenoreceptors in Complex Regional Pain Syndrome (CRPS) patients [21].

Psychological mechanisms can be involved in the pathogenesis of Chronic pain. Emotional factors can initiate pain. Chronic behavior patterns can be positively rewarded by others or maintained by escaping noxious stimuli through drugs or avoidance of activities like work. Psychological disorders like depression, anxiety, post-traumatic stress disorder (PTSD), schizophrenia and somatoform disorder can also affect the experience of pain.

## Biopsychosocial Experience of Pain

Pain is more than a physical symptom resulting from disease, pathology or tissue injury. There are psychosocial, cultural and environmental factors involved in addition to biology that affect the complex pain experience.

Once pain becomes chronic, sensory input plays a diminished role and affective and cognitive pathways play a more prominent role in the creation of painful perceptions [22]. Patient's cognition, mood, affect, beliefs, attitudes and psychological state affect the pain experience, as well as sleep, energy and level of function. Ethnicity, gender, occupation, social groups, and family can influence pain perception and response, which can in turn affect how pain is communicated and treated. This Biopsychosocial model of pain experience is important in designing effective chronic pain programs and including an interdisciplinary model in the evaluation and treatment of chronic pain.

### Key Points

1. Chronic pain is a complex disease process that lasts or recurs for longer than 3 months and has physiologic, neurologic and psychological components that lasts more than the anticipated usual course of an acute disease or injury.
2. The mechanisms and modulation of pain involves a complicated process involving peripheral nociceptors, spinal cord, central nervous system, ascending and descending pain pathways, pain producing chemical agents, neurotransmitters, immune cells, and sympathetic nervous system
3. **Nociceptive pain** is the most common cause of musculoskeletal pain which results from actual or threatened damage to non-neural tissue and is due to excitation of peripheral pain receptors which activates specific sensory nuclei and spinal cord pathways. Nociceptive pain can be either somatic or visceral type.
4. **Neuropathic pain** is caused by an injury or disease of the central and peripheral nervous system that is typically along a nerve distribution, described as burning in quality, electric-like, pins and needles sensation, numbness, or shooting and is commonly associated with sensory changes like hypo/hyperesthesia (decreased/increased sensitivity to stimulation), paresthesia (abnormal sensation), hypo/hyperalgesia (decrease/increase pain in response to a normally painful stimulus), and allodynia (pain from a stimulus that does not normally cause pain)
5. Nociceptive information is transmitted by two main pain nerve fibers: The myelinated fast A-delta fibers which convey sharp, easily localized pain signals to the cortical regions of the brain; and the unmyelinated slow C fibers conducts a more generalized, aching type of pain to the subcortical regions
6. Peripheral sensitization occurs after repeated stimulation to the nerve endings resulting to enhanced sensitivity, lowered threshold to stimulation and enhanced response to stimulation largely due to the action of several cells such as macrophages, mast cells, platelets, endothelial cells and other immune cells
7. **Central Sensitization or Facilitation** mechanisms involve the glutamate and N- methyl D-aspartate (NMDA) receptor—mediated sensitization, wind-up (perceived increase in intensity of pain over time when a painful stimulus is delivered repeatedly above a critical rate), disinhibition of spinal cord interneurons, receptive field expansion and hyperexcitability of reflexes and microglial activation
8. **Inhibition** of pain transmission can occur in the spinal cord by segmental activity in the cord itself through the inhibitory neurotransmittters Glycine and $\gamma$ Aminobutyric acid (GABA), and/or by descending pathways from supraspinal centers through Norepinephrine, Serotonin and the endogenous opioid system which act to inhibit the release of Substance P
9. The Sympathetic nervous system can play a role in patients with chronic pain like in the case of Complex Regional Pain Syndrome, although the exact pathophysiology remains unclear.
10. The Biopsychosocial model of pain experience, which recognizes that there are psychosocial, cultural and environmental factors involved in addition to biology that affect the complex pain experience, is important in

designing effective chronic pain programs and including an interdisciplinary model in the evaluation and treatment of chronic pain

**Questions**

1. A 65 year old male has a history of lumbar degenerative disc disease and poorly controlled diabetes. He presents with tingling and numbness in his hands and feet and his pain is often described as burning in quality, electric-like, pins and needles sensation. Based on pathophysiology, what kind of pain is he exhibiting?
   A. Nociceptive pain
   B. Somatic Pain
   C. Visceral Pain
   D. Neuropathic pain

   **Answer D:**

   **Explanation:** Neuropathic pain is caused by an injury or disease of the central and peripheral nervous system. The pain is typically along a nerve distribution and is commonly associated with sensory changes like hypo/hyperesthesia (decreased/increased sensitivity to stimulation), paresthesia (abnormal sensation), hypo/hyperalgesia (decrease/increase pain in response to a normally painful stimulus), and allodynia (pain from a stimulus that does not normally cause pain). Neuropathic pain is commonly described as burning in quality, electric-like, pins and needles sensation, numbness, or shooting.
2. A 50 year old patient with right shoulder pain from rotator cuff tear has progressive increasing pain in the shoulder over the years. He had this problem since he was 35 years old which he sustained after an aggressive tennis match. His pain started in a pinpoint region in the outside of his shoulder which has spread to the whole upper arm, side of the neck and shoulder blade. What is an important process involved in the pathophysiology of his increased pain?
   A. Repeated stimulation in the nerve endings can result to enhanced sensitivity, lowered threshold to stimulation and enhanced response to stimulation.
   B. Tissue damage caused by injury or disease can cause pain that stays localized to the site of injury.
   C. The host of inflammatory factors that cells release including cytokines, histamine, bradykinin, PGE2, ATP and others do not cause excitability of the peripheral sensory nerve endings and are not involved in Sensitization
   D. Sensitization only happens in the central nervous system where various processes occurs in the spinal cord leading to increased pain perception

   **Answer A.**

   **Explanation:** Repeated stimulation in the nerve endings can result to enhanced sensitivity, lowered threshold to stimulation and enhanced response to stimulation. Tissue damage caused by injury or disease can cause pain that can spread beyond the site of injury. Peripheral Sensitization involves the action of several cells such as macrophages, mast cells, platelets, endothelial cells and other immune cells. The host of inflammatory factors that cells release including cytokines, histamine, bradykinin, PGE2, ATP contribute to the process of Sensitization. Sensitization only happens peripherally as well as centrally in the spinal cord .
3. Chronic Musculoskeletal Pain affects approximately 37% of the United States population with an economic burden of up to $635 billion per year. Musculoskeletal conditions are also the biggest contributor to years lived with disability (YLDs) worldwide with approximately 149 million YLDs, accounting for 17% of all YLDs worldwide. What important mechanisms are involved in the pathophysiology of chronic pain?
   A. The pathophysiology of chronic pain involves a combination of peripheral, central and psychologic mechanisms.
   B. Peripheral mechanisms include sensitization, disinhibition, wind-up and re-organization of neural connections.
   C. The Sympathetic nervous system has clearly understood pathophysiology and is not involved in patients with chronic pain.
   D. Psychological mechanisms like emotional factors and psychologic disorders are separate conditions not involved in the pathogenesis of Chronic pain.

   **Answer A.**

   **Explanation:** The pathophysiology of chronic pain involves a combination of peripheral, central and psychologic mechanisms. Sensitization, disinhibition, wind-up and re-organization of neural connections are processes involved in Central sensitization. The Sympathetic nervous system can play a role in patients with chronic pain although the exact pathophysiology remains unclear. Psychological mechanisms can be involved in the pathogenesis of Chronic pain.
4. Which of the following is true about the Biopsychosocial experience of Chronic Pain?
   A. Chronic pain is simply a physical symptom resulting from disease, pathology or tissue injury and is not influenced by other factors.
   B. Patient's cognition and beliefs do not affect the pain experience.
   C. This Biopsychosocial model of pain experience is not important in designing effective chronic pain programs.
   D. There are psychosocial, cultural and environmental factors involved in addition to biology that affect the complex pain experience.

**Answer D.**

**Explanation:** Pain is more than a physical symptom resulting from disease, pathology or tissue injury. There are psychosocial, cultural and environmental factors involved in addition to biology that affect the complex pain experience. Patient's cognition, mood, affect, beliefs, attitudes and psychological state affect the pain experience, as well as sleep, energy and level of function. This Biopsychosocial model of pain experience is important in designing effective chronic pain programs and including an interdisciplinary model in the evaluation and treatment of chronic pain.

5. Which of the following is true about Nociceptors?
   A. A nociceptor is defined as a high-threshold sensory receptor of the peripheral somatosensory nervous system that is capable of transducing and encoding noxious stimuli
   B. The unmyelinated slow C fibers convey sharp, easily localized pain signals to the cortical regions of the brain
   C. The myelinated fast A-delta fibers conducts a more generalized, aching type of pain to the subcortical regions
   D. Muscles, joints, fascia and other deep somatic structures are supplied by A delta fibers only.

**Answer A.**

**Explanation:** A nociceptor is defined as a high-threshold sensory receptor of the peripheral somatosensory nervous system that is capable of transducing and encoding noxious stimuli. Nociceptive information is transmitted by two main pain nerve fibers. The myelinated fast A-delta fibers convey sharp, easily localized pain signals to the cortical regions of the brain; while the unmyelinated slow C fibers conducts a more generalized, aching type of pain to the subcortical regions. Muscles, joints, fascia and other deep somatic structures are supplied by both C and A delta fibers.

## References


1. Grace PM, Hutchinson M, Maier S, et al. Pathological pain and the neuroimmune interface. Nat Rev Immunol. 2014;14:217–31.
2. Pizzo PA, Clark NM. Alleviating suffering 101-pain relief in the United States. N Engl J Med. 2012;366:197–9.
3. Cieza A, Causey K, Kamenov K, et al. Global estimates of the need for rehabilitation based on the global burden of disease study 2019: a systematic analysis for the global burden of disease study 2019. Lancet. 2021;396:2006–17.
4. World Health Organization. Musculoskeletal Health. Accessed 23 Oct 2002. www.who.int/news-room/fact-sheets/detail/musculoskeletal-conditions
5. Raja S, Carr DB, Cohen M, et al. The revised International Association for the Study of Pain definition of pain: concepts, challenges, and compromises. Pain. 2020;161:1976–82.
6. Wlash NE, Dumitru D, Dchoenfield L, et al. Treatment of the patient with chronic pain. In: DeLisa JA, editor. Physical medicine and rehabilitation principles and practice. Philadelphia, PA: Lipppincott Williams & Wilkins; 2005. p. 493–530.
7. Treede RD, Rief W, Barke A, et al. Chronic pain as a symptom or a disease: the IASP classification of chronic pain for the international classification of diseases (ICD-11). Pain. 2019;160:19–27.
8. Treede RD, Rief W, Barke A, et al. A classification of chronic pain for ICD-11. Pain. 2015;156:1003–7.
9. Ambach MA, Rogers CJ, Lane JG. Osteochondral pathologies as effect of general diseases. In: Gobbi A, Lane JG, Long UG, et al., editors. Joint function preservation—a focus on the osteochondral unit. Cham, Switzerland: Springer; 2022. p. 181–90.
10. El-Tallawy SN, Nalamasu R, Salem GI, et al. Management of musculoskeletal pain: an update with emphasis on chronic musculoskeletal pain. Pain Ther. 2021;10:181–209.
11. Vrooman BM, Youngren KM. Chronic pain management. In: Butterworth IV JF, Mackey DC, Wasnick JD, editors. Morgan & Mikhail's clinical anesthesiology. 7th ed. New York: McGraw Hill; 2022.
12. Stucky C, Gold MS, Zhang X. Mechanisms of pain. Proc Natl Acad Sci. 2001;98:11845–6.
13. International Association for the Study of Pain. Terminology. Accessed 22 Oct 2022. www.iasp-pain.org/resources/terminology/.
14. Practical Pain Management. A Conceptual Model of Pain. Accessed 22 Oct 2022. www.practicalpainmanagement.com/conceptual-model-pain.
15. Raj P. Pain Mechanisms. In: Raj P, editor. Pain medicine a comprehensive review. Mosby-Year Book, Inc, St. Louis; 1992. p. 12–23.
16. Hanna VS, Hafez EA. Synopsis of arachidonic acid metabolism: a review. J Adv Res. 2018;13:23–32.
17. Almeida TF, Roizenblatt S, Tufik S. Afferent pain pathways: a neuro-anatomical review. Brain Res. 2004;1000:40–56.
18. Ossipov MH, Dussor GO, Porreca F. Central modulation of pain. J Clin Invest. 2010;120:3779–87.
19. Gerdle B, Ghafouri B, Ernberg M, et al. Chronic musculoskeletal pain: review of mechanisms and biochemical biomarkers as assessed by the microdialysis technique. J Pain Res. 2014;12:313–26.
20. Cheng J. Mechanisms of pathologic pain. In: Cheng J, Rosenquist R, editors. Fundamentals of pain medicine. Cham: Springer; 2018. p. 21–5.
21. PM&R Knowledge now. Autonomically mediated pain-autonomic pain syndromes. Accessed 8 Feb 2023. https://now.aapmr.org/autonomically-mediated-pain-autonomic-pain-syndromes/
22. McChesney KA, Weits G. Psychosocial and cultural aspects of pain. In: Yong R, Nguyen M, Nelson E, et al., editors. Pain medicine, An essential review. Cham: Springer; 2017. p. 129–32.

# Regenerative Medicine for the Shoulder

# 17

Alexandre Lavigne, Andrew Hiett, Kenneth Mautner, and Michael Khadavi

## Introduction

The shoulder is the joint with the most mobility in the human body [1], such that physicians often compare the shoulder joint to a golf ball sitting on a tee. Its great mobility allows a broad range of motion (ROM) in all directions to throw a ball, paint a wall or perform overhead activities. However, such mobility also increases the risk of soft tissue and cartilage injuries. Orthobiologics have been used thoroughly in the last decade to treat a variety of shoulder injuries. This chapter will explain the role of regenerative medicine in the non-operative management of shoulder injuries.

## Anatomy

The humeral head lies on the glenoid of the scapula, as shown in Fig. 17.1a, to form the glenohumeral (GH) joint, often called the shoulder joint [2]. The glenoid has an outer border of cartilage called the labrum that increases the stability of the shoulder joint. The GH ligaments (superior, middle and inferior) attach the humeral head to the glenoid and form the shoulder joint capsule. The rotator cuff is a group of four tendons (supraspinatus, infraspinatus, teres minor and subscapularis), as shown in Fig. 17.1b, that allow dynamic stability of the humeral head on the glenoid [3]. The long head of the biceps tendon also contributes to the stability of the GH joint. The subacromial bursa is a thin sac between the rotator cuff tendons (mostly the supraspinatus) and the acromion, decreasing friction between these structures. The junction of the acromion and clavicle forms the acromioclavicular joint which is supported by the acromioclavicular and coracoclavicular ligaments.

Injury can occur at each of the shoulder structures described above. However, only rotator cuff tendinopathy (including partial tears), GH osteoarthritis, and adhesive capsulitis will be discussed in this chapter since randomized controlled trials (RCTs) have examined the use of regenerative therapies to treat these three conditions. There is no high-quality literature on regenerative treatments of other shoulder pathologies such as labral tears, acromioclavicular arthropathy, acromioclavicular separations, and biceps tendinopathy.

A. Lavigne
Department of Physical Medicine and Rehabilitation, Faculty of Medicine, University of Montreal, Montreal, QC, Canada
e-mail: alexandre.lavigne.2@umontreal.ca

A. Hiett
Department of Physical Medicine and Rehabilitation, University of Kansas Medical Center, Kansas City, KS, USA

K. Mautner
Department of Rehabilitation Medicine, Emory University, Atlanta, GA, USA
e-mail: kmautne@emory.edu

M. Khadavi (✉)
Kansas City Orthopedic Alliance, Leawood, KS, USA

Kansas City Ballet, Sporting Kansas City, Kansas City, KS, USA


A. Navani et al. (eds.), *Essentials of Regenerative Medicine in Interventional Pain Management*,
https://doi.org/10.1007/978-3-031-50357-3_17

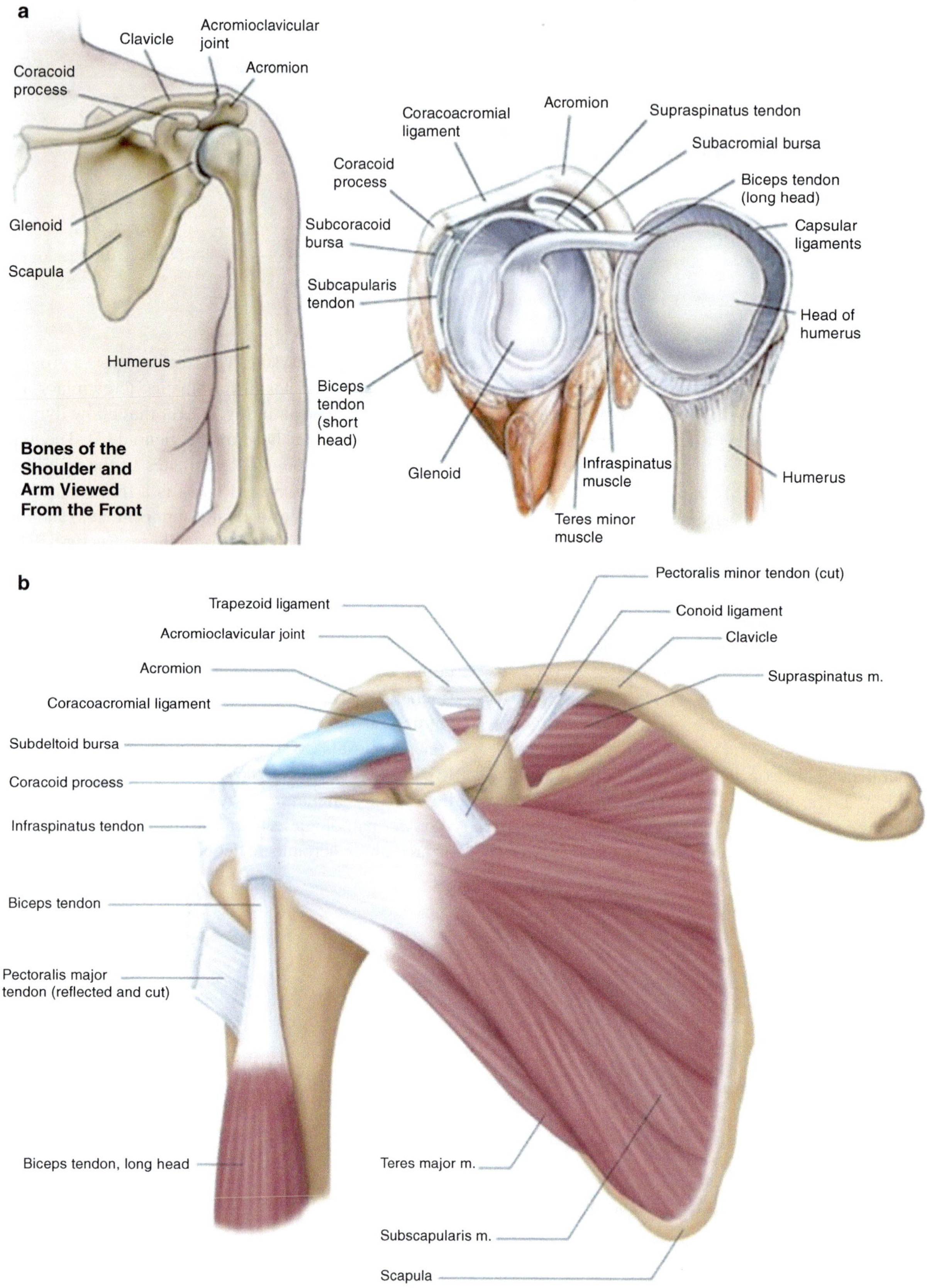


**Fig. 17.1** Shoulder anatomy images illustrating structures at risk of developing pathologies such as the glenohumeral joint and the rotator cuff tendons. (**a**). Bones of the shoulder. (**b**). Muscles and tendons of the shoulder

## Rotator Cuff Tendinopathy

### Brief Overview of Rotator Cuff Tendinopathy

Tendinopathy is an all-encompassing term that includes various pathologies such as tendinitis, tendinosis and partial tendon tear [4, 5]. Symptomatic rotator cuff tendinopathy occurs in 9.9–50% of adults and significantly increases in the elderly population [6]. Rotator cuff tendinopathies can be acute but are more often chronic injuries. They usually occur through repetitive tendon overload beyond the tendon's adaptive capacity. The high mechanical demand, chronic impingement between the acromion and the humeral head, and poor vascularization of the tendon footprint are leading factors for the high prevalence of rotator cuff tendinopathy [7]. Histological changes that occur with tendinopathy include micro-tearing, disorganized neovascularization, increased matrix metalloproteinases and lipid deposition [4, 8, 9]. Rotator cuff tendon tears can be partial or full-thickness [10]. Partial-thickness tears are often located on the articular side of the tendon, and less commonly on the bursal surface or within the interstitial portion. Rotator cuff tears are well visualized with ultrasound imaging and magnetic resonance imaging (MRI) [10–13].

Rotator cuff tendinopathies (including partial tears) are often treated with injections when physical therapy has failed to resolve the symptoms or functional impairments [10]. This chapter discusses these injections in depth. Full-thickness and massive tears more commonly require surgical treatments. However, massive tears can lead to secondary GH osteoarthritis, which can also be treated with GH injections [14].

There is high variability in the inclusion criteria of regenerative medicine studies on rotator cuff tendinopathy. Studies mentioned in this subchapter include either exclusively participants with rotator cuff partial tears or a mix of participants with partial tears and participants with rotator cuff tendinopathy alone. Most studies confirmed the presence of partial tears with either ultrasound imaging or MRI except when mentioned otherwise, and most studies used imaging to guide their injections.

### Platelet-Rich Plasma for Rotator Cuff Tendinopathy

Ten RCTs and one observational study assessed the efficacy of platelet rich plasma (PRP) in treating rotator cuff partial tears which are worth discussing. In these studies, PRP was injected into the rotator cuff lesion, the subacromial bursa, or at the injection sites mentioned below. They all used ultrasound guidance except Ilhanli et al. [15] and Shams et al. [16]. There have been several PRP classification systems put forth over the past 10 years, including Mautner et al. [17] in 2015, but most of the studies mentioned below did not include all the necessary detail, as shown in Table 17.1. This absence of important treatment variable measurements, like platelet and leukocyte counts, limits our ability to compare studies and draw conclusions such as which PRP product works better and why a negative study is so. A few participants in these PRP trials have reported a short-term increase in shoulder pain and a handful of participants developed adhesive capsulitis following PRP injected into the rotator cuff tendons. No tendon-specific complications nor major adverse events have been reported.

Five RCTs compared the efficacy of PRP against glucocorticoids.

First, Sari et al. [18] compared PRP, glucocorticoids, prolotherapy and lidocaine only. PRP injections had a platelet concentration five times higher than whole blood. They noticed that the 30 participants in the PRP group had statistically greater pain and functional improvement at 24 weeks compared to both the glucocorticoid and the prolotherapy groups. They did not discuss the specific types of rotator cuff tears each participant had.

Second, Thepsoparn et al. [19] injected leukocyte-poor PRP that had a platelet concentration two times higher than whole blood and noted that the PRP group (15 participants) had significantly better pain and functional improvements at the 6-month follow-up compared to the glucocorticoid group. They did not identify the types of tears.

Third, Shams et al. [16] found that 20 participants who received PRP had statistically greater pain and functional improvement at 12 weeks than those who received glucocorticoids, but there was no significant difference between the two groups at 6 months [16]. They did not report PRP characteristics such as the platelet count and the level of leukocytes. They did not identify the types of tears.

Fourth, Damjanov et al. [20] used four weekly injections of autologous conditioned serum compared with three weekly glucocorticoid injections and one saline injection in the fourth week. While autologous conditioned serum usually has a lesser platelet concentration than PRP and is leukocyte-poor, this study did not characterize their injectate content. They concluded that the autologous conditioned serum group (16 participants) had better pain and functional improvement at the 24-week follow-up than the glucocorticoid group. They did not identify the types of tears.

Fifth, Kwong et al. [21] used leukocyte-poor plasma with a platelet concentration 1.6 times greater than whole blood. Of note, this plasma is not considered PRP by some of the PRP classifications [22, 23]. Study results demonstrated that their PRP group (49 participants) had better outcomes than their glucocorticoid group regarding pain and function at 3

**Table 17.1** PRP characteristics and other clinical trial characteristics for each PRP study on rotator cuff tendinopathy

| Rotator cuff tendinopathy trials using PRP | Platelet count | Plasma volume (mL) | Platelet concentration | WBC count | Neutrophils count | RBC count | Activator | US-guided | Number of injections | Participants who received PRP | Comparison | Follow-up length (months) | Favorable outcome |
|---|---|---|---|---|---|---|---|---|---|---|---|---|---|
| Sari & Erogla, 2020 [18] | 5 folds | 5 | NR | NR | NR | NR | Yes (calcium chloride) | Yes | 1 | 30 | GCI | 6 | Yes |
| Thepsoparn et al., 2021 [19] | 2 folds [a] | 5 | NR | Poor | NR | Poor | NR | Yes | 1 | 15 | GCI | 6 | Yes |
| Shams et al., 2016 [16] | NR | 2–2.5 | NR | NR | NR | NR | NR | No | 1 | 20 | GCI | 6 | No |
| Damjanov et al., 2018 [20] | NR | 2 | NR | NR | NR | NR | NR | Yes | 1 | 16 | GCI | 6 | Yes |
| Kwong et al., 2021 [21] | 1.6 folds | 5 | NR | Poor | NR | NR | NR | Yes | 1 | 49 | GCI | 12 | No |
| Cai et al., 2019 [24] | $4 \times 10^9$ [b] | 4 | $1 \times 10^{12}$/L | Poor | NR | NR | NR | Yes | 1 | 45 | NS, HA, PRP + HA | 12 | Yes |
| Kesikburun et al., 2013 [25] | $5 \times 10^9$ [b] | 5 | $1 \times 10^{12}$/L | NR | NR | NR | No | Yes | 1 | 20 | NS | 12 | No |
| Schwitzguebel et al., 2019 [26] | 1.6 folds [a] | 2 | NR | Poor | NR | NR | No | Yes | 2 (1 month apart) | 41 | NS | 7 | No |
| Nejati et al., 2017 [27] | 3.6-6 $\times 10^9$ [b] | 4 | 0.9–1.5 $\times 10^{12}$/L | 20′000–40′000 | NR | NR | NR | Yes | 2 (2 months apart) | 31 | PT | 6 | No |
| Ilhanli et al., 2015 [15] | 2.1–2.5 folds | 6 | NR | 1.1–1.3 folds | NR | NR | Yes (calcium chloride) | No | 1 | 30 | PT | 12 | Yes (for functional scores) |
| Prodromos et al., 2021 [28] | NR | 4 | 0.75–1.25 $\times 10^{12}$/L | Rich | Rich | NR | No | Yes | 1 | 71 | N/A | 24 | Yes (but no comparison) |

*WBC* white blood cell, *RBC* red blood cell, *US* ultrasound, *NR* not reported, *GCI* glucocorticoid injection, *HA* hyaluronic acid, *NS* normal saline, *PT* physical therapy, *N/A* not applicable

[a] Data found on the merchant website

[b] Platelet count calculated from the reported PRP concentration and volume

months. Still, there was no significant difference at the 6-month and 12-month follow-ups. 11% of the participants had tendinopathy without a tendon tear, 23% had an articular-sided tear, 6% had an intrasubstance tear, 2% had a bursal-sided tear and 51% had a combination of tears. However, they did not identify what type of tendinopathy responded better to PRP and glucocorticoids.

An RCT by Cai et al. [24] compared the efficacy of four different treatment regimens. One group received four weekly PRP injections, another group received four weekly hyaluronic acid (HA) injections and another group received four weekly saline injections. The final group received both PRP and HA injections weekly for 4 weeks. Their PRP was leukocyte-poor and contained $4 \times 10^9$ platelets in the 4-mL injectate. All their participants had a bursal-sided tear. The group with PRP plus hyaluronic acid (48 participants) had statistically greater pain and functional improvement and 6 and 12 months compared with the PRP group (45 participants) and the hyaluronic acid group (44 participants). Moreover, participants with PRP alone had statistically greater pain and functional improvement at 6 and 12 months compared to those with hyaluronic acid alone. Finally, PRP plus hyaluronic acid, PRP alone and hyaluronic acid alone were all superior to saline injections at 3, 6 and 12 months in terms of pain and functional improvement.

Two additional RCTs compared PRP against saline. Kesikburun et al. [25] noted that PRP containing $5 \times 10^9$ platelets in a 5-mL injectate (20 participants) was equivalent to saline at the 1-year follow-up. Three-quarters of their participants had tendinopathy without a tendon tear, and among those who had a partial tear, the majority of tears were located either on the bursal side or intrasubstance. The authors did not mention if there was a link between the type of tear and pain improvement in either group.

Schwitzguebel et al. [26] found that two plasma injections 1 month apart (41 participants) had a similar effect as two saline injections regarding pain and function at the 7-month follow-up. As for Kwong et al's [21] injectate, the platelet concentration in this study is only 1.6 folds greater than whole blood, which is not considered PRP by some of the PRP classifications [22, 23]. They only included participants with an intrasubstance tear, and excluded candidates who had an articular-sided or a bursal-sided tear.

Additionally, two RCTs assessed the efficacy of PRP injections compared to physical therapy. Nejati et al. [27] used two PRP injections with a platelet count between $3.6\text{-}6 \times 10^9$ in each 4-mL injectate, injected 2 months apart. These PRP injections were compared to the effect of a 3-month physical therapy regimen with one weekly supervised session and at-home exercises during the other days of the week. In both groups, about 70% of participants had a partial tendon tear, and the rest had tendinopathy without a tendon tear. In the PRP group, only 77% showed up for their second injection, and in the physical therapy group, only 69% completed their 3 months of physical therapy. At the 1- and 3-month follow-ups, the physical therapy group had statistically greater pain improvement, but there was no statistically significant difference between the two groups at the 6-month follow-up. In addition, the physical therapy group had a statistically greater functional improvement than the PRP group at all three follow-up assessments. They did not indicate if participants with a rotator cuff partial tear progressed differently than those with tendinosis alone.

The second study by Ilhanli et al. [15] compared three weekly PRP injections (30 participants) with 3 weeks of intensive physical therapy, including five supervised weekly sessions (32 participants). They did not distinguish the location of their participants' partial tears. It is noteworthy that their PRP injections were intra-articular rather than into or around the rotator cuff tendons, compared to all the other studies mentioned in this chapter that evaluated PRP's effect on rotator cuff tendinopathy. Their PRP had a platelet concentration 2.1–2.5 times greater than whole blood, and a leukocyte concentration 1.1–1.3 times greater than whole blood. The study did not document the participants' adherence to physical therapy. The active shoulder ROM in most directions was statistically greater in the physical therapy group at the 12-month follow-up. The functional improvement was statistically greater in the PRP group. There was no statistically significant difference between the two groups for pain while doing activities and discomfort while sleeping, but the PRP group had less pain at rest.

One PRP observational study by Prodromos et al. [28] is worth mentioning even though it is not an RCT because it has the largest cohort (71 participants) and the longest follow-up time (24 months). They administered two PRP injections on the day of recruitment: the first one in the supraspinatus insertion zone and the second in the GH joint. There was statistically significant pain and functional improvement at the 6-month, 12-month and 24-month follow-ups. They did not differentiate the types of tears that were treated.

In conclusion, several high-quality RCTs show favorable outcomes when using PRP in the treatment of rotator cuff tendinopathy. However, there are significant outcome differences from one study to another, and this may be explained by the great diversity of these studies. The methodological variances include a lack of description of the types and severities of tendon tears and tendinosis, the concentration of leukocytes and platelets, the quantity and location of PRP delivery, and post-procedural rehabilitation protocols. Therefore, the possibility of pooling data is limited. The current literature does not allow precision in determining if one type of partial tear responds to PRP better than others. The ideal platelet and leukocyte concentrations for treating rotator cuff pathologies are also unclear. There is evidence to suggest that these variables influence PRP's effects on tendons [29, 30].

## Bone Marrow Aspirate Concentrate for Rotator Cuff Tendinopathy

Two studies have assessed the efficacy of bone marrow aspirate concentrate (BMAC) on rotator cuff tendinopathy. Injections in both studies were performed under ultrasound guidance.

First, in a RCT with cross-over, Centeno et al. compared the efficacy of a 2 mL injection containing 60% of BMAC, 20% of PRP and 20% of platelet lysate (14 participants) with physical therapy [13]. The BMAC+PRP group had statistically greater pain and functional improvement at the 3-month and 6-month assessments. By the sixth month, all participants in the exercise group had crossed over to the BMAC group. Moreover, pain and function in the BMAC+PRP group kept improving at 12 and 24 months.

Second, Kim et al. [31] led a non-randomized control trial to compare the outcomes of an injection containing 2 mL of BMAC and 1 mL of PRP (12 participants) with a physical therapy program, and noticed that the BMAC+PRP group had statistically greater pain and functional improvement at 3 months. They also kept track of the rotator cuff partial tear size and found no statistically significant difference in the tear size at 3 months.

While this chapter focuses on the non-operative management of shoulder pathologies, one surgical study is worth mentioning regarding cellular treatments for rotator cuff tears. Hernigou et al. [32] compared 45 participants suffering a complete tear who had a rotator cuff repair and an intraoperative BMAC injection with 45 participants who only had the rotator cuff repair. In the BMAC group, participants received a 4-mL intratendinous BMAC injection and an 8-mL intraosseous BMAC injection at the footprint of the rotator cuff tendon. At the 10-year follow-up, the BMAC group had a profoundly lower retear rate (13% in the BMAC group versus 56% in the control group), and faster and more prominent tendon healing. They also noted a correlation between the mesenchymal stem cell concentration and the quality of tendon healing.

In conclusion, BMAC injections seem promising for treating rotator cuff partial tears.

## Adipose Tissue-Derived Cells for Rotator Cuff Tendinopathy

Two trials have observed the effect of adipose tissue-derived cells on rotator cuff tendinopathy specifically.

Jo et al. observed the effect of a 3 mL intratendinous injection containing enzymatically-digested adipose tissue, also called the stromal vascular fraction, that was then culture-expanded (18 participants) [11]. The authors noticed that participants had statistically significant pain and functional improvement compared to baseline. At the 6-month follow-up, MRI demonstrated that the partial tear size decreased by 90% compared to baseline which reached statistical significance. Also, the investigator saw some regenerated tendon tissue in 100% of the 16 participants who underwent an arthroscopic second look at 6 months. None of the participants had a rotator cuff tear size increase.

Hogaboom et al. [33] studied a cohort of 8 wheelchair users with spinal cord injuries who had rotator cuff tendinopathy. They used a non-enzymatic isolation system that generated a micro-fragmented adipose tissue (MFAT) injectate using mechanical forces.

Then, they injected 2.5–6 mL of MFAT in the supraspinatus tendon lesion, 1–2 mL of MFAT in the subacromial bursa and 0–3 mL of MFAT in other shoulder structures causing pain according to the physical examination (infraspinatus tendon, subscapularis tendon, GH joint or acromioclavicular joint). There was statistically significant pain and functional improvement at 6 and 12 months compared to baseline, according to validated scales and physical examination maneuvers. In addition, they noticed some degree of supraspinatus tendinopathy healing in four participants according to ultrasound assessments at follow-ups. However, those pain, functional and biological improvements were not compared with a control group.

Furthermore, two case series were published on the efficacy of MFAT in participants who were either affected by rotator cuff tendinopathy and/or GH osteoarthritis, but the outcomes of those two shoulder pathologies were all mixed. They used the same technology as Hogaboom et al. [33] to harvest and prepare their MFAT injectates. Also, they both used ultrasound guidance to inject MFAT in the rotator cuff tendon lesions and the GH joint.

Striano et al. [34] assessed the efficacy of MFAT in 18 subjects, injecting 1–2 mL in the rotator cuff lesion and 4 mL in the GH joint occasionally associated with a 2-mL injection on the posterior surface of the glenoid labrum. Their participants had statistically significant pain and functional improvement at the 3, 6, and 12-month follow-ups compared to baseline. Interestingly, it was during the first 24 h that the most significant pain improvement occurred.

Robinson et al. [35] injected MFAT in 18 shoulders from 16 participants, averaging 1.4 mL in the rotator cuff tendon lesions and 4.9 mL in the GH joints. At the 6-month follow-up, their participants had statistically significant pain and functional improvement.

In conclusion, adipose tissue-derived cell injections also seem promising as a treatment for rotator cuff tendinopathy, just like BMAC. However, the efficacy of those two treatments has yet to be compared.

## Glenohumeral Osteoarthritis

### Brief Overview of Glenohumeral Osteoarthritis

Another prevalent shoulder pathology is GH osteoarthritis, which affects 4–26% of the population [36]. Clinicians often consider osteoarthritis a cartilage disorder, but it is a more complex issue that involves the surrounding structures such as the capsule, synovium, surrounding ligaments, and subchondral bone [37]. Cartilage thinning on the humeral head and glenoid is often associated with subchondral bone marrow lesions, sclerosis, cysts, osteophytes, and ultimately, humeral head dysplasia [38]. In later stages of osteoarthritis, subchondral bone marrow lesions contribute significantly to symptomatology because it is greatly innervated by nociceptive pain fibers [37, 39].

As for pathophysiology, GH osteoarthritis can be a primary or secondary disease [38]. Primary GH osteoarthritis accounts for 90% of cases and occurs due to the cumulative effect of mechanical stresses on the joint throughout an individual's lifetime. Cytokines such as interleukin 1-beta, tumor necrosis factor, and matrix metalloproteinases are thought to be involved in the biochemical destruction of cartilage [38, 40]. On the other hand, secondary osteoarthritis can be caused by GH dislocation, subluxation, or massive rotator cuff tear. It can also be caused by atraumatic osteonecrosis for which multiple risk factors have been identified, such as systemic glucocorticoid use, alcohol consumption and hematopoietic diseases among others [38]. A radiograph is the most common imaging modality to diagnose GH osteoarthritis. However, subchondral bone marrow lesions are better visualized with MRI [37].

Non-surgical treatments of GH osteoarthritis include activity modification, physical therapy modalities, nonsteroidal anti-inflammatory drugs (oral or intra-articular injection), glucocorticoid injection and orthobiologic injection.

### Platelet-Rich Plasma for Glenohumeral Osteoarthritis

While many RCTs have assessed the efficacy of PRP in treating osteoarthritis in other joints, no RCT or cohort study has evaluated the usefulness of PRP in treating GH osteoarthritis.

### Bone Marrow Aspirate Concentrate for Glenohumeral Osteoarthritis

Two studies assessed the efficacy of cellular treatments in treating GH osteoarthritis, neither with a control group.

Centeno et al. [41] led an observational study that included 34 participants who received a combination of BMAC, PRP, and platelet lysate. They followed participants' pain for an average of 8.3 months and participants' function for an average of 7.1 months and noticed that both pain and function were significantly improved at the latest follow-up.

Darrow et al. [42] also conducted an observational study of 32 participants, in which they followed participants who had received either BMAC (42%) or whole bone marrow (58%). Also, 12 of their participants received one injection, and 20 received two injections with an average of 22 days apart. Their results showed that the participants who received one and two injections significantly improved in pain and function at the follow-up (on average, 7 months post-injection) compared to baseline.

In conclusion, there is insufficient evidence for or against BMAC in treating GH osteoarthritis, and further high-quality evidence is needed.

## Adhesive Capsulitis

### Brief Overview of Adhesive Capsulitis

Adhesive capsulitis, often called frozen shoulder, is characterized by an insidious onset of diffuse shoulder pain that is later associated with an active and passive limitation of the shoulder ROM [43]. The prevalence of adhesive capsulitis is 0.75% in the general population and 13.4% in the diabetic population [44, 45]. It is also more prevalent in females, especially between 40 and 60 years old [46]. Numerous risk factors have been described, such as diabetes mellitus, thyroid dysfunction, immune diseases, stroke, myocardial infarction, previous breast cancer treatments and recent shoulder immobilization [46]. Adhesive capsulitis is a clinical diagnosis and does not need to be confirmed with imaging. However, adhesive capsulitis signs can be depicted with ultrasound imaging, such as a positive doppler signal at the rotator interval and increased thickness of the coracohumeral ligament [47]. Even though a radiograph does not help diagnose adhesive capsulitis, it is always recommended to rule out other etiologies that could influence management.

Several treatments have been described for adhesive capsulitis. The main treatments include physical therapy with mobilization exercises, oral analgesics, glucocorticoid injections with or without capsular distension, mobilization under anesthesia and arthroscopic capsular release [46, 48].

### Platelet-Rich Plasma for Shoulder Adhesive Capsulitis

Even though PRP is less commonly used for the treatment of shoulder adhesive capsulitis, three RCTs evaluated its potential in treating adhesive capsulitis (Table 17.2). All three research groups hypothesized that the anti-inflammatory

**Table 17.2** PRP characteristics and other clinical trial characteristics for each PRP study on shoulder adhesive capsulitis

| Adhesive capsulitis trials using PRP | Platelet count | Plasma volume (mL) | Platelet concentration | WBC count | Neutrophils count | RBC count | Activator | US-guided | Number of injections | Participants who received PRP | Comparison | Follow-up length (months) | Favorable outcome |
|---|---|---|---|---|---|---|---|---|---|---|---|---|---|
| Kothari et al. 2017 [49] | NR | 2 | NR | NR | NR | NR | NR | NR | 1 | 62 | GCI | 3 | Yes |
| Ünlü et al. 2021 [50] | NR | 2 | NR | NR | NR | NR | No | No | 1 | 17 | NS | 3 | Yes |
| Thu et al. 2020 [51] | NR | 4 | NR | NR | NR | NR | Yes (calcium chloride) | Yes | 1 | 31 | PT | 1.5 | No |

*WBC* white blood cell, *RBC* red blood cell, *US* ultrasound, *NR* not reported, *GCI* glucocorticoid injection, *NS* normal saline, *PT* physical therapy

properties of PRP could target the inflammation and fibrosis of the joint capsule of patients affected by adhesive capsulitis.

Kothari et al. [49] compared the efficacy of a GH PRP injection, glucocorticoid injection and ultrasonic therapy. PRP injections averaged platelet concentrations 6.1 times greater than whole blood. Those receiving ultrasonic therapy had seven sessions administered every other day. The use of image guidance was not mentioned. The 62 participants in the PRP group had significantly greater pain and functional improvement than the glucocorticoid and the ultrasonic therapy groups at the six- and 12-week follow-ups.

Ünlü et al. [50] compared the efficacy of three GH PRP injections administered every other week (17 participants) with that of three saline injections. Injections were performed under landmark guidance. The PRP concentration was not mentioned. Participants in the PRP group had significantly greater pain and functional improvement at 3 months post-intervention than in the saline group.

Thu et al. assessed the usefulness of one ultrasound-guided GH PRP plus lidocaine injection (31 participants) versus three weekly physical therapy sessions for 6 weeks that included diathermy and stretching exercises [51]. PRP concentration was not mentioned. They found no statistically significant pain or functional improvement between the two groups at weeks 1, 3 and 6.

In conclusion, evidence shows that GH PRP injections may have greater efficacy than glucocorticoid, saline injections or physical therapy in treating adhesive capsulitis. None of the aforementioned RCTs utilized GH hydrodistention with a high-volume injection containing PRP or glucocorticoids. There is debate about whether or not capsular distension may enhance outcomes in adhesive capsulitis [47, 52]. The mechanism by which intra-articular PRP injections without capsular distension could potentially be beneficial for adhesive capsulitis remains unclear.

## Cellular Treatments for Shoulder Adhesive Capsulitis

No RCT or cohort study has evaluated the usefulness of BMAC or MFAT in treating shoulder adhesive capsulitis.

## Technique Considerations for Orthobiologic Injections in the Shoulder

Patients should be offered a thorough explanation of orthobiologic procedures with realistic expectations before giving consent [53].

For orthobiologic injections in the shoulder, technique variables and the use of guidance likely influence the treatment's outcome [30]. While no orthobiologic shoulder study has yet specifically compared pain and functional improvements based upon injection technique, the accuracy of GH and subacromial injections is improved with the use of ultrasound guidance [54]. Ultrasound images of ultrasound-guided injections for rotator cuff tendinopathy and GH osteoarthritis are shown in Figs. 17.2 and 17.3 respectively.

For GH osteoarthritis, before injecting orthobiologics, any effusion should be aspirated to remove inflammatory cellular and acellular components and prevent dilution of the orthobiologic treatment [55]. For rotator cuff tendinopathy, injecting an orthobiologic solution into and around a tendon involves several considerations. Percutaneous tenotomy is often performed in association with orthobiologic injections [56, 57]. This ultrasound-guided procedure is done at the same time as the injection and consists of multiple fenestrations through the site of pathology, to stimulate the tendon healing mechanisms [56].

Another important consideration is the utilization of local or regional anesthesia for the tolerability of most intratendinous orthobiologic procedures [58].

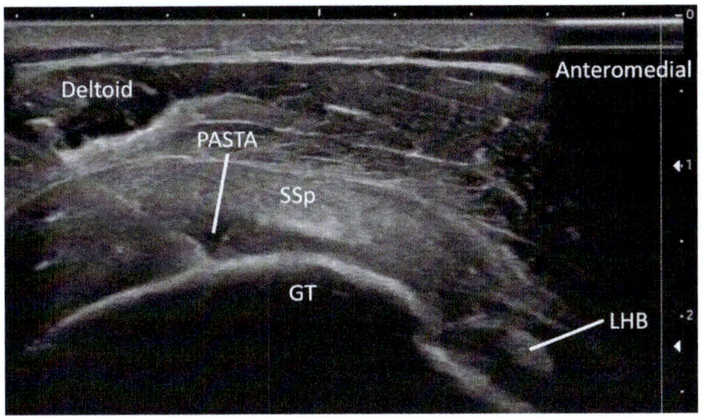


**Fig. 17.2** An in-plane ultrasound-guided injection of BMAC and PRP into a partial articular supraspinatus tendon avulsion (PASTA). The labels also show the supraspinatus tendon (SSp), the long-head biceps tendon (LHB) and the greater tuberosity (GT) of the humeral head. This tear is visualized as an anechoic portion of the tendon, which is seen here in short axis. Procedural pain control was achieved with a suprascapular nerve block 30 min before the BMAC and PRP injection. The BMAC and PRP were then injected with a peppering technique into and around the tear. (**Source:** The ultrasound image was provided by Dr. Michael Khadavi)

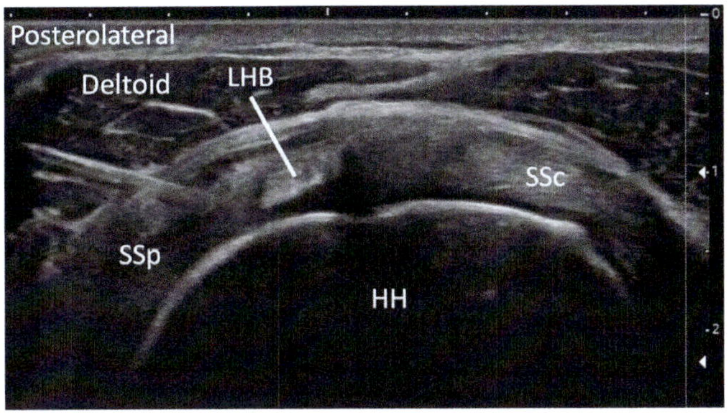


**Fig. 17.3** An in-plane ultrasound-guided glenohumeral injection of leukocyte-poor PRP from an anterior approach via the intra-articular long-head biceps tendon. The labels show the supraspinatus tendon (SSp), the long-head biceps tendon (LHB), the subscapular tendon (SSc) and the humeral head (HH). The glenohumeral injection is also commonly performed from a posterior approach with ultrasound guidance. (**Source:** The ultrasound image was provided by Dr. Michael Khadavi)

## Timing Considerations for Orthobiologic Injections in the Shoulder

Physical therapy should be an initial treatment for rotator cuff tendinopathy, rotator cuff partial tear and GH osteoarthritis, as it significantly decreases pain and increases function in most cases [10, 59]. It is also less invasive than regenerative medicine procedures.

The relative timing of regenerative medicine procedures remains controversial. Traditionally, clinicians have waited until physical therapy plateaus or fails to provide satisfactory pain and functional improvements. However, there may be cases in which orthobiologic injections should be performed sooner, alongside physical therapy, such as in professional athletes and others with a tight timeline.

Continued debate exists on the utility of glucocorticoid injections for shoulder pathologies. On the one hand, glucocorticoid injections are cheaper and often less painful [60], however, there are growing concerns about the long-term safety of glucocorticoids. These include local and systemic side effects such as tenotoxicity, chondrotoxicity, hyperglycemia, osteoporosis, and hypertension [10, 61–66]. In contrast, orthobiologics seem to be safe for tendons and cartilage, and do not cause any known systemic side effects [36, 67]. Therefore, the authors believe that regenerative medicine procedures could be offered before glucocorticoids for rotator cuff tendinopathies despite the associated costs.

## Conclusion

This chapter describes the literature on regenerative medicine for the most common shoulder pathologies. There is robust evidence on the efficacy and safety to support the use of PRP in treating rotator cuff tendinopathies when performed by trained physicians [53, 68]. There is also growing evidence that BMAC and adipose-derived mesenchymal stem cells (ADMSC) injections can be valuable treatment options for rotator cuff tendinopathies and GH osteoarthritis in selected cases where other treatments have failed. Nonetheless, more robust clinical trials are needed to routinely offer BMAC and ADMSC for those shoulder pathologies [68]. PRP seems to be beneficial for adhesive capsulitis as well, but again, the literature is still too scarce to fully understand its potential in treating adhesive capsulitis. Research gaps on the value of orthobiologics for other shoulder pathologies such as labral tears and acromioclavicular arthropathy need to be filled.

**Key Points**

1. Randomized controlled trials on orthobiologics for the shoulder have only focussed on three pathologies: rotator cuff tendinopathy, glenohumeral osteoarthritis, and adhesive capsulitis.
2. Platelet-rich plasma in the treatment of rotator cuff tendinopathy has demonstrated favorable outcomes in several high-quality randomized controlled trials.
3. The high variability in the inclusion criteria of regenerative medicine studies on rotator cuff tendinopathy and in the injectates being used limits the possibility of comparing studies together.

4. The efficacy of bone marrow aspirate concentrate and adipose tissue-derived cell injections in the treatment of rotator cuff tendinopathy has shown promise, however has yet to be compared to other treatments.
5. No randomized controlled trials have assessed the efficacy of platelet-rich plasma on glenohumeral osteoarthritis.
6. The literature on bone marrow aspirate concentrate in treating glenohumeral osteoarthritis is insufficient to provide recommendations.
7. While the mechanism of action remains incompletely understood, intra-articular platelet-rich plasma injections into the glenohumeral joint may have greater efficacy than glucocorticoid injections, saline injections, or physical therapy in treating adhesive capsulitis.
8. Ultrasound guidance improves the accuracy of orthobiologic injections into the rotator cuff tendons and glenohumeral joint.
9. While orthobiologic injections for shoulder pathologies are usually offered when conservative care plateaus, there may be exceptions in which it is appropriate to try orthobiologic procedures sooner.
10. In most shoulder studies, the cellular components of platelet-rich plasma and the post-procedural rehabilitation protocols are not clearly described or standardized, somewhat limiting our ability to draw comparative conclusions. Better characterization will allow us to more precisely tie a specific patient or orthobiologic variable to an outcome, and therefore prescribe the proper protocol for a particular patient.

**Questions**

1. Which shoulder pathology has the most robust evidence for being treated with PRP?
   A. Rotator cuff tendinopathy
   B. Glenohumeral osteoarthritis
   C. Adhesive capsulitis
   D. Acromioclavicular arthropathy
   **Answer: A.**
   **Explanation:** "There is robust evidence on the efficacy to support the use of PRP in treating rotator cuff tendinopathies." Moreover, "PRP seems to be beneficial for adhesive capsulitis as well", but the literature is not as robust as it is for rotator cuff tendinopathies. No RCT or cohort study has evaluated the usefulness of PRP in treating glenohumeral osteoarthritis and acromioclavicular arthropathy.
2. What is the preferred guidance for rotator cuff tendinopathy orthobiologic injections?
   A. Landmark
   B. Ultrasound
   C. Fluoroscopy
   D. Computerized tomography scan
   **Answer: B.**
   **Explanation:** The accuracy of rotator cuff tendinopathy injections is improved with the use of ultrasound guidance. Indeed, ultrasound guidance allows clinicians to visualize the tendon lesion and inject orthobiologics into and around the lesion. Landmark, fluoroscopic and computerized tomography guidances do not allow clinicians to visualize the tendon lesion.
3. Why may we prefer platelet-rich plasma over glucocorticoids for rotator cuff tendinopathy injections?
   A. Cheaper
   B. Easier to inject
   C. Fewer local and systemic adverse effects
   D. It requires drawing blood from the patient
   **Answer: C.**
   **Explanation:** PRP is safer for tendons and cartilage than glucocorticoids, and it does not cause any known systemic side effects. However, PRP is more expensive than glucocorticoids and requires drawing blood from the patient. The level of difficulty of ultrasound-guided PRP injections is generally more complex than ultrasound-guided glucocorticoid injection because it requires identifying the lesion and injecting PRP into the lesion, instead of simply injecting glucocorticoids in the subacromial bursa.
4. What injection characteristics need to be considered when performing a platelet-rich plasma injection for rotator cuff tendinopathy?
   A. Platelet count, leukocyte count, neutrophil count and red blood cell count.
   B. Bone marrow harvesting site.
   C. Molecular weight.
   D. Particulate versus nonparticulate preparation.
   **Answer: A.**
   **Explanation:** Cellular concentrations are important variables of PRP. The bone marrow harvesting site is relevant for BMAC injection. The molecular weight is relevant for hyaluronic injection. Glucocorticoids are classified as particulated or nonparticulated.
5. Which one is an important consideration when injecting orthobiologics for glenohumeral osteoarthritis?
   A. Immobilize the shoulder for a week following an orthobiologic injection.
   B. Perform a glenohumeral hydrodistension right after injecting orthobiologics.
   C. Proceed with a rotator cuff tenotomy following the glenohumeral injection.
   D. Aspirate the intra-articular effusion before injecting orthobiologics.
   **Answer: D.**

**Explanation:** "Any effusion should be aspirated to remove inflammatory cellular and acellular components and prevent dilution of the orthobiologic treatment" The shoulder should not be immobilized after a glenohumeral orthobiologic injection. No study has assessed the value of combining glenohumeral orthobiologic injection with either a hydrodistension or a rotator cuff tenotomy.

## References


1. Halder AM, Kuhl SG, Zobitz ME, et al. Effects of the glenoid labrum and glenohumeral abduction on stability of the shoulder joint through concavity-compression: an in vitro study. J Bone Joint Surg Am. 2001;83:1062–9.
2. Bakhsh W, Nicandri G. Anatomy and physical examination of the shoulder. Sports Med Arthrosc Rev. 2018;26:e10–22.
3. Lewis JS. Rotator cuff tendinopathy. Br J Sports Med. 2009;43:236–41.
4. Abadin AA, Orr JP, Lloyd AR, et al. An evidence-based approach to orthobiologics for tendon disorders. Phys Med Rehabil Clin N Am. 2023;34:83–103.
5. Hall MM, Allen GM, Allison S, et al. Recommended musculoskeletal and sports ultrasound terminology: a Delphi-based consensus statement. Br J Sports Med. 2022;56:310–9.
6. Teunis T, Lubberts B, Reilly BT, et al. A systematic review and pooled analysis of the prevalence of rotator cuff disease with increasing age. J Shoulder Elb Surg. 2014;23:1913–21.
7. Condron NB, Kester BS, Tokish JM, et al. Nonoperative and operative soft-tissue, cartilage, and bony regeneration and orthopaedic biologics of the shoulder: an Orthoregeneration network (ON) foundation review. Arthroscopy. 2021;37:3200–18.
8. Griffith KM, Hammer LC, Iannuzzi NP, et al. Review of human supraspinatus tendon mechanics. Part II: tendon healing response and characterization of tendon health. J Shoulder Elb Surg. 2022;31:2678–82.
9. Lewis JS, Raza SA, Pilcher J, et al. The prevalence of neovascularity in patients clinically diagnosed with rotator cuff tendinopathy. BMC Musculoskelet Disord. 2009;10:163.
10. Pasculli RM, Bowers RL. Evidence-based management of rotator cuff tears (acute and chronic). Curr Phys Med Rehabil Rep. 2022;10:239–47.
11. Jo CH, Chai JW, Jeong EC, et al. Intratendinous injection of autologous adipose tissue-derived mesenchymal stem cells for the treatment of rotator cuff disease: a first-in-human trial. Stem Cells. 2018;36:1441–50.
12. Bertrand H, Reeves KD, Bennett CJ, et al. Dextrose prolotherapy versus control injections in painful rotator cuff tendinopathy. Arch Phys Med Rehabil. 2016;97:17–25.
13. Centeno C, Fausel Z, Stemper I, et al. A randomized controlled trial of the treatment of rotator cuff tears with bone marrow concentrate and platelet products compared to exercise therapy: a midterm analysis. Stem Cells Int. 2020;2020:5962354–10.
14. Zingg PO, Jost B, Sukthankar A, et al. Clinical and structural outcomes of nonoperative management of massive rotator cuff tears. J Bone Joint Surg Am. 2007;89:1928–34.
15. Ilhanli I, Guder N, Gul M. Platelet-rich plasma treatment with physical therapy in chronic partial supraspinatus tears. Iran Red Crescent Med J. 2015;17:e23732.
16. Shams A, El-Sayed M, Gamal O, et al. Subacromial injection of autologous platelet-rich plasma versus corticosteroid for the treatment of symptomatic partial rotator cuff tears. Eur J Orthop Surg Traumatol. 2016;26:837–42.
17. Mautner K, Malanga GA, Smith J, et al. A call for a standard classification system for future biologic research: the rationale for new PRP nomenclature. PM R. 2015;7:S53–9.
18. Sari A, Eroglu A. Comparison of ultrasound-guided platelet-rich plasma, prolotherapy, and corticosteroid injections in rotator cuff lesions. J Back Musculoskelet Rehabil. 2020;33:387–96.
19. Thepsoparn M, Thanphraisan P, Tanpowpong T, et al. Comparison of a platelet-rich plasma injection and a conventional steroid injection for pain relief and functional improvement of partial supraspinatus tears. Orthop J Sports Med. 2021;9:23259671211024937.
20. Damjanov N, Barac B, Colic J, et al. The efficacy and safety of autologous conditioned serum (ACS) injections compared with betamethasone and placebo injections in the treatment of chronic shoulder joint pain due to supraspinatus tendinopathy: a prospective, randomized, double-blind, controlled study. Med Ultrason. 2018;20:335–41.
21. Kwong CA, Woodmass JM, Gusnowski EM, et al. Platelet-rich plasma in patients with partial-thickness rotator cuff tears or tendinopathy leads to significantly improved short-term pain relief and function compared with corticosteroid injection: a double-blind randomized controlled trial. Arthroscopy. 2021;37:510–7.
22. Lana JFSD, Purita J, Paulus C, et al. Contributions for classification of platelet rich plasma - proposal of a new classification: MARSPILL. Regen Med. 2017;12:565–74.
23. Rossi LA, Murray IR, Chu CR, et al. Classification systems for platelet-rich plasma. Bone Joint J. 2019;101-B:891–6.
24. Cai YU, Sun Z, Liao B, et al. Sodium hyaluronate and platelet-rich plasma for partial-thickness rotator cuff tears. Med Sci Sports Exerc. 2019;51:227–33.
25. Kesikburun S, Tan AK, Yilmaz B, et al. Platelet-rich plasma injections in the treatment of chronic rotator cuff tendinopathy: a randomized controlled trial with 1-year follow-up. Am J Sports Med. 2013;41:2609–16.
26. Schwitzguebel AJ, Kolo FC, Tirefort J, et al. Efficacy of platelet-rich plasma for the treatment of interstitial supraspinatus tears: a double-blinded, randomized controlled trial. Am J Sports Med. 2019;47:1885–92.
27. Nejati P, Ghahremaninia A, Naderi F, et al. Treatment of subacromial impingement syndrome: platelet-rich plasma or exercise therapy? A randomized controlled trial. Orthop J Sports Med. 2017;5:2325967117702366.
28. Prodromos CC, Finkle S, Prodromos A, et al. Treatment of rotator cuff tears with platelet rich plasma: a prospective study with 2 year follow-up. BMC Musculoskelet Disord. 2021;22:499.
29. Lin KY, Chen P, Chen AC, et al. Leukocyte-rich platelet-rich plasma has better stimulating effects on tenocyte proliferation compared with leukocyte-poor platelet-rich plasma. Orthop J Sports Med. 2022;10:23259671221084706.
30. Berger DR, Centeno CJ, Steinmetz NJ. Platelet lysates from aged donors promote human tenocyte proliferation and migration in a concentration-dependent manner. Bone Joint Res. 2019;8:32–40.
31. Kim SJ, Kim EK, Kim SJ, et al. Effects of bone marrow aspirate concentrate and platelet-rich plasma on patients with partial tear of the rotator cuff tendon. J Orthop Surg Res. 2018;13:1.
32. Hernigou P, Flouzat Lachaniette CH, Delambre J, et al. Biologic augmentation of rotator cuff repair with mesenchymal stem cells during arthroscopy improves healing and prevents further tears: a case-controlled study. Int Orthop. 2014;38:1811–8.
33. Hogaboom N, Malanga G, Cherian C, et al. A pilot study to evaluate micro-fragmented adipose tissue injection under ultrasound guidance for the treatment of refractory rotator cuff disease in wheelchair users with spinal cord injury. J Spinal Cord Med. 2021;44:886–95.
34. Striano RD, Malanga GA, Bilbool N, et al. Refractory shoulder pain with osteoarthritis, and rotator cuff tear, treated with micro-fragmented adipose tissue. J Orthop Spine Sports Med. 2018;2:014.

35. Robinson DM, Eng C, Mitchkash M, et al. Outcomes after micronized fat adipose transfer for glenohumeral joint arthritis and rotator cuff pathology: a case series of 18 shoulders. Muscle Ligaments Tendons J. 2020;10:393–8.
36. Robinson DM, Eng C, Makovitch S, et al. Non-operative orthobiologic use for rotator cuff disorders and glenohumeral osteoarthritis: a systematic review. J Back Musculoskelet Rehabil. 2021;34:17–32.
37. Pasculli RM, Kenyon CD, Berrigan WA, et al. Mesenchymal stem cells for subchondral bone marrow lesions: from bench to bedside. Bone Rep. 2022;17:101630.
38. Rossi LA, Piuzzi NS, Shapiro SA. Glenohumeral osteoarthritis: the role for orthobiologic therapies: platelet-rich plasma and cell therapies. JBJS Rev. 2020;8:e0075.
39. Madry H, van Dijk CN, Mueller-Gerbl M. The basic science of the subchondral bone. Knee Surg Sports Traumatol Arthrosc. 2010;18:419–33.
40. Kapoor M, Martel-Pelletier J, Lajeunesse D, et al. Role of proinflammatory cytokines in the pathophysiology of osteoarthritis. Nat Rev Rheumatol. 2011;7:33–42.
41. Centeno CJ, Al-Sayegh H, Bashir J, et al. A prospective multi-site registry study of a specific protocol of autologous bone marrow concentrate for the treatment of shoulder rotator cuff tears and osteoarthritis. J Pain Res. 2015;8:269–76.
42. Darrow M, Shaw B, Schmidt N, et al. Treatment of shoulder osteoarthritis and rotator cuff tears with bone marrow concentrate and whole bone marrow injections. Cogent Med. 2019;6:1628883.
43. Hsieh LF, Hsu WC, Lin YJ, et al. Addition of intra-articular hyaluronate injection to physical therapy program produces no extra benefits in patients with adhesive capsulitis of the shoulder: a randomized controlled trial. Arch Phys Med Rehabil. 2012;93:957–64.
44. Zreik NH, Malik RA, Charalambous CP. Adhesive capsulitis of the shoulder and diabetes: a meta-analysis of prevalence. Muscles Ligaments Tendons J. 2016;6:26–34.
45. Dyer BP, Rathod-Mistry T, Burton C, et al. Diabetes as a risk factor for the onset of frozen shoulder: a systematic review and meta-analysis. BMJ Open. 2023;13:e062377.
46. Lee LC, Lieu FK, Lee HL, et al. Effectiveness of hyaluronic acid administration in treating adhesive capsulitis of the shoulder: a systematic review of randomized controlled trials. Biomed Res Int. 2015;2015:1–8.
47. Fields BKK, Skalski MR, Patel DB, et al. Adhesive capsulitis: review of imaging findings, pathophysiology, clinical presentation, and treatment options. Skelet Radiol. 2019;48:1171–84.
48. Park KD, Nam HS, Lee JK, et al. Treatment effects of ultrasound-guided capsular distension with hyaluronic acid in adhesive capsulitis of the shoulder. Arch Phys Med Rehabil. 2013;94:264–70.
49. Kothari SY, Srikumar V, Singh N. Comparative efficacy of platelet rich plasma injection, corticosteroid injection and ultrasonic therapy in the treatment of periarthritis shoulder. J Clin Diagn Res. 2017;11:RC15–8.
50. Ünlü B, Çalış FA, Karapolat H, et al. Efficacy of platelet-rich plasma injections in patients with adhesive capsulitis of the shoulder. Int Orthop. 2021;45:181–90.
51. Thu AC, Kwak SG, Shein WN, et al. Comparison of ultrasound-guided platelet-rich plasma injection and conventional physical therapy for management of adhesive capsulitis: a randomized trial. J Int Med Res. 2020;48:300060520976032.
52. Maund E, Craig D, Suekarran S, et al. Management of frozen shoulder: a systematic review and cost-effectiveness analysis. Health Technol Assess. 2012;16:1–264.
53. Shapiro SA, Finnoff JT, Awan TM, et al. Highlights from the American medical Society for Sports Medicine position statement on responsible use of regenerative medicine and orthobiologics in sports medicine. Br J Sports Med. 2022;56:121–2.
54. Daley EL, Bajaj S, Bisson LJ, et al. Improving injection accuracy of the elbow, knee, and shoulder: does injection site and imaging make a difference? A systematic review. Am J Sports Med. 2011;39:656–62.
55. Khadavi M, Pourcho A, Podesta L. Protocols and techniques for orthobiologic procedures. Phys Med Rehabil Clin N Am. 2023;34:105–15.
56. Finnoff JT, Fowler SP, Lai JK, et al. Treatment of chronic tendinopathy with ultrasound-guided needle tenotomy and platelet-rich plasma injection. PM R. 2011;3:900–11.
57. Krey D, Borchers J, McCamey K. Tendon needling for treatment of tendinopathy: a systematic review. Phys Sportsmed. 2015;43:80–6.
58. Khadavi M, Rehor D, Roney A, et al. Regional anesthesia for orthobiologic procedures. Phys Med Rehabil Clin N Am. 2023;34:291–309.
59. Dubé MO, Desmeules F, Lewis J, et al. Rotator cuff-related shoulder pain: does the type of exercise influence the outcomes? Protocol of a randomised controlled trial. BMJ Open. 2020;10:e039976.
60. Andia I, Maffulli N. New biotechnologies for musculoskeletal injuries. Surgeon. 2019;17:244–55.
61. Stout A, Friedly J, Standaert CJ. Systemic absorption and side effects of locally injected glucocorticoids. PM R. 2019;11:409–19.
62. Aleem AW, Syed UAM, Nicholson T, et al. Blood glucose levels in diabetic patients following corticosteroid injections into the subacromial space of the shoulder. Arch Bone Jt Surg. 2017;5:315–21.
63. Brook EM, Hu CH, Kingston KA, et al. Corticosteroid injections: a review of sex-related side effects. Orthopedics. 2017;40:e211–5.
64. Haraldsson BT, Langberg H, Aagaard P, et al. Corticosteroids reduce the tensile strength of isolated collagen fascicles. Am J Sports Med. 2006;34:1992–7.
65. Wernecke C, Braun HJ, Dragoo JL. The effect of intra-articular corticosteroids on articular cartilage: a systematic review. Orthop J Sports Med. 2015;3:2325967115581163.
66. Lavigne A, Nguyen D, Chartrand-Oberoi O, et al. Incidence and risk factors of systemic side effects and complications of ultrasound- and fluoroscopy-guided glucocorticoid injections: a retrospective cohort study. Am J Phys Med Rehabil. 2023;103(1):31–7.
67. Annaniemi JA, Pere J, Giordano S. Platelet-rich plasma versus corticosteroid injections for rotator cuff tendinopathy: a comparative study with up to 18-month follow-up. Clin Shoulder Elb. 2022;25:28–35.
68. Shapiro SA, Master Z, Arthurs JR, et al. Tiered approach to considering orthobiologics for patients with musculoskeletal conditions. Br J Sports Med. 2023;57:179–80.

# 18 Regenerative Medicine for the Elbow

Andrew Hiett, Alexandre Lavigne, Luga Podesta, Kent C. Doan, and Michael Khadavi

## Introduction

The elbow is a unique joint which can experience a wide variety of pathologies that impact the nerves, muscles, tendons and ligaments regardless of activity level or occupation. Certain conditions such as ulnar collateral ligament (UCL) injury tend to occur in athletic populations [1] while disorders such as lateral epicondylitis (LE) occur more often in individuals with certain occupations requiring repetitive wrist and arm movement [2–4]. Regardless of causative etiology, there are a number of pathologies that can cause significant elbow pain and dysfunction for patients. Initial treatment for these conditions often consists of conservative measures such as rest, Nonsteroidal anti-inflammatory drugs (NSAIDs), bracing and corticosteroid injections [5, 6]. Often these modalities are effective in alleviating symptoms and avoiding operative interventions. Surgery inherently carries its own set of risks in addition to prolonged healing and recovery times depending on the intervention and demands of the individual. By capitalizing on the body's natural ability to heal using its own cells and growth factors, orthobiologics offer the possibility of shorter recovery times. In addition, orthobiologics are less invasive procedures with the promise of higher efficacy in treating certain pathologies about the elbow. As evidence for orthobiologic procedures has grown, these office-based procedures can limit the need for more invasive surgical options. Orthobiologics have been well studied for conditions such as LE and UCL injury with some studies investigating patients with medial epicondylitis. Elbow osteoarthritis, osteochondral injuries, nerve entrapments (cubital tunnel and radial tunnel syndrome), distal biceps tendinopathy and distal triceps tendinopathy are other conditions that can cause significant pain and dysfunction, but there is less evidence in the literature for treatment of these pathologies using orthobiologics.

The term orthobiologics encompasses many different types of injectates. Platelet-rich plasma (PRP) is defined as having a platelet concentration greater than autologous whole blood (AWB) [7]. However, most studies using PRP as an injectate have platelet concentrations that are 4–7 fold higher than those found in AWB. Autologous conditioned plasma (ACP) is an alternative injectate that consists of centrifuged plasma that contains platelet concentrations approximately twice those found in AWB. In addition, PRP can be subdivided into leukocyte-rich PRP (LR-PRP) and leukocyte-poor PRP (LP-PRP) based on the relative concentration of leukocytes, although definitions for these formulations vary. In the studies discussed below, platelet concentration, leukocyte concentration and use of ultrasound guidance are included where available. If these characteristics are not mentioned in this chapter then they were not published by the study authors. This continues to prove one of the most limiting factors in evidence for orthobiologics, with inconsistent quality control and reporting of harvest, processing, resultant product qualities, and delivery systems. Future research will help elucidate best practices for consistent and reliable procedures, with efficacious and durable results.

A. Hiett
Department of Physical Medicine and Rehabilitation, University of Kansas Medical Center, Kansas City, KS, USA

A. Lavigne
Department of Physical Medicine and Rehabilitation, Faculty of Medicine, University of Montreal, Montreal, QC, Canada
e-mail: alexandre.lavigne.2@umontreal.ca

L. Podesta
Regenerative Sports Medicine, Bluetail Medical Group, Naples, FL, USA

Podesta Sports Medicine and Orthopedics, Naples, FL, USA

Florida Everglades, Homestead, FL, USA

K. C. Doan
Kansas City Orthopedic Alliance, Overland Park, KS, USA
e-mail: kdoan@kcorthoalliance.com

M. Khadavi (✉)
Kansas City Orthopedic Alliance, Leawood, KS, USA

Kansas City Ballet, Sporting Kansas City, Kansas City, KS, USA


A. Navani et al. (eds.), *Essentials of Regenerative Medicine in Interventional Pain Management*,
https://doi.org/10.1007/978-3-031-50357-3_18

## Lateral Epicondylitis

Lateral Epicondylitis, also known as tennis elbow or perhaps more appropriately common extensor tendinopathy of the elbow, is a common cause of elbow pain that affects between 1–3% of the population [4, 8, 9]. Symptoms of LE include lateral elbow pain with reduced grip strength. Tennis elbow obtained this moniker as it was observed in athletes playing racquet or "swinging" sports. Nonathletes who perform repetitive gripping or wrist movements such as painters, carpenters, mechanics, plumbers and butchers are also at increased risk for this common condition.

The common extensor tendon is composed of the extensor carpi radialis brevis (ECRB), extensor carpi radialis longus (ECRL), extensor digitorum, extensor digiti minimi and extensor carpi ulnaris. In LE, pain is believed to be initially caused by overuse or microtears of the common extensor tendon—most commonly the ECRB—leading to an initial inflammatory reaction. After initial inflammation decreases, tissue healing begins with collagen reorganization and microvascular adaptations. Failure of this repair mechanism or repetitive insult may prevent adequate tendon healing leading to a chronic tendinosis with fibrovascular scar instead of robust tendon. Histologic samples from patients with chronic LE do not show the presence of inflammatory cells lending further evidence to the theory that the pathology is more of a tendinosis and degenerative process rather than a tendinitis and inflammatory process [10].

Treatment of LE initially consists of removal of the offending activity with rest, physical therapy, and possibly bracing and NSAIDs. While many patients typically respond well to conservative measures, there is a minority whose symptoms are refractory. It is important to consider a diagnosis of radial tunnel syndrome in patients who do not respond to initial treatment as it can closely mimic LE [11]. Historically, corticosteroids were often the next step in the treatment algorithm if initial conservative measures failed. However, more recent data suggests that corticosteroid injections have limited short-term benefits and may actually impair tendon healing in the long-term [12, 13]. In light of this evidence, PRP has been explored as a possible alternative. PRP contains a variety of growth factors and cytokines important for repair and healing, making PRP a theoretical remedy to diseased tendon that has undergone failure of normal healing and tissue remodeling.

Use of PRP for LE has been studied extensively after an initial pilot study by Mishra and Pavelko in 2006 which showed participants receiving PRP for LE demonstrated a superior improvement in pain and functional scores compared to those receiving lidocaine injections at 8 weeks and 6 months [14]. Participants received an average of 3.31 million platelets via 2–3 mL injection of PRP with mean increase of 539% compared to whole blood. Injections were administered at the point of maximal tenderness without ultrasound guidance. This small, 20-participant study sparked great interest in PRP for LE and other conditions.

## Randomized Controlled Trials

In 2011, Thanasas et al. [15] looked at participants with LE for at least 3 months who had never received any kind of injection. They compared pain and functional scores in 28 participants receiving either a single injection of 3–6 mL of LR-PRP or an injection of AWB under ultrasound guidance. The PRP group had improved visual analog scale (VAS) scores at 6-week timepoint ($p < 0.05$) but head-to-head group comparisons at 3 months and 6 months showed no significant differences. Platelet concentrations from 2 participants were measured and found to be 1.29 × 106/mL (5.5-fold increase).

In a double-blind, randomized trial, Peerbooms et al. [13] compared an injection with 3 mL LR-PRP or corticosteroids at the point of maximal tenderness in 100 patients with LE and found that a single injection of PRP led to greater rate of success at 52 weeks in terms of pain and functional scores compared to corticosteroids. Success was defined as at least a 25% reduction in pain or functional scores without need for repeat injection or surgery. For pain scores, PRP showed a success rate of 73% compared to 49% for corticosteroids ($p < 0.001$). For functional scores, PRP showed a success rate of 73% compared to 51% in the corticosteroid group ($p = 0.005$). Of note, patients receiving corticosteroid injections reported more rapid improvement in pain and function for the first 8 weeks compared to those receiving PRP, but this difference in benefit disappeared by 12 weeks. Patients receiving PRP continued to improve throughout the duration of follow-up (52 weeks) while those receiving corticosteroids had a decline in benefit after 8 weeks. Platelet concentrations were not specified for this study.

Using the same data set and definitions as Peerbooms et al. [13], Gosens et al. [12] followed the patients for a total of 2 years. Nearly twice as many participants receiving PRP (39/51) had a successful outcome at 2 years compared to participants receiving corticosteroid injections (21/49) ($p < 0.0001$) [12]. Platelet concentrations were not specified for this study.

In the largest randomized controlled trial to date, Mishra et al. [16] compared 2–3 mL of LR-PRP with needling injected without ultrasound guidance to needling alone in 230 participants with at least 3 months of elbow pain. The study examined pain scores during resisted elbow extension and found greater improvement in the group receiving a single PRP injection at 4, 8, 12 and 24 weeks with statistically signficant differences noted at 8 and 24 weeks. There was no difference in functional scores between groups at any time point. The study states that platelet concentrations increased at least five-fold, but no participants were tested to confirm platelet concentrations.

Another study examined 120 participants with symptoms of LE for more than 6 weeks to determine if ACP injected without ultrasound guidance was superior to betamethasone injections for functional improvement. The study showed that those receiving betamethasone had a greater improvement in functional scores at 6 weeks and 6 months, but ACP was superior at 12 months [17]. ACP was prepared using Double Syringe System, but no platelet concentrations were measured.

The aforementioned studies lend support to the use of PRP, ACP or autologous blood for use in LE, particularly over longer follow-up intervals. However, there have been a number of randomized controlled trials demonstrating equivocal results when comparing PRP to other injectates for the treatment of LE. Creaney et al. [18] treated 150 participants with symptoms of LE for greater than 6 months who had already failed conservative management with PT and stretching. Participants were given two separate injections of either 1.5 mL of LR-PRP or AWB using ultrasound guidance. Injections were separated by 4 weeks and functional scores were assessed with success defined as a functional improvement of at least 25 points. Investigators found no difference in success rates between PRP (66%) and AWB (72%) at 6 months. Of note, analysis was only performed on individuals who had not undergone surgery at 6 months which excluded a significant number of participants and artificially inflated the success rate of the remaining participants. Ten participants from each group had platelet counts measured with AWB group averaging 234 × 109/L and PRP averaging 652 × 109/L (2.8-fold increase).

Raeissadat et al. [19] performed a study similar to the one performed by Creaney et al. [18] in which they evaluated 61 participants with LE for at least 3 months who received an injection of 2 mL of LR-PRP or AWB without ultrasound guidance. Investigators found that both PRP and AWB showed significant improvements in pain and functional scores at 1, 2, 6 and 12 months compared to baseline. There were no significant differences noted between groups and it is interesting to note that both groups showed maximum benefit at 6 months with a slight worsening in both outcome variables at 12 months. Platelet concentrations averaged 2.5 × 105 ± 53,000/microliter for all participants at baseline and 1.23 × 106 ± 250,000/microliter for those receiving PRP (4.8-fold increase). Data from studies by Creaney and Raeissadat must be interpreted carefully as the "control" group in these trials was AWB rather than corticosteroid, saline, or lidocaine, meaning that one or both treatments could be superior to alternative controls.

One study completed in 2013 compared LR-PRP, corticosteroid, and saline with 20 participants in each of three arms. Injections were performed under ultrasound guidance and those receiving PRP received 3–3.5 mL of injectate. Pain and functional scores were superior in the corticosteroid group at 1 month, but all groups were equal by 3 months [20]. This study also evaluated ultrasound properties of the tendon and found that corticosteroids showed a statistically significant difference in tendon thickness with corticosteroid injection reducing tendon thickness at 3 months while the other therapies led to an increase in tendon thickness. Limitations of this study include the large dropout rate as participants were allowed to drop out at 3 months if they had an unsatisfactory result. The study stated that kits used for PRP preparation should produce an eight-fold increase in platelet concentrations, but participants were not tested for actual quantification.

Martin et al. [21] studied 71 participants with epicondylitis (60 lateral, 11 medial) to assess 25% improvements in pain and functional scores over 12 months. Participants were either given two injections of 5 mL LP-PRP or two injections of lidocaine with needle tenotomy using ultrasound guidance. Procedures were separated by 2 weeks. Investigators found the number of participants meeting 25% pain improvement threshold peaked for PRP at 6 months while lidocaine with tenotomy peaked at 3 months. For functional scores, a similar trend was noted. However, there were no statistically significant differences between treatment groups at 3, 6 and 12 months. Platelet concentrations were 2.3 ± 0.68 times the concentration of peripheral blood but no absolute quantification was provided.

Despite the promising nature of some of the studies noted, there is a lack of standardization across studies making it difficult to compare one study to another. Studies vary widely in terms of type of control, type of PRP used (LR vs LP, platelet concentration), use of image-guidance, location of injection (peritendinous vs. intratendinous), number of injections, and post-injection rehabilitation protocol. In fact, many studies make no mention of these important variables. Several meta-analyses have been conducted in an effort to make sense of the variety of trials that demonstrate conflicting results.

## Meta Analyses

Fitzpatrick et al. [22] analyzed 18 tendinopathy trials with 1066 participants to compare active interventions (PRP, AWB, ACP) against controls such as saline, anesthetic, dry needling or corticosteroid. The analysis included 13 studies examining LE including a number of studies that were previously mentioned [12–16, 18, 20] as well as several that have not been discussed previously [23–28]. To qualify for inclusion studies had to have a minimum of 3-month follow-up.

Trials were evaluated for change in pain or function. Standardized mean difference (SMD) showed that LR-PRP was superior to LP-PRP [22]. ACP showed a SMD greater than LP-PRP but less than LR-PRP. Controls (saline, anesthetic, dry needling, corticosteroid) were shown to be of benefit based on positive SMD's for each, with corticosteroid and dry needling showing greater benefit than saline and anesthetic. These differences in control groups were not statistically significant. While SMD's for active treatments were greater than those of controls, comparisons between the two groups were not performed due to the wide heterogeneity of control groups.

An additional meta-analysis looked at pain scores in 1031 participants receiving PRP for various conditions with 384 participants treated for LE. Follow-up intervals were defined by short term (<6.5 months) and long term (≥1 year). Among the 8 studies for LE with short term follow-up included in the meta-analysis, PRP showed a greater reduction in pain compared to controls weighted mean difference (WMD) = −1.14 (−1.85 to −0.43); ($p < 0.01$) [29]. Pooled data from the 5 trials included with long term follow-up also showed reduction in pain with PRP compared to controls with WMD = −1.39 (−2.49 to −0.29); ($P = 0.01$).

Chen et al. [30] examined nine level 1 randomized controlled trials comparing PRP to various injectates using VAS and Disabilities of the Arm, Shoulder, and Hand (DASH) scores. Four studies compared PRP to corticosteroids when assessing pain scores [12, 31–33] and found that there were no differences at 3 months, but by 6 months PRP showed superior pain relief. This pain relief surpassed both the 5% and 10% minimal clinically important difference (MCID) thresholds. Five studies examined functional outcomes when comparing PRP and corticosteroid injections [12, 17, 31, 32, 34]. The pooled data showed that PRP was superior at 6 months post injection and this improvement surpassed the 5% but not the 10% MCID threshold.

Not all meta-analyses have shown PRP to be superior to controls, highlighting the importance of identifying the outcome measures and methodology of each study. Simental-Mendia et al. [35] examined 5 RCT's with placebo control groups and follow-up times of at least 2 months. Among three studies reporting pain as the outcome measure, there was no significant difference between PRP and control [36–38]. The study also reported no difference in functional outcomes but each study included utilized a different functional outcome measure. Not surprisingly, several of the studies included noted a greater rate of post procedural pain in the participants receiving PRP compared to controls.

Studies have attempted to examine whether differences in PRP formulation (LR vs. LP) may impact the efficacy of injections for LE. A recent systematic review and meta-analysis by Shim et al. [39] examined 11 different studies which compared PRP to various alternative injectates including corticosteroids, whole blood, saline and local anesthetic. Studies were grouped into LR or LP-PRP based on whether the leukocyte concentration in the PRP was greater than or less than that of whole blood. Duration of follow-up was categorized as short term (<3 months) or long term (≥6 months). A total of eight studies (4 LR, 4 LP) evaluated pain using pain scores and found no difference between either type of PRP and controls at short term follow-up. At long term follow-up, LP-PRP was no different than controls but LR-PRP showed greater benefit compared to controls. The study also concluded that among five studies reporting success outcomes, LR-PRP was found to have a greater likelihood of success than control groups while LP-PRP was no more likely to reach predetermined success thresholds. Results related to success rate must be interpreted with caution as studies defined success differently based on symptoms [17], functional improvement [12, 34] or pain improvement [16, 19].

As part of a meta-analysis performed by Niemiec et al. [40] in 2022, 14 studies were grouped according to LR or LP-PRP. While both LR-PRP and LP-PRP were found to be efficacious, head-to-head comparisons between them were impossible because baseline values for all variables between LR-PRP and LP-PRP groups differed.

## Alternative Orthobiologics

While PRP is the most frequently studied orthobiologic for the treatment of LE, there have been small studies examining alternative injectates. One study by Wang et al. [41] examined 16 participants who received a single injection of autologous tenocytes harvested from the patellar tendon and found that participants had improved pain, function and grip strength at 1 and 12 months. Singh et al. [42] demonstrated that a single injection of bone marrow aspirate concentrate (BMAC) improved functional scores 2, 6 and 12 weeks compared to baseline values in 26 participants. Most recently, Khoury et al. [43] injected adipose derived stromal cells under ultrasound guidance and found that pain and functional scores were improved in 18 participants at 6 months compared to baseline. It is important to note that all of the aforementioned studies lacked control groups and used very small sample sizes, making it difficult to draw meaningful conclusions from these studies. Nonetheless, the benefit demonstrated by these studies opens the door for further, more rigorous clinical trials using alternative orthobiologics in the future.

## Medial Epicondylitis

While less common than LE, medial epicondylitis still occurs frequently and tends to affect those with repetitive hobbies and occupations that involve forearm pronation and wrist flexion. Despite being termed "golfer's elbow", the vast majority of medial epicondylitis cases occur in a nonathletic population [4]. Pain tends to occur at the medial epicondyle about the common flexor origin. The common flexor tendon includes pronator teres, flexor carpi ulnaris, flexor carpi radialis, flexor digitorum superficialis and palmaris longus. Given the common flexor tendon's proximity to the ulnar nerve, certain cases of medial epicondylitis may involve compression of the ulnar nerve. Most treatment algorithms for medial epicondylitis closely resemble those of LE with an emphasis on conservative measures initially with advancement to injections and possibly surgery if necessary.

There are far fewer studies involving the use of PRP in the treatment of medial epicondylitis. A study by Varshney et al. [44] studied participants with both medial and LE. The study found that an injection with 2 mL of LR-PRP showed greater improvements in pain and function scores compared to participants receiving corticosteroid injections at 6 months. Unfortunately, only 20/83 participants analyzed were being treated for medial epicondylitis and results were not stratified by pathology type. The study states the minimum platelet concentration was 1.5 × 106 platelets/mL to qualify for injection but no further quantification is provided. The lack of stratification along with a high dropout rate (29/112) makes it difficult to draw reliable conclusions from the study.

A randomized controlled trial by Martin et al. [21] compared two treatments of 5 mL of LP-PRP with tenotomy to lidocaine with tenotomy in participants with refractory epicondylitis as mentioned previously. Treatments were separated by 2 weeks and utilized ultrasound guidance. Pain and function scores were assessed at 6 and 12 months and the number of participants meeting the minimum 25% improvement for each outcome measure was calculated. There were no differences in the proportion of participants meeting the 25% improvement threshold between the two treatment groups at 3, 6 and 12 months. As mentioned previously, platelet concentrations were 2.3 ± 0.68 times the concentration of peripheral blood but no absolute quantification was provided. Similar to the previous study, the majority of the participants in this trial (60/71) had LE, making it difficult to draw meaningful conclusions for treatment of medial epicondylitis.

## Alternative Procedures

Several cohort studies have compared PRP to alternative interventions (percutaneous ultrasound tenotomy, surgery) in the treatment of medial epicondylitis. In 2019, Boden et al. [45] examined 62 participants with epicondylitis (medial or lateral) who had failed conservative management including corticosteroid injections for at least 3 months. Participants were allowed to choose between percutaneous ultrasound tenotomy and debridement (Tenex) or injection of 3 mL of LP-PRP under ultrasound guidance and then followed for 6 months with pain and functional outcome measures. There was a statistically significant difference in all outcome measures between baseline and follow-up for both groups, but there were no significant differences between groups. Quantification of components of the PRP was not performed. Of the 62 participants who completed follow-up, the majority were diagnosed with LE (52/62) with only 5 participants in each group diagnosed with medial epicondylitis.

Bohlen et al. [46] examined 33 participants with Type I medial epicondylitis who had failed conservative management for at least 3 months. Participants were given the choice to receive two 4–7 mL LR-PRP injections under ultrasound guidance or undergo surgery. 15/33 participants elected to have PRP injections with 2–3 weeks in between injections. Participants were followed at 2 weeks, 6 weeks, 3 months and 12 months and scored using the Nirschl grading system with pain and functional scores as secondary outcome measures. A patient was considered to have a successful outcome if the patient returned to full activity without pain while good outcomes required return to full activity with only occasional pain. Success was defined as participants with good or excellent outcomes. Overall, 12/15 participants in the PRP group and 17/18 participants in the surgery group achieved a successful outcome with no statistically significant difference between the two groups at any time point. Of note, there was a significant difference in the time to return of full range of motion (ROM) (42.3 vs. 96.1 days) and time to pain free status 56.2 vs. 108.0 days) favoring those who received PRP injections ($p < 0.01$). Platelet concentrations in the PRP were reported at 1500 × 109/L.

While no high-level publications examine utilization of BMAC in medial epicondylitis, unpublished data from the senior author suggests that outcomes differ if a tear is present with common flexor tendinosis, and that those with a partial tear may respond better to BMC than PRP.

## Ulnar Collateral Ligament

The UCL is critical for stabilization of the elbow during valgus stress, especially during overhead throwing [47–49]. Traditionally, injury of the UCL was predominantly diagnosed in high-level collegiate or professional throwing athletes; however, there has been a dramatic increase in the number of elbow UCL injuries diagnosed in younger adolescent, high school and collegiate athletes. The anterior bundle is particularly important for providing stabilization and injury to this structure can lead to chronic pain and instability which predisposes the elbow joint to further injury and significant loss of velocity and control for throwers.

UCL injury was previously viewed as career ending, but in 1974 Dr. Frank Jobe described a procedure known as Tommy John surgery or UCL reconstruction. In brief, Tommy John surgery involves grafting a tendon from elsewhere in the body (usually the forearm or hamstring) and using the graft to reconstruct the damaged native ligament. This procedure has allowed athletes to return to the field of play at their previous level of competition or continue advancing to higher levels of play in 53–90% of cases [50–57]. While this procedure has revolutionized the treatment of UCL injuries, it does carry a significant recovery period, often spanning 1–2 years in length [58–61]. Surgical options have expanded significantly since publication of the original article by Jobe et al. [62] and now include UCL reconstruction, repair or repair with augmentation [58, 63, 64].

Due to the inherent risks of surgery and prolonged recovery times, some athletes elect to trial conservative measures such as physical therapy prior to proceeding with surgical intervention. Rettig et al. [65] examined 31 participants with UCL injuries and found that 42% of participants returned to their previous level of play at an average of 24.5 weeks with conservative management and a two-phase therapy program. Subsequent studies have found 66–93% of participants avoided surgery with conservative management [66, 67]. Distal tears and high-grade changes were noted to be less responsive to conservative management [66].

Recently, PRP has become a promising alternative option to treat UCL injuries that may offer similar outcomes in terms of return to play while reducing the necessary recovery time. In 2013, Podesta et al. [68] examined 34 athletes who had failed at least 2 months of nonoperative treatment for partial UCL tear. Participants were given a single, 5 mL injection of LR-PRP under ultrasound guidance, followed by a strict rehabilitation protocol. 88% of participants returned to play at an average of 12 weeks with significant improvements noted in functional scores between baseline and final follow-up (70-week average follow-up). Mean platelet concentration was estimated to be 780.2 ± 246.5 × 103 platelets/microliter but no cell counts were actually measured.

Dines et al. [69] examined 44 participants with evidence of UCL injury on magnetic resonance imaging (MRI) who had failed conservative management. Participants were given 1–3 injections of 3 mL LP-PRP based on whether or not pain was still present 3 weeks after previous injection. Following the injection, participants completed relative rest for 2 weeks followed by 4 weeks of stretching and strengthening exercises. 34% of participants achieved an excellent outcome defined as returning to preinjury level of performance. An additional 39% of participants had a good outcome defined as return to play at a lower level of competition. The mean time to return to throwing was 5 weeks and the mean time to return to competition was 12 weeks. All 7 participants who had distal tears achieved a poor outcome defined as inability to return to previous sport. PRP was prepared using the ACP system (Arthrex) but cell counts were not measured.

An additional study by Deal et al. [58] examined UCL injured participants receiving 2 injections of 5–8 mL of LR PRP under ultrasound guidance spaced 2 weeks apart. 20/22 participants had full reconstitution of the medial UCL on MRI taken 6 weeks after the second injection. The average time to return to play was 82 days and the overall return to play rate was 96% in 25 participants. Platelet concentrations averaged 1228 ± 312 × 103 platelets/microliter with a 4.44-fold increase from baseline.

Mills and colleagues [70] examined 50 participants with varying degrees of UCL injury who received one 7 mL injection of LP PRP under ultrasound guidance. Following injection, athletes completed passive ROM for 2 weeks, active ROM with strengthening for 5 weeks and reintroduction to throwing at the 9-week mark. 52% of participants had successful outcomes defined as full return to preinjury level or return to preinjury level after additional PRP injection or further therapy. Participants with lower grade tears were 3.83 times more likely to return to play compared to higher grade tears. Cell counts were not measured.

In each of the studies noted above, there was no control group to compare to those receiving PRP injections. Further randomized trials are needed to determine if PRP is superior to standard conservative management with physical therapy, but preliminary results seem promising. It is important to take specific patient characteristics such as age, activity level and location of tear into account when determining the appropriate steps in treatment. Distal tears appear to predict a greater likelihood for surgery and variations in blood supply may help explain some of this predisposition [66, 71].

## Distal Biceps Tendinopathy

The biceps muscle is composed of two separate heads (long head, short head) that originate at different locations about the shoulder. These heads both function to flex the elbow and supinate the forearm. As the muscle bellies extend distally, they come together and twist 90 degrees to form the distal biceps tendon which inserts on the radial tuberosity just distal to the elbow joint as well as connecting to the antebrachial fascia in the form of the lacertous fibrosis. Irritation of the distal biceps tendon can occur with repeated lifting, pulling or supination of the elbow. Initial irritation to the tendon can typically be categorized as tendinitis while more chronic stresses may lead to tendinosis and potential rupture of the distal biceps tendon.

For the purposes of this review, attention will be focused on the distal biceps tendon and its insertion, rather than the proximal heads. Tendinopathy of the distal biceps tendon is uncommon but presents with pain in the antecubital fossa with elbow flexion and supination. Distal biceps tendinopathy is typically treated with activity modification, NSAIDs and physical therapy [72, 73]. Surgery is reserved for refractory chronic tendinopathy but is frequently recommended for complete ruptures. Several small studies have attempted to utilize PRP in the treatment of distal biceps tendinopathy to assess potential benefit in those that fail nonoperative measures. Similar to injections in other locations around the elbow, injection into the distal biceps tendon can prove difficult and accuracy can be improved with the use of ultrasound guidance [74].

Sanli et al. [75] assessed 12 participants with evidence of distal biceps tendinopathy on MRI that was refractory to physical therapy, NSAIDs or local corticosteroid injections. Participants received a single injection of 3 mL of LR-PRP under ultrasound guidance. The study showed significant reduction in pain and functional outcomes for all patients with a median follow-up of 47 months. Platelet concentration was stated to be an eight-fold increase from baseline but it is unclear if cell counts were actually measured.

Barker et al. [73] treated 6 participants with a diagnosis of distal biceps tendinopathy confirmed by MRI or ultrasound with a single 3 mL LR-PRP injection under ultrasound guidance. 4/6 patients had symptom resolution at six-week follow-up. 2/6 participants had improvement, but not resolution, of symptoms after the first injection and received a second injections 6 weeks after the first. Both patients who received repeat injections had symptom resolution 6 weeks after the second injection. The authors of the study stated that the platelet concentration was less than a five-fold increase from baseline.

In addition, unpublished data from the senior author's clinic shows high rates of success for tendinosis with intratendinous PRP injection. The senior author notes good outcomes with use of bone marrow concentrate for distal biceps partial tears.

While literature to support the use of orthobiologics is somewhat scarce, there is some encouraging results noted in the studies above. However, none of the studies mentioned featured a control group so drawing meaningful conclusions remains somewhat difficult.

## Elbow Injection Technique Considerations

When injecting orthobiologics into and around the elbow, technique variables and use of guidance likely influence the treatment's outcome. There is great diversity in the injection techniques performed in the aforementioned studies. Research examining pain and functional outcomes based upon specific use of orthobiologic technique variables is lacking.

Injecting an orthobiologic solution into and around a tendon or ligament involves several considerations. Use of local or regional anesthesia allows for improved tolerability of most intratendinous orthobiologic procedures. Similarly, ultrasound guidance is recommended for maximal accuracy and safety during these injections as shown in Figs. 18.1 and 18.2.

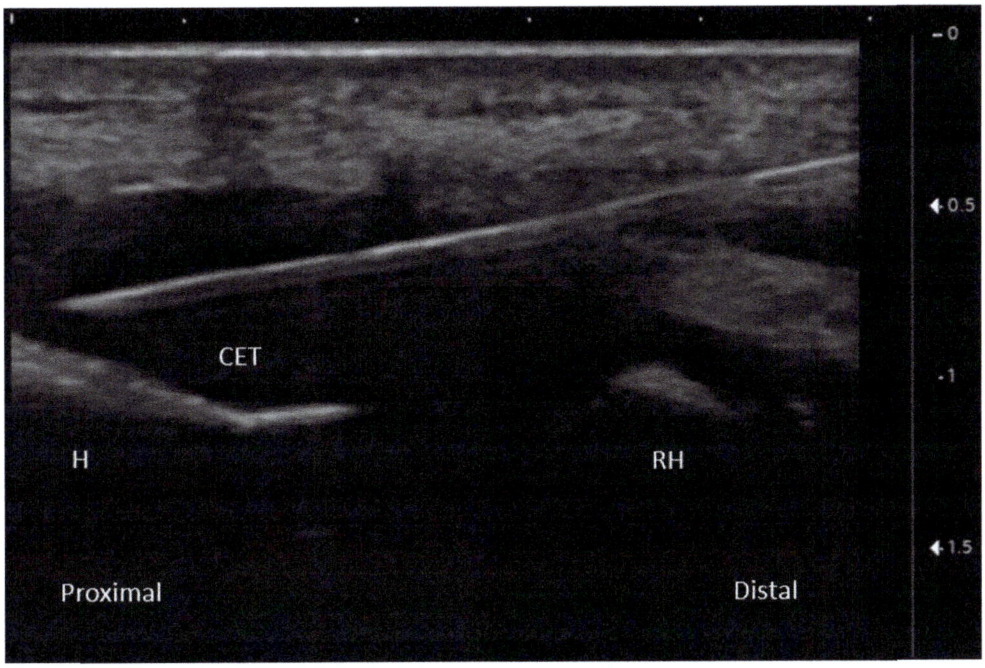


**Fig. 18.1** In-plane ultrasound-guided injection of 3 ml of LR-PRP into a common extensor tendon with severe tendinosis. Twenty minutes before the injection, the tendon was anesthetized with 3 mL 0.5% ropivacaine injected peri-tendinous and less than 1 mL injected into the tendon. The PRP was injected with a 22-gauge needle and a peppering technique, followed immediately by further tendon fenestration with approximately 20 more passes throughout the tendon with the same needle. The patient was sent home with a wrist splint and hot pack to use during the following days and a short opioid prescription post-procedural pain. *CET* Common extensor tendon, *H* humerus, *RH* radial head. (Image reproduced with permission from Michael Khadavi, MD)

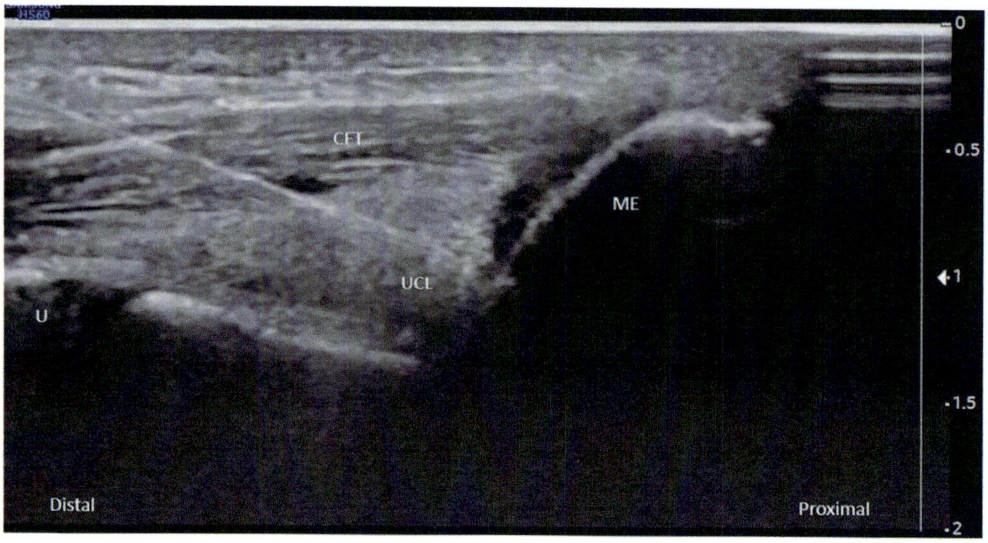


**Fig. 18.2** In-plane ultrasound-guided injection of 3 mL LR-PRP into the superficial fibers of a proximal UCL. The PRP was peppered with a 25-gauge needle into and around the partial tear, visualized as hypoechoic and anechoic ligament tissue at and just deep to location of the needle tip. Acetaminophen was advised for post-procedural pain. *CFT* common flexor tendon, *UCL* ulnar collateral ligament, *ME* medial epicondyle, *U* ulna. (Image reproduced with permission from Michael Khadavi, MD)

There is debate regarding the use of tendon fenestration, often called percutaneous tenotomy [76, 77]. While various techniques have been described, this procedure may be performed with a needle of between 17 and 23 gauge, either during or just before the orthobiologic injection, and with multiple fenestrations through the site of pathology, essentially creating a new injury. Platelet concentration may also be tailored to the patient's age for maximal tenocyte proliferation, and with a total volume proportional to the size of the tendon and area of pathology [78]. Use of activators such as autologous thrombin may decrease runoff, particularly in cases of tendon tearing, however outcomes with these techniques have yet to be evaluated [79].

## Conclusion

LR-PRP has a growing body of support as an evidence-based treatment for LE. Other pathologies about the elbow have also gained traction for PRP, such as proximal partial UCL injuries of the elbow. Image-guidance is recommended to visualize and inject into the injured tendon or ligament. Further research is needed for medial epicondylitis, elbow osteoarthritis, osteochondral injuries, distal biceps tendinopathy, distal triceps tendinopathy, and other conditions about the elbow. Future studies with quantification of orthobiologic "dose" variables, such as platelet and leukocyte concentration, and technique variables discussed above are essential [7].

**Key Points**

1. Injection of leukocyte rich platelet-rich plasma (LR-PRP) into the common extensor tendon is effective at improving symptoms and long-term outcomes in patients with lateral epicondylitis.
2. Following injection of LR-PRP for common extensor tendinosis, symptoms typically improve in 3–4 months, although longer and shorter healing periods may occur.
3. Radial tunnel syndrome is an important mimic of lateral epicondylitis and should be considered in the differential diagnosis of lateral and anterior elbow pain.
4. There are fewer studies investigating orthobiologic treatment of medial epicondylitis in comparison to lateral epicondylitis.
5. Ulnar Collateral Ligament (UCL) injuries tend to occur in overhead athletes, commonly presenting as a decrease in throwing velocity or medial elbow pain.
6. Location and severity of UCL injury are important characteristics as distal tears and complete tears are more likely to require surgical treatment.
7. LR-PRP appears to be an effective treatment of proximal partial UCL elbow tears and sprains, with an average 12-week return to throwing.
8. Procedural pain control needs consideration, especially for intra-tendinous orthobiologic injections, and may be achieved with a regional block or with pre-procedural local anesthetic injection at the target site.
9. Orthobiologic research has grown significantly in recent years, but standardization is needed with consistent

reporting of variables such as platelet concentration, leukocyte concentration, delivery method, and post-procedural rehabilitation protocols.

10. Orthobiologic injections about the elbow should be performed under ultrasound guidance to ensure proper delivery of the medication to the area of pathology.

**Questions**

1. A 45-year-old female with past medical history of hypothyroidism presents with a 3.5 months history of R arm pain along the lateral aspect of her elbow. Symptoms started insidiously several months ago after playing in a pickleball tournament with her husband. Initially, she took a few days off and symptoms improved so she returned to her typical routine playing in weekly volleyball, tennis and pickleball leagues. She saw her PCP about 6 weeks ago who sent her to physical therapy, provided her with a counterforce brace, and performed a corticosteroid injection with only 4 weeks relief. She reports symptoms are still bothersome and have prevented her from doing what she enjoys. On physical exam, she has tenderness to palpation over the common extensor tendon with pain during resisted wrist extension and finger extension. Which of the following treatment options should be discussed next?
   A. Complete rest for at least 4 weeks
   B. Corticosteroid Injection
   C. Leukocyte-Rich PRP injection
   D. Leukocyte-Poor PRP injection

   **Answer: C**

   **Explanation:** This patient has classic features of lateral epicondylitis with lateral elbow pain and pain with resisted wrist and finger extension. In addition, she has a history of multiple racquet sports which predisposes her to such pathology. While most patients will improve with conservative measures (rest, counterforce bracing, physical therapy), some may require further, more invasive interventions. There is no evidence that complete rest (choice A) will cure this patient. While her symptoms might improve with rest, they will likely return following resumption of her typical activity. Corticosteroid injections (Choice B) were previously a mainstay in the treatment of lateral epicondylitis but more recent evidence suggests that PRP (Choices C and D) may actually provide greater long-term symptom and pain improvement. Between the two classifications of PRP (LR vs. LP), there are several meta-analyses that suggest LR PRP (Choice C) has greater improvements in pain control and functional outcomes.

2. A 56-year-old male patient presents with a 6-month history of pain at the lateral elbow that is worse with gripping items. He is diagnosed with lateral epicondylitis and provided with a counterforce brace as well as physical therapy exercises to be completed at home. He returns 6 weeks later with no symptomatic improvement. He is given an injection of LR-PRP at the point of maximal tenderness. Symptoms initially worsened for several days following the procedure. He returns 8 weeks after the PRP injection reporting that his lateral elbow pain has improved, however anterolateral elbow and proximal forearm pain has become prominent. Which of the following is the next best step in management?
   A. Repeat LR-PRP injection
   B. Surgical Referral
   C. Ultrasound evaluation of the lateral elbow and surrounding structures
   D. Corticosteroid injection about the lateral epicondyle

   **Answer: C**

   **Explanation:** Ultrasound evaluation would be the most beneficial and most cost-effective way to further evaluate the patient's pathology. His initial PRP injection was palpation guided so the treating physician may have missed the opportunity to appreciate radial nerve entrapment. Surgical referral (Choice B) is not appropriate as only 8 weeks have passed since the PRP, and improvement often takes 3-4 months. Additionally alternative diagnoses need to be considered before pursuing referral. Corticosteroid injection (Choice D) is not appropriate until we have confirmed the reason for the patient's severe pain and allow sufficient time for the PRP to take effect. The patient has reported different symptoms since the LR-PRP injection, suggesting an alternative diagnosis, making a repeat injection (Choice A) inappropriate.

3. A 17-year-old, right-handed high school pitcher presents to a sports medicine clinic complaining of a 9-month history of right elbow pain. Initially, he was able to continue pitching despite the pain, but recently he has noted a drop in his velocity. He is not improved after a 6-week period of complete rest. Physical exam shows pain over the medial epicondyle with and pain and mild laxity with valgus stress testing of the elbow. Ultrasound examination reveals partial thickness tearing of the proximal portion of the UCL, and dynamic scanning shows a 1mm increase in medial joint space with valgus stress. The patient's senior season starts in two months and he would like to be able to return to the field as soon as possible in hopes of securing a college scholarship. Which of the following is the most appropriate initial intervention for this patient?
   A. Complete rest with gradual return to throwing in 4-6 week
   B. Platelet-rich plasma injection
   C. Corticosteroid Injection
   D. Surgical Referral

   **Answer: B**

   **Explanation:** The patient likely has an ulnar collateral ligament injury as a result of repetitive throwing which places stress on the ligament in overhead athletes. Given the proximal partial thickness tear, he will likely benefit from a PRP injection (Choice B). There are variable rates of success with conservative management (Choice A) and gradual progression through an extended rehabilitation program could cause him to miss a significant portion of his season. Surgery (Choice D) is not indicated at this time as this given its inherent risks and prolonged recovery. There is no evidence to support the use of corticosteroid injections (Choice C) in the treatment of UCL injuries.

4. A 59-year-old mechanic with a history of chronic low back pain, hypertension and type II diabetes presents with numbness and tingling in his medial hand. He denies acute injury and notes that symptoms have slowly worsened over the last few months. Recently, he has had difficulty holding on to objects like his coffee mug or tools. On physical exam he has tenderness to palpation over the majority of the medial elbow and pain with resisted pronation. He has mild numbness noted over the palmar aspect of the fourth and fifth digits with mild weakness with finger abduction on his R side when compared to the left. Which of the following is the most appropriate next step in management?
   A. Surgical Referral
   B. EMG
   C. Counterforce Brace
   D. Corticosteroid Injection

   **Answer: B**

   **Explanation:** The patient's symptoms are concerning for possible medial epicondylitis or ulnar neuropathy. Initially, it is unclear which disease the patient has as there is a fair amount of overlap between the two diseases. EMG is a useful first step in this patient to further elucidate the cause of the patient's symptoms (Choice B). If EMG is negative, then the patient may benefit from a counterforce brace (Choice C) and other conservative measures. If EMG is positive, the patient may need surgical referral (Choice A) for potential release of the ulnar nerve. Injection of corticosteroid (Choice D) is inappropriate as there is little evidence to support its use in medial epicondylitis.

5. A professional baseball pitcher presents to a sports medicine clinic following significant pain at his elbow which caused him to leave his last start after only 2 innings. He reports a dull aching sensation in his elbow for the last 3 months that acutely worsened after throwing a fastball in his game several days ago. He immediately left the game and has not attempted to throw since that time. Given your concern for a potential injury to the medial ulnar collateral ligament, you order an MRI arthrogram of the elbow. Which of the following factors, if noted on MRI, would increase the player's likelihood of needing Tommy John surgery?
   A. Proximal tear of the UCL
   B. Ulnar neuritis
   C. Distal tear of the UCL
   D. Low grade tear of the UCL

   **Answer: C**

   **Explanation:** The patient has evidence of UCL injury and there are certain characteristic findings on MRI such as distal tear (Choice C), complete tear, or excessive laxity on valgus stress that make surgical intervention more likely. Proximal partial tears (Choice A) and low-grade tears (Choice D) tend to recover better and there is evidence from multiple studies demonstrating success in these types of injuries with conservative management including platelet-rich plasma injection. Presence of ulnar neuritis does not suggest need for Tommy John surgery (Choice B).

## References

1. Chen FS, Rokito AS, Jobe FW. Medial elbow problems in the overhead-throwing athlete. J Am Acad Orthop Surg. 2001;9:99–113.
2. Kurppa K, Viikari-Juntura E, Kuosma E, et al. Incidence of tenosynovitis or peritendinitis and epicondylitis in a meat-processing factory. Scand J Work Environ Health. 1991;17:32–7.
3. Haahr JP, Andersen JH. Physical and psychosocial risk factors for lateral epicondylitis: a population based case-referent study. Occup Environ Med. 2003;60:322–9.
4. Shiri R, Viikari-Juntura E, Varonen H, et al. Prevalence and determinants of lateral and medial epicondylitis: a population study. Am J Epidemiol. 2006;164:1065–74.
5. Struijs PA, Kerkhoffs GM, Assendelft WJ, et al. Conservative treatment of lateral epicondylitis: brace versus physical therapy or a combination of both-a randomized clinical trial. Am J Sports Med. 2004;32:462–9.
6. Croisier JL, Foidart-Dessalle M, Tinant F, et al. An isokinetic eccentric programme for the management of chronic lateral epicondylar tendinopathy. Br J Sports Med. 2007;41:269–75.
7. Mautner K, Malanga GA, Smith J, et al. A call for a standard classification system for future biologic research: the rationale for new PRP nomenclature. PM R. 2015;7:S53–9.
8. Verhaar JA. Tennis elbow. Anatomical, epidemiological and therapeutic aspects. Int Orthop. 1994;18:263–7.
9. Allander E. Prevalence, incidence, and remission rates of some common rheumatic diseases or syndromes. Scand J Rheumatol. 1974;3:145–53.
10. Nirschl RP. Elbow tendinosis/tennis elbow. Clin Sports Med. 1992;11:851–70.
11. Gill B, Rahman R, Khadavi M. Ultrasound-guided hydrodissection provides complete symptom resolution in radial tunnel syndrome: a case series and scoping review on hydrodissection for radial nerve pathology. Curr Sports Med Rep. 2022;21:328–35.
12. Gosens T, Peerbooms JC, van Laar W, et al. Ongoing positive effect of platelet-rich plasma versus corticosteroid injection in lateral epicondylitis: a double-blind randomized controlled trial with 2-year follow-up. Am J Sports Med. 2011;39:1200–8.

13. Peerbooms JC, Sluimer J, Bruijn DJ, et al. Positive effect of an autologous platelet concentrate in lateral epicondylitis in a double-blind randomized controlled trial: platelet-rich plasma versus corticosteroid injection with a 1-year follow-up. Am J Sports Med. 2010;38:255–62.
14. Mishra A, Pavelko T. Treatment of chronic elbow tendinosis with buffered platelet-rich plasma. Am J Sports Med. 2006;34:1774–8.
15. Thanasas C, Papadimitriou G, Charalambidis C, et al. Platelet-rich plasma versus autologous whole blood for the treatment of chronic lateral elbow epicondylitis: a randomized controlled clinical trial. Am J Sports Med. 2011;39:2130–4.
16. Mishra AK, Skrepnik NV, Edwards SG, et al. Efficacy of platelet-rich plasma for chronic tennis elbow: a double-blind, prospective, multicenter, randomized controlled trial of 230 patients. Am J Sports Med. 2014;42:463–71.
17. Lebiedzinski R, Synder M, Buchcic P, et al. A randomized study of autologous conditioned plasma and steroid injections in the treatment of lateral epicondylitis. Int Orthop. 2015;39:2199–203.
18. Creaney L, Wallace A, Curtis M, et al. Growth factor-based therapies provide additional benefit beyond physical therapy in resistant elbow tendinopathy: a prospective, single-blind, randomised trial of autologous blood injections versus platelet-rich plasma injections. Br J Sports Med. 2011;45:966–71.
19. Raeissadat SA, Rayegani SM, Hassanabadi H, et al. Is platelet-rich plasma superior to whole blood in the management of chronic tennis elbow: one year randomized clinical trial. BMC Sports Sci Med Rehabil. 2014;6:12.
20. Krogh TP, Fredberg U, Stengaard-Pedersen K, et al. Treatment of lateral epicondylitis with platelet-rich plasma, glucocorticoid, or saline: a randomized, double-blind, placebo-controlled trial. Am J Sports Med. 2013;41:625–35.
21. Martin JI, Atilano L, Merino J, et al. Platelet-rich plasma versus lidocaine as tenotomy adjuvants in people with elbow epicondylopathy: a randomized controlled trial. J Orthop Surg Res. 2019;14:109.
22. Fitzpatrick J, Bulsara M, Zheng MH. The effectiveness of platelet-rich plasma in the treatment of tendinopathy: a meta-analysis of randomized controlled clinical trials. Am J Sports Med. 2017;45:226–33.
23. Wolf JM, Ozer K, Scott F, et al. Comparison of autologous blood, corticosteroid, and saline injection in the treatment of lateral epicondylitis: a prospective, randomized, controlled multicenter study. J Hand Surg Am. 2011;36:1269–72.
24. Ozturan KE, Yucel I, Cakici H, et al. Autologous blood and corticosteroid injection and extracoporeal shock wave therapy in the treatment of lateral epicondylitis. Orthopedics. 2010;33:84–91.
25. Behera P, Dhillon M, Aggarwal S, et al. Leukocyte-poor platelet-rich plasma versus bupivacaine for recalcitrant lateral epicondylar tendinopathy. J Orthop Surg (Hong Kong). 2015;23:6–10.
26. Stenhouse G, Sookur P, Watson M. Do blood growth factors offer additional benefit in refractory lateral epicondylitis? A prospective, randomized pilot trial of dry needling as a stand-alone procedure versus dry needling and autologous conditioned plasma. Skelet Radiol. 2013;42:1515–20.
27. Arik HO, Kose O, Guler F, et al. Injection of autologous blood versus corticosteroid for lateral epicondylitis: a randomised controlled study. J Orthop Surg (Hong Kong). 2014;22:333–7.
28. Kazemi M, Azma K, Tavana B, et al. Autologous blood versus corticosteroid local injection in the short-term treatment of lateral elbow tendinopathy: a randomized clinical trial of efficacy. Am J Phys Med Rehabil. 2010;89:660–7.
29. Chen X, Jones IA, Park C, et al. The efficacy of platelet-rich plasma on tendon and ligament healing: a systematic review and meta-analysis with bias assessment. Am J Sports Med. 2018;46:2020–32.
30. Chen XT, Fang W, Jones IA, et al. The efficacy of platelet-rich plasma for improving pain and function in lateral epicondylitis: a systematic review and meta-analysis with risk-of-bias assessment. Arthroscopy. 2021;37:2937–52.
31. Gautam VK, Verma S, Batra S, et al. Platelet-rich plasma versus corticosteroid injection for recalcitrant lateral epicondylitis: clinical and ultrasonographic evaluation. J Orthop Surg (Hong Kong). 2015;23:1–5.
32. Omar AS, Ibrahim ME, Ahmed AS, et al. Local injection of autologous platelet rich plasma and corticosteroid in treatment of lateral epicondylitis and plantar fasciitis: randomized clinical trial. Egypt Rheumatol. 2012;34:43–9.
33. Khaliq A, Khan I, Inam M, et al. Effectiveness of platelets rich plasma versus corticosteroids in lateral epicondylitis. J Pak Med Assoc. 2015;65:S100–4.
34. Palacio EP, Schiavetti RR, Kanematsu M, et al. Effects of platelet-rich plasma on lateral epicondylitis of the elbow: prospective randomized controlled trial. Rev Bras Ortop. 2016;51:90–5.
35. Simental-Mendia M, Vilchez-Cavazos F, Alvarez-Villalobos N, et al. Clinical efficacy of platelet-rich plasma in the treatment of lateral epicondylitis: a systematic review and meta-analysis of randomized placebo-controlled clinical trials. Clin Rheumatol. 2020;39:2255–65.
36. Yerlikaya M, Talay Calis H, Tomruk Sutbeyaz S, et al. Comparison of effects of leukocyte-rich and leukocyte-poor platelet-rich plasma on pain and functionality in patients with lateral epicondylitis. Arch Rheumatol. 2018;33:73–9.
37. Seetharamaiah VB, Gantaguru A, Basavarajanna S. A comparative study to evaluate the efficacy of platelet-rich plasma and triamcinolone to treat tennis elbow. Indian J Orthop. 2017;51:304–11.
38. Montalvan B, Le Goux P, Klouche S, et al. Inefficacy of ultrasound-guided local injections of autologous conditioned plasma for recent epicondylitis: results of a double-blind placebo-controlled randomized clinical trial with one-year follow-up. Rheumatology (Oxford). 2016;55:279–85.
39. Shim JW, Lee JS, Park YB, et al. The effect of leucocyte concentration of platelet-rich plasma on outcomes in patients with lateral epicondylitis: a systematic review and meta-analysis. J Shoulder Elb Surg. 2022;31:634–45.
40. Niemiec P, Szyluk K, Jarosz A, et al. Effectiveness of platelet-rich plasma for lateral epicondylitis: a systematic review and meta-analysis based on achievement of minimal clinically important difference. Orthop J Sports Med. 2022;10:23259671221086920.
41. Wang A, Breidahl W, Mackie KE, et al. Autologous tenocyte injection for the treatment of severe, chronic resistant lateral epicondylitis: a pilot study. Am J Sports Med. 2013;41:2925–32.
42. Singh A, Gangwar DS, Singh S. Bone marrow injection: a novel treatment for tennis elbow. J Nat Sci Biol Med. 2014;5:389–91.
43. Khoury M, Tabben M, Rolon AU, et al. Promising improvement of chronic lateral elbow tendinopathy by using adipose derived mesenchymal stromal cells: a pilot study. J Exp Orthop. 2021;8:6.
44. Varshney A, Maheshwari R, Juyal A, et al. Autologous platelet-rich plasma versus corticosteroid in the management of elbow epicondylitis: a randomized study. Int J Appl Basic Med Res. 2017;7:125–8.
45. Boden AL, Scott MT, Dalwadi PP, et al. Platelet-rich plasma versus Tenex in the treatment of medial and lateral epicondylitis. J Shoulder Elb Surg. 2019;28:112–9.
46. Bohlen HL, Schwartz ZE, Wu VJ, et al. Platelet-rich plasma is an equal alternative to surgery in the treatment of type 1 medial epicondylitis. Orthop J Sports Med. 2020;8:2325967120908952.
47. Morrey BF, An KN. Articular and ligamentous contributions to the stability of the elbow joint. Am J Sports Med. 1983;11:315–9.
48. Morrey BF, An KN. Functional anatomy of the ligaments of the elbow. Clin Orthop Relat Res. 1985;201:84–90.
49. Morrey BF, Tanaka S, An KN. Valgus stability of the elbow. A definition of primary and secondary constraints. Clin Orthop Relat Res. 1991;265:187–95.

50. Chahla J, Kennedy MI, Aman ZS, et al. Ortho-biologics for ligament repair and reconstruction. Clin Sports Med. 2019;38:97–107.
51. Erickson BJ, Gupta AK, Harris JD, et al. Rate of return to pitching and performance after Tommy John surgery in major league baseball pitchers. Am J Sports Med. 2014;42:536–43.
52. Makhni EC, Lee RW, Morrow ZS, et al. Performance, return to competition, and reinjury after Tommy John surgery in major league baseball pitchers: a review of 147 cases. Am J Sports Med. 2014;42:1323–32.
53. Osbahr DC, Cain EL, Raines BT, et al. Long-term outcomes after ulnar collateral ligament reconstruction in competitive baseball players: minimum 10-year follow-up. Am J Sports Med. 2014;42:1333–42.
54. Park JY, Oh KS, Bahng SC, et al. Does well maintained graft provide consistent return to play after medial ulnar collateral ligament reconstruction of the elbow joint in elite baseball players? Clin Orthop Surg. 2014;6:190–5.
55. Rohrbough JT, Altchek DW, Hyman J, et al. Medial collateral ligament reconstruction of the elbow using the docking technique. Am J Sports Med. 2002;30:541–8.
56. Conway JE, Jobe FW, Glousman RE, et al. Medial instability of the elbow in throwing athletes. Treatment by repair or reconstruction of the ulnar collateral ligament. J Bone Joint Surg Am. 1992;74:67–83.
57. Dodson CC, Thomas A, Dines JS, et al. Medial ulnar collateral ligament reconstruction of the elbow in throwing athletes. Am J Sports Med. 2006;34:1926–32.
58. Deal JB, Smith E, Heard W, et al. Platelet-rich plasma for primary treatment of partial ulnar collateral ligament tears: MRI correlation with results. Orthop J Sports Med. 2017;5:2325967117738238.
59. Azar FM, Andrews JR, Wilk KE, et al. Operative treatment of ulnar collateral ligament injuries of the elbow in athletes. Am J Sports Med. 2000;28:16–23.
60. Cain EL, Andrews JR, Dugas JR, et al. Outcome of ulnar collateral ligament reconstruction of the elbow in 1281 athletes: results in 743 athletes with minimum 2-year follow-up. Am J Sports Med. 2010;38:2426–34.
61. Vitale MA, Ahmad CS. The outcome of elbow ulnar collateral ligament reconstruction in overhead athletes: a systematic review. Am J Sports Med. 2008;36:1193–205.
62. Jobe FW, Stark H, Lombardo SJ. Reconstruction of the ulnar collateral ligament in athletes. J Bone Joint Surg Am. 1986;68(8):1158–63.
63. Savoie FH, Trenhaile SW, Roberts J, et al. Primary repair of ulnar collateral ligament injuries of the elbow in young athletes: a case series of injuries to the proximal and distal ends of the ligament. Am J Sports Med. 2008;36:1066–72.
64. Savoie FH, Morgan C, Yaste J, et al. Medial ulnar collateral ligament reconstruction using hamstring allograft in overhead throwing athletes. J Bone Joint Surg Am. 2013;95:1062–6.
65. Rettig AC, Sherrill C, Snead DS, et al. Nonoperative treatment of ulnar collateral ligament injuries in throwing athletes. Am J Sports Med. 2001;29:15–7.
66. Frangiamore SJ, Lynch TS, Vaughn MD, et al. Magnetic resonance imaging predictors of failure in the nonoperative management of ulnar collateral ligament injuries in professional baseball pitchers. Am J Sports Med. 2017;45:1783–9.
67. Ford GM, Genuario J, Kinkartz J, et al. Return-to-play outcomes in professional baseball players after medial ulnar collateral ligament injuries: comparison of operative versus nonoperative treatment based on magnetic resonance imaging findings. Am J Sports Med. 2016;44:723–8.
68. Podesta L, Crow SA, Volkmer D, et al. Treatment of partial ulnar collateral ligament tears in the elbow with platelet-rich plasma. Am J Sports Med. 2013;41:1689–94.
69. Dines JS, Williams PN, ElAttrache N, et al. Platelet-rich plasma can be used to successfully treat elbow ulnar collateral ligament insufficiency in high-level throwers. Am J Orthop (Belle Mead NJ). 2016;45:296–300.
70. Mills FB 4th, Misra AK, Goyeneche N, et al. Return to play after platelet-rich plasma injection for elbow UCL injury: outcomes based on injury severity. Orthop J Sports Med. 2021;9:2325967121991135.
71. Buckley PS, Morris ER, Robbins CM, et al. Variations in blood supply from proximal to distal in the ulnar collateral ligament of the elbow: a qualitative descriptive cadaveric study. Am J Sports Med. 2019;47:1117–23.
72. Simons S, Dixon JB. Biceps tendinopathy and tendon rupture. 2022. Accessed 23 Jan 2023. https://www.amboss.com/us/knowledge/Biceps_tendinopathy_and_biceps_tendon_rupture.
73. Barker SL, Bell SN, Connell D, et al. Ultrasound-guided platelet-rich plasma injection for distal biceps tendinopathy. Shoulder Elbow. 2015;7:110–4.
74. van der Vis J, Janssen SJ, Bleys RLAW, et al. Distal biceps tendon injection. Clin Shoulder Elb. 2021;24:93–7.
75. Sanli I, Morgan B, van Tilborg F, et al. Single injection of platelet-rich plasma (PRP) for the treatment of refractory distal biceps tendonitis: long-term results of a prospective multicenter cohort study. Knee Surg Sports Traumatol Arthrosc. 2016;24:2308–12.
76. Finnoff JT, Fowler SP, Lai JK, et al. Treatment of chronic tendinopathy with ultrasound-guided needle tenotomy and platelet-rich plasma injection. PM R. 2011;3:900–11.
77. Krey D, Borchers J, McCamey K. Tendon needling for treatment of tendinopathy: a systematic review. Phys Sportsmed. 2015;43:80–6.
78. Berger DR, Centeno CJ, Steinmetz NJ. Platelet lysates from aged donors promote human tenocyte proliferation and migration in a concentration-dependent manner. Bone Joint Res. 2019;8:32–40.
79. Matuska AM, Klimovich MK, Chapman JR. An ethanol-free autologous thrombin system. J Extra Corpor Technol. 2018;50:237–43.

# Regenerative Therapy of Hip and Related Joints, Ligaments, and Tendons

19

Mary A. Ambach, Mirielle Diaz-Molina, and Christopher Rogers

## Introduction

Pain in the hip region is commonly seen in the interventional pain management practice [1]. Thirty-six percent of adults 45 years and older reported significant pain or stiffness in the hip in a study on the prevalence of hip-related osteoarthritis [2]. In some countries, including the United States, the economic burden of osteoarthritis amounts to 1–2.5% of the gross national product. The United States spends an estimated $72 billion in direct medical costs related to osteoarthritis [3].

Hip conditions often present a diagnostic and therapeutic challenge due to multiple pathologies that can present with pain in the hip region. In addition to musculoskeletal etiology, other conditions, including neurologic, infectious, autoimmune, or genetic etiologies, can also present with hip pain. Musculoskeletal conditions causing hip pain include intra-articular and extra-articular causes [4, 5]. The intra-articular causes of hip pain include osteoarthritis, chondral damage, labral tears, femoroacetabular impingement, capsular laxity, ligamentum teres rupture, and avascular necrosis. Osteoarthritis is the most common cause of hip pain and is the most common joint disease, affecting more than 240 million people worldwide. The hip joint ranks as the second most affected joint after the knee [6]. Extra-articular causes of pain around the hip include iliopsoas tendinopathy, gluteal tendinopathy or greater trochanteric pain syndrome (GTPS), and iliotibial band (ITB) syndrome, among others. Other conditions can present similarly to those with hip-related etiology, such as athletic pubalgia, also known as sports hernia, and osteitis pubis [4].

M. A. Ambach (✉) · C. Rogers
San Diego Orthobiologics Medical Group, Carlsbad, CA, USA

Department of Orthopedic Surgery, University of California San Diego School of Medicine, San Diego, CA, USA
e-mail: ambach@sdomg.com; rogers@sdomg.com

M. Diaz-Molina
San Diego Orthobiologics Medical Group, Carlsbad, CA, USA
e-mail: mdiaz@sdomg.com

Conservative management consists of oral medications, physical therapy, and injections [4, 5, 7–10]. Pharmacologic treatment is mainly symptomatic and includes analgesics, non-steroidal anti-inflammatory drugs, and opioids. Intra-articular injections of corticosteroids and hyaluronic acid (HA) are palliative, but are short lived. Surgical interventions are associated with a high risk of morbidity and mortality, specifically in the elderly. Younger individuals may require surgical revision after 15–20 years [11]. Therefore, surgery should be reserved for patients that failed extensive conservative treatment.

The treatment of osteoarthritis of the hip and other conditions, including cartilage and tendon lesions, is complex due to their inherent limited healing potential. The regenerative capacity of cartilage is limited due to its isolation from systematic regulation and avascularity of chondrocytes. However, numerous basic science studies suggest that orthobiologics may have beneficial effects on tendon and cartilage pathology [12]. Regenerative medicine principles are being applied and provide promising options for managing hip pathology by promoting angiogenesis, collagen synthesis, and immune response regulation.

## Anatomy

Understanding the anatomy of the hip joint is crucial in the assessment and management of hip-related pathology.

### Osteology

The hip joint is a synovial joint. The head of the femur articulates with the acetabulum of the pelvis forming the femoral-acetabular joint (Fig. 19.1). The hip is a major weight-bearing joint that allows a significant level of mobility. The labrum is composed of fibrocartilage and increases the depth of the acetabulum at the outer rim. The hip bone is comprised of three different parts: the ischium, ilium, and pubis. The ilium


A. Navani et al. (eds.), *Essentials of Regenerative Medicine in Interventional Pain Management*,
https://doi.org/10.1007/978-3-031-50357-3_19

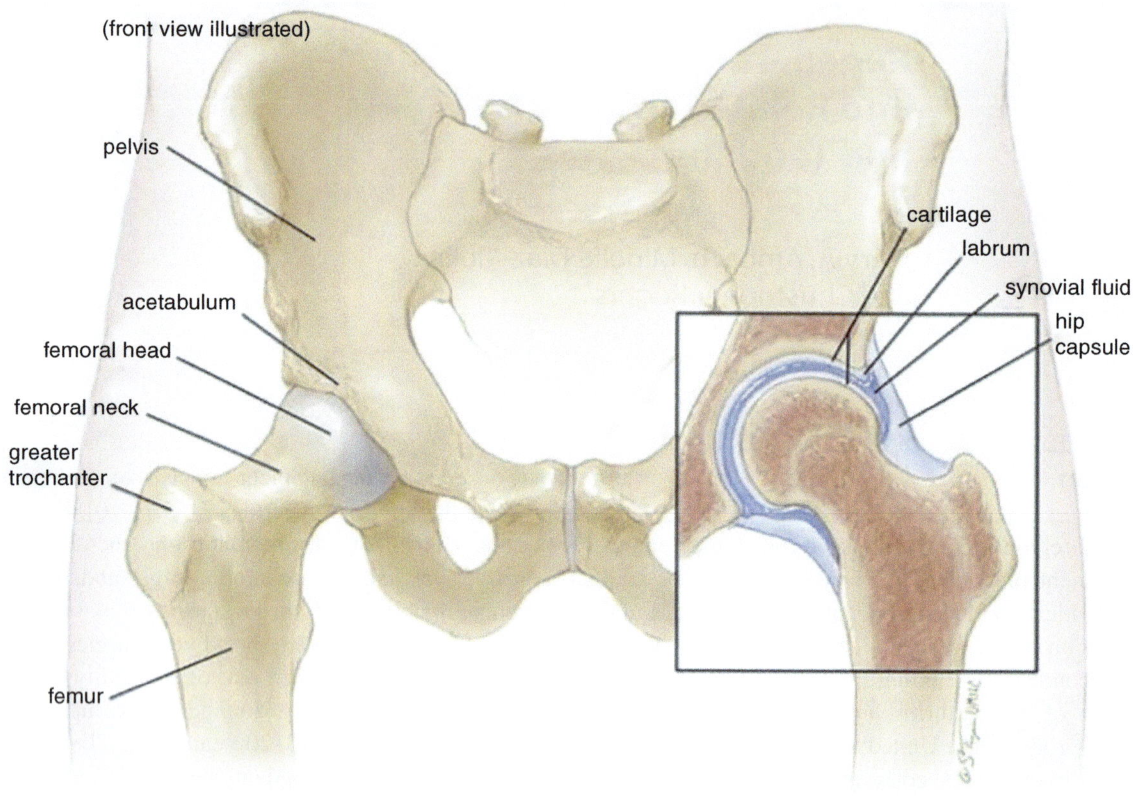


**Fig. 19.1** Anatomy of the hip. Chang Chien G, Kim B, Candido KD, Manchikanti L. Hip and related joints, ligaments and tendons. In: Manchikanti L, Navani A, Atluri S (eds). *Essentials of Regenerative Medicine in Interventional Pain Management*. ASIPP Publishing, Paducah, KY, 2019, pp 257–282 [1]

articulates with the sacrum posteriorly via the sacroiliac joint. The slight convexity of the iliac surface and the concavity of the sacral surface provide stability and limit its motion.

## Musculature

Anteriorly, the psoas and iliacus muscles originate from the lumbar spine transverse processes (T-12 and L1 through L5) and pelvis (superior anterior iliac crest). These muscles converge to form the iliopsoas muscle, the main hip flexor, which inserts onto the lesser trochanter of the proximal femur as the iliopsoas tendon. The pectineus muscle, a relatively weak hip flexor, arises from the superior pubic ramus and attaches to the pectineal line along the superior aspect of the femur. The rectus femoris muscle (part of the quadriceps muscles) originates from the anterior inferior iliac spine and attaches to the tibial tuberosity and the patella. It also plays a major role in flexing the hip as it crosses over the hip joint. The sartorius is a weak hip flexor that originates from the anterior-superior iliac spine and attaches to the pes anserine at the medial tibia.

The muscles located medially are the adductor longus, adductor brevis, adductor magnus, and gracilis. These muscles are the main adductors of the hip and thigh.

The gluteal muscles consist of the gluteus maximus, gluteus medius, and gluteus minimus muscles. Their main function is to facilitate abduction and extension of the hip joint, however, they can also assist in adduction, external rotation, and internal rotation of the thigh. The greater trochanter of the femur is the primary site of attachment for the gluteus medius and minimus, while the gluteus maximus muscle attaches to the gluteal tuberosity of the femur and iliotibial tract. The gluteus medius is a strong hip abductor, while the gluteus maximus is a major hip extender. The muscles deep to the gluteus maximus, which includes the piriformis, obturator internus, quadratus femoris, and superior/inferior gemellus, aid in stabilizing the hip during single leg stance and hip external rotation.

Posteriorly, the hamstring muscles, namely the biceps femoris, semimembranosus, and semitendinosus, originate from the ischial tuberosity and assist in hip extension and knee flexion. The muscles of the hip are illustrated in Fig. 19.2 and summarized in Table 19.1.

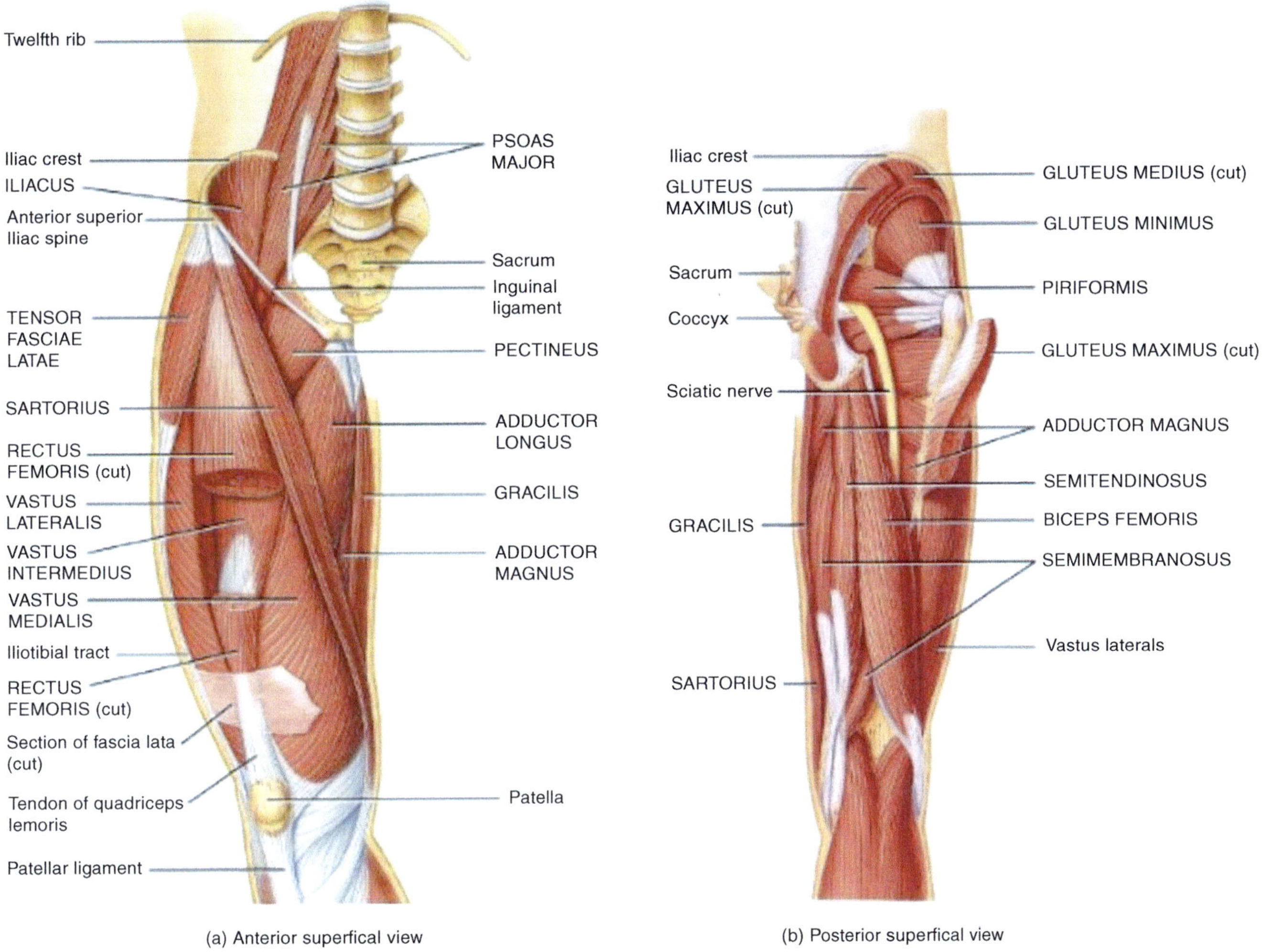


**Fig. 19.2** Muscles of the hip joint. *Courtesy: San Diego Mesa College*. Chang Chien G, Kim B, Candido KD, Manchikanti L. Hip and related joints, ligaments and tendons. In: Manchikanti L, Navani A, Atluri S (eds). *Essentials of Regenerative Medicine in Interventional Pain Management*. ASIPP Publishing, Paducah, KY, 2019, pp 257–282 [1]

## Tendons/Ligaments

Tendons serve as the strong, fibrous connection between muscle and bone, and allow energy transfer to provide tensile strength and motion. Tendinopathy is a clinical presentation characterized by chronic, localized pain due to hypervascularity, hypercellularity, and degenerative changes in the collagenous matrix. It is to be distinguished from tendinitis, which involves the presence of an acute inflammatory process. Inflammation is minimal in tendinopathy due to chronic overuse, as evidenced in histopathological findings of a wide variety of biopsied tendons [13, 14]. Several factors contribute to the generation of tendinopathy including the load intensity, duration, frequency, and direction of movement causing chronic microtrauma. Tendon rupture can occur following acute injuries or chronic, cumulative trauma from degenerative conditions. Studies have shown a remarkably increased risk for tendon rupture in degenerative conditions, illustrating the significance of treatment approaches for promoting tendon healing in addition to symptomatic management [13–15].

The hip ligaments consist of the iliofemoral ligament anteriorly, the ischiofemoral ligament posteriorly, and the pubofemoral ligament inferiorly. The iliofemoral and ischiofemoral ligaments originate from the anterior inferior iliac spine (AIIS) and acetabular rim and insert into the intertrochanteric line. The pubofemoral ligament originates from the acetabular rim inserting medially to the greater trochanter. The pubofemoral ligament prevents excess abduction and extension, ischiofemoral prevents excess extension, and the iliofemoral prevents hyperextension [16]. The ligament of the femoral head, also known as ligamentum teres, originates from the obturator crest and inserts into fibrous capsule of the hip joint. It prevents excessive hip abduction.

**Table 19.1** Muscles of the hip

| Muscle | Origin/insertion | Function | Innervation and root levels |
|---|---|---|---|
| Psoas Major | Transverse processes T12-L5/lesser trochanter | Hip flexion | Anterior rami of L1-L2 |
| Iliacus | Upper two-third of the iliac fossa/lesser trochanter joins Psoas to form Iliopsoas tendon | Hip flexion, anterior pelvic tilt | Femoral Nerve (L2-L4) |
| Pectineus | Pubic between iliopectineal eminence and pubic tubercle along pectineal line/lesser trochanter (linea aspera) | Hip adduction, flexion medial rotation | Femoral Nerve (L2-L4) |
| Rectus Femoris | Anterior inferior iliac spine/patellar tendon | Hip flexion, knee extension | Femoral Nerve (L2-L4) |
| Adductor Magnus | Ischial tuberosity, pubis/linear aspera, adductor tubercle | Hip adduction, hip flexion: proximal portion<br>Hip extension: distal portion | Obturator Nerve (L2-L4) |
| Adductor Longus | Anterior pubic between crest and symphysis/medial lip of linea aspera | Hip adduction, hip flexion assist | Obturator Nerve (L2-L4) |
| Adductor Brevis | Inferior pubic ramus/linea aspera | Hip adduction, hip flexion assist | Obturator Nerve (L2-L4) |
| Gracilis | Inferior pubic ramus/pes anserinus | Hip adduction, hip flexion/ internal rotation | Obturator Nerve (L2-L4) |
| Gluteus Maximus | Posterior gluteal line, sacrum, coccyx, sacrotuberous ligament/gluteal tuberosity and iliotibial tract | Hip extension. Assist in adduction and external rotation | Inferior Gluteal Nerve (L5-S2) |
| Gluteus Medius | Ilium from iliac crest and posterior luteal line to anterior gluteal line and aponeurosis/lateral greater trochanter | Hip abduction, hip internal rotation | Superior gluteal nerve (L4, L5, S1) |
| Gluteus Minimus | Outer ilium between anterior and inferior gluteal lines and margin of greater sciatic notch/greater trochanter | Hip abduction, hip internal rotation | Superior gluteal nerve (L4, L5, S1) |
| Piriformis | Sacrum/greater trochanter | Hip external rotation, hip abduction | Sacral plexus (S1-S2) |
| Quadratus Femoris | Ischial tuberosity/intertrochanteric crest | Hip external rotation, hip adduction | Lumbar plexus (L4-S1) |
| Obturator Externus | Obturator foramen and membrane/Trochanteric fossa of femur | External rotation, adduction | Posterior branch of obturator nerve (L3, L4) |
| Obturator Internus | Obturator foramen, membrane and fascia/greater trochanter | Hip abduction, hip external rotation | Nerve to obturator Internus (L5, S1, S2) |
| Gemellus Superior | Ischial spine/Greater trochanter | Hip external rotation | Nerve to obturator Internus (L5, S1, S2) |
| Gemellus Inferior | Ischial tuberosity/Greater trochanter | External rotation | Nerve to quadratus femoris (L4-S1) |
| Biceps Femoris | Long: Ischial tuberosity/fibular head<br>Short: Linea aspera/fibular head | Hip extension, knee flexion, external rotation | Long: Tibial nerve<br>Short: Common peroneal nerve |
| Semimembranosus | Ischial tuberosity/Medial condyle of tibia | Hip extension, knee flexion | Tibial nerve (L5, S1, S2) |
| Semitendinosis | Ischial tuberosity/Pes anserinus | Hip extension, knee flexion | Tibial nerve (L5, S1, S2) |

## Nerve Supply

The major nerves of the hip originate in the lumbosacral region, as shown in Fig. 19.3, which may make it difficult to distinguish between primary hip pain and radicular pain referred from the lumbar spinal nerves. A review of 17 different neuroanatomical studies of the hip found that the hip is supplied by the femoral, obturator, sciatic, and superior gluteal nerves, as well as the nerve to the quadratus femoris [17]. The obturator nerve (L2-L4) supplies the skin over the upper medial thigh. The lateral femoral cutaneous nerve (L2, L3) supplies sensation over the upper lateral thigh. The genitofemoral nerve (L1-L2) supplies the skin over the proximal anteromedial thigh. The densest innervation is found in the anterior aspect of the hip capsule and labrum. Despite significant literature on the innervation of the hip, there is limited data on the innervation of the greater trochanteric region of the femur [18].

## Vascular Supply

The vascular supply of the hip joint predominantly arises from the medial and lateral circumflex femoral arteries. The circumflex arteries are branches of the profunda femoris artery, which is a branch of the femoral artery (Fig. 19.4). These branches anastomose at the base of the femoral neck to form a ring from which smaller arteries arise to supply the joint. Most of the arterial supply is from the medial circumflex artery, thus, any damage to this artery can result in osteonecrosis of the femoral head.

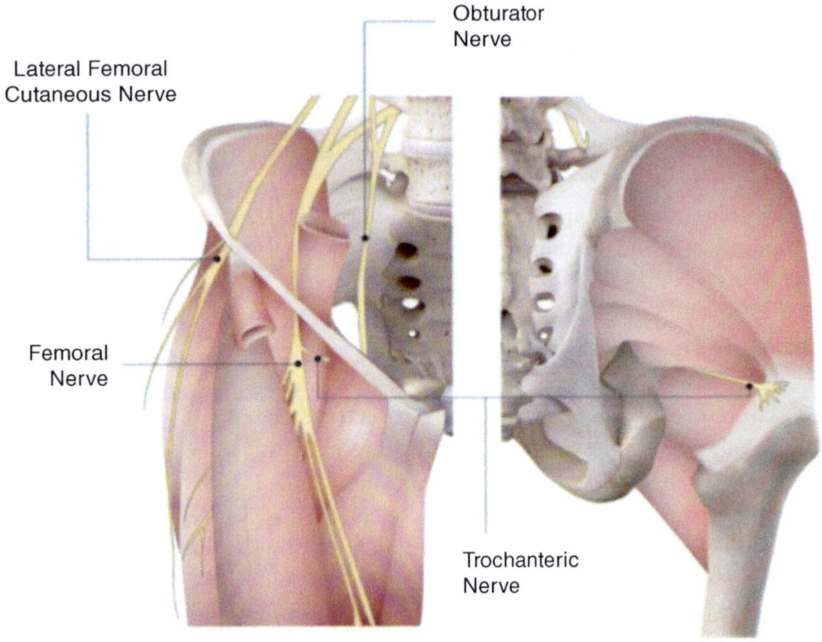


**Fig. 19.3** Anatomical diagram of the innervation of the hip. *Source: Amin NH, West JA, Farmer T, et al. Nerve blocks in the geriatric patient with hip fracture: a review of the current literature and relevant neuroanatomy. Geriatr Orthop Surg Rehabil 2017; 8:268–275* [18]. Chang Chien G, Kim B, Candido KD, Manchikanti L. Hip and related joints, ligaments and tendons. In: Manchikanti L, Navani A, Atluri S (eds). *Essentials of Regenerative Medicine in Interventional Pain Management*. ASIPP Publishing, Paducah, KY, 2019, pp 257–282 [1]

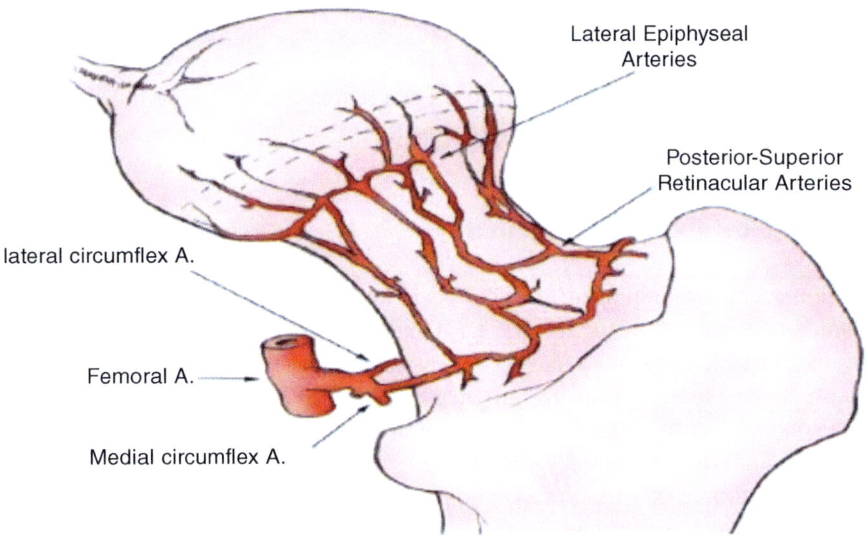


**Fig. 19.4** Blood supply of the hip. *Source:* SlideShare. Chang Chien G, Kim B, Candido KD, Manchikanti L. Hip and related joints, ligaments and tendons. In: Manchikanti L, Navani A, Atluri S (eds). *Essentials of Regenerative Medicine in Interventional Pain Management*. ASIPP Publishing, Paducah, KY, 2019, pp 257–282 [1]

## Pathophysiology

There has been a better understanding of the functional anatomy and physiology of the hip joint and the surrounding structures in recent years. Advances in magnetic resonance imaging (MRI) and the increasing use of musculoskeletal ultrasound (MSKUS) have improved the ability to diagnose causes of hip pain accurately and broadened the differential diagnosis of pain in and around the hip joint.

### Osteoarthritis

Hip osteoarthritis is a common disabling condition of the hip joint. The age-standardized incidence rate (ASIR) of hip osteoarthritis globally was 18.70 per 100,000 persons in 2019 [19]. The complex pathogenesis of osteoarthritis involves mechanical, inflammatory, and metabolic factors, leading to structural failure of the synovial joint. Osteoarthritis is a dynamic condition caused by an imbalance between the repair and destruction of joint tissues rather than a passive degenerative disease [20]. Intra-articular pro-inflammatory cytokines and proteinases in osteoarthritis interfere with the synthesis of HA, resulting in a significantly decreased molecular weight HA and reduced synovial fluid viscoelasticity, which leads to degradation of the articular cartilage and disruption of the mechanical homeostasis of the joint [21, 22].

Hip osteoarthritis most often presents with pain deep in the groin, which can radiate to the lateral or anteromedial thigh. It frequently involves the unilateral hip, unlike knee osteoarthritis which frequently presents with bilateral involvement. Hip osteoarthritis can cause restricted range in all planes of motion, however, internal rotation is usually the earliest affected in both active and passive motion. End-stage hip osteoarthritis often presents with an adducted, flexed, and externally rotated position. Major risk factors for hip osteoarthritis include obesity, advanced age, history of injury, anatomical alignment, and gender, with females more commonly affected than males [23–25].

## Labral Tear

The labrum is a dense fibrocartilage ring that lines the rim of the acetabulum and is responsible for joint stability while reducing mechanical stress on the hip joint. Patients with labral tears can present with a clicking or catching sensation in the affected joint. In a study of 100 patients with hip pain and mechanical symptoms such as locking or giving way, 66% had a labral tear on magnetic resonance arthrography [26]. Common complaints are groin pain when descending stairs or getting up from a sitting position. Risk factors for labral tears include trauma, structural abnormalities, or forceful mechanical stress from repetitive motions [27]. Labral tears are frequently associated with sports that require excessive rotational force, such as hockey, football, or soccer.

## Femoroacetabular Impingement

Femoroacetabular impingement (FAI) occurs when the anatomic abnormalities of the femoral head and/or acetabulum result in abnormal contact during hip motion, which can cause damage to the labrum and articular cartilage [28]. FAI is a common cause of hip pain, functional deficits, and progression to early osteoarthritis [29].

FAI syndrome is a triad of symptoms, signs, and radiographic findings. Symptoms include positional hip or groin pain, clicking, and catching. Signs include pain with flexion, adduction, and internal rotation (positive FADIR test) and decreased internal rotation. An anteroposterior (AP) radiograph of the pelvis and oblique view of the hip will show abnormal anatomy of the femoral head or acetabulum. CAM morphology is characterized by an aspherical femoral head referred to as having a "pistol grip" appearance. It is more likely seen in athletes and is associated with repetitive hip loading in maximal flexion [30]. Pincer morphology is an over coverage of the acetabulum around the femoral head. A mixed type with a combination of CAM and Pincer characteristics, is the most common type of impingement pathology.

## Osteonecrosis

Osteonecrosis of the femoral head (ONFH) is a destructive disease of the hip joint caused by a critical decrease in the vascular supply to the femoral head. ONFH causes progressive bone tissue necrosis. A majority of untreated lesions lead to femoral head collapse [31]. The exact pathophysiology is not clearly understood. Corticosteroid treatment is one of the most common causes of osteonecrosis [31, 32].

## Greater Trochanteric Pain Syndrome/Gluteal Tendinopathy

GTPS, formerly known as trochanteric bursitis, is one of the most common causes of hip pain. GTPS involves overuse tendinopathy of the gluteus medius and minimus tendons. Patients typically present with pain in the lateral hip adjacent to the greater trochanter. There is increased pain with ambulation, prolonged standing, and rising from a chair. The pain is reproduced by applying direct pressure on the greater trochanter. Risk factors for GTPS include obesity, female gender, low back pain, and ITB tenderness [33].

## Iliopsoas Tendinopathy

The iliopsoas muscle passes anterior to the pelvic brim and hip capsule. The iliopsoas tendon and bursa lie between the musculotendinous junction and the pelvic brim. They can become irritated by snapping over the anterior inferior iliac spine, the iliopectineal eminence, or the lesser trochanter. Internal snapping hip syndrome is most commonly a result of the iliopsoas tendon snapping over the iliopectineal eminence. This repeated snapping can cause tendinopathy and bursitis, which present with anterior hip pain [34, 35]. Internal snapping hip syndrome pain may be reproduced by bringing the hip from a flexed and externally oriented position to extension and internal rotation [36]. Iliopsoas impingement is characterized by increased friction or adhesions formed between the iliopsoas and the femoral head and anterior labrum [37]. This has been associated with internal snapping hip syndrome and presents with an audible snap or click in the hip or groin. Repeated hip flexion or external rotation can cause iliopsoas tendinopathy and overuse injury. Younger athletes at higher risk include jumpers, hurdlers, dancers, and uphill runners, who engage in dynamic movements and forceful hip flexion.

## Iliotibial Band Syndrome

The ITB is a fibrous sheath that extends along the entire lateral thigh. Proximally, it splits into superficial and deep layers and encloses the tensor fascia latae anchoring it to the iliac crest [38]. It receives contributions from the tendon of the gluteus maximus and tensor fascia latae at the greater trochanter [39]. In ITB syndrome, sometimes called external snapping hip syndrome, the ITB rubs over the greater trochanter of the femoral head during hip movements causing lateral hip pain. If the ITB pathology presents with pain along the iliac crest over the iliac tubercle, it is called proximal ITB syndrome. Symptoms include ITB tightness, pain, and clicking over the lateral hip, which is common in runners and cyclists.

### Hamstring Tendinopathy

Proximal hamstring tendinopathy presents primarily with pain in the lower gluteal region, sometimes radiating to the hamstrings and posterior thigh. The pain mainly manifests during activities where the hamstring tendon is overused or overloaded, such as running and jumping, or sitting for a prolonged time. On exam, there is palpable tenderness and pain over the ischial tuberosity and pain can be elicited with resisted knee flexion. Active stretching of the hamstrings can also recreate the pain at the ischial tuberosity.

### Differential Diagnosis

There are a vast number of other conditions that can cause hip pain. These conditions can be intra-articular or extra-articular etiologies, referred pain from other structures, traumatic injuries such as fractures, congenital conditions, and systemic conditions including infection, inflammatory, autoimmune, neurologic, neoplastic process, and vascular diseases [36]. The discussion of these conditions is out of the scope of this chapter however it is important to be aware of these various etiologies to have a comprehensive evaluation of the hip patient.

## Clinical Assessment

### History and Physical Examination

The evaluation of hip pain and identification of the exact pathology can be complex as etiological factors can be vast. A thorough clinical assessment, including a detailed orthopedic history and a comprehensive inquiry into all aspects of the patient's medical history, is essential in arriving at a solid diagnosis and treatment plan. The physical examination of the hips should include the identification of any abnormalities in alignment that can cause biomechanical dysfunction and overload to the hip region; assessment for gait instability, weakness, or pain; evaluation for leg-length discrepancy, range of motion testing; and a neurologic examination as appropriate [40–43]. Table 19.2 summarizes the provocative tests performed during the hip exam.

### Diagnostic Imaging Studies

Plain film radiographs are typically the initial imaging study utilized to evaluate the patient with hip pain. A weight-bearing AP x-ray of the pelvis is used to assess pelvic obliquity, sacroiliac joint, hip joint space, and overall bone density. Osteophyte formation, subchondral cyst formation, decreased

**Table 19.2** Special tests for examination of the hip

| Structure evaluated | Test | Positioning | Positive sign |
|---|---|---|---|
| Femoral-acetabular joint | Log roll | Patient supine<br>Practitioner places hand on the anterior knee/thigh and rotates the leg alternating between internal and external rotation | Pain elicited at the anterior groin suggests advanced osteoarthritis, or femoral-acetabular pathology<br>Clicking/popping suggests labral tear |
| | FABER<br>Flexion, Abduction, External rotation | Patient supine<br>Hip flexed to approximately 90°. The knee is flexed. The practitioner takes the hip into end ROM in ER | Pain elicited at the groin suggests femoral-acetabular pathology. Possibly osteoarthritis, or labrum |
| | FADIR<br>Flexed, Adduction, Internal Rotation | Patient supine<br>Hip flexed to approximately 90°. The knee is flexed. The practitioner takes the hip into end ROM in IR | Pain elicited at the groin suggests femoral-acetabular pathology. Possibly osteoarthritis, or labrum |
| Iliotibial band | Ober's Test | Patient placed in lateral position with affected side up and with hip in slight extension, abduct the leg then allow it to drop into adduction | If unable to adduct leg, suspect tight ITB |
| Gluteus medius | Trendelenburg sign | Patient to stand on both legs and alternate standing on a single leg with the goal to maintain staying upright | When standing on one leg, the pelvis drops on the side opposite to the stance leg. The muscle weakness is present on the side of the stance leg |
| | Tenderness over greater trochanter | Palpation over the greater trochanter of the femur | Tenderness that may increase with activation of the gluteus medius in hip abduction |

(continued)

**Table 19.2** (continued)

| Structure evaluated | Test | Positioning | Positive sign |
|---|---|---|---|
| Iliopsoas tendon | Snapping hip sign | Start with the affected hip in a flexed, abducted, and externally rotated position (knee is flexed for ease of testing), and passively move the hip into extension | An audible snap or palpable impulse over the inguinal region. Pain associated with this maneuver is highly suggestive of iliopsoas tendinitis or bursitis |
| | Ludloff sign | Patient sitting in a chair with affected leg elevated and knee extended. The patient elevates the heel/leg on the affected side | Pain caused by this maneuver is isolated to the iliopsoas which is the sole hip flexor activated in this position |
| Hamstring tendon | Resisted knee flexion | Palpate the muscle for a defect with the patient in a prone position and the knee flexed to 90° | Defect palpated, or pain at or near the hamstring insertion on the ischial tuberosity |

*ROM* Range of motion, *ER* External rotation, *IR* Internal rotation, *ITB* iliotibial band

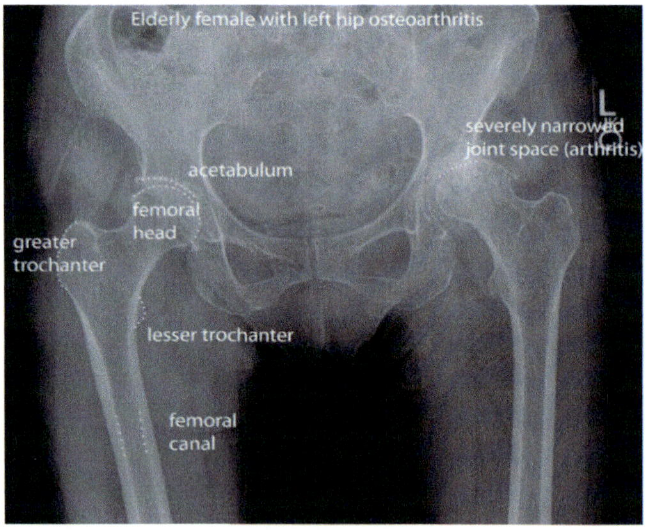


**Fig. 19.5** Radiological anatomy of the bilateral hips and pelvis demonstrate left hip osteoarthritis. Chang Chien G, Kim B, Candido KD, Manchikanti L. Hip and related joints, ligaments and tendons. In: Manchikanti L, Navani A, Atluri S (eds). *Essentials of Regenerative Medicine in Interventional Pain Management*. ASIPP Publishing, Paducah, KY, 2019, pp 257–282 [1]

joint space, and periarticular sclerosis are the findings of osteoarthritis seen on x-ray (Fig. 19.5).

Computed tomography (CT) is helpful to evaluate for fractures of the femur, pelvic ring, or sacrum in acute trauma settings. It is also a valuable tool in guiding percutaneous biopsies of tumors, preoperative evaluation for hip fractures and hip arthroplasty, and identification of intra-articular loose bodies.

MRI is the radiographic study of choice to evaluate the soft tissues of the pelvic girdle, including evaluation for tendinopathy, and labral tears in the form of MR arthrography. MRI can be used to identify suspected fractures not demonstrated by plain radiographs, for early diagnosis and accurate staging of osteonecrosis, as well as evaluation for infection and tumor. Osteoblastic activity can be localized when the MRI is performed with gadolinium.

A radionuclide bone scan is reserved for suspected fracture or osteonecrosis not demonstrated by plain film radiography or when MRI is unavailable or contraindicated. Increased activity is nonspecific and can be seen in fractures, osteonecrosis, acute and chronic arthritis, and metastatic bone lesions.

## Diagnostic Ultrasound

Diagnostic MSKUS is an indispensable tool for the interventional pain and regenerative medicine physician. It allows for the precise, safe and comfortable delivery of biologic therapies and other injection procedures. MSKUS provides real-time dynamic imaging capabilities, enabling examiners to identify dynamic impingement, ligament laxity, snapping iliopsoas or gluteal tendon, and other pathologies that are otherwise not appreciated with still images. A limitation of the US is its incomplete and limited evaluation of intra-articular pathologies, such as bone, cartilage and labrum injuries.

The sonographic examination of the hip includes anterior, lateral, and posterior evaluation. A linear array transducer with high frequency or a curvilinear transducer with low frequency can be utilized depending upon the patient body habitus.

The ultrasound evaluation of the hip joint (Fig. 19.6) typically starts anteriorly. The transducer is placed in the transverse plane at the mid-femur. It is then moved cranially until the greater and lesser trochanters are visualized. Once located, the transducer is rotated to the sagittal oblique plane parallel to the long axis of the femoral neck, distal and medial to the ASIS. It is then moved medially to identify the acetabulum, femoral head, and anterior recess. The hip joint is identified in long-axis to the femoral neck by the easily identifiable bony curvatures of the femoral head and acetabulum. The homogeneous hyperechoic fibrocartilaginous triangular structure between the acetabulum and the femoral head is the anterior acetabular labrum. Typical ultrasound appearance of the osteoarthritic joint is notable for osteophytes, an irregular contour of the femoral head, and at times, fluid within the anterior joint recess suggesting joint effusion.

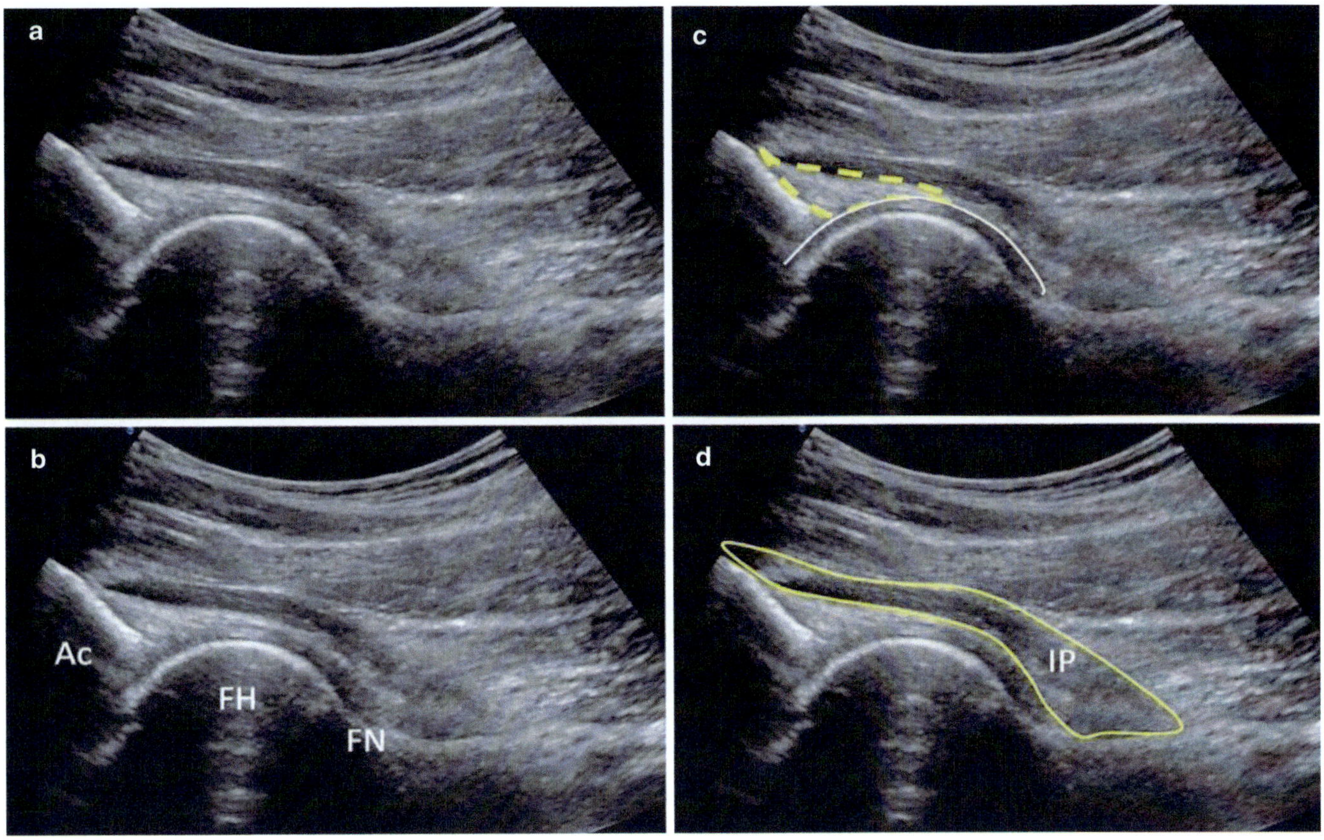


**Fig. 19.6** Ultrasound of the anterior femoral-acetabular joint in long axis view. (**a**) unlabeled, transducer oriented with left side of the image superior-medial towards the umbilicus, the right side inferior-lateral towards the greater trochanter. (**b**) Bone landmarks—*AC* Acetabulum, *FH* Femoral head, *FN* Femoral neck. (**c**) Dashed outline of the normal anterior labrum. white line demonstrating the outline over the anechoic cartilage. (**d**) Outline of the Iliopsoas tendon (IP). Chang Chien G, Kim B, Candido KD, Manchikanti L. Hip and related joints, ligaments and tendons. In: Manchikanti L, Navani A, Atluri S (eds). *Essentials of Regenerative Medicine in Interventional Pain Management*. ASIPP Publishing, Paducah, KY, 2019, pp 257–282 [1]

The iliopsoas tendon and its insertion point at the lesser trochanter can also be seen on the anterior sonographic view of the hip joint. The transducer is placed over the femoral head in the transverse plane and moved cranially to visualize the iliopsoas tendon which appears as an uniform hyperechoic density (Fig. 19.7). Tendon thickening, hypervascularity, and loss of hyperechogenicity, or heterogeneous echo signal, are signs of iliopsoas tendinopathy. To evaluate for "snapping hip syndrome", the transducer is placed over the iliopsoas tendon; then, the patient is instructed to flex and externally rotate the hip. Internal snapping of the iliopsoas tendon can be identified with the tendon snapping back and forth over the iliacus muscle, or more commonly, over the iliopectineal eminence. External snapping is typically caused by the ITB sliding over the greater trochanter. Dynamic ultrasound can be used to visualize the ITB snapping over the greater trochanter as the patient mobilizes the hip alternating between flexion and extension.

To evaluate the lateral hip, the patient is placed in a lateral decubitus position on the unaffected side. The transducer is placed longitudinally, parallel to the femoral shaft, and moved cranially to visualize the apex of the greater trochanter. The gluteus medius attaches to the posterosuperior and lateral facets. The gluteus minimus inserts into the anterior facet. The iliotibial tract is superficial to the gluteus medius tendon as a hyperechoic homogeneous band (Fig. 19.8).

The posterior hip is evaluated in a prone position where a careful assessment of the hamstring tendon, piriformis muscle, and hip abductor muscles is completed.

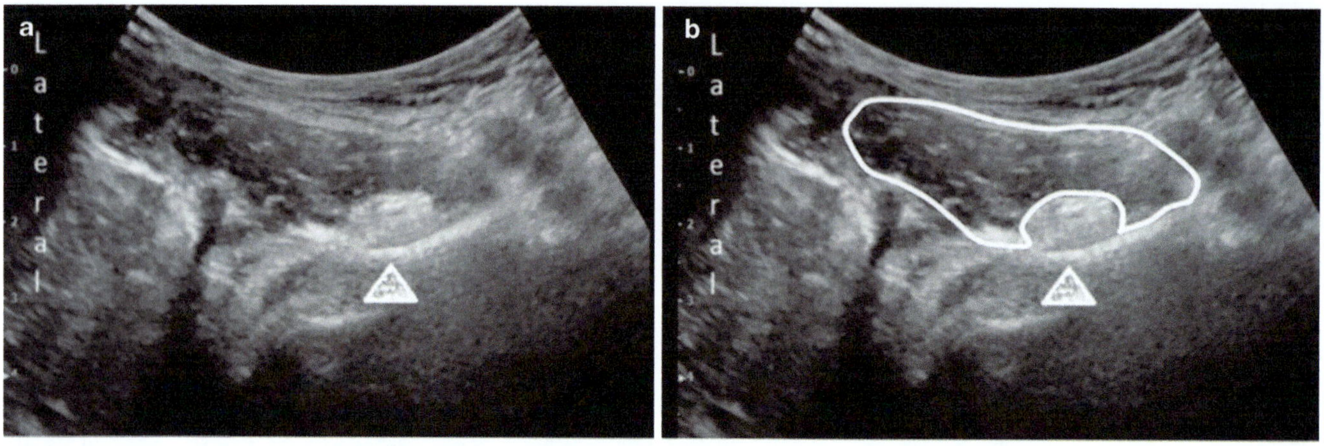


**Fig. 19.7** Ultrasound of the Iliopsoas tendon at the level of the iliopectineal line in short axis view. (**a**) Arrowhead pointing to the normal, hyperechoic psoas tendon. (**b**) The iliacus muscle (white outline) surrounds the psoas tendon. Chang Chien G, Kim B, Candido KD, Manchikanti L. Hip and related joints, ligaments and tendons. In: Manchikanti L, Navani A, Atluri S (eds). *Essentials of Regenerative Medicine in Interventional Pain Management*. ASIPP Publishing, Paducah, KY, 2019, pp 257–282 [1]

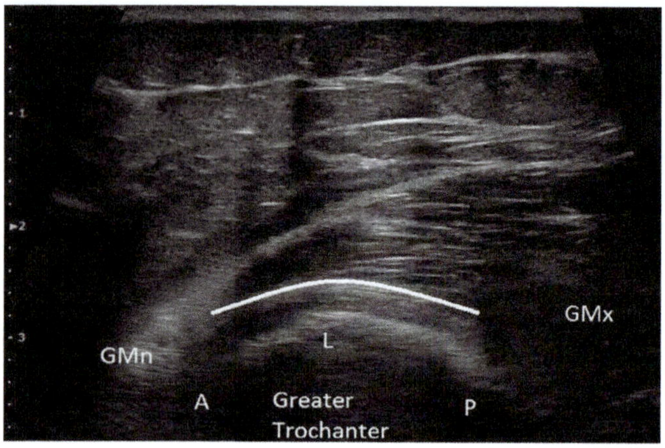


**Fig. 19.8** Ultrasound image of the greater trochanter in short axis view. Left side of the image is anterior, right side of the image is posterior. Gluteus minimus (GMn) insertion on anterior facet (A), Gluteus medius insertion on lateral facet (L), and gluteus maximus (Gmx) overlying the posterior facet (P). Chang Chien G, Kim B, Candido KD, Manchikanti L. Hip and related joints, ligaments and tendons. In: Manchikanti L, Navani A, Atluri S (eds). *Essentials of Regenerative Medicine in Interventional Pain Management*. ASIPP Publishing, Paducah, KY, 2019, pp 257–282 [1]

## Regenerative Medicine

There are currently limited studies showing the efficacy of cell-based therapies for hip conditions. However, there is increasing research on the use of regenerative medicine to manage hip pain with promising results and a good safety profile. The vast majority of the literature includes the study of platelet-rich plasma (PRP), with a fewer number of studies exploring the use of bone marrow aspiration concentrate (BMAC) and microfragmented adipose tissue (MFAT). A summary of the important clinical studies is shown in Tables 19.3 and 19.4.

### Joint OA

Osteoarthritis results from a complex interaction of biomechanical and biochemical factors leading to cartilage degeneration, synovial inflammation, and dysregulation of the bone. The cartilage has a limited regenerative capacity due to its isolation from systemic regulation and paucity of blood vessels and nerve supply. Chondrocytes cannot migrate from an intact healthy site into the site of injury, unlike other tissues [44–46]. PRP can be beneficial in treating osteoarthritis by attenuating catabolic and inflammatory events and by promoting anabolic responses through various mechanisms, primarily through the release of growth factors, namely, platelet-derived growth factors, transforming growth factor-ß, type I insulin-like growth factor, and vascular endothelial growth factor. Platelet rich plasma has been shown to inhibit chondrocyte apoptosis, stimulate bone and blood vessel remodeling, modulate inflammation, and support bone and collagen synthesis [47, 48].

In a prospective study of 40 patients with severe OA, Sanchez et al. showed that PRP significantly reduced Visual Analogue Scale (VAS), Western Ontario and McMaster Universities Osteoarthritis Index (WOMAC), and Harris Hip Scores (HHS) for up to 6 months [49]. In a randomized controlled trial (RCT) comparing intra-articular injections of PRP with HA, the PRP group had the lowest VAS scores for up to 12 months and had significantly better WOMAC scores than HA for up to 6 months [21]. Two other RCTs were recently published showing a superior effect of PRP when compared with HA. In a double-blind RCT pilot of 34 patients (36 hips), Kraeutler et al. [50] reported that leukocyte poor-PRP resulted in improved WOMAC scores and increased hip internal rotation at 6 months, with a delayed need for total hip arthroplasty when compared with low

**Table 19.3** Human clinical trials of platelet rich plasma for hip pathology

| Author/ year | Indication | Study design | Study methods | Outcome measures | Results/conclusions |
|---|---|---|---|---|---|
| **Joint osteoarthritis** | | | | | |
| Nouri et al. [51], 2022 | To compare the effectiveness of IA injections of PRP, HA, and their combination in patients with hip OA | Randomized controlled trial | 105 patients with mild to moderate hip OA were divided in 3 injection groups: PRP, HA, and PRP + HA. All groups received 2 US guided injections within a 2 weeks interval | VAS, WOMAC, and Lequesne questionnaires. | Although all three interventions were associated with improvement of pain and function, the therapeutic effect of PRP and PRP + HA injections lasted longer (6 months) and the effects of these two interventions on patients' performance, disabilities, and ADLs were superior to HA |
| Kraeutler et al. [50], 2021 | To compare efficacy of IA injections of LMW-HA and LP-PRP in patients with hip OA | Randomized controlled trial | 34 patients (36 hips) were randomized to receive 3 blinded weekly IA injections of either LP-PRP or LMW-HA | Conversion to THA or hip resurfacing, WOMAC, and hip range of motion | IA hip injections of LP-PRP resulted in improvement in WOMAC scores and internal rotation at 6 months and delayed the need for THA or resurfacing compared with treatment with LMW-HA |
| Sánchez et al. [49], 2012 | To assess the safety and symptomatic changes of IA injections of PRP in patients with unilateral severe hip OA | Prospective, non-controlled | 40 patients received 3 weekly IA joint PRP injections | HHS, WOMAC, and VAS at 6-months. | Statistically significant reductions in VAS, WOMAC, and HHS for pain and function were reported at 7 weeks and 6 months |
| Dallari et al. [21], 2016 | To compare the therapeutic efficacy of autologous PRP, HA, or a combination of both (PRP + HA) in hip OA | Prospective, double-blinded controlled randomized trial | 44 patients received PRP, 31 patients received PRP + HA, and 36 patients received HA Each group received 3 weekly US-guided IA joint injections | VAS at 2, 6, and 12 months after treatment. HHS, WOMAC, and concentration of growth factors in PRP and their correlation with clinical outcomes | PRP group had the lowest VAS scores. WOMAC scores of the PRP group was significantly better at 2-month follow-up and 6-month follow-up, but not at 12-month follow-up. The addition of PRP + HA did not lead to a significant improvement in pain symptoms |
| **Tendon tear and tendinopathy** | | | | | |
| Fitzpatrick et al. [63], 2019 | To determine whether there would be sustained long-term difference in the modified HHS at 2 years for a LR-PRP injection in the treatment of chronic gluteal tendinopathy | Double-blind randomized controlled trial | 80 patients were randomized 1:1 to receive intratendinous LR-PRP or CSI under US guidance | Modified HHS | Among patients with chronic gluteal tendinopathy and a length of symptoms >15 months, a single intratendinous LR-PRP injection resulted in greater improvement in pain and function than a single CSI. The improvement after LR-PRP was sustained at 2 years, whereas improvement from a CSI is maximal at 6 weeks and not maintained beyond 24 weeks |
| Jacobson et al. [74], 2016 | To compare US-guided percutaneous tendon fenestration to PRP injection for the treatment of greater trochanteric pain syndrome | Prospective randomized single blinded trial | 30 patients randomized to receive US-guided intratendinous PRP injections (15 patients) vs US-guided tendon fenestration (15 patients) | Mean pain score at 1 week, and 2 weeks post procedure | Both US-guided tendon fenestration and PRP injection are effective for treatment of gluteal tendinosis, with no significant difference between the treatments |

*IA* Intra-articular, *PRP* Platelet rich plasma, *HA* Hyaluronic acid, *OA* Osteoarthritis, *US* Ultrasound, *VAS* Visual Analogue Score, *WOMAC* Western Ontario and McMaster Universities Osteoarthritis Index, *ADL* Activities of Daily Living, *LMW* Low-Molecular Weight, *LP-PRP* Leukocyte Poor Platelet Rich Plasma, *LR-PRP* Leukocyte Rich Platelet Rich Plasma, *THA* Total Hip Arthroplasty, *HHS* Harris Hip Score, *CSI* Corticosteroid injection

**Table 19.4** Human clinical trials using MSC for osteonecrosis of the femoral head

| Author/year | Indication | Study design | Study methods | Outcome measures | Results/conclusions |
|---|---|---|---|---|---|
| Hernigou et al. [82], 2018 | Osteonecrosis | Prospective randomized study (30 year follow up) | 125 patients with bilateral hip osteonecrosis (stage I or II). Hip with smaller lesion was treated with decompression; contralateral hip with larger lesion was treated with BM-derived MSCs injection | HHS, WOMAC, VAS, plain radiograph, MRIs | Core decompression with BM injections had improved outcome compared to core decompression alone |
| Houdek et al. [81], 2018 | Osteonecrosis | Prospective case series | 22 patients (35 hips) with corticosteroid-induced osteonecrosis. Hip decompression was performed prior to injection of 6 cc of PRP and 12 cc of BMAC | Radiographs, MRIs, Modified Kerboul angle, HHS, conversion to THA | Core hip decompression with injection of BMAC plus PRP improved pain and function; >90% of hips were without collapse at a minimum of 2 years. Successful results were seen when nucleated cell count was high and modified Kerboul grade was low |
| Emadedin et al. [83], 2015 | Osteonecrosis of the femoral head | Prospective case series | BM-derived MSCs were isolated and cultured. Each patient received one MSC injection | VAS, WOMAC, MRI imaging at 2, 6, 12, and 30 months | All patients exhibited therapeutic benefits such as increased walking distance, decreased VAS, and total WOMAC scores |
| Gangji et al. [78], 2011 | Osteonecrosis | Prospective controlled study | 19 patients and 24 hips with early stage osteonecrosis of the femoral head received BM implantation under fluoroscopy | VAS, Lequesne index, WOMAC score at 3, 6, 12, 24, 36, 48, and 60 months | BM implantation resulted in a significant reduction in pain and reduced incidence of fractural stages. At 60 months, 8 of 11 hips in the control group had deteriorated to the fractural stage; only 3 of the 13 hips in the BM graft group progressed to that stage |
| Hernigou et al. [80], 2009 | Osteonecrosis | Prospective study | 342 patients (534 hips) with avascular osteonecrosis at early stages (stages I and II) treated with core decompression and autologous BM grafting under fluoroscopic guidance | HHS, progression and radiographic stages, change in volume (digitizing area necrosis), need for THA. Follow-up: 8–18 years | THA was necessary in 94 hips among 534 hips in stages I and II. 69 hips with stage I osteonecrosis demonstrated total resolution of osteonecrosis based on preoperative and postoperative MRI studies. For the 371 hips without collapse at 12-year follow-up, significant improvement was seen |
| Hernigou and Beaujean [77], 2002 | Osteonecrosis | Prospective study | 116 patients (184 hips) with avascular osteonecrosis underwent fluoroscopic guided core decompression and autologous BM grafting | HHS, progression and radiographic stages, need for THA Follow-up: 5–10 years | In patients operated on before collapse (stage I and stage II), THA was done in 9 of 145 hips. THA was necessary in 25 hips among the 44 hips operated on after collapse (stage III and stage IV) Patients that had a greater number of progenitor cells transplanted in their hips had better outcomes |

*BM* Bone Marrow, *MSC* Mesenchymal Stem Cell, *HHS* Harris Hip Score, *WOMAC* Western Ontario and McMaster Universities Osteoarthritis Index, *VAS* Visual Analogue Score, *MRI* magnetic resonance imaging, *BMAC* Bone Marrow Aspirate Concentrate, *THA* Total Hip Arthroplasty

molecular weight HA. Nouri et al. [51] conducted an RCT comparing intra-articular injections of HA versus PRP versus PRP combined with high molecular weight HA. When compared with HA, superior results and duration of benefit on patient performance, disability and activities of daily living (ADL) were seen with the PRP injection and with the PRP combined with HA injection. A few studies have not reported a significant difference between treatments with HA and PRP [52–54]. A meta-analysis of four RCTs comparing PRP vs HA indicated that PRP was associated with a significant reduction of VAS score at 2 months compared with HA, however it did not show significantly better outcomes at 6 and 12 months [55].

To address subchondral pathology that may not be impacted by intra-articular injections, intraosseous infiltration techniques have been developed. A pilot study of 40 patients with severe hip osteoarthritis that combined intra-articular and intraosseous infiltrations of PRP to the femoral head and acetabulum showed pain reduction and improved hip functionality for up to 12 months without any serious adverse effects [56].

There are a limited number of published clinical trials that have studied the use of culture expanded mesenchymal stem cells or medicinal signaling cells (MSCs) derived from bone marrow (BM-MSCs) or adipose tissue for the treatment of hip OA. A small number of case series reported decreased pain levels and functional improvement using BM-MSCs for hip osteoarthritis [57–59]. Two studies that used intra-articular injections of autologous MFAT achieved beneficial clinical results in treating early to moderate hip osteoarthritis [60, 61]. Adding PRP to the MFAT showed similar significant improvements in VAS and Oxford hip score compared with MFAT alone.

## Tendon Tear and Tendinopathy

Platelet rich plasma assists in tendon healing by releasing growth factors, cytokines, exosomes and other molecules that promote blood vessel formation, collagen proliferation, decreased inflammatory responses, and the recruitment of mesenchymal stem cells [62]. The use of PRP to treat tendon pathologies have shown mixed results in published studies. A double-blind RCT of 80 patients using a single intratendinous PRP injection to treat chronic gluteal tendinopathy (>15 months) showed that PRP resulted in greater improvement in pain and function than a single corticosteroid injection [63]. Improvement was sustained for up to 2 years in the PRP group, but less than 6 weeks in the corticosteroid group. A systematic review of the use of PRP to treat GTPS showed significant improvements in the first 3 months and persisted up to 12 months when PRP was injected into the gluteal tendon and/or surrounding bursa [64]. Several studies have evaluated the use of intratendinous and intramuscular hamstring PRP injections to promote soft tissue healing and accelerate return to play in athletes with mixed results [65–73]. Jacobson et al. [74, 75] showed that both US-guided tendon fenestration and PRP injection were effective for treating gluteal tendinosis.

## Femoral Head Osteonecrosis

There are three mechanisms by which orthobiologics can help treat ONFH: inducing angiogenesis and osteogenesis to accelerate bone healing, inhibiting inflammatory reactions in necrotic lesions, and preventing apoptosis induced by glucocorticoids [76]. The majority of the studies published showed positive outcomes in the use of PRP, BM-MSCs, or a combination of both [77–83]. Table 19.4 shows the clinical evidence of the application of MSCs in managing osteonecrosis of the femoral head. Many studies combined PRP or BM-MSC with core decompression procedures to remove necrotic tissue. The use of MSCs seems to be more beneficial for younger patients (<40 years) with early Association Research Circulation Osseous (ARCO) classification stage I or II. Better outcomes have been reported in post-traumatic ONFH [84, 85]. Studies have shown that MSCs could postpone the disease progression and increase the survival of the hip in the long term, decreasing the collapse of the femoral head and the need for surgery [84].

## Technical Considerations in Performing Interventional Procedures in the Hip

### Femoral-Acetabular Joint Injection

The most commonly used approach when injecting biologics into the hip joint is the anterior approach. The patient is placed in the supine position with the hip internally rotated to 15° to minimize the risk of damaging the iliopsoas tendon and neuromuscular structures. A 22 or 25 G needle with a 3.5 inch length is adequate for most hip injections.

When using ultrasound guidance, a low-frequency curvilinear probe is recommended. The transducer is positioned parallel to the long axis of the femoral neck (Fig. 19.9). Using an in-plane approach, the needle is inserted from a lateral to medial and inferior to superior trajectory until it approximates the femoral head and neck junction. Once inside the joint capsule, the injectate can be seen to flow superiorly over the head of the femur into the joint.

Using fluoroscopic guidance (Fig. 19.10), the needle is also directed to the neck of the femur towards the base of the femoral head and can also be performed between the acetabulum and the femoral head. Radiographic contrast media can be used to verify intra-articular injection, where contrast flow is seen around the femoral head and within the joint capsule. Fluoroscopy is useful when performing intraosse-

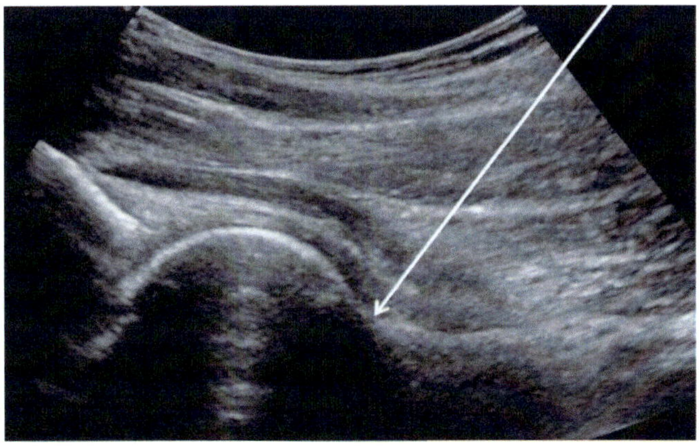

**Fig. 19.9** In-plane needle trajectory for anterior hip joint ultrasound guided injection at the junction of the femoral head and neck. Femur shown in long axis view. Cranial direction to the left and caudal direction to the right. Chang Chien G, Kim B, Candido KD, Manchikanti L. Hip and related joints, ligaments and tendons. In: Manchikanti L, Navani A, Atluri S (eds). *Essentials of Regenerative Medicine in Interventional Pain Management*. ASIPP Publishing, Paducah, KY, 2019, pp 257–282 [1]

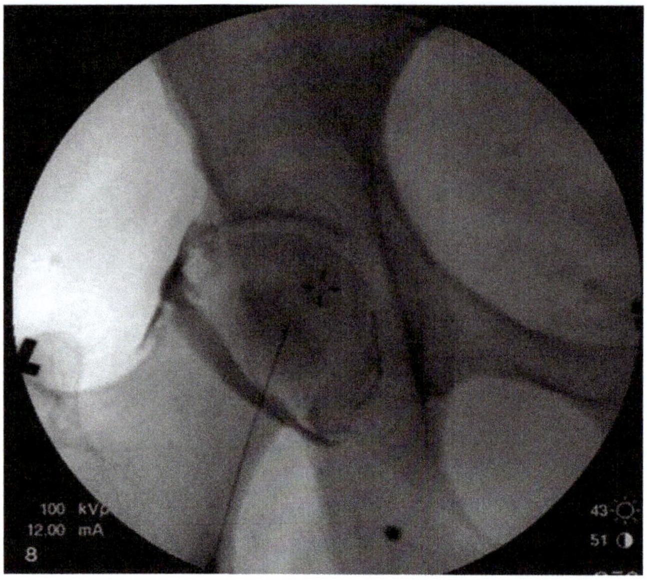


**Fig. 19.10** Fluoroscopic-guided femoral acetabular joint injection. Note the radiographic contrast within the joint capsule. Chang Chien G, Kim B, Candido KD, Manchikanti L. Hip and related joints, ligaments and tendons. In: Manchikanti L, Navani A, Atluri S (eds). *Essentials of Regenerative Medicine in Interventional Pain Management*. ASIPP Publishing, Paducah, KY, 2019, pp 257–282 [1]

ous infiltration of biologics. Fluoroscopically guided intraosseous infiltration of the subchondral bone can be performed in the acetabulum and femoral head. This technique has been combined with intra-articular injections of biologics in treating severe hip osteoarthritis and osteonecrosis. Fiz et al. [86] described an intraosseous acetabular infiltration approach by placing an 11-gauge bone biopsy trocar in the craniocaudal direction at an inclination of 20°. For infiltration of the femoral head, the trocar is introduced into the head of the femur 1 cm lateral to the sartorius muscle and 1 cm distal to the joint line at the union of the femoral neck and head. Intraosseous injections are typically performed under light conscious sedation for patient comfort.

## Tendon Injections

### Tendon Fenestration

Ultrasound-guided percutaneous fenestration has been used to treat tendinosis and tendon tears [74]. The act of perforating a ligament or tendon is termed fenestration or tenotomy, where inherent growth factors are stimulated with bleeding to promote tendon healing. The needle is passed into the target, withdrawn, and redirected multiple times [87] to promote bleeding into the injured region. The exact number of passes or re-directions necessary to achieve an optimal outcome is not known at this time [87–90]. With calcific tendinosis, fenestration is utilized to disrupt the calcium deposit and facilitate calcium reabsorption [90, 91]. Caution should be utilized to not create a larger tendon defect. A large injection into pathological tissue can lead to hydrodissection through the tissue [87–89]. Typically, 18 to 22-gauge needles are used for calcific tendinosis, while 20–25 gauge needles are used for fenestration of the interstitial tendon tissue.

### Iliopsoas Tendon and Bursa Injection

With the patient lying supine, the ultrasound transducer is placed on the transverse plane of the femoral shaft and then moved cranially to visualize the tendon in the short axis over the iliopectineal eminence. The bursa lies at the interface between the bone and the tendon. The femoral neurovascular bundle, which lies medial to the femoral head, should be avoided when performing the injection. Utilizing either an in-plane or out-of-plane approach, the needle is advanced through the iliacus muscle, with the target adjacent to the edge of the psoas tendon before touching the adjacent bone.

### Gluteus Medius Tendon Injection

The patient is positioned in a lateral decubitus position with the hips slightly flexed. A pillow may be placed between the patient's knees. The ultrasound probe is placed in the transverse plane, starting at the femoral shaft and moving cranially until the greater trochanter is located. The target site for injection is the areas of tendinopathy within the tendon, typically at its insertion point at the lateral facet of the greater trochanter (Fig. 19.11). Tendinopathy on ultrasound will show heterogenous or hypoechoic areas, fiber disruption/tears, and/or calcifications. Gentle fenestration and deposition of the biologic in small amounts is recommended.

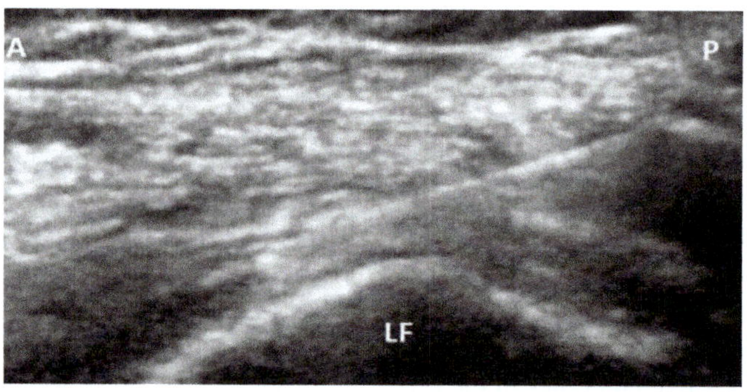


**Fig. 19.11** Ultrasound-guided In-plane approach to the gluteus medius tendon, at the insertion on the lateral facet (LF). A: anterior. P: posterior. Chang Chien G, Kim B, Candido KD, Manchikanti L. Hip and related joints, ligaments and tendons. In: Manchikanti L, Navani A, Atluri S (eds). *Essentials of Regenerative Medicine in Interventional Pain Management*. ASIPP Publishing, Paducah, KY, 2019, pp 257–282 [1]

## Conclusion

Hip pain is a common presentation to the regenerative medicine physician, requiring thoughtful evaluation and imaging studies to derive a complete and accurate diagnosis. Scientific evidence for the use of orthobiologics in the treatment of hip pathologies is limited, but promising. PRP, BMAC, MFAT and culture expanded MSCs have demonstrated the potential to heal hip injuries, providing long-lasting pain relief with a non-surgical treatment approach. However, high-quality randomized controlled trials are necessary to support the use of these innovative procedures. The lack of product characterization or standardized treatment protocols continues to be a significant limitation in our understanding of the best use of orthobiologic therapies. Proficiency in the use of fluoroscopic and ultrasound-guided interventions is vital to increase the diagnostic accuracy, safety and comfort of regenerative medicine procedures for the treatment of hip pain.

### Key Points

1. Management of hip conditions can be challenging, given the multiple sources of pathology including intra-articular and extra-articular musculoskeletal causes and other non-orthopedic conditions like neurologic, infectious, autoimmune, or genetic etiologies that can present with hip pain.
2. Hip osteoarthritis is the most prevalent cause of pain in the hip region affecting more than 240 million people worldwide causing significant disability and economic burden.
3. The complex pathogenesis of osteoarthritis involves mechanical, inflammatory, and metabolic factors, leading to structural failure of the synovial joint.
4. Musculoskeletal Ultrasound is an indispensable tool for the interventional pain and regenerative medicine physician allowing for the precise, safe and comfortable delivery of injection therapies and providing real-time dynamic imaging capabilities important for diagnosing dynamic pathologies.
5. There is increasing research on the use of regenerative medicine to manage hip pain with promising results and a good safety profile. The vast majority of the literature includes the study of platelet-rich plasma (PRP), with a fewer number of studies exploring the use of bone marrow aspiration concentrate (BMAC) and microfragmented adipose tissue (MFAT).
6. Platelet-rich plasma can be beneficial in treating osteoarthritis and tendon pathologies by releasing growth factors, cytokines, exosomes and other molecules that promote blood vessel formation, collagen proliferation, decreased inflammatory responses, recruitment of mesenchymal stem cells and immune response regulation.
7. Orthobiologics can help treat osteonecrosis of the femoral head by inducing angiogenesis and osteogenesis to accelerate bone healing, inhibiting inflammatory reactions in necrotic lesions, and preventing apoptosis induced by glucocorticoids.
8. Addressing subchondral pathology in osteoarthritis is a promising new treatment approach in treating osteoarthritis by Intraosseous infiltration of orthobiologics.
9. Scientific evidence for the use of orthobiologics in the treatment of hip pathologies is limited but promising. PRP, BMAC, MFAT, and culture expanded MSCs have demonstrated the potential to heal hip injuries, providing long-lasting pain relief with a non-surgical treatment approach.
10. Lack of product characterization and standardized treatment protocols has been a limitation in our understanding of the best use of orthobiologic therapies.

### Questions

Question 1

Which of the following hip flexor muscles also contributes to knee joint extension?

A. Iliopsoas muscle
B. Pectineus muscle
C. Rectus femoris muscle
D. Sartorius muscle

**Answer: C**

**Explanation:** The psoas and iliacus muscles originate from the T12 and L1 through L5 transverse processes and insert as the iliopsoas tendon on the lesser trochanter of the proximal femur. The pectineus muscle originates from the superior pubic ramus and inserts on the pectineal line along the superior aspect of the femur. These muscles do not cross the knee joint and, therefore, do not contribute to knee extension. The sartorius muscle originates from the anterior superior iliac spine and inserts on the medial tibia. As a result, contraction of the sartorius muscle results in hip flexion, external rotation and abduction while producing knee flexion and medial tibial rotation. The rectus femoris muscle originates from the anterior inferior iliac spine and inserts on the tibial tuberosity and patella. Contraction of the rectus femoris muscle results in hip flexion and knee extension.

Question 2

Which of the following common causes of hip pain is characterized by pain with hip flexion, adduction and internal rotation and often reveal radiographic signs of an aspherical femoral head or acetabular over-coverage?

A. Iliopsoas tendinopathy
B. Iliotibial band syndrome
C. Greater trochanteric pain syndrome (GTPS)
D. Femoroacetabular impingement (FAI) syndrome

**Answer: D**

**Explanation:** Femoroacetabular impingement (FAI) syndrome is a triad of symptoms, signs, and radiographic findings. Symptoms include positional hip or groin pain, clicking, and catching. Signs include pain with flexion, adduction, and internal rotation (positive FADIR test) and decreased internal rotation. Radiographs of the pelvis and oblique view of the hip will show abnormal anatomy of the femoral head or acetabulum. CAM morphology is characterized by an aspherical femoral head and "pincer" morphology is an over coverage of the acetabulum around the femoral head. Iliopsoas tendinopathy, iliotibial band syndrome and greater trochanteric pain syndrome (GTPS) are not typically associated with abnormal radiographic findings.

Question 3

A 70 year old female has a 5 year history of progressive right hip pain. He was assessed to have Hip Osteoarthritis (OA)after maging studies and clinical assessment by an orthopedic physician. Which if the following is true about her condition?

A. Males are commonly affected than females.
B. The complex pathogenesis of OA involves mechanical, inflammatory, and metabolic factors, leading to structural failure of the synovial joint
C. Hyaluronic acid synthesis is not affected in the inflammatory and degradative processes involved in OA.
D. Common presentation involves pain in the thigh or buttock.

**Answer: B**

**Explanation:** Hip osteoarthritis (OA) is a common disabling condition of the hip joint. The complex pathogenesis of OA involves mechanical, inflammatory, and metabolic factors, leading to structural failure of the synovial joint. Intra-articular pro-inflammatory cytokines and proteinases in OA interfere with the synthesis of hyaluronic acid (HA), resulting in a significantly decreased molecular weight HA and reduced synovial fluid viscoelasticity, which leads to degradation of the articular cartilage and disruption of the mechanical homeostasis of the joint. Hip OA most often presents with pain deep in the groin, which can radiate to the lateral or anteromedial thigh and frequently involves the unilateral hip. Hip OA can cause restricted range in all planes of motion, however, internal rotation is usually the earliest affected in both active and passive motion Major risk factors for hip OA include obesity, advanced age, history of injury, anatomical alignment, and gender, with females more commonly affected than males.

Question 4

When using image guidance for injection of the hip joint, what approach is most commonly recommended?

A. Either an ultrasound-guided or fluoroscopic-guided, anterior approach is recommended
B. Only an ultrasound-guided, anterior approach is recommended
C. Only an ultrasound-guided, posterior approach is recommended
D. Only a fluoroscopic-guided, anterior approach is recommended

**Answer: A**

**Explanation:** The use of fluoroscopic-guided and ultrasound-guided imaging for injection of the hip joint have advanced the safety, efficacy and comfort of these procedures. Each imaging modality has its own advantages. When using fluoroscopic-guidance for hip injections, the bony anatomy is well visualized and radiographic contrast can be seen within the hip joint capsule. Vascular uptake of contrast can be seen as well. When using ultrasound-guidance for hip injections, the soft tissues such as blood vessels, nerve, tendon and muscle are well visualized allowing the physician the ability to avoid inadvertent injection or injury of these structures. The use of both imaging modalities have been reported to result in the accurate and safe injection of the hip joint.

Question 5

A 55 year-old woman presents with progressive hip pain for more than 6 months. She notes progressive difficulty with walking and climbing stairs. A supervised weight loss program, hip mobilization and strengthening exercises and oral anti-inflammatory medications have not improved her condition. Physical examination reveals an antalgic gait and pain with hip flexion and internal rotation with normal hip range of motion in all planes and normal strength. There is no tenderness to palpation over the anterior or lateral hip. Radiographs of the hip reveal mild joint space narrowing, with small acetabular osteophytes, but no other bony lesions.

According to the published scientific literature, which of the following statements is most accurate regarding the use of platelet rich plasma (PRP) for this patient.

A. There are no published randomized clinical trials on the safety and efficacy of platelet rich plasma for the treatment of mild to moderate hip osteoarthritis.
B. There are no published randomized clinical trials on the safety and efficacy of platelet rich plasma for the treatment of mild to moderate gluteal tendinopathy.
C. Every published randomized clinical trial on the safety and efficacy of platelet rich plasma for the treatment of mild to moderate hip osteoarthritis has demonstrated superiority over hyaluronic acid injections.
D. Randomized clinical trials have demonstrated superiority of platelet rich plasma over hyaluronic acid injections and a delayed need for total hip arthroplasty.

**Answer D**

**Explanation:** Hip osteoarthritis is a common form of hip pain presenting in middle aged adults. Mild to moderate hip osteoarthritis is characterized by progressive hip pain with ambulation, and pain with hip flexion and internal rotation on physical examination. Radiographs often will reveal joint space narrowing and acetabular osteophytes without bony lesions. Randomized clinical trials on the use of platelet rich plasma (PRP) for hip osteoarthritis have demonstrated safety and efficacy. Krautler, et al. showed that leukocyte poor PRP resulted in improved WOMAC scores and improved hip internal rotation at 6 months with a delayed need for total hip arthroplasty when compared with low molecular weight hyaluronic acid.

**Acknowledgments** This book chapter is modified and updated from a previous book chapter, "Hip and Related Joints, Ligaments, and Tendons" by George Chang Chien, DO, Baruch Kim, DO, Kenneth D. Candido, MD, and Laxmaiah Manchikanti, MD, in *Essentials of Regenerative Medicine in Interventional Pain Management* published by ASIPP Publishing.

## References

1. Chang Chien G, Kim B, Candido KD, Manchikanti L. Hip and related joints, ligaments and tendons. In: Manchikanti L, Navani A, Atluri S, editors. Essentials of regenerative medicine in interventional pain management. Paducah: ASIPP Publishing; 2019. p. 257–82.
2. Jordan JM, Helmick CG, Renner JB, et al. Prevalence of hip symptoms and radiographic and symptomatic hip osteoarthritis in African Americans and Caucasians: the Johnston County Osteoarthritis Project. J Rheumatol. 2009;36:809–15.
3. Leifer VP, Katz JN, Losina E. The burden of OA-health services and economics. Osteoarthr Cartil. 2022;30:10–6.
4. Tibor LM, Sekiya JK. Differential diagnosis of pain around the hip joint. Arthroscopy. 2008;24:1407–21.
5. Fricka KB, Gortz S, Bugbee WD. The hip. In: Hochberg MC, Silman AJ, Smolen JS, Weinblatt ME, Weisman MH, editors. Rheumatology. 4th ed. New York: Mosby Publishing; 2008. p. 653–62.
6. Katz JN, Arant KR, Loeser RF. Diagnosis and treatment of hip and knee osteoarthritis: a review. JAMA. 2021;325:568–78.
7. Giordano BD. Comparison of two injection techniques for intra-articular hip injections. J Ultrasound Med. 2016;35:1259–67.
8. Clementi D, D'Ambrosi R, Bertocco P, et al. Efficacy of a single intra-articular injection of ultra-high molecular weight hyaluronic acid for hip osteoarthritis: a randomized controlled study. Eur J Orthop Surg Traumatol. 2018;28:915–22.
9. Wu B, Li YM, Liu YC. Efficacy of intra-articular hyaluronic acid injections in hip osteoarthritis: a meta-analysis of randomized controlled trials. Oncotarget. 2017;8:86865–76.
10. Sperati A, Picconi O, Tancioni V, et al. Outcomes of hip replacement: a hospital-based longitudinal study in Lazio region. Ann Ig. 2008;20:141–57.
11. Evans JT, Evans JP, Walker RW, et al. How long does a hip replacement last? A systematic review and meta-analysis of case series and national registry reports with more than 15 years of follow-up. Lancet. 2019;393:647–54.
12. Fice MP, Miller JC, Christian R, et al. The role of platelet-rich plasma in cartilage pathology: an updated systematic review of the basic science evidence. Arthroscopy. 2019;35:961–76.
13. Kragsnaes MS, Fredberg U, Stribolt K, et al. Stereological quantification of immune-competent cells in baseline biopsy specimens from achilles tendons: results from patients with chronic tendinopathy followed for more than 4 years. Am J Sports Med. 2014;42:2435–45.
14. Simon MJK, Yeoh J, Nevin J, et al. Histopathology of long head of biceps tendon removed during tenodesis demonstrates degenerative histopathology and not inflammatory changes. BMC Musculoskelet Disord. 2022;23:185.
15. Murrell G. Understanding tendinopathies. Br J Sports Med. 2002;36:392–3.
16. Gold M, Munjal A, Varacallo M. Anatomy, bony pelvis and lower limb, hip joint. 2022 Jul 25. In: StatPearls [Internet]. Treasure Island: StatPearls Publishing; 2023.
17. Simons MJ, Amin NH, Cushner FD, et al. Characterization of the neural anatomy in the hip joint to optimize periarticular regional anesthesia in total hip arthroplasty. J Surg Orthop Adv. 2014;24:221–4.
18. Amin NH, West JA, Farmer T, et al. Nerve blocks in the geriatric patient with hip fracture: a review of the current literature and relevant neuroanatomy. Geriatr Orthop Surg Rehabil. 2017;8:268–75.
19. Fu M, Zhou H, Li Y, et al. Global, regional, and national burdens of hip osteoarthritis from 1990 to 2019: estimates from the 2019 Global Burden of Disease Study. Arthritis Res Ther. 2022;24:8.
20. Hunter DJ, Bierma-Zeinstra S. Osteoarthritis. Lancet. 2019;393:1745–59.
21. Dallari D, Stagni C, Rani N, et al. Ultrasound-guided injection of platelet-rich plasma and hyaluronic acid, separately and in combination, for hip osteoarthritis: a randomized controlled study. Am J Sports Med. 2016;44:664–71.
22. Hunter DJ. Viscosupplementation for osteoarthritis of the knee. N Engl J Med. 2015;372:1040–7.
23. Chaganti RK, Lane NE. Risk factors for incident osteoarthritis of the hip and knee. Curr Rev Musculoskelet Med. 2011;4:99–104.
24. Prieto-Alhambra D, Judge A, Javaid MK, et al. Incidence and risk factors for clinically diagnosed knee, hip and hand osteoarthritis: influences of age, gender and osteoarthritis affecting other joints. Ann Rheum Dis. 2014;73:1659–64.
25. Brossi PM, Moreira JJ, Machado TS, et al. Platelet-rich plasma in orthopedic therapy: a comparative systematic review of clinical and experimental data in equine and human musculoskeletal lesions. BMC Vet Res. 2015;11:98.
26. Neumann G, Mendicuti AD, Zou KH, et al. Prevalence of labral tears and cartilage loss in patients with mechanical symptoms of the hip: evaluation using MR arthrography. Osteoarthr Cartil. 2007;15:909–17.
27. Freehill MT, Safran MR. The labrum of the hip: diagnosis and rationale for surgical correction. Clin Sports Med. 2011;30:293.
28. Pun S, Kumar D, Lane NE. Femoroacetabular impingement. Arthritis Rheumatol. 2015;67:17–27.
29. Menge TJ, Truex NW. Femoroacetabular impingement: a common cause of hip pain. Phys Sportsmed. 2018;46:139–44.
30. Trigg SD, Schroeder JD, Hulsopple C. Femoroacetabular Impingement Syndrome. Curr Sports Med Rep. 2020;19:360–6.
31. Mont MA, Salem HS, Piuzzi NS, et al. Nontraumatic osteonecrosis of the femoral head: where do we stand today?: a 5-year update. J Bone Joint Surg Am. 2020;102:1084–99.
32. Hernigou P, Guerin G, Homma Y, et al. History of concentrated or expanded mesenchymal stem cells for hip osteonecrosis: is there a target number for osteonecrosis repair? Int Orthop. 2018;42:1739–45.

33. Fearon AM, Scarvell JM, Neeman T, et al. Greater trochanteric pain syndrome: defining the clinical syndrome. Br J Sports Med. 2013;47:649.
34. Yen Y-M, Lewis CL, Kim Y-J. Understanding and treating the snapping hip. Sports Med Arthrosc Rev. 2015;23:194–9.
35. Deslandes M, Guillin R, Cardinal E, et al. The snapping iliopsoas tendon: new mechanisms using dynamic sonography. AJR Am J Roentgenol. 2008;190:576–81.
36. Hasan BA. The presenting symptoms, differential diagnosis, and physical examination of patients presenting with hip pain. Dis Mon. 2012;58:477–91.
37. Heiderscheit B, McClinton S. Evaluation and management of hip and pelvis injuries. Phys Med Rehabil Clin N Am. 2016;27:1–29.
38. Fairclough J, Hayashi K, Toumi H, et al. The functional anatomy of the iliotibial band during flexion and extension of the knee: implications for understanding iliotibial band syndrome. J Anat. 2006;208:309–16.
39. Sher I, Umans H, Downie S, et al. Proximal iliotibial band syndrome: what is it and where is it? Skelet Radiol. 2011;40:1553–6.
40. Martin HD, Kelly BT, Leunig M, et al. The pattern and technique in the clinical evaluation of the adult hip: the common physical examination tests of hip specialists. Arthroscopy. 2010;26:161–72.
41. Wilson JJ, Furukawa M. Evaluation of the patient with hip pain. Am Fam Physician. 2014;89:27–34.
42. Maslowski E, Sullivan W, Harwood JF, et al. The diagnostic validity of hip provocation maneuvers to detect intra-articular hip pathology. PM R. 2010;2:174–81.
43. Chamberlain R. Hip pain in adults: evaluation and differential diagnosis. Am Fam Physician. 2021;103:81–9.
44. Kon E, Mandelbaum B, Buda R, et al. Platelet-rich plasma intra-articular injection versus hyaluronic acid viscosupplementation as treatments for cartilage pathology: from early degeneration to osteoarthritis. Arthroscopy. 2011;27:1490–501.
45. Buckwalter JA, Brown TD. Joint injury, repair, and remodeling: roles in post-traumatic osteoarthritis. Clin Orthop Relat Res. 2004;423:7–16.
46. Ochi M, Uchio Y, Kawasaki K, et al. Transplantation of cartilage-like tissue made by tissue engineering in the treatment of cartilage defects of the knee. J Bone Joint Surg Br. 2002;84:571–8.
47. Bennell KL, Hunter DJ, Paterson KL. Platelet-rich plasma for the management of hip and knee osteoarthritis. Curr Rheumatol Rep. 2017;19:24.
48. Sundman EA, Cole BJ, Karas V, et al. The anti-inflammatory and matrix restorative mechanisms of platelet-rich plasma in osteoarthritis. Am J Sports Med. 2014;42:35–41.
49. Sánchez M, Guadilla J, Fiz N, et al. Ultrasound-guided platelet-rich plasma injections for the treatment of osteoarthritis of the hip. Rheumatology (Oxford). 2012;51:144–50.
50. Kraeutler MJ, Houck DA, Garabekyan T, et al. Comparing intra-articular injections of leukocyte-poor platelet-rich plasma versus low-molecular weight hyaluronic acid for the treatment of symptomatic osteoarthritis of the hip: a double-blind, randomized pilot study. Orthop J Sports Med. 2021;9:2325967120969210.
51. Nouri F, Babaee M, Peydayesh P, et al. Comparison between the effects of ultrasound guided intra-articular injections of platelet-rich plasma (PRP), high molecular weight hyaluronic acid, and their combination in hip osteoarthritis: a randomized clinical trial. BMC Musculoskelet Disord. 2022;23:856.
52. Battaglia M, Guaraldi F, Vannini F, et al. Efficacy of ultrasound-guided intra-articular injections of platelet-rich plasma versus hyaluronic acid for hip osteoarthritis. Orthopedics. 2013;36:e1501–8.
53. Di Sante L, Villani C, Santilli V, et al. Intra-articular hyaluronic acid vs platelet-rich plasma in the treatment of hip osteoarthritis. Med Ultrason. 2016;18:463–8.
54. Doria C, Mosele GR, Caggiari G, et al. Treatment of early hip osteoarthritis: ultrasound-guided platelet rich plasma versus hyaluronic acid injections in a randomized clinical trial. Joints. 2017;5:152–5.
55. Ye Y, Zhou X, Mao S, et al. Platelet rich plasma versus hyaluronic acid in patients with hip osteoarthritis: a meta-analysis of randomized controlled trials. Int J Surg. 2018;53:279–87.
56. Fiz N, Delgado D, Garate A, et al. Intraosseous infiltrations of platelet-rich plasma for severe hip osteoarthritis: a pilot study. J Clin Orthop Trauma. 2020;11:S585–90.
57. Darrow M, Shaw B, Darrow B, et al. Short-term outcomes of treatment of hip osteoarthritis with 4 bone marrow concentrate injections: a case series. Clin Med Insights Case Rep. 2018;11:1179547618791574.
58. Mardones R, Jofré CM, Tobar L, et al. Mesenchymal stem cell therapy in the treatment of hip osteoarthritis. J Hip Preserv Surg. 2017;4:159–63.
59. Rodriguez-Fontan F, Piuzzi NS, Kraeutler MJ, et al. Early clinical outcomes of intra-articular injections of bone marrow aspirate concentrate for the treatment of early osteoarthritis of the hip and knee: a cohort study. PM R. 2018;10:1353–9.
60. Dall'Oca C, Breda S, Elena N, et al. Mesenchymal stem cells injection in hip osteoarthritis: preliminary results. Acta Biomed. 2019;90:75–80.
61. Natali S, Screpis D, Romeo M, et al. Is intra-articular injection of autologous micro-fragmented adipose tissue effective in hip osteoarthritis? A three-year follow-up. Int Orthop. 2023;47(6):1487–92.
62. Ziegler CG, Van Sloun R, Gonzalez S, et al. characterization of growth factors, cytokines, and chemokines in bone marrow concentrate and platelet-rich plasma: a prospective analysis. Am J Sports Med. 2019;47:2174–87.
63. Fitzpatrick J, Bulsara MK, O'Donnell J, et al. Leucocyte-rich platelet-rich plasma treatment of gluteus medius and minimus tendinopathy: a double-blind randomized controlled trial with 2-year follow-up. Am J Sports Med. 2019;47:1130–7.
64. Ali M, Oderuth E, Atchia I, et al. The use of platelet-rich plasma in the treatment of greater trochanteric pain syndrome: a systematic literature review. J Hip Preserv Surg. 2018;5:209–19.
65. Stasinopoulos D, Manias P. Comparing two eccentric exercise programs for the management of Achilles tendonopathy: a pilot trial. J Bodyw Mov Ther. 2013;17:309–15.
66. Guillodo Y, Madouas G, Simon T, et al. Platelet-rich plasma (PRP) treatment of sports-related severe acute hamstring injuries. Muscles Ligaments Tendons J. 2016;5:284–48.
67. Davenport KL, Campos JS, Nguyen J, et al. Ultrasound-guided intratendinous injections with platelet-rich plasma or autologous whole blood for treatment of proximal hamstring tendinopathy: a double-blind randomized controlled trial. J Ultrasound Med. 2015;34:1455–63.
68. A Hamid MS, Mohamed Ali MR, Yusof A, et al. Platelet-rich plasma injections for the treatment of hamstring injuries: a randomized controlled trial. Am J Sports Med. 2014;42:2410–8.
69. Hamilton B, Tol JL, Almusa E, et al. Platelet-rich plasma does not enhance return to play in hamstring injuries: a randomized controlled trial. Br J Sports Med. 2015;49:943–50.
70. Fader RR, Mitchell JJ, Traub S, et al. Platelet-rich plasma treatment improves outcomes for chronic proximal hamstring injuries in an athletic population. Muscles Ligaments Tendons J. 2015;4:461–6.
71. Reurink G, Goudswaard GJ, Moen MH, et al. Dutch HIT-study Investigators. Rationale, secondary outcome scores and 1-year follow-up of a randomised trial of platelet-rich plasma injections in acute hamstring muscle injury: the Dutch Hamstring Injection Therapy study. Br J Sports Med. 2015;49:1206–12.

72. Zanon G, Combi F, Combi A, et al. Platelet-rich plasma in the treatment of acute hamstring injuries in professional football players. Joints. 2016;4:17–23.
73. Lee JJ, Harrison JR, Boachie-Adjei K, et al. Platelet-rich plasma injections with needle tenotomy for gluteus medius tendinopathy: a registry study with prospective follow-up. Orthop J Sports Med. 2016;4:2325967116671692.
74. Jacobson JA, Yablon CM, Henning PT, et al. Greater trochanteric pain syndrome: percutaneous tendon fenestration versus platelet-rich plasma injection for treatment of gluteal tendinosis. J Ultrasound Med. 2016;35:2413–20.
75. Jacobson JA, Rubin J, Yablon CM, et al. Ultrasound-guided fenestration of tendons about the hip and pelvis: clinical outcomes. J Ultrasound Med. 2015;34:2029–35.
76. Han J, Gao F, Li Y, et al. The use of platelet-rich plasma for the treatment of osteonecrosis of the femoral head: a systematic review. Biomed Res Int. 2020;2020:2642439.
77. Hernigou P, Beaujean F. Treatment of osteonecrosis with autologous bone marrow grafting. Clin Orthop Relat Res. 2002;405:14–23.
78. Gangji V, De Maertelaer V, Hauzeur JP. Autologous bone marrow cell implantation in the treatment of non-traumatic osteonecrosis of the femoral head: five-year follow-up of a prospective controlled study. Bone. 2011;49:1005–9.
79. Rastogi S, Sankineani SR, Nag HL, et al. Intralesional autologous mesenchymal stem cells in management of osteonecrosis of femur: a preliminary study. Musculoskelet Surg. 2013;97:223–8.
80. Hernigou P, Poignard A, Zilber S, et al. Cell therapy of hip osteonecrosis with autologous bone marrow grafting. Indian J Orthop. 2009;43:40–5.
81. Houdek MT, Wyles CC, Collins MS, et al. Stem cells combined with platelet-rich plasma effectively treat corticosteroid-induced osteonecrosis of the hip: a prospective study. Clin Orthop Relat Res. 2018;476:388–97.
82. Hernigou P, Dubory A, Homma Y, et al. Cell therapy versus simultaneous contralateral decompression in symptomatic corticosteroid osteonecrosis: a thirty-year follow-up prospective randomized study of one hundred and twenty five adult patients. Int Orthop. 2018;42:1639–49.
83. Emadedin M, Ghorbani Liastani M, Fazeli R, et al. Long-Term follow-up of intra-articular injection of autologous mesenchymal stem cells in patients with knee, ankle, or hip osteoarthritis. Arch Iran Med. 2015;18:336–44.
84. Mao L, Jiang P, Lei X, et al. Efficacy and safety of stem cell therapy for the early-stage osteonecrosis of femoral head: a systematic review and meta-analysis of randomized controlled trials. Stem Cell Res Ther. 2020;11:445.
85. Xu Y, Jiang Y, Xia C, et al. Stem cell therapy for osteonecrosis of femoral head: opportunities and challenges. Regen Ther. 2020;15:295–304.
86. Fiz N, Pérez JC, Guadilla J, et al. Intraosseous infiltration of platelet-rich plasma for severe hip osteoarthritis. Arthrosc Tech. 2017;6:e821–5.
87. Wilson JJ, Lee KS, Chamberlain C, et al. Intratendinous injections of platelet-rich plasma: feasibility and effect on tendon morphology and mechanics. J Exp Orthop. 2015;2:5.
88. Wiegerinck JI, Reilingh ML, de Jonge MC, et al. Injection techniques of platelet-rich plasma into and around the Achilles tendon: a cadaveric study. Am J Sports Med. 2011;39:1681–6.
89. Loftus ML, Endo Y, Adler RS. Retrospective analysis of postinjection ultrasound imaging after platelet-rich plasma or autologous blood: observational review of anatomic distribution of injected material. AJR Am J Roentgenol. 2012;199:W501–5.
90. Klontzas ME, Vassalou EE, Karantanas AH. Calcific tendinopathy of the shoulder with intraosseous extension: outcomes of ultrasound-guided percutaneous irrigation. Skelet Radiol. 2017;46:201–8.
91. Jo H, Kim G, Baek S, et al. Calcific tendinopathy of the gluteus medius mimicking lumbar radicular pain successfully treated with barbotage: a case report. Ann Rehabil Med. 2016;40:368–72.

# Regenerative Medicine for Knee Joint Disorders

# 20

Sairam Atluri and Navneet Boddu

## Introduction

The knee joint is the largest joint in the body and is most commonly injured as it takes a direct brunt of the body weight and is subjected to unusual torsion forces that may predispose the knee to early wearing out causing arthritis [1, 2]. Knee injuries may cause fracture of bones, damage to articular cartilage or more frequently cause damage to ligaments and/or meniscus. The knee ligaments involved in the injuries are anterior and/or posterior cruciate ligaments, medial and/or lateral collateral ligaments. The meniscus injuries may be due to medial and/or lateral meniscus. In addition, there are also several systemic diseases which cause disorders or degeneration of the knee joint and its structures.

Due to poor vascular supply and low cellularity of the tendons, ligaments, meniscus and articular cartilage of the knee joint the damage to these structures results in low healing potential [3]. In addition, the edema of the inflamed tissue and effusion increases pressure in the joint space, further reducing the blood supply. This combined with age related degeneration exacerbates the instability that further damages the joint. The articular cartilage of the knee joint is avascular with limited ability to repair and self-regenerate. Defects in the articular cartilage can cause debilitating degenerative joint disease. Lately numerous studies have shown that this poor healing of these structures can be augmented by using biological agents that provide nutrients and growth factors to improve vascular supply and induce cell growth, thus improving the quality of the healing in these structures [4]. Biological agents and stem cells have also been shown to improve and rebuild the damaged knee articular cartilage.

S. Atluri (✉)
Tri-State Spine Care Institute, Cincinnati, OH, USA

N. Boddu
Advanced Pain and Regenerative Specialists, Oceanside, CA, USA

## Treatment of Anterior Cruciate Ligament (ACL) with Biologics

Anterior cruciate ligament (ACL) originates on the posteromedial wall of the lateral femoral condyle, it courses anteriorly, inferiorly and medially and inserts into the intercondylar tibial eminence [5, 6]. The anteromedial bundle (AM) becomes tight in flexion and the posterolateral bundle (PL) becomes tight at extension (Fig. 20.1). *Even though this ligament is intra-articular because it is covered by synovium, they are considered extrasynovial* [7]. *This is important because intra-articular applications of biologics will not reach the ACL and in order to repair ACL tears, it has to be separately injected.* The main function of ACL is to prevent the tibia from sliding forward relative to the femur. The ACL also assists with preventing excessive knee extension, knee varus and valgus movements, and tibial rotation.

## Diagnosis of ACL Injury

The patient with an acute ACL tear typically presents with pain, knee effusion, hemarthrosis [6], reduction of knee motion and difficulty bearing weight. Often a "pop" is heard or felt by the athlete at the time of injury. The patient may not be able to fully flex the knee, but loss of hyperextension is more indicative of an ACL disruption [8]. Magnetic resonance imaging (MRI) is the primary study used to diagnose ACL injury in the United States with sensitivity of 87–94% and accuracy of 91–95% [4]. Partial ACL tears are more difficult to diagnose on MRI than complete tears because of various tear patterns. ACL injuries are graded as follows [9]:

- Grade 1 Sprain: the ligament is partially torn, with less than half of the ligament substance disrupted.
- Grade 2 sprain: the ligament is partially torn, with more than half of the ligament substance disrupted.
- Grade 3 sprain: the ligament is completely torn.


A. Navani et al. (eds.), *Essentials of Regenerative Medicine in Interventional Pain Management*,
https://doi.org/10.1007/978-3-031-50357-3_20

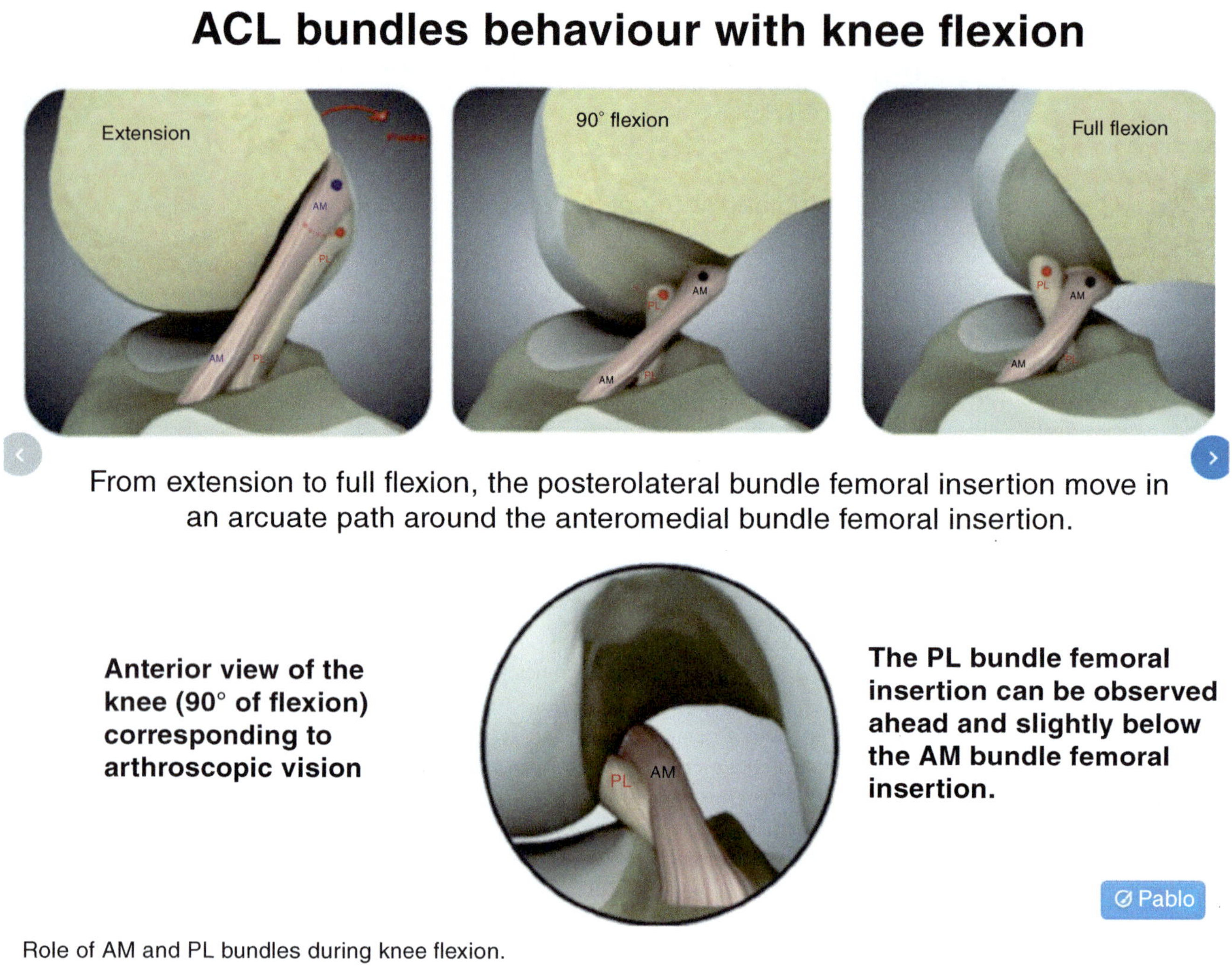


**Fig. 20.1** Movement of the ACL and PCL bundles during flexion and extension of the knee. **Source:** https://www.sciencedirect.com/science/article/pii/S1877056815003515?via%3Dihub

## Treatment

### Nonsurgical

Initial non-surgical treatment includes rehabilitation programs aimed at improving strength and balance along with bracing. With the conservative approach used in isolated ACL injury, a study showed only 31% of patients had excellent to good results upon 38 months follow up. 35% of patients in this study eventually required surgical reconstruction [10].

## Interventional Treatments

### Percutaneous Bone Marrow Concentrate (BMC) Infiltration to Treat ACL Injuries

Following ACL injury, the inability of ACL to regenerate itself makes interventions inevitable [11], which underscores the importance of regenerative biologics. Centeno et al. [12] reported a case series in which ten patients with grade 1, 2, or 3 ACL tears with less than 1 cm retraction in the subacute phase were treated with fluoroscopic percutaneous ligament infiltration and intra articular injection with autologous bone marrow concentrate (BMC) (average total nucleated count of 694 in 2.7 mL), platelet-rich plasma (PRP) and platelet lysate. The median time of follow up was 6 months. Most patients reported improvement in pain and functional scores. The majority of patients exhibited positive changes in the follow up MRI's reflecting regeneration of the ACL. The authors concluded that although more research is required, percutaneous BMC treatment of ACL injuries may be promising [12].

This group also published a second study of 23 patients with similar ACL injuries and the same treatment protocol as described above. This time, the average total nucleated cell count (TNCC) from the BMC was 690 million and the average volume of the injectate used for treatment was 3.7 mL. The mean time from injury to treatment was 22.6 weeks. Decreases in pain scores and improvements in functional scores were noted during an average follow-up time of 23 months. When MRI was repeated after an average duration of 8.7 months, positive changes in the ACL were seen, suggesting regeneration. Only 17% of patients eventually required surgery [13].

## Procedure for Percutaneous Infiltration of ACL

ACL is preferably approached under fluoroscopic guidance as it is difficult to visualize this ligament under ultrasound. Extreme flexion of the knee is required for ultrasound visualization of the ACL which most patients cannot tolerate.

Under fluoroscopic guidance the origin, midportion and the insertion sites of ACL can be targeted. The patient is in supine position with knee flexed. Flexion opens up the intercondylar fossa. The area is prepped with antiseptic solution and draped in a sterile fashion. Anteroposterior (AP) fluoroscopic views are taken and C-arm cranio-caudal angulation is done to maximize the opening of the intercondylar space (Fig. 20.2). The target is the medial eminence of the two tibial eminences. A 25 g, 3.5-inch needle is used to contact the target using the "hubshot" technique (Fig. 20.2). In the "hubshot" approach, the needle is placed in line with the fluoroscopy beam and in an ideal situation, mainly the hub is seen, and minimal needle shaft is seen in the fluoroscopic image. After contact is made with the medial tibial eminence, the needle is pulled back a mm and contrast is injected. Contrast spread outlining the ligament should be seen (Fig. 20.3). If ideal spread is not achieved, the needle is pulled back and contrast re-injected until appropriate dye spread is seen. If not, the needle has to be placed either slightly medially or laterally from the original target site until satisfactory dye spread is noted. Usually this will suffice, since most of the tears are midsubstance of the ligament, but if needed, the mid and superior part of the ligament can be sought. 2–3 mL of the biologic should be injected. Although separate AM and PL infiltration is desired (most tears involve both bundles), as these two bundles are covered by one synovial fold (Fig. 20.4), separate infiltration may not be required especially for stem cells as they are known to seek the inflamed tissue by the process known as "homing".

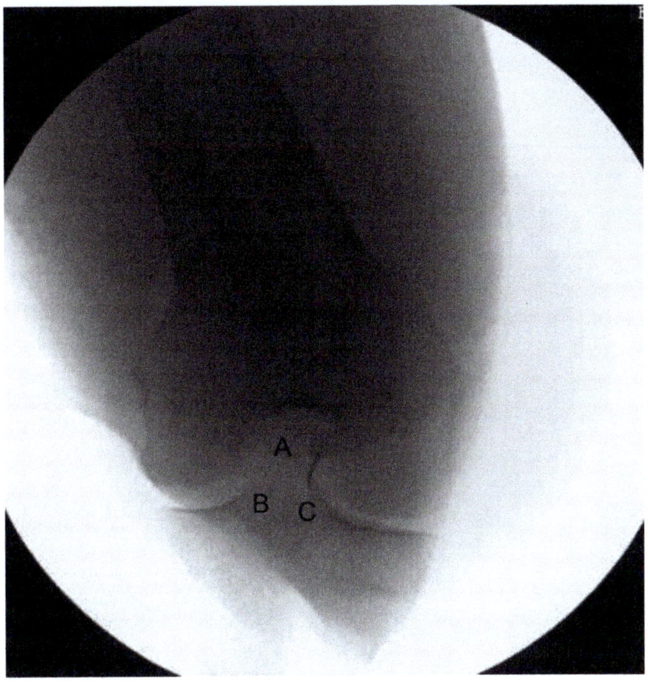


**Fig. 20.2** Anterior cruciate ligament injection under fluoroscopy. (A) Intercondylar notch maximized by knee flexion. (B) Lateral Tibial Eminence. (C) Medial Tibial Eminence. Needle contacting the medial tibial eminence. **Source:** Author

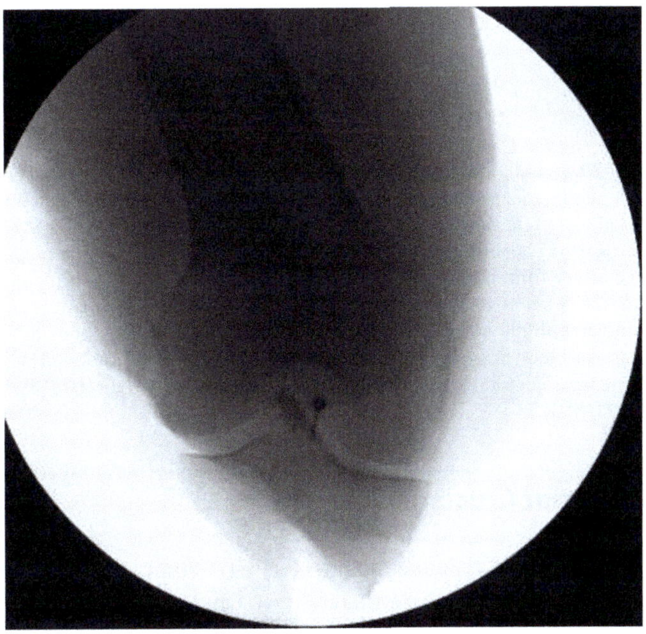

**Fig. 20.3** Needle pulled back slightly after contacting the medial tibial eminence and dye injected. Picture shows dye spread along the ACL. **Source:** Author

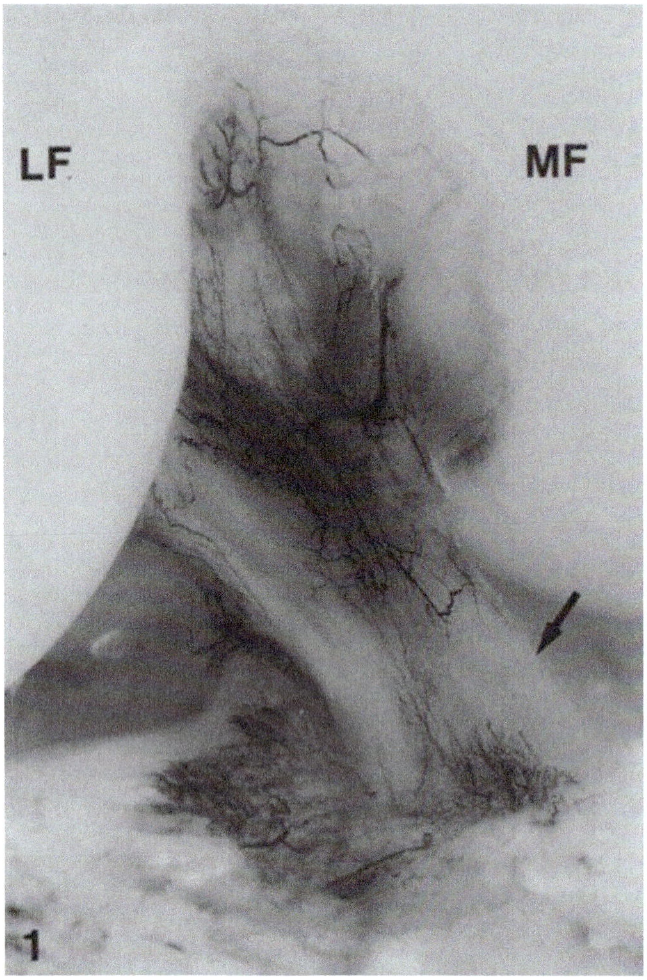


**Fig. 20.4** India ink injection. The major blood supply of the anterior cruciate ligament (ACL) arises from the middle geniculate artery. The distal part of the ACL is vascularized by branches of the inferior geniculate artery. The terminal branches of the middle and the inferior geniculate artery form a periligamentous network. In the anterior part of the ACL, 0.5 cm proximal to the tibial insertion where the ligament is in contact with the notch in full knee extension, the periligamentous network of blood vessels is absent (arrow). *MF* medial femoral condyle, *LF* lateral femoral condyle. https://link.springer.com/article/10.1007/BF02488915

## Posterior Cruciate Ligament (PCL)

The posterior cruciate ligament (PCL) complex is composed of the larger anterolateral (AL) and smaller posteromedial (PM) bundles and the meniscofemoral ligaments. Like ACL, the PCL is an intra-articular extrasynovial ligament. This is important because intra-articular applications of biologics will not reach the PCL and in order to repair PCL tears, it has to be separately injected. PCL is an important restraint of posterior tibial translation relation to the femur. In addition, PCL acts as a secondary restraint to resist varus, valgus and external rotation movements of the knee.

### PCL Injury

Only 3% of knee injuries have PCL disruption. PCL injuries are commonly produced as a result of high-energy trauma, such as sports and motor vehicle accidents.

### Diagnosis of PCL Injury

Physical examination of the acute conditions of the knee can be difficult due to pain and guarding [14]. An effusion is usually present after an acute injury. The majority of physical examination maneuvers attempt to demonstrate posterior instability and subluxation of the knee joint. The most frequently used tests are the posterior sag sign (patient is supine with knees flexed 90° and hips flexed approximately 45°, comparison prominence of tibial tuberosity bilaterally is done, the Godfrey's test, dial test (with the patient in prone or supine, an external rotation force is applied to both feet with knee at 30° and then 90° of flexion. A side-to-side difference of 10° or more is considered abnormal. The quadriceps active test, the patient lies supine, and the knee is flexed to 90°. While the examiner stabilizes the foot, the patient is asked to contract quadriceps isometrically.

MRI is highly accurate for diagnosing acute PCL tears. High sensitivity (nearly 100%) and specificity (near 97%) have been reported. PCL can be torn at its tibial or femoral attachments or more commonly through its substance.

### Treatment

#### Non-surgical Treatment

Nonsurgical treatment has been advocated for patients with isolated grade I or II PCL injuries or those with grade III injuries but have mild symptoms or for those who only participate in low-demand activities. Rehabilitation including improving range of motion (ROM), normalization of gait and reactivation of the quadriceps musculature [15]. Knee is immobilized for 2–4 weeks.

### Biologics for PCL Injuries

No human studies have been reported which treated PCL injuries with platelet products of stem cells.

### Procedure for Percutaneous Infiltration of PCL

PCL can be percutaneously accessed by fluoroscopy or ultrasound guidance, however it's difficult to visualize this ligament under ultrasound and is even more challenging to inject the origin and insertion sites. Under fluoroscopic guidance the origin, midportion and insertion sites can be targeted. The patient is in supine position with knee flexed. Flexion opens up the intercondylar fossa. The area is prepped and draped in a sterile fashion. AP fluoroscopic views are taken and C-arm cranio-caudal angulation is done to maximize the opening of the intercondylar space (Fig. 20.5). The target is the two tibial eminences. A 25 g, 3.5-inch needle is advanced using the "hubshot" technique, in between and couple of millimeters above the two tibial eminences (Fig. 20.5). In the "hubshot" approach, the needle is placed in line with the fluoroscopy beam and in an ideal situation, mainly the hub is seen, and minimal needle shaft is seen in the fluoroscopic image. Lateral fluoroscopic views are checked, and the needle is advanced until it is at the posterior border of the tibia (Fig. 20.6). It is imperative to see a "crisp" posterior tibial border either by manipulating the C-arm or by moving the knee. Dye is injected and the spread outlining the ligament should be seen (Fig. 20.7). If ideal spread is not achieved, the needle is pulled back and contrast re-injected until appropriate dye spread is seen. If not, the needle has to be placed either slightly medially or laterally from the original target site until satisfactory dye spread is noted. *Since the PCL dives down inferiorly from the joint line for its attachment into the posterior superior part of the tibia, the needle is close to the midpoint of the ligament.* Usually this gives a good spread but if needed, the superior part of the ligament can be sought. Two to three milliliters of the biologic should

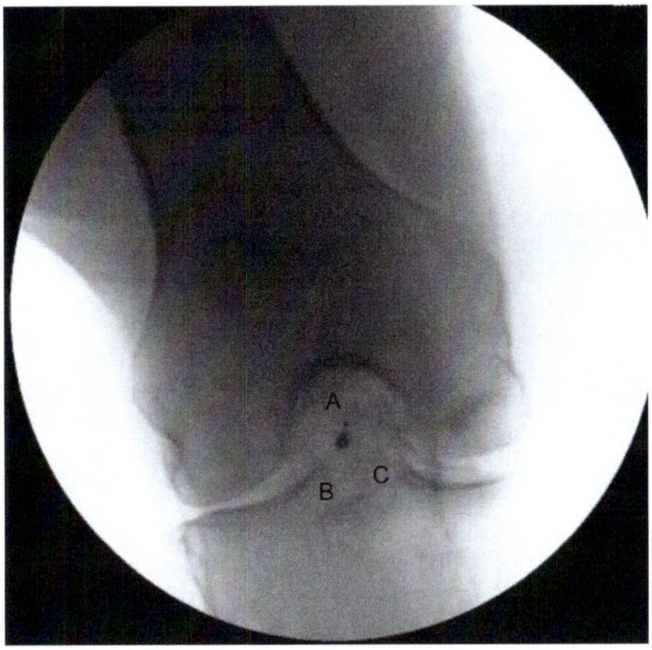


**Fig. 20.5** Needle placement into the ACL. (A) Intercondylar notch maximized by knee flexion. (B) Lateral Tibial Eminence. (C) Medial Tibial Eminence. Needle placed in between and couple of mm above the medial and lateral eminences

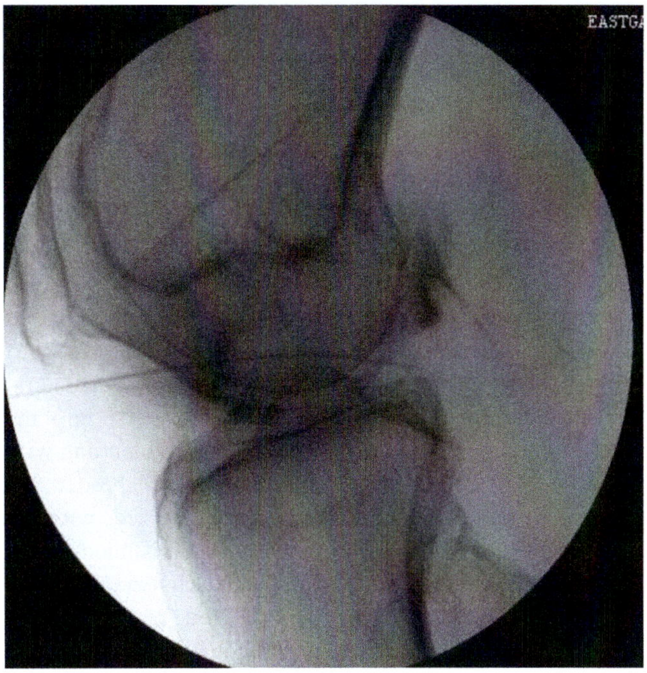

**Fig. 20.6** Needle is advanced in the lateral view until it reaches the posterior edge of the tibia

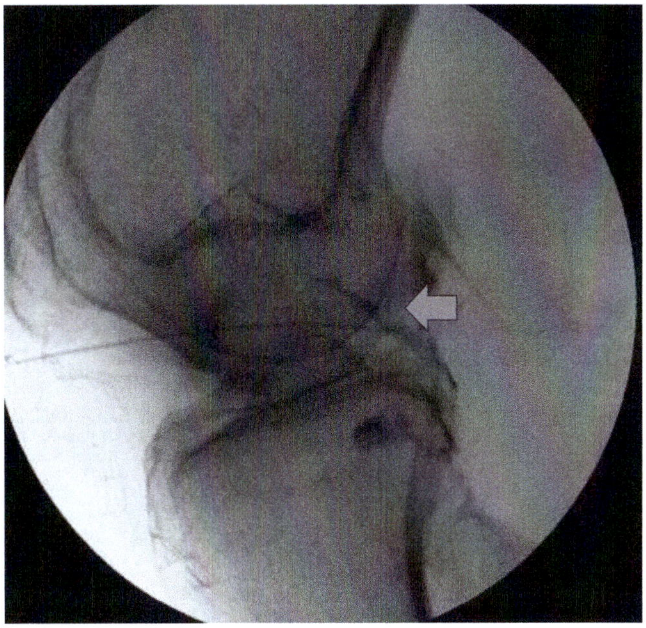

**Fig. 20.7** Arrow showing dye spread along the PCL

be injected. Although separate AL and PM infiltration is desired (most tears involve both bundles), as these two bundles are covered by one synovial fold, separate infiltration may not be required especially for stem cells as they are known to seek the inflamed tissue. Additionally, the two bundles are tightly together at the tibial insertion site and are challenging to do separate bundle infiltration.

## Knee Menisci

### Anatomy

The knee meniscus are crescent shaped wedges of fibrocartilage between the femoral condyles and the tibial plateaus. The outer edges of the menisci are broader and convex with attachments to the joint capsule and the inner edges taper to a concave free edge. The medial meniscus (MM) is C-shaped and covers approximately 60% of the medial compartment. The lateral meniscus (LM) is more circular than the MM and has been reported to cover as much as 80% of the lateral compartment surface. The anterior horn inserts into the anterior intercondylar area with its fibers also blending with the ACL. The peripheral 10–30% of the MM border and 10–25% of the LM border are well vascularized, with the remainder of the meniscus receiving nourishment from synovial fluid. This has led to meniscus zones being described in a radial direction as red-red, red-white, and white-white based on vascularity [16, 17] (Fig. 20.8). Knee menisci are important structures that provide lubrication, stability, joint congruity, load transmission, and functionally act as "shock absorbers" for joint preservation. Injuries to the menisci alter normal knee kinematics, increase peak contact stress, and can result in early degenerative changes of the knee [18].

### Classification of Meniscal Tears

Classification of meniscal injuries occurs depending on location, thickness, and resulting stability. Thus, tears in the peripheral vascularized portion are denoted as red-red tears, in the middle third portion as red-white or white-white tears, and in the inner avascular portion as white-white tears (see above).

#### Types of Tears [19] (Fig. 20.9)

1. Horizontal Tear: A horizontal tear runs parallel to the tibial plateau dividing the meniscus into superior and inferior halves.
2. Longitudinal Tear: Longitudinal tears run perpendicular to the tibial plateau and divide the meniscus into central and peripheral halves.
3. Radial Tear: A radial tear runs perpendicular to both the tibial plateau and the long axis of the meniscus dividing the meniscus into anterior and posterior halves.
4. Complex Tear: A complex tear includes a combination of radial, horizontal, and longitudinal components (any two or all three).
5. Displaced Tear: Displaced tears include free fragments, displaced flap tears, and bucket-handle tears.
6. Bucket-Handle Tear: A bucket-handle tear is a longitudinal tear with central migration of the inner "handle" fragment.

### Diagnosis

An assessment of gait should be performed first, looking for varus or valgus, quadriceps avoidance, and antalgic gaits. Substantial primary varus or valgus deformity may be the result of single-compartment osteoarthritis (OA). Excessive femoral anteversion may be identified by an inward-pointing or so-called squinting patella.

MR imaging demonstrates high sensitivity (93% for the MM and 79% for the LM) and specificity (88% for the MM and 96% for the LM) for detection of meniscal tears [19].

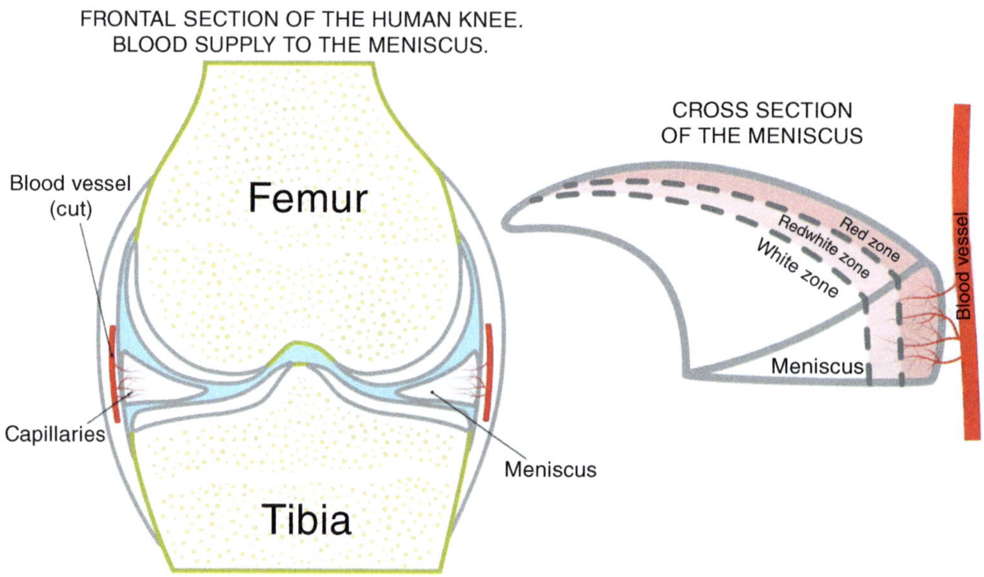


**Fig. 20.8** Vascular supply of the meniscus

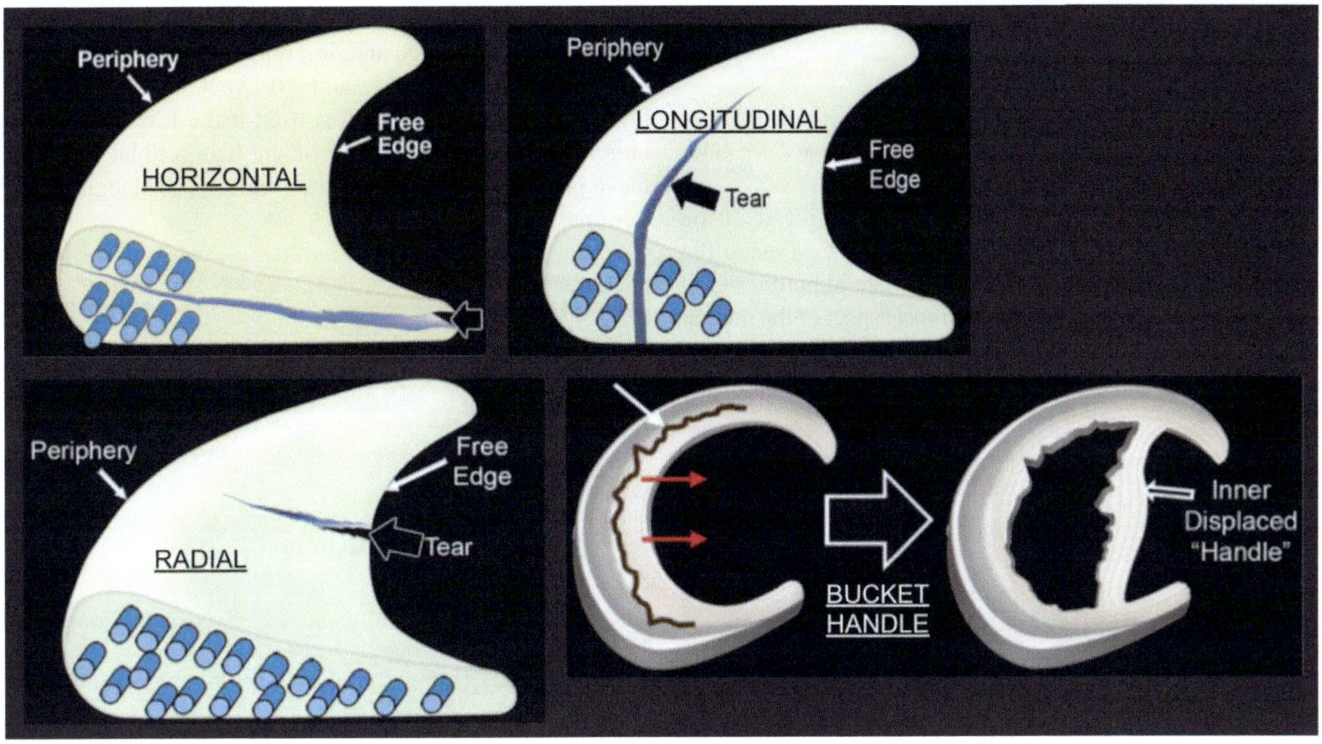


**Fig. 20.9** Classification of meniscal tears

## Treatment of Meniscal Injuries

There is evidence that treatment of older patients with degenerative meniscal tears with PT focusing on improving muscle strength, flexibility and proprioception for 8 weeks is beneficial and can possibly avoid surgery. However, there is no similar data for younger patients with acute meniscal injuries [20].

### Surgical Augmentation with PRP for Meniscus Tears

Two studies [21, 22] have demonstrated that augmenting arthroscopic meniscus repair with PRP resulted in better healing (on MRI) and functional outcomes.

### Percutaneous Infiltration of Meniscal Tear with PRP

In the only study available, ten recreational athletes with intrasubstance meniscal lesions (Grade II according to Reicher) proven by MR-Imaging (MRI) were treated by percutaneous injections of PRP in the affected meniscal area under fluoroscopic guidance. Three sequential injections in 7-day intervals were performed in every patient. Four of ten patients (40%) showed improvement of meniscal lesion in follow up MRI after 6 months. Six of ten patients (60%) showed improvement of Numeric Rating Scale (NRS) score at final follow up. Average NRS score improved significantly from 6.9 before injections to 4.5 6 months after treatment. Six of ten patients (60%) reported an increase of sports activity compared to the situation before injections. In four patients (40%), additional surgical treatment was necessary because of persistent knee pain or progression of meniscal lesion. *It is interesting to note that the PRP injection was done with only a 10 cc blood draw and was leukocyte poor* [23].

## Stem Cells in Meniscal Injuries

Vangsness et al. [24] performed a double-blind, randomized, controlled clinical study in which single superolateral knee injection was performed within 7–10 days after partial meniscectomy surgery. Fifty-five patients were randomized to one of three treatment groups: Group A (18 patients), in which patients received an injection of $50 \times 10^6$ allogeneic bone marrow mesenchymal stem cells (MSCs); Group B (18 patients), $150 \times 10^6$ allogeneic MSCs; and the control group (19 patients), a sodium hyaluronate (hyaluronic acid/hyaluronan) vehicle control. There was significantly increased meniscal volume (defined a priori as a 15% threshold) determined by quantitative MRI in 24% of patients in Group A and 6% in Group B at 12 months post meniscectomy. No patients in the control group met the 15% threshold for increased meniscal volume. Patients with OA changes who received MSCs experienced a significant reduction in pain compared with those who received the control, on the basis of Visual Analog Scale (VAS) assessments [24].

## Procedure for Percutaneous Meniscal Injury Infiltration

Medial and LM can be accessed either by ultrasound (US) or by fluoroscopy and the technique for the approach for both the medial and lateral menisci is similar.

In the fluoroscopic approach, AP view is utilized. If the MM is the target, 22 G 3.5-inch needle is placed using the "hubshot" approach to target the supero-lateral border of the tibia (Fig. 20.10) to enter the lateral most aspect of the medial femorotibial joint (Fig. 20.11). Depending on the location of the tear on the MRI, the needle should be advanced in the lateral view, using the superior part of the tibia as a guide, until the needle is in the anterior, mid or posterior part of the tibia (Figs. 20.11, 20.12, and 20.13). One to two cc is injected. A similar approach is used if the LM needs to be targeted. Severe degeneration of the femorotibial joint can make placing the needle into the meniscus challenging with this approach.

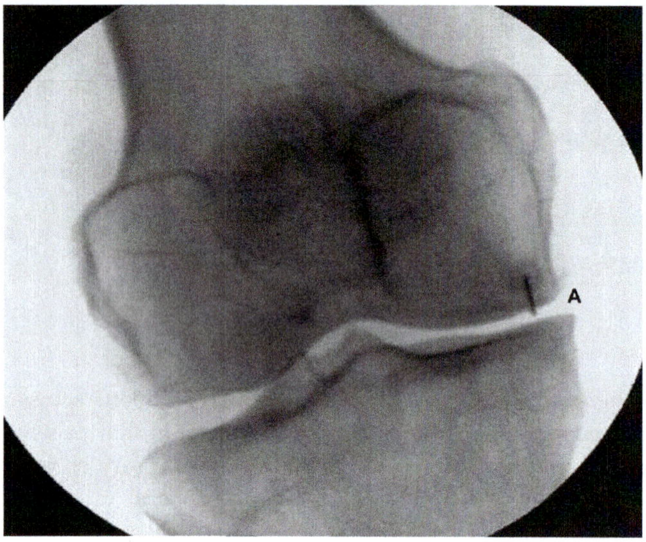


**Fig. 20.10** Needle placement at the lateral part of the femorotibial joint in the AP view

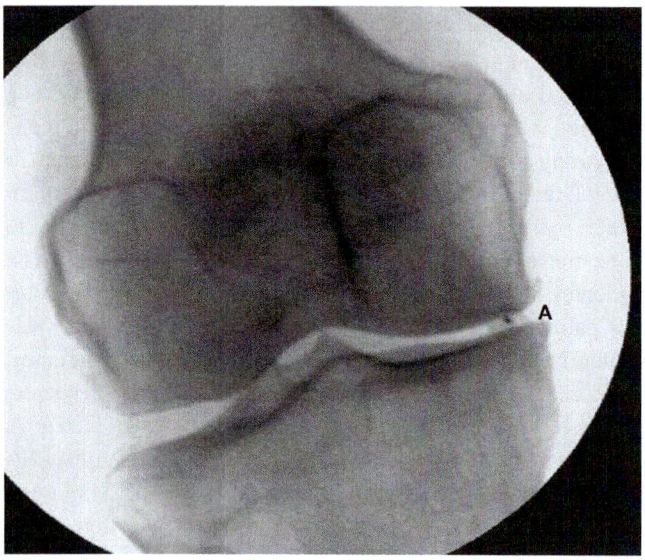


**Fig. 20.11** Needle placed inside the lateral aspect of the femorotibial joint

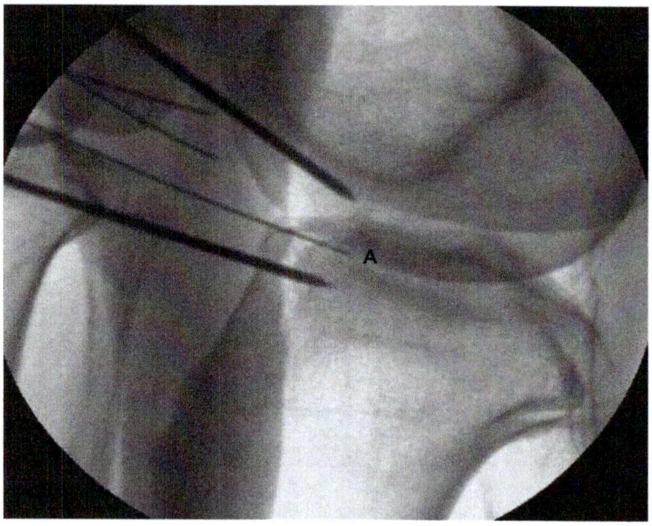


**Fig. 20.12** In the lateral view, needle placed in the anterior portion of the meniscus

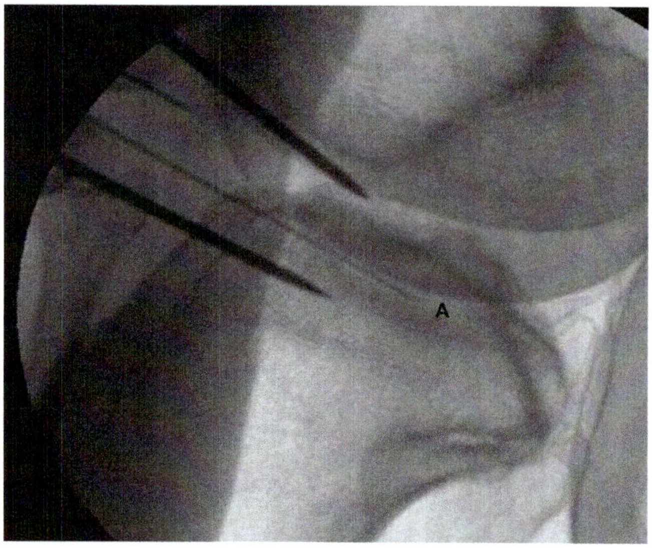


**Fig. 20.13** In the lateral view, needle placed in the mid portion of the meniscus

In the US technique, the knee is flexed 30–60° and placed sideways on the bed so that the painful side is facing the ceiling. The transducer is placed longitudinally so that the short axis view of the meniscus is seen (Fig. 20.14). In this view, the needle can be placed utilizing the "in plane" technique to enter

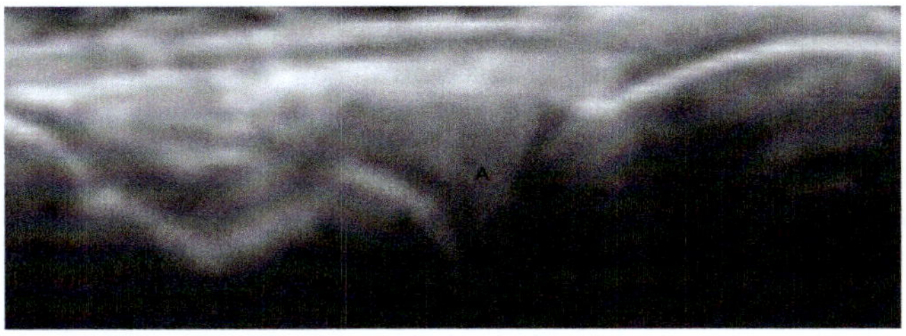


**Fig. 20.14** Wedge shaped meniscus seen in the long axis view under ultrasound

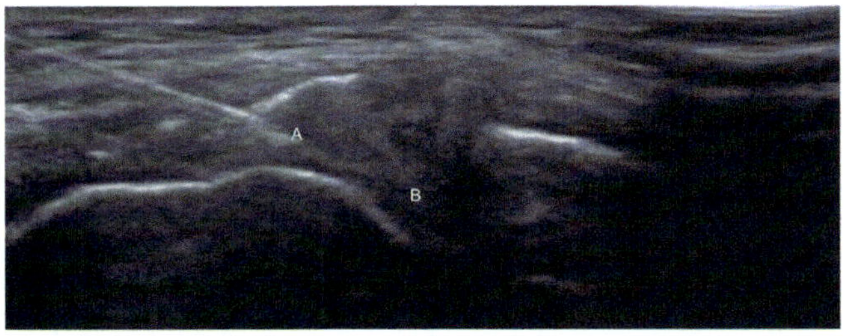


**Fig. 20.15** Needle tip seen in the long axis view using the "in plane" technique. Further advancement of the needle is required to enter the meniscus

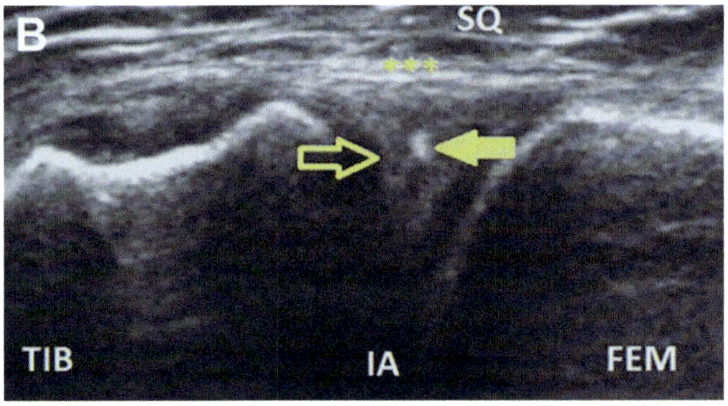


**Fig. 20.16** Needle tip in the body of the meniscus seen during the "out of plane" technique

the body of the meniscus (Fig. 20.15). Alternatively, using the "out of plane" technique, the 25G, 2-inch needle is placed into the meniscus (Fig. 20.16). 1–2 cc of the biologic is injected as the needle is withdrawn in order to inject the whole width of the meniscus. MRI will indicate which part of the meniscus has the tear (anterior, mid or posterior) and that part should be injected with the technique described above.

## Medial and Lateral Collateral Ligaments

As the name suggests, the medial and lateral collateral ligaments are on the medial and lateral sides stabilizing the knee joint. The medial collateral ligament (MCL) originates on the medial epicondyle of femur to the posteromedial crest of the tibia. The ligament is a broad and strong band whose main function is to stabilize the knee joint in the coronal plane on the medial side. MCL is also attached to the MM via meniscofemoral and meniscotibial ligament. MCL prevents rotational forces of the knee joint. The lateral collateral ligament originates on the lateral epicondyle of femur and inserts on the fibular head. The lateral collateral ligament's primary purpose is to prevent excess varus stress and posterior-lateral rotation of the knee. The lateral collateral ligament is the least frequently injured ligament. The collateral ligaments can be stretched, partial or complete tear. The MCL is injured in sports like football during a block to the lateral part of the knee, MCL injury is commonly associated with ACL injury. Common symptoms of collateral ligament injuries include knee swelling, locking or catching of the knee when you move it, Pain and tenderness along the ligament, knee that gives way or feels like it. Ultrasound and MRI can help diagnosis.

### Treatment

Common injury to these ligaments are treated with knee bracing, NSAIDs, cold compress therapy and physical therapy.

#### Interventional Treatment

Andia and Maffulli [25] showed that intra-articular injection of autologous PRP is effective for refractory pain after low grade MCL injury. Yoshida and Marumo [26] reported a case series of three patients with chronic injury of MCL with insufficient healing responded well with PRP injection, with all cases returning to sports activity at the previous level without symptoms [26] and showing complete healing of the proximal ligament was identified by the MRI imaging.

Ultrasound guided injections of biologics preferably into and around the tears after visualization can possibly help heal the injured medial and lateral collateral ligaments. In-plane or out-of-plane techniques can be used.

## Knee Osteoarthritis

Osteoarthritis (OA) is the most common joint disorder in the United States [27, 28]. Symptomatic knee OA occurs in 10% in men and 13% in women [28, 29] aged 60 years or older. The number of people affected with symptomatic OA is likely to increase due to the aging of the population and the obesity epidemic. OA has a multifactorial etiology and can be considered the product of an interplay between systemic and local factors. Old age, female gender, overweight and obesity, knee injury, repetitive use of joints, bone density, muscle weakness and joint laxity all play roles in the development of joint OA [28].

## Pathophysiology

OA in the knee has long been considered as a primary disorder of articular cartilage, however, lately, the contribution of subchondral bone to the physiopathology of OA is gaining traction [30, 31]. Subchondral bone deterioration is commonly associated with articular cartilage defects and together are widely considered as a hallmark of OA. In OA there is abnormally high bone turnover in the SC region leading to increase in the trabeculae and bone volume however it is hypomineralized and is of poor quality resulting in microdamage/fractures, bone marrow edema and bone cysts probably from poorly functioning native stem cells which are subjected to age-related and environmental degradation. Aging also decreases the number of stem cells.

Cartilage of the knee joint has distal non calcified cartilage, which is in communication with the synovial fluid. Proximal to this is the calcified cartilage layer (CL). Immediately next this layer is the subchondral bone (SC) bone which consists of distal SC bone plate (immediately beneath the calcified cartilage layer) and the more proximal SC trabecular bone. The SC bone plate which lies between the calcified cartilage (CC) and SC trabecular bone consists of channels (Fig. 20.17). It is through these channels blood vessels and nerves traverse from the SC trabecular bone to the CC. Hence, changes in CC translate to changes in SC trabecular bone and vice versa and because of this close connection, this functional unit is called ***osteochondral junction*** which consists of three structures. The more distal synovial membrane, middle cartilage (NC and CC) and the distal SC bone (SC bone plate and trabecular bone) [30]. The aneural and avascular cartilage lies functionally sandwiched between two highly vascularized and innervated tissues, namely, the synovial membrane and subchondral bone [32].

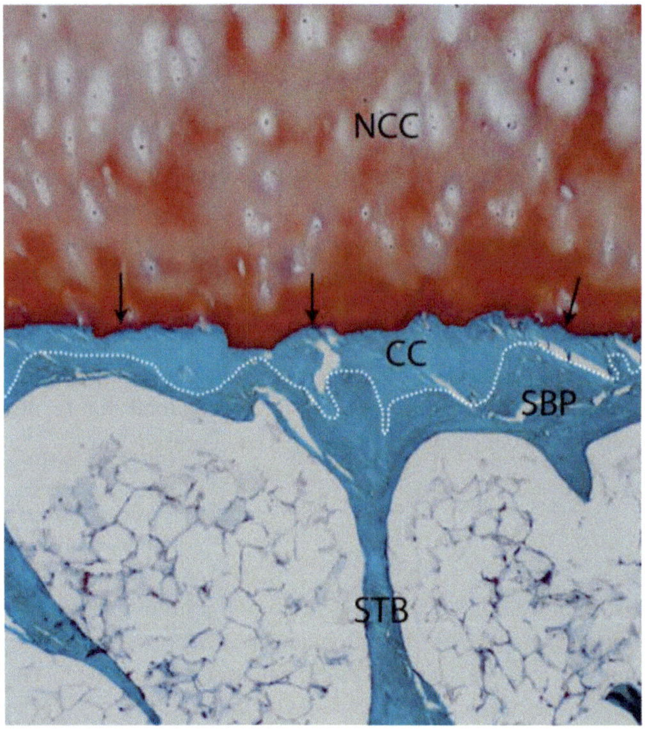


**Fig. 20.17** Histology of articular cartilage and subchondral bone. *NCC* Non calcified cartilage, *CC* Calcified cartilage, *SBP* Subchondral bone plate, *STB* Subchondral trabecular bone

Bone remodeling is an active and dynamic process that depends upon the tightly regulated balance between bone resorption and bone formation. Bone resorption is primarily through osteoclasts. The osteoclastic activity results in the release of TGF-β and IGF-1 which stimulates the MSCs to differentiate into osteoblasts which create new bone formation. This constant activity of the osteoclasts and osteoblasts results in bone remodeling [31]. Changes in MSC function and numbers may in fact play a major role in the failed joint repair responses in OA. Excessive activity of MSCs causes increases in bone volume, sclerosis and osteophytes in some areas and decreased activity leads to hypo-mineralization, bone cysts and bone marrow lesions.

Since the three layers of the osteochondral junction are intimately connected as described above, alterations in any of these three tissues will modulate the properties and functions of other parts of the osteochondral junction. *In vivo* studies showed that prostaglandins, leukotrienes and various growth factors released by osteoblasts during subchondral bone remodeling could reach overlying articular cartilage. Accordingly, inflammatory and osteoclast stimulation factors released by articular cartilage may also lead to subchondral bone deterioration through increased bone remodeling in OA [30]. In patients with severe OA, the subchondral bone undergoes changes which include microcracks, structural defects, vascularization of channels and nerve growth. This promotes crosstalk and a pathway for direct transport of growth factors such as transforming growth factor B (TGFβ) and nerve growth factor (NGF) and even for cells such as macrophages and MSCs between the subchondral bone and articular cartilage [32]. *If biologics are considered to treat OA, it seems that all three layers need to be targeted in order to maximize results.* The biologics can reach the cartilage and synovial membrane through intraarticular (IA) application and SC bone can be addressed by a SC injection. IA infiltrations of biologics are insufficient to reach the SB and deeper layers of cartilage, thereby limiting their efficacy in patients with moderate to severe OA.

Sanchez has conducted clinical trial injecting PRP into knee synovial membrane, synovial fluid, articular cartilage, and subchondral bone with intra-articular injections and intraosseous infiltrations of PRP reduces pain and MSCs in SF, besides significantly improving knee joint function in patients with severe knee OA.

Hernigou et al. [33] has shown in clinical trial that autologous BMC was an effective procedure for treating young patients with knee OA following secondary osteonecrosis of the knee related to corticosteroids with a lower complication rate and quicker recovery as compared with total knee arthroplasty (TKA).

## Diagnosis

In addition to a thorough history and physical, radiographic imaging is required to assess OA. The recommended views include standing AP, standing lateral in extension and a skyline view of the patella. A standing 45° posteroanterior (PA) view of the knee may be obtained. Radiographic findings of OA are joint space narrowing, osteophyte formation, subchondral sclerosis, and subchondral cysts. Semiquantitative Kellgren-Lawrence (KL) grading scale used to grade knee OA.

Treatment of knee OA begins with conservative methods and progresses to surgical treatment options. Non-surgical treatment options include patient education about weight loss and activity modification, knee bracing, physical therapy, weight loss, knee bracing, acetaminophen, NSAIDs, Glucosamine and chondroitin sulfate, corticosteroid injections, hyaluronic acid injections.

## Surgical Treatments

Unicompartmental knee arthroplasty and total knee arthroplasty.

## Interventional Treatments

There are several studies including randomized controlled trials (RCTs) showing the efficacy of platelet rich plasma and MSCs for the treatment of knee OA. The intra-articular injection into the knee joint showed a significant improvement in Western Ontario and McMaster Universities Osteoarthritis Index (WOMAC) score, quality of life, activities of daily living, knee range of motion, VAS pain scores and SF-36 (Table 20.1). Clinical trials of subchondral injection of stem cells into femoral condyles and tibial condyles are reporting functional improvement, pain control, improved range of motion, stimulation of intra-articular cartilage growth and seem superior to IA administration [32, 39].

Twenty-six RCTs assessed the efficiency of PRP in treating knee OA, which are discussed below. The PRP was injected into the knee joint and compared to control groups which include hyaluronic acid or corticosteroid or ozone or saline [34–38, 40–59].

Most of the studies have shown that intra-articular injection of PRP or plasma rich in growth factors (PRGF) into the knee joint has significantly superior results than intra-articular injection of hyaluronic acid products or corticosteroids or ozone or saline. PRP injection produced better pain control with lower VAS, decreased stiffness,

**Table 20.1** Publications on treatment of knee osteoarthritis with platelet rich plasma

| Author | Year | Number of pts | Type of biologic | Plt conc | Kl grade | Delivery | Los months | Outcome |
|---|---|---|---|---|---|---|---|---|
| Raeissadat et al. [34] | 2021 | 200(PRP59/ PRGF60/HA59/ O3-60) | PRP 2 mL/PRGF 2 mL/HA/Ozone | ×4–6/ PRGF ×5 | 2, 3 | HA 7 O3 IA × 1 q week, PRP &PRGF ×2 q 3 weeks | 12 | PRP, PRFG significant improvement, persisted for more than 12 months |
| Park et al. [35] | 2021 | 110 (PRP 55/HA 55) | PRP 3 mL/HA | INR | 1,2,3 | IA ×1 | 6 | PRP superior to HA, significant improvement of pain scores |
| Cole et al. [36] | 2017 | 111 PRP 49/ HA50) | PRP 4 mL/HA 16 mg | 1.73 | 1,2,3 | IA ×3, weekly | 12 | No significant difference between PRP and HA |
| Elksniņš-Finogejevs et al. [37] | 2020 | 40 (PRP 20/K 20) | PRP 8 mL/K40 | 2.5 times | 2, 3 | IA ×1 | 12 | No significant difference in the group short term. PRP significant improvement over CS at 1-year follow-up |
| Tucker et al. [38] | 2021 | 17 (PRP 11/NS6) | PRP/NS (INR) | | 2,3 | | 12 | PRP group has less pain, stiffness and improved function scores |
| Ke Su et al. [39] | 2018 | 86 (A 28/B26/C32) | IO+ IAPRP ×2, q2 weeks/IA PRP ×2 q 2 weeks/HA IA ×4 q 1 week | | 2 to 3 | IO + IA ×2 q 2 weeks | 18 | IO +IA is significantly superior to IA PRP which is better than IA HA |

(continued)

**Table 20.1** (continued)

| Author | Year | Number of pts | Type of biologic | Plt conc | Kl grade | Delivery | Los months | Outcome |
|---|---|---|---|---|---|---|---|---|
| Xu et al. [40] | 2021 | 122 (HA34/ PRP40/ PRP + HA48) | HA 2 mL/PRP 4 mL/PRP + HA 4 + 2 mL | 5.13 | 2,3 | IA ×1 | 12 | PRP and PRP + HA significantly superior to HA. At 24 months PRP + HA had better results than PRP alone |
| Rahimzadeh et al. [41] | 2018 | 42(PRP21/PRL21) | PRP7/PRL 7 (25% dextrose) | INR | 1,2 | IA ×1 | 6 | PRP more effective than PRL |
| Duymus et al. [42] | 2017 | 120(PRP40/HA40/ O340) | PRP 3–4 mL/HA 40 mg/ozone | ×9–13 | 2,3 | IA ×2/×1/×4 | 12 | PRP is significantly better than HA and O3 |
| Lin et al. [43] | 2019 | 87 (PRP 31/HA29/ NS27) | PRP31/HA29/ NS27 | 1.8 | 2,3 | IA ×3, weekly | 12 | Clinically significant function improvement atleast 1 year |
| Pishgahi et al. [44] | 2020 | 92 (PRL 32/ PRP30/ACS30) | PRL32 (25% dex)/PRP30/ ACS30 | ×4 | 2,3 | IA (prl ×3, PRP ×2, ACS ×2) | 6 | PRP and ACS treatments more effective than prolotherapy |
| Li et al. [45] | 2021 | 200(prp100/ HA100) | PRP 4–6 mL/HA | INR | INR | IA ×1 | 24 | PRP can offer more lasting outcome than HA |
| Filardo et al. [46] | 2012 | 109 (PRP55/ HA54) | PRP 5 mL/HA | 5 | 3 | IA ×3 q week | 12 | PRP is not better in moderate OA and middle age than HA. PRP group significantly improved 1 year follow-up |
| Cerza et al. [47] | 2012 | 120 (PRP60)/ HA60) | PRP5.5/HA 20 mg | | 1,2,3 | IA ×4 q week | 5.5 | ACP showed significantly better outcome than PRP |
| Ghai et al. [48] | 2019 | 20 (PRP 10/NS 10)) | PRP 8 mL/NS 8 mL | ~2 | 1,2 | IA ×1 | 6 | Improved physical function |
| Huang et al. [49] | 2019 | 120 (PRP40/ HA40/CS40) | PRP 4 mL/HA 2 mL/CS 1 mL | ×3 | 1,2 | IA 3×, q 3 weeks | 12 | All groups did good, no difference in 3 groups at 3 months. Long term efficiency PRP superior |
| Patel et al. [50] | 2013 | 78 pts (156 knees) (52 pts PRP ×1/50 PRP ×2/46NS ×1) | PRP 4 mL/NS 8 mL | 10 | 1,2 (Ahlback) | IA ×1/IA ×2/ IA ×1 | 6 | Single dose of PRP is as effective as 2 doses of PRP, both superior to NS |
| Chu et al. [51] | 2022 | 610 (PRP305/ NS305) | PRP/NS | | | IA ×3, q 1 week | 24 | PRP is superior to NS, PRP effective for 24 months with symptom relief and slowing process of KOA |
| Spakova et al. [52] | 2012 | 120 (PRP60/ HA60) | PRP 3 mL/HA | 4.5 | 1,2 and 3 | IA ×3 q week | 6 | PRP is significant better than HA |
| Paterson et al. [53] | 2016 | 23 (PRP12/HA11) | PRP 3 mL/HA 3 mL | | 2,3 | Ia ×3 q week | 6 | PRP improves pain, symptoms and function, no difference between groups |
| Forogh et al. [54] | 2016 | 48 knees (PRP24/ CS24) | PRP 5 mL/MP 40 mg | 4 | 2,3 | IA ×1 | 6 | PRP significantly better than corticosteroid and for long term, ADL and QOL significantly high |
| Montañez-Heredia et al. [55] | 2016 | 55 (28 PRP/27 HA) | PRP/HA 25 mg | INR | 1,2,3 | IA ×3 q 15 days | 6 | Both groups improved, statistical difference. PRP group improve quality of life at 3 months. PRP more effective than HA. KOOS, VAS, EUROQOL scale |
| Nunes-Tamashiro et al. [56] | 2022 | 100 | PRP/CS (T) 40 mg/NS | INR | 2,3 | IA ×1 | 12 | PRP and Triamcinolone superior to NS |
| Louis et al. [57] | 2018 | 54 (PRP 26/HA28) | PRP 3 mL/HA60 | ×2 | 2,3 | IA ×1 | 6 | Both groups similar knee function status and symptom relief (WOMAC, VAS) |
| Wang et al. [58] | 2022 | 110 (54 PRP/56 HA) | PRP 4 mL/HA 60 mg | INR | 4 | IA ×1 | 6 | PRP and HA no difference |
| Lisi et al. [59] | 2018 | 58 (PRP30/HA28) | PRP/HA 20 mg | INR | 2,3 | IA ×3 q 4 weeks | 6 | PRP significantly better |

*PRP* platelet rich plasma, *PRGF* platelet rich growth factors, *HA* Hyaluronate, *KL grade* Kellgren Lawrence grade, *O3* Ozone, *IA* intraarticular, *IO* intraosseous, *INR* information not recorded increased range of motion, improved quality of life. PRP treatment has also shown improved WOMAC and International Knee Documentation Committee (IKDC) scores. Intra-articular injection of PRP does not have side effects like osteonecrosis caused by intra-articular steroid injection and it can be repeated as needed in future.

Preparation of PRP plays a key role in predicting the outcomes of its healing effect. There is direct relation concentration of the platelets in the PRP, dose of the platelets in the PRP injections. Baria and Flanigan [60] illustrated that inadequate platelet concentration and dose were responsible for poor outcomes in many trials.

## Mesenchymal Stem Cell Injection

Multiple RCTs displayed the efficacy of mesenchymal stem cells for treatment of knee osteoarthritis and these are listed below in Table 20.2 [61–73]. The MSCs were injected into the knee joint and compared to control groups which include one of the following, PRP or hyaluronic acid or TKA or saline. Study of particular interest is done by Hernigou et al. [33] who published a study in which 30 patients with bilateral steroid induced knee osteonecrosis were randomly assigned to receive either the total knee replacement (TKR) in one knee and autologous bone marrow MSCs in the femoral and tibial condyles of the other knee. These patients were followed for an average of 12 years. In terms of pain relief, both groups had similar decreases in pain scores. However, in the stem cell group, there was about 4% increase in the knee cartilage and a 3 cm cubic area decrease in the bone arthritis. *Most importantly, 21 of the 30 patients were satisfied with stem cell therapy compared to only 9 of the 30 patients who had the knee surgery.* Six of the thirty knees which had the TKR required a second surgery compared to only three in the stem cell group who required TKR. Moreover, 30% of patients in the surgery group had blood transfusion versus none in the stem cell group. 15% of patients in the surgery group also developed deep vein thrombosis (DVT) in their legs. None of the patients in the stem cell group had any DVT.

**Table 20.2** Publications on treatment of knee osteoarthritis with stem cells

| Author | Year | Number of patients | Type/dose of biologic | OA KL stage | Delivery | Los month | Outcome |
|---|---|---|---|---|---|---|---|
| Vega et al. [61] | 2015 | 30 (15 MSC/15 HA) | Al-ex BM-MSC 40 × 10^6/HA60 | 2,3,4 | IA ×1 | 12 | MSC significant improvement |
| Gupta et al. [62] | 2016 | 60 (30 MSC/30 HA) | Al BMMSC 25, 50, 75, 150 ×10^6/HA20 | 2,3 | IA ×1 | 24 | MSC significant improvement |
| Kalifeh Soltani et al. [63] | 2019 | 20 (10 MSC/10 NS) | UCMSC 50–60 ×10^6/ NS | 2,3,4 | IA ×1 | 5.5 | MSC significant improvement |
| Lee et al. [64] | 2019 | 24 ad ex MSC 12/NS12 | au Ex adMSC 100 × 10^6 | 2,3 | IA ×1 | 6 | MSC significant improvement |
| Lu et al. [65] | 2019 | 53 (ad MSC26/HA27) | Auto Ex adMSC 50 × 10^6/HA 25 mg | 1,2,3 | IA MSC ×2, HA ×4 | 12 | MSC significant improvement |
| Anz et al. [66] | 2022 | 90 Leukocyte poor PRP/BMC | PRP 7 mL/BMC 7 mL | 1,23 | IA ×1 | 24 | No difference |
| Lamo Espinosa et al. [67] | 2020 | 60 Ex BMMSC + PRGF/ PRGF | Ex BMMSC 100 × 10^6/PRGF 8 mL | INR | IA ×3 q week (only one MSC application) | 12 | MSC better but not statistically significant |
| Hernigou et al. [68] | 2021 | 120 BMC/TKA | BMC (IO, 10 mL to each/TKA) | 2,3 | IO ×1 | 120 | IO MSC equivalent to TKA |
| Wang et al. [69] | 2016 | 36 Ex UCMSC/HA | UCMSC 20–30 × 10^6/ HA | INR | IA MSC ×2, HA ×5 | 6 | MSC significant improvement |
| Dulic et al. [70] | 2021 | 175 (111 BMC/34 PRP/30/ HA) | BMC 6.3 mL/PRP ×7/ HA 20 mg | 1,2 &3 | BMC and PRP IA ×1, HA IA ×3 q week | 12 | BMC could do better, PRP better than HA |
| Matas et al. [71] | 2019 | 29 pts, UCMSC 20 × 10^6/ UCMSC (20 × 10^6)2/HA ×2, 1:1:1 | UCMSC 20 × 10^6/ (20 × 10^6)2/HA | | IA ×1/IA ×2 | 12 | MSC ×2 significantly better than MSC and HA |
| Shapiro et al. [72] | 2017 | 25 pts (50 knees) BMAC in one knee/saline other | BMC 6 mL/NS | 1,2,3 | IA ×1/IA ×1 | 6 | Same relief |
| Hernigou et al. [73] | 2021 | 60 pts -120 knees (IA 60/ IO60) | IA BMC 20 mL/IO BMC 20 mL 910^4 in each group) | 1–4 | IA ×1/!0 ×1 | 24 and 180 | BMC IO significantly better than IA |

*UCMSC* umbilical cord mesenchymal stem cells, *BMC* bone marrow concentrate, *MSC* mesenchymal stem cells, *TKA* total knee arthroplasty

Hernigou et al. [73] conducted a double blind RCT study for knee arthritis treatment with bone marrow MSCs. There were 60 patients with bilateral knee arthritis, one knee was treated with intraarticular knee injection of BMC and the other treated with subchondral injection of BMC into femur and tibia. A 15-year follow-up showed 42 out of 60 knees (70%) had TKA in the intraarticular group, whereas 12 out of 60 knees (20%) had TKA in the subchondral group.

## Percutaneous Subchondral Knee Procedure

Although ultrasound can be used, fluoroscopic guidance is the authors' preferred approach to the subchondral femoral and tibial condyles. Since this can be a painful procedure, anesthesia in the form of epidural, femoral nerve block, adductor canal block or genicular nerves block is recommended. The knee is prepped and draped in a sterile fashion. The position of the knee is mild flexion. The oblique views of the C-arm are adjusted until the patella is in the center of the knee joint (Fig. 20.18). If the medial femoral condyle needs to accessed, AP fluoroscopic views are taken. In order to avoid the articular cartilage and the synovial joint, the target is in the middle of the outer border of the condyle in the AP view (Fig. 20.18). Once contact is made with this target point, using the "hubshot" technique, the needle is then angled medially and inferiorly to enter the condyle (Fig. 20.19). We find using a 15 g needle with a drill very useful. Final needle position is adjusted in the lateral view so that the needle is approximately in the mid part of the condyle (Fig. 20.20). 2–5 cc of the biologic is injected. Similar technique is used for the condyle. The technique to inject the

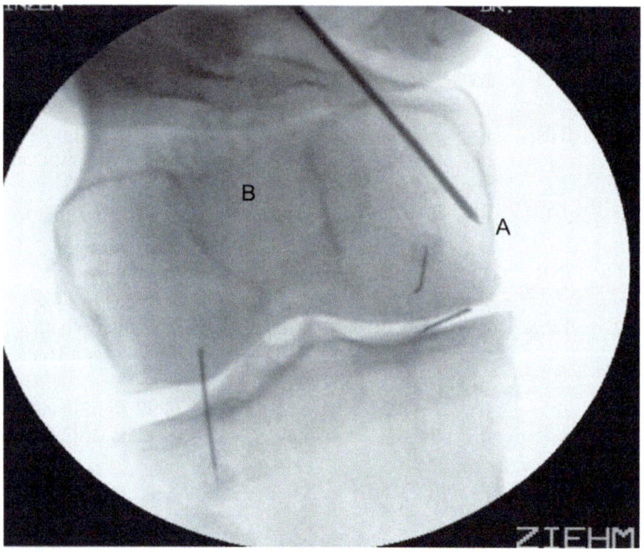


**Fig. 20.18** Subchondral needle insertion sites on femoral condyle. (A) Initial target point. The middle of the outer border of the femoral condyle. (B) Oblique views of CARM are adjusted until the patella is in the center of the knee

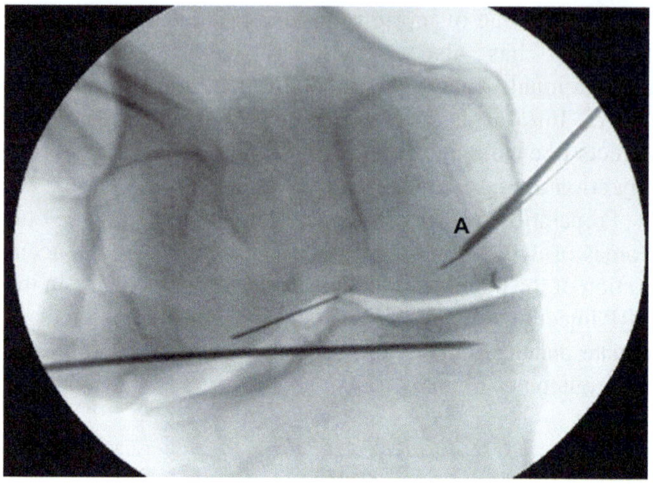


**Fig. 20.19** Subchondral needle entry points. (A) In the AP view the needle is advanced with a medial and inferior angulation to enter the femoral condyle

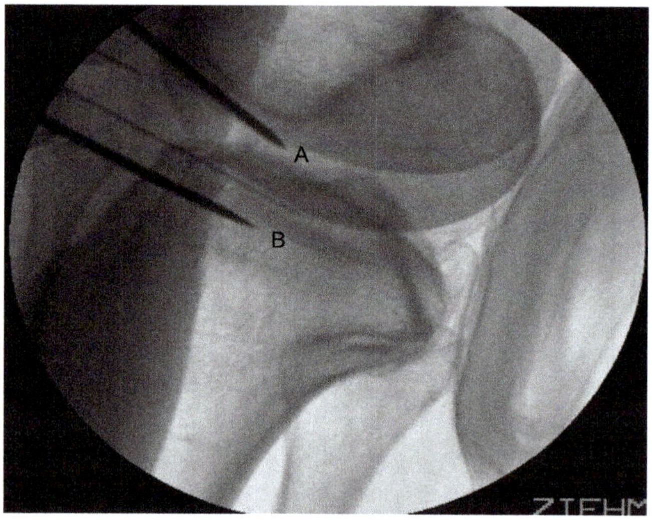


**Fig. 20.20** Lateral view of needle placement into subchondral bone for both femur and tibia. (A) Needle in the middle of the femoral condyle. (B) Needle in the middle of the tibial condyle

tibial condyles is similar for both the medial and lateral condyles. AP views are used and the target is the mid part of the condyle, just under the joint line with the patella in the midline (Fig. 20.21). Once contact is made with the bone, the needle is drilled in about a centimeter (Fig. 20.22). The needle is advanced under lateral view until it is in the middle of the condyle (Fig. 20.20). Two to five cc of the biologic is injected. There is a possibility of piercing through the iliotibial band on when accessing the lateral tibial condyle.

Another technique has been described by Sánchez et al. [74]. In this approach, the femoral condyles are accessed from their superolateral aspect initially and the needles are advanced into the inferomedial aspect. The final needle positions are in the middle of the condyles. a Similar approach is used for the tibial condyles except that they are accessed

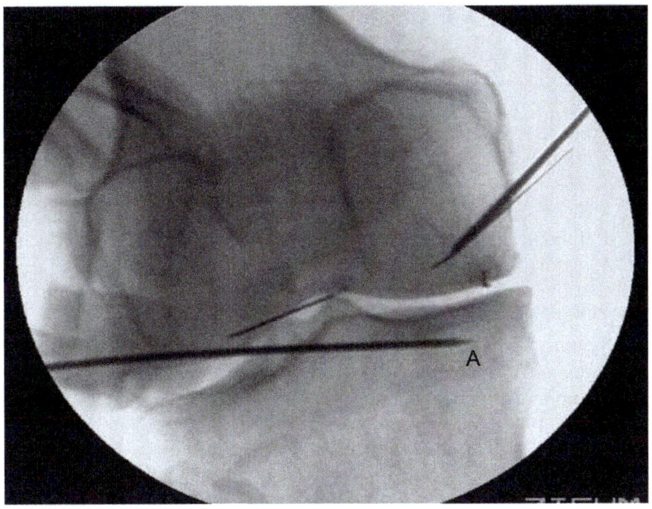


**Fig. 20.21** Entry point of the needle in the middle of the medial tibial tubercle

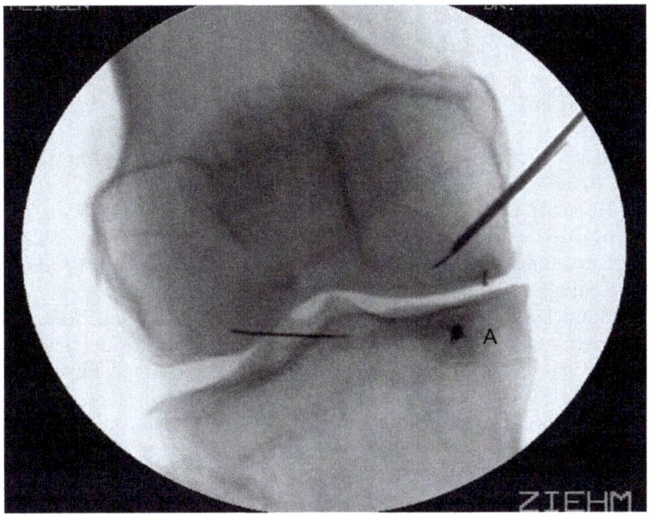


**Fig. 20.22** Needle entering the medial tibial tubercle in the AP view

from the inferolateral part and the needles are angled superomedially to enter the mid part of the condyles.

To minimize bleeding complications, particular attention should be paid to medications which increase this risk. Guidelines on anticoagulants and antithrombotics are described in Chap. 15 [75] and also on the ASIPP app, which is regularly updated.

## Conclusion

There is a tremendous growth in treatment with biologics of the musculoskeletal arena. Utilizing image guidance technology, biologics can be delivered with precision to the different structures of the knee joint. There are numerous double blinded controlled studies that have shown the effectiveness of treatment of different knee disorders with biologics. The outcomes arc comparable to or exceed that of surgical treatments. There needs to be standardization of the procurement, processing and injection of PRP and MSCs to obtain consistent outcomes. Treatment of knee joints with biologics preserves the anatomy and integrity of the knee joint. There are no long-lasting adverse effects with treatment of biologics.

### Key Points

1. Biologics have important role in management of knee injuries and knee arthritis.
2. Bone marrow concentrate seems to be longer lasting than platelet rich plasma
3. Proper dose and concentration is a prerequisite for good outcomes for PRP treatment of knee disorders
4. Biologic seem to be providing pain relief and improved function for mild to moderate knee arthritis.
5. If properly performed these procedures appear to be safe.

### Questions

Questions 1:

Which technique has better success for percutaneous anterior cruciate ligament injections?

A. Clinical anatomy
B. Ultrasound guidance
C. Fluoroscopic guidance
D. Blind technique

**Answer: C**

**Explanation:** Even though ACL injections can be performed using clinical anatomy and blind technique does not guarantee injection into ACL as there is no confirmation. Ultrasound can only visualize the insertion of ACL to Tibia, it is very difficult to visualize the mid portion and femoral attachment. Fluoroscopy guidance with dye study can visualize the insertion into Femur, Tibia and the body of the ACL.

Question 2:

Intra-articular injection of medications or biologics are inadequate for treatment ACL and PCL disorders because.

A. ACL and PCL are located outside the knee joint.
B. The blood supply of ACL and PCL is different blood supply from knee joint
C. ACL and PCL are covered with synovial sheath
D. ACL and PCL are non living tissues

**Answer: C.**

**Explanation:** Both ACL and PCL are intraarticular ligaments. However these ligaments are covered with their own synovial sheaths isolating them from the intraarticular joint space. Therefore both the ligaments have to be injected directly.

Question 3:

The blood supply to meniscus is

A. Uniform from outer margin to inner margin.
B. Inner margin has richer blood supply than outer margin
C. The middle third of the meniscus has a rich blood supply compared to outer and inner thirds of the meniscus.
D. The outer third of the meniscus has richer blood supply.

**Answer: D**

**Explanation:** Depending on the blood supply meniscus can be divided into three zones. Outer third red zone with rich blood supply, middle third red-white zone with marginal blood supply and inner third white zone with very little blood supply (Fig. 20.8).

Question 4:

Which technique shows better outcomes for treatment of knee osteoarthritis with biologics.

A. Intraarticular knee injections shows better improvement than subchondral Femur and tibia injections
B. Intraarticular knee injections and subchondral femur and tibia injections shows similar improvement
C. Subchondral femur and tibia injections shows better improvement than intraarticular knee injections
D. Both intraarticular injections and subchondral show similar improvement.
E. Neither Intraarticular nor subchondral show any improvement.

**Answer: C.**

**Explanation:** Subchondral injection of Femur and Tibia area at knee joint shows better improvement than intraarticular injection of knee joint shown in clinical trial by Hernigou [71] and Sanchez.

Question 5:

Most commonly injured intraarticular knee ligament is

A. PCL
B. Medial collateral ligament
C. Lateral collateral ligament
D. ACL
E. Patellar ligament.

**Answer: D**

**Explanation:** ACL is the most commonly injured ligament in the knee joint. Most common knee joint injury is injury to the meniscus. Very commonly ACL and PCL are associated with meniscus injuries.

**Acknowledgments** This book chapter is modified and updated from previous book chapters, "Regenerative Medicine for Knee Osteoarthritis" by Laxmaiah Manchikanti, MD, Sheri L. Albers, DO, Richard E. Latchaw, MD, and Sairam Atluri, MD, and **"Treatment of Knee Ligaments, Meniscus, and SubChondral Bone Disorders with Biologics"** by Sairam Atluri, MD, in *Essentials of Regenerative Medicine in Interventional Pain Management* published by ASIPP Publishing.

## References

1. Manchikanti L, Albers SL, Latchaw RE, Atluri S. Regenerative medicine for knee osteoarthritis. In: Manchikanti L, Navani A, Atluri S, editors. Essentials of regenerative medicine in interventional pain management. Paducah: ASIPP Publishing; 2019. p. 283–338.
2. Atluri S. Treatment of knee ligaments, meniscus, and subchondral bone disorders with biologics. In: Manchikanti L, Navani A, Atluri S, editors. Essentials of regenerative medicine in interventional pain management. Paducah: ASIPP Publishing; 2019. p. 339–68.
3. Neurath MF, Printz H, Stofft E. Cellular ultrastructure of the ruptured anterior cruciate ligament: a transmission electron microscopic and immunohistochemical study in 55 cases. Acta Orthop Scand. 1994;65:71–6.
4. LaPrade RF, Dragoo JL, Koh JL, Murray IR, Geeslin AG, Chu CR. AAOS research symposium updates and consensus: biologic treatment of orthopedic injuries. J Am Acad Orthop Surg. 2016;24:e62–78.
5. Flandry F, Hommel G. Normal anatomy and biomechanics of knee. Sports Med Arthrosc Rev. 2011;19:82–92.
6. Siegel L, Vandenakker-Albanese C, Siegel D. Anterior cruciate ligament injuries: anatomy, physiology, biomechanics, and management. Clin J Sport Med. 2012;22:349–55.
7. Markatos K, Kaseta MK, Lallos SN, Korres DS, Efstathopoulos N. The anatomy of the ACL and its importance in ACL reconstruction. Eur J Orthop Surg Traumatol. 2013;23:747–52.
8. Cimino F, Volk BS, Setter D. Anterior cruciate ligament injury: diagnosis, management, and prevention. Am Fam Physician. 2010;82:917–22.
9. Hong SH, Choi JY, Lee GK, Choi JA, Chung HW, Kang HS. Grading of anterior cruciate ligament injury. Diagnostic efficacy of oblique coronal magnetic resonance imaging of the knee. J Comput Assist Tomogr. 2003;27:814–9.
10. Barrack RL, Bruckner JD, Kneisl J, Inman WS, Alexander AH. The outcome of nonoperatively treated complete tears of the anterior cruciate ligament in active young adults. Clin Orthop Relat Res. 1990;259:192–9.
11. Negahi Shirazi A, Chrzanowski W, Khademhosseini A, Dehghani F. Anterior cruciate ligament: structure, injuries and regenerative treatments. Adv Exp Med Biol. 2015;881:161–86.
12. Centeno CJ, Pitts J, Al-Sayegh H, Freeman MD. Anterior cruciate ligament tears treated with percutaneous injection of autologous bone marrow nucleated cells: a case series. J Pain Res. 2015;8:437–847.
13. Centeno C, Markle J, Dodson E, et al. Symptomatic anterior cruciate ligament tears treated with percutaneous injection of autologous bone marrow concentrate and platelet products: a non-controlled registry study. J Transl Med. 2018;16:246.
14. Pache S, Aman ZS, Kennedy M, et al. Posterior cruciate ligament: current concepts review. Arch Bone Jt Surg. 2018;6:8–18.
15. Wang D, Graziano J, Williams RJ 3rd, Jones KJ. Nonoperative treatment of PCL injuries: goals of rehabilitation and the natural history of conservative care. Curr Rev Musculoskelet Med. 2018;11:290–7.
16. Chirichella PS, Jow S, Iacono S, et al. Treatment of knee meniscus pathology: rehabilitation, surgery, and orthobiologics. PM R. 2019;11:292–308.
17. Fox AJ, Bedi A, Rodeo SA. The basic science of human knee menisci. Structure, composition, and function. Sports Health. 2012;4:340–51.
18. Makris EA, Hadidi P, Athanasiou KA. The knee meniscus: structure-function, pathophysiology, current repair techniques, and prospects for regeneration. Biomaterials. 2011;32:7411–31.

19. Nguyen JC, De Smet AA, Graf BK, Rosas HG. MR imaging-based diagnosis and classification of meniscal tears. Radiographics. 2014;34:981–99.
20. Mordecai SC, Al-Hadithy N, Ware HE, Gupte CM. Treatment of meniscal tears: an evidence based approach. World J Orthop. 2014;5:233–41.
21. Kaminski R, Kulinski K, Kozar-Kaminska K, et al. A prospective, randomized, double-blind, parallel-group, placebo-controlled study evaluating meniscal healing, clinical outcomes, and safety in patients undergoing meniscal repair of unstable, complete vertical meniscal tears (bucket handle) augmented with platelet-rich plasma. Biomed Res Int. 2018;2018:9315815.
22. Pujol N, Salle De Chou E, Boisrenoult P, Beaufils P. Platelet-rich plasma for open meniscal repair in young patients: any benefit? Knee Surg Sports Traumatol Arthrosc. 2015;23:51–8.
23. Blanke F, Vavken P, Haenle M, von Wehren L, Pagenstert G, Majewski M. Percutaneous injections of platelet rich plasma for treatment of intrasubstance meniscal lesions. Muscles Ligaments Tendons J. 2015;5:162–6.
24. Vangsness CT Jr, Farr J 2nd, Boyd J, Dellaero DT, Mills CR, LeRoux-Williams M. Adult human mesenchymal stem cells delivered via intra-articular injection to the knee following partial medial meniscectomy: a randomized, double-blind, controlled study. J Bone Joint Surg Am. 2014;96:90–8.
25. Andia I, Maffulli N. Use of platelet-rich plasma for patellar tendon and medial collateral ligament injuries: best current clinical practice. J Knee Surg. 2015;28:11–8.
26. Yoshida M, Marumo K. An autologous leukocyte-reduced platelet-rich plasma therapy for chronic injury of the medial collateral ligament in the knee: a report of 3 successful cases. Clin J Sport Med. 2019;29:e4–6.
27. Zhang Y, Jordan JM. Epidemiology of osteoarthritis. Clin Geriatr Med. 2010;26:355–69.
28. Brooks PM. The burden of musculoskeletal disease—a global perspective. Clin Rheumatol. 2006;25:778–81.
29. Felson DT, Lawrence RC, Dieppe PA, et al. Osteoarthritis: new insights. Part 1: the disease and its risk factors. Ann Intern Med. 2000;133:635–46.
30. Li G, Yin J, Gao J, et al. Subchondral bone in osteoarthritis: insight into risk factors and microstructural changes. Arthritis Res Ther. 2013;15:223.
31. Ilas DC, Churchman SM, McGonagle D, Jones E. Targeting subchondral bone mesenchymal stem cell activities for intrinsic joint repair in osteoarthritis. Future Sci OA. 2017;3:FSO228.
32. Sánchez M, Delgado D, Sánchez P, et al. Combination of intra-articular and intraosseous injections of platelet rich plasma for severe knee osteoarthritis: a pilot study. Biomed Res Int. 2016;2016:4868613.
33. Hernigou P, Auergan JC, Dubory A, Flouzat-Lachaniette CH, Chevallier N, Rouard H. Subchondral stem cell therapy versus contralateral total knee arthroplasty for osteoarthritis following secondary osteonecrosis of the knee. Int Orthop. 2018;42:2563–71.
34. Raeissadat SA, Ghazi Hosseini P, Bahrami MH, et al. The comparison effects of intra-articular injection of platelet rich plasma, plasma rich in growth factor, hyaluronic acid, and ozone in knee osteoarthritis; A one year randomized clinical trial. BMC Musculoskelet Disord. 2021;22:134.
35. Park YB, Kim JH, Ha CW, Lee DH. Clinical efficacy of platelet rich plasma injection and its association with growth factors in the treatment of mild to moderate knee osteoarthritis: a randomized double blind controlled clinical trial as compared with hyaluronic acid. Am J Sports Med. 2021;49:487–96.
36. Cole BJ, Vasili K, Hussey K, Pilz K, Fortier LA. Hyaluronic acid versus platelet rich plasma: a prospective, double-blind randomized controlled trial comparing clinical outcomes and effects on intra-articular biology for the treatment of knee osteoarthritis. Am J Sports Med. 2017;45:339–46.
37. Elksniņš-Finogejevs A, Vidal L, Peredistijs A. Intra-articular platelet rich plasma vs corticosteroids in the treatment of moderate knee osteoarthritis: a single enter prospective randomized controlled study with a 1-year follow up. J Orthop Surg Res. 2020;15:257.
38. Tucker JD, Lance LL, Duncan MB, et al. Randomized placebo-controlled analysis of the knee synovial environment following platelet rich plasma treatment for knee osteoarthritis. PM R. 2021;13:707–19.
39. Su K, Bai Y, Wang J, Zhang H, Liu H, Ma S. Comparison of hyaluronic acid and PRP intra-articular injection with combined intra-articular and intraosseous PRP injections to treat patients with knee osteoarthritis. Clin Rheumatol. 2018;37:1341–50.
40. Xu Z, He Z, Shu L, Li X, Ma M, Ye C. Intra-articular platelet rich plasma combined with hyaluronic acid injection for knee osteoarthritis is superior to platelet rich plasma or hyaluronic acid alone in inhibiting inflammation and improving pain and function. Arthroscopy. 2021;37:903–15.
41. Rahimzadeh P, Imani F, Faiz SHR, Entezary SR, Zamanabadi MN, Alebouyeh MR. The effects of injecting intra-articular platelet rich plasma or prolotherapy on pain score and functioning knee osteoarthritis. Clin Interv Aging. 2018;13:73–9.
42. Duymus TM, Mutlu S, Dernek B, Komur B, Aydogmus S, Kesiktas FN. Choice of intra-articular injection in treatment of knee osteoarthritis: platelet rich plasma, hyaluronic acid or ozone options. Knee Surg Sports Traumatol Arthrosc. 2017;25:485–92.
43. Lin KY, Yang CC, Hsu CJ, Yeh ML, Renn JH. Intra-articular injection of platelet rich plasma is superior to hyaluronic acid or saline solution in the treatment of mild to moderate knee osteoarthritis: a randomized, double blind, triple parallel, placebo-controlled clinical trial. Arthroscopy. 2019;35:106–17.
44. Pishgahi A, Abolhasan R, Shakouri SK, et al. Effect of dextrose prolotherapy, platelet rich plasma and autologous conditioned serum on knee osteoarthritis: a randomized clinical trial. Iran J Allergy Asthma Immunol. 2020;19:243–52.
45. Li D, Wang Y, Shen Y. A randomized study protocol comparing the platelet rich plasma with hyaluronic acid in the treatment of symptomatic knee osteoarthritis. Medicine (Baltimore). 2021;100:e23881.
46. Filardo G, Kon E, Di Martino A, et al. Platelet rich plasma vs hyaluronic acid to treat knee degenerative pathology: study design and preliminary results of a randomized controlled trial. BMC Musculoskelet Disord. 2012;13:229.
47. Cerza F, Carni S, Carcangiu A, et al. Comparison between hyaluronic acid and platelet-rich plasma, intra-articular infiltration in the treatment of gonarthrosis. Am J Sports Med. 2012;40:2822–7.
48. Ghai B, Gupta V, Jain A, Goel N, Chouhan D, Batra YK. Effectiveness of platelet rich plasma in pain management of osteoarthritis knee: double blind, randomized comparative study. Braz J Anesthesiol. 2019;69:439–47.
49. Huang Y, Liu X, Xu X, Liu J. Intra-articular injections of platelet rich plasma, hyaluronic acid or corticosteroids for knee osteoarthritis: a prospective randomized controlled study. Orthopade. 2019;48:239–47.
50. Patel S, Dhillon M, Aggarwal S, Marwaha N, Jain A. Treatment with platelet rich plasma is more effective than placebo for knee osteoarthritis: a prospective, double blind, randomized trial. Am J Sports Med. 2013;41:356–64.
51. Chu J, Duan W, Yu Z, et al. Intra-articular injections of platelet rich plasma decrease pain and improve functional outcomes than sham saline in patients with knee osteoarthritis. Knee Surg Sports Traumatol Arthrosc. 2022;30:4063–71.
52. Spakova T, Rosocha J, Lacko M, Harvanová D, Gharaibeh A. Treatment of knee joint osteoarthritis with autologous platelet rich plasma in comparison with hyaluronic acid. Am J Phys Med Rehabil. 2012;91:411–7.
53. Paterson K, Nicholls M, Bennell KL, Bates D. Intra-articular injection of photo activated platelet rich plasma in patients with knee

osteoarthritis: a double blind, randomized controlled pilot study. BMC Musculoskelet Disord. 2016;17:67.
54. Forogh B, Mianehsaz E, Shoaee S, Ahadi T, Raissi GR, Sajadi S. Effect of single injection of platelet rich plasma in comparison with corticosteroid on knee osteoarthritis: a should blind randomized clinical trial. J Sports Med Phys Fitness. 2016;56:901–8.
55. Montañez-Heredia E, Irízar S, Huertas PJ, et al. Intra articular injections of platelet rich plasma versus hyaluronic acid in the treatment of osteoarthritic knee pain: a randomized clinical trial in the context of the Spanish national health care system. Int J Mol Sci. 2016;17:1064.
56. Nunes-Tamashiro J, Natour J, Natour J, et al. Intra-articular injection with platelet rich plasma compared to triamcinolone hexacetonide or slime solution in knee osteoarthritis: a double blinded randomized controlled trial with one year follow up. Clin Rehabil. 2022;36:900–15.
57. Louis ML, Magalon J, et al. Growth factors level determine efficacy of platelets rich plasma injection in knee osteoarthritis: a randomized double blind noninferiority trial compared with viscosupplementation. Arthroscopy. 2018;34:1530–40.
58. Wang YC, Lee CL, Chen YJ, et al. Comparing the efficacy of intra-articular single platelet rich plasma versus novel crosslinked hyaluronic acid for early-stage knee osteoarthritis: a prospective, double blind, randomized controlled trial. Medicina (Kaunas). 2022;58:1028.
59. Lisi C, Perotti C, Scudeller L, et al. Treatment of knee osteoarthritis: platelet derived growth factors vs hyaluronic acid. A randomized controlled trial. Clin Rehabil. 2018;32:330–9.
60. Baria MR, Flanigan DC. Intra-articular platelet-rich plasma vs placebo injection and pain and medial tibial cartilage volume in patients with knee osteoarthritis. JAMA. 2022;327:1184–5.
61. Vega A, Martín-Ferrero MA, Del Canto F, et al. Treatment of knee osteoarthritis with allogeneic bone marrow mesenchymal stem cells: a randomized controlled trial. Transplantation. 2015;99:1681–90.
62. Gupta PK, Chullikana A, et al. Efficacy and safety of adult human bone marrow derived, cultured, pooled, allogeneic mesenchymal stromal cells (Stempeucel®): preclinical and clinical trial osteoarthritis of the knee joint. Arthritis Res Ther. 2016;18:301.
63. Khalifeh Soltani S, Forogh B, Ahmadbeigi N, et al. Safety and efficacy of allogenic placental mesenchymal stem cells for treating knee osteoarthritis: a pilot study. Cytotherapy. 2019;21:54–63.
64. Lee WS, Kim HJ, Kim KI, Kim GB, Jin W. Intra-articular injection of autologous adipose tissue derived mesenchymal stem cells for the treatment of knee osteoarthritis: a phase IIb randomized placebo controlled clinical trial. Stem Cells Transl Med. 2019;8:504–11.
65. Lu L, Dai C, Zhang Z, et al. Treatment of knee osteoarthritis with intra-articular injection of autologous adipose derived mesenchymal progenitor cells: a prospective randomized double blind active controlled phase IIb clinical trial. Stem Cell Res Ther. 2019;10:143.
66. Anz AW, Plummer HA, Cohen A, Everts PA, Andrews JR, Hackel JG. Bone marrow aspirate concentrate is equivalent to platelet rich plasma for the treatment of knee osteoarthritis at 2 years: a prospective randomized trial. Am J Sports Med. 2022;50:618–29.
67. Lamo-Espinosa JM, Blanco JF, Sánchez M, et al. Phase II multicenter randomized controlled clinical trial on the efficacy of intra-articular injection of autologous bone marrow mesenchymal stem cells with platelet rich plasma for treatment of knee osteoarthritis. J Transl Med. 2020;18:356.
68. Hernigou P, Delambre J, Quiennec S, Poignard A. Human bone marrow mesenchymal stem cell injection in subchondral lesions of knee osteoarthritis: a prospective randomized study versus contralateral arthroplasty at a mean fifteen year follow up. Int Orthop. 2021;45:365–73.
69. Wang Y, Jin W, Liu H, et al. [Curative effect of human umbilical cord mesenchymal stem cells by intra-articular injection for degenerative knee osteoarthritis]. Zhongguo Xiu Fu Chong Jian Wai Ke Za Zhi. 2016;30:1472–7.
70. Dulic O, Rasovic P, Lalic I, et al. Marrow aspirate concentrate versus platelet rich plasma or hyaluronic acid for the treatment of knee osteoarthritis. Medicina (Kaunas). 2021;57:1193.
71. Matas J, Orrego M, Amenabar D, et al. Umbilical cord derived mesenchymal stromal cells for knee osteoarthritis: repeated MSC dosing is superior to a single MSC dose and to Hyaluronic acid in a controlled randomized phase I/II trial. Stem Cells Transl Med. 2019;8:215–24.
72. Shapiro SA, Kazmerchek SE, Heckman MG, Zubair AC, O'Connor MI. A prospective, single-blind, placebo-controlled trial of bone marrow aspirate concentrate for knee osteoarthritis. Am J Sports Med. 2017;45:82–90.
73. Hernigou P, Bouthors C, Bastard C, Flouzat Lachaniette CH, Rouard H, Dubory A. Subchondral bone or intra-articular injection of bone marrow concentrate mesenchymal stem cells in bilateral knee osteoarthritis: what better postpone knee arthroplasty at fifteen years? A randomized study. Int Orthop. 2021;45:391–9.
74. Sánchez M, Fiz N, Guadilla J, et al. Intraosseous infiltration of platelet-rich plasma for severe knee osteoarthritis. Arthrosc Tech. 2014;3:e713–7.
75. Manchikanti L, Sanapati M, Kaye AD, Kaye AM. Antiplatelet and anticoagulant therapy in regenerative medicine. In: Manchikanti L, Navani A, Atluri S, Sanapati M, editors. Manchikanti's essentials of regenerative medicine in interventional pain management. 2nd ed. New York: Springer; 2023.

# 21 Regenerative Therapy of Ankle and Foot and Related Joints, Ligaments and Tendons

Rowan Paul

## Introduction

Foot and ankle pain can be debilitating. Instability, sprains, resulting in tendinosis, fasciitis/fasciosis, arthritis, neuritis and osteochondral defects are common [1]. Appropriate dose, precisely guided orthobiologic injection therapy is a strategy that can offer significant relief for many conditions. Knowledge of functional anatomy, kinetic chain forces, injection treatment algorithms should focus on treating the functional unit. This chapter discusses anatomy, pathophysiology, physical exam, the use of fluoroscopy and ultrasound (US) guidance. Focus on the biologic basis for the use of regenerative and restorative therapies such as platelet-rich plasma (PRP), bone marrow concentrates (BMC), micronized adipose grafts will be discussed. The data, interventional techniques and complications will be discussed. Regulatory limitations will be considered. Regenerative orthobiologic injection therapy can be effective for chronic foot and ankle pain, particularly due to instability and poor local microcirculation. Novel injection techniques such as intraosseous/subchondral injections for severe joint arthritis with bone marrow lesions, osteonecrosis, as well as hydrodissection for neural entrapment, and percutaneous tarsal tunnel release under US guidance may be tried as an interventional treatment gap tool, prior to surgical alternative in select situations for pain and functional improvement.

## Anatomy

The anatomy of the ankle and foot is not well known to the interventional pain physician, and so we will spend time on these structures and anatomic terms. In the foot and ankle, there are 26 bones, 33 joints, and more than 100 tendons, muscles, and ligament attachments that create an intricate balance and inter-relationship for the biomechanics that support standing and locomotion. Pain-free status, strength, and flexibility is dependent on integrity of the ligaments and tendons that maintain the dynamic alignment of the interlocking puzzle of the bones.

## Bones, Joints, and Ligaments

There are seven bones that make up the ankle and hindfoot: the calcaneus (which makes up the majority of the heel region), talus, navicular, lateral cuneiform, intermediate cuneiform, medial cuneiform, and cuboid. The bones of the foot consist of the five metatarsal bones and then five proximal phalanges; the great toe is made up of a proximal and distal phalange, while the other four toes have a proximal, intermediate, and distal phalange (Figs. 21.1 and 21.2). The sinus tarsi (the "eye of the foot") is a small bony depression between the talus and calcaneus in the lateral portion of the ankle. The ankle joint is a diarthrodial articulation with three main articulations: the talocrural joint (also known as the tibiotalar joint or the mortise joint), the subtalar joint, and the tibiofibular joint. The talocrural joint consists of the rounded dome of the talus, the tibial plafond (the articular surface of the distal tibia, which is the "ceiling" of the joint), the medial malleolus (which is the medial portion of the tibia), and the lateral malleolus (which is the lateral fibula). This joint allows dorsiflexion and plantar flexion like a hinge joint. It is the only example of a true mortise (hinge) joint in the human body.

The subtalar joint is the articulation between the talus and the calcaneus, which acts like a dual ball and socket joint, and consists of two joint cavities:

- The anterior subtalar joint (talocalcaneonavicular joint), which consists of the head of the talus, the anterosuperior facets, the sustentaculum tali of the calcaneus, and the proximal concave surface of the tarsal navicular

R. Paul (✉)
Dartmouth Geisel School of Medicine, San Francisco and Palo Alto, CA, USA

Regen Core Method, San Francisco and Palo Alto, CA, USA
e-mail: Dr.Paul@regencoremethod.com


A. Navani et al. (eds.), *Essentials of Regenerative Medicine in Interventional Pain Management*,
https://doi.org/10.1007/978-3-031-50357-3_21

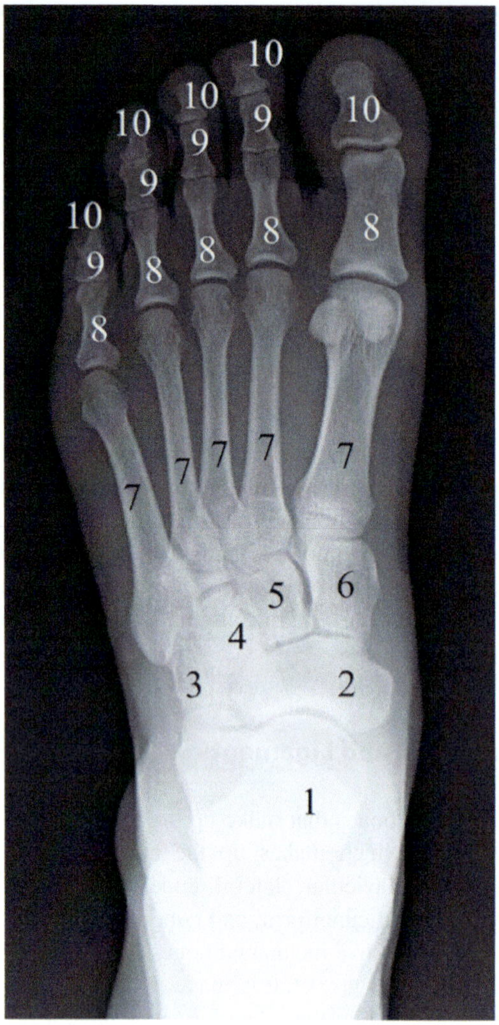


**Fig. 21.1** AP x-ray of the foot and ankle. 1 = talus; 2 = navicular; 3 = cuboid; 4 = lateral cuneiform; 5 = intermediate cuneiform; 6 = medial cuneiform; 7 = metatarsal; 8 = proximal phalanges; 9 = middle phalanges; 10 = distal phalanges

- The posterior subtalar joint, which consists of the inferoposterior facet of the talus and the superoposterior facet of the calcaneus

The talofibular joint consists of the distal tibia and fibula syndesmosis (the term for an "immovable" joint) and functions as the "roof" of the mortise. These bones and joints create three arches that collaboratively support and maintain the foot: the transverse arch, medial arch, and longitudinal arch. These bones are held together by the anteroinferior tibiofibular ligament (AITL) and posteroinferior tibial ligament (PITL). The tibia and fibula then articulate with the talus and the calcaneus and are held together by the calcaneofibular ligament (CFL) (sometimes also called fibulocalcaneal), which travels from the tip of the distal fibula inferiorly and slightly posteriorly to attach on a tubercle on the lateral calcaneus, as well as the anterior talofibular ligament (ATFL), and the posterior talofibular ligament (PTFL), which lie on either side of the CFL and travel from the fibula to lateral talus anteriorly and posteriorly, respectively. The deltoid ligament is on the medial aspect of the ankle, superficial to the medial talocalcaneal ligament, while the plantar calcaneonavicular ligament (spring ligament) supports the arch of the foot. The proximal tibiofibular joint (PTFJ) at the knee transfers the torsional stresses from the ankle to the knee (Figs. 21.3 and 21.4). The spring ligament is a ligament complex made up of a subgroup of the superomedial ligament, the medioplantar oblique ligament, and the inferoplantar longitudinal ligament.

## Muscles and Tendons

There are not many muscles in the ankle region, but there are a multitude of tendons that cross the ankle region. The foot has both plantar and dorsal muscle groups. The greatest complexity occurs in the plantar group with four distinct layers, while the dorsal foot has only one intrinsic muscle, the extensor digitorum brevis.

At the level of the ankle, pathology generally occurs from the multitude of tendons that cross the ankle region. Anteriorly, the tibialis anterior tendon (TAT), extensor hallucis longus tendon (EHLT), and extensor digitorum longus tendon (EDLT) extend to the toes dorsally. Coursing medially in the tarsal tunnel is posterior tibial tendon (PTT), the flexor digitorum longus tendons (FDLT) and flexor hallucis longus tendon (FHLT), which attach along the medial aspect of the arch to the toes ventrally to bend the toes and invert and plantar flex the foot. Laterally stabilization via eversion and plantar flexion occurs via the peroneus longus tendon (PLT) and peroneus brevis tendon (PBT), which travel together in the same sheath posterior to the fibula within the fibular groove to attach on to the head of the fifth metatarsal bone, with the PBT located anteromedial to the PLT; the superior and inferior peroneal retinaculum (SPR and IPR) holds the two tendons in place within the retro-malleolar groove. The plantar fascia is a dense aponeurosis that is attached to the calcaneal tuberosity proximally and the metatarsal heads distally, and it acts as the "bow" of a bow and arrow, to maintain the integrity of the metatarsal arc.

## Nerves

The nerves in this region include the common peroneal (fibular) nerve (which passes along the head of the fibula and innervates the PTFJ), which then divides into the superficial peroneal (fibular) nerve (which innervated the distal 2/3 of the lateral leg and most of the dorsum of the foot) [2], and the deep peroneal (fibular) nerve (which innervates the dorsum

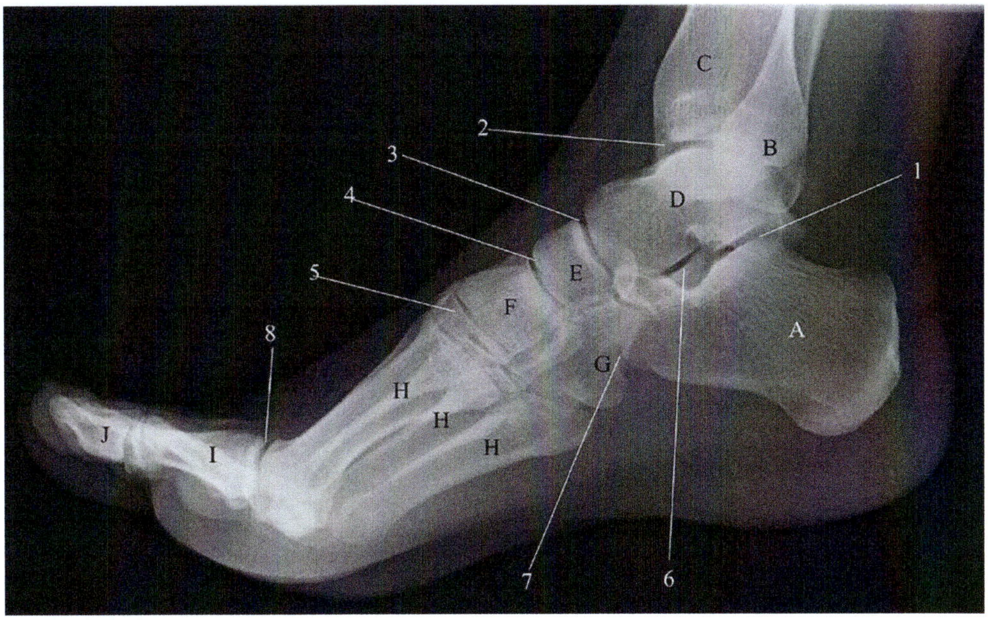


**Fig. 21.2** Lateral x-ray of the foot and ankle. A = calcaneus; B = fibula; C = tibia; D = talus; E = navicular; F = cuneiform; G = cuboid; H = metatarsals; I = phalanges. 1 = talocalcaneal joint; 2 = tibiotalar joint; 3 = talonavicular joint; 4 = naviculocuboid joint; 5 = tarsometatarsal joint; 6 = sinus tarsi; 7 = calcaneocuboid joint; metatarsophalangeal joint (MTP)

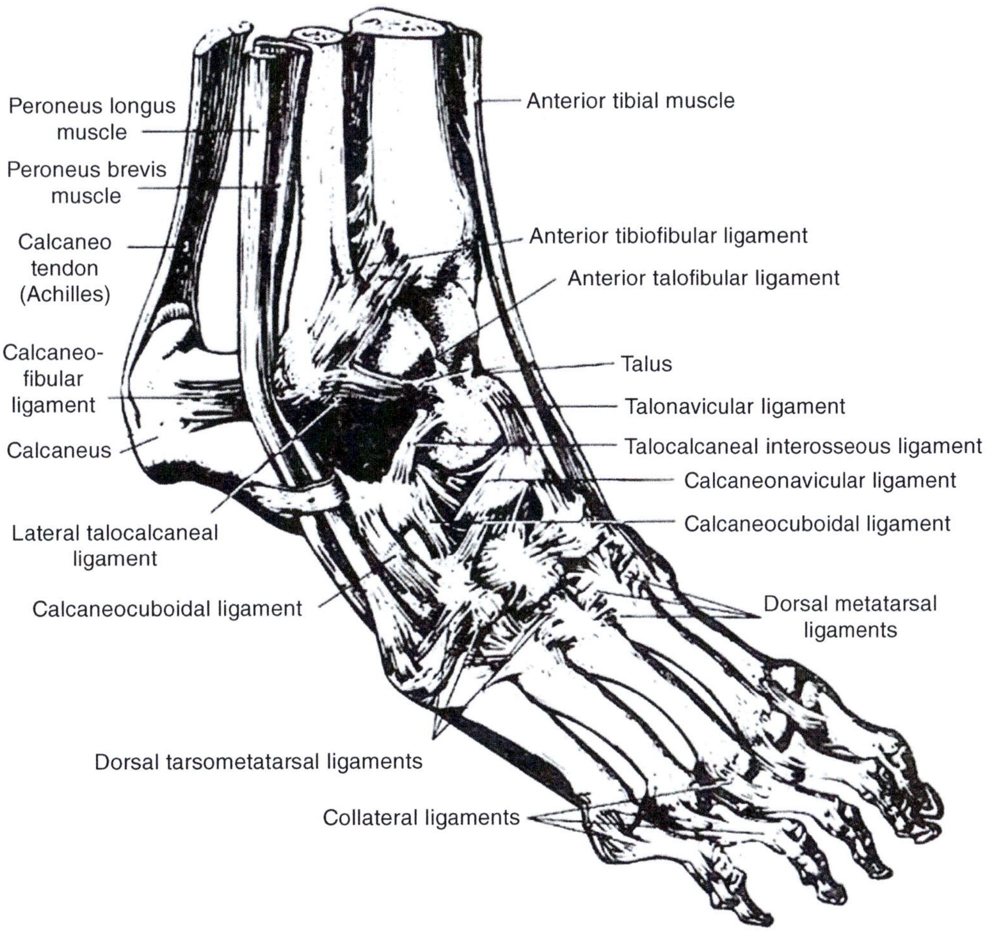


**Fig. 21.3** Lateral view of the ligaments of the foot. Image from Hackett GS. *Ligament & Tendon Relaxation—Treated by Prolotherapy*. 3rd ed. Charles C. Thomas, Springfield, IL, 1958. Image courtesy of the Hackett Hemwall Foundation, with permission

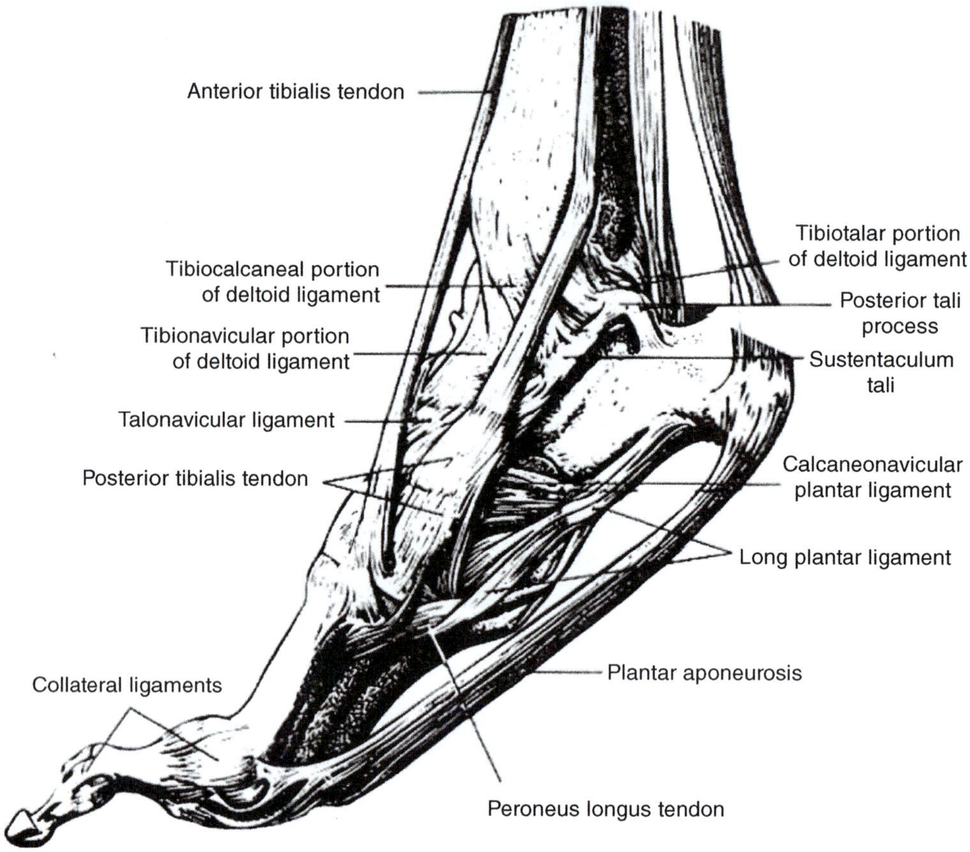


**Fig. 21.4** Medial view of the ligaments of the foot. **Image from** Hackett GS. *Ligament & Tendon Relaxation—Treated by Prolotherapy*. 3rd ed. Charles C. Thomas, Springfield, IL, 1958. Image courtesy of the Hackett Hemwall Foundation, with permission

of the foot and the space between the first and second toes) [3]. The superficial saphenous nerve innervates the medial anterior foot, frequently down to the great toe [4]. The sural nerve innervates the posterior and lateral aspect of the ankle and foot, and pathology of this nerve is often associated with paresthesia over the lateral ankle and the dorsum and lateral aspect of the foot [5].

The tarsal tunnel, on the medial side of the ankle, is a canal that serves neuromuscular, vascular, and tendon structures, composed of the calcaneus, talus, and the superior and inferior retinaculum. The mnemonic for the tarsal tunnel contents is "Tom, Dick, and Very Nervous Harry" (tibialis posterior tendon, flexor digitorum tendon, posterior tibial artery, vein(s) and nerve, and flexor hallucis tendon) (Fig. 21.5) [6].

The posterior tibial nerve then divides into the medial plantar nerve (MPN), the lateral plantar nerve (LPN), and medial calcaneal nerve (MCN) (Fig. 21.6). The MPN innervates the flexor digitorum brevis (FDB), abductor hallucis (AbH), flexor hallucis brevis (FHB), and the first lumbrical muscles; these plantar muscles flex the toes, especially at the metatarsophalangeal (MTP) joint of the great toe. Patients with MPN pathology complain of burning pain, dysesthesias, and aching in the medial arch of the foot that increases with activity and may radiate to the plantar aspect of the first and second toes or to the heel [7, 8]. The LPN innervates the flexor digiti minimi brevis muscle (FDMB), the quadratus plantae muscle (QP), and the adductor hallucis muscle (AdH), which together help to maintain the arch of the foot, flex the toes at the MTP joints, and adduct the toes. Patients with LPN pathology usually have burning pain, paresthesia, and numbness in the lateral side of the sole that can extend to the lateral toes, and the first branch off the LPN, known as the inferio-calcaneal or Baxter's nerve may contribute to heel pain due to innervation of the calcaneal periosteum [9–13]. The MCN innervates the AbH muscle; and these patients complain of sharp "neuralgic" heel pain with burning and occasional tingling or sensory disturbances in the medial heel that worsens during the day, or the longer they stand or walk. The interdigital nerves, which are the terminal branches of the MPN and LPN nerves, innervate either side of each toe; pathology of these nerves cause burning and sharp pain that radiates to the toes [14, 15].

## Vascular

The main arteries to the foot generally arise from the popliteal artery and include the posterior tibial artery, which runs behind the medial malleolus within the tarsal tunnel, and the

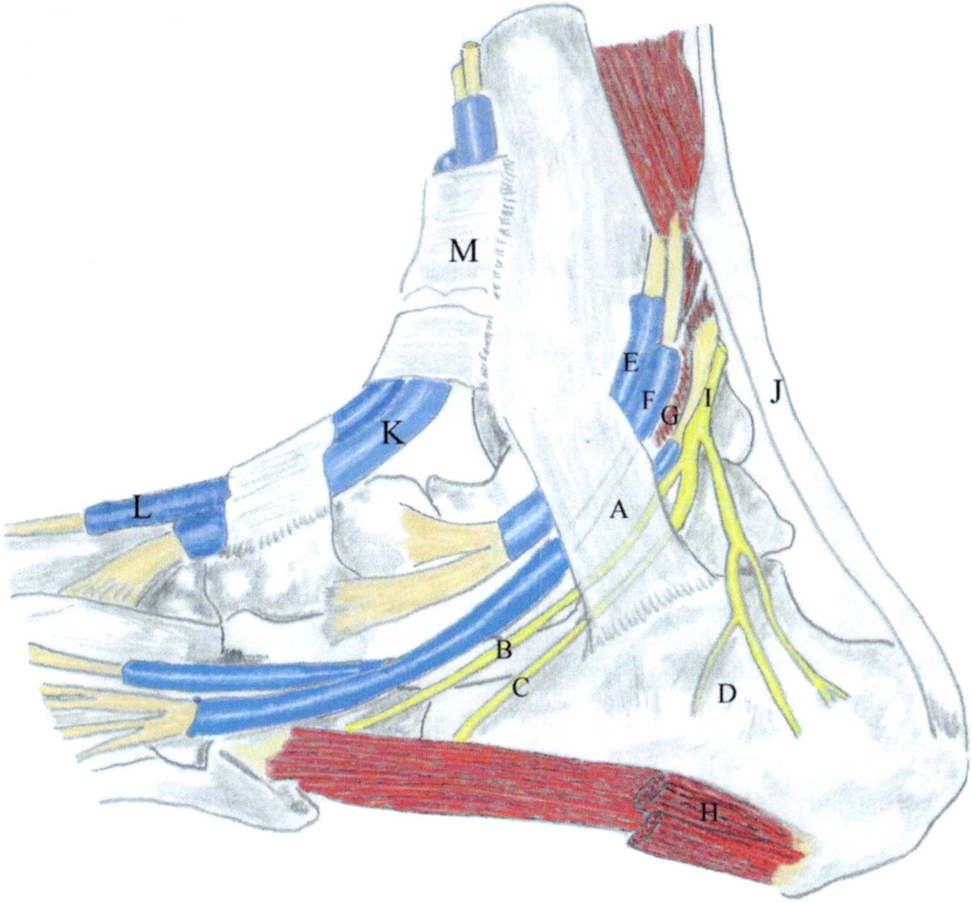


**Fig. 21.5** Anatomy of the tarsal tunnel and the medial foot. A = flexor retinaculum; B = medial plantar nerve; C = lateral plantar nerve; D = medial calcaneal nerve; E = tibialis posterior tendon; F = flexor digitorum longus tendon; G = posterior tibial artery; H = adductor hallicus muscle; I = posterior tibial nerve; J = Achilles tendon; K = extensor hallicus longus tendon; L = extensor digitorum tendon; M = extensor retinaculum. Image courtesy of Michael Brown, MD, with permission

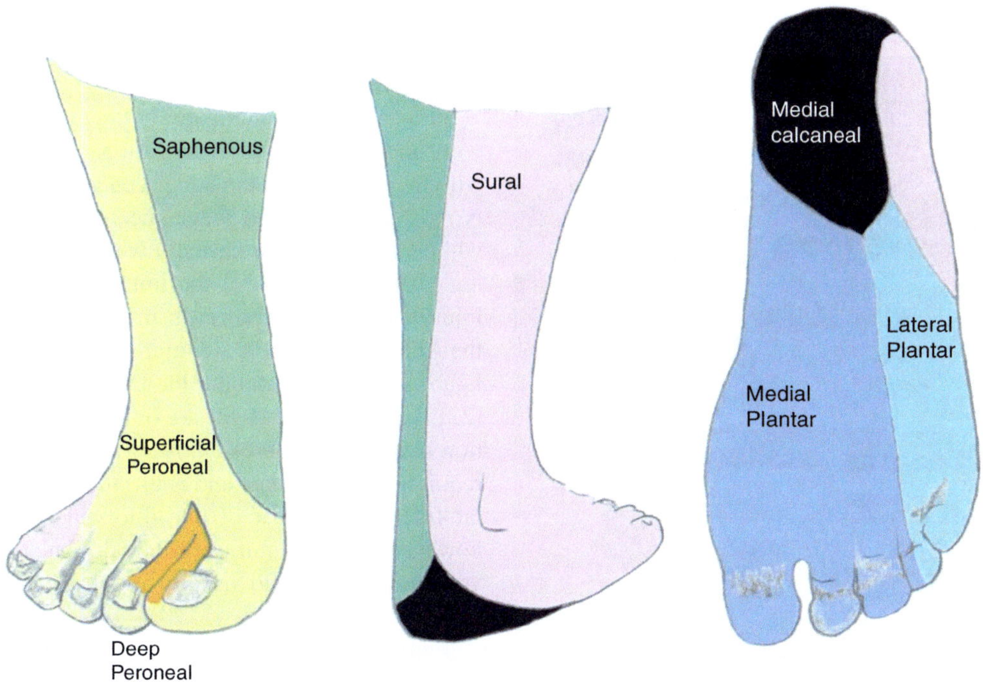


**Fig. 21.6** Distribution of the nerves of the foot. Image courtesy of Michael Brown, MD

anterior tibial artery, which becomes superficial at the ankle midway between the malleoli and later terminates anteriorly as the dorsalis pedis artery, and the fibular (or peroneal) artery, which courses postero-laterally.

### Movement

Dorsiflexion and plantar flexion describe the motion of the tibiotalar joint, inversion and eversion involve the movement of the subtalar joint, and internal rotation and external rotation are the terms used to describe the combined tibiotalar and talocalcaneal movements, while pronation and supination refer to the movements of the midfoot and forefoot [16].

## Pathophysiology

The commonly experienced chronic conditions of pain result from the extremes in the biomechanics of the foot. Pes planus and pes cavus cause a stress overload reaction with degenerative syndromes rather than the inflammatory outcomes of sport or accident pathologies. The resultant pain pattern will parallel the unresolved acute syndromes [17].

Hypermobility of the joints of the foot can be the primary factor in chronic sprains, recurrent capsulitis, heel pain, metatarsalgia, and bunions [18]. Ankle sprains are caused by an unexpected twisting movement of the ankle, either by supination (inversion injury causing a lateral ankle sprain) (Fig. 21.7), which is much more common, or by pronation

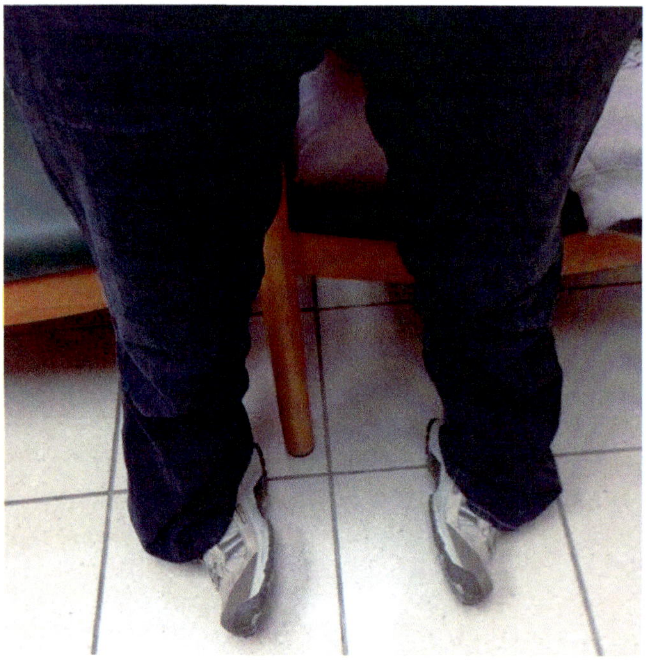

**Fig. 21.7** Inversion movement in a patient with ligament laxity

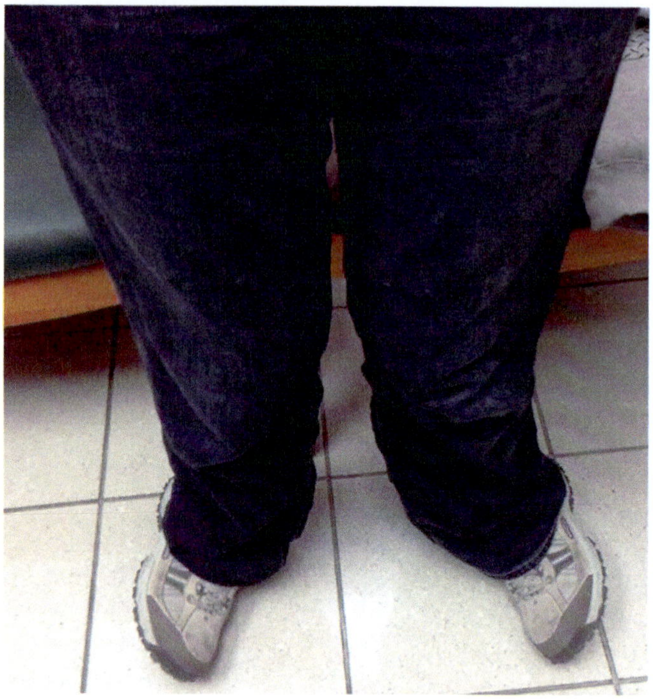

**Fig. 21.8** Eversion movement in a patient with ligament laxity

(eversion injury causing a medial ankle sprain) forces (Fig. 21.8). Because the fibula extends more inferiorly than the tibia does, there is a physical block to eversion. Since inversion is more common, the injuries of the lateral ligament complex (ATFL, PTFL, and CFL) are by far more common [19]. The third type of ankle injury is the high ankle sprain (syndesmotic sprain), which is an injury to the ligaments holding the fibula onto the tibia at the ankle, as well as at the knee.

In a supination lateral sprain, there is an ankle inversion with internal twisting of the foot as well as plantar flexion with the subtalar joint in adduction and inversion. The ATFL is most likely injured when the foot is in plantar flexion, while CFL injuries are usually when the foot is in dorsiflexion [20]. About 65% of the time, there is only an isolated injury to the ATFL; however, in 20% of ankle sprains, both the ATFL and the CFL are injured [21]. The CFL is the only ligament that bridges both the talocrural and subtalar joints [22]; injuries to the PTFL are only seen with severe dislocation of the ankle joint [23]. A sprain of the deltoid ligament requires severe trauma and is usually associated with a fibular fracture or a tear of the tibiofibular syndesmosis [24]; however, isolated injuries of the talofibular syndesmosis are very uncommon and usually require a hyper-dorsiflexion injury [25].

Plantar heel pain is a common problem affecting about 10% of the population at some time in their life [26]. The most common causes are plantar fasciitis, heel spurs (which may represent the consequence of the heel pain and not the cause), fat pad dysfunction, stress fractures, biomechanical dysfunction [26], and gastrocnemius equinus [27]. However, another cause of heel pain can be entrapment of the posterior tibial nerve in the tarsal tunnel, which causes pain and tingling in a large area of the medial and plantar foot. The MCN, a branch of the tibial nerve (or rarely the lateral plantar nerve), causes pain in the anterior-medial plantar heel region, and is often misdiagnosed as plantar fasciitis [26].

Achilles tendinopathies, one of the most common overuse injuries in sports medicine, is similarly a source of chronic pain and dysfunction [28]. Unless acute and requiring surgical repair, it can be a vexing condition recalcitrant to most classic conservative interventions. The importance of accurate assessment cannot be overstated. Posterior heel pain can also be from insertional enthesopathy, bony ostosis from a Haglund's deformity, or from retrocalcaneal bursitis, which all must be addressed differently.

Common foot injuries include lateral talar fractures, trigonum injuries, Lisfranc injuries, "turf toe," peroneus syndrome, and navicular stress fractures [29]. Lateral talar fractures, also known as "snowboarder's fractures," can be difficult to detect on plain x-rays, but leaving it untreated can lead to osteoarthritis. About 5–10% of the world's population has the trigonum (an accessory bone) on the posterior-lateral aspect of the talus, which can be misdiagnosed as Achilles tendonitis or flexor hallucis longus injuries. Injury to the Lisfranc joint (also known as the tarsometatarsal complex) can be subtle, but misdiagnosis can lead to chronic pain. Chronic mild injury can lead to micro-instability, which can progress to traumatic arthritis of the Lisfranc joint. "Turf toe" describes a sprained great toe, usually caused by hyperdorsiflexion of the first MTP, which, untreated, can lead to joint arthritis. Pain from a non-healing navicular fracture is nonspecific and therefore easily missed. Painful peroneum syndrome similarly can be a frustrating clinical variant of lateral hindfoot pain, with localized tenderness along the distal course of the PLT at the cuboid tunnel.

Sinus tarsi syndrome presents with localized pain at the sinus tarsi with a feeling of instability on uneven surfaces and pain with weight bearing. This commonly results after severe pronation in patients with chronic lateral ankle instability or as seen in the chronic lateral overload of a cavus foot [30].

There are also a group of patients with congenital hypermobility syndromes, such as benign joint hypermobility syndrome (BJHS), Ehlers-Danlos syndrome (EDS), and Marfan's syndrome. BJHS is estimated to range in from 10 to 30% of adults, which is clearly not insignificant. Less common is EDS, which affects at least one per 5000 patients, while Marfan's syndrome also affects one in 5000 patients [31]. These hypermobile patients may be at increased risk of ligament laxity and progressive instability.

## Physical Examination

It is the physical examination that guides and directs treatment, by identifying the areas of tenderness to palpation. Structures that are not tender are unlikely to be pain generators, though the corollary is not always true, since an area could be tender secondary to a different pain generator. For instance, the plantar heel might hurt due to medial calcaneal neuralgia, but the plantar fascia calcaneal attachment might be tender due to excessive toe walking to avoid contact with the nerve. The history of the onset of pain then becomes important to differentiate the initial trigger from the subsequent conditions.

The clinician should specifically palpate the ligaments at the medial malleolus, the lateral malleolus, the base of the fifth metatarsal, the tibialis posterior and peroneal tendons, the plantar fascial attachments, the Achilles attachment, the first MCP, and the metatarsal bones. In the medial hindfoot, particularly at the sustentaculum tali, it is important to examine the spring ligament, deltoid, and medial talocalcaneal ligament sites; these are often the source of pain in the flat foot, with or without PTT symptoms.

The maximal points of tenderness for lateral ankle ligament pathology should be at the ATFL and/or CFL areas, although a patient with a medial sprain would be expected to be tender over the deltoid ligament. To evaluate lateral ankle ligament pathology, passive inversion of the ankle with plantar flexion (which puts strain on the ATFL) will cause pain. Stressing the ankle in inversion (the talar tilt test) will also stress the ATFL; place the patient supine with the gastrocnemius muscle relaxed by flexing the knee, and tilt the talus from side to side into adduction and abduction. If a firm end point cannot be felt, suspect additional damage to the CFL [20]. If a high ankle or fibular injury is suspected, use the fibular compression test, or squeeze test, squeezing the tibia and fibula together at the midpoint of the lower leg. Pain down the length of the fibula suggests a high ankle sprain or a Maisonneuve fracture (spiral fracture of the proximal fibula) [32].

Subluxation and bowing of the peroneal tendons occurs when the SPR becomes detached from its attachment on the posterior border of the fibula and/or the lateral surface of the fibula. Patients complain of a painful snapping sensation along the lateral aspect of the ankle associated with a sense of instability, especially in patients with high arches [33].

The sinus tarsi syndrome presents with tenderness to palpation over the sinus tarsi on the lateral aspect of the ankle and pain on forced inversion of the foot. Clinical diagnosis of peroneum syndrome is often best identified with single stance heel rise and varus inversion stress along with resisting plantar flexion of the first ray. Tenderness will be localized along the distal course of the PLT at the cuboid tunnel along with positive radiographic evidence of the accessory bone.

Intermetatarsal bursitis occurs when a fluid filled sac between the metatarsal heads, or sometimes more proximal just distal to the tarsometatarsal joint, becomes inflamed. Morton's neuroma occurs when the interdigital nerve becomes inflamed, and sometimes fibrotic, usually due to compression at the metatarsal heads. Both intermetatarsal bursitis and Morton's neuroma can present similarly, with pain in the sole of the foot as if walking on a pebble, sometimes with decreased sensation, tingling or burning in the web between the toes, worse with walking, particularly with narrow or compressive shoes [34].

## Imaging (Fluoroscopy and Ultrasound)

The use of fluoroscopy can aid in the localization of the joint lines for intraarticular joint injections as well as ligamentous attachments. US may be useful to examine the tendons, ligaments, nerves of the foot and ankle, and is particularly useful for correlating pain with anatomy, as well as dynamic stress testing. Careful diagnostic high resolution musculoskeletal US may pick up pathology missed on even high resolution (1.5–3 T) magnetic resonance imaging (MRI) [35]. Musculoskeletal elastography is a relatively novel way of looking qualitatively and quantitatively at the mechanical properties of tissues [36].

## Diagnosis

X-rays can show broken bones of the ankle and foot, though they hold only clues to ligamentous injuries. For instance, for the high ankle sprain, x-rays may show widening of the tibia and fibula mortise or a fracture of the medial malleolus, which would suggest instability, though normal x-rays do not rule out significant ligamentous injuries. One study found no association between the tibiofibular gap seen on radiographs with syndesmotic injury found on MRI [37]. MRIs may give the greatest detail of ligamentous injuries and enthesopathies [38]. Electromyographic examinations are rarely diagnostic but may be useful as part of the evaluation of individuals who have suffered severe ankle sprains. They have shown that 80% of these patients have some degree of peroneal nerve injury [33].

Milz and colleagues [38] evaluated 20 patients with ankle injuries using US with MRI as the "gold standard"; US had a 92% sensitivity and an 83% specificity for injury and rupture of the ATFL.

For the evaluation of plantar heel pain, the history is important, and many conditions mimic the presumptive diagnosis of plantar fasciitis. Pain from plantar fasciitis and heel spurs is insidious; although spurs are often present, the etiology is less likely to cause pain from the bone prominence. Heel spurs pain is located specifically in the anterior plantar heel, while plantar fasciitis may extend to the MTP joints of the first four toes [26]. On the other hand, MCN pathology usually starts with an extrinsic trauma along the medial border of the calcaneus [26]. Patients with plantar fasciitis usually have pain on initial weight bearing ("first step pain") and then improve, while MCN pain usually gets worse gradually with continued weight bearing.

Sometimes the etiology of the ankle or foot pain is unclear. Tendon injections, by landmark or US guidance, with small volumes of local anesthetic with or without steroids, can confirm the ligamentous injury. Intra-articular joint injections under fluoroscopy or US can confirm the integrity of the joint capsule. If a nerve has been injured along with the ligament, there can be a persistent pain that does not get better as the ligament heals. These nerve entrapments may result in a complex regional pain syndrome (CRPS). These nerves can be identified by physical examination (Fig. 21.9) and confirmed by injection of small volumes of local anesthetic and steroids directed by landmarks, fluoroscopy, or US.

Hydrodissection is a novel injection technique that involves injecting local anesthetic, saline, 5% dextrose or an orthobiologic like PRP to separate the nerve from surrounding fascia, tissue or surrounding structures. Hydrodissection is used for both diagnostic and therapeutic purposes to look for and potentially release regions of entrapped or compressed nerves [39].

## Biologic Basis

A variety of materials have been used for regenerative treatment of the ankle. The oldest of the techniques, fibroproliferative therapy or prolotherapy, involves the injection of a deliberately irritating medication, usually 50% dextrose plus local anesthetic, into the fibro-osseous junction to stimulate the proliferation of fibroblasts to lay down new tissue [40]. A variety of concentrations of dextrose have been used (typically 12.5–25%) with local anesthetic (usually lidocaine). Hauser and Feister [41] described using 10 mL of 12.5% dextrose with 1 IU of human growth hormone (HGH).

Although multiple theories regarding prolotherapy exist, most accept that dextrose acts as a biologically inflammatory agent that stimulates repair cascade without tissue deformation, and along with the technique of needling, takes a relatively avascular fibrous structure and causes an acute inflammatory response [42]. In addition, Linetsky et al. [43] proposed that the high concentration of dextrose acts as a neurolytic to treat the periosteal nerve and blood vessel in-

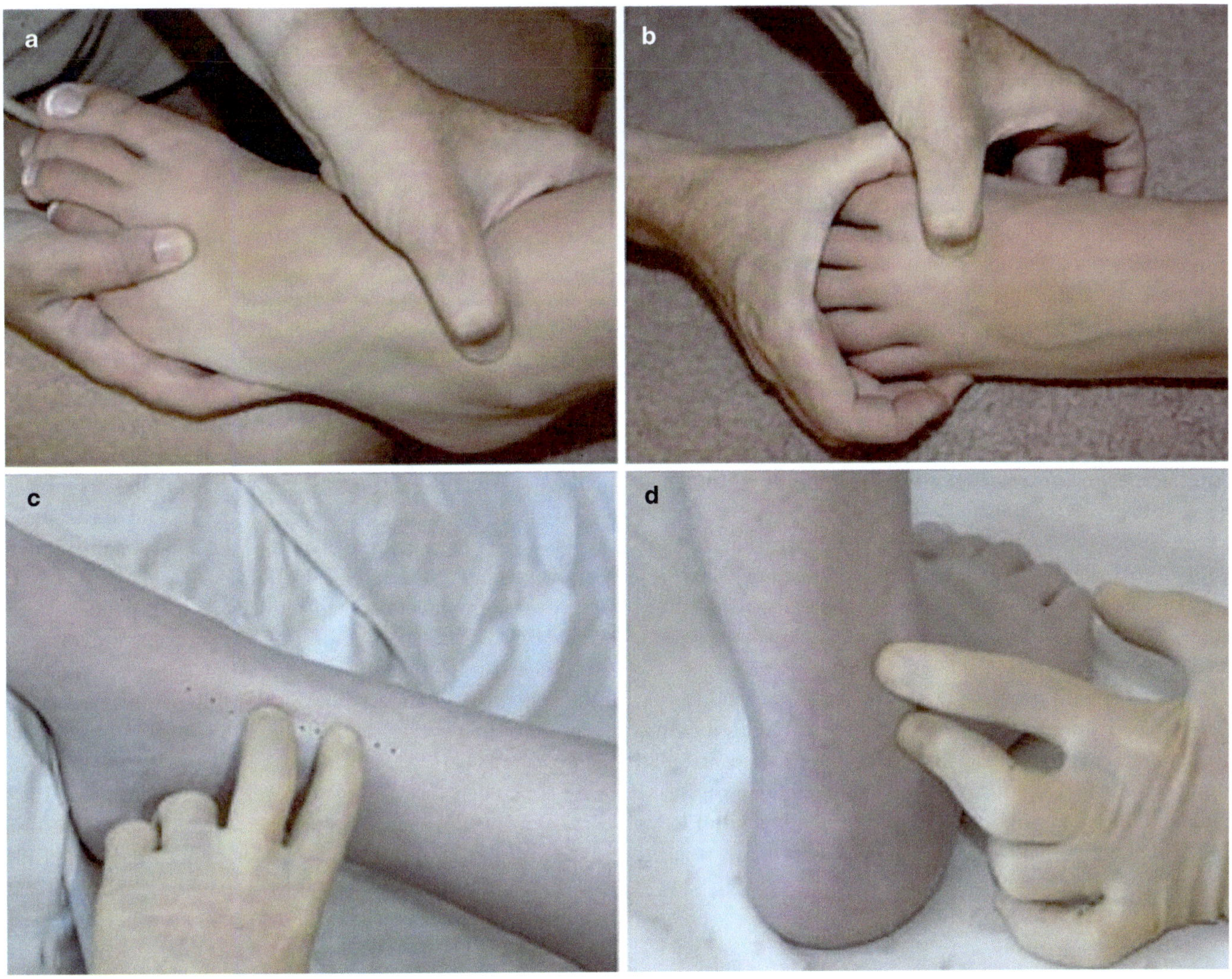


**Fig. 21.9** Physical exam of the nerves of the foot. (**a**) superficial peroneal nerve; (**b**) interdigital nerve; (**c**) superficial saphenous nerve; (**d**) sural nerve

growth, which may explain the rapid pain relief after the injections.

PRP involves the harvesting of autologous blood, which is then spun down, and the platelets are injected into the fibro-osseous junction. Autologous mesenchymal and hematopoietic stem cells, growth factors, cytokines and extracellular matrix can be collected from adipose tissue or bone marrow concentrate (BMC) and then injected to the same fibro-osseous junction, tendon sheath, intra-tendinous, intra-osseous [44]. Perineural orthobiologics injections [39], or even intra-neural which may be an emerging technique [45].

Gellan gum, allogenic dermal fibroblasts, epidermal keratinocytes, acellular dermal grafts, biodegradable biomaterials, bone morphogenetic proteins, and cellular silk fibroin-based scaffolds have all been used to treat ankle ligamentous and cartilaginous pathologies [46]. Human placenta, umbilical, and amniotic derived tissues, particularly the HCHA/PTX3 (Heavy Chain Hyaluronic Acid/Pentraxin 3) compounds appear to function effectively to halt the cycle of inflammation and actually re-program host healthy cells to resolve pathology [47].

Each of these therapies is used primarily for ligament or tendon injuries at the fibro-osseous junction, which function as "anchors" of the ligament and tendons. It is therefore important to consider the muscle-nerve-tendon-bone complex as a unit [17], since muscle spasms can occur because of joint instability, and nerve entrapment can occur by this same instability, and can be treated by some of the same modalities [39–47].

These regenerative orthobiologics may function via the paracrine effect modulating inflammation, proliferation and migration, angiogenesis, hemostasis, remodeling, contraction, as well as the differentiation of cells like medicinal signaling cells, also known as mesenchymal stem cells, or mesenchymal stromal cells [48–50].

## Evidence

### Achilles Tendinosis

A study by Maxwell et al. [51] used US to treat Achilles tendinosis with prolotherapy and showed 78% improvement in the structure after prolotherapy; they also noted no adverse effects or complications. Yelland et al. [52] reported a single-blind study of 43 patients with Achilles tendinosis, using exercise, prolotherapy, or both treatments combined. Although all three groups improved at 6 weeks and 12 months, these increases were significantly better for the combined treatment. Gross et al. [53] performed a systematic review of injections for Achilles tendinosis; they found 9 studies that fit the inclusion criteria, involving a total of 312 Achilles tendons. There were 54 injections of PRP, 40 injections of autologous blood, 72 tendons injected with sclerosing agents, 26 injected with protease inhibitors, 60 with hemodialysate, 52 with corticosteroids, and 20 with prolotherapy. However, only one study met the criteria for a high quality randomized controlled trial, and they were not able to compare the studies because of the disparate techniques used.

An additional systematic review by Sanderson and Bryant [54] showed 5 moderate to good studies of Achilles tendinosis treated with prolotherapy involving 338 participants; all 5 studies reported a significant reduction in pain. However, the treatment of the non-insertional Achilles tendon with PRP is controversial with at least two recent systematic reviews showing no benefit over controls.

### Bunion

Hauser and Feister [41] retrospectively studied 12 patients with severe hallux rigidus and bunion deformities; all had pain scores of at least 4/10, and 5 of the 12 patients preoperatively could not walk 50 feet without pain. After an average of four treatments of Dextrose based prolotherapy (range of 3–6) every 4 weeks, the patients were contacted by phone 10–12 months after the completion of therapy. After treatment, all but one patient reported no restrictions in walking at all, with an average visual analog scale (VAS) score of 1.17 and a greater than 75% improvement in activities of daily living.

### Plantar Fasciitis

For plantar fasciitis/fasciosis, Sanderson and Bryant [54] found 2 good studies involving 41 patients, including one study by Ryan et al. [55] using US to direct injections of the plantar fascia and one study by Kim and Lee [56] comparing prolotherapy and PRP. Both studies showed improvement in symptoms; 80% of the patients in Ryan et al.'s study [55] noted improvement while Kim and Lee [56] showed a greater improvement in pain and Foot Function Index with PRP than prolotherapy. No complications were noted in either study.

### Osteoarthritis and Osteochondral Lesions

Hauser and Orlofsky [57] described the use of whole tibial or iliac bone marrow aspirate to treat degenerative joint disease of the hip, knee, and ankle in four new patients and three prior prolotherapy patients. The hips were injected with 50 mL of aspirate in 38 sites, while knees were injected with a total of 40 mL at 30 locations. All of the patients noted "significant gains" and 5 of the 7 patients noted "complete relief or strong functional improvement," and no adverse events were noted. Yausep et al. [58] did a systematic review of level 1 and 2 I/O talus injections and found that, "PRP improves joint function, and reduces pain in patients with OLT regardless of the method of implementation. In addition, inter-study comparison demonstrated that patients that received surgery along with PRP injections improved more than those that received PRP only" [58]. Based on a meta-analysis review by Migliorini et al. [59] looking at 30 RCTs, they concluded that there was level one evidence that PRP was more efficacious for knee osteoarthritis compared to corticosteroid, hyaluronic acid and placebo. Abas et al. [60] did a retrospective cohort study for ankle osteochondral lesions over 3 years with BMC and concluded that BMC is a safe, efficacious treatment both as a primary as well as if patients fail primary treatment. Vannini et al. [61] showed that osteochondral lesions can be successfully treated with BMC seeding into a scaffold in a single stage surgery with 10-year follow-up [61]. A retrospective study by Kim et al. [62] showed improvement of pain and functional scoring with augmentation of marrow release, and osteotomy surgery in the adipose derived stem cell group. It should be noted that between 2020 and 2022, there have been 365 clinical trials registered for the treatment of chondral lesion and arthritis on www.ClinicalTrials.gov [63]. The results of these publications should be quite elucidating in the near future.

### Tendinosis/Fasciitis

Jia et al. [64] described the use of PRP and BMC in the management of foot and ankle pathologies in a total of 634 patients. They chose PRP for tendons such as Achilles, tibialis posterior, tibialis anterior, peroneal, extensor digitorum longus, and metatarsal phalangeal joint ligaments. BMC was used primarily for complex lower extremity surgical reconstructions in high-risk patients; they were able to confirm bone union in 88% of the patients, compared to a 35% non-union rate for these surgeries, even with bone stimulators. They noted only one patient with prolonged (10 days) pain at the BMC harvest site, that resolved spontaneously, and there were no infections or wound complications related to the PRP or BMC. Sanderson and Bryant [54] performed a systematic review of prolotherapy for lower limb tendinopathy and fasciopathy. They reviewed 203 studies and found 3 randomized controlled trials and 5 prospective case-series studies that met their criteria. Their review suggested that prolotherapy injections appeared to be superior to eccentric loading exercises for Achilles tendinopathy, and PRP was effective for plantar fasciopathy and for Osgood-Schlatter disease. No complications were identified.

## Intraosseous Injections

There are emerging techniques and evidence for the use of intraosseous (I/O) injections or sub-chondral injections. Most of the evidence is in the knee and hip, with emerging data on I/O injections into the talus. There is one systematic review of only four studies, by Chahla et al. [65] for osteochondral lesions of the talus, that acknowledged that there is a paucity of data but given that, the "evidence available showed varying degrees of beneficial results of bone marrow aspirate concentrate (BMAC) for the treatment of ankle cartilage defects". It must be noted that the use of orthobiologics is very patient dependent, dose dependent as well as technique dependent, making studying the efficacy of these injections quite difficult, prompting a call for standardization of study protocols so that therapeutic differences can be more accurately measured [66]. Several studies show no difference over control, used sub-standard equipment, dosing, and sometimes without the use of U/S or fluoroscopic guidance.

## Technical Considerations

Precise image guidance with musculoskeletal ultrasound, fluoroscopy is greatly preferred over landmark based injection, but clinicians must get adequately trained in their use. In addition, it is becoming increasing clear that dosing is critically important to an excellent outcome. The author uses no less than 30 mL's of drawn blood, and more often 60 mL's or even 120 mL's for more severe cases, particularly in older patients or those that have had prior exposure within the past year to injectable or oral corticosteroids. For landmark-guided intra-articular injections of the ankle, feel the anterior joint line while flexing and extending the foot, and then feel for the dorsalis pedis artery (and the accompanying superficial peroneal nerve); inject during extension of the ankle to open up the joint space, depositing injectate intra- and peri-articularly. Under fluoroscopy or musculoskeletal US, identify the anterior joint line (Fig. 21.9); under real-time, high resolution musculoskeletal US guidance with image documentation, the capsule, joint line, and neurovascular structures are visible, and the use of color or power doppler can be used as a contrast corollary, to assess fluid flow patterns, and confirm proper placement of injectate.

The sites of injection are dictated by the sites of tenderness, and directed by landmarks (Fig. 21.10), x-rays, or US to the fibro-osseous attachments of each site, and in regions of interstitial degeneration or suspected split tearing (Figs. 21.11 and 21.12). In addition, given the dynamic nature and frequent instability in the foot and ankle, particularly after injury, the authors recommend considering the biomechanical kinetic chain, and treating up and down the chain as appropriate, even above the foot and ankle.

Although the structures of the ankle and foot are superficial and usually easy to palpate and inject, there are several areas where imaging is useful, if not critical. The ligaments around the medial malleolus are more technically difficult, because of the vessels and nerves of the tarsal tunnel, and so

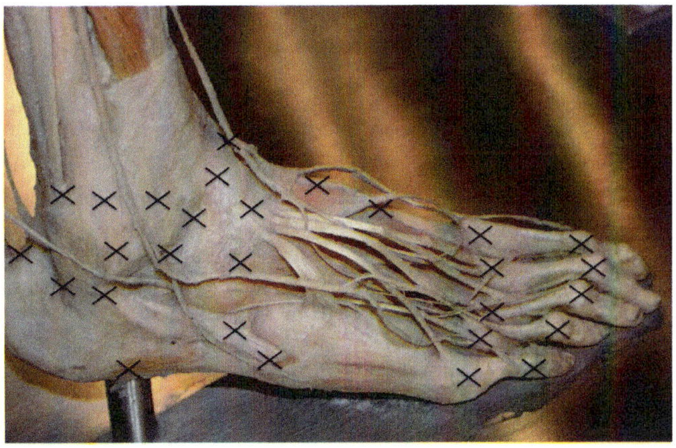

**Fig. 21.10** Regenerative injection sites

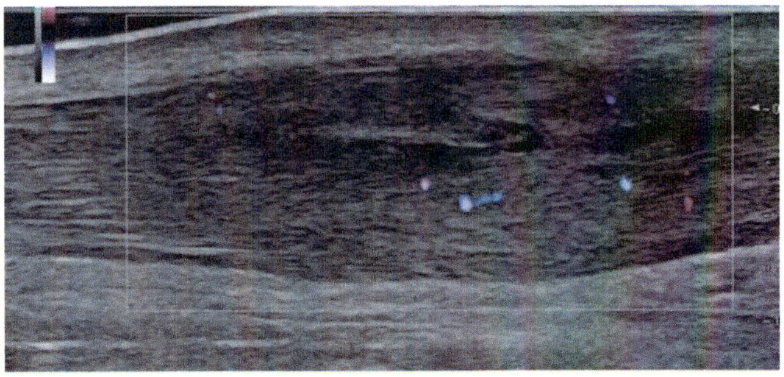

**Fig. 21.11** Interstitial tearing of the Achilles with superimposes severe tendinopathy. **Source:** Courtesy of ultrasound cases https://www.ultrasoundcases.info/partial-ruptures-7286/#gallery-3

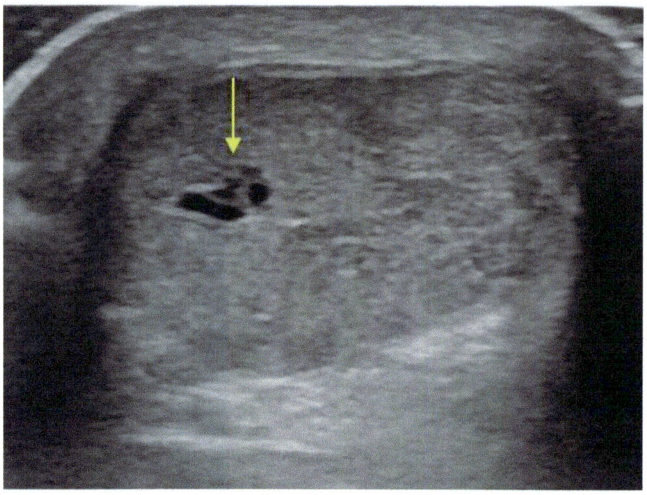

**Fig. 21.12** Short axis view of the Achilles tendon interstitial tear. **Source:** Courtesy of ultrasound cases. https://www.ultrasoundcases.info/partial-ruptures-7286/#gallery-6

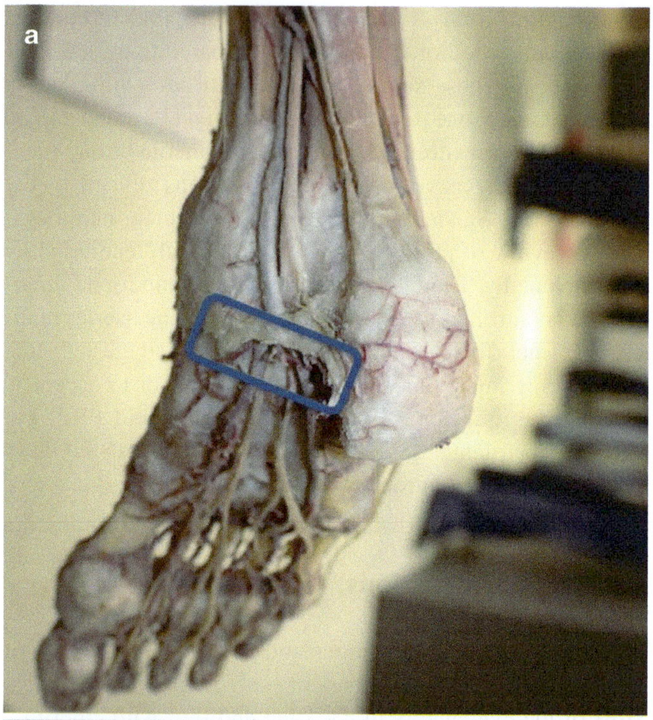


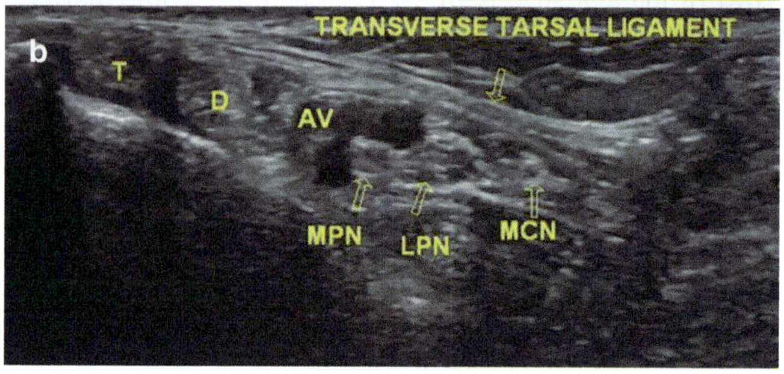


**Fig. 21.13** Vascular and neurologic structures of the tarsal tunnel and planar arch of the foot. (**a**) Anatomic dissection; (**b**) ultrasound image. *T* tibialis posterior tendon, *D* flexor digitorum longus tendon, *AV* posterior tibial artery and veins, *MPN* medial plantar nerve, *LPN* lateral plantar nerve, *MCN* medial calcaneal nerve. Image courtesy Michael N. Brown, MD and Andrea Trescot, MD

US is very useful in this region to identify the nerves and vessels. Identifying and injecting interstitial split tears must be done under real-time US guidance with visualization both long and short axis to the structure being injected. A linear transducer of at least 12 Mhz, preferably 15 Mhz should be used. Occasionally, a test injection under US with anesthetic can help determine pain generators, and help highlight interstitial, delamination tears. The spring ligament lies very close to the plantar nerves and vessels (Fig. 21.13) and is strongly recommended to be injected under US. Intra-articular ankle joint injections are usually best performed under fluoroscopy or high-resolution US with color/power doppler, especially at the talotibial, talonavicular, tarsometatarsal and naviculocalcaneal joints (Figs. 21.13, 21.14, 21.15, 21.16, and 21.17). Intraosseous injections must be done under US and/or fluoroscopic guidance. Fluoroscopic contrast can help visualize sub-chondral intravascular needle placement (Fig. 21.18a, b). Hydrodissection of the tibial nerve through the tarsal tunnel, or any branch of the innervating nerves in the foot and ankle must be done under real-time US guidance, usually with linear probe with at least 12Mhz, preferably higher, in both long and short axis to the nerve, both in and out of plane to the needle. The author usually starts above the tarsal tunnel and aims distally (Fig. 21.19). There is an emerging technique for US guided tarsal tunnel release as a potential alternative to endoscopic or open decompression [67]. Another emerging technique is calcaneal fat grafting whereby a patient's abdominal, flank or gluteal fat, harvested via liposuction, processed then injected into the macrochamber of the calcaneal fat pad.

An RCT by James et al. [68] showed statistically significant improvements in foot pain and appearance at 6 months. A 20-gauge spinal needle, two V-shaped straight curettes no. 1 and no. 2, a blunt dissector, a 3-mm retrograde hook knife and high-resolution U/S of 8–17 Mhz probe was used. A total of 76.54% [62] patients obtained excellent results, 13.58% [11] good results, and 9.87% [8] poor results on the Takakura scale (Fig. 21.3) [69].

When utilizing US for diagnostic purposes, it is important to use the appropriate probe, frequencies and depth as well as include dynamic stress and instability testing, and power/color doppler imaging for a more full clinical pathophysiologic picture. Musculoskeletal US and strain or shear wave elastography may give additional insight into the mechanical properties of tendons, ligaments and fascia, even where MRI is reportedly normal and is an emerging diagnostic tool that

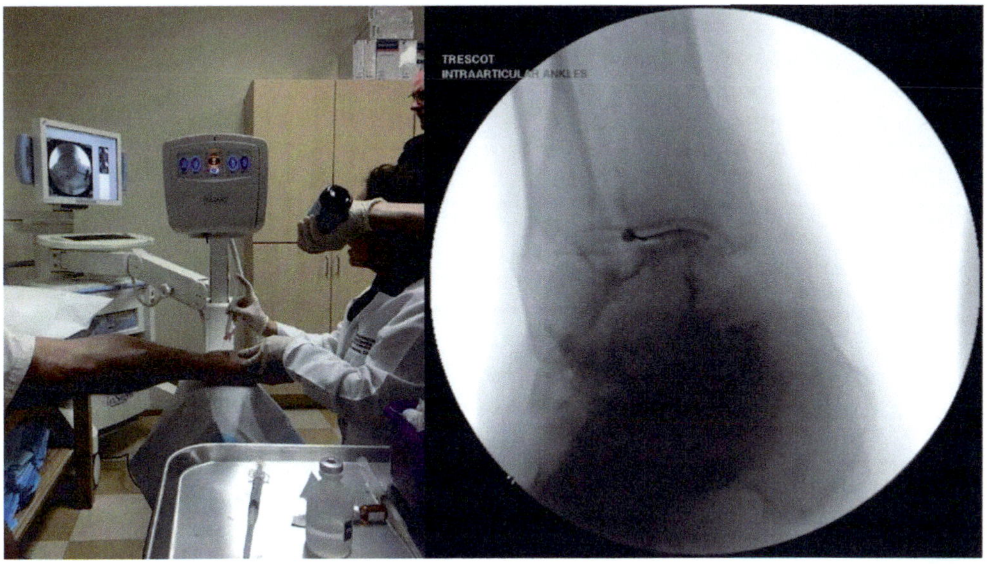


**Fig. 21.14** Fluoroscopic ankle injection

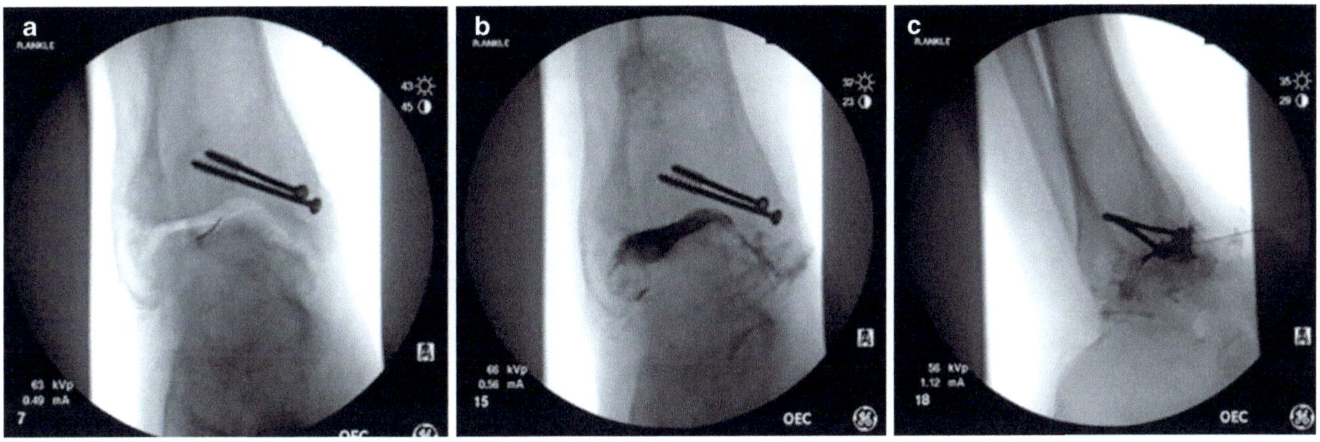


**Fig. 21.15** Fluoroscopic ankle injection. (**a**) Needle placement. (**b**) Contrast display pattern (AP View). (**c**) Contrast display pattern (Lateral view)

can give both subjective and objective tracking information on healing. When possible, pre-procedural mapping correlation of US with MRI imaging is recommended.

Regional nerve blocks can be quite useful in reducing or eliminating procedural pain, as well as for diagnostic and therapeutic purposes. For instance, the use of tibial nerve blocks can be useful for medial, or plantar injections of the Posterior Tibialis, Flexor Hallucis Longus, Lisfranc ligaments, sural nerve blocks can be useful for Achilles, lateral foot and ankle injections, and superficial peroneal blocks can be used for targets in the dorsal foot and ankle and into the first webspace. The author will often do LPN (Baxter's nerve), or inferior calcaneal blocks for chronic heel pain, particularly if the plantar fascia itself appears normal or clinical findings point against a traditional plantar fasciitis/fasciosis. The clinician can use hydrodissection of these nerves with PRP, if the diagnosis of neuritis, with or without neural entrapment is clear, and conservative care has failed [39].

Another re-emerging treatment option is the use of focused (ECSWT-f) and radial shockwave therapy (ECSWT-r) for the use of chronic Achilles tendinosis, plantar fasciitis and posterior tibial tendinosis. Al-Siyabi et al. [69] published a recent systematic review showing benefit of shockwave therapy over US therapy. Asheghan et al. [70] published an RCT showing that both dextrose based prolotherapy and Radial shockwave therapy were effective in improving pain and function in patients with plantar fasciosis. Xu et al. [71] showed that both corticosteroids and shockwave helped plantar fasciitis pain, though shockwave was superior for duration of improvement.

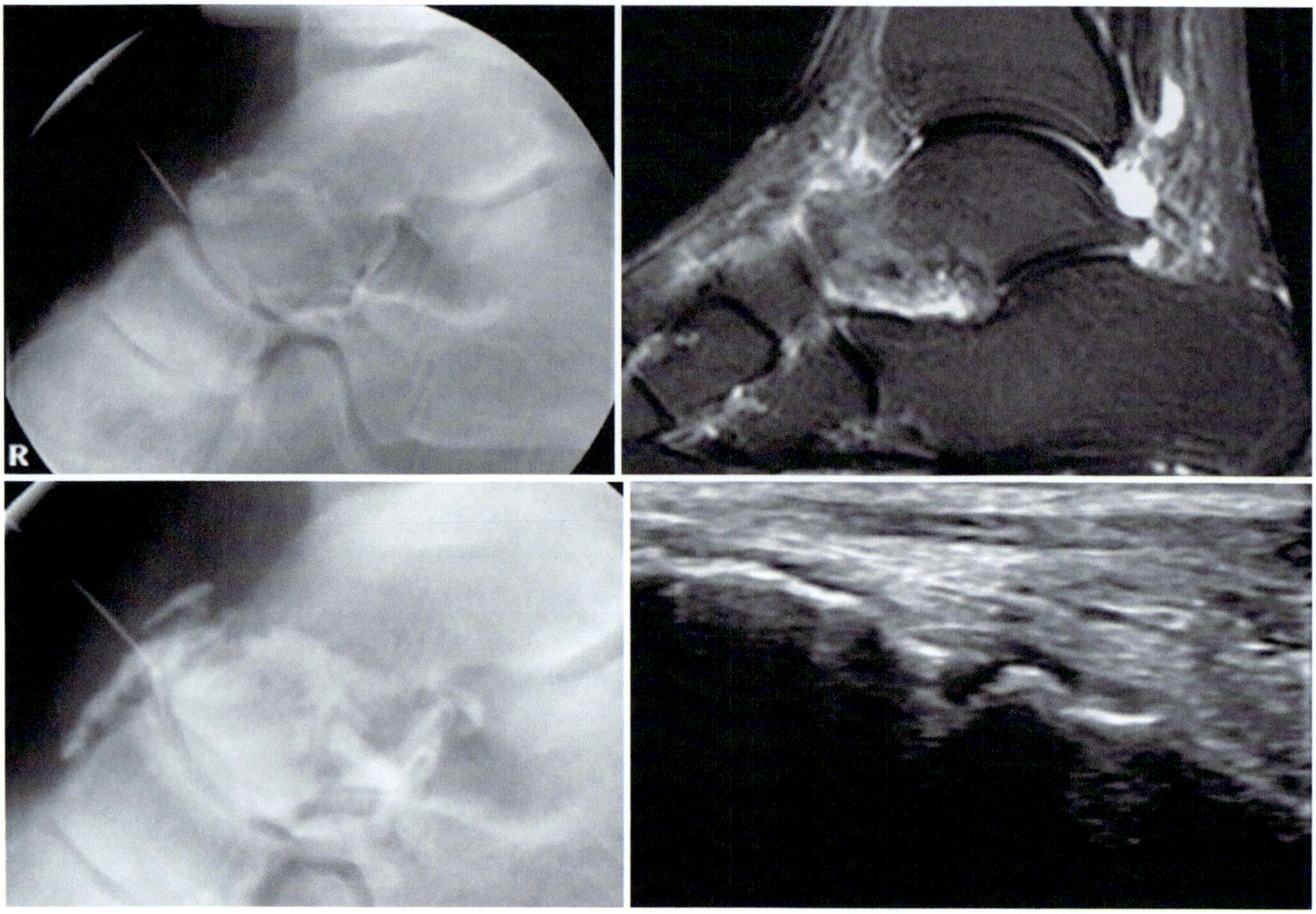


**Fig. 21.16** Tibiotalar joint injection

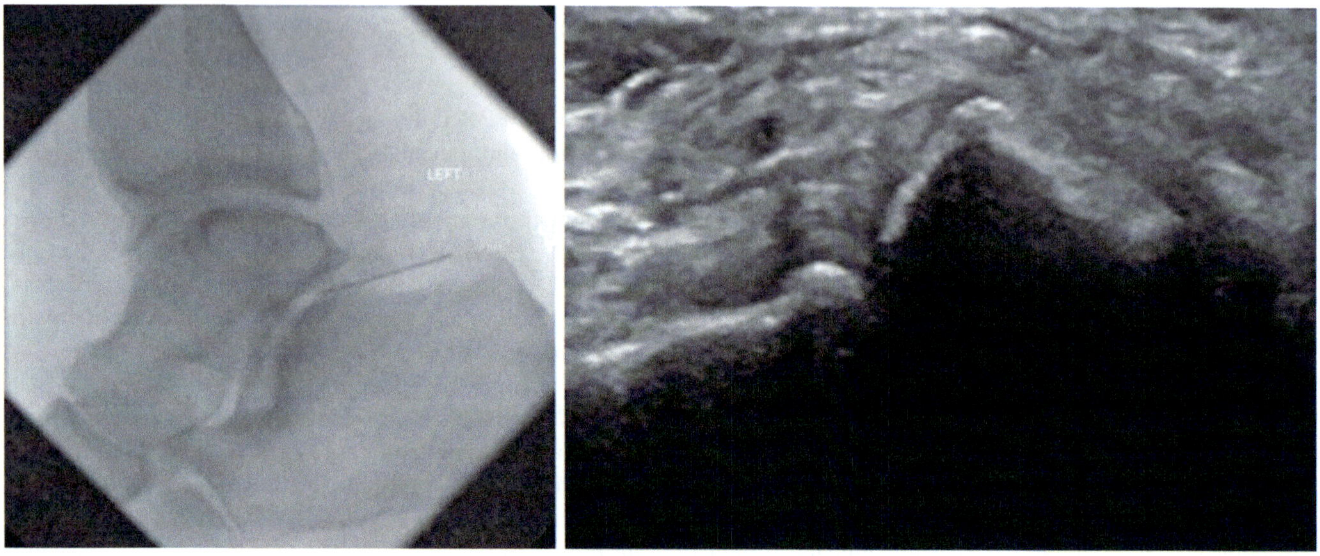


**Fig. 21.17** Posterior subtalar joint injection

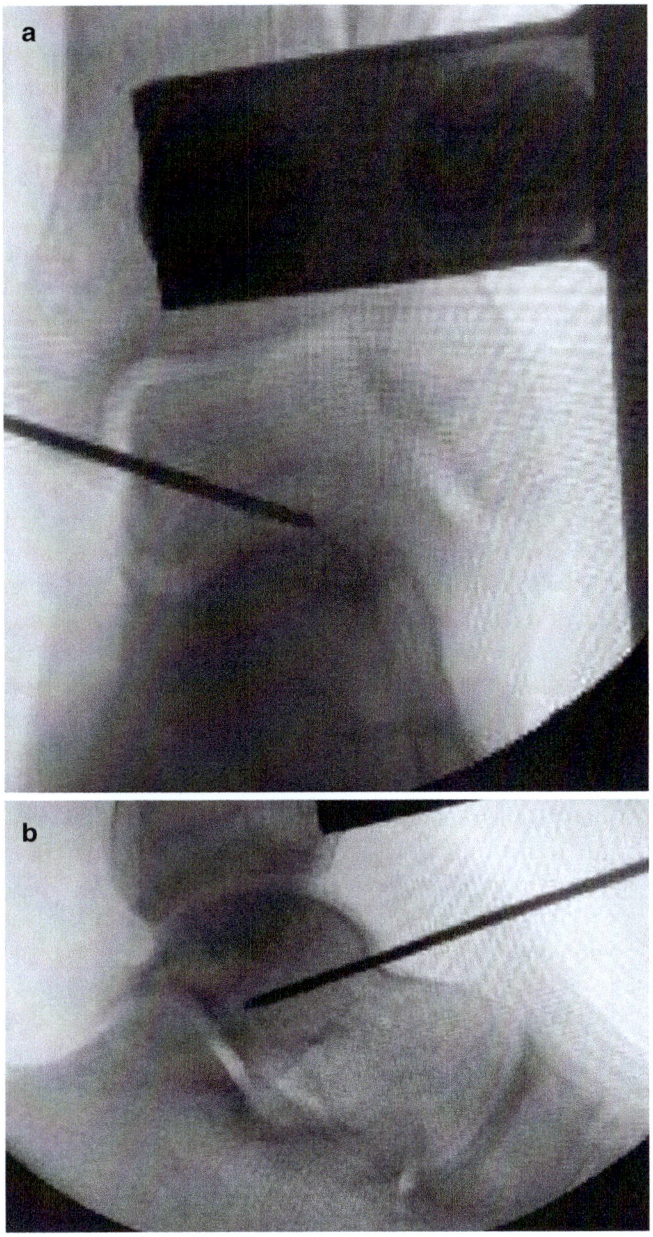


**Fig. 21.18** Intraosseous injection of talus. **Source:** Orgiu A, Choufani C, Litrico L, et al. Intraosseous platelet-rich plasma injection in treatment of a subtalar osteochondral lesion: A case report. *Foot & Ankle Surgery: Techniques, Reports & Cases*, Volume 2, Issue 1, 2022. https://doi.org/10.1016/j.fastrc.2021.100116 (Open Access Creative Commons)

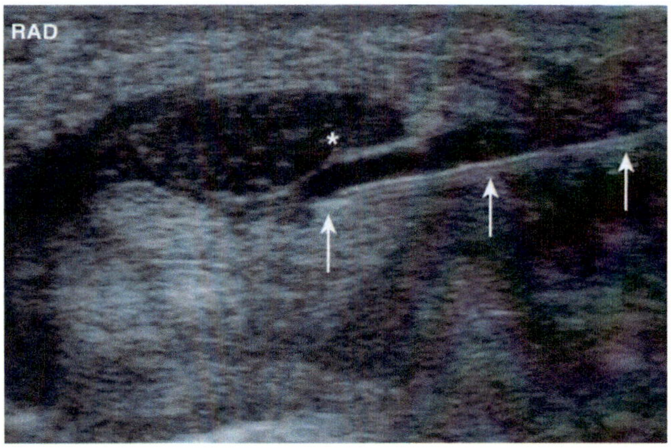


**Fig. 21.19** Nerve hydrodissection carpal tunnel median nerve under ultrasound. **Source:** https://www.researchgate.net/figure/18-Ultrasound-guided-carpal-tunnel-injection-ulnar-approach-Arrows-depict-needle-Note_fig5_235953694 (Courtesy of Jay Smith)

## Complications

There are relatively few complications that have been reported after regenerative orthobiologic injections. There may be allergies to anesthetics. There is a potential for nerve damage from the injections, especially around the tarsal tunnel, the medial arch, and the dorsum of the foot. The skin in these areas is very thin, and therefore potentially prone to damage. For intraosseous injections, there is an added rare risk of osteomyelitis requiring IV antibiotics, subchondral fracture that could result in the need for surgical fixation or joint replacement, as well as rare embolic events that would be managed with anticoagulants and could result in pulmonary embolism (PE) and cerebrovascular accident (CVA). These risks can be minimized with sterile technique, careful patient selection, precise pre-procedural mapping with Radiographs, MRI, computed tomography (CT) and high-resolution musculoskeletal US. During the procedure, precise trained guidance with US and/or fluoroscopy is critical. In extremely rare circumstances, underlying autoimmune disease can be unroofed, and complex regional pain syndrome can be triggered.

## Regulatory Considerations

Consideration must be given to local regulatory environments in the reader's country of practice. While a full regulatory analysis is out of scope for this chapter, the reader is urged to review US FDA-21 CFR 1271.15(b) same surgical procedure exemption, as well as what is considered processed tissue (361), vs a cell drug (351), particularly for U.S. based clinicians [72, 73]. Key points to understand are the need to stay minimally manipulated, homologous use, and avoid the combination of substances prior to injection.

## Conclusion

Regenerative orthobiologic injection therapy can be very useful for chronic foot and ankle pain, particularly if the diagnosed underlying pathology is related to instability, relative poor local microcirculation, resulting in tendinopathy, ligament laxity, fasciosis, arthritis, osteochondral damage and neuritis. There are multiple sites of pathology in the foot and ankle, and many potential injection sites. Prolotherapy, PRP, BMC and micronized Adipose grafts all appear to be effective in the treatment of these debilitating conditions, although larger clinical trials are necessary. Novel injection techniques such as intraosseous/subchondral injections for severe joint arthritis with bone marrow lesions, osteonecrosis, as well as hydrodissection for neural entrapment may potentially be tried as an interventional treatment gap tool, prior to surgical alternative in select situations. With the advent of high-resolution US, microinvasive techniques such as percutaneous tarsal tunnel decompression can be utilized as an alternative to endoscopic or open surgery.

**Key Points**

1. Structural integrity and dynamic alignment of the foot and ankle ligaments and are key to proper functioning and preventing pain.
2. Neurovascular Irritation or injury may be a common under-diagnosed contributor to foot and ankle pain.
3. Clear knowledge of mechanism of injury allows the clinician to diagnose and treat different biomechanically related structures.
4. Dynamic imaging with U/S and weightbearing imaging with Xray fluoroscopy or CT are important to evaluate instability.
5. Hydrodissection of nerves and tendons or ligaments can be used both for diagnosis of entrapment, adhesion and to deliver therapeutic agents such as PRP or dilute Dextrose solutions.
6. Evidence is mixed but generally shows good reduction in pain and improvement for Achilles Tendinosis.
7. There is positive evidence for the use of Dextrose based Prolotherapy for bunions.
8. Both PRP and Dextrose prolotherapy have been shown to be helpful for plantar fasciosis/fasciitis and tendinosis with PRP being more effective in the literature.
9. PRP, BMAC and Adipose therapies show positive benefit in ankle arthritis, and augmentation of surgical interventions with orthobiologics show early evidence in support their use.
10. BMAC has been shown to be beneficial for tendinosis and for delayed unions in the foot and ankle in emerging evidence.
11. There is emerging evidence in favor of Intraosseous (I/O)/subchondral injections for bone marrow lesions (BMLs), and osteochondral defects, particularly in later stages of Osteoarthritis.
12. The use of ultrasound and fluoroscopy for precise guided injections and nerve blocks are highly preferred over landmark or blind injections.
13. While infections are rare with orthobiologics, early identification and treatment is important.
14. The use of machine energy tools like shockwave therapy (ECSWT), high field magnetic therapy (EMTT), and high dose Class IV lasers are emerging as helpful non-invasive, therapeutic and diagnostic options.
15. Regulatory considerations must be explicitly discussed and followed with patients.

**Questions**

1. An 18 year old male basketball player develops acute left foot pain after landing in dorsiflexion and eversion during a game. Initially, he was able to painfully weight-bear. Initial radiographs including weight bearing were negative for fracture. He was diagnosed with a foot sprain. He was placed in a long leg cam boot for 6 weeks, then weaned out after an additional 3 weeks. 4 months after initial injury the man continues to complain about dorsal, medial foot pain at the first and second tarsometatarsal joints while running and jumping despite NSAIDS, taping, orthotics, footwear changes, and topical as well as prescription pain relieving creams. He failed 8 weeks of excellent physical therapy. MRI was reportedly normal with the exception of mild T2 edema of the Lisfranc ligaments suggestive of chronic sprain. Exam shows no gross instability but tenderness in the first and second Tarsometatarsal joints with shear and torsional testing, and trace laxity and 1st TMT effusion on dynamic ultrasound imaging. You offer the following treatment options:
   A. A corticosteroid injection into the Lisfranc joint complex
   B. A Leukocyte Rich Platelet Rich Plasma injection
   C. Further immobilization in a boot for 6 weeks
   D. Refer to a foot and ankle surgeon for Lisfranc surgical ligament repair

   **Answer: B**

   Key Concept: Midfoot sprains can be minor, non-surgical yet cause significant disability due to post-traumatic ligamentous laxity, joint articulation synovitis, and can result in progression to traumatic arthritis, which may result in the need for fusion later in life. The use of Platelet Rich Plasma, possibly Leukocyte Rich, at least 5 times over baseline concentration can help with pain, potentially promote ligament healing, reduce chronic inflammation potentially preventing progression to traumatic arthritis. While corticosteroid may offer temporary benefit for weeks to months, it may result in inhibition of healing, and further laxity of ligaments.

References
1. Burroughs KE, Reimer CD, Fields KB. Lisfranc injury of the foot: a commonly missed diagnosis. American Family Physician. 1998 Jul;58(1):118–124.
2. DeOrio M, Erickson M, Usuelli FG, Easley M. Lisfranc injuries in sport. Foot Ankle Clin. 2009 Jun;14(2):169–86. Doi: https://doi.org/10.1016/j.fcl.2009.03.008. PMID: 19501801.
3. Lewis JS Jr, Anderson RB. Lisfranc Injuries in the Athlete. Foot Ankle Int. 2016 Dec;37(12):1374–1380. doi: https://doi.org/10.1177/1071100716675293. PMID: 27899721.
4. Website: "Platelet Rich Plasma." Arthrex, Inc., www.arthrex.com/orthobiologics/platelet-rich-plasma."Platelet-Rich Plasma (PRP) Treatment."
5. Website: Johns Hopkins Medicine: www.hopkinsmedicine.org/health/treatment-tests-and-therapies/plateletrich-plasma-prp-treatment.
6. Solan MC, Moorman CT 3rd, Miyamoto RG, et al. Ligamentous restraints of the second tarsometatarsal joint: a biomechanical evaluation. Foot Ankle Int 2001; 22:637.
7. Watson, Troy S. MD; Shurnas, Paul S. MD; Denker, Jacques Treatment of Lisfranc Joint Injury: Current Concepts, American Academy of Orthopaedic Surgeon: December 2010—Volume 18—Issue 12—p 718–728.
8. Koehler, L, BR Waterman , NA Kusnezov , JA Blair , PJ Belmont, J D Orr.Occupational Outcomes and Return to Running After Operative Management of Lisfranc Injuries in a High-Demand Population. Foot Ankle Spec. 2020 Jun 9.

2. A 68 year old overweight woman, comes into your clinic with 3 years of progressive right ankle joint pain worse with prolonged hiking and walking, relieved with rest and non-weight bearing, and associated with stiffness and mild occasional self-resolving effusions. She has no major pain with activities of daily living beyond stiffness. She notes numerous ankle sprains in her teens and twenties as a gymnast. She has tried 2 Cox-1 NSAIDS at appropriate doses and duration courses but complains of gastritis with more than a week on both NSAIDS and has a recent history of mild myocardial infarction 6 months ago. She has tried modest weight loss, ankle bracing, orthotics, 6 weeks of physical therapy with only mild temporary benefit. 3 months ago, she had a corticosteroid which helped moderately for 6 weeks, but now she feels "worse than she ever felt before." Radiographs show grade 2 tibiotalar joint space narrowing and marginal osteophytes, and MRI shows ATFL and CFL degeneration, Grade 2 chondral thinning and effusion, with a small 2 mm Osteochondral defect of the anterolateral talus. She is interested in further interventions. The next appropriate step would be:

A. A 2 week course of Cox-2 inhibitor NSAID Celecoxib
B. Surgical Microfracture and arthroscopic debridement
C. Platelet Rich Plasma injection for her ankle ligaments and intra-articular, followed by 3–6 weeks of an ASO ankle brace, and a structured foot and ankle home exercise program.
D. Hyaluronic Acid injection intra-articularly along with an ASO brace, and a foot and ankle home exercise program.

**Answer: C**

Key Concept: Intra-articular PRP injection demonstrated significantly enhanced recovery of function, and decreased pain scores compared to HA in a systematic review by Yausep, et al. with level 1 and 2 evidence. This patient has a small osteochondral defect with grade 2 traumatic arthritis, has relatively low disability and has failed conservative care, as well as corticosteroid injection. While surgical intervention may help, the results in this type of patient are inconsistent, and could worsen her arthritic symptoms and course of the disease. Therefore, an interventional orthobiologic approach with PRP would be quite reasonable as a next step with a goal of treating appropriate ligaments, and intra-articularly followed by another good course of comprehensive physical therapy from the spine down to the foot and ankle. Cox 2 inhibitors are not recommended in patients with recent myocardial infarctions.

References
1. Yausep OE, Madhi I, Trigkilidas D. Platelet rich plasma for treatment of osteochondral lesions of the talus: A systematic review of clinical trials. J Orthop. 2020 Jan 30;18:218–225. doi: https://doi.org/10.1016/j.jor.2020.01.046. PMID: 32071508; PMCID: PMC7013135.
2. Patrono C. Cardiovascular effects of cyclooxygenase-2 inhibitors: a mechanistic and clinical perspective. Br J Clin Pharmacol. 2016 Oct;82(4):957-64. doi: https://doi.org/10.1111/bcp.13048. Epub 2016 Jul 18. PMID: 27317138; PMCID: PMC5137820.

3. A 32-year-old female runner comes into your clinic with right medial plantar pain, worse with the first few steps of the day, worse with poorly supportive footwear, better with calf and heel stretching. She is training for her first marathon in 9 months. She has tried several rounds of Ibuprofen without long term benefit. She has failed 6 weeks of rest, a night splint, long leg cam boot, running shoe changes, over the counter and semi-custom orthotics, foot and calf stretches, and rolling her foot with a frozen water bottle. 6 months ago, she received an ultrasound guided corticosteroid injection with 6 weeks of moderate benefit, then the pain returned. On exam she is

directly tender to palpation at the calcaneal insertion of the plantar fascia with medial calcaneal heel pain walking barefoot. Seated-slump test is negative. Diagnostic musculoskeletal ultrasound imaging confirms a thickened degenerative plantar fascia with no evidence of tear. She is interested in further interventions. The next appropriate step would be:

A. A second corticosteroid injection into the plantar fascia under ultrasound guidance followed by continued stretching, and foot and ankle physical therapy.
B. A Leukocyte Rich Platelet Rich Plasma injection followed by 3–6 weeks of rest, stretching and continued foot and ankle physical therapy.
C. Prolotherapy injections followed by 3–6 weeks of rest, stretching and continued foot and ankle physical therapy.
D. Surgical Plantar Fascia release

**Answer: B**

Key Concepts: Platelet Rich Plasma has been shown to be quite effective in relieving plantar fasciitis pain when conservative care has failed. PRP has been shown to be superior to prolotherapy and corticosteroid for plantar fasciitis/plantar fasciosis. Surgical plantar fascia release, while palliative, is not indicated in a younger runner, particularly one that has a goal to train and run for a marathon in 9 months. A second corticosteroid injection is not encouraged due to risk of rupture of the plantar fascia with repetitive corticosteroid injections, only marginal clinical benefit vs placebo when the Minimal Clinical Important Difference is considered.

References

1. Sanderson LM, Bryant A. Effectiveness and safety of prolotherapy injections for management of lower limb tendinopathy and fasciopathy: A systematic review. *J Foot Ankle Res* 2015; 8:57.
2. Ryan M, Wong A, Rabago D, et al. Ultrasound-guided injections of hyperosmolar dextrose for overuse patellar tendinopathy: A pilot study. *Br J Sports Med* 2011; 45:972–977.
3. Kim E, Lee JH. Autologous platelet-rich plasma versus dextrose prolotherapy for the treatment of chronic recalcitrant plantar fasciitis. *PM R* 2014; 6:152–158.
4. David JA, Sankarapandian V, Christopher PR, Chatterjee A, Macaden AS. Injected corticosteroids for treating plantar heel pain in adults. Cochrane Database Syst Rev. 2017 Jun 11;6(6):CD009348. doi: https://doi.org/10.1002/14651858.CD009348.pub2. PMID: 28602048; PMCID: PMC6481652.

4. A 42 year old male marathon runner with pes planus and moderate pronation comes in with 6 months of moderate persistent medial ankle pain, aching, and occasional tingling with radiation into the plantar aspect of the foot, worse with prolonged running, better with rest. He has tried an appropriate course of NSAIDS, and has changed to a higher stability shoe, with modest benefit only. He has done 3 months of rest followed by 6 weeks of PT for his core, kinetic chain, foot and ankle with moderate benefit. He notes no back pain or radiculopathy symptoms and has had a recent MRI with no evidence of nerve root impingement. He has seen a foot and ankle surgeon who diagnosed tarsal tunnel syndrome. Nerve conduction and Electromyography studies are diagnostic of a mild tibial neuritis across the tarsal tunnel with no axonal damage. The surgeon did a tarsal tunnel injection 8 weeks ago with corticosteroid which was tremendously effective but only lasted 4 weeks. He is interested in further interventions but is unwilling to do anything surgical due to an upcoming race in 3 months. The next appropriate step would be:

A. A second corticosteroid injection in the tarsal tunnel
B. A Platelet Rich Plasma hydro-dissection of the tibial nerve through the tarsal tunnel
C. Rest for another 3 months
D. Tarsal tunnel release

**Answer: B**

Key Concepts: Tarsal tunnel syndrome affects many runners with pes planus and pronation. While conservative care, corticosteroids can be helpful as first line treatments, many present with recurrent refractory symptoms limiting ability to exercise. PRP with a hydro-dissection technique holds excellent promise for the tarsal tunnel syndrome. Its upper extremity corollary, the carpal tunnel has significant emerging data for the use of PRP hydro-dissection. In this relatively young patient, it may be useful to try a PRP hydro-dissection of the tarsal tunnel before committing this patient to tarsal tunnel release given the change in biomechanics that can result after release, especially given existing pes planus and moderate baseline pronation with ambulation. Another 3 months of rest may help temporarily but is likely to recur upon resumption of marathon training, and the patient has already failed this once.

References

1. Lam KHS, Hung CY, Chiang YP, Onishi K, Su DCJ, Clark TB, Reeves KD. Ultrasound-Guided Nerve Hydrodissection for Pain Management: Rationale, Methods, Current Literature, and Theoretical Mechanisms. J Pain Res. 2020 Aug 4;13:1957-1968. doi: https://doi.org/10.2147/JPR.S247208. PMID: 32801851; PMCID: PMC7414936.
2. Wu YT, Ho TY, Chou YC, Ke MJ, Li TY, Huang GS, Chen LC. Six-month efficacy of platelet-rich plasma for carpal tunnel syndrome: A prospective randomized, single-blind controlled trial. Sci Rep. 2017 Dec;7(1):94. doi: https://doi.org/10.1038/s41598-017-00224-6. Epub 2017 Mar 7. PMID: 28273894; PMCID: PMC5427966.

**Acknowledgments** This book chapter is modified and updated from a previous book chapter, "Ankle and Foot, and Related Joints, Ligaments and Tendons" by Andrea Trescot, MD, and Beth S. Pearce, DPM, in *Essentials of Regenerative Medicine in Interventional Pain Management* published by ASIPP Publishing.

## References

1. Trescot A, Pearce BS. Ankle and foot, and related joints, ligaments and tendons. In: Manchikanti L, Navani A, Atluri S, editors. Essentials of regenerative medicine in interventional pain management. Paducah: ASIPP Publishing; 2019. p. 369–84.
2. Pacha D, Carrera A, Llusa M, et al. Clinical anatomy of the superficial peroneal nerve in the leg. Eur J Anat. 2003;7:15–20.
3. Dellon AL. Deep peroneal nerve entrapment on the dorsum of the foot. Foot Ankle. 1990;11:73–80.
4. Flanigan RM, DiGiovanni BF. Peripheral nerve entrapments of the lower leg, ankle, and foot. Foot Ankle Clin. 2011;16:255–74.
5. Pringle RM, Protheroe K, Mukherjee SK. Entrapment neuropathy of the sural nerve. J Bone Joint Surg Br. 1974;56B:465–8.
6. Horwitz MT. Normal anatomy and variations of the peripheral nerves of the leg and foot. Application in operations for vascular diseases: study of one hundred specimens. Arch Surg. 1938;36:626–36.
7. Rask MR. Medial plantar neurapraxia (jogger's foot): report of 3 cases. Clin Orthop Relat Res. 1978;134:193–5.
8. Oh SJ, Lee KW. Medial plantar neuropathy. Neurology. 1987;37:1408–10.
9. Oh SJ, Kwon KH, Hah JS, et al. Lateral plantar neuropathy. Muscle Nerve. 1999;22:1234–8.
10. Rao SB, Crawford AH. Lateral plantar nerve entrapment following Rush rod fixation for congenital pseudarthrosis of the tibia. Orthopedics. 1996;19:1043–6.
11. Coulter KR, Gerbert J, Shea TP. Neurilemmoma of the lateral plantar nerve: a case report. J Am Podiatry Assoc. 1978;68:721–4.
12. Kwok KB, Lui TH, Lo WN. Neurilemmoma of the first branch of the lateral plantar nerve causing tarsal tunnel syndrome. Foot Ankle Spec. 2009;2:287–90.
13. Belding RH. Neurilemoma of the lateral plantar nerve producing tarsal tunnel syndrome: a case report. Foot Ankle. 1993;14:289–91.
14. McCrory P, Bell S, Bradshaw C. Nerve entrapments of the lower leg, ankle and foot in sport. Sports Med. 2002;32:371–91.
15. Schon LC, Baxter DE. Neuropathies of the foot and ankle in athletes. Clin Sports Med. 1990;9:489–509.
16. Hyman GS, Solomon J, Dahm D. Physical exam of the foot and ankle. In: Malanga GA, Nadler S, editors. Musculoskeletal physical examination: an evidence-based approach. Philadelphia: Elsevier Mosby; 2006. p. 315–43.
17. Benjamin M, Toumi H, Ralphs JR, et al. Where tendons and ligaments meet bone: attachment sites ('entheses') in relation to exercise and/or mechanical load. J Anat. 2006;208:471–90.
18. Tsatsos G, Mandal R. Prolotherapy in the treatment of foot problems. J Am Podiatr Med Assoc. 2002;92:366–8.
19. Ferran NA, Maffulli N. Epidemiology of sprains of the lateral ankle ligament complex. Foot Ankle Clin. 2006;11:659–62.
20. Fong DT, Chan YY, Mok KM, et al. Understanding acute ankle ligamentous sprain injury in sports. Sports Med Arthrosc Rehabil Ther Technol. 2009;1:14.
21. Brostroem L. Sprained ankle. Anatomic lesions in recent sprains. Acta Chir Scand. 1964;128:483–95.
22. Golanó P, Vega J, de Leeuw PA, et al. Anatomy of the ankle ligaments: a pictorial essay. Knee Surg Sports Traumatol Arthrosc. 2016;24:944–56.
23. Kerkhoffs GM, Rowe BH, Assendelft WJ, et al. Immobilisation and functional treatment for acute lateral ankle ligament injuries in adults. Cochrane Database Syst Rev. 2002;3:CD003762.
24. Linklater J. Ligamentous, chondral, and osteochondral ankle injuries in athletes. Semin Musculoskelet Radiol. 2004;8:81–98.
25. Xenos JS, Hopkinson WJ, Mulligan ME, et al. The tibiofibular syndesmosis. Evaluation of the ligamentous structures, methods of fixation, and radiographic assessment. J Bone Joint Surg Am. 1995;77:847–56.
26. Diers DJ. Medial calcaneal nerve entrapment as a cause for chronic heel pain. Physiother Theory Pract. 2008;24:291–8.
27. Wenzel EM, Kajgana Z, Kelley KD, et al. Prevalence of equinus in patients diagnosed with plantar fasciitis. Foot Ankle Online J. 2009;2:1.
28. Gerdesmeyer L, Mittermayr R, Fuerst M, et al. Current evidence of extracorporeal shock wave therapy in chronic Achilles tendinopathy. Int J Surg. 2015;24:154–9.
29. Reissig J, Bitterman A, Lee S. Common foot and ankle injuries: what not to miss and how best to manage. J Am Osteopath Assoc. 2017;117:98–104.
30. Helgeson K. Examination and intervention for sinus tarsi syndrome. N Am J Sports Phys Ther. 2009;4:29–37.
31. Jacobs JW, da Silva JA. Hypermobility syndromes from the clinician's perspective: an overview. Acta Reumatol Port. 2014;39:124–36.
32. de César PC, Avila EM, de Abreu MR. Comparison of magnetic resonance imaging to physical examination for syndesmotic injury after lateral ankle sprain. Foot Ankle Int. 2011;32:1110–4.
33. Young CC. Ankle sprain clinical presentation. Medscape; 14 Jan 2019. https://emedicine.medscape.com/article/1907229-clinical. Accessed 03/02/2023.
34. Bhatia M, Thomson L. Morton's neuroma—current concepts review. J Clin Orthop Trauma. 2020;11:406–9.
35. Mei-Dan O, Kots E, Barchilon V, et al. A dynamic ultrasound examination for the diagnosis of ankle syndesmotic injury in professional athletes: a preliminary study. Am J Sports Med. 2009;37:1009–16.
36. Drakonaki EE, Allen GM, Wilson DJ. Ultrasound elastography for musculoskeletal applications. Br J Radiol. 2012;85:1435–45.
37. Nielson JH, Gardner MJ, Peterson MG, et al. Radiographic measurements do not predict syndesmotic injury in ankle fractures: an MRI study. Clin Orthop Relat Res. 2005;436:216–21.
38. Milz P, Milz S, Steinborn M, et al. [13-MHc high frequency ultrasound of the lateral ligaments of the ankle joint and the anterior tibia-fibular ligament. Comparison and results of MRI in 64 patients]. Radiologe. 1999;39:34–40.
39. Lam KHS, Hung CY, Chiang YP, et al. Ultrasound-guided nerve hydrodissection for pain management: rationale, methods, current literature, and theoretical mechanisms. J Pain Res. 2020;13:1957–68.
40. Hackett GS. Ligament and tendon relaxation—treated with prolotherapy. 5th ed. Gustav, Oak Park: Hemwall; 1991.
41. Hauser RA, Feister WA. Dextrose prolotherapy with human growth hormone to treat chronic first metatarsophalangeal joint pain. Foot Ankle Online J. 2012;5:9.
42. Banks A. A rationale for prolotherapy. J Orthop Med. 1991;13:55–9.
43. Linetsky FS, Miguel R, Torres F. Treatment of cervicothoracic pain and cervicogenic headaches with regenerative injection therapy. Curr Pain Headache Rep. 2004;8:41–8.
44. Correia SI, Pereira H, Silva-Correia J, et al. Current concepts: tissue engineering and regenerative medicine applications in the ankle joint. J R Soc Interface. 2013;11:20130784.
45. García de Cortázar U, Padilla S, Lobato E, et al. Intraneural platelet-rich plasma injections for the treatment of radial nerve section: a case report. J Clin Med. 2018;7:13.

46. Tseng SC. HC-HA/PTX3 purified from amniotic membrane as novel regenerative matrix: insight into relationship between inflammation and regeneration. Invest Ophthalmol Vis Sci. 2016;57:ORSFh1-8.
47. DeChellis DM, Cortazzo MH. Regenerative medicine in the field of pain medicine: prolotherapy, platelet-rich plasma therapy, and stem cell therapy—theory and evidence. Tech Reg Anesth Pain Manage. 2011;15:74–80.
48. Caplan AI. Mesenchymal stem cells: time to change the name! Stem Cells Transl Med. 2017;6(6):1445–51. https://doi.org/10.1002/sctm.17-0051. Accessed 03/02/2023.
49. Pybus PL. In: di Fabio A, editor. Intraneural injections for rheumatoid arthritis and osteoarthritis with control of pain in arthritis in the knee. Birmingham: Rheumatoid Disease Foundation; 1989. p. 9–22.
50. di Summa PG, Kingham PJ, Raffoul W, et al. Adipose-derived stem cells enhance peripheral nerve regeneration. J Plast Reconstr Aesthet Surg. 2010;63:1544–52.
51. Maxwell NJ, Ryan MB, Taunton JE, et al. Sonographically guided intratendinous injection of hyperosmolar dextrose to treat chronic tendinosis of the Achilles tendon: a pilot study. AJR Am J Roentgenol. 2007;189:W215–20.
52. Yelland MJ, Sweeting KR, Lyftogt JA, et al. Prolotherapy injections and eccentric loading exercises for painful Achilles tendinosis: a randomised trial. Br J Sports Med. 2011;45:421–8.
53. Gross CE, Hsu AR, Chahal J, et al. Injectable treatments for noninsertional achilles tendinosis: a systematic review. Foot Ankle Int. 2013;34:619–28.
54. Sanderson LM, Bryant A. Effectiveness and safety of prolotherapy injections for management of lower limb tendinopathy and fasciopathy: a systematic review. J Foot Ankle Res. 2015;8:57.
55. Ryan M, Wong A, Rabago D, et al. Ultrasound-guided injections of hyperosmolar dextrose for overuse patellar tendinopathy: a pilot study. Br J Sports Med. 2011;45:972–7.
56. Kim E, Lee JH. Autologous platelet-rich plasma versus dextrose prolotherapy for the treatment of chronic recalcitrant plantar fasciitis. PM R. 2014;6:152–8.
57. Hauser RA, Orlofsky A. Regenerative injection therapy with whole bone marrow aspirate for degenerative joint disease: a case series. Clin Med Insights Arthritis Musculoskelet Disord. 2013;6:65–72.
58. Yausep OE, Madhi I, Trigkilidas D. Platelet-rich plasma for treatment of osteochondral lesions of the talus: a systematic review of clinical trials. J Orthop. 2020;18:218–25.
59. Migliorini F, Driessen A, Quack V, et al. Comparison between intra-articular infiltrations of placebo, steroids, hyaluronic and PRP for knee osteoarthritis: a Bayesian network meta-analysis. Arch Orthop Trauma Surg. 2021;141:1473–90.
60. Abas S, Kuiper JH, Roberts S, et al. Osteochondral lesions of the ankle treated with bone marrow concentrate with hyaluronan and fibrin: a single-centre study. Cells. 2022;11:629.
61. Vannini F, Filardo G, Altamura SA, et al. Bone marrow aspirate concentrate and scaffold for osteochondral lesions of the talus in ankle osteoarthritis: satisfactory clinical outcome at 10 years. Knee Surg Sports Traumatol Arthrosc. 2021;29:2504–10.
62. Kim YS, Lee M, Koh YG. Additional mesenchymal stem cell injection improves the outcomes of marrow stimulation combined with supramalleolar osteotomy in varus ankle osteoarthritis: short-term clinical results with second-look arthroscopic evaluation. J Exp Orthop. 2016;3:12.
63. Zhang Z, Schon L. The current status of clinical trials on biologics for cartilage repair and osteoarthritis treatment: an analysis of clinicaltrials.gov data. Cartilage. 2022;13:19476035221093065.
64. Jia X, Peters PG, Schon L. The use of platelet-rich plasma in the management of foot and ankle conditions. Oper Tech Sports Med. 2011;19:177–84.
65. Chahla J, Cinque ME, Shon JM, et al. Bone marrow aspirate concentrate for the treatment of osteochondral lesions of the talus: a systematic review of outcomes. J Exp Orthop. 2016;3:33. Erratum in: J Exp Orthop. 2016;3:38.
66. Mautner K, Malanga GA, Smith J, et al. A call for a standard classification system for future biologic research: the rationale for new PRP nomenclature. PM R. 2015;7:S53–9.
67. Iborra A, Villanueva M, Sanz-Ruiz P. Results of ultrasound-guided release of tarsal tunnel syndrome: a review of 81 cases with a minimum follow-up of 18 months. J Orthop Surg Res. 2020;15:30.
68. James I, Gusenoff B, Wang SS, et al. Abstract 137: fat grafting improves foot pain associated with fat pad atrophy of the heel: early findings from a randomized controlled clinical trial. Plast Reconstr Surg Glob Open. 2018;6:107–8.
69. Al-Siyabi Z, Karam M, Al-Hajri E, et al. Extracorporeal shockwave therapy versus ultrasound therapy for plantar fasciitis: a systematic review and meta-analysis. Cureus. 2022;14:e20871.
70. Asheghan M, Hashemi SE, Hollisaz MT, et al. Dextrose prolotherapy versus radial extracorporeal shock wave therapy in the treatment of chronic plantar fasciitis: a randomized, controlled clinical trial. Foot Ankle Surg. 2021;27:643–9.
71. Xu D, Jiang W, Huang D, et al. Comparison between extracorporeal shock wave therapy and local corticosteroid injection for plantar fasciitis. Foot Ankle Int. 2020;41:200–5.
72. US District Court USA vs California Stem Cell Treatment Center, Inc. https://s3.documentcloud.org/documents/22272935/berman-bernal-ruling.pdf. Accessed 03/02/2023.
73. U.S. Food and Drug Administration. CFR—Code of Federal Regulations Title 21. https://www.accessdata.fda.gov/scripts/cdrh/cfdocs/cfcfr/cfrsearch.cfm?fr=1271.15. Accessed 03/02/2023.

# Part III

# Regenerative Medicine in the Spine

# Radiology of the Spine for the Interventionalist

# 22

Sheri L. Albers and Richard E. Latchaw

## Introduction

The role of this chapter is to demonstrate and discuss pertinent imaging findings, both normal and pathological, that will aid the spinal interventionalist in providing therapy to the patient. This chapter is divided according to presentation; for example, cervical pain conditions are discussed separately from those in the lumbosacral (L-S) spine; both are subdivided into disc disease, facet disease, focal nerve root compression for which nerve root block may be an option, and other causes of spinal and radicular pain. Phrases are used in the text to highlight key features. Important points regarding the imaging characteristics of both normal and abnormal conditions are discussed in the legends to the many illustrations. Inserts have been added that stress the important findings to observe when treating a given condition, and the advantages and disadvantages of the modalities used to image the spine.

The appearances of many of the conditions affecting the spine and spinal cord discussed in this chapter, including the various techniques to image them, have been presented in previous books and chapters [1–15]. Because of this, references for a given topic have been chosen that discuss findings on imaging that ensure the specificity of the diagnosis or that will affect treatment.

## Imaging Modalities

### Plain X-Rays and Fluoroscopy

- X-rays generated by a cathode ray tube pass through body [2].
- X-rays detected by small packets (pixels) of rare earth metals, photomultiplied, and computer-processed.
- Quick, provides overview of body part, but overlap of structures makes evaluation of a specific bony structure difficult, limited differentiation of soft tissues [2].
- Continuous or pulsed beam of x-rays, similar to plain x-ray technique, but allows for continual visualization of one or more body parts over time [2].
- Electronic subtraction technique subtracts the initial image (scout) from the sequential images, allows for visualization of tiny amounts of infused contrast agent without its being masked by overlying tissues [2].
- Advantages
  - Plain films (radiographs) provide an overview of an entire body part.
  - Plain films are a good screening technique for gross fractures of a large bone.

S. L. Albers (✉)
Radiology Research and Consultation, Sacramento, CA, USA

R. E. Latchaw
Department of Radiology, Neuroradiology Section, University of California Medical Center, Sacramento, CA, USA


A. Navani et al. (eds.), *Essentials of Regenerative Medicine in Interventional Pain Management*,
https://doi.org/10.1007/978-3-031-50357-3_22

  - Plain films are acquired rapidly; multiple body parts can be imaged in the same x-ray room.
  - Fluoroscopy is active, can be additive to CT and MR examinations.
- Disadvantages
  - Overlap of many bony structures makes detection of an abnormality at a specific site difficult.
  - Not a good technique to view small bony structures or subtle abnormalities.
  - Does not define the various components of soft tissues, so only nonspecific soft tissue "swelling" is seen.
  - Uses ionizing radiation, with the long-term risk of cancer induction.
  - Fluoroscopy exposes physicians and technologists to direct and indirect (scatter) radiation exposure.

## Nuclear Medicine

- Either a planar gamma camera or single photon emission computed tomography (SPECT) system used to detect isotopes that emit gamma rays, such as 99mTechnesium diphosphonate for three-phase bone scanning, gallium-67 citrate and Indium-111 for infection [2, 14].
- Positron emission tomography (PET) scanning is used to study the uptake of a positron emitter such as 18fluorodexoglucose (18FDG); used to study glucose-dependent processes; better anatomical localization achieved with PET/CT.
- Most common indications:
  - Determine sites of primary or metastatic disease
  - Detect avascular necrosis
  - Evaluate patient who has undergone diffuse trauma, including pediatric non-accidental trauma.
  - Study complex regional pain syndrome
  - Evaluate periprosthetic complications
- Role in differentiating acute from subacute/chronic spinal fractures: bone scan remains positive in 41% at 1 year, 10% at 2 years, and 3% at 3 years [14]; MRI with non-FS T1 before and after contrast enhancement and short Tau inversion recovery (STIR) sequences more accurately show the decrease in edema and inflammation over time.

## Computed Tomography (CT)

- Cathode ray tube moves in arc opposite ring of rare earth detectors, producing thousands of points of x-ray detection; sophisticated computer programs determine the density of the tissue at a given point that absorbs some of the beam energy [2].
- Matrix of relative densities converted into picture of body part with much improved soft tissue contrast resolution over plain x-rays.
- Best technique for cortical and trabecular bone detail and for lung parenchyma
- Advantages
  - Ability to evaluate soft tissues such as the brain, unlike plain films which only show the calvarium, or the internal aspects of abdominal and pelvic organs, unlike plain films which only show the organs' external contours.
  - Very high resolution of small highly dense structures, so excellent for detecting subtle fractures, small osteophytes, and tiny calcifications.
  - Much improved soft tissue discrimination relative to plain films, so hematoma, abscess, or edema can be differentiated from surrounding normal soft tissues; however, this improved soft tissue contrast is more limited within a small space surrounded by bone, such as the spinal canal.
  - Use of an infused iodinated contrast agent permits differentiation of vascular structures and anomalies such as malformations and aneurysms from surrounding tissues, and visualization of leakage of the contrast through a damaged blood-tissue barrier typically seen with tumors and infections.
  - Thin slice CT angiography (CTA) with processing that produces multiple projections is almost or as accurate in diagnosis as catheter-bases angiography.
  - Thin slices permit the computer production of data in other planes, e.g., data acquired in the axial plan, but sagittal and coronal images constructed.
  - Rapid, can be used to evaluate most body parts.
- Disadvantages
  - Less than optimal differentiation of the components of a soft tissue mass, so diagnosis is usually "mass present", without ability to be more specific, i.e., tissue contrast resolution less than optimum.
  - Adequate multiple projections require very thin cuts which are time consuming, requiring patient cooperation.
  - Technique delivers a significant amount of radiation to a given body part. Potential for cancer induction not a trivial issue.
  - Much more expensive than plain films (but less expensive than MR studies).
  - Enhancement requires administration of a nonionic contrast agent, which has a low but significant risk of allergenic response, varying from rash to anaphylaxis.

## Magnetic Resonance Imaging (MRI)

- Patient's body part within a magnetic field, causing many protons to align north to south [2].
  - Instead of an x-ray beam as in CT, a beam of RF energy is pulsed into tissues within the magnetic field; that beam rebounds to a receiver in proportion to the chemical constitution of the tissue(s) studied. No ionizing radiation and its potential long term ill effects.
- MRI uses a variety of RF beams with varying amplitudes and frequencies, thus producing many pictures of the tissue studied.
- MRI is far superior to CT in tissue characterization (contrast resolution), but lesser spatial resolution than CT. Example: tiny calcifications well seen with CT, but will be "averaged" with surrounding tissue on MRI and are thus invisible.
- Paramagnetic IV contrast infusion interacts with tissue protons, producing relative enhancement, especially if tissue is pathologic or hypervascular.
- RF beam with specific frequency used to evaluate flowing blood, producing images of blood vessels (MR Angiography, MRA); different RF techniques can differentiate arteries (MRA) from veins (MRV).
- RF beam specific to a chemical element can determine chemical nature of tissue (MR Spectroscopy, MRS).
- Problem: most MR sequences require more time than a CT scan, and they are motion sensitive. So, a long MR examination may require sedation/anesthesia.
- Another problem: MRI is contraindicated with most pacemakers, spinal cord stimulators, cochlear implants, peripheral nerve stimulators, and other electromagnetic devices.
- Advantages
  - Superb contrast resolution, so able to characterize the nature of soft tissues and their diseases; e.g., can not only determine the make-up of a soft tissue abnormality, but able to propose a histological diagnosis that is accurate in the majority of cases.
  - Multiplanar imaging easy, does not require thin slices for multiplanar reformations as with CT.
  - No ionizing radiation and no soft tissue injury via imaging.
  - MRI is more sensitive to subtle contrast enhancement than CT.
  - Able to detect abnormalities next to bony structures missed by CT because cortical bone is relatively less visible on MRI.
  - While standard MRI sequences may not demonstrate a subtle fracture, sequences such as fat-saturated (FS) T2 or STIR may show increased signal indicative of a subtle compression fracture.
  - Use of STIR and FS T2 may also show abnormal signal representing foci of inflammation or tumor, not seen with other MRI sequences or CT images that have higher spatial but not contrast resolution.
  - Propeller software techniquc removes much of the movement artifact from T2 sequences when scanning an uncooperative patient.
  - Other MR-based techniques are MR angiography (MRA) which has less spatial accuracy than CTA; MR venography (MRV) which is more accurate than CT venography for evaluation of dural sinuses because of the absence of contiguous bone; perfusion MRI, which determines the relative quantity of blood flow into a volume of soft tissue such as the brain; diffusion MRI, which detects the movement of hydrogen molecules in a volume of soft tissue; functional MRI (fMRI) which evaluates the physiological activity of tissues based upon its blood flow patterns (usually venous at this time); and MR spectroscopy (MRS) which permits evaluation of the chemical nature of tissue.
- Disadvantages
  - Some patients cannot be in a magnetic field (e.g., those with most pacemakers and most subcutaneous stimulators), and others cannot tolerate the claustrophobic environment.
  - Hard to monitor a patient that is deep inside an imaging cylinder.
  - Slow technique compared to CT, so patient must be cooperative or under sedation/anesthesia.
  - Spatial resolution not equal to CT, and cortical bony structures are less visible, so subtle fractures may be missed.
  - MRI is insensitive to small or fine calcifications, which may be critical to the proper diagnosis; CT is very sensitive to such findings.
  - Significant artifacts produced by iron-containing structures in the scanned volume.
  - Expensive, sophisticated equipment required, space needed away from sources of interfering radiofrequencies that produce artifacts.
  - Gadolinium relatively contraindicated in patients with glomerular filtration rate (GFR) $< 30$ due to greater potential to acquire nephrogenic systemic fibrosis (NSF).
  - MRI cannot be performed in patients with metal in an eye, having non-MRI compatible cerebral aneurysm clips, shrapnel near vascular structures.

## Myelography

- Puncture of the thecal sac, instillation of a small volume of an iodinated contrast agent or air into the subarachnoid space, followed by plain films. Patients can subsequently undergo CT scanning (post-myelographic CT) for better

definition of the spinal canal, cord, nerve roots, and intracranial subarachnoid spaces [2].
- Problem: ionizing radiation used for both the myelography and CT scanning.
- Problem: cannot characterize the specific cord tissues, only size, configuration.

## Degenerative Lumbosacral Spine

### Intervertebral Disc Disease

- Types [16, 17]
  - Bulging disc: increased transverse disc diameter over more than 180° of disc circumference (Fig. 22.1).
  - Disc protrusion: focal disc projection with obtuse borders relative to parent disc (Fig. 22.1).
  - Disc extrusion: disc fragment with acute borders relative to parent disc (Fig. 22.2).
  - Free fragment (sequestration): disc material separated from parent disc (Fig. 22.3).
- Annular fissure: high intensity focus of discontinuity in annulus on T2 images allowing for disc protrusion, eventually disc extrusion (Fig. 22.4).
- Acute vs. chronic extrusion: hyperintensity within the disc fragment indicates a high water content, and thus probably an acute/early subacute extrusion (Fig. 22.2a) vs. isointensity to the rest of the disc indicating a more chronic extrusion (Figs. 22.2b and 22.3b).
- Relationship of disc protrusion/extrusion to nerve root(s):
  - Touching: protruding or extruding disc may just touch nerve root without producing radiculopathy, or radiculopathy only in certain body positions.
  - Compressing/displacing nerve root: usually produces radiculopathy.
  - Central: midline, below posterior longitudinal ligament (PLL), does not usually affect nerve roots (Fig. 22.1).
  - Paracentral: slightly off midline, still below PLL, may affect nerve root(s) if large.
  - Into lateral recess: anterolateral corner of disc space between disc and ligamentum flavum overlying facets; descending nerve root runs in this space as shown in Figs. 22.2, 22.3b, and 22.5) [18].
  - Into a neural foramen: sagittal MR views are essential to determine if the disc actually touches the nerve root which is higher than the usual location of the disc (Figs. 22.1a and 22.6).
  - Disc disease is one component of central spinal canal stenosis compressing all roots at a given level.

### Inflammatory Reaction to Intervertebral Disc Disease

- Modic Classification [19]
  - Modic Type 1: low signal intensity on MR non-FS T1 sequence, high on T2 with or without FS, indicative of edema and thus represents acute to subacute reaction as shown in Fig. 22.7 [20–22].

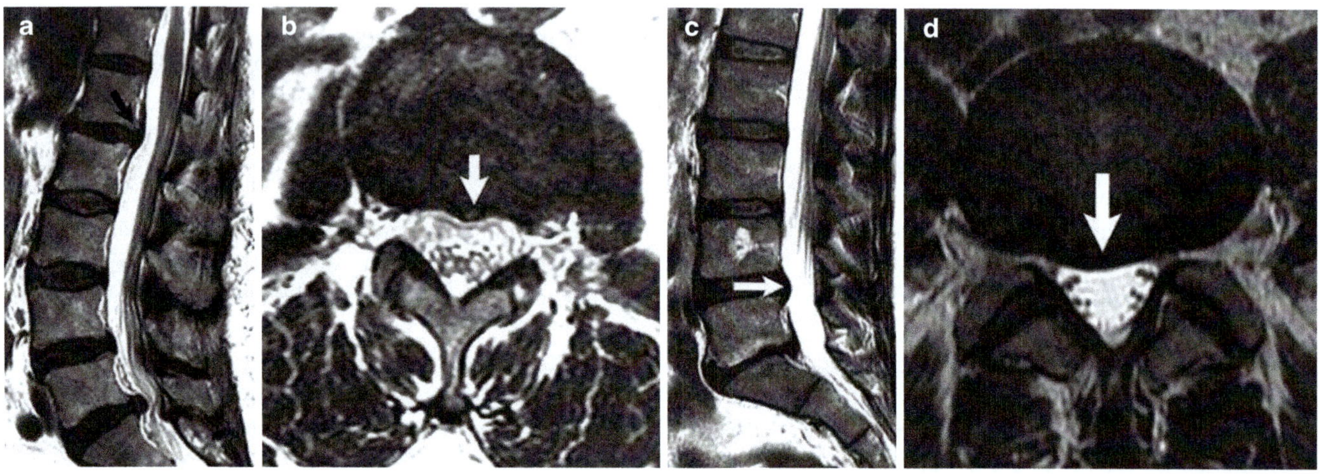


**Fig. 22.1** Bulging lumbar disc with small focal protrusion through an annular fissure, and mild disc protrusion at second level. At L1–L2, the sagittal view (**a**) demonstrates the disc to project both anterior and posterior to the spinal column. By definition, a diffusely bulging must encompass 270–360° of disc circumference. Between 90 and 270 it is an asymmetric bulge, while 90° or less is a protrusion. The axial view (**b**) demonstrates a small disc protrusion in the midline, having obtuse margination with the rest of the disc, typical of a protrusion. There is a small annular fissure (arrow) through which the nucleus pulposus projects through the annulus. (**c**, **d**) The L4–L5 disc projects posteriorly in the midline (arrows, **c**, **d**). The obtuse margins of this projection relative to the rest of the disc and its focality, less than 90° of the disc circumference, make this a protrusion

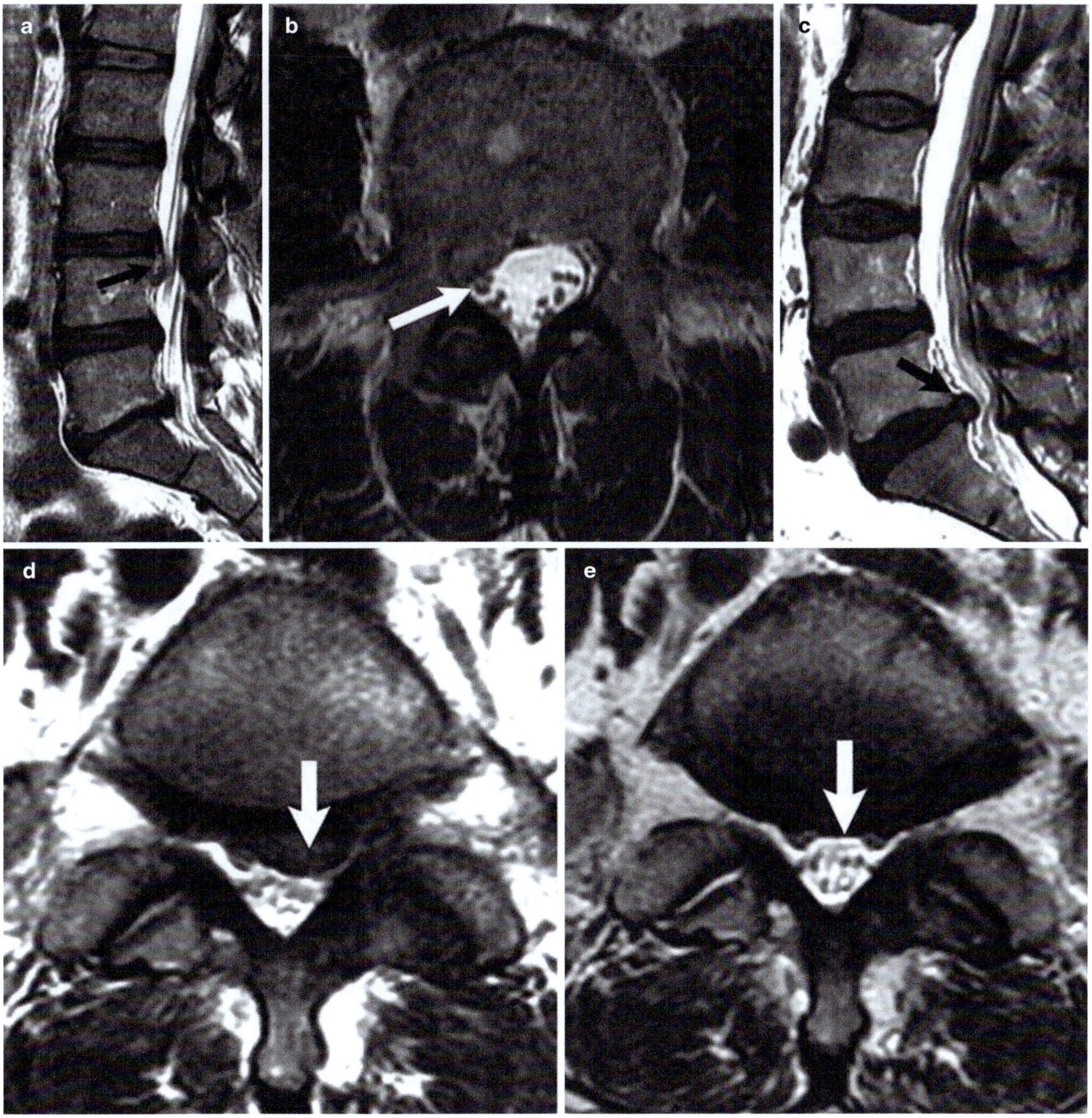


**Fig. 22.2** Disc extrusion with recent inferior migration and L4 root compression, and second level of disc extrusion and resorption. The sagittal T2 image (**a**) shows an extrusion from the L3–L4 disc extending over the superior margin of L4. Part of this extrusion has a higher water content (arrow), indicating its recent herniation. The axial view (**b**) demonstrates another component extending into the lateral recess at the disc space level, posteriorly displacing the right L4 nerve root (arrow). The initial sagittal (**c**) and axial (**d**) T2 sequences demonstrate a large disc extrusion centrally and to the left at L5–S1 (arrows), extending into the left lateral recess to compress the left S1 root. The patient opted for no surgery, and 13 months later, a repeat MRI (**e**) demonstrates that the extruded segment has resorbed, leaving only an asymmetrically bulging disc (arrow) which touches but does not compress or displace the adjacent S1 roots

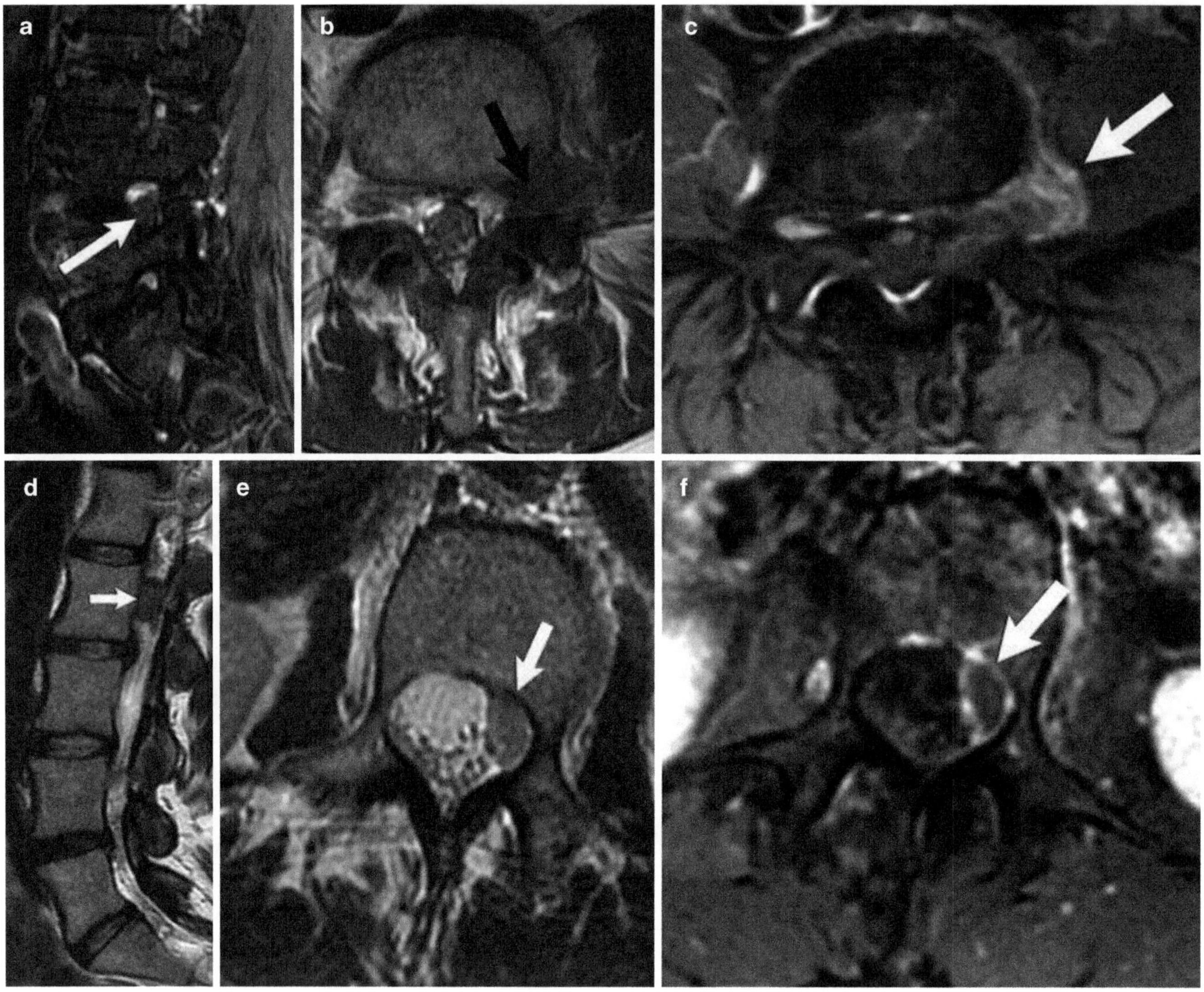


**Fig. 22.3** MRIs of an enhancing free lumbar disc fragment within a neural foramen on, and a second case of a migratory free disc fragment with peripheral enhancement. The sagittal IR (**a**) and unenhanced axial T1 (**b**) images demonstrate a mass in the L4 left neural foramen (arrows). Considerations are a nerve sheath tumor such as a schwannoma and an extruded or free disc fragment. The mass has an intensity like that of disc, being slightly lighter than muscle. The enhanced fat suppressed axial T1 image (**c**) shows peripheral enhancement (arrow), typical of the inflammatory component around a disc fragment. A schwannoma would enhance homogeneously unless there was central cyst formation, which the unenhanced sequences do not demonstrate. The sagittal T2 view (**d**) demonstrates a mass (arrow) at the mid-L2 level, with bulging of the L2–L3 disc below it. The axial T2 image (**e**) shows that this mass (arrow) is epidural in location, displacing the thecal sac to the right. An enhanced axial image (**f**) demonstrates the classic peripheral ring of enhancement typically seen surrounding a disc fragment (arrow). While the location on the sagittal image (**d**) might have suggested a meningioma or schwannoma, both possible tumors would have been intradural and shown homogenous enhancement, plus a schwannoma would have been hyperintense on the T2 images. This is a free disc fragment which migrated from the L2–L3 disc space

  - Modic Type 2: hyperintense on both T1 and T2 sequences, indicating fat deposition and thus a subacute to chronic reaction (Fig. 22.8).
  - Modic Type 3: sclerotic bone producing low intensity on both T1 and T2, chronic.
- Hyperintensity on STIR and fat-suppressed (FS) T2 MR sequences: represents edema and thus acute/subacute inflammation (Fig. 22.9).
- Contrast enhancement: acute inflammation
- Chronic noninfectious inflammation [23] may lead to Charcot spine (Fig. 22.10).
- Bones but not disc hot on bone scan (resolution may not allow differentiation).

## Facet Disease

- Moderate in degree, but no spondylolisthesis [23].
- Severe permits spondylolisthesis (slippage) of the upper vertebral body on the lower, plus deformity of the neural foramina, potentially compressing the exiting nerve roots within as shown in Fig. 22.11 [23].

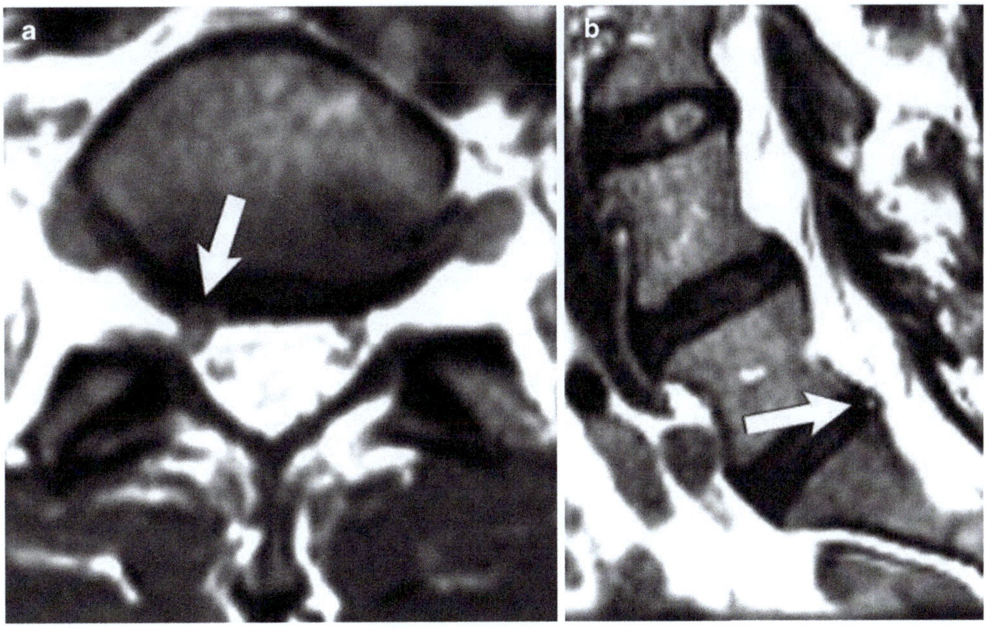

**Fig. 22.4** Classical annular fissure. The axial (**a**) and sagittal (**b**) T2 MR images demonstrate a small hyperintense focus in the annulus (arrows), overlying the right S1 nerve root. This is an annular fissure, through which disc material may herniate. The fissure is usually asymptomatic, although there may be irritating substances secreted through it during its initial formation

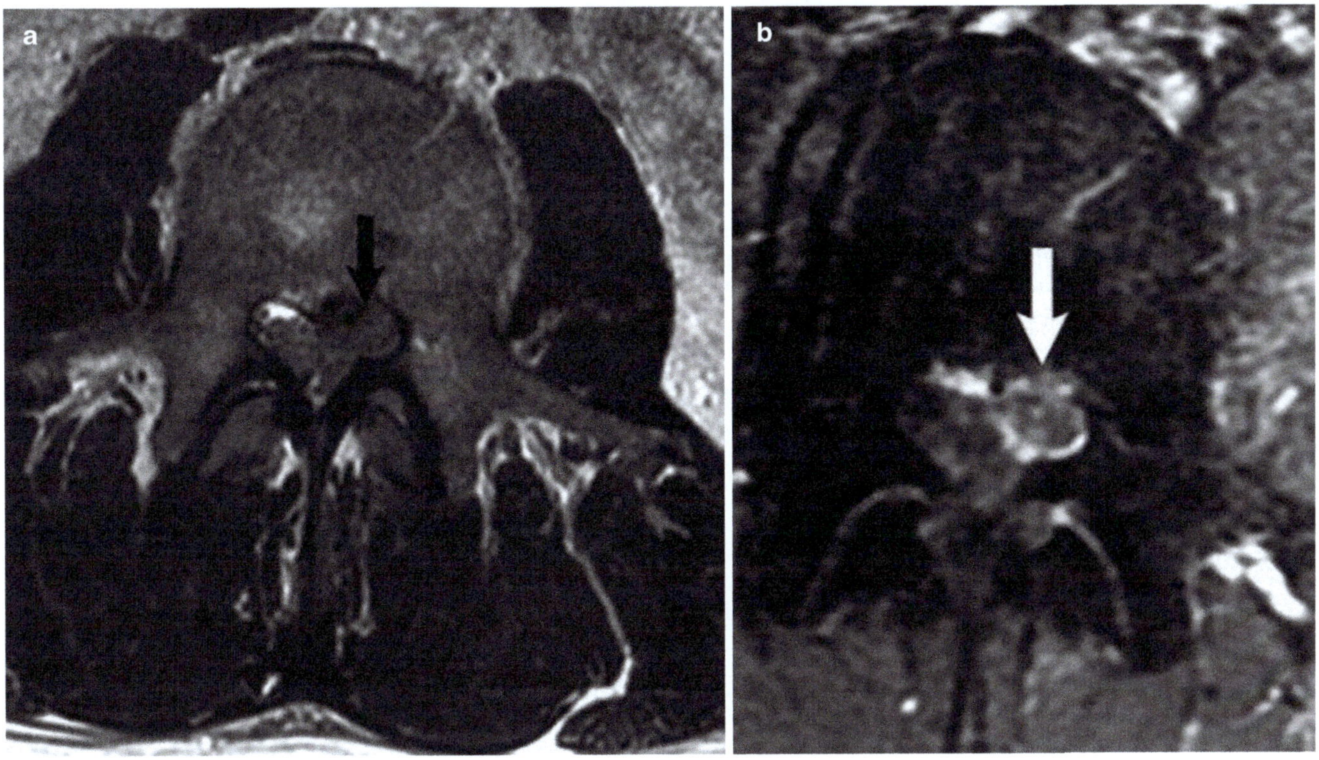

**Fig. 22.5** Large enhancing extruded lumbar disc fragment displacing the thecal sac laterally. The axial unenhanced T1-weighted MR image (**a**) shows a complicated appearance of two masses, one extending from the disc space into the left lateral recess (arrow) and a second along the right ligamentum flavum. Differential diagnosis would include an extruded disc, a schwannoma, and some other type of intraspinal tumor. The enhanced fat suppressed T1 image (**b**) demonstrates peripheral enhancement of the left lateral recess mass (arrow), typical of a disc fragment, with the thecal sac being the "second mass," displaced to the right

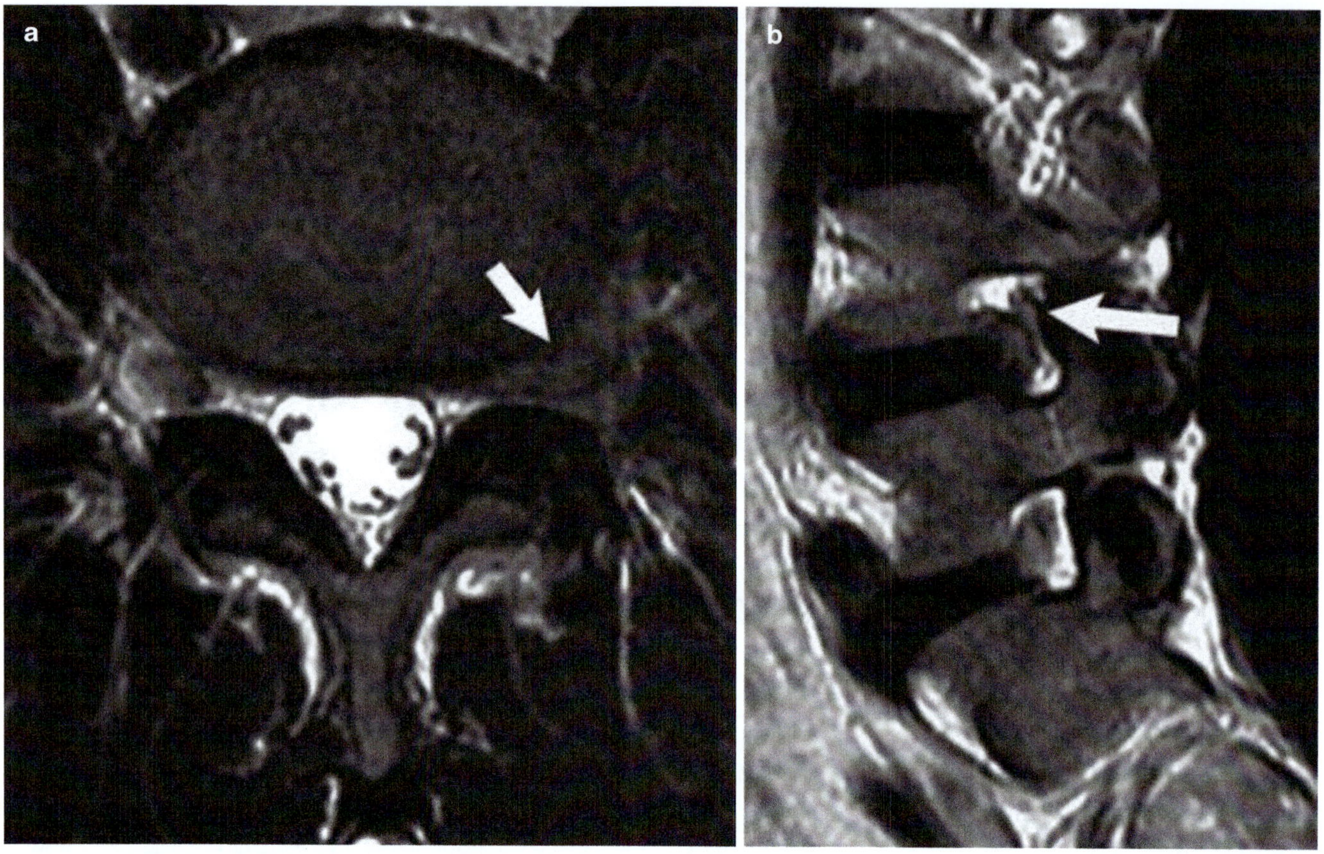


**Fig. 22.6** Disc extrusion into the left L4 neural foramen, compressing the exiting nerve root. The axial T2 MR image (**a**) shows an extruded disc (arrow) into the lower aspect of the left L4 neural foramen. The root courses in the foramen superior to the disc plane, and so the sagittal view (**b**) is needed to determine if the root is compressed by this disc extrusion, which it is (arrow), or if the root courses above the disc extrusion

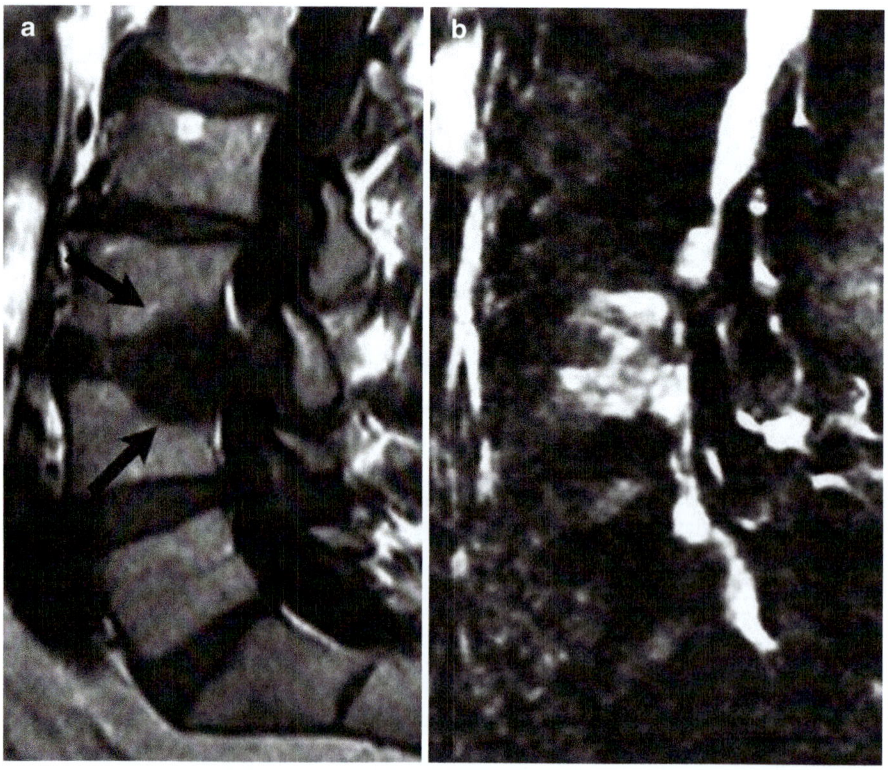


**Fig. 22.7** Endplate edema from acute degenerative disc disease, not infection. The sagittal T1 (**a**) MR sequence demonstrates low density within the L3 and L4 vertebral bodies (arrows) on either side of the disc, typical of edema (also called Modic type 1 changes). This finding is confirmed on the STIR sequence (**b**), which is very sensitive to the presence of edema, a nonspecific finding. While osteomyelitis/discitis can produce edema in the bone, there was no edema in the disc and no disc enhancement (not shown), typical of such a process. An acute discogenic event in this patient with early degenerative disc disease was responsible

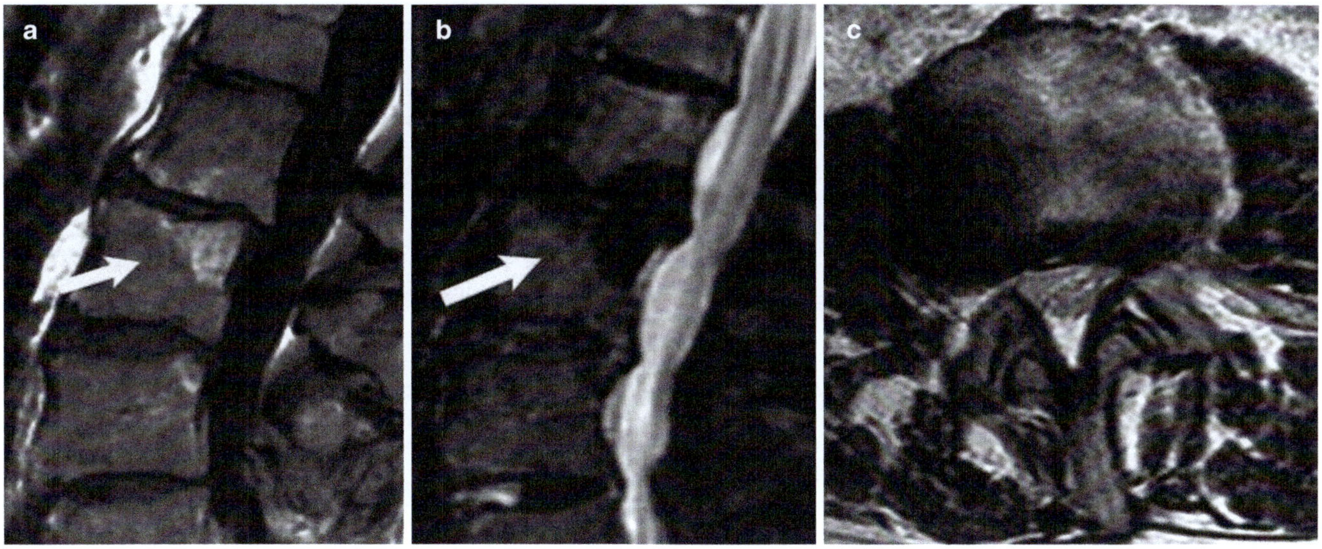


**Fig. 22.8** Chronic endplate changes (Modic type 2) and moderate central spinal canal stenosis. The MR sagittal T1-weighted image (**a**) shows hyperintense changes (arrow) within both the L2 and L3 vertebral bodies on either side of a narrowed, bulging, degenerated disc. This signal indicates that fatty marrow is present, a chronic response. The STIR image (**b**) suppresses the fat signal and excludes the presence of edema, typical of acute, Modic type 1 changes. The axial T2-weighted image (**c**) shows a bulging disc with an annular fissure and moderate facet hypertrophy, producing a moderate degree of central canal spinal stenosis

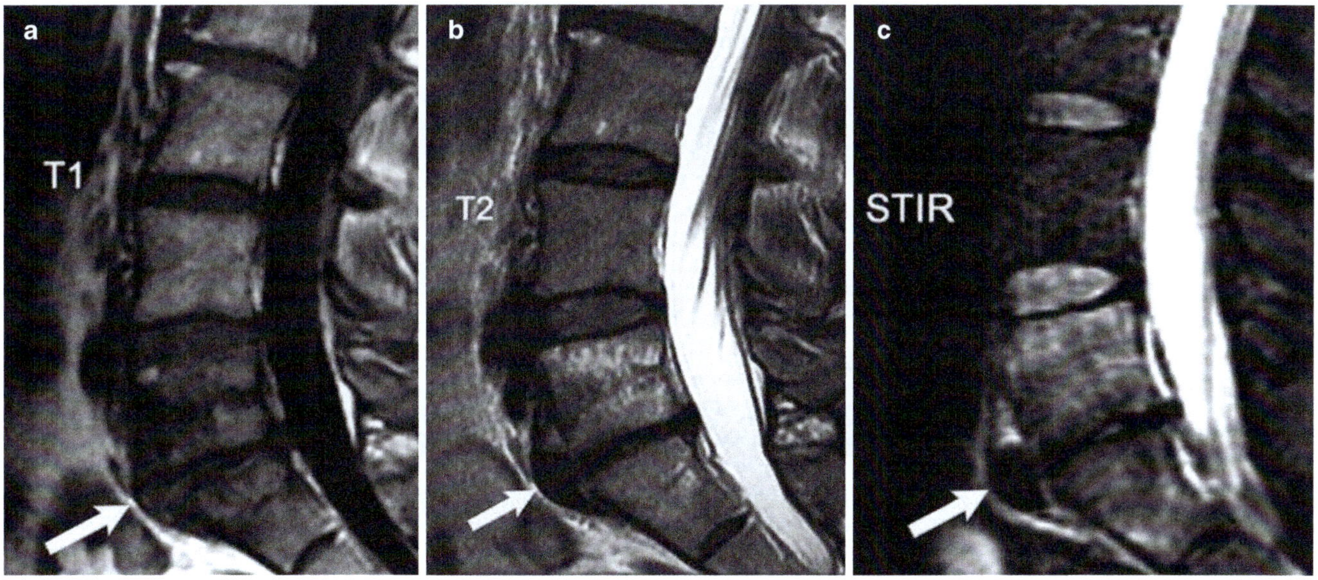


**Fig. 22.9** Edema of vertebral bodies from acute traumatic anterior disc extrusion. The sagittal T1 image (**a**) demonstrates low signal intensity in the L5 and S1 bodies representing edema, between which is a large anterior extrusion (arrow) from the L5–S1 disc. The T2 fat-saturated sagittal image (**b**) shows this edema to be hyperintense, as it is on the STIR sequence (**c**), which shows the extent of the edema better than the other sequences because of its inherent ability to suppress fat. The anterior longitudinal ligament was acutely stripped away from the vertebrae by the extrusion

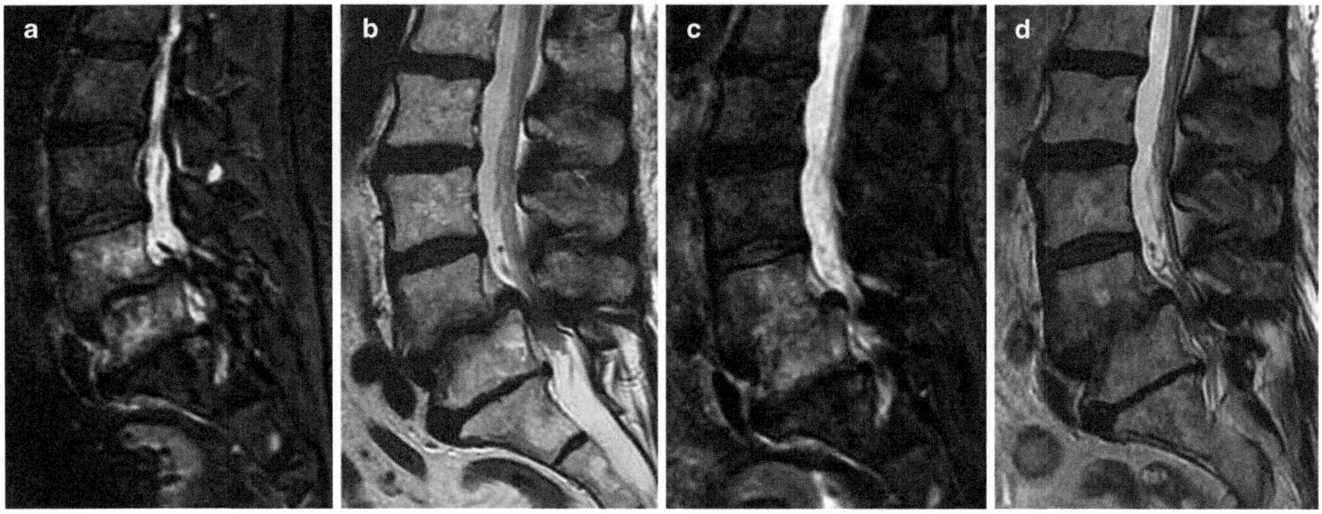


**Fig. 22.10** Rapid progression of vertebral inflammation and fusion (Charcot spine). The initial sagittal STIR (**a**) and T2 (**b**) images show edema throughout the L4 and L5 vertebral bodies, sparing the disc which is degenerated and protruding. The lack of disc edema is against an infectious etiology of this process. In only 9 months, there is a significant increase in noninfectious inflammation throughout the L4 and L5 vertebrae, as seen on the sagittal STIR (**c**) and T2 (**d**) images, but now involving the disc; this process will rapidly progress to fusion

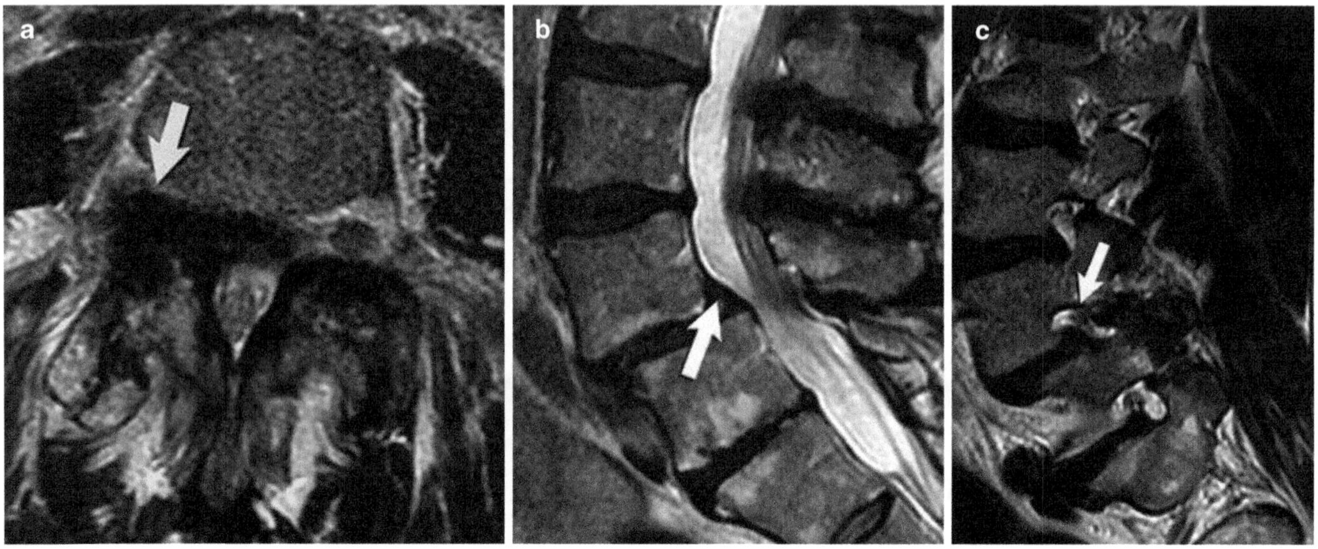


**Fig. 22.11** Severe degenerative facet disease producing grade 2 spondylolisthesis (25–50% slippage), marked osteophyte formation, and compression of bilateral nerve roots. The axial image (**a**) at the L4–L5 level demonstrates severe degenerative facet disease with a marked degree of osteophyte formation along the right vertebral body margin (arrow, very low intensity), extending into the right lateral recess to produce significant compression of the right L5 nerve root. The sagittal view (**b**) demonstrates a grade 2 spondylolisthesis (arrow) secondary to the severe facet disease. The left neural foramen is deformed by the slippage (**c**), and the exiting left L4 root (arrow) is compressed between the osteophytic facets and the degenerated disc

- Severe, with osteophytic spurs extending into a neural foramen and/or lateral recess to produce nerve root compression.
- Thickening of the ligamenta flava [24], a response to facet disease, best seen on MRI (Fig. 22.12a).
- Hypertrophy of the laminae, frequently a component of congenital spinal stenosis (Fig. 22.12b).

## Combination: Central Canal Stenosis

- Central lumbar spinal canal stenosis from disc and facet disease, plus thickened ligamenta flava with or without anteroposterior slippage, may be predisposed due to congenital spinal stenosis (Fig. 22.13).
- Degree of stenosis relative, measurements can be misleading.

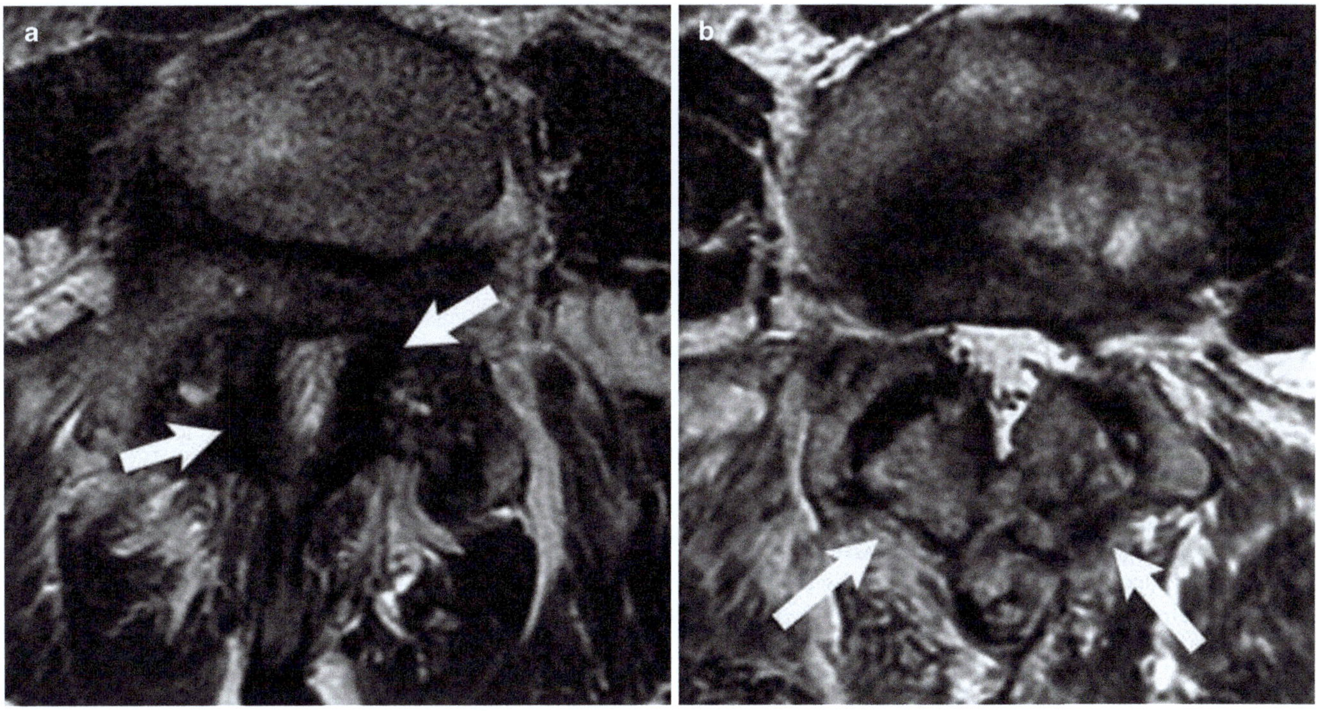


**Fig. 22.12** Marked thickening of the ligamenta flava secondary to severe degenerative facet disease and marked hypertrophy of laminae. (**a**) The axial image demonstrates the severely degenerative facets with very low-intensity osteophytes, and the marked thickening of the ligamenta flava (arrows), which are an attempt to "buttress" this level which is prone to abnormal motion and spondylolisthesis because of the facet degeneration. (**b**) This axial MR T2 image at the L5 level shows marked hypertrophy of the laminae bilaterally (arrows), producing mild central canal stenosis

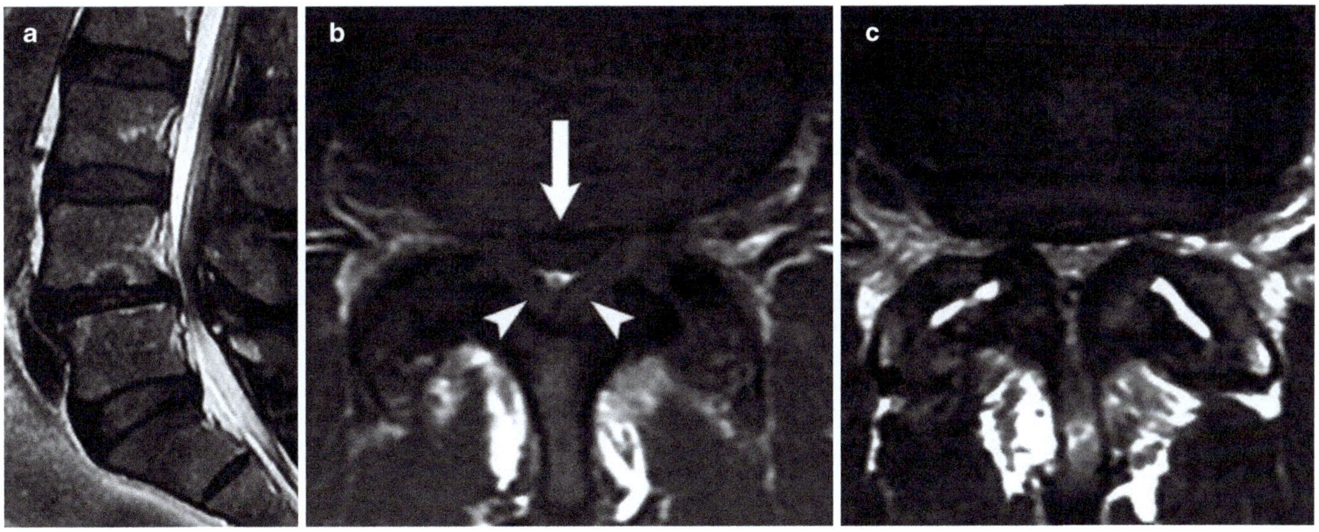


**Fig. 22.13** Multifactorial severe L4–L5 central canal stenosis. The sagittal T2 sequence (**a**) shows the central canal throughout the lumbar spine to be congenitally narrow. There is diffuse bulging of the L4–L5 disc and slight anterior slippage of L4 on L5 due to the severe degenerative facet disease that is better seen on the axial images. The axial T1 image (**b**) better demonstrates the thickening of the ligamenta flava (arrowheads) secondary to the severe facet disease and the focal protrusion (arrow) superimposed on the bulging disc. The T2 axial image (**c**) demonstrates fluid within the degenerated facet joints and the severe compression of the thecal sac. The result is a severe degree of central canal stenosis which can be decompressed with laminectomies, while fusion will be necessary because of the facet disease

- Mild: Some cerebrospinal fluid (CSF) surrounds each crowded nerve root.
- Moderate: Minimal or no CSF remaining but roots still definable.
- Severe: No CSF within compressed sac, roots cannot be defined.

- Thecal sac stenosis: increased amount of epidural fat in obese patients or due to iatrogenic steroids (epidural lipomatosis), which may or may not be reversible [25].

## Neural Foraminal Narrowing

- Due to bulging disc which does not interact with nerve root which exits above the disc [26, 27].
- Due to osteophytic spurs from facets, possibly with superimposed foraminal disc protrusion/extrusion, touching or compressing exiting nerve root.

## Degenerative Sacroiliac (SI) Joint Disease

- CT is excellent for showing the erosive and degenerative changes of the joint, including fusions, while the STIR, FS T2, and contrast enhanced sequences of MRI show the inflammation in the joints and peri-articular bone [28, 29].
- Must be differentiated from the benign osteitis condenses ilii.
- Inflammatory SI joint disease discussed under non-infectious inflammatory diseases.

## Post-Operative Spine

- Scar formation involving a nerve root: best examination is enhanced MRI with fat-suppression, looking for relatively uniform enhancing scar touching and/or encircling nerve root as shown in Fig. 22.14) [5, 11].
- Must look for recurrent disc extrusion/retained fragment, differentiated from scar by its lower signal intensity, also encircled with enhancing scar (Fig. 22.15).
- Other, new levels of disc protrusion above/below that of fusion (Fig. 22.16).

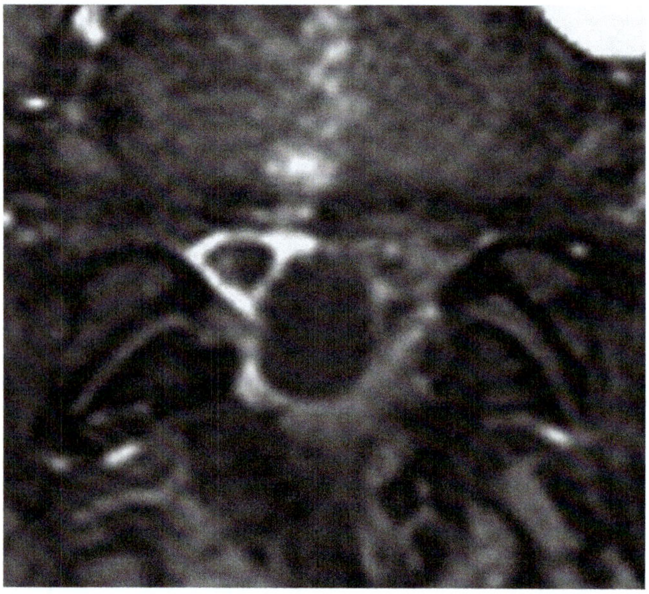

**Fig. 22.14** Scarring surround the right L5 nerve root. A contrast-enhanced axial MR image through the L4–L5 disc space shows bright enhancement surrounding the descending L5 nerve root in the right lateral recess, indicative of scarring. There is no recurrent disc extrusion

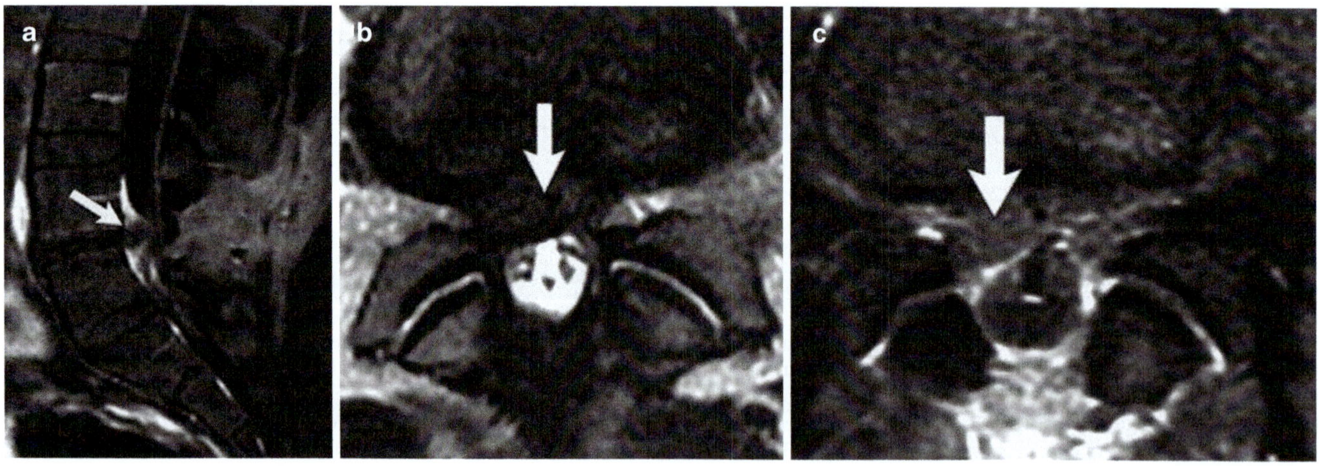


**Fig. 22.15** Postoperative retained disc fragment surrounded by enhancing scar on MRI. This patient complained of persisting pain in the right leg for days after surgery and finally had an MRI. The sagittal enhanced image (**a**) shows an epidural mass (arrow) extending from the disc space, surrounded by enhanced epidural tissue. The unenhanced (**b**) and enhanced (**c**) images at the same L4–L5 disc space level demonstrate the mass (arrows) to be a large disc fragment surrounded by enhancing scar tissue. There is extensive enhancing scar tissue surrounding the thecal sac in this patient who had undergone bilateral laminectomies

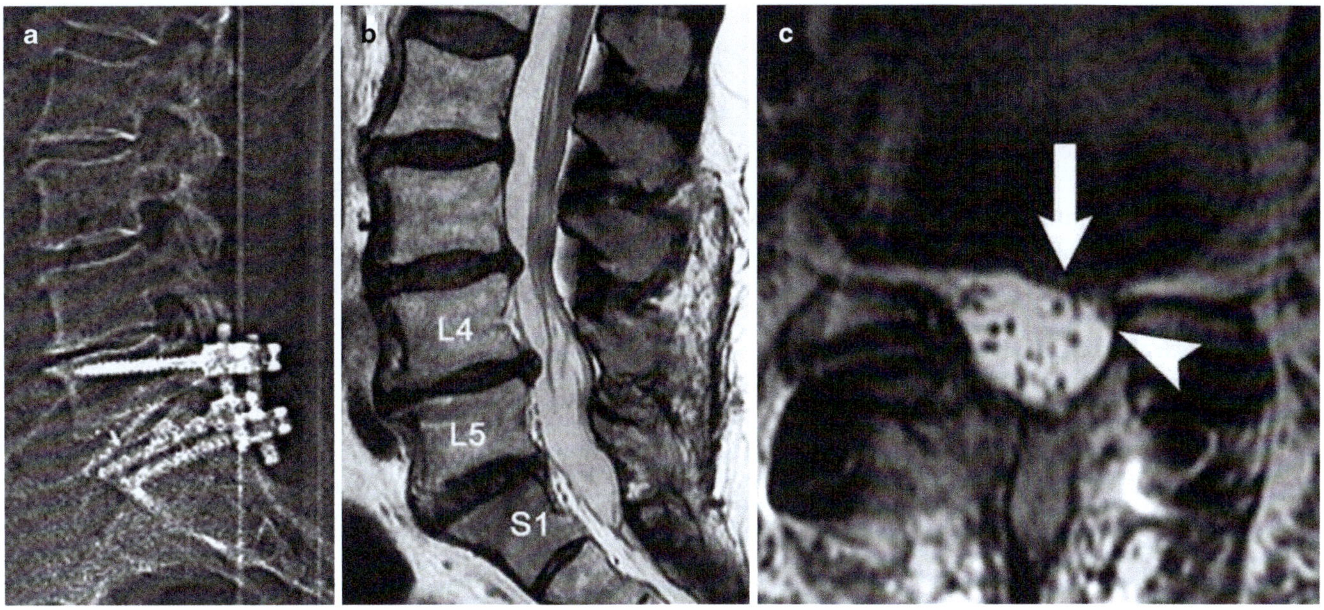


**Fig. 22.16** Disc degeneration and protrusion above a level of fusion. The scout image for a CT scan (**a**) shows that the L5 and S1 levels have been fused with pedicle screws. The fusion has resulted in altered pressures within the nearby discs, with subsequent degeneration. The sagittal T2 MR image (**b**) demonstrates posteriorly protruding discs at both the L4–L5 and L3–L4 levels. At L4–L5, there is protrusion of the disc (**c**, arrow) into the left lateral recess, compressing and displacing the left L5 nerve root (**c**, arrowhead)

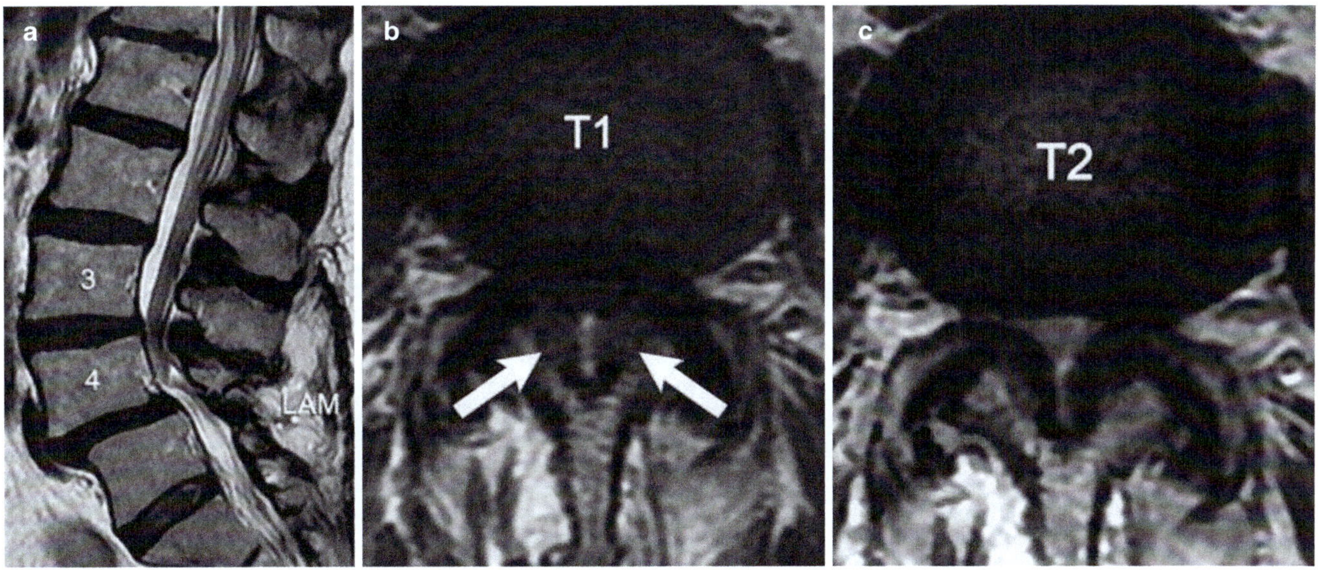


**Fig. 22.17** Development of central canal stenosis above the level of previous laminectomies and fusion for spondylolisthesis. Laminectomies and bilateral pedicle screw fusion were performed at the L4–L5 level for a grade 2 spondylolisthesis (slightly >25% slippage) due to degenerative facet disease (**a**). A few years later, the patient had symptoms of central canal stenosis. The axial T1 scan at the L3–L4 (**b**) level shows the hypertrophied ligamenta flava (arrows), while the T2 scan (**c**) better shows the degenerated facets and disc, all combining to produce a moderately severe degree of central canal stenosis. The postoperative relocation of new sites impacted by the body's weight is responsible for the acceleration of these degenerative changes

- New levels of bony inflammation, disc degeneration, facet disease with synovitis, thickened ligamenta flava, leading to central canal stenosis, all possibly secondary to abnormal motion following fusion (Fig. 22.17).
- Scar vs disc inadequately evaluated with any CT (plain, enhanced, post-myelo).
- Spatial resolution of RN scan inferior to MRI; may see persistent postoperative activity with fusion or disc prosthesis up to 2–3 years even though no complication.

## Degenerative Thoracic Spine

### Disc Protrusion

- Degenerated thoracic disc frequently calcified, best seen on CT (Fig. 22.18).
- Disc protrusion/extrusion best demonstrated on MRI or myelography (Fig. 22.18).

### Costochondral Joint Abnormalities

- Degenerative changes are best seen on CT.
- Acute or subacute inflammation best seen on enhanced MRI.

### Kyphosis

- Plain radiographs demonstrate the degree and location of kyphosis, endplate irregularities.
- Scheuermann's disease, thoracic and upper lumbar endplate irregularities, with wedge deformities producing kyphosis, two types:
  - Upper thoracic: less common, mild but painful endplate irregularities, more marked kyphosis.
  - Lower thoracic/upper lumbar: more common, more irregular endplates, less kyphosis.
- MRI is best for cord or root compression.

## Degenerative Cervical Spine Disease

### Disc Protrusion

- Obtuse margins, characterized as midline, paramedian, extending into lateral recess or neural foramen.
- Location determines whether there will be pain only, myelopathy from cord compression, or radiculopathy from nerve root compression.
- MRI best screening test

### Disc Extrusion

- MRI best screening test, myelography + CT excellent but invasive; if CT alone, without myelography, recommend use of intravenous contrast to enhance the epidural venous plexus.
- Acute extrusion denoted by hyperintensity (edema) in disc fragment (Fig. 22.19).
- Large extrusion, or extrusion into a narrow canal, may produce cord compression and major neurological injury (Fig. 22.20).

### Chronic Disc/Osteophyte Complex

- Chronic disc bulging or protrusion stimulates osteoblasts to form an osteophytic ridge (disc/osteophyte complex) that protects against further disc displacement.

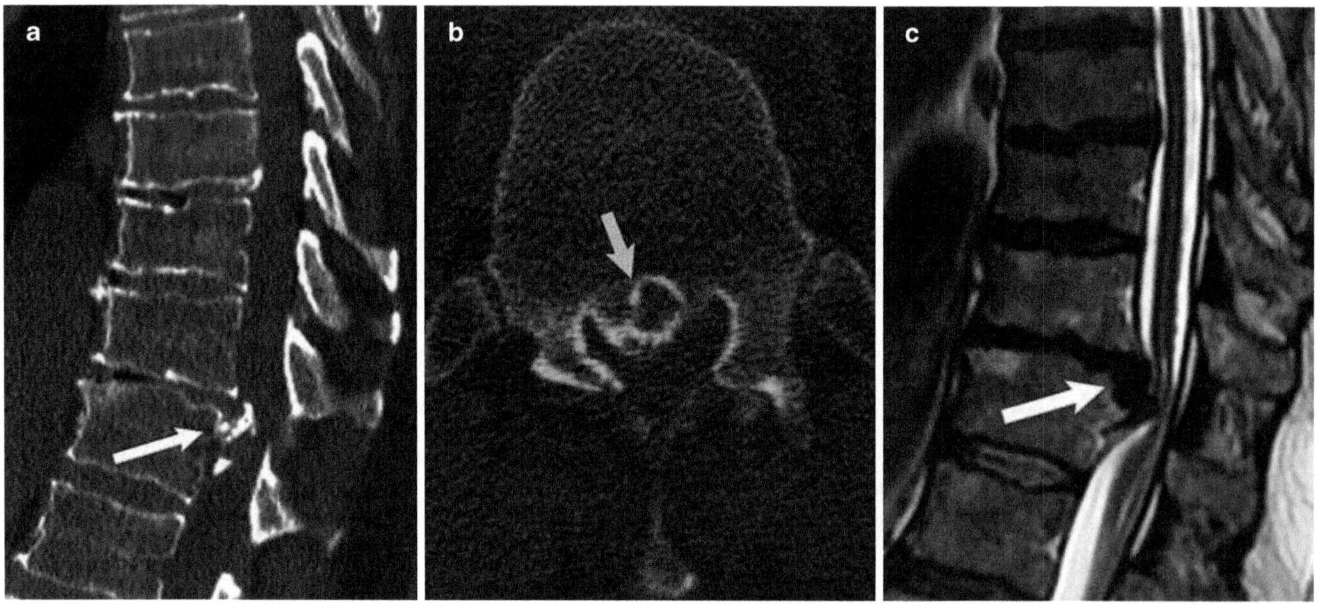


**Fig. 22.18** Annular calcification of an extruded thoracic disc. The mid-thoracic disc is markedly degenerated, denoted by its narrowing and the gas within it, and there is calcification of its annulus (arrows) as seen on the sagittal (**a**) and axial (**b**) CT images. The sagittal T2-weighted MR image (**c**) shows the extent of the extrusion (arrow) which severely compresses the spinal cord. The increased signal within the cord is either edema, if acute, or gliosis, if chronic

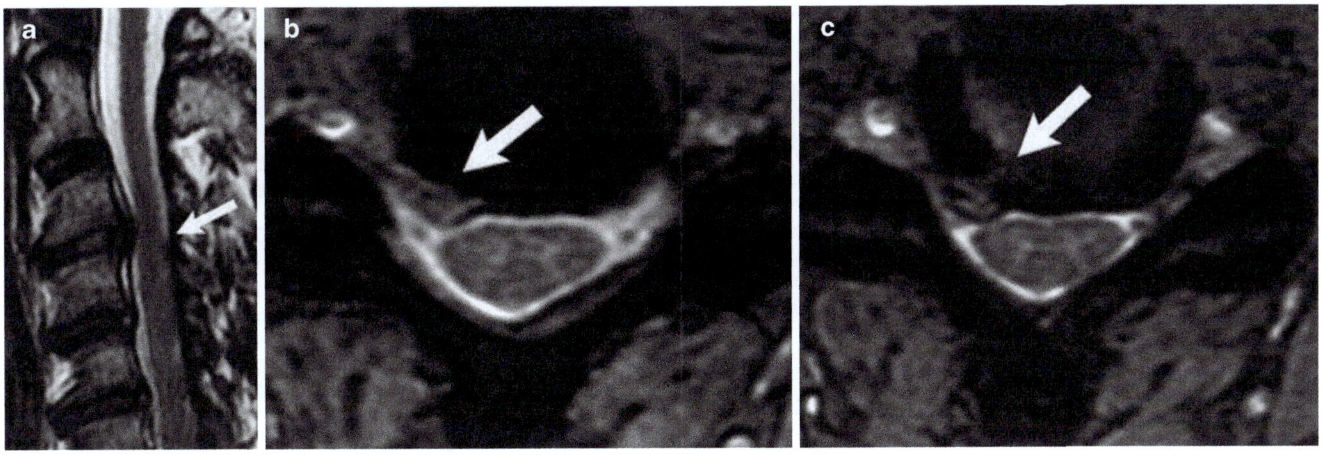


**Fig. 22.19** Acute cervical disc extrusion. The sagittal T2-weighted image (**a**) shows a disc extrusion at the C3–C4 level (arrow), elevating the posterior longitudinal ligament. Thin T2 gradient echo images at the C3–C4 neural foraminal level (**b**, **c**) demonstrate a hyperintense disc extrusion (arrows) into the right lateral recess and proximal right neural foramen, obliterating visualization of the right C4 root, while that on the left is well seen. The increased signal within the disc fragment is indicative of its relatively high water content and thus its acute extrusion

**Fig. 22.20** Acute quadriplegia during a football game due to acute protrusion of a disc into a small spinal canal. The sagittal view (**a**) demonstrates a relatively narrow cervical spinal canal, disc material (arrow) protruding through an annular fissure at the C3–C4 disc space, and hyperintensity within the cord at and below the disc space level. The axial view (**b**) shows the disc protrusion to be broad-based; the annular fissure (arrow) to be quite irregular, suggesting recent tearing; and the edematous spinal cord to be significantly compressed

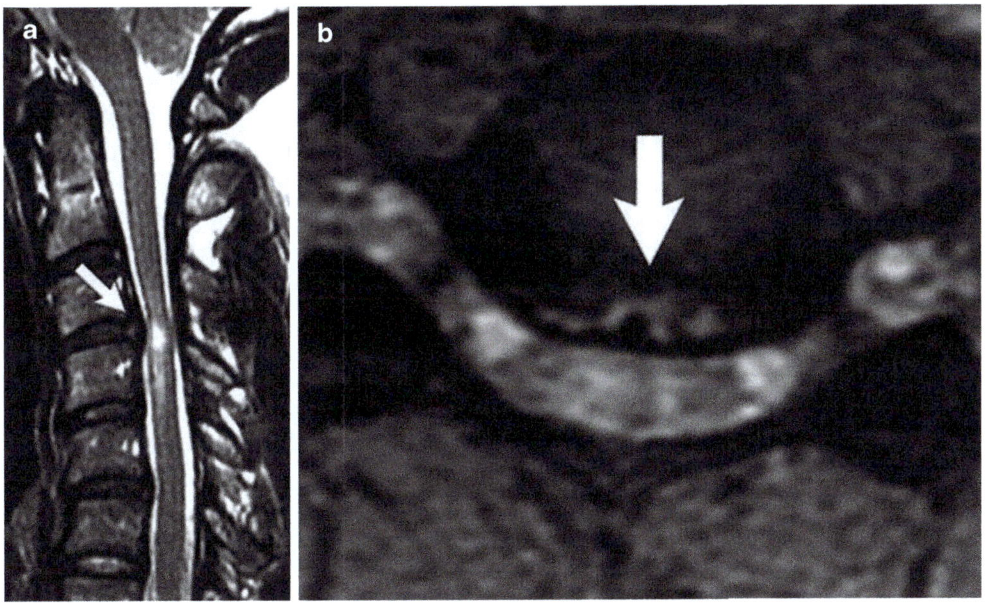


- However, the combination produces canal narrowing with potential cord compression, the degree depending upon the sizes of the antecedent canal and the disc/osteophyte complex.
  - Large chronic disc/osteophyte complexes (Fig. 22.21).
  - Ossification of the posterior longitudinal ligament (OPLL).
  - Multiple congenital and degenerative factors in combination.

## Central Cervical Spinal Canal Stenosis

- Central cervical spinal canal stenosis may occur from a number of factors:
  - Congenitally small canal, defined as a significantly decreased amount of CSF around cord; measurements can be misleading [30].

## Facet Joint Degenerative Disease with Osteophytic Spurring

- Facet joint degenerative disease with osteophytic spurring from chronic low-level trauma narrows the foramen from the lateral position of the joint.

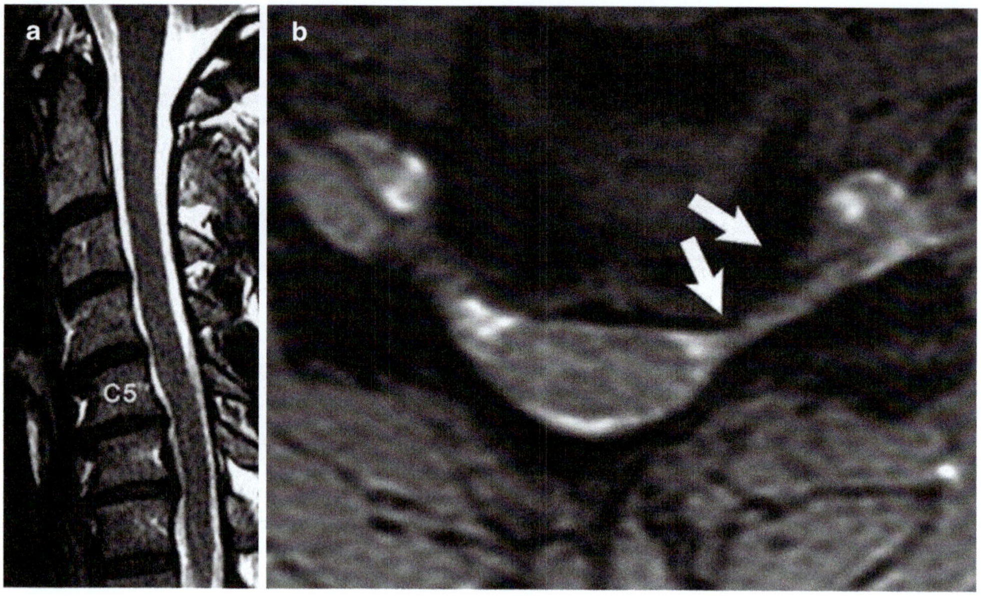


**Fig. 22.21** Disc/osteophyte complex at C5–C6 narrowing the canal and flattening the spinal cord. The sagittal (**a**) and axial (**b**) T2 MR images show diffuse disc bulging and secondary osteophyte formation (arrows) along the posterior vertebral body margins. This produces mild flattening of the cord. The left neural foramen is narrower than the right, secondary to osteophyte formation along the uncinate process. The left C6 root is compressed, resulting in left arm pain in the C6 distribution

### Narrowed Neural Foramen from Disc/ Osteophyte Complex Affecting the Uncinate Process

- Superior projection of the vertebral body around the disc that protects against acute lateral disc extrusion that would traumatize the exiting nerve root or vertebral artery with or without concomitant facet disease compresses the exiting nerve root, a common cause of upper extremity radiculopathy (Fig. 22.21) [31].

## Traumatic Injuries

- Acute Fractures
  - Lumbar [32]
    - Acute compression fracture
      - Superior to inferior loading forces
      - Superior endplate compressed, especially anteriorly (Fig. 22.22).
      - The degree of trauma may be only mild to moderate if osteopenia present.
      - CT has highest spatial resolution to show fracture line.
      - Fracture may be occult on CT; MR STIR sequence most sensitive for showing bony edema, indicative of an occult injury to trabecular bone (Fig. 22.23).
      - Best differentiated from a chronic compression fracture using MRI with STIR or FS T2 sequences to show edema, and/or enhanced T1 to show inflammation.
      - RN bone scan can be misleading: 40% remain positive at 1 year, 10% at 2 years [14].
      - 18FDG PET scanning: high standardized uptake value (SUV) in acute fracture; chronic fracture has SUV $< 2$; pathologic fracture has SUV $>2$, frequently up to 4 [14].
    - Burst—similar superior to inferior loading forces, only more severe:
      - Involvement of anterior and middle columns (Denis Classification), variable involvement of posterior columns.
      - Displacement of middle column into spinal canal leads to neurological injury.
  - Type A: midportion of vertebral body middle column projects into canal.
    - Type B: superior-posterior edge of vertebral body middle column projects into canal (Fig. 22.24).
    - Facet and transverse process fractures—frequently accompanied by intra-abdominal injuries.
    - Flexion-distraction: "seat belt" injuries (rapid flexion over a fulcrum)
      - Type A (Chance fracture)—horizontal fracture through vertebral body anteriorly and laminae/ facets posteriorly.
      - Type B ("Naked facets")—horizontal fracture through the intervertebral disc, and separation of the facets posteriorly via ligamentous disruption.
      - Type C = bony fracture anteriorly, ligamentous disruption posteriorly.
      - Type D = fractured disc anteriorly, bony fractures posteriorly
  - Thoracic
    - Acute compression fracture—as above in lumbar
    - Flexion-distraction—as above in lumbar
    - Other types of fractures—Unusual unless a high velocity injury is sustained. The forces required for acute fracture are much greater than for an acute cervical or lumbo-sacral fracture because the rib cage holds the vertebrae in position; when there are

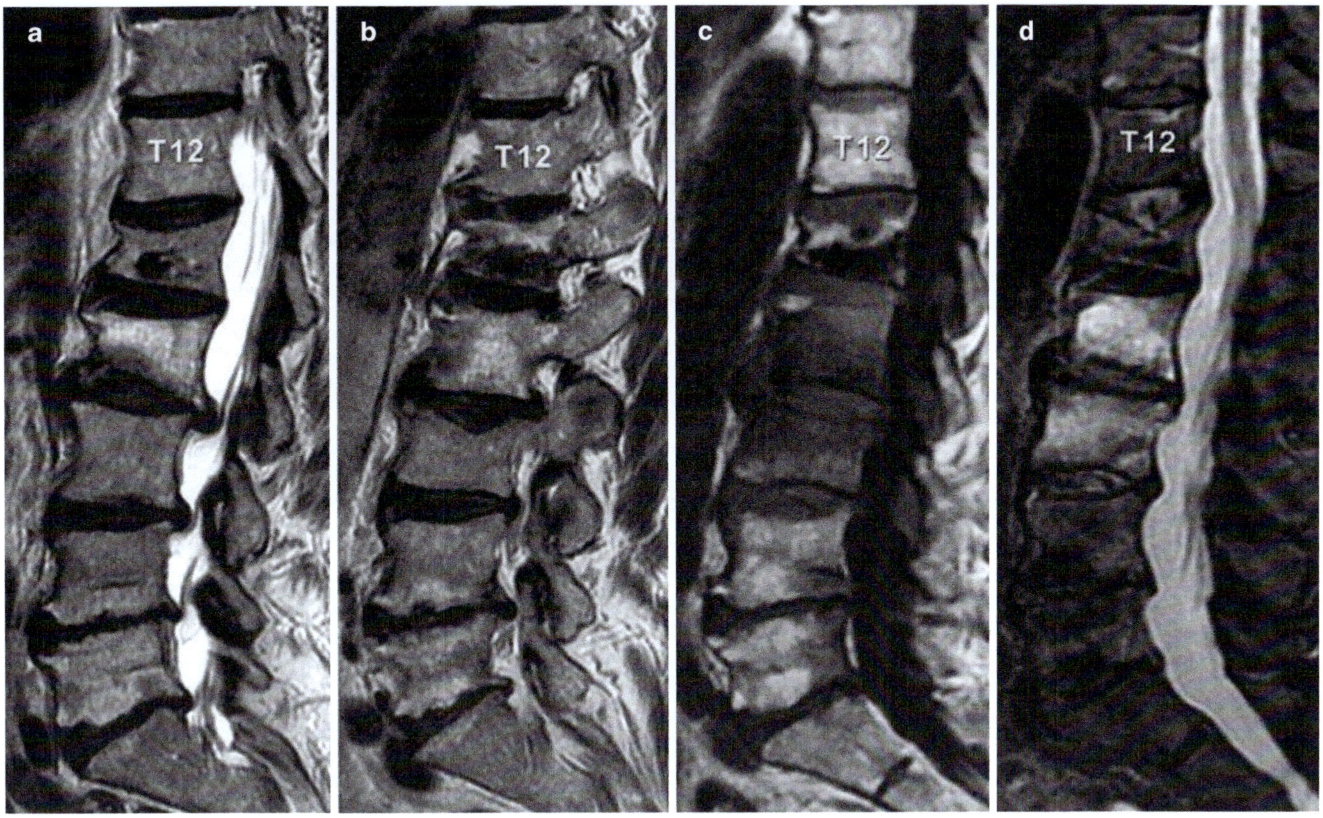


**Fig. 22.22** Acute compression fractures of L2 and L3. The two T2-weighted sagittal MR images (**a**, **b**) demonstrate deformities of the L2 and L3 vertebral bodies, but the age of these compression fractures cannot be determined because edema is difficult to see on non-fat suppressed T2 images. The hypointensity within the L1 body represents methylmethacrylate infused during previous vertebroplasty. The T1 (**c**) and STIR (**d**) sagittal sequences demonstrate edema (gray/black on T1, white on IR) within the L2 and L3 bodies, indicative of the acuteness of the injuries

**Fig. 22.23** STIR MR sequence shows occult vertebral body fractures. The initial CT scan (**a**) after acute trauma does not show definite fractures of the upper thoracic spine in this patient with upper back pain. However, the STIR sequence naturally suppresses the fat of the yellow marrow, allowing the traumatic edema to be seen within the T4 and T5 vertebral bodies (**b**, arrows)

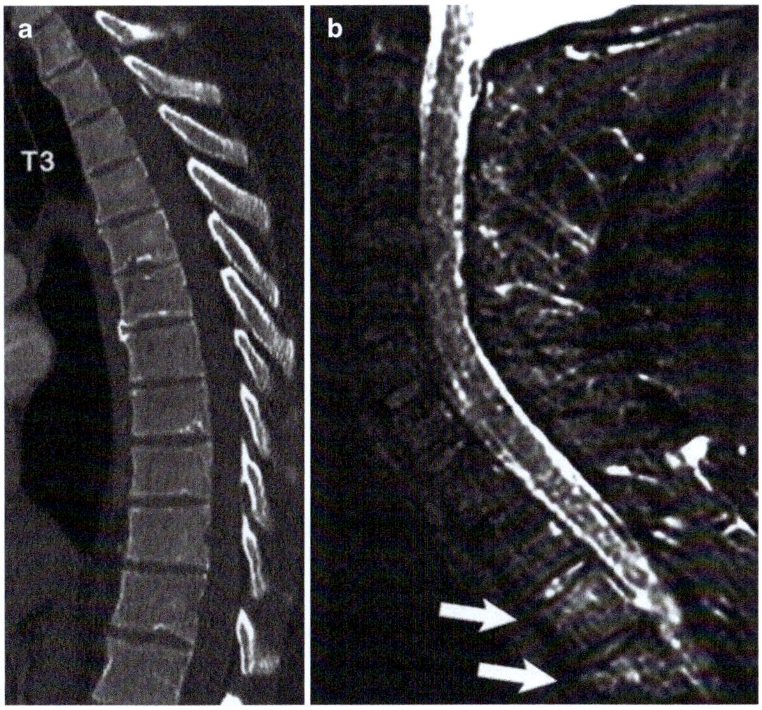

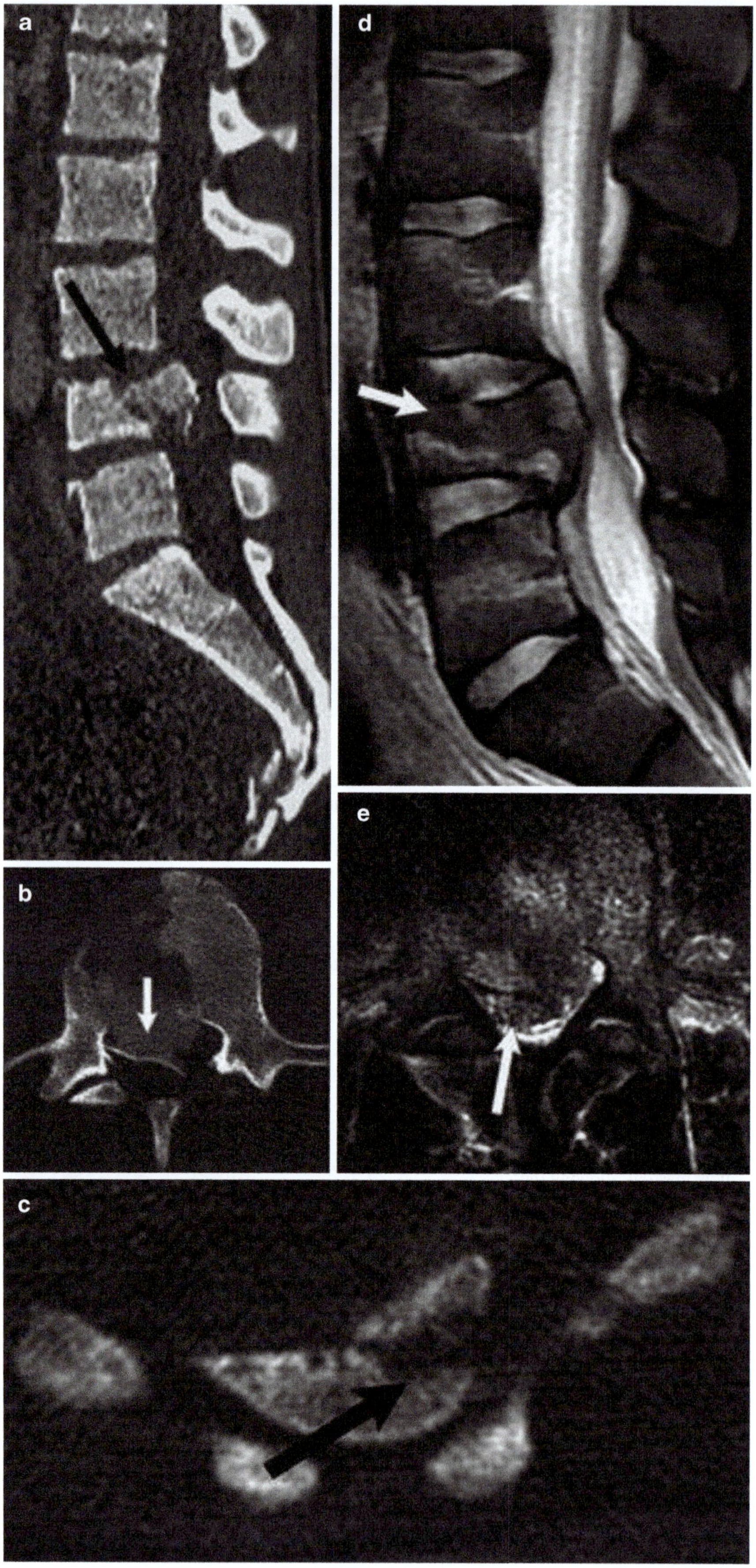


**Fig. 22.24** CT and MR demonstration of an L4 burst fracture and S4 fracture-dislocation. The sagittal reformatted CT image (**a**) shows the pronounced posterior dislocation of the posterior-superior aspect of the L4 body into the spinal canal. The axial image (**b**) through that level demonstrates bony retropulsion (arrow) secondary to cephalocaudal (loading) forces, characteristic of a burst fracture. A lower axial slice (**c**) shows the shattered S4 segment (arrow). The sagittal (**d**) and axial (**e**) MR images show the significant mass effect of the fractured L4 body (**d**, arrow) on the thecal sac (**e**, arrow)

such forces, fracture-dislocation and severe neurological injury result.
- Cervical
  Jefferson—explosive fractures of C1 ring due to loading forces from superior to inferior (landing on head), usually no cord compression.
  Odontoid—cord compression variable, depending upon degree of displacement of the odontoid process and the size of the spinal canal.
  Teardrop—while the name is from the displacement of the anteroinferior corner of the vertebral body, the posteroinferior corner is also fractured and is displaced into the cord.
  Burst—as above
  Facet—usually due to rapid lateral flexion, may produce chronic pain.
  Spinous process (clay shovelers')—rapid head/neck rotation.
- Subacute fractures
  - If producing pain and no impingement on cord or nerve root, may be a candidate for bony stabilization, such as vertebroplasty or kyphoplasty.
- Chronic fractures
  - Best differentiated from acute fractures by the presence of osteophytes or negative MRI STIR or FS T2 sequences [14, 32].
  - RN scans misleading.
  - Pseudarthroses are fractures through a fusion mass producing chronic pain.
- Operative trauma and complications
  - Intraspinal hematoma following spinal surgery (Fig. 22.25).
  - Impingement of nerve roots during pedicle screw fusion may be seen.

## Infectious Diseases of the Spine and Cord

- Infectious diseases of the spine and cord [14, 33–37]
  - Origins and modes of spread:
    Primary pulmonary or soft tissue infection with hematogenous spread; metaanalysis of non-TB origins: Staphylococcus aureus most common pathogen, case volume: average = 50%, range = 20–84%).

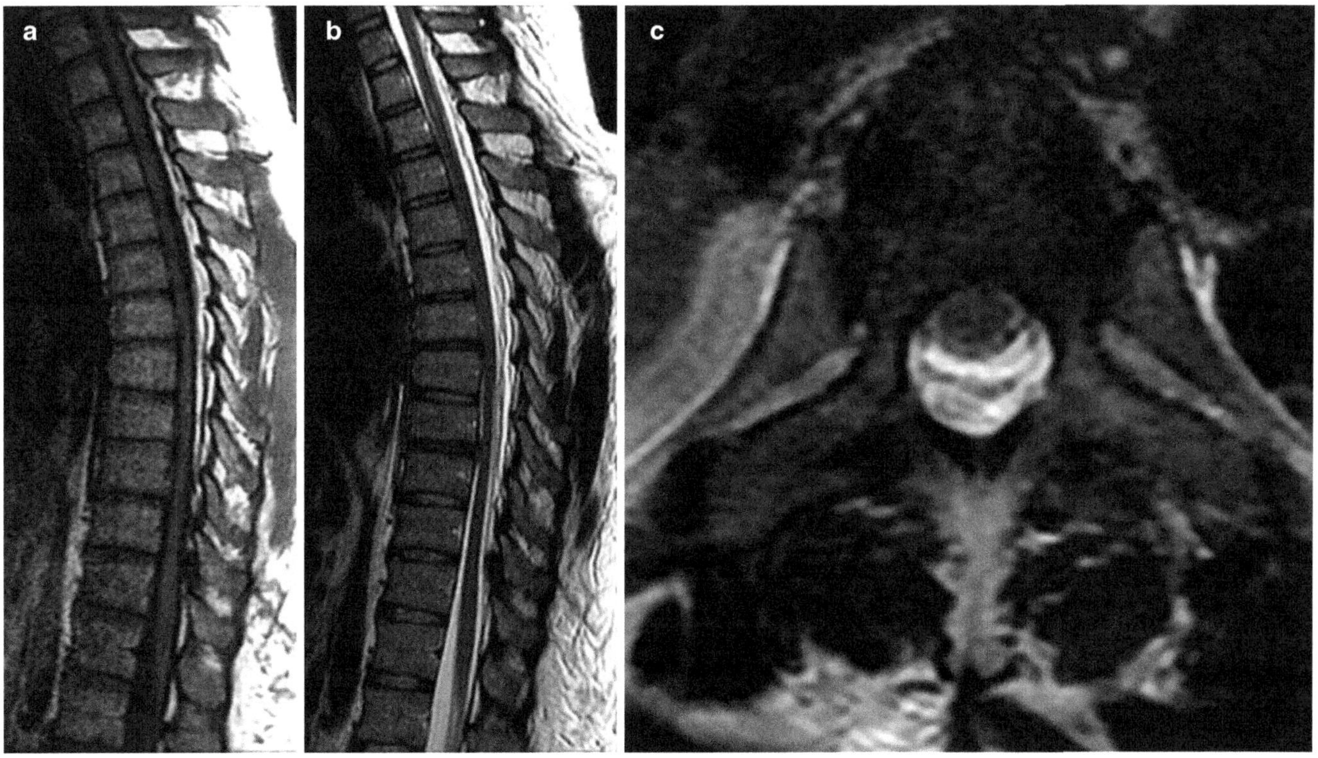


**Fig. 22.25** Acute thoracic spinal canal hematoma producing paraplegia following dural opening for tumor removal. There is a long extra-axial mass posterior to the compressed spinal cord; once the dura is opened, there is no longer a distinction between epidural and subdural. On the sagittal T1 sequence (**a**), the mass is a combination of gray and white, while on the sagittal (**b**) and axial (**c**) T2 sequences, it also has combined white and gray colors, but in different locations than on the T1 sequence. The evolution from oxyhemoglobin to deoxyhemoglobin and intracellular methemoglobin is much faster than in a closed environment such as the brain. Once the hemorrhage has been exposed to room air, this conversion can occur within hours, not days. On T1, the gray of oxy- and deoxyhemoglobin is combined with the white of early conversion to intracellular methemoglobin. On T2, the white of oxyhemoglobin is combined with the gray of deoxyhemoglobin

Genitourinary sites of primary infection with local invasion or transvenous spread (via Batson's plexus).

The disc in a child under 10 is vascular, initial infection may be discitis only.

Age > 10, infection goes to vascular zone deep to endplate, producing spondylitis.

Spread through destroyed endplate into avascular disc leads to spondylodiscitis; spread to paraspinal tissues and to epidural space common (Fig. 22.26).

Cavitation of epidural phlegmon produces epidural abscess (Figs. 22.26 and 22.27).

SI joint infection may be accompanied by a pelvic abscess (Fig. 22.28).

Extension through dura leads to inflammation of the leptomeninges (meningitis) with spread to the cauda equina.

Imaging best with enhanced MRI, using fat suppression to detect subtleties; enhanced CT has lesser contrast resolution of soft tissue abnormalities.

18FDG PET 95% sensitive, 85% specific for infection; SPECT scanning 80% sensitive, 80% specific.

Gallium-67 RN scan not dependent upon pyogenic response, so can be used in immunocompromised patient; indium-111 and 99mTc-labeled leukocytes very sensitive in rest of body, but not spine due to increased pressure in disc.

Spondylodiscitis from tuberculosis (TB) [38], atypical mycobacterium, and fungus.

TB infection commonly produces relative "disc sparing" (Fig. 22.27).

Spreads to another vertebral level ("skip lesions") via paraspinal soft tissues, skipping the intervening disc, while producing caseating psoas abscesses (Fig. 22.27).

The patient is usually immunocompromised.

Involvement of posterior elements common with TB (Fig. 22.27).

Other similar agents include atypical mycobacterium, Brucella.

Fungal spondylodiscitis usually occurs in the immunocompromised patient.

Meningitis without spondylodiscitis

Viral meningitis may be mild to moderate in severity.

Pneumococcus and Neisseria are the most common bacterial pathogens.

Meningococcal meningitis feared for its rapid spread in a susceptible group, such as school students or persons in close contact (e.g., military forces).

E. coli meningitis is typically found in infancy; death is common if in an adult.

Coccidiomycosis and cryptococcosis found in immunocompromised patients.

Arachnoiditis may follow meningitis from any pathogen, or from a non-infectious inflammatory process, resulting in similar findings on imaging (Figs. 22.29 and 22.30).

Spinal cord abscess

Distinctly rare entity, usually via hematogenous dissemination.

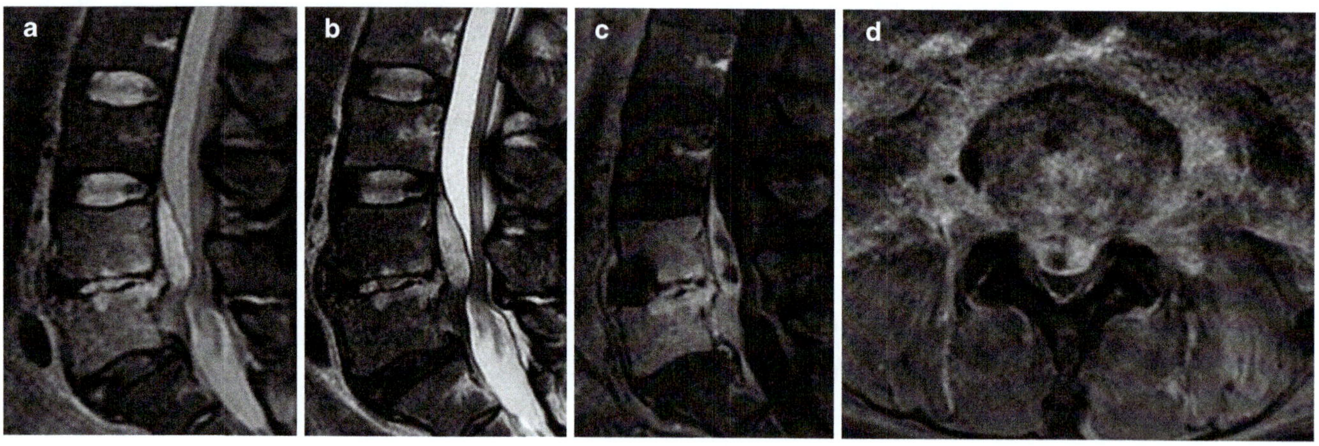


**Fig. 22.26** E. coli spondylodiscitis and epidural abscess at L4–L5. The IR (**a**) and T2 (**b**) sagittal MR sequences demonstrate edema throughout the L4 and L5 vertebral bodies, including the intervening disc space. There is a large epidural abscess and paraspinal inflammation. All of these inflamed structures enhance dramatically on the enhanced fat-suppressed sagittal (**c**) and axial (**d**) sequences, with the epidural abscess seen to markedly compress the thecal sac (**d**)

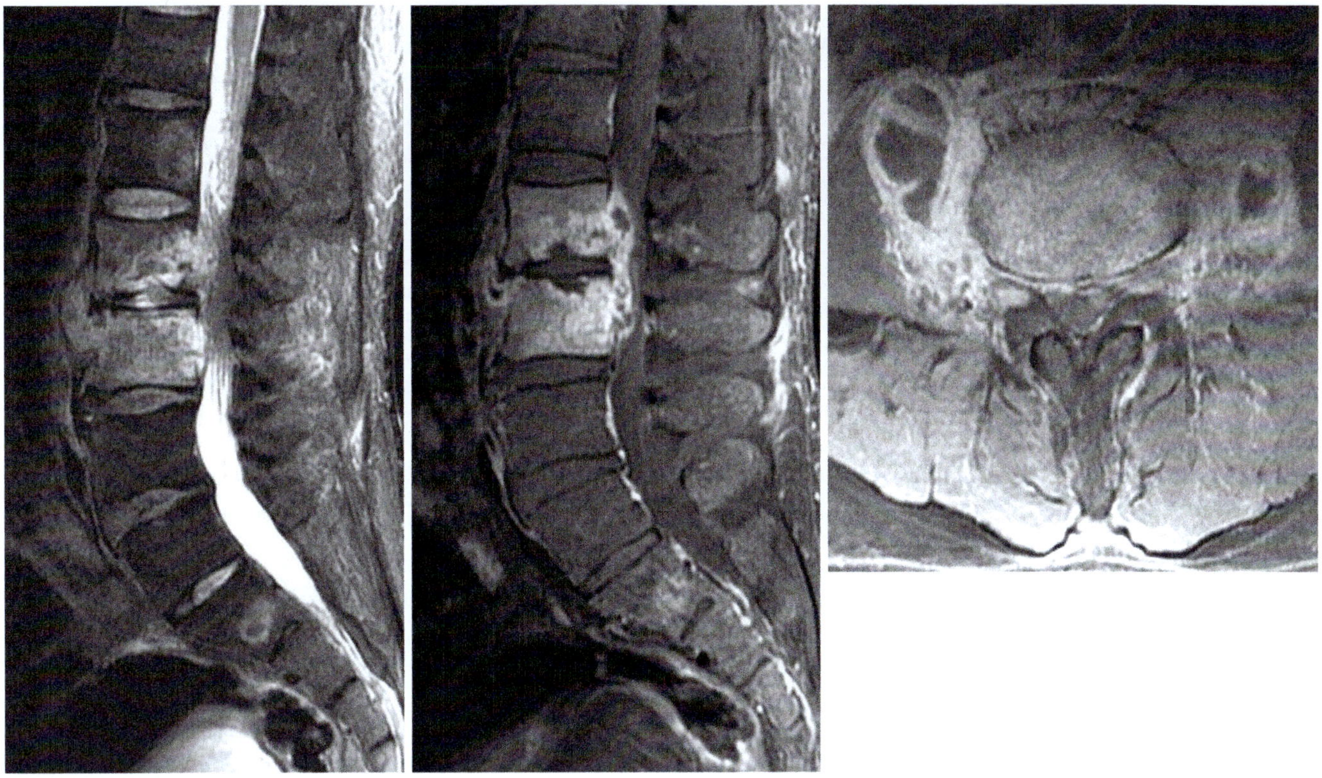

**Fig. 22.27** Tuberculous spondylodiscitis, with epidural, paraspinal, and psoas abscesses. At L3–L4 the sagittal T2 and enhanced fat-suppressed T1 images demonstrate infection through the vertebral bodies, but significantly less enhancement and endplate destruction relative to that seen with a bacterial infection (Fig. 22.26). There is also much more involvement in the posterior elements and facets than typically seen with bacterial infection. Non-contiguous involvement at L2 and throughout the sacrum ("skip lesions") are characteristic of TB. There are large epidural and paraspinal abscesses, which are nonspecific, but the large psoas abscesses (axial view) are typical of TB

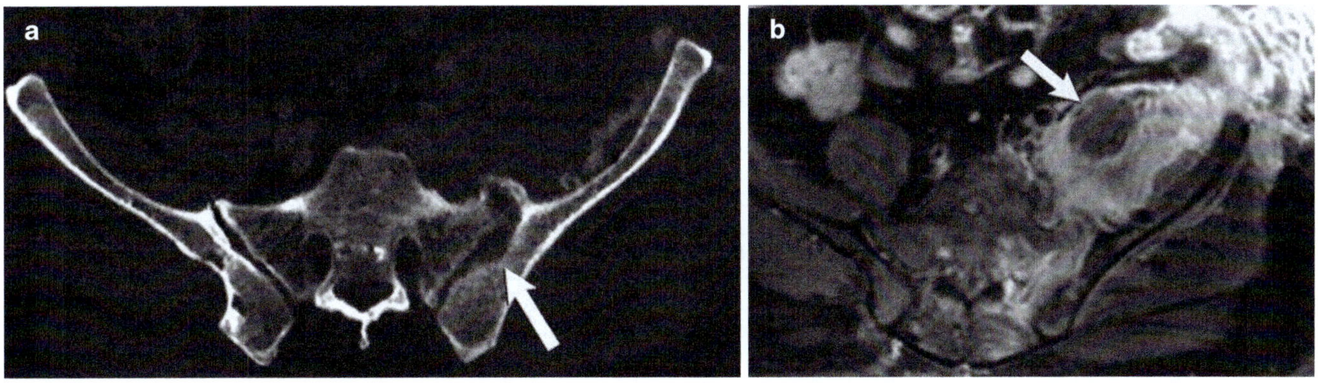


**Fig. 22.28** Unilateral SI joint infection. The CT scan (**a**) demonstrates destruction of bone on either side of the left sacroiliac joint and periosteal new bone along the inner margin of the left ileum (arrow). The axial enhanced scan (**b**) shows the extensive enhancing soft tissue sur rounding an abscess (arrow), extending into the left sacroiliac joint, S1 bony structures, and left facet joint

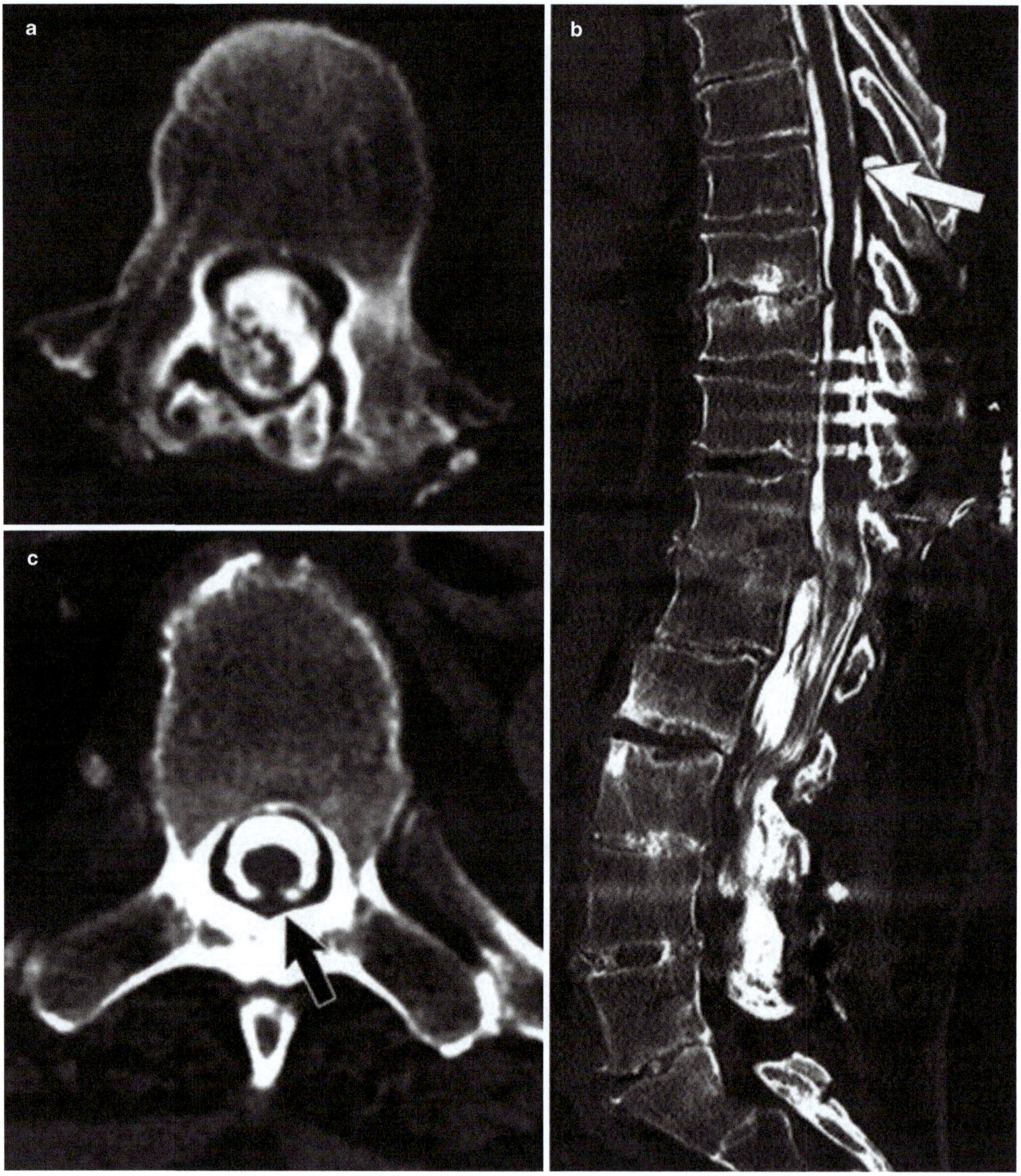


**Fig. 22.29** Adhesions and arachnoiditis requiring a stimulator for pain relief. Post-myelographic CT imaging of the lumbar spine (**a**) shows clumping of the nerve roots indicative of arachnoiditis. Posteriorly at the T8 level, there is an adhesion between the cord and the meninges, seen on the sagittal T2 sequence (**b**, arrow), but better demonstrated on the post-myelography CT scan (**c**, arrow)

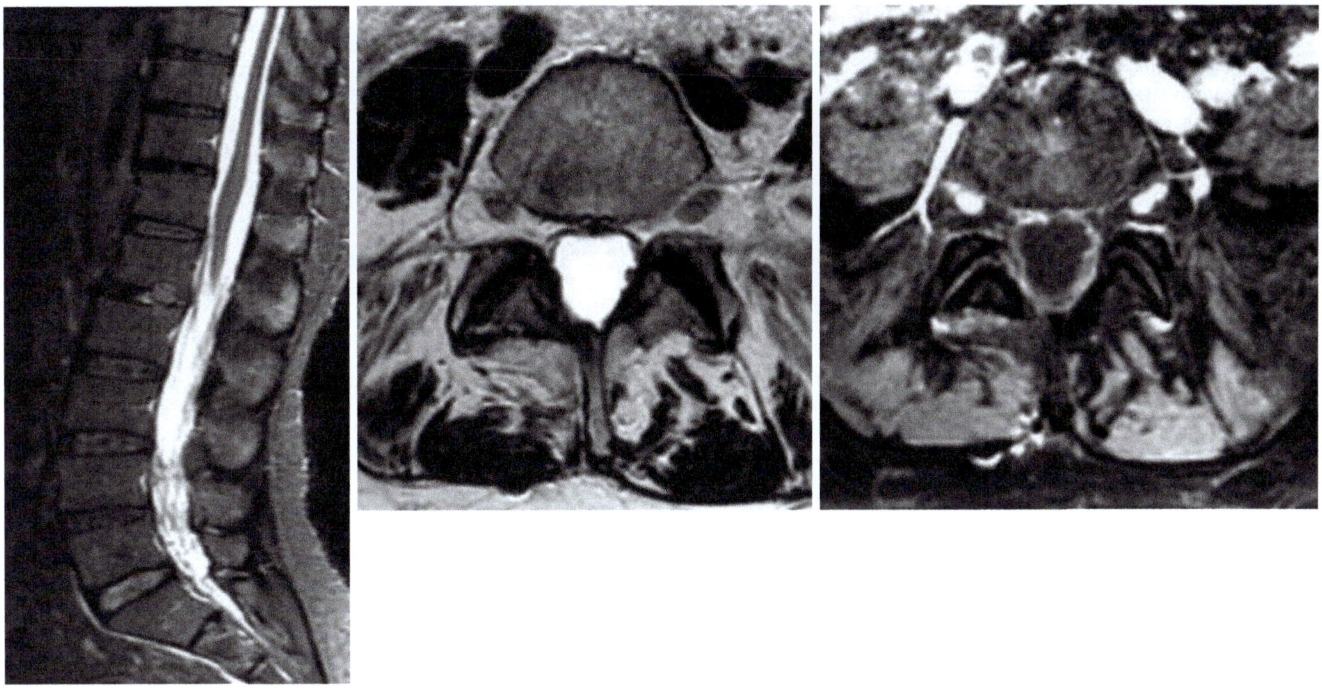

**Fig. 22.30** Arachnoiditis on MRI. The thecal sac is relatively "empty" as seen on the sagittal and axial T2-weighted images, due to the adhesions of the nerve roots along the inner margins of the thecal sac. The axial enhanced fat-suppressed T1 sequence shows significantly thickened meninges from previous inflammation

Tuberculoma (caseous center having low signal on T2 because of the low water content within caseation), rare; tuberculous abscess exceptionally rare.

## Regenerative Medicine and the Intervertebral Disc

Low back pain affects up to 80% of the population at some point in their life [39]. Low back pain creates a major social and economic burden [40]. For those patients with chronic low back pain, discogenic pain is the cause in over 1/3 of cases. It is important therefore to understand the basic physical properties of the healthy intervertebral disc (IVD) as well as the pathophysiology of the degenerating disc, if we are to advance IVD regenerative therapy. The following is a brief discussion of the complex anatomy of the normal IVD and the deleterious complexity of the degenerative cascade affecting the IVD.

## Disc Anatomy [41]

The IVD has three components:

- The annulus fibrosus (AF)
- The nucleus pulposus (NP)
- The cartilaginous vertebral endplates (CEP)

In adults, both the NP and AF are primarily avascular and aneural.

Annulus fibrosus (AF) [42]:

- Fibrocartilaginous structure that houses the nucleus pulposus
- Withstands circumferential loads
- Limits torsional rotation and bending
- Two distinct layers
  - Inner annulus (primarily Collagen Type II)
  - Outer annulus (primarily Collagen Type I, thus it is stiffer and stronger)
- Up to 25 distinct layers (lamellae)
  - Variable in thickness (increasing with age).
  - Alternating collagen fiber angles (20–65°) producing a "zig-zag" or crimped structure which allows for stretch/expansion [43] (Fig. 22.31).

Extracellular membrane (ECM):

- Composed of Proteoglycans (glycosylated proteins)
  - Aggrecan
- Also composed of attached glycosaminoglycans (GAG) [44].
- This combination promotes hydration due to water binding, produces the swell pressure and provides resistance to compressive stress [44, 45].

Nucleus pulposus (NP) [42]:

- 80–90% water, making it soft and gelatinous [46, 47].
- Center of the disc

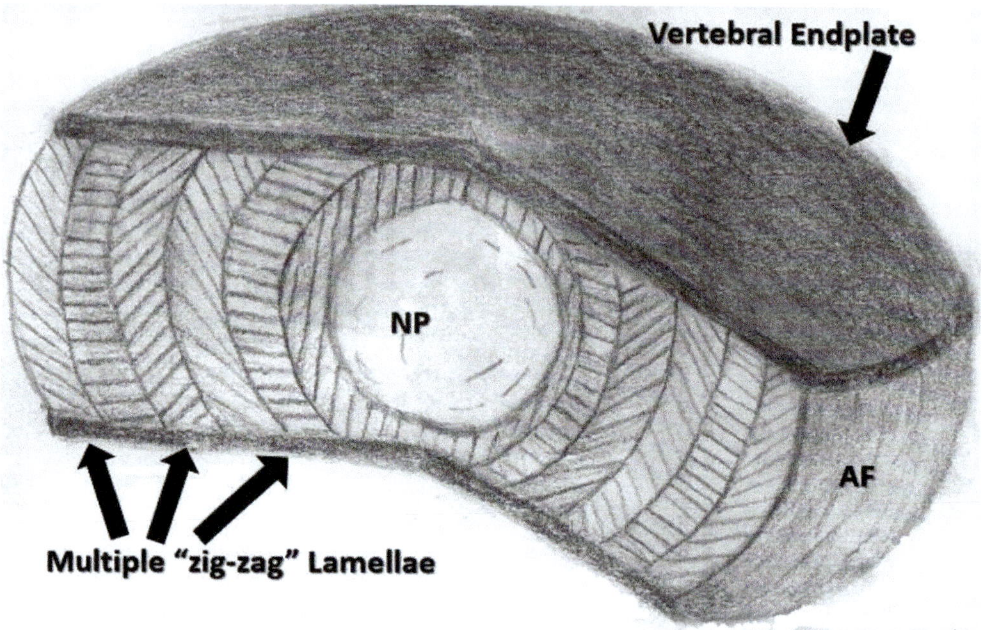


**Fig. 22.31** Normal disc with wedge removed to display the central nucleus pulposus (NP) and the multiple lamellae of the annulus fibrosus (AF). Notice the "zig-zag" appearance of the lamellae which allow the disc to stretch/expand and return to its pre-stressed shape. There can be up to 25 lamellae

- Natural swell pressure increases against compressive loads [46].
- Four primary components [48]
  - Loose network of collagen Type II fibrils giving some firmness to the nucleus.
  - Elastin fibers which confer mechanical stability and limit elasticity.
  - Proteoglycans
  - Chondrocyte-like cells which replace the notochord cells that diminish in number from birth to adulthood. Most are gone by age [49]. These cells turn over a large volume of ECM, which allows the disc to withstand high compressive loads [50].

Cartilaginous vertebral endplates (CEP) [42]:

- Strongest part of disc [51]
  - Interface between the dense cortical bone of the vertebral body and the AF/NP. Keeps the NP pressurized and from bulging into the cancellous trabecular vertebral body bone [52].
  - Embedded with chondrocytes
  - Primarily Collagen Type II
  - Bony component has a porous layer of fused trabecular bone which allows nutrients and disc waste products to pass through [52, 53].
  - Strongest where thick (1.0 mm), porous and curved.
    Thinnest centrally

Potential resident stem/progenitor cells

- Reside in a niche at the AF/CEP border and nearby perichondrium.

Blood vessels:

- Up to 12 months of age, blood and lymph vessels extend through the endplate into the annulus and nucleus.
- These vessels recede after 12 months to 20 years until they are only found within the outer annulus and within the endplates [54, 55].
- Without significant central blood supply, the NP is relatively hypoxic.
  - Energy for the NP is from anaerobic glycolysis.
  - Generates lactic acid which further lowers the pH [42, 56, 57].

Nerve supply:

- Numerous branches
  - Recurrent sinuvertebral nerves, ventral primary rami and grey rami communicantes, recurrent branches of the rami communicantes.
  - In young adults, sensory nerves extend only 3 mm into the disc, thereby involving only the superficial layers of the outer annulus [58].
  - With age and injury, neoinnervation occurs within granulation tissue and extends into the mid and inner annulus and even into the NP.
    As neoinnervation progresses, pressure on the disc also produces pressure on the nerves [59].

## Pathomechanism of Disc Degeneration

Discs are among the most avascular tissues in the body. They have a low cell population which limits the proliferation potential resulting in an inability to self-regenerate.

- Many causes of disc degeneration
  - Genetics is a primary cause (50–70%) leading to increased susceptibility to environmental factors including sequalae from smoking [60–63].

- Wear and tear, aging, obesity, Diabetes Mellitus I & II, trauma, altered enzymes, cell senescence (aging), neurovascular ingrowth, chronic infection, gene polymorphisms [64–67].
- Ultimately degeneration leads to:
  - Increased catabolic activity [68]
  - Decreased anabolic activity [41]
  - Changes in ECM
  - NP: collagen changes from Type II to Type I
    Increased cross-linkage increases stiffness.
    Decrease in proteoglycans, so water content decreases as does the swell pressure [42].
    Increase in metal metalloproteinases
    NP cannot tolerate compressive forces and becomes narrowed, bulges and eventually cannot be differentiated from the AF.
  - AF: Lamellae become disrupted (possibly the start of degenerative disc disease) [41].
    Fissures develop. Depending on severity, can allow the NP to protrude or extrude [69].
    Increased vascularization and innervation which with stretching and compression lead to discogenic pain.
    Increase in number of immune cells with increased vascularization [42] (macrophages, mast cells, T-lymphocytes).
  - CEP:
    Become structurally disorganized, cracked, thinned and mineralized (usually calcium).
    Subchondral bone develops microfractures and sclerosis [41].
    These changes alter the ability for nutrient and disc waste products to pass through the endplates.
    Loses their normal curvature, becoming flattened with surface irregularity [70, 71].
    Weaken, and may develop Schmorl's nodes

## Pfirrmann Classification: Disc Degenerative Scoring Based on T2-Weighted MRI Changes

The grades are based on disc structure, signal intensity, distinction between the nucleus and annulus, and the height of the disc [72].

- Grade 1: Nucleus is homogeneous, bright white with a clear distinction between the NP and AF. Signal is hyperintense and equal to that of CSF. Disc height is normal. The NP appears bright white on T2-weighted images because it is predominantly water. The AF appears dark due to its significant fibrous composition (Fig. 22.32).

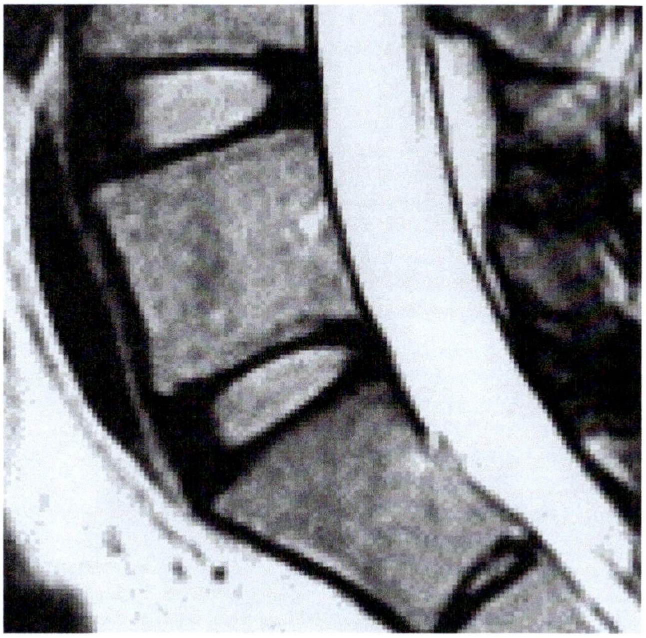

**Fig. 22.32** Pfirrmann Grade 1 discs. Nucleus pulposus (NP) of L4/5 and L5/S1 are bright white. There is a clear distinction between the NP and the annulus fibrosus (AF). Both the NP and AF appear normal in signal and demonstrate normal disc height. The NP is 80–90% water leading to its high signal on T2-weighted images

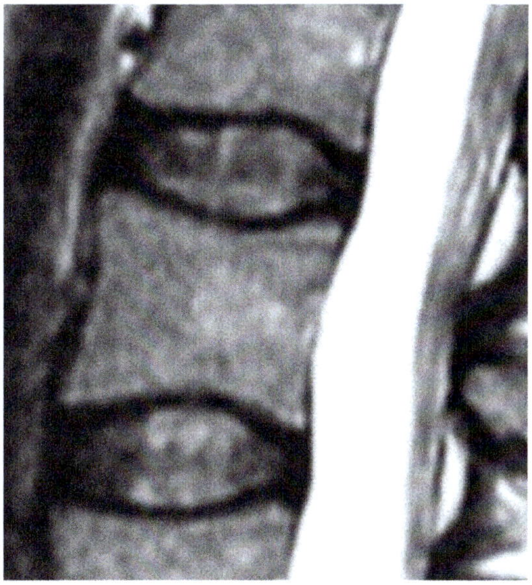

**Fig. 22.33** Pfirrmann Grade 2 discs. Both disc demonstrate inhomogeneous signal within the NP. Both vertical and horizontal bands can be seen. There is still distinction between the NP and AF, and the disc height remains normal

- Grade 2: Now the NP is inhomogeneous with or without horizontal bands. The remainder is the same as for Grade 1 (Fig. 22.33).
- Grade 3: NP is inhomogeneous and gray. The distinction between the NP and AF is unclear. The NP signal is inter-

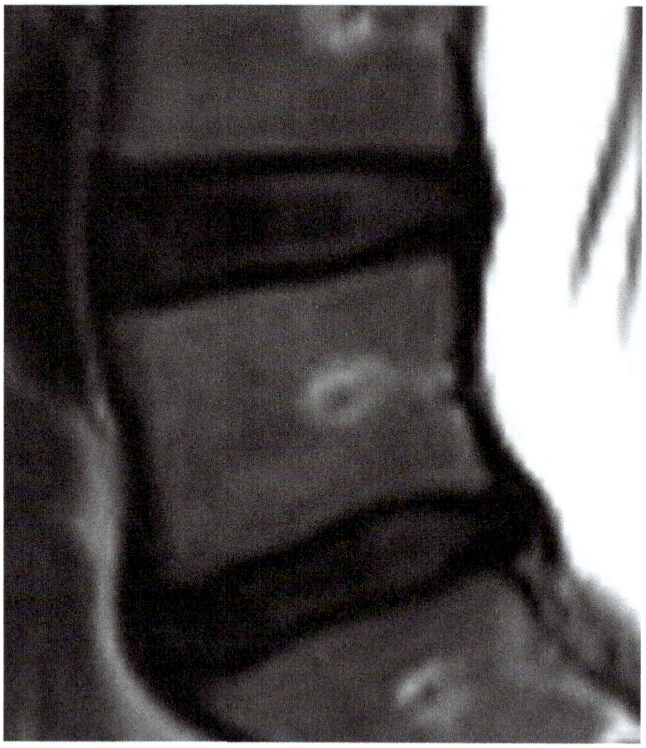

**Fig. 22.34** Pfirrmann Grade 3 discs. The NP of both discs are inhomogeneous and gray. The distinction between the NP and AF is no longer clear. The L4/5 disc height is normal but the L5/S1 disc height is slightly decreased

mediate. The disc height is normal to slightly decreased (Fig. 22.34).

- Grade 4: The NP is inhomogeneous gray to black. The distinction between the NP and AF is lost. The NP signal is intermediate to hypointense. The disc height is normal to moderately decreased. Commonly demonstrates Modic Type 1 endplate changes (Figs. 22.35, 22.36, and 22.37).
- Grade 5: The NP/disc is inhomogeneous and black. The distinction between the NP and AF is lost. The signal is hypointense. The disc space is collapsed. Commonly demonstrate Modic Type 1 endplate changes (Fig. 22.36).
- A Modified Pfirrmann Score was created which divides the classic Pfirrmann Grade 4 into a modified Grade 4 making the classic Grade 4 into a new Grade 5 [73]. This modification also adds Grades 6–8. The modification produces greater specificity as to the degree of disc degeneration. The classic Pfirrmann Grades are still widely used.
- Modified Pfirrmann Grade 4: The NP has mild hyperintensity, greater than the outer AF signal. The disc height is normal to moderately decreased.
- Modified Pfirrmann Grade 5: The NP is inhomogeneous gray to black. The distinction between the NP and AF is lost. The NP is intermediate to hypointense in signal. The disc height is normal to moderately decreased (original Pfirrmann Grade 4).
- Modified Pfirrmann Grade 6: Modified Pfirrmann Grade 5 plus there is <30% decrease in disc height. Commonly demonstrates Modic Type 1 endplate changes (Fig. 22.36).

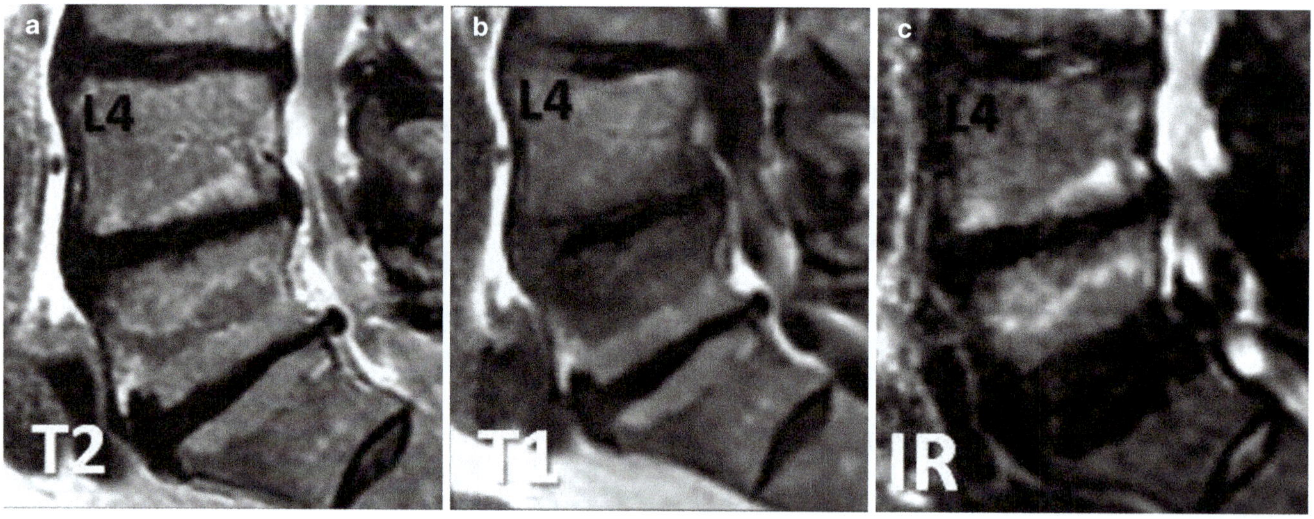


**Fig. 22.35** Pfirrmann Grades 4 and 5 with Modic endplate changes. (**a**) is a T1-weighted image demonstrating increased signal involving the L4/L5 and L5/S1 endplates. (**b**) is a T2-weighted image. On T2, signal at the L4/5 endplates is decreased. Comparing (**a**) and (**b**), the findings are consistent with Modic type 1 change, which represents edema/inflammation. Modic type 1 change is associated with back pain. It is common in Pfirrmann grades 4 and 5. Notice how much better IR imaging (**c**) demonstrates the endplate edema at L4/5. Findings at L4/5 also demonstrate a Pfirrmann Grade 4 disc. The nucleus is inhomogeneous gray to black. There is no distinction between the NP and AF. Disc height is moderately decreased. The L5/S1 disc demonstrates high signal on T2-weighted imaging (**b**). Comparing (**a**) and (**b**) the findings represent Modic type 2 change. This process represents fatty marrow replacement and is chronic. The L5/S1 disc also demonstrates Pfirrmann Grade 5 change. The NP is inhomogeneous gray to black. There is no distinction between the NP and AF. The L5/S1 disc space is collapsed

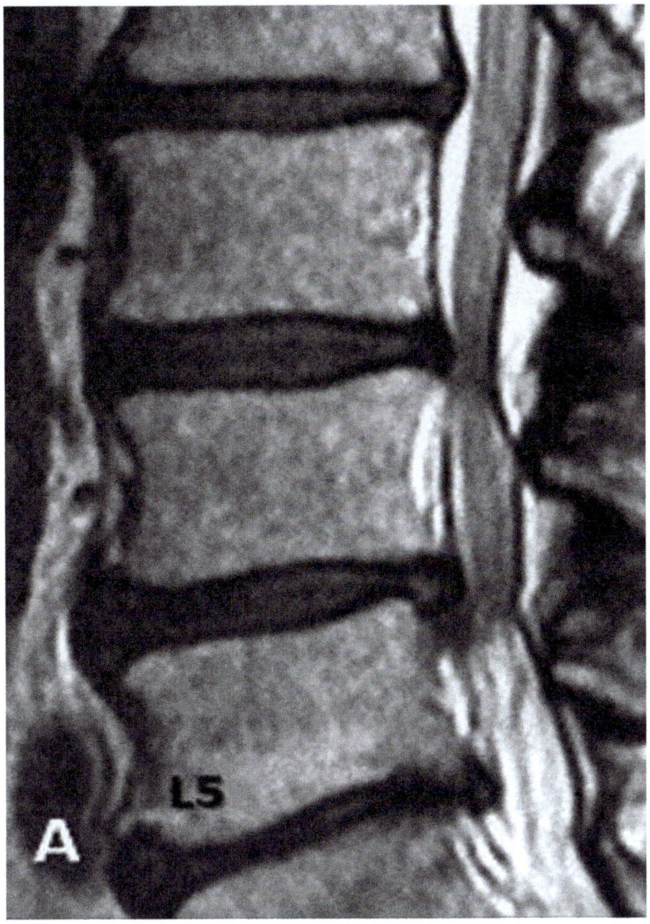


**Fig. 22.36** L2/3 through L4/5 demonstrate Pfirrmann Grade 4 changes. The NP for each disc is inhomogeneous gray to black. The NP cannot be distinguished from the AF. Comparatively, the disc heights vary slightly. This is an advantage of the Modified Pfirrman scoring system which assesses the degree of disc height loss in addition to the classic Pfirrmann changes. The L3/4 disc represents a Modified Pfirrmann grade 6. Grade 6 discs demonstrate <30% decrease in disc height but are otherwise consistent with a classic Pfirrmann grade 5. The L2/3 and L4/5 discs represent a Modified Pfirrmann grade 7. For Modified Pfirrmann grade 7, the disc must be decreased in height 30–60%. The L5/S1 disc represents a Modified Pfirrmann grade 8 where the disc height is decreased >60%, while the remaining disc characteristics are consistent with a classic Pfirrmann grade 5

- Modified Pfirrmann Grade 7: Modified Pfirrmann Grade 5 plus there is 30–60% decrease in disc height. Commonly demonstrates Modic Type 1 endplate changes (Fig. 22.36).
- Modified Pfirrmann Grade 8: Modified Pfirrmann Grade 5 plus there is >60% decrease in disc height (Fig. 22.36).

## Regenerative Therapies

There are many regenerative strategies for the stabilization/restoration of the IVD being used in pre-clinical and/or clinical studies [74]. The following are a few examples:

- Scaffolds and other biomaterials for the restoration of both the AF and NP.
- Reimplantation of primary cells such as NP and AF progenitor cells.
- Implantation of autologous IVD cells and/or chondrocytes.
- Intradiscal gene therapies
- Direct IVD injection of growth factors.
- Targeting of IL-1β using IL-1βRa and n IL-1β (n = neutralizing) therapies

The two most commonly used IVD biologics in clinical practice are intradiscal platelet rich plasma (PRP) and intradiscal mesenchymal stromal/stem cell (MSC) therapies [39, 75].

## MSC Therapy for IVD Regeneration

- Most common clinically evaluated cell type.
- FDA approved for use in musculoskeletal diseases.
- Can differentiate into any lineage of mesenchymal and mesodermal origins (osteoblasts, chondroblasts, and adipocytes) [75].
- Can be harvested from a wide variety of vascularized tissues.
  - Most common sources are bone marrow and adipose tissue. Bone marrow derived cells can be used clinically for intradiscal therapy [75] (Fig. 22.37).
- Studies have shown MSCs decrease pain and improve function, promote tissue repair, reduce inflammatory damage, inhibit the inflammatory cascade, can increase disc height and a few studies have demonstrated an increase in disc hydration (Fig. 22.38) [76, 77].
- The iliac crest is the most common aspiration site due to its high concentration of MSC, hematopoietic stem cells and endothelial progenitor cells, IL-1ra (receptor antagonist), as well as numerous growth factors.
  - Benefits of growth factors include their anabolic and anti-inflammatory effects.
  - IL-1ra inhibits catabolism and may be partially responsible for pain relief.
- Bone marrow aspiration technique is crucial for optimum results [77].
  - Minimally invasive, although as with any procedure, there are risks and complications [78].

    Procedural pain, hematoma, seroma, infection, lateral femoral cutaneous nerve injury, neuroma formation, chronic numbness, and chronic pain.

    Majority of complications are minor and self-limiting.

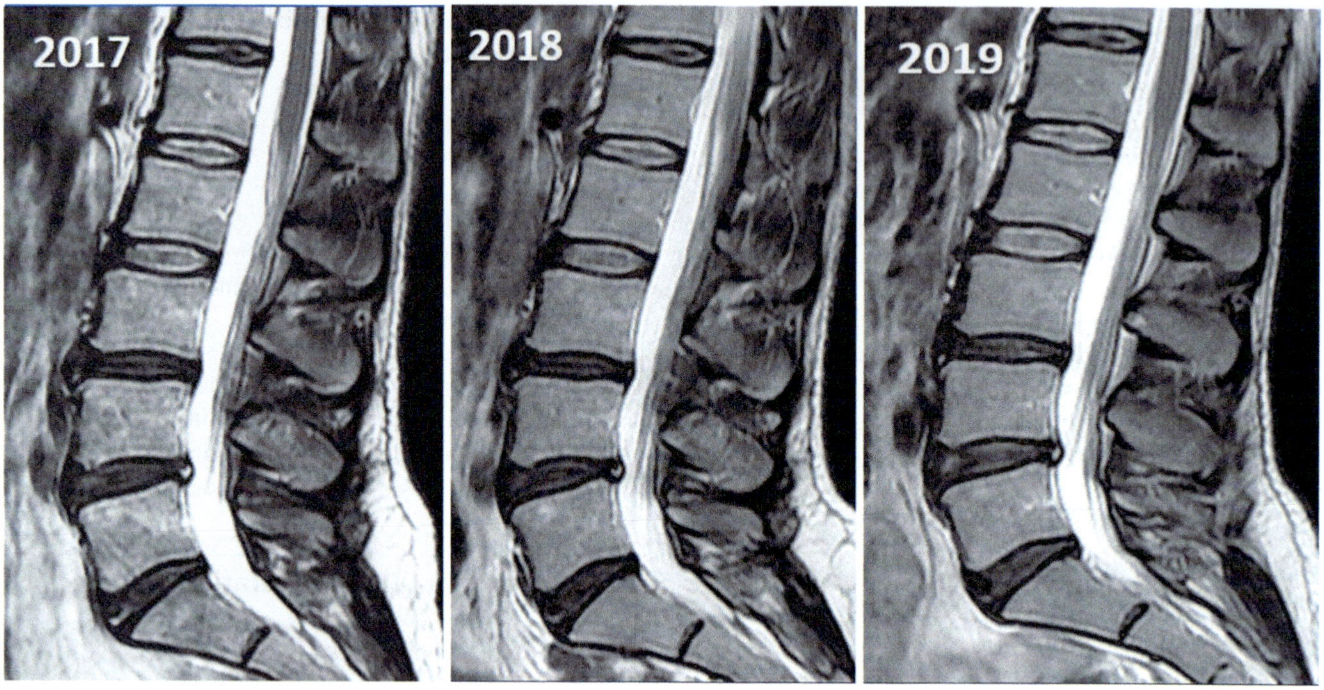


**Fig. 22.37** Intradiscal therapy. These are T2-weighted sagittal lumbar MRIs of a 42 yo male with low back pain at L5/S1, that were obtained over a 2 year period. The patient's initial evaluation was in 2017. In 2018, the patient was treated with intradiscal MSCs at L5/S1 due to worsening low back pain. The 2018 MRI demonstrates a continued decrease in disc signal (Pfirrmann grade 4) without significant interval decrease in disc height (which is only slightly diminished). A small anterior annular fissure remains visible. The MRI obtain at the 1 year follow-up demonstrates a subtle increase in disc height and decrease in the size and signal intensity of the anterior annular fissure. The patient was asymptomatic at his 1-year follow-up

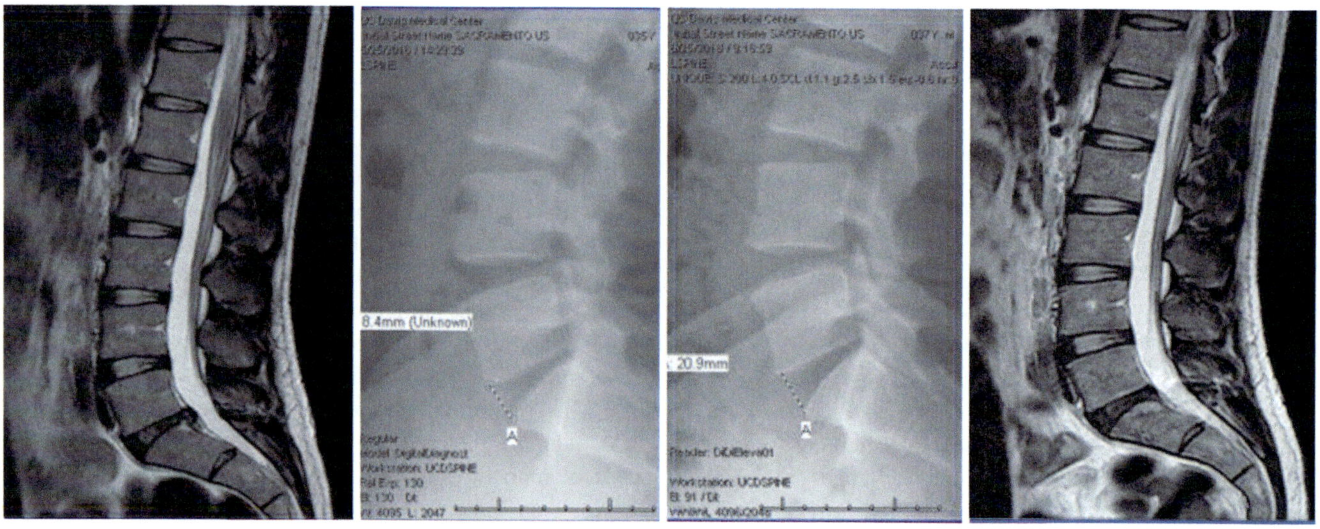


**Fig. 22.38** Intradiscal therapy. This patient is a 35 year old pharmacist with low back pain for 8 months. The patient reports his pain as averaging 5/10, but he states that his pain can be 7/10 depending on his activities. The patient is very active (kite boarding, motorcross sports). The patient's pain is centered at L5/S1. The first two images on the left represent a pre-treatment sagittal T2-weighted lumbar spine MRI and lateral standing lumbar spine x-ray. The anterior disc height was measured at 18.4 mm on the lateral x-ray. In June of 2016 the patient received intradiscal MSC therapy. At his 3 month follow-up he reported improved back pain, and at his 6 month follow-up he stated that he no longer had back pain. A 2 year follow-up standing lateral lumbar spine x-ray and sagittal T2-weighted lumbar spine MRI (the two studies on the right) demonstrate an increase in the anterior disc height now measuring 20.9 mm. The L5/S1 disc also appears increased in height on the post-treatment sagittal MRI. The patient remained asymptomatic on his 2 year follow-up evaluation

**Key Points**

1. Imaging the lumbosacral spine for radiculopathy
   (a) Need to detect nerve root compression.
   (b) MRI is superior to other techniques for demonstrating disc disease such as disc protrusion or extrusion.
   (c) MRI is superior for showing the narrowing of a lateral recess on axial view by disc, hypertrophied ligamentum flavum, synovial cyst, etc., producing compression of a descending (traversing) nerve root.
   (d) MRI is superior to CT for demonstrating a compressed nerve root within a neural foramen. While CT can demonstrate spurs projecting from a facet, the exiting nerve root is not well seen on CT.
   (e) Enhanced MRI is superior for demonstrating a tumor compressing a nerve root in multiple projections.
   (f) Spondylolisthesis is well demonstrated by plain films, MRI, and CT, but CT is best for demonstrating spondylolysis as the cause. Either MRI or CT can demonstrate facet degeneration as another cause.
   (g) Narrowing of a neural foramen with nerve root compression from spondylolisthesis plus marked disc narrowing best seen on MRI
   (h) Evaluation of S1 compression by the pyriformis muscle requires MRI
   (i) Paraspinal lumbosacral plexus evaluation requires enhanced fat suppressed coronal and axial MRI
2. Imaging the thoracic spine for radiculopathy
   (a) MRI is superior to CT for seeing the nerve roots within the spinal canal and neural foramina
   (b) MRI is superior to CT for detecting a protruded or extruded disc, or other mass within the canal, neural foramen, or paraspinal soft tissues. CT is better than MRI for showing disc calcification indicative of metabolic disorders and degeneration.
3. Imaging the cervical spine for radiculopathy
   (a) Disc disease better evaluated with MRI than with CT, given MR's better contrast resolution and multiprojectional capabilities
   (b) Neural foraminal stenosis due to uncinate or facet spurs is much better seen with CT than MRI, which overcalls the stenosis
   (c) The nerve root within a foramen is much better seen on MRI than CT
   (d) Post-myelographic CT only demonstrates the nerve root sleeve to the proximal aspect of the neural foramen. The rest of the neural foramen is seen as on a non-myelographic CT study, with the nerve root within the foramen poorly seen. Thus, myelography may not offer much additional information beyond plain CT; it may not be worth the stress on the patient, the cost, and the complications.
4. Imaging the spine for myelopathy
   (a) MRI is far superior to all other modalities for demonstrating cord compression, expansion, cyst formation, enhancement from neoplastic or infectious disease, or abnormal signal without enhancement which may indicate non-infectious inflammation, demyelination, or infarction
   (b) CT after myelography excellent for demonstrating a block to CSF flow and adequate for demonstrating cord expansion but not its etiology.
   (c) MRI is superior to CT and CT after myelography for demonstrating the level of the tip of the conus, filum thickening, and possible tethering of the cord
   (d) MRI is superior to CT and nuclear studies for demonstrating ischemic cord changes due to a dural arteriovenous fistula (AVF). Abnormal veins are seen along the posterior cord surface before and after enhancement. Specialized MRA can demonstrate the probable site of the fistula, while catheter angiography will definitively show the fistula and can be used for endovascular occlusion of the AVF.
   (e) MRI is superior to all other techniques for evaluating a spinal cord arteriovenous malformation (AVM). Catheter angiography is required to understand the inflow to and outflow from the AVM as an aid to endovascular or surgical treatment. MRI is the only reliable technique for detecting a cavernous angioma.
5. Imaging the spine to determine the age of a compression fracture
   (a) STIR is the most sensitive MR sequence for detecting subtle edema and small fluid collections.
   (b) The STIR sequence, and frequently a fat-saturated (FS) T2 sequence, will show increased signal representing inflammation and thus indicate that a fracture is acute to subacute.
   (c) A bone scan demonstrates increased radionuclide uptake indicating bone turnover in an acute to subacute fracture. However, in an osteoporotic patient, such uptake may be delayed. Only 30% of acute spinal and pelvic fractures are positive on bone scan, while they are always positive on STIR and non-FS/non-enhanced T1 MRI.
   (d) Bone scans can be misleading when attempting to differentiate acute from subacute/chronic compression fractures. The bone scan remains positive in 41% of patients at 1 year, 10% at 2 years, and 3% at 3 years. MRI, especially the STIR sequence, more accurately shows decreased edema and inflammation over time.

6. Evaluation for spondylodiscitis and/or epidural abscess
   (a) CT is better than plain films for showing early endplate erosion.
   (b) MRI is best for demonstrating the early signs of inflammation, and for showing enhancement of the disc, meninges, synovial lining of facet joints, and paraspinal soft tissues.
   (c) A bone scan can demonstrate the resolution of infection when enhancement on MRI is still present due to enhancing inflamed but sterile scar tissue.

**Acknowledgments** This book chapter is modified and updated from a previous book chapter, "Radiology of the Spine for Interventionalist" by Sheri L Albers, DO, and Richard E. Latchaw, MD, in *Essentials of Interventional Techniques in Managing Chronic Spinal Pain* published by Springer.

## References

1. Albers SL, Latchaw RE. Radiology of the spine for the interventionalist. In: Manchikanti L, Kaye AD, Falco FJE, Hirsch JA, editors. Essentials of interventional techniques in managing chronic spinal pain. New York: Springer; 2018. p. 89–108.
2. Bushberg JT, Seibert JA, Leidholdt EM, et al. The essential physics of medical imaging. 3rd ed. Philadelphia: Lippincott Williams & Wilkins; 2012.
3. Ross JS. Degenerative disease of the spine. In: Latchaw RE, Kucharczyk J, Moseley ME, editors. Imaging of the nervous system: diagnostic and therapeutic applications. Philadelphia: Elsevier Mosby; 2005. p. 1501–36.
4. Ross JS, Brant-Zawadzki M, Moore KR, et al. Diagnostic imaging: spine. Salt Lake City: Amirsys; 2004.
5. Ross JS. Specialty imaging: postoperative spine. Salt Lake City: Amirsys; 2011.
6. Ross JS, Moore K. Diagnostic imaging: spine. Salt Lake City: Amirsys; 2015.
7. Harnsberger HR, Osborn AG, Ross JS, et al. Diagnostic and surgical imaging anatomy: brain, head and neck, spine. Salt Lake City: Amirsys; 2006.
8. Czervionke L, Fenton D. Imaging painful spine disorders. Philadelphia: Saunders; 2011.
9. Naidich T, Castillo M, Soonmee C, et al. Imaging of the spine. Philadelphia: Saunders; 2010.
10. Waldman S, Campbell RSD. Imaging of pain. Philadelphia: Saunders; 2010. Expert Consult Series.
11. Oritz O. Imaging of the postoperative spine. Philadelphia: Elsevier; 2014. Neuroimaging Clinics of North America.
12. Resnick D. Internal derangement of joints. In: Resnick D, Kransdorf M, editors. Bone and joint imaging. 3rd ed. Philadelphia: Elsevier; 2005. p. 905–94.
13. Zagzebski JA. Essentials of ultrasound physics. St. Louis: Mosby-Year Book; 1996.
14. Mettler FA, Guiberteau MJ. Essentials of nuclear medicine imaging. 6th ed. Philadelphia: Elsevier; 2012.
15. Jahan R, Gobin YP. Angiography and embolization of vascular lesions of the spine. In: Latchaw RE, Kucharczyk J, Moseley ME, editors. Imaging of the nervous system: diagnostic and therapeutic applications. Philadelphia: Elsevier Mosby; 2005. p. 1723–44.
16. Adams A, Roche O, Mazumder A, et al. Imaging of degenerative lumbar intervertebral discs; linking anatomy, pathology and imaging. Postgrad Med J. 2014;90:511–9.
17. Fardon DF, Williams AL, Dohring EJ, et al. Lumbar disc nomenclature: version 2.0: recommendations of the combined task forces of the North American Spine Society, the American Society of Spine Radiology, and the American Society of Neuroradiology. Spine J. 2014;39:2525–45.
18. Hoch B, Hermann G. Migrated herniated disc mimicking a neoplasm. Skelet Radiol. 2010;39:1245–9.
19. Modic MT, Masaryk TJ, Ross JS, et al. Imaging of degenerative disk disease. Radiology. 1988;168:177–86.
20. Bailly F, Maigne J-Y, Genevay S, et al. Inflammatory pain pattern and pain with lumbar extension associated with Modic 1 changes on MRI: a prospective case-controlled study of 120 patients. Eur Spine J. 2014;23:493–7.
21. Brayda-Bruno M, Schistad MEI, Espeland A, et al. The association between Modic changes and pain during 1-year follow-up in patients with lumbar radicular pain. Skelet Radiol. 2014;43:1271–9.
22. Charran AK, Tony G, Lalam R, et al. Destructive discovertebral degenerative disease of the lumbar spine. Skelet Radiol. 2012;41:1213–21.
23. Varlotta GV, Lowitz TR, Schweitzer M, et al. The lumbar facet joint: a review of current knowledge: part 1: anatomy, biomechanics, and grading. Skelet Radiol. 2011;40:13–23.
24. Viejo-Fuertes D, Liguoro D, Rivel J, et al. Morphologic and histologic study of the ligamentum flavum in the thoraco-lumbar region. Surg Radiol Anat. 1998;20:171–6.
25. Tok CH, Kaur S, Gangi A. Symptomatic spinal epidural lipomatosis after a single local epidural steroid injection. Cardiovasc Intervent Radiol. 2011;34:250–5.
26. Maric DL, Krstonosic B, Eric M, et al. An anatomical study of the lumbar external foraminal ligaments: appearance at MR imaging. Surg Radiol Anat. 2015;37:87–91.
27. Khiami F, Aziria S-A, Ragot S, et al. Reliability and validity of a new measurement of lumbar foraminal volume using computed tomography. Surg Radiol Anat. 2015;37:93–9.
28. Vleemin Q, Schuenke MD, Masi AT, et al. The sacroiliac joint: an overview of its anatomy, function and potential clinical implications. J Anat. 2012;221:537–67.
29. Muche B, Bollow M, Francois RJ. Anatomic structures involved in early- and late-stage sacroiliitis in spondyloarthritis: a detailed analysis by contrast-enhanced magnetic resonance imaging. Arthritis Rheum. 2003;48:1374–84.
30. Morishita Y, Naito M, Hymanson H, et al. The relationship between the cervical spinal canal diameter and the pathological changes in the cervical spine. Eur Spine J. 2009;18:877–83.
31. Hartman J. Anatomy and clinical significant of the uncinate process and uncovertebral joint: a comprehensive review. Clin Anat. 2014;27:431–40.
32. Latchaw RE, Falcone S, Saraf-Lavi E. Imaging of spinal trauma. In: Latchaw RE, Kucharczyk J, Moseley ME, editors. Imaging of the nervous system: diagnostic and therapeutic applications. Philadelphia: Elsevier Mosby; 2005. p. 1567–625.
33. Holz AJ. Infectious diseases. In: Latchaw RE, Kucharczyk J, Moseley ME, editors. Imaging of the nervous system: diagnostic and therapeutic applications. Philadelphia: Elsevier Mosby; 2005. p. 1661–73.
34. Fuster D, Tomas X, Mayoral M, et al. Prospective comparison of whole-body (18)F-FDG PET CT and MRI of the spine in the diagnosis of haematogenous spondylodiscitis. Eur J Nucl Med Mol Imaging. 2015;42:264–71.
35. An HS, Seldomridge JA. Spinal infections: diagnostic tests and imaging studies. Clin Orthop Relat Res. 2006;444:27–33.
36. Hong SH, Choi JY, Lee JW, et al. MR imaging assessment of the spine: infection or an imitation? Radiographics. 2009;29:599–612.
37. Kucera T, Brtkova J, Sponer P, et al. Pyogenic sacroiliitis: diagnosis, management and clinical outcome. Skelet Radiol. 2015;44:63–71.

38. Jung NY, Jee WH, Ha KY, et al. Discrimination of tuberculous spondylitis from pyogenic spondylitis on MRI. AJR Am J Roentgenol. 2004;182(6):1405–10.
39. Kaye AD, Edinoff AN, Rosen YE, et al. Regenerative medicine: pharmacological considerations and clinical role in pain management. Curr Pain Headache Rep. 2022;26:751–65.
40. Dieleman JL, Squires E, Bui AL, et al. Factors associated with increase in US health care spending, 1996–2013. JAMA. 2017;318:1668–78.
41. Kirnaz S, Capadona C, Wong T, et al. Fundamentals of intervertebral disc degeneration. World Neurosurg. 2022;157:264–73.
42. Frost BA, Camarero-Espinosa S, Foster EJ. Materials for the spine: anatomy, problems, and solutions. Materials. 2019;12:253.
43. Smith LJ, Fazzalari NL. The elastic fibre network of the human lumbar anulus fibrosus: architecture, mechanical function and potential role in the progression of intervertebral disc degeneration. Eur Spine J. 2009;18:439–48.
44. Mouw JK, Ou G, Weaver VM. Extracellular matrix assembly: a multiscale deconstruction. Nat Rev Mol Cell Biol. 2014;15:771–85.
45. Yanagishita M. Function of proteoglycans in the extracellular matrix. Pathol Int. 1993;43:283–93.
46. Iatridis JC, Setton LA, Weidenbaum M, et al. Alterations in the mechanical behavior of the human lumbar nucleus pulposus with degeneration and aging. J Orthop Res. 1997;15:318–22.
47. Calve J, Galland M. The intervertebral nucleus pulposus—its anatomy, its physiology, its pathology. J Bone Joint Surg. 1930;12:555–78.
48. Trout JJ, Buckwalter JA, Moore KC. Ultrastructure of the human intervertebral disc: II. Cells of the nucleus pulposus. Anat Rec. 1982;204:307–14.
49. Purmessur D, Cornejo MC, Cho SK, et al. Notochordal cell-derived therapeutic strategies for discogenic back pain. Global Spine J. 2013;3:201–18.
50. Akkiraju H, Nohe A. Role of chondrocytes in cartilage formation, progression of osteoarthritis and cartilage regeneration. J Dev Biol. 2015;3:177–92.
51. Adams MA. Intervertebral disc tissues. In: Derby B, Akhtar R, editors. Mechanical properties of aging soft tissues. Cham: Springer International Publishing; 2015. p. 7–35.
52. Moore RJ. The vertebral endplate: disc degeneration, disc regeneration. Eur Spine J. 2006;15:S333–7.
53. Raj PP. Intervertebral disc: anatomy-physiology-pathophysiology-treatment. Pain Pract. 2008;8:18–44.
54. Nerlich AG, Schaaf R, Walchli B, et al. Temporo-spatial distribution of blood vessels in human lumbar intervertebral discs. Eur Spine J. 2007;16:547–55.
55. Rudert M, Tillmann B. Lymph and blood-supply of the human intervertebral disc—cadaver study of correlations to discitis. Acta Orthop Scand. 1993;64:37–40.
56. Horner HA, Urban JP. 2001 Volvo award winner in basic science studies: effect of nutrient supply on the viability of cells from the nucleus pulposus of the intervertebral disc. Spine (Phila Pa 1976). 2001;26:2543–9.
57. Grunhagen T, Shirazi-Adl A, Fairbank JC, et al. Intervertebral disk nutrition: a review of factors influencing concentrations of nutrients and metabolites. Orthop Clin North Am. 2011;42:465–77, vii.
58. Edgar MA. The nerve supply of the lumbar intervertebral disc. J Bone Joint Surg Br. 2007;89:1135–9.
59. Freemont AJ, Peacock TE, Goupille P, et al. Nerve ingrowth into diseased intervertebral disc in chronic back pain. Lancet. 1997;350:178–81.
60. Battie MC, Videman T, Levalahti E, et al. Genetic and environmental effects on disc degeneration by phenotype and spinal level: a multivariate twin study. Spine (Phila Pa 1976). 2008;33:2801–8.
61. Buckwalter JA. Aging and degeneration of the human intervertebral disc. Spine (Phila Pa 1976). 1995;20:1307–14.
62. Kalichman L, Hunter DJ. The genetics of intervertebral disc degeneration. Familial predisposition and heritability estimation. Joint Bone Spine. 2008;75:383–7.
63. Cong L, Pang H, Xuan D, et al. The interaction between aggrecan gene VNTR polymorphism and cigarette smoking in predicting incident symptomatic intervertebral disc degeneration. Connect Tissue Res. 2010;51:397–403.
64. Cong L, Pang H, Xuan D, et al. Association between the expression of aggrecan and the distribution of aggrecan gene variable number of tandem repeats with symptomatic lumbar disc herniation in Chinese Han of Northern China. Spine (Phila Pa 1976). 2010;35:1371–6.
65. Shiri R, Karppinen J, Leino-Arjas P, et al. The association between obesity and low back pain: a meta-analysis. Am J Epidemiol. 2010;171:135–54.
66. Liuke M, Solovieva S, Lamminen A, et al. Disc degeneration of the lumbar spine in relation to overweight. Int J Obes (Lond). 2005;29:903–8.
67. Navone SE, Marfia G, Canzi L, et al. Expression of neural and neurotrophic markers in nucleus pulposus cells isolated from degenerated intervertebral disc. J Orthop Res. 2012;30:1470–7.
68. Oichi T, Taniguchi Y, Oshima Y, et al. Pathomechanism of intervertebral disc degeneration. JOR Spine. 2020;3:e1076.
69. Inoue N, Espinoza Orias AA. Biomechanics of intervertebral disk degeneration. Orthop Clin North Am. 2011;42:487–99.
70. Lotz JC, Fields AJ, Liebenberg EC. The role of the vertebral end plate in low back pain. Global Spine J. 2013;3:153–63.
71. Rodriguez AG, Rodriguez-Soto AE, Burghardt AJ, et al. Morphology of the human vertebral endplate. J Orthop Res. 2012;30:280–7.
72. Pfirrmann CW, Metzdorf A, Zanetti M, et al. Magnetic resonance classification of lumbar intervertebral disc degeneration. Spine (Phila Pa 1976). 2001;26:1873–8.
73. Griffith JF, Wang YX, Antonio GE, et al. Modified Pfirrmann grading system for lumbar intervertebral disc degeneration. Spine (Phila Pa 1976). 2007;32(24):E708–12.
74. Yamada K, Iwasaki N, Sudo H. Biomaterials and cell-based regenerative therapies for intervertebral disc degeneration with a focus on biological and biomechanical functional repair: targeting treatments for disc herniation. Cells. 2022;11:602.
75. Navani A, Manchikanti L, Albers S, et al. Responsible, safe, and effective use of biologics in the management of low back pain: American Society of Interventional Pain Physicians (ASIPP) guidelines. Pain Physician. 2019;22:S1–S74.
76. Beall DP, Davis T, DePalma MJ, et al. Viable disc tissue allograft supplementation; one- and two-level treatment of degenerated intervertebral discs in patients with chronic discogenic low back pain: one year results of the VAST randomized controlled trial. Pain Physician. 2021;24:465–77.
77. Atluri S, Murphy MB, Dregalia R, et al. Evaluation of the effectiveness of autologous bone marrow mesenchymal stem cells in the treatment of chronic low back pain due to severe lumbar spinal degeneration: a 12-month, open-label, prospective controlled trial. Pain Physician. 2022;25:193–207.
78. Elattar O, Christophersen CM, Farber D. Pain and complications following bone marrow aspirate concentrate harvested from the iliac crest in foot and ankle surgery. Foot Ankle Spec. 2021:19386400211017379.

# Intradiscal Injections of Biologics

# 23

Laxmaiah Manchikanti, Mahendra Sanapati,
Annu Navani, Vijay Singh, Frank J. E. Falco,
and Joshua A. Hirsch

## Introduction

The increasing socioeconomic burden imposed by chronic spinal pain, the resources utilized in managing chronic spinal pain with the growing number of modalities applied together with alleged low quality and the high cost of interventions has led to multiple negative health policy implications [1–10]. Dieleman et al. [9, 10], in an assessment of US spending on personal health care in public health from 1996 to 2016 showed an estimated spending of $134.5 billion in 2016, with a 53.5% increase from $87.6 billion spent in 2013, in managing low back and neck pain, accounting for the highest amount of various disease categories. They also showed that for musculoskeletal disorders, the spending increased from $183.5 billion in 2013 to $263.3 billion in 2016 with an increase of 43.5%. The intervertebral discs, the zygapophysial (facet) joints, and sacroiliac joints have all been demonstrated, with controlled diagnostic techniques, to be common causes of chronic spinal pain [5–8]. Based on extensive evaluations, the prevalence of discogenic pain was shown to be variable from 16.9 to 26% without internal disc disruption and 16.9 to 42% for the diagnosis of internal disc disruption. Further, the cervical discogenic pain prevalence ranged from 16 to 53%. However, the prevalence of discogenic pain in the thoracic spine has not been available.

The intervertebral disc has been implicated as a source of spinal pain based on decades of pre-clinical, clinical, and epidemiological research, though the precise mechanisms continue to be debated as the literature evolves [11–20]. The pain emanating from pathologic changes within the disc itself, without disc herniation [11–13, 21], was described before the classic description of disc herniation by Mixter and Barr in 1934 [22]. Thus, chronic spinal pain with or without extremity pain from an intervertebral disc has been described with two inter-related, but distinct etiologies, namely disc herniation and discogenic pain. Discogenic pain without disc herniation is variably termed discogenic pain, internal disc disruption, and painful degenerative disc disease [11–13, 21]. Despite numerous advances, diagnosis based on history, physical exam, and radiological imaging has low sensitivity and specificity in determining whether or not the disc is a primary source of spinal pain [11].

Numerous modalities of treatments have been described in managing discogenic pain, including conservative management with physical therapy, pharmacological therapy [11–13, 23–26], interventional techniques, intradiscal therapies, and surgical interventions with fusion or disc replacement. It is clear that in disc degeneration, inflammatory cytokines produced by macrophages or disc cells play an important role [27]. Currently, in addition to the above treatments described, there are also treatments such as anti-nerve growth factor therapy with multiple available drugs [27–30].

Frank J.E. Falco has died before the publication of this book.

L. Manchikanti (✉)
Pain Management Centers of America, Paducah, KY, USA
e-mail: drlm@thepainmd.com

M. Sanapati
Pain Management Centers of America, Evansville, IN, USA

A. Navani
LeReve Regenerative Wellness, Campbell, CA, USA

Boomerang Healthcare, Walnut Creek, CA, USA

Stanford University School of Medicine, Stanford, CA, USA
e-mail: anavani@boomeranghc.com

V. Singh
Spine Pain Diagnostics Associates, Niagara, WI, USA
e-mail: vj@secure.paindiagnostics.net

F. J. E. Falco
Mid Atlantic Spine and Pain Physicians, Newark, DE, USA

J. A. Hirsch
Massachusetts General Hospital and Harvard Medical School, Boston, MA, USA
e-mail: jahirsch@mgh.harvard.edu


A. Navani et al. (eds.), *Essentials of Regenerative Medicine in Interventional Pain Management*,
https://doi.org/10.1007/978-3-031-50357-3_23

In addition, several cell-based therapies have been proposed in recent years [6, 27]. These therapies include mesenchymal cell therapy and platelet-rich plasma (PRP) injection [6, 31–34]. Consequently, evidence is emerging with significant literature published with basic and clinical science leading to controlled trials and systematic reviews.

## History

The lumbar disc was not considered an important cause of low back pain or radiculopathy until Mixter and Barr in 1934 [22] reported the surgical removal of a posterior disc prolapse leading to relief of these symptoms. In 1940, Roofe [35] ascertained the nerve supply to the annulus that laid down a foundation for understanding that the intervertebral disc alone could be a pain generator. Lindblom [36], in 1944, revealed the existence of annular pathology within the cadaveric discs after injecting red dye into the disc nucleus that extended into annular tears. Hirsch [37] noted in 1948 that the injection of procaine into the lumbar disc provided pain relief confirming that the disc was a source for axial low back pain. Subsequently, Lindblom [38], in 1948 reported the clinical use of lumbar disc injections to identify disc(s) responsible for low back pain and sciatica.

Cloward [39] and Smith and Robinson [40] initially pioneered cervical discography and in 1958 separately published papers regarding cervical discography in the evaluation of chronic neck pain, headache, and shoulder pain. Holt [41] later published a report in 1964 that brought into question the validity and role of cervical discography. Schellhas et al. [42] found that normal discs were not painful in either asymptomatic or symptomatic groups and that abnormal discs produced pain similar to the presenting complaint.

Thoracic discography was not implemented as a diagnostic tool until approximately 30 years after lumbar discography. In 1975, Simmons and Segil [43] described thoracic discography and nucleography in the evaluation of a man with mid-thoracic radicular pain. In 1995, Schellhas et al. [44] published a retrospective review of 100 outpatient thoracic discographies performed on patients whose magnetic resonance imaging (MRI) findings revealed thoracic disc degeneration.

Over the years, numerous treatments have been described in managing discogenic pain ranging from over-the-counter medications to complex surgical fusions, and more recently, injection of biologics into the discs with significant and emerging evidence of effectiveness.

## Pathophysiology

Two of the most common problems associated with the intervertebral disc are intervertebral disc degeneration and intervertebral disc protrusion or extrusion; however, the processes are not mutually exclusive (Figs. 23.1 and 23.2) [11, 45]. Extruded nucleus pulposus spontaneously produces increased amounts of matrix metalloproteinases, nitric oxide, interleukin-6, prostaglandin E2, and other chemical mediators (Fig. 23.1) [11, 45]. These products may be intimately involved in both the biochemistry of disc degeneration and the pathophysiology of radiculopathy.

Intervertebral disc degeneration is difficult to define and is also difficult to separate from the normal aging process of the intervertebral disc. However, there is a wide variation in the aging and degenerative process of the disc with some septuagenarians having the intervertebral discs of 30-year-olds and vice versa [11, 46–48]. The hallmarks of disc degeneration are loss of fluid and fluid pressure, disruption, or breakdown of collagen (including tearing of the annulus fibrosus) and proteoglycans, and sclerosis of the cartilaginous endplate and adjacent subchondral bone [11, 45–48].

The normal aging and degeneration of the intervertebral disc is closely related to the number of vessels that reach the intervertebral disc, especially the cartilaginous endplate. As the number of vessels decrease and the nutrition provided and waste removed by these vessels decrease, the changes associated with intervertebral disc aging and degeneration increase [11, 45].

Injuries to the disc may evoke pain by activation of sensory nociceptors. Injury of the annulus can lead to the release of significant amounts of inflammatory chemicals that may in turn irritate or sensitize the annulus and adjacent structures in the spinal canal and intervertebral foramina. Inflammatory and immunologic reaction to herniated disc material is well recognized as a pain generating mechanism [11]. Even though the initiating factors in disc degeneration are not fully understood, it is hypothesized to be a cell driven process [27, 33]. There is an increase in multiple proinflammatory cytokines, occurring independently of cytokine antagonist, resulting in a shift towards an increased inflammatory response and matrix catabolism [27, 33]. This increase in cytokine profile stimulates the expression of matrix degrading enzymes without a concomitant increase in their endogenous inhibitors, resulting in elevated levels of activated enzymes [33]. Alongside the catabolic changes, the excess of cytokines within the discs, particularly IL-1, alters the expression of anabolic genes and matrix proteins [49]. This structural and mechanical compromise can potentially lead to microtrauma and herniation of nucleus pulposus tissue through the damaged annular fibrosis. As the degeneration progresses, there is also ingrowth of blood vessels and nociceptive nerve fibers into the normally avascular disc through the AF and extending into the nucleus pulposus enabling the infiltration of immune cells resulting in increased pain [33].

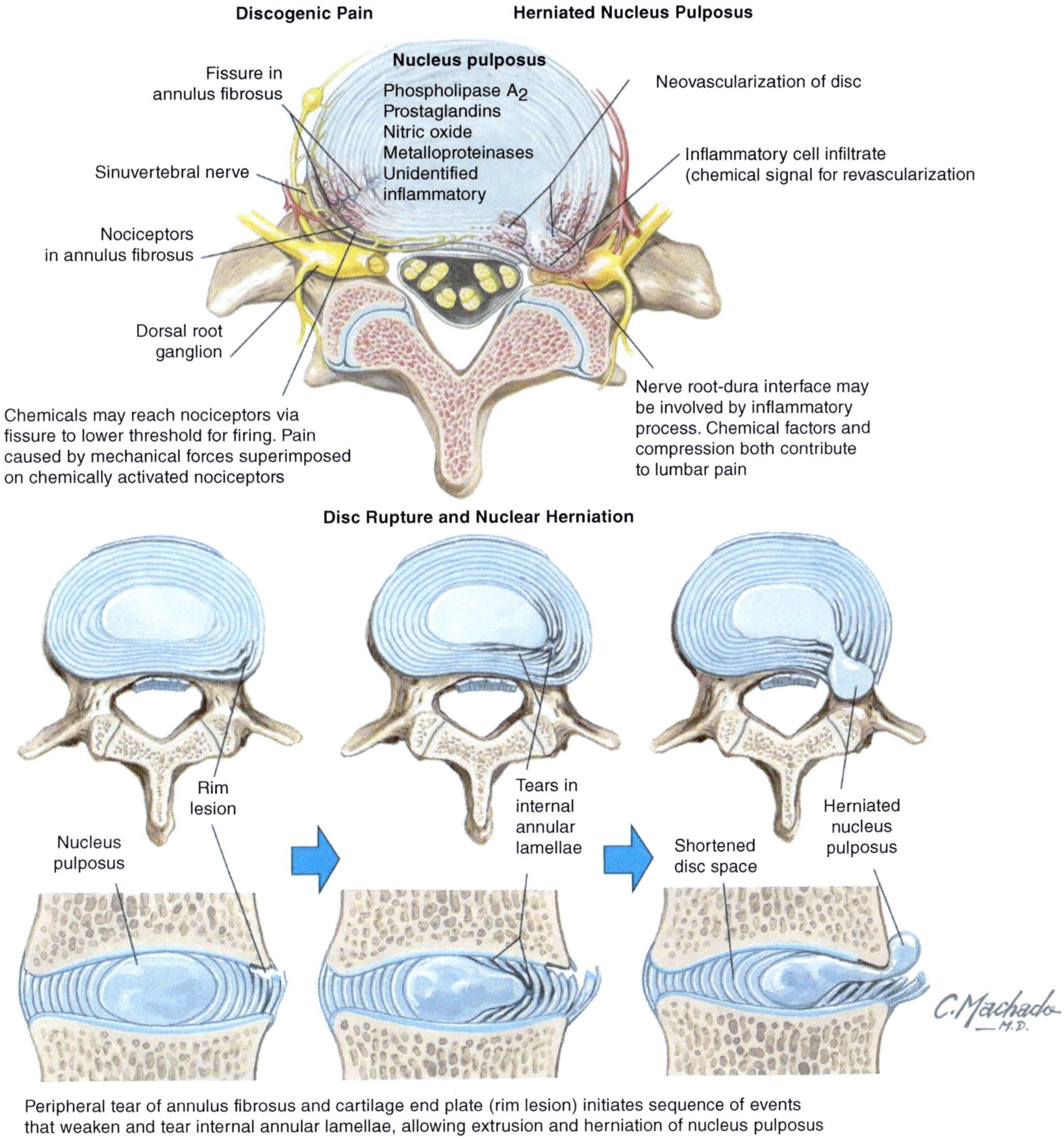


**Fig. 23.1** Pathophysiology of disc related pain. Netter medical illustration used with permission of Elsevier. 

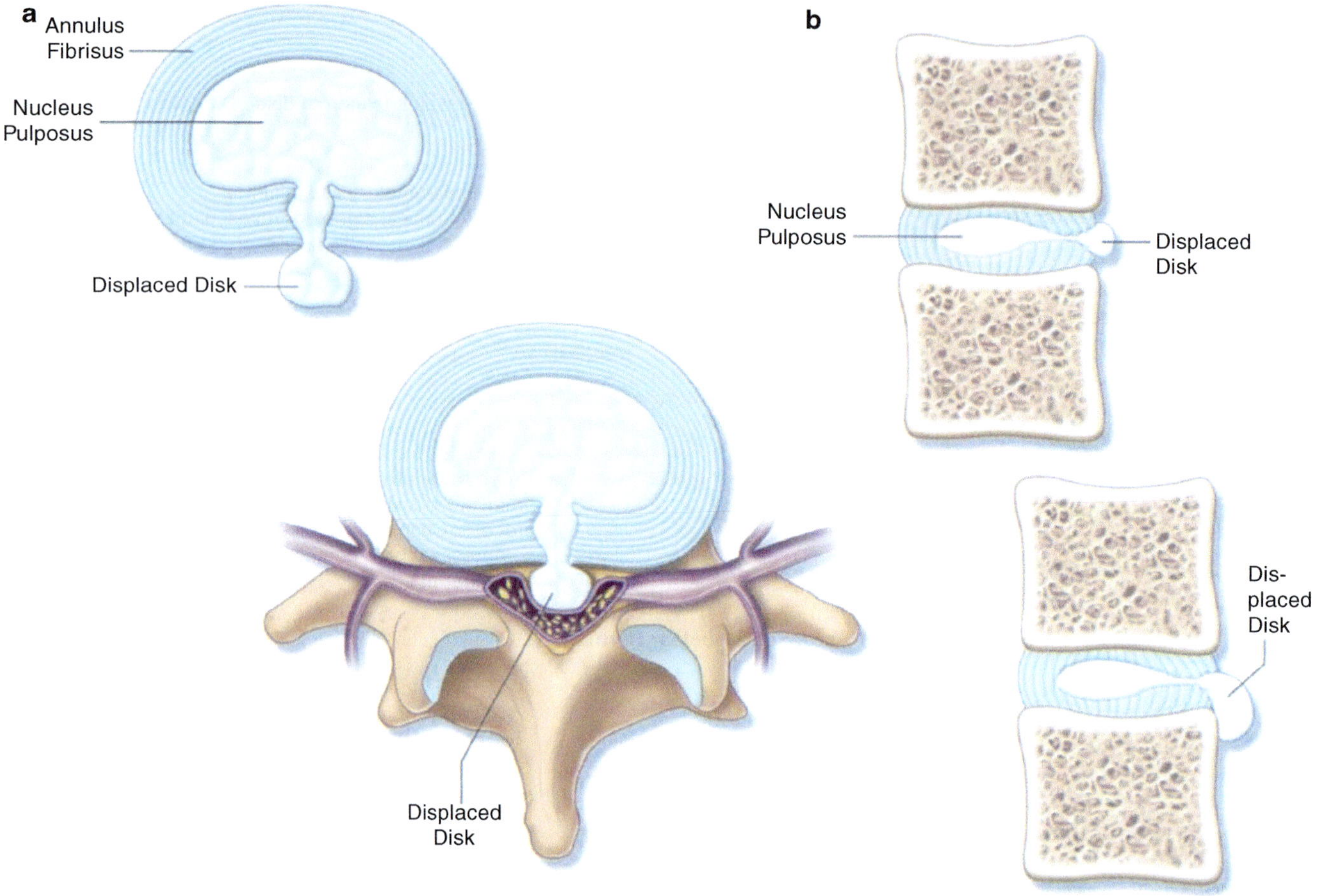


**Fig. 23.2** Schematic illustration of disk extrusion. (**a**) Axial and (**b**) sagittal images demonstrate that the greatest measure of the displaced disk material is greater than the base of the displaced disk material at the disk space of origin when measured in the same plane. **Source:** Manchikanti L, Albers SL, Hirsh JA, Boswell, MV. Lumbar disk herniation. In: Kaye AD (ed). Scientific American Pain Management. Hamilton: Decker; September 2017. DOI: https://doi.org/10.2310/7900.15047. www.DeckerIP.com [46]

## Regenerative Medicine Therapy in Disc Repair

Multiple treatment options for intervertebral disc degeneration or for discogenic pain have been described. In view of the lack of significant progress with the current treatments, there remains a growing need for therapies capable of restoring cell population, intervertebral disc function, and alleviation of the pain with improvement in function. Increasing efforts have therefore been directed towards investigating the potential of cell-based therapies for treatment of intervertebral disc degeneration [6, 8].

The versatility of biologic therapy allows its use in spinal disc degeneration, as well as disc protrusion. The two biologics most commonly utilized and well-studied are PRP and mesenchymal stem cells (MSCs) [6]. In the United States, PRP is approved by the Food and Drug Administration (FDA) for use with ligament grafting and bony matrices during reconstructive procedures. This benefit occurs by increased concentration of growth factors that are secreted by a platelet in an inflammatory environment.

PRP is a concentrate of whole blood that is centrifuged to obtain a concentrate of plasma rich in platelets and hence growth factors. PRP provides multiple benefits with disc repair by the increased concentration of growth factors that are secreted by a platelet in an inflammatory environment, i.e., platelet-derived growth factors (PDGF) [6, 50–53].

These growth factors are essential to the healing process, as they increase fibroblast or osteoblast metabolic activity while reducing cell apoptosis; promote angiogenesis, thereby increasing blood flow and circulation to the new-forming tissues; and increase the expression of the pro-collagen gene and collagen-derived growth factor, which increases the tensile strength of the new tissue [6, 27, 34, 50–56]. Even though, components of PRP include platelets, leukocytes, and red blood cells, platelets are central to mediating the anabolic effects of PRP by virtue of releasing growth factors stored in their alpha granules [6, 53]. Notable growth factors released from platelets that are involved in the healing process include PDGF, TGF-β (transforming growth factor-β), VEGF (vascular endothelial growth factor), EGF (epidermal growth factor), bFGF (basic fibroblast growth factor), IGF-1 (insulin-like growth factor).

An in vitro study of PRP-infused human intervertebral disc cultures supported the hypothesis of inducing natural healing cascade and exhibited nucleus pulposus proliferation and differential, as well as upregulated proteoglycan synthesis [56, 57]. Further, in animal models of experimentally injured intervertebral discs, treatments with intradiscal PRP have demonstrated restoration of normal cellular architecture and disc height [58, 59]. Further, PRP also has been postulated to have an antiinflammatory effect which was demonstrated in in vitro study showing cytokine (TNF-α and IL-1) induced proinflammatory degrading enzymes and mediators showing separation of mediators with the addition of PRP into a collagen matrix of human nucleus pulposus cells [60]. In addition, PRP-impregnated gelatin hydrogel injection in a rabbit model with degenerated intervertebral discs results in significantly higher water content determined by MRI, which corresponded with increased intradiscal proteoglycan content, up regulated micro-RNA precursors for Type II collagen, and significantly reduced apoptotic nucleus pulposus cells [61]. Additionally, in a percutaneous annulus puncture induced degenerated disc rat model, discs treated with PRP had fewer inflammatory cells, higher preservation of normal morphology, and higher fluid content in T2 MRI compared with sham at 4 weeks post injection [58]. According to the International Society of Cellular Therapy (ISCT), MSCs are defined by their ability to adhere to plastic under standard culture conditions; express cell surface markers; lack expression of other markers; and differentiate into osteoblasts, adipocytes, and chondrocytes under in vitro conditions [51]. In addition to displaying multi-potentiality and having self-renewal capacity, MSCs have immunomodularity and antiinflammatory properties [34] and have the capacity to modulate the activity of microphages [54, 55].

Multiple research activities have intensified in tissue engineering and regenerative medicine [31, 62], with multiple preclinical studies demonstrating encouraging results [63–66]. The preclinical research was focused on one of the three biological approaches addressing the degenerative cascade with stimulation of anabolic processes, modulation of catabolic processes, or provision of new cells [64]. Thus, focus has been on tissue engineered cellular therapy on chondrocytes [65] and stem cell replacement therapy. In fact, a systematic review of MSCs in spinal cord injury, intervertebral disc repair, and spinal fusion showed 28 eligible articles identified that assessed the role of MSCs in intervertebral disc degeneration [6]. Minimally invasive strategies were adapted in different studies using a fine needle microinjection to induce degeneration and MSC implantation with needle sizes of 33 gauge for mice and 30 gauge for larger animals such as rats, and a 21-gauge needle in humans. Multiple studies reported better results when the MSCs were used in coculture with other cells or used in scaffolds. MSCs were also found to have an immune-modulatory role, which can improve surgical outcome [50].

## Evidence Base of Intradiscal Biologics

PRP injections and injections of MSCs have been utilized with increasing frequency in disc-related disorders of the spine. However, stem cells are used more commonly for intervertebral disc degeneration and disc repair.

### Platelet Rich Plasma

Basso et al. [67] in a systematic review of regenerative therapies in intervertebral disc degeneration identified three publications relating to PRP, without randomized controlled trials (RCTs) [68–73]. The number of patients included in these studies varied from 8 to 86, with variable follow-ups from 8 weeks to 2 years. The majority were very small studies with only two studies with large sample sizes of 47 and 86.

Sanapati et al. [66] also assessed the role of PRP with lumbar disc injections. Based on the available evidence, including one high quality RCT [71], multiple moderate quality observational studies [68–70, 72, 73], and a single arm meta-analysis for injection of PRP [31]. The qualitative evidence was assessed to be Level III on a scale of Level I to V using a qualitative modified approach to grading of evidence based on best evidence synthesis, with I indicating strong evidence and V indicating evidence derived from consensus. Table 23.1 shows in a tabular format the characteristics and outcomes of studies of PRP in intervertebral disc degeneration utilized in the analysis by Sanapati et al. [66]. Figures 23.3 and 23.4 show the results of PRP injections into lumbar disc.

Since the publication of these systematic reviews [60, 67], Cheng et al. [74] presented data on 5–9-year follow-up study in a small study with statistically significant improvement in pain and function compared to baseline in the 19 patients that completed the survey. However, six patients (29%) had undergone surgery on average 4 years from injection. All patients received injections of intradiscal PRP in a previous RCT. Data on pain, function, satisfaction, and need for surgery were collected at one time point of 5–9 years postinjection and compiled with existing data. The results showed in comparison to baseline, there were statistically significant improvements in pain and function ($p < 0.001$). All improvements were clinically significant. Six patients had undergone surgery during the follow-up period. They concluded this subset of patients demonstrated statistically and clinically significant improvements in pain and function at 5–9 years postinjection.

**Table 23.1** Characteristics and outcomes of studies of PRP in intervertebral disc degeneration

| Study details | Chronicity of injury and biologic used | Follow-up period | Conclusions |
|---|---|---|---|
| Tuakli-Wosornu et al., 2016 [71]<br>Lumbar discogenic pain<br>Prospective, double-blind, randomized controlled study, n = 47 | Chronic PRP injections | 1 year | Intradiscal injections of PRP ×1 showed significant improvement at 8-week follow-up, with maintained improvement compared to controls at 1-year follow-up |
| Monfett et al., 2016 [70]<br>Lumbar discogenic pain, lumbar disc degeneration<br>Prospective trial, n = 29 | Chronic PRP injections | 2 years | Intradiscal PRP injections show continued safety and improvements in pain and function at 2 years post-procedure |
| Navani et al., 2018 [68]<br>Lumbar discogenic pain<br>Prospective case series n = 20 | Chronic PRP, single injection, 2 mL injected up to three disc levels | 18 months | At 18 months, 15 patients remained for survey compared to 18 patients surveyed at 6 months: >50% relief in VAS in 93% of patients at 18 months (n = 14/15) and in 94% of patients (n = 17/18) at 6 months [2]. Improvement in SF-36 scores in 93% of patients at 18 months (n = 14/15) compared to 100% (n = 18/18) at 6 months |
| Akeda et al., 2017 [73]<br>Lumbar discogenic pain<br>Preliminary clinical trial, n = 14 | Chronic PRP injections | 12 months | Intradiscal injection of autologous PRP releasate in patients with low back pain was safe with no adverse events observed during follow-up<br>The results showed reduction in mean pain scores at 1 month, sustained throughout the observation periods of 6 months and 12 months |
| Levi et al., 2016 [69]<br>Lumbar discogenic pain<br>Prospective trial, n = 8 | Chronic PRP, single injection | 6 months | Single or multiple levels (up to 5) of discogenic pain injected with PRP showed encouraging improvement, with more patients developing improvement over time. Cohort up to 6 months |
| Kirchner and Anitua, 2016 [72]<br>Lumbar disc degeneration<br>Observational retrospective pilot study, n = 86 | Chronic PRGF-Endoret | 6 months | Fluoroscopy-guided infiltrations of intervertebral discs and facet joints with PRGF in patients with chronic LBP resulted in significant pain reduction assessed by VAS<br>The results showed reduction of the VAS over time. The study ended at 6 months with 91% of the patients showing an excellent score, 8.1% showing moderate improvement, and 1.2% showing lack of response |

*PRP* platelet rich plasma, *VAS* Visual Analog Scale, *SF-36* Short-form 36

**Adapted from:** Sanapati J, Manchikanti L, Atluri S, et al. Do regenerative medicine therapies provide long-term relief in chronic low back pain: A systematic review and metaanalysis. Pain Physician 2018; 21:515–550 [66]

Schepers et al. [75] published a study of intradiscal PRP for discogenic low back pain in patients without modic changes in an RCT. Provocation discography was performed to confirm the suspected disc was the source of pain. Participants were randomized to receive 1.0 cc intradiscal PRP (intervention) or 1.0 cc Saline with 0.2 g Kefzol (control). Data on pain (Numeric Rating Scale [NRS]), physical function (Roland Morris Disability Questionnaire [RMDQ]), and participants' general perceived health (12-Item Short Form Survey [SF-12]) were collected at 1 week, 4 weeks, 2 months, 6 months, 9 months and 1 year. A repeated-measures analysis (mixed model) was used for comparing the outcomes of the groups. The results show of the initial 98 (49 intervention, 49 control) patients randomized, 89 (91%) (44 intervention, 45 control) with complete outcome data were analyzed. Groups were balanced at baseline. After 12 months no differences between groups were found in the average pain (improved 21/44 in intervention vs 16/45 in control, p = 0.244), the disability scores (RMDQ minimal 3 points improvement 22/44 in intervention vs 24/45 in control, p = 0.753) and the SF-12 (mean difference physical health −1.19, 95% CI −5.39 to 2.99, p = 0.721, and mental health −0.34, 95% CI −3.99 to 3.29, p = 0.834). One serious adverse event occurred (spondylodiscitis) after intervention.

**Study name** — **Statistics for each day**

| Study name | Difference in means | Standard error | Variance | Lower limit | Upper limit | Z-Value | p-Value |
|---|---|---|---|---|---|---|---|
| Akeda et al 2017 | -43.000 | 7.691 | 59.152 | -58.074 | -27.926 | -5.591 | 0.000 |
| Monfett et al 2016 | -16.600 | 4.888 | 23.890 | -26.180 | -7.020 | -3.396 | 0.001 |
| Navani et al 2018 | -40.120 | 5.235 | 27.406 | -50.381 | -29.859 | -7.664 | 0.000 |
| Kirchner and Anitua 2016 | -76.000 | 2.183 | 4.767 | -80.279 | -71.721 | -34.807 | 0.000 |
| Levi et al 2016 | -26.600 | 6.217 | 38.647 | -38.784 | -14.416 | -4.279 | 0.000 |
| | -40.631 | 14.001 | 196.032 | -68.073 | -13.190 | -2.902 | 0.004 |

Heterogeneity

| Q-Value | df(Q) | p-value | I-squared | Tau squared | Tau |
|---|---|---|---|---|---|
| 187.85 | 4 | 0.0001 | 97.782 | 949.782 | 30.817 |

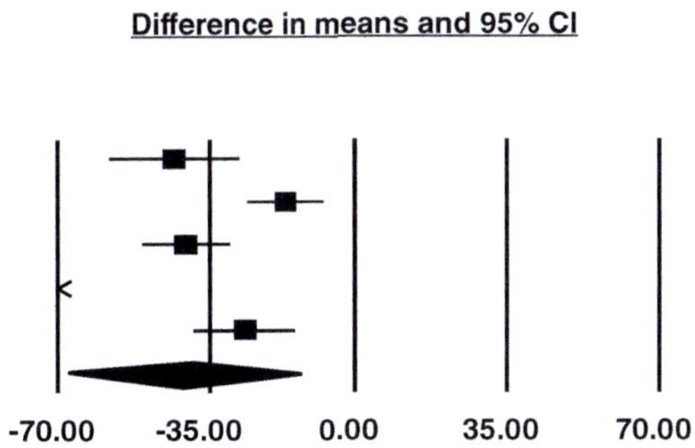


**Fig. 23.3** Decreased pain score (numerical rating scale or visual analog scale, 0–100) after treatment (6-month follow-up data) of lumbar disc injections of PRP. **Source:** Sanapati J, Manchikanti L, Atluri S, et al. Do regenerative medicine therapies provide long-term relief in chronic low back pain: A systematic review and metaanalysis. *Pain Physician* 2018; 21:515–550 [66]

**Study name** — **Statistics for each day**

| Study name | Difference in means | Standard error | Variance | Lower limit | Upper limit | Z-Value | p-Value |
|---|---|---|---|---|---|---|---|
| Akeda et al 2017 | -46.000 | 8.580 | 73.616 | -62.816 | -29.184 | -5.361 | 0.000 |
| Monfett et al 2016 | -21.900 | 4.936 | 24.629 | -31.627 | -12.173 | -4.413 | 0.000 |
| Navani et al 2018 | -43.230 | 4.674 | 21.847 | -52.391 | -34.069 | -9.249 | 0.000 |
| | -36.408 | 8.114 | 65.839 | -20.505 | -20.505 | -4.487 | 0.000 |

Heterogeneity

| Q-Value | df(Q) | p-value | I-squared | Tau squared | Tau |
|---|---|---|---|---|---|
| 11.711 | 2 | 0.003 | 77.166 | 82.923 | 12.653 |

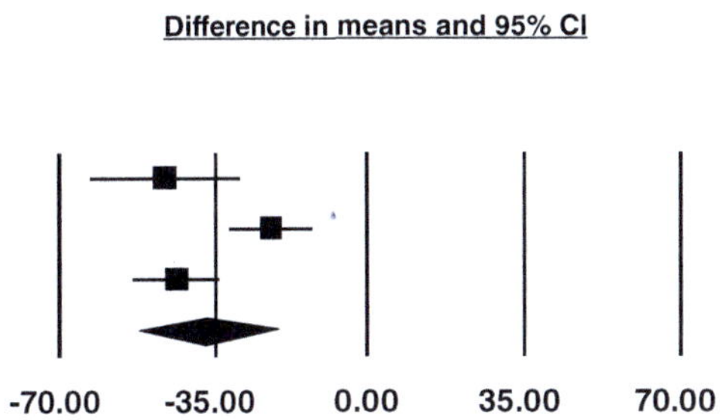


**Fig. 23.4** Pain scores (numerical rating scale or visual analog scale, 0–100) after treatment (12-month follow-up data) with lumbar disc PRP injections. **Source:** Sanapati J, Manchikanti L, Atluri S, et al. Do regenerative medicine therapies provide long-term relief in chronic low back pain: A systematic review and metaanalysis. *Pain Physician* 2018; 21:515–550 [66]

The concluded the participants who received intradiscal PRP showed no significant improvement in pain or functionality compared to the control group at 1 year follow up.

In another prospective multicenter randomized double-blind study, Zielinski et al. [76], 26 (12 men, 14 women) with a diagnosis of chronic lumbar discogenic pain, were randomly assigned to active (PRP) or control (saline) groups in a ratio of 2 active to 1 control. Baseline and follow-up Oswestry Disability Index (ODI) and Numeric Pain Rating Scale (NRS) questionnaires were obtained to track patient outcomes at 8 weeks postoperatively. The results showed within group assessment showed clinically significant improvement in 17% of PRP patients and clinically significant decline in 5% (one patient) of the active group. Clinically significant improvement was seen in 13% of placebo group patients and no placebo patients had clinically significant decline secondary to the procedure. They concluded these findings are markedly different than the highly promising results of the 2016 PRP study. This study posits necessary caution for researchers who wish to administer PRP for therapeutic benefit and may ultimately point to necessary redirection of interventional research for discogenic pain populations.

Lutz et al. [77] published a report of retrospective study of 37 patients 37 patients who received intradiscal injections of higher-concentration (>10×) PRP and had post-procedure outcomes data (visual numerical scale pain score, Functional Rating Index [FRI], and North American Spine Society [NASS] Patient Satisfaction Index). Outcomes were compared to a historical cohort of 29 patients who received intradiscal injections of <5× PRP. The results showed pain and FRI scores significantly improved by $3.4 \pm 2.5$ and $46.4 \pm 27.6$, respectively, at $18.3 \pm 13.3$ months following intradiscal injections of >10× PRP ($p < 0.001$). These improvements were greater than those reported by the historical cohort ($1.7 \pm 1.6$ and $33.7 \pm 12.3$; $p = 0.004$ and $0.016$, respectively). Additionally, the satisfaction rate was higher in patients receiving >10× PRP compared to those receiving <5× PRP (81% vs. 55%; $p = 0.032$). They concluded that the findings from this study suggest that clinical outcomes can be optimized by using PRP preparations that contain a higher concentration of platelets. Further research is needed to continue to optimize the composition of PRP used to treat patients with lumbar disc disease.

## Intradiscal Autologous Bone Marrow Cells

Stem cell therapy with bone marrow aspirate (BMA) has been studied more frequently than PRP with multiple publication including preclinical studies [31, 63–66, 78–89].

Amirdelfan et al. [87] in a multicenter, randomized, controlled study evaluated the safety and efficacy of a single intradiscal injection of STRO-3+ adult allogeneic mesenchymal precursor cells (MPCs) combined with hyaluronic acid (HA) in subjects with chronic low back pain associated with degenerative disc disease through 36-month follow-up. The results showed there were significant differences between the control and MPC groups for improvement in Visual Analog Scale (VAS) and ODI. The post-treatment interventions (PTI)-corrected VAS and ODI Means and Mean Change analyses; the proportion of subjects with VAS ≥30% and ≥50% improvement from baseline; absolute VAS score ≤20; and ODI reduction ≥10 and ≥15 points from baseline showed MPC therapy superior to controls at various time points through 36 months. Additionally, the proportion of subjects achieving the minimally important change and clinically significant change composite endpoints for the MPC groups was also superior compared with controls at various time points from baseline to 36 months. There were no significant differences in change in MP score from baseline across the groups. There were also no statistically significant differences in change in modified MP score at the level above or below the level treated between study arms. Both the procedure and treatment were well tolerated and there were no clinical symptoms of immune reaction to allogeneic MPCs. They concluded that the results provide evidence that intradiscal injection of MPCs could be a safe, effective, durable, and minimally invasive therapy for subjects who have chronic low back pain associated with moderate degenerative disc disease.

Atluri et al. [88] evaluated the effectiveness of autologous bone marrow MSCs in the treatment of chronic low back pain due to severe lumbar spinal degeneration with involvement of multiple structures. The treatment group patients received a one-time bone marrow concentrate injection into spinal structures (i.e., discs, facets, spinal nerves, and sacroiliac joints), along with conventional treatment, whereas the control group received conventional treatment with nonsteroid antiinflammatory drugs, over-the-counter drugs, structured exercise programs, physical therapy, spinal injections and opioids, etc., as indicated. Outcomes were assessed at 1, 3, 6, and 12 months. The results showed significant improvement was achieved in functional status measured by ODI, pain relief measured by NRS-11, and other parameters measured by European Quality of Life 5 Dimensions 3 Level Version (EQ-5D-3L), global mental health (GMH), and global physical health (GPH), in the study group relative to the control group at all time periods. The results showed significant improvements at 12-month follow-up with 67% of the patients in the study group achieving minimal clinically important difference (MCID) utilizing ODI when compared to 8% in the control group. Greater than 2-point pain reduction was seen in 74% of the patients at 3 months, 66% of the patients at 6 months, and 56% of the patients at 12 months. Both MCID and pain relief of 2 points were significantly different compared to the control group. Opioid use decreased in the investigational group, whereas, there was a slight increase in the control group.

Wolff et al. [89] in a retrospective analysis assessed injections of concentrated bone marrow aspirate (cBMA) as treatment for discogenic pain. Thirty-three patients who received intradiscal injections of cBMA to relieve low back pain were followed up based on NRS, Oswestry Low Back Pain Disability Index (ODI), and Short Form-36 Health Survey (SF-36) scores. Patients were also subdivided into those with a pre-injection NRS > 5 and pre-injection NRS ≤ 5. The proportion of patients demonstrating at least 50% improvement (and 95% confidence intervals) from baseline at five follow-up visits for each outcome was evaluated. The results showed at least 50% improvement in NRS was observed for 13.8, 45.8, 41.1, 23.5, and 38.9% of patients across five follow-up visits, out to 1 year. When stratified by high (>5) versus low (≤5) baseline NRS scores, the values were 14.3, 45.5, 71.4, 22.2, and 44.4% among those with high baseline pain, and 13.3, 46.2, 20.0, 25.0, and 33.3% among those with low baseline pain. The 50% improvement rates across visits were 4.3, 28.6, 30.0, 22.2, and 30.8% for SF36, and 4.2, 26.7, 36.4, 55.6, and 30.8% for ODI. They concluded that intradiscal cBMA injections may be effective to reduce pain and improve function. Patients with relatively higher initial pain may have potential for greatest improvement.

MSCs have been studied in multiple clinical studies, three systematic reviews [31, 66, 67], and guidelines [6, 31, 32, 66, 67, 78–86]. In a systematic review and a single or meta-analysis, Wu et al. [31] included six studies which were eligible for the review. They reported the pooled mean difference in pain score from baseline to follow-up points of 44.2 points decreased. Meanwhile, the pooled mean difference in ODI from baseline to follow-up points was 32.2 points. There were no adverse effects. They concluded that cell-based therapy is for patients who have discogenic low back pain associated with improved pain relief and disability scores. However, they recommended that most stringently designed randomized double blind clinical trials with appropriately determined sample sizes will be needed to confirm their clinical efficacy and safety. Wu et al. [31] showed that three studies [78, 80, 84] used stem cells and another three studies used chondrocytes

[82, 83, 85]. They also showed that five studies used expanded cells [78, 82–85] and one used unexpanded cells [80]. Cells were injected into the lumbar disc with a range of from 1 to 23 ± 5 million cell dose. The mean follow-up among the six trials [78, 80, 82–85] was 22 months with a range of 12–6 months.

Clinical outcomes were assessed with NRS and VAS after treatment in six studies [78, 80, 82–85]. The pooled mean difference in pain scores from baseline to follow-up points was 44.2 points decrease. Similarly, ODI scoring was performed in only five studies [78, 80, 83–85]. The pooled mean difference in ODI from baseline to follow-up points was 32.2 points decrease.

In this systematic review, they also assessed MRI evaluation. Four trials performed pathology grading of the MRI data after intervention [82–85]. The results showed that improvements were observed in some patients with improvements in disc contour or height in one study [83]. In another study, the ratio of fluid content of the affected segments to healthy segments was low at the beginning of the treatment, and it increased at 12 months, but not at 6 months [84]. In contrast to the pain relief, autologous cultured disc-derived chondrocyte treated group showed a substantially higher normalization as 41% normal fluid content compared with only 25% normal content in the control group [85]. Another study [82] showed no worsening of intervertebral disc up to the final follow-up at 3 years.

Sanapati et al. [66] in a systematic review with single arm meta-analysis effectiveness of MSCs for disc injection. The evaluation showed Level III evidence on a scale of I to V based on best evidence synthesis with inclusion of one RCT [86] and multiple observational studies [32, 80, 83–86] as shown in Table 23.2, Figs. 23.5 and 23.6 show the changes in pain scores and ODI with injection of MSCs into the discs.

**Table 23.2** The characteristics and outcomes of the studies of stem cell therapy in disc degeneration

| Study details | Population | Cell/solution type | Cell or solution dose and delivery pathway | Outcome parameters | Results | Conclusion |
|---|---|---|---|---|---|---|
| Noriega et al., 2017 [86]<br>Sample size = 24<br>Follow-up = 12 months<br>RCT | 24 patients with chronic low back pain with lumbar disc degeneration and unresponsive to conservative treatments were randomized into two groups<br>Patient age (years) mean age ± SE = 38 ± 2 s | Allogeneic bone marrow MSCs by intradiscal injection or a sham infiltration of paravertebral musculature with anesthetic | The intervention group received allogeneic bone marrow MSCs by intradiscal injection of 25 × 10, 6 cells per segment under local anesthesia | VAS, ODI, MRI, SF-12 | • MSC-treated patients displayed a quick and significant improvement in all algofunctional indices versus the controls<br>• Both lumbar pain and disability were significantly reduced at 3 months and improvement was maintained at 6 and 12 months. Overall there was an average 28% improvement in pain and disability 1-year after the intervention<br>• Only 5 of the 12 outcomes in patients (40%) receiving MSCs were described as perfect treatment with 100% improvement | • 28% improvement in all patients<br>• 40% of patients perfect result<br>• Positive result |

(continued)

**Table 23.2** (continued)

| Study details | Population | Cell/solution type | Cell or solution dose and delivery pathway | Outcome parameters | Results | Conclusion |
|---|---|---|---|---|---|---|
| Pettine et al., 2015, 2016, 2017 [79–81]<br>Sample size = 26<br>Follow-up = 3 years<br>Prospective, open-label, non-randomized, two-arm study | 26 patients presented with symptomatic moderate to severe discogenic low back pain<br>Patient age (years) = 18–61 years (median 40) | Autologous bone marrow concentration (nonexpanded) | 2–3 mL of bone marrow concentrate was injected in lumbar disc (1.66_106/mL) | ODI, VAS, and MRI | • The average ODI and VAS scores were reduced to 22.8 and 24.4 at 3 months. After 36 months, 6 patients proceeded to surgery<br>• After 36 months, 20 of the 26 patients reported average ODI and VAS improvement to 17.5 ± 32 and 21.9 ± 4.4 respectively<br>• One year MRI indicated 40% of patients improved one modified Pfirrmann Grade and no patient worsened radiographically | • At 36-month follow-up, 6 of 26 patients progressed to surgery. The remaining 20 patients (77%) reported significant ODI and VAS improvements<br>• Authors concluded that there were no adverse effects and the study provided evidence of safety and feasibility of intradiscal BMC therapy |
| Coric et al., 2013 [83]<br>Sample size = 15<br>Follow-up = 1 year<br>Prospective cohort | 15 patients with single-level, symptomatic lumbar DDD from L-3 to S-1 and medically refractory low back pain<br>Patient age (years) = 19–47 years (median 40) | Expanded allogeneic juvenile chondrocyte cells | Mean 1.3 mL (1–2 mL, 107/mL) cells solution was injected in the center of the disc space | ODI and NRS scores, 36-item Short Form Health Survey and MRI | • The mean ODI, NRS, and SF-36 physical component summary scores all improved significantly from baseline<br>• Ten (77%) of these 13 patients exhibited improvements on MRI. Of these, the HIZ was either absent or improved in 8 patients (89%) by 6 months<br>• Of the 10 patients who exhibited radiological improvement at 6 months, findings continued to improve or were sustained in 8 patients at the 12-month follow-up<br>• Only 3 patients (20%) underwent total disc replacement by the 12-month follow-up due to persistent, but not worse than baseline, LBP | • The results of this prospective cohort are promising with 77% of patients improving<br>• Positive result |

**Table 23.2** (continued)

| Study details | Population | Cell/solution type | Cell or solution dose and delivery pathway | Outcome parameters | Results | Conclusion |
|---|---|---|---|---|---|---|
| Orozco et al., 2011 [84]<br>Sample size = 10<br>Follow-up = 1 year<br>Pilot phase 1 trial | 10 patients with degenerative disc disease and persistent low-back pain (>6 months; decrease of disc height >50%;)<br>Patient age (years) = 35_7 (mean SD) | Autologous expanded bone marrow–derived mesenchymal stem cells | 23 ± 5 × 106 autologous expanded BMSCs was injected into the nucleus pulposus area | ODI and VAS scores and MRI | • Patients exhibited rapid improvement of pain and disability (85% of maximum in 3 months) that approached 71% of optimal efficacy<br>• This study confirmed feasibility and safety with identification of strong indications of clinical efficacy | • Authors concluded that MSC therapy may be a valid alternative treatment for chronic back pain caused by degenerative disc disease<br>• They also concluded that advantages over current gold standards include simpler and more conservative intervention without surgery, preservation of normal biomechanics, and same or better pain relief |
| Kumar et al., 2017 [32]<br>Sample size = 10<br>Follow-up = 1 year<br>Phase 1 study | 10 patients with chronic low back pain lasting for more than 3 months with a minimum intensity of 4/10 on a visual analog scale and disability level ≥30% on the Oswestry Disability Index<br>Patient age (years) = between 19 and 70 | Combined hyaluronic acid derivative and adipose-tissue derived mesenchymal stem cells (AT-MSCs) | A single intradiscal injection at a dose of 2 × 107 cells/disc (N = 5) or 4 × 107 cells/disc (N = 5) | VAS, ODI, Short-form 36, lumbar spine x-ray, MRI | • VAS, ODI, and SF-36 scores significantly improved in both groups receiving both low and high cell doses, and did not differ signi ficantly between the two groups<br>• At 12-month follow-up 7 patients reported 50% or greater im provement in VAS<br>• 6 patients achieved treatment success with pain reduction of 50% or greater and improvement on disability scores on ODI<br>• Among 6 patients who achieved significant improvement in VAS, ODI, and SF-36, 3 patients were determined to have increased water content based on an increased apparent diffusion coefficient on diffusion MRI | • 60% significant improvement with no adverse effect<br>• Authors concluded that combined implantation of AT-MSCs and hyaluronic acid derivative in chronic discogenic low back pain is safe and tolerable<br>• Positive result |

(continued)

**Table 23.2** (continued)

| Study details | Population | Cell/solution type | Cell or solution dose and delivery pathway | Outcome parameters | Results | Conclusion |
|---|---|---|---|---|---|---|
| Mochida et al. [82]<br>Sample size = 9<br>Follow-up = 3 years<br>Prospective clinical study | 9 patients with Pfirrmann grade III disc degeneration and posterior lumbar intervertebral fusion<br>Patient age (years) = 20–29 years | Autologous cultured nucleus pulposus chondrocytes that cocultured with MSCs | One million activated autologous NP cells were injected into the degenerated disc 7 days after fusion surgery | JOA and MRI | • Clinical outcomes based on Japanese Orthopedic Association (JOA) scoring system for low back pain showed significant improvement from 14.2 ± 4.8 points preoperatively to 27.2 ± 1.6 points at 3 years after transplantation of the activated NP cells (maximum possible score of 29 points)<br>• The JOA scoring system also showed improvement in low back pain subscale from 1.2 ± 0.5 points preoperatively to 2.7 ± 0.2 points at 3 years after the transplantation with maximum possible score of 3 points for no pain<br>• No adverse effects were observed during the 3-year follow-up period | • Significant improvement in function and pain scores was reported<br>• This study confirmed the safety of activated NP cell transplantation, and the findings suggest the minimal efficacy of this treatment to slow the further degeneration of human intervertebral discs |

**Table 23.2** (continued)

| Study details | Population | Cell/solution type | Cell or solution dose and delivery pathway | Outcome parameters | Results | Conclusion |
|---|---|---|---|---|---|---|
| Meisel et al., 2006 [85]<br>Sample size = 12<br>Follow-up = 2 years | Patients with discogenic pain after repeat discograms. Patients were treated with cell therapy at least 3 months post the endoscopic<br>Patient age (years) = 18–75 years | Autologous cultured disc-derived chondrocytes (from surgical treatment of their disc prolapse) | Cells are injected into disc approximately 12 weeks following discectomy. The cell dose was not mentioned | ODI and VAS scores and MRI | • The median total Oswestry Score was 2 in the autologous disc chondrocyte transplantation (ADCT) group compared with 6 in the control group. Decreases in the Disability index in autologous disc chondrocyte transplantation (ADCT)-treated patients correlated with the reduction of low back pain<br>• Decreases in disc height over time were only found in the control group, and of potential significance, intervertebral discs in adjacent segments appeared to retain hydration when compared to those adjacent to levels that had undergone discectomy without cell intervention | • Significant improvement<br>• Positive result |

*BMSCs* Bone marrow derived stem cells, *HUCMSC* human umbilical cord tissue–derived mesenchymal stem cells, *MSCs* mesenchymal stem cells, *JOA* Japanese Orthopedic Association, *MRI* magnetic resonance imaging, *NP* nucleus pulposus, *DDD* degenerative disc disease, *BMC* bone marrow concentrate, *LBP* low back pain, *IND* investigational new drug, *NRS* Numerical Rating Scale, *ODI* Oswestry Disability Index, *VAS* Visual Analog Scale, *SD* standard deviation

| Study name | Statistics for each day | | | | | | |
|---|---|---|---|---|---|---|---|
| | Difference in means | Standard error | Variance | Lower limit | Upper limit | Z-Value | p-Value |
| Coroc D 2013 | -22.330 | 1.030 | 1.061 | -24.349 | -20.311 | -21.680 | 0.000 |
| Kumar H 2017 | -36.000 | 6.609 | 43.679 | -48.953 | -23.047 | -5.447 | 0.000 |
| Meisel HJ 2006 | -41.450 | 8.509 | 72.403 | -58.127 | -24.773 | -4.871 | 0.000 |
| Noreiga DC 2017 | -20.000 | 11.768 | 138.486 | -43.065 | 3.065 | -1.700 | 0.089 |
| Orozco L 2011 | -48.900 | 7.288 | 53.115 | -63.184 | -34.616 | -6.710 | 0.000 |
| Pettine K, 2015 | -53.400 | 7.920 | 62.726 | -68.923 | -37.877 | -6.742 | 0.000 |
| | -36.943 | 6.588 | 43.407 | -49.855 | -24.030 | -5.608 | 0.000 |

| Heterogeneity Q-Value | df(Q) | p-value | I-squared | Tau squared | Tau |
|---|---|---|---|---|---|
| 35.678 | 5 | 0.000 | 85.989 | 204.543 | 14.302 |

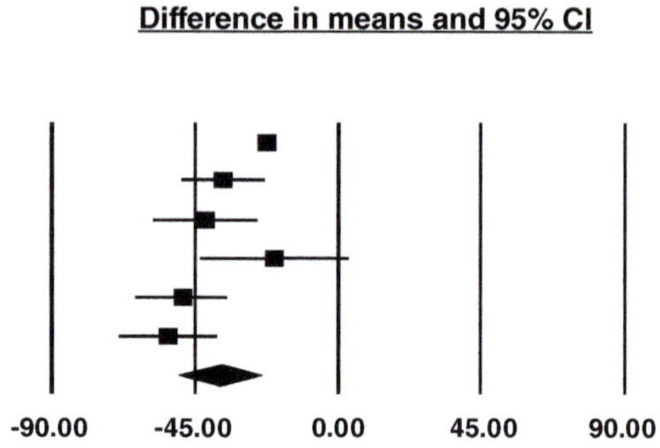


**Fig. 23.5** Changes in pain score (numerical rating scale or visual analog scale, 0–100) after treatment (12 months follow data) of cell therapy of lumbar disc. **Source:** Sanapati J, Manchikanti L, Atluri S, et al. Do regenerative medicine therapies provide long-term relief in chronic low back pain: A systematic review and metaanalysis. *Pain Physician* 2018; 21:515–550 [66]

| Study name | Statistics for each study | | | | | | |
|---|---|---|---|---|---|---|---|
| | Difference in means | Standard error | Variance | Lower limit | Upper limit | Z-Value | p-Value |
| Coric D 2013 | -27.649 | 0.555 | 0.308 | -28.737 | -26.561 | -49.818 | 0.000 |
| Kumar H 2017 | -26.000 | 5.672 | 32.172 | -37.117 | -14.883 | -4,584 | 0.000 |
| Meisel HJ 2006 | -41.190 | 7.438 | 55.324 | -55.768 | -26.612 | -5.538 | 0.000 |
| Noriega DC 2017 | -12.000 | 9.523 | 90.688 | -30.665 | 6.665 | -1.260 | 0.208 |
| Orozco L 2011 | -17.600 | 4.699 | 22.081 | -26.810 | -8.390 | -3.745 | 0.000 |
| Pettine KA 2015 | -33.900 | 8.393 | 70.442 | -50.350 | -17 450 | -4.039 | 0.000 |
| | -26.342 | 3.070 | 9.425 | -32.359 | -20.325 | -8.580 | 0.000 |

| Heterogeneity Q-Value | df(Q) | p-value | I-squared | Tau squared | Tau |
|---|---|---|---|---|---|
| 11.1922 | 5 | 0.048 | 55.324 | 26.461 | 5.144 |

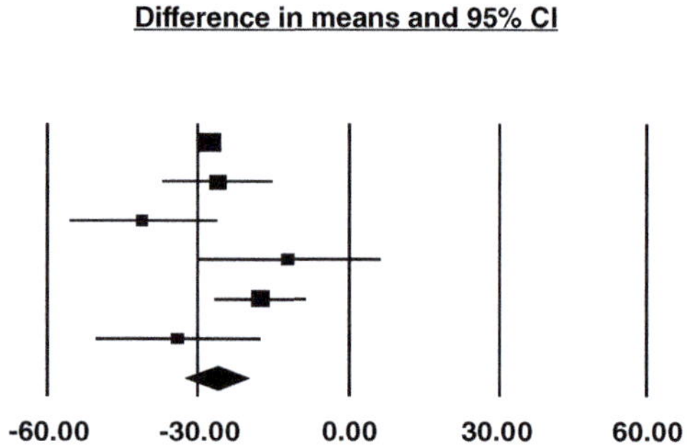


**Fig. 23.6** Changes in Oswestry Disability Index (ODI) after treatment (12 months follow data) of cell therapy of lumbar disc. **Source:** Sanapati J, Manchikanti L, Atluri S, et al. Do regenerative medicine therapies provide long-term relief in chronic low back pain: A systematic review and metaanalysis. *Pain Physician* 2018; 21:515–550 [66]

## Indications

Biologic or cell-based therapy is recommended in patients after failure of conservative management as follows:

- Unremitting spinal pain, with or without extremity pain, of greater than 3 months duration.
- Pain unresponsive to all appropriate methods of conservative therapy, including fluoroscopically directed epidural injections.
- Before cell-based disc injections, affected patients should have undergone investigations with other modalities that have failed to explain the source of pain; such modalities should include, but not be limited to, either computed tomography (CT) scanning, MRI, or myelography. In these circumstances, discography may be the only study capable of providing a diagnosis or permitting a precise description of the internal anatomy of the disc.

## Anatomy

The vertebral column is flexible because it consists of many vertebrae that are separated by resilient intervertebral discs [11, 90–95]. The 25 cervical, thoracic, lumbar, and first sacral vertebrae also articulate at synovial zygapophyseal joints. The spinal column typically consists of 33 vertebrae. The mobile section of the spine comprises 7 cervical, 12 thoracic, and 5 lumbar vertebrae, whereas inflexible sacrum which offers a relatively rigid connection to the innominate bones is formed by 5 fused vertebrae (Fig. 23.7). Below the sacrum, the coccyx is composed of four or five irregular ossicles [94].

Even though each of the 33 vertebrae is unique, most of the vertebrae demonstrate characteristic features identifying them as belonging to one of the five regions of the vertebral column. The articulating surfaces of the adjacent vertebrae are connected by intervertebral discs and ligaments. The intervertebral discs provide strong attachments

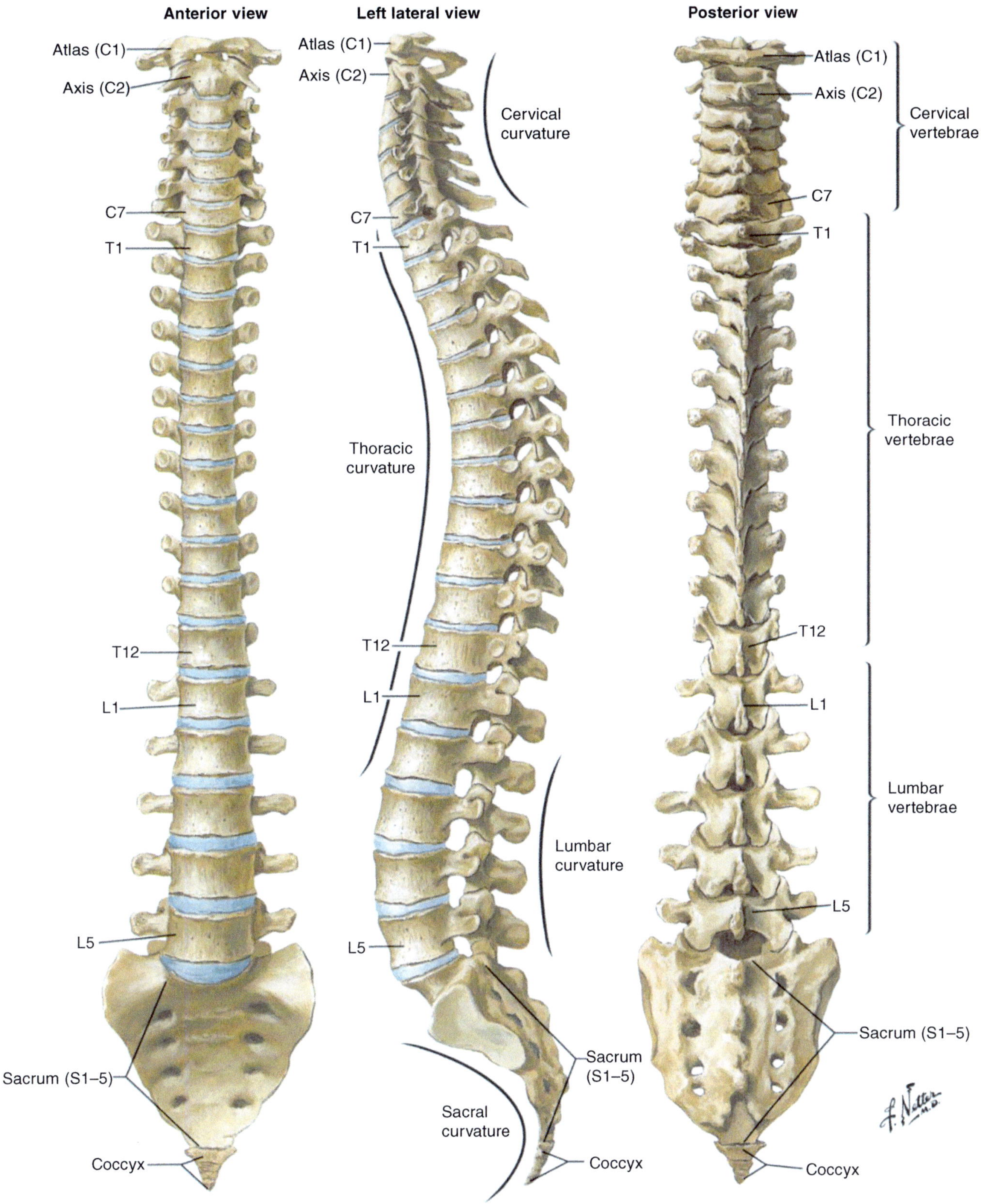


**Fig. 23.7** Illustrative anatomy of the vertebral column. Netter medical illustration used with permission of Elsevier. All rights reserved

between the vertebral bodies, uniting them into a continuous semi rigid column and forming the inferior half of the anterior border of the intervertebral foramen. Overall, the discs account for 20–25% of the length (height of the vertebral column).

Intervertebral discs also serve as shock absorbers. Each intervertebral disc consists of an annulus fibrosus, an outer fibrous part which is composed of concentric lamellae of fibrocartilage forming the circumference of the intervertebral disc, and a gelatinous central mass called the nucleus pulposus. The annulus fibrosis is a fibrous ring consisting of concentric lamellae of fibrocartilage forming the circumference of the intervertebral disc. The nucleus pulposus is the central core of the intervertebral disc. The nucleus pulposus is not centered in the disc, it is more posteriorly placed. The nucleus pulposus is avascular; it receives its nourishment by diffusion from blood vessels at the periphery of the annulus fibrosis and vertebral body. There is no intervertebral disc between C1 and C2 vertebrae, whereas the most inferior functional disc is between L5 and S1 vertebrae.

The intervertebral disc consists of three main parts: the outer annular fibrosus (AF), which consists of a series of fibrocartilaginous rings; the inner gelatinous nucleus pulposus; and the cartilaginous endplates (CEP) of hyaline-like cartilage (Fig. 23.8). The endplates are located between the bony vertebrae and other parts of the intervertebral disc.

## Technical Aspects

Discography and disc injections may be performed in lumbar, thoracic, and cervical regions [11, 91–95]. However, lumbar discography and lumbar cell-based therapies are the most studied and published. To perform cell-based disc injections, it is essential to be proficient with discography and it may be also essential to perform discography prior to injection of biologics. However, some may inject at the same time as discography without contrast injection.

## Lumbar Discography and Disc Injections

The two techniques that have been used for lumbar discography and injections are the transdural and extraspinal approaches. The extraspinal technique is the more common of the two methods used in performing lumbar discography and is described in this chapter. The transdural technique was the original way in which discography was performed but is now generally limited to patients who have undergone posterior instrumented fusions with bone grafts that prevent the extraspinal route.

The procedure is performed under absolute aseptic conditions. The patient is positioned in a prone position and is prepped and draped in a sterile manner. Intravenous sedation can be given to relax the patient as long as there is no compromise to the patient's ability to participate during the pro-

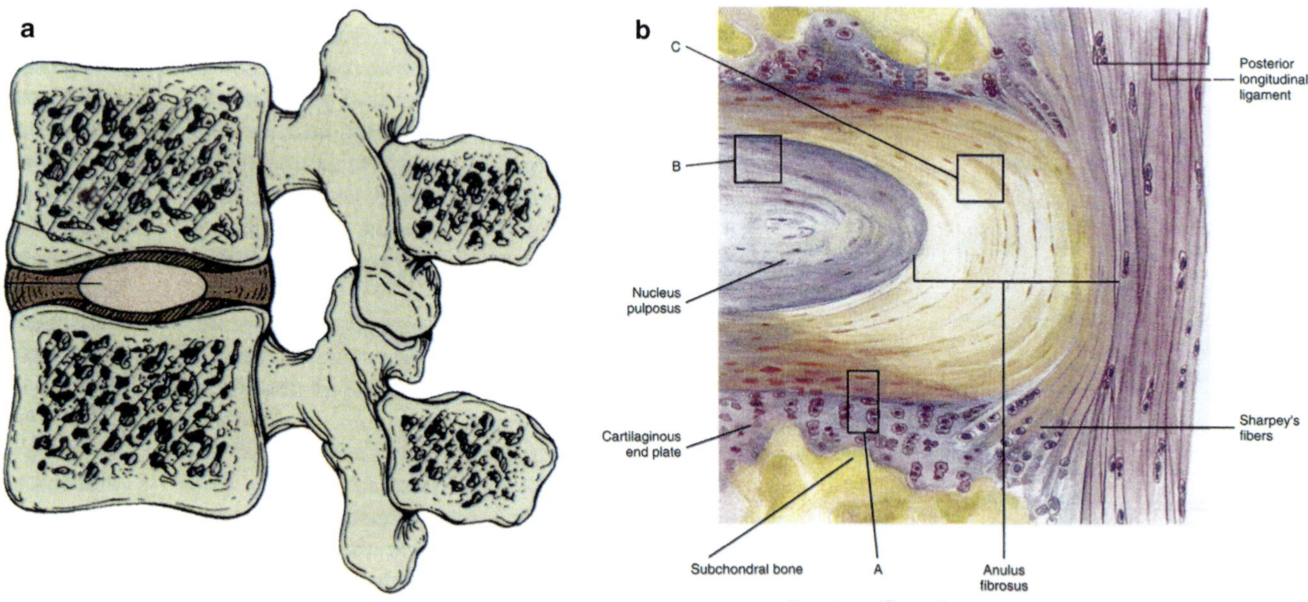


**Fig. 23.8** Anatomy of intervertebral disc. (**a**) Midsagittal section of two adjacent lumbar vertebrae and the intervertebral disc separating the two vertebral bodies. Notice the components of the intervertebral disc: annulus fibrosus, nucleus pulposus, and vertebral (cartilaginous) endplate. (**b**) Higher magnification (magnification = ×15) view showing the annulus fibrosus, nucleus pulposus, and cartilaginous endplate, the vertebral body and posterior longitudinal ligament. **Adapted from:** Manchikanti L, et al. Discography In: Manchikanti L, Kaye AD, Falco FJE, et al. (eds). *Essentials of Interventional Techniques in Managing Chronic Pain*. Springer, 2018, pp. 273–300 [91]

cedure. Heart rate, pulse oximetry, blood pressure, and respiration are monitored during the procedure. The administration of intravenous antibiotics prior to the procedure is standard as a prophylaxis against infection, particularly discitis [11, 96, 97]. However, intradiscal and oral antibiotics can be given at the discretion of the physician. With the injection of biologics, contrast injection and intradiscal antibiotic administration is discouraged because of the toxicity of these materials to biologics.

The patient may be positioned in a modified lateral decubitus position with the symptomatic side down to avoid any confusion between needle induced annular pain and a provocative pain response. This position also facilitates optimal fluoroscopic imaging and mobilizes the bowel away from the needle path.

The C-arm is adjusted to achieve a posterior oblique position with the superior articular process (SAP) dividing the vertebral body in half (Fig. 23.9).

A two-needle system is used with a longer, small gauge procedure needle that is passed through a shorter, larger gauge needle into the disc (Fig. 23.10). The introducer needle tip is positioned just anterior to the SAP and superior to the subjacent endplate. While being aware of any paresthesias from contact with the exiting nerve root at that level, the introducer needle is advanced to the outer annulus. Under biplanar fluoroscopic guidance, the procedure needle is then advanced through the introducer needle into the central third of the disc.

An alternative approach would be to use a single needle technique for all needle placements with the possible exception of the L5/S1 disc level where a two-needle method facilitates navigating past the iliac crest.

### Disc Injection

Regardless of the technique, after the needles are positioned in the disc(s), each disc is evaluated by injecting saline, or contrast, or both, in sequence. For diagnostic discography, data may be recorded during the study as to the injected volume, discometry (end point or recorded pressure), pain level, analgesia, pain quality, and nucleogram. A firm endpoint reflects a normal intact annulus. A spongy endpoint is a reflection of annular pathology with an intact annulus, and a complete annular tear through the outer annular fibers is characterized by no endpoint. Depending on the size of the patient, a normal lumbar disc accepts from 0.5 mL up to 3 mL with a firm endpoint, or high discometric pressure.

Figure 23.11 illustrates normal discograms. Figure 23.12 illustrates a step-by-step approach of lumbar discography. Figure 23.13 illustrates approach to discography at L4/L5 level. Figure 23.14 illustrates discography at L5/S1 level. However, Fig. 23.15 illustrates abnormal discographic patterns. A herniated disc with an intact annulus or a degenerative disc typically accepts 1.5–4 mL with intermediate pressure described as a spongy endpoint. Discs with complete rupture of the outer annulus can accept 3 mL or more as the injectate spreads into the epidural space with very low or varying pressure and no endpoint.

Lumbar intradiscal pressure can also be directly measured with a pressure gauge in pounds per square inch (psi) at the onset of pain or with a firm endpoint.

The disc pressure (discometry) at the onset of pain during lumbar discography can also be measured and recorded with a pressure gauge in psi. Painful lumbar discs are defined as chemically sensitive (<15 psi), mechanically sensitive (15–50 psi) or intermediate (>50 psi) depending on the measured pressure above the opening pressure [11, 91, 92].

The pain level is a reflection of the pain intensity experienced by the patient during the injection regardless if the pain is similar or not (concordant or disconcordant). The intensity is graded verbally on a scale, often using 10 as the greatest degree of pain and 0 as no pain at all. In the case of two or more contiguously painful discs, injecting discs with lidocaine is helpful in determining truly painful discs. The discs are evaluated again by re-injection to determine the lidocaine response. If a painful disc is no longer painful upon re-injection after being anesthetized, then it is a true pain generator.

Once a painful disc is anesthetized and "taken out of the equation," nearby discs can be re-injected to determine if they are really a source of pain or were actually referring pain by stimulating the previously painful disc. The pain quality is most important since it establishes whether or not pain produced during the procedure is exactly what the patient clinically experiences. Patients relate whether they have any pain at all and if the pain quality is the same as "their" pain.

The pain can be vague and discordant, just a component of their typical pain, or an exact reproduction of their concordant pain.

Upon completion of the injections, x-rays of the lumbar spine are obtained in the posteroanterior (PA) and lateral views. A normal lumbar disc nucleogram appears as a rounded or bilobular-contained component of the lumbar disc. Annular disruption appears as contrast that spreads beyond the nuclear border, typically in a radial fashion. Contrast is commonly recognized on lateral view as spreading within the posterior portion of the disc (but can also spread into the anterior annulus). Nucleogram patterns can range from normal (cotton ball and lobular) to abnormal (irregular, fissured, ruptured, and degenerative).

The Dallas Discogram classification scheme assists in assessing patients with lumbar spine pain and in the selection of patients for surgical and intradiscal procedures (Fig. 23.16). Figure 23.17 illustrates multiple images of computerized tomography discograms showing multiple grades of annular tear. Figure 23.18 illustrates discographic patterns of normal degeneration, moderate degeneration and

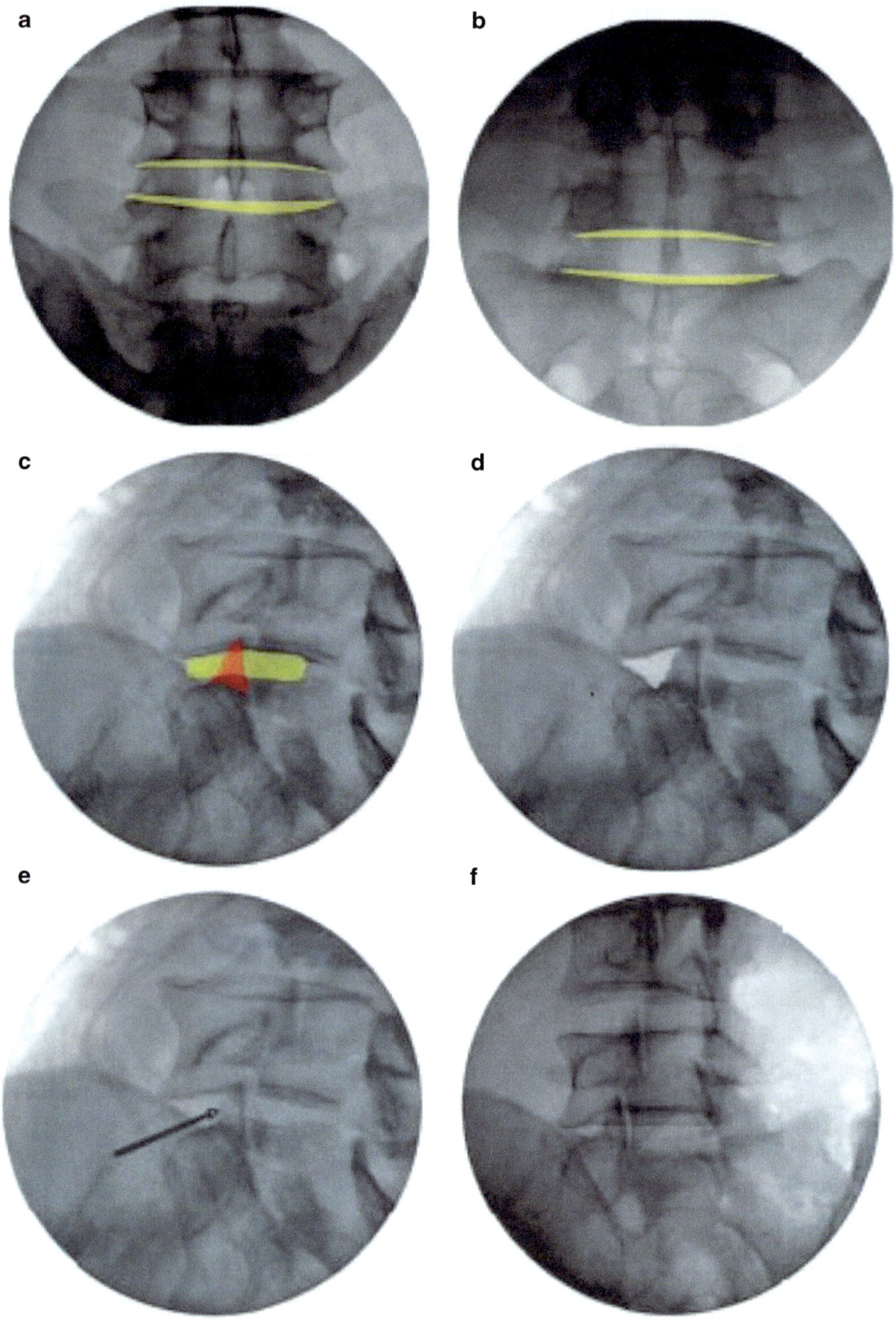


**Fig. 23.9** Fluoroscopic visualization of disc area for needle entry in performing lumbar discography. (**a**) Endplate (squared) PA view of the L4/5 disc space (endplates outlined in yellow) taken with the C-arm in a vertical position. (**b**) Endplate (squared) PA view of the L5/S1 disc space (endplates outlined in yellow) taken with the C-arm in a 45 degree cephalad tilt. (**c**) L5/S1 disc space after an oblique orientation of the disc space (yellow) with the C-arm to position the tip of the S1 superior articular process (red) in order to divide the disc space in half. (**d**) The disc space anterior to the superior articular process (white) represents the area for needle insertion when performing lumbar discography. (**e**) Needle entry position for entering the L5/S1 disc is just superior to the S1 endplate and anterior to the base of the S1 superior articular process. (**f**) Needle entry position for entering the L4/5 disc is just superior to the L5 endplate and anterior to the base of the L5 superior articular process. **Adapted from:** Manchikanti L, et al. Discography In: Manchikanti L, Kaye AD, Falco FJE, et al. (eds). *Essentials of Interventional Techniques in Managing Chronic Pain*. Springer, 2018, pp. 273–300 [91]

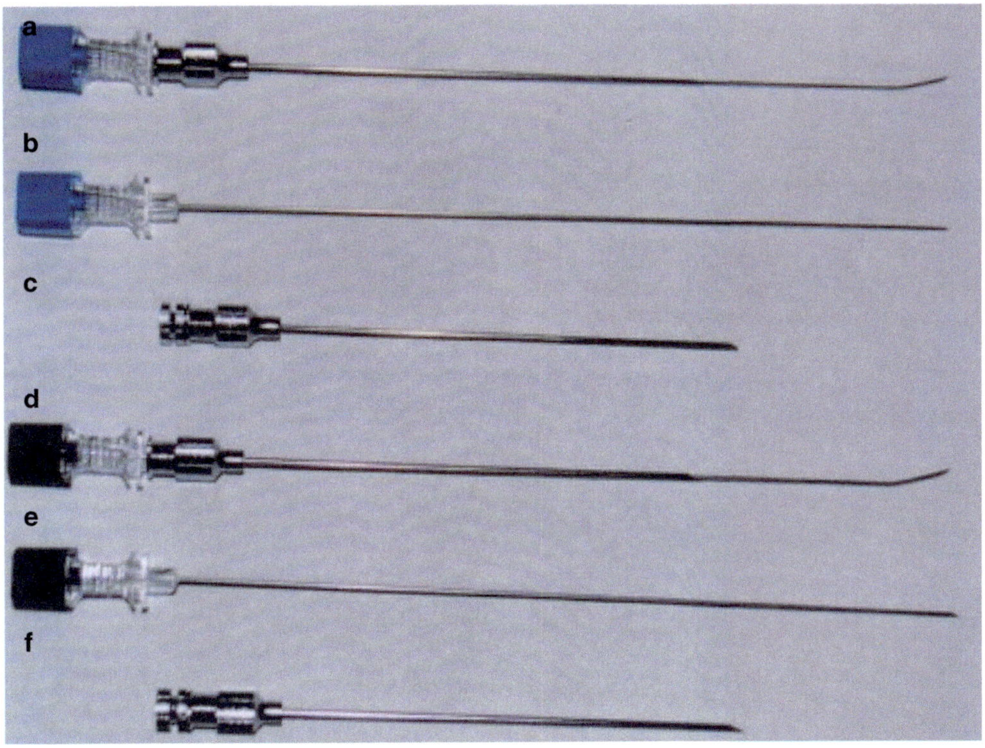


**Fig. 23.10** Discography needles. (**a**) Two needle set with a 6-inch, 25-gauge procedure needle with a distal 10° bend within a 3½-inch, 20-gauge introducer needle used for lumbar discography. (**b**) 6-inch, 25-gauge procedure needle. (**c**) 3½-inch, 20-gauge introducer needle. (**d**) Two needle set with a 6-inch, 22-gauge procedure needle with a distal 10° bend within a 3½-inch, 18-gauge introducer needle used for lumbar discography. (**e**) 6-inch, 22-gauge procedure needle. (**f**) 3½-inch, 22-gauge introducer needle. **Adapted from:** Manchikanti L, et al. Discography In: Manchikanti L, Kaye AD, Falco FJE, et al. (eds). *Essentials of Interventional Techniques in Managing Chronic Pain*. Springer, 2018, pp. 273–300 [91]

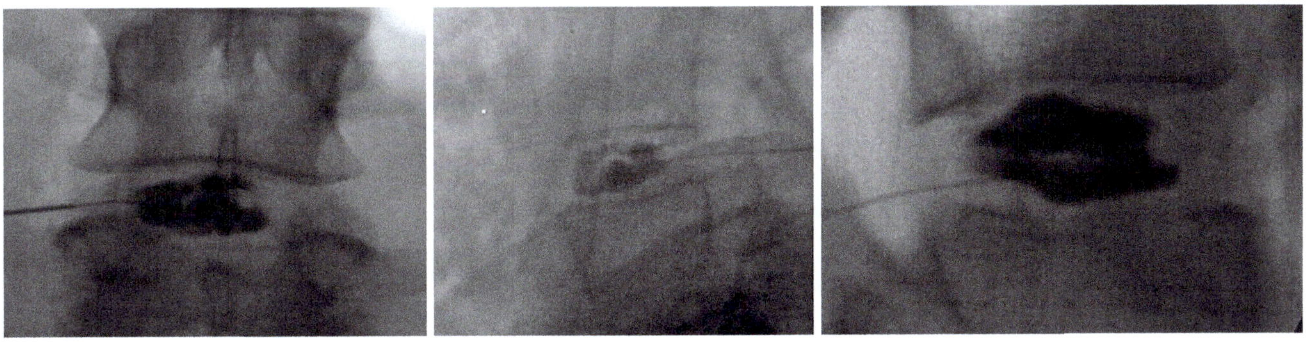

**Fig. 23.11** Illustration of normal L4/5 and L5/S1 discograms. (**a**–**c**) L4/5 Discogram (left) AP view and (right) lateral view. **Adapted from:** Manchikanti L, et al. Discography In: Manchikanti L, Kaye AD, Falco FJE, et al. (eds). *Essentials of Interventional Techniques in Managing Chronic Pain*. Springer, 2018, pp. 273–300 [91]

severe degeneration. Figure 23.19 illustrates posterior annular tears (PAT) disruption.

## Intradiscal Injection of Biologics

Injection of biologics may be performed in a separate session after diagnostic discography during which contrast and antibiotic solutions are avoided. At this time, a total of 2 mL of biologic solution may be injected. The needle may be left inside the disc for a few minutes after the injection so that the entire solution will be absorbed into the disc rather than leaking into the subarachnoid space.

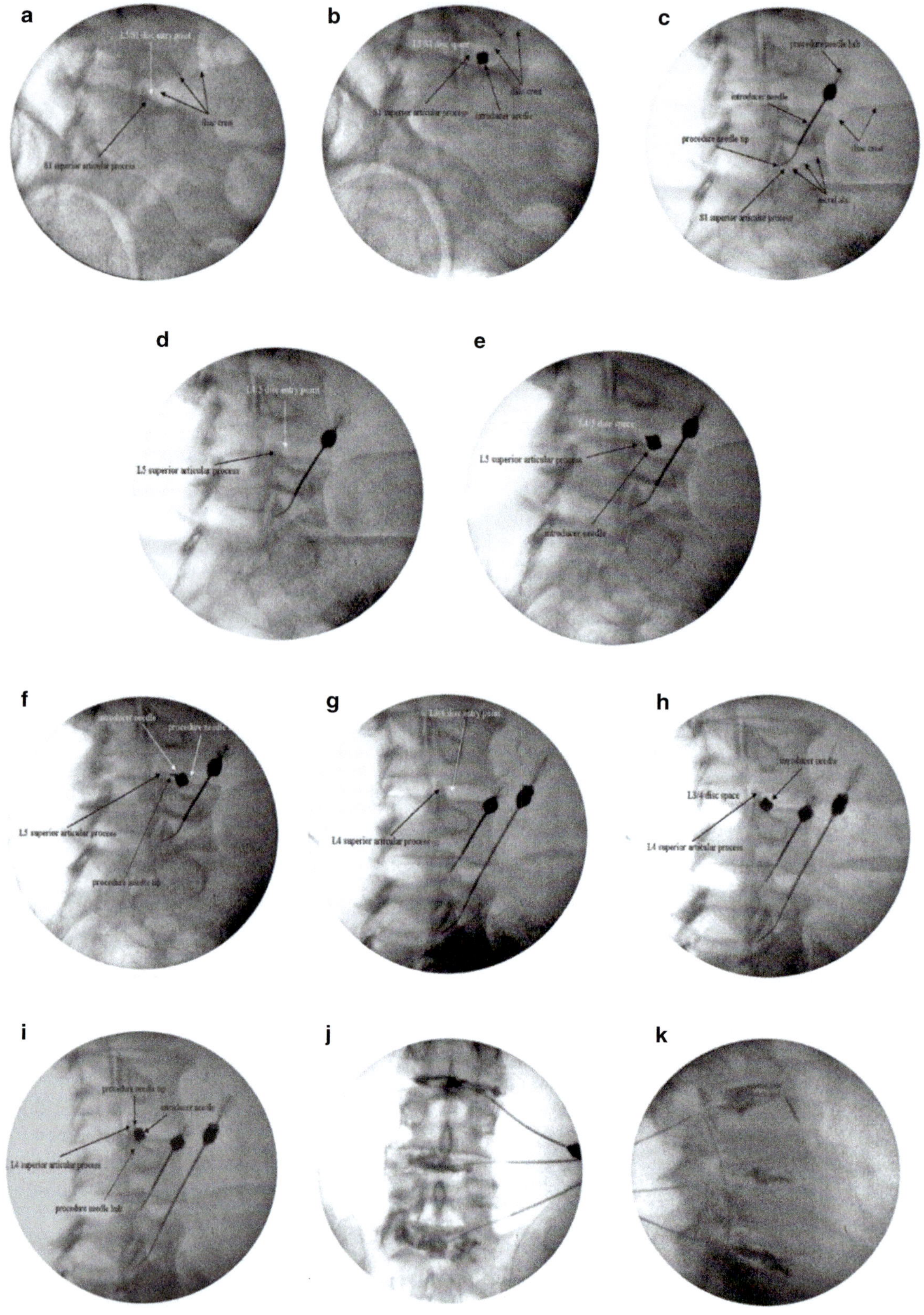

**Fig. 23.12** A step-by-step approach of lumbar discography. This figure illustrates needle placement into L3/4, L4/5 and L5/S1 disc spaces with an oblique approach. (**a**) Oblique approach to the L5/S1 disc space. (**b**) Introducer needle position for L5/S1 disc space. (**c**) Oblique projection of the lumbar spine. (**d**) Oblique approach to the L4/5 disc space. (**e**) Introducer needle position for L4/5 disc space. (**f**) Procedure needle position within L4/5 disc space. (**g**) Oblique approach to the L3/4 disc space. (**h**) Introducer needle position for L3/4 disc space. (**i**) Procedure needle position within L3/4 disc space. (**j**) Post discography anteroposterior view. (**k**) Post discography lateral view. **Adapted from:** Manchikanti L, et al. Discography In: Manchikanti L, Kaye AD, Falco FJE, et al. (eds). *Essentials of Interventional Techniques in Managing Chronic Pain*. Springer, 2018, pp. 273–300 [91]

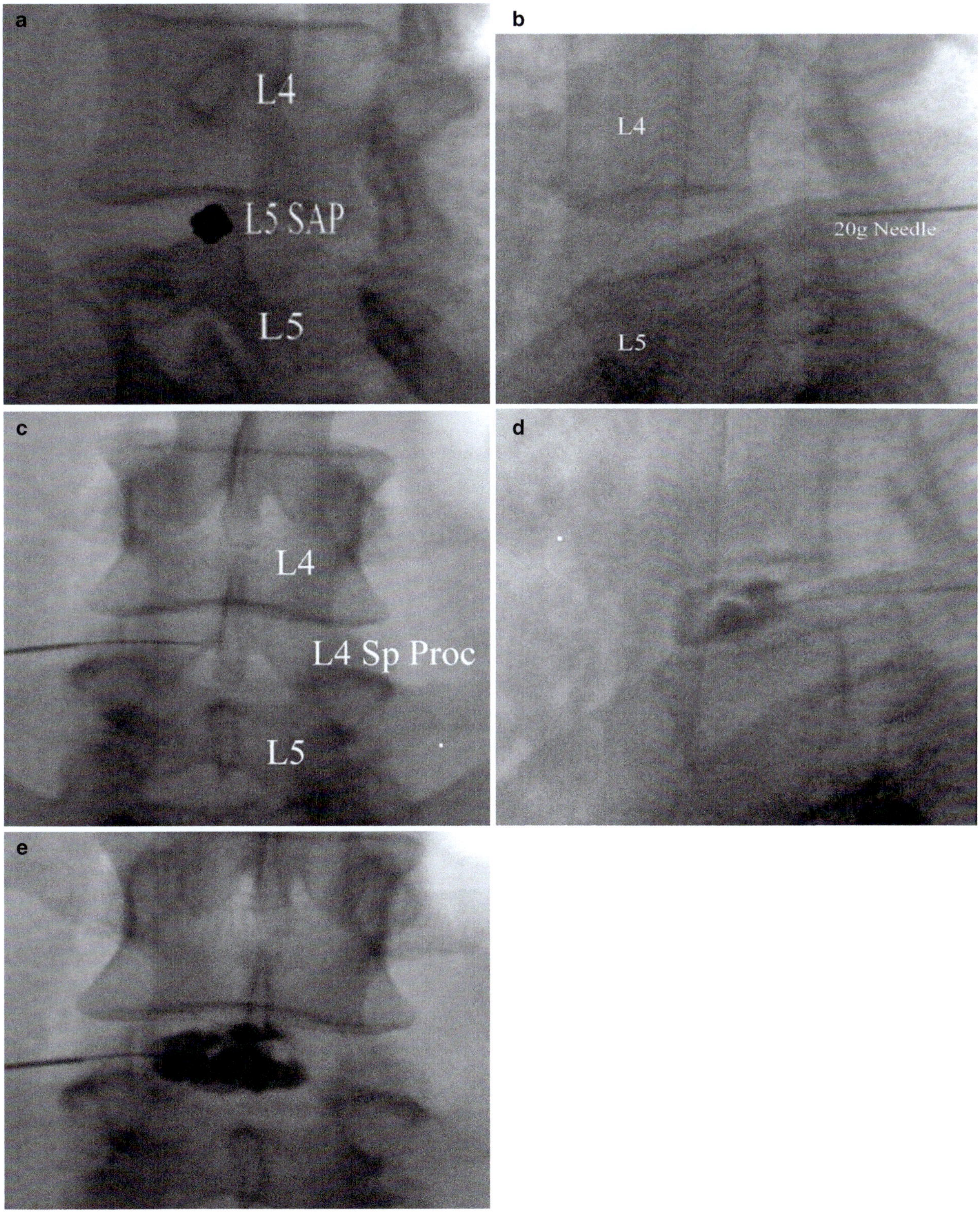


**Fig. 23.13** Illustrated approach for discography at the L4/5 disc level. (**a**) Oblique view of the L4/5 disc space with the L5 superior articular process (SAP) dividing the disc space in half and the introducer needle positioned superior to the L5 endplate and anterior to the L5 SAP. (**b**) Lateral view of the introducer needle position for the L4/5 disc at the level of the L5 SAP in route to the level of the posterolateral L5 superior endplate. (**c**) Anteroposterior view of the L4/5 disc space after the procedure needle has been passed through the introducer needle into the center of the L4/5 nucleus pulposus. (**d**) Lateral view of the L4/5 disc after contrast injection demonstrating a normal nucleus pulposus without annular disruption. (**e**) Anteroposterior view of the L4/5 disc after injection again showing a normal nucleus pulposus without annular disruption. **Adapted from:** Manchikanti L, et al. Discography In: Manchikanti L, Kaye AD, Falco FJE, et al. (eds). *Essentials of Interventional Techniques in Managing Chronic Pain*. Springer, 2018, pp. 273–300 [91]

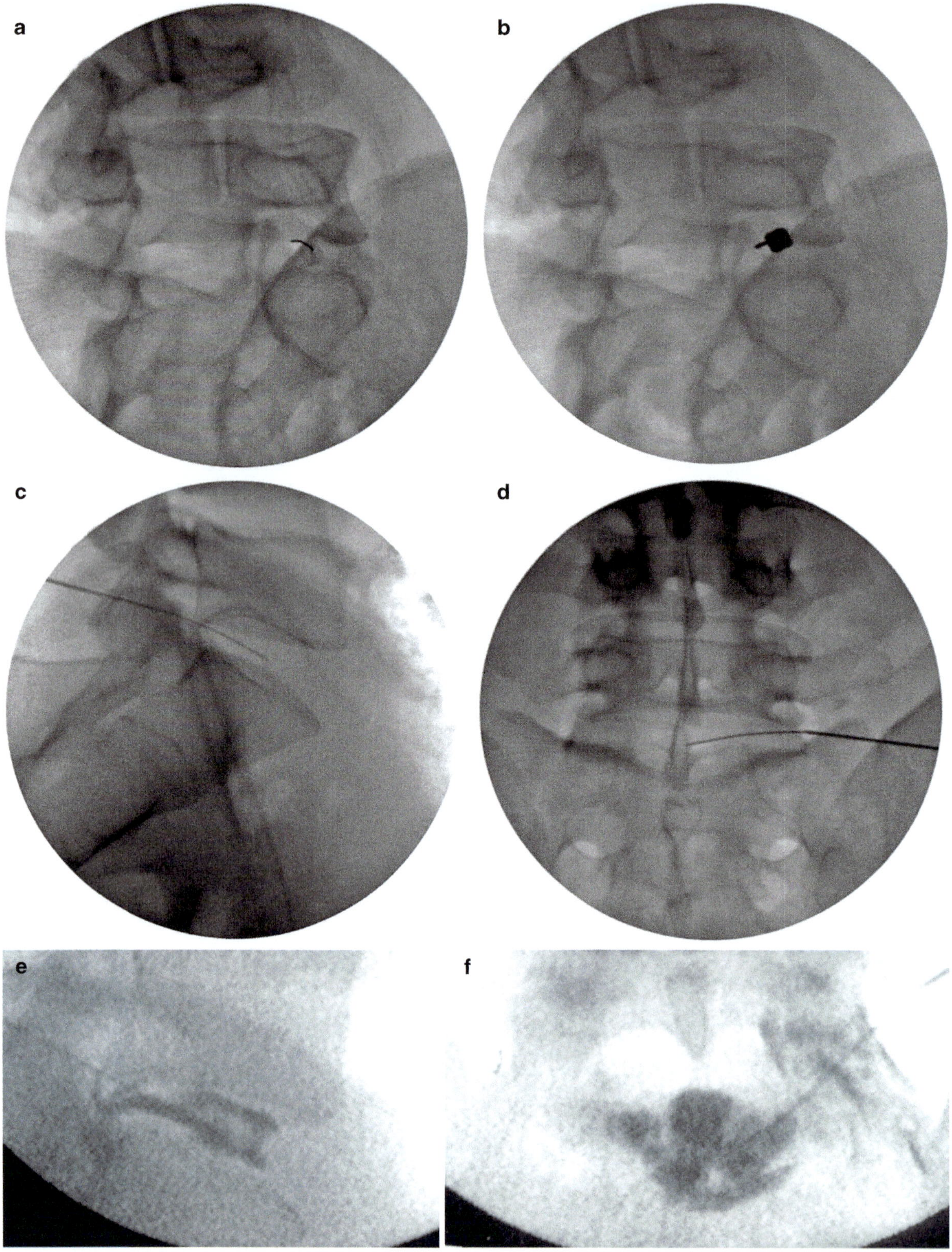

**Fig. 23.14** Illustrated approach for discography at the L5/S1 disc level. (**a**) Oblique view of the L5/S1 disc space with the L5 superior articular process (SAP) dividing the disc space in half. (**b**) Introducer needle positioned superior to the S1 endplate and anterior to the S1 SAP. (**c**) L5/S1 disc space after the procedure needle has been passed through the introducer needle into the center of the L5/S1 nucleus pulposus. (**d**) Anteroposterior view of the L5/S1 disc space after the procedure needle has been passed through the introducer needle into the center of the L5/S1 nucleus pulposus. (**e**) Anteroposterior (AP) view of the L5/S1 disc after injection shows a normal nucleus pulposus without any apparent annular disruption. (**f**) Lateral view of the L5/S1 disc after contrast injection demonstrates a normal nucleus pulposus and a centroposterior radial annular fissure/tear that was not evident on AP view due to the overlying contrast within the nucleus pulposus. **Adapted from:** Manchikanti L, et al. Discography In: Manchikanti L, Kaye AD, Falco FJE, et al. (eds). *Essentials of Interventional Techniques in Managing Chronic Pain*. Springer, 2018, pp. 273–300 [91]

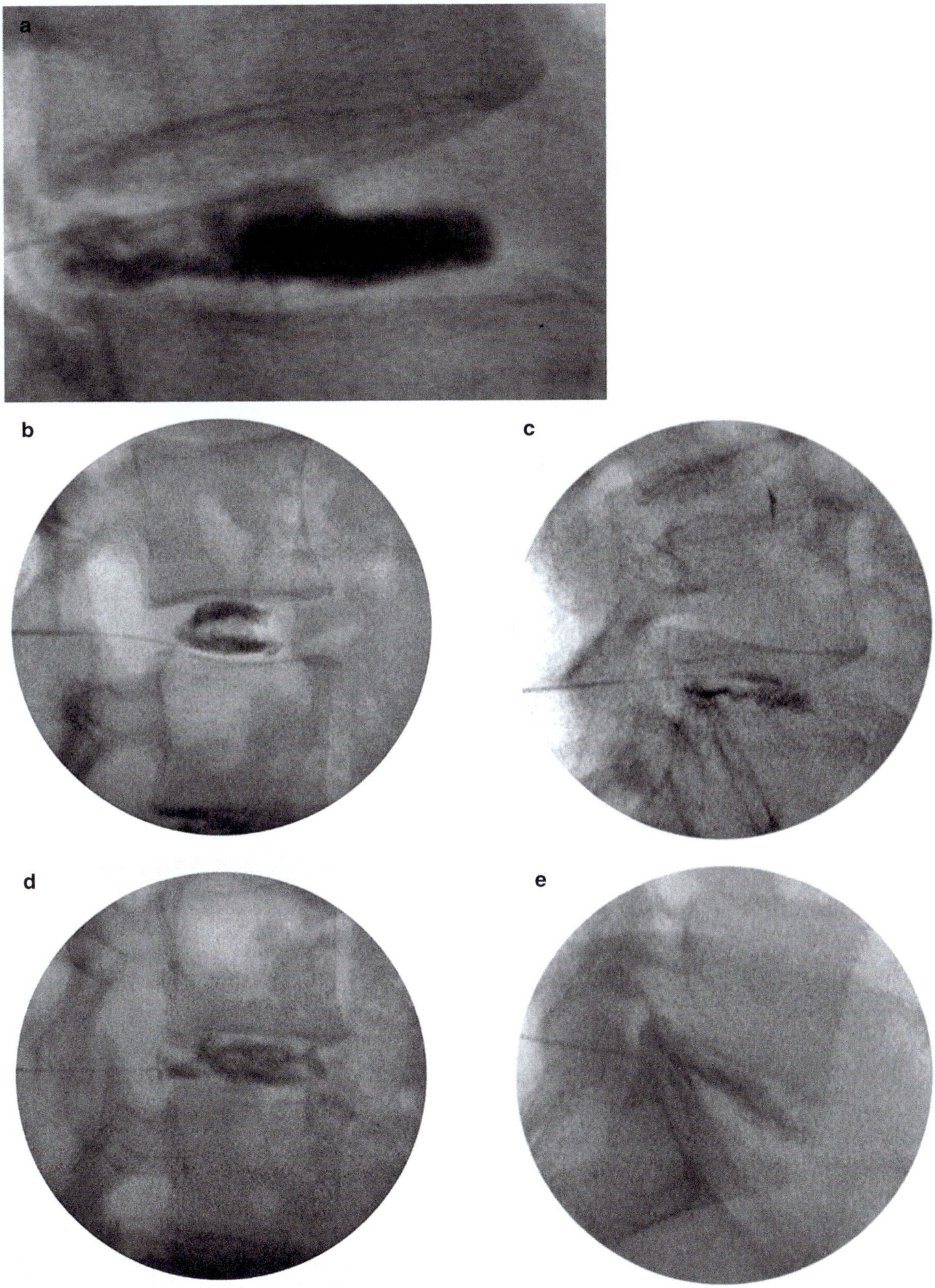


**Fig. 23.15** Illustration of abnormal discographic patterns. (**a**) Disrupted disc with posterior radial tear. (**b**) Normal bi-lobed discogram. (**c**) Grade four posterior radial tear. (**d**) Grade 5 posterior radial tear. (**e**) Severly disrupted L5/S1 disc. (**f**) L5/S1 disrupted disc. (**g**) L4/5 disrupted disc. (**h**) Grade 5 posterior radial tear. (**i**) Severe disruption. (**j**) Disrupted L4/5 with a normal L5/S1 disc. **Adapted from:** Manchikanti L, et al. Discography In: Manchikanti L, Kaye AD, Falco FJE, et al. (eds). *Essentials of Interventional Techniques in Managing Chronic Pain*. Springer, 2018, pp. 273–300 [91]

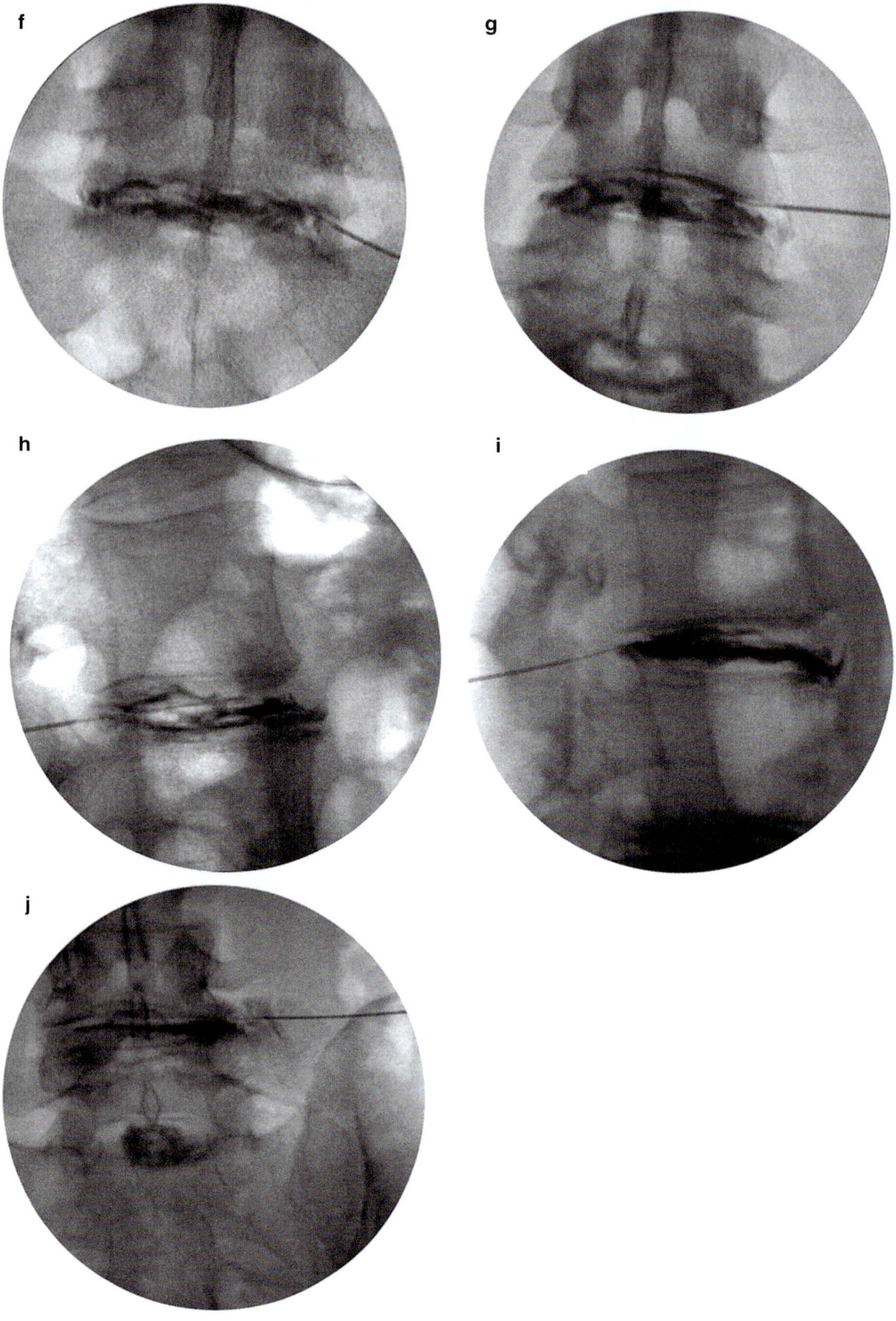

**Fig. 23.15** (continued)

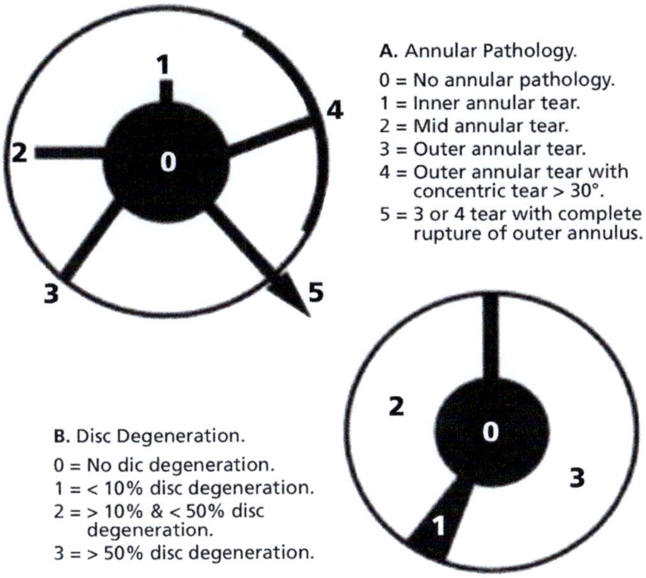


**Fig. 23.16** Modified Dallas discogram classification. (**a**) Annular pathology. 0 = No annular pathology. 1 = Inner annular tear. 2 = Mid annular tear. 3 = Outer annular tear. 4 = Outer annular tear with concentric tear >30°. 5 = 3 or 4 tear with complete rupture of outer annulus. (**b**) Disc degeneration. 0 = No disc degeneration. 1 = <10% disc degeneration. 2 = >10% and <50% disc degeneration. 3 = >50% disc degeneration. **Adapted from:** Manchikanti L, et al. Discography In: Manchikanti L, Kaye AD, Falco FJE, et al. (eds). *Essentials of Interventional Techniques in Managing Chronic Pain*. Springer, 2018, pp. 273–300 [91]

## Cervical Discography and Disc Injection

Cervical discography may be performed utilizing an original anterior approach or modified anterolateral approach [11, 91, 93]. However, the anterolateral approach is the most commonly utilized technique associated with less risk.

### Anterior Approach

The original anterior technique for performing cervical discography involves placing the patient in the supine position. The procedure is performed in a sterile atmosphere with all precautions. Intravenous antibiotics and light intravenous sedation is given to the patient prior to starting the procedure.

The C-arm fluoroscope or an overhead fluoroscopy unit is employed to visualize the cervical spine. The head and neck of the patient is placed in extension so as to widen the anterior disc space for easier access into the disc.

The patient or the fluoroscope is positioned so the spinous processes are midline and there is clear visualization of the vertebral endplates and uncinate processes.

A single 22- or 25-gauge spinal needle is used to enter each disc level for the study. A right sided approach when performing cervical discography can lessen the risk of puncturing the esophagus at the lower levels of the cervical spine where it typically is located left of midline. Barium contrast can be swallowed prior to the procedure in order to outline and avoid the esophagus during the study. The needle should enter more laterally at the C2/3 and C3/4 disc levels to avoid the hypopharynx and more medially at the C7/T1 level to avoid the apex of the lung.

The first needle is typically inserted into the most cephalad disc, followed by needle insertion into the remaining discs in a sequential and caudad direction. A metallic marker is used to locate the chosen intervertebral disc level for needle placement. The non-dominate index finger compresses the soft tissue structure underlying the marker, and at the same time displaces the great vessels laterally, and the trachea and larynx medially. The dominant hand then places the needle into the disc under continuous fluoroscopy by making contact with the endplate below the disc before "walking" the needle into the disc space. The needle tip is then carefully adjusted into the center of the disc using biplanar fluoroscopy.

The needle tip is only advanced under lateral fluoroscopy in order to prevent injury to the spinal cord.

### Anterolateral Approach

The anterolateral approach for cervical discography positions the patient supine with positioning the fluoroscope obliquely positioned to produce an optimal view of the foramen at the intended disc level (Fig. 23.20a, b). This allows for the advancement of the needle tip at an oblique angle to the disc and posterior to the great vessels, trachea, and esophagus. The needle is advanced from this orientation with intermittent fluoroscopy until contact is made with the uncinate process (Fig. 23.20c). The needle tip is then "walked off" carefully into the disc (Fig. 23.20d) and centered within the disc using biplanar fluoroscopy (Fig. 23.21a). Again, the needle tip is only advanced under lateral fluoroscopy in order to prevent spinal cord injury (Fig. 23.21c, d). The advantages of this technique are a significant reduction of x-ray exposure to the physician's hands and potentially a lower risk of disc infection for the patient.

### Disc Injection

Each disc is evaluated by injecting either saline or contrast, or both, in sequence after the needles are positioned in the discs (Fig. 23.21). Data may be recorded during the study about the injected volume, discometry (end point), pain level, analgesia, pain quality, and nucleogram [98]. However, discographic pressures have not been well studied in the cervical spine.

In the cervical spine iatrogenic disc injury may be caused at significantly lower pressures and volumes infused than in the lumbar spine.

A normal cervical disc accepts less than 0.5 mL with a firm endpoint or high discometric pressure [98]. Cervical discs with posterolateral uncinate seepage, but an intact annulus, generally accept up to 1.5 mL. Discs with outer

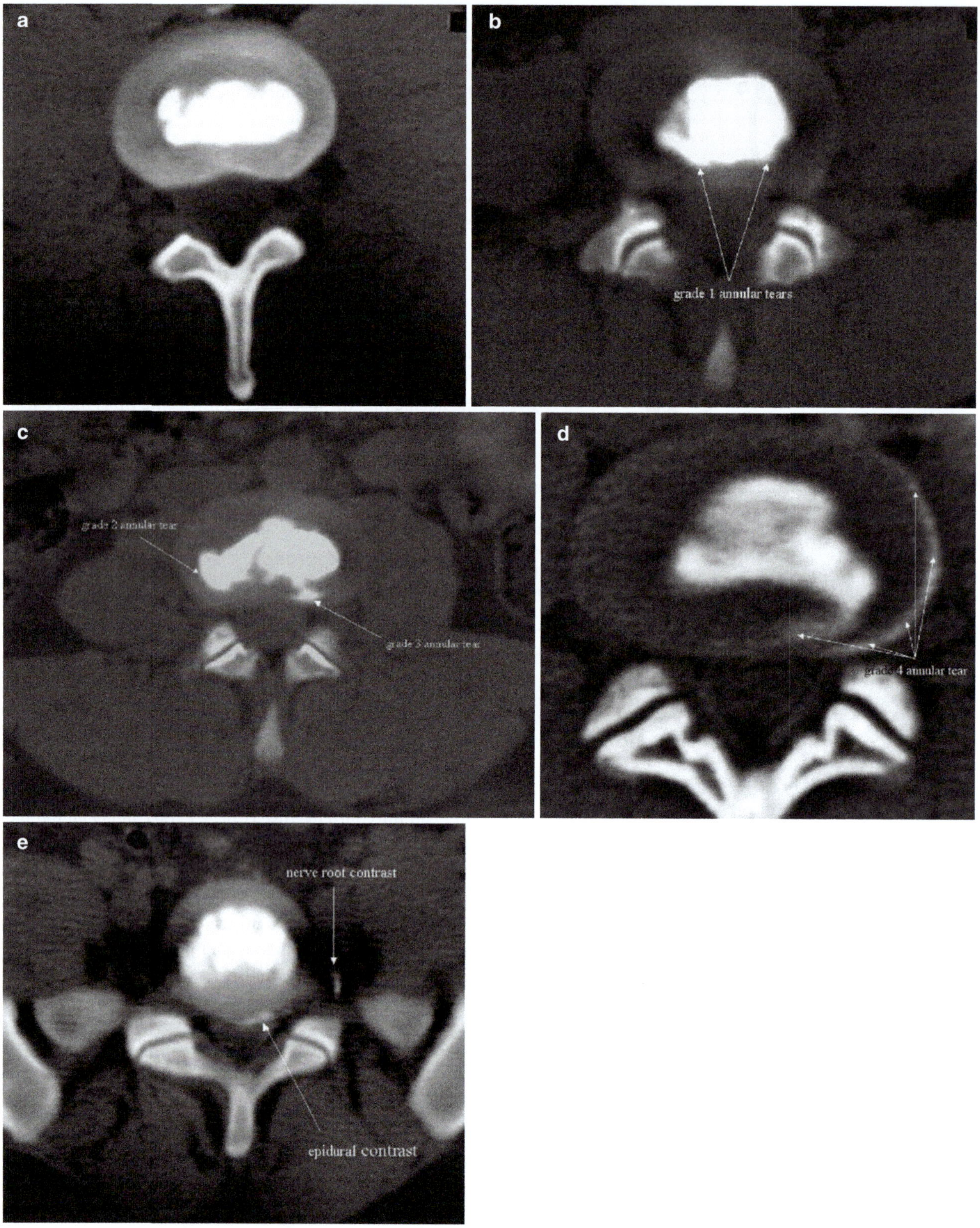


**Fig. 23.17** Illustration of various types of annular tears. (**a**) CT discogram showing a normal grade 0 annulus without any annular disruption. (**b**) CT discogram presenting grade 1 annular tears within the inner annular region. (**c**) CT discogram demonstrates grade 2 and grade 3 annulus fissures/tears. (**d**) CT-Discogram demonstrating a grade 4 annular tear. (**e**) CT discogram displaying a grade 5 annular tear with contrast outside the outer annular fibers within the anterior epidural space and along the exiting left L5 nerve root. **Adapted from:** Manchikanti L, et al. Discography In: Manchikanti L, Kaye AD, Falco FJE, et al. (eds). *Essentials of Interventional Techniques in Managing Chronic Pain*. Springer, 2018, pp. 273–300 [91]

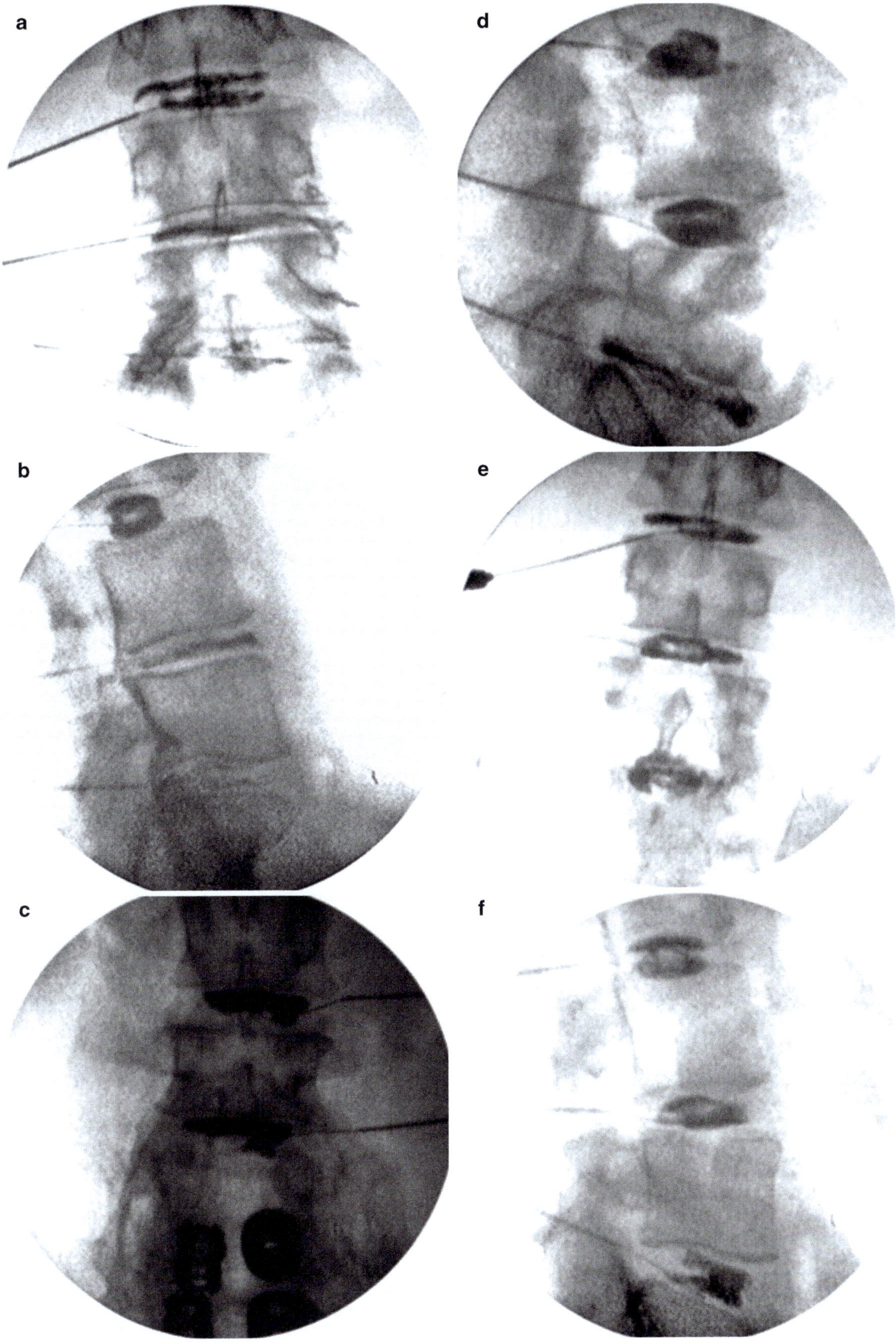

**Fig. 23.18** Illustration of discographic patterns of normal degeneration, moderate degeneration and severe degeneration. (**a**) Normal—degenerative. (**b**) Normal—degenerative. (**c**) Moderate—degenerative. (**d**) Moderate—degenerative. (**e**) Moderate—degenerative. (**f**) Moderate—degenerative. (**g**) Severe—degenerative. (**h**) Severe—degenerative. (**i**) Severe—degenerative. (**j**) Severe—degenerative. (**k**) Severe—degenerative. (**l**) Severe—degenerative. **Adapted from:** Manchikanti L, et al. Discography In: Manchikanti L, Kaye AD, Falco FJE, et al. (eds). *Essentials of Interventional Techniques in Managing Chronic Pain*. Springer, 2018, pp. 273–300 [91]

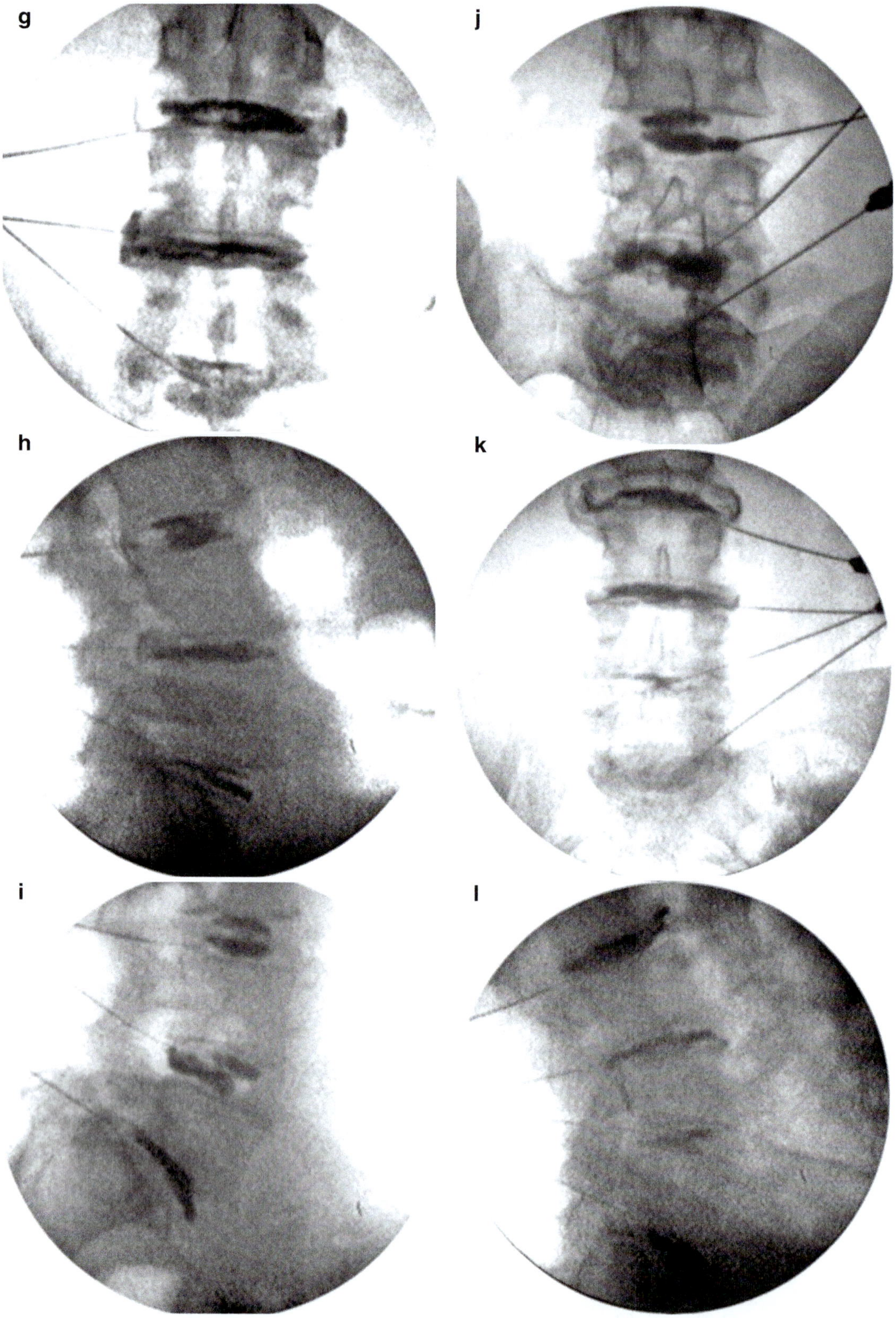


**Fig. 23.18** (continued)

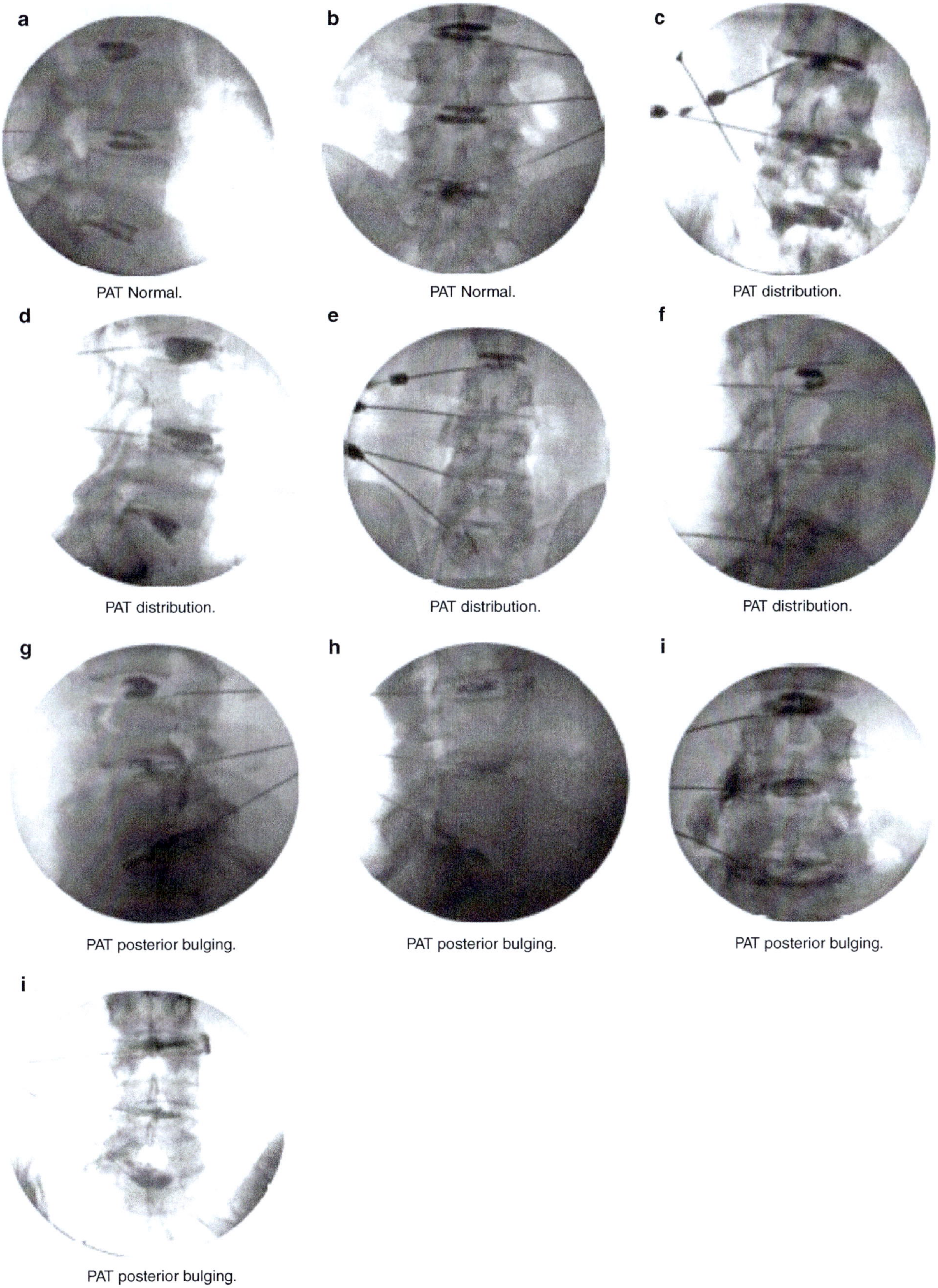


**Fig. 23.19** Illustration of posterior annular tears (PAT) disruption. (**a**, **b**) PAT normal. (**c**–**f**) PAT distribution. (**g**–**j**) PAT posterior bulging. **Adapted from:** Manchikanti L, et al. Discography In: Manchikanti L, Kaye AD, Falco FJE, et al. (eds). *Essentials of Interventional Techniques in Managing Chronic Pain*. Springer, 2018, pp. 273–300 [91]

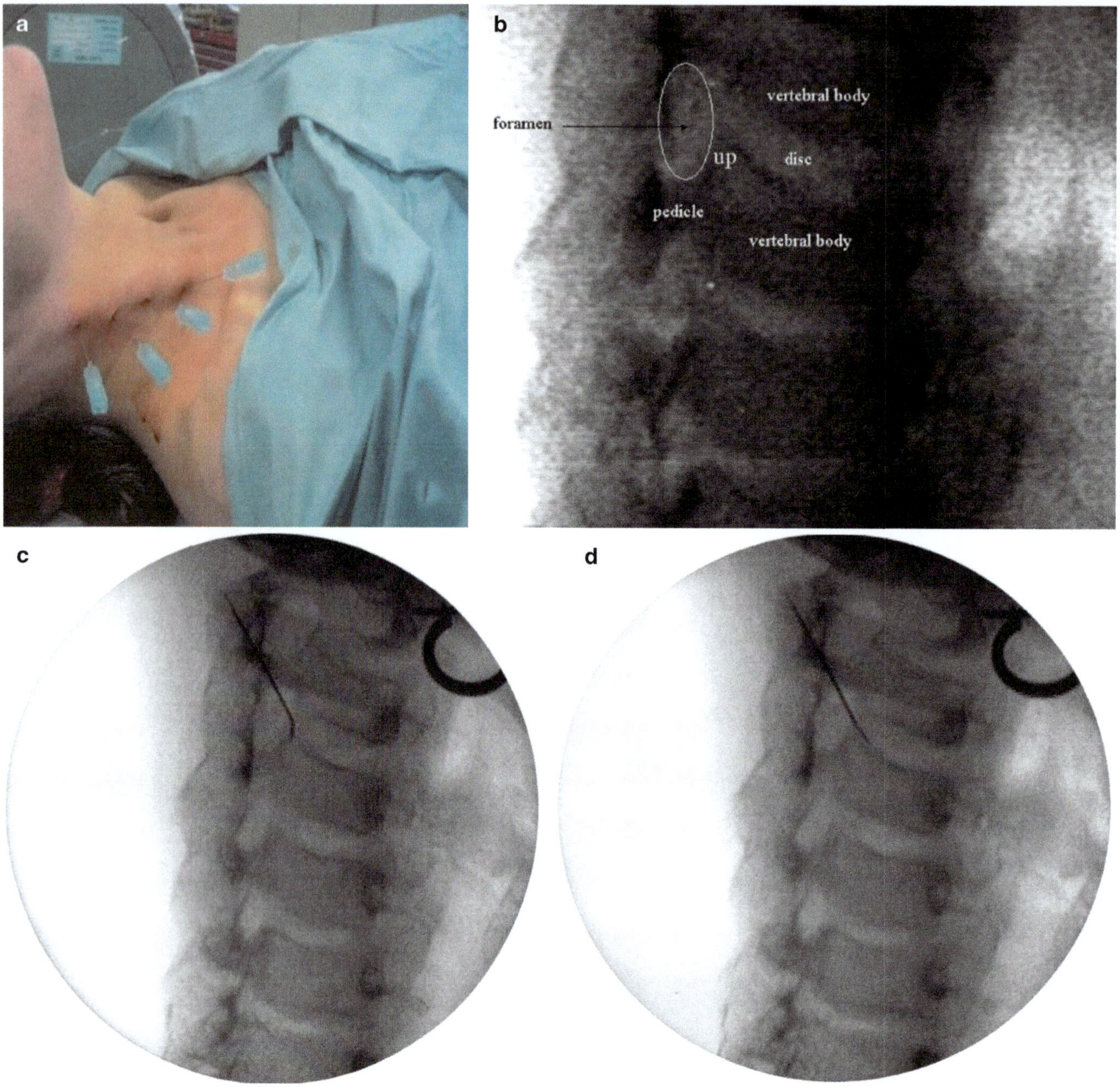


**Fig. 23.20** Anterolateral cervical discography technique. (**a**) Patient view. *up* uncinate process. (**b**) The patient and/or C-arm are placed in an oblique position to optimize the foraminal view of the cervical spine. (**c**) Spinal needle contacts with the uncinate process to establish needle depth, avoid the vertebral artery, and to serve as launching point into the disc. (**d**) The spinal needle is re-positioned into the outer annulus of the disc. **Adapted from:** Manchikanti L, et al. Discography In: Manchikanti L, Kaye AD, Falco FJE, et al. (eds). *Essentials of Interventional Techniques in Managing Chronic Pain*. Springer, 2018, pp. 273–300 [91]

annular disruption can accept greater than 1.5 mL and more as the injectate spreads into the epidural space with very low or varying pressure and no endpoint. A degenerative (Fig. 23.21) or herniated disc with an intact annulus (Fig. 23.20) typically accepts 0.5–1.5 mL with intermediate pressure described as a spongy endpoint. A complete annular tear is characterized by no endpoint (Fig. 23.21).

The pain level is a reflection of the pain intensity experienced by the patient during the injection, regardless if the pain is similar to pre-injection pain or not. The intensity is graded verbally on a scale, often using 10 as the greatest degree of pain and 0 as no pain at all. Injecting discs with lidocaine are helpful in determining truly painful discs in the case of two or more contiguously painful discs. The discs are evaluated again by re-injection to determine the lidocaine response. If a painful disc is no longer painful upon re-injection after being anesthetized, then it is a true pain generator.

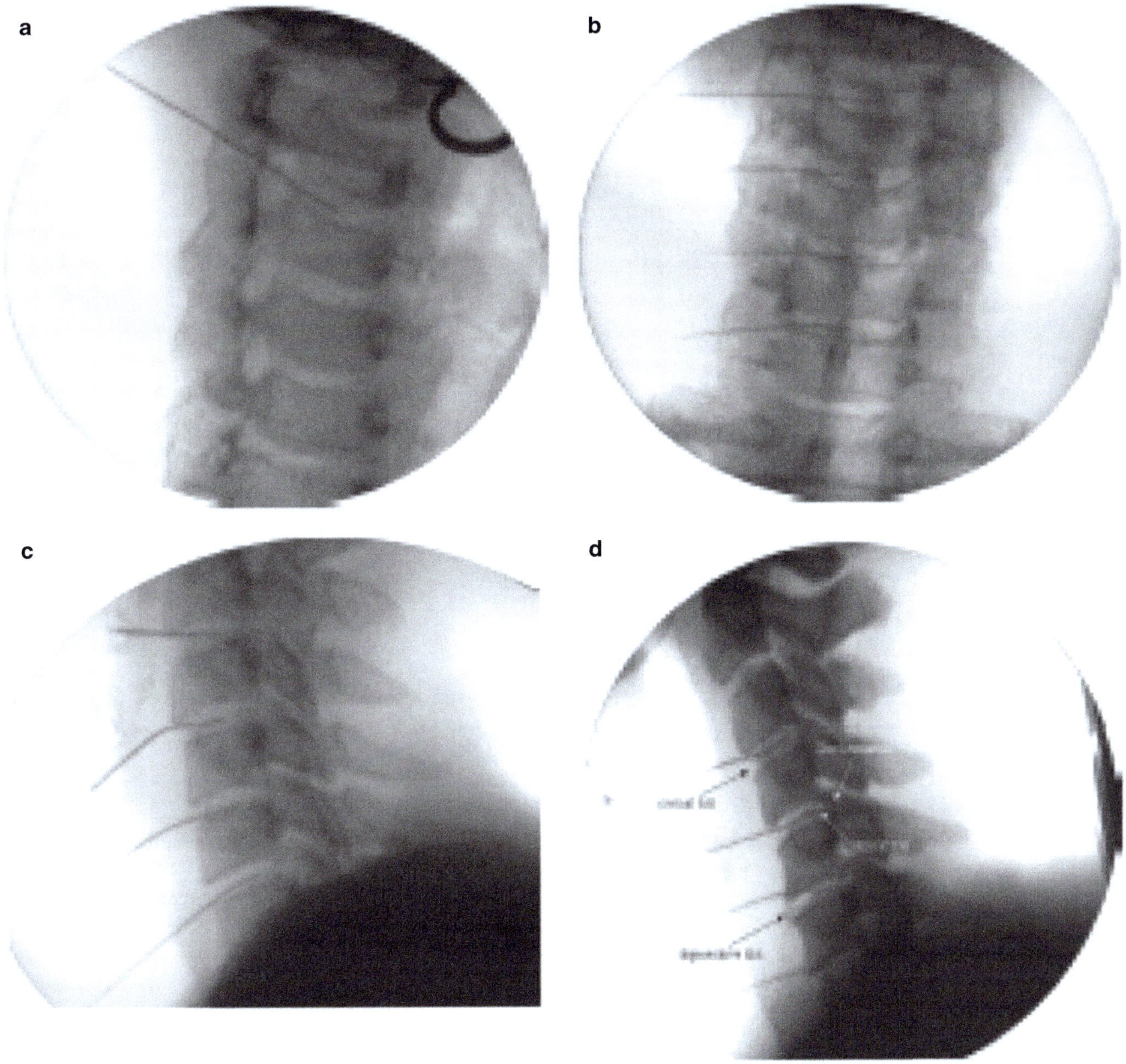


**Fig. 23.21** (**a–d**) Anterolateral cervical discography technique. Final needle tip position in the cervical discs prior to injection. **Adapted from:** Manchikanti L, et al. Discography In: Manchikanti L, Kaye AD, Falco FJE, et al. (eds). *Essentials of Interventional Techniques in Managing Chronic Pain*. Springer, 2018, pp. 273–300 [91]

Once a painful disc is anesthetized and "taken out of the equation," nearby discs can be injected to determine if they are really a source of pain or were actually referring pain by stimulating the previously painful disc.

The pain quality is most important since it establishes whether or not that any pain produced during the procedure is exactly what the patient clinically experiences. The patient relates whether he has any pain at all and if the pain quality is the same as the baseline pain. The pain can be vague and discordant, just a component of their typical pain, or an exact reproduction of their concordant pain.

X-rays of the cervical spine are obtained in the PA and lateral views upon completion of the injections (Fig. 23.22). A normal cervical disc nucleogram appears as a rounded or bilobular-contained component of the cervical disc. Annular disruption (Fig. 23.23) appears as contrast that spreads beyond the nuclear borders. On a lateral view, the contrast is commonly recognized as spreading within the posterior portion of the disc, but it can also spread into the anterior annulus. Nucleogram patterns can range from normal (cotton ball and lobular) to abnormal (irregular, fissured, ruptured, and degenerative). A PA view must be checked to rule out leak-

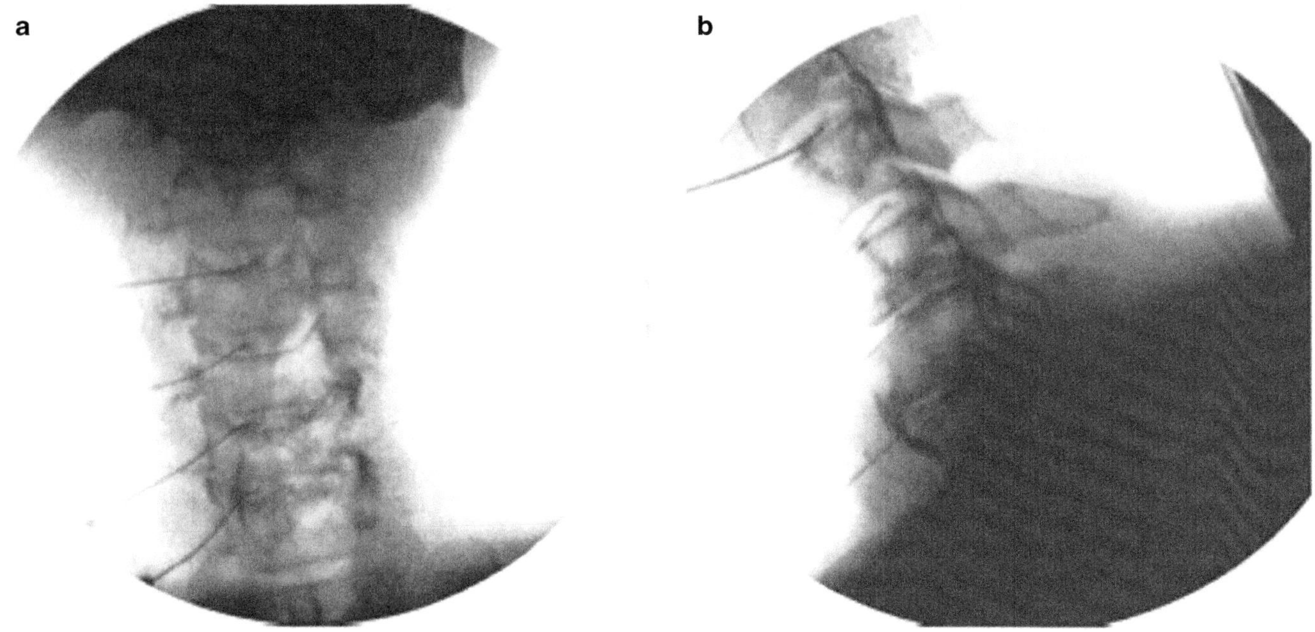


**Fig. 23.22** Cervical discogram. Cervical spine x-rays obtained after completing the cervical discography study with contrast. (**a**) AP view. (**b**) Lateral view. **Adapted from:** Manchikanti L, et al. Discography In: Manchikanti L, Kaye AD, Falco FJE, et al. (eds). *Essentials of Interventional Techniques in Managing Chronic Pain*. Springer, 2018, pp. 273–300 [91]

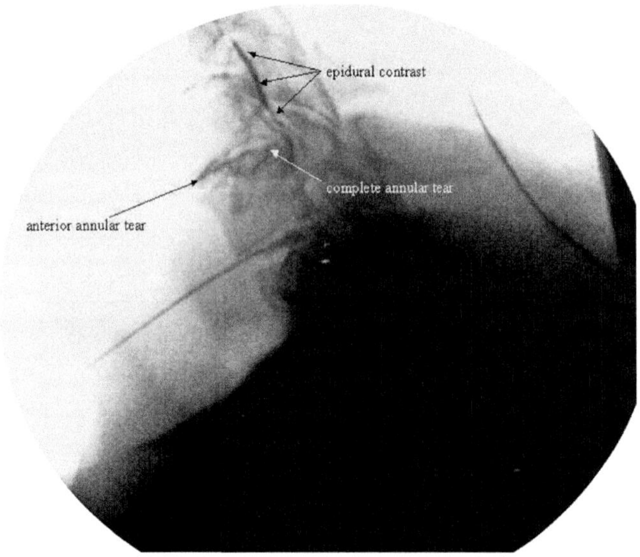


**Fig. 23.23** Lateral view of the cervical spine status post disc injection with contrast revealing a disc with an anterior and posterior annular tear. The posterior annular tear extends through the outer annulus with contrast extending into the epidural space. **Adapted from:** Manchikanti L, et al. Discography In: Manchikanti L, Kaye AD, Falco FJE, et al. (eds). *Essentials of Interventional Techniques in Managing Chronic Pain*. Springer, 2018, pp. 273–300 [91]

age toward the uncinate recess. There is decreased thickness of posterolateral annulus due to the presence of the uncovertebral articulation. This is typically recognized only in adults and not in children.

The size of the uncovertebral cleft is variable from side to side and thus fills with contrast in an irregular fashion. This is recognized as a maturation of the intervertebral disc and not as a form of disc degeneration.

### Intradiscal Injection of Biologics

In the cervical spine, intradiscal injection of biologics must be performed with more caution than in lumbar spine or thoracic spine. The volume injected is approximately 1 mL. Once again, contrast injection and antibiotic injection is discouraged. The needle may be left for 30 s to 1 min to avoid leakage.

## Thoracic Discography and Injection of Biologics

The technique of thoracic discography is similar to lumbar discography. However, the extraspinal technique is the recommended approach, which is safe considering the location of the spinal cord in the thoracic spine [11, 91, 94].

An extraspinal technique similar to that used for lumbar discography is utilized in performing thoracic discography.

This is a safe approach as long as the physician employs the technique with the required attention to detail, precision, and a thorough understanding of the surrounding anatomical structures.

The procedure must be performed under aseptic conditions. The patient is prepped and draped in a sterile manner in the prone position. Intravenous sedation can be given to relax the patient as long as there is no compromise to their ability to participate during the procedure. Heart rate, pulse oximetry, blood pressure, and respirations are monitored during the procedure. Administration of intravenous antibiotics prior to the procedure is standard as a prophylaxis against infection, particularly discitis [96, 97, 99–103]. Intradiscal antibiotics may be given at the discretion of the physician.

The patient is approached from the asymptomatic side to avoid any confusion between needle-induced annular pain and a provocative pain response. Either a single or a two-needle system is used with a longer, small gauge procedure needle being passed through a shorter, larger gauge needle into the disc (Fig. 23.10). A two-needle approach can be more difficult in thinner individuals who lack soft tissue to hold the introducer needle. In thinner subjects, a single needle technique can be accomplished with a shorter 3½-inch spinal needle.

The C-arm is adjusted to achieve a posterior oblique position with the SAP dividing the vertebral body in half and squaring off the endplates (Fig. 23.24). Described by Schellhas, this view provides a "box" configuration formed by the endplates, the superior articular process, and the rib head (Fig. 23.24). The box defines a safe pathway into the annulus while avoiding the spinal cord medially and the lung laterally.

Figure 23.25 illustrates important landmarks in cross-section of a cadaver [92, 104]. The introducer needle tip is positioned and advanced within the confines of the box to the outer annulus. The procedure needle is then advanced through the introducer needle into the central third of the disc under biplanar fluoroscopic guidance.

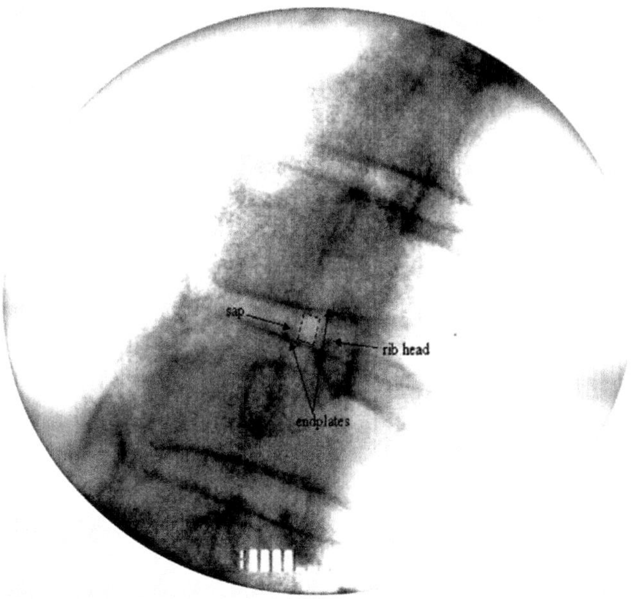


**Fig. 23.24** Thoracic discography view of "box" configuration. This oblique starting point allows for safe entry of spinal needles into the thoracic disc. **Adapted from:** Manchikanti L, et al. Discography In: Manchikanti L, Kaye AD, Falco FJE, et al. (eds). *Essentials of Interventional Techniques in Managing Chronic Pain*. Springer, 2018, pp. 273–300 [91]

## Disc Injection

Each disc is evaluated by injecting saline, or contrast, or both in sequence after the needles are positioned in the discs. Data recorded during the study may include the injected volume, discometry (end point), pain level, analgesia, pain quality, and nucleogram. A firm endpoint reflects a normal intact annulus. A spongy endpoint is a reflection of annular pathology with an intact annulus, and a complete annular tear through the outer annular fibers is characterized by no endpoint. A normal thoracic disc accepts from 0.5 mL up to 1 mL with a firm endpoint and an even larger volume depending on the size of the patient [43]. A herniated disc with an intact annulus, or a degenerative disc, typically accepts more volume with intermediate pressure described as a spongy endpoint. Discs with complete rupture of the outer annulus can accept 3 mL or more as the injectate spreads into the epidural space with no endpoint.

The pain level is a reflection of the pain intensity experienced by the patient during the injection, regardless of whether the pain is similar or not. The intensity is graded verbally on a scale, often using 10 as the greatest degree of pain and 0 as no pain at all. In the case of two or more contiguously painful discs, injecting discs with lidocaine is helpful in determining the truly painful disc(s). The discs are evaluated again by re-injection to determine the lidocaine response. If a painful disc is no longer painful upon re-injection after being anesthetized, then it is a true pain generator. Once a painful disc is anesthetized and "taken out of the equation," nearby discs can be re-injected to determine if they are really a source of pain or were actually referring pain by stimulating the previously painful disc.

The pain quality is most important since it establishes whether or not any pain produced during the procedure is exactly what the patient clinically experiences. The patient relates whether they have pain at all and if the pain quality is the same as "their" pain. The pain can be vague and discordant, just a component of their typical pain, or an exact reproduction of their concordant pain.

Upon completion of the injections, x-rays of the thoracic spine are obtained in the PA and lateral views (Fig. 23.26). A normal nucleogram appears as a rounded or bilobular-contained component of the disc. Annular disruption appears as contrast that spreads beyond the nuclear border typically in a radial fashion. On lateral view, contrast is commonly recognized as spreading within the posterior portion of the disc, but it can also spread into the anterior annulus. Nucleogram patterns can range from normal (cotton ball and lobular) to abnormal (irregular, fissured, ruptured and degenerative).

Figure 23.27 illustrates multiple thoracic discograms.

## Intradiscal Injection of Biologics

Injection of biologics may be performed in a separate session after diagnostic discography during which contrast and anti-

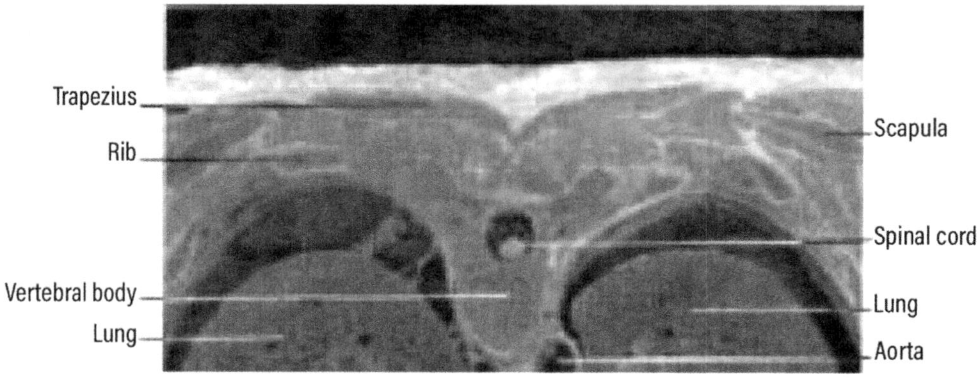


**Fig. 23.25** Important landmarks in cross-section of cadaver thoracic cavity. **Reproduced from:** Singh V. Thoracic discography. *Pain Physician* 2004; 7:451–458 [104] with permission from the author and ASIPP

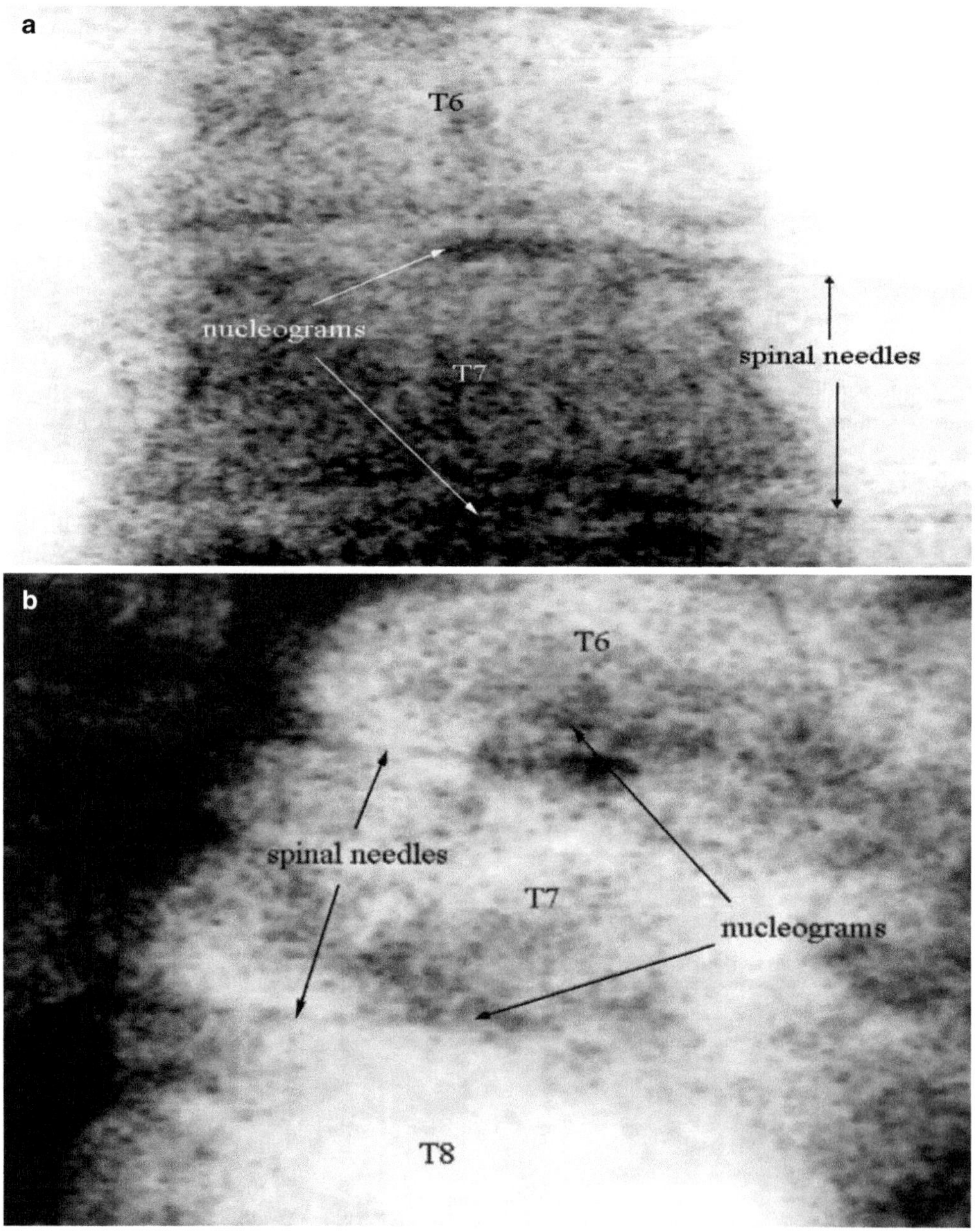


**Fig. 23.26** Thoracic discogram. (**a**) AP view. (**b**) Lateral view. T6 = T6 vertebra, T7 = T7 vertebra, T8 = T8 vertebra. **Adapted from:** Manchikanti L, et al. Discography In: Manchikanti L, Kaye AD, Falco FJE, et al. (eds). *Essentials of Interventional Techniques in Managing Chronic Pain.* Springer, 2018, pp. 273–300 [91]

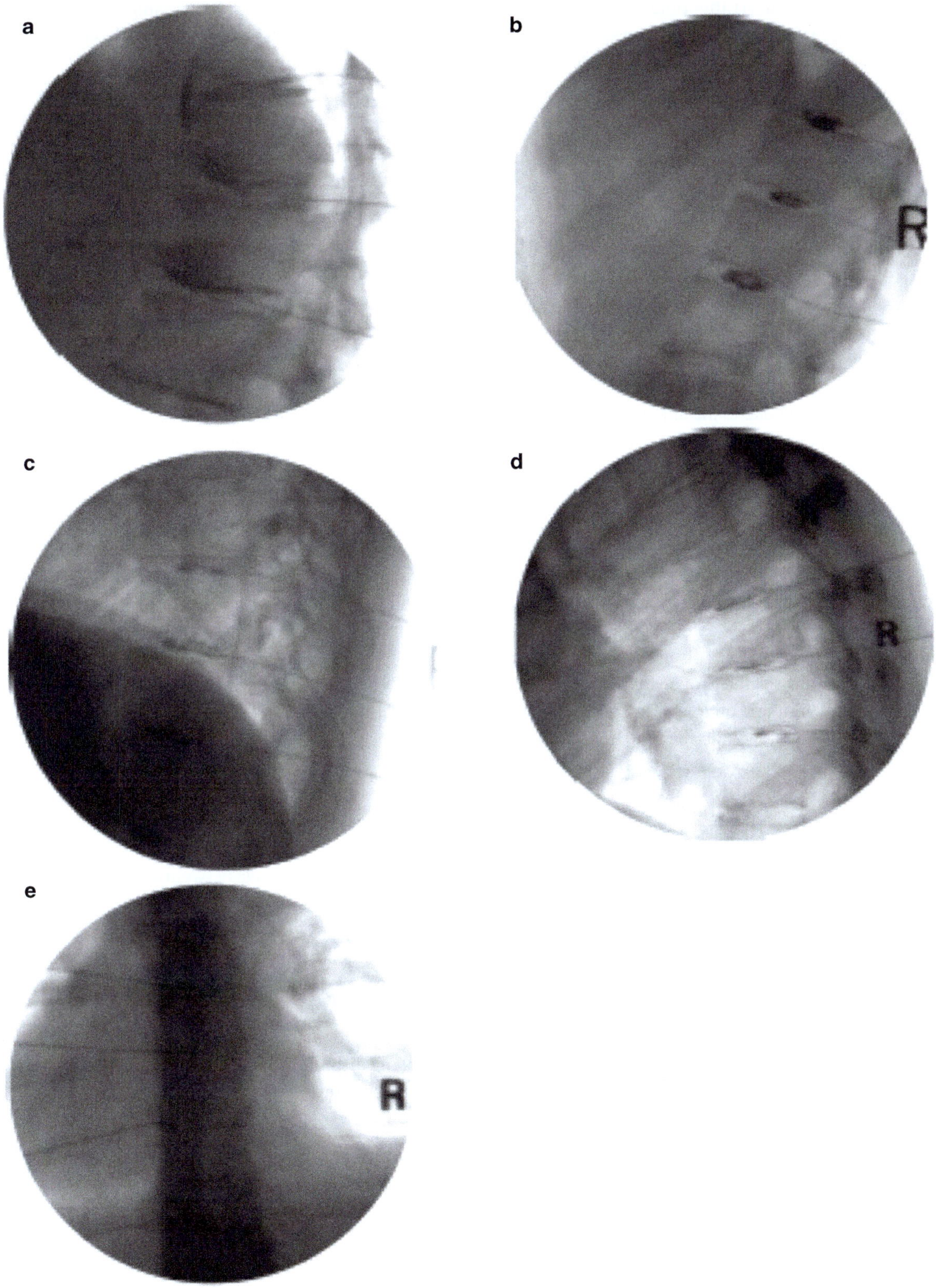


**Fig. 23.27** Illustration of thoracic discograms. (**a**) Lateral view thoracic discograms, T9/10, T10/11, and T11/T12. Note contrast leaking anteriorly at T9/10. Severe degeneration at T10/11 and T11/12, with contrast highlighting posterior bulge at T11/12. (**b**) Example of wider disc spaces in the lower thoracic and upper lumbar spine. Contrast visible at four levels T10/T11, T11/T12, T12/L1, and L1/L2. (**c**) Later view of thoracic discogram in mid-thoracic region. Disc spaces are slightly narrower in the region. (**d**) Lateral view thoracic discography. T6/T7, T7/T8, and T8/T9, showing quite narrow disc spaces at these levels. (**e**) AP view of narrow disc spaces at T6/T7, T7/T8, and T8/T9. **Adapted from:** Manchikanti L, et al. Discography In: Manchikanti L, Kaye AD, Falco FJE, et al. (eds). *Essentials of Interventional Techniques in Managing Chronic Pain*. Springer, 2018, pp. 273–300 [91]

biotic solutions are avoided. At this time, 0.5–1 mL of biologic solution may be injected. The needle may be left inside the disc for a few minutes after the injection so that the entire solution will be absorbed into the disc rather than leaking into the subarachnoid space.

## Side Effects and Complications

Potential complications from discography include discitis, injury to nerve roots, bleeding, epidural abscess, allergic contrast reaction, subarachnoid puncture, and chemical meningitis. The most recognized complication of a discography procedure is bacterial discitis with a reported incidence that is typically less than 1% [97, 99]. Pain associated with disc injection is often unrecognized by the patient due to the amount of pain involved during and after the procedure. A common microbe in discitis is Staphylococcus epidermidis, but streptococcus and Escherichia coli are also commonly involved. Escherichia coli can be inoculated from the hypopharynx in cervical discography. A face mask for the patient might help prevent the patient's own oral flora from contaminating the sterile field as the result of an inadvertent cough or sneeze during the performance of a cervical discography.

Despite adequate treatment of discitis with intravenous antibiotics, disc infection has the propensity to cause further disc degeneration and vertebral endplate destruction. Further complications from discitis include the development of epidural or retropharyngeal abscesses [101–103]. Changes in pain or quality of symptoms after discography should raise the suspicion for discitis. Screening is performed by erythrocyte sedimentation rate. If elevated, MRI is the current gold standard for detection of discitis [105, 106].

A contrast allergy is a relative contraindication that can be circumvented by pre-medication with antihistamines and corticosteroids, or by substituting gadolinium for nonionic contrast.

Further complications include a vasovagal response, a hematoma that can include neural compromise within the spinal canal, an allergic drug reaction, headache, herniated cervical disc, quadriplegia, pneumothorax with lower cervical disc injections, thecal sac puncture, and arachnoiditis [98, 99].

## Precautions

Prior to discography, patients should have completed a full clinical assessment with a history, physical, neurological examination, and have undergone MRI scanning.

Relative contraindications to interventional techniques have been described in patients receiving treatment with antithrombotics and anticoagulants [6, 7, 107, 108]. Guidelines on anticoagulants and antithrombotics are described in Chap. 15 [108] and also on the ASIPP app, which is regularly updated.

## Conclusion

Intervertebral discs have been shown to be generators for spinal pain. Intervertebral discs are innervated by branches from the ventral rami, the grey rami communicans, and the sinuvertebral nerve. The prevalence of chronic discogenic low back pain ranges from 26 to 39% among those with an unremarkable diagnostic workup and 20 to 61% in the cervical spine, with an unknown proportion in the thoracic spine.

Discography can determine disc pain through provocation and analgesia. Discography is the only means to correctly identify whether a disc that appears normal or abnormal on imaging is painful or not. Discography and CT-discography are more accurate than MRI or plain CT in detecting annular pathology.

Discography plays a key role in determining whether or not a biologic injection is indicated for discogenic low back pain.

Although rare, discitis is the most common risk of discography. Only physicians with adequate experience and appropriate training should perform discography using sterile and precise technique.

**Key Points**

1. The understanding of the lumbar disc as a significant cause of low back pain and radiculopathy began with pioneering work in the early twentieth century, leading to the development of diagnostic techniques like discography.
2. The intervertebral disc can suffer from two common issues: intervertebral disc degeneration and intervertebral disc protrusion or extrusion. These processes are not mutually exclusive and involve the release of various inflammatory chemicals that contribute to the biochemistry of disc degeneration and the pathophysiology of radiculopathy.
3. Injuries to the disc can activate sensory nociceptors and trigger an inflammatory and immunologic reaction, leading to pain generation. The degeneration process is associated with a decrease in vascular supply, altered cytokine profiles, and changes in matrix proteins, culminating in structural compromise and potential herniation of the nucleus pulposus tissue, which further contributes to pain.
4. With the limitations of current treatments for intervertebral disc degeneration and discogenic pain, there is a growing need for therapies capable of restoring cell

population, disc function, and alleviating pain. Cell-based therapies, such as platelet-rich plasma (PRP) and mesenchymal stem cells (MSCs), have emerged as promising approaches to address these challenges. PRP, containing growth factors released by platelets, promotes tissue healing, angiogenesis, and collagen synthesis. In vitro and animal studies have demonstrated the potential of PRP in restoring disc architecture and function.

5. Platelet-rich plasma (PRP) injections and mesenchymal stem cell (MSC) injections have been increasingly used for disc-related spine disorders. While the evidence for both treatments is still evolving, there have been systematic reviews and studies reporting positive outcomes with PRP injections, including improvements in pain and function over a 5–9 year follow-up period.
6. Biologic or cell-based therapy is recommended for patients who have experienced unremitting spinal pain, with or without extremity pain, lasting for more than 3 months and have not responded to conservative treatments, including epidural injections.
7. Discography and disc injections can be performed in the lumbar, thoracic, and cervical regions. The most studied and published approach is lumbar discography and lumbar cell-based therapies. Proficiency in discography is essential before performing cell-based disc injections, and in some cases, discography may be necessary to obtain a precise diagnosis of the internal anatomy of the disc before injection of biologics.
8. The larger and more robust randomized controlled trials are needed to establish the efficacy and long-term benefits of these regenerative therapies for disc-related conditions.

**Acknowledgments** This book chapter is modified and updated from a previous book chapter, "Intradiscal Injections of Biologics" by Laxmaiah Manchikanti, MD, Sheri L. Albers, DO, Annu Navani, MD, Vijay Singh, MD, Frank J.E. Falco, MD, and Joshua A. Hirsch, MD, in *Essentials of Regenerative Medicine in Interventional Pain Management* published by ASIPP Publishing.

## References

1. Manchikanti L, Albers SL, Navani A, Singh V, Falco FJE, Hirsch JA. Intradiscal injections of biologics. In: Manchikanti L, Navani A, Atluri S, editors. Essentials of regenerative medicine in interventional pain management. Paducah: ASIPP Publishing; 2019. p. 385–420.
2. Hoy D, March L, Brooks P, et al. The global burden of low back pain: estimates from the Global Burden of Disease 2010 study. Ann Rheum Dis. 2014;73:968–74.
3. Hoy D, March L, Woolf A, et al. The global burden of neck pain: estimates from the global burden of disease 2010 study. Ann Rheum Dis. 2014;73:1309–15.
4. U.S. Burden of Disease Collaborators. The state of US health, 1990–2010: burden of diseases, injuries, and risk factors. JAMA. 2013;310:591–608.
5. Manchikanti L, Kaye AD, Soin A, et al. Comprehensive evidence-based guidelines for facet joint interventions in the management of chronic spinal pain: American Society of Interventional Pain Physicians (ASIPP) guidelines. Pain Physician. 2020;23:S1–S127.
6. Navani A, Manchikanti L, Albers SL, et al. Responsible, safe, and effective use of biologics in the management of low back pain: American Society of Interventional Pain Physicians (ASIPP) guidelines. Pain Physician. 2019;22:S1–S74.
7. Manchikanti L, Knezevic NN, Navani A, et al. Epidural interventions in the management of chronic spinal pain: American Society of Interventional Pain Physicians (ASIPP) comprehensive evidence-based guidelines. Pain Physician. 2021;24:S27–S208.
8. Manchikanti L, Centeno CJ, Atluri S, et al. Bone marrow concentrate (BMC) therapy in musculoskeletal disorders: evidence-based policy position statement of American Society of Interventional Pain Physicians (ASIPP). Pain Physician. 2020;23:E85–E131.
9. Dieleman JL, Baral R, Birger M, et al. US spending on personal health care and public health, 1996–2013. JAMA. 2016;316:2627–46.
10. Dieleman JL, Cao J, Chapin A, et al. US health care spending by payer and health condition, 1996–2016. JAMA. 2020;323:863–84.
11. Manchikanti L, Soin A, Benyamin RM, et al. An update of the systematic appraisal of the accuracy and utility of discography in chronic spinal pain. Pain Physician. 2018;21:91–110.
12. Manchikanti L, Hirsch JA. An update on the management of chronic lumbar discogenic pain. Pain Manag. 2015;5:373–86.
13. Manchikanti L, Hirsch JA. Clinical management of radicular pain. Expert Rev Neurother. 2015;17:681–93.
14. Bogduk N, Aprill C, Derby R. Lumbar discogenic pain: state-of-the-art review. Pain Med. 2013;14:813–36.
15. Zhao L, Manchikanti L, Kaye AD, Abd-Elsayed A. Treatment of discogenic low back pain: current treatment strategies and future options—a literature review. Curr Pain Headache Rep. 2019;23:86.
16. Cohen SP, Larkin TM, Barna SA, et al. Lumbar discography: a comprehensive review of outcome studies, diagnostic accuracy, and principles. Reg Anesth Pain Med. 2005;30:163–83.
17. Hancock MJ, Maher CG, Latimer J, et al. Systematic review of tests to identify the disc, SIJ or facet joint as the source of low back pain. Eur Spine J. 2007;16:1539–50.
18. Willems P. Decision making in surgical treatment of chronic low back pain: the performance of prognostic tests to select patients for lumbar spinal fusion. Acta Orthop Suppl. 2013;84:1–35.
19. Peng BG. Pathophysiology, diagnosis, and treatment of discogenic low back pain. World J Orthop. 2013;4:42–52.
20. Brayda-Bruno M, Tibiletti M, et al. Advances in the diagnosis of degenerated lumbar discs and their possible clinical application. Eur Spine J. 2014;23:S315–23.
21. Keyes DC, Compere EL. The normal and pathological physiology of the nucleus pulposus of the intervertebral disc: an anatomical, clinical, and experimental study. J Bone Joint Surg Am. 1932;14:897–938.
22. Mixter WJ, Barr JS. Ruptures of the intervertebral disc with involvement of the spinal canal. N Engl J Med. 1934;211:210–5.
23. Manchikanti L, Staats PS, Nampiaparampil DE, Hirsch JA. What is the role of epidural injections in the treatment of lumbar discogenic pain: a systematic review of comparative analysis with fusion and disc arthroplasty. Korean J Pain. 2015;28:75–87.
24. Helm S 2nd, Simopoulos TT, Stojanovic MP, et al. Effectiveness of thermal annular procedures in treating discogenic low back pain. Pain Physician. 2017;20:447–70.

25. Kaye AD, Manchikanti L, Abdi S, et al. Efficacy of epidural injections in managing chronic spinal pain: a best evidence synthesis. Pain Physician. 2015;18:E939–E1004.
26. Manchikanti L, Nampiaparampil DE, Manchikanti KN, et al. Comparison of the efficacy of saline, local anesthetics, and steroids in epidural and facet joint injections for the management of spinal pain: a systematic review of randomized controlled trials. Surg Neurol Int. 2015;6:S194–235.
27. Miguélez-Rivera L, Pérez-Castrillo S, González-Fernández ML, et al. Immunomodulation of mesenchymal stem cells in discogenic pain. Spine J. 2018;18:330–42.
28. Pimentel DC, Abd OE, El Abd O, et al. Anti-tumor necrosis factor antagonists in the treatment of low back pain and radiculopathy: a systematic review and meta-analysis. Pain Physician. 2014;17:E27–44.
29. Leite VF, Buehler AM, El Abd O, et al. Anti-nerve growth factor in the treatment of low back pain and radiculopathy: a systematic review and a meta-analysis. Pain Physician. 2014;17:E45–60.
30. Sainoh T, Sakuma Y, Miyagi M, et al. Efficacy of anti-nerve growth factor therapy for discogenic neck pain in rats. Spine (Phila Pa 1976). 2014;39:E757–62.
31. Wu T, Song HX, Dong Y, Li J. Cell-based therapies for lumbar discogenic low back pain: systematic review and single-arm meta-analysis. Spine (Phila Pa 1976). 2018;43:49–57.
32. Kumar H, Ha DH, Lee EJ, et al. Safety and tolerability of intradiscal implantation of combined autologous adipose-derived mesenchymal stem cells and hyaluronic acid in patients with chronic discogenic low back pain: 1-year follow-up of a phase I study. Stem Cell Res Ther. 2017;8:262.
33. Gilbert HT, Hoyland JA, Richardson SM. Stem cell regeneration of degenerated intervertebral discs: current status (update). Curr Pain Headache Rep. 2013;17:377.
34. Zeckser J, Wolff M, Tucker J, Goodwin J. Multipotent mesenchymal stem cell treatment for discogenic low back pain and disc degeneration. Stem Cells Int. 2016;2016:3908389.
35. Roofe PG. Innervation of the annulus and posterior longitudinal ligament. Arch Neurol Psychiatry. 1940;44:100–3.
36. Lindblom K. Protrusions of the discs and nerve compression in the lumbar region. Acta Radiol Scand. 1944;25:195–212.
37. Hirsch C. Attempt to diagnose the level of disc lesion clinically by disc puncture. Acta Orthop Scand. 1948;18:131–40.
38. Lindblom K. Diagnostic puncture of the intervertebral disc in sciatica. Acta Orthop Scand. 1948;17:231–9.
39. Cloward RB. The anterior approach for removal of ruptured disks. J Neurosurg. 1958;15:602–17.
40. Smith GW, Robinson RA. The treatment of certain cervical-spine disorders by anterior removal of the intervertebral disc and interbody fusion. J Bone Joint Surg. 1958;40:607–23.
41. Holt EP Jr. Fallacy of cervical discography. Report of 50 cases in normal subjects. JAMA. 1964;188:799–801.
42. Schellhas KP, Smith MD, Gundry CR, et al. Cervical discogenic pain. Prospective correlation of magnetic resonance imaging and discography in asymptomatic subjects and pain sufferers. Spine (Phila Pa 1976). 1996;21:300–12.
43. Simmons EH, Segil CM. An evaluation of discography in the localization of symptomatic levels in discogenic disease of the spine. Clin Orthop Relat Res. 1975;108:57–69.
44. Schellhas KP, Pollei SR, Dorwart RH. Thoracic discography. A safe and reliable technique. Spine (Phila Pa 1976). 1994;19:2103–9.
45. Cramer GD, Darby SA, Frysztak RJ. Pain of spinal origin. In: Cramer GD, Darby SA, editors. Basic and clinical anatomy of the spine, spinal cord and ANS. 2nd ed. St. Louis: Elsevier Mosby; 2005. p. 480–520.
46. Manchikanti L, Albers SL, Hirsh JA, Boswell MV. Lumbar disk herniation. In: Kaye AD, editor. Scientific American pain management. Hamilton: Decker; 2017. https://doi.org/10.2310/7900.15047. www.DeckerIP.com.
47. Peng B, Fu X, Pang X, et al. Prospective clinical study on natural history of discogenic low back pain at 4 years of follow-up. Pain Physician. 2012;15:525–32.
48. Cramer GD, Bakkum BW, Stathopoulos PC. Microscopic anatomy of the zygapophysial joints, intervertebral discs, and other major tissues of the back. In: Cramer GD, Darby SA, editors. Basic and clinical anatomy of the spine, spinal cord and ANS. 2nd ed. St. Louis: Elsevier Mosby; 2005. p. 569–619.
49. Le Maitre CL, Pockert A, Buttle DJ, et al. Matrix synthesis and degradation in human intervertebral disc degeneration. Biochem Soc Trans. 2007;35:652–5.
50. Chen WH, Lo WC, Lee JJ, et al. Tissue-engineered intervertebral disc and chondrogenesis using human nucleus pulposus regulated through TGF-beta1 in platelet-rich plasma. J Cell Physiol. 2006;209:744–54.
51. Dominici M, Le Blanc K, Mueller I, et al. Minimal criteria for defining multipotent mesenchymal stromal cells. The International Society for Cellular Therapy position statement. Cytotherapy. 2006;8:315–7.
52. Hunt CL, Her YF, Law LA, et al. Five generations of cell preparation: a translational framework for categorizing regenerative stem cell therapies. J Am Acad Reg Med. 2017;1:7239.
53. Lana JF, Purita J, Silva Rava CA, et al. Bone marrow concentrate. In: Manchikanti L, Navani A, Atluri S, editors. Essentials of regenerative medicine in interventional pain management. Paducah: ASIPP Publishing; 2019. p. 83–98.
54. Moresco EM, LaVine D, Beutler B. Toll-like receptors. Curr Biol. 2011;21:R488–93.
55. Watters TM, Kenny EF, O'Neill LA. Structure, function and regulation of the toll/IL-1 receptor adaptor proteins. Immunol Cell Biol. 2007;85:411–9.
56. Bodor M, Dregalla R, Uribe Y. Platelet rich plasma. In: Manchikanti L, Navani A, Atluri S, editors. Essentials of regenerative medicine in interventional pain management. Paducah: ASIPP Publishing; 2018. p. 67–82.
57. Mascarinas A, Harrison J, Boachie-Adjei K, Lutz G. Regenerative treatments for spinal conditions. Phys Med Rehabil Clin N Am. 2016;27:1003–17.
58. Gullung GB, Woodall JW, Tucci MA, James J, Black DA, McGuire RA. Platelet-rich plasma effects on degenerative disc disease: analysis of histology and imaging in an animal model. Evid Based Spine Care J. 2011;2:13–8.
59. Obata S, Akeda K, Imanishi T, et al. Effect of autologous platelet-rich plasma-releasate on intervertebral disc degeneration in the rabbit anular puncture model: a preclinical study. Arthritis Res Ther. 2012;14:R241.
60. Kim HJ, Yeom JS, Koh YG, et al. Anti-inflammatory effect of platelet-rich plasma on nucleus pulposus cells with response of TNF-a and IL-1. J Orthop Res. 2014;32:551–6.
61. Sawamura K, Ikeda T, Nagae M, et al. Characterization of in vivo effects of platelet-rich plasma and biodegradable gelatin hydrogel microspheres on degenerated intervertebral discs. Tissue Eng Part A. 2009;15:3719–27.
62. Khan S, Mafi P, Mafi R, Khan W. A systematic review of mesenchymal stem cells in spinal cord injury, intervertebral disc repair and spinal fusion. Curr Stem Cell Res Ther. 2018;13:316–23.
63. Wang Z, Perez-Terzic CM, Smith J, et al. Efficacy of intervertebral disc regeneration with stem cells—a systematic review and meta-analysis of animal controlled trials. Gene. 2015;564:1–8.
64. Benneker LM, Andersson G, Iatridis JC, et al. Cell therapy for intervertebral disc repair: advancing cell therapy from bench to clinics. Eur Cell Mater. 2014;27:5–11.

65. Acosta FL Jr, Metz L, Adkisson HD, et al. Porcine intervertebral disc repair using allogeneic juvenile articular chondrocytes or mesenchymal stem cells. Tissue Eng Part A. 2011;17:3045–55.
66. Sanapati J, Manchikanti L, Atluri S, et al. Do regenerative medicine therapies provide long-term relief in chronic low back pain: a systematic review and metaanalysis. Pain Physician. 2018;21:515–50.
67. Basso M, Cavagnaro L, Zanirato A, et al. What is the clinical evidence on regenerative medicine in intervertebral disc degeneration? Musculoskelet Surg. 2017;101:93–104.
68. Navani A, Ambach MA, Navani R, Wei J. Biologics and lumbar discogenic pain: 18 month follow-up for safety and efficacy. IPM Reports. 2018;3:111–8.
69. Levi D, Horn S, Tyszko S, Levin J, Hecht-Leavitt C, Walko E. Intradiscal platelet-rich plasma injection for chronic discogenic low back pain: preliminary results from a prospective trial. Pain Med. 2016;17:1010–22.
70. Monfett M, Harrison J, Boachie-Adjei K, Lutz G. Intradiscal platelet-rich plasma (PRP) injections for discogenic low back pain: an update. Int Orthop. 2016;40:1321–8.
71. Tuakli-Wosornu YA, Terry A, Boachie-Adjei K, et al. Lumbar intradiskal platelet-rich plasma (PRP) injections: a prospective, double-blind, randomized controlled study. PM R. 2016;8:1–10.
72. Kirchner F, Anitua E. Intradiscal and intra-articular facet infiltrations with plasma rich in growth factors reduce pain in patients with chronic low back pain. J Craniovertebr Junction Spine. 2016;7:250–6.
73. Akeda K, Ohishi K, Masuda K, et al. Intradiscal injection of autologous platelet-rich plasma releasate to treat discogenic low back pain: a preliminary clinical trial. Asian Spine J. 2017;11:380–9.
74. Cheng J, Santiago KA, Nguyen JT, Solomon JL, Lutz GE. Treatment of symptomatic degenerative intervertebral discs with autologous platelet-rich plasma: follow-up at 5–9 years. Regen Med. 2019;14:831–40.
75. Schepers MO, Groot D, Kleinjan MM, Mylenbusch H, Klopper-Kes AHJ. Effectiveness of intradiscal platelet rich plasma for discogenic low back pain without Modic changes: a randomized controlled trial. Int Pain Med. 2022;1:100011.
76. Zielinski MA, Evans NE, Bae H, et al. Safety and efficacy of platelet rich plasma for treatment of lumbar discogenic pain: a prospective, multicenter, randomized, double-blind study. Pain Physician. 2022;25:29–34.
77. Lutz C, Cheng J, Prysak M, Zukofsky T, Rothman R, Lutz G. Clinical outcomes following intradiscal injections of higher-concentration platelet-rich plasma in patients with chronic lumbar discogenic pain. Int Orthop. 2022;46:1381–5.
78. Pang X, Yang H, Peng B. Human umbilical cord mesenchymal stem cell transplantation for the treatment of chronic discogenic low back pain. Pain Physician. 2014;17:E525–30.
79. Pettine KA, Suzuki RK, Sand TT, Murphy MB. Autologous bone marrow concentrate intradiscal injection for the treatment of degenerative disc disease with three-year follow-up. Int Orthop. 2017;41:2097–103.
80. Pettine K, Suzuki R, Sand T, Murphy M. Treatment of discogenic back pain with autologous bone marrow concentrate injection with minimum two year follow-up. Int Orthop. 2016;40:135–40.
81. Pettine KA, Murphy MB, Suzuki RK, Sand TT. Percutaneous injection of autologous bone marrow concentrate cells significantly reduces lumbar discogenic pain through 12 months. Stem Cells. 2015;33:146–56.
82. Mochida J, Sakai D, Nakamura Y, et al. Intervertebral disc repair with activated nucleus pulposus cell transplantation: a three-year, prospective clinical study of its safety. Eur Cell Mater. 2015;29:202–12.
83. Coric D, Pettine K, Sumich A, Boltes MO. Prospective study of disc repair with allogeneic chondrocytes presented at the 2012 Joint Spine Section Meeting. J Neurosurg Spine. 2013;18:85–95.
84. Orozco L, Soler R, Morera C, et al. Intervertebral disc repair by autologous mesenchymal bone marrow cells: a pilot study. Transplantation. 2011;92:822–8.
85. Meisel HJ, Ganey T, Hutton WC, et al. Clinical experience in cell based therapeutics: intervention and outcome. Eur Spine J. 2006;15:S397–405.
86. Noriega DC, Ardura F, Hernández-Ramajo R, et al. Intervertebral disc repair by allogeneic mesenchymal bone marrow cells: a randomized controlled trial. Transplantation. 2017;101:1945–51.
87. Amirdelfan K, Bae H, McJunkin T, et al. Allogeneic mesenchymal precursor cells treatment for chronic low back pain associated with degenerative disc disease: a prospective randomized, placebo-controlled 36-month study of safety and efficacy. Spine J. 2021;21:212–30.
88. Atluri S, Murphy MB, Dragella R, et al. Evaluation of the effectiveness of autologous bone marrow mesenchymal stem cells in the treatment of chronic low back pain due to severe lumbar spinal degeneration: a 12-month, open-label, prospective controlled trial. Pain Physician. 2022;25:193–207.
89. Wolff M, Shillington JM, Rathbone C, Piasecki SK, Barnes B. Injections of concentrated bone marrow aspirate as treatment for Discogenic pain: a retrospective analysis. BMC Musculoskelet Disord. 2020;21:135.
90. Manchikanti L, Falco FJE, Singh V, Hirsch JA. Discography. In: Manchikanti L, Singh V, Falco FJE, Kaye AD, Soin A, Hirsch JA, editors. Manchikanti's essentials of interventional techniques in managing chronic pain. 2nd ed. New York: Springer; 2023.
91. Manchikanti L, Falco FJE, Singh V, Hirsch JA. Discography. In: Manchikanti L, Kaye AD, Falco FJE, et al., editors. Essentials of interventional techniques in managing chronic pain. Springer; 2018. p. 273–300.
92. Falco FJE, Zhu J, Irwin L, et al. Lumbar discography. In: Manchikanti L, Singh V, editors. Interventional techniques in chronic spinal pain. Paducah: ASIPP Publishing; 2007. p. 527–52.
93. Falco FJE, Irwin L, Zhu J, et al. Cervical discography. In: Manchikanti L, Singh V, editors. Interventional techniques in chronic spinal pain. Paducah: ASIPP Publishing; 2007. p. 567–80.
94. Falco FJE, Zhu J, Irwin L, et al. Thoracic discography. In: Manchikanti L, Singh V, editors. Interventional techniques in chronic spinal pain. Paducah: ASIPP Publishing; 2007. p. 553–66.
95. Parke WW, Bono CM, Garfin SR. Applied anatomy of the spine. In: Herkowitze HN, Garfin SR, Eismont FJ, Bell GR, Balderston RA, editors. The spine. 5th ed. Philadelphia: Saunders Elsevier; 2006. p. 16–54.
96. Fraser HF, Osti OL, Vernon-Roberts B. Discitis after discography. J Bone Joint Surg Br. 1987;69:26–35.
97. Osti OL, Fraser HF, Vernon-Roberts B. Discitis after discography. The role of prophylactic antibiotics. J Bone Joint Surg Br. 1990;72:271–4.
98. Kambin P, Abda S, Kurpicki F. Intradiskal pressure and volume recording: evaluation of normal and abnormal cervical disks. Clin Orthop. 1980;146:144–7.
99. Laun A, Lorenz R, Agnoli AL. Complications of cervical discography. J Neurosurg Sci. 1981;25:17–20.
100. van Heeswijk VM, Thambyah A, Robertson PA, Broom ND. Does an annular puncture influence the herniation path? Spine (Phila Pa 1976). 2018;43:467–76.
101. Connor PM, Darden BV II. Cervical discography complications and clinical efficacy. Spine (Phila Pa 1976). 1993;18:2035–8.
102. Zeidman SM, Thompson K, Ducker TB. Complications of cervical discography: analysis of 4400 diagnostic disc injections. Neurosurgery. 1995;37:414–7.

103. Guyer RD, Ohnmeiss DD, Mason SL, et al. Complications of cervical discography: findings in a larger series. J Spinal Disord. 1997;10:95–101.
104. Singh V. Thoracic discography. Pain Physician. 2004;7:451–8.
105. Szypryt E, Hardy J, Hinton C, et al. A comparison between magnetic resonance imaging and scintigraphic bone imaging in the diagnosis of disc space infection in an animal model. Spine (Phila Pa 1976). 1988;13:1042–8.
106. Arrington JA, Murtagh FR, Silbiger ML, et al. Magnetic resonance imaging of post discogram discitis and osteomyelitis in the lumbar spine: case report. J Fla Med Assoc. 1986;73:192–4.
107. Kaye AD, Manchikanti Lm Novitch MB, et al. Responsible, safe, and effective use of antithrombotics and anticoagulants in patients undergoing interventional techniques: American Society of Interventional Pain Physicians (ASIPP) guidelines. Pain Physician. 2019;22:S75–S128.
108. Manchikanti L, Sanapati M, Kaye AD, Kaye AM. Antiplatelet and anticoagulant therapy in regenerative medicine. In: Manchikanti L, Navani A, Atluri S, Sanapati M, editors. Manchikanti's essentials of regenerative medicine in interventional pain management. 2nd ed. New York: Springer; 2023.

# Epidural Administration of Biologics

# 24

Laxmaiah Manchikanti, Sairam Atluri,
Mahendra Sanapati, and Joshua A. Hirsch

## Introduction

Spinal pain and related disabilities are common [1–5]. Spinal and extremity pain secondary to degenerative disc disease with disc herniation, disc protrusion, or disc extrusion; central or foraminal stenosis; discogenic pain without disc herniation; and post lumbar surgery syndrome have been amenable to surgical and non-surgical interventions, including epidural injections [6–13].

Access to the epidural space is available by interlaminar and transforaminal approaches in the cervical and thoracic spine; both approaches and caudal access are used in the lumbar spine [6–13]. Substantial differences in techniques and outcomes exist among the three approaches, spinal levels, and disease patterns [6–13].

Epidural injections are performed using local anesthetic with or without steroids, or with sodium chloride solution and steroids. In recent years, the value of epidural administration of autologous products including platelet-rich plasma (PRP) and medicinal signaling cells or mesenchymal stem cells (MSCs) has been described [14–18]. However, the literature for PRP and stem cell therapy with epidural administration lags behind their use in the peripheral joints, for a variety of musculoskeletal conditions, and into degenerated discs. Recognition of the role of inflammation producing radiculitis following disc herniation instead of simple mechanical compression of lumbar nerve roots by the herniated nucleus pulposus has led to the development of antiinflammatory therapies, including solutions based on regenerative medicine principles [14–16].

New concepts regarding the use of biological agents, and access to the target, continue to emerge. For example, there is a need to minimize the use of contrast agents, which are toxic to biological tissues. Another is target localization and administration of biologicals with the caudal approach, specifically in the post-surgery syndrome, which has to be performed with adhesiolysis and catheterization [6–13].

## History

The first description of epidural injections was in 1901 with caudal epidural injections. In 1933, interlaminar epidural injections using the loss of resistance technique was described, followed by transforaminal epidural injections in 1952 [10–12], and adhesiolysis in the 1980s [13].

## Pathophysiology

Tissues in the lower back capable of transmitting pain include the disc, nerve roots dura, muscle, ligaments, fascia, and facet joints [19]. Pain from lumbar disc herniation can arise from nerve root compression and stimulation of nociceptors in the anulus or posterior longitudinal ligament (PLL) (Fig. 24.1). Mixter and Barr [20] in 1934 described intervertebral disc herniation, which led many practitioners to assume that intervertebral disc herniation is the most common cause of back problems. The simple concept of compression, or mass effect, is insufficient to explain the variety of presentations [21].

A multitude of mechanisms have been proposed to explain radicular pain, which include partial axonal damage, neuroma formation, focal demyelination, intraneural edema, impaired microcirculation, chemical irritation, and inflammation [6–13, 21–23]. Inflammatory reactions between the

L. Manchikanti (✉)
Pain Management Centers of America, Paducah, KY, USA
e-mail: drlm@thepainmd.com

S. Atluri
Tri-State Spine Care Institute, Cincinnati, OH, USA

M. Sanapati
Pain Management Centers of America, Evansville, IN, USA

J. A. Hirsch
Massachusetts General Hospital and Harvard Medical School, Boston, MA, USA
e-mail: jahirsch@mgh.harvard.edu


A. Navani et al. (eds.), *Essentials of Regenerative Medicine in Interventional Pain Management*,
https://doi.org/10.1007/978-3-031-50357-3_24

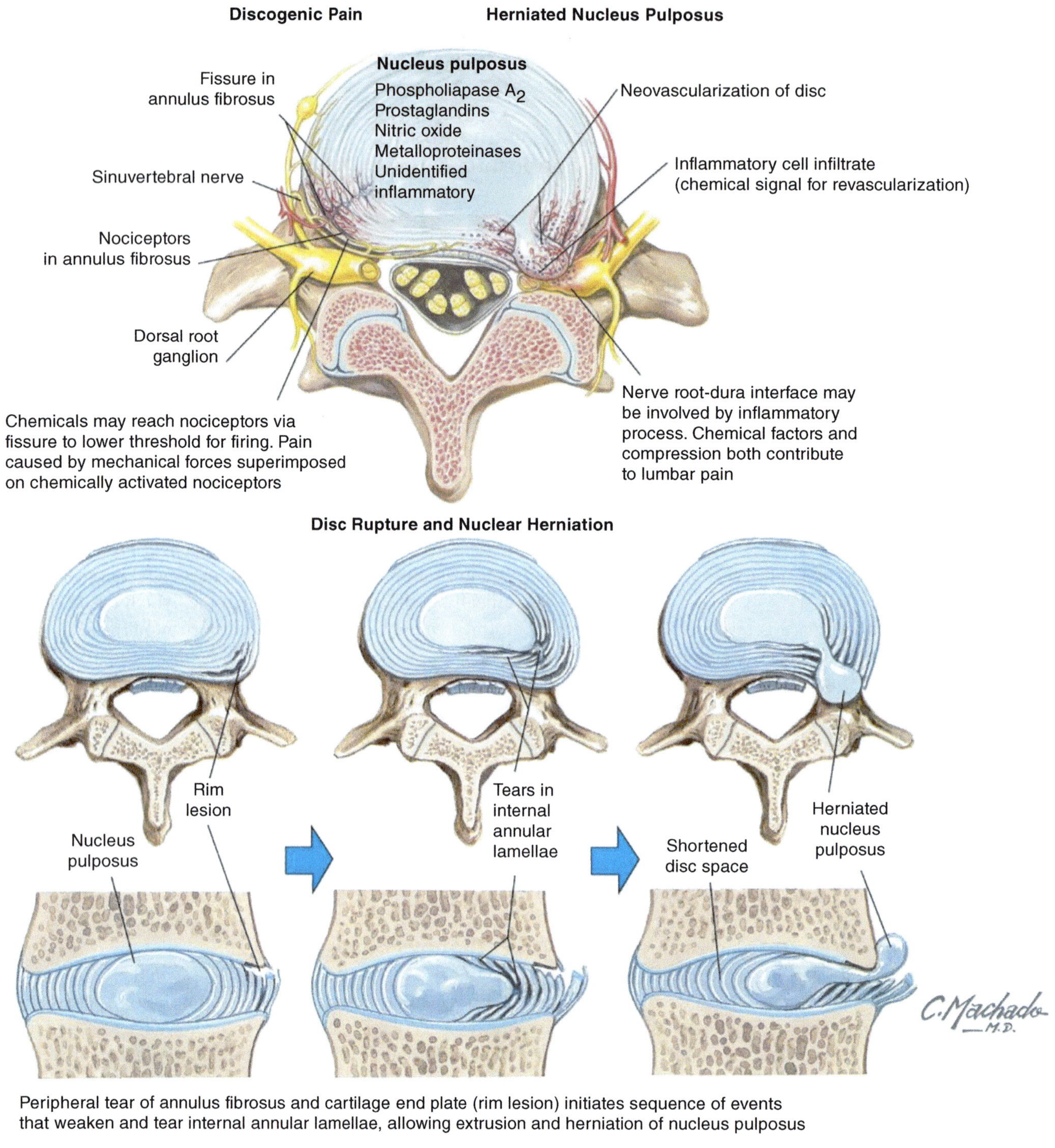


**Fig. 24.1** Pathophysiology of lumbar disc degeneration, disc herniation, and discogenic pain. (**Reproduced from:** Netter Medical Illustrated. Used with permission of Elsevier)

nucleus pulposus and nerve roots have been suggested as playing an important role in the production of sciatica from disc herniation [6–13, 21–23].

Intervertebral disc disease without herniation has been implicated as a source of spinal pain based on decades of pre-clinical, clinical, and epidemiological research, though the precise mechanisms still continue to be debated [6–13, 21–23]. Low back pain without disc herniation was described by Mixter and Ayers [22] soon after the description of disc herniation [20]. In fact, diagnostic discography showed the prevalence of discogenic pain without disc herniation in 16.9–26% of patients with low back pain, whereas it showed internal disc disruption in 16.9–42% with cervical discogenic pain prevalence ranging from 16% to 53% [23].

The term "spinal stenosis" is used by some to indicate any type of narrowing of the central canal, with the term "foraminal stenosis" used for narrowing of the lateral canals containing the exiting nerve roots. Others use the term "spinal stenosis" for both the central and the foraminal forms. It may be local, segmental, or generalized, congenital or acquired [7]. Central spinal stenosis may be congenital, or may result from disc bulging, protrusion, extrusion (protrusion and extrusion are more specific and have largely replaced "herniation"), ossification and thickening of ligamenta flava, ossification of the PLL, osteophytosis, and arthritic changes of facet joints. Further, spinal stenosis is a multifactorial disorder with the pathogenesis of neurogenic claudication in lumbar spinal canal stenosis explained by mechanical forces (compression of both nerve roots and the subarachnoid spaces), circulatory disturbances with venous congestion, and injury to the nerves with Wallerian degeneration.

Lumbar post-surgery syndrome is one of the causes of continued persistent pain and disability with low back and lower extremity pain, requiring a re-operation rate of 9.5–25% [13, 24, 25]. The unremitting pain and disability in the low back and lower extremities following lumbar spine surgery has multiple causes including epidural fibrosis, sacroiliac joint pain, disc herniation, discogenic pain, spinal stenosis, arachnoiditis, and facet joint pain [13, 24, 25].

The underlying mechanisms of action for epidurally administered local anesthetics and steroids have been described [6–13]. The various modes of action of corticosteroids include membrane stabilization, inhibition of neural peptide synthesis or action, blockade of phospholipase A2 activity, prolonged suppression of ongoing neuronal discharge, and suppression of sensitization of dorsal horn neurons. Local anesthetics also have been shown to have a significant effect in relieving low back and lower extremity pain, both in experimental and clinical settings [6–13]. Local anesthetics have been postulated to provide relief by multiple mechanisms which include suppression of nociceptive discharge, the blockade of the sympathetic reflex arc, the blockade of axonal transport, the blockade of sensitization, and anti-inflammatory effects [6–13].

## Epidural Biologics

The most pressing area of investigation involves the role that epidural PRP or MSCs injections play in the treatment of disc herniation, any associated radiculopathy, radiculopathy without disc herniation, and in other biochemical and mechanical disorders [26–28]. The nucleus pulposus contains a variety of inflammatory pain mediators, including phospholipase A2, nitric oxide, and prostaglandin E. In addition, cytokines such as interleukin 1 (IL-1) have been identified as mediators of inflammatory and degenerative changes [14–23, 26, 27, 29]. It has been hypothesized that the disc material, with inflammatory substances, causes direct toxic injury to the nerve root by chemical mediation which subsequently amplifies intra- and extraneural inflammation. This results in venous congestion and conduction block [28, 30]. Of the multiple cytokines identified within the disc, IL-1—is of special interest regarding its role in the causation of lower back pain [31]. Strategies for inhibiting the biological activities of IL-1 include the use of IL-1 receptor antagonist (-ra), soluble forms of IL-1 receptors, and Type-1 cytokines such as IL-4, IL-10, and IL-13 that inhibit synthesis of IL-1 and/or increase the synthesis of IL-1ra [15]. The therapeutic use of cytokine inhibitors and growth factors was proposed in the late 1970s and early 1980s [14–23]. In fact, some of the proponents of epidural injections believe that epidural injections produce anti-inflammatory effects with or without the use of steroids [8–13, 32–34]. The role of steroids in epidural injections for managing discogenic pain continues to be debated and remains controversial [34].

With significant developments in biologicals as an evolutionary model, PRP and its derivatives, along with MSCs, have been proposed for epidural administration. Autologous condition serum (ACS) preparations have been described as a source of anti-inflammatory cytokines, including IL-4, IL-10, IL-13, and IL-1ra, and also contain elevated concentrations of growth factors like fibroblast growth factor (FGF-2), hepatocyte growth factor (HGF), and transforming growth factor beta (TGF-β1) [31]. ACS contains high concentrations of IL-1ra, an antagonist to IL-1 that is a "biochemical sensitizer" of nerve roots in radiculopathy [35, 36].

## Evidence Base

Evidence of effectiveness is determined based on best evidence synthesis ranging from Levels I to V, with Level I evidence being the highest level of evidence obtained from multiple relevant high-quality, randomized controlled trials (RCTs) [37], whereas Level IV and V evidence is based on observational studies and consensus as shown in Table 24.1.

**Table 24.1** Qualitative modified approach to grading of evidence

| | | |
|---|---|---|
| Level I | Strong | Evidence obtained from multiple relevant high-quality randomized controlled trials |
| Level II | Moderate | Evidence obtained from at least one relevant high-quality randomized controlled trial or multiple relevant moderate or low-quality randomized controlled trials |
| Level III | Fair | Evidence obtained from at least one relevant moderate or low-quality randomized trial<br>or<br>Evidence obtained from at least one relevant high-quality non-randomized trial or observational study with multiple moderate or low-quality observational studies |
| Level IV | Limited | Evidence obtained from multiple moderate or low-quality relevant observational studies |
| Level V | Consensus based | Opinion or consensus of large group of clinicians and/or scientists |

**Modified from:** Manchikanti L, Falco FJE, Benyamin RM, Kaye AD, Boswell MV, Hirsch JA. A modified approach to grading of evidence. *Pain Physician* 2014; 17:E319–E325 [37]

ACS has been considered as a promising new treatment option for patients with radicular pain. ACS has been studied in one RCT [36] and in a prospective assessment [35]. Similarly, the epidural injection of PRP with its multiple growth factors has also been studied [38]. In addition, multiple other innovations including plasma lysate [39] and plasma rich in growth factors (PRGF)-Endoret [40] have been studied for epidural use. However, the literature has been tainted with flawed studies of spinal injections providing inadequate evidence [14–18, 36, 38–46]. Despite this, single arm meta-analysis did show moderate results with Level IV (on a scale of Level I–V) evidence using lumbar epidural injections of PRP or its derivatives.

Table 24.2 and Fig. 24.2 show summary characteristics of studies of lumbar epidural injections of PRP and changes in pain scores after treatment.

There was only one study assessing the role of biologics in the cervical spine, which described the efficacy of epidural perineural injection of ACS in unilateral cervical radiculopathy in the form of a pilot study [18]. They also

**Table 24.2** Summary of lumbar epidural injection PRP studies published to date

| Study details | Population | Cell/solution type | Cell or solution dose and delivery pathway | Outcome parameters | Results | Conclusion |
|---|---|---|---|---|---|---|
| Becker et al. 2007 [36]<br>Sample size = 84<br>Follow-up = 6 months<br>Randomized controlled trial | 84 patients with chronic lumbar radicular pain<br>Patient age (years) = 29–81 (mean 53.9) | IL-1 RA-enriched autologous conditioned serum (Orthokine; Autologous conditioned serum [ACS]) | Transforaminal or perineural injection of ACS once per week for 3 consecutive weeks. Perineural space was approached through interlaminar space with injection of 1 mL of local anesthetic and autologous serum in Group 3. Groups 1 and 2 received 1 mL of local anesthetic plus 10 mg of triamcinolone or 5 mg of triamcinolone | VAS, ODI | • Patients in all 3 groups improved comparably. Results were significant within each treatment group from week 12 to the final evaluation at week 26, injections with ACA showed a consistent pattern of superiority over both triamcinolone groups with VAS, with no significant difference for ODI among the groups | • ACS showed similar response to triamcinolone injection with local anesthetic initially; however, with significant improvement at 26 weeks with pain relief, but not with function. Results similar to local anesthetic and steroid injections<br>• One of the limitations in this trial is that of multiple injections |
| Centeno et al. 2017 [39]<br>Sample size = 470<br>Follow-up = 2 years<br>Prospective registry | Patients suffering with lumbar radicular pain and MRI findings that were consistent with symptoms | Platelet lysate | • Transforaminal or interlaminar epidural injection under fluoroscopic guidance. The final injectate consisted of plasma lysate 50% by volume, 4% lidocaine at 25% by volume, and compounded preservative free 100–200 ng/mL of hydrocortisone at 25% by volume<br>• For transforaminal and interlaminar epidural injections, 3–5 $cm^3$ volume was injected | Numeric pain score (NPS), modified single assessment numeric evaluation (SANE) rating, functional rating index (FRI) | • Patients treated with platelet lysate epidurals reported significantly lower NPS and FRI change scores at all time points compared to baseline<br>• Posttreatment FRI change score means exceeded the minimal clinically important difference beyond 1 month<br>• Average modified SANE rating showed 49.7% improvement at 24 months posttreatment<br>• 29% or 6.3% patients reported mild adverse events related to treatment | • Positive results with registry presenting real world data<br>• A large proportion of patients withdrawn<br>• 36% of the patients failed to provide baseline NPS data<br>• Information on number of procedures is unknown<br>• 26% of the patients failed to show any significant improvement |

**Table 24.2** (continued)

| Study details | Population | Cell/solution type | Cell or solution dose and delivery pathway | Outcome parameters | Results | Conclusion |
|---|---|---|---|---|---|---|
| Kumar et al. 2015 [35]<br>Sample size = 20 patients<br>Follow-up = 6 months<br>Prospective evaluation | 20 patients with unilateral lumbar radiculopathy | Autologous conditioned serum (ACS) | • Interlaminar injection with 2 mL of contrast followed by 2 mL of ACS<br>• A maximum of 3 injections at 7 days interval were given based on the clinical response<br>• 20 patients received number of injections ranging from 1 to 3 | VAS, ODI | • There was significant improvement in pain relief and function with VAS improving from 6.95 to 2.0 and ODI improving from 27.9 to 8.5<br>• 5 patients (20%) had complications which were immediate and systematic rather than local and were of short duration lasting less than 30 min | • Positive results with injection of ACS with disease modifying course with reduction of pain and disability<br>• Small sample size with injection of contrast, which may diminish the effect of ACS |
| Kirchner and Anitua 2016 [40]<br>Sample size = 86 patients<br>Follow-up = 6 months<br>Observational retrospective pilot study | 86 patients<br>47 females: age range 22–81, median age 58<br>39 males: age range 29–79 years, median age 55 | 86 patients each received one intradiscal, one intraarticular facet and one transforaminal epidural injection of plasma rich in growth factors | Transforaminal injection of 2 mL of activated PRGF after intradiscal injection of 4 mL of PRGF and facet joint infiltration | VAS | • At the end of the study (6 months), 90.7% of the patients showed an excellent score, 8.1% showed a moderate VAS score, and 1.2% of patients were included in the ineffective score group<br>• There were no major adverse events | • Approximately 90% of the patients improved after 6 months<br>• Positive study<br>• Results are unreliable because of observational study and multiple injections into the disc, facet joint and with transforaminal application<br>• Number of injections are unknown |
| Bhatia and Chopra 2016 [38]<br>Sample size = 10 patients<br>Follow-up = 3 months<br>Prospective evaluation | 10 patients with findings of lumbar disc herniation/ prolapse in MRI, or either sex with age less than 65 years | Platelet-rich plasma | Interlaminar epidural injection with 2 mL of PRP | VAS and ODI | Patients who had received epidural injections of autologous PRP showed improvements in their scores of evaluation tools. Improvement was sustained during the 3 month study period and was not associated with any complications | Autologous PRP can be considered as a good alternative to epidural steroids and surgery in management of patients with chronic prolapsed intervertebral disc |

*ODI* Oswestry Disability Index; *VAS* Visual Analog Scale; *PRGF* platelet-rich growth factor; *PRP* platelet-rich plasma

**Adapted from:** Sanapati J, Manchikanti L, Atluri S, et al. Do regenerative medicine therapies provide long-term relief in chronic low back pain: A systematic review and metaanalysis. *Pain Physician* 2018; 21:515–540 [14]

| Study name | Difference in means | Standard error | Variance | Lower limit | Upper limit | Z-Value | p-Value |
|---|---|---|---|---|---|---|---|
| | **Statistics for each day** | | | | | | |
| HS Kumar 1976 | -49.500 | 3.852 | 14.835 | -57.049 | -41.941 | -12.852 | 0.000 |
| Becker C 2007 | -54.500 | 5.256 | 27.625 | -64.801 | -44.199 | -10.369 | 0.000 |
| Kirchner 2016 | -76.000 | 2.183 | 4.767 | -80.279 | -71.721 | -34.807 | 0.000 |
| | -60.278 | 9.691 | 93.906 | -79.272 | -41.285 | -6.220 | 0.000 |

| Heterogeneity Q-Value | df (Q) | p-value | I-squared | Tau squared | Tau |
|---|---|---|---|---|---|
| 43.082 | 2 | 0.0001 | 95.358 | 266.285 | 16.318 |

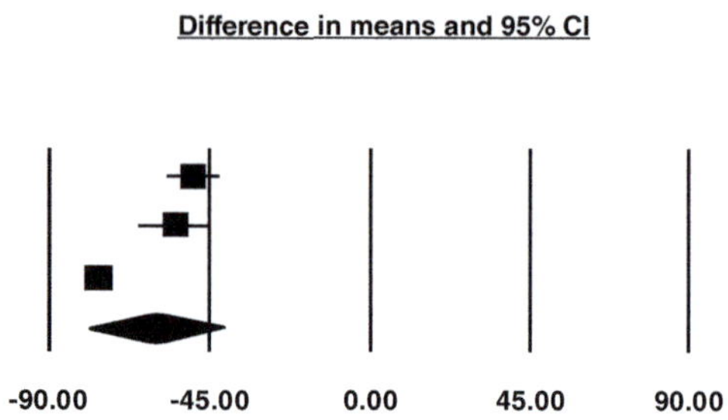


**Fig. 24.2** Changes in pain scores (0–100) after treatment (6 months follow data) of epidural PRP injections. (**Adapted from:** Sanapati J, Manchikanti L, Atluri S, et al**.** Do regenerative medicine therapies provide long-term relief in chronic low back pain: A systematic review and metaanalysis. *Pain Physician* 2018; 21:515–540 [14])

showed that patients receiving perineural injections of ACS or methylprednisolone both had improvements in the evaluation scores. However, the improvement in the ACS patients was gradual and sustained during the entire study period, whereas that in the methylprednisolone group had some deterioration over time. The authors concluded that ACS can be considered an equally good or better modality of nonoperative management in patients with unilateral cervical radiculopathy as methylprednisolone. Considering numerous complications related to cervical transforaminal epidural injections, the safety profile may be somewhat better with ACS compared to methylprednisolone injections [12].

## Indications

The indications for epidural injections (caudal, interlaminar, or transforaminal) with biologics are chronic moderate to severe spinal and extremity pain of at least 3 months duration which has failed to respond or poorly responded to nonsurgical conservative management resulting from:

- Disc herniation/lumbar radiculitis
- Lumbar spinal stenosis
- Axial or discogenic neck or low back pain
- Post lumbar surgery syndrome

Interlaminar injections may be performed in the post-surgery syndrome only if access to the epidural space is obtained above or below the scar. Caudal epidural access is the technique of choice for the post-surgery syndrome based on the level of pathology, with adhesiolysis and catheterization.

Finally, while the lumbar transforaminal approach has been utilized extensively for the treatment of disc herniation and spinal stenosis, its effectiveness has only been demonstrated with disc herniation, and the indications have not been established for using transforaminal epidural injections for axial or discogenic pain. Cervical and thoracic transforaminal epidural injections are associated with high risk [6].

## Anatomy

The human spine consists of 33 vertebrae, with 7 cervical vertebrae between the thorax and the skull, 12 thoracic vertebrae between the cervical spine and lumbar spine, 5 lumbar vertebrae inferior to the thoracic vertebrae, along with 5 sacral vertebrae fused into 1 single bone, the sacrum, followed by 4 coccygeal vertebrae fused into a single small triangular bone called the coccyx. Figures 24.3 and 24.4 show the spinal vertebrae and the regional differences between cervical, thoracic and lumbar vertebrae. In the intact spine, the central canal of each vertebral segment are aligned to form a continuous channel, which is referred to as the vertebral or spinal canal (Figs. 24.3 and 24.4).

The spine is often divided anatomically into the anterior, neuraxial, and posterior compartments (Fig. 24.5). The anterior compartment consists of the vertebral body and intervertebral disc. The neuraxial compartment includes all structures within the osseous and ligamentous boundaries of the spinal canal, including the PLL, ligamenta flava, dura, and arachnoid. The posterior compartment consists of facet joints and associated bony vertebral arch ligaments.

The epidural space lies between the dural sac and the osseo-ligamentous boundaries of the vertebral canal and is a narrow space (Fig. 24.5). It surrounds the dural sac and is bordered posteriorly by a right and a left ligamentum flavum which are joined in the midline posteriorly, and underlying periosteum, anteriorly by the PLL and vertebral bodies, and laterally by the pedicles and intervertebral foramina. The epidural space is widest in the midline underneath the junction of the laminae and narrows laterally beneath the zygapophysial joints. The actual size and shape of the epidural space is determined by the manner of attachment of the dural sac to the walls of the spinal canal, as well as the shape of the

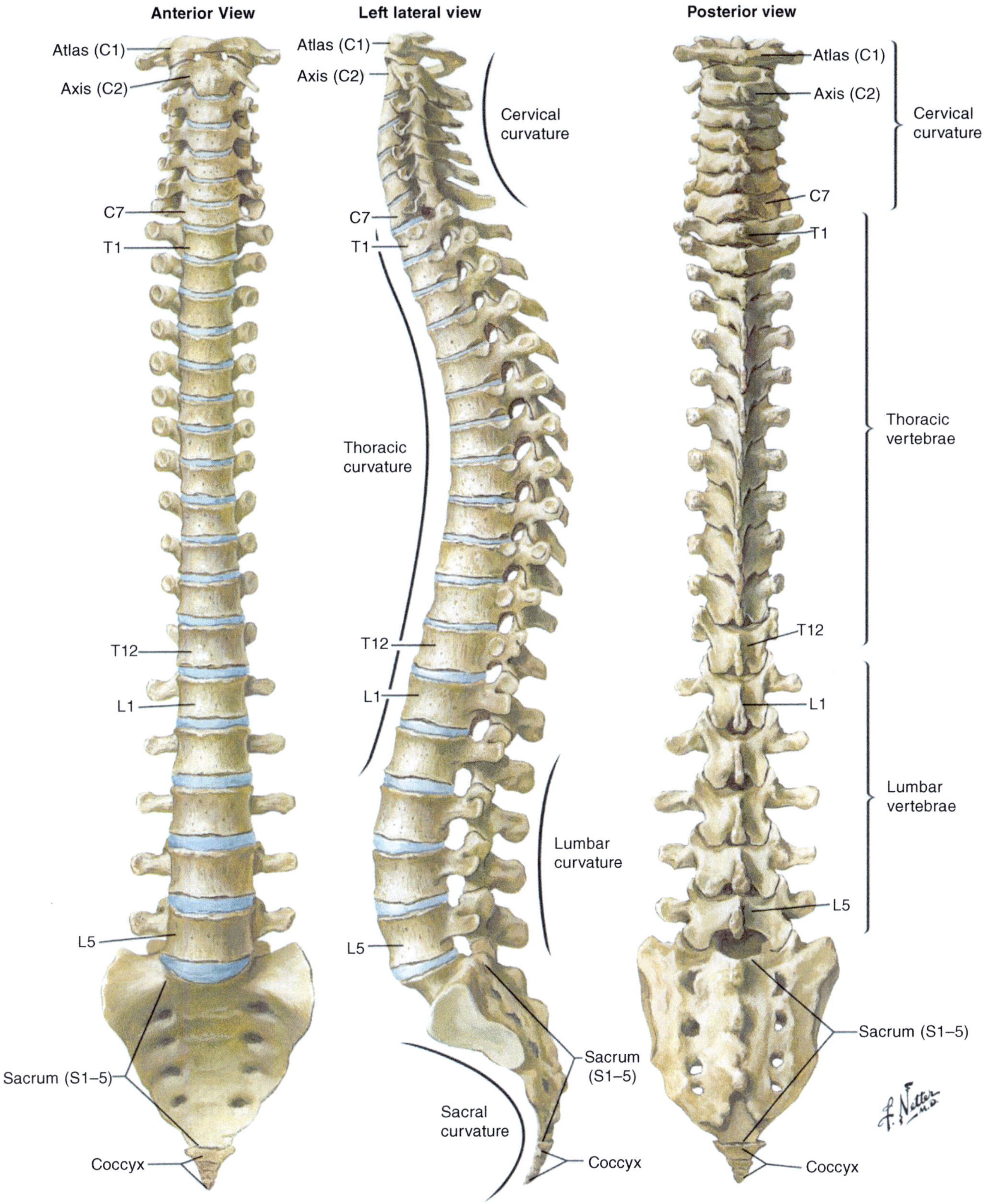


**Fig. 24.3** Anatomy of vertebral column. (**Reproduced from:** Netter Medical Illustrated. Used with permission of Elsevier)

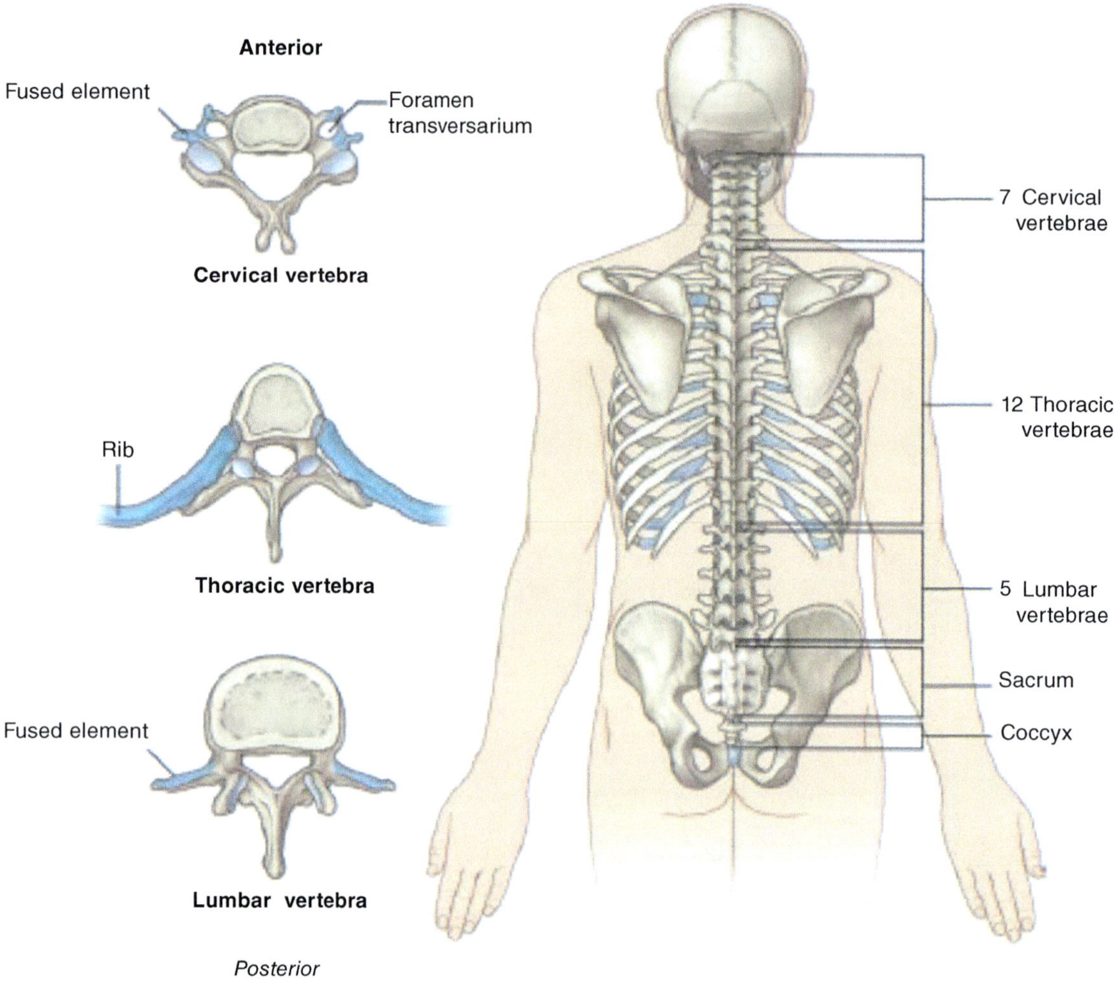


**Fig. 24.4** Anatomy of vertebral column with cranium, ribs, and pelvis, with descriptive characteristics of cervical, thoracic, and lumbar vertebrae. (**Reproduced from:** Gray's Anatomy for Students, Drake, © 2004, with permission from Elsevier)

spinal canal at different levels. The size of the lumbar epidural space is 4.0–6.0 mm and is 3.0 mm in the sacral canal. Contents of the epidural space include the vertebral venous plexus, the spinal branches of the segmental arteries, the lymphatics, and the dural and arachnoidal projections that surround the spinal nerve roots, along with abundant fat.

Cryomicrotome sections of the epidural space have shown that there is a variable degree of fusion of the ligamenta flava in the midline. The ligamenta flava in the lumbar spine are thicker than in the cervical and thoracic spine.

The anatomy of the neural foramina which contains the spinal nerves and of the vertebral (central or spinal) canal is of crucial importance. Ventral and dorsal segmental roots join to form the segmental spinal nerve that traverses the neural foramen, the spinal nerves dividing into the dorsal and ventral rami outside the foramen (Fig. 24.6). Spinal nerve roots leave the dural sac just above the level of each intervertebral foramen by penetrating the dural sac at variable angles in an inferolateral direction, taking with them an extension of dura and of arachnoid referred to as the dural sleeve. Immediately proximal to the formation of the spinal nerve, the dorsal root forms an enlargement, the dorsal root ganglion, which contains the cell bodies of sensory fibers in the dorsal root.

Epidural arteries are present in the epidural space and supply the surrounding bony and ligamentous structures as well as the spinal cord. Segmental radicular vessels enter the epidural space through the intervertebral foramina. These segmental arteries are derived from the intercostal, lumbar, and internal iliac arteries in the thoracolumbar spine, and from the vertebral, ascending cervical, and deep cervical arteries in the cervical spine. The spinal cord receives its blood supply through both longitudinal and segmental vessels, which produce an anastomotic longitudinal network of arterial vessels which surround the cord. Many of the arterial vessels that supply blood to the cord are end arteries and collateral blood supplies are lacking throughout much of the cord. Epidural veins are located in the epidural space. The vertebral venous plexus drains through multiple segmental epidural veins, which exit the spinal canal through the intervertebral foramina, thus allowing venous return to the inferior cava and azygous vein via the thoracic and abdominal veins.

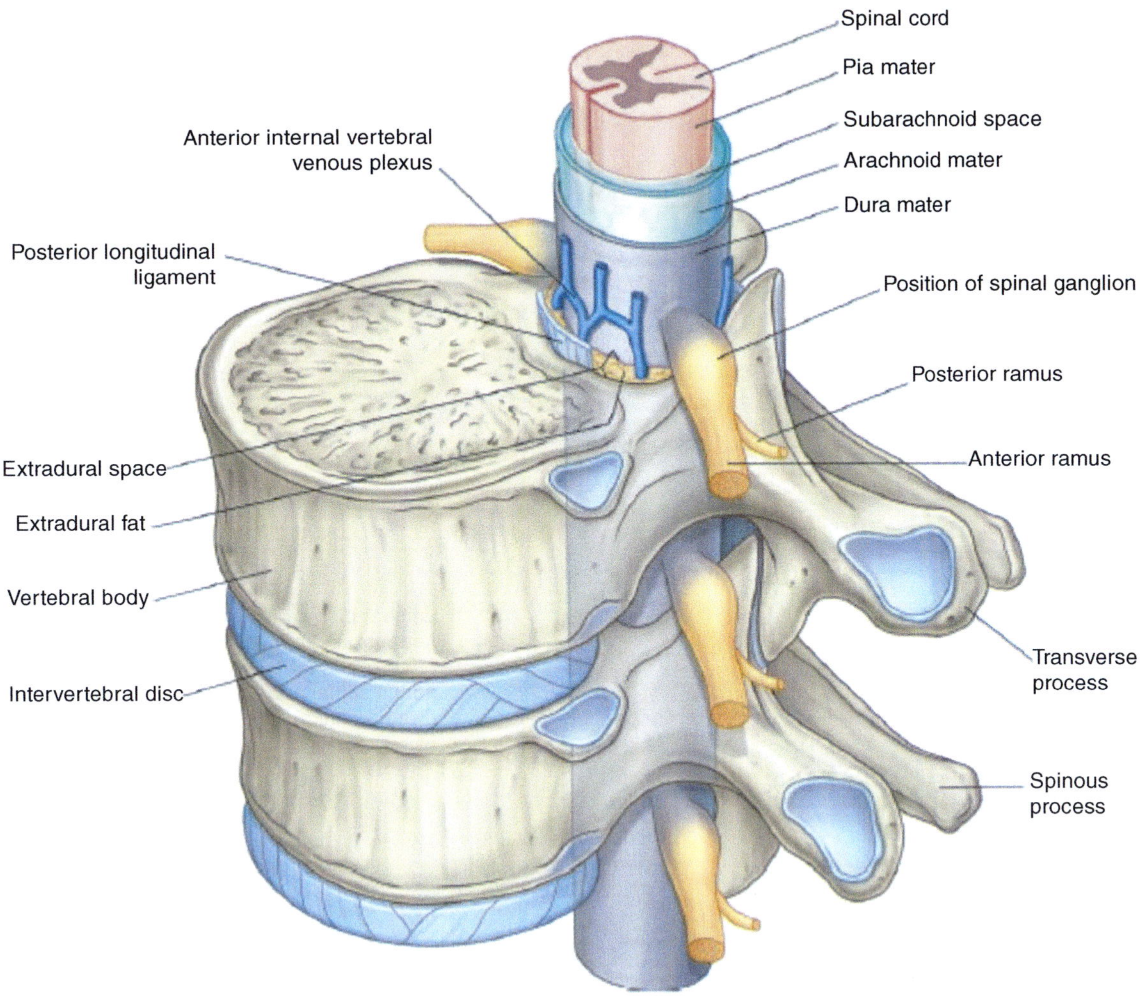


**Fig. 24.5** Lumbar vertebral canal and its contents, showing anterior, neuraxial, and posterior compartments

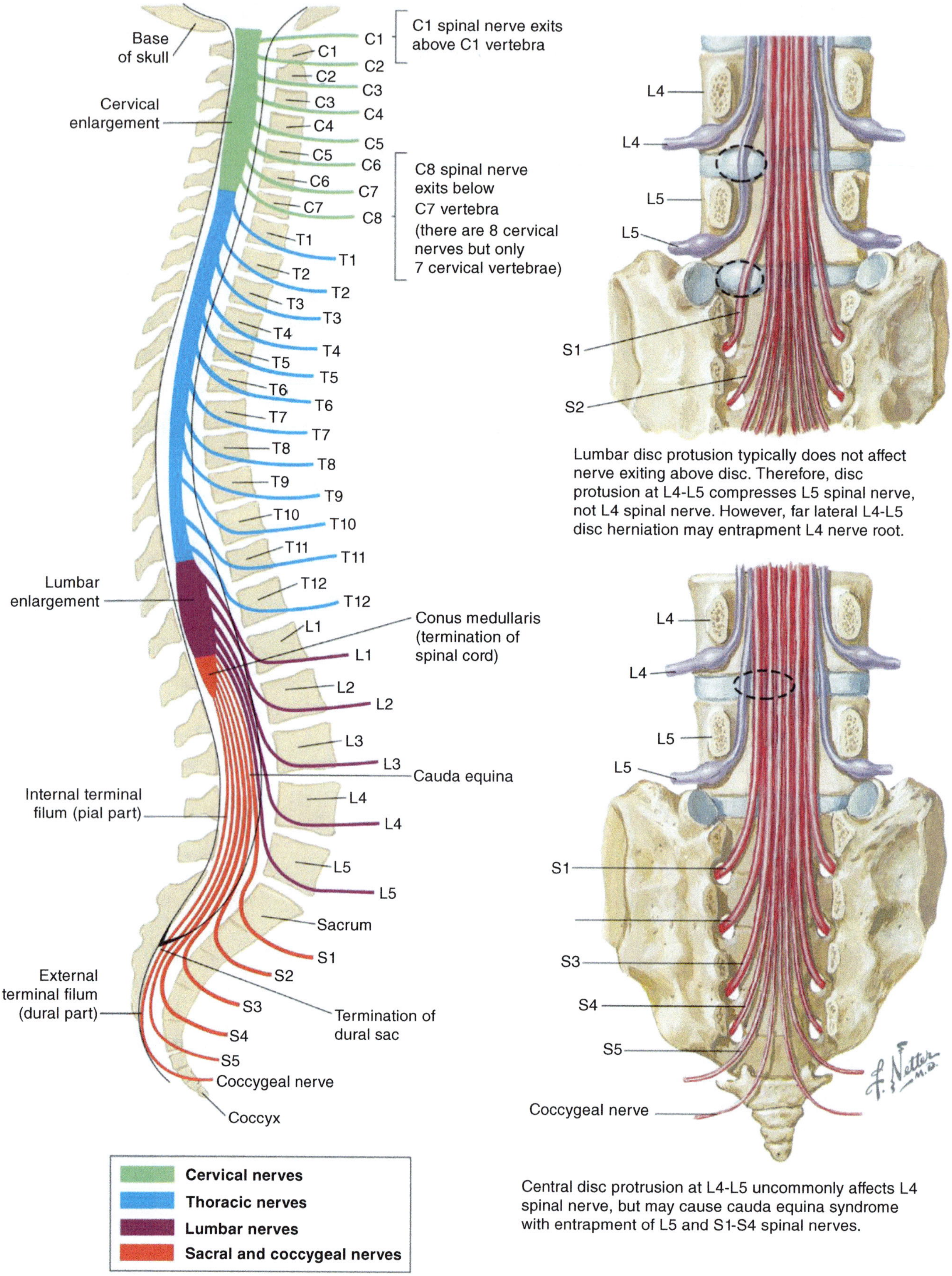


**Fig. 24.6** Illustration of vertebral canal, spinal cord, and relation to spinal nerve roots to vertebrae. (**Reproduced from:** Netter Medical Illustration. Used with permission of Elsevier)

## Technical Considerations

Lumbar epidural injections are administered under fluoroscopy by three approaches—interlaminar, transforaminal, and caudal, each different with its own technical considerations [11]. Cervical and thoracic epidural injections are safely administered via interlaminar access [12, 46].

### Lumbar Interlaminar Epidural Injections

The patient is placed in prone position with firm padding under the abdomen with appropriate sterile preparation. The desired interlaminar space is identified in the PA view, with the rotation of the C-arm toward the patient's right side or left side until the spinous process is exactly equidistant from the right and the left pedicles, and with cranial or caudal angulation until the desired interlaminar space is maximally opened. In many cases, cranial or caudal angulation is not required, especially if the L5-S1 space is chosen. Typically, the L5-S1 interlaminar space is the widest, and they are smaller—at L4-L5 and L3-L4. T12-L1 and L1-L2 are again easily accessible. The epidural space may be accessed with a midline, paramedian, or parasagittal approach at the interspace most closely associated with the patient's level of pain.

The skin is anesthetized with 1% preservative-free lidocaine. A Tuohy needle with or without a wing tip is placed in the midline or in a parasagittal or paramedian position. A parasagittal or paramedian approach is better for patients with unilateral radicular pain syndromes. Intermittent pressure is applied on the syringe plunger with advancement of the needle, with intermittent fluoroscopic visualization, until loss of resistance is felt, or needle position is observed in the epidural space. As the needle is advanced through the interspinous ligament and is noted to be in a direct midline or a paramedian position, the fluoroscope is positioned in a lateral view in order to ascertain the needle depth. The needle is then advanced into the ligamentum flavum, and subsequently into the epidural space, while intermittent lateral fluoroscopic views are obtained.

For a parasagittal approach, using the fluoroscope, the needle may be advanced towards the upper edge of the inferior lamina at the target interspace. The needle is advanced superiorly into the ligamentum flavum, and subsequently into the epidural space, utilizing the loss of resistance technique.

Once the needle has reached the epidural space with appropriate loss of resistance to either air or saline, 3–5 mL of nonionic contrast, (Omnipaque or Isovue), is injected to confirm epidural placement. Posteroanterior (PA) and lateral fluoroscopic images are obtained. The contrast spread may have an areolar appearance (Fig. 24.7). However, for injection of biologicals into the epidural space, injection of contrast must be minimized or eliminated. Subsequently, 3–5 mL of biologic solution is injected.

Subarachnoid puncture and injection or subarachnoid myelographic contrast patterns are observed in 1% with lumbar epidural injections [11, 47, 48]. Subdural contrast patterns are very rare but are more commonly seen with the interlaminar approach [11, 47, 49].

### Lumbar Transforaminal Epidural Injections

Lumbar transforaminal epidural injections are considered target-specific because the injectate is placed over the relevant nerve root which may maximize drug concentration and possibly diffuse toward the ventral epidural space of the nerve root and dorsal root ganglion related to the symptoms.

### Clinical Anatomy

Understanding of the anatomy of the nerve root and its relation to the vasculature is critically important to avoid major neurovascular complications [11]. The segmental radiculomedullary arteries are variable in number and location. They course through various neural foramina and carry oxygenated blood from various cervical, thoracic, and lumbosacral arteries into the longitudinal anterior and posterior spinal artery systems and ultimately into the parenchyma of the spinal cord (Fig. 24.8). Generally, the size of the radicular artery is less than 0.2 mm, whereas the size of the medullary artery is 0.2–0.8 mm. The largest of these arteries is called the artery of Adamkiewicz and is found in the thoracolumbar region where it carries blood from an intercostal or lumbar artery through a thoracic or lumbar neural foramen to supply a large portion of the middle and lower spinal cord [11]. There are multiple smaller radicular arteries from the intercostal, lumbar, and internal iliac arteries that vary in location and number which also contribute to the blood supply of the spinal cord. Figures 24.8 and 24.9 show the relationships of various structures in the intervertebral foramen. The main stem of the lumbar artery courses backwards, with anterior and posterior radicular arteries running upward along the nerve root before penetrating the dural sleeve just medial to the vertebral pedicle. Anatomical studies show that the radicular artery is located predominantly in the superior part of the neural foramen [11, 50]. Murthy et al. [50] in a retrospective review showed 97% of the radiculomedullary arteries are located in the superior one-half of a neural foramen as opposed to only 9% and 2% for mid and inferior thirds, respectively, as shown in Fig. 24.10. In fact, the artery of Adamkiewicz was located in the upper one-half of a foramen 97% of the time (110/113) and was never seen in the most inferior one-fifth of the foramen.

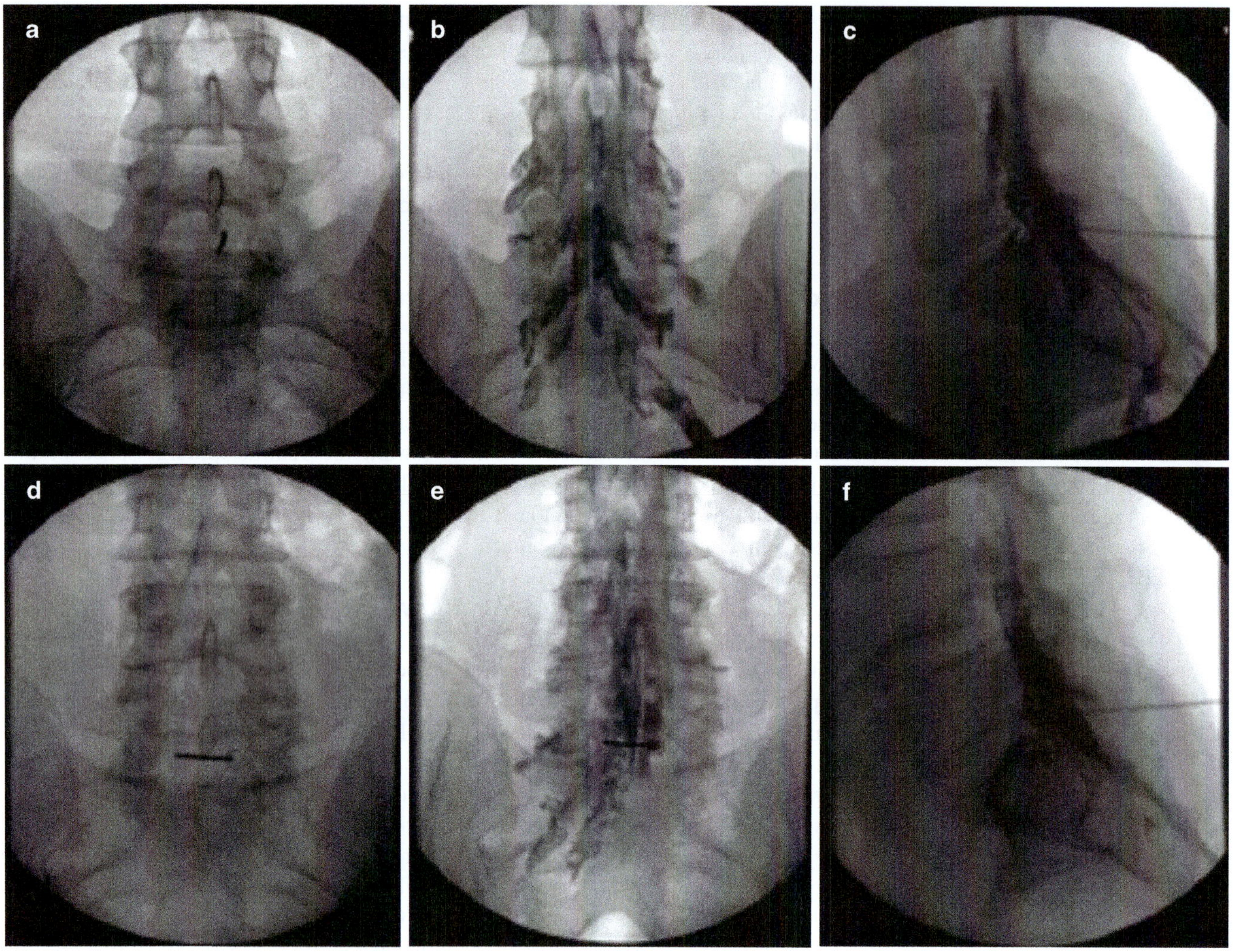


**Fig. 24.7** Lumbar interlaminar epidural injections with midline (**a**–**c**) and paramedian or parasagittal (**d**–**f**) approach. (**a**) Needle tip in PA view with placement in midline between L5 and S1 and slightly lateral to L5 spinous process. (**b**) PA view with injection of contrast with bilateral filling with typical illustration of nerve roots and air bubbles with a Christmas tree appearance. (**c**) Lateral view of the epidural filling pattern illustrating double-line railroad pattern showing ventral as well as dorsal epidural space. (**d**) PA view with needle position to paramedian or parasagittal between L5 and S1 interspace. (**e**) Contrast filling pattern observed in the PA view with typical nerve root filling at multiple levels predominantly on the left side with filling of epidural space and multiple nerve roots including L5, S1, and S2, with partial filling noted on the right side. (**f**) Lateral view with contrast filling predominantly with dorsal filling; however with railroad track pattern, also with ventral filling

## Technical Implications

A safe triangle was described by Bogduk corresponding to three sides corresponding to the horizontal base of the pedicle, the outer vertical border of the intervertebral foramen, and the connecting diagonal nerve root and dorsal ganglion (Figs. 24.8, 24.9, and 24.11). Consequently, a needle placed into the safe triangle will lie above and lateral to the nerve root. In this method, the injection needle is moved towards the safe triangle under the inferior surface of the pedicle to stop at the superolateral aspect of the spinal nerve. This approach has been favored because agents can be injected into the ventral epidural space, the inflammatory site at the junction of the herniated intervertebral disc, and the anterior nerve root dural sleeve. It also reduces the risk of damaging the dura, as the injection needle goes through the border of the lateral upper intervertebral foramen. However, 92% of radiculomedullary arteries are located in this location [11].

Multiple alternate approaches have been described, including the infraneural placement of the needle called Kambin's triangle approach, an inferior triangle approach, an inferior ganglion approach, and a retrodiscal approach, as shown in Fig. 24.12 [51]. Based on the analysis of the available anatomical studies and radiological studies, Atluri et al. [52] have identified the "inferior triangle." The boundaries of the inferior triangle in the oblique fluoroscopic view include the lateral border of the superior articular process (SAP) forming one side of the triangle, the transverse process (TP)

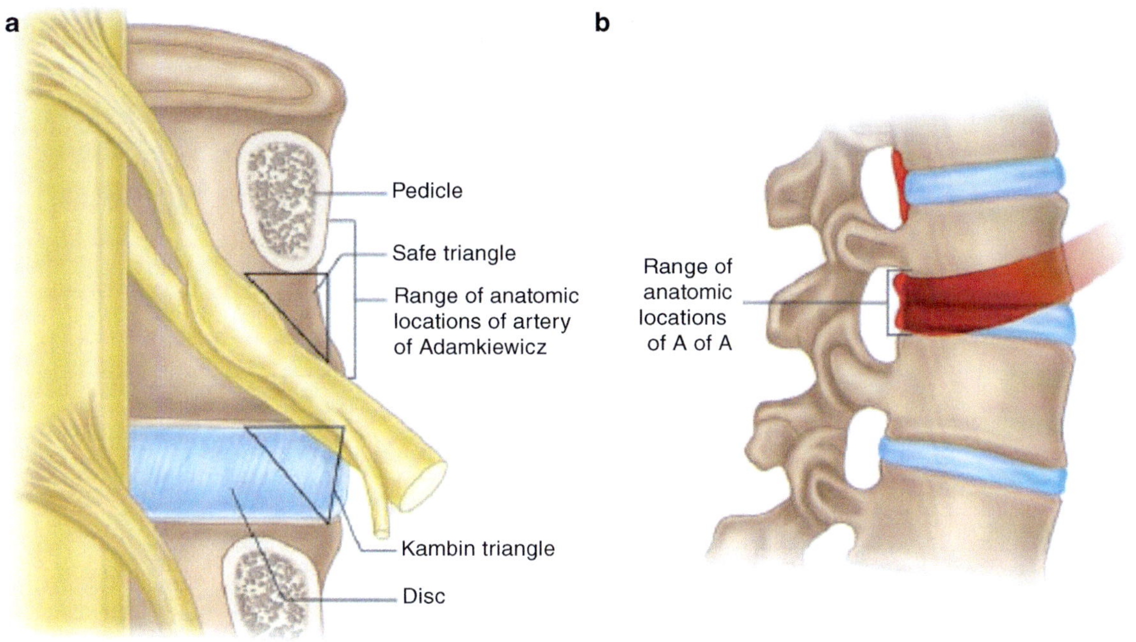


**Fig. 24.8** (**a**, **b**) Relationship of arteries supplying the spinal cord with nerve roots. A of A—artery of Adamkiewicz

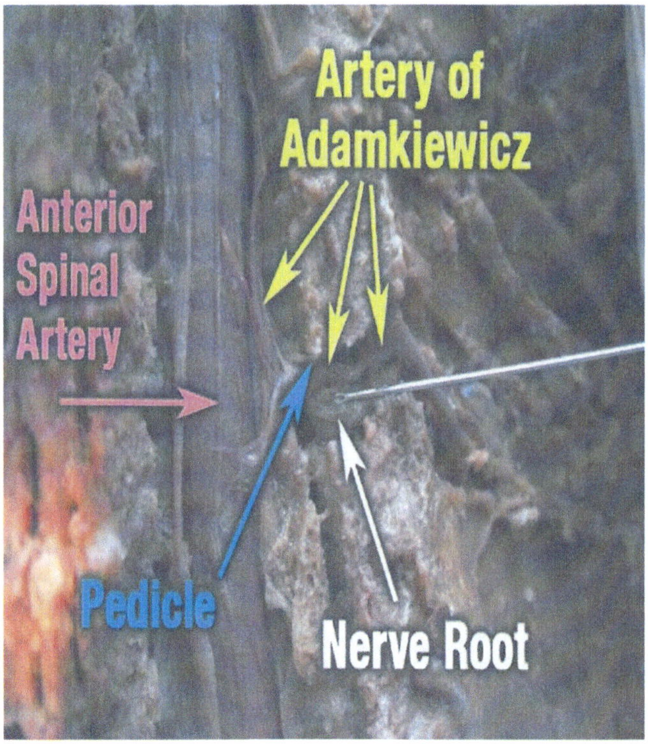


**Fig. 24.9** Relationship of Artery of Adamkiewicz to nerve root in safe triangle

forming the base, and the traversing nerve root forming the other side of the triangle. In a PA view for the posterior approach, the inferior two-thirds or middle half of the triangle is formed by the inferior margin of the superior pedicle and the superior margin of the inferior pedicle (Figs. 24.10 and 24.11). Thus, these approaches incorporate similar approaches when directing the needle placement in the inferior aspect of a foramen, thus avoiding a radicular artery (Figs. 24.10 and 24.11).

### Safe Triangle Approach

The safe triangle approach has been widely used with oblique or posterior approaches [11]. For the posterior approach, the patient is placed in the prone position with firm padding under the abdomen and the fluoroscopy unit positioned with the spinous process in the center of the spine. A diamond tipped, Quincke, Whitacre, Bella-D Coudé® or blunt needle is inserted into the skin over the lateral border between the two adjacent TPs at the target interspace, closer to the upper TP. The needle is advanced toward the lower edge of the TP, near its junction with the SAP. Alternatively, it may be directed towards the edge of the TP, at which time the needle may be retracted slightly and redirected toward the base of the appropriate pedicle and advanced very slowly to the final position, or the needle may also be directed towards the pedicle without this intermediate step. The needle position may be observed in lateral view, a small volume of contrast is injected, and the pattern of dispersion into the nerve root is noted. If the needle has penetrated the epiradicular membrane surrounding the nerve root, an appropriate and positive image of the nerve root will be seen on fluoroscopy, with appropriate dispersion of the contrast, as shown in Fig. 24.13.

In an oblique approach, the patient and the fluoroscopy unit are rotated as needed to provide an oblique projection of the

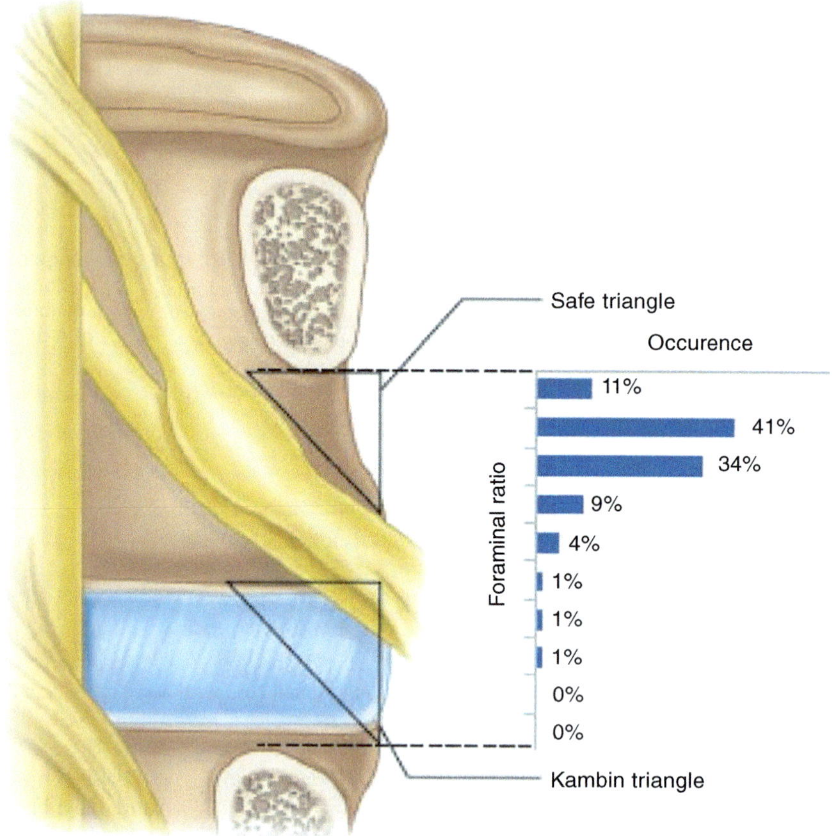


**Fig. 24.10** Location of radicular medullary artery as determined at the mid-pedicular plane, is graphically depicted using the foraminal ratio, in 0.10 increments

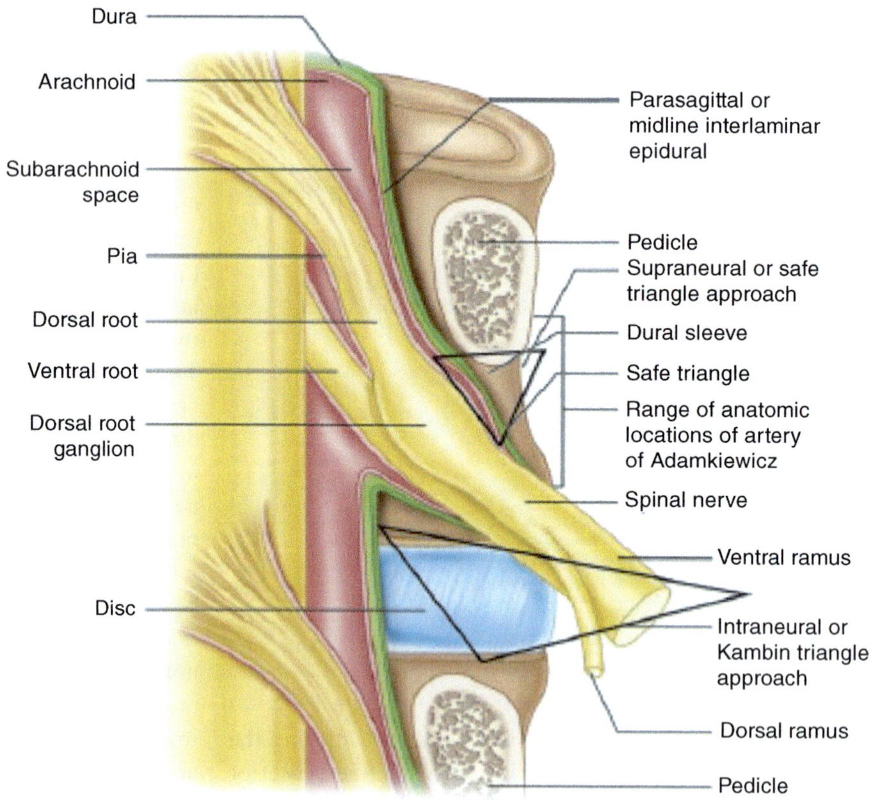


**Fig. 24.11** Nerve root anatomy and implications of positioning needles for transforaminal and interlaminar epidural injections

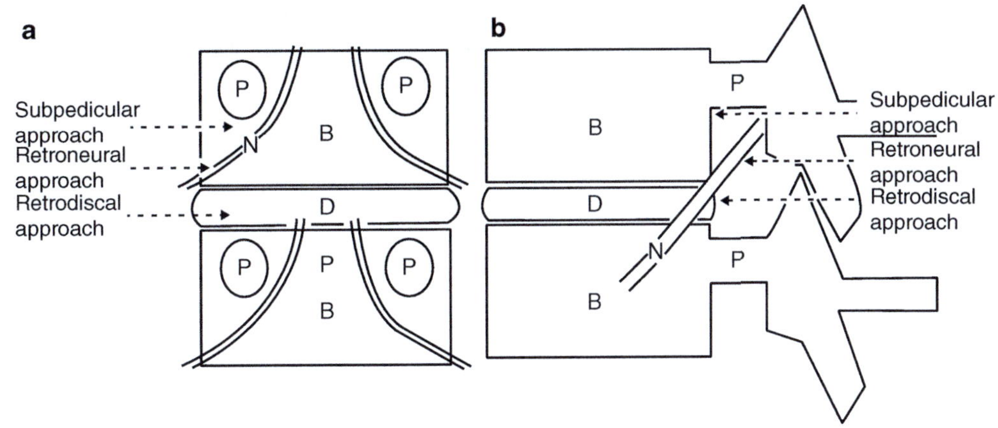


**Fig. 24.12** Three transforaminal approaches. (**a**) Anteroposterior view, (**b**) lateral view. B: vertebral body, D: intervertebral disc, N: spinal nerve, P: pedicle. (**Adapted from:** Kim WJ, et al. Comparison of epidural spreading patterns and clinical outcomes of transforaminal epidural steroid injection with high-volume injectate via the subpedicular versus the retrodiscal approach. *Pain Physician* 2018; 21:269–278 [51])

pedicle on the side of the targeted nerve root (Fig. 24.12). The oblique position is achieved by rotating the fluoroscopic C-arm until the SAP is seen between the anterior and posterior edges of the vertebral body and the base of the articular process is in line with the pedicle. A needle is inserted slightly above the SAP and directed towards the base of the pedicle, advancing slowly until contact is made with the bone below the pedicle (Fig. 24.12). Following the placement of the needle, a very low dose of contrast is injected slowly and the dispersal pattern around the nerve root is assessed. This may be followed by an injection of sodium chloride solution prior to 1 mL of biologic injected at each level.

### Infraneural or Kambin Triangle Approach

The procedure may be performed using the Kambin's triangle approach, either with a posterior or an oblique approach (Figs. 24.9, 24.10, and 24.11). Atluri et al. [52] suggested the needle be placed as inferior and posterior in the neural foramen as possible, which corresponds to the inferomedial part of this inferior triangle.

For the posterior approach, the patient is placed in the prone position and the fluoroscopy unit is positioned with the spinous process in the center of the spine (Fig. 24.14). A diamond tipped, Quincke, Whitacre, Bella-D Coudé or a blunt needle is inserted into the skin over the lateral border between the two adjacent TPs at the target interspace, two-thirds below the upper TP. The needle is advanced toward the inferior aspect of the foramen. Observing the needle position in a posterior or a lateral view, the injection of a small volume of contrast and the pattern of dispersion into the nerve root is noted. If the needle has penetrated the epiradicular membrane surrounding the nerve root, an appropriate and positive image of the nerve root will be seen on fluoroscopy, with appropriate dispersion of the contrast, as shown in Fig. 24.14. However, a classic contrast pattern with a dispersion showing a neurogram is often not achieved with this approach.

For the oblique approach, the patient is placed in the prone position and the fluoroscopy unit is positioned into the oblique position ipsilaterally until the "target point" is identified. The target point is the junction of the SAP and the TP (Fig. 24.14). If this landmark is not clearly visualized, the inferolateral part of the SAP can be targeted. Craniocaudal angulation of the C-arm may be required to "crisp up" these target points. Targeting either of these points is critical as it will ensure placement of the needle into the inferior aspect of the foramen. If not, the likelihood of needle placement in the midzone of the foramen increases. This is not desirable because the radicular artery lays in the midzone of the foramen up to 23% of the time [50].

For the injection of biologicals, extremely low volumes of nonionic contrast is recommended or contrast is avoided. One to 2 mL of biologic may be injected at each level.

### First Sacroiliac Transforaminal

For the transforaminal approach to the S1 nerve root, the patient is in the prone position, and the S1 foramen is visualized under fluoroscopy, which appears as a small radiolucent circle just below the oval S1 pedicle (Fig. 24.15). It may be necessary to direct the fluoroscopic beam in a cephalocaudad direction for the alignment of the anterior and posterior aspects of the foramina, often to 30°. The needle is inserted slightly lateral and inferior to the S1 pedicle and advanced slowly through the posterior foramen to the medial edge of the pedicle. Extreme caution must be exercised to avoid advancing the needle through both the posterior and anterior aspects of the S1 foramen and into the pelvis. By first contacting the posterior sacral bone prior to entering the S1 foramen, the depth and direction of the needle is provided. Following appropriate placement of the needle, once again dispersion of the contrast is observed. If it is appropriate, local anesthetic and corticosteroid solution is injected, similar to the lumbar transforaminal epidural injection technique.

## Caudal Epidural Injections

The technical aspects of caudal epidural injections are distinctly different from lumbar interlaminar and transforami-

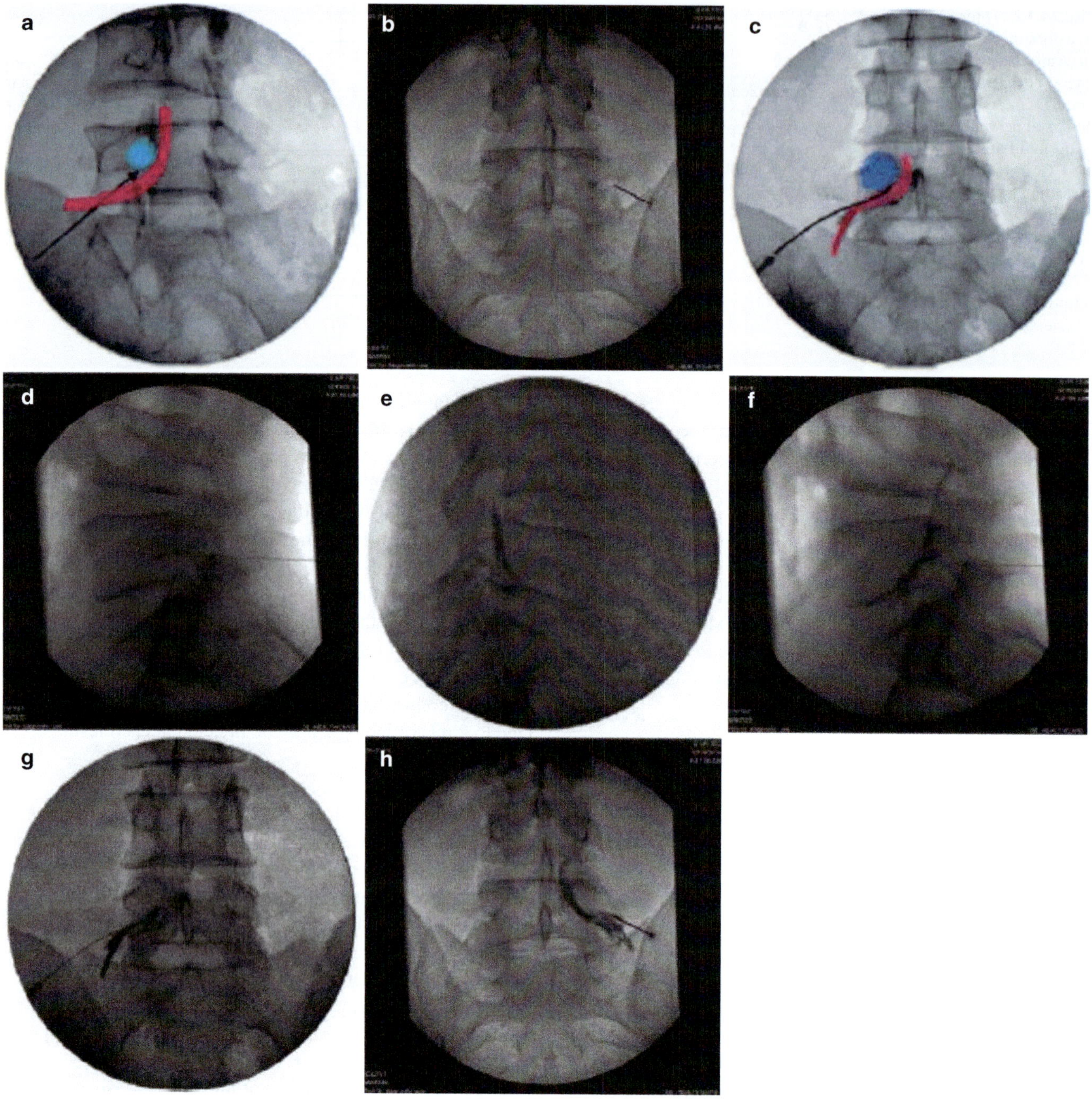


**Fig. 24.13** Transforaminal epidural at L5, utilizing safe triangle approach with oblique (**a**, **c**, **e**, **g**) or posterior (**b**, **d**, **f**, **h**) approach. (**a**) "Scotty dog" and location of needle placement with illustrative anatomy of L5—oblique approach. (**b**) Placement of needle at L5 with posterior approach in PA view. (**c**) Illustration of nerve root, pedicle, and needle in PA view. (**d**) Lateral view of L5 needle placement. (**e**) Lateral view of nerve root filling partially into epidural space. (**f**) Contrast display pattern at L5 in lateral view. (**g**) Nerve root filling noticed after contrast injection at L5 in PA view contrast into the nerve root. (**h**) Contrast display pattern of L5 nerve root in PA view

nal epidural injections. In addition, for delivery of injectate with biologics, target-specific catheter placement is required with percutaneous adhesiolysis. In patients without previous surgical interventions, catheterization may be simply performed without adhesiolysis: A catheter may be placed to reach the appropriate area in patients without adhesions. To place the catheter appropriately, a large 16-gauge RK needle is utilized, along with a Racz catheter.

The caudal epidural procedure is performed with the patient in the prone position with placement of firm padding under the pelvis, with the patient's head turned away from the operator and with appropriate tilting of the pelvis to make the sacral hiatus more prominent. Following the preparation of a wide area of skin with an antiseptic solution followed by appropriate draping, a C-arm is brought over the lumbosacral area and is positioned in either a PA or a lateral position.

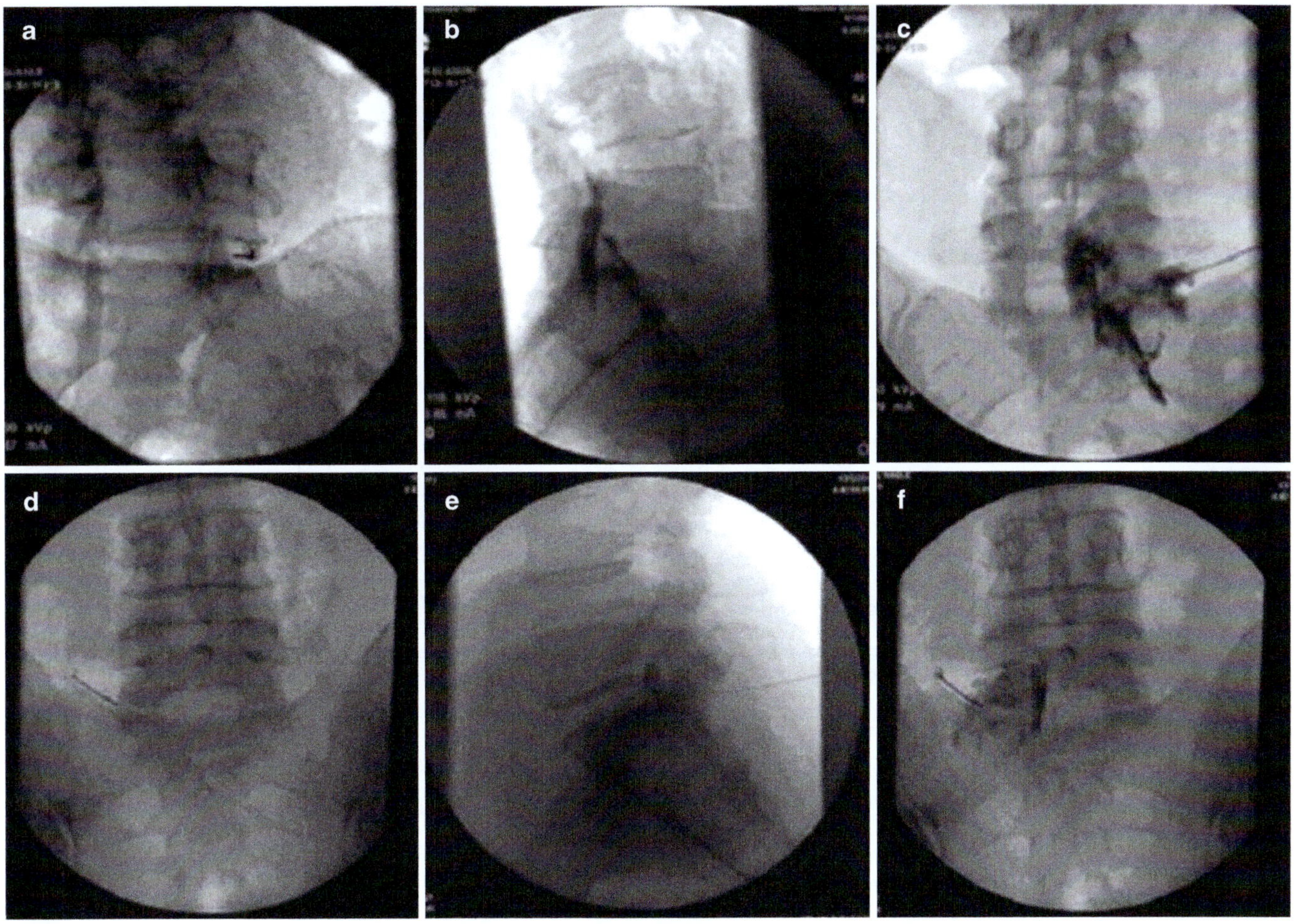


**Fig. 24.14** Transforaminal epidural injection with Kambin's triangle approach with oblique (**a–c**) or posterior (**d–f**) approach at L5. (**a**) Oblique L5 transforaminal epidural injection. (**b**) Lateral with contrast L5 transforaminal epidural injection. (**c**) PA view of L5 transforaminal epidural injection with contrast. (**d**) PA view with needle placement at L5—posterior approach. (**e**) Lateral view of L5 with contrast injection. (**f**) PA view with contrast injection

Understanding of the anatomy of the sacrum is crucial for successful placement of the needle.

The sacrum is a triangular bone, dorsally convex, that consists of the fused five sacral vertebrae articulating cephalad with the fifth lumbar vertebra and caudad with the coccyx, as shown in Figs. 24.3, 24.4, and 24.16. In the midline, there is a median crest with three or more, commonly four, prominent tubercles, representing the sacral spinous processes. The remnants of the S5 inferior articular processes are prominent and palpable through the skin and constitute the sacral cornua and, together with the adjacent coccygeal cornua, which they abut, are key landmarks for identification of the sacral hiatus and successful caudal blockade.

The coccyx is a small triangular bone consisting of three to five fused rudimentary vertebrae (Figs. 24.2, 24.3, and 24.16). The coccyx attaches to the lower part of the sacrum, and its tip is an important landmark.

The sacral hiatus is a defect in the lower part of the posterior wall of the sacrum, formed by the failure of the S5 laminae, and usually part of those of S4, to meet and fuse in the median plane. Thus, a variable space is left which is described as an inverted U or V, covered by the thick fibrous posterior sacrococcygeal ligament. Direct access to the caudal canal is obtained by penetration of the sacrococcygeal ligament. There is significant variation in the normal anatomy in this area because the sacral hiatus is highly variable in size and shape. The hiatus lies higher than the lower one-third of S4 in about 50% of specimens. The distance between the tip of the dural sac and the apex of the hiatus is also highly variable, from 20 to 45 mm. The coccyx and sacral hiatus are palpated, a needle used as the marker (Fig. 24.16), and appropriate local anesthetic infiltration is carried out.

Subsequently, an epidural needle (either 18- or 20-gauge Tuohy) is inserted, as shown in Fig. 24.17, with injection of contrast and observation under fluoroscopy. In a PA view of

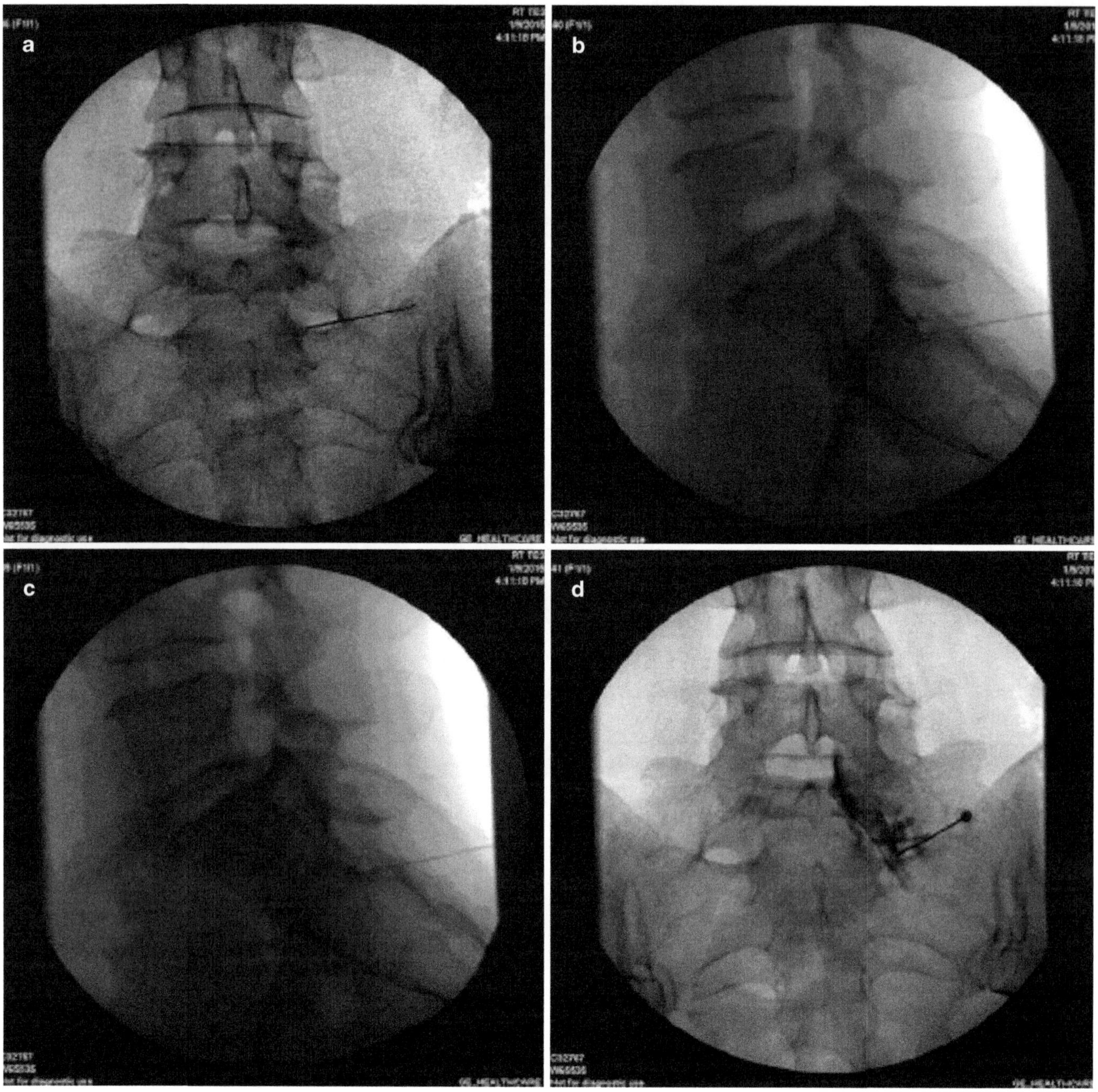

**Fig. 24.15** First sacral transforaminal epidural injection with postero-anterior approach on right side. (**a**) Needle placement at right S1 for transforaminal in a PA view. (**b**) Needle placement at right S1 with contrast display in lateral view. (**c**) Needle position in lateral view prior to contrast injection. (**d**) Contrast display at S1 in PA view

the epidurogram, the nerve roots are clearly outlined by contrast as they exit the intervertebral foramina, and in a lateral view, the contrast extends along both the anterior and posterior aspects of the epidural space.

## Percutaneous Adhesiolysis and Catheterization

The procedure is performed in a sterile operating room with appropriate sterile precautions utilizing a fluoroscope, an RK needle, and a stainless-steel spiral-tipped catheter (Fig. 24.18).

Procedural steps are illustrated in Fig. 24.19. In the prone position, sterile preparation is carried out with povidone-iodine (Betadine) prep or chlorhexidine. Draping is carried out to cover the patient, extending into the midthoracic or cervical region. Appropriate monitoring of blood pressure, pulse, and pulse oximetry is carried out, and sedation may be slowly administered if necessary. Under fluoroscopic visualization, an RK needle is introduced into the epidural space

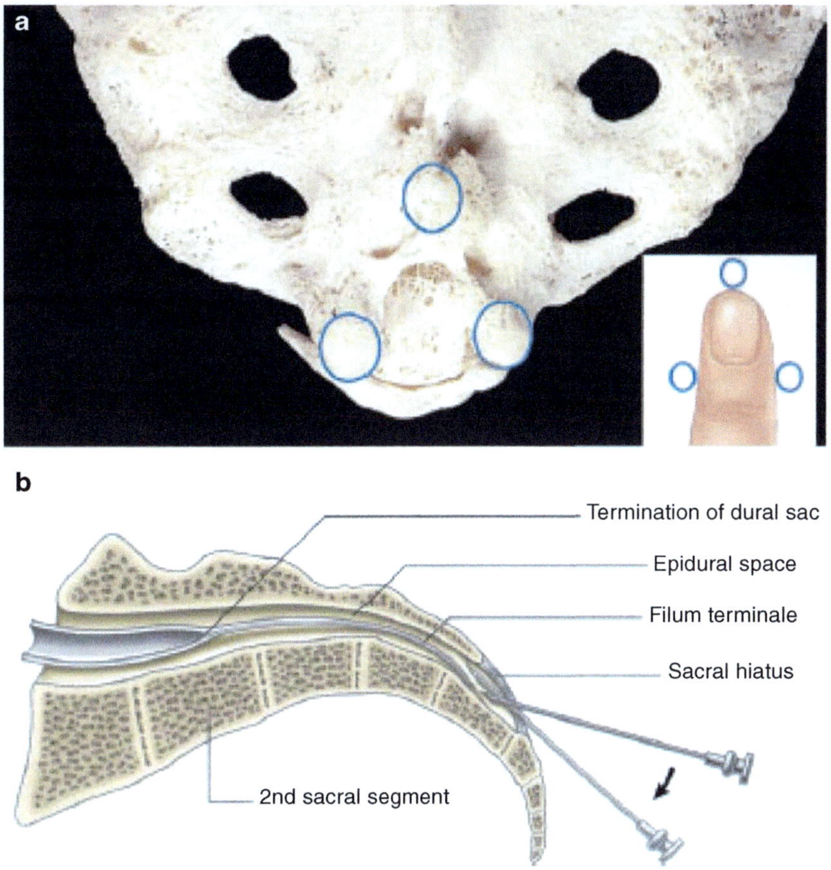


**Fig. 24.16** Illustration of palpation of sacral cornua (**a**) and needle position for caudal epidural (**b**)

after local anesthetic infiltration. Once the needle placement is confirmed to be in the epidural space, a lumbar epidurogram is carried out utilizing approximately 2–5 mL of contrast medium. Finding the filling defects by examining the contrast medium flow into the nerve roots is the purpose of the epidurogram. After visualizing the contrast medium spread by epidurography, a Racz catheter, which is a spring-guided, reinforced catheter, is slowly passed through the RK needle to the area of the filling defect or the site of pathology, as previously determined by magnetic resonance imaging, computed tomography, or patient symptoms. Following the positioning of the catheter into the appropriate area by mechanical means, adhesiolysis is carried out. After the adhesiolysis is completed, a repeat epidurogram is carried out by injecting additional contrast medium. If appropriate adhesiolysis is achieved, nerve root filling, as well as epidural filling, will be noted.

Figure 24.20 illustrates percutaneous adhesiolysis at L5 and S1 on the right side. Filling defects can be seen on lumbar epidurography, followed by adhesiolysis and appropriate nerve root filling seen on the PA view and ventral epidural filling seen on the lateral view.

Figure 24.21 shows left-sided filling defects. Attempted catheter placement from the left was difficult. The problem was resolved by entering the catheter from the right side and crossing to the left side, followed by successful adhesiolysis. Follow-up epidurography demonstrates excellent filling patterns in the left L5 and S1 nerve roots with excellent ventral epidural filling.

Figure 24.22 shows a case of post-fusion syndrome with left L5 radiculitis and filling defects. With an appropriate adhesiolysis, the filling defects were removed. Excellent L5 nerve root filling is observed, along with ventral filling as seen on lateral epidurography.

Following the adhesiolysis, 1–2 mL of biologic is injected at each nerve root level or 5 mL to cover multiple nerve roots. As described earlier, if contrast is injected it should be washed away with sodium chloride solution and then the biologic is injected after a waiting period of at least 5 min so that the sodium chloride solution diffuses from the area. Following completion of the injection, the catheter is removed.

## Cervical Epidural Injections

Cervical interlaminar epidural injections also may be performed transforaminally; however, because of the numerous risks associated with the injection procedure itself, in addition to the injection of particulate steroids, transforaminal injections are not advised in the cervical spine.

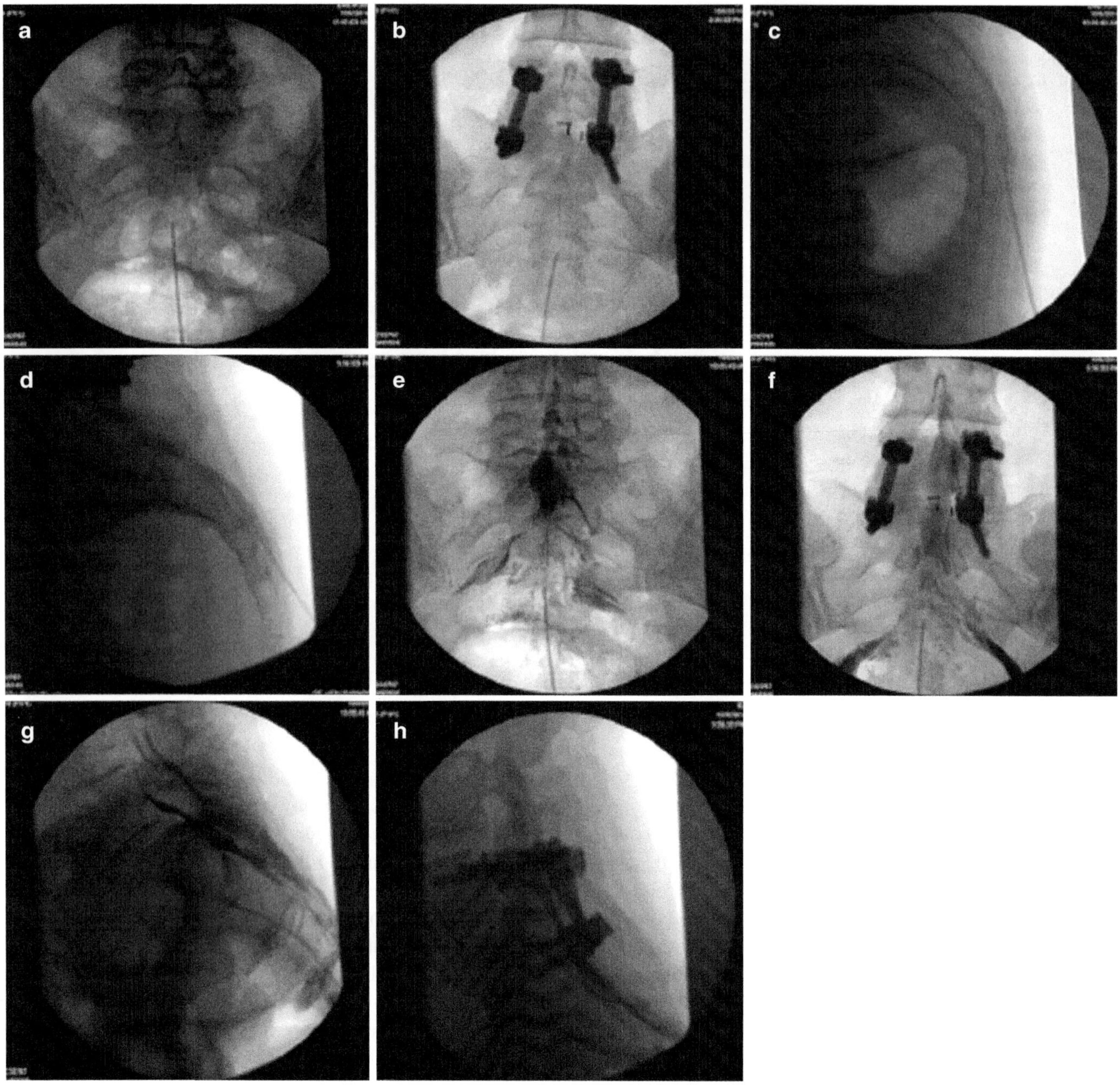


**Fig. 24.17** (**a**–**h**) Steps in performing caudal epidural injections in patients with (**b**, **d**, **f**, **h**) and without (**a**, **c**, **e**, **g**) fusion. (**a**, **b**) PA view of the placement of the needle into sacral hiatus. (**c**, **d**) Lateral view of needle placement into the sacral hiatus. (**e**) Contrast injection into the epidural space showing epidural filling pattern, along with outlining of multiple nerve roots with a typical Christmas tree pattern. (**f**) PA view with good epidural filling pattern despite surgical intervention with no significant nerve root filling noted. (**g**) Lateral view showing a double-line pattern with double lines with ventral and dorsal filling pattern. (**h**) Lateral view of the caudal epidural filling with predominantly dorsal epidural filling pattern with some ventral epidural filling pattern

## Interlaminar Epidural Injections

Fluoroscopic guidance for cervical epidural injections for chronic pain is mandatory in order to assure that medications reach the appropriate and desired intervertebral space either with the midline approach or a paramedian approach.

Goodman et al. [53] described the optimal patient positioning and fluoroscopic imaging for the performance of cervical interlaminar epidural steroid injections.

Compared to the lumbar spine, the epidural space in the cervical spine is smaller [54–56]. In addition, there is also reported to be a higher incidence of discontinuity in the ligamenta flava. Consequently, there is a higher rate of dural puncture during cervical interlaminar epidural injections, which can lead to rare but potentially serious complications [48].

Among the multiple types of needles utilized in performing cervical interlaminar epidural injections, the Tuohy nee-

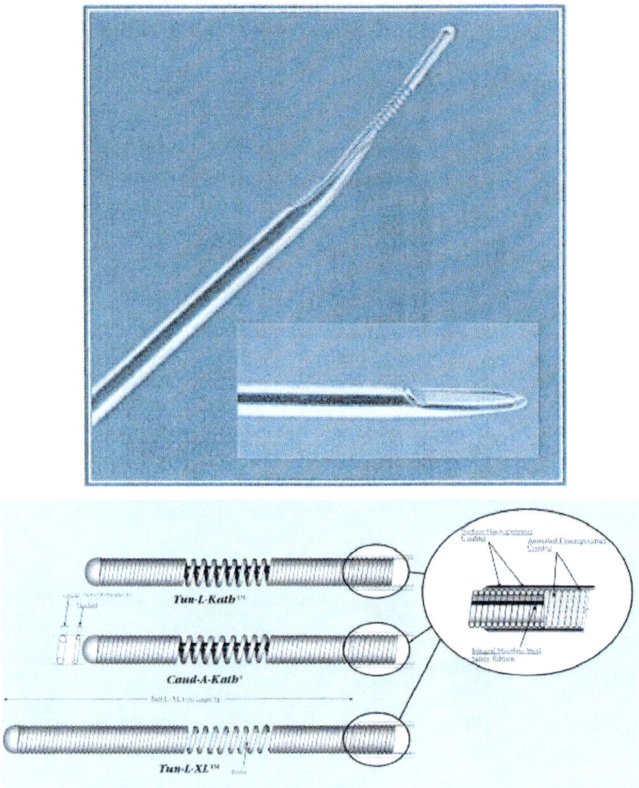


**Fig. 24.18** Special needle and catheter for adhesiolysis. (Top) R.K. epidural needle (close-up). (Bottom) Tun-L-Kath and Tun-L-XL/24 epidural catheters. (**Courtesy:** Epimed International, Gloversville, NY)

dle is the most common, using a loss of resistance technique. The needle's orifices align nearly perpendicular to the shaft to direct the needle, catheter, or solutions through the needle along the plane of the epidural space.

Utilizing fluoroscopy, the epidural space is entered by loss of resistance technique introduced by Dogliotti [57] in 1933, with a lubricated glass syringe partially filled with air or saline [58]. Potential complications such as pneumocephalus, subcutaneous emphysema, or venous air embolism can occur with this loss of resistance to air technique [59, 60].

The patient is placed in the prone position with firm padding under the chest, followed by appropriate sterile preparation. The desired interlaminar space is identified in the PA view, with the rotation of the C-arm toward the patient's right side or left side until the spinous process is exactly equidistant from the right and the left pedicles, and with cranial or caudal angulation until the desired interlaminar space is maximally opened. The midline or paramedian technique is performed at the interspace most closely associated with the patient's level of pain.

Access to the epidural space is obtained between the C7 and T1 interspace or, if necessary, at C6–7 or C5–6. Entry between C5 and C6 provides ease of manipulation of the C-arm to obtain oblique or lateral views, since the shoulders do not prevent C-arm positioning at this level.

The International Spine Intervention Society (SIS), the Multi-Society Pain Workgroup (MPW), and the American Society of Anesthesiology (ASA) have provided recommendations to the Food and Drug Administration (FDA) that cervical interlaminar epidural injections should not be performed above C5 and C6; however, this has not been established by evidence in the literature and has not been approved by the FDA Advisory Committee [61–65].

The skin is anesthetized with 1% preservative-free lidocaine. A Tuohy needle with or without a wing tip is placed in the midline or in a paramedian position (Figs. 24.23, 24.24, and 24.25). A paramedian approach is best utilized in patients with unilateral radicular pain syndromes. Intermittent pressure is applied on the syringe plunger with advancement of the needle, with intermittent fluoroscopic visualization, until loss of resistance is felt, or needle position is observed in the epidural space (Fig. 24.25). As the needle is advanced through the interspinous ligament and is noted to be in the direct midline or intended paramedian location, the fluoroscope may be positioned in an oblique or a lateral view in order to ascertain the needle depth. In the lateral view, with appropriate positioning, the needle tip lies just across the spinal laminar line and the contrast agent smoothly fills the posterior epidural space (Fig. 24.25).

During the cervical interlaminar epidural injection, the landmark for the posterior epidural space is the spinal laminar line in the fluoroscopic lateral view. If the needle is just posterior to the spinal laminar line in the lateral view, it should be slowly advanced very carefully through the ligamentum flavum with intermittent injections of the contrast.

Once the needle has reached the epidural space with appropriate loss of resistance to either air or saline, 0.5–2 mL of nonionic contrast (Omnipaque or Isovue) is injected to confirm epidural placement. PA fluoroscopic images with or without an oblique or lateral fluoroscopic image(s) are then obtained (Fig. 24.25). The classic contrast spread appears with or without an areolar appearance or the filling of nerve roots (Fig. 24.25).

The drainage of fluid back from the needle—whether contrast media, local anesthetic, or saline—indicates a dural puncture may have occurred but can also occur due to poor compliance of the epidural space which can worsen by a rapid rate or increased volume of injection. Dripping back of cerebrospinal fluid (CSF) related to subarachnoid puncture will usually cease within 30 s as the epidural pressure re-establishes equilibrium.

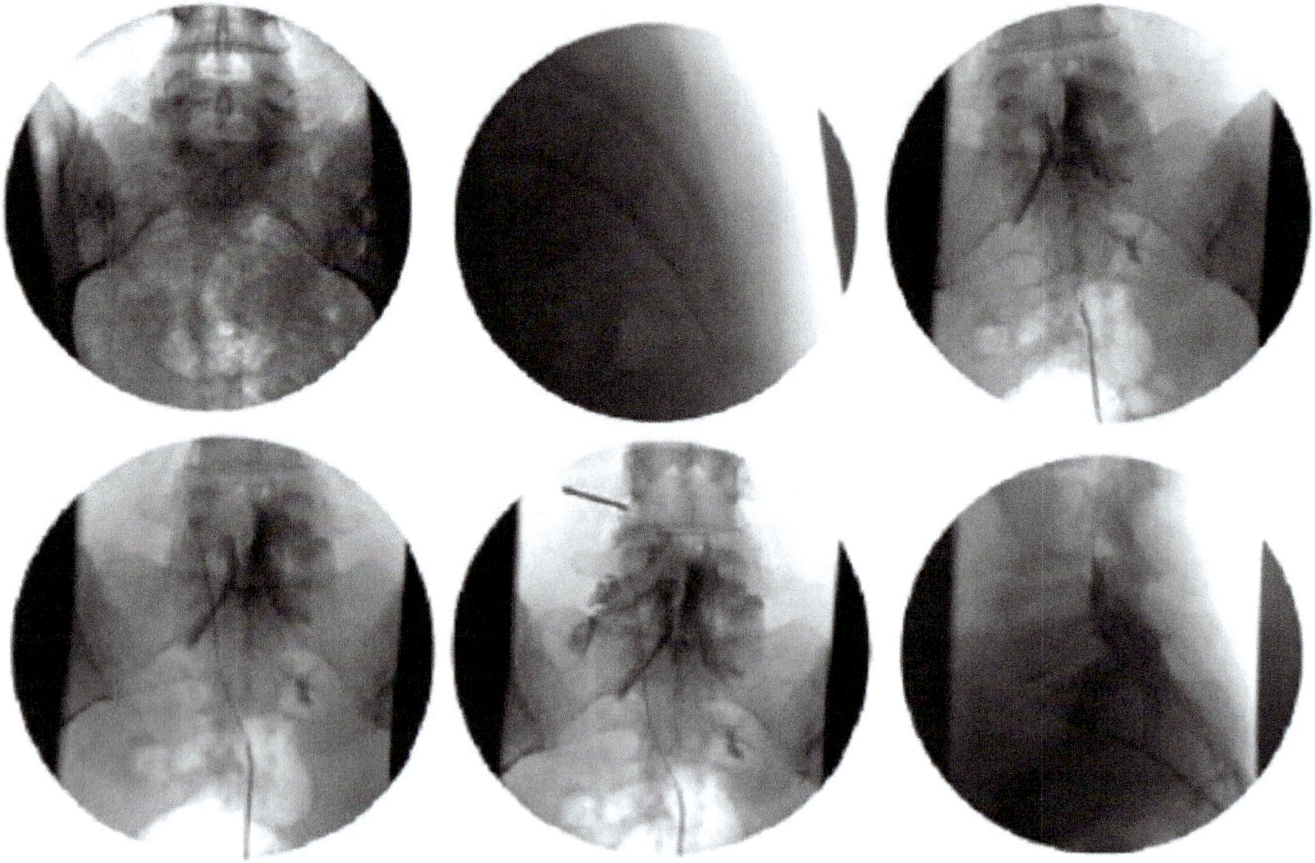

**Fig. 24.19** Percutaneous adhesiolysis with the caudal approach—positioning, preparation, fluoroscopy, and adhesiolysis. (Top left) Fluoroscopic view in AP position. (Top middle) Fluoroscopic view in lateral position. (Top right) AP fluoroscopic view after injection of 5 mL of contrast medium showing lack of filling on the left side in L5 distribution. (Bottom left) Placement of Racz catheter and adhesiolysis and positioning of the catheter over the L5 body to the left laterally. (Bottom middle) Contrast medium injection after placement of the catheter following adhesiolysis showing excellent filling of the L5 nerve root. (Bottom right) Lateral fluoroscopic view showing nerve root filling along with ventral and dorsal filling of the epidural space

If no intravascular or soft tissue contrast pattern is seen, sodium chloride solution is injected to wash out the contrast, followed by injection of the biologic solution.

Figure 24.26 illustrates cervical epidural filling patterns with a midline approach. Figure 24.27 illustrates epidurographic patterns of cervical epidural contrast injections in postsurgical patients.

## Extra-Epidural Placement

Dural puncture with cervical interlaminar epidural puncture under fluoroscopy has been shown to occur in approximately 1% of procedures, in contrast to 0.8% of lumbar interlaminar epidural procedures [48]. Dural puncture may be diagnosed with return of fluid; however, if there is no return of CSF, the contrast display pattern may illustrate a subarachnoid (or myelographic) pattern. A typical myelogram demonstrates contrast within the thecal sac as shown in Fig. 24.28. A myelogram has a significantly identifiable contrast display pattern compared to an unintended dural puncture and subdural injection of the contrast as shown in Figs. 24.28 and 24.29. It may be confusing to differentiate an excellent epidural filling pattern in the lateral view with its railroad track pattern from a myelographic (subarachnoid) pattern. The differentiation may be made by nerve root filling, areolar appearance, lateral filling of the epidural space and nerve roots, and dissipation of the contrast, which occurs rapidly, as shown in Figs. 24.28 and 24.30. A PA radiographic view of the cervical spine after intrathecal placement of contrast media outlines the cervical nerve roots within the thecal sac as they travel laterally toward the intervertebral foramina. The nerve roots appear radiolucent as they are surrounded by contrast within the CSF. A lateral radiographic view of the cervical spine after intrathecal placement of contrast media may show outlining of the cervical nerve roots within the thecal sac as they travel laterally toward the neural foramina (Fig. 24.30).

Subdural placement of the needle is rarely seen with cervical interlaminar epidural injections. The subdural compart-

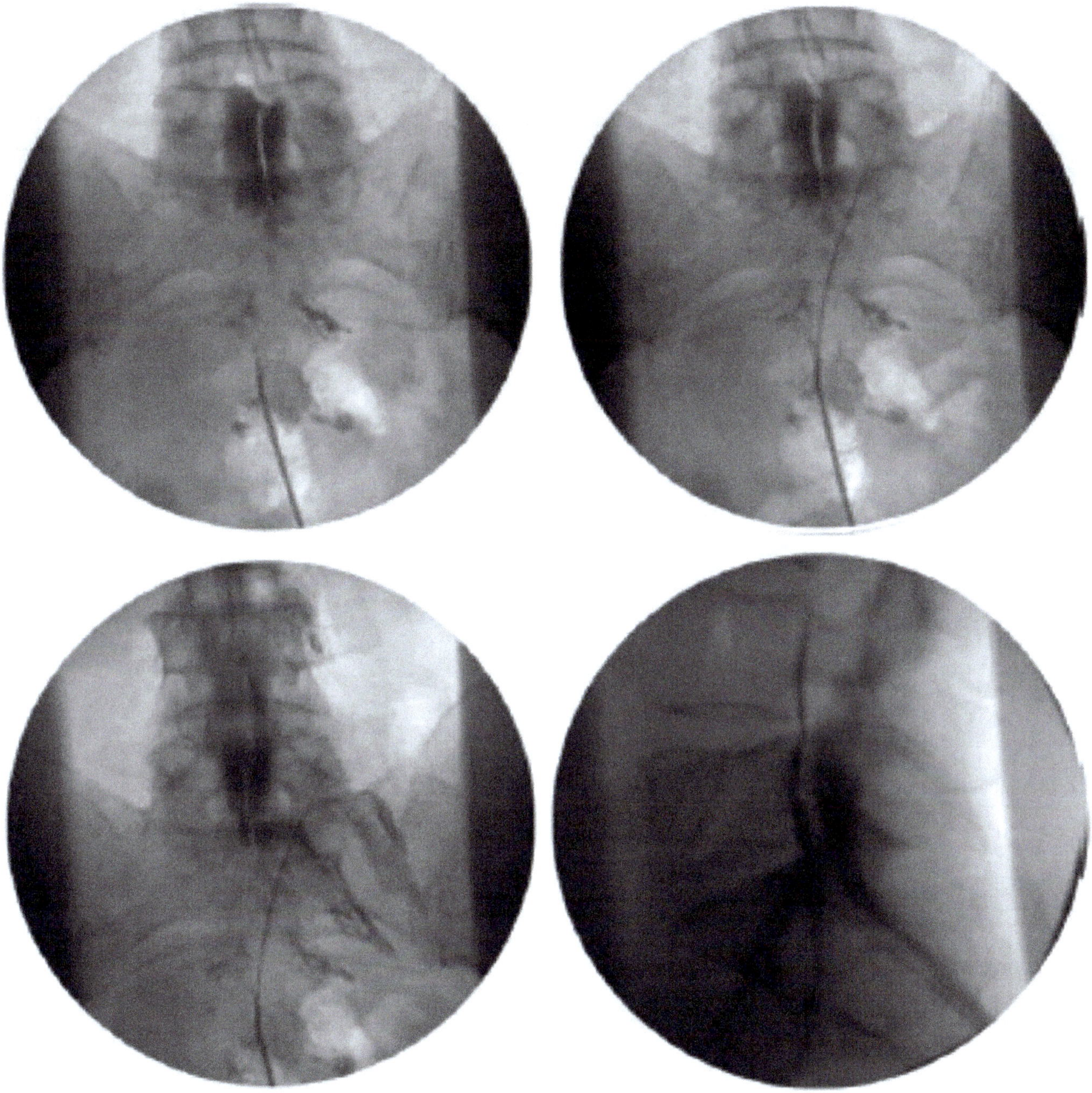

**Fig. 24.20** Illustration of caudal epidural adhesiolysis for right L5 and S1 radiculitis. (Top left) AP view of lumbar epidurography with no nerve root filling seen on either side. (Top right) AP view with Racz catheter positioned at L5 after adhesiolysis. (Bottom left) AP view of post-adhesiolysis with nerve root filling pattern seen at L5 and S1 on the right side. (Bottom right) Lateral view with lumbar epidurography with ventral and dorsal filling

ment lies between the dura and the arachnoid, extending from the lumbar spine into the cranium. This space contains only a small volume of serous fluid between opposing membranes [49, 66]. The subdural compartment is widest in the cervical region, narrowest in the lumbar region, and has more potential capacity posteriorly and laterally at all vertebral levels, present both ventrally and posteriorly [49, 66]. The subdural space does not communicate directly with the subarachnoid space but does extend laterally over the exiting nerve roots and dorsal root ganglia. Unlike the epidural space, the subdural compartment extends intracranially.

Injections of anesthetics and biologics into the subdural space can result in physical symptoms which are variable. The different symptoms are secondary to the anterior and

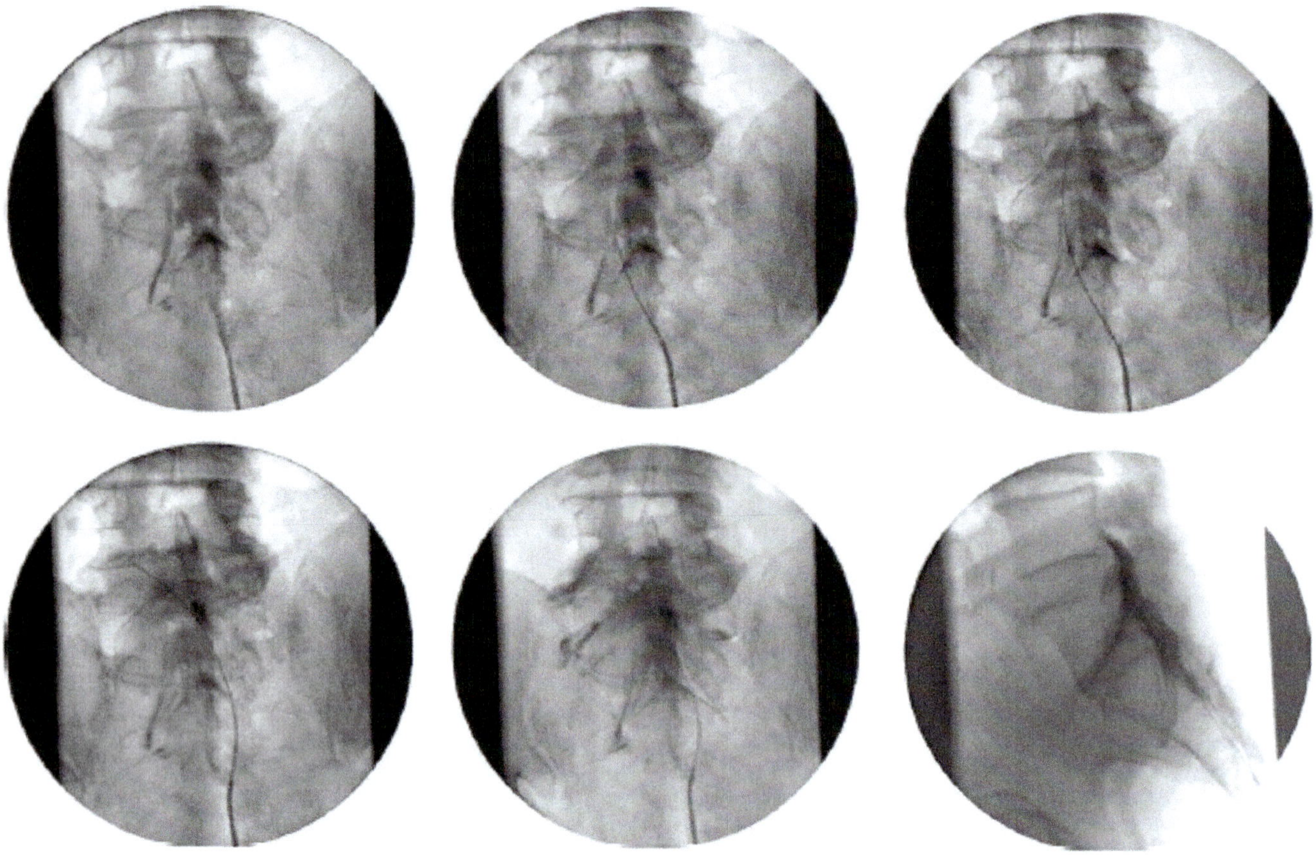

**Fig. 24.21** Epidural adhesiolysis. (Top left) AP view showing fibrosis and filling defects at L5 and S1 on the right side. (Top middle) Attempted adhesiolysis. (Top right) Adhesiolysis difficult as catheter goes in path of least resistance. (Bottom left) Catheter positioning on the left side accessing from right to left. (Bottom middle) Adhesiolysis with excellent filling of left L5 and S1 nerve roots. (Bottom right) Lateral view with epidural filling

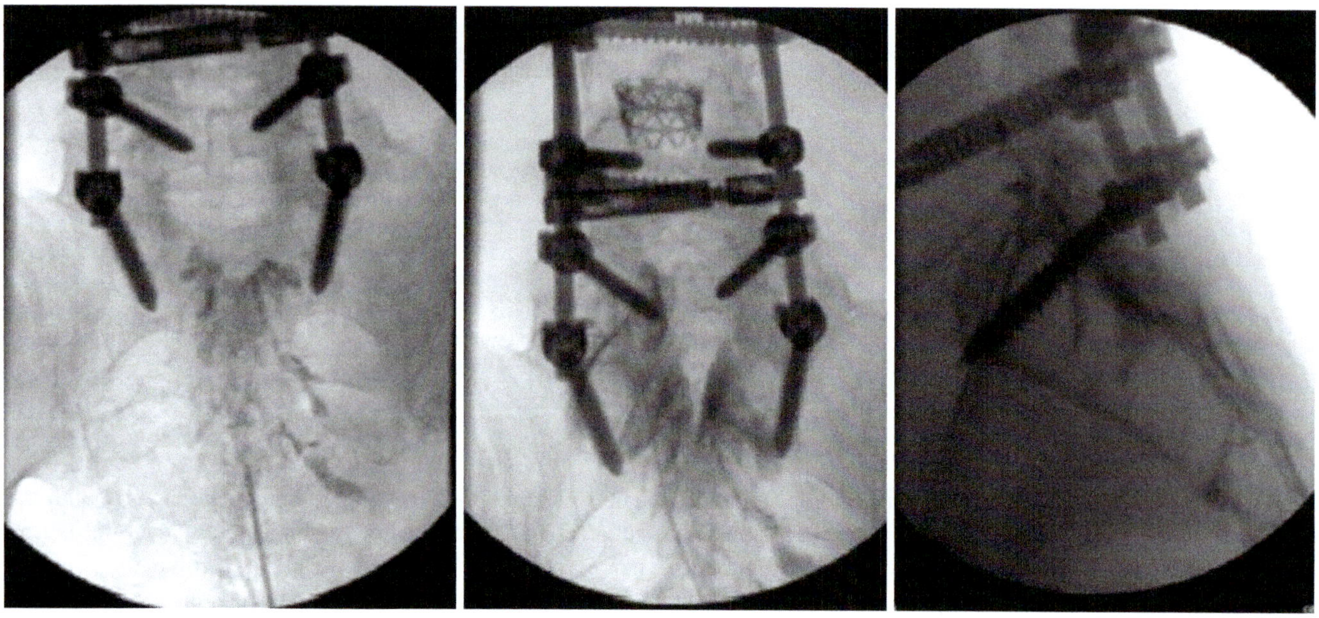

**Fig. 24.22** Caudal epidural adhesiolysis in a post-fusion patient with left L5 radiculitis

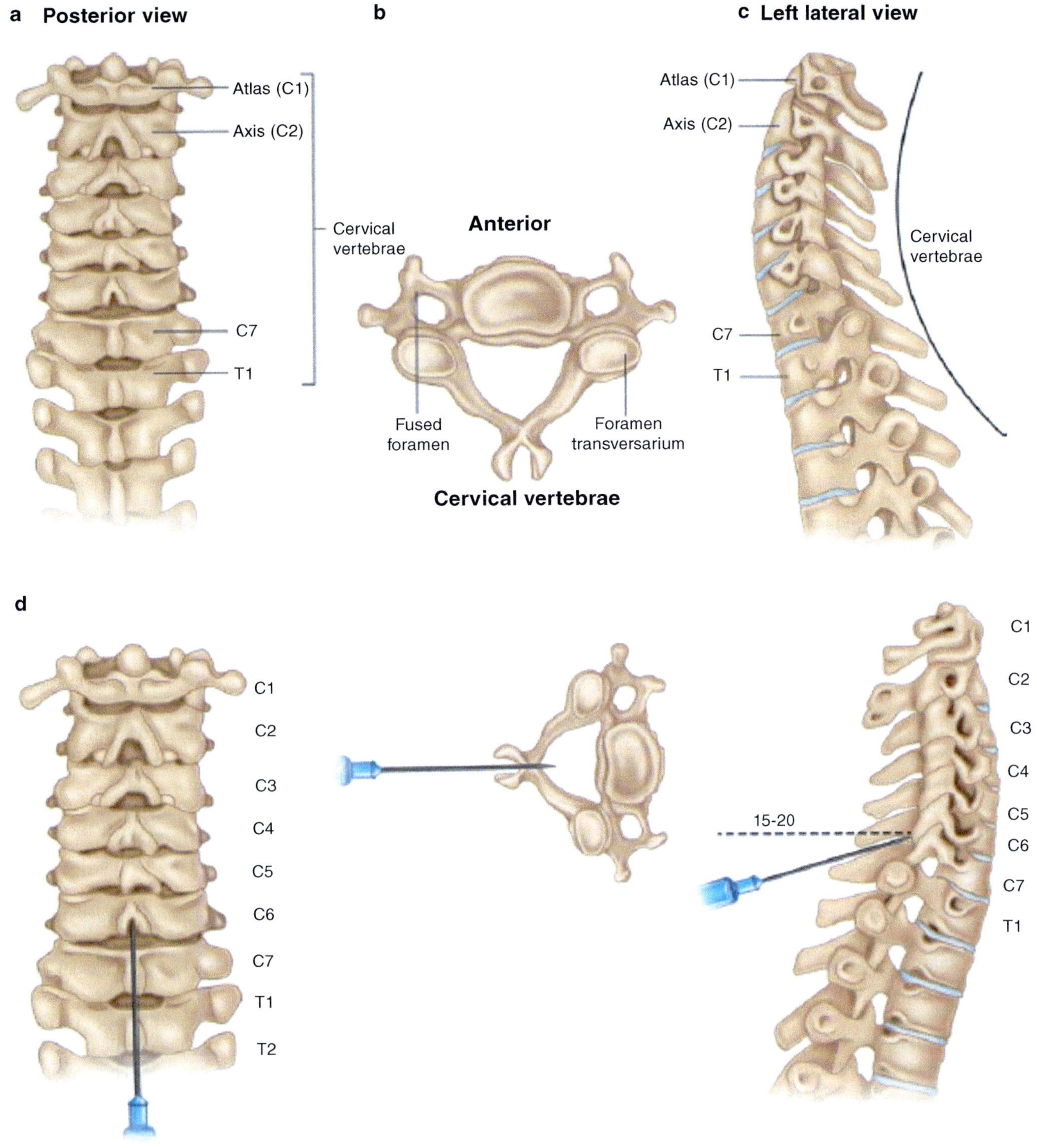


**Fig. 24.23** (**a**) Posterior view. (**b**) Anterior view. (**c**) Left lateral view. (**d**) Position of angle of needle entry for cervical interlaminar epidural injection

posterior compartmentalization of the subdural space. Most commonly, posterior subdural injections are performed, but anterior subdural injections may also be performed. The symptoms depend on the extent of subdural compartmentalization and the volume injected. Injection into the anterior subdural space may be associated with high motor and sensory blocks and even loss of consciousness [66], while posterior subdural injection usually presents as a failed spinal

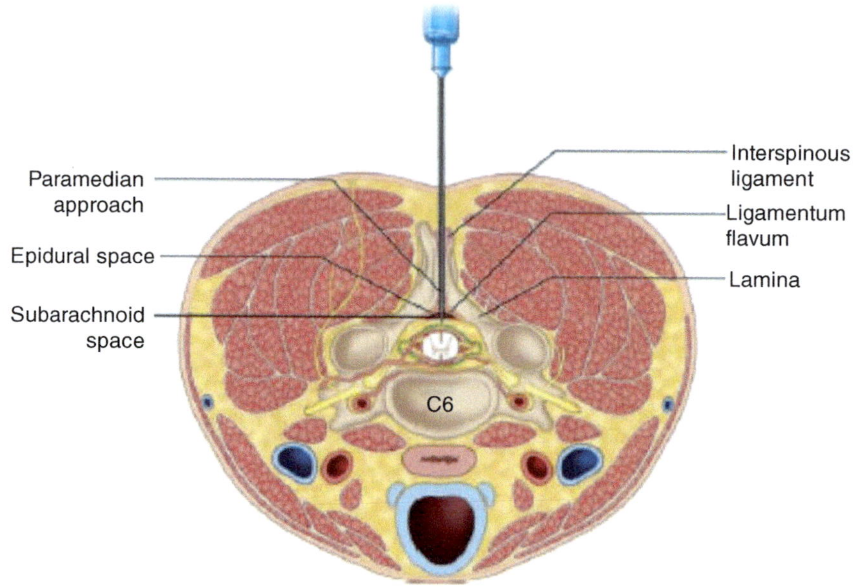


**Fig. 24.24** Axial diagram of interlaminar cervical epidural injection

block or a patchy epidural block of slow onset, requiring large volumes of local anesthetic for correction. In a PA view of the spine, subdural contrast media appears as a dense collection of contrast media confined to the central portion of the spinal canal. It does not extend to outline the exiting spinal roots laterally or the inferior portion of the thecal sac (Fig. 24.31). It may be difficult to diagnose subdural patterns on the lateral view. The pattern illustrated in Fig. 24.31c shows subdural placement of contrast and injectate. This was not identified at the time of the injection. Thirty-five minutes after an injection of 3 mL of 0.5% preservative-free Xylocaine, the patient complained of tingling in her fingertips. Subsequently she developed numbness below T4 followed by motor weakness bilaterally, more so on the left than the right, with loss of deep reflexes on the left side.

Intravascular entry with interlaminar epidural injections is less frequent than in transforaminal epidural injections; however, intravascular entry of the needle has been shown in 4.2% of epidural procedures.

### Cervical Transforaminal Epidural Injections

Cervical transforaminal epidural injections are performed infrequently and are associated with significant risk. They have been the subject of controversy because of a number of severe patient injuries and deaths [61–65, 67, 68]. In fact, the FDA has issued a warning against cervical transforaminal epidural injections [61].

These procedures have been performed with modified techniques, under continuous fluoroscopy, utilizing digital subtraction angiography (DSA) and injection of only non-particulate steroids [12].

The superiority or even equivalency of cervical transforaminal epidural injections compared to interlaminar epidural injections has not been proven. Adding the differences in risk factors strongly warrants against cervical transforaminal injections.

## Thoracic Epidural Injections

Thoracic epidural injections are performed using two approaches, interlaminar and transforaminal, with distinctly different technical issues. Fluoroscopy must be used for both approaches in chronic pain management settings.

### Thoracic Interlaminar Epidural Injections

In the thoracic spine, the angulation of the spinous processes is variable. The angulation is mild from T1 to T4 and T9 to T12, whereas angulation is more marked downward from T5 to T8, influencing the ease of entry into the epidural space, with increased technical difficulty between T5 and T8. Both a midline and a paramedian approach may be used.

Among the multiple types of needles utilized in performing thoracic epidural injections, the Tuohy needle is the most common using a loss of resistance technique. The needle orifices align nearly perpendicular to the shaft to direct the needle, catheter, or solutions through the needle along the plane of the epidural space.

Utilizing fluoroscopy, the epidural space is identified by loss of resistance technique introduced by Dogliotti [57] in 1933, with a lubricated glass syringe partially filled with air or saline [58]. Potential complications such as pneumocephalus, subcutaneous emphysema, or venous air embolism can occur with the loss of resistance to air technique [59, 60].

The patient is placed in the prone position with firm padding under the chest. The desired space is identified in a PA

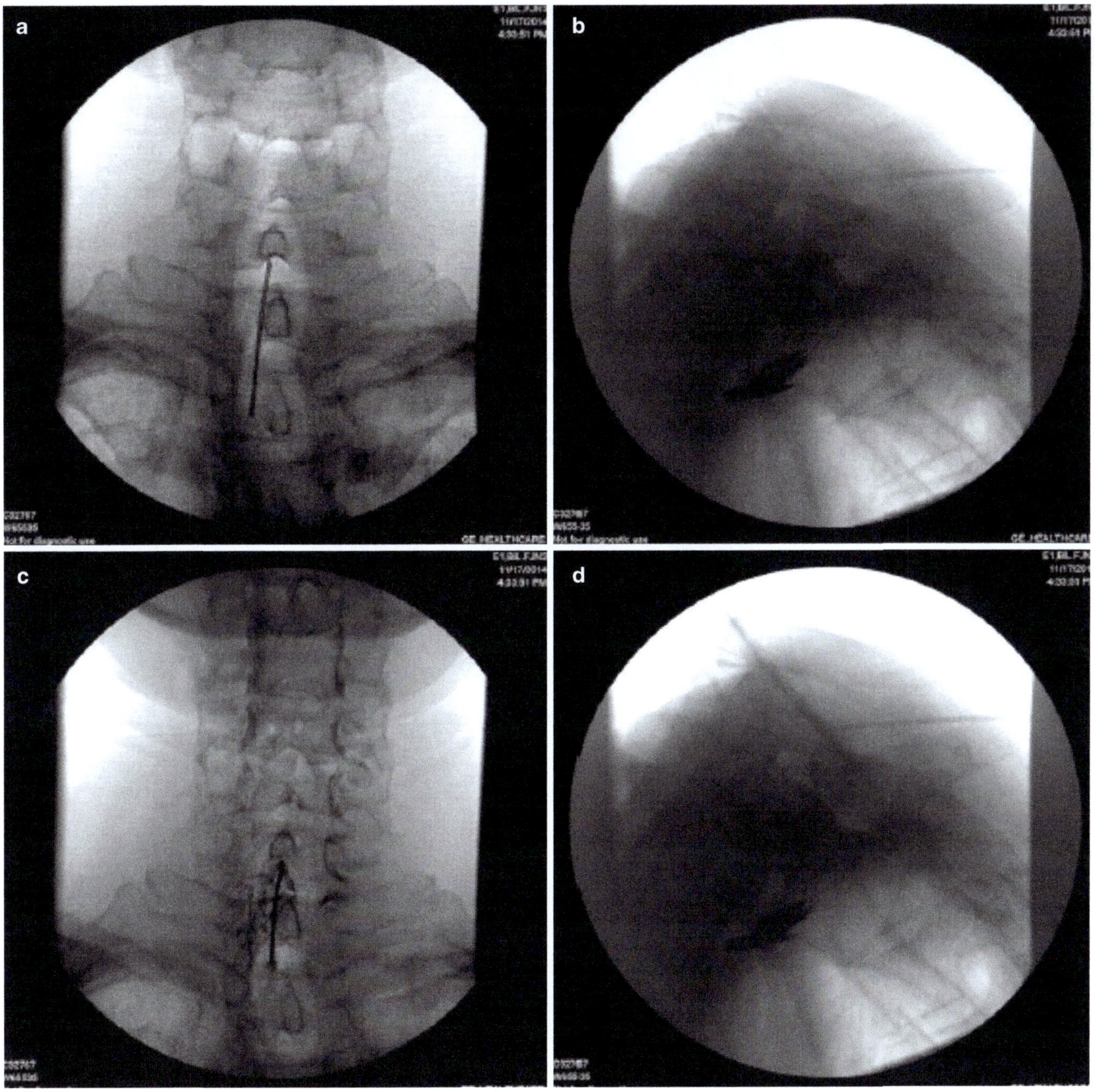

**Fig. 24.25** Illustration of steps of performing a typical cervical interlaminar epidural injection with midline approach. (**a**) Placement of needle in PA view between C6 and C7 interspace. (**b**) Lateral view showing the needle placement at the laminar line in the posterior epidural space identified with loss of resistance technique. (**c**) PA view of epidural needle placement at C6/7 with injection of contrast showing extensive epidural filling pattern bilaterally. (**d**) Lateral view of the epidural filling pattern demonstrated as a single line extending into cervical and thoracic epidural space

view with a rotation of the C-arm of the fluoroscope toward the patients' side until the spinous process is equidistant from the right and left pedicles, with cranial or caudal angulation until the desired interlaminar space is maximally opened. In many cases, cranial or caudal angulation is not required. Typically, T1–T4 and T9–T12 are the widest interspaces, whereas other interspaces are smaller, as shown in Figs. 24.3, 24.4, 24.5, and 24.6. The needle placement differs for midline and paramedian approaches (Fig. 24.32).

The skin is anesthetized with 1% preservative-free lidocaine. The Tuohy needle with or without a wing tip is placed in the midline or paramedian position. A paramedian approach is best utilized in patients with unilateral radicular pain syndromes. Intermittent pressure is applied on the

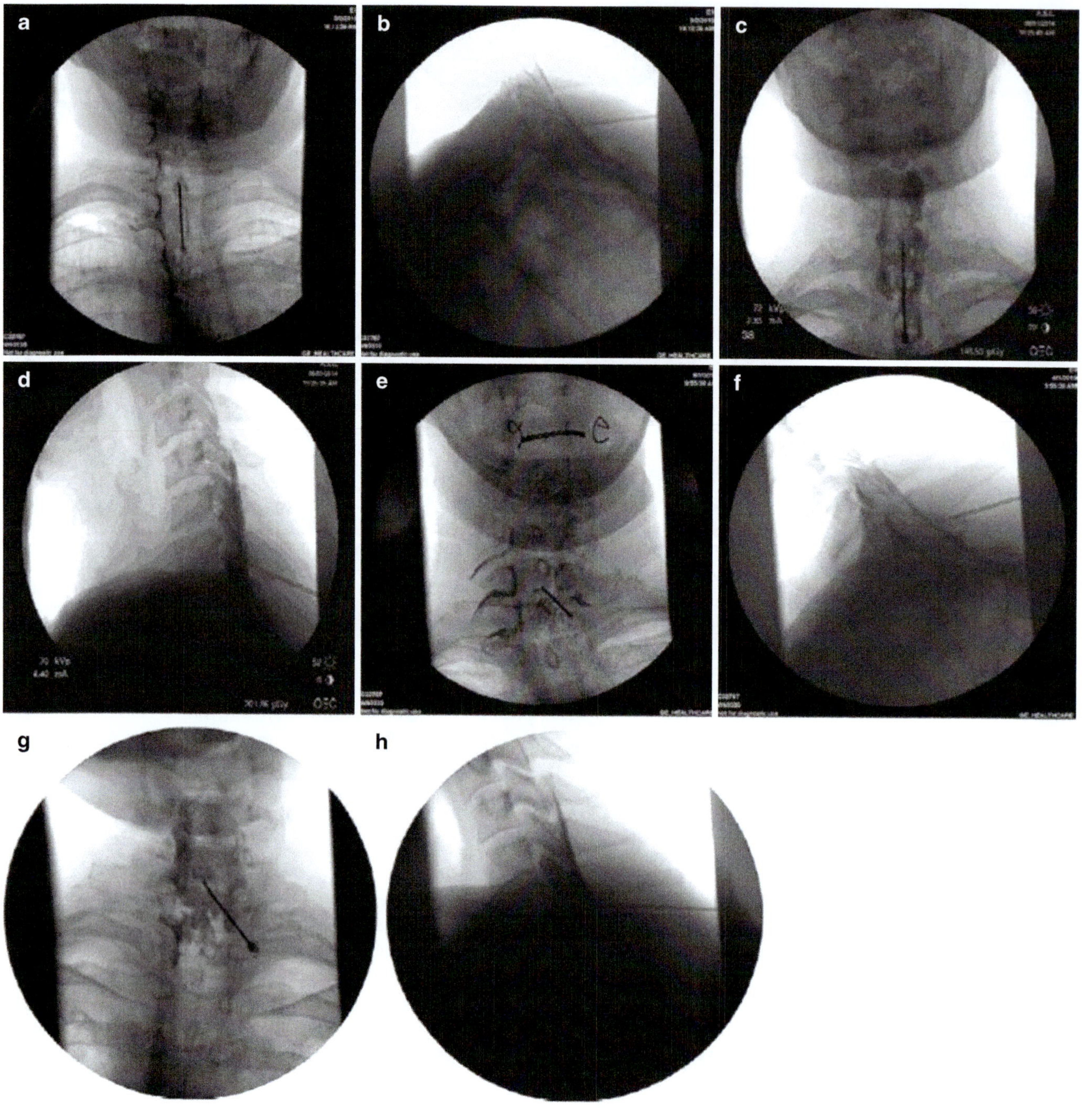

**Fig. 24.26** Cervical epidurographic contrast display patterns. (**a**) Cervical epidural needle placement in PA view with contrast injection with needle placed between C7 and T1 with contrast filling pattern illustrating filling pattern into multiple nerve roots. (**b**) The lateral view with typical epidural filling pattern noted. (**c**) PA view with entry into the cervical epidural space with loss of resistance followed by injection of contrast at C7 and T1. (**d**) Lateral view showing appropriate dispersion of contrast. (**e**) Epidural placement of the needle between C7 and T1 in midline approach showing good filling pattern into the epidural space, predominantly on the left side, with some on the right side, along with display of nerve root filling. (**f**) Lateral view of epidural filling with needle placement between C7 and T1 illustrating cervical and thoracic epidural filling pattern. (**g**) Entry into the cervical epidural space between C6 and C7 with loss-of-resistance technique followed by injection of contrast. (**h**) Lateral view with appropriate dispersion of contrast using midline approach

syringe plunger during advancement of the needle with intermittent fluoroscopic visualization, until loss of resistance is felt, or the needle position is observed to be in the epidural space (Fig. 24.33). As the needle is advanced through the interspinous ligament and is noted to be in the direct midline or in a paramedian position, the fluoroscope is positioned in

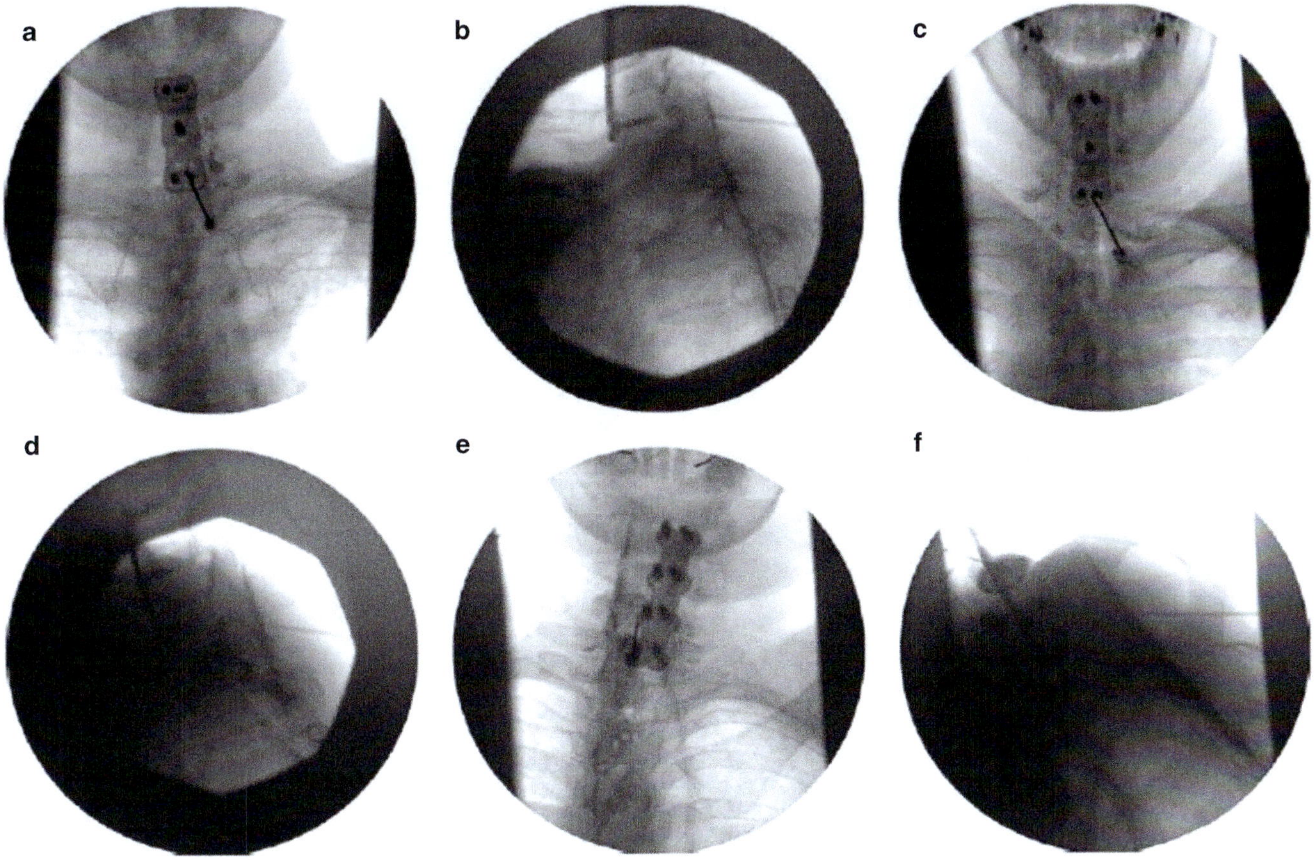


**Fig. 24.27** Illustration of cervical epidurographic patterns in postsurgical patients. (**a**) Needle placement in patients with anterior cervical fusion with injection of contrast showing filling in the cervical and thoracic epidural space. (**b**) Lateral view illustrating predominantly thoracic and some cervical epidural filling pattern. (**c**) Epidural placement of the needle between C6 and C7 in a patient with anterior cervical fusion. Contrast injection shows good filling pattern into the cervical epidural space with filling also noted in the thoracic epidural space. (**d**) Lateral view showing the hardware and appropriate filling of the contrast. (**e**) Entry into the cervical epidural space between C6 and C7 in a patient with anterior cervical fusion. Contrast injection showing excellent dispersion of the contrast, specifically on the left side, extending into the thoracic spine. (**f**) Lateral view illustrating appropriate contrast dispersion

the lateral view in order to ascertain the needle placement (Fig. 24.33).

Once the needle has reached the epidural space with appropriate loss of resistance to either air or saline, 3–5 mL of nonionic contrast (Omnipaque or Isovue) is injected to confirm epidural placement. PA and lateral fluoroscopic images are then obtained. With the paramedian approach using the fluoroscope, the needle may be advanced toward the upper edge of the inferior lamina at the targeted space. The needle is advanced superiorly into the ligamentum flavum and, subsequently, into the epidural space, utilizing the loss of resistance technique. The contrast spread generally has an areolar appearance on the PA view. If no non-epidural (intravascular, subarachnoid, or soft tissue) contrast pattern is appreciated following the injection of 1–2 mL of contrast, sodium chloride is used for washout, and an injectate of biologic solution is injected into the epidural space.

In a typical PA epidurogram, the nerve roots are clearly outlined by contrast as they exit the intervertebral foramina, and in a lateral view, contrast extends along both the anterior and posterior aspects of the epidural space on the lateral radiograph producing a double line or railroad track appearance. The "railroad track" appearance is characteristic of epidural localization of the contrast (Fig. 24.33).

Figure 24.34 shows multiple filling patterns of thoracic transforaminal epidural injections.

## Extra-Epidural Placement

Dural puncture has been observed with thoracic epidural injections in 1.3% of the procedures [48]. This may be identified either with return of CSF or after injection of contrast. Contrast injection may reveal a subarachnoid pattern typical of a myelogram. Figure 24.35 illustrates typical myelographic patterns (Fig. 24.35a, b) and subarachnoid injection of contrast patterns (Fig. 24.35c, d). A typical myelogram demonstrates contrast within the thecal sac. An AP radiographic view of the thoracic spine after intrathecal placement of contrast media may outline the thoracic nerve roots within

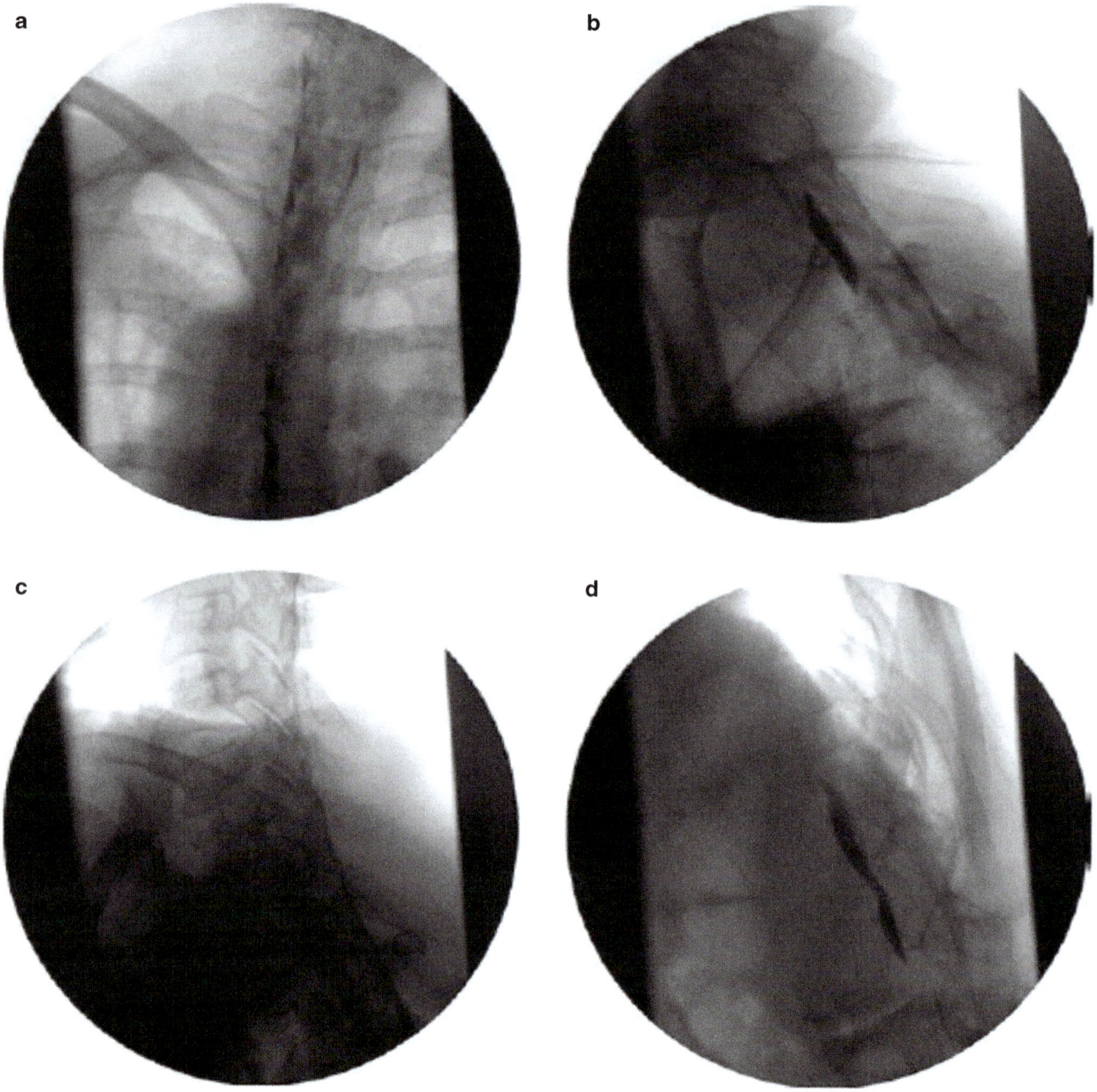


**Fig. 24.28** Unintended placement of epidural needle into the subarachnoid space showing subarachnoid filling patterns during an epidural procedure. (**a**) PA view of subarachnoid contrast pattern of cervical and thoracic spine. (**b**) Lateral view of cervical and thoracic subarachnoid contrast. (**c**) Lateral view of cervical subarachnoid contrast. (**d**) Subarachnoid contrast in the lumbar region in lateral view illustrating a different pattern compared to the cervical subarachnoid contrast

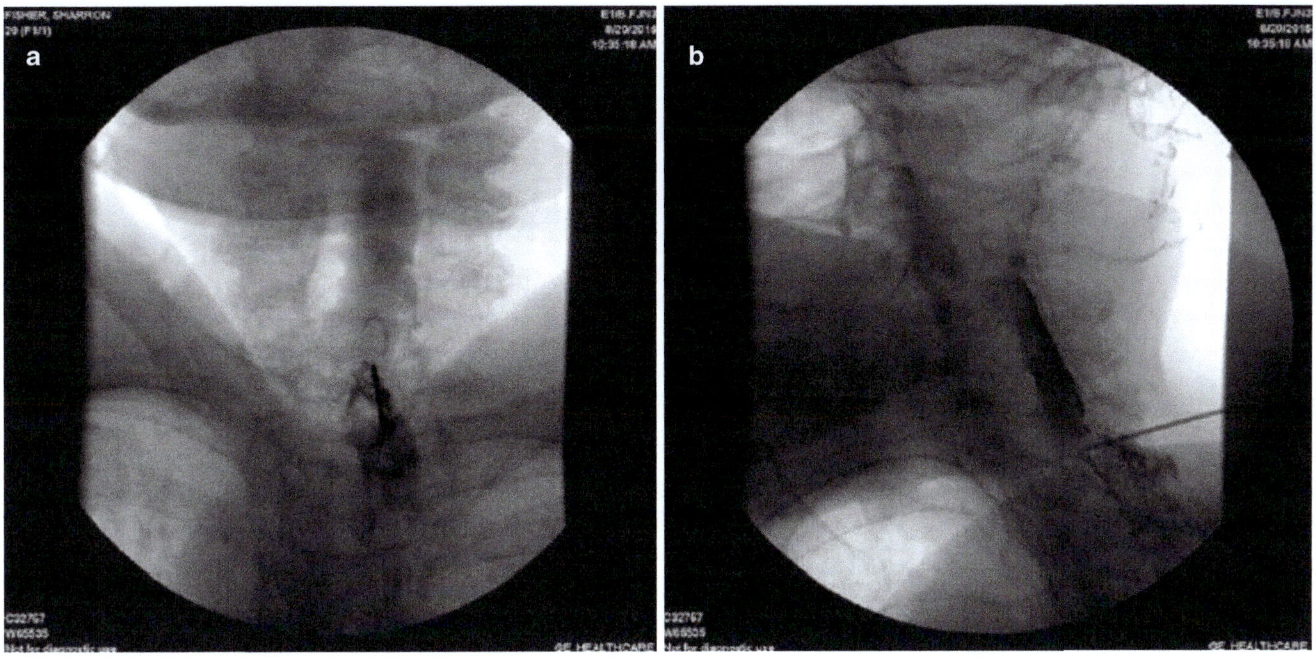


**Fig. 24.29** Unintended subarachnoid placement in the cervical spine. (**a**) PA view of subarachnoid contrast. (**b**) Oblique view of subarachnoid contrast

the thecal sac as they travel laterally toward the intervertebral foramina. The nerve roots appear radiolucent as they are surrounded by contrast within the CSF. A lateral radiograph of the thoracic spine after intrathecal placement of contrast media may show outlining of the thoracic nerve roots within the thecal sac as they travel laterally toward the neural foramina. However, this feature is much easier to see with myelography than following the unintended placement of subarachnoid contrast during an epidural injection (Fig. 24.35).

Subdural compartment entry is more commonly seen with the interlaminar approach rather than the transforaminal approach and more commonly in the lumbar region than the thoracic or cervical regions (Fig. 24.36).

## Thoracic Transforaminal Epidural Injections

Thoracic transforaminal epidural injections are performed infrequently. Due to the significant risk associated with thoracic transforaminal epidural injections and the lack of evidence of improved efficacy using the thoracic transforaminal approach compared to the interlaminar approach, they are performed infrequently. Nevertheless, they have been described in the literature along with their associated complications [10].

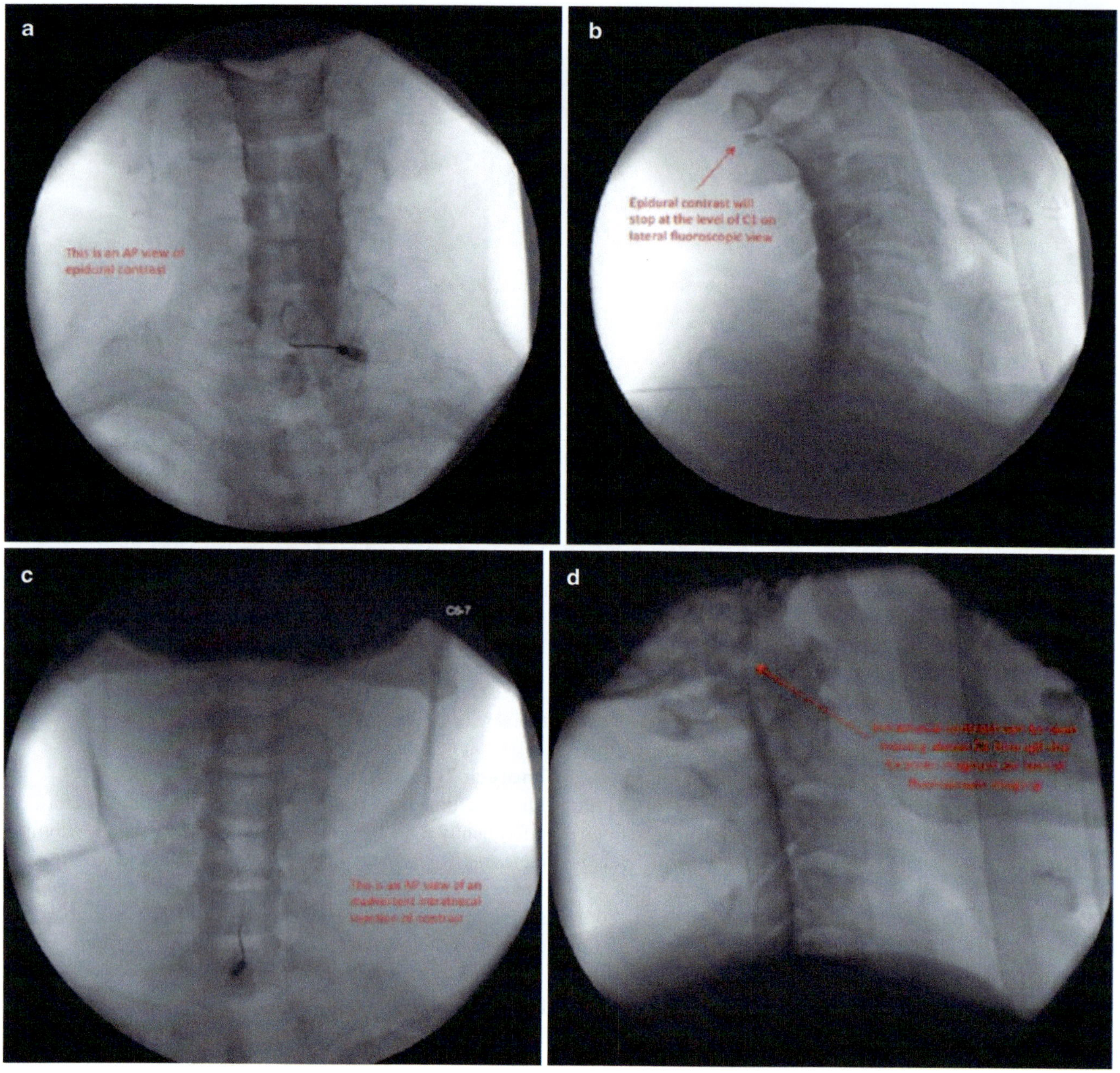


**Fig. 24.30** Differentiation of epidural and intrathecal contrast dispersion. (**a**) PA view of epidural contrast. (**b**) Lateral view of epidural contrast showing cephalad stop at C1. (**c**) PA view of inadvertent intrathecal contrast injection. (**d**) Intrathecal contrast can be seen moving through the foramen magnum cephalad of C1

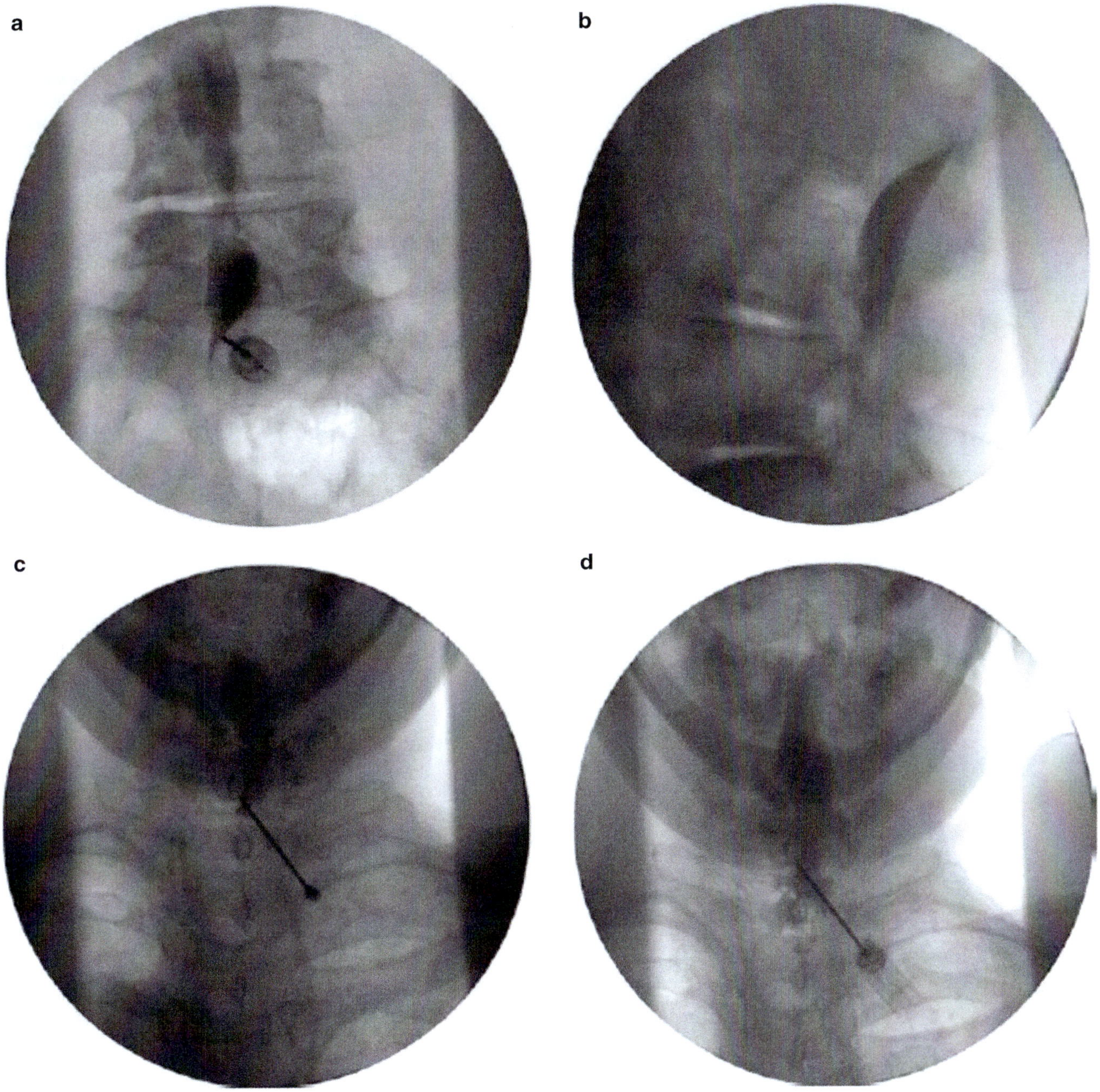


**Fig. 24.31** Illustration of subdural filling patterns during lumbar and cervical interlaminar epidural injections. (**a**) Contrast pattern of subdural injection in an AP view of lumbar spine. (**b**) Lateral view of subdural contrast pattern in lumbar spine. (**c**) Contrast injection into the epidural space shows an appearance without areolar pattern. (**d**) Injection of local anesthetic shows some enlargement of the contrast pattern, however, without nerve root distribution

**Fig. 24.32** Axial diagram of interlaminar thoracic epidural injection

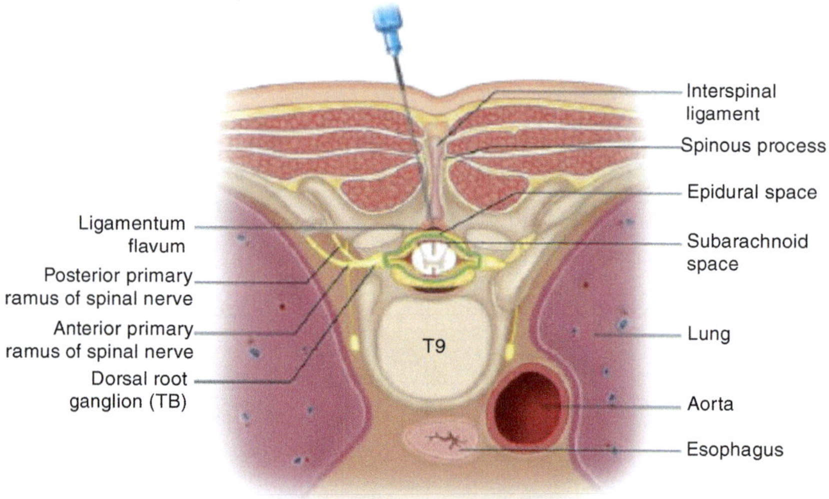


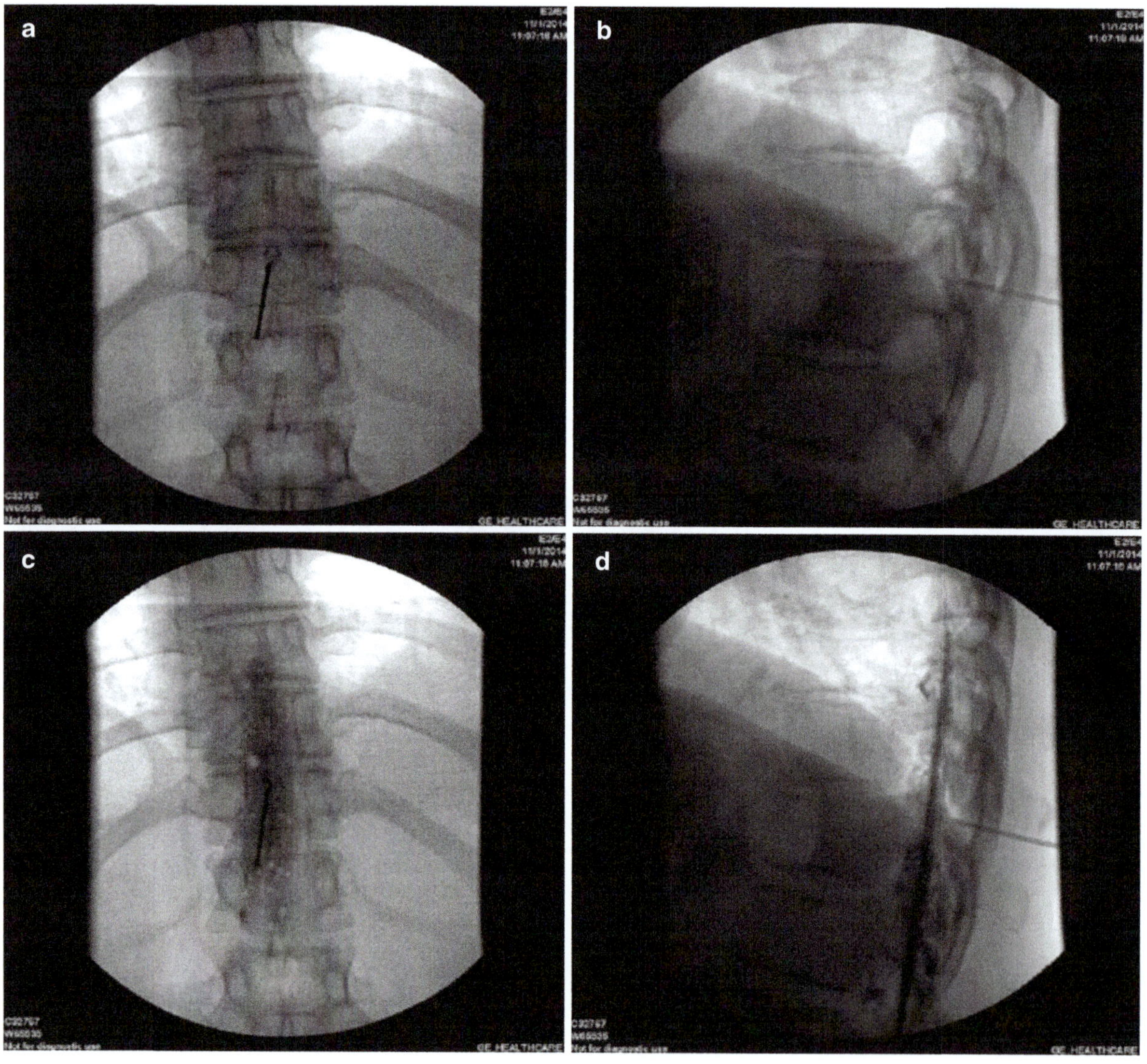


**Fig. 24.33** Typical steps in performing a thoracic epidural injection. (**a**) Placement of the needle between T10 and T11 epidural space in the PA view showing the needle tip in the midline. (**b**) Lateral view of the needle placement in the epidural space after passing the ligamentum flavum with loss of resistance technique inferior to T10–T11 disc space. (**c**) Epidural filling pattern with typical air bubbles extending into thoracic and lumbar epidural space with no significant nerve root outlining in PA view. (**d**) Epidural contrast display pattern in the lateral view extending into the thoracic and lumbar epidural space

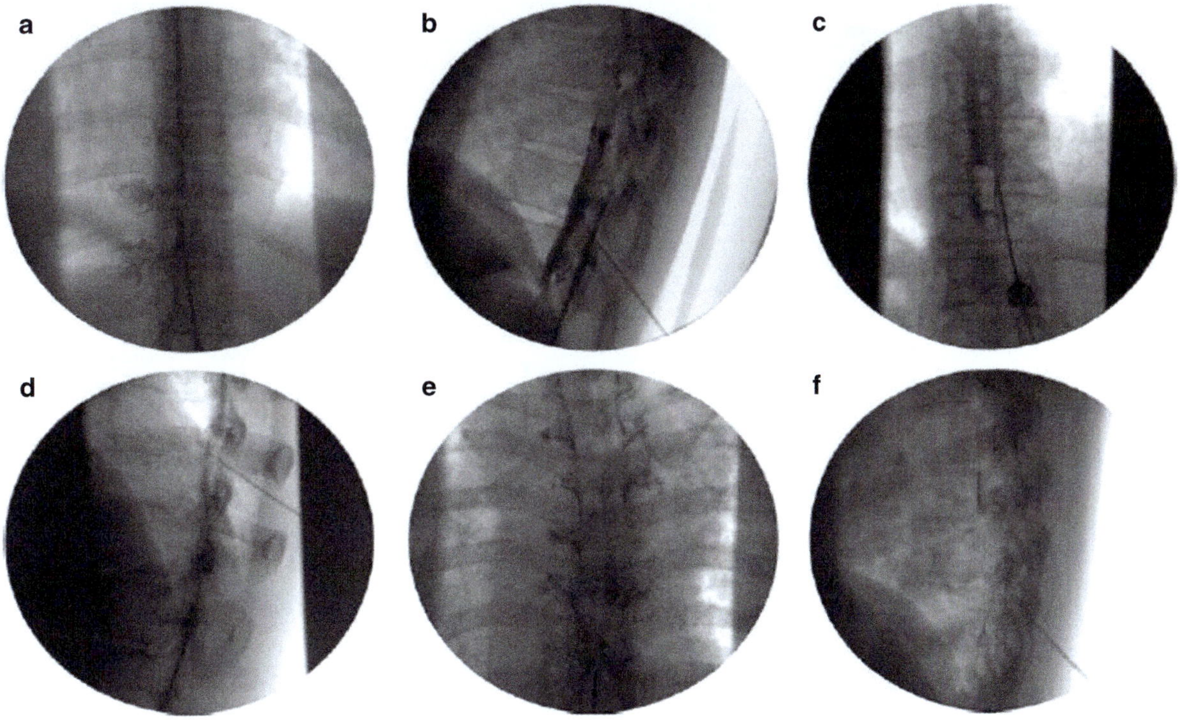


**Fig. 24.34** Typical thoracic epidural filling patterns. (**a**) PA view of contrast injection showing epidural filling accompanied with nerve root filling, both cephalad and caudad. (**b**) Lateral view with contrast extending along the anterior and posterior aspects of the epidural space producing a "double line" or "railroad track" appearance characteristic of epidural localization of the contrast. (**c**) PA View of thoracic epidurogram showing a typical bilateral pattern. (**d**) Lateral view with appropriate epidural filling pattern. (**e**) PA view with excellent nerve root filling patterns. (**f**) Lateral view of double line pattern

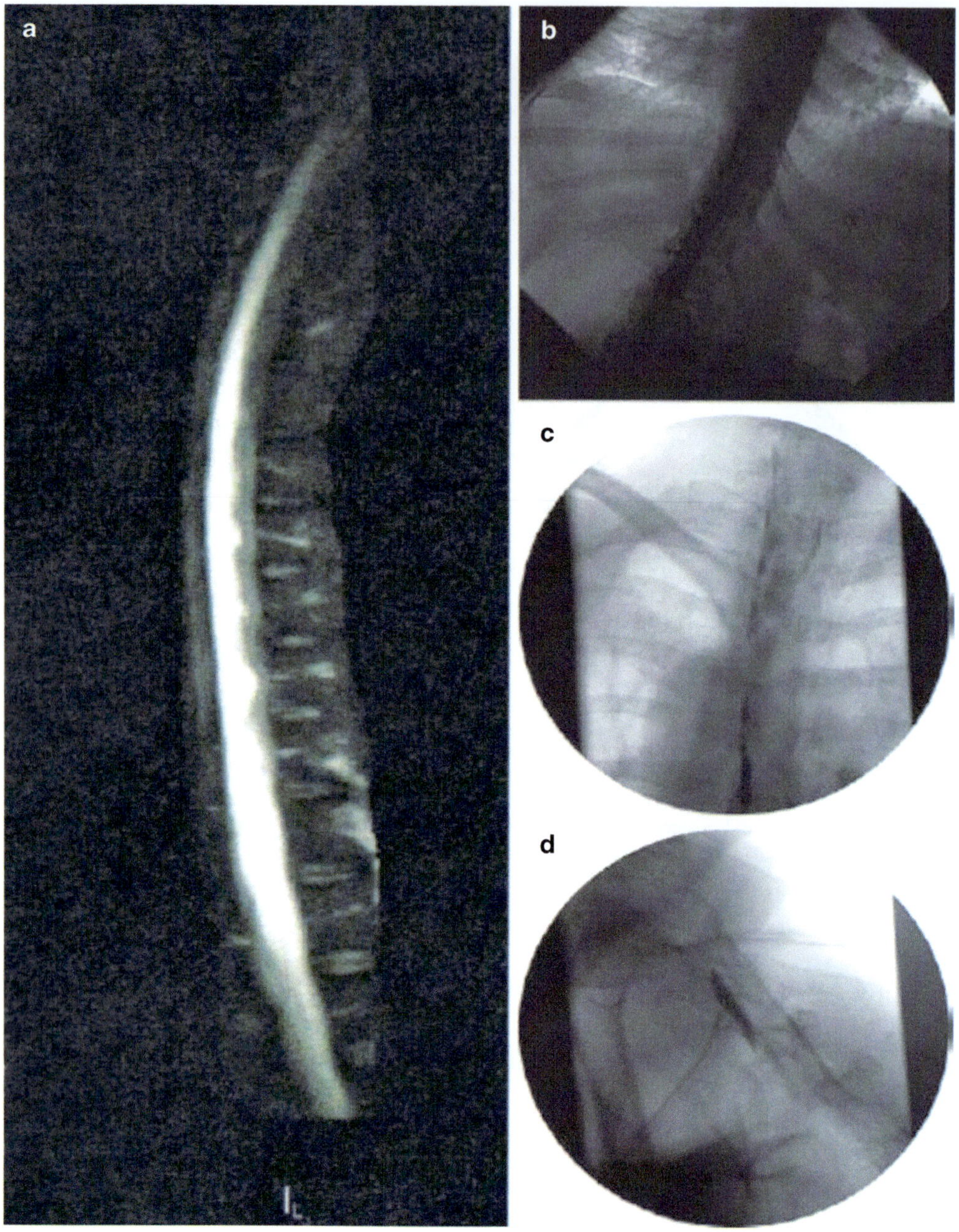


**Fig. 24.35** Myelographic and subarachnoid patterns of contrast. (**a**) MR myelography of the thoracic spine. (**b**) An AP view of myelography of the thoracic spine. (**c**) PA view of a subarachnoid injection of contrast of the thoracic spine. (**d**) Lateral view of a subarachnoid injection of contrast of the thoracic spine

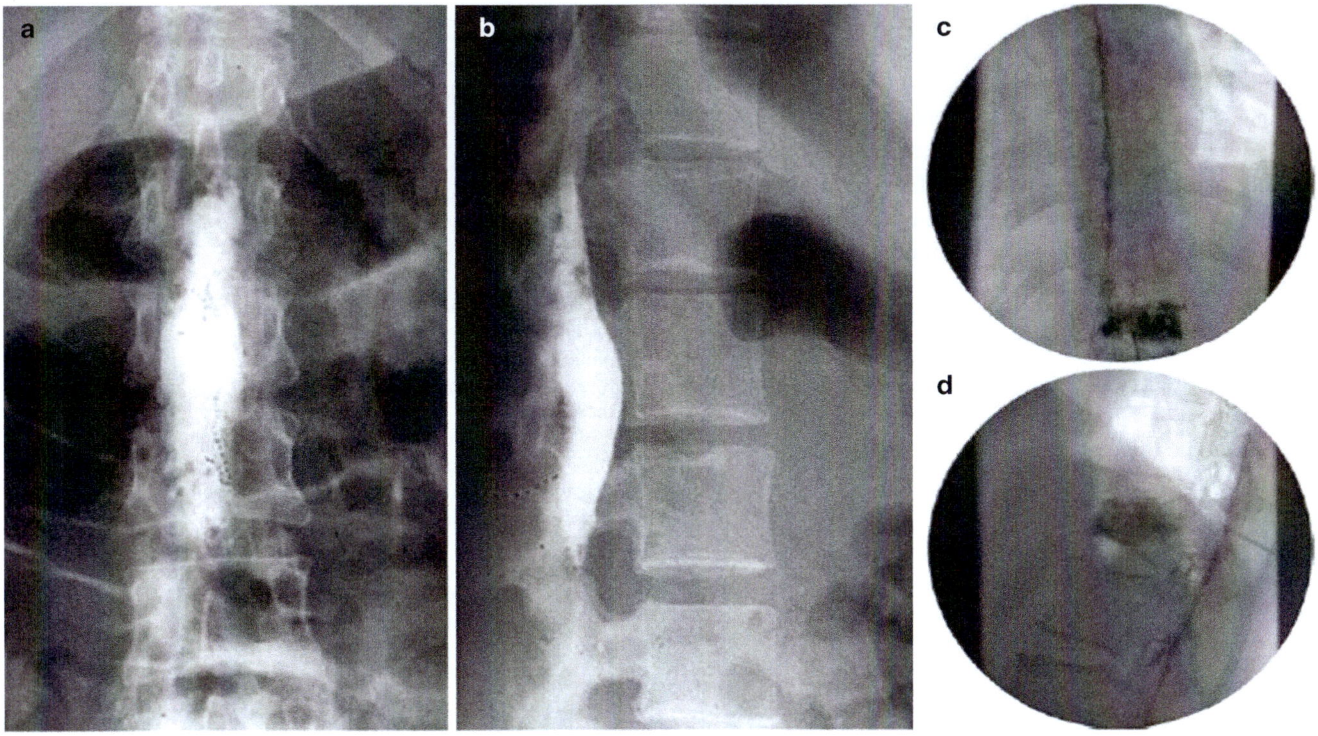


**Fig. 24.36** Subdural contrast patterns in the thoracic and lumbar spine. (**a**) AP view of subdural contrast, L2–4. (**b**) Lateral view of subdural contrast, with marked anterior bulge. (**c**) Subdural placement of 1 mL of contrast with the interlaminar approach—PA view. (**d**) Subdural placement of 1 mL of contrast with interlaminar approach—lateral view

## Side Effects and Complications

Side effects and complications related to epidural injections are either related to the needle placement or the drug administration and are rare. However, occasional complications occur specifically with transforaminal epidural injections because of possible neural trauma, vascular trauma, or intravascular injection leading to paralysis and/or death. Complications related to epidural injections have been described extensively [10–13].

## Precautions

Relative contraindications to interventional techniques, specifically epidural injections, have been described in patients receiving treatment with antithrombotics and anticoagulants [10–13, 69, 70].

Guidelines on anticoagulants and antithrombotics are described in Chap. 15 [71] and also on the ASIPP app, which is regularly updated.

**Key Points**

1. Various tissues in the lower back can transmit pain, including the disc, nerve roots, dura, muscle, ligaments, fascia, and facet joints. The compression or mass effect alone cannot explain the diverse presentations of back problems, suggesting the involvement of other mechanisms.
2. Radicular pain caused by lumbar disc herniation is thought to involve inflammatory reactions between the nucleus pulposus and nerve roots. Inflammatory mechanisms, along with other factors like partial axonal damage, neuroma formation, and impaired microcirculation, contribute to the production of sciatica from disc herniation.
3. Inflammatory pain mediators and cytokines within the disc, such as interleukin 1 (IL-1), have been implicated in inflammatory and degenerative changes, making them potential targets for intervention. Strategies to inhibit the biological activities of IL-1, including the use of IL-1 receptor antagonist and certain cytokines, have been proposed.
4. Inflammation, rather than mechanical compression alone, has been recognized as a contributor to radiculitis following disc herniation, leading to the development of anti-inflammatory therapies based on regenerative medicine principles.
5. Lumbar epidural injections can be performed through three approaches: interlaminar, transforaminal, and caudal. The interlaminar approach is commonly used for cervical and thoracic epidural injections, while the transforaminal approach allows for target-specific injection over the relevant nerve root.

6. Epidural injections, including the use of autologous products like platelet-rich plasma (PRP) and mesenchymal stem cells (MSCs), have shown promise in the treatment of spinal pain and related disabilities, particularly in cases of degenerative disc disease, disc herniation, and discogenic pain.
7. The investigation into the role of epidural platelet-rich plasma (PRP) or mesenchymal stem cell (MSC) injections in the treatment of disc herniation, associated radiculopathy, radiculopathy without disc herniation, and other biochemical and mechanical disorders is a pressing area of research.
8. Epidural injections, including the use of PRP and MSCs, are believed to produce anti-inflammatory effects, with or without the use of steroids. The role of steroids in managing discogenic pain through epidural injections remains a topic of debate and controversy.

**Acknowledgments** This book chapter is modified and updated from a previous book chapter, "Epidural Administration of Platelet Rich Plasma and Stem Cells" by Laxmaiah Manchikanti, MD, Sairam Atluri, MD, Sheri L. Albers, DO, and Richard E. Latchaw, MD, in *Essentials of Regenerative Medicine in Interventional Pain Management* published by ASIPP Publishing.

## References

1. US Burden of Disease Collaborators. The state of US health, 1990-2016: burden of diseases, injuries, and risk factors among US states. JAMA. 2018;319:1444–72.
2. Dieleman JL, Baral R, Birger M, et al. US spending on personal health care and public health, 1996-2013. JAMA. 2016;316:2627–46.
3. Dieleman JL, Cao J, Chapin A, et al. US health care spending by payer and health condition, 1996-2016. JAMA. 2020;323:863–84.
4. Navani A, Manchikanti L, Albers SL, et al. Responsible, safe, and effective use of biologics in the management of low back pain: American Society of Interventional Pain Physicians (ASIPP) guidelines. Pain Physician. 2019;22:S1–S74.
5. Manchikanti L, Centeno CJ, Atluri S, et al. Bone marrow concentrate (BMC) therapy in musculoskeletal disorders: evidence-based policy position statement of American Society of Interventional Pain Physicians (ASIPP). Pain Physician. 2020;23:E85–E131.
6. Manchikanti L, Atluri A, Albers SL, Latchaw RE. Epidural administration of platelet rich plasma and stem cells. In: Manchikanti L, Navani A, Atluri S, editors. Essentials of regenerative medicine in interventional pain management. Paducah, KY: ASIPP Publishing; 2019. p. 421–58.
7. Manchikanti L, Knezevic NN, Navani A, et al. Epidural interventions in the management of chronic spinal pain: American Society of Interventional Pain Physicians (ASIPP) comprehensive evidence-based guidelines. Pain Physician. 2021;24:S27–S208.
8. Knezevic N, Manchikanti L, Urits I, et al. Lack of superiority of epidural injections with lidocaine with steroids compared to without steroids in spinal pain: a systematic review and meta-analysis. Pain Physician. 2020;23:S239–70.
9. Manchikanti L, Kosanovic R, Vanaparthy R, et al. Steroid distancing in interventional pain management during COVID-19 and beyond: safe, effective and practical approach. Pain Physician. 2020;23:S319–52.
10. Manchikanti L, Schultz DM, Glaser SE, Falco FJE. Thoracic epidural injections. In: Manchikanti L, Singh V, Falco FJE, Kaye AD, Soin A, Hirsch JA, editors. Manchikanti's essentials of interventional techniques in managing chronic pain. 2nd ed. New York, NY: Springer; 2023.
11. Manchikanti L, Schultz DM, Atluri S, et al. Lumbar epidural injections. In: Manchikanti L, Kaye AD, Falco FJE, Hirsch JA, editors. Essentials of interventional techniques in managing chronic spinal pain. New York, NY: Springer; 2018. p. 141–86.
12. Manchikanti L, Schultz DM, Falco FJE. Cervical epidural injections. In: Manchikanti L, Kaye AD, Falco FJE, Hirsch JA, editors. Essentials of interventional techniques in managing chronic spinal pain. New York, NY: Springer; 2018. p. 209–40.
13. Manchikanti L, Heavner JE, Racz GB. Percutaneous adhesiolysis. In: Manchikanti L, Kaye AD, Falco FJE, Hirsch JA, editors. Essentials of interventional techniques in managing chronic spinal pain. New York, NY: Springer; 2018. p. 241–72.
14. Sanapati J, Manchikanti L, Atluri S, et al. Do regenerative medicine therapies provide long-term relief in chronic low back pain: a systematic review and metaanalysis. Pain Physician. 2018;21:515–40.
15. Basso M, Cavagnaro L, Zanirato A, et al. What is the clinical evidence on regenerative medicine in intervertebral disc degeneration? Musculoskelet Surg. 2017;101:93–104.
16. Wu T, Song HX, Dong Y, et al. Cell-based therapies for lumbar discogenic low back pain: systematic review and single-arm meta-analysis. Spine (Phila Pa 1976). 2018;43:49–57.
17. Khan S, Mafi P, Mafi R, et al. A systematic review of mesenchymal stem cells in spinal cord injury, intervertebral disc repair and spinal fusion. Curr Stem Cell Res Ther. 2018;13:316–23.
18. Goni VG, Singh Jhala S, Gopinathan NR, et al. Efficacy of epidural perineural injection of autologous conditioned serum in unilateral cervical radiculopathy: a pilot study. Spine (Phila Pa 1976). 2015;40:E915–21.
19. Kuslich SD, Ulstrom CL, Michael CJ. The tissue origin of low back pain and sciatica: a report of pain response to tissue stimulation during operation on the lumbar spine using local anesthesia. Orthop Clin North Am. 1991;22:181–7.
20. Mixter WJ, Barr JS. Rupture of the intervertebral disc with involvement of the spinal canal. N Engl J Med. 1934;211:210–5.
21. Manchikanti L, Boswell MV, Singh V, et al. Comprehensive review of neurophysiologic basis and diagnostic interventions in managing chronic spinal pain. Pain Physician. 2009;12:E71–E120.
22. Mixter WJ, Ayers JB. Herniation or rupture of the intervertebral disc into the spinal canal. N Engl J Med. 1935;213:385–95.
23. Manchikanti L, Falco FJE, Singh V, et al. Discography. In: Manchikanti L, Kaye AD, Falco FJE, Hirsch JA, editors. Essentials of interventional techniques in managing chronic spinal pain. New York, NY: Springer; 2018. p. 273–300.
24. Manchikanti L, Knezevic E, Knezevic NN, et al. The role of percutaneous neurolysis in lumbar disc herniation: systematic review and meta-analysis. Korean J Pain. 2021;34:346–68.
25. Manchikanti L, Knezevic NN, Knezevic E, et al. Efficacy of percutaneous adhesiolysis in managing low back and lower extremity pain: a systematic review and meta-analysis of randomized controlled trials. Pain Ther. 2023;12:903–37.
26. Manchikanti L, Hirsch JA. An update on the management of chronic lumbar discogenic pain. Pain Manag. 2015;5:373–86.
27. Manchikanti L, Hirsch JA. Clinical management of radicular pain. Expert Rev Neurother. 2015;17:681–93.
28. Spangfort EV. The lumbar disc herniation: a computer-aided analysis of 2,504 operations. Acta Orthop Scand Suppl. 1972;142:1–95.
29. Bogduk N, Aprill C, Derby R. Lumbar discogenic pain: state-of-the-art review. Pain Med. 2013;14:813–36.
30. Saal JA. Natural history and nonoperative treatment of lumbar disc herniation. Spine (Phila Pa 1976). 1996;21:2S–9S.
31. Wright-Carpenter T, Klein P, Schaferhoff P, Appell HJ, Mir LM, Wehling P. Treatment of muscle injuries by local administration of autologous conditioned serum: a pilot study on sportsmen with muscle strains. Int J Sports Med. 2004;25:588–93.
32. Manchikanti L, Staats PS, Nampiaparampil DE, Hirsch JA. What is the role of epidural injections in the treatment of lumbar discogenic

pain: a systematic review of comparative analysis with fusion and disc arthroplasty. Korean J Pain. 2015;28:75–87.
33. Bogduk N. Epidural steroids. Spine (Phila Pa 1976). 1995;20:845–8.
34. Shanthanna H, Busse JW, Thabane L, et al. Local anesthetic injections with or without steroid for chronic non-cancer pain: a protocol for a systematic review and meta-analysis of randomized controlled trials. Syst Rev. 2016;5:18.
35. Kumar R, Goni VG, Batra YK. Autologous conditioned serum as a novel alternative option in the treatment of unilateral lumbar radiculopathy: a prospective study. Asian Spine J. 2015;9:916–22.
36. Becker C, Heidersdorf S, Drewlo S, et al. Efficacy of epidural perineural injections with autologous conditioned serum for lumbar radicular compression: an investigator-initiated, prospective, double-blind, reference-controlled study. Spine (Phila Pa 1976). 2007;32:1803–8.
37. Manchikanti L, Falco FJE, Benyamin RM, Kaye AD, Boswell MV, Hirsch JA. A modified approach to grading of evidence. Pain Physician. 2014;17:E319–25.
38. Bhatia R, Chopra G. Efficacy of platelet rich plasma via lumbar epidural route in chronic prolapsed intervertebral disc patients—a pilot study. J Clin Diagn Res. 2016;10:UC05–7.
39. Centeno C, Markle J, Dodson E, et al. The use of lumbar epidural injection of platelet lysate for treatment of radicular pain. J Exp Orthop. 2017;4:38.
40. Kirchner F, Anitua E. Intradiscal and intra-articular facet infiltrations with plasma rich in growth factors reduce pain in patients with chronic low back pain. J Craniovertebr Junction Spine. 2016;7:250–6.
41. Noriega DC, Ardura F, Hernández-Ramajo R, et al. Intervertebral disc repair by allogeneic mesenchymal bone marrow cells: a randomized controlled trial. Transplantation. 2017;101:1945–51.
42. Tuakli-Wosornu YA, Terry A, Boachie-Adjei K, et al. Lumbar intradiskal platelet-rich plasma (PRP) injections: a prospective, double-blind, randomized controlled study. PM R. 2016;8:1–10.
43. Wu J, Zhou J, Liu C, et al. A prospective study comparing platelet-rich plasma and local anesthetic (LA)/corticosteroid in intra-articular injection for the treatment of lumbar facet joint syndrome. Pain Pract. 2017;17:914–24.
44. Singla V, Batra YK, Bharti N, et al. Steroid vs. platelet-rich plasma in ultrasound-guided sacroiliac joint injection for chronic low back pain. Pain Pract. 2017;17:782–91.
45. Ko GD, Mindra S, Lawson GE, Whitmore S, Arseneau L. Case series of ultrasound-guided platelet-rich plasma injections for sacroiliac joint dysfunction. J Back Musculoskelet Rehabil. 2017;30:363–70.
46. Peng B, Wu W, Hou S, Li P, Zhang C, Yang Y. The pathogenesis of discogenic low back pain. J Bone Joint Surg Br. 2005;87:62–7.
47. Botwin KP. Lumbar interlaminar epidural steroid injections. In: Manchikanti L, Singh V, editors. Interventional techniques in chronic spinal pain. Paducah, KY: ASIPP Publishing; 2007. p. 355–82.
48. Manchikanti L, Malla Y, Wargo BW, et al. A prospective evaluation of complications of 10,000 fluoroscopically directed epidural injections. Pain Physician. 2012;15:131–40.
49. Ajar AH, Rathmell LP, Mukherji SK. The subdural compartment. Reg Anesth Pain Med. 2002;27:72–6.
50. Murthy NS, Maus TP, Behrns CL. Intraforaminal location of the great anterior radiculomedullary artery (artery of Adamkiewicz): a retrospective review. Pain Med. 2010;11:1756–64.
51. Kim WJ, Shin HY, Yoo SH, et al. Comparison of epidural spreading patterns and clinical outcomes of transforaminal epidural steroid injection with high-volume injectate via the subpedicular versus the retrodiscal approach. Pain Physician. 2018;21:269–78.
52. Atluri S, Glaser S, Shah R, et al. Needle position analysis in cases of paralysis from transforaminal epidurals: consider alternative approaches to traditional technique. Pain Physician. 2013;16:321–34.
53. Goodman B, Petalcorin JS, Mallempati S. Optimizing patient positioning and fluoroscopic imaging for the performance of cervical interlaminar epidural steroid injections. PM R. 2010;2:783–6.
54. Yoon SP, Kim HJ, Choi YS. Anatomic variations of cervical and high thoracic ligamentum flavum. Korean J Pain. 2014;27:321–5.
55. Ho PS, Yu SW, Sether LA, et al. Ligamentum flavum: appearance on sagittal and coronal MR images. Radiology. 1988;168:469–72.
56. Lirk P, Kolbitsch C, Putz G, et al. Cervical and high thoracic ligamentum flavum frequently fails to fuse in the midline. Anesthesiology. 2003;99:1387–90.
57. Dogliotti AM. Segmental peridural anesthesia. Am J Surg. 1933;20:107–18.
58. Shenouda P, Cunningham B. Assessing the superiority of saline versus air for use in the epidural loss of resistance technique: a literature review. Reg Anesth Pain Med. 2003;28:48–53.
59. Mehta S, Khalil AA, Alshekhlee A. Air myelopathy following a cervical epidural injection. Pain Med. 2010;11:1678–9.
60. Kim TS, Shin SS, Kim JR, et al. Air bubbles mimic disc herniation in MRI after cervical epidural block. Korean J Pain. 2010;23:202–6.
61. Friedly JL, Comstock BA, Turner JA, et al. A randomized trial of epidural glucocorticoid injections for spinal stenosis. N Engl J Med. 2014;371:11–21.
62. Manchikanti L, Candido KD, Singh V, et al. Epidural steroid warning controversy still dogging FDA. Pain Physician. 2014;17:E451–74.
63. Manchikanti L, Falco FJE, Benyamin RM, et al. Epidural steroid injections safety recommendations by the Multi-Society Pain Workgroup (MPW): more regulations without evidence or clarification. Pain Physician. 2014;17:E575–88.
64. Rathmell JP, Benzon HT, Dreyfuss P, et al. Safeguards to prevent neurologic complications after epidural steroid injections: consensus opinions from a multidisciplinary working group and national organizations. Anesthesiology. 2015;122:974–84.
65. Manchikanti L, Falco FJE. Safeguards to prevent neurologic complications after epidural steroid injections: analysis of evidence and lack of applicability of controversial policies. Pain Physician. 2015;18:E129–38.
66. Collier CB. Accidental subdural injection during attempted lumbar epidural block may present as a failed or inadequate block: radiographic evidence. Reg Anesth Pain Med. 2004;29:45–51.
67. Wallace MA, Fukui MB, William RL, et al. Complications of cervical selective nerve root blocks performed with fluoroscopic guidance. AJR Am J Roentgenol. 2007;188:1218–21.
68. Scanlon GC, Moeller-Bertram T, Romanowsky SM, et al. Cervical transforaminal epidural steroid injections. More dangerous than we think? Spine (Phila Pa 1976). 2007;32:1249–56.
69. Netter FH. Atlas of human anatomy. 4th ed. Philadelphia, PA: Elsevier (Saunders); 2006.
70. Kaye AD, Manchikanti L, Novitch MB, et al. Responsible, safe, and effective use of antithrombotics and anticoagulants in patients undergoing interventional techniques: American Society of Interventional Pain Physicians (ASIPP) guidelines. Pain Physician. 2019;22:S75–S128.
71. Manchikanti L, Sanapati M, Kaye AD, Kaye AM. Antiplatelet and anticoagulant therapy in regenerative medicine. In: Manchikanti L, Navani A, Atluri S, Sanapati M, editors. Manchikanti's essentials of regenerative medicine in interventional pain management. 2nd ed. New York, NY: Springer; 2023.

# Lumbar Facet Joints and Intraarticular Injection of Biologics

# 25

Annu Navani, Laxmaiah Manchikanti, and Mahendra Sanapati

## Introduction

Low back pain is the most common of all disabling chronic spinal pain problems [1–7]. Structures capable of transmitting pain in the lumbar spine, resulting in symptoms of low back pain and lower extremity pain, include lumbar facet joints, lumbar intervertebral discs, sacroiliac joints, ligaments, fascia, muscles, and nerve root dura.

Utilizing controlled diagnostic blocks, the prevalence of persistent low back pain secondary to the involvement of lumbosacral facet joints has been established as varying from 16% to 41% depending on the type of population and setting studied [7–12]. Lumbar facet joint pain has been managed utilizing intraarticular injections, facet joint nerve blocks, and/or radiofrequency neurotomy [7, 13].

The evidence for therapeutic modalities continues to be debated with moderate to good evidence for therapeutic facet joint nerve blocks of average 3 months and with radiofrequency neurotomy of an average 9 months with limited evidence for intraarticular injections [7, 13]. Consequently, various other types of interventions are considered. Regenerative medicine appears to be an attractive option, based on the results of effectiveness in knee joint and other joints with repair and restoration of injuries [14–22]. Even though there is paucity of literature and its clinical applications and research interest in managing facet joint pain with platelet-rich plasma (PRP) and medicinal signaling cells or mesenchymal stem cells (MSCs), there is a significant potential for their usage in managing lumbar facet joint pain.

A. Navani
LeReve Regenerative Wellness, Campbell, CA, USA
Stanford University School of Medicine, Stanford, CA, USA
Boomerang Healthcare, Walnut Creek, CA, USA
e-mail: anavani@boomeranghc.com

L. Manchikanti (✉)
Pain Management Centers of America, Paducah, KY, USA
e-mail: drlm@thepainmd.com

M. Sanapati
Pain Management Centers of America, Evansville, IN, USA

## History

Lumbar facet joints were recognized as potential sources of back pain in 1911 by Goldthwait [23]. A few years later, Ghormley coined the term facet syndrome, describing the lumbosacral pain [24]. Subsequently, in 1941 Badgley [25] proposed that facet joints themselves could be a primary source of pain separate from nerve root compression based on his observations of large numbers of patients with low back pain without disc herniation. Twenty years later, Hirsch et al. [26] experimented with injection of hypertonic saline into the facet joints, with reproduction of the pain generation in the low back, hip, and sacroiliac area, thus providing a physiologic basis. Subsequently, among many authors attempting to map pain of facet joint origin, Mooney and Robertson [27] and McCall et al. [28] injected hypertonic saline into the facet joints of asymptomatic volunteers under fluoroscopic guidance demonstrating resultant low back and lower extremity pain. Multiple investigators [27–31] described the pattern of pain caused by facet joint stimulation. Kaplan et al. [32] and Dreyfuss et al. [33] showed that facet joint pain could be relieved by anesthetizing the facet joints responsible for low back pain.

In addition to the conventional management [7, 13, 34], multiple biologics have been injected into the facet joints, including platelet-rich plasma (PRP) and medicinal signaling cells or mesenchymal stem cells (MSCs).

## Pathophysiology

The innervation for lumbar facet joints has been well described [7, 8, 12, 13, 27–36]. Recently, biologic solutions have been injected into the facet joints [15–18, 21, 22]. The


A. Navani et al. (eds.), *Essentials of Regenerative Medicine in Interventional Pain Management*,
https://doi.org/10.1007/978-3-031-50357-3_25

presence of free and encapsulated nerve endings in the lumbar facet joints, as well as nerves containing substance P and calcitonin gene-related peptide have been demonstrated by neuroanatomic, neurophysiologic, and biomechanical studies [37–40]. It is well recognized that lumbar facet joints are a source of low back and referred pain in the lower extremities in patients with chronic low back pain [7–13, 23–34].

Facet joint pain has been proposed to be induced by both mechanical injury and inflammation of the facet joints resulting in persistent pain in otherwise normal rats [41]. It has been shown that inflammatory mediators, such as cytokines, prostaglandins, and neuropeptides increase within the joint and the dorsal root ganglia in joint inflammation and arthritis [42, 43].

Even though osteoarthritis is a common condition to be treated with biologicals, radiologically present arthritis in facet joints and their relationship to chronic pain has not been established [6–8, 12, 13, 34]. Overall, facet joint pain has been reported in a large proportion of patients, specifically in the elderly.

## Regenerative Medicine

The biologic primarily referred to with regards to lumbar facet joints is PRP. PRP includes concentration of growth factors that are secreted by a platelet in an inflammatory environment, i.e., platelet-derived growth factors. These growth factors are essential to the healing process, as they increase fibroblast or osteoblast metabolic activity while reducing cell apoptosis; promote angiogenesis, thereby increasing blood flow and circulation to the new-forming tissues; and increase the expression of the pro-collagen gene and collagen-derived growth factor, which increases the tensile strength of the new tissue [44, 45]. Proteins present in PRP that are proliferative in nature include endothelial growth factor, platelet-derived growth factor, vascular endothelial growth factor, and basic fibroblast growth factor (b-FGF). PRP is prepared from whole blood with the addition of only an acidifying agent and a platelet anticoagulant added for stabilization.

The other biologics that may be relevant in this regard is medicinal signaling cells or mesenchymal stem cells, and other commercially available biologic preparations however they are currently no published studies supporting their use for lumbar facet joint pain. Interestingly, Kristjánsson and colleagues [46] reported the MSCs harvested from the facet joints and interspinous ligaments of ten patients had the trilineage potential to be differentiated into osteogenic, adipogenic, and chondrogenic cells under appropriate inductions. They concluded that the facet joints and interspinous ligaments may provide alternative sources of MSCs for tissue engineering applications.

## Evidence Base

The literature supporting the use of PRP for lumbar facet joint pain is limited. Sanapati et al. [47] summarized the literature pertaining to lumbar facet intraarticular injections, along with a single arm meta-analysis. Their study yielded three studies assessing the role of PRP in lumbar facet joint injections [16–18]. Of the three available studies, only one was randomized comparing PRP comparing to a local anesthetic combined with a corticosteroid [17]. Of the 3 studies, 2 were performed by 1 group of authors with a sample size of 19 [16] and 46 [17]. The third study by Kirchner and Anitua [18] was a complicated study with multiple injections (intradiscal facet joint as well as transforaminal) and reported excellent results. Consequently, Sanapati et al. [47] based on one high quality randomized controlled trial [17] and two

**Table 25.1** Summary of lumbar facet joint PRP studies published to date

| Study details | Results | Conclusions |
|---|---|---|
| Wu et al. 2016 [16] Prospective clinical evaluation, n = 19 humans Chronic facet joint pain | 19 patients with lumbar facet syndrome given intra-articular injections of PRP | Combined measures of pain relief and functional improvement that were "good" or "excellent" were: 9/19 patients immediately after treatment; 14/19 patients 7 days post-injections, 15/19 patients 4 weeks, 8 weeks, and 3 months after injections |
| Wu et al. 2017 [17] Prospective randomized trial, n = 46 humans Chronic facet joint pain | 46 patients with lumbar facet syndrome were randomized to intra-articular injections of PRP versus LA/ corticosteroid | VAS score at rest or during flexion, the RMQ, and the ODI ($P < 0.01$) improved in both groups ($P < 0.05$). For PRP group, subjective satisfaction based on the modified MacNab criteria and objective success rate increased over time until 6-month duration of study |
| Kirchner et al. 2016 [18] Observational retrospective pilot study, n = 86 humans Facet joint syndrome | One intradiscal, one intra-articular facet, and one transforaminal epidural injection of PRGF under fluoroscopic guidance-control were carried out in 86 patients with chronic LBP | VAS showed a statistically significant drop at 1, 3, and 6 months after the treatment ($P < 0.0001$) except for the pain reduction between the third and sixth month whose signification was lower ($P < 0.05$) |

moderate quality observational studies [16, 18], showed the qualitative evidence for facet joint injections with PRP as Level IV on a scale of Level I to V using qualitative modified approach to grading of evidence based on best evidence synthesis. Table 25.1 summarizes lumbar facet joint PRP studies.

A study has been published by El-Kadiry et al. [19] not specifically dedicated to facet joints in a nonrandomized comparative case series of 18 patients. There were only five patients with facet joint injections. It was extremely difficult to interpret the data in this study.

The results of the study by Atluri et al. [20] showed significant improvement was achieved in functional status measured by ODI, pain relief measured by NRS-11, and other parameters measured by EQ-5D-3L, GMH, and GPH, in the study group relative to the control group at all time periods. The results showed significant improvements at 12-month follow-up with 67% of the patients in the study group achieving MCID utilizing ODI when compared to 8% in the control group. Greater than 2-point pain reduction was seen in 74% of the patients at 3 months, 66% of the patients at 6 months, and 56% of the patients at 12 months. Both MCID and pain relief of two points were significantly different compared to the control group. Opioid use decreased in the investigational group, whereas, there was a slight increase in the control group. Age, gender, opioid use, and body mass index did not affect the outcomes in the stem cell group. Conclusions: The first available controlled study utilizing BM-MSCs in severe degenerative spinal disease with interventions into multiple structures simultaneously, including disc, facet joints, nerve roots, and sacroiliac joint based on symptomatology, showed promising results.

## Clinical Presentation

The diagnosis of facet-joint pain is established with detailed history, comprehensive physical examination, review of diagnostic studies, and lastly corroboration of all the facts obtained from these sources. It is imperative to first rule out all the red flag causes of back pain, such as infection, tumors, vascular or metabolic diseases thorough history and examination, laboratory tests, and medical imaging if necessary. Figure 25.1 shows pain patterns.

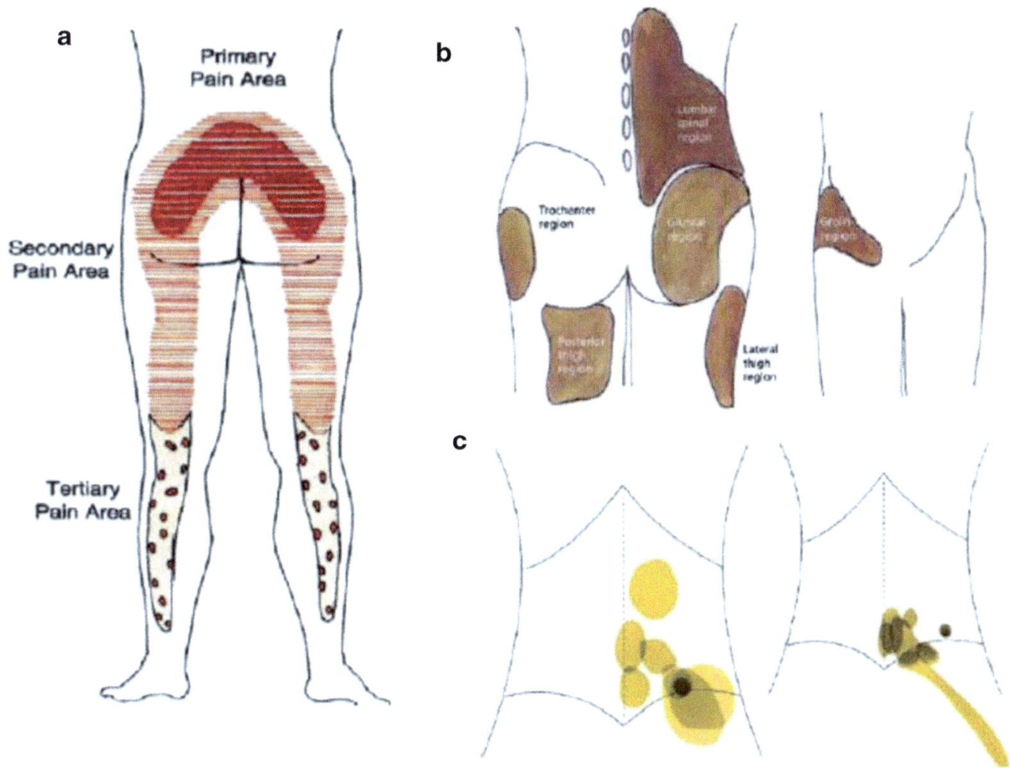


**Fig. 25.1** Patterns of lumbar facet joint pain based on descriptions of multiple authors [26–31]. (*Source: Manchikanti L, Schultz DM, Singh V, et al. Lumbar facet joint interventions. In: Manchikanti L, Singh V (eds). Interventional Techniques in Chronic Spinal Pain. ASIPP Publishing, Paducah, KY, 2007, pp. 253–276* [50]). (**a**) Illustration of distribution of lumbar facet joint pain [5]. (**b**) Referred pain distribution. (1) Lumbar spinal region; (2) gluteal region; (3) trochanter region; (4) lateral thigh region; (5) posterior thigh region; (6) groin region [8]. (**c**) Composite drawing of the referral zones of all eight subjects derived from the minimal threshold stimulation of their right L3 (left) and right L4 (right) medial branches [9]

Although there are no clinical features that allow a physician to predict with confidence that the patient with back pain will respond positively to lumbar facet joint nerve blocks, there are certain features that increase the likelihood of zygapophyseal joint pain, such as age greater than 65 years, pain relieved by recumbency and absence of aggravation of pain by coughing, forward flexion or rising from flexion. Typically, the lumbar facet pain presents as low-back pain that is aggravated by hyperextension or extension-rotation [6–12, 34]. The diagnosis of facet joint pain is reliably confirmed with controlled diagnostic blocks with concordant response with short-acting and long-acting local anesthetics [6–12, 34].

## Anatomy

The facet joint is a true synovium-lined joint located in the posterior compartment of the spine that allows the spine to flex, extend, and rotate.

The lumbar spine can be divided into two to three compartments—anterior, middle, and posterior. The anterior compartment consists of vertebral bodies and disk spaces, the middle compartment consists of epidural space and neural foramina, and the posterior compartment consists of facets joints. The vertebrae articulate anteriorly through the disks and posteriorly through the two facet joints, one on each side. The facet joints are paired diarthrodial synovial joints formed by inferior articular process of one vertebra and the superior articular process of the subjacent vertebrae below (Fig. 25.2).

Biomechanically, this produces complex tripod structure that has evolved into distinctive patterns at each spinal level. The facet joints share the same general characteristics of the synovial joint. The articular surface is covered by cartilage and the synovial membrane bridges the margins of the cartilage of two facets in each joint. Covering the synovial membrane superiorly, posteriorly, and inferiorly is a thick fibrous joint capsule, however, anteriorly the synovial membrane lacks true fibrous capsule and instead is in direct contact with the ligamentum flavum. Superiorly and inferiorly, the fibrous joint capsule forms a pocket connected to the lumbar facet joint between the capsule pocket and synovium are two structures. The first has been referred to a meniscoid. There are several descriptions of these meniscoids, but probably they are best understood as a fold of synovium that contains adipose tissue, collagen, and some blood vessels. The normal function of these meniscoids is not clear although in the diseased facet joints, they have been implicated as a source of damage to the articular cartilage causing the facet joints to lock. The second structure is simply adipose tissue, which occupy any leftover space beneath the capsule. The superior and inferior parts of the capsule each have a hole. The superior hole connects the intracapsular fat to the epidural space. The capsule is attached to the margins of articular process of the adjacent vertebra, accessory ligaments unite the lamina, transverse process, and spinous process is to help stabilize the joint.

The zygapophyseal joints are innervated by articular branches that arrive from the medial branches (MBs) of the dorsal primary rami of the spinal nerves as these nerves pass posteriorly and inferiorly and lie in grooves on the posterior surfaces of medial parts of the transverse process (Figs. 25.3 and 25.4) [48]. Each articular branch supplies two adjacent joints; therefore, each joint is supplied by two nerves, the nerve at the same level and the level above, e.g., L4-L5 joint

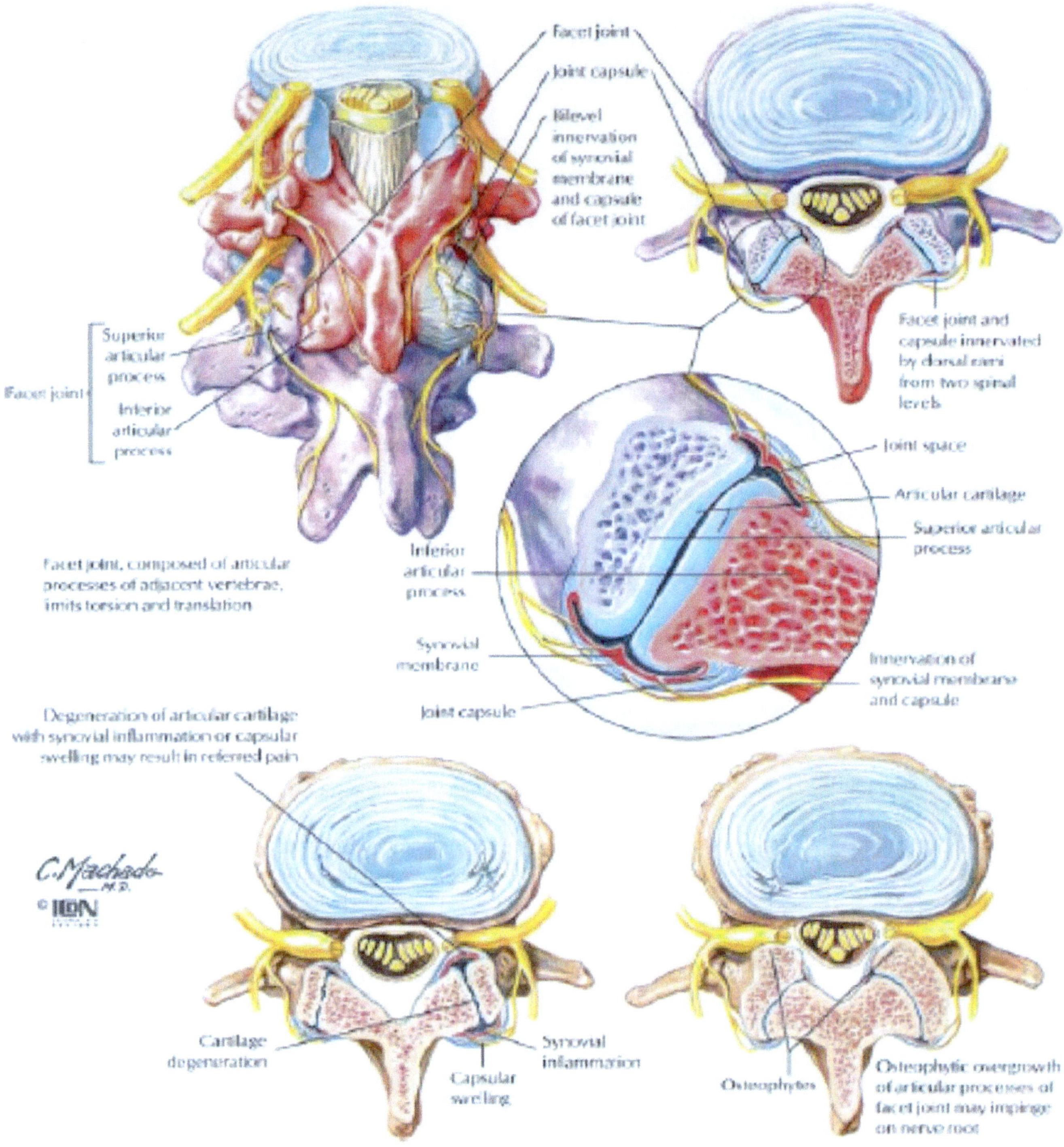


**Fig. 25.2** Anatomic and pathophysiologic aspects of facet joint pain. (*Source: Reproduced from Netter Medical Illustrated. Used with permission of Elsevier*)

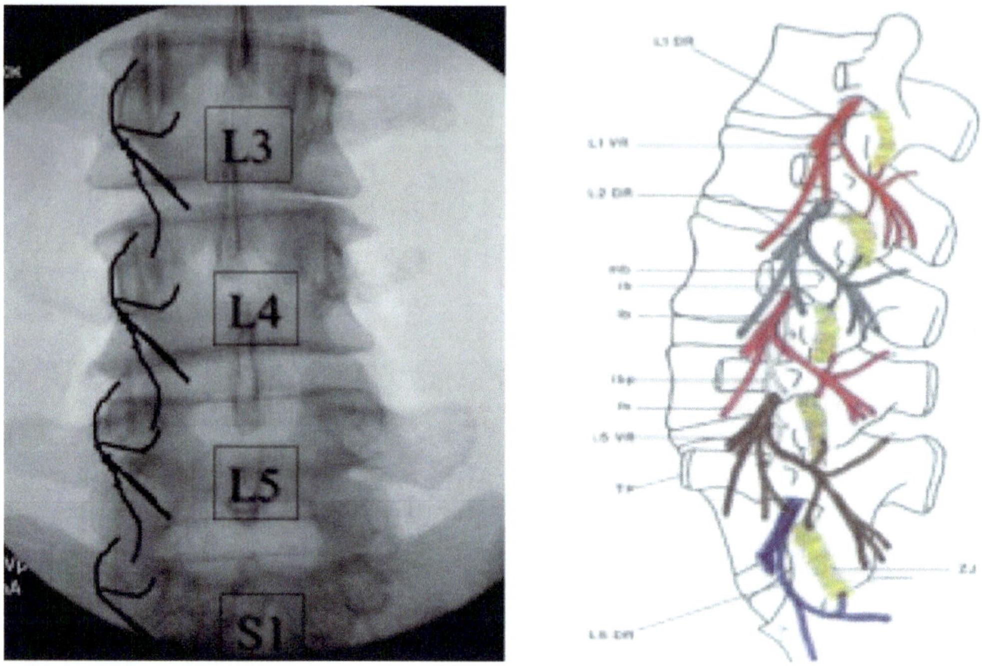


**Fig. 25.3** Innervation of lumbar facet joints

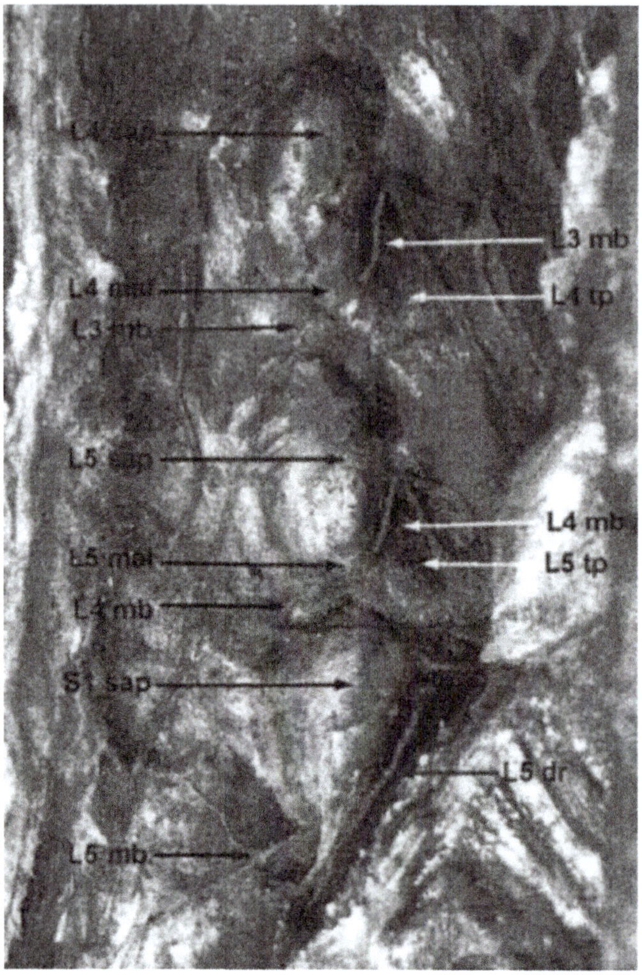


**Fig. 25.4** A close-up view of a dissection of the left L3, L4, and L5 medial branches and the L5 dorsal ramus. (**Source:** Bogduk N. Lumbar Medial Branch Blocks. In: Practice Guidelines for Spinal Diagnostic and Treatment Procedures. 2nd edition. International Spine Intervention Society (ISIS), San Francisco, 2013, pp. 559–600 [48])

is innervated by L3 and L4 MBs, whereas L5-S1 joint is innervated by L4 MB and L5 dorsal ramus.

## Technical Considerations

Lumbar facet joint interventions in regenerative medicine include intraarticular injections and facet joint nerve blocks [6, 49].

It is imperative to understand radiographic/fluoroscopic anatomy prior to attempting facet interventions. Figure 25.5 illustrates the radiographic anatomy of the lumbar spine.

### Lumbar Facet Joint Intraarticular Injections

The patient is placed prone on the fluoroscopy table. A rolled towel or pillow may be placed under the abdomen to facilitate easier entry into the joint. As with any lumbar intervention, a baseline anteroposterior (AP) fluoroscopic view of the lumbar spine is obtained and the fluoroscope is oriented. Facet joints may or may not be visible on AP fluoroscopy depending on the specific anatomy of the patient and projection of fluoroscopy (Fig. 25.6).

Fluoroscopic views of the target joint are obtained. The fluoroscopy beam is rotated from the AP view toward the oblique view until the target facet joint is visible. The upper lumbar facet joints are typically aligned toward the sagittal plane and may be visible on AP imaging, whereas the lower joints are typically and increasingly oriented toward the coronal plane so that oblique imaging is necessary to identify joint lines (Fig. 25.6). The target joint is then visualized under fluoroscopic guidance and the skin may be marked. A

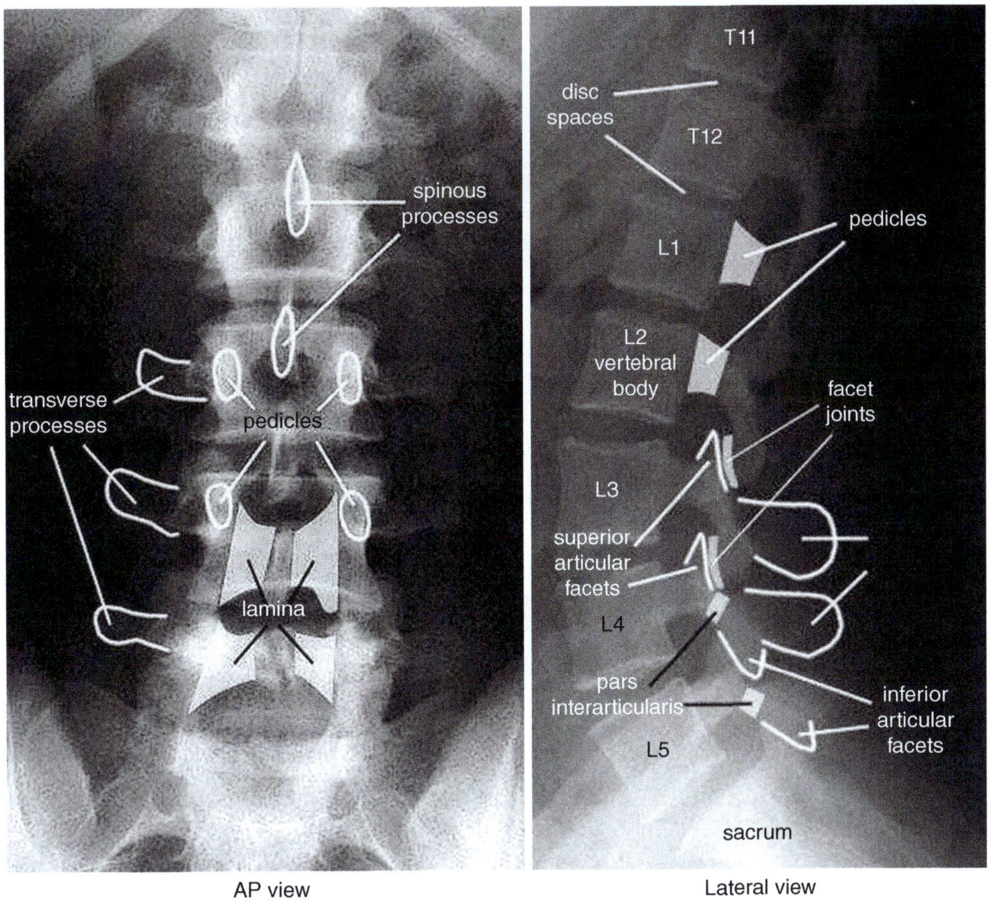


**Fig. 25.5** Radiographic anatomy of the lumbar spine. (*Source: Atlas of Rheumatology, Volume EU, Chapter 9*)

22- to 25-gauge, 3½-7-in. spinal needle is then inserted through the anesthetized area. Figures 25.7 and 25.8 show intraarticular injections at the L3/4, L4/5, and L5/S1 levels. The needle is directed downward and obliquely (from lateral to medial) toward the selected joint under direct fluoroscopic visualization. Contact is made with the inferior articular process. The needle is then withdrawn slightly and redirected to enter the target facet joint.

As the needle is felt to penetrate the joint, advancement is stopped to prevent any potential damage to the articular cartilage. If there is difficulty in obtaining capsular penetration, one may try to access the articular recesses by redirecting the needle just off the margins of the inferior articular process.

Another method of gaining intracapsular entry is to redirect the needle slightly medial or lateral to the posterior joint line so that the needle gains access via its medial placement to the insertion of the capsule on the articular process. Once the needle is in an appropriate position, 0.2–0.5 mL of contrast medium is then injected into the joint to confirm proper placement. Arthrogram is omitted with injection of PRP and stem cells. An arthrographic image should then be obtained. During contrast medium injection, an outline of the oval-shaped joint capsule should be visualized with a lack of vascular uptake and/or epidural spread. After contrast medium confirmation of the intraarticular needle tip placement, the joint is injected with 1 mL or less of an anesthetic agent to complete the diagnostic block.

## Facet Joint Nerve Branch Blocks

To block the sensory innervation to a lumbar facet joint, it is necessary to block the two medial branch nerves that supply the joint. It is simple to remember which nerves need to be blocked to anesthetize a particular joint: Block the medial branch at the transverse process at the same level as the joint. Block the medial branch at the transverse process at the level below the joint. To block the L3/4 facet joint, block the L2 medial branch at the transverse process of L3 and the L3 medial branch at the transverse process of L4. To block the L5/S1 facet joint, block the L4 medial branch at the L5 transverse process and the L5 dorsal ramus at the sacral ala.

Lumbar facet joint nerve blocks have essentially replaced, or should replace, intraarticular injections for diagnosing lumbar facet joint pain. Facet joint nerve blocks are relatively easier to perform and safer. Intraarticular injections of lumbar facet joints may lack significant therapeutic utility [7,

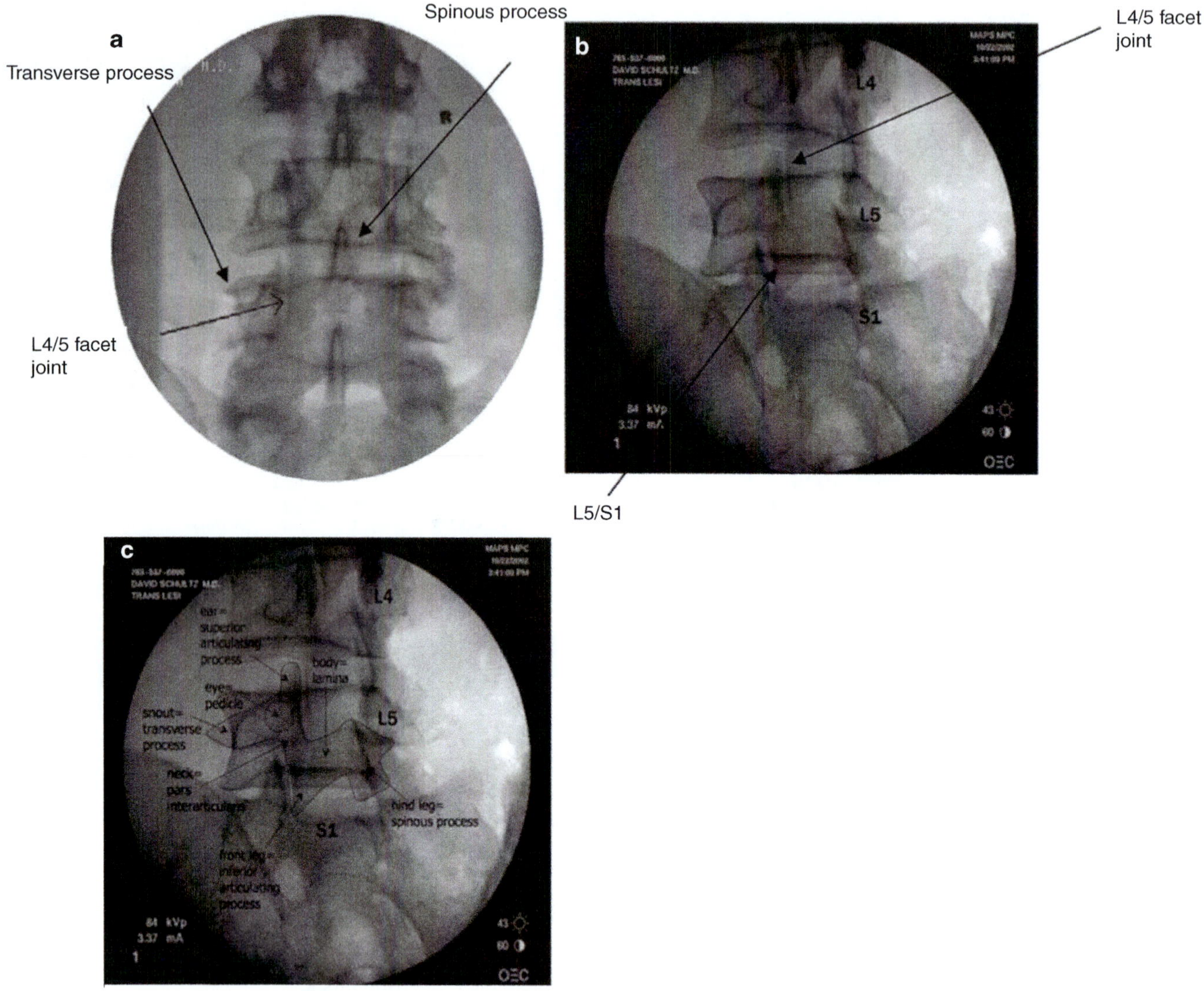


**Fig. 25.6** Fluoroscopic view of lumbar facet joints. (**a**) AP image of lumbar spine with visible, sagittally oriented L4/5 facet joint lines shown by an arrow. (**b**) Oblique view with joint lines visible (arrows). (**c**) Oblique view with "Scotty dog"

12, 13]. Facet joint nerve blocks have therapeutic utility. There is evidence that facet joint nerve block therapy and radiofrequency neurotomy of lumbar facet joint nerve blocks are effective in managing chronic low back pain [7, 12, 13].

When the posterior primary ramus of the exiting segmental nerve exits the foramen, it divides into lateral and medial branches. The medial branch runs in a dorsal and caudal direction reaching the junction of the superior articular process of the vertebra below the level of the exiting segmental nerve. Therefore, the medial branch of L4 reaches the junction of the L5 process. Facet joint nerve block or radiofrequency neurotomy is performed at this junction between the superior articular and transverse processes. Anatomy may be quite variable with the target point hidden from sight when an AP projection is used because the posterior articular process is blocking the view. In some cases access to the target point can only be obtained by using an oblique projection. For an L5 dorsal ramus block, at the junction between the sacral articular process and upper sacral border where the L5 dorsal ramus runs there is no pedicle to serve as a radiological landmark.

## L1 to L4 Facet Joint Nerve Blocks

### Prone Position Oblique

The patient is placed prone with a pillow under the abdomen and fluoroscopic imaging is obtained (Fig. 25.9). The C-arm must be adjusted to an oblique position. To maximally visualize the landmarks of the "Scotty dog" configu-

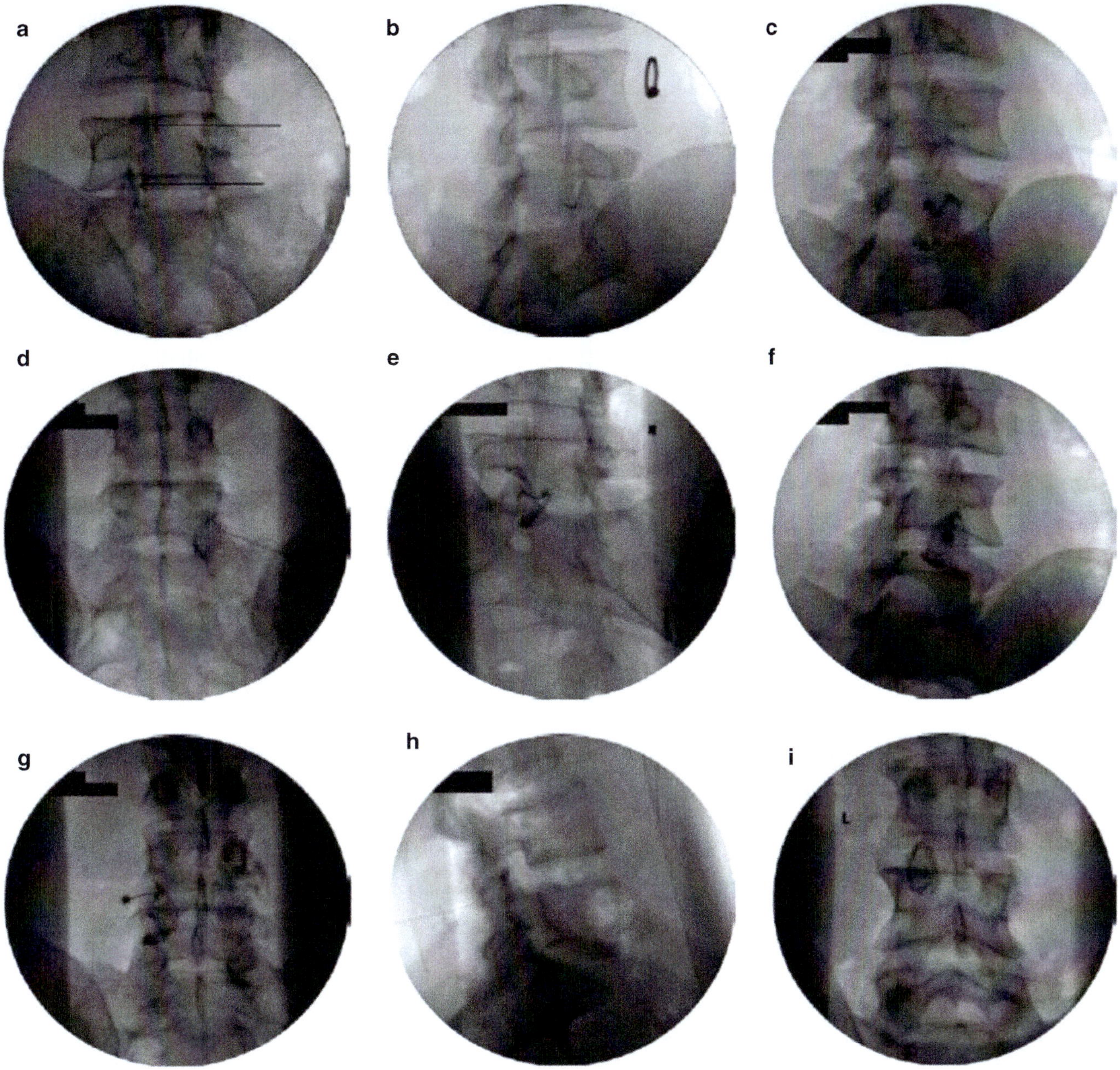


**Fig. 25.7** Intraarticular facet injections in the lumbosacral spine. (**a**) With "down the beam" needle placement, the skin insertion point is directly over the target. (**b**) 25 G spinal needle curving into L5/S1 facet joint. (**c**) L5/S1 arthrogram. (**d**) AP view of L5/S1 facet arthrogram. (**e**) Contrast exiting superior and inferior poles of L5/S1 joint. (**f**) L4/5 arthrogram. (**g**) AP view of L4/5 facet arthrogram. (**h**) Lateral view of L4/5 facet arthrogram. (**i**) AP view of L3/4 facet arthrogram

ration, an approximate 25–30° angle is necessary depending on the specific level from the L1–L4 medial branches (Fig. 25.7).

The needle is advanced toward the dorsal aspect of the root of the transverse process to ensure safe needle depth away from the ventral ramus. Utilizing an oblique view with the "Scotty dog" identified, the needle is advanced "down the beam" toward the target using a slightly superior starting position to the final target (Fig. 25.7). The needle is directed anterior, medial, and caudad to reach the target location, utilizing a "Scotty dog" view.

Needles are typically placed down to contact with the bony endpoint utilizing oblique fluoroscopic imaging. Prior to injection, however, the final needle tip position should be

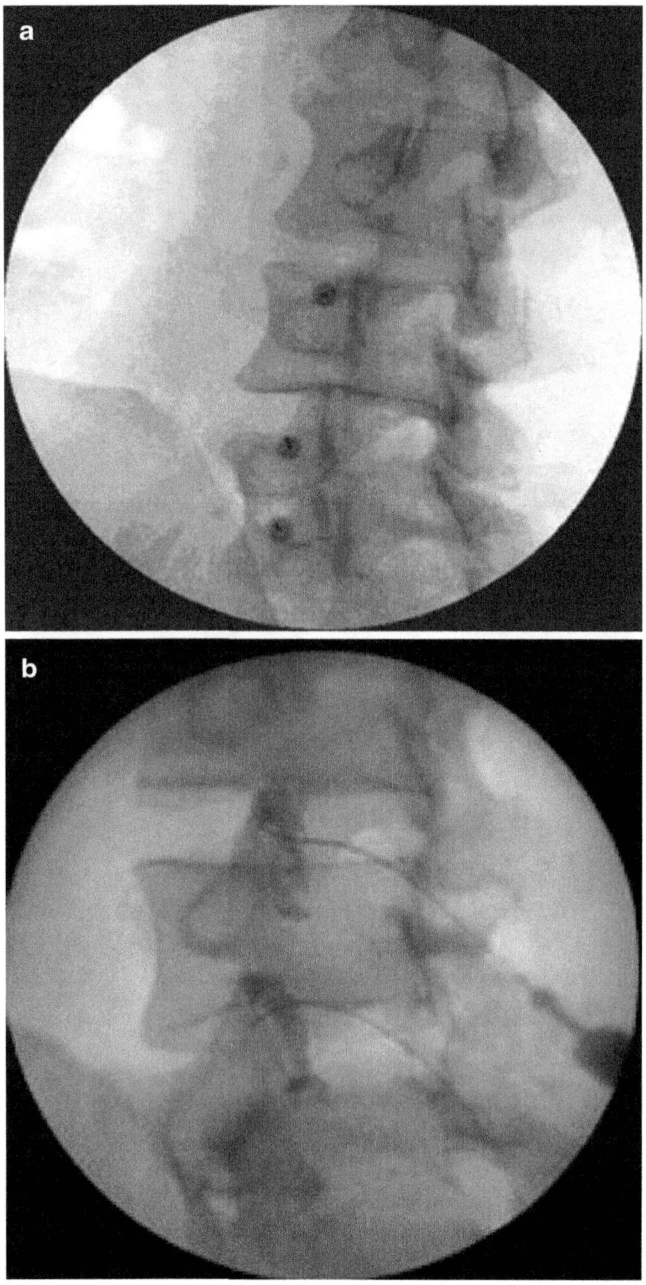


**Fig. 25.8** Needle placement into facet joints. (**a**) The left oblique fluoroscopy image demonstrating needle placement for left L3, L4 and L5 medial branches at the junction of superior articular and transverse process of L4, L5 for L3 and L4 and at the junction of superior articular process and sacral ala for L5 dorsal ramus. (**b**) The left oblique fluoroscopic image demonstrating arthrogram at left L3–4, L4–5 facet joint after needle placement via intra-articular approach

confirmed using both AP and lateral imaging or by arthrography to ensure that the needle tip is neither too deep nor too medial.

On AP imaging, the tip of the needle should at least be in line with the lateral margin of the silhouette of the superior articular process and, if possible, medial to this margin. On lateral imaging, the needle tip should be within the confines of the shadow of the dorsal elements and not protruding into the foramen. The superior articular process frequently bulges laterally, overlapping the target point dorsally. If the needle appears lateral to this point, it has contacted a thick transverse process instead of the superior articular process. In this case, the needle usually needs to be adjusted dorsally until the correct position is obtained on both AP and oblique or lateral views.

Before injection, the bevel opening should be medial and slightly inferior to reduce lateral and superior flow into the intervertebral foramen, especially if the needle is placed inadvertently higher than the target position.

### Prone Position Anteroposterior

The patient is positioned prone with or without a pillow under the abdomen and fluoroscopic imaging is obtained [6, 50]. The C-arm must be adjusted straight AP from the skin entry point laterally using posteroanterior (PA) imaging, which is usually just above the tip of the target transverse process. With axial rotation of the C-arm in the sagittal plane, the angle in the sagittal plane is changed to eliminate the double contour in the L4-endplate.

The needle is advanced toward the back of the root of the transverse process to ensure a safe needle depth away from the ventral ramus (Fig. 25.10). On AP imaging, the tip of the needle should at least be in line with the lateral margin of the silhouette of the superior articular process and, if possible, medial to this margin. The superior articular process frequently bulges laterally, overlapping the target point dorsally. If the needle appears lateral to this point, it has contacted a thick transverse process instead of the superior articular process. In this case, the needle usually needs to be adjusted dorsally until the correct position is obtained. Before injection, the bevel opening should be medial and slightly inferior to reduce lateral and superior flow to the intervertebral foramen, especially if the needle is placed inadvertently higher than the target position.

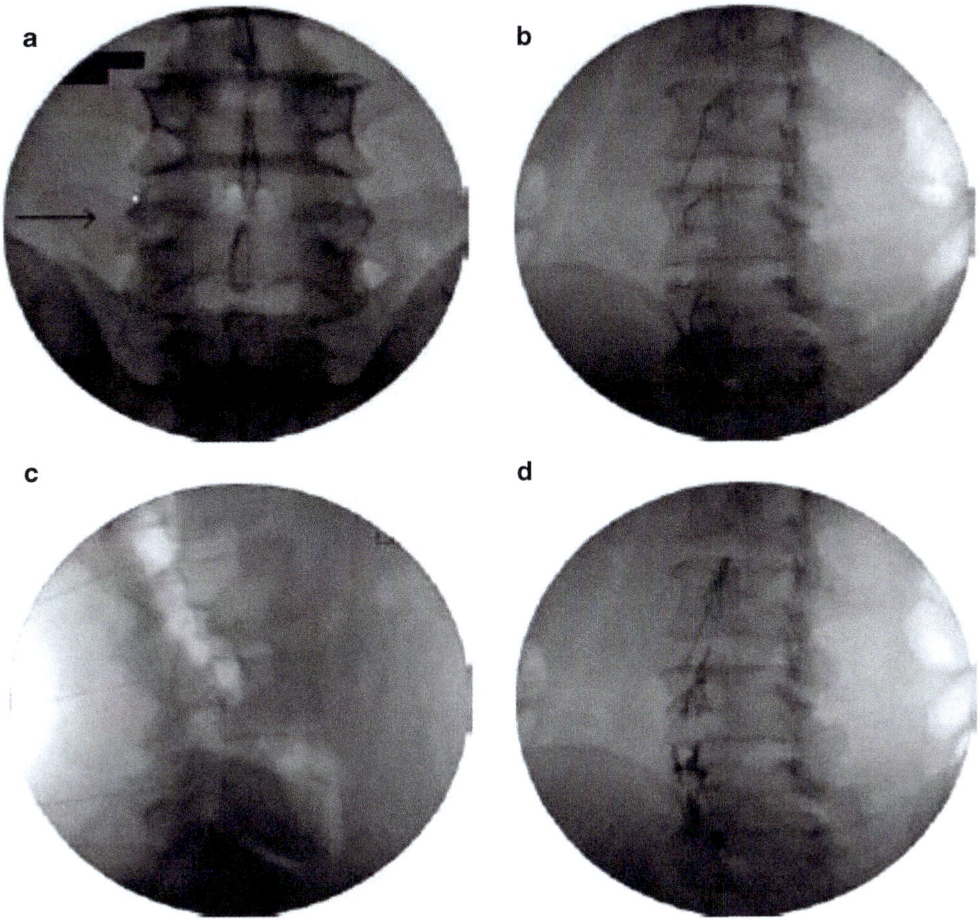


**Fig. 25.9** Lumbar facet joint nerve blocks—oblique approach. (**a**) Arrow points to L5 transverse process; white dot is on L4 medial branch. (**b**) Needles placed obliquely onto the medial branches of L2, L3, L4, and L5. (**c**) Lateral view of same needle placement. (**d**) 0.5 mL of contrast injected, oblique view

## Dorsal Ramus Blocks

Position the patient prone with or without a pillow under the abdomen and begin with an AP view of the L5/S1 segment (Fig. 25.10). Rotate the fluoroscope approximately 10–15° oblique toward the side to be blocked to view the junction of the sacral ala and the superior articular process of S1.

Further obliquity usually places the medial iliac crest in front of the trajectory to the target position. If the ilium obscures the view of the target point as the C-arm is rotated obliquely, a cephalad tilt of the C-arm will usually bring the target into clear view by moving the ilium caudad within the fluoroscopic image. An angled needle tip will allow for adequate steering of the needle to the target despite a cephalad-to-caudad needle trajectory.

Once a clear path to the target point for the L5 dorsal ramus is identified, a skin insertion point is chosen. The target point is recognized as a notch between the sacral ala and the superior articular process of S1. The target point lies opposite the middle of the base of the superior articular process and thus slightly below the silhouette at the top of the sacral ala. Higher placement is associated with spread into the L5/S1 epidural space and lower placement with spread to the S1 posterior sacral foramina. The needle is advanced directly "down the beam" to the target position. AP imaging is then obtained to verify that the needle is placed at, or preferably medial to, the lateral silhouette of the S1 superior articular process. Lateral imaging should confirm that the needle tip lies within the shadow of the posterior spinal elements and does not protrude into the L5/S1 neuroforamen. After the needle tip is confirmed to be in the proper location, the bevel opening should be rotated medially. This has been shown to reduce inadver-

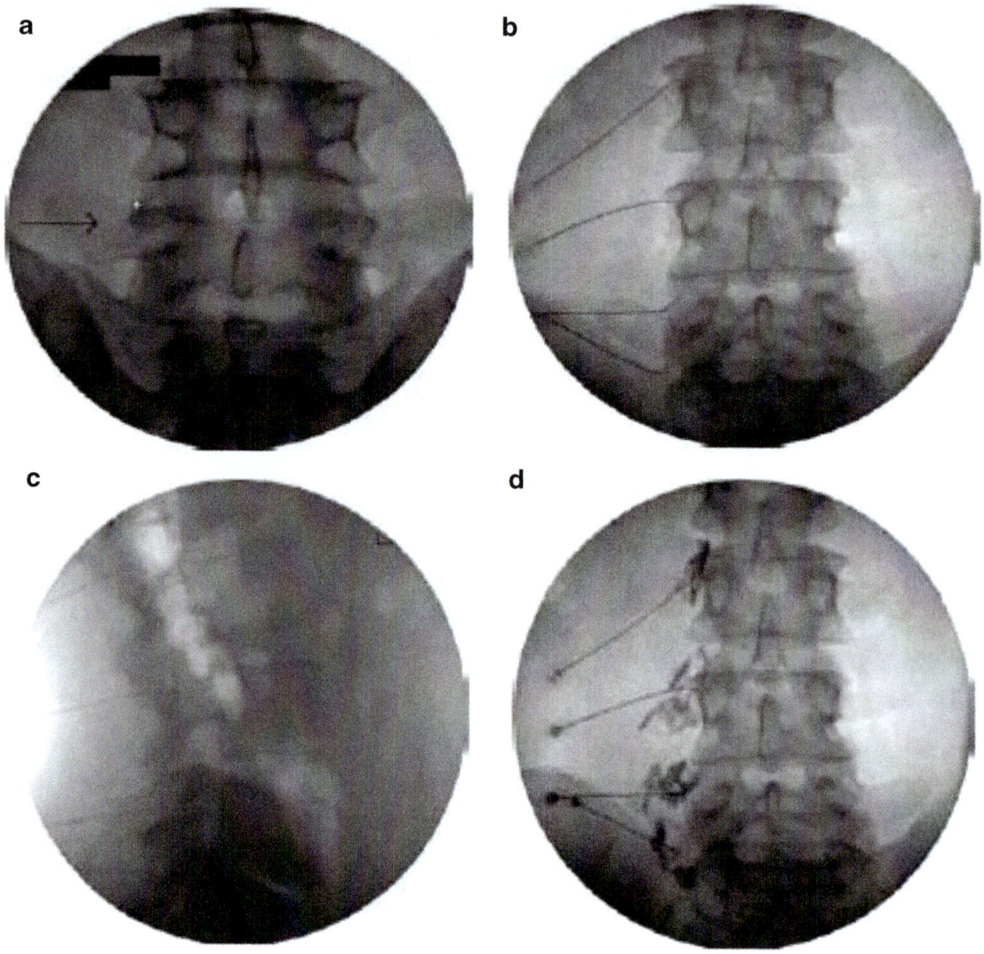


**Fig. 25.10** Facet joint nerve blocks with a posterior approach. (**a**) Arrow points to L5 transverse process; white dot is on L4 medial branch. (**b**) PA view of needle placement at multiple levels. (**c**) Lateral view of needle placement at multiple levels. (**d**) PA view with contrast injection

tent spread to the S1 posterior foramen or the L5 vertebral foramen [6, 50].

## Indications

Indications for diagnostic facet joint nerve blocks are as follows: somatic or nonradicular low back pain or lower extremity pain of at least 3 months duration, moderate to severe pain causing functional disability, failure to respond to more conservative management, including physical therapy, exercise, chiropractic management, and nonsteroidal anti-inflammatory drugs (NSAIDs), lack of evidence, for either discogenic or sacroiliac joint pain and lack of painful disc herniation or evidence of radiculitis.

Indications for therapeutic facet joint interventions are as follows: therapeutic modalities are applied based on a diagnosis established with a positive response to controlled diagnostic blocks, either placebo or comparative local anesthetic blocks, with a criterion standard of 75% or 80% pain relief and the ability to perform prior painful movements without significant pain.

The fundamental invitation for lumbar facet joint injections is to diagnose origin of low back pain if it is from facet joints.

Regenerative medicine solutions are applied in managing facet joint pain after failure of conservative modalities of treatments including therapeutic facet joint interventions.

## Side Effects and Complications

Complications from facet joint nerve blocks, intraarticular injections, with biologic solutions in the lumbar spine are exceedingly rare [6, 7, 12, 13, 15–18, 21, 22, 49, 50]. The most common complications of facet joint interventions are twofold: complications related to the placement of the needle and complications related to the administration of various drugs including biologics.

Most problems, such as local swelling, pain at the site of the needle insertion, and pain in the low back, are short-lived and self-limited. More serious complications may include dural puncture, spinal cord trauma, subdural injection, neural trauma, injection into the intervertebral foramen, and hematoma formation; infectious complications including epidural abscess and bacterial meningitis; and side effects related to the administration of steroids, local anesthetics, and other drugs.

## Precautions

Relative contraindications to interventional techniques have been described in patients receiving treatment with antithrombotics and anticoagulants [6, 7]. Guidelines on anticoagulants and antithrombotics are described in Chap. 15 [51] and also on the ASIPP app, which is regularly updated.

## Conclusion

Lumbar facet joints have been determined to be the cause for chronic low back pain in 16–41% of patients who have chronic low back pain. Facet joint pain cannot be diagnosed by demographic features, pain characteristics, physical findings, electrodiagnostic studies, or radiological evaluation.

Diagnostic facet joint nerve blocks, either by means of comparative local anesthetic blocks or placebo injections, are the only reliable diagnostic measures available according to today's literature. The validity, specificity, and sensitivity of lumbar facet joint nerve blocks are considered strong for the diagnosis of facet joint pain.

Comparative local anesthetic blocks or placebo-controlled blocks are essential, since single block false-positive rates of 25–44% have been reported in the lumbar spine. Facet joint blocks are performed reliably only under fluoroscopic visualization. Chronic low back pain may be treated therapeutically with facet joint nerve blocks or intraarticular injections with administration of PRP and/or MSCs.

**Key Points**

1. Low back pain is a prevalent and disabling chronic spinal pain problem, with the involvement of lumbosacral facet joints accounting for a significant proportion of cases.
2. The recognition of lumbar facet joints as potential sources of back pain dates back to the early twentieth century, and subsequent studies have provided evidence of the physiological basis and pain patterns associated with facet joint involvement.
3. The facet joint is a synovium-lined joint in the posterior compartment of the spine, allowing for spinal movement. It is innervated by articular branches of the medial branches of dorsal primary rami.
4. Lumbar facet joints are well-innervated structures associated with the presence of nerve endings containing substance P and calcitonin gene-related peptide. They have been identified as a source of both low back pain and referred pain in patients with chronic low back pain.
5. Facet joint pain can be induced by mechanical injury and inflammation, with inflammatory mediators playing a role in joint inflammation and arthritis. While the presence of radiological arthritis in facet joints is common, its direct relationship to chronic pain has not been established. Facet joint pain is frequently reported, particularly in the elderly population.
6. Therapeutic interventions such as facet joint nerve blocks and radiofrequency neurotomy have shown moderate to good evidence for pain relief.
7. In addition to conventional management approaches, biologics such as platelet-rich plasma (PRP) and medicinal signaling cells or mesenchymal stem cells (MSCs) have been used in facet joint injections as alternative treatment options.
8. Lumbar facet joint interventions in regenerative medicine include intraarticular injections and facet joint nerve blocks. These procedures require proper radiographic/fluoroscopic anatomy understanding and precise needle placement guided by imaging.

**Acknowledgments** This book chapter is modified and updated from a previous book chapter, "Lumbar Facet Joints and Regenerative Medicine" by Annu Navani, MD, and Laxmaiah Manchikanti, MD, in *Essentials of Regenerative Medicine in Interventional Pain Management* published by ASIPP Publishing.

## References

1. Navani A, Manchikanti L. Lumbar facet joints and regenerative medicine. In: Manchikanti L, Navani A, Atluri S, editors. Essentials of regenerative medicine in interventional pain management. Paducah, KY: ASIPP Publishing; 2019. p. 459–72.
2. US Burden of Disease Collaborators. The state of US health, 1990-2016: burden of diseases, injuries, and risk factors among US states. JAMA. 2018;319:1444–72.
3. Dieleman JL, Cao J, Chapin A, et al. US health care spending by payer and health condition, 1996-2016. JAMA. 2020;323:863–84.
4. Manchikanti L, Centeno CJ, Atluri S, et al. Bone marrow concentrate (BMC) therapy in musculoskeletal disorders: evidence-based policy position statement of American Society of Interventional Pain Physicians (ASIPP). Pain Physician. 2020;23:E85–E131.
5. Hoy D, March L, Brooks P, et al. The global burden of low back pain: estimates from the Global Burden of Disease 2010 study. Ann Rheum Dis. 2014;73:968–74.
6. Manchikanti L, Schultz DM, Falco FJE, Singh V. Lumbar facet joint interventions. In: Manchikanti L, Singh V, Falco FJE, Kaye AD, Soin A, Hirsch JA, editors. Manchikanti's essentials of inter-

ventional techniques in managing chronic pain. 2nd ed. New York, NY: Springer; 2023.
7. Manchikanti L, Kaye AD, Soin A, et al. Comprehensive evidence-based guidelines for facet joint interventions in the management of chronic spinal pain: American Society of Interventional Pain Physicians (ASIPP) guidelines. Pain Physician. 2020;23:S1–S127.
8. Bogduk N. Low back pain. Clinical anatomy of lumbar spine and sacrum. 4th ed. New York: Churchill Livingstone; 2005. p. 183–216.
9. DePalma MJ, Ketchum JM, Saullo TR. Multivariable analyses of the relationships between age, gender, and body mass index and the source of chronic low back pain. Pain Med. 2012;13:498–506.
10. Pang WW, Mok MS, Lin ML, et al. Application of spinal pain mapping in the diagnosis of low back pain—analysis of 104 cases. Acta Anaesthesiol Sin. 1998;36:71–4.
11. Manchikanti L, Singh V, Pampati V, et al. Evaluation of the relative contributions of various structures in chronic low back pain. Pain Physician. 2001;4:308–16.
12. Manchikanti L, Kosanovic R, Pampati V, et al. Low back pain and diagnostic lumbar facet joint nerve blocks: assessment of prevalence, false-positive rates, and a philosophical paradigm shift from an acute to a chronic pain model. Pain Physician. 2020;23:519–30.
13. Manchikanti L, Kosanovic R, Pampati V, et al. Equivalent outcomes of lumbar therapeutic facet joint nerve blocks and radiofrequency neurotomy: comparative evaluation of clinical outcomes and cost utility. Pain Physician. 2022;25:179–92.
14. Patel A, Koushik S, Schwartz R, et al. Platelet-rich plasma in the treatment of facet mediated low back pain: a comprehensive review. Orthop Rev (Pavia). 2022;14:37076.
15. Navani A, Benyamin RM, Boswell MV, et al. Responsible, safe, and effective use of biologics in the management of low back pain: American Society of Interventional Pain Physicians (ASIPP) Guidelines. Pain Physician. 2019;22:S1–S74.
16. Wu J, Du Z, Lv Y, et al. A new technique for the treatment of lumbar facet joint syndrome using intra-articular injection with autologous platelet rich plasma. Pain Physician. 2016;19:617–25.
17. Wu J, Zhou J, Liu C, et al. A prospective study comparing platelet-rich plasma and local anesthetic (LA)/corticosteroid in intra-articular injection for the treatment of lumbar facet joint syndrome. Pain Pract. 2017;17:914–24.
18. Kirchner F, Anitua E. Intradiscal and intra-articular facet infiltrations with plasma rich in growth factors reduce pain in patients with chronic low back pain. J Craniovertebr Junction Spine. 2016;7:250–6.
19. El-Kadiry AE, Lumbao C, Rafei M, Shammaa R. Autologous BMAC therapy improves spinal degenerative joint disease in lower back pain patients. Front Med (Lausanne). 2021;8:622573.
20. Atluri S, Murphy MB, Dragella R, et al. Evaluation of the effectiveness of autologous bone marrow mesenchymal stem cells in the treatment of chronic low back pain due to severe lumbar spinal degeneration: a 12-month, open-label, prospective controlled trial. Pain Physician. 2022;25:193–207.
21. Navani A, Li G, Chrystal J. Platelet rich plasma in musculoskeletal pathology: a necessary rescue or a lost cause? Pain Physician. 2017;20:E345–56.
22. Chakravarthy K, Chen Y, He C, Christo PJ. Stem cell therapy for chronic pain management: review of uses, advances, and adverse effects. Pain Physician. 2017;20:293–305.
23. Goldthwait JE. The lumbosacral articulation: an explanation of many cases of lumbago, sciatica, and paraplegia. Boston Med Surg J. 1911;164:365–72.
24. Ghormley RK. Low back pain. With special reference to the articular facets, with presentation of an operative procedure. JAMA. 1933;101:1773–7.
25. Badgley CE. The articular facets in relation to low back pain and sciatic radiation. J Bone Joint Surg. 1941;23:481.
26. Hirsch D, Inglemark B, Miller M. The anatomical basis for low back pain. Acta Orthop Scand. 1963;33:1–17.
27. Mooney V, Robertson J. The facet syndrome. Clin Orthop. 1976;115:149–56.
28. McCall IW, Park WM, O'Brien JP. Induced pain referral from posterior elements in normal subjects. Spine (Phila Pa 1976). 1979;4:441–6.
29. Marks R. Distribution of pain provoked from lumbar facet joints and related structures during diagnostic spinal infiltration. Pain. 1989;39:37–40.
30. Fukui S, Ohseto K, Shiotani M, et al. Distribution of referral pain from the lumbar zygapophyseal joints and dorsal rami. Clin J Pain. 1997;13:303–7.
31. Windsor RE, King FJ, Roman SJ, et al. Electrical stimulation induced lumbar medial branch referral patterns. Pain Physician. 2002;5:347–54.
32. Kaplan M, Dreyfuss P, Halbrook B, et al. The ability of lumbar medial branch blocks to anesthetize the zygapophysial joint. Spine (Phila Pa 1976). 1998;23:1847–52.
33. Dreyfuss P, Schwarzer AC, Lau P, et al. Specificity of lumbar medial branch and L5 dorsal ramus blocks: a computed tomography study. Spine (Phila Pa 1976). 1997;22:895–902.
34. Manchikanti L, Hirsch JA, Falco FJ, Boswell MV. Management of lumbar zygapophysial (facet) joint pain. World J Orthop. 2016;7:315–37.
35. Bogduk N, Wilson AS, Tynan W. The human lumbar dorsal rami. J Anat. 1982;134:383–97.
36. Cavanaugh JM, Lu Y, Chen C, et al. Pain generation in lumbar and cervical facet joints. J Bone Joint Surg Am. 2006;88:63–7.
37. Ishikawa T, Miyagi M, Ohtori S, et al. Characteristics of sensory DRG neurons innervating the lumbar facet joints in rats. Eur Spine J. 2005;14:559–64.
38. Gong K, Shao W, Chen H, et al. Rat model of lumbar facet joint osteoarthritis associated with facet-mediated mechanical hyperalgesia induced by intra-articular injection of monosodium iodoacetate. J Formos Med Assoc. 2011;110:145–52.
39. Igarashi A, Kikuchi S, Konno S, et al. Inflammatory cytokines released from the facet joint tissue in degenerative lumbar spinal disorders. Spine (Phila Pa 1976). 2004;29:2091–5.
40. Sakuma Y, Ohtori S, Miyagi M, et al. Up-regulation of p55 TNF alpha-receptor in dorsal root ganglia neurons following lumbar facet joint injury in rats. Eur Spine J. 2007;16:1273–8.
41. Lee KE, Davis MB, Winkelstein BA. Capsular ligament involvement in the development of mechanical hyperalgesia after facet joint loading: behavioral and inflammatory outcomes in a rodent model of pain. J Neurotrauma. 2008;25:1383–93.
42. Dong L, Guarino BB, Jordan-Sciutto KL, et al. Activating transcription factor 4, a mediator of the integrated stress response, is increased in the dorsal root ganglia following painful facet joint distraction. Neuroscience. 2011;193:377–86.
43. Quinn JH, Bazan NG. Identification of prostaglandin E2 and leukotriene B4 in the synovial fluid of painful, dysfunctional temporomandibular joints. J Oral Maxillofac Surg. 1990;48:968–71.
44. Yoshida R, Cheng M, Murray MM. Increasing platelet concentration in platelet-rich plasma inhibits anterior cruciate ligament cell function in three-dimensional culture. J Orthop Res. 2014;32:291–5.
45. Wasterlain AS, Braun HJ, Harris AH, et al. The systemic effects of platelet-rich plasma injection. Am J Sports Med. 2013;41:186–93.
46. Kristjánsson B, Limthongkul W, Yingsakmongkol W, et al. Isolation and characterization of human mesenchymal stem cells

from facet joints and interspinous ligaments. Spine (Phila Pa 1976). 2016;41:E1–7.
47. Sanapati J, Manchikanti L, Atluri S, et al. Do regenerative medicine therapies provide long-term relief in chronic low back pain: a systematic review and metaanalysis. Pain Physician. 2018;21:515–40.
48. Bogduk N. Lumbar medial branch blocks. In: Practice guidelines for spinal diagnostic and treatment procedures. 2nd ed. San Francisco: International Spine Intervention Society (ISIS); 2013. p. 559–600.
49. Navani A, Melnik I, Derby R, Jeong-Eun LJ. Lumbar facet and median branch block. In: Baheti DK, Bakshi S, Gupta S, Gehdoo RS, editors. Interventional pain management: a practical approach. 2nd ed. New Delhi: Jaypee Brothers Medical Publishers; 2016. p. 366–85.
50. Manchikanti L, Schultz DM, Singh V, et al. Lumbar facet joint interventions. In: Manchikanti L, Singh V, editors. Interventional techniques in chronic spinal pain. Paducah, KY: ASIPP Publishing; 2007. p. 253–76.
51. Manchikanti L, Sanapati M, Kaye AD, Kaye AM. Antiplatelet and anticoagulant therapy in regenerative medicine. In: Manchikanti L, Navani A, Atluri S, Sanapati M, editors. Manchikanti's essentials of regenerative medicine in interventional pain management. 2nd ed. New York, NY: Springer; 2023.

# Cervical Facet Joints and Intraarticular Injection of Biologics

# 26

Laxmaiah Manchikanti, Annu Navani, David M. Schultz, Sairam Atluri, Frank J. E. Falco, and Mahendra Sanapati

## Introduction

Chronic neck pain in the general population, although is less common than low back pain, it is the third leading cause of disability [1–5]. Multiple structures are capable of transmitting pain in the cervical spine into neck, head, and upper extremities. Intervertebral discs, cervical facet joints, ligaments, fascia, muscles, and nerve root dura have been described as pain generators [6–9]. Symptoms of cervical spine disorders include neck pain, upper extremity pain, upper back pain, and headache. Cervical facet joints may cause chronic neck pain, which may also refer pain into the head and extremities.

The prevalence of persistent neck pain, with or without upper extremity pain, and with or without headache secondary to the involvement of cervical facet joints, has been established as varying from 36% to 67% with a false-positive rate of 27–63% based on the type of population and setting studied utilizing controlled diagnostic blocks [6–8, 10–13]. Pain of cervical facet joint origin has been treated with intraarticular injections, facet joint nerve blocks, and radiofrequency neurotomy [6–8]. The evolution of regenerative medicine, its expanding clinical applications in managing chronic pain has led to application of regenerative medicine principles in managing various spinal disorders including pain of cervical facet joint origin [14–16].

## History

Pawl [17] reported reproduction of pain in the neck head in 1977, after injecting hypertonic saline into cervical facet joints. The role of cervical facet joints in causing idiopathic neck pain was studied by Bogduk and Marsland [18] by utilizing diagnostic cervical medial branch blocks and facet joint injections. Mapping of specific locations of referred neck pain was studied by Dwyer et al. [19] by performing facet joint injections in asymptomatic volunteers. The pain chart developed by Dwyer et al. [19] was confirmed by anesthetizing the medial branches of the dorsal rami above and below the symptomatic joint. The referred pain distribution of cervical facet joints and cervical dorsal rami were also studied by Aprill et al. [20] and Fukui et al. [21] with similar results. Electrical stimulation-induced cervical medial branch pain referral patterns were described by Windsor et al. [22]. Finally, Cooper et al. [23] provided pain maps in patients with cervical zygapophysial joint pain.

Okada [24] described intraarticular cervical facet joint injections using a lateral approach in 1981, and a posterior approach was described by Dory in 1983 [25]. The cervical medial branch blocks were described by Bogduk and Marsland [18] in 1988.

Mesenchymal stem cells are medicinal signaling cells known as MSCs harvested from the facet joints and interspinous ligaments add the trilineage potential to be differenti-

Frank J. E. Falco has died before the publication of this book.

L. Manchikanti (✉)
Pain Management Centers of America, Paducah, KY, USA
e-mail: drlm@thepainmd.com

A. Navani
LeReve Regenerative Wellness, Campbell, CA, USA

Boomerang Healthcare, Walnut Creek, CA, USA

Stanford University School of Medicine, Stanford, CA, USA
e-mail: anavani@boomeranghc.com

D. M. Schultz
Nura Precision Pain Management, Minneapolis, MN, USA
e-mail: dschultz@nuraclinics.com

S. Atluri
Tri-State Spine Care Institute, Cincinnati, OH, USA

F. J. E. Falco (Deceased)
Mid Atlantic Spine & Pain Physicians, Newark, DE, USA

M. Sanapati
Pain Management Centers of America, Evansville, IN, USA


A. Navani et al. (eds.), *Essentials of Regenerative Medicine in Interventional Pain Management*,
https://doi.org/10.1007/978-3-031-50357-3_26

ated into osteogenic, adipogenic, and chondrogenic cells under appropriate conditions [16]. Thus, facet joints and interspinous ligaments may provide alternative sources of MSCs for tissue engineering applications. Conversely, facet joint pain may be treated with MSCs. In addition, there is also literature supporting utilization of platelet-rich plasma (PRP) in managing lumbar facet joint pain [15].

## Pathophysiology

Cervical facet joints are a recognized source of neck pain and referred pain into the upper extremities, head and upper back in patients with chronic neck pain [6–9, 26, 27]. They are innervated by the medial branches of the dorsal primary rami of spinal nerves [6–9, 26, 27]. Neuroanatomic, neurophysiologic, and biomechanical studies have demonstrated free and encapsulated nerve endings in the cervical facet joints, as well as nerves containing substance *P* and calcitonin gene-related peptides [6–9, 26–35].

Multiple factors have been shown to induce facet joint pain. Both mechanical injury and inflammation in facet joints can produce persistent pain in otherwise normal rats [6–9, 26, 27, 36–44]. Mechanical injury of the facet joints was shown to increase cytokine messenger RNA in the dorsal root ganglion (DRG) [37]. In fact, an intraarticular injection of a nonsteroidal anti-inflammatory drug (NSAID) alleviated injury-induced pain in the rat model [38], suggesting the crucial role of inflammation in the pain response after mechanical joint insult. Further, the increased expression of two inflammatory markers (IL-1a and $PGE_2$) in the spinal cord in a rat model at 1 day after painful joint injury suggests that spinal inflammation may contribute to the initiation of pain after cervical facet joint injury. In addition, inflammatory mediators, such as cytokines, prostaglandins, and neuropeptides were shown to increase within the joint and the DRG in joint inflammation and arthritis [39, 40].

Chua et al. [43] showed the differences in sensory processing between chronic facet joint pain patients with and without cervicogenic headache. However, facet joint pain is seen in patients with or without arthritis. Arthritis denotes joint inflammation, whereas osteoarthritis indicates arthritis characterized by erosions of articular cartilage, which mainly affect weight-bearing joints.

The classic radiographic hallmarks of facet joint osteoarthritis include both degenerative and proliferative features, with narrowing of the facet joint space, subarticular bone erosions, subchondral cysts, osteophyte formation, and hypertrophy of the articular processes [45, 46]. The degenerative process in the cervical spine is characterized by intradiscal tears, synovial joint cartilage loss, and later by paradiscal and facet joint osteophytes and endplate irregularities [45, 47]. Cervical facet joint osteoarthritis was prevalent in 19% of adults age 45–65 years, and 57% of adults age 65 years and older, in a community based U.S. population [48]. Proposed pain mechanisms of pain in osteoarthritis include capsular stretch, entrapment of synovial villi between the articular surfaces, nerve impingement by osteophytes, and release of inflammatory factors [49, 50].

Pain in the neck, head, and upper extremities is commonly described following cervical spinal injuries, predominately whiplash type of injuries; however, multiple injuries of facet joints have been described in the literature. Of all the injuries, whiplash is the common cause of neck and upper extremity pain, specifically associated with headaches. Whiplash may cause increased laxity of the cervical capsular ligament [51]. In fact, Bogduk [52] described the biological features of whiplash from motor vehicle accidents, based on the results of postmortem studies, illustrating a spectrum of injuries that could occur in cervical facet joints. Further, Curatolo et al. [53] discussed the role of tissue damage in whiplash-associated disorders. Their results demonstrated that numerous investigations conducted in animals, cadavers, healthy volunteers, and patients have documented lesions in various tissues.

## Regenerative Medicine

The biologic primarily referred to with regards to cervical facet joints is PRP. PRP includes concentration of growth factors that are secreted by a platelet in an inflammatory environment, i.e., platelet-derived growth factors. These growth factors are essential to the healing process, as they increase fibroblast or osteoblast metabolic activity while reducing cell apoptosis promoting angiogenesis, thereby increasing blood flow and circulation to the new-forming tissues [54–56]. These processes increase the expression of the pro-collagen gene and collagen-derived growth factors, which increase the tensile strength of the new tissue [54–56]. Proteins present in PRP that are proliferative in nature include endothelial growth factor, platelet-derived growth factor, vascular endothelial growth factor, and basic fibroblast growth factor (b-FGF).

The other biologic that may be relevant in this regard is MSCs, and other commercially available biologic preparations. However, they are currently no published studies supporting their use for facet joint pain. Interestingly, Kristjánsson and colleagues [16] reported the MSCs harvested from the facet joints and interspinous ligaments of ten patients had the tri-lineage potential to be differentiated into osteogenic, adipogenic, and chondrogenic cells under appropriate conditions. They concluded that the facet joints and interspinous ligaments may provide alternative sources of MSCs for tissue engineering applications.

## Evidence Base

The literature supporting the use of PRP for lumbar facet joint pain in limited with none available for thoracic and cervical regions [15]. However, therapeutic facet joint interventions available including intraarticular facet joint injections, facet joint nerve blocks, and radiofrequency neurotomy often present with limited to moderate evidence along with paucity of literature. Consequently, regenerative medicine with intraarticular injections or occasionally medial branch blocks is attractive in managing chronic neck pain.

## Clinical Presentation

Conventional clinical features are unreliable in diagnosing cervical facet joint pain [6–9, 26, 27]. The features of somatic or referred pain secondary to facet joints, and radicular pain secondary to disc pathology, may not be able to distinguish facet joint pain from radicular pain as shown in Fig. 26.1 [19, 21, 22, 57–60]. Since there is no gold standard for diagnosing neck pain, the recommended reference standards typically involve anesthetic or provocative injections, regardless of whether the suspected source is a facet joint or an intervertebral disc.

The published literature to screen patients suitable for diagnostic cervical facet joint blocks suggests that negative findings on the manual spinal examination and/or segmental palpation may inform clinicians that facet joint blocks may not be an optimal procedure for their patients with persistent neck pain. Published radiological investigations report no correlation between the clinical symptoms of neck pain and degenerative spinal changes observed on radiological imaging studies [6–9, 26, 27].

Many of the maneuvers used in physical examination of the cervical spine simultaneously stress multiple structures in addition to facet joints and therefore these maneuvers fail to provide useful diagnostic criteria.

Level I evidence has been presented demonstrating the reliability of cervical facet joint nerve blocks for diagnosing neck pain, upper extremity pain, and headache [6–9, 26, 27]. Significant false-positive rates with single facet joint blocks in the cervical spine have been documented (27–63%) [6–9, 26, 27].

## Anatomy

The cervical facet joints are paired, diarthrodial, synovial joints located between the superior and inferior articular pillars in the posterior cervical column extending from C2/3 to

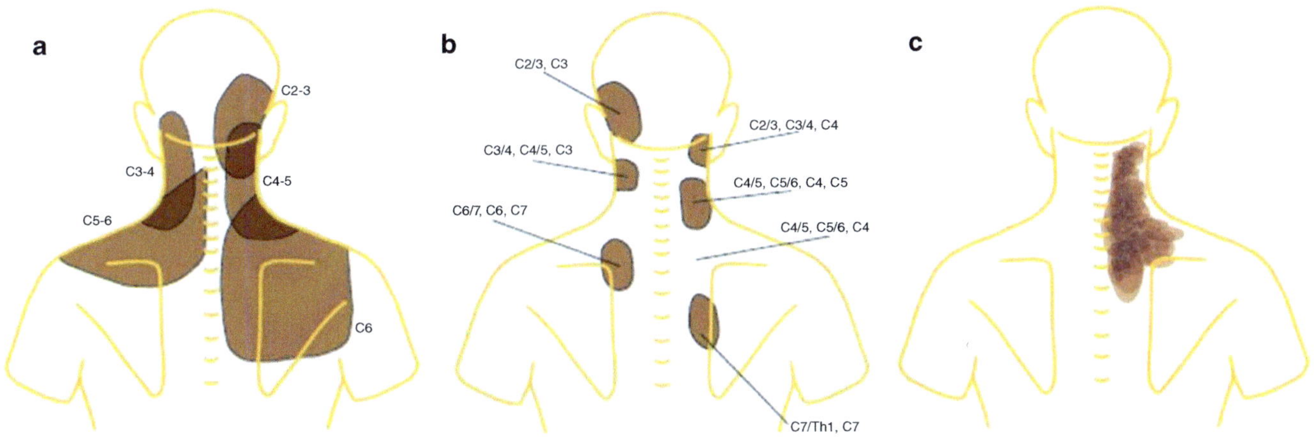


**Fig. 26.1** Referral patterns for cervical facet joint pains, as described by various investigators [19, 21, 22]. (**a**) Diagram of cervical zygapophysial joint pain distribution in volunteers. (Reproduced from Dwyer et al. [19]). Cervical zygapophysial joint pain patterns I: a study in normal volunteers. Spine 1990; 6:453–457. (**b**) Main referred pain distributions for the zygapophyseal joints from C0/1 to C7/T1 and the dorsal rami C3 to C7. (Reproduced from Fukui et al. [21]). Referred pain distribution of the cervical zygapophysial joints and cervical dorsal rami. Pain 1996; 68:79–83. (**c**) A composite drawing of the referral patterns of all subjects derived from the minimal threshold stimulation of their right third occipital nerve and C3 to C8 medial branches. (Reproduced from Windsor et al. [22]). Electrical stimulation induced cervical medical branch referral patterns. Pain Physician 2003; 6:411–418

C7/T1 [26, 27, 61]. The atlanto-occipital and atlanto-axial synovial joints are also present in the cervical spine as two paired joints but are not considered to be facet joints, as they are anterior rather than posterior spinal structures.

The cervical facet joints are formed by the inferior articular process of the superior vertebral segment and superior articular process of the inferior vertebral segment. The superior aspect of the joint faces forward and downward at 45°, whereas the inferior aspect of the joint faces backward and upward at 45° (Fig. 26.2).

The facet joints in the cervical spine form paired vertebral columns or "pillars," which together with the discover-

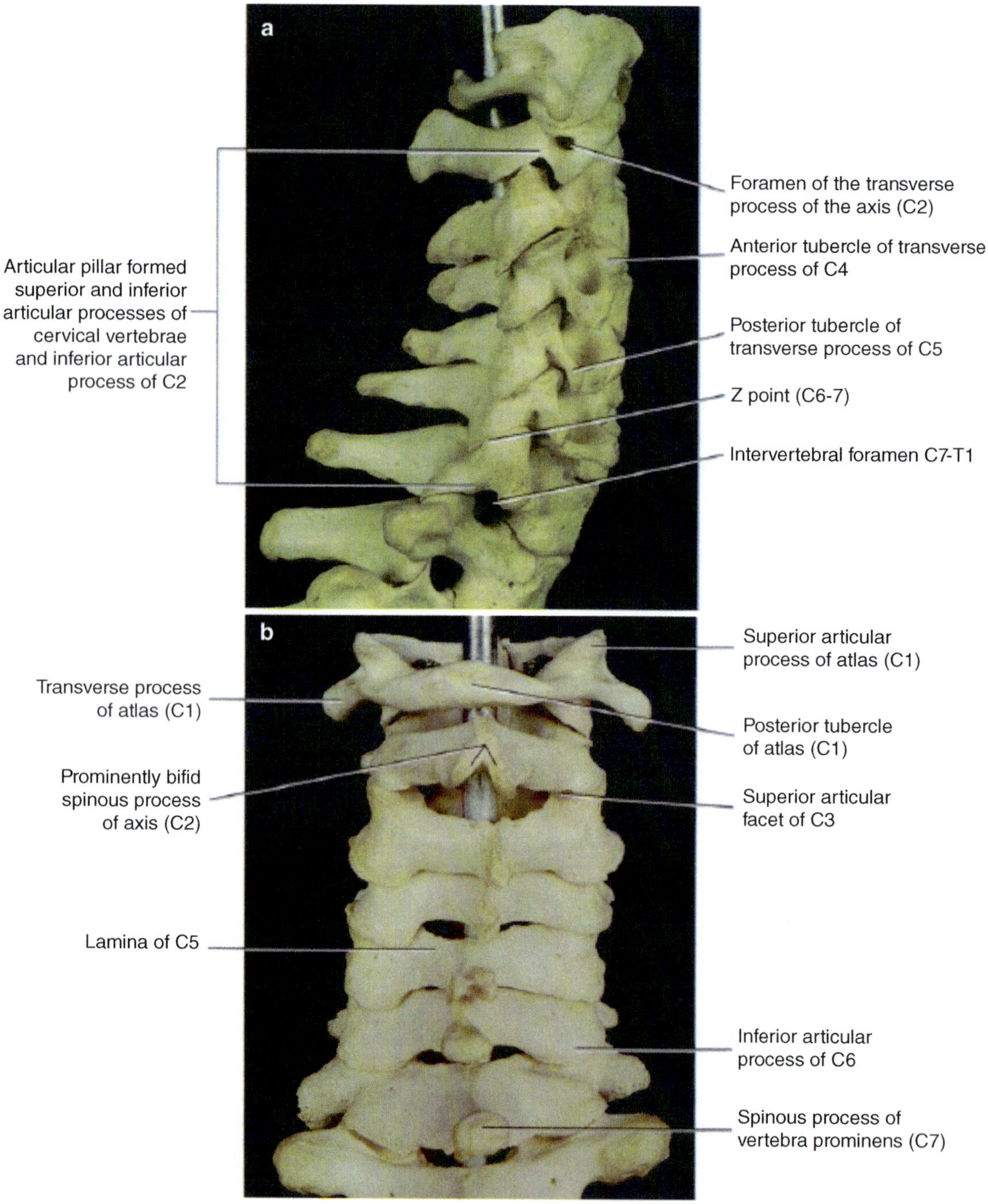


**Fig. 26.2** Lateral (**a**) and posterior (**b**) views of the cervical portion of the vertebral column. (*Reprinted with permission from Manchikanti et al. "Cervical Facet Joint Interventions" in Interventional Techniques in Chronic Spinal Pain. Paducah, KY. ASIPP Publishing. 2007*)

tebral unit provide the three-pronged structural support for the cervical segment. The obliquity of the facet joints allows degrees of flexion, extension, and rotation. The articular facets are covered by articular cartilage, and a synovial membrane bridges the margins of the articular cartilage of the two facets in each joint. The average joint volume is less than 1 mL.

The cervical facet joints are innervated by the medial branches of the dorsal rami (Figs. 26.3 and 26.4). The fibrous joint capsule is richly innervated with mechanoreceptors, as well as nociceptors [6–8, 26, 27, 62, 63]. The cervical facet joints below C2/3 are supplied by medial branches of the cervical dorsal rami above and below the joint. The C2/3 joint is supplied by the third occipital nerve [62, 63]. However, the innervation of the atlanto-occipital and atlanto-axial joints is derived from the C1 and C2 root, respectively.

Each C3 to C7 dorsal ramus crosses the same segments' transverse processes and divides into lateral and medial branches [35]. The medial branches curve around the waist of the articular pillars of the same numbered vertebra. They are bound by fascia, held against the articular pillars, and covered by the tendinous slips of the origin of the semispinalis capitis. As the nerve approaches the posterior aspect of the articular pillar, articular branches arise with an ascending branch innervating the joint above and a descending branch innervating the joint below. But at C7, in contrast to C3 to C6, the medial branches are located at a higher level due to the transverse processes. At C7, the bases of the transverse processes occupy most of the lateral aspect of the articular pillars, pushing the medial branches higher. The majority (70%) of C7 medial branches are located high on the C7 articular pillars and cross the C6/7 facet joint. However, a few may be lower on the C7 transverse processes. In contrast, the course of the C4 and C5 medial branch nerves has been shown to be relatively constant following the waist of their respected articular pillars in cadavers. Thus, C3, C6, and C7 show more variation compared to C4 and C5. The C3 medial branch nerves, with their more superior location at the upper third of the C3 articular pillars, often overlap the third occipital nerve, with the third occipital nerve being rostral to the C3 medial branches. The C3 medial branches and the third occipital nerve have a common origin in the C3 dorsal ramus. Finally, the C6 medial branches course around the waist of the articular pillars or above them, between the waists and the superior articular processes. The distance between the nerves and bone varies from close proximity to separation by 2–3 mm. The C2/3 facet joints are largely innervated from the third occipital nerve, which is the superficial medial branch of the C3 dorsal ramus [35]. The deep medial branch of the C3 dorsal ramus is referred to as the C3 medial branch. Articular branches may also arise from a communicating loop that crosses the back of the joint

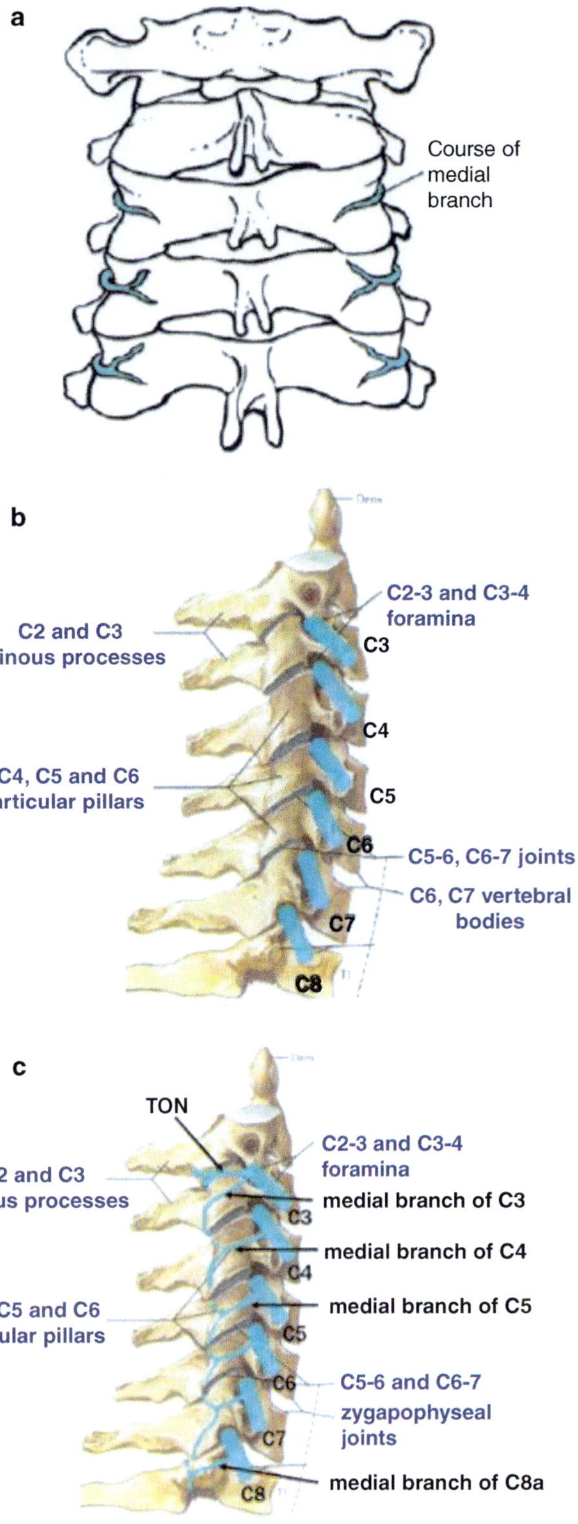


**Fig. 26.3** Schematic drawings of the innervation of cervical facet joints. Each facet joint below C2/3 receives dual innervation from the medial branch of the dorsal primary ramus above and below the joint. (**a**) Posterior view. (**b**) Lateral view. (**c**) Cervical medial branch nerves (TON is third occipital nerve)

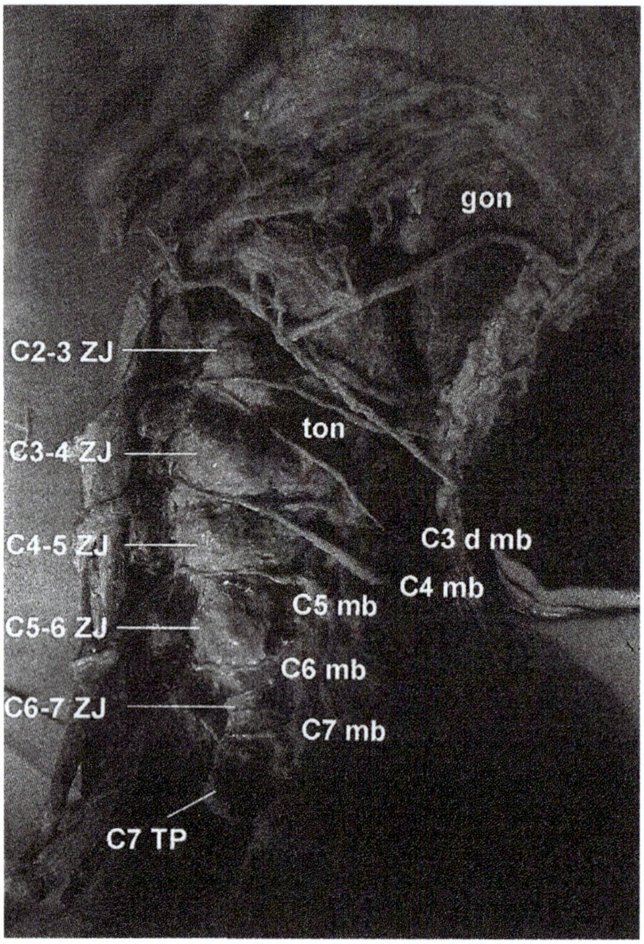


**Fig. 26.4** A dissection of the left cervical dorsal rami showing the relationship between the nerves and the cervical zygapophysial joints. *ZJ* zygapophysial joint; *mb* medial branch; *dmb* deep medial branch; *ton* third occipital nerve; *gon* greater occipital nerve. (*From Bogduk N. Practice Guidelines for Spinal Diagnostic and Treatment Procedures, 2nd edition. International Spine Intervention Society (ISIS), San Francisco, 2013* [63])

between the third occipital nerve and the C2 dorsal ramus. The third occipital nerve continues around the lower lateral and dorsal surface of the C2/3 joint embedded in the connective tissue that invests the joint capsule.

Relation to vascular structures is crucial. The vertebral artery ascends through the cervical transverse foramina of C1 to C6, which is located anterolaterally (Fig. 26.5). The vertebral artery at C2 to C7 is located anterior to the facet joints from both posterior and lateral injection approaches. It passes directly superior in the neck until the transverse process of the axis where it courses upward and laterally to the transverse foramen of the atlas. The vertebral artery courses medially and superiorly from its lateral position at C1 to the medial foramen magnum. Thus, the course of the vertebral artery from the transverse foramen of the atlas to the foramen magnum may be tortuous and variable (Fig. 26.5).

The anterior spinal canal branches arise from single fine branches of the vertebral arteries just outside the cervical intervertebral foramina. They are closely related to the pedicle of the cervical vertebrae from C2 downward. The posterior spinal canal arteries derived from the vertebral arteries

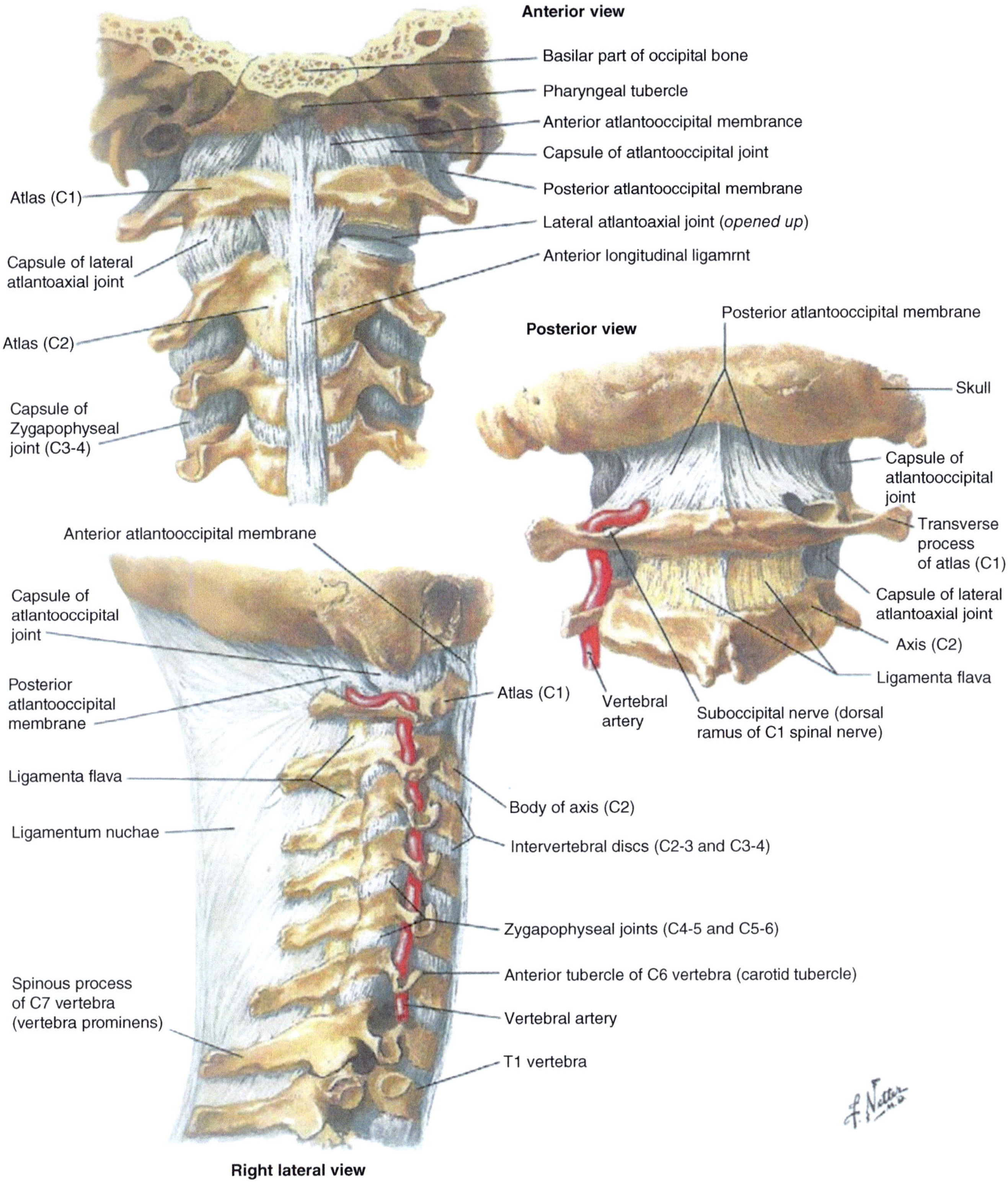


**Fig. 26.5** Anatomy of cervical spine. (*Netter medical illustration used with permission of Elsevier. All rights reserved*)

supply the posterior spinal elements, the outer surfaces of which are also covered by plexuses derived from the ascending and deep cervical arteries and by descending branches of the occipital arteries.

## Technical Considerations

Cervical facet joint interventions in regenerative medicine include intraarticular injections and medial branch blocks. It is imperative to understand radiographic and fluoroscopic anatomy prior to proceeding with facet joint interventions.

### Intraarticular Cervical Facet Joint Injections

Intraarticular cervical facet joint injections may be performed utilizing a posterior approach or a lateral approach.

#### Posterior Approach

The posterior approach for a cervical intraarticular facet joint block can be performed with the patient placed prone (Fig. 26.6a), or if required, sitting. It involves introducing a 22- or a 25-gauge needle into the target joint from behind, along an oblique trajectory that coincides with the plane of the joint (Fig. 26.6a).

The patient is placed prone with a cushion under the chest and a completely prone head and neck or with the neck rotated to the side opposite the needle insertion. Under aseptic conditions, skin entry is carried out approximately two or more segments below the target joint. Skin entry is determined by identifying the lateral aspect of the facet joint column on prone fluoroscopic viewing (Fig. 26.6b). The needle is passed upward and ventrally through the posterior neck muscles until it makes contact with the posterior surface of the articular pillar below the target joint. Following the above step, the needle may be readjusted until it enters the joint cavity. However, this may step require repeated posteroanterior and lateral fluoroscopic visualization to ensure the needle stays on course. Caution is required as directing the needle medially toward the interlaminar space may result in epidural or intrathecal puncture and/or spinal cord trauma.

After satisfactory localization of the needle into the joint, water-soluble contrast medium is injected to obtain an arthrogram and verify accurate placement. Then corticosteroid and/or local anesthetic is injected for diagnostic or therapeutic purposes (Fig. 26.7). Due to the risk of a pneumothorax and the proximity of other neurovascular structures, the posterior approach is often preferred for the C7/T1 facet joint. However, the C2/3 joint may be technically more difficult to visualize and enter due to its anatomic features. The C2/3 joint is more angulated vertically and medially and is not clearly evident on lateral views. For C2/3 the posterior approach may be modified by rotating the patient's head to bring the cavity of the C2/3 joint into view as it rotates forward on the long axis of the vertebral column (Fig. 26.8).

The benefits of the posterior approach are many. The posterior approach is considered safe since the needle penetrates only the skin and posterior neck muscles, with the deep cervical artery being the only structure at risk of inadvertent puncture. Posterior cervical arteries pose minimal risk of morbidity, as they supply no major structures.

a

b

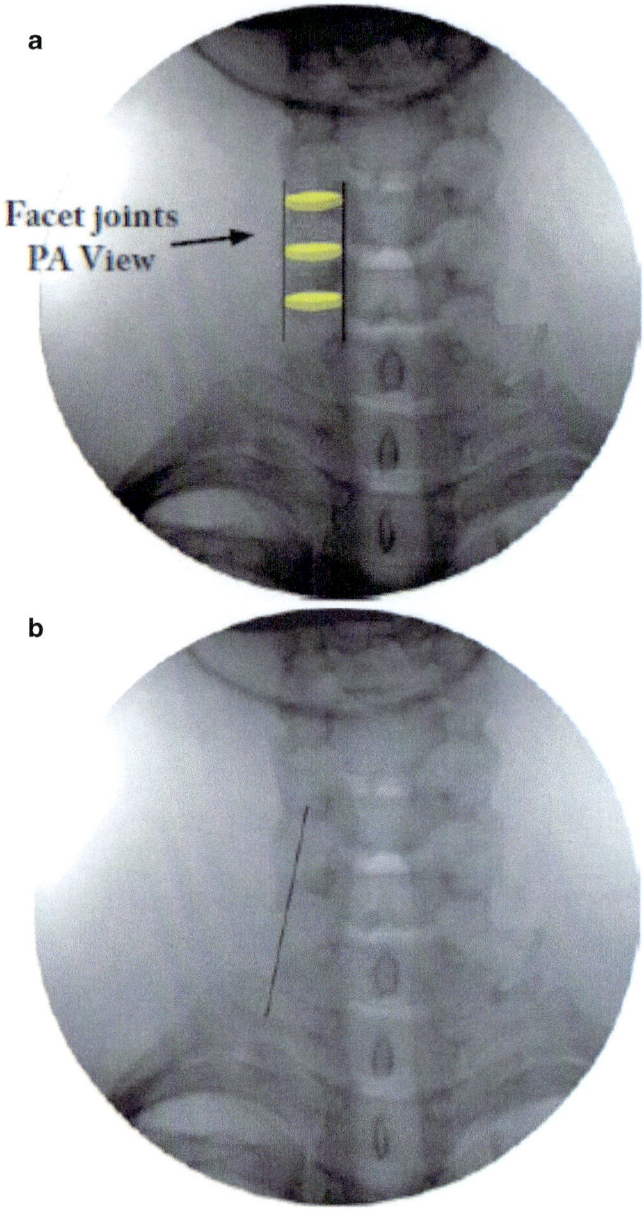


**Fig. 26.6** Posterior approach to intraarticular injections. (**a**) Posterior fluoroscopic view of facet joint column. (**b**) Radiographic and fluoroscopic anatomy of cervical spine

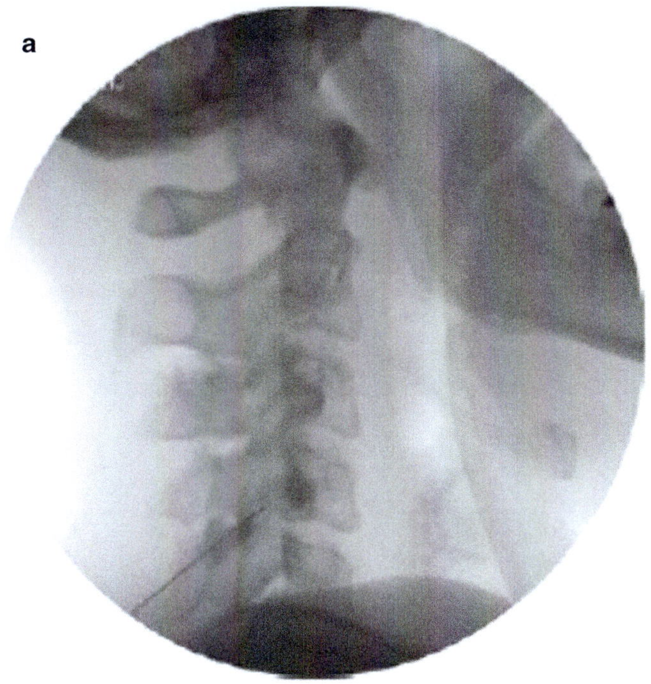

**Fig. 26.7** (**a**, **b**) Lateral views of C4/5 facet joint injections as confirmed with contrast medium

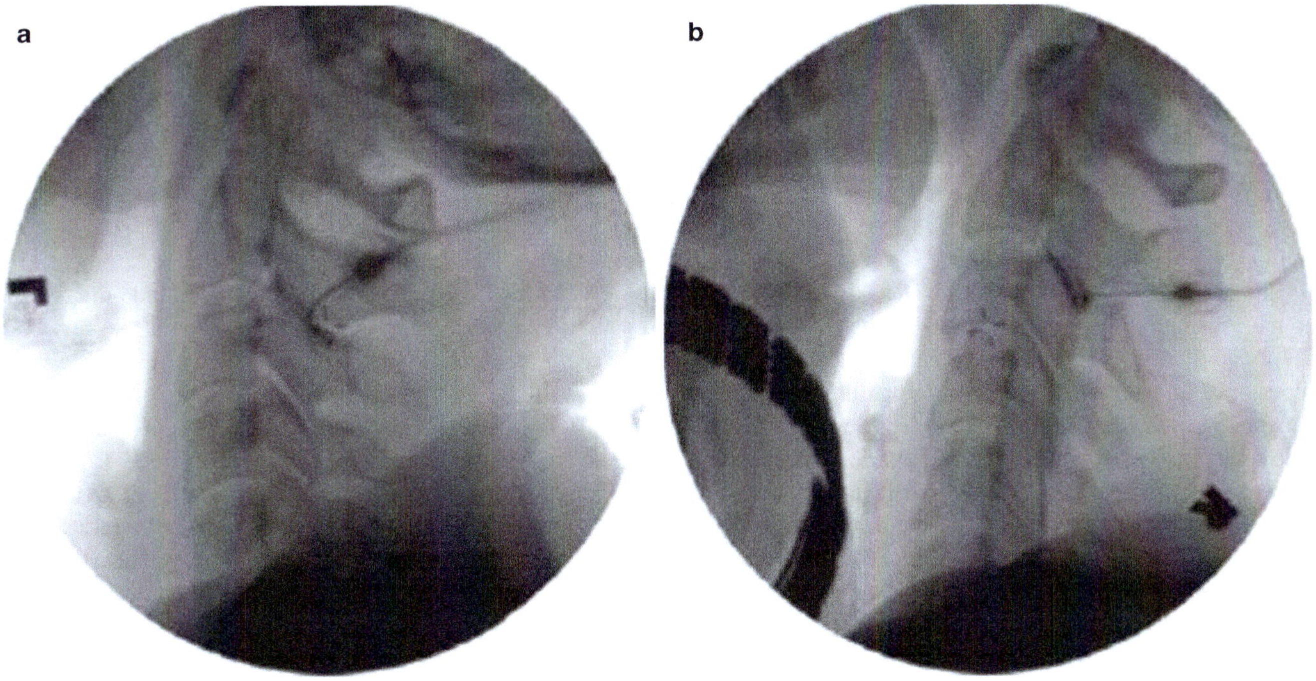

**Fig. 26.8** (**a**, **b**) Lateral views of C2/3 facet joint injections as confirmed with contrast medium

Risks of the posterior approach include penetration of the anterior joint capsule moving into the neural foramen and into the vicinity of the DRG, the cervical radicular artery, and/or the vertebral artery, epidural space, or spinal cord. Further, leakage of local anesthetic and steroid onto the DRG may cause issues.

**Lateral Approach**

For the lateral, the patient is positioned lying on his or her side with the target side up. The shoulders are pulled down to avoid obscuring the joints under fluoroscopy and rotated slightly posterior, about 25° into the plane of the upper torso and shoulders [64]. The target joint is identified on lateral

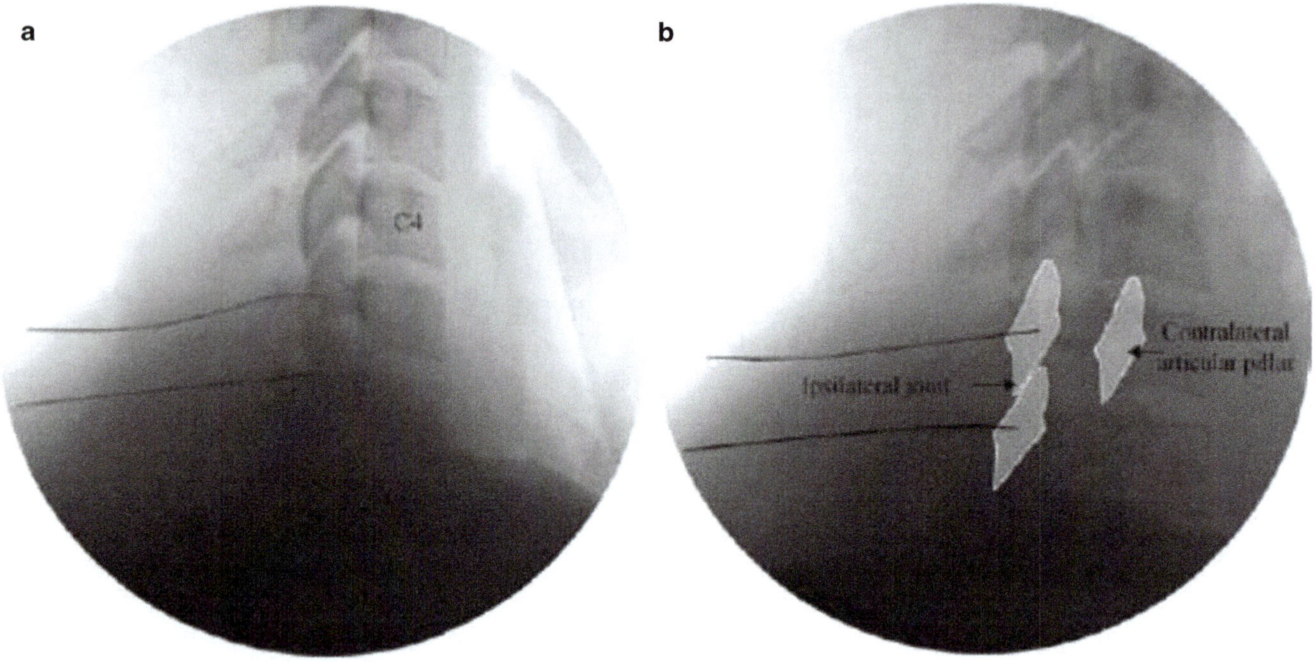


**Fig. 26.9** Illustration of lateral approach. (**a**) True lateral fluoroscopy view with superimposed articular pillars. (**b**) Oblique view with offset articular pillars

imaging of the neck (Fig. 26.9). Lateral fluoroscopic imaging must appropriately identify both joints so that the uppermost target joint is differentiated from the downside contralateral joint. This can be done by precisely superimposing one joint over the other (true lateral view in Fig. 26.9a) or by offsetting the joints adequately to view them independently (oblique view, Fig. 26.9b).

The needle is introduced through the skin over the midpoint of the joint for lateral approach. The needle is advanced deeply until it makes contact with the bone of either the superior or inferior articular process. Rotation of the intensifier may help to identify both joints. It is absolutely crucial that right and left joints be identified properly so that the uppermost joint is targeted. After the identification of the joint, the needle is advanced until the superior articular process is contacted just above the joint line. Then the needle is then directed and advanced through the joint capsule. At this time, the needle may be felt to pierce the capsule and to enter the joint space. However, only minimal penetration is required. The appropriate final position into the joint is confirmed by injecting a small dose of contrast medium to obtain an arthrogram.

The C2/3 joint may be technically more difficult to visualize and enter due to its anatomic features. The C2/3 joint is more angulated vertically and medially and is not clearly evident on lateral views. Even then, multiple benefits of the lateral approach have been described. These include less demanding technique and may be performed with smaller gauge needles [24]. In contrast, a lateral approach may be more comfortable for the patient because less soft tissue is traversed. Finally, the risks of the lateral approach includes the needle into the epidural space or spinal cord due to aggressive maneuvering or over-penetration.

Precautions with the lateral approach include the needle to remain posterior to the ventral ramus and the vertebral and radicular arteries. Further, in a patient with a large neck and shoulders, C7/T1 may not be able to be reached from a lateral approach and may require the posterior approach. While C3/4, C4/5, and C6/7 joint injections are easily performed using the lateral approach with no disadvantages compared to the posterior approach, the C7/T1 joint injection may be more easily performed using the posterior approach.

Low volumes must be injected into the cervical facet joints. If volumes of more than 1 mL are injected, or the injection is carried out rapidly or forcefully, the joint capsule may rupture and/or medication may spread onto nearby structures. Even with smaller volumes, extraarticular leaks have been observed in up to 7% of cases as shown in Fig. 26.10 [23, 24, 65]. Finally, the extraarticular spread will compromise the specificity of the block.

## Cervical Medial Branch Blocks

To block the nerve supply to a single cervical facet joint, two medial branches must be blocked due to the dual innervation of each facet joint.

The target points for medial branches, other than the third occipital nerve, are the crossing points of the waists of the articular pillars—points proximal to the origin of the articular branches at which the nerves have a constant relationship to the bone (Figs. 26.11 and 26.12). These points may be reached by needles using either a posterior, lateral, or anterior approach. Posterior and lateral approaches are most commonly utilized.

The medial branches of the cervical dorsal rami do not lie at unique points in the cervical spine. Consequently, the con-

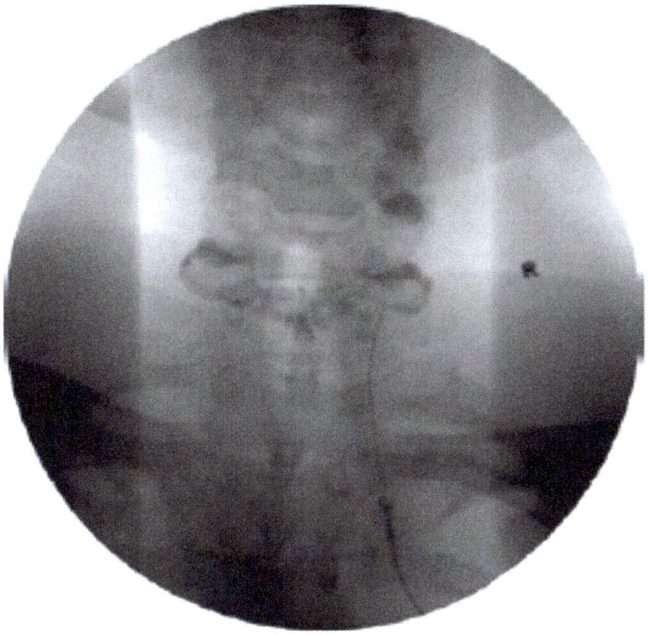

**Fig. 26.10** Filling of bilateral C6/7 facet joint capsules from injection into a single right-sided C6/7 joint capsule

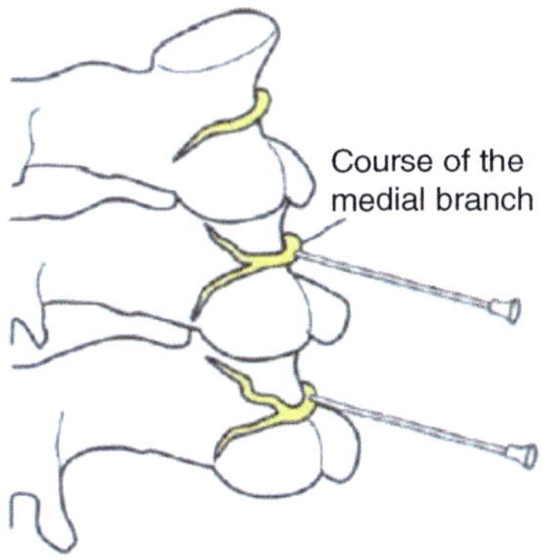


**Fig. 26.12** Diagram showing the needle positioning for cervical medial branch block. (*Reproduced from Mathis, Image-Guided Spine Interventions, 2004, with permission from Springer-Verlag, New York*)

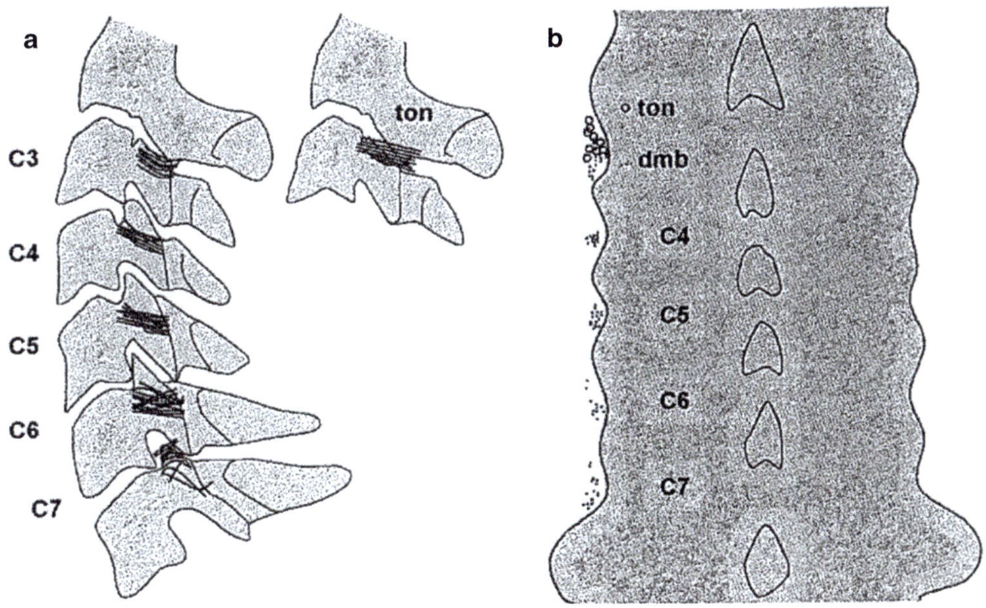


**Fig. 26.11** Composite tracings of radiographs of cadavers in which wires were applied to the medial branches of the cervical dorsal rami in order to show the variation in their location. (**a**) Lateral view. At the C3 level, the locations of the C3 deep medial branch are shown. Note that the C5 medial branches are fairly constantly located over the middle fifth of the C5 articular pillar, but the medial branches are located progressively higher on their respective articular pillars at levels progressively further away from the C5 level. The inset shows the locations of the third occipital nerve. (**b**) AP view. The nerves are seen as dots passing from front to back. (From: Bogduk N. Cervical medial branch thermal radiofrequency neurotomy. In: Practice Guidelines for Spinal Diagnostic and Treatment Procedures. 2nd edition. International Spine Intervention Society (ISIS), San Francisco, 2013, pp. 165–218)

cept of target points it not always accurate. At typical cervical segments, the medial branches of the cervical dorsal rami assume a curved course around the anterolateral, lateral, and posterolateral aspects of their ipsisegmental articular pillars [65]. The typical relationship between nerve and bone has been shown at the C5 level. At the C5 level, as shown in the lateral view, the C5 medial branch runs transversally across the centroid of the C5 articular pillar (Fig. 26.11a). In anteroposterior (AP) views (Fig. 26.11b), the C5 medial branch is located in the lateral concavity of the articular pillar. But at other levels, the medial branches differ and vary slightly in their location from their typical patterns. Thus, beyond C5, medial branches are located at higher levels on the articular pillar, and the third occipital nerve and the C7 medial branches leave their articular pillar and actually cross the zygapophysial joint above. Consequently, at all levels except for C5, medial branches run slightly to significantly higher across the articular pillars.

At the C3 level, two medial branches are present with C3 deep medial branch innervating C3/4 facet joint, and the C3 superficial medial branch or third occipital nerve innervating the C2/3 facet joint. Thus, the C3 deep medial branch is usually found running across the upper half of the C3 articular pillar; however, sometimes it crosses the lower half of the C2/3 facet joint. The third occipital nerve crosses the C2/3 zygapophysial joint, and may be located anywhere from the opposite effects of the C3 superior articular process to opposite the bottom of the C2/3 intervertebral foramen. In AP views as shown in Fig. 26.11b, either the deep medial branch or the third occipital nerve may occupy the concavity of the C3 articular pillar, but C3 deep medial branches tend to spill caudally onto the upper half of the convexity of the C3/4 facet joint, whereas the third occipital nerves tend to spill onto the lower half of the convexity of the C2/3 facet joint [65].

### Posterior Approach

For the posterior approach, the patient is placed prone with a pillow under the chest with head maintained looking straight downward in a neutral position or is turned to the side opposite the needle insertion. A posteroanterior view is obtained to identify the posterior aspect of the waists of the articular pillars from C3 to C7 (Fig. 26.13a). After identifying the waists of the articular pillars at the levels to be blocked, a 22- or 25-gauge, 2- to 3-in. spinal needle is inserted through the skin and posterior neck muscles, aiming first for the dorsal aspect of the articular pillar medial to its lateral concavity (Fig. 26.13a). Once the needle has made contact with the bone, it is readjusted laterally to the deepest of this concavity where the C3 to C7 medial branches lie (Fig. 26.13a). However, directing the needle medially to bone ensures that the needle is not placed too deeply. The needle is then directed laterally until the tip reaches the lateral margin of the articular pillar's waist. The needle should be felt to barely slip off the bone laterally in a ventral direction at the deepest point of the articular pillar's concavity [64]. Lateral images may ensure the needle tip rests at the centroid of the articular pillar. The centroid is defined as the intersection of the two diagonals of the diamond-shaped pillar (Fig. 26.13b) [64].

The C8 medial branch is blocked by placing the needle onto the transverse process of T1 and then directing it until it lies at the superolateral border of the transverse process. However, lateral viewing of the needle tip location may be difficult below C6. Moving the C-arm to an oblique projection may help to identify needle depth by moving the shoulders out of the line of the fluoroscopic beam (Fig. 26.13c).

Figure 26.14 illustrates cervical medial branch blocks at multiple levels. Figure 26.15 shows medial branch blocks at C4, C5, and C6 nerves. Figure 26.16 shows medial branch blocks at C3, C4, and C5, C6.

### Lateral Approach

The lateral approach was described and utilized during the 1980s and early 1990s, until Lord et al. [66] introduced the posterior approach with the patient placed prone. The posterior approach was based on the argument that while using this technique the electrode ends up in a position that is parallel to the medial branch. The lateral approach with the patient placed supine is reported by proponents to allow for much greater patient tolerance. Thus, the lateral approach is described as a fast, well-tolerated, effective procedure [64]. However, this is not an easy technique to learn as it is not a tunnel vision procedure. The direction of the X-rays is used in this procedure as a technique to avoid contact with the exiting segmental nerves.

For the lateral approach, the patient is placed laterally, target side up, or supine. The articular pillars are identified by lateral fluoroscopy (Fig. 26.9a). The uppermost articular pillar can be distinguished from the opposite side by moving the fluoroscope (Fig. 26.9b). The needle will be seen to travel with the uppermost articular pillar as the two articular pillars separate on the fluoroscopic image. This approach is well suited for C3 to C6 medial branches. The needle is directed through the skin and posterolateral neck muscles toward the centroid of the articular pillar as seen on a true lateral radiograph.

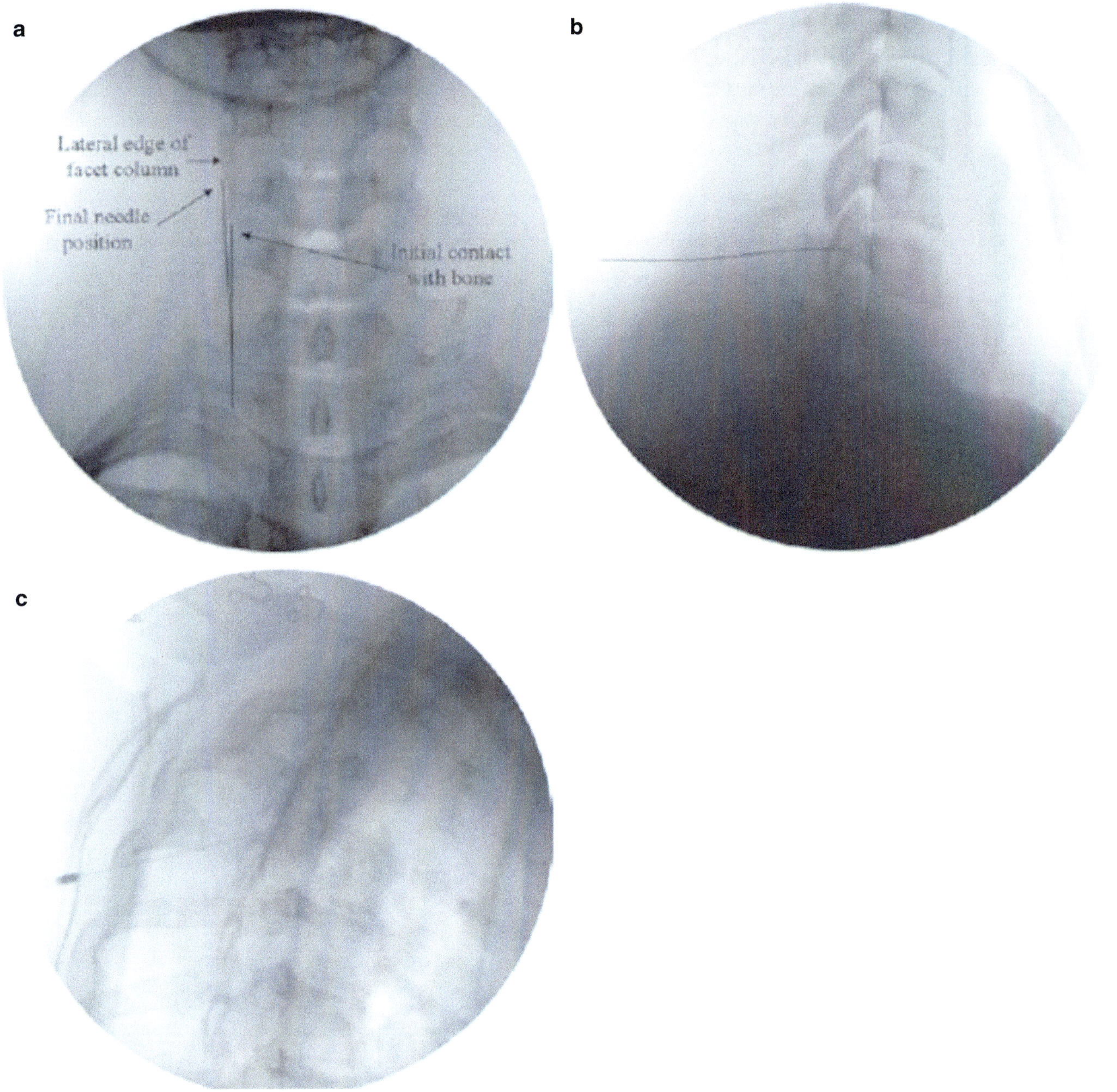


**Fig. 26.13** Illustration of posterior approach to medial branch blocks. (**a**) Posterior placement of needles for medial branch block. (**b**) Needle tip near the center of the articular pillar centroid on lateral view. (**c**) Oblique fluoroscopic image allows for better viewing of upper thoracic and lower cervical levels

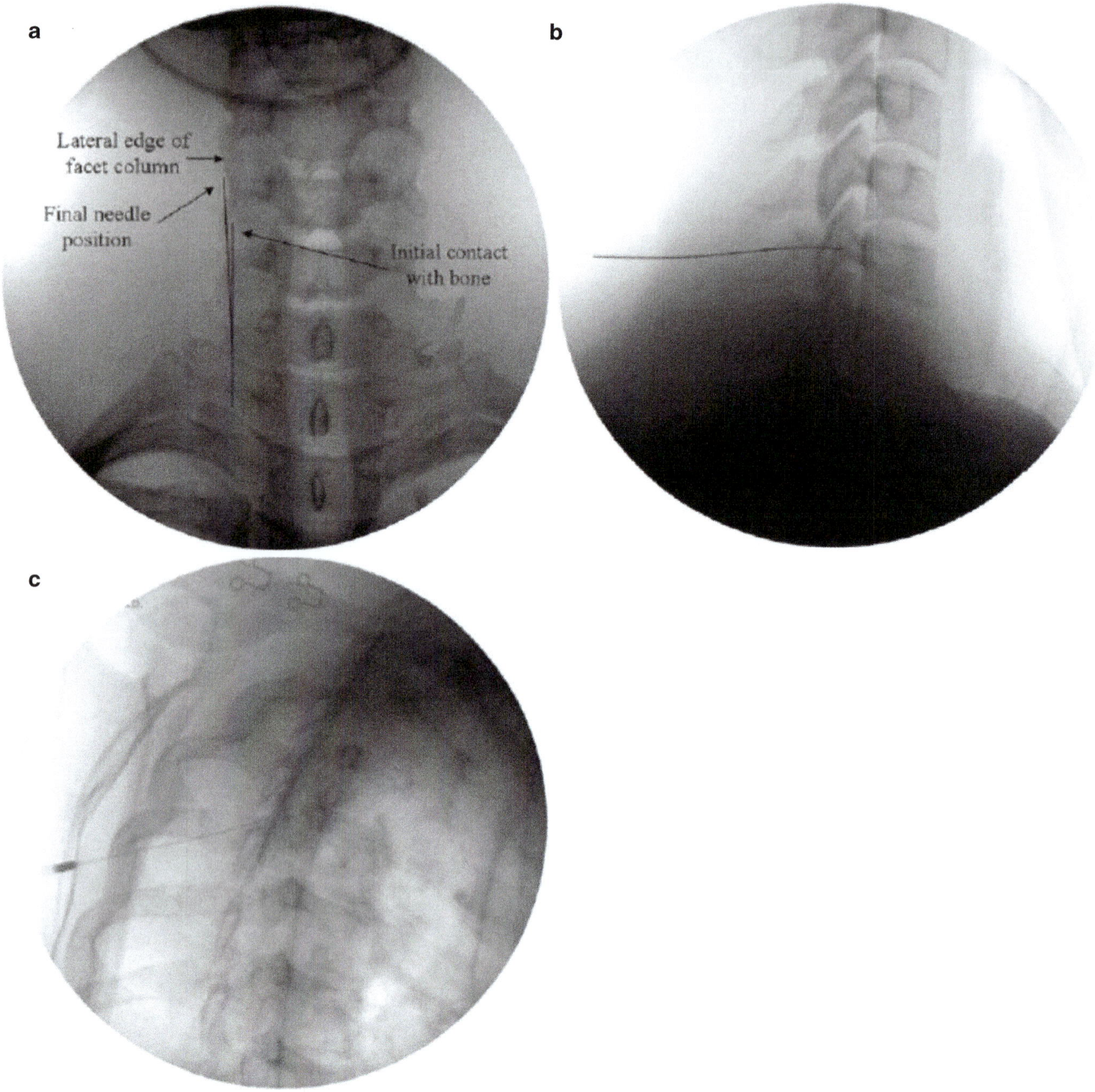


**Fig. 26.14** Illustration of posterior approach to medial branch blocks. (**a**) Posterior placement of needles for medial branch block. (**b**) Needle tip near the center of the articular pillar centroid on lateral view. (**c**) Oblique fluoroscopic image allows for better viewing of upper thoracic and lower cervical levels

To block the C7 medial branch with the lateral approach, the needle is advanced so that it stays within the confines of the C7 superior articular process, which prevents excessive advancement into the C8 foramen and toward the vertebral artery. Once the superior articular process is contacted, AP imaging should verify that the needle lies against the lateral aspect of the superior articular process. With posterior and lateral approaches, contrast medium in doses of 0.1–0.2 mL may be injected to confirm appropriate needle placement. For injection of biologics contrast may be withheld. After confirming the needle position, local anesthetic with or without steroid is injected incrementally around the nerve.

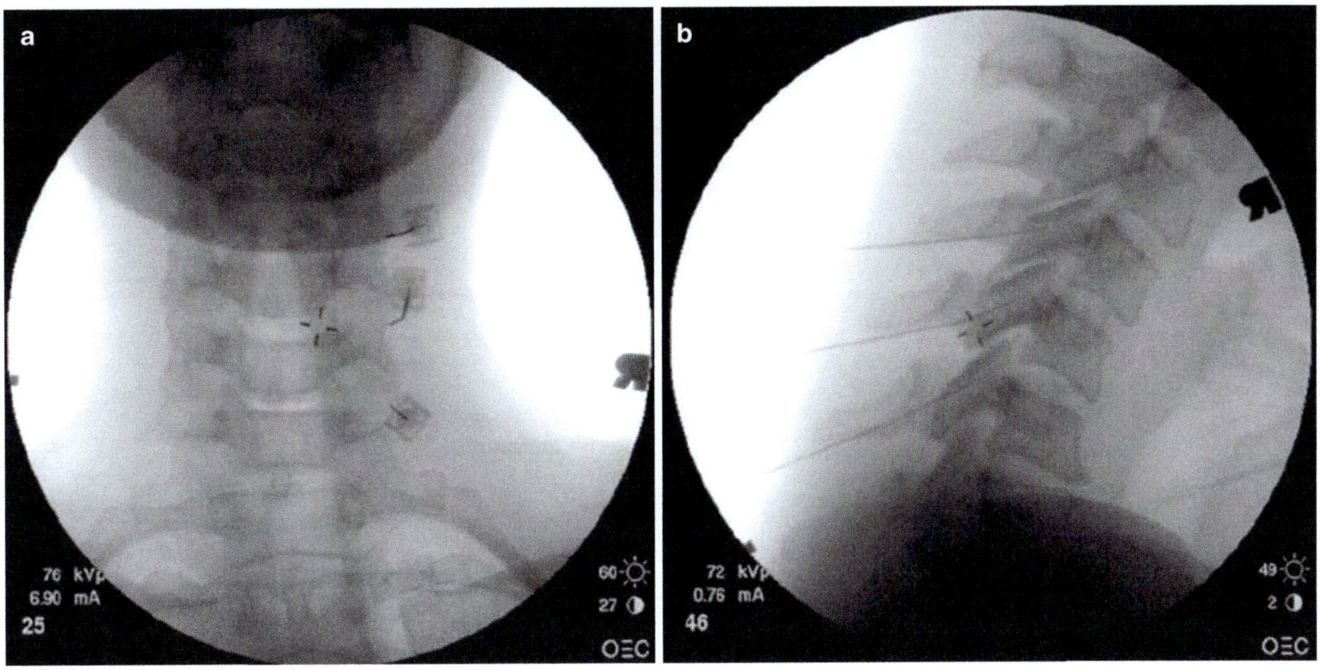


**Fig. 26.15** Medial branch blocks at C4, C5, and C6 levels with posterior approach under fluoroscopy. (**a**) PA view. (**b**) Lateral view

For a patient placed supine, a different trajectory of the needle is used from the direction of the X-rays [64]. The entry point cannot simply be found by marking the skin right over the target point, as is the case for tunnel vision procedures. One therefore has to do some stereoscopic thinking, and not everybody finds that easy [67]. As an additional difficulty, the distance from the skin to the target point is quite variable. This has very important consequences for the changes in the direction of the needle that have to be made to effectuate a correction of position. Remember, the shorter the trajectory, the larger the change in direction must be. That is why the medial branch procedure is technically easier if the patient has a large neck. In addition, the headrest assists so that the C-arm can be moved freely around the head and neck. The C-arm should be positioned in line with the operating table, not perpendicular to it. This facilitates handling of the C-arm and prevents less than optimal projections for different levels of the cervical spine. The head should be very slightly hyperextended; however, if the neck is short, the patient's shoulder may be dropped or an assistant may apply gentle traction to the arm [67]. A moderately oblique projection is used for this procedure. The degree of obliquity is judged by observing the relative positions of the contralateral pedicles and the anterior border of the vertebral bodies. The pedicles should be projected a little anterior to 50% of the vertebral bodies. This is different from the more oblique projection that is optimal for transforaminal epidural injections or DRG lesioning. In this projection the segmental nerves exit approximately in a plane that is perpendicular to the monitor screen. Thus, contact of the needle tip with the segmental nerves can at all times be avoided by keeping the needle posterior to the imaginary line connecting the posterior aspects of the foramina.

In the frontal plane, there should be a slight angle with the transverse plane, so that the intervertebral discs and foramina are clearly visible. In the projection that has now been obtained, the medial branch runs at the base of the superior articular process. This is the superior articular process and the target point. Further, the target point is also at the same level as the inferior aspect of the intervertebral foramen. This also helps to identify the target point if the superior articular process is not clearly identifiable. The entry point of the needle should be somewhat posterior and caudal to the target point as it is seen on the monitor. The distance that should be taken depends on the distance from the skin to the target point. Figure 26.17 shows the normal distance, and depending on the size of the neck, the distances should be varied proportionately. The depth of the needle should not be more than 2 cm. Before advancing farther, the position should first be checked. The needle should be in line with the target point. If the needle is pointing too far cephalad or too far caudad, it should be corrected first without inserting farther. Figure 26.18 shows four needles in place. The position is also checked anteroposteriorly. The tip should be adjacent to the spine, in the concavity between the articular processes.

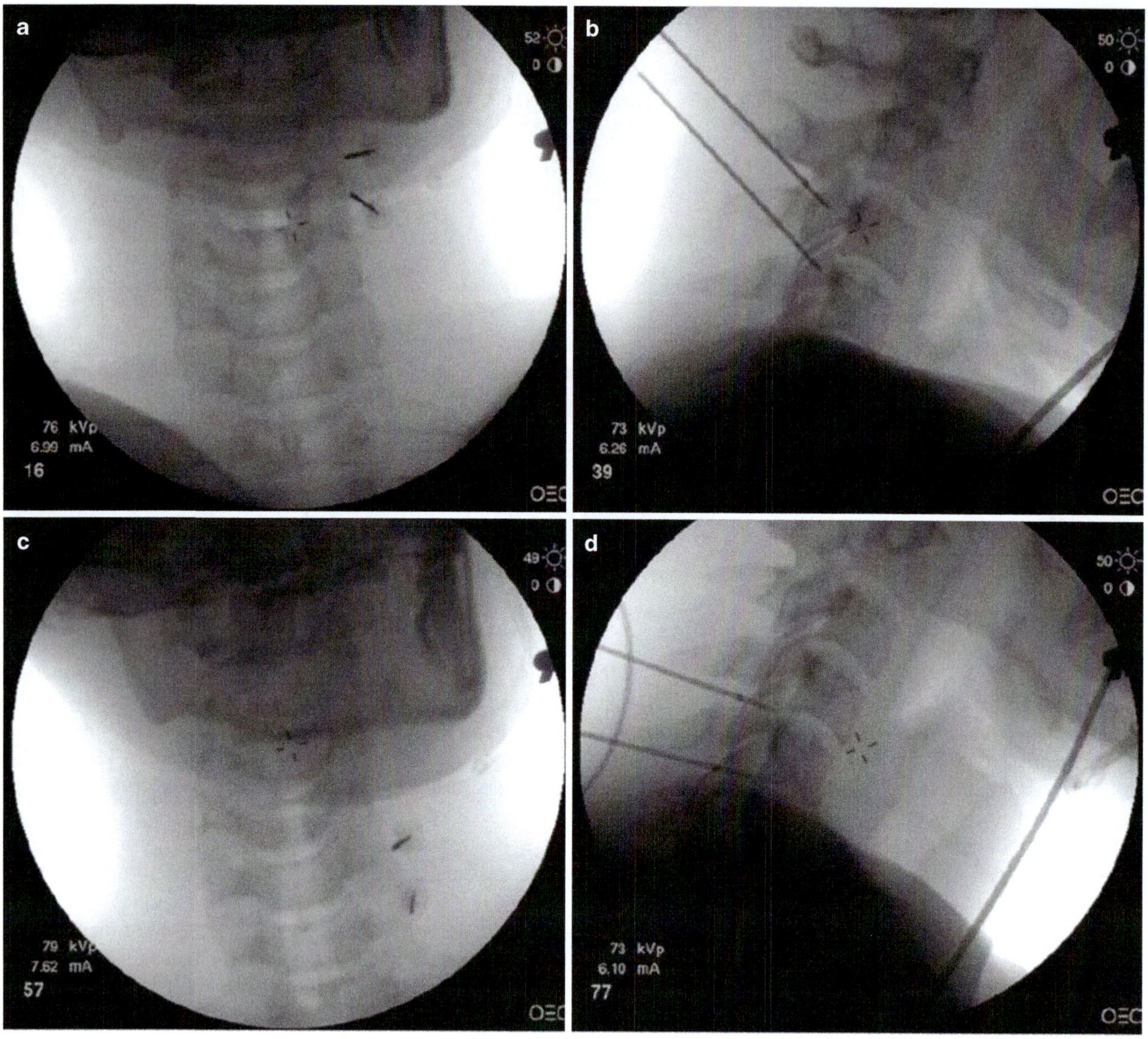


**Fig. 26.16** Needle placement for medial branch neurotomy. (**a**, **b**) at C3, C4 medial branches and (**c**, **d**) C5, C6 medial branches

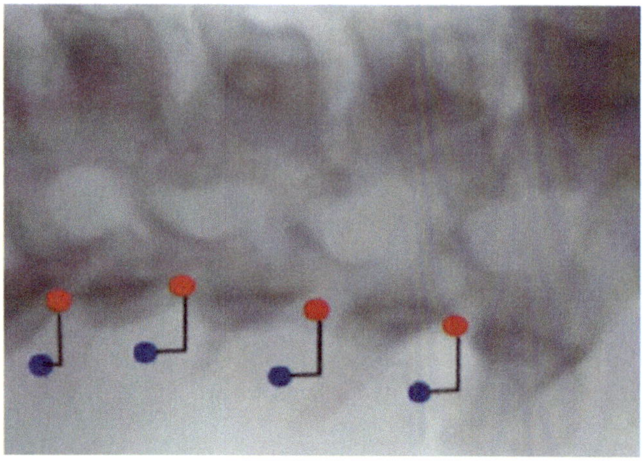

**Fig. 26.17** Relation of the entry points (blue) to the target points

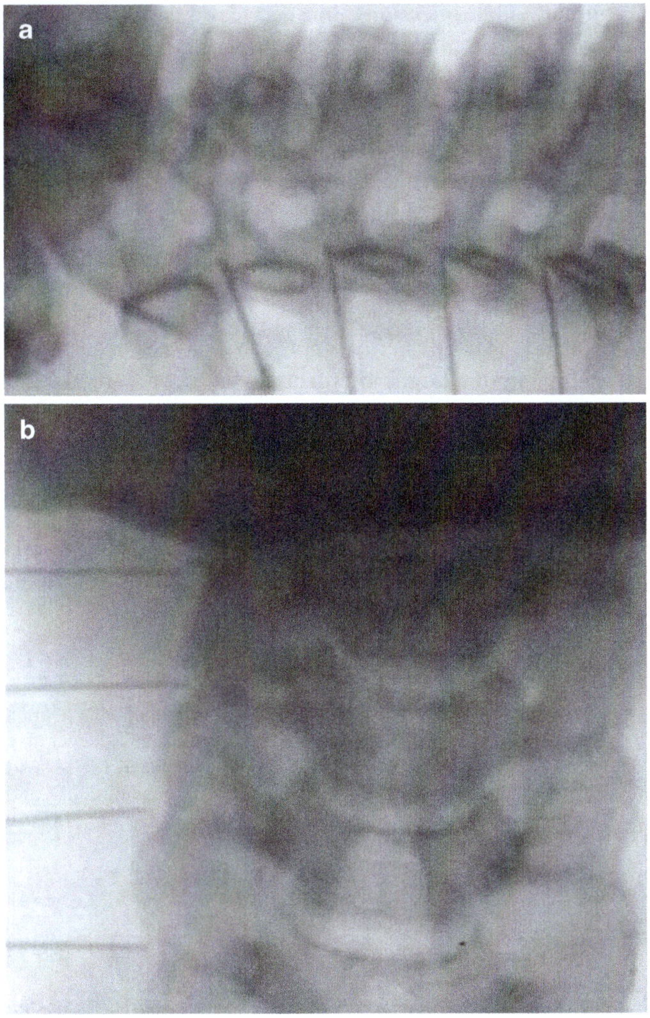

Fig. 26.18 Positions for a medial branch block. (a) Oblique projection. (b) Anteroposterior projection

## Indications

Indications for cervical diagnostic facet joint interventions are persistent and disabling axial neck and/or upper back pain or suspected cervicogenic headache of at least 3 months. In addition, average pain levels greater than 5 on a scale of 0–10 with intermittent or continuous pain causing functional disability. Indications also show failure to respond to more conservative management, including medication management, physical therapy exercise, and/or chiropractic treatment. Further, lack of evidence for disc herniation and radiculitis causing pain and lack of obvious evidence for discogenic pain must be shown. However, some patients with facet joint pain also have asymptomatic disc herniation or discogenic and facet joint involvement.

Indications for therapeutic facet joint interventions are established diagnosis with a positive response to controlled diagnostic blocks, either placebo or comparative local anesthetic blocks, with a criterion standard of 80% pain relief and the ability to perform prior painful movements without significant pain.

## Side Effects and Complications

Complications from intraarticular injections or medial branch blocks in the cervical spine are exceedingly rare [6–9, 26, 27]. However, serious complications with cervical facet joint injections may occur. Complications include those related to placement of the needle and those related to the administration of various drugs. The needle's proximity to the vertebral artery, spinal cord, and nerve root creates risk for injury and makes precise and accurate needle placement exceedingly important.

Complications may include dural puncture, spinal cord trauma, subdural injection, neural trauma, injection into the intervertebral foramen and intravertebral arteries; intravascular injection into veins or vertebral arteries; infectious complications including epidural abscess and bacterial meningitis; and side effects related to the administration of steroids, local anesthetics, and other drugs (Table 26.1).

**Table 26.1** Potential complications of cervical facet joint interventions

| | |
|---|---|
| • Pain<br>– Pain at the site of the needle insertion<br>– Exacerbation of existing pain<br>– Pain in the spine | • Trauma<br>– Soft tissue<br>– Medial branch<br>– Nerve root<br>– Spinal cord |
| • Infection<br>– Soft tissue abscess<br>– Epidural abscess<br>– Facet joint abscess<br>– Meningitis<br>– Encephalitis | • Inadvertent injection<br>– Dural puncture<br>– Subdural injection<br>– Epidural injection<br>– Foraminal injection<br>– Intravascular injection |
| • Bleeding<br>– Soft tissue hematoma<br>– Epidural hematoma<br>– Spinal cord hematoma<br>– Nerve root sheath hematoma | • Hypoesthesia |
| Steroid effects | Local anesthetic effects |

## Precautions

Relative contraindications to interventional techniques have been described in patients receiving treatment with antithrombotics and anticoagulants [6, 68]. Guidelines on anticoagulants and antithrombotics are described in Chap. 15 [69] and also on the ASIPP app, which is regularly updated.

## Conclusions

Cervical facet, or zygapophysial, joints have been shown to be a source of pain in the neck and referred pain in the head and upper extremities. Cervical facet joints were recognized as a potential source for neck pain in 1977. Based on controlled diagnostic blocks, the prevalence of persistent neck pain secondary to the involvement of cervical facet joints has been established as varying from 36% to 67%. Cervical facet joint pain may be managed with intraarticular injections, facet joint nerve blocks, and/or radiofrequency neurotomy and regenerative therapy with PRP and MSCs.

### Key Points

1. Chronic neck pain is a significant cause of disability, with multiple structures in the cervical spine contributing to pain generation. Cervical facet joints play a role in chronic neck pain and can refer pain to the head and extremities.
2. Cervical facet joints are a significant source of chronic neck pain, with innervation by dorsal primary rami. Mechanical injury and inflammation play a role in facet joint pain, and radiographic features of osteoarthritis are commonly observed. Whiplash injuries can also affect facet joints, contributing to neck and upper extremity pain.
3. The cervical facet joints are paired, synovial joints located between the superior and inferior articular pillars in the posterior cervical column. They provide structural support and allow for flexion, extension, and rotation.
4. The facet joints are innervated by the medial branches of the dorsal rami, and their fibrous capsules contain mechanoreceptors and nociceptors.
5. Conventional clinical features and radiological imaging are unreliable in diagnosing cervical facet joint pain. Physical examination maneuvers and radiological findings often do not correlate with the presence of facet joint pain. Diagnostic cervical facet joint blocks are considered a reliable reference standard for diagnosing neck pain, upper extremity pain, and headache, despite the significant false-positive rates associated with single facet joint blocks.
6. Intraarticular injections for cervical facet joints were described using lateral and posterior approaches.
7. Platelet-rich plasma (PRP) is a biologic treatment option for cervical facet joint pain, containing growth factors that promote tissue healing and regeneration. Show potential for tissue engineering applications. However, there is currently limited published evidence supporting their use specifically for facet joint pain.
8. Intraarticular cervical facet joint injections can be performed using either a posterior approach or a lateral approach. The posterior approach involves inserting a needle into the target joint from behind, while the lateral approach is performed with the patient positioned laterally, targeting the articular pillar through the skin and posterolateral neck muscles.
9. Complications from cervical facet joint injections or medial branch blocks are rare, but they can include serious risks such as dural puncture, spinal cord trauma, neural trauma, and injection into critical structures.

**Acknowledgments** This book chapter is modified and updated from a previous book chapter, "Cervical Facet Joints and Regenerative Medicine" by Laxmaiah Manchikanti, MD, Annu Navani, MD, David M. Schultz, MD, Frank FJ Falco, MD, and Ramsin Benyamin, MD, and in *Essentials of Regenerative Medicine in Interventional Pain Management* published by ASIPP Publishing.

## References

1. Manchikanti L, Navani A, Schultz DM, Falco FJE, Benyamin RM. Cervical facet joints and regenerative medicine. In: Manchikanti L, Navani A, Atluri S, editors. Essentials of regenerative medicine in interventional pain management. Paducah, KY: ASIPP Publishing; 2019. p. 473–94.
2. US Burden of Disease Collaborators. The state of US health, 1990-2016: burden of diseases, injuries, and risk factors among US states. JAMA. 2018;319:1444–72.
3. Hoy D, March L, Woolf A, et al. The global burden of neck pain: estimates from the global burden of disease 2010 study. Ann Rheum Dis. 2014;73:1309–15.
4. Dieleman JL, Baral R, Birger M, et al. US spending on personal health care and public health, 1996-2013. JAMA. 2016;316:2627–46.
5. Dieleman JL, Cao J, Chapin A, et al. US health care spending by payer and health condition, 1996-2016. JAMA. 2020;323:863–84.
6. Manchikanti L, Kaye AD, Soin A, et al. Comprehensive evidence-based guidelines for facet joint interventions in the management of chronic spinal pain: American Society of Interventional Pain Physicians (ASIPP) guidelines. Pain Physician. 2020;23:S1–S127.
7. Manchikanti L, Hirsch JA, Kaye AD, Boswell MV. Cervical zygapophysial (facet) joint pain: effectiveness of interventional management strategies. Postgrad Med. 2016;128:54–68.
8. Manchikanti L, Kosanovic R, Cash KA, et al. Assessment of prevalence of cervical facet joint pain with diagnostic cervical medial branch blocks: analysis based on chronic pain model. Pain Physician. 2020;23:531–40.

9. Manchikanti L, Knezevic NN, Knezevic E, et al. A systematic review and meta-analysis of the effectiveness of radiofrequency neurotomy in managing chronic neck pain. Pain Ther. 2023;12:19–66.
10. Manchikanti L, Kosanovic R, Pampati V, Sanapati MR, Hirsch JA. Outcomes of cervical therapeutic medial branch blocks and radiofrequency neurotomy: clinical outcomes and cost utility are equivalent. Pain Physician. 2022;25:35–47.
11. Manchikanti L, Boswell MV, Singh V, et al. Prevalence of facet joint pain in chronic spinal pain of cervical, thoracic, and lumbar regions. BMC Musculoskelet Disord. 2004;5:15.
12. Manchukonda R, Manchikanti KN, Cash KA, et al. Facet joint pain in chronic spinal pain: an evaluation of prevalence and false-positive rate of diagnostic blocks. J Spinal Disord Tech. 2007;20:539–45.
13. Yin W, Bogduk N. The nature of neck pain in a private pain clinic in the United States. Pain Med. 2008;9:196–203.
14. Manchikanti L, Navani A, Sanapati M. Evolution of regenerative medicine in managing musculoskeletal and spinal disorders. In: Manchikanti L, Navani A, Atluri S, Sanapati M, editors. Manchikanti's essentials of regenerative medicine in interventional pain management. 2nd ed. New York, NY: Springer; 2023.
15. Navani A, Manchikanti L, Albers SL, et al. Responsible, safe, and effective use of biologics in the management of low back pain: American Society of Interventional Pain Physicians (ASIPP) guidelines. Pain Physician. 2019;22:S1–S74.
16. Kristjánsson B, Limthongkul W, Yingsakmongkol W, et al. Isolation and characterization of human mesenchymal stem cells from facet joints and interspinous ligaments. Spine (Phila Pa 1976). 2016;41:E1–7.
17. Pawl RP. Headache, cervical spondylosis, and anterior cervical fusion. Surg Ann. 1977;9:391–408.
18. Bogduk N, Marsland A. The cervical zygapophysial joints as a source of neck pain. Spine (Phila Pa 1976). 1988;13:610–7.
19. Dwyer A, Aprill C, Bogduk N. Cervical zygapophysial joint pain patterns I: a study in normal volunteers. Spine (Phila Pa 1976). 1990;6:453–7.
20. Aprill C, Dwyer A, Bogduk N. Cervical zygapophysial joint pain patterns II: a clinical evaluation. Spine (Phila Pa 1976). 1990;6:458–61.
21. Fukui S, Ohseto K, Shiotani M, et al. Referred pain distribution of the cervical zygapophysial joints and cervical dorsal rami. Pain. 1996;68:79–83.
22. Windsor RE, Nagula D, Storm S, et al. Electrical stimulation induced cervical medial branch referral patterns. Pain Physician. 2003;6:411–8.
23. Cooper G, Bailey B, Bogduk N. Cervical zygapophysial joint pain maps. Pain Med. 2007;8:344–53.
24. Okada K. Studies on the cervical facet joints using arthrography of the cervical facet joint. J Jpn Orthop Assoc. 1981;55:563–80.
25. Dory MA. Arthrography of the cervical facet joints. Radiology. 1983;148:379–82.
26. Manchikanti L, Schultz DM, Falco FJE, et al. Cervical facet joint interventions. In: Manchikanti L, Singh V, editors. Interventional techniques in chronic spinal pain. Paducah, KY: ASIPP Publishing; 2007. p. 295–320.
27. Manchikanti L, Schultz DM, Falco FJE, Singh V. Cervical facet joint interventions. In: Manchikanti L, Kaye AD, Falco FJE, Hirsch JA, editors. Essentials of interventional techniques in managing chronic spinal pain. New York, NY: Springer; 2018. p. 387–412.
28. Cavanaugh JM, Lu Y, Chen C, et al. Pain generation in lumbar and cervical facet joints. J Bone Joint Surg Am. 2006;88:63–7.
29. Inami S, Shiga T, Tsujino A, et al. Immunohistochemical demonstration of nerve fibers in the synovial fold of the human cervical facet joint. J Orthop Res. 2001;19:593–6.
30. Kallakuri S, Singh A, Chen C, et al. Demonstration of substance P, calcitonin gene-related peptide, and protein gene product 9.5 containing nerve fibers in human cervical facet joint capsules. Spine (Phila Pa 1976). 2004;29:1182–6.
31. Ohtori S, Takahashi K, Moriya H. Calcitonin gene-related peptide immunoreactive DRG neurons innervating the cervical facet joints show phenotypic switch in cervical facet injury in rats. Eur Spine J. 2003;12:211–5.
32. Ohtori S, Moriya H, Takahashi K. Calcitonin gene-related peptide immunoreactive sensory DRG neurons innervating the cervical facet joints in rats. J Orthop Sci. 2002;7:258–61.
33. McClain RF. Mechanoreceptor endings in human cervical facet joints. Spine (Phila Pa 1976). 1994;19:495–501.
34. Chen C, Lu Y, Kallakuri S, et al. Distribution of A-delta and C-fiber receptors in the cervical facet joint capsule and their response to stretch. J Bone Joint Surg Am. 2006;88:1087–816.
35. Lord SM. Cervical zygapophysial joint pain after whiplash injury, precision diagnosis, prevalence, and evaluation of treatment by percutaneous radiofrequency. Neurotomy [doctoral thesis]. Newcastle: University of Newcastle; 1996.
36. Dong L, Odeleye AO, Jordan-Sciutto KL, et al. Painful facet joint injury induces neuronal stress activation in the DRG: implications for cellular mechanisms of pain. Neurosci Lett. 2008;443:90–4.
37. Lee KE, Davis MB, Winkelstein BA. Capsular ligament involvement in the development of mechanical hyperalgesia after facet joint loading: behavioral and inflammatory outcomes in a rodent model of pain. J Neurotrauma. 2008;25:1383–93.
38. Dong L, Guarino BB, Jordan-Sciutto KL, et al. Activating transcription factor 4, a mediator of the integrated stress response, is increased in the dorsal root ganglia following painful facet joint distraction. Neuroscience. 2011;193:377–86.
39. Quinn JH, Bazan NG. Identification of prostaglandin E2 and leukotriene B4 in the synovial fluid of painful, dysfunctional temporomandibular joints. J Oral Maxillofac Surg. 1990;48:968–71.
40. Winkelstein BA, Santos DG. An intact facet capsular ligament modulates behavioral sensitivity and spinal glial activation produced by cervical facet joint tension. Spine (Phila Pa 1976). 2008;33:856–62.
41. Lee KE, Davis MB, Mejilla RM, et al. In vivo cervical facet capsule distraction: mechanical implications for whiplash and neck pain. Stapp Car Crash J. 2004;48:373–95.
42. Dong L, Winkelstein BA. Simulated whiplash modulates expression of the glutamatergic system in the spinal cord suggesting spinal plasticity is associated with painful dynamic cervical facet loading. J Neurotrauma. 2010;27:163–74.
43. Chua NH, van Suijlekom HA, Vissers KC, et al. Differences in sensory processing between chronic cervical zygapophysial joint pain patients with and without cervicogenic headache. Cephalalgia. 2011;31:953–63.
44. Kras JV, Dong L, Winkelstein BA. Increased interleukin-1α and prostaglandin E2 expression in the spinal cord at 1 day after painful facet joint injury: evidence of early spinal inflammation. Spine (Phila Pa 1976). 2014;39:207–12.
45. Laplante BL, DePalma MJ. Spine osteoarthritis. PM&R. 2012;4:S28–36.
46. Gellhorn AC, Katz JN, Suri P. Osteoarthritis of the spine: the facet joints. Nat Rev Rheumatol. 2013;9:216–24.
47. Lee MJ, Riew KD. The prevalence cervical facet arthrosis: an osseous study in a cadaveric population. Spine J. 2009;9:711–4.
48. Weiler C, Schietzsch M, Kirchner T, Nerlich AG, Boos N, Wuertz K. Age-related changes in human cervical, thoracal and lumbar intervertebral disc exhibit a strong intra-individual correlation. Eur Spine J. 2012;6:S810–8.
49. Kettler A, Werner K, Wilke HJ. Morphological changes of cervical facet joints in elderly individuals. Eur Spine J. 2007;16:987–92.

50. Morishita K, Kasai Y, Uchida A. Hypertrophic change of facet joint in the cervical spine. Med Sci Monit. 2008;14:CR62–4.
51. Ivancic PC, Ito S, Tominaga Y, et al. Whiplash causes increased laxity of cervical capsular ligament. Clin Biomech (Bristol, Avon). 2008;23:159–65.
52. Bogduk N. On cervical zygapophysial joint pain after whiplash. Spine (Phila Pa 1976). 2011;36:S194–9.
53. Curatolo M, Bogduk N, Ivancic PC, et al. The role of tissue damage in whiplash-associated disorders: discussion paper 1. Spine (Phila Pa 1976). 2011;36:S309–15.
54. Yoshida R, Cheng M, Murray MM. Increasing platelet concentration in platelet-rich plasma inhibits anterior cruciate ligament cell function in three-dimensional culture. J Orthop Res. 2014;32:291–5.
55. Wasterlain AS, Braun HJ, Harris AH, et al. The systemic effects of platelet-rich plasma injection. Am J Sports Med. 2013;41:186–93.
56. Manchikanti L, Centeno CJ, Atluri S, et al. Bone marrow concentrate (BMC) therapy in musculoskeletal disorders: evidence-based policy position statement of American Society of Interventional Pain Physicians (ASIPP). Pain Physician. 2020;23:E85–E131.
57. Bogduk N, Lord S. Cervical zygapophysial joint pain. Neurosurgery. 1998;8:107–17.
58. King W, Lau P, Lees R, et al. The validity of manual examination in assessing patients with neck pain. Spine J. 2007;7:22–6.
59. Grubb SA, Kelly CK. Cervical discography: clinical implications for 12 years of experience. Spine (Phila Pa 1976). 2000;25:1382–9.
60. DeBoer KF, Harmon R, Tuttle CD, et al. Reliability study of detection of somatic dysfunctions in the cervical spine. J Manip Physiol Ther. 1985;8:9–16.
61. Cramer GD. The cervical region. In: Cramer GD, Darby SA, editors. Basic and clinical anatomy of the spine, spinal cord, and ANS. 2nd ed. St. Louis, MO: Elsevier; 2005. p. 142–209.
62. Johnson GM. The sensory and sympathetic nerve supply within the cervical spine: review of recent observations. Man Ther. 2004;9:71–6.
63. Bogduk N. Practice guidelines for spinal diagnostic and treatment procedures. 2nd ed. San Francisco: International Spine Intervention Society (ISIS); 2013.
64. Sluijter ME. The medial branch procedure. In: Radiofrequency, part 2. Meggen (LU): Flivo-Press SA; 2003. p. 99–110.
65. Wedel DJ, Wilson PR. Cervical facet arthrography. Reg Anesth. 1985;10:7–11.
66. Lord S, Barnsley L, Wallis B, et al. Percutaneous radio-frequency neurotomy for chronic cervical zygapophyseal-joint pain. N Engl J Med. 1996;335:1721–6.
67. Rathmell JP. Facet injection: intraarticular injection, medial branch block, and radiofrequency treatment. In: Atlas of image-guided intervention in regional anesthesia and pain medicine. Philadelphia, PA: Lippincott Williams & Wilkins; 2006. p. 65–92.
68. Kaye AD, Manchikanti L, Novitch MB, et al. Responsible, safe, and effective use of antithrombotics and anticoagulants in patients undergoing interventional techniques: American Society of Interventional Pain Physicians (ASIPP) guidelines. Pain Physician. 2019;22:S75–S128.
69. Manchikanti L, Sanapati M, Kaye AD, Kaye AM. Antiplatelet and anticoagulant therapy in regenerative medicine. In: Manchikanti L, Navani A, Atluri S, Sanapati M, editors. Manchikanti's essentials of regenerative medicine in interventional pain management. 2nd ed. New York, NY: Springer; 2023.

# Thoracic Facet Joints and Intraarticular Injection of Biologics

27

Laxmaiah Manchikanti, Annu Navani, Mahendra Sanapati, and Vijay Singh

## Introduction

The mid back or upper back pain from the thoracic spine is considered the least common of chronic spinal problems with a prevalence of 13% compared to 32% for neck pain and 43% for low back pain [1–3]. However, thoracic pain can be as disabling as neck or low back pain, which are considered to be the number 2 and number 1 leading causes of disability [4–8]. Multiple structures of thoracic spine, including intervertebral discs, facet joints, ligaments, fascia, muscles, and nerve root dura, have been shown to be capable of transmitting pain resulting in symptoms of mid back and upper back pain with radiation into the chest wall or abdominal wall [9–14].

Persistent thoracic pain in the mid back or upper back secondary to the involvement of thoracic facet joints has been established to range from 34% to 48%, with a false-positive rate ranging from 42% to 58% using controlled diagnostic medial branch blocks [10–17]. Chronic thoracic facet joint pain has been treated with either intraarticular facet joint injections or nerve blocks or radiofrequency neurotomy [11–14]. Platelet-rich plasma (PRP) and medicinal signaling cells or mesenchymal stem cells (MSCs) appear to have significant potential for their usage in managing spinal pain of discogenic origin, facet joint origin, and ligamentous origin. However, there is no available literature in reference to injection of biologics in the thoracic spine. However, there is significant emerging literature in the lumbar spine in managing discogenic pain, and to some extent, facet joint pain [18–21].

L. Manchikanti (✉)
Pain Management Centers of America, Paducah, KY, USA
e-mail: drlm@thepainmd.com

A. Navani
LeReve Regenerative Wellness, Campbell, CA, USA

Boomerang Healthcare, Walnut Creek, CA, USA

Stanford University School of Medicine, Stanford, CA, USA
e-mail: anavani@boomeranghc.com

M. Sanapati
Pain Management Centers of America, Evansville, IN, USA

V. Singh
Spine Pain Diagnostics Associates, Niagara, WI, USA
e-mail: vj@secure.paindiagnostics.net

## History

Descriptions and studies of thoracic facet joints in chronic upper or mid back pain are relatively recent with thoracic facet syndrome described first in 1987 [22], with its characterization in 1991 [23]. Dreyfuss et al. in 1994 [24] and Fukui et al. in 1997 [25] described thoracic zygapophysial joint pain patterns. Thoracic intraarticular facet joint blocks, essentially diagnostic and therapeutic interventions including thoracic medial branch blocks and radiofrequency thermoneurolysis were developed as analogues to cervical and lumbar interventions.

## Pathophysiology

Thoracic facet joints are an established source of mid back, upper back, and referred pain into the chest wall and abdominal wall in patients with chronic thoracic pain [10–17]. Facet joints in the thoracic spine are well innervated by the medial branches of the dorsal rami [10–17, 26–30]. Neuroanatomic, neurophysiologic, and biomechanical studies have demonstrated free and encapsulated nerve endings in the lumbar and cervical facet joints along with the nerves containing substance P and calcitonin gene-related peptide [10, 11].

Multiple factors including mechanical injury and inflammation of facet joints have been shown to produce persistent pain in otherwise normal rats [31, 32]. Mechanical injury of the facet joint increases cytokine messenger mRNA in the dorsal root ganglion (DRG); an intraarticular injection of a


A. Navani et al. (eds.), *Essentials of Regenerative Medicine in Interventional Pain Management*,
https://doi.org/10.1007/978-3-031-50357-3_27

nonsteroidal anti-inflammatory agent alleviates injury-induced pain in the same model [32], suggesting that inflammation has a role in the pain response after mechanical joint insult. Inflammatory mediators, including cytokines, prostaglandins, and neuropeptides, have been shown to increase within the facet joint and the DRG as a result of joint inflammation and arthritis [33, 34].

The role of facet joint arthritis as a causative factor has been studied in spinal pain. It has described arthritis to denote joint inflammation, whereas osteoarthritis to indicate arthritis characterized by erosions of articular cartilage which mainly affect weight bearing joints. In fact, the classic radiographic hallmarks of facet joint osteoarthritis involve both degenerative and proliferative features, including narrowing of the facet joint space, subarticular bone erosions, subchondral cysts, osteophyte formation, and hypertrophy of the articular process [35]. Thus, facet joint osteoarthritis is a clinical and pathological construct that involves the functional failure of the synovial facet joints. The thoracic facet joints can be affected by osteoarthritis, rheumatoid arthritis, and spondylitis. Disc degeneration is assumed to lead to associated facet joint degeneration and subsequent spinal pain. In the majority of patients, facet joint pain has been shown to be independent of facet joint arthritis.

## Regenerative Medicine

The biologic primarily referred to with regards to spinal facet joints is PRP and MSCs. PRP includes concentration of growth factors that are secreted by a platelet in an inflammatory environment, i.e., platelet-derived growth factors [18, 19, 21, 36, 37]. These growth factors are essential to the healing process, as they increase fibroblast or osteoblast metabolic activity while reducing cell apoptosis promoting angiogenesis, thereby increasing blood flow and circulation to the new-forming tissues [34, 35]. These processes increase the expression of the pro-collagen gene and collagen-derived growth factors, which increase the tensile strength of the new tissue [36–38]. Proteins present in PRP that are proliferative in nature include endothelial growth factor, platelet-derived growth factor, vascular endothelial growth factor, and basic fibroblast growth factor (b-FGF). PRP is prepared by centrifugation of whole blood to precipitate predictable components of the blood. There is only an acidifying agent and a platelet anticoagulant added for stabilization [39].

The other biologics that may be relevant in this regard are medicinal signaling cells or mesenchymal stem cells (MSCs), and other commercially available biologic preparations. Despite lack of published studies supporting their use for facet joint pain, Kristjánsson and colleagues [40] reported the MSCs harvested from the facet joints and interspinous ligaments of ten patients had the tri-lineage potential to be differentiated into osteogenic, adipogenic, and chondrogenic cells under appropriate inductions. Consequently, they hypothesized that the facet joints and interspinous ligaments may provide alternative sources of MSCs for tissue engineering applications. Preparation of bone marrow concentrate and stromal vascular fraction have been well described [18, 19, 21, 41, 42].

## Evidence Base

The literature supporting the use of biologics for thoracic facet joint is not yet available. Similarly, therapeutic facet joint interventions available including intraarticular facet joint injections, facet joint nerve blocks, and radiofrequency neurotomy often suffer from the paucity of literature. The effectiveness of intraarticular injections in the thoracic spine is limited, whereas, effectiveness of medial branch blocks is Level 2 or moderate [9, 11, 43]. Further, the literature continues to be limited. Radiofrequency neurotomy also provides limited evidence, once again with a paucity of literature [9, 11]. Consequently, regenerative medicine with intraarticular injections or even medial branch blocks is attractive in managing chronic thoracic pain.

## Clinical Presentation

It is general consensus that conventional clinical features are unreliable in diagnosing thoracic facet joint pain [9–14]. In addition, the features of somatic or referred pain secondary to facet joints and radicular pain secondary to disc pathology may not be able to distinguish facet joint pain from radicular pain as shown in Fig. 27.1, the diagrams of facet joint pain may mimic discogenic pain and/or disc herniation.

Published literature of radiological investigations report no correlation between the clinical symptoms of thoracic pain and degenerative spinal changes observed on radiological imaging studies as in cervical and lumbar spine [9–14]. Plain radiographs, magnetic resonance imaging, computed tomography scanning, single photon emission computed tomography, and radionuclide bone scanning have not identified patients with facet joint pain based on degenerative changes or other features. The association between degenerative changes in thoracic facet joints and symptomatic mid and upper back pain remains unclear.

Thus, without a gold standard for diagnosing thoracic pain, the recommended reference standards typically involve anesthetic or provocative injections, regardless of whether the suspected source is the facet joint or intervertebral disc. Significant false-positive rates (42–48%) with single facet joint blocks, with a prevalence of 34–48%, have been documented with controlled diagnostic blocks in the thoracic spine [9–17].

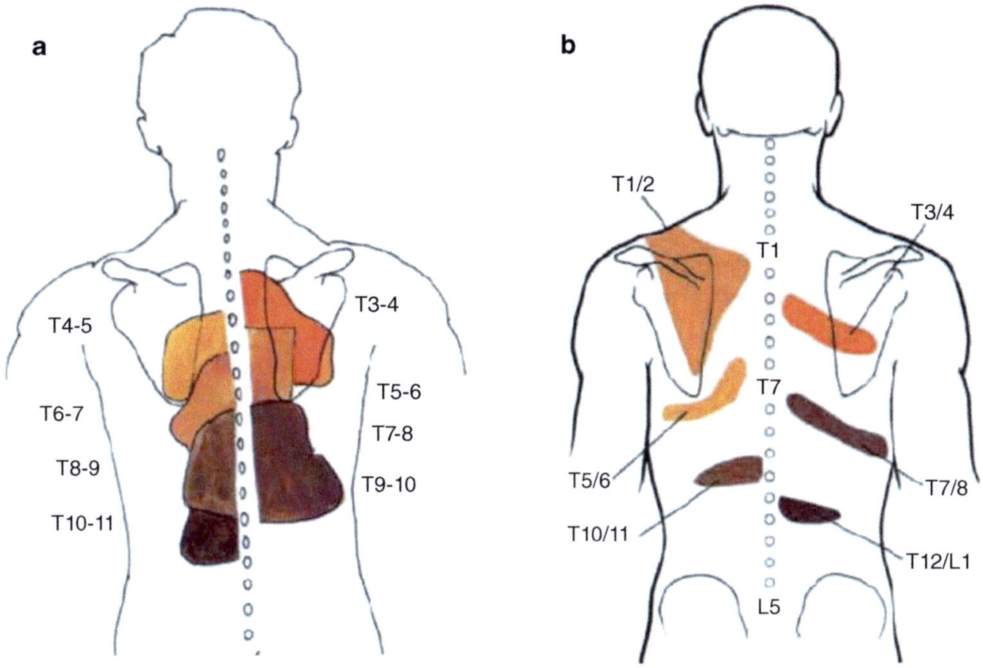


**Fig. 27.1** (a-b) Thoracic facet joint referral pain patterns. (A. Adapted from Dreyfus et al. [25]) (B. Adapted from Fukui et al. [24])

## Anatomy

Even though the anatomy of the thoracic spine has distinct features and marked differences, it has many features in common with the lumbar spine. Each thoracic vertebra is distinguished by costal facets on the side of the body and, in all but the last two or three segments, by articular facets on the transverse processes, articulating respectively with the heads and tubercles of the ribs [44]. The laminae are short, thick, broad, and overlap each other like roof tiles from above downwards. In addition, the typical spinous process slants back and downwards, overlapping between T5 and T8 and being less oblique above and below this level, projecting upwards at the junction of the laminae and pedicles are the superior articulating processes, facing backwards, upwards, and laterally. The articular processes of adjoining thoracic vertebrae form synovial joints that permit a limited degree of movement, primarily facilitating axial rotation [44]. The primary axis of rotation for the mid-thoracic segments is located close to the anterior region of the thoracic intervertebral disc.

The lower thoracic facet joints provide specialized features characteristic of this region. The first is the greater development of the mamillary processes, which originate as extensions of the superior articular processes [44]. The mamillary processes provide attachment for the multifidus muscle as it traverses medially to attach to the spinous processes of the segments above [44].

The anatomic features of thoracic vertebrae are illustrated in Fig. 27.2. The articular surface of thoracic facet joints is inclined 60° from the horizontal to the frontal plane and rotated 20° from the frontal to the sagittal plane in a medial direction. The lateral aspect of a thoracic facet joint is anterior; the medial aspect of the joint is posterior. In addition, the superior articular facet from the inferior vertebrae is almost flat and faces posterior, superior, and slightly lateral, whereas the inferior articular facet is oriented in a reciprocal manner. Further, there is also variation in the inclination of the joints by region, with the mid-thoracic level approximately 60° off the horizontal plane, while the upper segments achieve a more vertical orientation. In contrast to the lumbar spine, the superior and inferior articular processes of the thoracic spine cannot be identified separately because the joints are mostly in the frontal plane and do not face sideward [45]. The articular pillar is wider in the thoracic region than in the lumbar region and extends farther laterally in relation to the vertebral bodies.

The articular facets are covered by articular cartilage; a synovial membrane bridges the margins of the articular cartilage of the two facets in each joint. However, ribs may make fluoroscopic visualization more difficult. Finally, differences in the thoracic spine are related to the anatomy of the pedicles pointing straight backward and considerably upward, compared to those in the lumbar region which point slightly laterally, recognizable as a circular structure forming the eye of the "Scotty dog" configuration [45].

As facet joints in the lumbar and cervical spine, thoracic facet joints exhibit features typical of a synovial joint. Present on the posterolateral aspect of the facet joint is a tough fibrous capsule composed of several layers of fibrous tissue and a synovial membrane separated by a layer of loose alveolar tissue. However, there is no fibrous capsule on the ventral aspect of the joints. Instead, in its place, the ligamentum flavum is in

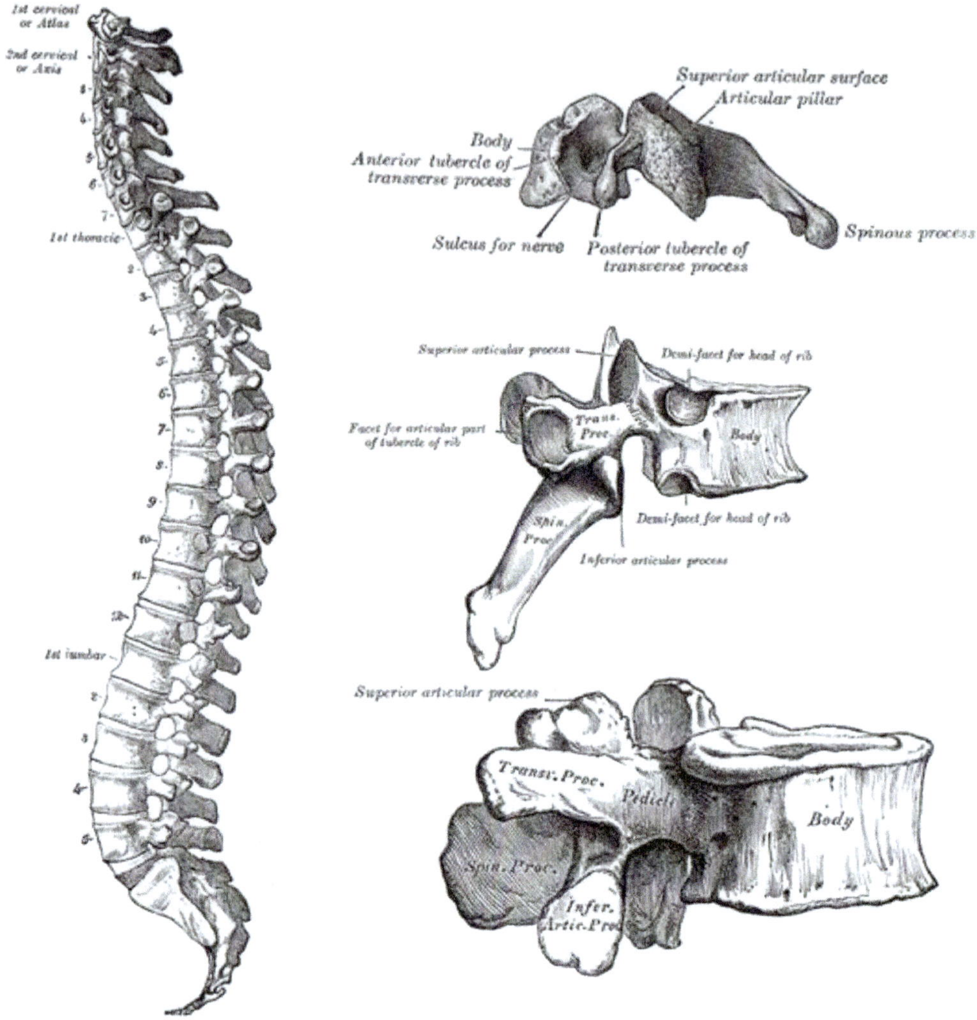


**Fig. 27.2** Illustrative anatomy of thoracic vertebrae

direct contact with the synovial membrane. Facet joints appear to be anatomically designed to restrain excessive mobility and distribute axial loading over a broad area.

Spinous processes T1–T3 and T11–T12 are located approximately at the level of the facet joint of the adjoining level and the one below; the T2 spinous process relates to the T2/3 facet joint [46–49]. Spinous processes T4–T10 are located halfway between the adjoining facet joint and the joint below; the T6 spinous process lies halfway between facet joints T6/7 and T7/8.

The thoracic facet joints are well innervated by the medial branches of the dorsal rami [26–30]. The fibrosus capsule is richly innervated by mechanoreceptors and nociceptors. Analogous to the innervation of the cervical and lumbar facet joints, the thoracic facet joints receive bi-segmental innervation from the medial branches of the dorsal ramus of the upper segment and one more cephalad level [26–30, 50–52]. Chua and Bogduk [30] described the surgical anatomy of thoracic facet denervation, whereas Ishizuka et al. [50] described the topographic anatomy of the posterior ramus of the thoracic spinal nerve and surrounding structures. The medial branches of the thoracic dorsal rami assume a reasonably constant course except at midthoracic levels from T5 to T8 as shown in Figs. 27.3 and 27.4 [30, 52]. The medial branches of the thoracic dorsal rami at midthoracic levels do not run on bone. Instead, they are suspended in the intertransverse space. Thoracic medial branches are not very close to the facet joint as they swing laterally to circumvent the multifidus muscle (Fig. 27.3). The dorsal ramus is separated from the ventral ramus (i.e., the intercostal nerve) by the anterior part of the superior costotransverse ligament. The archetypical course of the medial branches is consistently exhibited by nerves at the upper thoracic levels T1–T4 and lower levels T9–T10. The course of the T11 and T12 medial branches differ because of the different osseous anatomy. The T12 transverse process is much shorter than the typical transverse process. Consequently, the T11 medial branch runs across the lateral surface of the root of the superior articular process of T12 (Figs. 27.3 and 27.4). The T12 medial branch assumes a course analogous to that of lumbar medial branches crossing the junction of the superior articular process and the base of the transverse process.

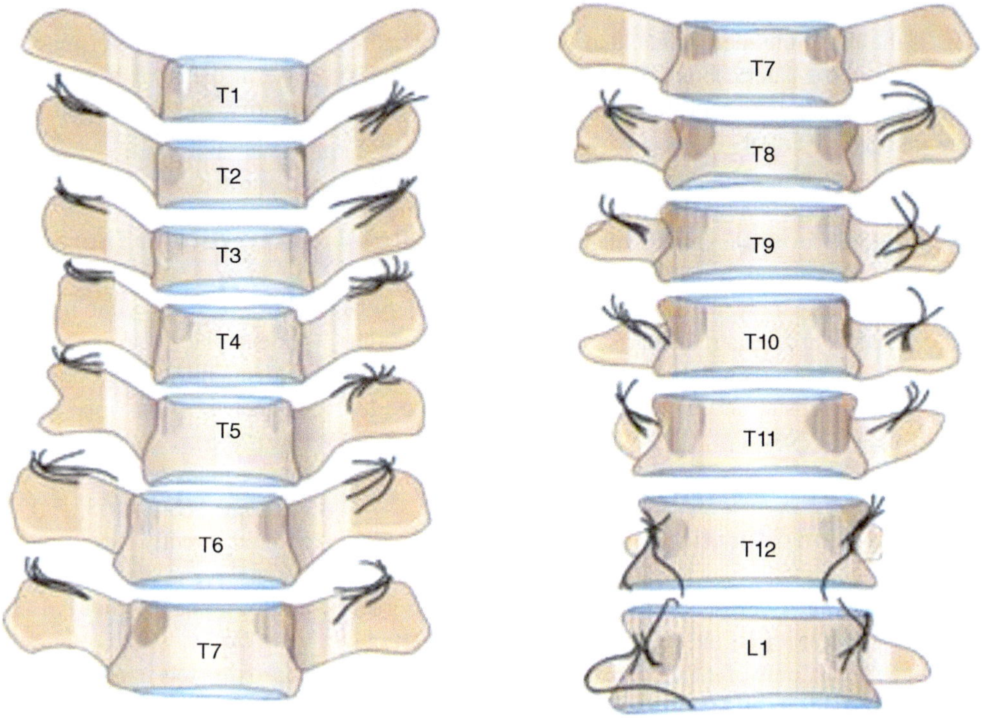


**Fig. 27.3** A composite sketch of a radiograph of cadaveric thoracic spines in which the medial branches of the thoracic dorsal rami were marked with wires in order to depict their relationship to the transverse processes. Note how at mid-thoracic levels, the nerves are suspended in the intertransverse space, whereas at other levels they cross, and lie on, the transverse process. (Adapted from: Chua WH, Bogduk N. The surgical anatomy of thoracic facet denervation. *Acta Neurochir* 1995; 136:140–144 [30])

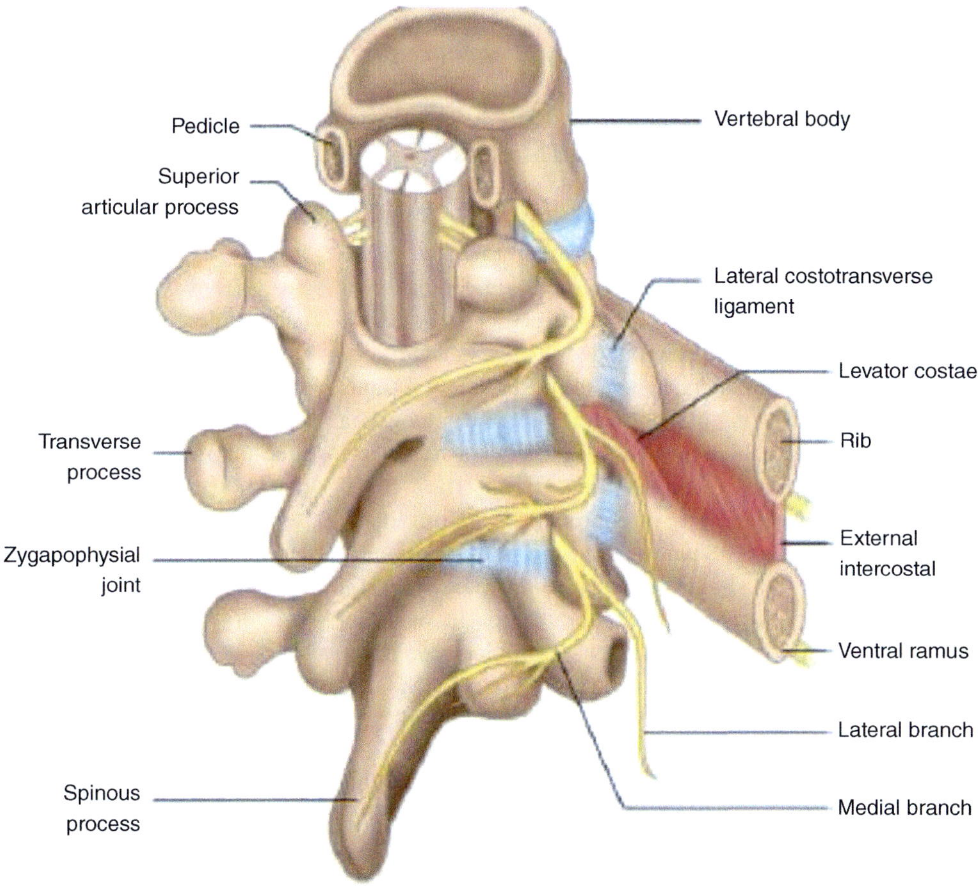


**Fig. 27.4** A sketch of the archetypical course and relations of the thoracic dorsal rami viewed from a right superior aspect. (Adapted from: Chua WH, Bogduk N. The surgical anatomy of thoracic facet denervation. *Acta Neurochir* 1995; 136:140–144 [30])

The medial branches supply the adjacent intrinsic muscles of the back, the facet joints, and skin in the upper thoracic region; the lateral branches supply intrinsic muscles, costotransverse joints, and the skin in the lower thoracic region [30, 44, 45, 50].

The arterial blood supply to the thoracic spinal cord is derived from large radicular arteries [46–49]. At the mid and lower thoracic levels, large radicular arteries with a diameter >400 mm originate from spinal branches of the intercostal arteries and supply the spinal cord [47]. The artery of Adamkiewicz, or arteria radicularis magna, the largest radicular artery, originates in the majority of cases (>80%) from the lower intercostal and upper lumbar arteries on the left side (T9-L2) [49].

## Technical Considerations

Radiographic and fluoroscopic understanding of the thoracic spine is essential for all interventions. At first glance, it may look like the anatomy of the thoracic spine has many features in common with the lumbar region [45]. However, the ribs may make radiology readings somewhat more difficult, even though the thoracic vertebrae have well-defined transverse and articular processes and laminae like a lumbar vertebra. In addition, there are several marked differences that make working in the thoracic region more difficult, compared to the lumbar spine. In the lumbar region, the pedicles are very important structures since they serve as a beacon during fluoroscopy [45] and are easily recognizable as a roughly circular, typical structure, forming the eye of the Scotty dog. In the thoracic region, the pedicles point straight backward and considerably upward (Fig. 27.2), a direction far different from the direction of the X-rays, with the pedicles not visible on many patients' X-rays. The rounded shadows that are visible close to the upper endplates are the costovertebral joints . Further, the transverse process in the thoracic region is a true transverse process, in contrast to the lumbar transverse process, which in an evolutionary sense is mostly a costal element. The thoracic transverse process is a formidable structure, with a broad connection to the inferior articular process pointing in a slightly cephalad and markedly posterior direction, which is important for the image when an oblique projection is used. Obliqueness provides for better visibility of the contralateral transverse process, whereas, ipsilateral transverse processes are more difficult to identify since they are more or less in tunnel vision [45].

The anatomy of the facet joints is quite different from the lumbar region because of the inability to identify superior and inferior articular processes separately because the joints are mostly in the frontal plane and do not face sideward. The articular pillar is wider in the thoracic region and extends farther laterally in relation to the vertebral bodies. In addition, the spinous processes in the thoracic region also are not suitable to determine the degree of obliquity, as is the case in the lumbar region. They are sometimes difficult to identify and they often deviate, to some degree, from the median line [45].

## Intraarticular Thoracic Facet Joint Injections

Thoracic facet joint interventions are performed with the patient placed prone. Thoracic facet joints are not clearly evident in an anteroposterior (AP) view. Their location is estimated from the location of the thoracic pedicles (Fig. 27.5). The target joint can be gauged to lie between the two pedicles bearing the same segmental number as the target joint [12–14, 53, 54]. Thoracic facet joints cannot be entered directly from behind due to their orientation in the coronal plane, and they cannot be entered laterally due to the risk of a pneumothorax [12–14, 53, 54]. The safest entry is from below, which requires an initial insertion of the needle approximately 1–2 segments below the target joint.

With intermittent fluoroscopic visualization, the needle is inserted through the skin cephalically, pointing toward the superior articular process. The needle should remain on an imaginary vertical line connecting the mid portion of the targeted facet joint and the one below. If the needle stays in this line without deviation, either medial or lateral, it is safe. However, risks with deviation of the needle include entering the epidural space and spinal cord, or the pleural space and lung. After inserting the needle approximately 4–5 cm, or once the needle tip is seen to lie at the mid to inferior aspect of the pedicle, the fluoroscope is rotated away from the side being injected until the outline of the joint is clearly visible (almost a lateral position). Following this, the needle is advanced through the capsule into the inferior aspect of the joint (Fig. 27.5). After the needle is inserted into the inferior aspect of the joint, contrast medium is injected. After confirmation, steroid and/or local anesthetic may be injected.

For superior levels from T1/2 to T5/6, a more perpendicular approach to the facet joints is needed. At these levels on AP imaging, skin entry is usually at the mid portion of the vertebral body rather than at its inferior aspect.

### Medial Branch Blocks

To block the nerves supplying a thoracic facet joint, two medial branches must be blocked due to the dual innervation of each facet joint. Problems related to a medial branch procedure arise not so much from the osseous anatomy, but from the innervation [45]. It was originally thought that in the thoracic region, placement of the electrode for radiofrequency neurotomy would be best performed analogous to the lumbar region, with the needle tip ending at the junction between the

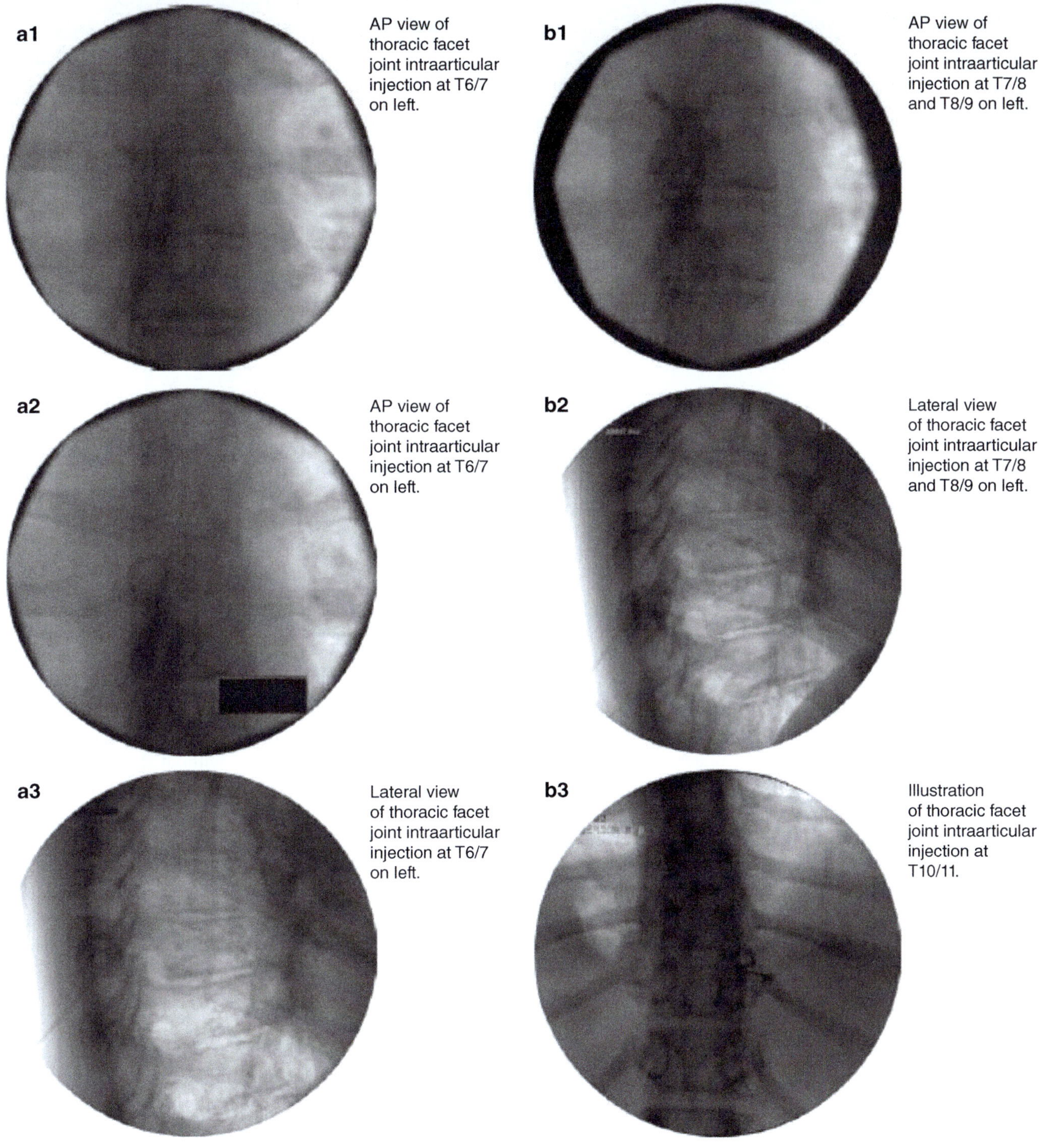


**Fig. 27.5** Illustration of thoracic facet joint intraarticular injections at multiple levels. (**a1**) AP view of thoracic facet joint intraarticular injection at T6/7 on left. (**a2**) AP view of thoracic facet joint intraarticular injection at T6/7 on left. (**a3**) Lateral view of thoracic facet joint intraarticular injection at T6/7 on left. (**b1**) AP view of thoracic facet joint intraarticular injection at T7/8 and T8/9 on left. (**b2**) Lateral view of thoracic facet joint intraarticular injection at T7/8 and T8/9 on left. (**b3**) Illustration of thoracic facet joint intraarticular injection at T10/11

superior articular and transverse processes. The optimal location for a medial branch procedure in the thoracic region would be the superolateral corner of the transverse process (Fig. 27.3).

Bogduk [12–14, 53, 54] articulated that for needle placement from T1 to T3 and T8 to T10, the needle is positioned directly overlying the target point of the nerve. The needle must be advanced until contact is made with the target trans-

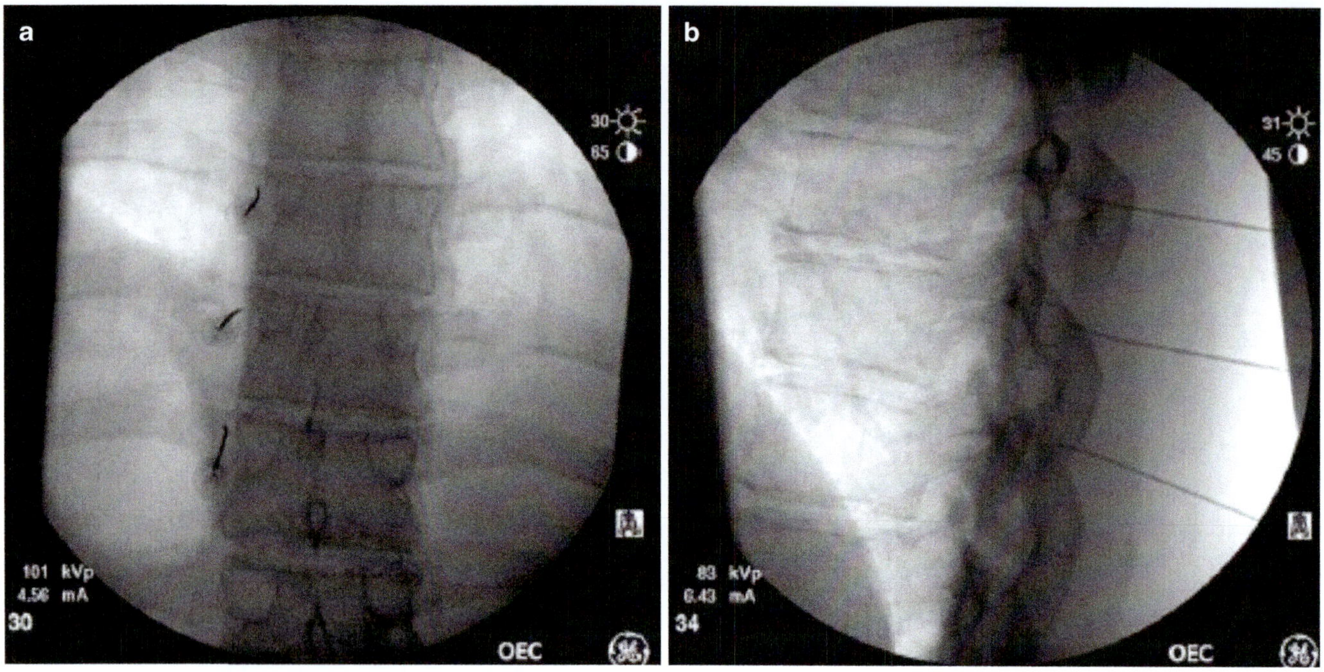


**Fig. 27.6** (**a**, **b**) Needle placement covering medial branches at left T9, T10, and T11 levels for medial branch blocks

verse process, with subsequent adjustment of the needle to rest on the back of the superolateral corner of the transverse process.

For medial branch blocks from T4 to T8, Bogduk [54] described to aim to position the needle onto the nerve that passes dorsally and caudally just above and slightly dorsal to the typical target joint on the superolateral corner of the transverse process. The depth of the needle placement is the same as the depth of the transverse process.

Medial branch blocks of T11 and T12 are performed similarly to medial branch blocks of the lumbar spine [53, 54]. The target location for the T11 and T12 medial branches is at the junction of the superior articular process and the transverse process that the nerve crosses, midway between the transverse process and the location of the mamilloaccessory notch. On AP imaging, the tip of the needle should at least be in line with the lateral margin of the silhouette of the superior articular process and, if possible, medial to this margin. After confirming the proper positioning of the needle, contrast medium and drugs may be injected. Figure 27.6 shows placement of needles for medial branches.

## Indications

Indications for diagnostic thoracic facet joint nerve blocks are somatic or nonradicular mid back or upper back pain with or without chest wall pain of at least 3-months duration. In addition, pain must be causing functional disability and has failed to respond to more conservative management, including nonsteroidal anti-inflammatory agents and physical therapy exercises and/or chiropractic treatment.

Therapeutic modalities are applied based on the diagnosis established with a positive response to controlled diagnostic blocks, either placebo or comparative local anesthetic blocks, with a criterion standard of 80% pain relief and the ability to perform prior painful movements without significant pain.

## Side Effects and Complications

Complications from facet joint nerve blocks, intraarticular injections, or radiofrequency thermoneurolysis in the thoracic spine are rare [10–17]. The most common complications of thoracic facet joint interventions are related to the placement of the needle and to the administration of various drugs. Serious complications may include dural puncture, spinal cord trauma, pneumothorax, subdural injection, neural trauma, injection into the intervertebral foramen, and hematoma formation; infectious complications including epidural abscess and bacterial meningitis; and side effects related to the administration of steroids, local anesthetics, and other drugs.

## Precautions

Relative contraindications to interventional techniques have been described in patients receiving treatment with antithrombotics and anticoagulants [9, 55]. Guidelines on anticoagulants and antithrombotics are described in Chap. 15 [56] and also on the ASIPP app, which is regularly updated.

## Conclusion

Thoracic facet joints have been shown to be capable of being a source of pain in the thoracic spine and referred pain in the chest wall. They are well innervated by the medial branches of the dorsal rami. Based on responses to controlled diagnostic blocks of thoracic facet joints, in accordance with the criteria established by the International Association for the Study of Pain (IASP), the prevalence of thoracic facet joint pain has been determined to be 34–48%. Further, to maintain the validity of diagnostic blocks, either comparative local anesthetic blocks or placebo-controlled blocks must be performed because single blocks carry a false-positive rate of 42–58%.

Multiple modalities are available for managing thoracic facet joint pain, including intraarticular injections, medial branch blocks, and radiofrequency neurotomy. However, there are no reports of application of biologics in managing thoracic facet joint pain. Even then, the evidence is limited for traditional management modalities of thoracic facet joint pain. Consequently, the application of biologics in managing thoracic facet joint pain, along with discogenic pain, as well as pain of soft tissue origin is evolving.

**Key Points**

1. Thoracic spine pain is less common than neck or low back pain but can be equally disabling. Multiple structures in the thoracic spine can transmit pain, with thoracic facet joints being a significant source of chronic pain. Treatments such as facet joint injections and radiofrequency neurotomy are used, but the use of biologics like PRP and MSCs in the thoracic spine requires further investigation.
2. Thoracic facet joints are a known source of chronic mid back, upper back, and referred pain to the chest and abdomen. Inflammation and mechanical injury of these joints play a role in persistent pain, with inflammatory mediators and cytokines implicated in the pain response. While facet joint osteoarthritis can occur in the thoracic spine, facet joint pain is often independent of arthritis in most patients.
3. The thoracic facet joints are well innervated by the medial branches of the dorsal rami, receiving bi-segmental innervation from the upper and one more cephalad level. The arterial blood supply to the thoracic spinal cord is derived from large radicular arteries, with the artery of Adamkiewicz being the largest radicular artery supplying the majority of cases on the left side.
4. Diagnosing thoracic facet joint pain is challenging, as conventional clinical features and radiological imaging do not reliably correlate with the presence of facet joint pain. Anesthetic or provocative injections are recommended as the reference standard for diagnosis, despite significant false-positive rates with single facet joint blocks.
5. Platelet-rich plasma (PRP) is the primary biologic used for spinal facet joints, containing growth factors that promote tissue healing and angiogenesis. Mesenchymal stem cells (MSCs) for facet joint pain, also show potential for tissue engineering applications.
6. Understanding the radiographic and fluoroscopic anatomy of the thoracic spine is essential for interventions. While there are similarities with the lumbar region, the presence of ribs makes radiology readings more challenging. The orientation of pedicles, articular facets, and transverse processes in the thoracic spine differs from the lumbar spine, requiring specific approaches for facet joint injections and medial branch blocks.
7. Careful needle placement is crucial to avoid complications and ensure accurate targeting of the nerves and facet joints.
8. Complications from thoracic facet joint interventions, such as nerve blocks, intraarticular injections, or radiofrequency thermoneurolysis, are rare. Potential complications include needle placement issues, drug-related side effects, infectious complications, and serious events like dural puncture, spinal cord trauma, and pneumothorax.

**Acknowledgments** This book chapter is modified and updated from a previous book chapter, "Thoracic Facet Joints and Regenerative Medicine" by Laxmaiah Manchikanti, MD, Annu Navani, MD, Frank J.E. Falce, MD, David M. Schultz, MD, and Vijay Singh, MD, in *Essentials of Regenerative Medicine in Interventional Pain Management* published by ASIPP Publishing.

## References

1. Manchikanti L, Navani A, Falco FJE, Schultz DM, Singh V. Thoracic facet joints and regenerative medicine. In: Manchikanti L, Navani A, Atluri S, editors. Essentials of regenerative medicine in interventional pain management. Paducah, KY: ASIPP Publishing; 2019. p. 495–508.
2. Leboeuf-Yde C, Fejer R, Nielsen J, et al. Pain in the three spinal regions: the same disorder? Data from a population-based sample of 34,902 Danish adults. Chiropr Man Therap. 2012;20:11.
3. Leboeuf-Yde C, Nielsen J, Kyvik KO, et al. Pain in the lumbar, thoracic or cervical regions: do age or gender matter? A population-based study of 34,902 Danish twins 20–71 years of age. BMC Musculoskelet Disord. 2009;10:39.
4. US Burden of Disease Collaborators. The state of US health, 1990-2016: burden of diseases, injuries, and risk factors among US states. JAMA. 2018;319:1444–72.

5. Hoy DG, Protani M, De R, Buchbinder R. The epidemiology of neck pain. Best Pract Res Clin Rheumatol. 2010;24:783–92.
6. Hoy D, Brooks P, Blyth F, Buchbinder R. The epidemiology of low back pain. Best Pract Res Clin Rheumatol. 2010;24:769–81.
7. Dieleman JL, Baral R, Birger M, et al. US spending on personal health care and public health, 1996-2013. JAMA. 2016;316:2627–46.
8. Dieleman JL, Cao J, Chapin A, et al. US health care spending by payer and health condition, 1996-2016. JAMA. 2020;323:863–84.
9. Manchikanti L, Kaye AD, Soin A, et al. Comprehensive evidence-based guidelines for facet joint interventions in the management of chronic spinal pain: American Society of Interventional Pain Physicians (ASIPP) guidelines. Pain Physician. 2020;23:S1–S127.
10. Boswell MV, Manchikanti L, Kaye AD, et al. A best-evidence systematic appraisal of the diagnostic accuracy and utility of facet (zygapophysial) joint injections in chronic spinal pain. Pain Physician. 2015;18:E497–533.
11. Manchikanti L, Kaye AD, Boswell MV, et al. A systematic review and best evidence synthesis of the effectiveness of therapeutic facet joint interventions in managing chronic spinal pain. Pain Physician. 2015;18:E535–82.
12. Manchikanti L, Schultz DM, Falco FJE. Thoracic facet joint interventions. In: Manchikanti L, Singh V, Falco FJE, Kaye AD, Soin A, Hirsch JA, editors. Manchikanti's essentials of interventional techniques in managing chronic pain. 2nd ed. New York, NY: Springer; 2023.
13. Manchikanti L, Schultz DM, Benyamin RM, Falco FJE. Thoracic facet joint interventions. In: Manchikanti L, Kaye AD, Falco FJE, Hirsch JA, editors. Essentials of interventional techniques in managing chronic spinal pain. New York, NY: Springer; 2018. p. 369–85.
14. Manchikanti L, Schultz DM, Falco FJE, Singh V. Thoracic facet joint interventions. In: Manchikanti L, Singh V, editors. Interventional techniques in chronic spinal pain. Paducah, KY: ASIPP Publishing; 2007. p. 277–94.
15. Manchikanti L, Singh V, Pampati VS, et al. Evaluation of the prevalence of facet joint pain in chronic thoracic pain. Pain Physician. 2002;5:354–9.
16. Manchikanti L, Boswell MV, Singh V, et al. Prevalence of facet joint pain in chronic spinal pain of cervical, thoracic, and lumbar regions. BMC Musculoskelet Disord. 2004;5:15.
17. Manchukonda R, Manchikanti KN, Cash KA, et al. Facet joint pain in chronic spinal pain: an evaluation of prevalence and false-positive rate of diagnostic blocks. J Spinal Disord Tech. 2007;20:539–45.
18. Manchikanti L, Navani A, Sanapati M. Evolution of regenerative medicine in managing musculoskeletal and spinal disorders. In: Manchikanti L, Navani A, Atluri S, Sanapati M, editors. Manchikanti's essentials of regenerative medicine in interventional pain management. 2nd ed. New York, NY: Springer; 2023.
19. Navani A, Manchikanti L, Albers SL, et al. Responsible, safe, and effective use of biologics in the management of low back pain: American Society of Interventional Pain Physicians (ASIPP) guidelines. Pain Physician. 2019;22:S1–S74.
20. Manchikanti L, Sanapati M, Navani A, Singh V, Falco FJE, Hirsch JA. Intradiscal injection of biologics. In: Manchikanti L, Navani A, Atluri S, Sanapati M, editors. Manchikanti's essentials of regenerative medicine in interventional pain management. 2nd ed. New York, NY: Springer; 2023.
21. Navani A, Manchikanti L, Sanapati M. Lumbar facet joints and intraarticular injection of biologics. In: Manchikanti L, Navani A, Atluri S, Sanapati M, editors. Manchikanti's essentials of regenerative medicine in interventional pain management. 2nd ed. New York, NY: Springer; 2023.
22. Wilson PR. Thoracic facet joint syndrome—a clinical entity? Pain Suppl. 1987;4:S87.
23. Skubic JW, Kostuik JP. Thoracic pain syndromes and thoracic disc herniation. In: Frymoyer JW, editor. The adult spine. New York, NY: Raven Press; 1991. p. 1443–61.
24. Dreyfuss P, Tibiletti C, Dreyer SJ. Thoracic zygapophyseal joint pain patterns: a study in normal volunteers. Spine (Phila Pa 1976). 1994;19:807–11.
25. Fukui S, Ohseto K, Shiotani M. Patterns of pain induced by distending the thoracic zygapophyseal joints. Reg Anesth. 1997;22:332–6.
26. Stolker RJ, Vervest AC, Groen GJ. Percutaneous facet denervation in chronic thoracic spinal pain. Acta Neurochir. 1993;122:82–90.
27. Stolker RJ, Vervest AC, Groen GJ. Parameters in electrode positioning in thoracic percutaneous facet denervation: an anatomical study. Acta Neurochir. 1994;128:32–9.
28. Stolker RJ, Vervest AC, Groen GJ. The treatment of chronic thoracic segmental pain by radiofrequency percutaneous partial rhizotomy. J Neurosurg. 1994;80:986–92.
29. Stolker RJ, Vervest AC, Ramos LM, et al. Electrode positioning in thoracic percutaneous partial rhizotomy: an anatomical study. Pain. 1994;57:241–51.
30. Chua WH, Bogduk N. The surgical anatomy of thoracic facet denervation. Acta Neurochir. 1995;136:140–4.
31. Dong L, Odeleye AO, Jordan-Sciutto KL, et al. Painful facet joint injury induces neuronal stress activation in the DRG: implications for cellular mechanisms of pain. Neurosci Lett. 2008;443:90–4.
32. Lee KE, Davis MB, Winkelstein BA. Capsular ligament involvement in the development of mechanical hyperalgesia after facet joint loading: behavioral and inflammatory outcomes in a rodent model of pain. J Neurotrauma. 2008;25:1383–93.
33. Dong L, Guarino BB, Jordan-Sciutto KL, et al. Activating transcription factor 4, a mediator of the integrated stress response, is increased in the dorsal root ganglia following painful facet joint distraction. Neuroscience. 2011;193:377–86.
34. Quinn JH, Bazan NG. Identification of prostaglandin E2 and leukotriene B4 in the synovial fluid of painful, dysfunctional temporomandibular joints. J Oral Maxillofac Surg. 1990;48:968–71.
35. Gellhorn AC, Katz JN, Suri P. Osteoarthritis of the spine: the facet joints. Nat Rev Rheumatol. 2013;9:216–24.
36. Yoshida R, Cheng M, Murray MM. Increasing platelet concentration in platelet-rich plasma inhibits anterior cruciate ligament cell function in three-dimensional culture. J Orthop Res. 2014;32:291–5.
37. Manchikanti L, Centeno CJ, Atluri S, et al. Bone marrow concentrate (BMC) therapy in musculoskeletal disorders: evidence-based policy position statement of American Society of Interventional Pain Physicians (ASIPP). Pain Physician. 2020;23:E85–E131.
38. Wasterlain AS, Braun HJ, Harris AH, et al. The systemic effects of platelet-rich plasma injection. Am J Sports Med. 2013;41:186–93.
39. Bodor M, Dregalla R, Uribe Y. Platelet rich plasma. In: Manchikanti L, Navani A, editors. Essentials of regenerative medicine in interventional pain management. Paducah, KY: ASIPP Publishing; 2018. p. 67–81.
40. Kristjánsson B, Limthongkul W, Yingsakmongkol W, et al. Isolation and characterization of human mesenchymal stem cells from facet joints and interspinous ligaments. Spine (Phila Pa 1976). 2016;41:E1–7.
41. Karli DC, Sand T. Platelet rich plasma, bone marrow concentrate and adipose derived cells: quantitative analysis and quality control—what does it all mean? In: Manchikanti L, Singh V, Falco FJE, Kaye AD, Soin A, Hirsch JA, editors. Manchikanti's essentials of interventional techniques in managing chronic pain. 2nd ed. New York, NY: Springer; 2023.
42. Sand TT. Adipose tissue-derived stromal vascular fraction. In: Manchikanti L, Singh V, Falco FJE, Kaye AD, Soin A, Hirsch JA, editors. Manchikanti's essentials of interventional techniques in managing chronic pain. 2nd ed. New York, NY: Springer; 2023.

43. Manchikanti L, Singh V, Falco FJE, Cash KA, Pampati V, Fellows B. The role of thoracic medial branch blocks in managing chronic mid and upper back pain: a randomized, double-blind, active-control trial with a 2-year follow-up. Anesthesiol Res Pract. 2012;2012:585806.
44. Singer KP, Goh S. Anatomy of the thoracic spine. In: Giles LGF, Singer KP, editors. Clinical anatomy and management of thoracic spine pain. Oxford: Butterworth-Heinemann; 2000. p. 17–33.
45. Sluijter ME. Thoracic pain and thoracic procedures. In: Radiofrequency part 2: thoracic and cervical region, headache and facial pain. Meggan (LU): FlivoPress SA; 2003. p. 143–58.
46. Piscol K. Die Blutversorgung Des Rückenmarkes Und Ihre Klinische Relevanz. Schriftenreihe der Neurologie. Band 8. Berlin: Springer; 1972. p. 21–57.
47. Gao L, Wang L, Su B, et al. The vascular supply to the spinal cord and its relationship to anterior spine surgical approaches. Spine J. 2013;13:966–73.
48. Tomioka H, Aomi S, Saito S, et al. Scanning the Adamkiewicz artery and collateral supply via dorsal thoracic artery. Ann Thorac Surg. 2012;93:1733.
49. Vrettos XC, Wyke BD. Articular reflexogenic systems in the costovertebral joints. J Bone Joint Surg. 1974;56B:382.
50. Ishizuka K, Sakai H, Tsuzuki N, Nagashima M. Topographic anatomy of the posterior ramus of thoracic spinal nerve and surrounding structures. Spine (Phila Pa 1976). 2012;37:E817–22.
51. Groen GJ, Stolker RJ. Thoracic neural anatomy. In: Giles GF, Singer KP, editors. Clinical anatomy and management of thoracic spine pain. Boston, MA: Butterworth-Heinemann; 2000. p. 114–41.
52. Chua WH. Clinical anatomy of the thoracic dorsal rami. BmedSc thesis. Newcastle: University of Newcastle; 1994.
53. Bogduk N. International spinal injection society guidelines for the performance of spinal injection procedures. Part 1: zygapophyseal joint blocks. Clin J Pain. 1997;13:285–302.
54. Bogduk N. Thoracic medial branch blocks. In: Practice guidelines for spinal diagnostic and treatment procedures. 2nd ed. San Francisco: International Spine Intervention Society (ISIS); 2013. p. 363–72.
55. Kaye AD, Manchikanti LM, Novitch MB, et al. Responsible, safe, and effective use of antithrombotics and anticoagulants in patients undergoing interventional techniques: American Society of Interventional Pain Physicians (ASIPP) guidelines. Pain Physician. 2019;22:S75–S128.
56. Manchikanti L, Sanapati M, Kaye AD, Kaye AM. Antiplatelet and anticoagulant therapy in regenerative medicine. In: Manchikanti L, Navani A, Atluri S, Sanapati M, editors. Manchikanti's essentials of regenerative medicine in interventional pain management. 2nd ed. New York, NY: Springer; 2023.

# Regenerative Medicine for Sacroiliac Joint Dysfunction

28

Joseph A. Cabaret, Laxmaiah Manchikanti, and Aaron K. Calodney

## Introduction

Injection of biologics have been successfully used to treat various painful joints in the extremities and occasionally in the spine. Among the spinal joints, the sacroiliac joint is a common source of low back pain and disability that can be difficult to treat and also may be effectively managed with application of regenerative medicine techniques utilizing injections of intraarticular biologics [1–3]. Even though the evidence based from randomized controlled trials (RCTs) and based on clinical trials is preliminary, it continues to evolve to demonstrate the effectiveness of sacroiliac joint injections of biologics [4–7].

Chronic low back pain with or without lower extremity pain continues to be the number 1 cause of disability among the 30 leading diseases and injuries in the United States with increasing healthcare costs [8–11]. In reference to the expenditures in the United States, Dieleman et al. [10, 11] estimated spending of $134 billion in 2016, with a 53.5% increase from $87.6 billion spent in 2013, in managing low back and neck pain, which essentially accounted for the highest amount of various disease categories.

The intervertebral discs, the zygapophysial (facet) joints, and sacroiliac joints have all been demonstrated, with controlled diagnostic techniques, to be common causes of chronic spinal pain [1–3, 12–15].

J. A. Cabaret
Genesis Pain Specialists, Camarillo, CA, USA

L. Manchikanti (✉)
Pain Management Centers of America, Paducah, KY, USA
e-mail: drlm@thepainmd.com

A. K. Calodney
Clinical Research, Precision Spine Care, Tyler, TX, USA

Department of Anesthesiology, Louisiana State University, Shreveport, LA, USA

## Pathophysiology

Sacroiliac joint dysfunction generally refers to pain in the sacroiliac joint region that is caused by either hypermobility or hypomobility because of weakened, injured, or sprained ligaments [1–3]. While the joint itself may initially appear structurally normal and healthy, abnormal motion can eventually result in sacroiliac joint arthritis (sacroiliitis), often not until after many years. The joint can be slightly subluxed and even locked positions can occur. The muscles can be involved or spasm with a painful and dysfunctional sacroiliac joint. Patients can develop tightness and dysfunction in the hamstring, quadriceps, iliotibial tract, and hip flexors.

Sacroiliac joint pain diagnosed by intra-articular blocks accounts for about 10–27% of chronic non-radicular low back pain [3]. Symptoms are non-specific and can easily be confused with other sources of low back pain so the diagnosis is frequently overlooked. Additionally, pathology often does not show up on X-ray or magnetic resonance imaging (MRI), making detection difficult [16]. Sacroiliac joint pain can be disabling. It is more common in females and pain referral patterns are highly variable [1–3].

Trauma may cause sacroiliac joint pain, usually due to a lifting and twisting injury or fall onto the buttocks. With an injury to the sacroiliac joint, pain tends to be unilateral and can refer to the posterior thigh, iliac fossa, and buttocks. Sprains of the iliolumbar ligaments can also result in referred pain into the groin and genitalia. Stress across the sacroiliac joint following lumbar fusion, especially at the L5-S1 level, appears to be a common cause of sacroiliac joint pain, and may be considered a form of adjacent segment disorder. It has been reported that the sacroiliac joint is a common source of post-fusion low back pain [3, 17, 18]. Other causes and predisposing factors of sacroiliac joint pain include joint hypermobility syndromes (e.g., Ehlers-Danlos syndrome and pregnancy), hypomobility syndromes (advanced age and inflammatory arthritis, such as ankylosing spondylitis, Reiter's syndrome, rheumatoid arthritis, and psoriatic arthri-


A. Navani et al. (eds.), *Essentials of Regenerative Medicine in Interventional Pain Management*,
https://doi.org/10.1007/978-3-031-50357-3_28

tis), leg or muscle length discrepancy, altered gait patterns, repetitive stress (marathoners), scoliosis, lumbosacral transitional vertebrae, and hip arthropathy [1–3, 17, 18]. Hormone imbalances, particularly those associated with pregnancy and the hormone relaxin, can also cause a ligamentous laxity resulting in the weakening of the sacroiliac structure.

## Regenerative Medicine

Most sacroiliac joint patients stabilize with conservative care of 6–8 weeks. This consists of ice, anti-inflammatory medicines, physiotherapy (stabilization and/or manual manipulation), self-mobilization techniques, sacroiliac joint belt or tape (external fixation), psychological counseling, steroid injections, and radiofrequency denervation. Surgical fusion is an option with some favorable 2-year outcomes data for patients who have failed conservative care. The anti-inflammatory effect of injection therapy is often not permanent, and many patients wish to avoid surgical solutions. In these cases, prolotherapy, platelet rich plasma, and stem cell therapies are an option.

Sacroiliac joint pain can be disabling and there are few good solutions if simple therapies don't work. Many patients with confirmed sacroiliac joint pain fail to obtain adequate sustained relief with conservative care or steroid injection therapy and wish to avoid invasive sacroiliac joint fusion surgery due to its less than perfect risk to benefit ratio. Prolotherapy and/or intraarticular injections with platelet rich plasma (PRP) and or stem cells is a relatively new and increasingly popular option. Effective application in patients with sacroiliac joint pain has been documented with benefits lasting for many years. Instead of merely masking symptoms, regenerative medicine treatments aim at healing the underlying pathology that is causing pain and disability. There have been published reports of dramatic results but not all studies support the use of regenerative medicine for musculoskeletal indications. However, the existing body of literature is largely encouraging and merits further investigation [19–21].

The decision whether to use PRP, bone marrow aspirate stem cell concentrate (BMAC), stromal vascular fraction (SVF) or some combination thereof is still open for debate [22–26]. There is little evidence that one is superior to the others or for combining various substrates. Experienced regenerative medicine clinicians often recommend stem cell concentrates when injecting into joints. At this point, this is largely conjectural and related to the belief that stem cell concentrates are better able to regenerate cartilage than platelet concentrates alone. Further high-quality laboratory and clinical studies are required to provide evidence-based answers to these questions. For now, regenerative medicine for sacroiliac joint pain remains a promising option.

The versatility of biologic therapy allows its use in spinal disc degeneration, as well as disc protrusion. The two biologics most commonly utilized and well studied are PRP and mesenchymal stem cells (MSCs) [22]. In the United States, PRP is approved by the Food and Drug Administration (FDA) for use with ligament grafting and bony matrices during reconstructive procedures. This benefit occurs by increased concentration of growth factors that are secreted by a platelet in an inflammatory environment.

PRP is a concentrate of whole blood that is centrifuged to obtain a concentrate of plasma rich in platelets and hence growth factors. PRP provides multiple benefits with disc repair by the increased concentration of growth factors that are secreted by a platelet in an inflammatory environment, i.e., platelet-derived growth factors (PDGF) [22, 24].

These growth factors are essential to the healing process, as they increase fibroblast or osteoblast metabolic activity while reducing cell apoptosis; promote angiogenesis, thereby increasing blood flow and circulation to the new-forming tissues; and increase the expression of the pro-collagen gene and collagen-derived growth factor, which increases the tensile strength of the new tissue [22, 24, 27–34]. Even though components of PRP include platelets, leukocytes, and red blood cells, platelets are central to mediating the anabolic effects of PRP by virtue of releasing growth factors stored in their alpha granules [22, 24]. Notable growth factors released from platelets that are involved in the healing process include PDGF, transforming growth factor-β (TGF-β), vascular endothelial growth factor (VEGF), epidermal growth factor (EGF), basic fibroblast growth factor (bFGF), insulin-like growth factor (IGF-1).

## Evidence-Base

The analysis was conducted using five levels of evidence ranging from strong to opinion- or consensus-based as shown in Table 28.1 [35].

Multiple research activities have intensified in tissue engineering and regenerative medicine [22–26], with multiple preclinical studies demonstrating encouraging results in multiple spinal conditions [22, 23, 36–38].

There have not been many studies assessing the role of PRP or MSCs with sacroiliac joint injections. However, Sanapati et al. [39] performed systematic assessment of the evidence, with identification of one RCT [5] and two observational studies [4, 40]. Since the systematic review by Sanapati et al. [39], a double blind RCT [7] and a prospective nonrandomized study [6] were published. In the RCT, Chen et al. [7] compared 26 patients with a positive diagnostic block of 80% pain relief with intraarticular platelet-rich plasma (PRP) injections with intraarticular steroids. The study showed improvement in both groups;

however, the results in the steroid group was superior to PRP group. In a prospective nonrandomized interventional study, Wallace et al. [6] studied 50 patients with sacroiliac joint dysfunction with injection of PRP under ultrasound guidance. They showed significant improvement at the 6-month follow-up after injection. However, overall improvement tapered off after 4 weeks with no statistically significant reduction from 4 weeks. The results of available studies are shown in Table 28.2. Sanapati et al. [39], based on one high quality RCT [5], one moderate quality observational study [40] and one low quality case report [4], presented the qualitative evidence as Level IV on scale of Level I–V using a qualitative modified approach to grading of evidence based on best evidence synthesis. Addition of these two new studies has not changed the level of evidence [6, 7].

In recent years, there has been significant criticism of the quality parameters of injected PRP, which may contribute for lackluster results [41, 42]. There have not been any studies with injection of bone marrow aspirate.

**Table 28.1** Qualitative modified approach to grading of evidence of therapeutic effectiveness studies

| | | |
|---|---|---|
| Level I | Strong | Evidence obtained from multiple relevant high-quality randomized controlled trials |
| Level II | Moderate | Evidence obtained from at least one relevant high-quality randomized controlled trial or multiple relevant moderate or low-quality randomized controlled trials |
| Level III | Fair | Evidence obtained from at least one relevant moderate or low-quality randomized trial<br>or<br>Evidence obtained from at least one relevant high-quality non-randomized trial or observational study with multiple moderate or low-quality observational studies |
| Level IV | Limited | Evidence obtained from multiple moderate or low-quality relevant observational studies |
| Level V | Consensus based | Opinion or consensus of large group of clinicians and/or scientists |

**Modified from:** Manchikanti L, et al. A modified approach to grading of evidence. *Pain Physician* 2014; 17:E319–E325 [35]

**Table 28.2** Summary of sacroiliac joint injection PRP studies published to date

| Study details | Methods | Results | Conclusion |
|---|---|---|---|
| Chen et al. 2022 [7]<br>Double-blind, randomized, clinical trial<br>Fluoroscopically guided injection<br>Sample size = 26 patients<br>Follow-up = 6 months<br>Chronic unilateral low back pain | • Patients were selected with positive diagnostic blocks<br>• PRP injections were compared with steroid injections<br>• Pain rating scales and functional disability score with NRS and ODI | • Positive diagnostic blocks in patients with provocative signs with steroid injection with sacroiliac joint injection with 0.5–1 mL contrast medium followed by combination of 1 mL of betamethasone 6 mg/mL, and 1 mL of 2% lidocaine<br>• The PRP group received a single injection of 2 mL PRP<br>• Both groups improved at 1, 3, and 6 months<br>• Patients who received steroid injections reported lower pain scores than did subjects who received PRP<br>• Significant pain relief defined by 50% or more from baseline was higher at 1 month and 3 months in the group who received steroids than in the group that received PRP | • PRP injections inferior to steroid injections<br>• Even though both groups showed improvement, the steroid group had significantly greater response and significantly more responders than did the PRP group |
| Singla et al. 2017 [5]<br>Prospective, randomized open blinded endpoint study<br>Ultrasound guided injection<br>Sample size = 40<br>Follow-up = 3 months<br>Chronic low back pain with sacroiliac joint pathology | • Clinical diagnosis with no diagnostic blocks<br>• Patients were randomized into two groups with one group receiving 1.5 mL of methylprednisolone 40 mg/mL and 1.5 mL of 2% lidocaine with 0.5 mL of saline, whereas, PRP group receiving 3 mL of leukocyte free PRP with 0.5 mL of calcium chloride<br>• Ultrasound guided sacroiliac joint injection<br>• Outcomes were assessed with VAS scores, ODI, short Form-12 | • At 3-month follow-up, 90% of the patients reported satisfactory relief with PRP; whereas, satisfactory relief was observed in 25% of the patients receiving steroids<br>• A strong association was observed in patients receiving PRP and showing a reduction of VAS of greater than 50% from baseline | • Positive first prospective, randomized study for PRP<br>• Small number of patients |

(continued)

**Table 28.2** (continued)

| Study details | Methods | Results | Conclusion |
|---|---|---|---|
| Wallace et al. 2020 [6]<br>Prospective, nonrandomized, interventional study<br>Ultrasound guided injection<br>Sample size = 50 patients<br>Follow-up = 6 months<br>Chronic low back pain | • Diagnosis confirmed by intraarticular injection with steroid. Patients were also positive for provocative tests<br>• 50 patients with chronic low back pain of 12 months duration<br>• 10 mL of PRP was prepared with platelet concentrations ranging from 5 to 10 times greater than the baseline level of platelets in the average patient's systemic circulation injected under ultrasound guidance at point maximal tenderness and posterior ligaments<br>• Outcomes included ODI and NRS | • NRS and ODI were reduced at 6 months after injection compared with baseline values<br>• NRS pain scores decreased a mean of 1.94, and ODI decreased with a mean change of 9.79% | • Positive for PRP injection<br>• PRP injections in the sacroiliac joint are effective at reducing disability and pain<br>• Maximum improvement was seen at 4 weeks after the injection and with sustained reduction of pain and disability at 6 months |
| Navani and Gupta 2015 [40]<br>Case review<br>Fluoroscopic guidance<br>Sample size = 10 (4 males, 6 females) with sacroiliac joint pain of greater than 6 months duration<br>Age distribution = 5 patients below 40 and 5 patients over 40<br>Sacroiliac joint pain | • 10 patients<br>• Positive clinical criteria and at least single block with 50% relief<br>• Sacroiliac joint injection under fluoroscopic guidance with PRP<br>• PRP was obtained using 60 mL of whole blood to yield 7 mL of PRP using a double spin technique at 3800 rpm for 1.5 min in the first cycle followed by 5 min in the second<br>• Approximately 4 mL of PRP was injected into each joint until increased pressure was felt in the plunger and solution could not be injected easily<br>• Patients were followed at 1-month, 3 months, 6 months, and 12-months post injection<br>• Post operative treatments included 2 weeks of limited activity with bracing followed by progressive sacroiliac joint stabilization and mobilization exercises until they reached their full functional potential | • All patients improved 3 months post injection and maintained low pain levels not requiring any additional treatment up to 6 months post injection<br>• SF-36 demonstrated improvement in both physical component summary scores and mental component summary scores in all patients<br>• No adverse events | • A positive case series of 10 patients |
| Ko et al. 2017 [4]<br>Case series<br>Ultrasound guidance<br>Sample size = 4<br>Follow-up = 2 years | • Clinical criteria<br>• Sacroiliac joint injection with PRP under ultrasound<br>• Outcomes were assessed with short form, McGill pain questionnaire, NRS, ODI | • At 12-month follow-up there was marked improvement in joint stability, a statistically significant reduction in pain, and improvement in quality of life<br>• The clinical benefits of PRP were still significant at 4 years post treatment | • PRP showed long lasting positive results in this short case series of 4 |

*PRP* platelet rich plasma, *NRS* numeric rating scale, *ODI* Oswestry Disability Index, *VAS* Visual Analog Scale

## Clinical Presentation

### Symptoms and Signs

The pain diagram usually centers around the Venus Dimple or posterior superior iliac spine (PSIS) region below the L5 level (pelvis, buttock, hip, and groin) with radiation down the leg (Fig. 28.1) sometimes with transient numbness, tingling, weakness, instability (leg giving way), and even urinary frequency [1–3, 43, 44]. For this reason, it can be mistaken for sciatica, but it rarely extends below the knee. Sacroiliac joint pain is usually a dull ache or stiffness but can be sharp and stabbing and is often aggravated with prolonged immobility (e.g., prolonged standing, sitting, or lying down). Pain is usually unilateral and worse with turning in bed, getting out of bed, standing up from a seated position, or stepping up with the affected leg (stair and hill climbing), and often results in disturbed sitting patterns. It can be exacerbated by sexual intercourse but this is not specific to the sacroiliac joint. Pain is reported to increase during menstruation in women. Patients with severe and disabling sacroiliac joint dysfunction can suffer from insomnia and depression.

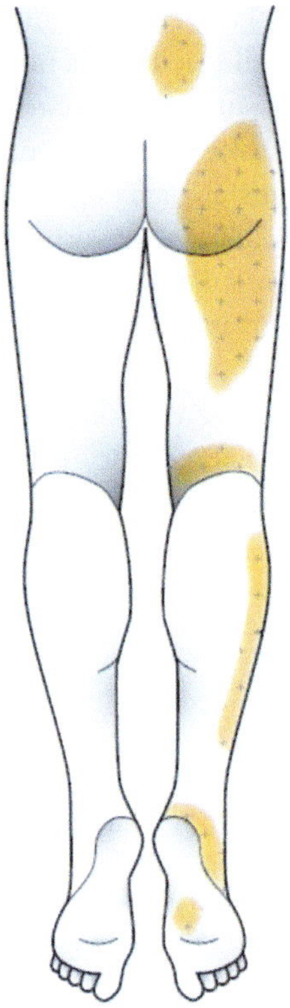

**Fig. 28.1** Illustration of pain referral patterns of the sacroiliac joint. (Adapted and modified from: Fortin JD, et al. Sacroiliac joint: Pain referral maps upon applying a new injection/arthrography technique. Part I: Asymptomatic volunteers. Spine (Phila Pa 1976) 1994; 19:1475–1482 [43] and Fortin JD, et al. Sacroiliac joints: Pain referral maps upon applying a new injection/arthrography technique. Part II: Clinical evaluation. *Spine (Phila Pa 1976)* 1994; 19:1483–1489 [44])

## Diagnosis

Diagnosis is established through patient history, physical examination, and other diagnostic tests [1–3, 17, 18, 43–50].

History of prior trauma, prior lumbar fusion, prior pregnancy, or inflammatory arthritis is often reported. Differential diagnosis includes hip and lumbar spine (low back pain should be attributed to the lumbar spine until proven otherwise).

Physical exam is non-specific and no single test is considered diagnostic [1, 3, 47–50]. The Fortin Test and provocative tests (Iliac Compression Test, Iliac Distraction Test, Thigh Thrust, Patrick's FABER Test, Gaenslen's Test) are 85% sensitive and 76% specific. For diagnostic purposes, positive results from three of five tests including Patrick's FABER Test, Gaenslen's Test, Distraction Test, Thigh Thrust, and Compression Test should be positive with inclusion of Iliac Compression Test or Thigh Thrust as shown in Fig. 28.2. In all the tests, pain along the typical area raises suspicion for sacroiliac joint dysfunction. Neurogenic weakness, numbness, or loss of reflex should alert clinicians to consider nerve root pathology. A new scale to diagnose sacroiliac joint instability that responds to prolotherapy has been recently

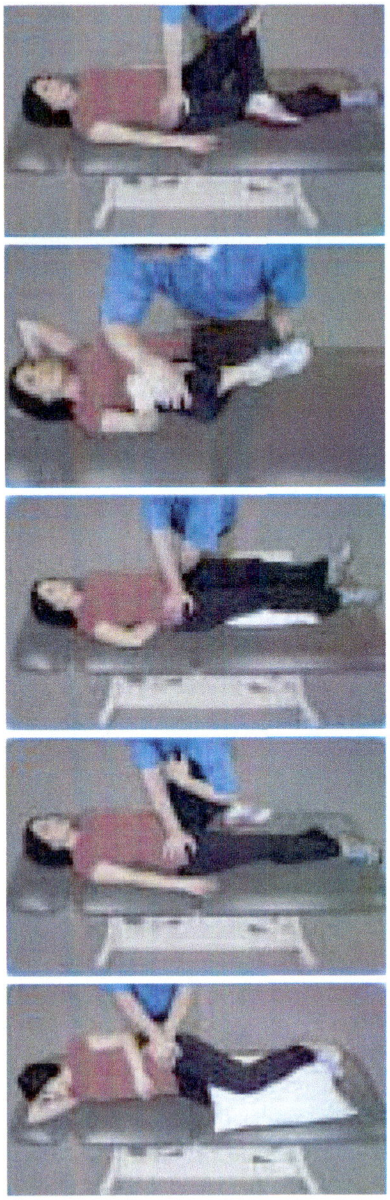

**Fig. 28.2** Sacroiliac joint stress tests. **Patrick's FABER Test** (flexion, abduction, external rotation test) The patient is in the supine position, and the examiner has the leg of the affected side bent at the hip and knee so that the foot is positioned just under the opposite knee. The examiner applies downward force on the knee to provoke the pain. **Gaenslen Test** (pelvic torsion test) The patient's affected side is placed at the edge of the exam table. The patient holds the unaffected limb pulling the flexed hip to the abdomen. The contralateral leg is allowed to hyperextend as it is permitted to move off the exam table towards the floor. Pain may then be expected to escalate on the affected side. **Distraction Test (gapping test)** The patient is placed in the supine position as the examiner moves to the affected side. The examines applies pressure in a dorsolateral direction with both hands on the ipsilateral anterior superior iliac spine. **Thigh Thrust (posterior shear test)** The supine position is assumed by the patient and the examiner stands on the affected side. The patient places the hip in a 90° flexed position with slight adduction. A downward pressure is the applied to the flexed knee to provoke the sacroiliac joint. **Compression/Approximation Test** The patient is placed in the lateral decubitus position with the painful side up and the hips and knees flexed at 45° and 90° respectively. The examiner stands behind the patient and exerts a downward and medial force after placing both hands on the front side of the iliac crest so as to replicate/provoke pain

**Table 28.3** Whitmore-Gordons sacroiliac instability tool

| Please circle ONE number corresponding to the statement in EACH question that BEST describes your low back pain | | | | | | |
|---|---|---|---|---|---|---|
| | Never | Rarely | Sometimes | Occasional | Often | Always |
| 1. I have pain in my buttock | | | | | | |
| | 0 | 1 | 2 | 3 | 4 | 5 |
| 2. My back feels unstable | | | | | | |
| | 0 | 1 | 2 | 3 | 4 | 5 |
| 3. I get pain (in the low back-buttock area) when I turn in bed | | | | | | |
| | 0 | 1 | 2 | 3 | 4 | 5 |
| 4. I get pain (same area) when I get out of a low chair | | | | | | |
| | 0 | 1 | 2 | 3 | 4 | 5 |
| 5. I get pains (same area) when I bend the leg up (such as to put on my sock) | | | | | | |
| | 0 | 1 | 2 | 3 | 4 | 5 |
| 6. I get pain (same area) when going up or down stairs | | | | | | |
| | 0 | 1 | 2 | 3 | 4 | 5 |
| 7. I feel clicking/clunking/popping (same area) when I move | | | | | | |
| | 0 | 1 | 2 | 3 | 4 | 5 |
| 8. I have had the following injuries to my low back: | | | | | | |
| __(a) motor vehicle accident with foot on brake at time of impact (score as 5) | | | | | | |
| __(b) fall on the same buttock or hip (score as 5) | | | | | | |
| __(c) women: pregnancy-related pelvic pain (score as 5) | | | | | | |
| __(c) men: unexplained pain (normal urology studies) referred into the testicle (score as 5) | | | | | | |
| __(d) temporary relief with spinal manipulation or sacroiliac belt (score as 5) | | | | | | |
| __(e) sports or work-related twisting-lifting injury—e.g., bowling, figure skating, etc. (score as 3) | | | | | | |
| __(f) history of hypermobile "loose" joints (score as 2) | | | | | | |
| Total Score __/60 | | | | | | |

co-developed and is undergoing validity and reliability testing (Whitmore-Gordons Sacroiliac Instability Tool) as shown in Table 28.3.

The current standard for diagnosis is sacroiliac joint injection confirmed under fluoroscopy or computed tomography (CT)-guidance using controlled comparative local anesthetic or placebo-controlled blocks, with greater than 75% visual analog scale (VAS) reduction confirming the diagnosis [3].

## Anatomy

The sacroiliac joints are paired C-shaped or L-shaped diarthrodial synovial joints formed between the articular surfaces of the sacrum and the ilium bones, one on the left and one on the right [1, 2, 17, 51]. The joints are covered by two different kinds of cartilage; the sacral surface has hyaline cartilage and the ilial surface has fibrocartilage (Fig. 28.3). The sacroiliac joint is like a keystone designed to take compressive loads and transfer loads from the upper body to the lower body, acting as a shock absorber and relieving forces on the spine. It has irregular elevations and depressions that produce interlocking of the two bones, which together with strong ligaments, resist shear loading. It is still capable of small movements that allow it to act like a shock absorber and torque converter. The primary motion of the sacroiliac joint is nutation and counternutation (nodding type of motion). This limited motion is very important when we walk or transition from sitting to standing.

### Ligaments

The ligaments of the sacroiliac joint (Fig. 28.4) include [1, 2, 17, 51, 52] the:

- Anterior Sacroiliac Ligament
- Dorsal Interosseous Sacroiliac Ligament
- Dorsal Posterior Sacroiliac Ligament
- Sacrotuberous Ligament
- Sacrospinous Ligament

The anterior ligament, in most cases, is just a slight thickening of the anterior joint capsule and not as well defined as the posterior sacroiliac ligaments. The dorsal interosseous ligaments are very short and strong ligaments that run perpendicular from the iliac surface to the sacrum, they keep the articular surfaces from abducting or opening/distracting. The long dorsal posterior sacroiliac ligaments run in an oblique vertical direction in a basket weave pattern. The sacrotuberous and sacrospinous ligaments (also known as the extrinsic sacroiliac joint ligaments) limit the amount the sacrum flexes.

### Muscles

Many large and small muscles have relationships with the ligaments of the sacroiliac joint including the piriformis (piriformis syndrome is often related with sacroiliac joint dys-

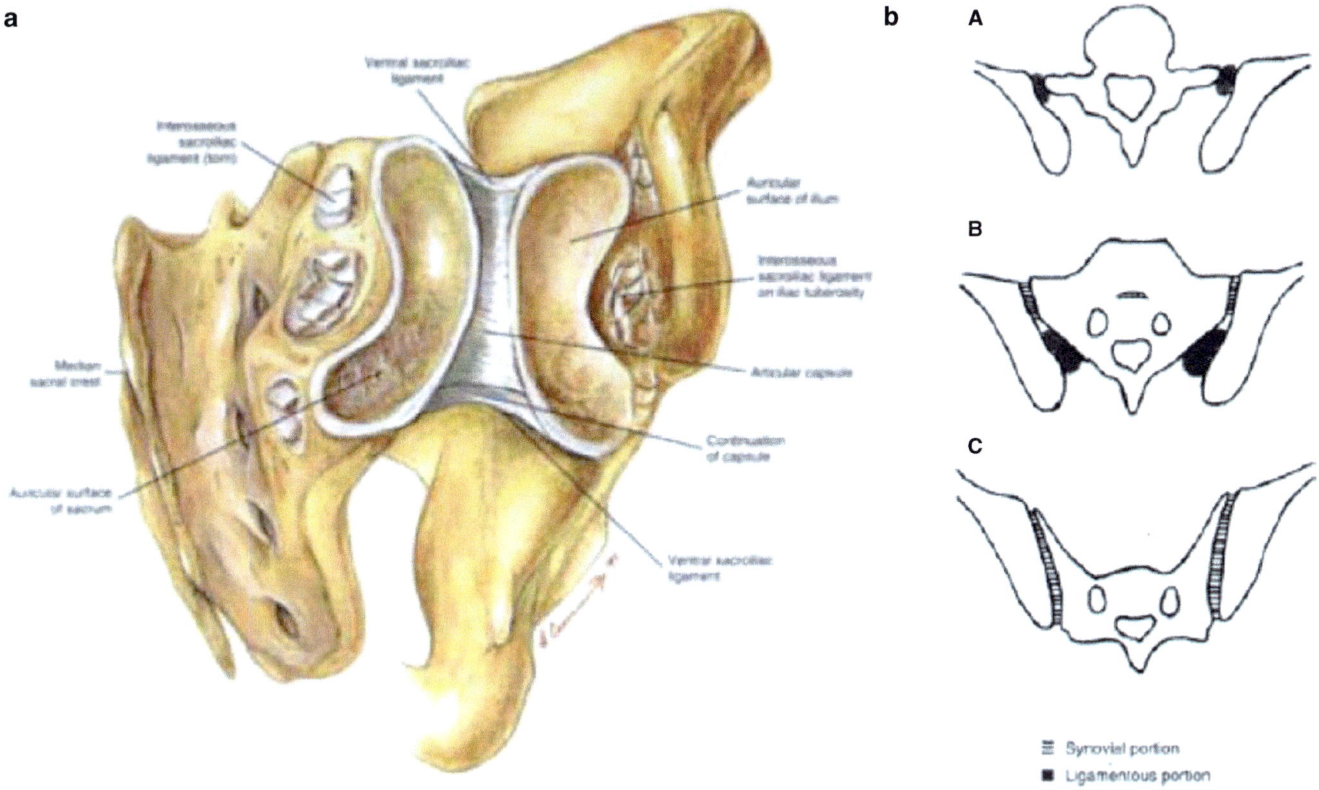


**Fig. 28.3** An illustrative anatomy of the sacroiliac joint. (**a**) Posterior view of an opened right sacroiliac joint demonstrating some of the important bony and soft tissue components. (**b**) Horizontal sections through A, the upper, B, middle, and C, lower portions of the sacroiliac joint. Diagram also demonstrates the synovial versus fibrous portions of the sacroiliac joint. (Reproduced from Basic and Clinical Anatomy of the Spine, Spinal Cord and ANS, 2nd ed. Cramer and Darby, ©2005, with permission from Elsevier [51])

function) rectus femoris, gluteus maximus and minimus, erector spinae, latissimus dorsi, thoracolumbar fascia, and iliacus.

## Innervation

The sacroiliac joint is a pain-sensitive structure richly innervated by a combination of unmyelinated free nerve endings of the posterior primary rami of spinal segments L2-S3 (Fig. 28.5). The wide possibility of innervation may explain why pain originating from the joint can manifest in so many various ways, with different and unique referral patterns for individual patients [1, 2, 17, 43, 44, 53–60]. Cox and Fortin [53] described the anatomy of the lateral branches of the sacral dorsal rami and summarized the innervation as shown in Table 28.4 with the descriptions from various authors [54–60].

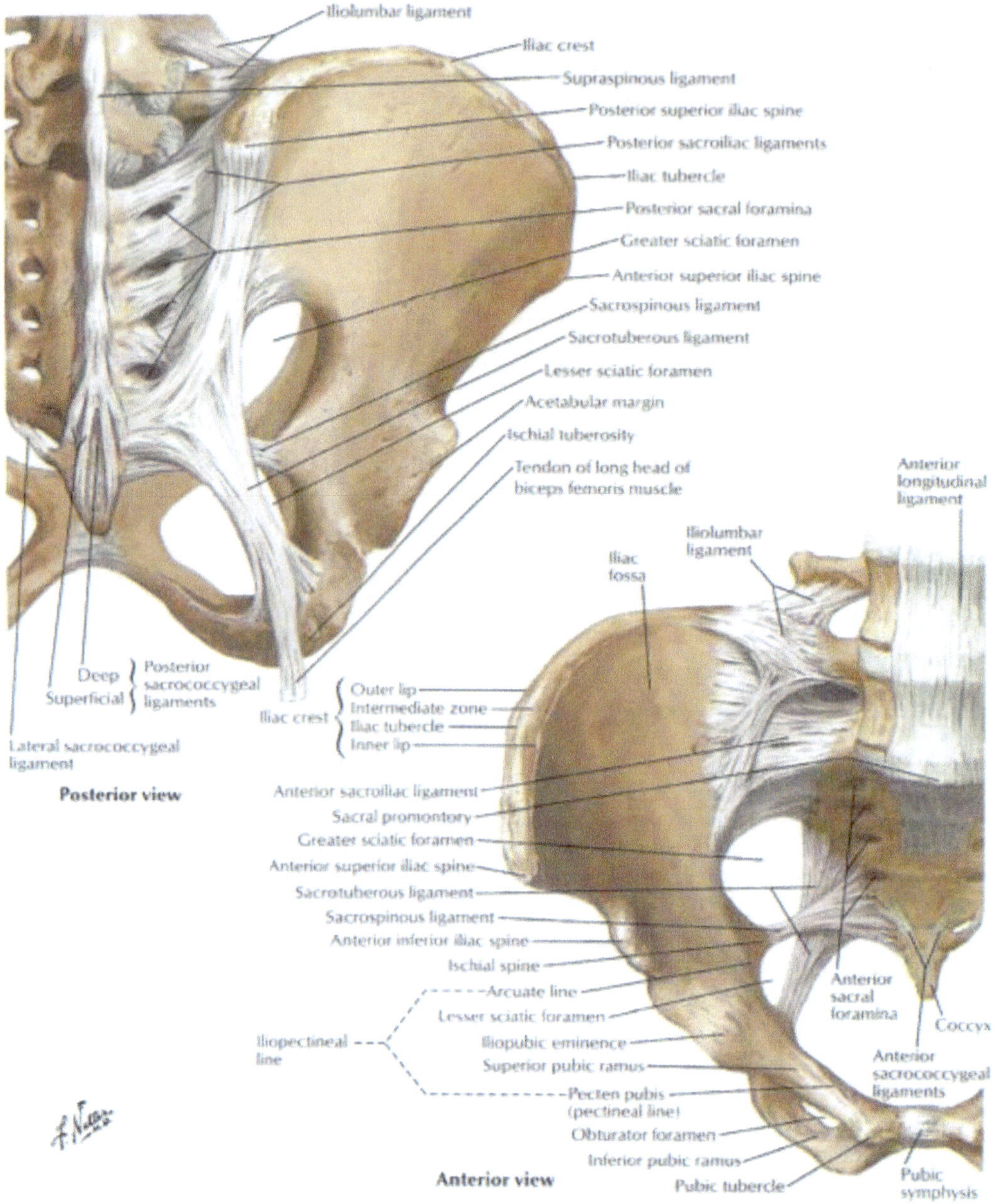


**Fig. 28.4** Ligaments in both the anterior and posterior aspect of the sacroiliac joint. (Reproduced with permission from: Netter FH. Atlas of Human Anatomy, 4th edition, Saunders, El*sevier, 2006* [52])

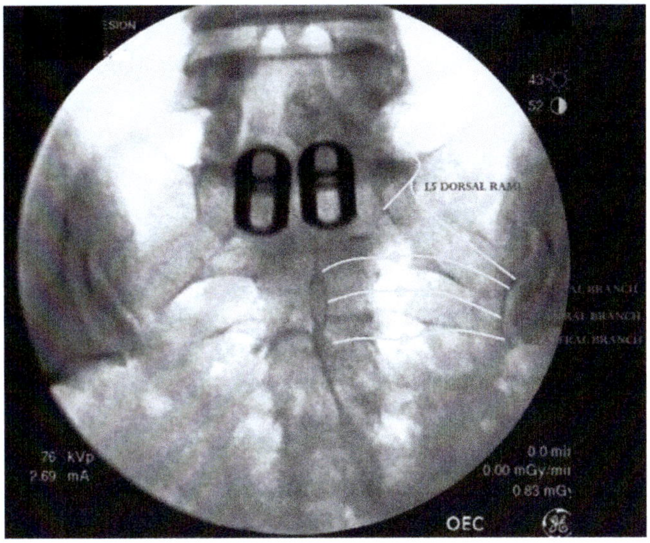


**Fig. 28.5** Neuroanatomical schematic in a previously instrumented spine at L5/S1, demonstrating the approximate locations of the L5 dorsal ramus and the sacral lateral branch which contribute to the majority of posterior sacroiliac joint innervation

**Table 28.4** Studies of sacroiliac joint innervation

| Author | Dorsal Rami contribution | Ventral Rami contribution | L5 contribution |
|---|---|---|---|
| Solonen [54], 1957 | S1 and S2 | Yes (L4, L5 and S1) | Yes |
| Ikeda [55], 1991 | L5, S1–S4 | Yes (L5 and S2) | Yes |
| Grob et al. [56], 1995 | S1–S4 | No | No |
| Yin et al. [60], 2003 | L5, S1–S3 | Not studied | Yes |
| McGrath and Zhang [57], 2005[a] | S2–S4 | Not studied | Not studied |
| Szadek et al. [58], 2008[b] | Not studied | Yes (L4 and L5) | Yes |
| Willard et al. [59], 2010 | L5, S1–S4 | Not studied | Yes |

[a] McGrath and Zhang [57] studied the innervation to the long posterior sacroiliac ligament only

[b] Szadek et al. [58] studied the innervation to the anterior sacroiliac ligaments

## Technical Considerations

Appropriate preparation with monitoring, intravenous access, and sedation as required, are indicated. While physiological monitoring is required, intravenous access or sedation are optional and are based on medical necessity. Multiple techniques have been described [19–21, 47–50, 55, 61–72].

## Classic Technique

The patient is positioned prone. A lateral position with the affected joint superiorly may be used if the patient is unable to lie in prone position.

A sterile technique must be applied. Positioning and radiographic anatomy of the sacroiliac joint are illustrated in Fig. 28.6. The sacroiliac joint is difficult to enter consistently. The difficulty arises because the joint does not present a single radiographic silhouette of its articular cavity [43]. The joint is sinuous, both in caudocephalad direction and in a dorso-ventral direction [71].

When viewed radiographically from behind, the joint presents multiple lines and shadows that represent its articular surfaces, but these lines are two-dimensional projections of a sinuous three-dimensional structure. Lines and spaces that may be evident may be located ventrally in the joint and overlaid by bone posteriorly, which prevents access to the apparent space [43]. Thus, selecting a target point for entry into the joint involves an exercise, both in interpreting these multiple shadows correctly, and in obtaining a view of the joint in which the required target point is not overlaid by bone. When multiple lines are seen, the medial lines indicate the posterior or dorsal joint margin, whereas lateral lines indicate ventral or anterior joint margins.

Under image intensifier control, a 25- or 22-gauge, 3.5-in. spinal needle is directed into the inferior aspect of the sacroiliac joint using a posterior approach. The usual border of entry is the inferior one-third of the joint. There is often a lucency in the inferior aspect of the joint, which allows the least resistance upon needle passage. There may be two or more "limbs" of the joint, because the joint is laterally divergent from its posterior to anterior borders and has interdigitations. In this instance, the border has interdigitations, wherein the medial or posterior division is the most amenable to needle placement [61].

If the inferior aspect of the joint cannot be entered, the joint is accessed through the deeper, more rostral aspect of the joint. Rotating the C-arm side to side in an axial plane will permit a better three-dimensional perspective of the joint and enable selection of the "window" for optimal needle trajectory. With few exceptions, a direct posterior approach is used [48]. Once the dorsal sacroiliac and interosseus ligaments are engaged, the needle often takes a characteristic bend because it conforms with the interdigitating contours of the diarthrodial joint. This phenomenon is often preceded by a subtle tactile sensation of a 'giving away' or loss of resistance as the needle penetrates through the ligaments to enter the joint.

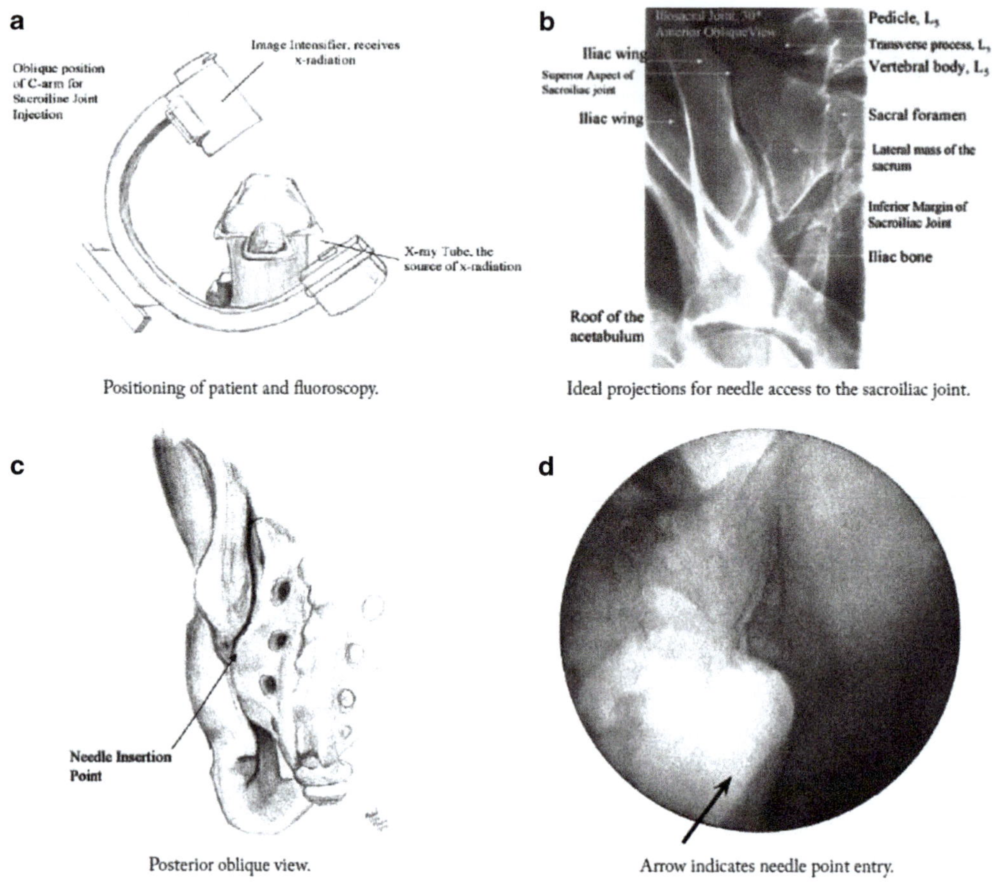


**Fig. 28.6** Fluoroscopy of the sacroiliac joint. (**a**) Positioning of patient and fluoroscopy. (**b**) Ideal projections for needle access to the sacroiliac joint. (**c**) Posterior oblique view. (**d**) Arrow indicates needle point entry

If bony resistance is met after the ligaments are engaged and the needle is not yet within the joint margin, the needle should be withdrawn slightly without becoming disengaged. Subsequent needle advancement, while simultaneously rotating it around its own longitudinal axis, will allow it to deflect and conform to the joint margins. Once the needle has entered the joint space, intraarticular placement is confirmed with an injection of contrast medium. If the needle has been correctly placed, injection of the contrast medium will outline the joint space. Only a minimum volume of contrast medium (0.3–0.5 mL) is required to establish intraarticular injection. As shown in Fig. 28.7, in posteroanterior view, the contrast medium should be seen to travel cephalad along the joint line. In lateral views, the contrast medium most densely outlines the parameter of the joint.

## Technique Without Oblique Angulation

Sacroiliac joint injection may be performed without angulation with the C-arm positioned in straight anteroposterior (AP) view. Khuba et al. [62] assessed fluoroscopic sacroiliac joint injection without oblique angulation and lack of necessity of oblique angulation. They successfully injected 60 sacroiliac joints of 58 patients under AP fluoroscopic view. Seventy percent of sacroiliac joints were seen as 2 separate medial and lateral joint spaces and were entered in distal 1 cm of the medial joint space. In 30%, the joints were seen as straight line rather than two separate spaces, so the image intensifier of the fluoroscopy was tilted cranially to elongate the image of the lower part of the posterior sacroiliac joint and then the sacroiliac joint was entered in its distal 1 cm.

Technique was as follows: The patients were positioned prone on the operating table. First an AP image of the respective SI joints was taken. If the AP image showed the anterior and posterior joint spaces as separate lateral and medial joint lines, respectively (Fig. 28.8a), then the lower part of medial joint line was entered with a 22 G, 10 cm spinal needle under a gun barrel view with the fluoroscope (Fig. 28.8b). The distal end of the spinal needle was slightly bent so as to facilitate easy manipulation of the needle into the joint space. If the SI joint was seen as a straight line rather than two joint spaces in the AP view (Fig. 28.8c) then the image intensifier of the fluoroscope was tilted cranially to elongate the image of the lower part of the posterior sacroiliac joint thus facilitating entry into the posterior SI joint space. This was confirmed by administering 0.3–0.5 mL of radiopaque contrast medium (Omnipaque 240) and seeing contrast flow in the

a

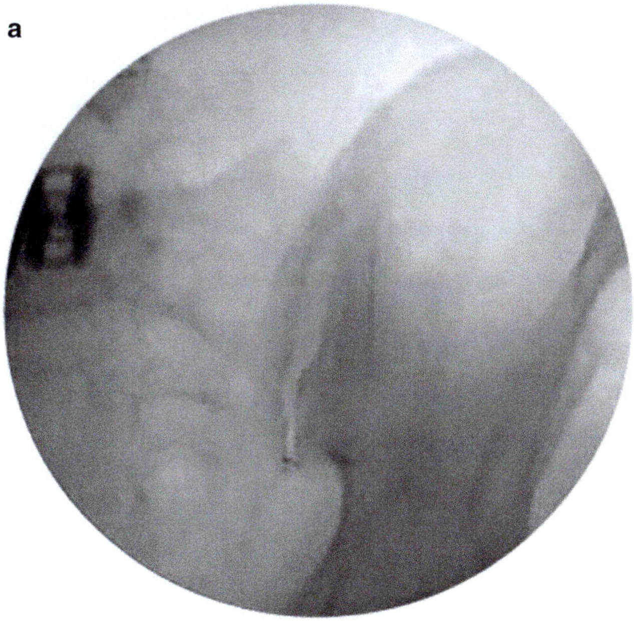

Joint space highlighted.

b

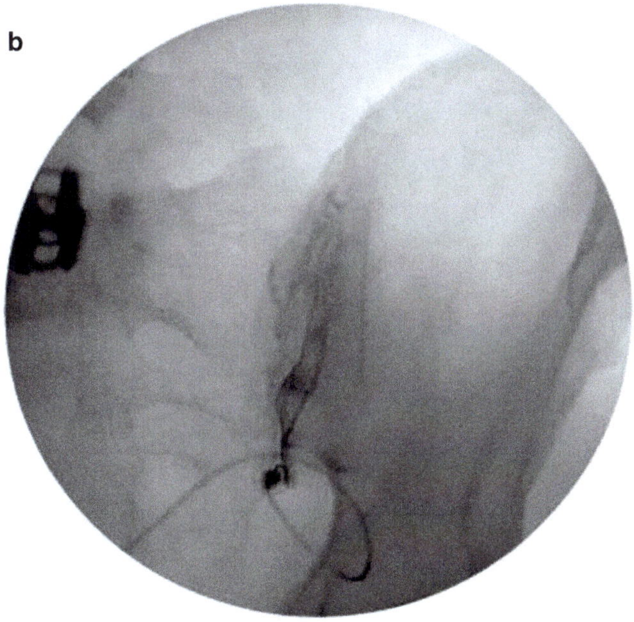

Contrast injection.

**Fig. 28.7** Needle placement and contrast injection. (**a**) Joint space highlighted. (**b**) Contrast injection

superior or anterior part of the joint (Fig. 28.8d). Following confirmation of the contrast in the sacroiliac joint, injection was administered into the joint.

With this technique, the procedure is performed in an AP view without C-arm angulation. The accuracy of the intraarticular injection is confirmed by using intraarticular contrast material. The authors report a successful intraarticular contrast spread in all sacroiliac joint injections. Figure 28.9 shows intraarticular injection in true AP view.

## Benyamin Technique

Benyamin [63] described an alternate technique to enter the sacroiliac joint under fluoroscopic guidance. The technique is as follows: the procedure is started with straight AP fluoro positioning as shown in Fig. 28.10a1, a2 showing two different cases. Following this, the fluoroscopy is tilted caudally instead of the traditional cephalad tilt until the bony ridges of the sacrum and ilium present as a contiguous line (Fig. 28.10b1, b2). Following this, fluoro is rotated to contralateral oblique until a lucency appears at midpoint of the joint (Fig. 28.10c1, c2).

As shown in Fig. 28.11a1, a2, lucency is seen at the midpoint of the right sacroiliac joint. After confirmation of lucency at midpoint of the sacroiliac joint, the needle is placed into the lucent spot using the tunnel vision and it is advanced deeply into the joint as shown in Fig. 28.11b1, b2 [63].

The next step is to inject contrast after confirming the needle position with its deep placement as shown in Fig. 28.12a1, a2. As shown in Fig. 28.12b1, b2, lateral fluoroscopic view further confirming correct needle placement and any leakages as seen with right sacroiliac joint injection in Fig. 28.12b2 [63].

## Double Needle Technique

Gupta [64] described an alternate double needle technique. After obtaining appropriate fluoroscopic images, the tip of a 3.5-in. long, 22-gauge curved tip spinal needle is advanced into the SIJ (Fig. 28.13). Once the tip of the needle is correctly placed, its position is checked under continuous fluoroscopy while moving the C-arm in the right and left oblique directions (dynamic fluoroscopy). On dynamic fluoroscopy, the tip of the needle should remain within the joint line and not appear to be on the bone. If the tip of the needle appears to be on the bone (Fig. 28.13b), a new joint line will need to be identified (the most translucent area through the joint) by dynamic fluoroscopy and another needle advanced into the newly identified joint line (Fig. 28.13c). Dynamic fluoroscopy is repeated again to confirm that the tip of the second needle remains in the joint line. Once both needles are in place, contrast is injected through the needle that is most likely to be in the joint (Fig. 28.13d). If the contrast dye spread is not satisfactory, then contrast is injected through the other needle.

## Modified Technique by Daitch et al.

Daitch et al. [65] described a modified sacroiliac joint injection technique. To aide in penetrating the potentially thick

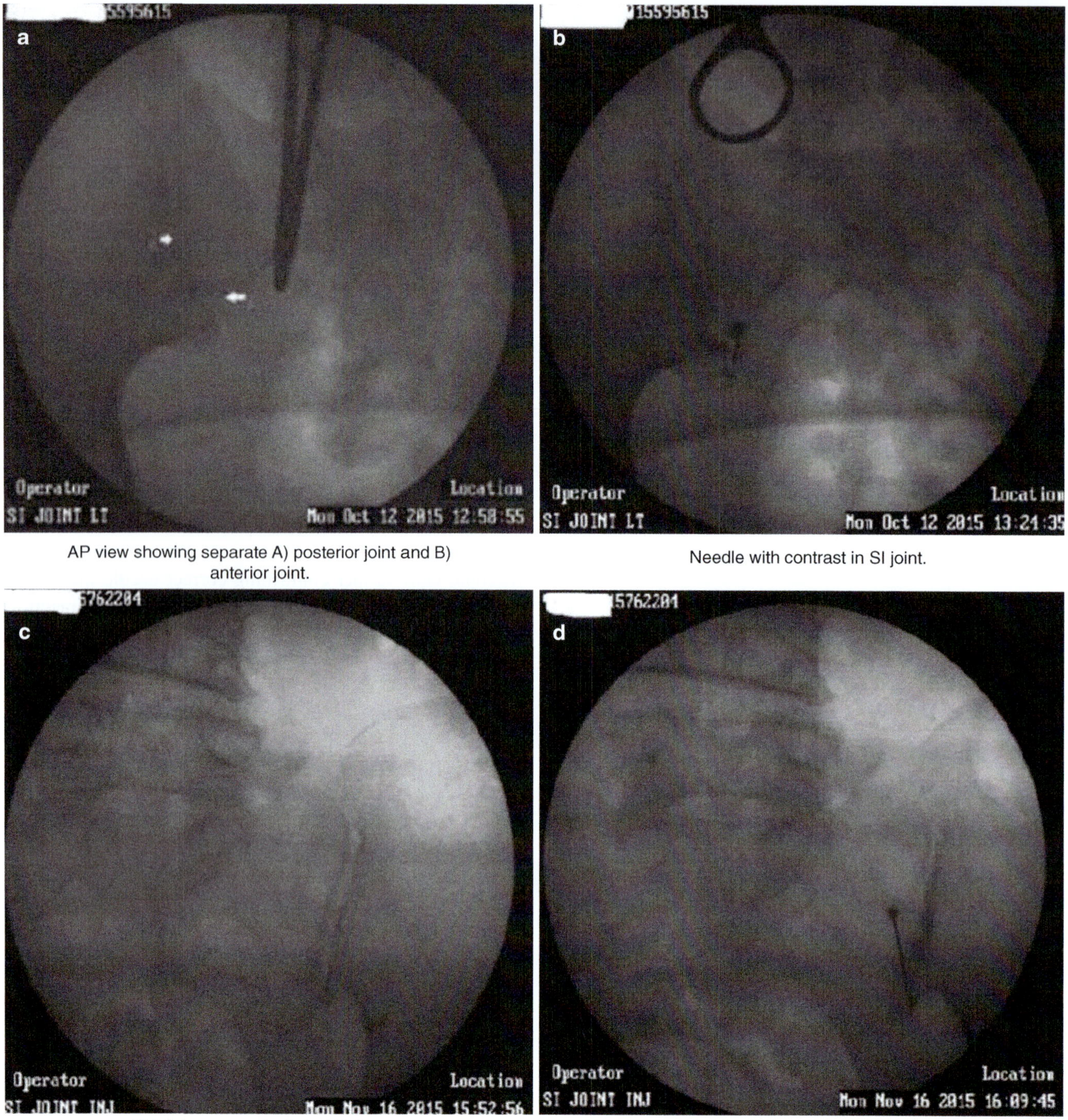


**Fig. 28.8** Illustration of sacroiliac joint injection without oblique angulation. (**a**) AP view showing separate (A) posterior joint and (B) anterior joint. (**b**) Needle with contrast in SI joint. (**c**) SI joint seen as single in AP view. (**d**) Needle with contrast in AP view

capsule in the lowest portion of the joint, it is recommended to use a 22-gauge spinal needle with a slight bend at the tip. The curved tip enhances maneuverability during the entry phase while the 22-gauge size is more easily manipulated once the joint is penetrated. The C-arm is positioned in a PA projection with slight cephalocaudal tilt and an oblique view 5° ipsilaterally. The C-arm is then adjusted until the medial view of the sacroiliac joint line is clearly visible (Fig. 28.14a). Once the medial joint line is identified, the C-arm is maneuvered until the medial lip of the joint and the edge of the

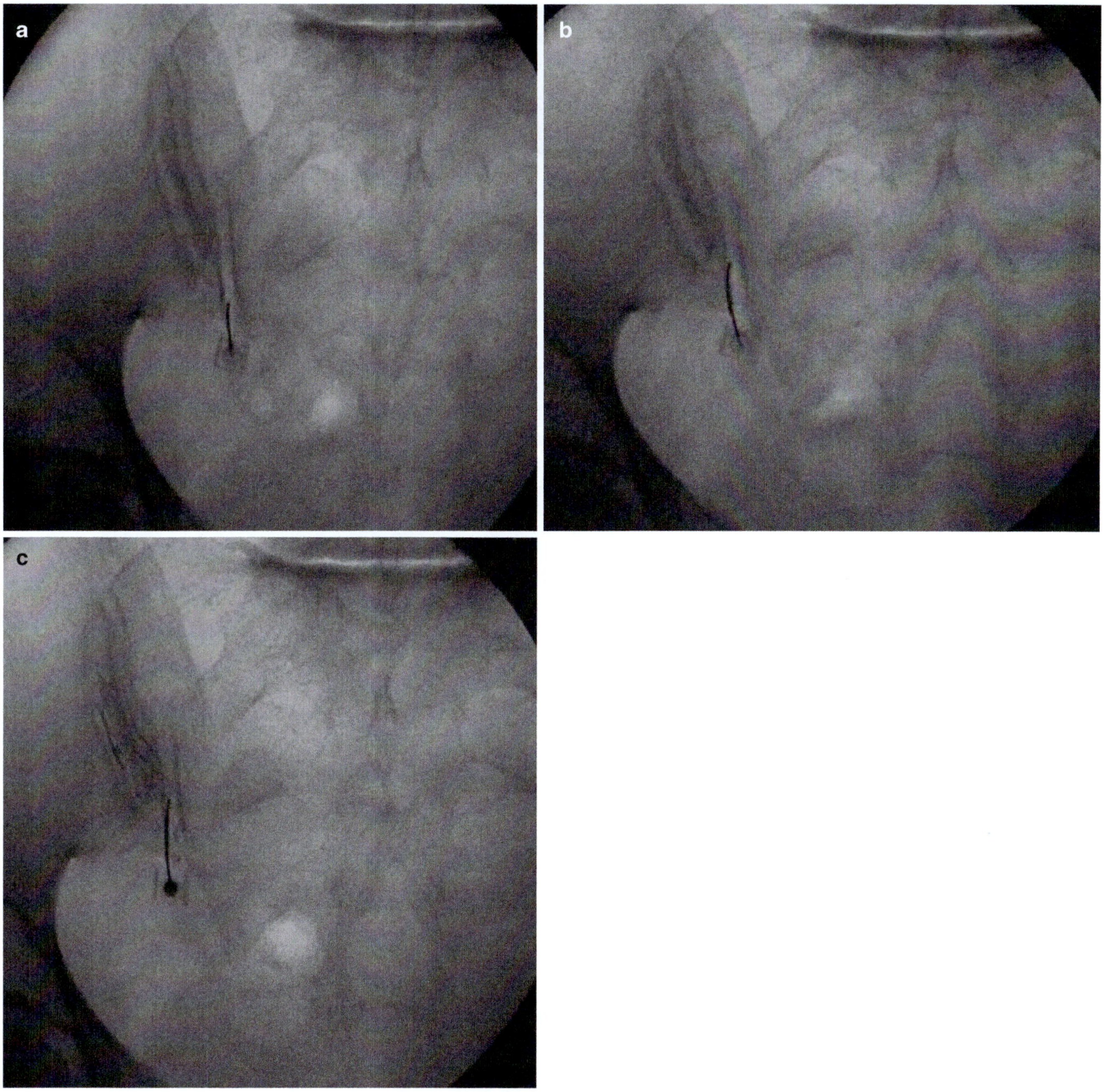


**Fig. 28.9** Intraarticular injection in true AP view. (**a**) Needle entering the joint in nontunnel view. (**b**) Needle in the joint and C-arm angulation 3-degree to right that certifies that the tip of the needle was in lucent area of joint space. (**c**) Final image with contrast material in sacroiliac joint injection

sacrum are clearly identified (Fig. 28.14b). Obtaining a clear view of this junction significantly improves success rate. The 22-gauge needle is guided into the most inferior and medial aspect of the sacroiliac joint (Fig. 28.14c). The joint is most easily accessed if the needle entry is slightly inferior to this aspect and guided upward into the base of the joint. Figure 28.14d shows the flow of contrast into both medial and lateral sacroiliac joint lines. The authors claim that this technique usually allows entry of the sacroiliac joint in less than 30 s. The key elements of this approach are entering the medial aspect of the joint and obtaining a clear view of the edge of the sacrum.

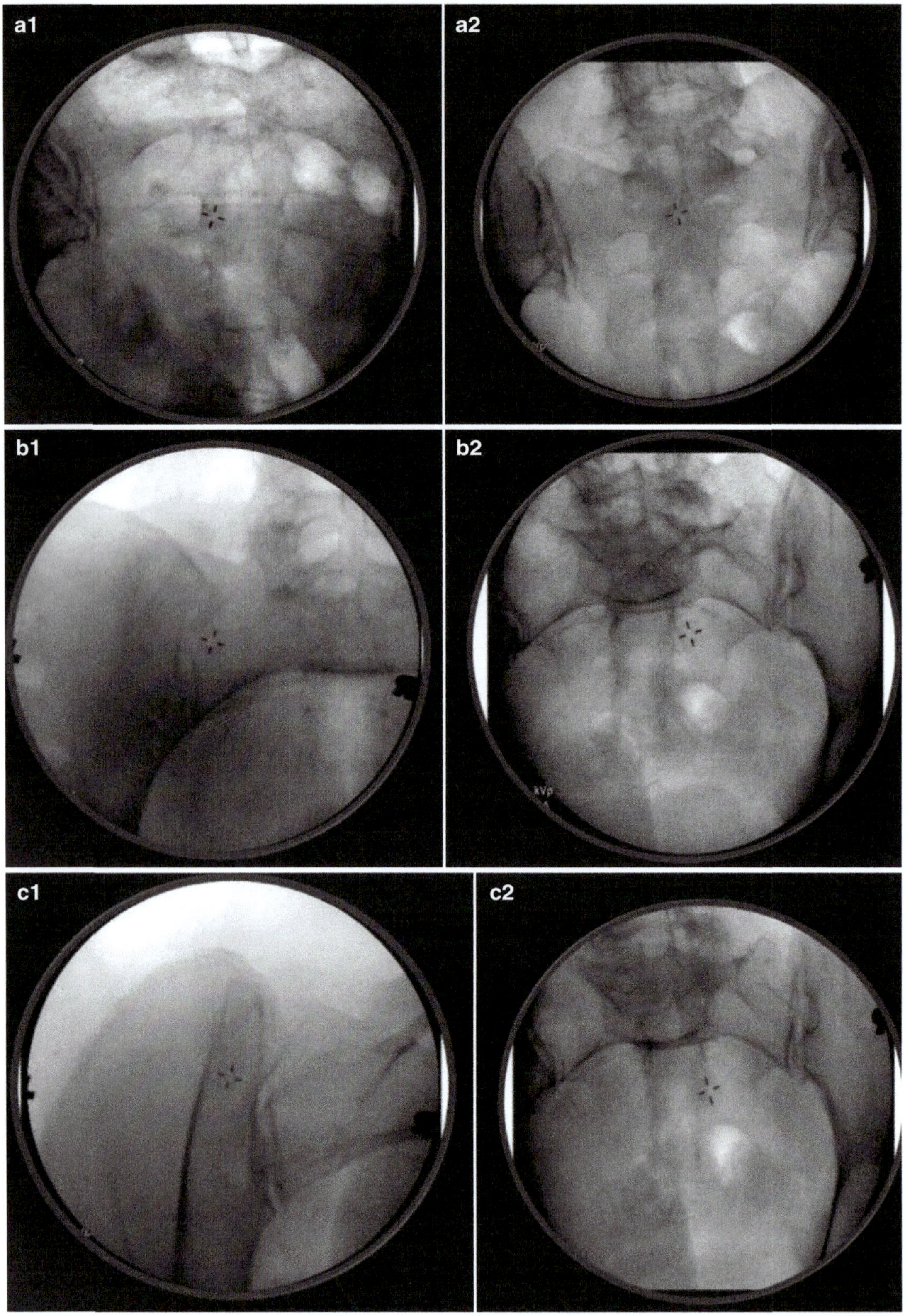


**Fig. 28.10** a1: Straight AP fluoro position: 2 different cases. a2: Straight AP fluoro position: 2 different cases. b1: Straight AP fluoro position: 2 different cases. Tilt fluoro caudally (instead of the traditional cephalad tilt) until the bony ridges of sacrum and iliac present as a contiguous line. b2: Straight AP fluoro position: 2 different cases. Tilt fluoro caudally (instead of the traditional cephalad tilt) until the bony ridges of sacrum and iliac present as a contiguous line. c1: Rotate the fluoro to contralateral oblique until a lucency appears at mid-point of the joint. c2: Rotate the fluoro to contralateral oblique until a lucency appears at mid-point of the joint

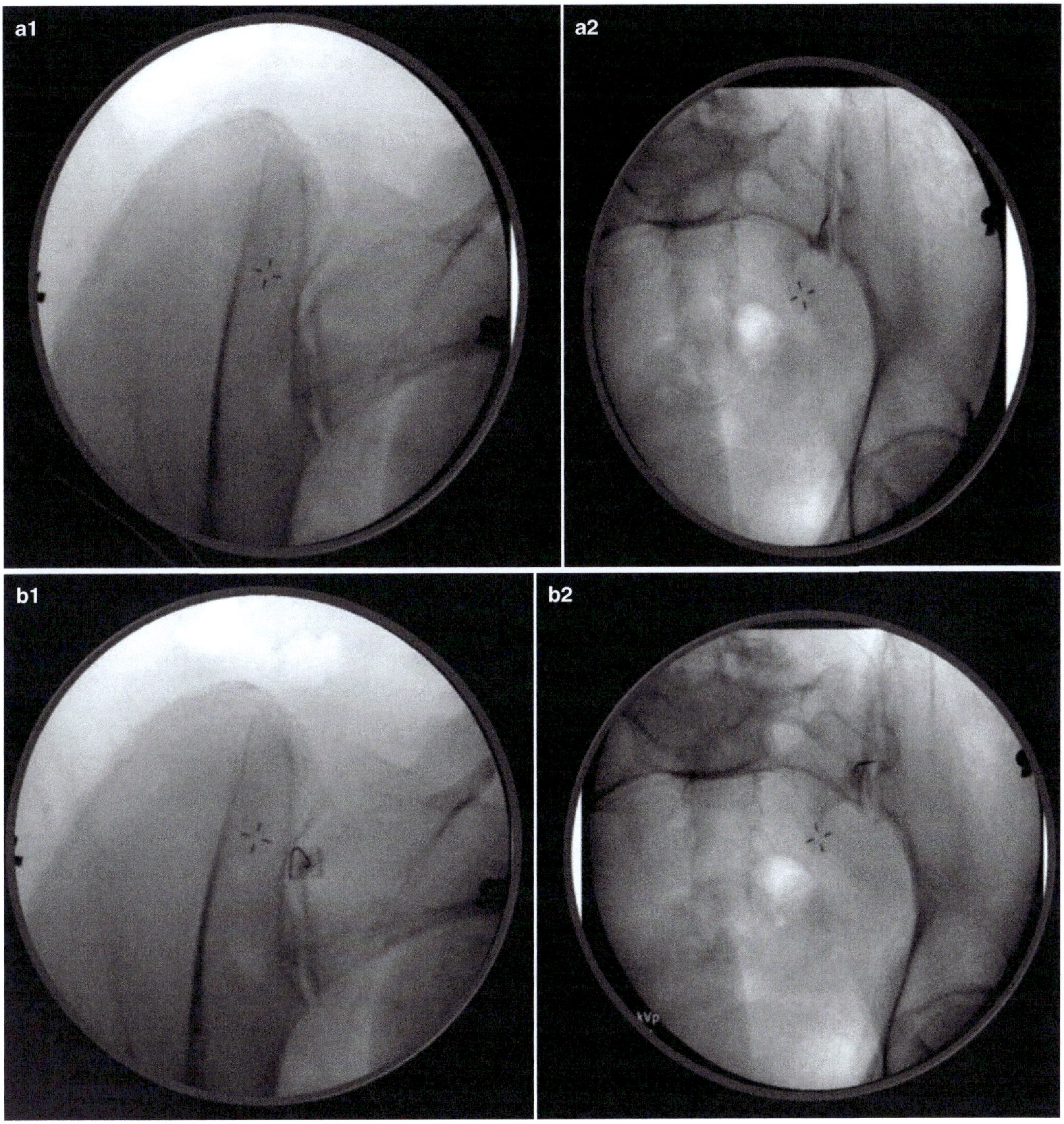


**Fig. 28.11** a1-2. Lucency seen at mid-point of right SI joint. b1-2: The needle is placed into the lucent spot using the tunnel vision and is advanced deeply into the joint

## Upper One-Third Joint Injection

Multiple techniques have been described for entering the sacroiliac joint and placement of the injectate. However, the majority of the techniques are limited to middle third and lower third, with lower one third joint technique being the most commonly used method. Consequently, it is also important to focus on multiple other techniques and also reach the upper third of the joint space. Park et al. [66] described the following a technique: "The patients were prepared in a prone position with a pillow under the abdomen. The procedure area was prepared and draped in the usual sterile fash-

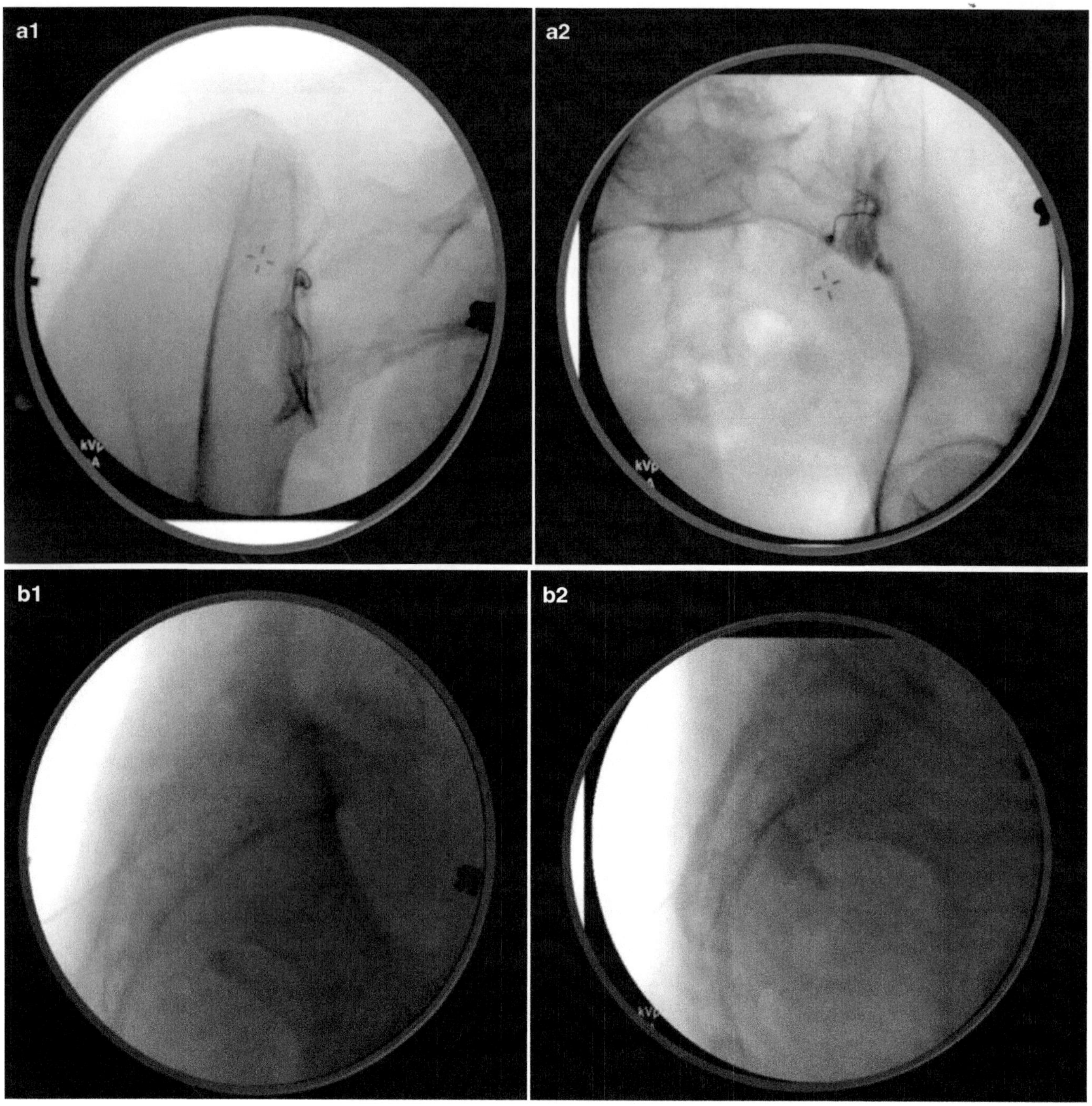


**Fig. 28.12** a1-2: Once the needle is embedded deep in the joint, a small volume of contrast is injected to confirm intra-articular needle placement. b1-2: Lateral fluoroscopic view further confirming correct needle placement (and in b2: small anterior leakage of contrast resulting from anterior SI ligament damage)

ion. Before insertion of the needle, the subcutaneous tissue was infiltrated with 1% lidocaine in the midline of the L5-S1 interspinous space. A 10 cm long, 22-gauge curved-tip spinal needle was advanced into the skin and directed towards the upper one-third joint at about a 45° angle. As the needle hit a firm tissue on the silhouette of the iliac crest in the AP fluoroscopic view, we could distinguish between the sacrum and the iliac crest by rotating the curved needle. The curved-tip needle was advanced beyond the line of the iliac crest until the needle reached the joint with a pop sensation. The intraarticular position was confirmed by injecting 0.2–0.5 mL contrast material through the needle. The injected contrast material spread throughout the SIJ in a cephalocaudal fashion. After the contrast material outlined the sacroiliac joint without vascular runoff, a solution of choice may be injected (Fig. 28.15)."

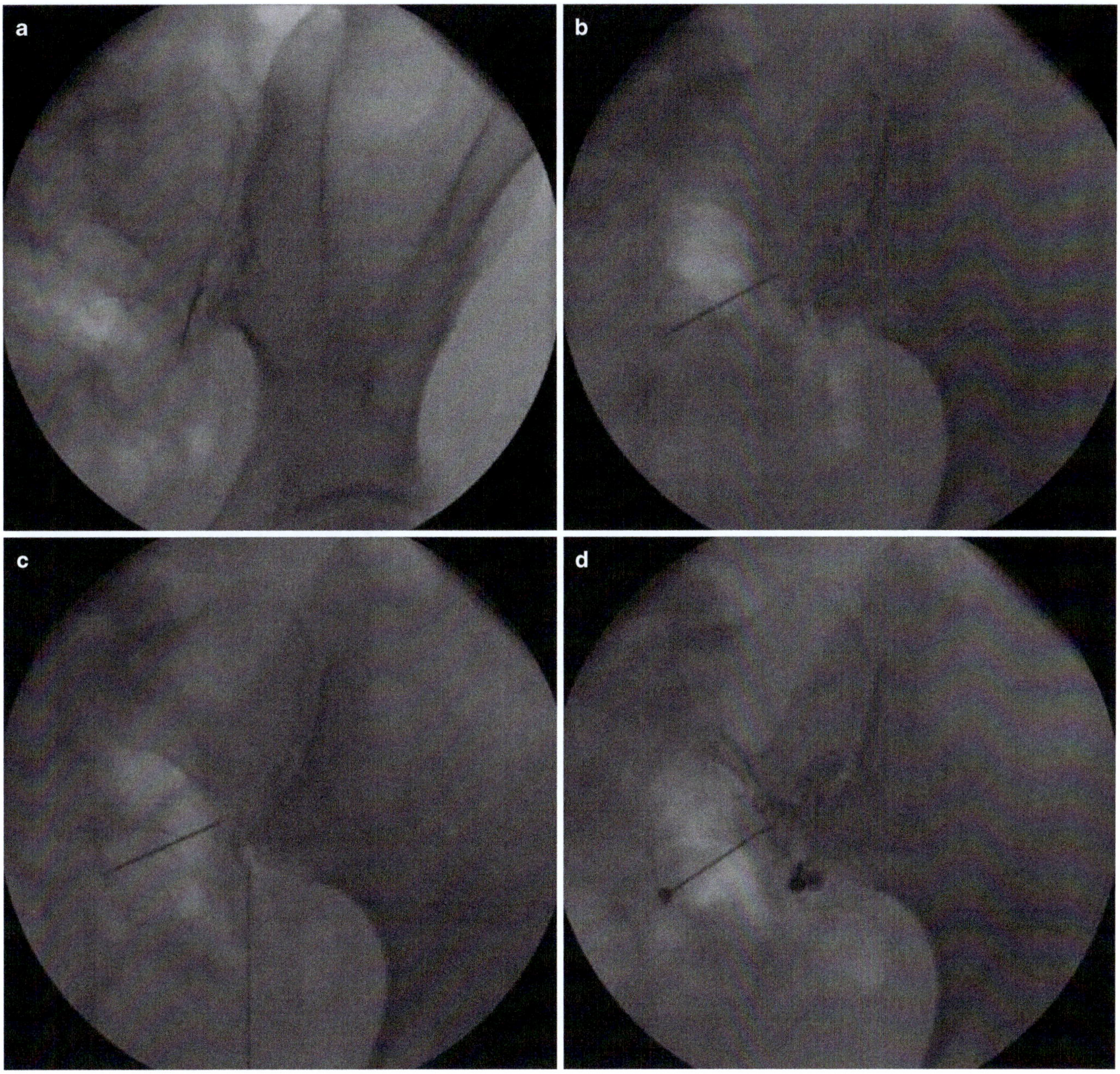


**Fig. 28.13** SI joint injection technique as described by Gupta [64]. (**a**) A curved tip needle is advanced into the right sacroiliac joint. (**b**) On dynamic fluoroscopy the tip of the needle appears to be on bone rather than in the joint. (**c**) On dynamic fluoroscopy another translucent joint is identified and a second needle is advanced into the joint. (**d**) Contrast is injected through the first needle which shows contrast spreading medially—possibly a vascular spread. (Source: Gupta S. Double needle technique: an alternative method for performing difficult sacroiliac joint injections. *Pain Physician* 2011; 14:281–284 [64])

## Injection of the Middle Portion of the Joint

As described earlier, the majority of injections are performed into the lower third of the sacroiliac joint. However, Kurosawa et al. [67] have described fluoroscopy-guided sacroiliac intraarticular injections via the middle portion of the joint as well. Injection of the middle portion of the joint was also described by Benyamin [63] and Ikeda et al. [55]. Potentially injection of lower third may be difficult and also may not reach the entire joint space. Consequently, targeting the middle third of the joint may be advantageous. Kurosawa et al. [67] described this technique in 69 consecutive patients showing an 80% success rate of entering the joint. Further, in 20 cases where this technique was unsuccessful, the conventional method was also unsuccessful in 17 of these cases. The described technique of Kurosawa et al. [67] is as follows:

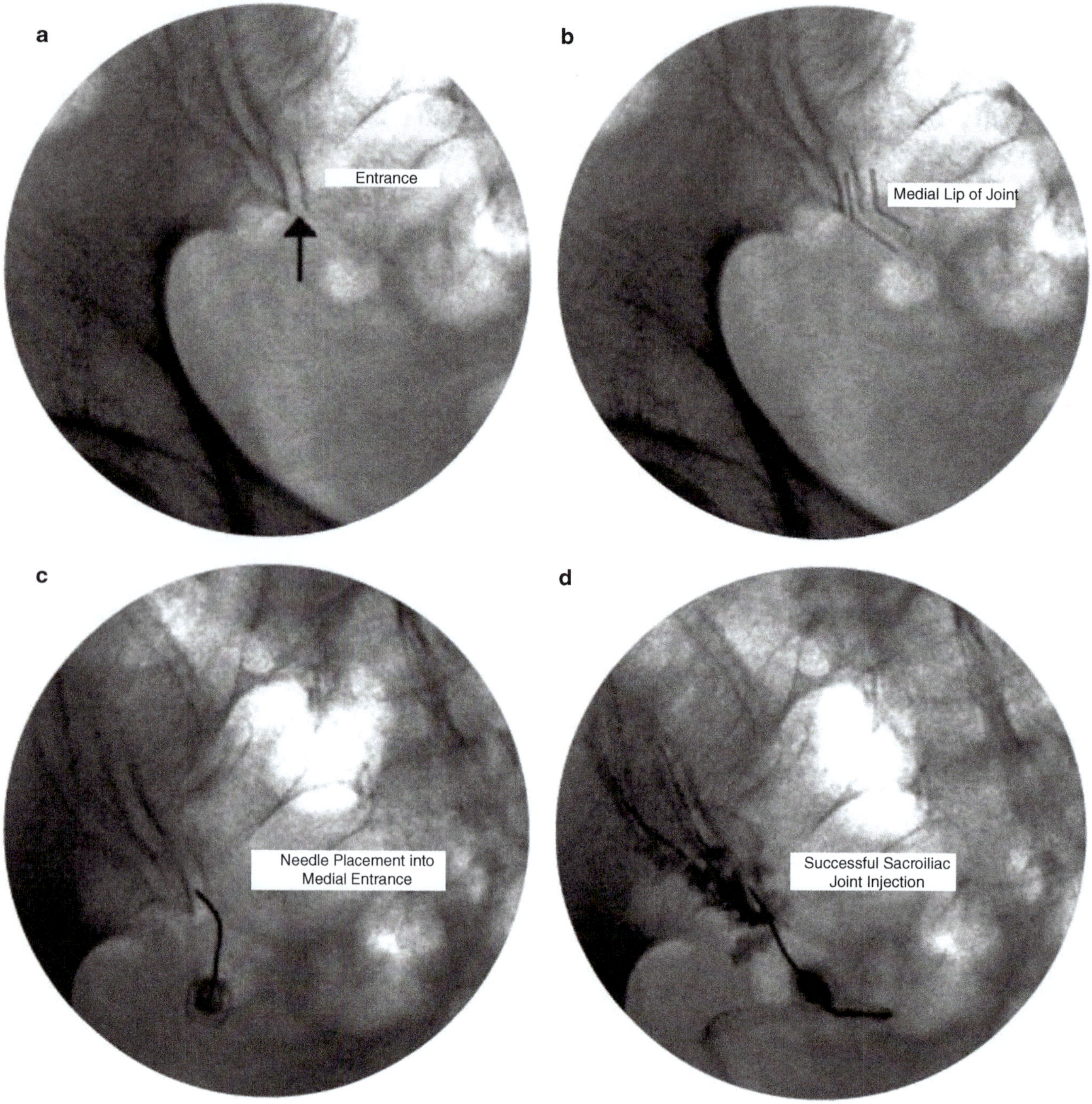


**Fig. 28.14** Modified sacroiliac joint injection techniques as described by Daitch et al. [65]. (**a**) Medial view of the sacroiliac joint line. (**b**) Medial tip of the joint and edge of the sacrum. (**c**) A 22-gauge needle guided into the most inferior and medial aspect of the sacroiliac joint. (**d**) Flow of contrast into medial and lateral sacroiliac joint lines. (Reproduced from Daitch et al. [65] with permission from the authors and the American Society of Interventional Pain Physicians)

"Patients are positioned prone on the fluoroscopy table. With the patient lying prone-oblique with the painful side down on a fluoroscopy table, the posterior sacroiliac joint line is divided into four sections (Fig. 28.16). After administering local anesthetics at the needle entry point, a 23-gauge or 22-gauge, 90-mm straight spinal needle is inserted into the middle portion of the joint. The needle is advanced nearly perpendicular to the fluoroscopic beam angle. After the needle is inserted and advanced until the bone tissue, the fluoroscopy tube was angled at a caudal tilt of 25–30°. The image clearly showed the recess between the ilium and sacrum. The needle direction could be adjusted under fluoroscopic guid-

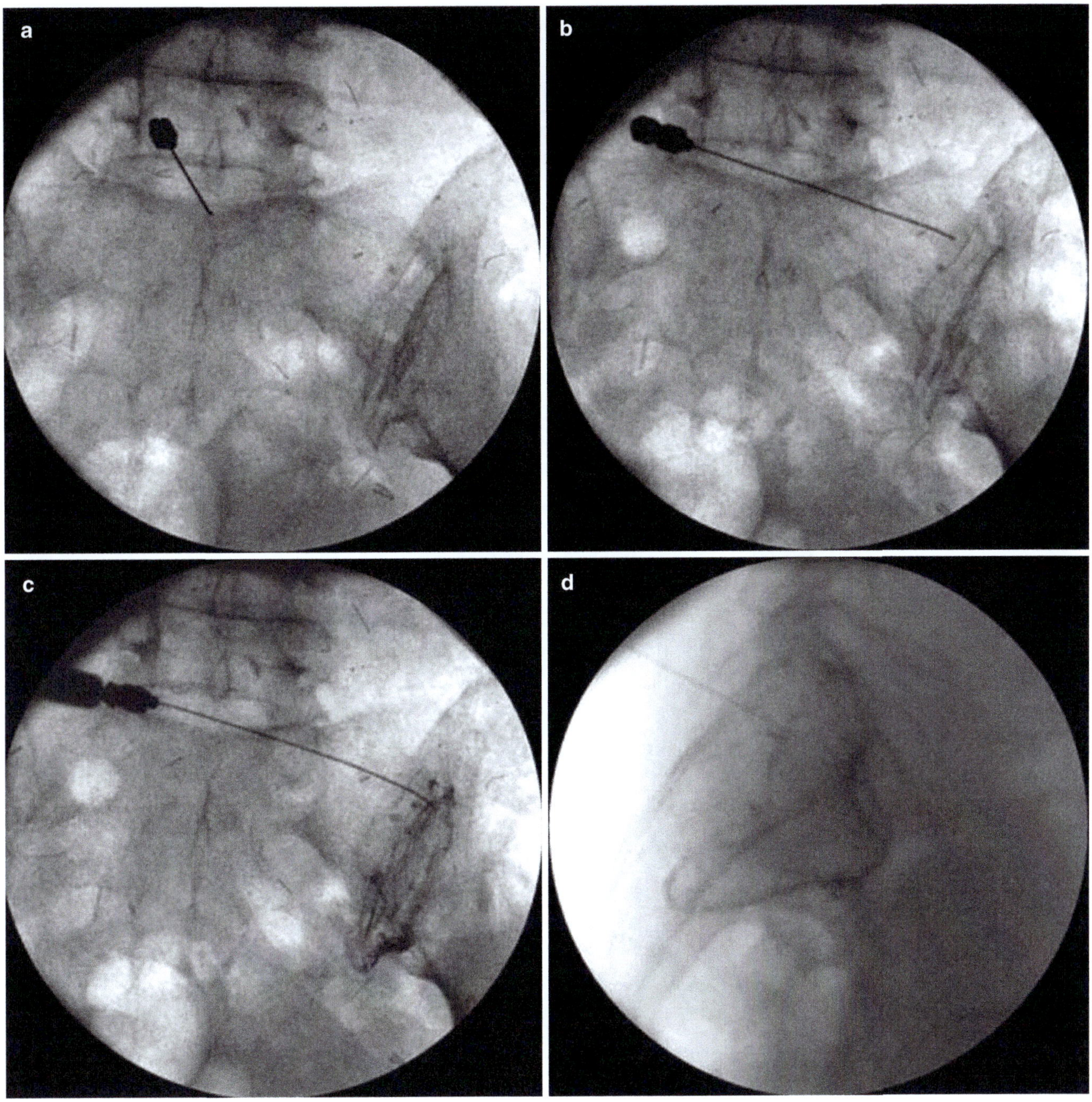


**Fig. 28.15** The upper one-third joint technique for fluoroscopically guided sacroiliac joint injection. (**a**) Note the placement of the needle in the midline of the L5-S1 interspinous space. (**b**) The needle directed toward the SIJ. (**c**) Injection of the contrast material spread throughout the SIJ in an superior to inferior fashion. (**d**) In lateral fluoroscopic view, the contrast material reached lower one-third joint. (Source: Park J, Park HJ, Moon DE, et al. Radiologic analysis and clinical study of the upper one-third joint technique for fluoroscopically guided sacroiliac joint injection. *Pain Physician* 2015; 18:495–503 [66])

ance to the obviously detected posterior margin of the joint. When the needle is advanced until it reached the ilium bone wall in the middle portion, contrast medium was injected. However, because of the presence of ilium cartilage, the tip of the needle was occasionally blocked; the needle, then, had to be slightly retracted while simultaneously injecting the contrast medium. This slight retraction allowed the contrast medium to flow freely into the joint cavity. After the joint was outlined by contrast medium, injectate is placed" (Fig. 28.17).

Depending on the needs of each case, the retro-oblique view was useful for confirming whether the needle tip had reached into the middle portion of the joint. When the needle

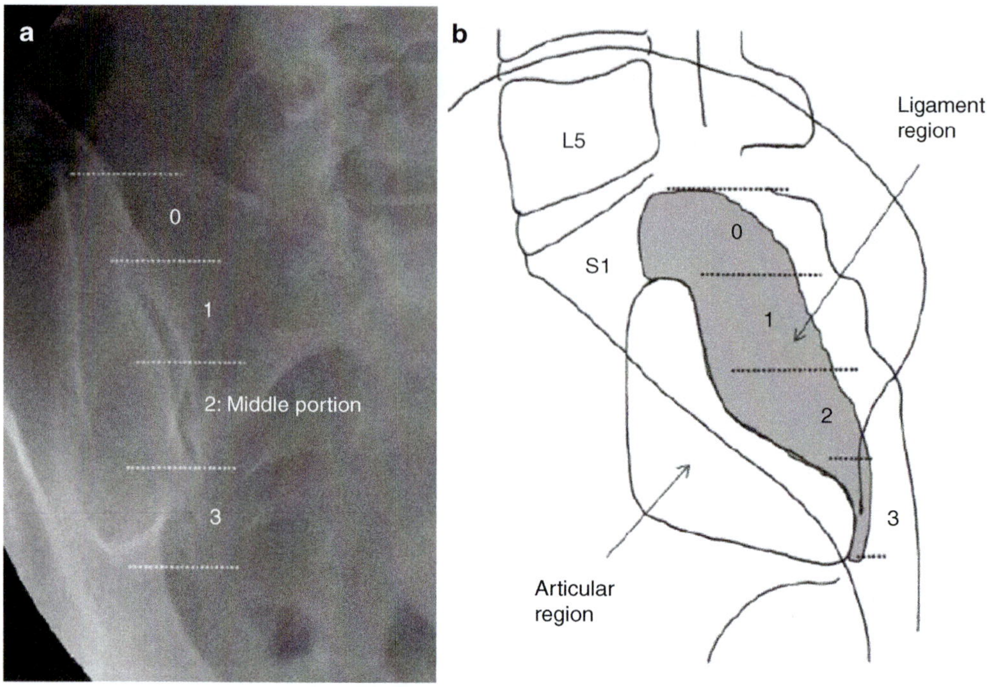


**Fig. 28.16** The posterior sacroiliac joint line is divided into four equal sections (0–3). Section 2 is the middle portion of the joint. (**a**) Prone-oblique view of the sacroiliac joint with the painful side down under fluoroscopy. (**b**) Diagram of four sections in the posterior sacroiliac ligament region

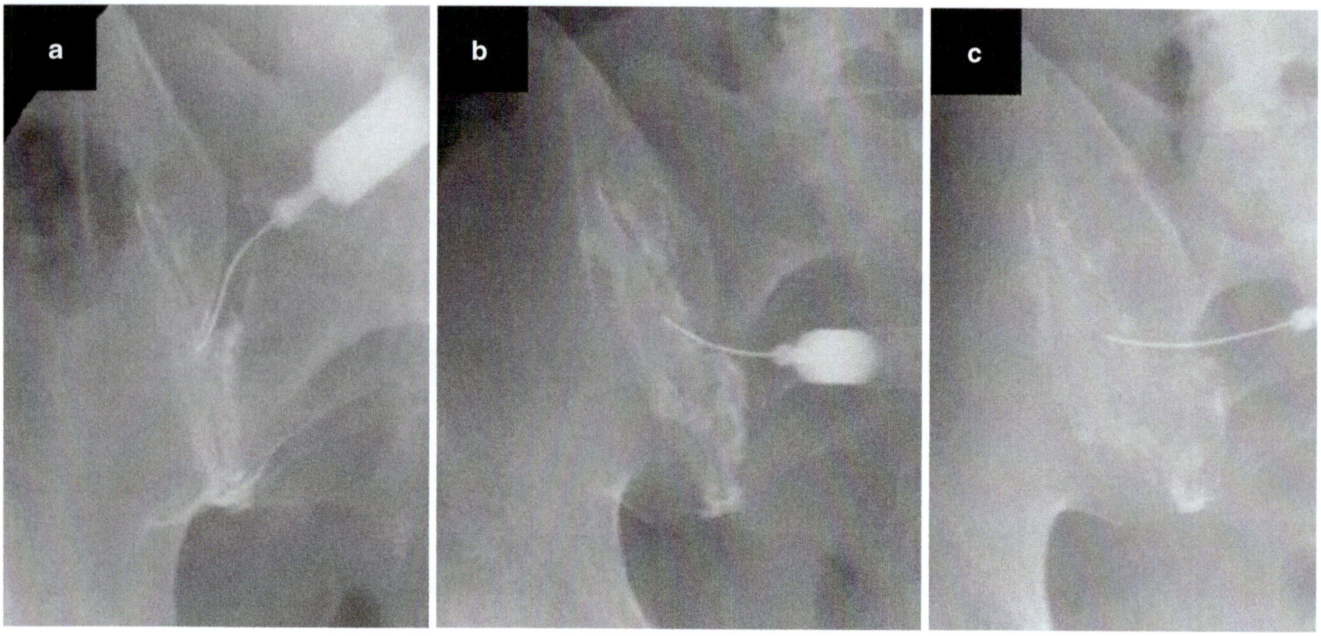


**Fig. 28.17** Injection of contrast medium to outline the joint. (**a**) Injection of contrast medium with the fluoroscopic tube angled caudally at 25–30°. The contrast medium outlines the joint. (**b**) Posteroanterior view after contrast medium injection. (**c**) Contra-oblique view. The needle is inserted into the joint via the middle portion

tip did not insert into the posterior margin of the joint with the posterior and anterior joint lines superimposed, we rotated the patient's body in the retro-oblique direction, such that the anterior and posterior joint lines were separated, to confirm that the needle tip had reached the ilium bone line in the middle portion of the joint (Fig. 28.18).

## True Anteroposterior View

Taheri et al. [72] described a technique without posterior angulation similar to the technique in true AP view similar to the technique described by Khuba et al. [62] without oblique angulation.

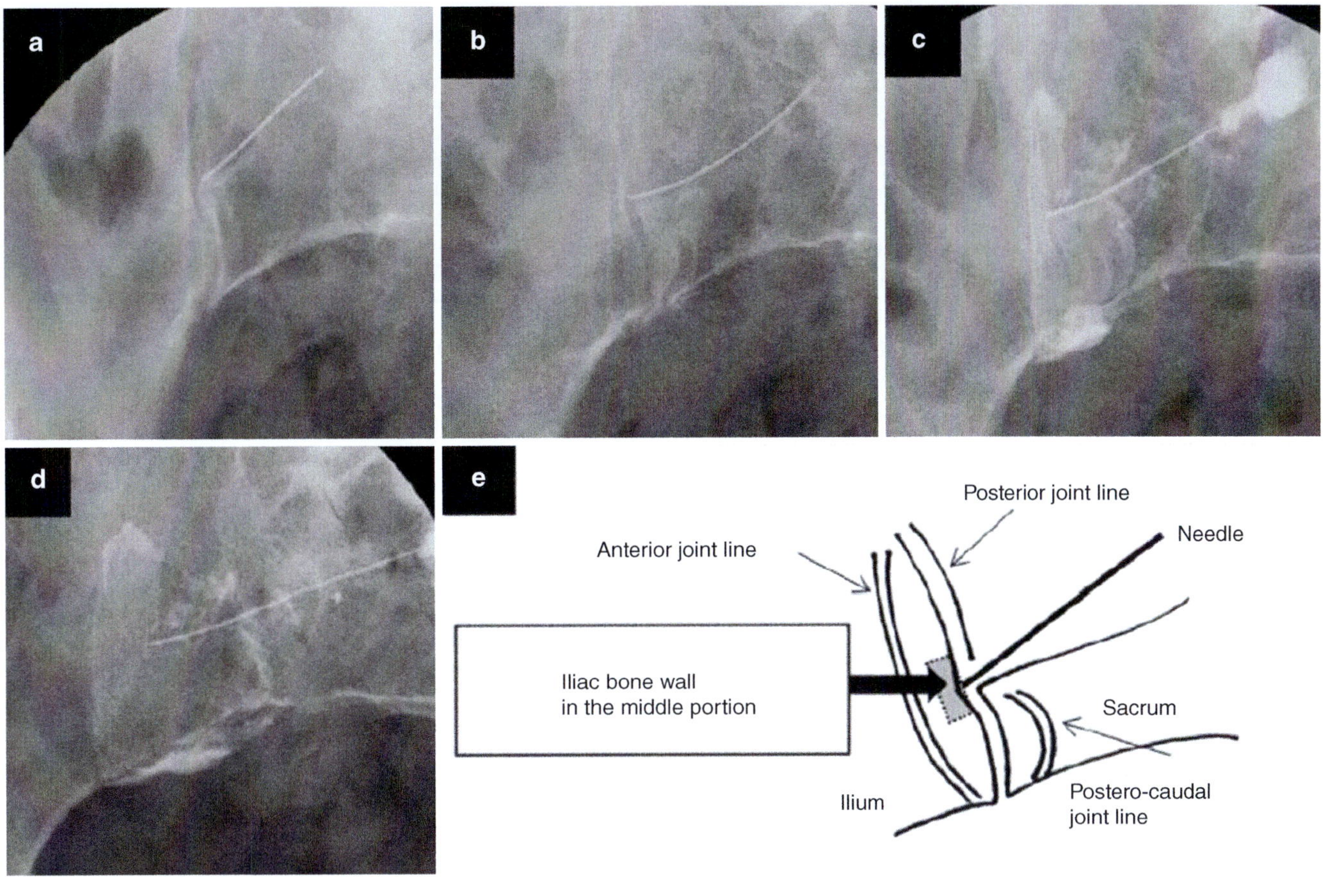


**Fig. 28.18** Retro-oblique view technique. (**a**) Image shows the patient lying prone-oblique with the painful side down and with the fluoroscopic tube angled caudally at 30°. The posterior and anterior joint line are superimposed. The needle seems to be blocked and cannot reach the posterior margin of the joint. (**b**) Radiograph after rotating the patient's body in a retro-oblique direction shows separation of the anterior and postero-caudal joint lines. Usually, the anterior joint line comes to be observed laterally and posterocaudal joint line moves medially. It can be confirmed that the needle tip has access to the ilium bone line beyond the sacrum boundary in the middle portion. (**c**) Radiograph displays contrast medium outlining the joint. Initially, the needle is advanced until it reaches the ilium bone wall. Subsequently, the needle is slightly retracted while contrast medium is simultaneously injected. This slight retraction allows the contrast medium to flow freely into the joint cavity. (**d**) Radiograph displays the contra-oblique view. The needle has been inserted into the joint via the middle portion. (**e**) Diagram of the retro-oblique view technique

## Precautions

- As with most procedures, a practitioner must be mindful of patient specific issues:
  - There should be careful consideration of risk versus benefit of SIJ interventions in patients with either acquired or iatrogenic coagulopathy.
  - Individuals with severely immunocompromised states may not be ideal for steroid injections, which have the potential to worsen any underlying infections.
- Guidelines on anticoagulants and antithrombotics are described in Chap. 15 [73] and also on the American Society of Interventional Pain Physicians (ASIPP) app, which is regularly updated.

## Conclusion

PRP and stem cell injections can potentially represent a viable treatment option for sacroiliac joint pain. Double-blind RCTs have shown both PRP and prolotherapy to be effective in treating varying regions of the body. It is therefore reasonable to consider that PRP could be similarly effective in treating pain related to the sacroiliac joints. At the very least, PRP is autologous and has limited side effects other than a temporary increase in pain. Further investigation with RCTs, larger patient numbers, and longer follow-up periods are needed.

**Key Points**

1. Sacroiliac joint dysfunction refers to pain in the sacroiliac joint caused by ligament issues, leading to hypermobility or hypomobility.
2. Diagnosis of sacroiliac joint dysfunction can be challenging due to non-specific symptoms and lack of detectable pathology on imaging tests.
3. The sacroiliac joints are synovial joints formed between the sacrum and ilium bones, acting as shock absorbers and allowing small movements like nutation and counternutation. Ligaments, including the anterior, dorsal interosseous, dorsal posterior, sacrotuberous, and sacrospinous ligaments, provide stability to the joint. The sacroiliac joint is innervated by a rich supply of nerves from spinal segments L2-S3, explaining the varied referral patterns of pain experienced by patients.
4. Clinical presentation of sacroiliac joint dysfunction includes pain in the pelvis, buttock, hip, and groin regions, with radiation down the leg. Symptoms may include numbness, tingling, weakness, instability, and urinary frequency. The pain is usually unilateral and worsens with movement, especially turning in bed, standing up, and climbing stairs. Diagnosis involves patient history, physical examination, and diagnostic tests such as provocative tests and sacroiliac joint injections under guidance, with significant pain reduction confirming the diagnosis.
5. Conservative care, including ice, anti-inflammatory medicines, physiotherapy, and injections, is the initial approach for sacroiliac joint patients, with surgical fusion as an option for those who fail conservative care. However, the anti-inflammatory effect of injection therapy may not be permanent, leading some patients to explore regenerative medicine therapies such as platelet-rich plasma (PRP), and stem cell therapies.
6. PRP is a concentrate of plasma rich in platelets and growth factors released from platelets in PRP, such as PDGF, TGF-β, VEGF, EGF, bFGF, and IGF-1, play a crucial role in the healing process by increasing metabolic activity, promoting angiogenesis, and enhancing the strength of new tissue.
7. Various techniques for sacroiliac joint injection under fluoroscopic guidance have been described.
8. These techniques aim to accurately enter the sacroiliac joint, confirm needle position with contrast medium, and facilitate successful intraarticular injection.

**Acknowledgments** This book chapter is modified and updated from a previous book chapter, "Regenerative Medicine for Sacroiliac Joint Dysfunction" by Joseph Cabaret, MD, Laxmaiah Manchikanti, MD, and Aaron K. Calodney, MD, in *Essentials of Regenerative Medicine in Interventional Pain Management* published by ASIPP Publishing.

## References

1. Cabaret J, Manchikanti L, Calodney AK. Regenerative medicine for sacroiliac joint dysfunction. In: Manchikanti L, Navani A, Atluri S, editors. Essentials of regenerative medicine in interventional pain management. Paducah, KY: ASIPP Publishing; 2019. p. 509–34.
2. Jani S, Simopoulos TT. Sacroiliac joint interventions. In: Manchikanti L, Kaye AD, Falco FJE, Hirsch JA, editors. Essentials of interventional techniques in managing chronic spinal pain. New York, NY: Springer; 2018. p. 337–48.
3. Simopoulos TT, Manchikanti L, Gupta S, et al. Systematic review of the diagnostic accuracy and therapeutic effectiveness of sacroiliac joint interventions. Pain Physician. 2015;18:E713–56.
4. Ko GD, Mindra S, Lawson GE, et al. Case series of ultrasound-guided platelet-rich plasma injections for sacroiliac joint dysfunction. J Back Musculoskelet Rehabil. 2017;30:363–70.
5. Singla V, Batra YK, Bharti N, et al. Steroid vs. platelet-rich plasma in ultrasound-guided sacroiliac joint injection for chronic low back pain. Pain Pract. 2017;17:782–91.
6. Wallace P, Bezjian Wallace L, Tamura S, Prochnio K, Morgan K, Hemler D. Effectiveness of ultrasound-guided platelet-rich plasma injections in relieving sacroiliac joint dysfunction. Am J Phys Med Rehabil. 2020;99:689–93.
7. Chen AS, Solberg J, Smith C, et al. Intra-articular platelet rich plasma vs corticosteroid injections for sacroiliac joint pain: a double-blinded, randomized clinical trial. Pain Med. 2022;23:1266–71.
8. Hoy D, March L, Brooks P, et al. The global burden of low back pain: estimates from the global burden of disease 2010 study. Ann Rheum Dis. 2014;73:968–74.
9. US Burden of Disease Collaborators. The state of US health, 1990-2016: burden of diseases, injuries, and risk factors among US states. JAMA. 2018;319:1444–72.
10. Dieleman JL, Baral R, Birger M, et al. US spending on personal health care and public health, 1996-2013. JAMA. 2016;316:2627–46.
11. Dieleman JL, Cao J, Chapin A, et al. US health care spending by payer and health condition, 1996-2016. JAMA. 2020;323:863–84.
12. Manchikanti L, Kaye AD, Soin A, et al. Comprehensive evidence-based guidelines for facet joint interventions in the management of chronic spinal pain: American Society of Interventional Pain Physicians (ASIPP) guidelines. Pain Physician. 2020;23:S1–S127.
13. Manchikanti L, Soin A, Benyamin RM, et al. An update of the systematic appraisal of the accuracy and utility of discography in chronic spinal pain. Pain Physician. 2018;21:91–110.
14. Manchikanti L, Hirsch JA. Clinical management of radicular pain. Expert Rev Neurother. 2015;17:681–93.
15. Manchikanti L, Knezevic NN, Navani A, et al. Epidural interventions in the management of chronic spinal pain: American Society of Interventional Pain Physicians (ASIPP) comprehensive evidence-based guidelines. Pain Physician. 2021;24:S27–S208.
16. Peterson C, Hodler J. Evidence-based radiology (part 1): is there sufficient research to support the use of therapeutic injections for the spine and sacroiliac joints? Skelet Radiol. 2010;39:5–9.
17. Goldthwait JE, Osgood RB. A consideration of the pelvic articulations from an anatomical, pathological and clinical standpoint. Boston Med Surg J. 1905;152:593–601.
18. Mixter WJ, Barr JS. Rupture of the intervertebral disc with involvement of the spinal cord. N Engl J Med. 1934;211:210–4.
19. Smyth NA, Murawski CD, Fortier LA, et al. Platelet-rich plasma in the pathologic processes of cartilage: review of basic science evidence. Arthroscopy. 2013;29:1399–409.
20. Kon E, Filardo G, Di Matteo B, et al. PRP for the treatment of cartilage pathology. Open Orthop J. 2013;7:120–8.

21. Lee KS, Wilson JJ, Rabago DP, et al. Musculoskeletal applications of platelet-rich plasma: fad or future? AJR Am J Roentgenol. 2011;196:628–36.
22. Navani A, Manchikanti L, Albers SL, et al. Responsible, safe, and effective use of biologics in the management of low back pain: American Society of Interventional Pain Physicians (ASIPP) guidelines. Pain Physician. 2019;22:S1–S74.
23. Manchikanti L, Navani A. Evolution of regenerative medicine in managing chronic pain. In: Manchikanti L, Navani A, Atluri S, editors. Essentials of regenerative medicine in interventional pain management. Paducah, KY: ASIPP Publishing; 2018. p. 1–38.
24. Bodor M, Dregella R, Uribe Y. Platelet rich plasma. In: Manchikanti L, Navani A, Atluri S, editors. Essentials of regenerative medicine in interventional pain management. Paducah, KY: ASIPP Publishing; 2018. p. 67–81.
25. Lana JF, Purita J, Rava CA, et al. Bone marrow concentrate. In: Manchikanti L, Navani A, Atluri S, editors. Essentials of regenerative medicine in interventional pain management. Paducah, KY: ASIPP Publishing; 2018. p. 83–98.
26. Lander EB. Adipose cells or stromal vascular fraction. In: Manchikanti L, Navani A, Atluri S, editors. Essentials of regenerative medicine in interventional pain management. Paducah, KY: ASIPP Publishing; 2018. p. 99–106.
27. Miguélez-Rivera L, Pérez-Castrillo S, González-Fernández ML, et al. Immunomodulation of mesenchymal stem cells in discogenic pain. Spine J. 2018;18:330–42.
28. Zeckser J, Wolff M, Tucker J, Goodwin J. Multipotent mesenchymal stem cell treatment for discogenic low back pain and disc degeneration. Stem Cells Int. 2016;2016:3908389.
29. Chen WH, Lo WC, Lee JJ, et al. Tissue-engineered intervertebral disc and chondrogenesis using human nucleus pulposus regulated through TGF-beta1 in platelet-rich plasma. J Cell Physiol. 2006;209:744–54.
30. Dominici M, Le Blanc K, Mueller I, et al. Minimal criteria for defining multipotent mesenchymal stromal cells. The International Society for Cellular Therapy position statement. Cytotherapy. 2006;8:315–7.
31. Hunt CL, Her YF, Law LA, et al. Five generations of cell preparation: a translational framework for categorizing regenerative stem cell therapies. J Am Acad Reg Med. 2017;1:7239.
32. Moresco EM, LaVine D, Beutler B. Toll-like receptors. Curr Biol. 2011;21:R488–93.
33. Watters TM, Kenny EF, O'Neill LA. Structure, function and regulation of the toll/IL-1 receptor adaptor proteins. Immunol Cell Biol. 2007;85:411–9.
34. Mascarinas A, Harrison J, Boachie-Adjei K, Lutz G. Regenerative treatments for spinal conditions. Phys Med Rehabil Clin N Am. 2016;27:1003–17.
35. Manchikanti L, Falco FJE, Benyamin RM, Kaye AD, Boswell MV, Hirsch JA. A modified approach to grading of evidence. Pain Physician. 2014;17:E319–25.
36. Manchikanti L, Navani A, Singh V, et al. Intradiscal injection of biologics. In: Manchikanti L, Navani A, Atluri S, editors. Essentials of regenerative medicine in interventional pain management. Paducah, KY: ASIPP Publishing; 2018. p. 385–420.
37. Manchikanti L, Atluri S, Albers SL, et al. Epidural administration of PRP and stem cells. In: Manchikanti L, Navani A, Atluri S, editors. Essentials of regenerative medicine in interventional pain management. Paducah, KY: ASIPP Publishing; 2018. p. 421–57.
38. Navani A, Manchikanti L. Lumbar facet joints and regenerative medicine. In: Manchikanti L, Navani A, Atluri S, editors. Essentials of regenerative medicine in interventional pain management. Paducah, KY: ASIPP Publishing; 2018. p. 459–72.
39. Sanapati J, Manchikanti L, Atluri S, et al. Do regenerative medicine therapies provide long-term relief in chronic low back pain: a systematic review and metaanalysis. Pain Physician. 2018;2018(21):515–40.
40. Navani A, Gupta D. Role of intra-articular platelet-rich plasma in sacroiliac joint pain. Tech Reg Anesth Pain Manag. 2015;19:54–9.
41. Atluri S, Murphy MB, Dragella R, et al. Evaluation of the effectiveness of autologous bone marrow mesenchymal stem cells in the treatment of chronic low back pain due to severe lumbar spinal degeneration: a 12-month, open-label, prospective controlled trial. Pain Physician. 2022;25:193–207.
42. Manchikanti L, Atluri S, Hirsch JA. Comment on "intra-articular platelet-rich plasma vs placebo injection on pain and medial tibial cartilage volume in patients with knee osteoarthritis: the restore randomized clinical trial". Pain Physician. 2022;25:E698.
43. Fortin JD, Dwyer AP, West S, et al. Sacroiliac joint: pain referral maps upon applying a new injection/arthrography technique. Part I: asymptomatic volunteers. Spine (Phila Pa 1976). 1994;19:1475–82.
44. Fortin JD, Aprill CN, Ponthieux B, et al. Sacroiliac joints: pain referral maps upon applying a new injection/arthrography technique. Part II: clinical evaluation. Spine (Phila Pa 1976). 1994;19:1483–9.
45. Fortin JD. Sacroiliac joint dysfunction: a new perspective. J Back Musculoskelet Rehabil. 1993;3:31–43.
46. Forst SL, Wheeler MT, Fortin JD, et al. The sacroiliac joint: anatomy, physiology, and clinical significance. Pain Physician. 2006;9:61–7.
47. Simpson R, Gemmell H. Accuracy of spinal orthopedic tests: a systematic review. Chiropr Osteopat. 2006;14:26.
48. Young S, April C, Laslett M. Correlation of clinical examination characteristics with three sources of chronic low back pain. Spine J. 2003;3:460–5.
49. Borowsky CD, Fagen G. Sources of sacroiliac region pain: insights gained from a study comparing standard intra-articular injection with a technique combining intra- and peri-articular injection. Arch Phys Med Rehabil. 2008;89:2048–56.
50. Fortin JD, Falco FJ. The Fortin finger test: an indicator of sacroiliac pain. Am J Orthop (Belle Mead NJ). 1997;26:477–80.
51. Cramer G, Darby S. Basic and clinical anatomy of the spine, spinal cord and ANS. 2nd ed. St. Louis, MO: Elsevier; 2005.
52. Netter FH. Atlas of human anatomy. 4th ed. Philadelphia, PA: Elsevier (Saunders); 2006.
53. Cox RC, Fortin JD. The anatomy of the lateral branches of the sacral dorsal rami: implications for radiofrequency ablation. Pain Physician. 2014;17:459–64.
54. Solonen K. The sacroiliac joint in the light of anatomical, roentgenological and clinical studies. Acta Orthop Scand. 1957;27:1–127.
55. Ikeda R. Innervation of the sacroiliac joint. Macroscopical and histological studies article in Japanese. Nihon Ika Daigaku Zasshi. 1991;58:587–96.
56. Grob K, Neuhuber W, Kissling R. [Die innervation des sacroiliacalgelenkes beim menschen] article in German. Z Rheumatol. 1995;54:117–22.
57. McGrath MC, Zhang M. Lateral branches of dorsal sacral nerve plexus and the long posterior sacroiliac ligament. Surg Radiol Anat. 2005;27:327–30.
58. Szadek K, Hoogland P, Zuurmond W, de Lange J, Perez R. Nociceptive nerve fibers in the sacroiliac joint in humans. Reg Anesth Pain Med. 2008;33:36–43.
59. Willard F, Carreiro J, Manko W. The long posterior interosseous ligament and the sacrococcygeal plexus. In: Third interdisciplinary world congress on low back and pelvic pain. Rotterdam: ECO; 1998. p. 207–9.

60. Yin W, Willard F, Carreiro J, Dreyfuss P. Sensory stimulation-guided sacroiliac joint radiofrequency neurotomy: technique based on neuroanatomy of the dorsal sacral plexus. Spine (Phila Pa 1976). 2003;28:2419–25.
61. Fortin JD, Sehgal N. Sacroiliac joint injection and arthrography with imaging correlation. In: Lennard TA, editor. Pain procedures in clinical practice. 2nd ed. Philadelphia, PA: Hanley and Belfus, Inc.; 2000. p. 265–75.
62. Khuba S, Agarwal A, Gautam S, Kumar S. Fluoroscopic sacroiliac joint injection: is oblique angulation really necessary? Pain Physician. 2016;19:E1135–8.
63. Benyamin, personal communication.
64. Gupta S. Double needle technique: an alternative method for performing difficult sacroiliac joint injections. Pain Physician. 2011;14:281–4.
65. Daitch J, Frey M, Snyder K. Modified sacroiliac joint injection technique. Pain Physician. 2006;9:367–8.
66. Park J, Park HJ, Moon DE, et al. Radiologic analysis and clinical study of the upper one-third joint technique for fluoroscopically guided sacroiliac joint injection. Pain Physician. 2015;18:495–503.
67. Kurosawa D, Murakami E, Aizawa T. Fluoroscopy-guided sacroiliac intraarticular injection via the middle portion of the joint. Pain Med. 2017;18:1642–8.
68. Kasliwal PJ, Kasliwal S. Fluoroscopy-guided sacroiliac joint injection: description of a modified technique. Pain Physician. 2016;19:E329–38.
69. Do KH, Ahn SH, Jones R, et al. A new sacroiliac joint injection technique and its short-term effect on chronic sacroiliac region pain. Pain Med. 2016;17:1809–13.
70. Liliang PC, Liang CL, Lu K, et al. Modified fluoroscopy-guided sacroiliac joint injection: a technical report. Pain Med. 2014;15:1477–80.
71. Maigne JY, Aivakiklis A, Pfefer F. Results of sacroiliac joint double block and value of sacroiliac pain provocation test in 54 patients with low back pain. Spine (Phila Pa 1976). 1996;21:1889–92.
72. Taheri A, Lajevardi M, Abad M, et al. Sacroiliac joint intraarticular injection in true anteroposterior view: description of a new C-arm guided method. Pain Physician. 2018;21:61–6.
73. Manchikanti L, Sanapati M, Kaye AD, Kaye AM. Antiplatelet and anticoagulant therapy in regenerative medicine. In: Manchikanti L, Navani A, Atluri S, Sanapati M, editors. Manchikanti's essentials of regenerative medicine in interventional pain management. 2nd ed. New York, NY: Springer; 2023.

# Part IV

# Miscellaneous Applications

# Regenerative Therapy for Chronic Pelvic Pain Syndromes

# 29

Sheldon Jordan, Andrea Trescot, and Laxmaiah Manchikanti

## Introduction

Chronic pelvic pain (CPP), a specific subset of chronic pain, is an important yet underreported multifactorial condition for both genders [1–5]. Rather than a single objective diagnosis, CPP represents a number of complex, potentially debilitating, disorders defined as, "A nonmalignant pain perceived in the pelvis in either men or women … in the case of documented nociceptive pain that becomes chronic, the pain must have been continuous for at least 6 months [6]." A classic well-known definition is that CPPS is pain located in the pelvic area that lasts over 6 months and is severe enough to limit functioning, unrelated to menstrual cycle, pregnancy, local trauma or pelvic operations in women [5]. The National Institute of Health (NIH) defined CPPS as, "Lower genitourinary symptoms, particularly pain in the perineum or genitalia, with voiding symptoms, such as dysuria or frequency, and sexual dysfunction… [7]".

Although the prevalence of CPP in both genders of the general population is not fully established, estimated prevalence ranges from 5.7% to 26.6% in females and 2–16% in males [8–11].

Based on the above definitions, it is clear that CPP is a nonspecific, all-inclusive term that includes a variety of diagnosis ranging from pudendal neuralgia and coccydynia to prostatodynia, vulvodynia, and painful bladder syndromes. In patients seen in interventional pain management settings, the usual symptoms of CPP include neuropathic symptoms such as paresthesia, numbness, burning, or lancinating pain in the pelvis, anus, and/or genitals. Pain is often initiated or exacerbated with sitting, urinating, defecating, intercourse, and ejaculation [3].

In a comprehensive review, Grinberg et al. [5] provided an overview of indications regarding the management of chronic pelvic pain syndrome (CPPS) patients, with management requiring a holistic approach, with inclusion of either interventional techniques or regenerative medicine techniques. They considered painful bladder syndrome (PBS) and provoked vestibulodynia (PVD) as subgroups [11] (Fig. 29.1). The management strategies include cognitive behavioral therapy, medications, surgery as a last resort, and myofascial physical therapy.

In addition to these described approaches, interventional techniques with various types of nerve blocks, including trigger point injections and Botox injections, superior hypogastric block, ganglion impar block, peripheral nerve blocks, pudendal nerve blocks, sacral nerve blocks, along with ablation in neuromodulation of these nerves including dorsal column and spinal cord stimulation. There is significant paucity of literature with application of regenerative techniques in managing CPP.

This chapter will focus on approaches that would fall under the rubric of regenerative medicine. Regenerative medicine may be defined as optimizing the body's endogenous mechanisms of self-repair. Regeneration involves delivering specific types of cells or cell products to diseased tissues or organs, where they will ultimately restore tissue and organ function. This can be done through cell-based therapy or by using cell products, such as growth factors. Other naturally occurring substances such as hormones may be considered in the armamentarium. Furthermore, the ability of exercise, dietary modifications, and mindfulness strategies to optimize natural functions may be considered part of the therapeutic approach.

S. Jordan
Neurological Associates of West Los Angeles, Santa Monica, CA, USA
e-mail: sjordan@theNeuroAssociates.com

A. Trescot
Physician Partners of America (PPOA), Orange Park, FL, USA

Curonix, Pompano Beach, FL, USA

L. Manchikanti (✉)
Pain Management Centers of America, Paducah, KY, USA
e-mail: drlm@thepainmd.com


A. Navani et al. (eds.), *Essentials of Regenerative Medicine in Interventional Pain Management*,
https://doi.org/10.1007/978-3-031-50357-3_29

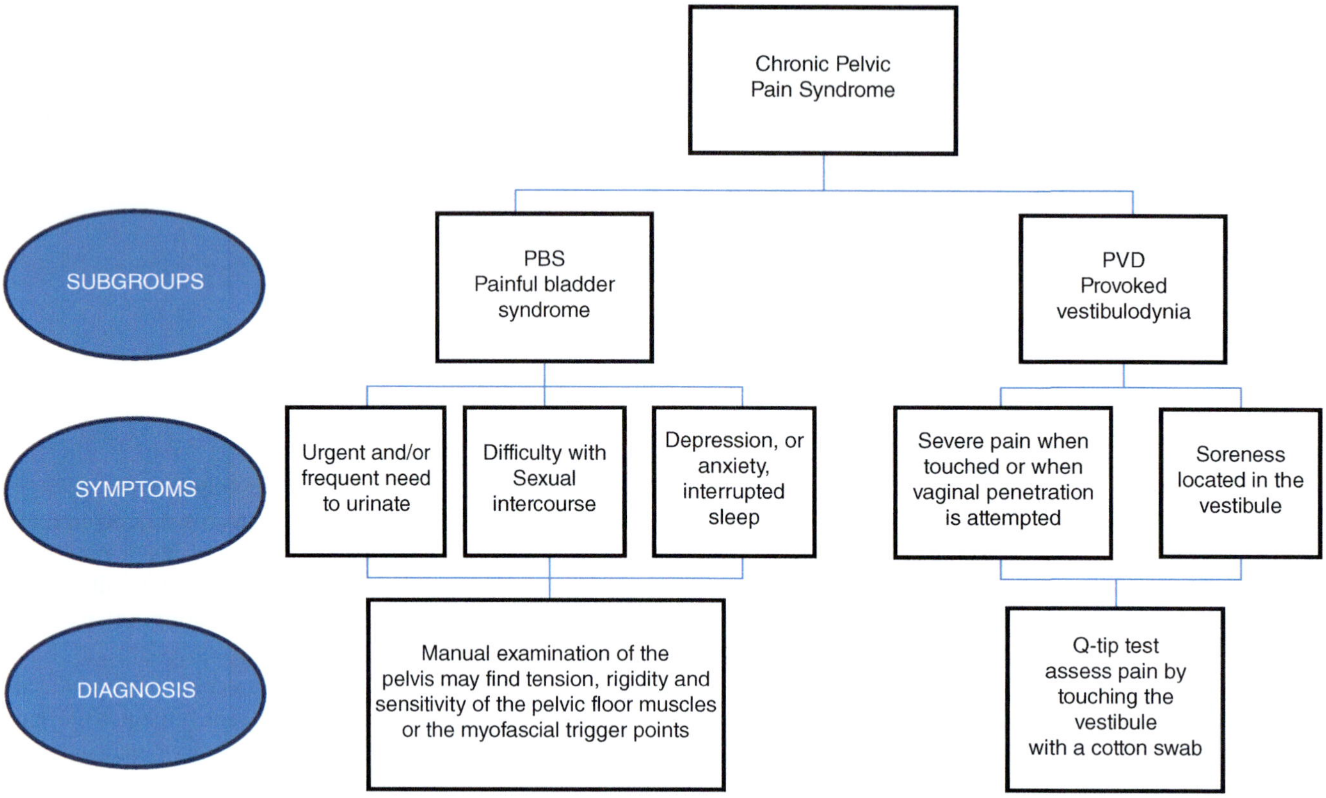


**Fig. 29.1** Symptoms and diagnosis of subgroups of chronic pelvic pain syndrome (CPPS). (**Adapted from:** Grinberg K, Sela Y, Nissanholtz-Gannot R. New insights about chronic pelvic pain syndrome (CPPS). *Int J Environ Res Public Health* 2020; 17:3005 [5])

## Pathophysiology

Multiple mechanisms have been suggested as the pathophysiologic basis of CPPS [5]:

1. Infection process, but no conclusive evidence has been found and antibiotic treatment is ineffective [12];
2. A neurogenic inflammation that includes local changes [8, 11];
3. Hypoxia. A disrupted blood flow to the pelvic area, reduced microvascular density of the bladder's submucosa layer, followed by decreased perfusion [13] is supported by clinical improvement following hyperbaric treatment [14–16]; and
4. Weakness or cramps of the pelvic floor muscles [17, 18].

However, none of the mechanisms have been described as the sole cause of the syndrome, and it is estimated that these mechanisms interact. Further, a common explanation is that, for an unknown reason, the *glycosaminoglycan* (GAG) layer that coats the mucous of the bladder and vagina is damaged. This damage leads to a chain of nerve cell processes that culminates in neurogenic inflammation and mast cell activation [19]. The GAG layer, which is impermeable barrier of urine solutes, becomes permeable. The infiltration of solutes into the submucosa irritates the nerve endings and creates heightened inflammation agents such as vasoactive intestinal polypeptides, substance P, and acetylcholine [19, 20]. Consequently, activated mast cells secrete histamines that widen blood vessels and generate local inflammation [21]. The inflammation irritates the C fibers and causes release of more inflammation agents, which in turn causes additional damage and the inflammation of fibrosus on the bladder [15, 21] or in the vagina [20–22].

Grinberg et al. [5] described PBS and PVD, as subgroups of CPPS [11]. These involve the dysfunction of the general pain system, not just pelvic dysfunction, with research findings expressing allodynia and hyperalgesia in the projected pain area above the pelvis [5, 15, 23, 24]. Women with PVD, in comparison to healthy women, presented increased sensitivity to pain in body areas distant from the genital area [5] conveyed by allodynia and hyperalgesia. It was also described that women suffering with PVD demonstrated disruptions of pain modulation, as well as increased central sensitization [25]. Thus far, two imaging studies examined central pain process in PVD in women and both found high activation response to painful experimental stimulation [26, 27]. The studies also showed reduced grey matter in anterior cingulate cortex in these individuals [28]. Similar neuroanatomic changes found in chronic pain conditions such as chronic back pain, inflammatory bowel syndrome (IBS), migraines, and phantom pains have been shown with relative decrease of grey matter volume [5]. In addition, structural changes related to depression and other emotional aspects that accompany chronic pain may have a significant role in processing emotional information in general and that related to chronic pain specifically. Overall, these findings indicate hypersensitive pain system among women with PBS and PVD, raising the possibility that dysfunction of the central pain system is at the core of CPPS. Coexistence of conditions such as fibromyalgia and IBS in approximately 19% meeting criteria for chronic pain in fibromyalgia indicates hypersensitivity syndrome.

Associated CPP symptoms in females may include pain with sexual intercourse, unexplained genital pain, and/or itching, menstrual pain and pelvic cramping. In males, chronic prostatitis/CPPS, also known as NIH Category 3 prostatitis, is defined by the NIH as, "The presence of genitourinary pain in the absence of uropathogenic bacteria detected by standard microbiological methodology [29]". The main CPP domains in males include urogenic discomfort, sexual dysfunction, urinary tract symptoms (i.e., pain with voiding and perineal pain). Consequently, it has been postulated that CPP is closely linked with sexual dysfunction in both males and females, impacting emotional status and overall quality of life [1].

CPP and substance use disorders have been described to share considerable psychosocial factors, such as post-traumatic stress disorder (PTSD) history, trauma, sleep disorders, anxiety, and depression [30–32]. Further, CPP and substance use disorders also intersect neurobiologically, especially when considering neural pain referred pathways [33]. With overlapping factors contributing to CPP and opioid use disorder (OUD), predictable comorbidity between these two disorders has been proposed, thus complicating their management by one another [2]. Reichmann et al. [2] proposed that poorly controlled CPP pain may lead to instability among patients receiving medication treatment for OUD, potentially manifesting as increased opioid cravings and relapsed risk. However, existing data on substance use among CPP patients is limited, but when reported, suggest that substance use may be more prevalent among these patients [34, 35]. This is also complicated by the fact that OUD and OUD medication treatment have been linked with sexual dysfunction, which may further complicate recovery and other health outcomes for these patients [2].

A variety of gender and organ-specific conditions have been described with risk for development of CPP as shown in Tables 29.1 and 29.2 [10, 36]. Consequently, CPP is a diagnostic exclusion likely encompassing other pathologic states, and often comprises and evolution of those states into a clearly chronic neuropathic pain condition. Thus, it has been compared to CRPS due to the many similarities the two conditions share, along with fibromyalgia and IBS, which are also described as neuropathic conditions [5, 6,

**Table 29.1** Gender-specific concerns [10, 36]

| Women | Men |
|---|---|
| Infection | Prostatitis |
| Endometriosis | Chronic orchalgia |
| Dysmenorrhea | Prostadynia |
| Dysparenia | Interstitial cystitis |
| Myofascial pain syndrome | Scrotal pain |
| Vulvodynia | Penile pain |
| Vulvitis | Ureteral obstruction |
| Cystitis | Irritable bowel syndrome |
| Ovarian remnant congestion | Sexual/physical abuse |
| Sympathetically mediated pain | Cancer pain |
| Pelvic congestion | Psychiatric disorders |
| Pelvic fibrosis | Proctalgia fugax |
| Pelvis neurodystonica | Radiation proctitis |
| Irritable bowel syndrome | Surgical procedures (adhesions) |
| Sexual/physical abuse | |
| Cancer pain | |
| Psychiatric disorders | |
| Surgical procedures (adhesions) | |

**Table 29.2** Organ-specific concerns [10, 36]

| Organ system | Organ-specific pathologies |
|---|---|
| Reproductive | Visceral: uterus, ovaries, bladder, urethra |
| | Somatic: skin, vulva, clitoris, vaginal canal |
| | Adhesions, endometriosis, salpingo-oophoritis |
| Vascular | Dilated pelvic vein/pelvic congestions theory |
| Musculocutaneous | Ligamentous structures, muscular (iliopsoas, piriformis, quadrates lumborum, sacroiliac joint, obturator internus, pubococcygeus) |
| | Skeletal (referred pain) |
| | Pelvic floor muscle tension/spasm |
| Spinal | Degenerative joint disease, disc herniation, spondylosis, neoplasm of spinal cord/sacral nerve, coccydynia, degenerative disease |
| Neurologic | Neuralgia/cutaneous nerve entrapment (surgical scar in the lower part of the abdomen), iliohypogastric, ilioingiunal, genitofemoral, lateral femoral cutaneous nerve, PN, shingles (herpes zoster infection), spine-related nerve compressions |
| Gastrointestinal | Irritable bowel syndrome, abdominal epilepsy, abdominal migraine, recurrent small bowel obstruction, hernia |
| Urologic | Bladder dysfunction, chronic (nonbacterial) prostatitis, chronic orchalgia, and prostatodynia |
| Psychological (psychosocial/sexual) | Anxiety, depression, somatization, physical or sexual abuse, drug addiction, dependence, family problems, sexual dysfunction |

10, 28–38]. In reference to the similarities to CRPS, allodynia and hyperesthesia in the affected areas with extreme pain from normally non-noxious stimulation like urination, sensation of a full bladder, ovulation, or sexual intercourse are found. Central sensitization [39], dysfunction, autonomic dysregulation (pelvic congestion), and a significant sympathetic component has been described at least in early stages. Responding sympathetic blocks similar to other CRPS conditions with ganglion impar or hypogastric plexus blocks [40].

In many cases of CPP, there is a history of injury, related surgical procedure or medical condition (i.e., endometriosis, ectopic pregnancy infection, etc.) that one can point to, predating the development of CPP, which may have precipitated a cascade of events leading to this neuropathic syndrome [39]. Similar to CRPS, these events may be innocuous or routine surgical procedures resulting in a nerve injury that fails to heal [41]. As described above, CPP also shares considerable similarities with substance use disorders with psychosocial factors, including PTSD history, trauma, sleep disorders, anxiety, and depression [5, 31, 32].

## Anatomy

Pelvic bone and articular structures such as the symphysis pubis and sacroiliac joints may sometimes be present in the setting of pelvic regional pain. During examination and in the process of trying to identify painful structures, it would be helpful to refer to relevant anatomy. The pelvic bony structures include the lower lumbar vertebrae, the innominate bones (joined anteriorly at the symphysis pubis and posteriorly on the sacrum at the sacroiliac joint), the fused sacrum, and one to four coccygeal bones connected to the sacrum by the fibrous sacrococcygeal joint, eight sacral foramina (four anterior, four posterior), and as sacral hiatus [3, 4, 42–56] (Fig. 29.2).

Nerves that are frequently involved in pelvic pain syndromes include the genitofemoral, ilioinguinal, pudendal and posterior femoral cutaneous nerves. Genitofemoral and ilioinguinal neuralgia may be due to intrinsic entrapment but many more are related to surgical injuries including those from mesh placements for treatment of hernias. Posterior femoral cutaneous neuropathy is caused by nerve injury at the ischial tuberosity and is associated with painful sitting.

Painful sitting would also suggest pudendal neuropathy; however, the latter may be associated with penile pain radiation, perianal pain, perineal paresthesias and a Tinel's sign

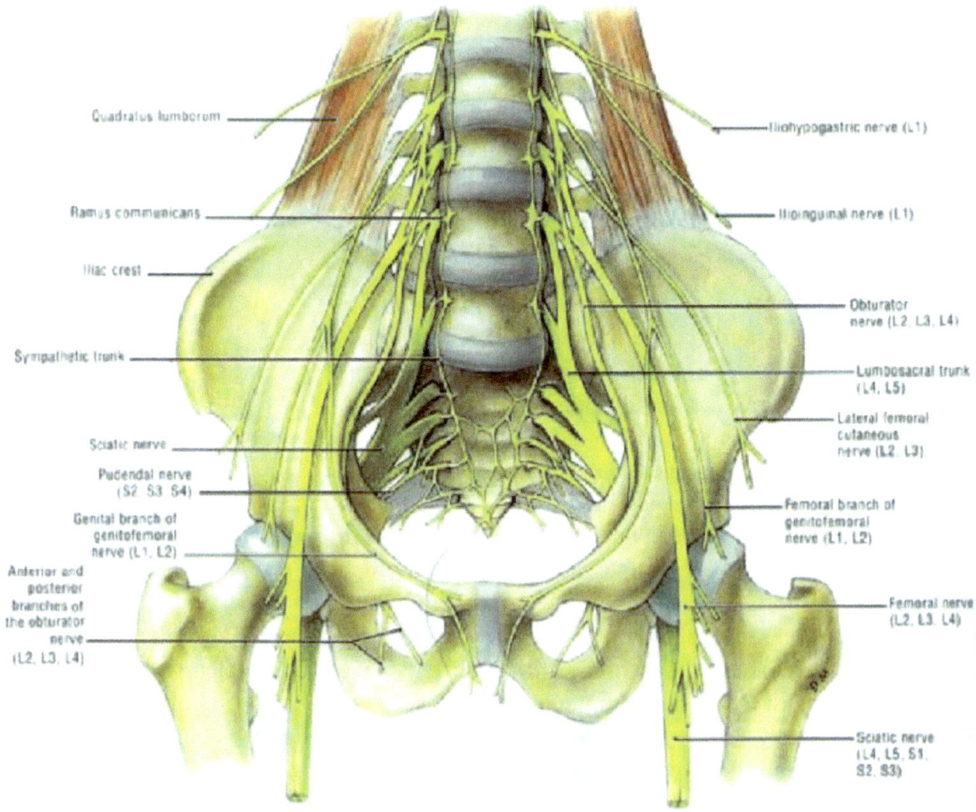


**Fig. 29.2** Illustration of anatomy of pelvic nerves. (**Source:** Netter medical illustration used with permission from Elsevier. All rights reserved)

over the pudendal canal. The high frequency of anomalous changes along the course of this nerve in cadaver dissections makes it perilous to establish a diagnosis based on anatomical imaging alone. Although there has been some demonstrated value of physiological confirmation with pudendal nerve latency testing [47], there is little in the literature to demonstrate that this diagnostic test predicts the outcome of targeted therapy. As with many regional pain syndromes, numerous factors degrade the predictive value of diagnostic testing including the common presence of comorbid conditions, the presence of overall pain sensitivity as well as the non-specificity of imaging and physical exam findings. Selective nerve blockade with proper controls for placebo effects and examiner biases would be a reasonable step for improving the prediction of targeted treatment.

Occasionally other nerves may be involved in patients presenting with pelvic pain. In order to provide a more detailed picture the following discussion includes a more comprehensive discussion of the relevant anatomy.

The somatic and visceral nerves of the pelvis involve the dermatomes at the T12-L3 and S2–4 levels. This includes the iliohypogastric, the ilioinguinal, genitofemoral, lateral femoral cutaneous, and the pudendal nerves. Except for the iliohypogastric and pudendal nerves, these nerves are afferent (sensory) with no significant efferent (motor) component [3, 4, 42–46].

## Anterior (Abdominal) Cutaneous Nerve

From superficial to deep, the abdominal wall is composed of skin, subcutaneous tissues, the external abdominal oblique, the internal abdominal oblique and the transversus abdominis muscles [44, 45]. Anterior divisions of the intercostal nerves T7-L1 traverse the abdominal wall extending toward midline. The intramuscular plane between the internal abdominal oblique and transversus abdominis muscle is termed transversus abdominis plane (TAP) (Fig. 29.3).

## Ilioinguinal, Iliohypogastric and Genitofemoral Nerves

Iliohypogastric, ilioinguinal, and genitofemoral nerves are known as "border nerves" because these nerves supply the skin that borders the area between the abdomen and thigh [4, 48].

Both the iliohypogastric and ilioinguinal nerves arise from the anterior rami of L1 with contributing filaments from the T12 (Figs. 29.4 and 29.5). Emerging from the lateral border of the psoas major muscle, both nerves run subperitoneally in front of quadratus lumborum before piercing the transverse abdominis muscle above the iliac crest [4, 49].

The iliohypogastric nerve runs downward and forward in the neurovascular plane between the internal and external obliques, and pierces and innervates the internal oblique muscle above the anterior superior iliac spine (ASIS). The iliohypogastric nerve then pierces the aponeurosis of the external oblique muscle an inch or so above the superficial inguinal ring and provides sensory fibers to the skin over the lower part of the rectus abdominis, including the skin of mons.

The ilioinguinal nerve runs parallel and below the iliohypogastric nerve and pierces the lower border of internal oblique muscle. It passes between the crura of the superficial inguinal ring in front of the spermatic cord. It supplies the

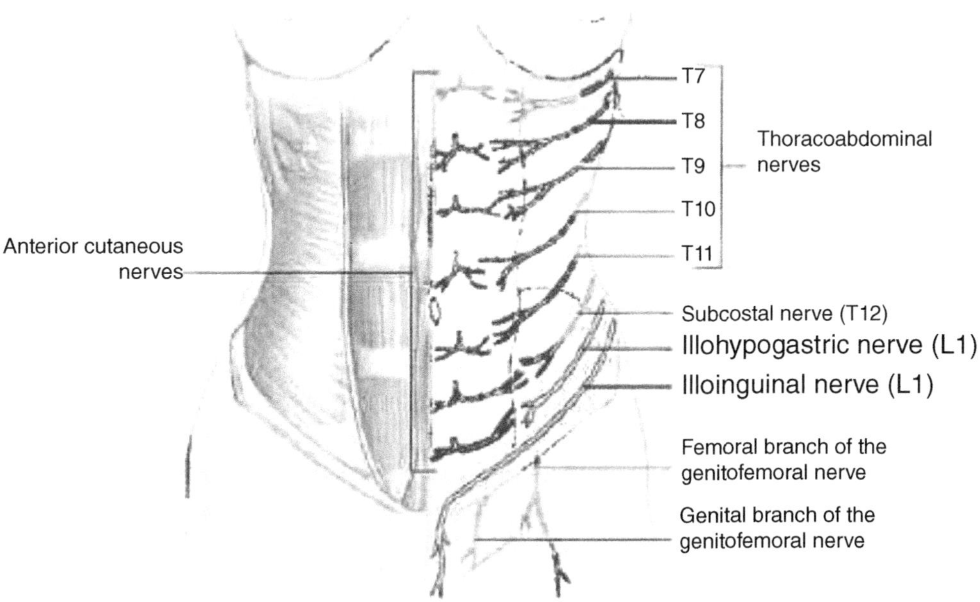


**Fig. 29.3** Anatomy of the abdominal wall. (**Source:** Amin N, Krashin D, Trescot AM. Ilioinguinal and iliohypogastric nerve entrapment: Abdominal. In: Trescot AM (ed). *Peripheral Nerve Entrapments: Clinical Diagnosis and Management.* Springer, Switzerland, 2016, pp. 413–424 [45])

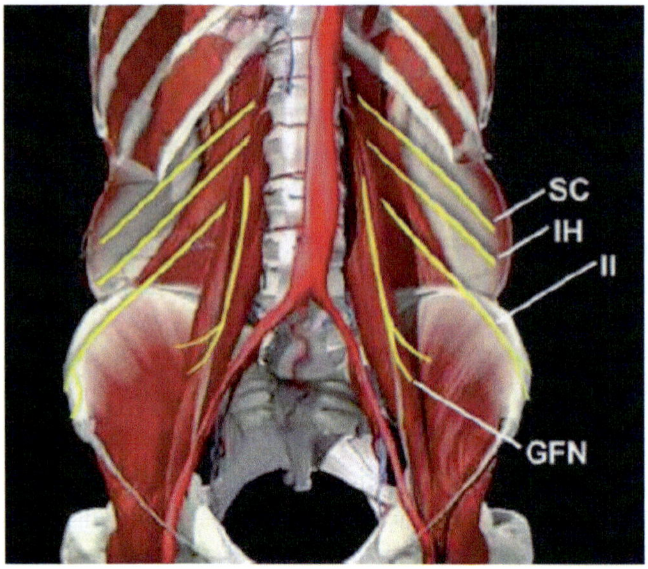


**Fig. 29.4** Genitofemoral nerve, ilioinguinal and iliohypogastric nerves all arise from the lumbar plexus. *SC* subcostal nerve; *IH* iliohypogastric nerve; *II* ilioinguinal nerve, *GFN* genitofemoral nerve

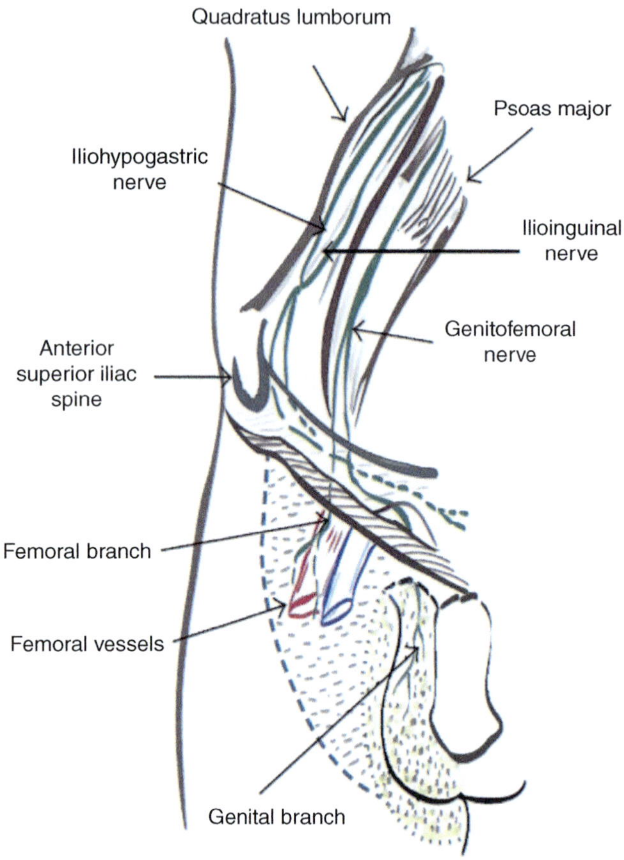


**Fig. 29.5** Anatomy of nerves around the inguinal region

skin of the superomedial area of the thigh, the skin over the root of the penis and anterior scrotum (the mons pubis and labium majus in the female).

The genitofemoral nerve arises from the first and second lumbar nerve roots. The nerve penetrates the psoas muscle at the level of the L3–4 intervertebral disc and comes to lie on its anterior surface, either as a single trunk or separate genital and femoral branch (Figs. 29.4 and 29.5) [48]. It then divides into a femoral and genital branch at a variable distance above the level of the inguinal ligament.

The femoral branch of the genitofemoral nerve follows the external iliac artery and passes with it under the inguinal ligament. It also penetrates the fascia lata in order to supply skin sensations to the femoral triangle. The genital branch of the genitofemoral nerve passes through the internal inguinal ring of the transversalis fascia and then continues into the inguinal canal. The relationship of the genital branch to the spermatic cord in the inguinal canal is highly variable. It can either run outside the spermatic cord in the ventral [4, 48], dorsal, and inferior locations [4, 50], or incorporate with the cremaster muscle [49].

It is important to note that there are many variations of sensory nerves within the inguinal region, with free communication between the branches of the genitofemoral, ilioinguinal, and iliohypogastric nerves. In one cadaver study, it was shown that the Ilioinguinal nerve was solely responsible for cutaneous innervation of the genital branch of genitofemoral nerve in 28% of the dissections and shared innervation with the genital branch of genitofemoral nerve in 8% [4, 48]. The site where the ilioinguinal and iliohypogastric nerves penetrate the different layers of abdominal muscle is also highly variable [51]. The size of the ilioinguinal nerve is inversely proportional to the iliohypogastric nerve. In some patients, the ilioinguinal nerve joins with the iliohypogastric nerve, or one of the nerves is entirely absent.

## Pudendal Nerve

The pudendal nerve is formed from the anterior rami of the second, third, and fourth sacral nerves [4]. Emerging from the anterior sacral foramina, the pudendal nerve is accompanied by the internal pudendal artery and exits the pelvis through the greater sciatic notch [4]. Subsequently, the nerve runs posterior to the sacrospinous ligament at the level of the ischial spine and passes between the sacrospinous and sacrotuberous ligament (interligamentous plane) [4, 52, 53] as shown in Fig. 29.6, with the pudendal artery on the lateral side [54]. The nerve then swings anteriorly to enter the pelvis through the lesser sciatic notch and Alcock's canal [55]. The Alcock's canal is the fascia tunnel formed by the duplication of the obturator internus muscle under the plane of the levator ani muscle on the lateral wall of the ischiorectal fossa [56]. The pudendal nerve subsequently splits into three terminal branches: the dorsal nerve of the penis (or clitoris), the inferior rectal nerve, and the perineal nerve, providing the sensory branches to the skin of the penis (or clitoris), the perianal area, and the posterior surface of the scrotum or labia majora as shown in Figs. 29.7, 29.8, and 29.9. It also innervates the external anal sphincter (inferior rectal nerve) and deep muscles of the urogenital triangle (perineal nerve) [4, 57].

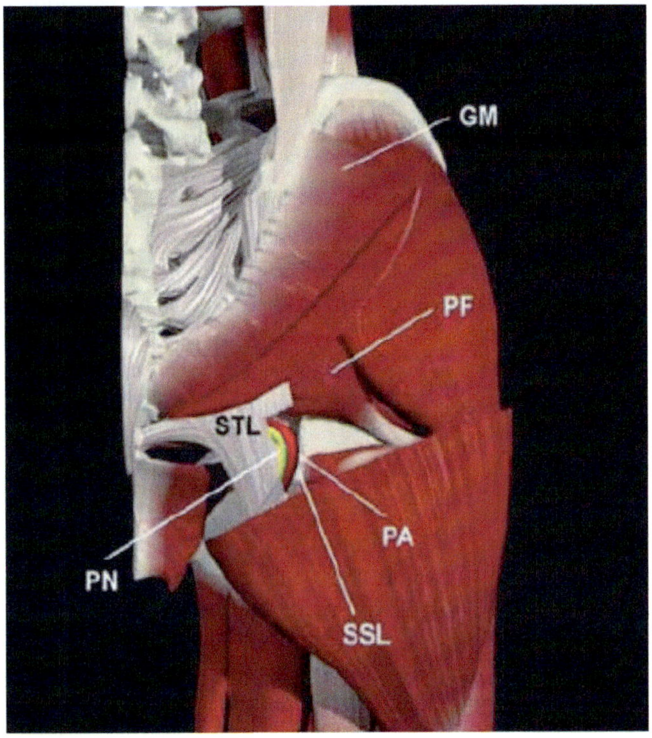


**Fig. 29.6** Anatomy of pelvic region. Gluteus maximus (GM) was partially removed to expose the piriformis muscle (PF), sacrospinous ligament (SSL), sacrotuberous ligament (STL), pudendal artery and nerve (PA and PN)

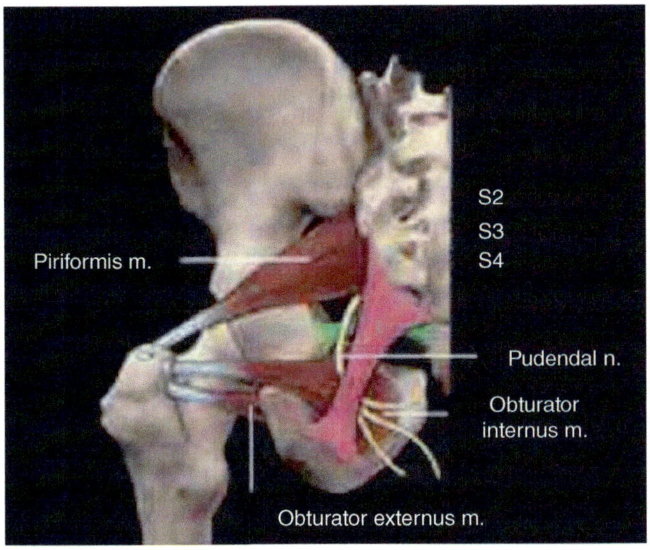


**Fig. 29.7** Illustration of pudendal nerve anatomy

## Piriformis Muscle

The piriformis muscle originates from the anterior surface of the second, third, and fourth sacral vertebrae and the capsule of the sacroiliac joint [4]. It runs laterally and exits the pelvis through the greater sciatic foramen, becomes tendinous, and then inserts into the upper border of the greater trochanter (Figs. 29.6, 29.7, and 29.10). The piriformis acts as an external rotator in the erect position and as an abductor in the supine position.

There are six possible anatomical relationships between the sciatic nerve and the piriformis muscle. The most common arrangement is found when the undivided nerve passes below the piriformis muscle (78–84%) [58, 59]. The second most common arrangement is found when the divided nerve passes through and below the muscle (12–21%). Other combinations include where the undivided nerve passes through the muscle, or the peroneal division (but not the tibial division) passes through the muscle. The aberrant course of the sciatic nerve through the piriformis muscle can cause sciatica and may suggest the important role of a nerve stimulator for the purposes of localization while trying to inject the piriformis muscle [4].

## Regenerative Medicine

Specific evidence demonstrating the effectiveness of regenerative techniques for treatment of pelvic pain syndromes is minimal at present. The best available evidence for utilizing orthobiological techniques for treatment of pelvic pain, for the time being, will depend on considering experiences obtained from treating analogous musculoskeletal targets in other body regions.

Blood and tissue components, characterized by enrichment of anti-inflammatory factors and growth factors, have been extensively studied with sources that include amniotic tissue, autologous blood, autologous adipose and bone marrow, and engineered biologicals. Improvements of pain control, functional and tissue integrity have been demonstrated. Since there is no clear front runner in terms of availability the choice of injectant type may depend on known side effects, invasiveness, and cost. Alpha 2 macroglobulin enhanced plasma fraction is easily obtained from a peripheral vein source [60], it is relatively inexpensive and, as an anti-inflammatory agent, it does not tend to produce any potential irritative effect that may otherwise characterize the use of platelet-rich plasma (PRP). The relative lack of short-term irritating effects would likely be a good choice in the hypersensitized patients with neuropathic pain, opioid hyperalgesia or fibromyalgia.

PRP can be considered in the treatment of pelvic disorders [61–63]. Harvesting of platelets starts with simple veni-

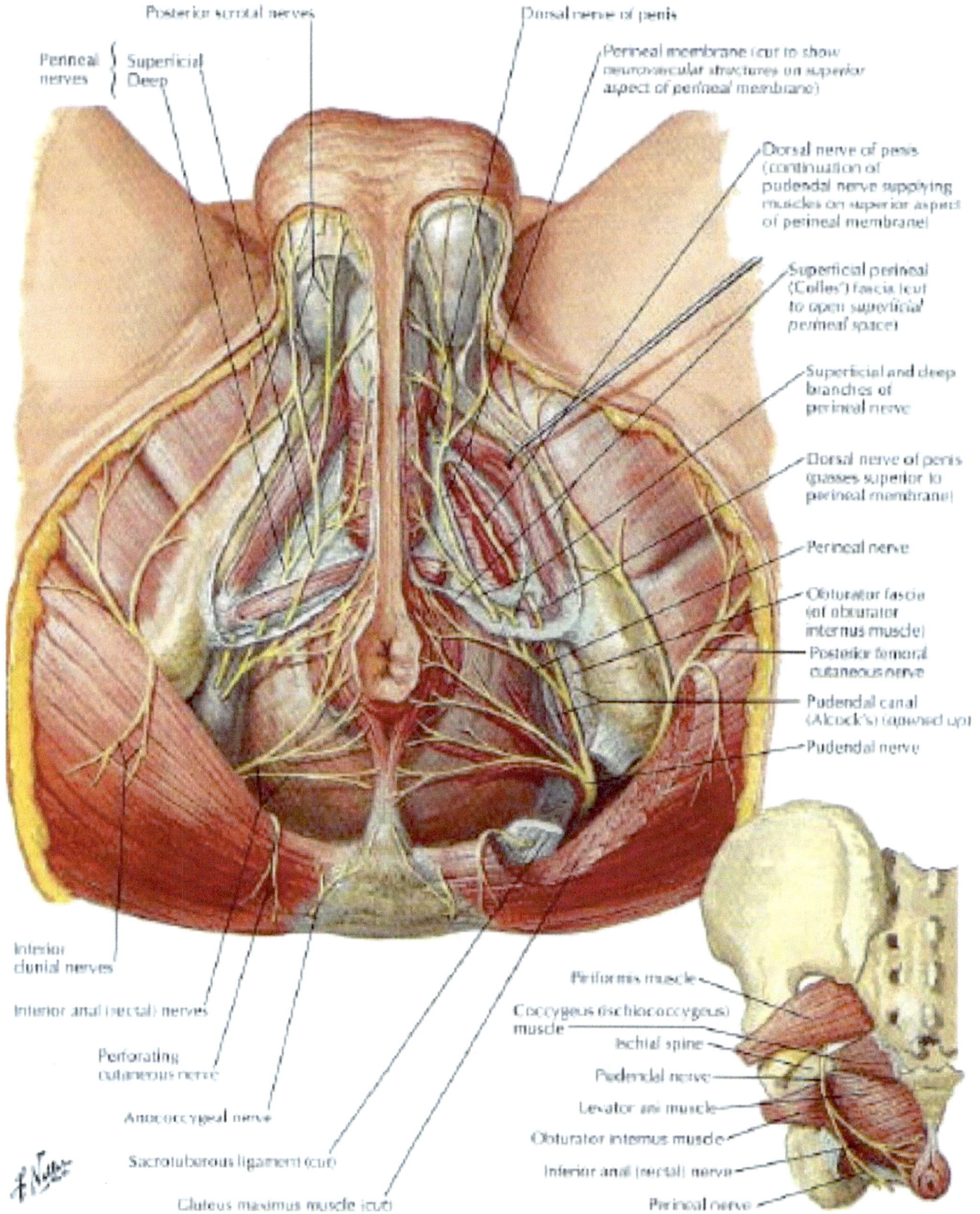


**Fig. 29.8** Nerves of perineum: male. (**Source:** Netter medical illustration used with permission of Elsevier. )

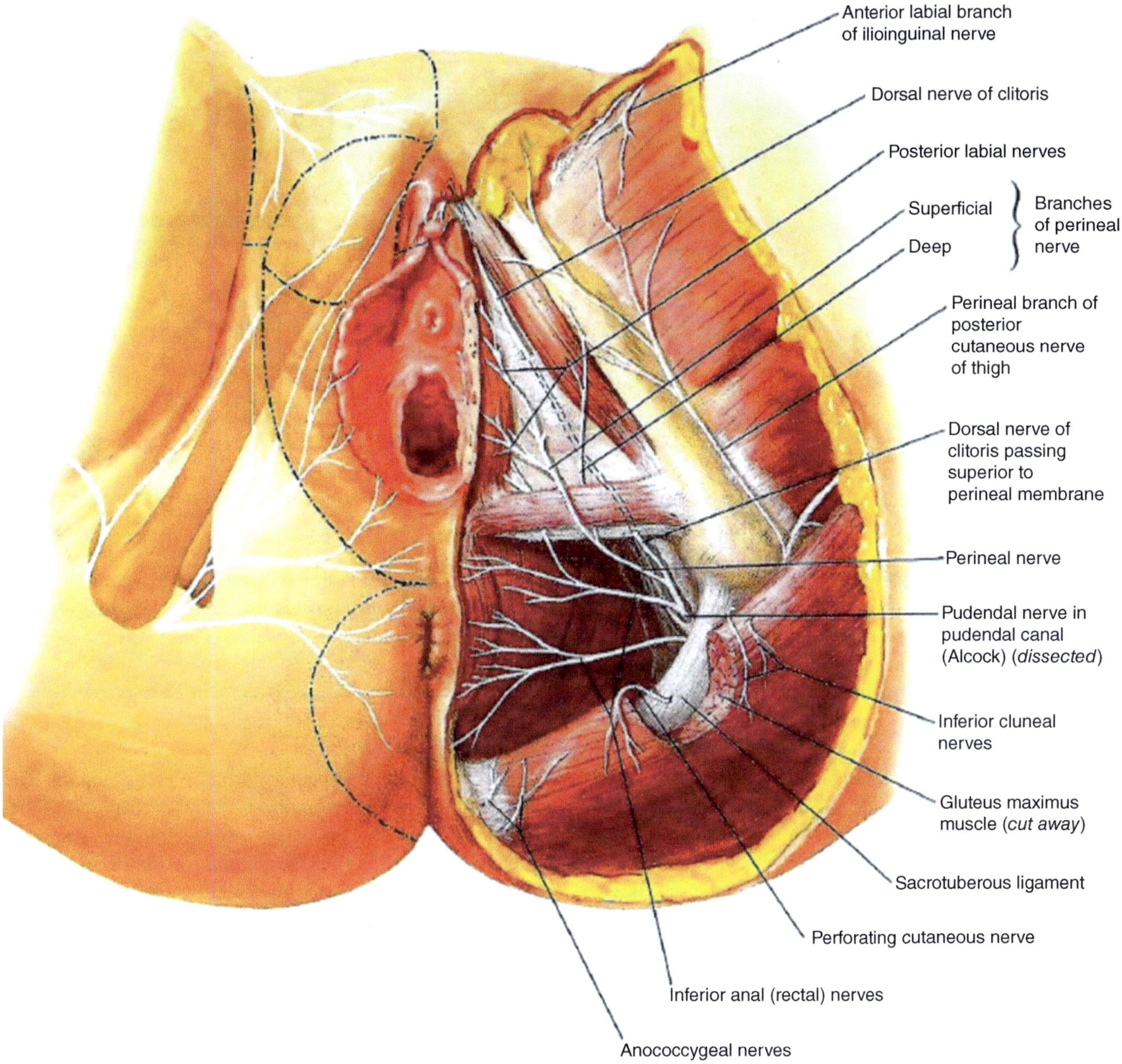


**Fig. 29.9** Illustration of pudendal branches. (**Source:** Netter medical illustration used with permission of Elsevier. All rights reserve)

puncture followed by centrifugation and selective pipetting of the obtained layered components, which separate based on density differences. The relative concentration of red cells, leukocytes, platelets and growth factors will vary widely with the system and protocol used for processing and may have consequences in terms of outcomes. Patients need to be warned that improvements may be delayed by up to 8 weeks and the early post-operative period may result in a temporary worsening of pain. The latter must be considered when using this approach in patients with chronic regional sensitization.

Autologous bone marrow and adipose tissue derived injectants enriched for mesenchymal stem cells (MSCs) may also be considered for treating pelvic disorders [64–68]. The harvesting techniques are more invasive than those described for PRP since they require either bone marrow aspiration or limited liposuction. Processing protocols, processing times and costs are substantially higher than with the other regenerative techniques described above. It is beyond the scope of this chapter to discuss regulatory issues that apply to these autologous transplants, however, an understanding of up-to-date guidelines by federal and state agencies should be

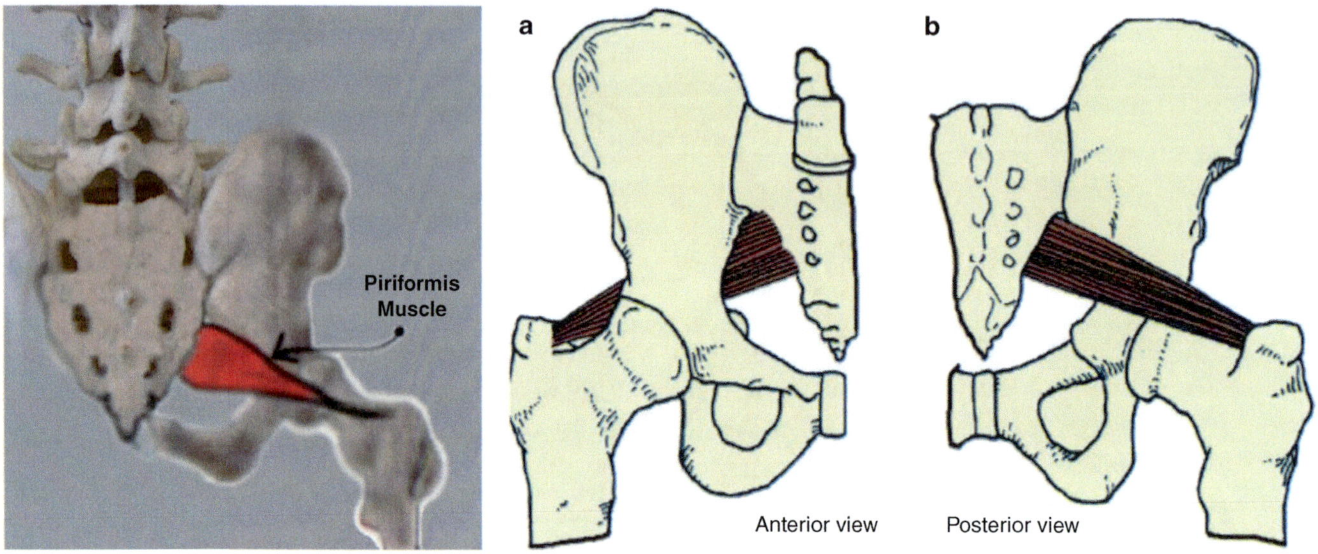


**Fig. 29.10** Illustrative anatomy of piriformis muscle. (**a**) Anterior view, (**b**) posterior view

understood by the physician prior to engaging in this approach. At the time of this writing, guidelines from federal agencies and state medical boards are being clarified; however, it appears at present that use of bone marrow derived injectants with minimal manipulation including centrifugation and filtration would likely continue to be considered acceptable for clinical use.

The concentration of MSCs varies widely from sample to sample even with continuing use of a given protocol [69]. The numbers of hematopoietic stem cells and nucleated blood components, including leukocytes, in bone marrow derivatives, dwarfs the number of actual MSCs [70]. Any potential disadvantage engendered by limited stem cell components or excess leukocyte concentrations may be overshadowed by the presence of growth factors which are better represented than what is commonly encountered with PRP preparations [71]. There is limited high level evidence for the use of stem cell preparations for pelvic disorders, however, the emerging experience suggests an apparent safety when injected in various bodily tissues (except the eye) [72]. In the experience of the present writers with stem cell preparations, there does not appear to be the same level of post-operative pain that may be seen with PRP injection.

In patients with CPP, the coexistence of multiple pain generators and regional comorbidities is the rule rather than the exception. Pelvic pain with myotendinous elements, nerve entrapment, bladder and/or bowel comorbidities, and locally problematic trophic tissue changes will be frequently encountered, and successful treatment will likely require a multitargeted approach. Systemic and psychiatric comorbidities also require attention including coexisting depression, "total body pain" syndromes, deconditioning, sexual and social dysfunctions, and endocrine imbalance. A multifaceted approach requires the skill sets of physical therapists, endocrinologists, urologists, gastroenterologists, psychiatrists, sexual dysfunction counselors and life coaches.

## Diagnosis

The diagnosis of pelvic pain syndromes is challenging. The main point that would be emphasized in this chapter is that in patients with CPP, there may have been an initiating pain generator such as anal fissures, vestibulitis, prostatitis or a surgical intervention which, over time, recruits other painful areas and engages psychosocial factors and produces central sensitization that must all be addressed for a good clinical outcome. Local nerve blocks may be helpful in determining the relative contribution of a particular pain generator and would be helpful in selecting targeted longer-term treatments.

Referred and radiating pain into the pelvis occurs with some degree of frequency. Athletic pubalgia, or so-called sports hernia, is due to tendinopathy of the abdominal and adductor muscle insertions around the pubic bone. A sometimes confusing presentation is that a herniated nucleus pulposus can be associated with inguinal, testicular and pubic pain; the relationship to back mechanics and straining maneuvers as well as localized palpation pain over the lumbar spine along with collaborating imaging tests will help in specifying the diagnosis.

Hip arthritis and labral tears commonly radiate into the inguinal region; furthermore, hip pathology may be associated with tendinopathy of the iliopsoas, adductors and gluteal muscles and their insertions at the greater trochanter. Reproduction of concordant pain with hip rotation is often helpful in making a diagnosis in these circumstances.

## Examination

In order to properly target injection treatments and to assess for possible hormonal interventions, a systematic performed pelvic examination will be necessary. Pelvic examination in females should be reserved to experienced physicians or gynecologists.

The bedside examination of the pelvis starts with an inspection of the skin for changes in coloration, atrophy or fissures that would suggest the need for referral to a dermatologist.

In women, the clitoris should be the size of the head of a cotton swab, the clitoral hood should move easily, without pain, to expose the organ. Reduction in the size of the head of the clitoris can be indicative of hormonal insufficiencies from oral contraceptive pill use, hormonal suppressive therapies, or menopause. Issues with mobility of the clitoral hood can stem from dermatologic diseases, infection, or connective tissue issues.

Next, check for perineal movement with concentric contraction (squeeze) and Valsalva movement (push). Pelvic floor muscles of normal length should shorten with an attempted contraction and relax or 'let go' at a similar speed. If the muscles do not shorten it could be because they are already in a shortened position because of a contracture or because the patient lacks motor control. Impaired muscles show little movement, they may not relax after contracting or they may relax slowly.

If no movement occurs during the Valsalva or push motion, the patient may have impaired pelvic floor motor control. It is not common that muscles in a nulliparous woman with pain are at normal or 'overlengthened' positions, though this can occur in women with chronic constipation, vaginal deliveries, and advanced age.

Inspect the vestibule for erythema and tissue integrity. The cotton swab test is done by lightly touching the vestibule and documenting location and severity of pain levels. This test is often painful for the patient and can sensitize them for the remainder of the examination. If you notice severe redness and they have unprovoked pain it may make sense to limit the number of areas touched or skip the cotton swab test because it is obvious they have pain. Check for sacral reflex integrity by eliciting an anal wink in both men and women.

During the digital internal examination, a single, gloved lubricated finger is used. For the sake of a physician screening of pelvic floor muscles, the examiner is feeling for tone and elasticity while also asking for reports of tenderness and pain. Healthy muscles do not hurt when they are palpated. Examine for motor control. Can the patient contract the muscles? Are they already contacted in a state and unable to be volitionally relaxed? Is the patient able to Valsalva and move the pelvic floor into a lengthened position if they cannot volitionally relax the muscles? These are all factors to keep in mind as the examiner palpates the different pelvic floor muscles.

The obturator internus is an external rotator of the hip. This makes it easy to identify by placing a finger in the vagina to a depth roughly past the second knuckle at 3 or 9 o'clock on the patient's left or right side, respectively, using the ventral aspect of a flat finger. The external hand can be placed on the patient's outer knee, asking the patient to lightly press into your hand. When the patient moves in this manner, the obturator internus will contract under the examiner's finger, confirming proper placement. This muscle often causes pain at the ischial tuberosities and tailbone. Importantly, the pudendal nerve travels through Alcock's canal, which is partially comprised of the aponeurosis of the obturator internus. This nerve can be examined for irritability by performing Tinel's Sign; light palpation causes sharp, shooting, stabbing pain. In some, but in all cases, pudendal neuropathy is associated with dysesthesia to light touch along the distribution of the nerve. Similar dermatomal dysesthesias may be noted in other cases of genitofemoral neuropathy and ilioinguinal neuropathy.

The urogenital diaphragm contains the bulbospongiosus, ischiocavernosus, and superficial transverse perineum; they may play an important role in vulvodynia in women and ejaculation pain in men. These muscles can be palpated using a pincer grasp by lightly using the index finger internally and the thumb of the same hand externally up and down the length of the bulbospongiosus, ischiocavernosus, and superficial transverse perineum in women. In men the bulbospongiosus can be palpated externally at the very base of the corpus spongiosum.

In women, the pubococcygeus can be found about 1 in. into the vagina between 7 and 11 o'clock on the right and 1 and 5 o'clock on the left. The puborectalis can be palpated going slightly deeper and straight down towards the table as it swings around the rectum, at 6 o'clock. The ischiococcygeus can be palpated by going slightly further into the vagina between 4 and 8 o'clock. The overall coordination of this muscle group can be tested by asking the patient to squeeze or do a Kegel exercise. The muscles are considered impaired if palpation is painful, if they cannot contract, or if the muscles do not relax after a concentric contraction. In men, the levator ani components can be palpated at rest and with Kegel exercises through the anus.

After muscular tone has been assessed, evaluation for endometriosis commences. Typically thickened uterosacral ligaments are present in women with endometriosis causing a deviation of the cervix from the midline towards either the right or left vaginal fornix. Palpation of the thickened, often nodular uterosacral ligaments produces pain. Also, the uterus is typically retroverted in women with endometriosis who have adhesions and scarring in the rectovaginal septum and uterosacral ligaments.

The following is a discussion of various localized conditions that may be targeted in an attempt to identify potentially treatable components of a regional pelvic pain condition.

## Anterior (Abdominal) Cutaneous Neuralgia

Persistent anterior abdominal wall pain [73], which is initially well localized, but eventually may become diffuse. It is usually unilateral but may be bilateral.

The symptoms may present after shingles (postherpetic neuralgia), or during pregnancy [74]. It may clinically manifest as sudden, acute abdominal pain resulting from mononeuritis multiplex of the abdominal nerves in patients with diabetes or vasculitis [75], which also may mimic endometriosis, interstitial cystitis, and abdominal adhesions [76].

Pain is reproduced by palpation with one finger at the lateral border of the rectus abdominis muscle. Careful, gentle palpation will often reveal a knot-like aponeurotic opening.

The diagnosis may be made by injection of the anterior cutaneous nerve with a local anesthetic and deposteroid, with a total volume of less than 2 mL.

## Ilioinguinal, Iliohypogastric and Genitofemoral Neuralgia

Because of the course of these nerves, they are at risk to injury from the lower abdominal incision (Pfannenstiel incision, appendectomy, inguinal herniorrhaphy) or trocar insertion performed in laparoscopic surgery [4]. Patients with neuropathy following injury to these nerves will present with groin pain that may extend to the scrotum or testicle in men, the labia majora in women, and the medial aspect of the thigh. These nerves may be injured by mechanical trauma of seatbelt injuries, large belts, holsters, and tight clothing. Accurate diagnostic block of those nerves is important in understanding the etiology of the clinical problem.

## Pudendal Neuralgia

Patients complain of peroneal, vaginal or perirectal pain. Robert et al. [52] described pudendal neuralgia as perineal pain that is aggravated by sitting, reduced by recumbence, and relieved sitting on a toilet seat. Pain is increased by flexion of the hip from jogging, abdominal crunches, cycling, leg presses, etc. [43]. Pain can be triggered by prolonged sitting at work, while driving, or on transoceanic air flights. Other causes include radiation therapy [77], direct trauma (bicycling) [78], hard chairs, surgery [79], straining from chronic constipation, difficult vaginal delivery [80] sacroiliac joint instability, neurofibroma entrapment [81] and postherpetic neuralgia [82]. The etiology in males is typically secondary to repetitive microtrauma to the nerve from common physical activities including high school sports and adult exercise [83].

Pathophysiology correlates with the restricted pathway of the pudendal nerve between the sacrospinous and sacrotuberous ligaments and the proximity of the ischial spine. Compression may occur in the pudendal canal (Alcock's canal). In bicyclists, the penile branch may be compressed at the pubic arch. Because compression sites in patients with pudendal neuropathy are not consistent, any branch or all branches may be affected, resulting in daily variations in sensory, motor, and autonomic symptoms, so clinicians should expect pudendal neuropathy to have protean manifestations. Although pathologic specimens of damaged pudendal nerves are unavailable, neuropathy appears to include neurapraxia, axon damage, and demyelinization, and is associated with denervation and reinnervation [84, 85].

Examination of sensory testing with pinprick often identifies hyperalgesia or hypalgesia or analgesia in one or more branches of the pudendal distribution (glans, posterior scrotum, posterior perianal skin). Pressure on the pudendal nerve medial to the ischial spine may reproduce subjective pains (the Valleix phenomenon) [86].

Neurophysiological tests are available to confirm the clinical diagnosis. Two useful quantitative sensory tests include the warm detection threshold (WDT) [87] and the pudendal nerve terminal motor latency test (PNTMLT) [85]. Each is simple to perform. The WDT is sensitive, being elevated in over 90% of patients with pudendal neuropathy. The test is useful for measuring other neuropathies [88]. A prolonged PNTMLT is absolutely specific but is abnormal in only 40% of patients.

Examination is carried out with patient in a lateral position, flexed at the hips over the examination table by palpating through the perineum to the medial ischium, pushing laterally and cephalad. Rectal examination may allow more cephalad palpation but is usually not necessary for vaginal/urethral pain. Patient may also be positioned in supine position on the examination table, in a modified lithotomy position with heels brought toward the buttocks, knees apart. Palpate through the perineum to the medial ischium, whereas vaginal examination may allow for more cephalad palpation it is usually not necessary.

The diagnosis may be made by injection of local anesthetic on the nerve (under fluoroscopic guidance at the ischial spine, under ultrasound guidance at the sacrospinatus ligament, or landmark guidance through the perineum at

Adcock's canal) with assessment of the subsequent loss of sensation in the pudendal nerve supply area.

## Piriformis Syndrome

The diagnosis of piriformis syndrome would be considered in a patient with gluteal pain that is typically worse with sitting and with walking up stairs. Examination would detect pain in the sciatic notch. Figure 29.11 shows trigger points and attachments of the piriformis muscle.

Some reports suggest 6:1 female to male predominance [89]. Bernard et al. [90] cited an incidence of 0.33% in a review of 1293 patients that were referred for treatment. Gluteal pain with or without radiation to the ipsilateral thigh that may sometimes extend below the knee may occur. Pain past the posterior knee into the calves may be due to the involvement of posterior femoral cutaneous nerve [91]. Low back pain is seen in approximately 50% of patients. Dyspareunia and pain with bowel movements is seen in 23% of patients.

During the physical examination, the gluteal area should be inspected for any obvious signs of atrophy or evidence of trauma. Tenderness in the region of the greater sciatic foramen to the greater trochanter is a frequent finding. A spindle-shaped mass or swelling can sometimes be palpated in thin patients. Pain is often elicited with specific signs voluntary hip flexion, internal rotation, and adduction of the hip joint.

- Straight leg raising test (referred to as Lasègue's sign) is an inconsistent finding [91].
- Pace sign [89]: With patient in the sitting position, pain is elicited on resisted abduction of the hip.
- Frieberg test: With patient in the standing or prone position, buttock pain may be elicited with forceful internal rotation of the thigh [58].
- Neurologic signs are usually negative unless there is involvement of the sciatic nerve.

In the investigation of patients with possible piriformis syndrome diagnostic tests may be utilized to exclude various spinal and extraspinal causes. Computed tomography (CT) and magnetic resonance imaging (MRI) findings of abnormal uptake by the muscle and enlarged piriformis muscle have been reported in the literature [92]. Nerve conduction studies may demonstrate delayed F waves and H reflexes [93].

The diagnosis may be made by injection of local anesthetic with a small volume of less than 2 mL.

## Associated Syndromes Characterized by Hypersensitivity and Psychological Distress

Patients with CPP often have comorbidities and multiple pain sources. Painful bladder and painful bowel syndromes are often encountered. Due to sleep disturbance and perhaps other factors that come into play with chronic regional pain, many pelvic pain patients will otherwise fulfill criteria for fibromyalgia. Depression, anxiety and obsessive-compulsive disorders may coexist and will require therapy. It has become more commonplace to avoid opioid medication, however, hypersensitivity and other psychological effects of pain med-

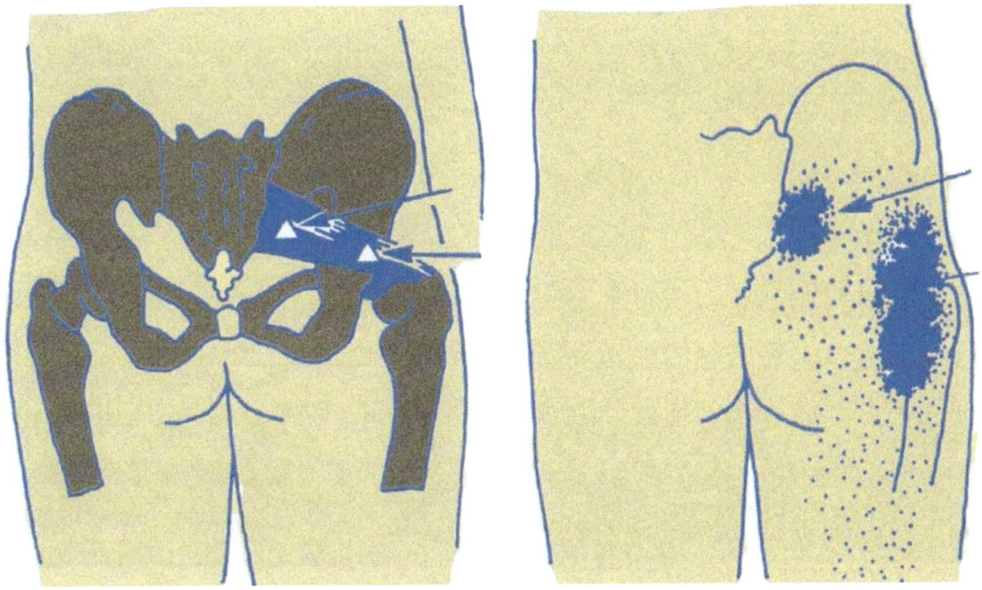

**Fig. 29.11** Trigger points and attachments of the piriformis muscle with pain zones

ications and sedative hypnotics may require artful pharmaceutical management and the substitution of therapies that don't rely on medications.

## Technical Considerations

### Conventional Therapy

Pudendal nerve blocks and genitofemoral blocks have been used for diagnosis and treatment of neuropathic pain in the pelvis. Longer duration effects may also result from pulsed radiofrequency of these nerves. Diagnostic and therapeutic muscle injections for levator ani, piriformis and other pelvic muscles are also described in the literature. Longer duration effects may occur with botulinum chemodenervation. For sympathetically mediated pain, inferior hypogastric blocks and ganglion impar blocks are often considered in treating patients with pelvic pain. For neuropathic pain, not responding to other techniques, spinal stimulation has also been used.

For pain thought to originate from the bladder, which can exist along with urinary symptoms such as frequency and urgency, intravesical instillations including alkalinized lidocaine, heparin, pentosan polysulfate sodium, chondroitin sulfate, and hyaluronic acid have been administered [94–96]. The efficacy and durability of these treatments differ greatly in the published literature. Newer intravesical treatments have focused on investigational drugs or botulinum toxin. Intravesical botulinum toxin Type A injections have shown some efficacy in reducing pain and irritative voiding symptoms in non-ulcerative chronic bladder pain patients [97]. Many of these intravesical treatments are thought to work by influencing chronic inflammation and apoptotic signaling molecules and down regulating of vascular endothelial growth factors [98, 99]. For those with complicated or medical refractory chronic bladder pain, reconstructive procedures such as cystectomy, urinary diversion without cystectomy, and augmentation ileocystoplasty have also been used with varying success [100, 101]. Many of these more radical treatments are reserved as last resort solutions to "end-stage" cases of chronic bladder pain.

### Imaging

For many years, fluoroscopic guidance has been used for guiding pelvic injection; in particular, piriformis injections and pudendal nerve block injections have been performed with this technology. Even when relying more on ultrasonic guidance, fluoroscopy skill may be required, particularly for the successful targeting with those more generously proportioned subjects. Figure 29.12 shows radiographic image of male pelvis.

Ultrasound imaging has gained significant common utilization in performing multiple nerve blocks and injection

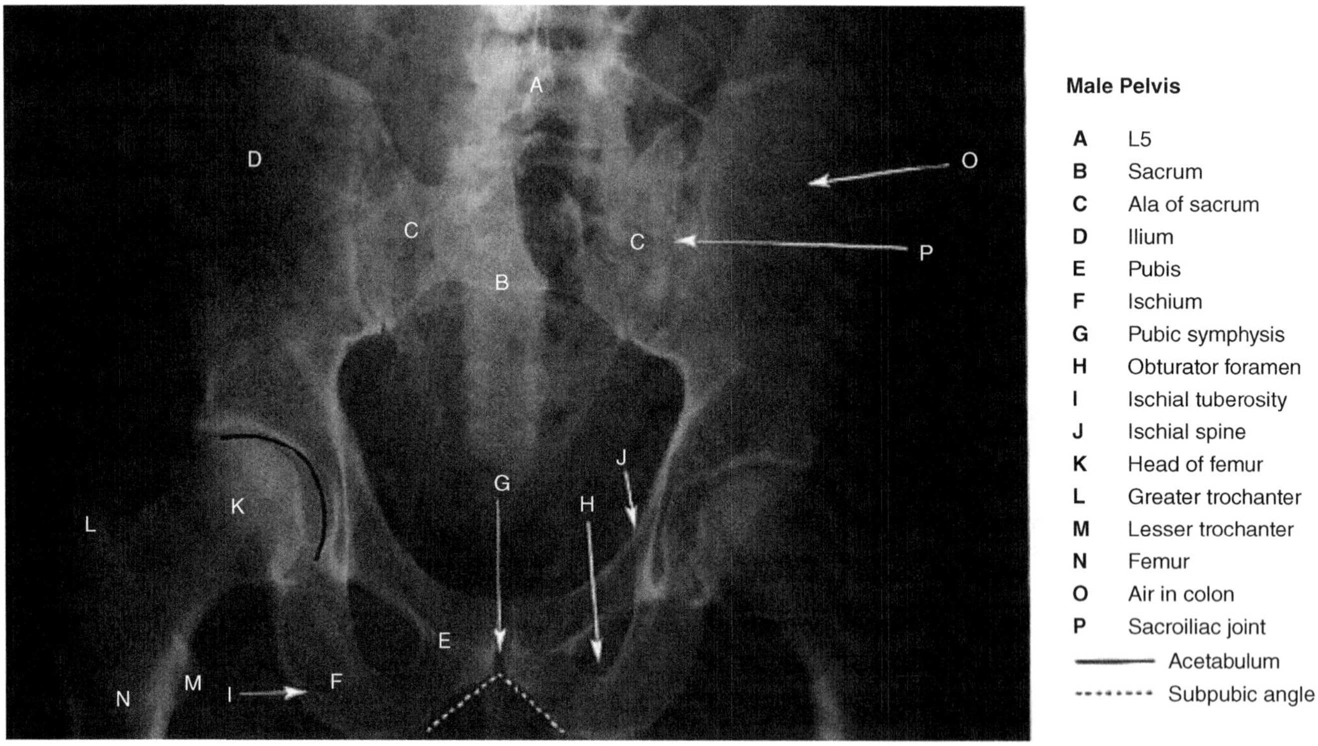


**Fig. 29.12** Radiographic image of male pelvis. (**Source:** Netter medical illustration used with permission of Elsevier. All rights reserved)

techniques in abdomen and pelvis. Figures 29.13 and 29.14 show ultrasound imaging.

## Anterior (Abdominal) Cutaneous Nerve Blocks

The injection of the anterior abdominal cutaneous nerves is performed by holding the rectus border by the non-injecting hand, with the tips of the fingers at the rectus border (Figs. 29.2 and 29.15). A 25- or 27-gauge needle is advanced from lateral to medial, feeling the "pop" as the needle pierces the abdominal fascia. These blocks may be performed blindly with palpation or under ultrasonic guidance. A peripheral nerve stimulator may increase the accuracy of needle placement. Multiple types of injections may be performed with local anesthetics with or without steroids, dextrose 5%, with biologics with PRP and stem cells, or neurolytic solutions; ablation may be performed with cryoneuroablation or pulsed radiofrequency.

## Ilioinguinal, Iliohypogastric, and Genitofemoral Nerve Blocks

Ilioinguinal, iliohypogastric, and genitofemoral nerve blocks in the past most commonly have been performed with landmark guidance, or rarely, they also have been performed under fluoroscopic or CT guidance. More recently, they have been performed with ultrasound-guided techniques.

### Landmark Guided Techniques

Peng et al. [4] described that, due to variable anatomy between the three nerves (Figs. 29.4 and 29.16), the existing techniques, based on landmarks, are confusing [42, 44, 102–105]. Most techniques requiring injection of the ilioinguinal and iliohypogastric nerves use the ASIS as a landmark (Figs. 29.16, 29.17, and 29.18).

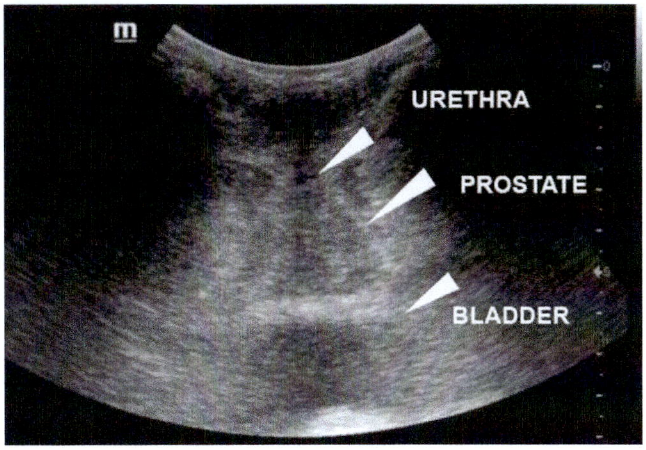


**Fig. 29.13** Coronal ultrasound of male pelvis

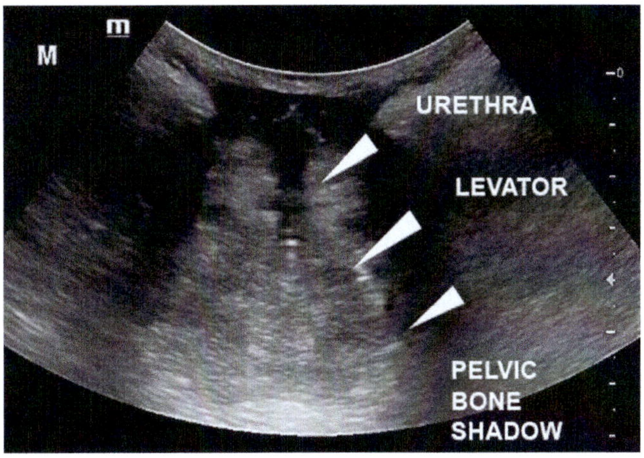


**Fig. 29.14** Coronal ultrasound of female pelvis

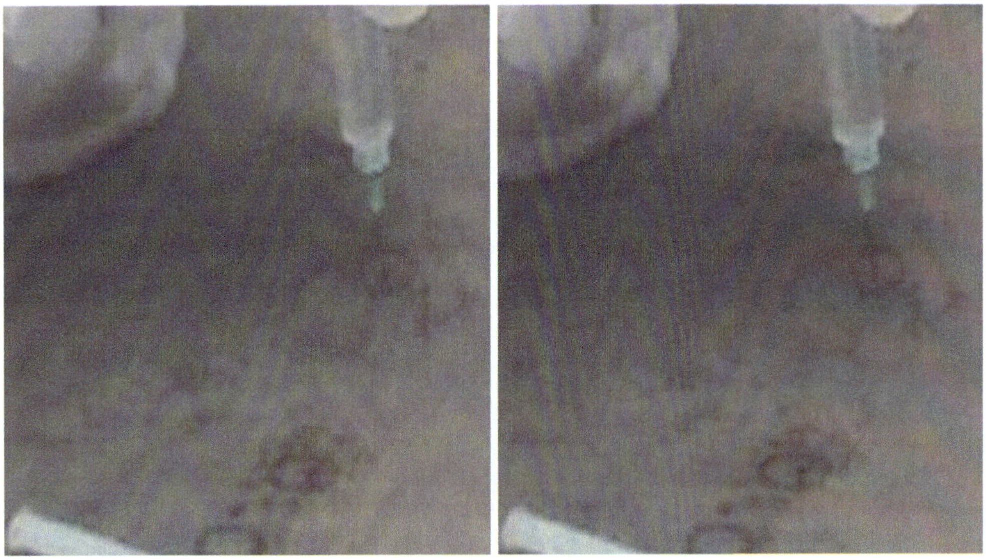

**Fig. 29.15** Infiltrating several abdominal cutaneous nerves at the border of the left rectus muscle

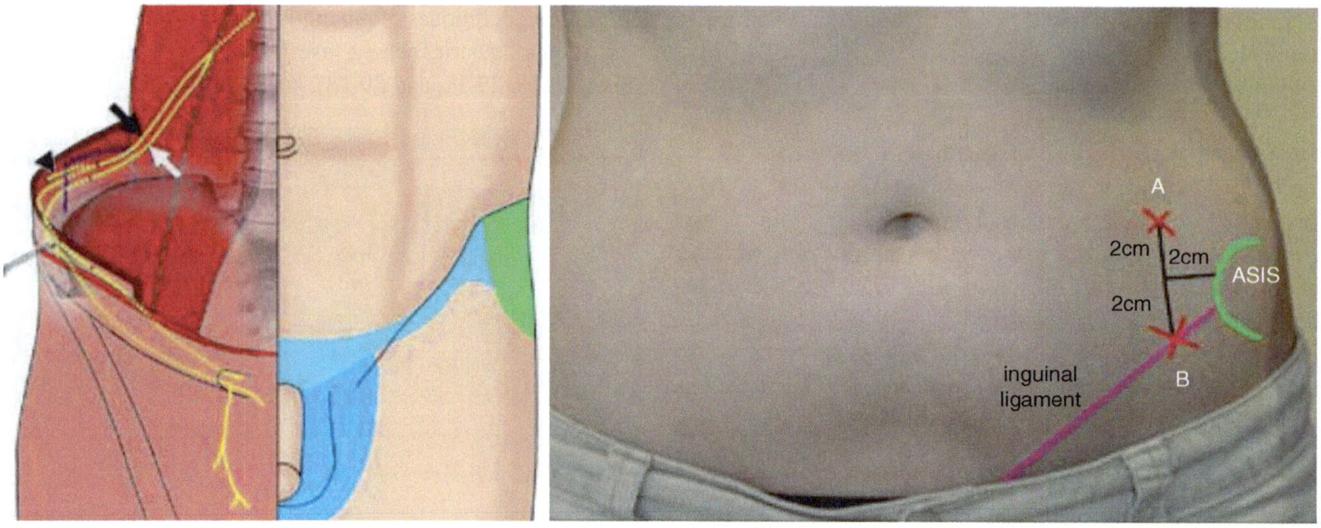


**Fig. 29.16** Landmark-based blocks of ilioinguinal and iliohypogastric nerve blocks. (**a**) Anatomical routes of both ilioinguinal (white arrow) and iliohypogastric (black arrow) nerves and their regions of distribution. Blue: Common to both nerves: light blue is inguinal region, dark blue is scrotal region. Green: Lateral cutaneous branch (black arrowhead) of iliohypogastric nerve distributing in upper and lateral part of thigh. (**b**) Site for landmark-guided ilioinguinal/iliohypogastric injection at ASIS. A site 2 cm superior, B site 2 cm inferior. (Image courtesy of Andrea Trescot, MD)

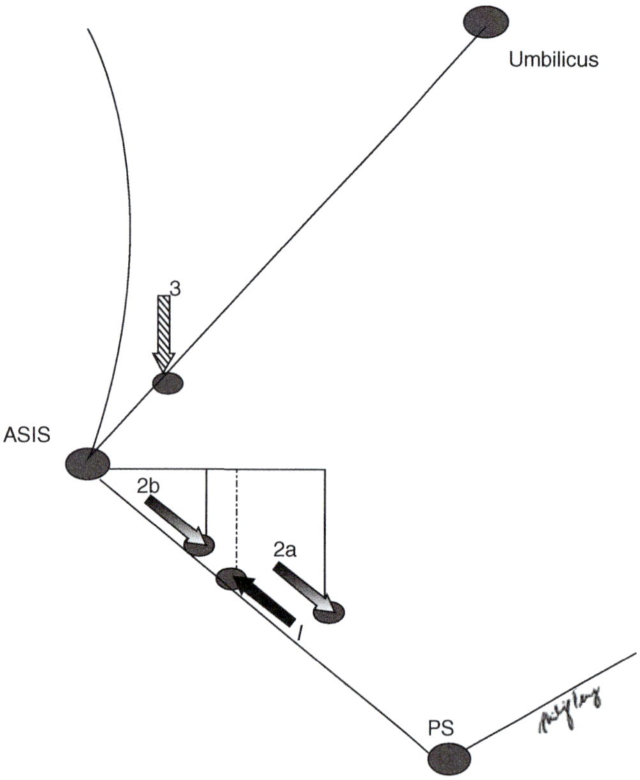


**Fig. 29.17** The three methods (four landmarks) described for ilioinguinal and iliohypogastric nerves injection. Methods 1, 2 and 3 were described by Brown [104], Waldman [105], and Katz and Brown [103]. *PS* pubic symphysis; *ASIS* anterior superior iliac spine

One technique described the needle insertion point as 1 in. medial to the ASIS on a line joining the ASIS and umbilicus. The needle was suggested to "pierce the fascia of oblique muscles" using "fanwise and up-and-down infiltration" [4, 102].

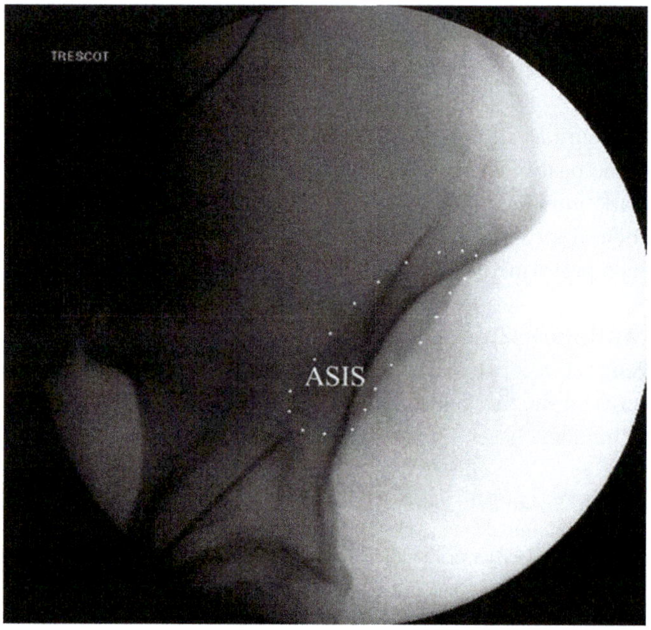


**Fig. 29.18** Needle marker at the anterior superior iliac spine

Another technique described the landmarks as 3 cm medial and inferior to the ASIS. It suggested the needle insertion to be in the cephalolateral direction until contacting the inner surface of the ilium. The investigator of this technique suggested infiltration of local anesthetic as the needle is withdrawn "through the layers of abdominal wall." Repeated needle entry at a steeper angle was also suggested to penetrate all three layers of abdominal muscles [4, 104].

Another investigator, who described separate procedures for Ilioinguinal and iliohypogastric nerves, suggested a completely different approach. The needle entry point for the ilioinguinal nerve was 2 in. medial and inferior to the ASIS. The needle was then directed "towards the pubic symphysis in a fan-like manner," piercing through the fascia of the external oblique muscle. The needle entry point for the iliohypogastric is 1 in. medial and inferior to the ASIS. The direction and depth of needle insertion was similar to that of the ilioinguinal blockade [4, 105].

Because the ilioinguinal and iliohypogastric nerves can be located at different fascia planes among the three muscles (IO, EO and TA), these blind techniques may have lower success rates [4, 106].

The description of a genitofemoral nerve block mainly refers to the injection of the genital branch of genitofemoral nerve. The landmark described is the pubic tubercle. The needle is suggested to be directed to a point 1 cm superior and lateral to the tubercle with a field block. Although the philosophy of this landmark is not clear, the needle is likely directed toward the spermatic cord in the inguinal canal. With a blind technique, important structures of the spermatic cord (testicular artery and vas deferens) or the peritoneum are at risk.

## Fluoroscopic and CT-Guided Technique for Ilioinguinal, Iliohypogastric and Genitofemoral Nerve

Fluoroscopic-guided and CT-guided techniques are based on the location of the iliac crest in supine position (Figs. 29.18 and 29.19). CT-guided neurolysis procedures as shown in Fig. 29.19 illustrate the fluoroscopic approach.

Another technique is performed in prone position similar to transforaminal epidural injections at T12, L1, or L2 (the origin of the ilioinguinal, iliohypogastric, and genitofemoral nerves). However, transforaminal epidural injections at this level are associated with high risk [107].

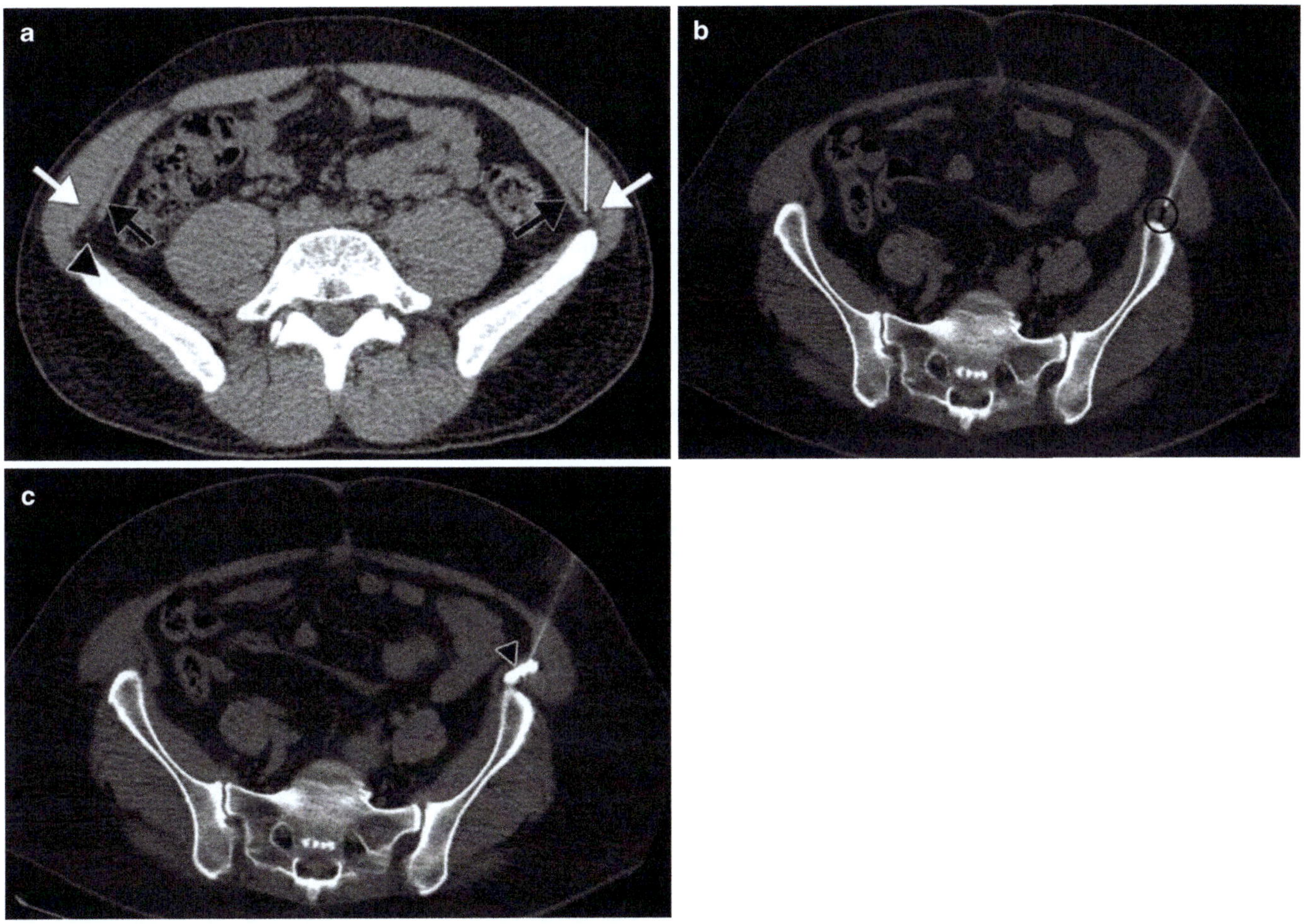


**Fig. 29.19** CT-guided anatomic visualization of ilioinguinal and iliohypogastric nerves. (**a**) CT-guided anatomic visualization of ilioinguinal and iliohypogastric nerves. Axial CT scan showing target location prior to needle insertion. Black arrow: transverse abdominal muscle. White arrows: internal oblique muscle. Arrowhead: ilioinguinal-iliohypogastric nerves. (**b**) CT-guided anatomic visualization of ilioinguinal and iliohypogastric nerves. CT slice showing needle tip artifact (black circle) at defined target. (**c**) CT slice after injection of contrast media (arrowhead) at defined target

Distal genitofemoral nerve blocks are performed in supine position under fluoroscopy with identification of symphysis pubis as shown in Fig. 29.20.

### Ultrasound-Guided Technique for Ilioinguinal, Iliohypogastric and Genitofemoral Nerves

The technique for ilioinguinal and iliohypogastric injections, under ultrasound guidance in adults has been described in the literature [4, 45, 103, 104]. Because of the superficial nature of the nerve, a high frequency linear probe will be sufficient. The lateral end of the probe should be placed just above ASIS. The orientation of the probe should be perpendicular to the inguinal line. Once the probe is placed in this position, the hyperechoic shadow of ASIS can be visualized. The probe is then tilted until all three layers of muscles (TA, IO, EO) are visualized (Fig. 29.21a, b) [4]. The peritoneum can be seen as the fascia layer underneath the TA muscles. Between the layers of TA and IO muscle, splitting of the fascia layer is usually observed. It is on this plane where ilioinguinal and iliohypogastric nerves pass through [4]. Sometimes, both nerves pierce the IO and appear between the IO and EO muscles. Both nerves can run together or run at a distance of approximately 1 cm apart. In this case, more than one fascia split will be seen.

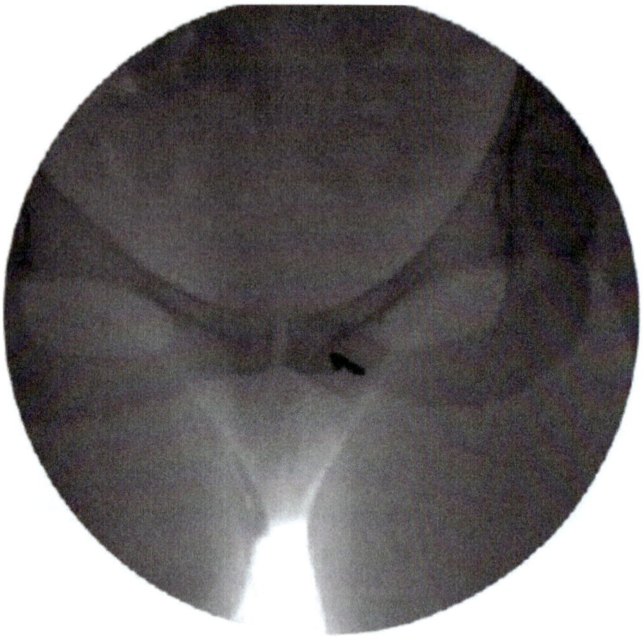

**Fig. 29.20** Illustration of genitofemoral nerve block under fluoroscopy

The nerve can be approached by in-plane or out-of-plane techniques [4]. With in-plane techniques, a nerve-stimulating needle is inserted toward the splitting fascia from the lateral end of the probe [4]. Stimulation of the nerve may then produce paresthesia of the groin area [4].

Bolash and Cheng [44] described the following ultrasound-guided technique: utilizing ultrasound, the lower anterior abdominopelvic wall is prepped in a sterile fashion. A linear transducer is positioned transversely on the abdomen with the lateral aspect of the transducer overlying the ASIS and the medial aspect directed toward the umbilicus (Fig. 29.22). Real-time sonography is utilized to visualize the external abdominal oblique, the internal abdominal oblique and the transversus abdominis muscles [108].

The ultrasonic image shows iliac crest at the lateral aspect, whereas bowel is seen below the muscular layers of the abdominal wall (Fig. 29.23).

The technique for blocking the genital branch of the genitofemoral nerve under ultrasound guidance has been described by Peng et al. [4] and others [46]. Peng et al. [4] described the following technique: similar to the ilioinguinal

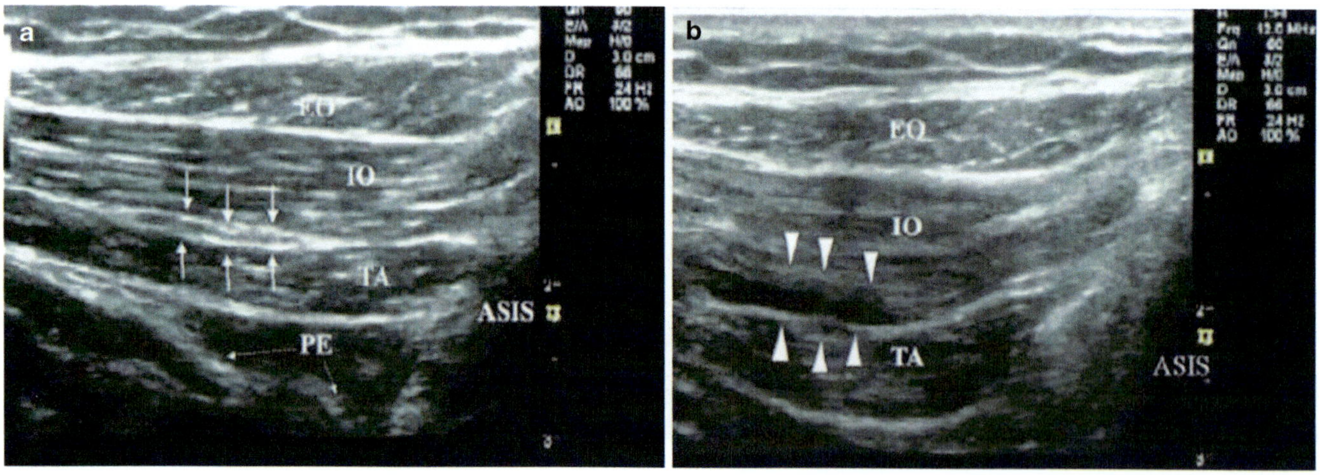


**Fig. 29.21** (**a**, **b**) Ultrasound pictures showing the fascia split before and after injection of a local anesthetic. *EO* external oblique muscle; *IO* internal oblique muscle; *TA* transversus abdominus muscle; *PE* peritoneum; *ASIS* anterior superior iliac spine. (**Source:** Peng PW, Tumber PS. Ultrasound-guided interventional procedures for patients with chronic pelvic pain - a description of techniques and review of literature. *Pain Physician* 2008; 11:215–224 [4])

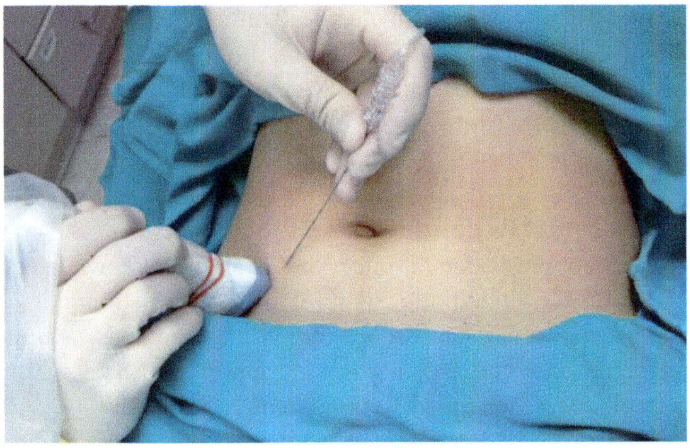

**Fig. 29.22** Ilioinguinal and iliohypogastric nerve block: the ultrasound is positioned the probe directed toward the umbilicus at the level of the anterior superior iliac spine. The needle enters medial to the probe permitting shallow advancement in-plane with the probe

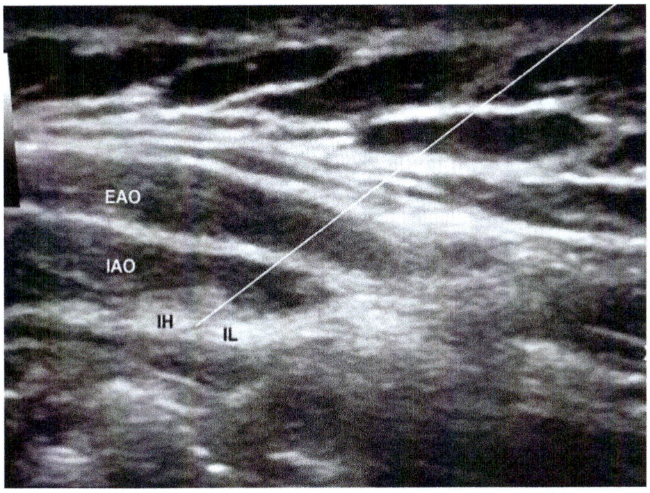


**Fig. 29.23** Ilioinguinal and iliohypogastric nerve block: simulated needle path as it courses through both the external abdominal oblique and internal abdominal oblique muscles with the final position located adjacent to the ilioinguinal and iliohypogastric nerves. *EAO* external abdominal oblique; *IAO* internal abdominal oblique; *IH* iliohypogastric nerve; *IL* ilioinguinal nerve

nerve block, a high frequency linear probe is used. The orientation of the probe is perpendicular to the inguinal ligament. The final position of the probe is about one finger-breadth lateral to the pubic tubercle [4]. However, it may be difficult to identify the spermatic cord, which is oval or circular in shape (Fig. 29.24a) with one or two arteries within it (the testicular artery and the artery to vas deferens). The vas deferens is often seen as a thick tubular structure within the spermatic cord. It is suggested to start scanning with the probe at the internal inguinal ring, where the femoral artery can be visualized in the longitudinal scan (along the length of the femoral artery) [4]. By moving the probe in the cephalad direction, the artery is seen diving deep toward the inguinal ligament [4]. At this point, an oval or circular structure can be seen superficial to the femoral artery. The probe is then moved slightly in the medial direction away from the femoral artery. An out-of-plane technique can also be used, with the needle approaching the skin from the lateral aspect of the probe. Because of the anatomical anomalies found around the location of the genital branch in the genitofemoral nerve, we suggest depositing 5 mL of local anesthetic inside and another 5 mL outside the spermatic cord (Fig. 29.24b).

With ultrasonography, it has been shown that ilioinguinal and iliohypogastric nerves are seen in 95% of adults [109]. In addition, with developments in ultrasonography and color Doppler, the deep circumflex iliac artery may be seen as an oval structure in proximity to the nerves. When passing the needle through the fascia enveloping the external abdominal oblique muscle, a "pop" may be felt. The needle courses in plane with ultrasound probe passing through the external abdominal oblique muscle after passing through subcutaneous fat.

The needle is advanced through the deep layer of the internal oblique muscle and through the muscle's posterior facial covering. At this time, a second "pop" may be felt. The final position of the needle tip lies posterior to the internal abdominal oblique muscle and anterior to the transverse

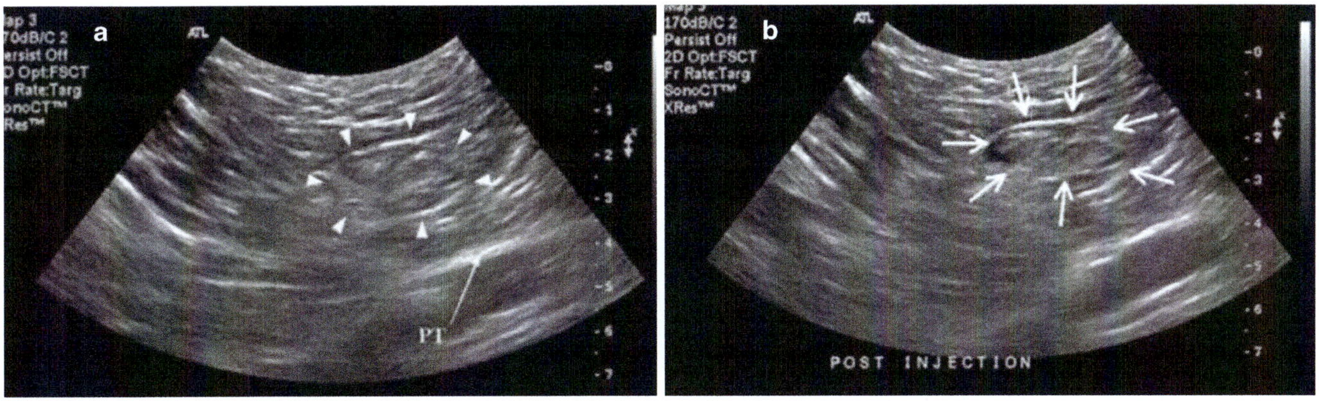


**Fig. 29.24** (**a**, **b**) Ultrasound pictures showing pre- and post-injection into spermatic cord. *PT* pubic tubercle

abdominis. However, if needle tip visualization is uncertain, a small dose of local anesthetic (1 mL) may be placed under real time sonography to confirm the location of the needle tip. During this injection, intramuscular tissue disruption indicates that the needle tip is at ideal location, and it should be advanced further toward the intramuscular plane.

Since ilioinguinal and iliohypogastric nerves travel either directly adjacent to each other or up to 1 cm apart, an injectate of 10 mL of local anesthetic may be needed. However, injection of biologic or neurolytic solutions must be performed with much smaller volumes of 1–3 mL.

At least two cases of retroperitoneal hematoma have been reported following an ilioinguinal and iliohypogastric nerve block, likely owing to inadvertent puncture of the deep circumflex iliac artery [110, 111]. Real-time ultrasound with Color Doppler adds an additional layer of safety when performing the ilioinguinal and iliohypogastric nerve blocks, allowing the practitioner to differentiate between vascular and neural structures.

## Pudendal Nerve Block

Pudendal neuropathy is a "tunnel syndrome", in that the nerve travels through several functional canals [112]. It is treated in a manner analogous to carpal tunnel syndrome; namely, splinting, injections, and surgical decompression.

Sequential treatment of pudendal neuropathy is as follows:

- Self-care, a nerve protection program
- Restrict sitting
- Use of a perineal suspension pad to prevent pressure on the perineum so that the body weight is supported by the ischial tuberosities
- Avoid bending, squatting, and lifting
- Cease hip flexion exercises
- Amitriptyline 10–50 mg at hs
- Pudendal nerve perineal injection (PNPI) using bupivacaine and steroid
  - Two at the ischial spine
  - One into the pudendal canal (Alcock's canal).
- Surgical decompression via a transgluteal approach.

Physical examination is the major method of diagnosis. Pinprick sensation is a rapid means of confirming suspected neuropathy. Few neurophysiologic tests are available to the practicing physician for objective confirmation PNPIs have shown good results. Amarenco et al. [113] reported long-term pain relief with pudendal nerve blocks in patients failing to respond to multiple interventional techniques including facet blocks, sacroiliac joint blocks, trigger point injections, ganglion of Impar blocks, and hypogastric plexus blocks.

As with any neuropathy, confounding adjacent neuropathies must be addressed. In addition, ilioinguinal and iliohypogastric neuropathies and the Maigne syndrome may exist concurrently causing vexatious aggravation of pudendal symptoms.

### Transgluteal Pudendal Nerve Block

While multiple pudendal injection techniques are available, including CT-guided, ultrasound-guided, nerve stimulation-guided, and other fluoroscopically-guided methods, a simple, rapid, and effective fluoroscopic approach also known as transgluteal pudendal nerve blocks can be utilized.

The interligamentous plane is the key location for needle placement. The use of CT scan allows the visualization of the interligamentous plane [114]. However, CT scan is not usually available to interventional pain specialists. Thus, a fluoroscopy-guided technique has become popular [52]; however, fluoroscopy cannot be used to visualize the interligamentous plane to identify all the important structures (Fig. 29.25). Instead, it involves the use of a surrogate landmark, the ischial spine (Fig. 29.26). Since the pudendal nerve is principally found medial to the pudendal artery (76–100% of the time) at the level of the ischial spine [115, 116], injection at this bony landmark may result in failure to disperse solution medially to the pudendal nerve.

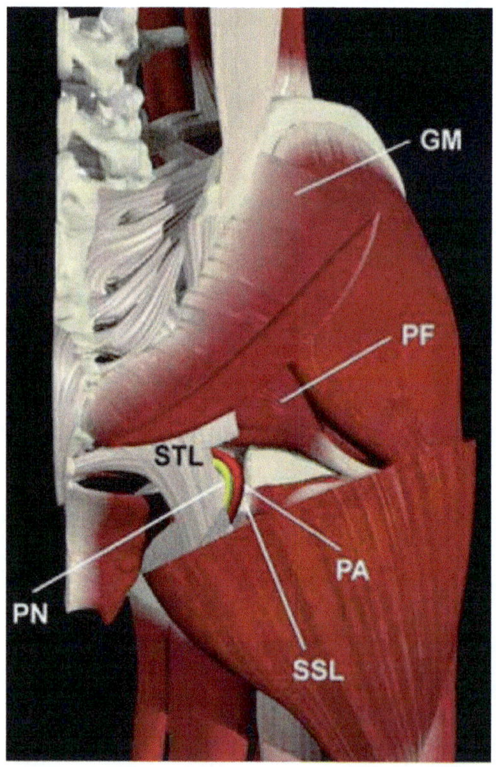


**Fig. 29.25** Anatomy of pudendal nerve with pudendal artery. *GM* gluteus maximus; *PF* piriformis muscle; *SSL* sacrospinous ligament; *STL* sacrotuberous ligament; *PA* pudendal artery; *PN* pudendal nerve

### Technique

For fluoroscopically-guided PNPIs, the patient is placed in prone position on a fluoroscopy table. A pillow is placed under the ipsilateral hip, raising the pelvis approximately 30° (or the C-arm is rotated ipsilaterally approximately 30°). This allows visualization of the ischial spine (Fig. 29.26). The nerve pathway is usually immediately medial to the tip of the ischial spine. The tip of the ischial spine is marked on the skin of the buttock. After antiseptic preparation, a 22-gauge 12 cm needle is passed through the gluteus muscle until the ischial spine is gently touched. Then the needle is retracted several millimeters and is angled about 1–2 mm medial to the tip of the bone (Fig. 29.27). It is advanced to the sacrotuberous ligament, which is identified by palpable resistance to needle passage. This is usually 3–5 mm shallower than the bony ischial spine. Slow penetration of the

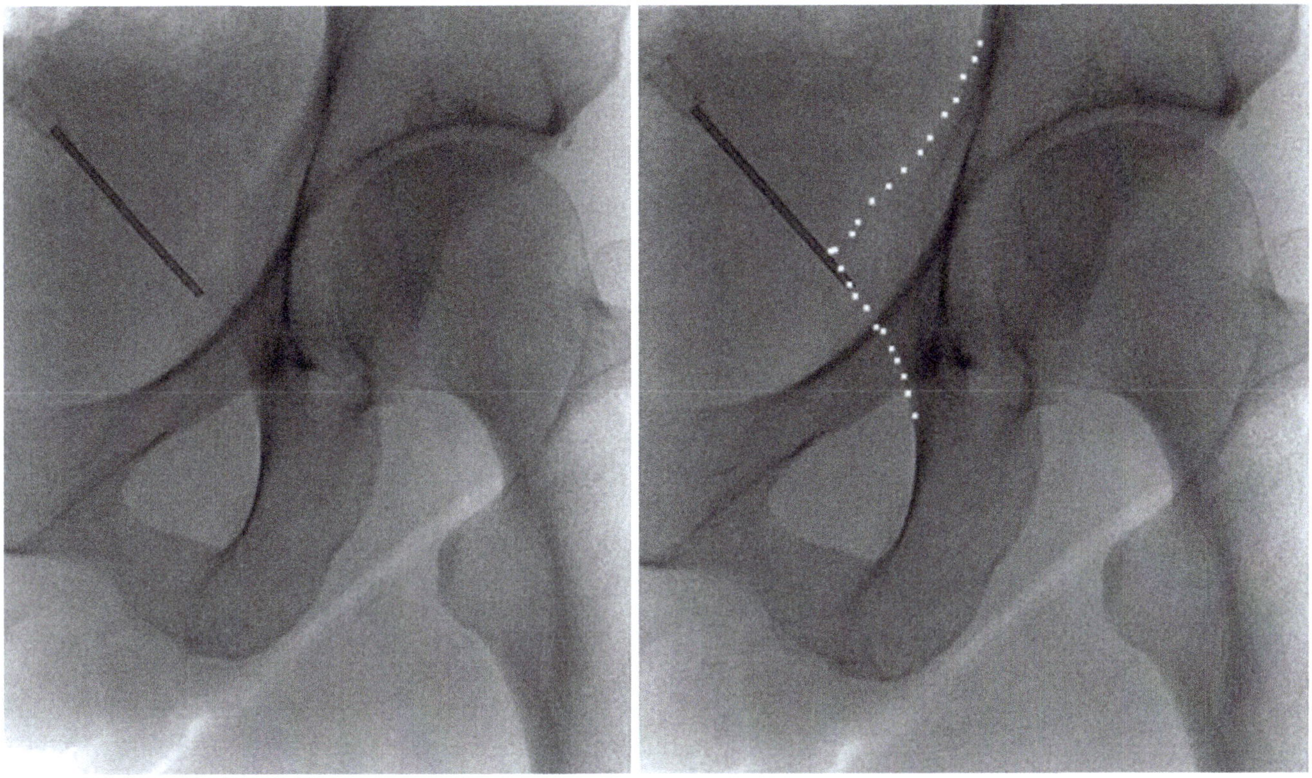

**Fig. 29.26** Fluoroscopic view of the ischial spine, the target in a fluoroscopically-guided pudendal injection

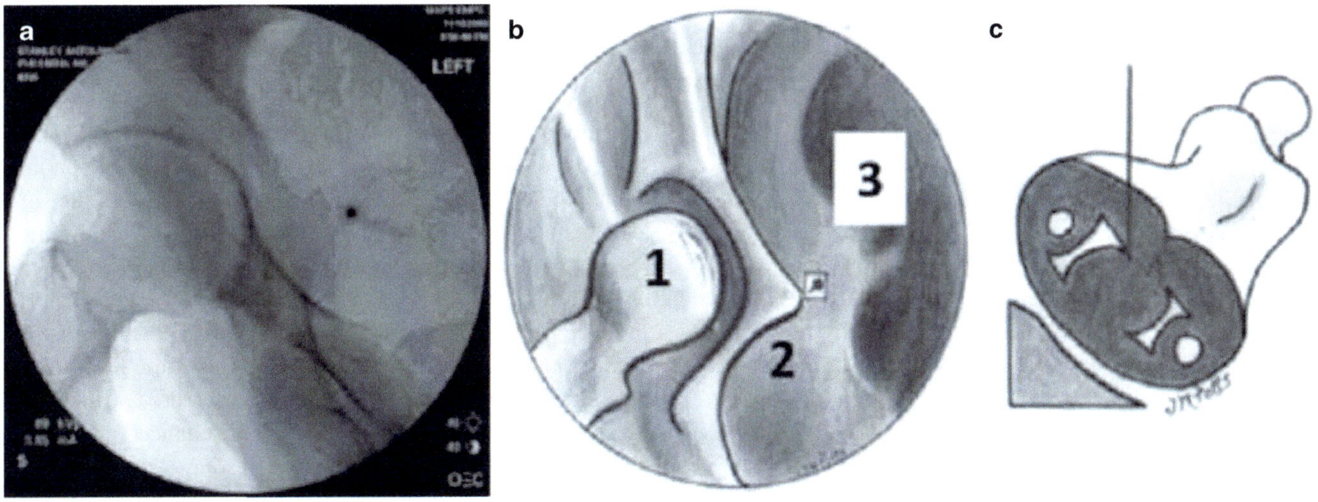


**Fig. 29.27** (**a**) Fluoroscopic guided PNPI. Hub of needle medial to ischial spine. (**b**) (1) Femoral head. (2) Ischial spine with hub of needle medially 1–2 mm. (3) Greater sciatic notch. (c) Position for PNPI. Ipsilateral hip elevated. Needle is inserted perpendicular to floor

sacrotuberous ligament is made into the interligamentary space. At this point, the needle stimulates paresthesias or pains in the pudendal territory. The needle is withdrawn 1–2 mm and medications are infiltrated into the interligamentary space.

### Dual Site Block

Kastler et al. [117] also have described a dual site pudendal nerve infiltration, which showed that resulted in better success. In this study, 95 patients underwent 155 procedures. Clinical success at 1 month was in 63.2% of the patients, at 3 months it was 50.5% of the patients, and at 6 months effectiveness decreased to 25.2%, with mean VA scores decreasing to 2.07 at 3 months, 2.90 at 6 months, and 3.2 at 6 months. However, efficacy assessment was based on 50% decrease in VAS scores. Figure 29.28 shows the dual site pudendal nerve blocks. The procedure was performed under CT guidance as shown in Fig. 29.29. The targets were at ischial spine and Alcock's canal. Figure 29.28 shows site 1 and site 2 for infiltration of the pudendal nerve. They described that at this second level, the pudendal nerve and the accompanying vessel are seen as a small bulge or a linear structure within the split in the aponeurosis of the internal obturator muscle (Fig. 29.30).

### Ultrasound-Guided Pudendal Nerve Block

There are only three reports on the use of ultrasound in the visualization of the pudendal nerve [4, 54, 115, 116]. Of these, only one described the ultrasound-guided injection technique [54]. Ultrasonography allows the visualization of

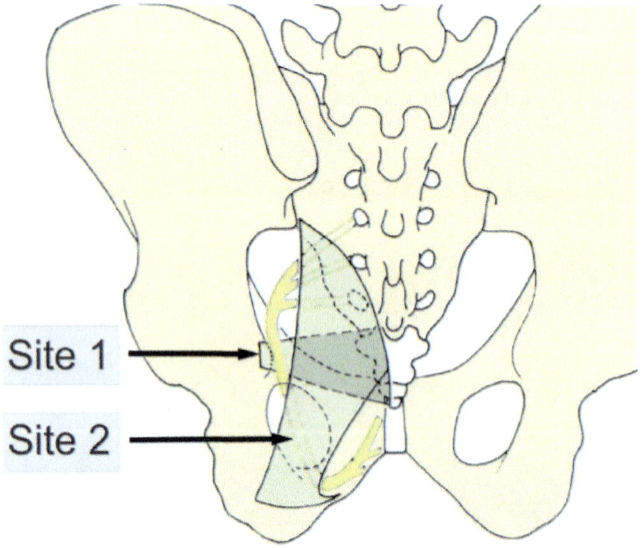


**Fig. 29.28** Pudendal nerve anatomical representation. This anatomical drawing represents the two potential conflicting pudendal nerve entrapment sites (black arrow head): site 1 corresponds to the segment of the pudendal nerve passing between the sacrotuberous and sacrospinous ligaments; site 2 corresponds to the pudendal nerve segment passing in the Alcock's canal alongside the internal obturator muscle

important landmarks: the ischial spine, pudendal artery, sacrospinous ligament, sacrotuberous ligament, and pudendal nerve (Figs. 29.25 and 29.31). It also allows real-time needle advancement and confirmation of injectate spread within the interligamentous plane.

A low frequency 2–5 MHz curved array ultrasound probe is used. After a sterile skin and sterile probe preparation, scanning is performed in transverse planes on the prone patient to visualize the ischium forming the lateral border of the sciatic notch. By moving the ultrasound probe in a cephalad-caudal direction, the ischium appears as a progressively lengthening hyperechoic line that is widest at the ischial spine level. The ischium is initially seen as a curved line as it forms the posterior aspect of the acetabulum.

When the probe is at the ischial spine level, the ischium will appear as a straight line. At this level, a color Doppler is used to localize the internal pudendal artery pulsations in close proximity to the ischial spine. Another arterial pulsation is often seen lateral to the tip of the ischial spine and is accompanied by the sciatic nerve. This is the inferior gluteal artery. Mistaking this artery for the pudendal artery will result in sciatic nerve block. The sacrospinous ligament appears as a hyperechoic line in continuity with the ischial spine, with lower echogenicity than bone. Similarly, the sacrotuberous ligament is seen as a light hyperechoic line deep within gluteus maximus muscle and appears parallel and superior to the sacrospinous ligament in ultrasound images.

Localization of the pudendal nerve is targeted in the plane between these two ligaments (Fig. 29.32). Under ultrasound guidance, a 22-gauge, 12 cm insulated peripheral nerve stimulating needle is inserted from the medial aspect of the probe. It is advanced in line with the ultrasound probe to the medial aspect of the internal pudendal artery (Fig. 29.33). Once the needle passes through the sacrotuberous ligament, a "click" is usually felt and a small volume (1–2 mL) of D5W is injected. The solution appears as a hypoechoic collection, in order to identify the plane between the sacrotuberous and the sacrospinous ligaments and to accentuate the pudendal nerve appearance (Fig. 29.33).

In a recent feasibility study, it was shown that the pudendal nerve could only be clearly visualized in 30% of the subjects tested [54]. This limited visibility may be due to a few reasons. The average diameter of the pudendal nerve at the level of the ischial spine is approximately 4–6 mm [53, 115, 118]. Nerves of this size are generally difficult to detect with an ultrasound at a depth of 5.2 ± 1.1 cm. At the level of the ischial spine, 30–40% of pudendal nerves are 2-trunked or 3-trunked [53, 119]. This reduces the chance of a direct depiction of the nerve with an ultrasound and may also account for the poor response to the nerve stimulator. Lastly, the pudendal nerve itself may be embedded in dense connective or fatty tissue [115], which may also limit direct

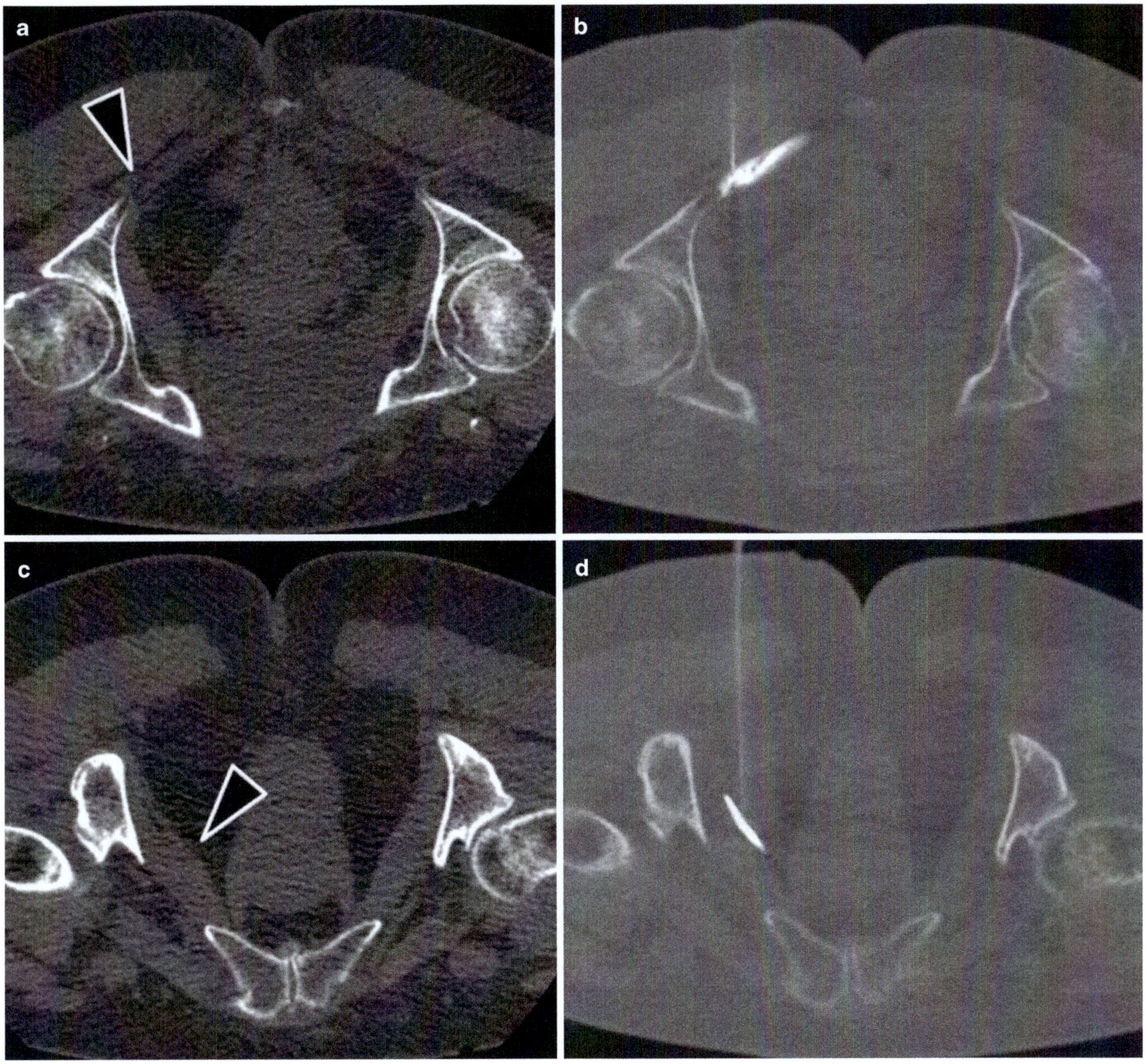


**Fig. 29.29** Intraoperative computed tomography images of dual-site pudendal nerve infiltration. Images (**a**, **c**) represent a planning CT slice at Site 1 (Image **a**) and Site 2 (Image **c**). Images (**b**, **d**) show accurate needle positioning with the needle tip artifact located at target sites (white arrowhead)

delineation of the nerve. Although visualization of the pudendal nerve was not possible in all cases, the two ligaments and internal pudendal artery can be easily identified. The needle is inserted medially toward the pudendal artery, since the pudendal nerve is principally located medial to this artery (76–100%) [115, 116]. With this technique, Rofaeel et al. [54] were able to produce a sensory block in 100% of the patients tested.

Nerve stimulation is used once the needle is positioned in the interligamentous plane medial to the pudendal artery, since the pudendal nerve is in proximity to the sciatic nerve. Absence of muscle stimulation in the sciatic distribution avoids subsequent blockade of the sciatic. Following satisfactory positioning of the needle tip, an admixture of local anesthetic (5 mL) and steroid (or dextrose 5% or biologic) is injected and the adequacy of local anesthetic dispersion around the nerve during injection is reassessed.

## Piriformis Injection

Piriformis syndrome is an infrequent (or under-diagnosed) occurrence in interventional pain management settings. The diagnosis is made by medical history and physical examination.

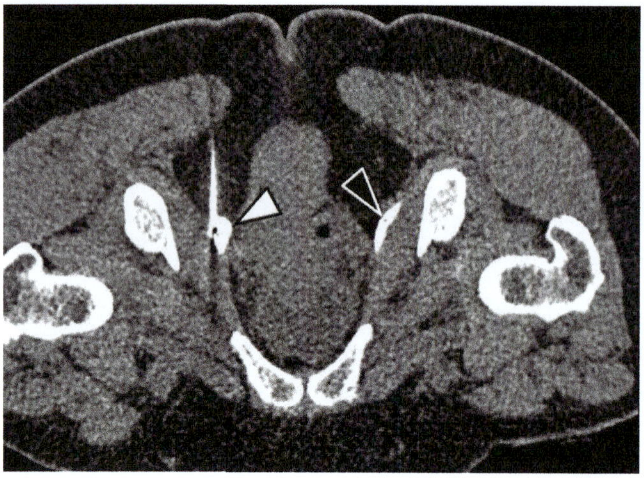

**Fig. 29.30** Intraprocedural CT image after Alcock's canal infiltration. This image depicts an accurate intra-Alcock's canal contrast media injection on the right side (black arrowhead) and inadequate contrast media injection on the left side (white arrowhead). Note how the product diffusion is contained in the canal on the left side creating a biconcave lentil as opposed to a round shaped product diffusion on the right, in favor of para-rectal fact location

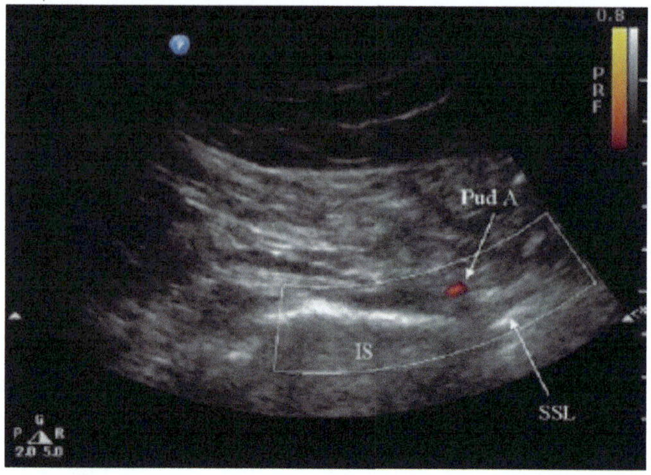


**Fig. 29.32** Ultrasound picture of the pudendal nerve before injections. SSL: sacrospinous ligament; Pud A: pudendal artery; IS: ischium at ischial spine level

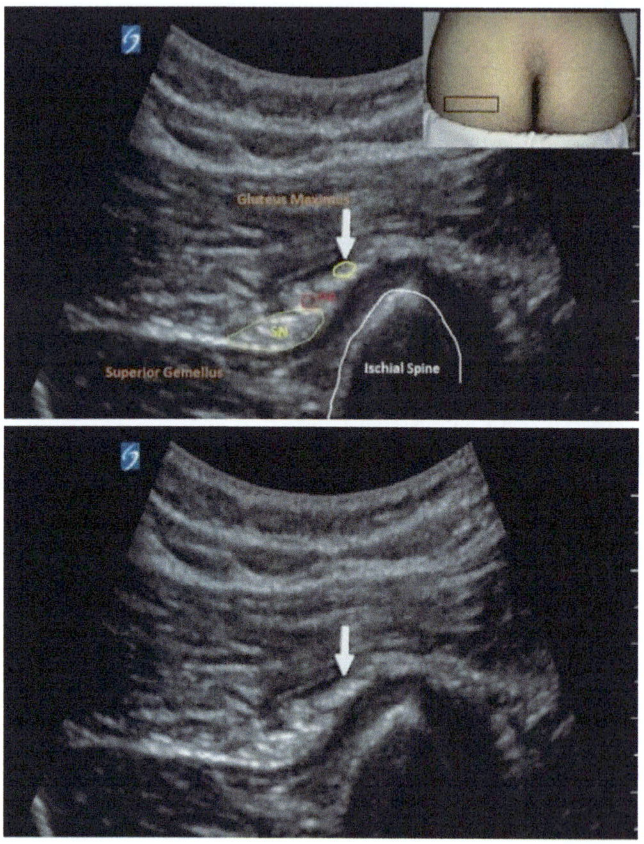


**Fig. 29.31** Ultrasound image of the pudendal nerve. SN = sciatic nerve; PA = pudendal artery; white arrow = pudendal nerve. (Image courtesy of Agnes Stogicza, MD with permission)

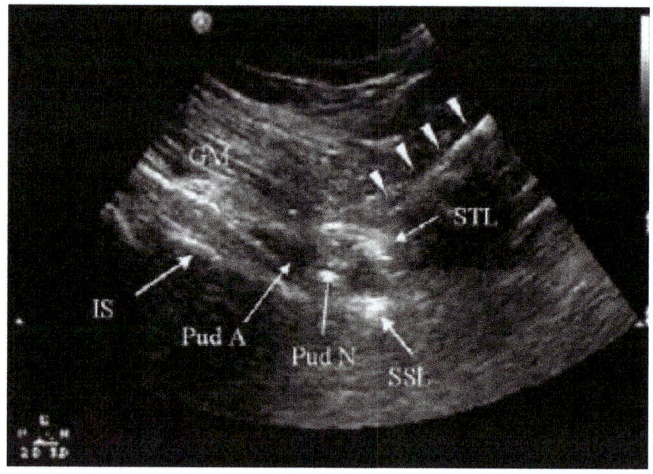


**Fig. 29.33** Ultrasound picture of the pudendal nerve during injection. Ultrasound picture showing the interligamentous plane and the needle insertion following injection of local anesthetic and steroid. The needle is outlined by the solid arrows. STL—sacrotuberous ligament; SSL—sacrospinous ligament; Pud A—pudendal artery; Pud N—pudendal nerve, IS—ischium at ischial spine level; GM—gluteus maximus. The needle was outlined by the solid arrows

Figure 29.11 shows trigger points and attachments of the piriformis muscle. CT and MRI findings of the abnormal uptake by the muscle and enlarged piriformis muscle have been reported in the literature. Nerve conduction studies may demonstrate delayed F waves and H reflexes [120]. There have been some observational studies reporting the effectiveness of piriformis injections. However, there are no randomized controlled trials. Further, there are no trials with injection of biologicals.

Fishman et al. [121] described a technique of piriformis injection using fluoroscopy and myelography. Correct placement into the muscle was confirmed using EMG and injection of contrast dye under fluoroscopy. With patients in the prone position the piriformis muscle was identified using the greater trochanter, SI joint, and the lateral border of the sacrum as landmarks.

Hanania and Ketain [122] described the technique of perisciatic injection in which the patient usually lies in the lateral or semiprone position. The dependent extremity is straight, and the nondependent extremity is flexed at the knees. The sciatic nerve is then located using a nerve stimulator and a mixture of local anesthetic with steroid is injected after withdrawing the needle a few centimeters.

Benzon et al. [123] described a technique in which the steroid is injected into the piriformis muscle and the area near the sciatic nerve using fluoroscopic guidance and a nerve stimulator. In this description the authors used steroids to reduce inflammation. However, they used only very dilute local anesthetic or no local anesthetic to minimize the motor blockade.

A commonly utilized technique involves injection of the piriformis muscle under fluoroscopic guidance. The patient is placed prone on the procedure table and sterile technique is used. Fluoroscopic view is obtained so that the lateral border of the sacrum as well as the greater trochanter of the femur can be visualized in the same view. The point of entry is marked over the superior-medial aspect of the trochanter. Although this site is probably slightly more lateral than the one described by various authors, it provides a margin of safety against accidental entry into the pelvic area. The 22G pencil point needle is advanced perpendicular to the skin, and in most cases a distinct "pop" is felt as the pencil-point needle pierces the posterior muscle fascia. A 3.5-in. needle is usually sufficient, but, in a larger patient a longer needle may be needed. As the needle is slightly lateral to the location of the sciatic nerve, paresthesia is usually uncommon. In the lateral fluoroscopic views, the needle tip may be found located approximately halfway between the posterior border of the sacrum and the greater trochanter (Fig. 29.34). Injection of radio-opaque contrast would then confirm the placement of the needle, and distinct striated spread of the contrast is noted in an oblique direction (Fig. 29.34). Usually, a total volume of 2–5 mL of 0.5% of lidocaine is injected.

Chronic piriformis spasm can result in a chronic enthesopathy (and subsequent calcification) at the distal piriformis attachment, which may trigger recurrent piriformis spasm. Local anesthetic with or without steroids injected at this site, followed by biologics if the steroid relief is only temporary, may give long-term relief.

## Pelvic Floor Myotendinous Injections

For pelvic floor myotendinous pain generators, the levator ani, piriformis and obturator internus are commonly targeted. The piriformis and obturator internus muscle is easily targeted with ultrasound guidance with a patient prone and with a 5 mHz probe oriented in transaxial dimension. With live ultrasound imaging, internal and external hip rotation will help identify the piriformis and obturator internus muscles. The levator ani muscle is best targeted while the patient is in a dorsal lithotomy position and the 5 mHz probe is angled coronally for transperineal injection on each side. The levator is seen as an obliquely oriented muscle originating from the boney pelvic structures and obturator internus and traversing towards the prostate and anus in men and towards the bladder outlet, vagina and anus in women. For all three muscles, with regards to the orientation of the ultrasound probe, an out of plane injection technique is used. Care must be taken to keep the needle tip in view at all times and avoid penetration of the rectum, bladder and urethra. In men, for the levator ani target, it is important to use an initial path between the corpus spongiosum and corpus cavernosum. Live ultrasound must be used during the actual injection to be sure there is no vascular uptake. A stimulating needle may be used or voluntary pelvic floor contractions may be elicited to help identify the location of the levator ani, and electromyographic monitoring is particularly useful to confirm intramuscular positioning for this muscle, since it is often thin or obliquely oriented, which can make accurate targeting challenging.

## Hormone Therapy

Hormonal evaluation should be considered in patients with pelvic pain originating in or involving the genital tissues. In women reporting superficial or deep dyspareunia, hormonal dysfunction is not uncommon. Vaginal dryness can often coexist with and contribute to dyspareunia and may be the result of hormonal deficiency. Deficiencies in both estradiol and testosterone can bring about a variety of problems in the genital tissues including decreased lubrication, provoked pain, atrophy of the clitoris, labia and vagina, and loss of vaginal rugae [124]. Each of these can be pain generators.

One example highlighting the relationship between hormones and pain is PVD in the female. PVD is a common cause of superficial dyspareunia. Because hormonal abnormalities are often found in patients with provoked vestibulodynia, blood should be collected to measure serum estradiol, total testosterone, free testosterone, albumin, sex hormone binding globulin (SHBG), follicle-stimulating hormone (FSH), and prolactin [125]. Increased SHBG and deficiencies in free testosterone and estradiol are commonly seen in women with PVD who have been exposed to hormonal contraceptives [125–127]. Often female patients with recognized hormonal abnormalities can achieve significant improvements in their pain when these abnormalities are

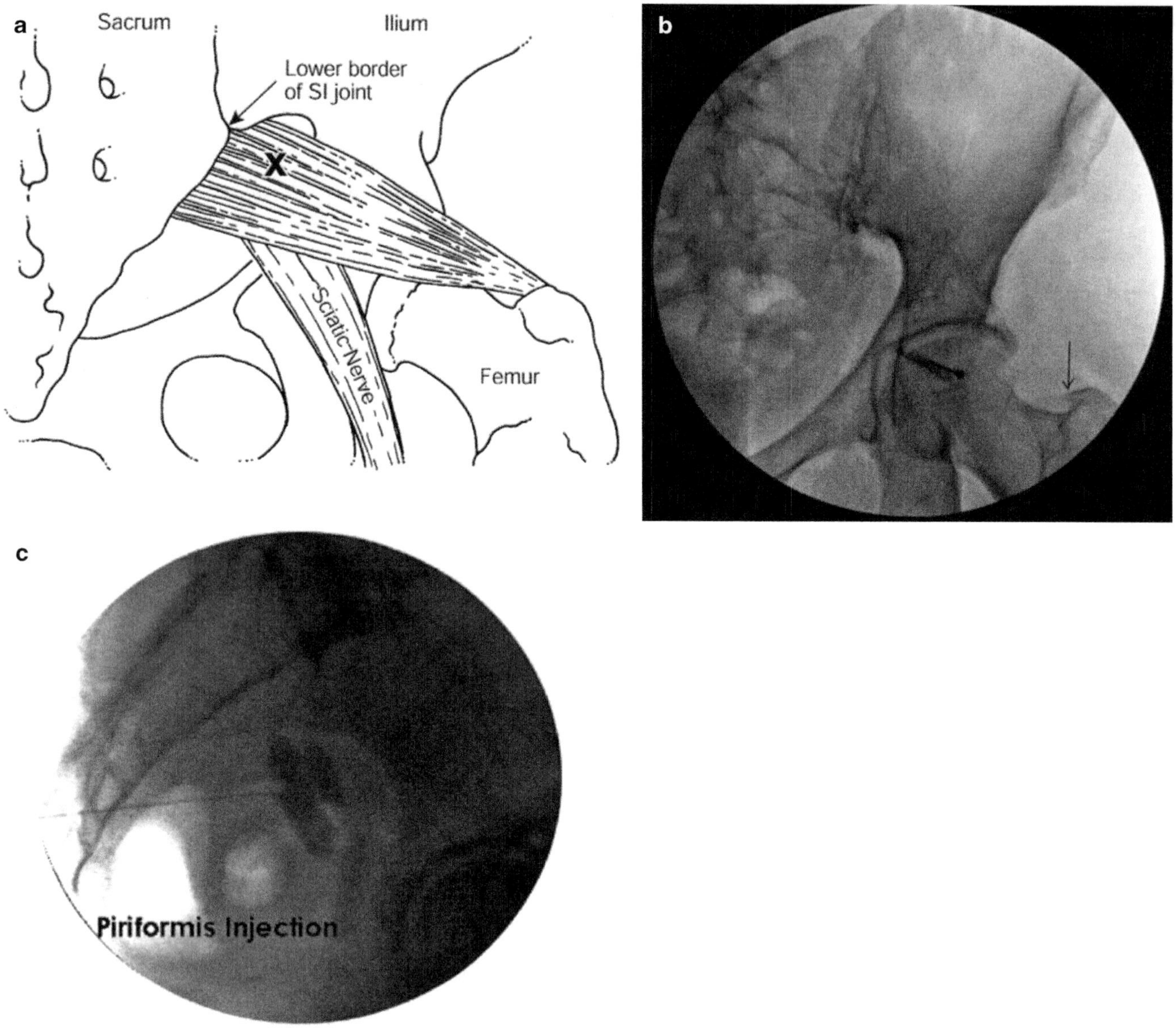


**Fig. 29.34** Fluoroscopic piriformis injection. (**a**) Illustration of radiopaque contrast illustrating distinct pattern. (**b**) Needle with contrast in the piriformis muscle, black arrow marking the calcified distal piriformis. (**c**) In the lateral fluoroscopic views, the needle tip may be found located approximately halfway between the posterior border of the sacrum and the greater trochanter

corrected with either local or systemic hormonal therapies [128].

There is an important relationship between hormones and inflammation that is also beginning to be appreciated. Returning to vestibulodynia as an example, one study reported that inflammatory infiltrates of lymphoid and mast cells were greater in menopausal cases compared with premenopausal cases [129]. Their results suggest that a hormonal deficient state may promote localized inflammation and pain [130].

Male patients can also have pain associated with hormonal abnormalities. In those presenting with chronic testicular pain especially when associated with atrophy or varicocele, an evaluation of testosterone should be considered. Both atrophy and varicocele can be sources of chronic pain in the male and often result in lower levels of serum testosterone.

## Conclusion

Regenerative techniques that have been applied to musculoskeletal conditions are becoming increasingly applied to pelvic pain syndromes. The present chapter discussed various clinical conditions that may present as pelvic pain. An improved knowledge of the relevant anatomy and how to examine these structures as part of the bedside examination should be helpful effective targeting of pain generators in the pelvis.

**Key Points**

1. Chronic pelvic pain (CPP) is a complex and multifactorial condition that affects both men and women. It is characterized by nonmalignant pain perceived in the pelvic area, lasting at least 6 months, and is severe enough to limit functioning.
2. CPPS is a multifactorial condition with various suggested mechanisms, including infection, neurogenic inflammation, hypoxia, and pelvic floor muscle weakness or cramps. These mechanisms often interact rather than being the sole cause of the syndrome.
3. CPP encompasses various disorders, including pudendal neuralgia, coccydynia, prostatodynia, vulvodynia, and painful bladder syndromes. The estimated prevalence of CPP ranges from 5.7% to 26.6% in females and 2–16% in males in the general population.
4. CPPS is closely linked with sexual dysfunction in both males and females, impacting emotional status and overall quality of life. There is a significant comorbidity between CPP and substance use disorders, sharing psychosocial factors and neurobiological pathways.
5. Pelvic regional pain may involve bony and articular structures, such as the symphysis pubis and sacroiliac joints. Nerves frequently involved in pelvic pain syndromes include the genitofemoral, ilioinguinal, pudendal, and posterior femoral cutaneous nerves.
6. Management of chronic pelvic pain syndrome (CPPS) requires a holistic approach and may involve cognitive behavioral therapy, medications, surgery as a last resort, and myofascial physical therapy. Additionally, interventional techniques and regenerative medicine techniques can be considered for management. Subgroups of CPPS, such as painful bladder syndrome (PBS) and provoked vestibulodynia (PVD), are also addressed in the management plan.
7. The iliohypogastric, ilioinguinal, and genitofemoral nerves are known as "border nerves" because they supply the skin bordering the area between the abdomen and thigh. The iliohypogastric and ilioinguinal nerves arise from the anterior rami of L1, whereas, genitofemoral nerve arises from the first and second lumbar nerve roots.
8. The pudendal nerve is formed from the anterior rami of the second, third, and fourth sacral nerves, and exits the pelvis through the greater sciatic notch, passing posterior to the sacrospinous ligament.
9. Specific evidence demonstrating the effectiveness of regenerative techniques for the treatment of pelvic pain syndromes is limited at present. Orthobiological techniques involving blood and tissue components enriched with anti-inflammatory factors and growth factors have been studied, including amniotic tissue, autologous blood, autologous adipose, and bone marrow-derived products.
10. While improvements in pain control, function, and tissue integrity have been demonstrated, the choice of injectant type may depend on factors such as side effects, invasiveness, and cost.

**Acknowledgments** This book chapter is modified and updated from a previous book chapter, "Regenerative Therapy for Chronic Pelvic Pain Syndrome" by Sheldon Jordan, MD, Stephanie Prendergast, MPT, Joshua Gonzalez, MD, Iris Orbuch, MD, Andrea Trescot, MD, and Laxmaiah Manchikanti, MD, in *Essentials of Regenerative Medicine in Interventional Pain Management* published by ASIPP Publishing.

## References

1. Jordan S, Prendergast S, Gonzalez J, Orbuch I, Trescot A, Manchikanti L. Regenerative therapy for chronic pelvic pain syndromes. In: Manchikanti L, Navani A, Atluri S, editors. Essentials of regenerative medicine in interventional pain management. Paducah, KY: ASIPP Publishing; 2019. p. 561–96.
2. Reichmann G, Parlier-Ahmad AB, Beck L, et al. Chronic pelvic pain and sexual dysfunction among females and males receiving treatment for opioid use disorder. Front Pain Res (Lausanne). 2022;2:787559.
3. Hunter CW, Stovall B, Chen G, Carlson J, Levy R. Anatomy, pathophysiology and interventional therapies for chronic pelvic pain: a review. Pain Physician. 2018;21:147–67.
4. Peng PW, Tumber PS. Ultrasound-guided interventional procedures for patients with chronic pelvic pain—a description of techniques and review of literature. Pain Physician. 2008;11:215–24.
5. Grinberg K, Sela Y, Nissanholtz-Gannot R. New insights about chronic pelvic pain syndrome (CPPS). Int J Environ Res Public Health. 2020;17:3005.
6. Kothari S. Neuromodulatory approaches to chronic pelvic pain and coccygodynia. Acta Neurochir Suppl. 2007;97:365–71.
7. Krieger JN, Nyberg L, Nickel CJ. NIH consensus definition and classification of prostatitis. JAMA. 1999;282:236–7.
8. Ahangari A. Prevalence of chronic pelvic pain among women: an updated review. Pain Physician. 2014;17:E141–7.
9. Smith CP. Male chronic pelvic pain: an update. Indian J Urol. 2016;32:34–9.
10. Hunter C, Dave N, Diwan S, Deer T. Neuromodulation of pelvic visceral pain: review of the literature and case series of potential novel targets for treatment. Pain Pract. 2013;13:3–17.
11. Grinberg K, Granot M, Lowenstein L, Abramov L, Weissman-Fogel I. A common pro-nociceptive pain modulation profile typifying subgroups of chronic pelvic pain syndromes is interrelated with enhanced clinical pain. Pain. 2017;158:1021–9.
12. Klotz SGR, Ketels G, Richardsen B, Lowe B, Brumahi CA. Physiotherapeutic assessment of chronic pelvic pain syndrome. Development of a standardized physiotherapeutic assessment instrument for interprofessional cooperation in patients with chronic pelvic pain syndrome. Man Med. 2019;57:181–7.
13. Andersson KE, Nomiya M, Sawada N, Yamaguchi O. Pharmacological treatment of chronic pelvic ischemia. Ther Adv Urol. 2014;6:105–14.

14. Hanno P, Nordling J, Fall M. Bladder pain syndrome. Med Clin N Am. 2011;95:55–73.
15. Homma Y, Udea T, Tomoe H, et al. Clinical guidelines for interstitial cystitis and hypersensitive bladder syndrome. Int J Urol. 2009;16:597–615.
16. Stones RW, Mountfield J. Interventions for treating chronic pelvic pain in women. Cochrane Database Syst Rev. 2000;4:CD000387.
17. Naess I, Bo K. Pelvic floor muscle function in women with provoked vestibulodynia and asymptomatic controls. Int Urogenecol J. 2015;26:1467–73.
18. Dumoulin C, Bourbonnais D, Lemieux MC. Development of a dynamometer for measuring the isometric force of the pelvic floor musculature. Neurourol Urodyn. 2003;22:648–53.
19. van Lankveld JJ, Granot M, Weijmar-Schultz WC, et al. Women's sexual pain disorders. J Sex Med. 2010;7:615–31.
20. Farage MA, Miller KW, Ledger W. Localized provoked vestibulodynia: do multiple etiologic triggers manifest as neurogenic vulvar pain? J Genit Syst Disord. 2014;3:1–6.
21. Forrest JB, Mishell DR. Breaking the cycle of pain in interstitial cystitis/painful bladder syndrome: toward standardization of early diagnosis and treatment-consensus panel recommendations. J Reprod Med. 2009;54:3–14.
22. Bohm-Starke N, Hilliges M, Brodda-Jansen G, Rylander E, Torebjork E. Psychophysical evidence of nociceptor sensitization in vulvar vestibulitis syndrome. Pain. 2001;94:177–83.
23. Lowenstein L, Kenton K, Mueller ER, et al. Patients with painful bladder syndrome have altered response to thermal stimuli and catastrophic reaction to painful experiences. Neurourol Urodyn. 2009;28:400–4.
24. Stav K. Painful bladder syndrome—it is worth knowing. Harefuah. 2011;150:168–74.
25. Harlow BL, Stewart BG. A population-based assessment of chronic unexplained vulvar pain: have we underestimated the prevalence of vulvodynia? J Am Med Women Assoc. 2003;58:82–8.
26. Pukall CF, Baron M, Amsel R, Khalife S, Binik YM. Tender point examination in women with vulvar vestibulitis syndrome. Clin J Pain. 2006;22:601–9.
27. Hampson JP, Reed BD, Clauw DJ, et al. Augmented central pain processing in vulvodynia. J Pain. 2013;14:579–89.
28. As-Sanie S, Harris RE, Napadow V, et al. Changes in regional gray matter volume in woman with chronic pelvic pain: a voxel-based morphometry study. Pain. 2012;153:1006–14.
29. Nickel JC, Alexander RB, Anderson R, et al. Category III chronic prostatitis/chronic pelvic pain syndrome: insights from the National Institutes of Health Chronic Prostatitis Collaborative Research Network studies. Curr Urol Rep. 2008;9:320–7.
30. Lamvu G, Carrillo J, Ouyang C, Rapkin A. Chronic pelvic pain in women: a review. JAMA. 2021;325:2381–91.
31. Bilevicius E, Sommer JL, Asmundson GJG, El-Gabalawy R. Posttraumatic stress disorder and chronic pain are associated with opioid use disorder: results from a 2012-2013 American Nationally Representative Survey. Drug Alcohol Depend. 2018;188:119–25.
32. Ross S, Peselow E. Co-occurring psychotic and addictive disorders: neurobiology and diagnosis. Clin Neuropharmacol. 2012;35:235–43.
33. Leknes S, Tracey I. A common neurobiology for pain and pleasure. Nat Rev Neurosci. 2008;9:314–20.
34. Walker E, Katon W, Harrop-Griffiths J, Holm L, Russo J, Hickok LR. Relationship of chronic pelvic pain to psychiatric diagnoses and childhood sexual abuse. Am J Psychiatry. 1988;145:75–80.
35. Lamvu G, Soliman AM, Manthena SR, Gordon K, Knight J, Taylor HS. Patterns of prescription opioid use in women with endometriosis: evaluating prolonged use, daily dose, and concomitant use with benzodiazepines. Obstet Gynecol. 2019;133:1120–30.
36. Benzon HT, Raja SN, Fishman SM, Liu SS, Cohen SP. Essentials of pain medicine. 3rd ed. Philadelphia, PA: Elsevier/Saunders; 2011.
37. Janicki TI. Chronic pelvic pain as a form of complex regional pain syndrome. Clin Obstet Gynecol. 2003;46:797–803.
38. Baranowski AP. Chronic pelvic pain. Best Pract Res Clin Gastroenterol. 2009;23:593–610.
39. Ploteau S, Labat JJ, Riant T, Levesque A, Robert R, Nizard J. New concepts on functional chronic pelvic and perineal pain: pathophysiology and multidisciplinary management. Discov Med. 2015;19:185–92.
40. Lee RB, Stone K, Magelssen D, Belts RP, Benson WL. Presacral neurectomy for chronic pelvic pain. Obstet Gynecol. 1986;68:517–21.
41. Ingber MS, Peters KM, Killinger KA, Carrico DJ, Ibrahim IA, Diokno AC. Dilemmas in diagnosing pelvic pain: multiple pelvic surgeries common in women with interstitial cystitis. Int Urogynecol J Pelvic Floor Dysfunct. 2008;19:341–5.
42. Trescot AM. Interventions in managing female pelvic pain. In: Interventional techniques in chronic non-spinal pain. Paducah, KY: ASIPP Publishing; 2009. p. 319–32.
43. Antolak SJ Jr. Interventions in managing male pelvic pain. In: Interventional techniques in chronic non-spinal pain. Paducah, KY: ASIPP Publishing; 2009. p. 303–18.
44. Bolash R, Cheng J. Abdominal wall blocks and neurolysis. In: Manchikanti L, Kaye AD, Falco FJE, Hirsch JA, editors. Essentials of interventional techniques in managing chronic spinal pain. Berlin: Springer International Publishing; 2018. p. 489–96.
45. Amin N, Krashin D, Trescot AM. Ilioinguinal and iliohypogastric nerve entrapment: abdominal. In: Trescot AM, editor. Peripheral nerve entrapments: clinical diagnosis and management. Cham: Springer; 2016. p. 413–24.
46. Shanthanna H. Successful treatment of genitofemoral neuralgia using ultrasound guided injection: a case report and short review of literature. Case Rep Anesthesiol. 2014;2014:371703.
47. Vodusek DB. Electromyogram, evoked sensory and motor potentials in neurourology. Neurophysiol Clin. 1997;27:204–10.
48. Rab M, Ebmer J, Dellon AL. Anatomic variability of the ilioinguinal and genitofemoral nerve: implications for the treatment of groin pain. Plast Reconstr Surg. 2001;108:1618–23.
49. Liu WC, Chen TH, Shyu JF, et al. Applied anatomy of the genital branch of the genitofemoral nerve in open inguinal herniorrhaphy. Eur J Surg. 2002;168:145–9.
50. Ducic I, Dellon AL. Testicular pain after inguinal hernia repair: an approach to resection of the genital branch of genitofemoral nerve. J Am Coll Surg. 2004;198:181–4.
51. Whiteside JL, Barber MD, Walters MD, Falcone T. Anatomy of ilioinguinal and iliohypogastric nerves in relation to trocar placement and low transverse incisions. Am J Obstet Gynecol. 2003;189:1574–8.
52. Robert R, Prat-Pradal D, Labat JJ, et al. Anatomic basis of chronic perineal pain: role of the pudendal nerve. Surg Radiol Anat. 1998;20:93–8.
53. Mahakkanukrauh P, Surin P, Vaidhayakarn P. Anatomical study of the pudendal nerve adjacent to the sacrospinous ligament. Clin Anat. 2005;18:200–5.
54. Rofaeel A, Peng P, Louis I, Chan V. Feasibility of real-time ultrasound for pudendal nerve block in patients with chronic perineal pain. Reg Anesth Pain Med. 2008;33:139–45.
55. Thompson JR, Gibb JS, Genadry R, et al. Anatomy of pelvic arteries adjacent to the sacrospinous ligament: importance of the coccygeal branch of the inferior gluteal artery. Obstet Gynecol. 1999;94:973–7.
56. Warrick R, Williams PL, editors. Myology: fascia of the trunk. In: Gray's anatomy. 36th ed. Edinburgh: Churchill Livingstone; 1980. p. 542–64.

57. Schraffordt SE, Tjandra JJ, Eizenberg N, Dwyer PL. Anatomy of the pudendal nerve and its terminal branches: a cadaver study. ANZ J Surg. 2004;74:23–6.
58. Pecina M. Contribution to the etiological explanation of the piriformis syndrome. Acta Anat. 1979;105:181–7.
59. Beason LE, Anson BJ. The relation of the sciatic nerve and its subdivisions to the piriformis muscle. Anat Record. 1937;70:1–5.
60. Cuéllar JM, Cuéllar VG, Scuderi GJ. α2-macroglobulin: autologous protease inhibition technology. Phys Med Rehabil Clin N Am. 2016;27:909–18.
61. Scholten PM, Massimi S, Dahmen N, et al. Successful treatment of athletic pubalgia in a lacrosse player with ultrasound-guided needle tenotomy and platelet-rich plasma injection: a case report. PM R. 2015;7:79–83.
62. Dönmez Mİ, İnci K, Zeybek ND, et al. The early histological effects of intravesical instillation of platelet-rich plasma in cystitis models. Int Neurourol J. 2016;20:188–96.
63. Ozyuvali E, Yildirim ME, Yaman T, et al. Protective effect of intravesical platelet-rich plasma on cyclophosphamide-induced hemorrhagic cystitis. Clin Invest Med. 2016;39:27514.
64. Kim A, Shin DM, Choo MS. Stem cell therapy for interstitial cystitis/bladder pain syndrome. Curr Urol Rep. 2016;17:1.
65. Kim A, Yu HY, Lim J, et al. Improved efficacy and in vivo cellular properties of human embryonic stem cell derivative in a preclinical model of bladder pain syndrome. Sci Rep. 2017;7:8872.
66. Ulukus M. Stem cells in endometrium and endometriosis. Womens Health (Lond). 2015;11:587–95.
67. Song M, Lim J, Yu HY, et al. Mesenchymal stem cell therapy alleviates interstitial cystitis by activating Wnt signaling pathway. Stem Cells Dev. 2015;24:1648–57.
68. Forte A, Cipollaro M, Galderisi U. Genetic, epigenetic and stem cell alterations in endometriosis: new insights and potential therapeutic perspectives. Clin Sci. 2014;126:123–38.
69. Hernigou P, Flouzat Lachaniette CH, et al. Biologic augmentation of rotator cuff repair with mesenchymal stem cells during arthroscopy improves healing and prevents further tears: a case-controlled study. Int Orthop. 2014;38:1811–8.
70. Hegde V, Shonuga O, Ellis S, et al. A prospective comparison of 3 approved systems for autologous bone marrow concentration demonstrated nonequivalency in progenitor cell number and concentration. J Orthop Trauma. 2014;28:591–8.
71. Cassano JM, Kennedy JG, Ross KA, et al. Bone marrow concentrate and platelet-rich plasma differ in cell distribution and interleukin 1 receptor antagonist protein concentration. Knee Surg Sports Traumatol Arthrosc. 2018;26:333–42.
72. Berman M, Lander E. A prospective safety study of autologous adipose-derived stromal vascular fraction using a specialized surgical processing system. Am J Cosmet Surg. 2017;34:129–42.
73. Hall PN, Lee AP. Rectus nerve entrapment causing abdominal pain. Br J Surg. 1988;75:917.
74. Peleg R, Gohar J, Koretz M, Peleg A. Abdominal wall pain in pregnant women caused by thoracic lateral cutaneous nerve entrapment. Eur J Obstet Gynecol Reprod Biol. 1997;74:169–71.
75. Hughes GS, Treadwell EL, Miller J. Syndrome of the rectus abdominis muscle mimicking the acute abdomen. Ann Emerg Med. 1985;14:694–5.
76. Stultz P. Peripheral nerve injuries resulting from common surgical procedures in the lower abdomen. Arch Surg. 1982;117:324–7.
77. Hagen NA. Sharp, shooting pain in the rectum or genitals: pudendal neuralgia. J Pain Symptom Manag. 1993;8:496–501.
78. Anderson KV, Bovim G. Impotence and nerve entrapment in long-distance amateur cyclists. Acta Neurol Scand. 1997;95:233–40.
79. Alevizon SJ, Finan MA. Sacrospinous colpopexy: management of postoperative pudendal nerve entrapment. Obstet Gynecol. 1996;88:713–5.
80. Tetzschner T, Sorensen M, Lose G, Christiansen J. Pudendal nerve function during pregnancy and after delivery. Int Urogynecol J Pelvic Floor Dysfunct. 1997;8:66–8.
81. Tagnetti F, Poppi M, Gaist G, et al. Pudendal neuralgia due to solitary neurofibroma: case report. J Neurosurg. 1982;56:732–3.
82. Howard EJ. Postherpetic pudendal neuralgia. JAMA. 1985;253:2196.
83. Antolak S, Hough D, Pawlina W, Spinner R. Anatomical basis of chronic pelvic pain syndrome: the ischial spine and pudendal nerve entrapment. Med Hypotheses. 2002;59:349–53.
84. Amarenco G, Ismael SS, Bayle B, et al. Electrophysiological analysis of pudendal neuropathy following traction. Muscle Nerve. 2001;24:116–9.
85. Popeney C, Ansell V, Renney K. Pudendal entrapment as an etiology of chronic perineal pain: diagnosis and treatment. Neurourol Urodyn. 2007;26:820–7.
86. Fall M, Baranowski AP, Fowler CJ, et al. EAU guidelines on chronic pelvic pain. Eur Urol. 2004;46:681–9.
87. Bleustein CB, Eckholdt E, Arezzo JC, Melman A. Quantitative somatosensory testing of the penis: optimizing the clinical neurological examination. J Urol. 2003;169:2266–9.
88. Jaaskelainen SK, Teerijoki-Oska T, Virtanen A, Tenovuo O, Forsell H. Sensory regeneration following intraoperatively verified trigeminal nerve injury. Neurology. 2004;62:1951–7.
89. Pace JB, Nagle D. Piriformis syndrome. West J Med. 1976;124:435–9.
90. Bernard TN Jr, Kirkaldy-Willis WH. Recognizing specific characteristics of nonspecific low back pain. Clin Orthop. 1987;217:266–80.
91. Papadopoulos E, Khan S. Piriformis syndrome and low back pain: a new classification and review of literature. Orthop Clin North Am. 2004;35:65–71.
92. Jankiewicz JT, Hennrikus WL, Houkom JA. The appearance of piriformis muscle in CT and MRI: a case report and review of literature. Clin Orthop. 1991;262:205–9.
93. Synek VM. The piriformis syndrome: a review and case presentation. Clin Exp Neurol. 1987;23:31–7.
94. Lv YS, Zhou HL, Mao HP, et al. Intravesical hyaluronic acid and alkalinized lidocaine for the treatment of severe painful bladder syndrome/interstitial cystitis. Int Urogynecol J. 2012;23:1715–20.
95. Nickel JC, Herschorn S, Whitmore KE, Forrest JB, Hu P, Friedman AJ, Baseman AS. Pentosan polysulfate sodium for treatment of interstitial cystitis/bladder pain syndrome: insights from a randomized, double-blind, placebo-controlled study. J Urol. 2015;193(3):857–62.
96. Porru D, Leva F, Parmigiani A, et al. Impact of intravesical hyaluronic acid and chondroitin sulfate on bladder pain syndrome/interstitial cystitis. Int Urogynecol J. 2012;23:1193–9.
97. Jhang JF, Jiang YH, Kuo HC. Potential therapeutic effect of intravesical botulinum toxin type A on bladder pain syndrome/interstitial cystitis. Int J Urol. 2014;21(Suppl 1):49–55.
98. Shie JH, Liu HT, Wang YS, Kuo HC. Immunohistochemical evidence suggests repeated intravesical application of botulinum toxin A injections may improve treatment efficacy of interstitial cystitis/bladder pain syndrome. BJU Int. 2013;111:638–46.
99. Peng CH, Jhang JF, Shie JH, Kuo HC. Down regulation of vascular endothelial growth factor is associated with decreased inflammation after intravesical OnabotulinumtoxinA injections combined with hydrodistention for patients with interstitial cystitis—clinical results and immunohistochemistry analysis. Urology. 2013;82(1452):e1–6.
100. Rössberger J, Fall M, Jonsson O, Peeker R. Long-term results of reconstructive surgery in patients with bladder pain syndrome/interstitial cystitis: subtyping is imperative. Urology. 2007;70:638–42.

101. Kim HJ, Lee JS, Cho WJ, et al. Efficacy and safety of augmentation ileocystoplasty combined with supratrigonal cystectomy for the treatment of refractory bladder pain syndrome/interstitial cystitis with Hunner's lesion. Int J Urol. 2014;21:69–73.
102. Kastler A, Aubry S, Piccand V, et al. Radiofrequency neurolysis versus local nerve infiltration in 42 patients with refractory chronic inguinal neuralgia. Pain Physician. 2012;15:237–44.
103. Katz J, Brown JW. Atlas of regional anesthesia. Norwalk, CT: Appleton-Century-Crofts and Fleschner Publishing Company; 1985.
104. Brown DL. Atlas of regional anesthesia. Philadelphia, PA: WB Saunders; 1999.
105. Waldman SD. Atlas of interventional pain management. Philadelphia, PA: Saunders; 2004.
106. van Schoor AN, Boon JM, Bosenberg AT, et al. Anatomical considerations of the pediatric ilioinguinal/iliohypogastric nerve block. Paediatr Anaesth. 2005;15:371–7.
107. Manchikanti L, Schultz DM, Atluri S, et al. Lumbar epidural injections. In: Manchikanti L, Kaye AD, Falco FJE, Hirsch JA, editors. Essentials of interventional techniques in managing chronic spinal pain. Berlin: Springer International Publishing; 2018. p. 141–86.
108. Willschke H, Marhofer P, Bösenberg A, et al. Ultrasonography for ilioinguinal/iliohypogastric nerve blocks in children. Br J Anaesth. 2005;95:226–30.
109. Eichenberger U, Greher M, Kirchmair L, et al. Ultrasound-guided blocks of the ilioinguinal and iliohypogastric nerve: accuracy of a selective new technique confirmed by anatomical dissection. Br J Anaesth. 2006;97:238–43.
110. Vaisman J. Pelvic hematoma after an ilioinguinal nerve block for orchialgia. Anesth Analg. 2001;92:1048–9.
111. Parvaiz MA, Korwar V, McArthur D, et al. Large retroperitoneal haematoma: an unexpected complication of ilioinguinal nerve block for inguinal hernia repair. Anaesthesia. 2012;67:80–1.
112. Pecina MM, Krmpotic-Nemanic J, Markiewitz AD. Pudendal nerve syndrome (syndrome of Alcock's tunnel). In: Tunnel syndromes: peripheral nerve compression syndromes. 3rd ed. Boca Raton, FL: CRC Press; 2001. p. 191–4.
113. Amarenco G, Kerdraon J, Bouju P, et al. Efficacy and safety of different treatments of perineal neuralgia due to compression of the pudendal nerve within the ischiorectal fossa or by ischiatic spine. Rev Neurol. 1993;153:331–4.
114. Hough DM, Wittenberg KH, Pawlina W, et al. Chronic perineal pain caused by pudendal nerve entrapment: anatomy and CT-guided perineural injection technique. Am J Roentgenol. 2003;181:561–7.
115. Gruber H, Kovacs P, Peigger J, Brenner E. New, simple ultrasound-guided infiltration of the pudendal nerve: topographic basics. Dis Colon Rectum. 2001;44:1376–80.
116. Kovacs P, Gruber H, Piegger J, Bodner G. New, simple, ultrasound-guided infiltration of the pudendal nerve: ultrasonographic technique. Dis Colon Rectum. 2001;44:1381–5.
117. Kastler A, Puget J, Tiberghien F, et al. Dual site pudendal nerve infiltration: more than just a diagnostic test? Pain Physician. 2018;21:83–90.
118. O'Bichere A, Green C, Phillips RK. New, simple approach for maximal pudendal nerve exposure: anomalies and prospects for functional reconstruction. Dis Colon Rectum. 2000;43:956–60.
119. Shafik A, el-Sherif M, Youssef A, Olfat ES. Surgical anatomy of the pudendal nerve and its clinical implications. Clin Anat. 1995;8:110–5.
120. Bahethi S, Patel V. Pyriformis syndrome and interventional therapy. In: Interventional techniques in chronic non-spinal pain. Paducah, KY: ASIPP Publishing; 2009. p. 411–8.
121. Fishman SM, Caneris OA, Bandman TB, et al. Injection of the piriformis muscle by fluoroscopic and electromyographic guidance. Reg Anesth Pain Med. 1998;23:554–9.
122. Hanania M, Ketain E. Perisciatic injection of steroid for the treatment of sciatica due to piriformis syndrome. Reg Anesth Pain Med. 1998;23:223–8.
123. Benzon HT, Katz JA, Benzon HA, Iqbal MS. Piriformis syndrome: anatomic consideration, a new injection technique, and a review of the literature. Anesthesiology. 2003;98:1442–8.
124. Goetsch MF, Lim JY, Caughey AB. Locating pain in breast cancer survivors experiencing dyspareunia: a randomized controlled trial. Obstet Gynecol. 2014;123:1231–6.
125. Goldstein AT, Belkin ZR, Krapf JM, et al. Polymorphisms of the androgen receptor gene and hormonal contraceptive induced provoked vestibulodynia. J Sex Med. 2014;11:2764–71.
126. Burrows LJ, Basha M, Goldstein AT. The effects of hormonal contraceptives on female sexuality: a review. J Sex Med. 2012;9:2213–23.
127. Bouchard C, Brisson J, Fortier M, et al. Use of oral contraceptive pills and vulvar vestibulitis: a case-control study. Am J Epidemiol. 2002;156:254–61.
128. Burrows LJ, Goldstein AT. The treatment of vestibulodynia with topical estradiol and testosterone. Sex Med. 2013;1:30–3.
129. Leclair CM, Goetsch MF, Li H, et al. Histopathologic characteristics of menopausal vestibulodynia. Obstet Gynecol. 2013;122:787–93.
130. Pukall CF, Goldstein AT, Bergeron S, et al. Vulvodynia: definition, prevalence, impact, and pathophysiological factors. J Sex Med. 2016;13:291–304.

# 30 The ABC of Stem Cell Therapies for Neurological Disorders

Miguel A. Pappolla, Ping Wu, and Kumar Sambamurti

## Basic Concepts in Cell Therapy for Neurological Disease

Stem cells are progenitors capable of cell self-renewal/differentiation and carrying critical functions during development and repair [1]. They can undergo symmetric or asymmetric divisions into two daughter cells. In symmetric cell division, both daughter cells are also stem cells. In asymmetric division, a stem cell produces one stem cell and a specialized somatic cell. Neural stem cells (NSCs), one of the cell types of interest to this chapter, can generate mature neurons, astrocytes, and oligodendrocytes; these are the main cell types of the developed adult brain. Stem cells can be categorized into two main groups, embryonic stem cells (ESCs) and tissue stem cells; the latter includes fetal stem cells (FSCs) or adult stem cells (ASCs) [2]. Depending on the extent of differentiation, they can also be classified as 1-totipotent, 2-pluripotent, or 3-multipotent. Totipotent cells can produce all other cell types; pluripotent cells are also known as ESCs and can give rise to cells of any of the three germ layers: ectoderm, mesoderm, or endoderm. Multipotent cells can differentiate into the cells of the tissues in which they reside, and these are cells that typically engage in physiological tissue repair after injury. "The stem cell niche" modulates the functions and division potential of stem cells, a term that encompasses the tissue microenvironment in which these cells reside. ASCs relevant to neurological disease include mesenchymal stem cells (MSCs) and NSCs (Fig. 30.1). Some investigators are exploring the use of adipose-derived mesenchymal stem cells (ADSCs) [3].

Also of interest are laboratory-modified stem cells that have been produced under certain experimental conditions. These include induced pluripotent stem cells (iPSCs), consisting of genetically engineered adult somatic cells that have regained many features of pluripotent embryonic stem cells. The use of iPSCs has circumvented the ethical concerns of obtaining embryonic cells but some of the potential complications i.e., teratoma formation, remain in place [2]. There also is a growing interest in another laboratory modified type of stem cell; i.e., transgenic stem cells; these cells can overexpress certain growth factors to promote growth of other cell types in target tissues.

## Neurogenesis

The generation of new neurons is known as neurogenesis, a process that mainly occurs during development but continues at a much slower pace throughout adult life [2, 4]. NSCs are neuroepithelial cells localized around the neural tube during early development. After fetal maturity, they are localized primarily in the subventricular zone of the lateral and third ventricles and the subgranular layer in the hippocampus. During development, numerous NSCs are produced and migrate to their definitive places while the brain is growing. This process continues to a much lesser extent in the adult brain, where new neurons are mainly integrated into pre-existing circuitries. After differentiation, NSCs can generate new synaptic connections in a process known as plasticity [5]. Neurogenesis in the adult brain is a process that can be activated after an injury as shown in many experimental models [6].

In neurological disease, potential benefits have been observed with the administration of both, NSCs and MSCs

M. A. Pappolla (✉)
Department of Neurology, University of Texas Medical Branch, Galveston, TX, USA
e-mail: mpappoll@utmb.edu

P. Wu
Department of Neuroscience and Cell Biology, University of Texas Medical Branch, Galveston, TX, USA
e-mail: piwu@utmb.edu

K. Sambamurti
Neurosciences, Medical University of South Carolina, Charleston, SC, USA
e-mail: Sambak@musc.edu


A. Navani et al. (eds.), *Essentials of Regenerative Medicine in Interventional Pain Management*,
https://doi.org/10.1007/978-3-031-50357-3_30

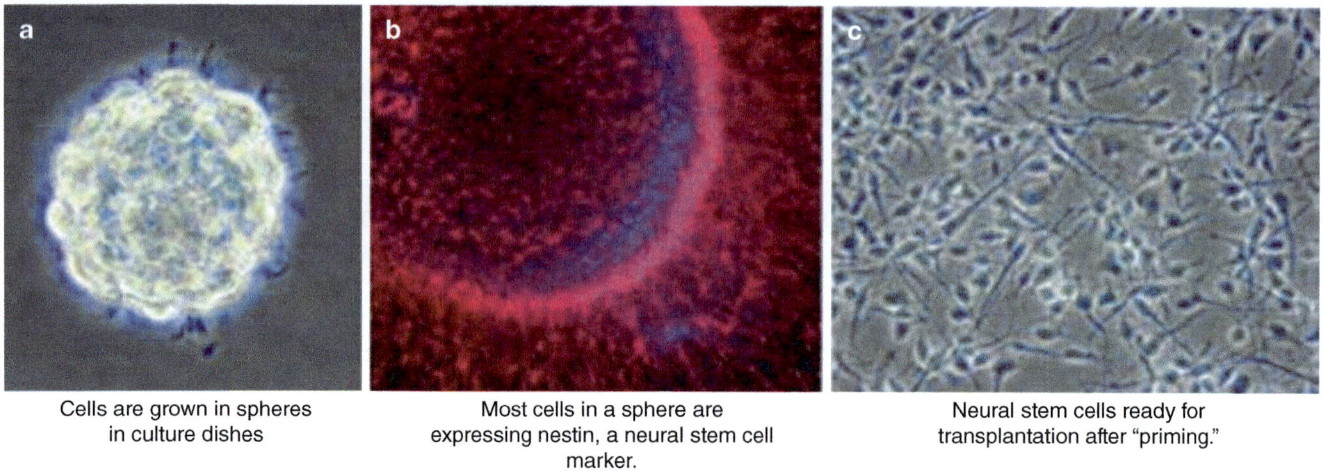


**Fig. 30.1** Human fetal brain-derived neural stem cells

[7]. NSCs have the capacity to differentiate into Schwann cells, oligodendrocytes (myelin-forming) and neurons [8]. In addition to their engrafting properties, stem cells can exert distant effects in target tissues by secreting many growth factors and cytokines (i.e., brain-derived neurotrophic factor (BDNF), nerve growth factor (NGF) and several others) [9]. This process is known as paracrine effect. Paracrine effects are important in tissue repair, and mediated by exosomes secreted by the transplanted cells. This mechanism include stimulation of endogenous stem cell proliferation and modulation of inflammatory responses via decreased production of pro-inflammatory cytokines and adhesion molecules. Improvement of blood-brain barrier function has also been observed following NSC transplantation, and this is also mediated through paracrine effects [10, 11].

## Mesenchymal Stem Cells and the Brain

In addition to NSCs, MSCs have often been employed therapeutically in animal models of neurological disease because they can also differentiate into various resident cells and participate in immunomodulation [10]. Three criteria have been proposed to identify MSCs by the Mesenchymal and Tissue Stem Cell Committee of the International Society for Cellular Therapy. They include the plastic adherence of the isolated cells in culture, the expression of CD105, CD73, and CD90 in more than 95% of the culture, and the lack of expression of certain markers including CD34, CD45, and CD14 or CD11b, CD79a or CD19 [7, 12, 13]. It is generally agreed that pericytes and adventitial cells surrounding cerebral blood vessels have stem cell properties that can differentiate into other specialized cells of the brain (i.e., dendritic cells). Despite some controversy, MSCs appear to have the potential to differentiate into neurons. As mentioned with NSCs, distant paracrine effects are also exerted by MSCs; these account for a significant extent of their therapeutic actions [11].

Stringent criteria to determine if there is genuine trans-differentiation of MSCs into neurons has been developed [14]. These include morphological changes, expression of neural biomarkers and electrophysiological activity. It has been described that after induction, trans-differentiation of perivascular MSCs is followed by loss of pericyte markers and acquisition of neuronal markers such as HUC/D, neuron specific enolase (NSE), Map2, synaptophysin, nestin, and others. Induction of a glial phenotype with expression of glial fibrillary acid protein (GFAP) and S100b has also been demonstrated [14]. Experimentally, glial or neuronal trans-differentiation requires specific induction protocols causing epigenetic changes in the MSCs. MSCs from various sources, including brain, bone marrow and adipose tissue, can promote endogenous neuronal growth, decrease apoptosis and downregulate the inflammatory response. In that manner, MSCs can facilitate repair and neuroprotection [15].

## Stem Cell Therapy for Stroke

Stroke is a major cause of death following cardiovascular disease and cancer. Stroke remains a major concern in neurology because of the irreversible damage to brain tissue. The risk of suffering a stroke depends on a wide variety of factors, many of which are modifiable. Existing methods for treating stroke are constrained by the time elapsed from onset to treatment. The current approaches are centered on secondary prevention and rehabilitative care, neither of which promote clinically meaningful tissue regeneration.

The two main types of stroke are ischemic and hemorrhagic; about 80% are deemed to be ischemic, that is, caused by lack of blood flow to a brain region. Standard treatments include thrombolytic therapy, percutaneous intravascular interventions, rehabilitation, anti-platelet agents and anti-coagulants, depending on the clinical setting. Other pharmacological agents along with treatment of associated risk factors and lifestyle modifications are frequently used to improve clinical outcomes after a stroke and to prevent recurrences.

The development of stem cell therapy involves understanding endogenous processes of brain regeneration leading to the identification of potential therapeutic targets in regenerative medicine. For example, the number of precursor cells gradually increases by around 30% within 2 weeks after a stroke, peaks at 2 weeks, and declines during the subsequent 3 to 5 weeks [16]. Intrinsic NSCs respond to ischemic stroke by proliferating and migrating toward the lesion, as shown in several studies [17–19].

Stem cell therapies have been investigated by several laboratories for the potential treatment of stroke. Many animal studies have been conducted in rats and mice using MSCs, (ADSCs; a variant of MSCs), NSCs and iPSCs [12]. A review of many of these studies in experimental animals shows, with a few exceptions, that administration of either MSCs or NSCs results in variable improvement of neurological functions following middle cerebral artery occlusion (MCAO) [20–22]. Direct injection of iPSCs into damaged areas of the rat cortex or subdural transplantation of iPSCs in fibrin glue was found to decrease infarct size significantly and cause an improvement of motor functions in rats with MCAO. These findings are accompanied by a reduction of proinflammatory cytokines and an increase in anti-inflammatory cytokines [23]. iPSCs derived from adult human fibroblasts transplanted into the rat MCAO model migrate to the infarcted areas, a phenomenon that correlates with motor and sensory improvements [24]. Synaptogenesis and functional electrophysiological integration of exogenous stem cells into the neuronal circuitry of the host brain has also been demonstrated [12]. However, several investigators have concluded that the number of engrafting neurons after stem cell transplantation is relatively insignificant to account for the extent of the benefits following this type of therapy. Instead, as previously mentioned, cytokines and growth factors secreted by the stem cells appear to underlie the observed proliferation and differentiation of endogenous NSCs during repair. This concept, repeated in numerous experimental settings and as mentioned throughout this chapter for different disorders, has also been supported by a paradigm in which suppression of endogenous adult neurogenesis in mice, despite engrafting of transplanted cells, has been found to worsen stroke outcomes [25]. Detection of active cell proliferation after a stroke in the ipsilateral subventricular zone was demonstrated unequivocally in humans. Several studies have shown that multiple neurotrophic signals, such as autocrine or paracrine, are involved in the regulation of neurogenesis.

As with NSCs transplantation, MSCs -have also shown promising therapeutic outcomes in pre-clinical stroke models [12, 15]. Injections of human or rat MSCs after experimental strokes were followed by functional improvement. As observed with other types of stem cells, the therapeutic effects of MSCs also reflect the activity of secreted factors on endogenous neurogenesis [26]. However, one substantive issue on the use of rodent studies is the poor correlation between the outcomes following cell transplantation in these animals versus the results in human clinical trials. For example, erythropoietin (EPO) has been shown to promote endogenous neurogenesis and angiogenesis in the rat stroke model [27]. In the rat, functional recovery significantly improved, along with increases in the density of cerebral microvasculature at the edges of the ischemic lesion. These exciting outcomes were accompanied by increased BrdU-labeled cells, doublecortin, and nestin immunoreactive cells in the subventricular zone [27]. In sharp contrast with these animal studies, the overall death rate of ischemic stroke patients treated with EPO in human trials was disappointingly higher (16.4%) than in the placebo control group (9%), leading to early termination of the trial due to safety concerns [28]. A possible partial explanation of these differences may reside in rodents' expression patterns of the genes involved in the inflammatory response compared to humans. However, other mechanisms are also probably involved.

Recent studies continue to show that the injection of ADSCs improves motor function and neurological state in animals after an ischemic stroke [3]. According to a meta-analysis, there is substantial evidence that transplanting ADSCs after an ischemic stroke improves neurological and motor function in animal models. With increasing knowledge of the mechanisms involved, there is hope that impediments to success in humans will be surmounted and efforts to test stem cells in human clinical trials continue at the time of this writing [29].

Another promising preclinical area of research involves the potential therapeutic use of exosomes. Exosomes from transplanted stem cells can activate several signaling pathways in target endogenous stem cells relevant to tissue repair (Akt, ERK, and STAT3) [30, 31]. These particles regulate intercellular communication and contribute to brain remodeling by transferring growth promoting factors to target cells [32]. The transfer of exosomes to target cells play essential roles in restorative events and inflammation after stroke and in other forms of neuronal injury [32]. Lately, microRNA-containing cell-derived exosomes (miRNAs) have gained attention as a possible treatment for cerebral ischemia-reperfusion injury (CIRI). However, their mode of action and therapeutic relevance are not fully understood. Recently, several investigators examined the potential therapeutic benefits of exosomes modified with miRNA in pre-clinical animal models of CIRI. According to a meta-analysis of 11 studies published in 2022, MiRNA-modified exosomal treatment can reduce infarct size and neurobehavioral function in preclinical animal models of CIRI [33]. The successful use of exosomes in human clinical trials has not yet been determined.

In conclusion, clinical trials using regenerative medicine in stroke patients are in early phases and have not yet excluded placebo effects or natural recovery of stroke; however, they provided evidence that transplantation of autologous cells is feasible and safe [34, 35]. As of 2022, efforts to translate the use of stem cells to the human population continue unabated [35]. A phase IIa, randomized, double-blind, placebo-controlled, single-center in patients aged ≥60 years with moderate to severe ischemic stroke of ADSCs published in 2022 was shown to be safe but showed no clinically meaningful differences between ADSCs and placebo as evaluated with a modified Rankin scale [35]. Despite the current shortcomings, efforts continue, and there are high hopes that a better understanding of the mechanisms involved in human patients will translate into positive outcomes in the foreseeable future.

## Stem Cell Therapy in Multiple Sclerosis

Among the chronic inflammatory disorders affecting the central nervous system, multiple sclerosis (MS) is a relatively common cause of disability in young people. This disorder is characterized by an inflammatory destruction of myelin in the brain and in the spinal cord by abnormally activated T helper cells [36, 37]. There are different clinical forms of MS; the most common is characterized by multiple relapses and remissions. Another more aggressive variant exhibits a relentless progression of myelin destruction without remissions (primary progressive form). With some exceptions, all forms of MS progress to substantive disability if left untreated. Many hypotheses have been proposed, including post-infection autoimmune mechanisms against myelin antigens, "slow" infections, and vitamin deficiencies; however, the cause of MS remains unknown. The typical neuropathological lesion in MS shows infiltration by inflammatory cells, demyelination, secondary axonal degeneration and cell loss [36]. Current treatments available are aimed at reducing the rate of relapses by decreasing the inflammatory response to myelin antigens. The prospect of employing cell therapies in MS has been investigated in many laboratories due to the anti-inflammatory and immunomodulatory effects of stem cells [7, 15]. In an experimental allergic encephalomyelitis (EAE) model of MS, it has been shown that rodents that received an intraventricular infusion of MSCs have almost twice the number of axons as control animals [15]. NSCs and MSCs have both been tried in the EAE model.

Following the administration of MSCs, remyelination and decreased inflammation were observed in preclinical studies [38]. However, results using NCSs in the same model are mixed [39, 40]. Gao et al. [41] reported that the anti-inflammatory effect of stem cells can be markedly enhanced by pretreating EAE mice with osthole, a natural coumarin analog with a broad spectrum of pharmacological activities, including anti-inflammatory properties, regulation of the immune-response, and neuroprotection. Yang et al. [42] engineered NSCs to secrete IL-10, an immunoregulatory cytokine, and induced a stronger recovery from ongoing EAE than that with control NSCs. Similar outcomes were also obtained by transfecting NSCs with neurotrophin 3 (NT-3), a potent factor that is both neuroprotective and immunomodulatory [43]. iPSCs were shown to differentiate into oligodendrocyte lineage in vitro and successful repopulate oligodendrocyte loss in animal models [7]. All things considered, the mechanisms of actions stem cells in MS may include immunomodulation primarily via paracrine effects and, secondarily, cell repopulation by engrafting of oligodendrocytes.

Clinical trials in human subjects for MS are undergoing earlier phases. In a randomized, double-blind, placebo-controlled, crossover phase II study, Llufriu et al. [44] investigated the safety and efficacy of relapsing-remitting MS patients. Efficacy was evaluated in terms of a cumulative number of gadolinium-enhancing lesions on magnetic resonance imaging (MRI) at 6 months and the end of the study. At baseline, nine patients were randomized to receive MSCs ($n = 5$) or placebo ($n = 4$). The authors did not identify any serious adverse events. At 6 months, patients treated with MSCs had a trend to a lower cumulative mean number of gadolinium-enhancing lesions ($p = 0.064$), and at the end of the study ($p = 0.075$). The investigators showed a non-significant decrease in the frequency of Th1 (CD4+ IFN-$\gamma$+) cells in the blood of MSCs treated patients. In another human study, Lublin et al. tested infusions of PDA-001, a preparation of mesenchymal-like stem cells derived from full-term human placenta [45]. Their objective was to evaluate safety, to rule out the possibility of paradoxical exacerbation of disease activity by PDA-001.

The infusions were safe and well tolerated in relapsing-remitting MS and secondary progressive MS patients. No paradoxical worsening of lesion counts was noted with either dose [45]. In an open-label study, Connick et al. recruited 10 patients with secondary progressive MS involving the visual pathways. Participants received an intravenous infusion of autologous bone marrow-derived MSCs. The primary objective was to assess feasibility and safety; adverse events were analyzed up to 10 months after the treatment. Secondary outcomes were efficacy measures to assess the anterior visual pathway. The administration of MSCs was safe in these patients, and data analysis showed some evidence of protection [46]. In a recent phase 2 clinical trial of intrathecal administration of MSCs, that included eighteen participants with non-relapsing progressive MS, there were no documented MS-related fatalities or adverse events [47]. Two participants experienced low back and leg pain symptoms that were compatible with an arachnoiditis diagnosis during one of three intrathecal treatments. At 28 weeks following treatment, 19% of the treated participants met the pre-specified 25% improvements in timed 25-foot walk speed relative to baseline and showed consistent efficacy signals for the pre-specified response criteria across various secondary efficacy outcomes. Following therapy, cerebrospinal fluid neuroprotective factors rose, and inflammatory indicators decreased, in line with the suggested mechanism of action involved in stem cell therapy of MS.

As can be appreciated in the above discussion, there is an overall trend favoring clinical benefits following the use of cell-based therapy for MS. Nevertheless; additional clinical studies are necessary to optimize these strategies.

## Stem Cell Therapy in Amyotrophic Lateral Sclerosis

Amyotrophic lateral sclerosis (ALS) is characterized by the progressive loss of motor neurons. Most cases of ALS (about 90%) are sporadic, and the rest are inherited. Patients with this disorder exhibit spastic weakness from the loss of upper motor neurons and atrophy of muscles with non-spastic weakness from the loss of spinal cord motor neurons. As the disease progresses, the tendon reflexes become hyperactive, and the muscles exhibit characteristic fasciculations. Speaking, swallowing, and breathing are gradually more difficult in the advanced stages. Patients with ALS eventually lose the ability to initiate and control most voluntary movements. Bowel and bladder control and ocular movements remain spared until the advanced stages of the disease. Death occurs inexorably in all patients from respiratory failure.

Enhancing the survival of motor neurons has been one of the main objectives of several experimental therapies in ALS. Therefore, cell-based therapies are attractive strategies for replacing dying motor neurons increasing their survival capacity.

Marcetto et al. [48] investigated the mechanisms leading to neuronal degeneration in a model of ALS. They showed that spinal motor neurons derived from human ESCs are selectively sensitive to the toxic effects of glial cells carrying a familial ALS-causing mutation in the SOD1 gene. Comparable results have been obtained from other investigators [48]. These studies suggest that non-neuronal cells may be involved in the pathogenesis of the disease by contributing to a toxic microenvironment for motor neurons. In a study illustrating this scenario, stem cells were transplanted into transgenic rats expressing a mutant form of human superoxide dismutase 1 (hSOD1(G93A)) associated with inherited ALS. This treatment resulted in transient improvement of the symptoms. However, there was an accelerated death of the transplanted cells with a prompt resumption of the progressive neurological decline typical of the disease [49]. Similar results were also reported by grafting human neural stem cell-derived motor neurons into the spinal cord of the ALS transgenic rats [50]. These results show that reducing the toxic microenvironment detrimental to motor neurons is an important therapeutic target.

In contrast, Serio et al. [51] reported that astrocytes derived from iPSCs cells carrying another ALS-causing mutation (TDP-43 mutation) showed no toxicity. Still, the astrocytes died earlier than control wild-type astrocytes derived from iPSCs. Their data suggests, in contrast with the SOD1 studies, that there is a loss of neuroprotection normally exerted by astrocytes. Therefore, these results highlight the existing gaps in our understanding of cell death in ALS [52]. Data from a meta-analysis study using the SOD1 (G93A) transgenic mouse model of ALS, however, showed that transplanted NSCs (both mouse and human) can slow the onset and progression of the disease and prolong survival of the ALS mice [52, 53].

MSCs transplantation has also been tried in ALS. In one human phase 1 clinical study, trMSCs delivered to in the spinal cord of ALS patients were shown to be safe and feasible [54]. An encouraging study using modified MSCs injected intrathecally showed evidence of possible clinically meaningful reductions in the disease rate of disease progression in 12 ALS patients [55]. The results reported by the investigators were mainly characterized by a decreased rate of progression of the forced vital capacity and of the ALS Functional Rating Scale-Revised score in patients receiving the modified MSCs [55]. In an open-label clinical trial to evaluate the safety and efficacy of repeated intrathecal administrations of autologous MSCs in ALS patients, the treatments were safe and well tolerated (primary endpoints). This trial also provided indications of potential clinical benefits [56]. Based on these results, efforts to introduce cell-based therapies for ALS are worth further investigation.

## Cell Therapies for Parkinson's Disease

Parkinson's disease (PD) is a degenerative disorder affecting the extrapyramidal motor system, characterized by tremors, rigidity, and slowness of movements (bradykinesia). Other regions are also involved resulting in a spectrum of non-motor symptoms, including autonomic dysfunction, cognitive difficulties, and psychiatric problems. Major neuropathological changes include the death of dopaminergic neurons in the brain stem and other affected nuclei and the accumulation of synuclein within the cytoplasm of some affected cells in the form of inclusion bodies called Lewy bodies (Fig. 30.2).

For the last 5 decades, dopaminergic therapies have been used to treat PD. While these therapies are very effective, especially in the early stages of the disease, they are ultimately constrained by the emergence of side effects related to the modified stimulation of dopamine receptors, including those in non-striatal areas. Therefore, there have been many attempts at treating the symptoms resulting from dying dopaminergic cells using cell transplantation techniques over the last 30 years [57, 58].

Proof of principle that such a strategy can deliver long-term clinical improvements with a reduced dependency on any oral dopaminergic medications has been demonstrated by the implantation of human fetal cells to the striatum of PD patients. Fetal tissue, however, presents several ethical and practical issues that make it unlikely to be used in the future for the regenerative treatment of PD. As a result, different cell sources have been investigated, and the techniques for creating real midbrain dopaminergic neurons from pluripotent cells have now developed to the point where it is possible to produce dopaminergic progenitors effectively and repeatedly at a much higher yield than is possible from human fetal tissue. Stem cell-based treatment for PD has the potential to get beyond many of the issues now connected to fetal cells.

Historically, Perlow et al. [58] were among the first investigators to implant fetal rat dopamine-containing neurons adjacent to the caudate nucleus of adult animals in which their endogenous dopaminergic afferents had been destroyed. The grafts showed good survival, axonal outgrowth, and reduced motor abnormalities induced by the destruction of the endogenous dopaminergic input to the caudate. Since then, experiments in many pre-clinical models of PD have demonstrated that cell replacement can be accomplished with functional improvement and reduction of motor deficits. Clinical trials, however, had an initially rocky start showing a lack of efficacy in sham surgery-controlled trials, development of dyskinesias in some treated patients, and lack of efficacy in non-motor symptoms. Despite these initial shortcomings, advances in the generation of dopaminergic neurons from stem cells and a better understanding of the mechanisms of graft-induced dyskinesias are now starting to show better long-term motor outcomes in experimentally treated patients with PD [59, 60]. One of the initially performed trials included a patient receiving an embryonic nigral transplant into the putamen; the grafts restored basal and drug-induced dopamine release to normal levels. Grafted neurons continued to function for a decade, giving substantial symptomatic relief [61]. In another study, 3 patients with PD received grafts with dopaminergic cells and showed long-term improvements in motor symptoms due to the replacement of dopaminergic neurons. Unfortunately, these patients continued to suffer from gradually progressive non-motor symptoms, including depression, fatigue, visual hallucinations, and sleep problems [62]. The non-motor symptomatology reflects the progressive degeneration of serotonergic raphe nuclei and their projections to regions involved in regulating sleep, arousal, feeding, satiety, mood,

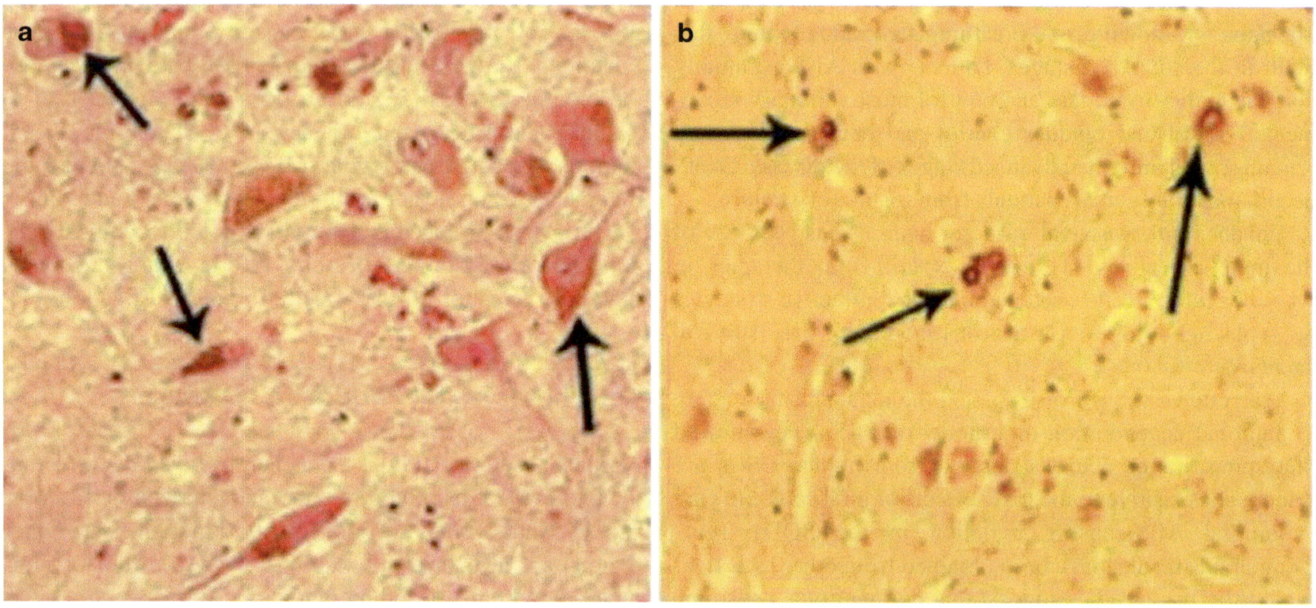


**Fig. 30.2** (**a**) Normal density of dopaminergic neurons (arrows) in the substantia nigra. (**b**) Alpha-synuclein immunostaining in tissue sections from the substantia nigra shows Lewy bodies in affected neurons (arrows); notice the paucity of nigral neurons in B compared to A. The loss of these cells is largely responsible for the motor symptoms of PD

and emotion [62]. Lack of improvement of the latter symptomatology highlights one of the unresolved problems of cell therapies for PD.

Currently, efforts are directed at producing dopaminergic cells from human iPSCs and their optimization for transplantation. Experimental data and clinical studies show that cells must be of human origin and have the properties of substantia nigra neurons to be clinically effective [60]. In this regard, there have been substantive advances in the protocols used to generate clinically effective dopaminergic grafts derived from human cells [63]. Currently, clinical trials are ongoing testing the efficacy of dopaminergic cell replacement in PD. The primary focus of these trials is to optimize tissue preparation and delivery, patient selection, and immunosuppressive treatment of candidate patients (for examples, see http://www.transeuro.org.uk/pages/objectives.html [64]).

Several investigators utilize various approaches using iPSCs to circumvent the need for immunosuppression. One such strategy consists of matching transplant recipients with cells that are less likely to be rejected. In this regard, a team at Kyoto University in Japan [65] is attempting to reduce the immunological response by producing cells matched for major histocompatibility Complexes (MHCs) for each transplant as opposed to using frozen cell lines [66].

Another approach is being attempted by Loring et al. at the Scripts Institute, employing iPSCs derived from a patient's autologous skin [67]. These cells have been induced to return to the stem-cell stage and differentiate into dopamine neuron progenitors. It is a scheme aiming at preventing cell rejection and the requirement for immunosuppressive medications. Another method used by Studer and collaborators is based on identifying and modulation two novel signaling inhibitors (Noggin and SB431542), which allows up to 80% of ESCs to develop into neural progenitors [68]. From that stage, the investigators use various protocols to induce the cells to differentiate into dopamine neurons.

Most trials using these approaches are in their early stages. However, initial outcomes of ongoing trials yielded promising data in a handful of patients for the reduction of the motor abnormalities characteristic of PD [65].

## Cell-Based Therapy in Traumatic Brain Injury

In the USA alone, nearly 1.5 million patients yearly are new victims of traumatic brain injury (TBI), adding to the six million patients permanently disabled by irreversible neurological deficits due to prior TBIs [69]. Despite considerable progress, clinical treatments are still limited to supportive care. Stem cell therapy has received much attention as a possible means of promoting endogenous neurogenesis and clinical recovery after TBI. For almost two decades, multiple investigators have studied the effect of various stem and progenitor cell transplantation in this injury environment to minimize the clinical impact of TBI. About 16 years ago, Lu et al. [70] administered MSCs intravenously 24 h after TBI to rats. The investigators reported significantly reduced motor and neurological deficits compared with control groups by day 15 after the injury. The cells preferentially entered and migrated into the parenchyma of the injured brain and expressed the neuronal and astrocytic markers. Work by other investigators has also shown, as discussed previously for other neurological conditions, that following MSCs transplantation, the brain is stimulated to amplify its endogenous restorative processes by several cytokines and growth factors secreted by the injected cells [71]. Following proliferation, NSCs migrate from the sub-ventricular zone (SVZ) to the injury site and differentiate into neuronal and glial cells [26]. In animal models, SVZ proliferation increases several-fold after TBI. However, the ability of endogenous NSCs to replace neuronal loss is limited [26]. In addition, the repair is frequently impaired by the severity of the inflammatory reaction associated with the injury. Although stem cell-mediated neural cell replacement via engrafting may be one beneficial effect of stem cell-based therapies, other therapeutic mechanisms are involved, as discussed in the preceding paragraphs, including trophic support [28] and suppression of inflammation [72–74]. Various stem cell types have shown beneficial therapeutic effects in animal models of TBI [75–78]. In neuroinflammation, it has been shown that transplanted human fetal brain-derived NSCs can attenuate microglia/macrophage activation, leading them to preferential differentiation into the neuroprotective M2 phenotype, which may contribute to neural repair [74]. Also, NSC grafts and treatment with glial derived neurotrophic factor reduce acute traumatic axonal injury and promote neurite outgrowth [79].

Currently, several groups are exploring the safety and feasibility of these approaches in phase 1 clinical trials. Cox et al. [80] recently measured several functional and cognitive outcomes in Stem cell treated patients with TBI and correlated them with imaging data in a pediatric population. Inflammatory cytokine arrays were measured in pretreatment plasma,—at post-treatment, and at 1- and 6-months follow-up. The treatment showed no serious adverse events and the structural preservation of critical regions of interest correlated with functional outcomes and downregulation of inflammatory cytokines [80]. These investigators demonstrated that treating TBI using intravenously delivered autologous bone marrow MSCs is safe. In another study, Wang et al. [81] examined the feasibility of MSCs transplantation from the umbilical cord via lumbar puncture. The investigators evaluated the safety and functional outcomes before and 6 months after stem cell transplantation. Preliminary data suggested an improvement in upper extremity and lower extremities motor and sensation sub-scores in the scales used for the assessments and balance sub-score in the patients receiving the stem cell transplantation 6 months after the intervention. The control group, which received no treatment, showed no improvements after 6 months [81]. A phase 2 clinical trial of intravenously delivered autologous bone marrow mononuclear cell infusion in TBI showed that the approach is safe and may improve central nervous system structural preservation and reduction of inflammatory biomarkers [80].

These initial results suggest that stem cell therapies are safe in human trials and may be associated with improved outcomes. Questions remain regarding the mechanisms, safety, dose, route, and timing of administration, which will eventually determine their optimal clinical use [26].

In summary, TBI is a major cause of mortality and morbidity, especially among young adults, and lifelong disability is common in those who survive. Although we are at an early stage of developing these therapies, the initial results are encouraging and worth pursuing aggressively. Several exciting studies are ongoing (clinicaltrials.gov), with results not yet available.

## Stem Cell Therapies as Potential Approaches to Reverse Aging

The capacity of stem cells for renewal declines with aging, and they experience changes in their potential to differentiate into various cell types. As a result, it is hypothesized that the pathophysiology of numerous age-related illnesses may be significantly influenced by abnormal stem cells. Understanding the pathophysiology of aging-associated disorders and creating novel, highly effective stem cell-based therapeutics to treat aging-associated diseases depend heavily on our understanding of how the aging process affects stem cell function.

Aging in vertebrates, including humans, is characterized by higher vulnerability to disease and increased mortality risk. Among the former, aging is by far the most important risk factor for many neurodegenerative disorders, including AD, PD, age-related frailty, and a host of other ailments ranging from age-related sarcopenia to osteoarthritis. Also, the health costs associated with frailty, a syndrome whose incidence rises with aging, are very high, which increases a person's vulnerability to poor health outcomes and mortality. Notably, the mechanisms involved in age-associated frialty, such as inflammation and stem cell failure, may differ from those causing biological aging [82, 83].

A review of the literature shows two groups of studies intended to either extend lifespan or promote healthspan. On the one hand, there are studies targeting changes in tissues' milieu, which may be associated with endogenous stem cell rejuvenation. The other group is intended to determine whether the administration of allogenic stem cells may enhance health span or ameliorate age-related frailty.

The first group of studies would require a better understanding of the aging process that is currently available. However, despite large gaps in our knowledge, we are beginning to unravel some of the basic pathways involved in the aging process so that aging itself, including the rejuvenation of stem cells, could be experimentally targeted [84]. Many genetic and environmental interventions have been shown to extend the life span of model organisms such as nematodes, Drosophila melanogaster, and mice [85–89]. For instance, we have discovered a new generation of indole-related molecules (i.e., indole-propionic acid; indole propionamide), which decrease the electron leak from aging mitochondria. Decreases Exposure to cell protective indoles results in decreased generation of reactive oxygen species by mitochondria and by prolongation of the maximum life span [90, 91]. Interestingly, recent epidemiological studies have demonstrated that higher serum levels of such molecules (i.e., indole propionic acid) strongly correlate to a lower incidence of certain age-related chronic disorders such as diabetes mellitus type 2 [92, 93]. Some of these effects have been attributed to stem cell protective mechanisms, such as age-related mitochondrial aging and reduction of oxidative stress.

Other pathways involved in the aging process that may be potential targets for intervention include insulin-like receptors, mechanistic target of rapamycin (mTOR) signaling, and AMPK. Daf-2, a gene involved in the endocrine signaling pathway, encodes an insulin-like receptor family member. Decreased Daf-2 signaling leads to an increase in lifespan [94]. The nutrient-sensing TOR pathway has also been identified as a critically important pathway affecting the longevity effects of dietary restriction, a powerful environmental method of lifespan extension by fasting [89, 94]. Activation of AMPK is another mechanism leading to lifespan extension in model organisms [95].

One other intervention yielding dramatic results in slowing down the aging process in vertebrate animals is heterochronic parabiosis, whereby the systemic circulations of two animals are joined together, i.e., young-to-old [96]. Historically, Paul Bert performed the earliest recorded parabiosis experiments in 1864, stitching together pairs of rats to share their circulatory systems [96]. Bert's research led to a revolutionary paradigm that has been used extensively to disentangle the nature of numerous and diverse biological phenomena, such as the roles of hormones or the way sugars cause dental decay [96]. McCay was the first investigator to test whether such animal pairs could influence aging [97–99]. In McCay's studies, after old and young rats were joined together for up to one and a half years, many traits associated with aging were reversed, including the abnormally age-related decreases in bone density and the loss of gloss in their furs (Fig. 30.3).

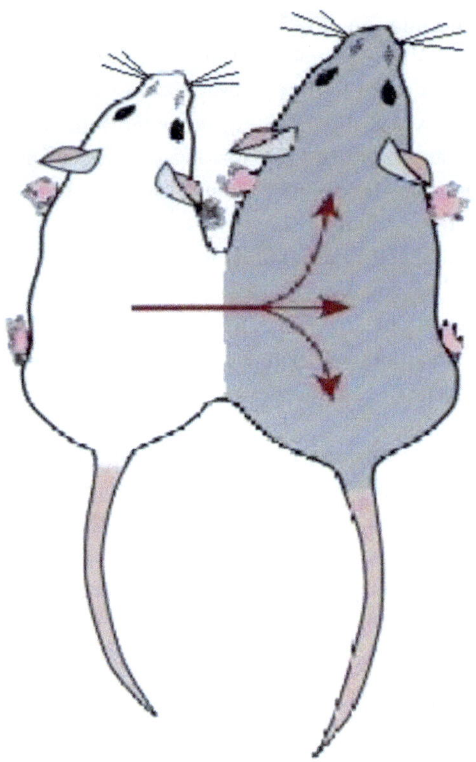

**Fig. 30.3** Heterochronic parabiosis

Heterochronic parabiosis is currently employed to identify anti-aging factors present in the circulation of young animals.

Following McCay's discovery, several investigators have demonstrated that the age-related loss in the capacity of tissues to regenerate after an injury can be restored in heterochronic old parabionts. More importantly, such a phenomenon results from the reactivation of endogenous stem cells in aged animals [100, 101]. This dramatic recovery in regenerative capacity to youthful levels was observed in many tissues, including muscle and liver [100–103]. Germane to this chapter, researchers have also used young-old mouse pairs to show restoration of neuronal stem activation and synapses in old mice. These results were also confirmed in individual (separate) old animals by exposing them to young blood or plasma derived from young, healthy donors [104, 105]. One unexpected finding in these experiments was that the stem cells in the young animals, when paired with old ones, also developed molecular and functional features typical of older cells. Therefore, heterochronic parabiosis experiments highlighted the bi-directional impact of environmental influences on cellular aging. It is essential to mention that stem cells "rejuvenated" in such a manner have substantive differences with younger cells. This is because, in addition to epigenetic changes, there are nuclear and mitochondrial age-related DNA mutations, not readily reversible by changes in the environmental milieu, which have been correlated with cellular aging [84]. However, there is no direct evidence that DNA mutations are the proximal cause of aging or that a reversal in such mutations may result in life span extension [84]. Therefore, considerable research is focused on the role of epigenetic processes as mediators of the aging process.

Together, these studies demonstrate that endogenous stem cells are profoundly influenced by their environment and imply that blood-borne factors may be responsible for at least some age-associated declines in stem cell function. Some factors involved are cytokines and immune system-associated molecules, which have been shown to change with age and either promote or inhibit aging phenotypes. The overall trend is that the increase in pro-inflammatory factors and the concomitant decrease in factors that promote tissue repair likely contribute to some aging phenotypes in many tissues [84]. The determination that plasma derived from young donors contains youth-promoting substances has led to a search for some of these factors, a potential "fountain of youth" within us. It is hoped that as our knowledge of the aging process advances, manipulation of these pathways will impact stem cell and tissue rejuvenation.

The second group of studies involves direct stem cell transplantation aiming to prevent or modify age-related disorders such as age-related frailty. Some initial encouraging data are beginning to emerge. A clinical trial of intravenous administration of allogenic human MSCs to patients afflicted by age-related fragility showed that this approach is safe and may be associated with reducing inflammatory biomarkers and enhancing physical performance [106]. While we are still in the beginning stages of developing these treatments, the preliminary findings are promising and warrant more investigation.

## Stem Cell Therapy in Alzheimer's Disease

Alzheimer's disease (AD) is the most frequent form of dementia affecting older adults and is characterized by gradual and progressive cognitive decline. It often (but not always) begins with memory deficits, but as the disease spreads to multiple brain regions, other cognitive domains become gradually affected, culminating in full-blown dementia. Neuropathologically, AD is characterized by progressive accumulation of the amyloid beta protein and the phosphorylated microtubule-associated protein tau (Fig. 30.4).

Historically, our laboratory was the first to identify oxidative stress in the human brain tissue of patients with AD [105]. We and others have confirmed these findings in transgenic mice and identified these mechanisms as critical events leading to neuronal loss [107, 108]. Oxidative distress has been proposed to impact stem cell aging [83].

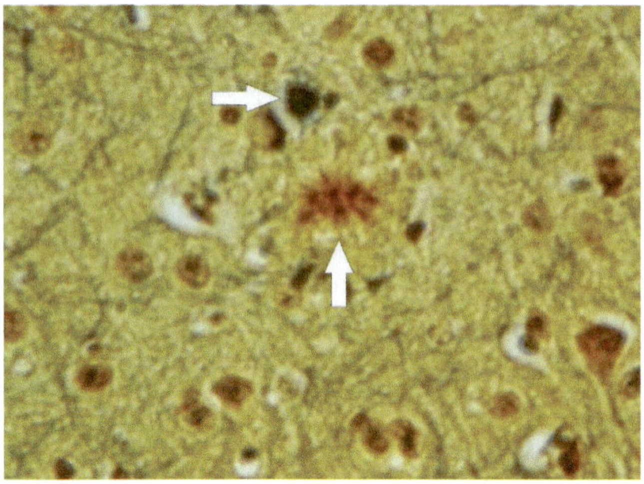

**Fig. 30.4** Intraneuronal accumulation of the abnormal microtubule-associated protein tau (horizontal arrow) and amyloid (vertical arrow) in a triple transgenic model of AD. A profuse loss of neurons and synaptic connections accompanies these changes. Oxidative stress and inflammation caused by oligomers of the amyloid protein are believed to be involved the cell loss

Currently, there are no disease-modifying therapies for AD. Pharmacological agents available only alleviate its symptoms temporarily but do not alter the progression of AD. Since aging is the strongest risk factor in AD, recent work has explored the possibility of using stem cell transplantation to slow down the cognitive decline associated with this disorder. Interestingly, the mechanisms that mediate these effects in animal models of AD seem to involve neuroprotection (such as against oxidative stress) and trophic support rather than a neuronal replacement [109]. Several phase 1 clinical trials in humans using MSCs and NSCs, similar to those shown for other neurological disorders, have demonstrated that these cell-based therapies are safe in patients with AD [110]. Some early trials involved "off-the-shelf" commercial stem cell preparations [2]. However, these trials are early and only beginning, and meaningful results await further investigation. Clinical trials and other research are warranted based on the early experiments with parabiosis and stem cells.

## Conclusions

This chapter illustrates, in a rather concise manner, the therapeutic potential of cell-based treatments for several neurological disorders. Because many diseases affecting the brain are either incurable or associated with permanent neurological damage, stem cell approaches have become appealing alternatives to supportive care. The benefits of these treatments include stimulation of repair, replacement of cells that have been lost or irreversibly damaged by the disease process, modulation of the associated inflammation, and protection of blood-brain barrier integrity. Although not all the disorders amenable to these therapies have been comprehensively reviewed in this chapter, the general therapeutic principles discussed are of widespread application in stem cell transplantation. There are many unanswered questions regarding the mechanisms by which stem cell administration leads to enhanced functional recovery and structural reorganization of neuronal circuitries [111]. The answers to many of these questions are critical to optimizing therapies and are being actively pursued by many laboratories worldwide.

**Key Points**

1. Stem cells, classified by their differentiation potential and origin, play a vital role in development and repair and offer significant potential for cell therapy in neurological diseases.
2. Neurogenesis, primarily driven by Neural Stem Cells (NSCs), is a continuous process that can be activated post-injury and leveraged for therapeutic benefits in neurological diseases.
3. Mesenchymal Stem Cells (MSCs) offer therapeutic potential in neurological disease models by differentiating into various resident brain cells, participating in immunomodulation, and facilitating repair and neuroprotection.
4. Various forms of stem cell therapies and microRNA-modified exosomal treatments have demonstrated promising results in animal models for stroke treatment, showcasing improved neurological and motor functions and potential for tissue repair. However, translating these results into successful human clinical trials remains a challenge.
5. Early research and clinical trials show stem cells might safely reduce inflammation and slow MS progression, but more studies are needed for optimization.
6. Preclinical and clinical studies suggest stem cells could enhance motor neuron survival and slow ALS progression, but further research is necessary to confirm these results.
7. Stem cell therapies, using dopaminergic neurons derived from induced pluripotent stem cells (iPSCs), offer potential for long-term improvement in PD symptoms. Ongoing research focuses on optimizing these methods to increase efficacy and manage non-motor symptoms.
8. Stem cell therapies are emerging as potential treatments for TBI, aiding in neurogenesis and reducing neurological deficits. Current clinical trials indicate safety, but optimal treatment protocols remain under study.
9. Stem cell therapies show promise in combating aging effects by rejuvenating stem cells and reducing age-related frailty. However, understanding aging's impact on stem cells is crucial for advancing these treatments.

10. Treatment of Alzheimer's Disease with mesenchymal stem cells and neural stem cells aimed at providing neuroprotection and trophic support has been deemed safe in initial human trials, although further is needed to evaluate its effectiveness.

**Acknowledgments** This book chapter is modified and updated from a previous book chapter, "Stem Cell Therapies for Neurological Disorders" by Miguel Pappolla, MD, PhD, Kumar Sambamurti, PhD, and Ping Wu, MD, PhD, in *Essentials of Regenerative Medicine in Interventional Pain Management* published by ASIPP Publishing.

## References

1. Pappolla M, Sambamurti K, Wu P. Stem cell therapies for neurological disorders. In: Manchikanti L, Navani A, Atluri S, editors. Essentials of regenerative medicine in interventional pain management. Paducah, KY: ASIPP Publishing; 2019. p. 605–19.
2. Batista CE, Mariano ED, Marie SK, et al. Stem cells in neurology—current perspectives. Arq Neuropsiquiatr. 2014;72:457–65.
3. Yousefifard M, Shamseddin J, Babahajian A, Sarveazad A. Efficacy of adipose derived stem cells on functional and neurological improvement following ischemic stroke: a systematic review and meta-analysis. BMC Neurol. 2020;20:294.
4. Christian KM, Song H, Ming GL. Functions and dysfunctions of adult hippocampal neurogenesis. Annu Rev Neurosci. 2014;37:243–62.
5. Ming GL, Song H. Adult neurogenesis in the mammalian brain: significant answers and significant questions. Neuron. 2011;70:687–702.
6. Opendak M, Gould E. Adult neurogenesis: a substrate for experience-dependent change. Trends Cogn Sci. 2015;19:151–61.
7. Xie C, Liu Y-Q, Guan Y-T, Zhang GX. Induced stem cells as a novel multiple sclerosis therapy. Curr Stem Cell Res Ther. 2015;11:313–20.
8. Pluchino S, Quattrini A, Brambilla E, et al. Injection of adult neurospheres induces recovery in a chronic model of multiple sclerosis. Nature. 2003;422:688–94.
9. Pluchino S, Zanotti L, Rossi B, et al. Neurosphere-derived multipotent precursors promote neuroprotection by an immunomodulatory mechanism. Nature. 2005;436:266–71.
10. Einstein O, Fainstein N, Vaknin I, et al. Neural precursors attenuate autoimmune encephalomyelitis by peripheral immunosuppression. Ann Neurol. 2007;61:209–18.
11. Appaix F, Nissou MF, van der Sanden B, et al. Brain mesenchymal stem cells: the other stem cells of the brain? World J Stem Cells. 2014;6:134–43.
12. Hao L, Zou Z, Tian H, Zhang Y, Zhou H, Liu L. Stem cell-based therapies for ischemic stroke. Biomed Res Int. 2014;2014:468748.
13. Nistor-Cseppento DC, Jurcau MC, Jurcau A, Andronie-Cioara FL, Marcu F. Stem cell- and cell-based therapies for ischemic stroke. Bioengineer (Basel). 2022;9:717.
14. Paul G, Ozen I, Christophersen NS, et al. The adult human brain harbors multipotent perivascular mesenchymal stem cells. PLoS One. 2012;7:e35577.
15. Joyce N, Annett G, Wirthlin L, Olson S, Bauer G, Nolta JA. Mesenchymal stem cells for the treatment of neurodegenerative disease. Regen Med. 2010;5:933–46.
16. Ejma M, Madetko N, Brzecka A, et al. The role of stem cells in the therapy of stroke. Curr Neuropharmacol. 2022;20:630–47.
17. Zhang RL, Chopp M, Roberts C, et al. Stroke increases neural stem cells and angiogenesis in the neurogenic niche of the adult mouse. PLoS One. 2014;9:e113972.
18. Hermann DM, Peruzzotti-Jametti L, Schlechter J, Bernstock JD, Doeppner TR, Pluchino S. Neural precursor cells in the ischemic brain—integration, cellular crosstalk, and consequences for stroke recovery. Front Cell Neurosci. 2014;8:291.
19. Marques BL, Carvalho GA, Freitas EMM, et al. The role of neurogenesis in neurorepair after ischemic stroke. Semin Cell Dev Biol. 2019;95:98–110.
20. Cao H, Cheng Y, Zhang J, Xu M, Ge L. The effect of umbilical cord mesenchymal stem cells combined with tetramethylpyrazine therapy on ischemic brain injury: a histological study. J Stroke Cerebrovasc Dis. 2020;29:105298.
21. Nazarian S, Abdolmaleki Z, Torfeh A, Shirazi Beheshtiha SH. Mesenchymal stem cells with modafinil (gold nanoparticles) significantly improves neurological deficits in rats after middle cerebral artery occlusion. Exp Brain Res. 2020;238:2589–601.
22. Salehi MS, Safari A, Pandamooz S, et al. The beneficial potential of genetically modified stem cells in the treatment of stroke: a review. Stem Cell Rev Rep. 2022;18:412–40.
23. Chen SJ, Chang CM, Tsai SK, et al. Functional improvement of focal cerebral ischemia injury by subdural transplantation of induced pluripotent stem cells with fibrin glue. Stem Cells Dev. 2010;19:1757–67.
24. Jiang M, Lv L, Ji H, et al. Induction of pluripotent stem cells transplantation therapy for ischemic stroke. Mol Cell Biochem. 2011;354:67–75.
25. Jin K, Wang X, Xie L, Mao XO, Greenberg DA. Transgenic ablation of doublecortin-expressing cells suppresses adult neurogenesis and worsens stroke outcome in mice. Proc Natl Acad Sci U S A. 2010;107:7993–8.
26. Gennai S, Monsel A, Hao Q, et al. Cell-based therapy for traumatic brain injury. Br J Anaesth. 2015;115:203–12.
27. Wang L, Zhang Z, Wang Y, Zhang R, Chopp M. Treatment of stroke with erythropoietin enhances neurogenesis and angiogenesis and improves neurological function in rats. Stroke. 2004;35:1732–7.
28. Batista PJ, Molinie B, Wang J, et al. M(6)a RNA modification controls cell fate transition in mammalian embryonic stem cells. Cell Stem Cell. 2014;15:707–19.
29. de Celis-Ruiz E, Fuentes B, Moniche F, et al. Allogeneic adipose tissue-derived mesenchymal stem cells in ischaemic stroke (AMASCIS-02): a phase IIb, multicentre, double-blind, placebo-controlled clinical trial protocol. BMJ Open. 2021;11:e051790.
30. Taverna S, Pucci M, Alessandro R. Extracellular vesicles: small bricks for tissue repair/regeneration. Ann Transl Med. 2017;5:83.
31. Silva AM, Teixeira JH, Almeida MI, Goncalves RM, Barbosa MA, Santos SG. Extracellular vesicles: immunomodulatory messengers in the context of tissue repair/regeneration. Eur J Pharm Sci. 2017;98:86–95.
32. Zhang ZG, Chopp M. Exosomes in stroke pathogenesis and therapy. J Clin Invest. 2016;126:1190–7.
33. Yu B, Zhang Y, Tian Y, Gong S. Effects of miRNA-modified exosomes alleviate cerebral ischemic reperfusion injury in pre-clinical studies: a meta-analysis. World Neurosurg. 2022;168:278–86.
34. Trounson A, McDonald C. Stem cell therapies in clinical trials: progress and challenges. Cell Stem Cell. 2015;17:11–22.
35. de Celis-Ruiz E, Fuentes B, Alonso de Lecinana M, et al. Final results of allogeneic adipose tissue-derived mesenchymal stem cells in acute ischemic stroke (AMASCIS): a phase II, randomized, double-blind, placebo-controlled, single-center, pilot clinical trial. Cell Transplant. 2022;31:9636897221083863.
36. Popescu BF, Lucchinetti CF. Pathology of demyelinating diseases. Annu Rev Pathol. 2012;7:185–217.

37. Engelhardt B, Ransohoff RM. The ins and outs of T-lymphocyte trafficking to the CNS: anatomical sites and molecular mechanisms. Trends Immunol. 2005;26:485–95.
38. Gerdoni E, Gallo B, Casazza S, et al. Mesenchymal stem cells effectively modulate pathogenic immune response in experimental autoimmune encephalomyelitis. Ann Neurol. 2007;61:219–27.
39. Reekmans KP, Praet J, De Vocht N, et al. Clinical potential of intravenous neural stem cell delivery for treatment of neuroinflammatory disease in mice? Cell Transplant. 2011;20:851–69.
40. Sher F, Amor S, Gerritsen W, et al. Intraventricularly injected Olig2-NSCs attenuate established relapsing-remitting EAE in mice. Cell Transplant. 2012;21:1883–97.
41. Gao Z, Wen Q, Xia Y, et al. Osthole augments therapeutic efficiency of neural stem cells-based therapy in experimental autoimmune encephalomyelitis. J Pharmacol Sci. 2014;124:54–65.
42. Yang J, Jiang Z, Fitzgerald DC, et al. Adult neural stem cells expressing IL-10 confer potent immunomodulation and remyelination in experimental autoimmune encephalitis. J Clin Invest. 2009;119:3678–91.
43. Yang J, Yan Y, Xia Y, et al. Neurotrophin 3 transduction augments remyelinating and immunomodulatory capacity of neural stem cells. Mol Ther. 2014;22:440–50.
44. Llufriu S, Sepulveda M, Blanco Y, et al. Randomized placebo-controlled phase II trial of autologous mesenchymal stem cells in multiple sclerosis. PLoS One. 2014;9:e113936.
45. Lublin FD, Bowen JD, Huddlestone J, et al. Human placenta-derived cells (PDA-001) for the treatment of adults with multiple sclerosis: a randomized, placebo-controlled, multiple-dose study. Mult Scler Relat Disord. 2014;3:696–704.
46. Connick P, Kolappan M, Crawley C, et al. Autologous mesenchymal stem cells for the treatment of secondary progressive multiple sclerosis: an open-label phase 2a proof-of-concept study. Lancet Neurol. 2012;11:150–6.
47. Cohen JA, Lublin FD, Lock C, et al. Evaluation of neurotrophic factor secreting mesenchymal stem cells in progressive multiple sclerosis. Mult Scler. 2023;29:92–106.
48. Marchetto MC, Muotri AR, Mu Y, Smith AM, Cezar GG, Gage FH. Non-cell-autonomous effect of human SOD1 G37R astrocytes on motor neurons derived from human embryonic stem cells. Cell Stem Cell. 2008;3:649–57.
49. Lopez-Gonzalez R, Kunckles P, Velasco I. Transient recovery in a rat model of familial amyotrophic lateral sclerosis after transplantation of motor neurons derived from mouse embryonic stem cells. Cell Transplant. 2009;18:1171–81.
50. Thonhoff JR, Gao J, Dunn TJ, Ojeda L, Wu P. Mutant SOD1 microglia-generated nitroxidative stress promotes toxicity to human fetal neural stem cell-derived motor neurons through direct damage and noxious interactions with astrocytes. Am J Stem Cells. 2012;1:2–21.
51. Serio A, Bilican B, Barmada SJ, et al. Astrocyte pathology and the absence of non-cell autonomy in an induced pluripotent stem cell model of TDP-43 proteinopathy. Proc Natl Acad Sci U S A. 2013;110:4697–702.
52. Coatti GC, Beccari MS, Olavio TR, Mitne-Neto M, Okamoto OK, Zatz M. Stem cells for amyotrophic lateral sclerosis modeling and therapy: myth or fact? Cytometry A. 2015;87:197–211.
53. Teng YD, Benn SC, Kalkanis SN, et al. Multimodal actions of neural stem cells in a mouse model of ALS: a meta-analysis. Sci Transl Med. 2012;4:165ra4.
54. Mazzini L, Ferrero I, Luparello V, et al. Mesenchymal stem cell transplantation in amyotrophic lateral sclerosis: a phase I clinical trial. Exp Neurol. 2010;223:229–37.
55. Petrou P, Gothelf Y, Argov Z, et al. Safety and clinical effects of mesenchymal stem cells secreting neurotrophic factor transplantation in patients with amyotrophic lateral sclerosis: results of phase 1/2 and 2a clinical trials. JAMA Neurol. 2016;73:337–44.
56. Petrou P, Kassis I, Yaghmour NE, Ginzberg A, Karussis D. A phase II clinical trial with repeated intrathecal injections of autologous mesenchymal stem cells in patients with amyotrophic lateral sclerosis. Front Biosci (Landmark Ed). 2021;26:693–706.
57. Bjorklund A, Stenevi U. Reconstruction of the nigrostriatal dopamine pathway by intracerebral nigral transplants. Brain Res. 1979;177:555–60.
58. Perlow MJ, Freed WJ, Hoffer BJ, Seiger A, Olson L, Wyatt RJ. Brain grafts reduce motor abnormalities produced by destruction of nigrostriatal dopamine system. Science. 1979;204:643–7.
59. Lindvall O. Developing dopaminergic cell therapy for Parkinson's disease—give up or move forward? Mov Disord. 2013;28:268–73.
60. Lindvall O. Treatment of Parkinson's disease using cell transplantation. Philos Trans R Soc Lond Ser B Biol Sci. 2015;370:20140370.
61. Piccini P, Brooks DJ, Bjorklund A, et al. Dopamine release from nigral transplants visualized in vivo in a Parkinson's patient. Nat Neurosci. 1999;2:1137–40.
62. Politis M, Wu K, Loane C, et al. Serotonin neuron loss and non-motor symptoms continue in Parkinson's patients treated with dopamine grafts. Sci Transl Med. 2012;4:128ra41.
63. Kriks S, Shim JW, Piao J, et al. Dopamine neurons derived from human ES cells efficiently engraft in animal models of Parkinson's disease. Nature. 2011;480:547–51.
64. Barker PR. Innovative approach for the treatment of Parkinson's Disease: Transeuro; 2023; http://www.transeuro.org.uk/pages/objectives.html. Accessed 9 May 2023.
65. Cyranoski D. 'Reprogrammed' stem cells implanted into patient with Parkinson's disease. Nature. 2018;
66. Morizane A, Kikuchi T, Hayashi T, et al. MHC matching improves engraftment of iPSC-derived neurons in non-human primates. Nat Commun. 2017;8:385.
67. Loring JF. Autologous induced pluripotent stem cell-derived neurons to treat Parkinson's disease. Stem Cells Dev. 2018;27:958–9.
68. Studer L. Strategies for bringing stem cell-derived dopamine neurons to the clinic-the NYSTEM trial. Prog Brain Res. 2017;230:191–212.
69. Walker PA, Shah SK, Harting MT, Cox CS Jr. Progenitor cell therapies for traumatic brain injury: barriers and opportunities in translation. Dis Model Mech. 2009;2:23–38.
70. Lu D, Mahmood A, Wang L, Li Y, Lu M, Chopp M. Adult bone marrow stromal cells administered intravenously to rats after traumatic brain injury migrate into brain and improve neurological outcome. Neuroreport. 2001;12:559–63.
71. Li Y, Chopp M. Marrow stromal cell transplantation in stroke and traumatic brain injury. Neurosci Lett. 2009;456:120–3.
72. Snyder EY, Teng YD. Stem cells and spinal cord repair. N Engl J Med. 2012;366:1940–2.
73. Pluchino S, Cossetti C. How stem cells speak with host immune cells in inflammatory brain diseases. Glia. 2013;61:1379–401.
74. Gao J, Grill RJ, Dunn TJ, et al. Human neural stem cell transplantation-mediated alteration of microglial/macrophage phenotypes after traumatic brain injury. Cell Transplant. 2016;25:1863–77.
75. Kulbatski I, Mothe AJ, Nomura H, Tator CH. Endogenous and exogenous CNS derived stem/progenitor cell approaches for neurotrauma. Curr Drug Targets. 2005;6:111–26.
76. Maegele M, Schaefer U. Stem cell-based cellular replacement strategies following traumatic brain injury (TBI). Minim Invasive Ther Allied Technol. 2008;17:119–31.
77. Richardson RM, Singh A, Sun D, Fillmore HL, Dietrich DW 3rd, Bullock MR. Stem cell biology in traumatic brain injury: effects of injury and strategies for repair. J Neurosurg. 2010;112:1125–38.
78. Schouten JW, Fulp CT, Royo NC, et al. A review and rationale for the use of cellular transplantation as a therapeutic strategy for traumatic brain injury. J Neurotrauma. 2004;21:1501–38.

79. Wang E, Gao J, Yang Q, et al. Molecular mechanisms underlying effects of neural stem cells against traumatic axonal injury. J Neurotrauma. 2012;29:295–312.
80. Cox CS Jr, Hetz RA, Liao GP, et al. Treatment of severe adult traumatic brain injury using bone marrow mononuclear cells. Stem Cells. 2017;35:1065–79.
81. Wang S, Cheng H, Dai G, et al. Umbilical cord mesenchymal stem cell transplantation significantly improves neurological function in patients with sequelae of traumatic brain injury. Brain Res. 2013;1532:76–84.
82. Clegg A, Young J, Iliffe S, Rikkert MO, Rockwood K. Frailty in elderly people. Lancet. 2013;381:752–62.
83. Ahmed AS, Sheng MH, Wasnik S, Baylink DJ, Lau KW. Effect of aging on stem cells. World J Exp Med. 2017;7:1–10.
84. Rando TA, Chang HY. Aging, rejuvenation, and epigenetic reprogramming: resetting the aging clock. Cell. 2012;148:46–57.
85. Guarente L, Kenyon C. Genetic pathways that regulate ageing in model organisms. Nature. 2000;408:255–62.
86. Gupta VK, Scheunemann L, Eisenberg T, et al. Restoring polyamines protects from age-induced memory impairment in an autophagy-dependent manner. Nat Neurosci. 2013;16:1453–60.
87. McCarroll SA, Murphy CT, Zou S, et al. Comparing genomic expression patterns across species identifies shared transcriptional profile in aging. Nat Genet. 2004;36:197–204.
88. Longo VD, Antebi A, Bartke A, et al. Interventions to slow aging in humans: are we ready? Aging Cell. 2015;14:497–510.
89. Fontana L, Partridge L. Promoting health and longevity through diet: from model organisms to humans. Cell. 2015;161:106–18.
90. Poeggeler B, Sambamurti K, Siedlak SL, Perry G, Smith MA, Pappolla MA. A novel endogenous indole protects rodent mitochondria and extends rotifer lifespan. PLoS One. 2010;5:e10206.
91. Chyan YJ, Poeggeler B, Omar RA, et al. Potent neuroprotective properties against the Alzheimer beta-amyloid by an endogenous melatonin-related indole structure, indole-3-propionic acid. J Biol Chem. 1999;274:21937–42.
92. de Mello VD, Paananen J, Lindstrom J, et al. Indolepropionic acid and novel lipid metabolites are associated with a lower risk of type 2 diabetes in the finnish diabetes prevention study. Sci Rep. 2017;7:46337.
93. Kimura KD, Tissenbaum HA, Liu Y, Ruvkun G. Daf-2, an insulin receptor-like gene that regulates longevity and diapause in Caenorhabditis elegans. Science. 1997;277:942–6.
94. Kapahi P, Kaeberlein M, Hansen M. Dietary restriction and lifespan: lessons from invertebrate models. Ageing Res Rev. 2017;39:3–14.
95. Apfeld J, O'Connor G, McDonagh T, DiStefano PS, Curtis R. The AMP-activated protein kinase AAK-2 links energy levels and insulin-like signals to lifespan in C. elegans. Genes Dev. 2004;18:3004–9.
96. Scudellari M. Ageing research: blood to blood. Nature. 2015;517:426–9.
97. Lunsford WR, Mc CC, Lupien PJ, Pope FE, Sperling G. Parabiosis as a method for studying factors which affect aging in rats. Gerontologia. 1963;7:1–8.
98. McCay CM, Pope F, Lunsford W, Sperling G, Sambhavaphol P. Parabiosis between old and young rats. Gerontologia. 1957;1:7–17.
99. Horrington EM, Pope F, Lunsford W, Mc CC. Age changes in the bones, blood pressure, and diseases of rats in parabiosis. Gerontologia. 1960;4:21–31.
100. Conboy IM, Rando TA. Heterochronic parabiosis for the study of the effects of aging on stem cells and their niches. Cell Cycle. 2012;11:2260–7.
101. Conboy MJ, Conboy IM, Rando TA. Heterochronic parabiosis: historical perspective and methodological considerations for studies of aging and longevity. Aging Cell. 2013;12:525–30.
102. Carlson ME, Suetta C, Conboy MJ, et al. Molecular aging and rejuvenation of human muscle stem cells. EMBO Mol Med. 2009;1:381–91.
103. Conboy IM, Conboy MJ, Wagers AJ, Girma ER, Weissman IL, Rando TA. Rejuvenation of aged progenitor cells by exposure to a young systemic environment. Nature. 2005;433:760–4.
104. Castellano JM, Kirby ED, Wyss-Coray T. Blood-borne revitalization of the aged brain. JAMA Neurol. 2015;72:1191–4.
105. Pappolla MA, Omar RA, Kim KS, Robakis NK. Immunohistochemical evidence of oxidative [corrected] stress in Alzheimer's disease. Am J Pathol. 1992;140:621–8.
106. Tompkins BA, DiFede DL, Khan A, et al. Allogeneic mesenchymal stem cells ameliorate aging frailty: a phase II randomized, double-blind, placebo-controlled clinical trial. J Gerontol A Biol Sci Med Sci. 2017;72:1513–22.
107. Pappolla MA, Chyan YJ, Omar RA, et al. Evidence of oxidative stress and in vivo neurotoxicity of beta-amyloid in a transgenic mouse model of Alzheimer's disease: a chronic oxidative paradigm for testing antioxidant therapies in vivo. Am J Pathol. 1998;152:871–7.
108. Siedlak SL, Casadesus G, Webber KM, et al. Chronic antioxidant therapy reduces oxidative stress in a mouse model of Alzheimer's disease. Free Radic Res. 2009;43:156–64.
109. Chen WW, Blurton-Jones M. Concise review: can stem cells be used to treat or model Alzheimer's disease? Stem Cells. 2012;30:2612–8.
110. Kang JM, Yeon BK, Cho SJ, Suh YH. Stem cell therapy for Alzheimer's disease: a review of recent clinical trials. J Alzheimers Dis. 2016;54:879–89.
111. Kim SU, de Vellis J. Stem cell-based cell therapy in neurological diseases: a review. J Neurosci Res. 2009;87:2183–200.

# Nerve Blocks for Regional Anesthesia for Regenerative Interventional Treatment

# 31

Michael Khadavi, Danielle Rehor, Alex Roney, Luga Podesta, and David R. Smith

## Introduction

The field of orthobiologics has a rapidly growing evidence base and frequency of use, and these procedures are often performed on an outpatient basis. Many of these procedures, such as into a tendon or bone, may be challenging due to injection discomfort and consequent patient movement without some form of anesthesia. While both intravenous (IV) sedation or target site local anesthetic (LA) injection are commonly used, their use is not without risk of complication or adverse effects. In the case of LA use into or around the target tissue, all commercially available LAs have negative effects on tenocytes, chondrocytes, and platelets, with ropivacaine the least and bupivacaine the most harmful to these tissues [1–6]. Similarly, many providers utilize IV sedation, which comes with a host of mild complications such as nausea, vomiting, flu-like symptoms, hypotension, dysesthesia, and headache [7–9]. Life-threatening risks include airway obstruction, cardiac arrythmias, pulmonary embolus, and aspiration [9–13]. Furthermore, heavily sedated patients can become disoriented and agitated, which can further challenge the procedure.

These situations can be avoided by blocking the sensory nerve supply to our orthobiologic target via a peripheral nerve block (PNB). Nerve blocks, also termed regional anesthesia, have been used since the late 1800's [14], and were in fact preferred due to the morbidity and mortality of general anesthesia in that period. More recently, they have been used with great success in a wide range of surgeries, including both large and small orthopedic surgeries [12, 15–19]. When used for perioperative pain control, PNBs are associated with dramatic decreases in complications [17], even compared to neuraxial nerve blocks [20].

The literature is robust and formal educational curricula are well-established for PNB techniques in many orthopedic surgical procedures, however, any mention of regional blockade before an orthobiologic injection is exceedingly rare in the literature. Yet, anesthetic goals for orthobiologic procedures differ from those in surgery. A complete block of both sensory and motor fibers is necessary in surgical settings; however, a sensory-only block is not only sufficient but preferred in orthobiologic settings.

These procedures have the potential to simultaneously minimize procedural pain and eliminate LA's detrimental effects on target tissues. Furthermore, limiting patient movement and anxiety while allowing for a more rapid procedure reduces the risk of infection and other complications [21–24]. In our experience, PNBs may allow even intraosseous injections to be performed in a clinic setting without the risks and adverse effects of heavy IV sedation.

While many orthobiologic providers are already performing PNBs on some occasions, most subspecialists who perform these treatments receive no formal training in PNB techniques and risks.

This chapter aims to summarize the basic concepts of regional anesthetic usage as pertains to the field of

M. Khadavi (✉)
Kansas City Orthopedic Alliance, Leawood, KS, USA

Kansas City Ballet, Sporting Kansas City, Kansas City, KS, USA

D. Rehor
University of Kansas Medical Center, Kansas City, KS, USA

A. Roney
Liberty University College of Osteopathic Medicine, Lynchburg, VA, USA

L. Podesta
Regenerative Sports Medicine, Bluetail Medical Group, Naples, FL, USA

Podesta Sports Medicine and Orthopedics, Naples, FL, USA

Florida Everglades, Homestead, FL, USA

D. R. Smith
Division of Anesthesia, Virginia Mason Clinic, Kingston Orthopedic Pain Institute, Queens University, Kingston, ON, Canada
e-mail: dsmith@kopi.ca


A. Navani et al. (eds.), *Essentials of Regenerative Medicine in Interventional Pain Management*,
https://doi.org/10.1007/978-3-031-50357-3_31

orthobiologics, including the risks of PNBs, strategies to minimize these risks, and specific PNB techniques for several commonly performed orthobiologic procedures.

## Risks and Precautions of Peripheral Nerve Blocks for Orthobiologic Procedures

The decision to perform a PNB primarily hinges on the balance between the potential for enhanced efficacy, safety, and comfort and the local and systemic risks of a PNB. These risks can be separated into two main categories: local anesthetic systemic toxicity (LAST) and target nerve damage from either LA chemical toxicity or physical injury from the needle and injection.

LAST is usually the result of LA toxicity to the central nervous system (CNS) or heart, but can also cause hematologic, allergic, and local tissue reactions. It typically manifests within 5 min of injection but can appear several hours after anesthetic administration; the former associated with intravascular injection and the latter with slower uptake of LA from the soft tissues where they have been deposited. Strategies to mitigate both situations are discussed below [25]. CNS symptoms include confusion, metallic taste, dizziness, tinnitus, circumoral numbness, muscle twitching, seizure, loss of consciousness, coma, respiratory depression, and death. Cardiovascular symptoms include hypertension, tachycardia, bradycardia, arrhythmias, and asystole. However, an estimated 40% of patients with LAST present atypically. The incidence of LAST has been estimated at 3 in 10,000 PNBs [26, 27], and half are caused by non-anesthesia providers. While self-limiting symptoms are common, the overall mortality rate of those who experience LAST is between 4–10% [25].

Risk of LA systemic absorption in PNBs is lower than for plexus blocks, especially in the neck [27], but higher than the subcutaneous and musculoskeletal injections that physiatrists are most experienced with, because of:

- Higher volumes of LA, and
- The highly vascularized anatomy that is more prone to indirect absorption or unintentional intravascular injection [25, 28].

Low muscle mass rather than low total body weight is a risk factor for LAST, since systemically circulating LAs get stored in muscle. Use of epinephrine with LA in PNBs can both decrease intravascular uptake and serve as an intravascular marker but carries additional risks that may not be outweighed by the benefits [29]. Other risk factors for LAST include the high and low extremes of age, and comorbidities, particularly heart and liver disease but also kidney disease and diabetes mellitus [25–27]. Ultrasound-guidance reduces LAST by 65% [25, 30, 31].

Among LAs, bupivacaine poses the greatest risk of LAST, compared to lidocaine or ropivacaine [11, 32]. Its uniquely high cardiac toxicity leads to more cardiac arrest and thus carries the highest morbidity and mortality in LAST among LAs [27, 33]. However, the American Society of Regional Anesthesia (ASRA) advisory panel on LAST opines that ropivacaine may have similar toxicity to bupivacaine [27]. Higher LA dose is a risk factor for LAST; however, even low doses have been implicated [25]. Because miscalculation of maximum LA dose is a common reason for overdose and LAST [34], maximum volumes corresponding to maximum doses of recommended LAs in PNBs are presented in Table 31.1 for three sample lean body weights [34, 35]. However, the risk of systemic toxicity increases continuously with escalating dose, rather than with a sudden increase

**Table 31.1** Concentration and maximum doses of lidocaine and ropivacaine

| | Concentration (mg/mL) | Max Dose (mg/kg) | Max Volume per kg of IBM (mL/kg) | Max LA Volume for 50 kg person (110 lb) | Max LA Volume per 70 kg person (154 lb) | Max LA Volume per 85 kg person (187 lb) |
|---|---|---|---|---|---|---|
| Lidocaine 1% | 10 | 5 | 0.5 | 25 | 35 | 42.5 |
| Lidocaine 2% | 20 | 5 | 0.25 | 12.5 | 17.5 | 21.25 |
| Lidocaine 1% c epi | 10 | 7 | 0.7 | 35 | 49 | 59.5 |
| Lidocaine 2% c epi | 20 | 7 | 0.35 | 17.5 | 24.5 | 29.75 |
| Ropivacaine 0.5% | 5 | 3 | 0.6 | 30 | 42 | 51 |
| Ropivacaine 0.2% | 2 | 3 | 1.5 | 75 | 105 | 127.5 |

$$\max \textit{Volume per kg IBM} = \frac{\max \textit{Dose}\left(\frac{mg}{kg}\right)}{\textit{Concentration}\left(\frac{mg}{ml}\right)}$$

*Max doses and volumes must be calculated using ideal body weight and not total body weight. *Max* maximum, *c* with, *Epi* epinephrine once the maximum dose is reached. No more than the lowest possible LA dose should be used [25, 27, 35].

If signs or symptoms of LAST occur, quick recognition and treatment is crucial for patient outcomes, beginning with immediate cessation of LA injection once any cardiovascular or mental status change is noticed. Treatment continues with respiratory and cardiac support, administration of 100% oxygen, benzodiazepines for seizures, and IV access to allow treatment with lipid emulsion therapy which acts as a "sink" for Las [25, 26, 36, 37]. Pharmacology, indications, and dosing of lipid emulsion therapy is outside the scope of this chapter.

A common procedural complication with often similar presentation to LAST is the vasovagal reaction. This presents as dizziness, light-headedness, weakness, nausea, sweating, or syncope and is known to occur in 1–8% of outpatient injections [38, 39]. In a sitting position, venous blood will pool in the legs and diminish the body's ability to augment preload to the heart when cardiac output drops because of the bradycardia seen in vasovagal reactions [40, 41]. Moreover, when possible, it is recommended to avoid performing painful injections in a sitting position. If a vasovagal response occurs, treatment begins with halting the procedure, placing the patient's·legs above their heart (Trendelenburg position), and applying a cold compress on their forehead. Oxygen, IV fluids, and vasoactive medications may be needed in the rare, severe case [42]. Atropine and ephedrine can be used to support heart rate, blood pressure, and cardiac output as described in Advanced Cardiovascular Life Support (ACLS) guidelines [42].

Local risks of PNBs include toxicity from the LA to the target nerve and mechanical trauma to the nerve from intraneural penetration or injection. These can present as motor or sensory deficits and pain in the sensory distribution of that peripheral nerve. Although transient deficits lasting up to 2 weeks occur in 8–15% of PNBs, incidence of long-lasting nerve injury is 2–4 per 10,000 PNBs [28].

This can be prevented by retracting and repositioning the needle if pain or paresthesia occurs during the injection, a situation that can only occur if the patient is awake. Furthermore, it is important that the physician instruct the patient before the injection on the importance of reporting paresthesia and how to recognize them before performing the block, and again just prior to starting. We suggest a statement such as "when I get near the nerve, you may feel a "zing" down the arm. Please tell me if you experience this and I will reposition my needle and it will go away." Then when injecting around the nerve, it is helpful to ask, "are you having any discomfort right now? Any uncomfortable sensations going down the arm?" Anxiolytic medications may be considered as long as the patient is still able to effectively communicate.

An indirect risk from some PNBs is injury that can occur secondary to the expected, temporary weakness after a block [43, 44]. These injuries may be prevented by proper warning before the procedure and precautions such as arm slings for upper extremity weakness and crutches or an ankle boot for lower extremity weakness.

Contraindications to PNBs include an inability to cooperate with the injection, pre-existing neurologic deficit, active local or systemic infection, and obesity obscuring optimal ultrasonographic visualization. Coagulopathy, anticoagulants, and heart and liver disease represent relative contraindications to PNBs [12, 27].

Lastly, an "incomplete block" is not uncommon and should be counseled. Several common reasons for this include inability of the LA to diffuse into the deeper nerve fibers, insufficient time for the block to take effect, and target innervation by nerves additional to those blocked. The latter scenario may knowingly occur in cases in which a portion of innervation is intentionally left unblocked because of challenges in blocking the remaining innervation branches. However, nerve anomalies are common and may explain some fraction of these incomplete nerve blockade phenomena [45]. In these scenarios, consideration may be given to injecting a few milliliters of ropivacaine around the target structure, with an even smaller volume into the target structure itself. PNB success rates are improved when expected onset time of the LA is observed [46].

## General Nerve Block Techniques

As the local anatomy and volume of LA for PNBs may be unfamiliar to even the most experienced ultrasound-guided proceduralist, the aforementioned risks and several unique techniques must be considered. When first learning PNB techniques, more superficial PNBs and those not part of a vascular plexus may be ideal—such as the suprascapular nerve, tibial nerve at the ankle or radial nerve. These procedures should only be performed by a skilled ultrasound proceduralist.

Ultrasound guidance improves the success rate [47–50] and safety [25, 30, 31] and decreases the required volume of LA in PNBs [51]; however, these risks are not completely eliminated. A brief anatomic scan with particular attention to nearby nerve branches and vessels is essential to prevent inadvertent puncture or intravascular injection. A "straight-line view" setup (Fig. 31.1) with the provider, body part, and ultrasound monitor in a line is recommended to enhance provider comfort and accuracy. Upon approaching the target nerve, aspiration followed by a "test" injection of a small volume of 1 mL LA may be injected just before reaching the nerve. If no "halo" occurs, advancing farther with another test dose is recommended [27]. Multiple small volume LA

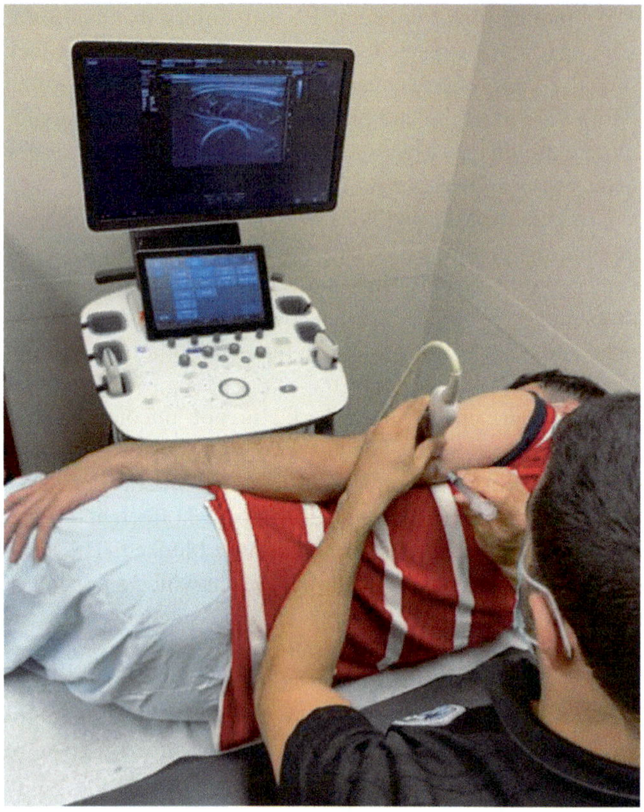


**Fig. 31.1** Physician, patient, and ultrasound oriented in a "straight line view," allowing the physician a comfortable position with an easy view back and forth between the ultrasound monitor and injection site. (Photo courtesy of Michael Khadavi, MD)

perineural injections in different locations around the nerve will reduce risk of concomitant intravascular uptake. We suggest a volume of 2–3 mL and no more than 5 mL of LA in one needle position to create a "halo" about the nerve before the needle is repositioned and another small volume is injected. Even if a "halo" occurs around the nerve, inadvertent or concomitant intravascular injection is possible. Each injection should be slow, without resistance, and with a "halo" flow pattern. If any of these do not occur, immediate repositioning is necessary to prevent injection into an unintended structure.

While optimal LA concentration and dose for orthobiologic procedures have not been studied, the surgical anesthesiology literature suggests that low volumes may often be sufficient, with volumes as low as 0.11 mL/mm$^2$ nerve cross-sectional area advocated [52]. Surgical literature suggests that at the same total dose, a lower injected volume of higher concentration provides more complete block than the converse [53–56]. There is also evidence that at the same total LA dose, higher injected volumes may result in greater systemic absorption and thus greater potential of LAST. In any case, the lowest possible LA dose should always be used [25, 27].

Further education and training through guidelines, publications, and organizations such as the ASRA and New York School of Regional Anesthesia (NYSORA) are recommended.

## Common Nerve Block Techniques

### Peripheral Nerve Block for Rotator Cuff Interventions

While interscalene blocks remain the gold standard for shoulder surgery regional anesthesia, this block is riskier than PNBs, with an expected phrenic nerve blockade and significant risks of respiratory distress [15, 57] and LAST [26, 27]. Fortunately, two of the most commonly diseased rotator cuff tendons, the supraspinatus and infraspinatus [58] are innervated by the suprascapular nerve (SSN), which may be targeted with a simple and less risky SSN block.

SSN blocks are widely studied and clinically used in the treatment of scapular fractures, various shoulder pathologies such as frozen shoulder, rotator cuff tendinopathy, glenohumeral arthritis, shoulder dislocations [59–63], radiofrequency neurotomy [64], and surgical and post-surgical regional anesthesia [16]. The SSN is a mixed sensory and motor nerve that branches from the superior trunk of the brachial plexus, travels toward the scapula and through the suprascapular notch to the supraspinatus fossa to provide innervation to the supraspinatus via its first branch, then the acromioclavicular joint, subacromial bursa, and then the infraspinatus and 70% of the glenohumeral capsule [65]. It is important to be aware that a SSN block will not anesthetize the subscapularis, teres minor, or long head biceps tendons.

**Technique**: Suprascapular Nerve Block.

**Indication**: Supraspinatus and infraspinatus intratendinous procedures.

**Positioning**: Sitting, facing away from physician.

**Suggested Needle**: 2 ½ or 3 ½ inch Quincke, 22 or 25-gauge.

**Needle Approach**: Inferomedial to superolateral, in-plane.

**LA Volume**: 5–10 mL.

Several SSN block techniques have been described, each with differing risks, ease of performance, and completeness of SSN blockade [58, 66–69], including direct SSN blockade proximal to the suprascapular notch from an anterior approach as it courses deep to the omohyoid and indirect blocks in which the nerve is not visualized and require larger LA volumes. The former has greater risk of proximal spread and LAST and as such our preferred technique is just distal to the suprascapular notch at the lateral floor of the supraspinatus fossa.

This block may be performed with the patient sitting upright (Fig. 31.2). Shoulder adduction and forward flexion

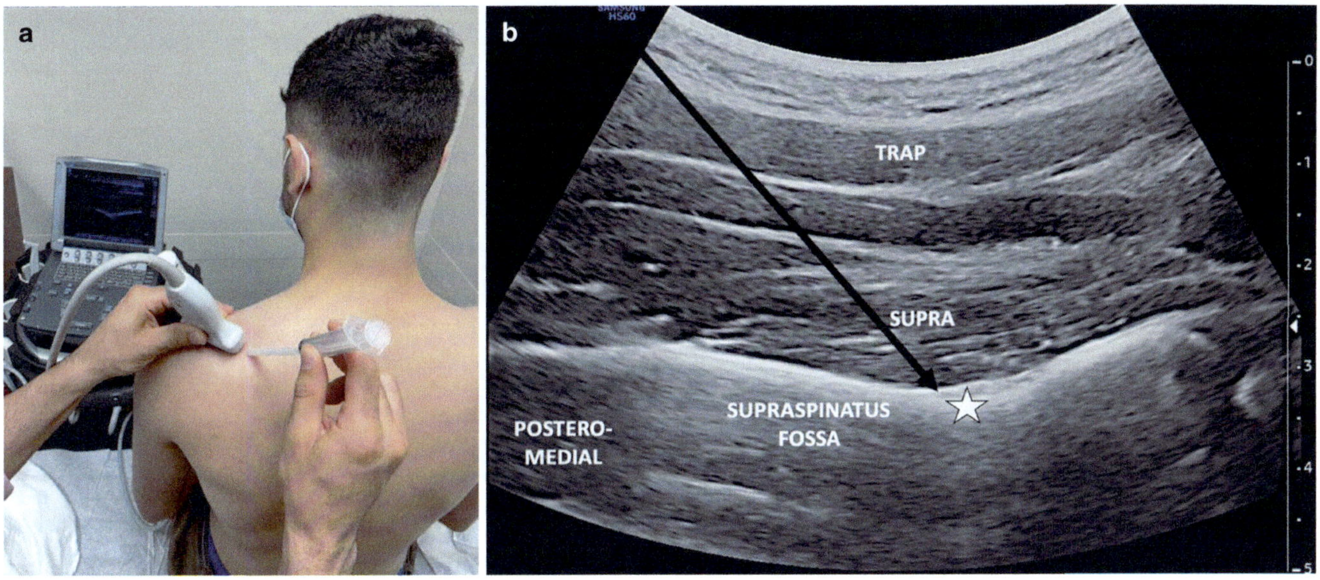


**Fig. 31.2** (**a**) Probe positioning for a suprascapular block within the supraspinatus fossa with (**b**) corresponding US image of an in-plane suprascapular nerve injection, short-axis to the nerve. *Trap* trapezius, *Supra* supraspinatus, ☆ suprascapular nerve, ➔ needle path. (Photo courtesy of Michael Khadavi, MD)

may assist by bringing the nerve into a more superficial position. Scanning the lateral supraspinatus fossa will find the suprascapular notch on its anterior border with the SSN and artery running deep and in a posterolateral direction in a shallow canal known as the suprascapular canal. In our experience, color doppler often misses the small accompanying artery, but the nerve can most easily be found just as it exits anteriorly from the suprascapular notch. With the US probe placed short-axis to the SSN and artery, the needle is directed from an inferomedial approach, in-plane through the overlying trapezius, supraspinatus, and finally the supraspinatus fascia overlying the suprascapular canal. In some larger patients the suprascapular canal is difficult to visualize, and the needle may be directed beneath the supraspinatus fascia in the lateral suprascapular fossa without direct nerve visualization [70].

Before the SSN block, the patient should be advised of anticipated rotator cuff weakness. If a sling is given, we suggest range of motion exercises such as pendulums and wall-crawls to reduce the risk of immobility-associated adhesive capsulitis.

## Peripheral Nerve Block for Elbow Common Extensor Tendon Procedures

Regional anesthesia literature describes proximal interscalene and axillary plexus blocks [17–19, 71] for most elbow surgery regional anesthesia, however these blocks are associated with higher respiratory and systemic risks [72]. Thus, PNB of the radial nerve may be considered.

While the proximal tendon of the extensor carpi radialis brevis is the most relevant in lateral epicondylitis, blockade of the entire common extensor tendon (CET) is necessary for optimal anesthesia. Variations commonly occur in nerve branching patterns to the components of the wrist extensors, including branching directly from the radial nerve, the posterior antebrachial cutaneous nerve, the posterior interosseous nerve (PIN), and even the radial sensory nerve in over 20% of individuals [73–75]. Therefore, to maximize efficacy of this blockade, the authors suggest blocking the radial nerve proximal to branching of the posterior antebrachial cutaneous nerve at the intermuscular septum of the arm. Consequent wrist-drop and lateral forearm and thumb numbness should be explained beforehand and a wrist splint may be considered.

**Technique Description**: Radial nerve block.

**Indication**: Elbow common extensor intratendinous procedures.

**Positioning**: Side-lying with affected arm on top, facing away from the physician.

**Suggested Needle**: 1 ½ inch, 22 or 25-gauge.

**Needle Approach**: Posterior to anterior, in-plane.

**LA Volume**: 5–10 mL.

With the patient side-lying, facing away from the physician and with the affected arm on top (Fig. 31.3), the radial nerve may be found either from proximal to distal, posterolateral along the spiral groove. In an axial plane, the nerve will be visualized coursing the intermuscular septum between the triceps brachii and brachialis after which the posterior antebrachial cutaneous nerve diverges. Just proximal to this location, the block is performed. Either an anterior or posterior approach may be performed, and the latter is our preference.

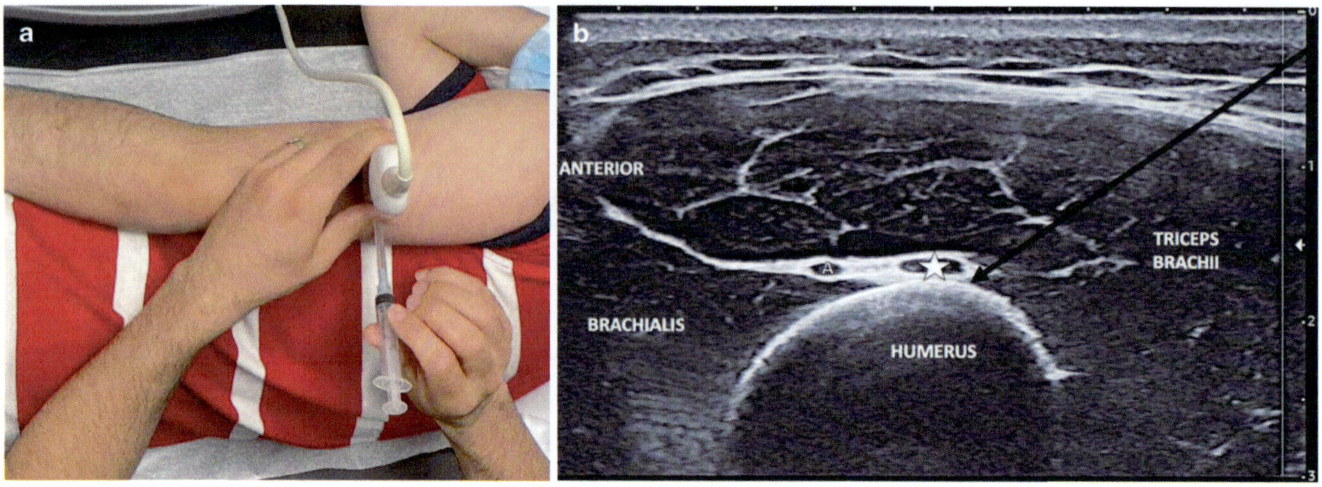


**Fig. 31.3** (**a**) Probe positioning for radial nerve block just proximal to its passing through the intermuscular septum with (**b**) corresponding US image of an in-plane injection in short-axis to the radial nerve. A radial artery, ☆ radial nerve, ➔ needle path. (Photo courtesy of Michael Khadavi, MD)

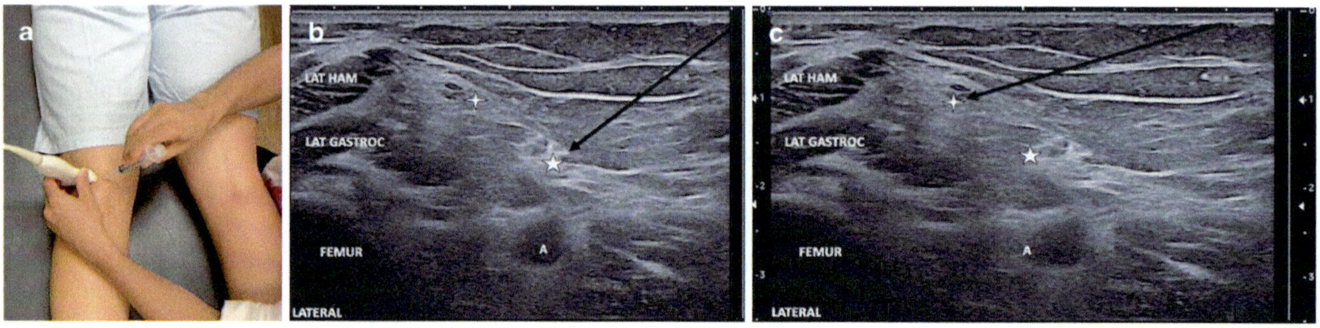


**Fig. 31.4** (**a**) Probe positioning for popliteal block with corresponding US image of an in-plane tibial nerve (**b**) and common peroneal nerve (**c**) injection, both in short-axis to the nerves. *Lat Ham* lateral hamstrings, *Lat Gastroc* lateral gastrocnemius, *A* popliteal artery, ☆ tibial nerve, ✦ common peroneal nerve, ➔ needle path. (Photo courtesy of Michael Khadavi, MD)

## Peripheral Nerve Block for Ankle Interosseus and Achilles Tendon Procedures

The ankle joint and capsule receives its innervation from the tibial nerve (TN), saphenous nerve, sural nerve (SUN), as well as superficial peroneal nerve (SPN) and deep peroneal nerve (DPN) [76]. Blocking the TN and common peroneal nerve (CPN) just past their bifurcation within the popliteal fossa provides superior anesthesia compared to blocking the sciatic nerve before this bifurcation [77] and is thus our preferred PNB technique for the ankle joint and capsule for interosseus procedures [78]. Subsequent weakness in plantarflexion and dorsiflexion is anticipated [79] and a well-fitting orthopedic boot is necessary after TN and CPN blocks to avoid tripping and falls.

**Technique**: Tibial and common peroneal nerve blocks at the knee ("Popliteal Block").

**Indication**: Achilles intratendinous procedures; intraosseous injection of distal tibia and talus.

**Positioning**: Prone, treatment leg away from physician.

**Suggested Needle**: 2 ½ or 3 ½ inch, 22 or 25-gauge Quincke.

**Needle Approach**: Medial to lateral, in-plane.

**LA Volume**: 5–10 mL for TN; 5 mL for CPN.

With the patient in prone and the US probe in axial position over the posterior knee, the sciatic nerve and its TN and CPN branches can be traces from proximal to distal or distal to proximal. The popliteal artery and its branches should be identified without and with color doppler. The block location is found by moving the transducer until just past the bifurcation of TN and CPN (Fig. 31.4). A medial approach avoids unintentional puncture of the more superficial CPN. With meticulous, small volume injection technique, the perineural injection about the TN is performed, followed by the CPN. Two separate needle entrance sites will sometimes be required depending on patient specific anatomy.

## Peripheral Nerve Blocks for Medial Knee Intraosseous Procedures

The innervation of the boney and capsular anatomy of the knee is complex, with over a dozen nerve branches that originate from the CPN, TN, sciatic nerve, femoral nerve, and obturator nerves [80, 81]. Furthermore, several common variant innervation patterns have been described [81]. Because of these complex and overlapping innervation patterns to the knee, incomplete blocks are unfortunately common when performing intraosseous knee injections.

No studies have yet been aimed at regional anesthesia techniques specifically for intra-osseous injections of the knee, however, extensive innervation anatomy of the knee has been studied over the past 80 years [82]. More recently, knee innervation pattern literature has focused on regional anesthesia techniques for knee surgeries [83–86] and for knee radiofrequency neurotomy procedures [81, 87]. Within the medial knee, the medial femoral condyle and capsule are most commonly innervated by the:

- Nerve to vastus medialis (NVM)—a branch of the femoral nerve,
- Superior medial geniculate nerve (SMGN)—an articular branch of the tibial nerve,
- Posterior division of the obturator nerve, and the medial tibial plateau is innervated primarily by the:
- Inferior medial geniculate nerve (IMGN)—a branch off the anastomosis of the articular branches of the tibial nerve and the posterior division of the obturator nerve
- Infrapatellar branch of the saphenous nerve (IPBSN)

Historically a femoral nerve block was utilized for post-operative pain management after knee surgeries, however the immediate post-operative period was challenged by quadriceps weakness [83, 84] and incomplete post-operative analgesia for more extensive surgeries such as total knee arthroplasty [85]. Adductor canal block (ACB) has become popular and widely utilized for major knee surgery with the advantage of motor sparing due to lack of blockade of several components of the quadriceps muscle [83].

The adductor canal begins at the apex of the femoral triangle and ends at the adductor hiatus, just posterior to the sartorius and carries the saphenous nerve and NVM. The NVM enters the adductor canal at the apex of the femoral triangle and then exits the adductor canal and courses medially and parallel to it midway down the thigh, separated by aponeurosis. The NVM has been found to provide more afferent innervation than previously thought to the medial knee. Therefore, for maximal medial knee blockade, an ACB should be performed midway or just proximal to midway down the thigh to include the NVM. The other nerves to the medial knee are the SMGN, IMGN, and posterior division of the obturator nerve. While not always necessary, these nerves can be blocked simultaneously as they course posteriorly through the popliteal plexus of Rudinger, via the IPACK (Interspace between the Popliteal Artery and Capsule of the Knee) block [85, 86]. However, for anyone not highly experienced in PNBs who desires blockade beyond the ACB, we suggest SMGN and IMGN blocks. It is important to be aware that the lateral knee remains sensate as the nerve to the vastus intermedius, nerve to the vastus lateralis, superior lateral geniculate nerve, inferior lateral geniculate nerve, and recurrent fibular nerve are spared. However, since osteoarthritis effecting primarily the medial compartment is very common, these PNBs for the medial knee remain highly useful in clinical orthobiologics practice.

**Technique**: Adductor Canal Block.

**Indication**: Intraosseous injection to the medial femoral condyle and medial tibial plateau of the knee, medial meniscus and medial collateral ligament injection.

**Positioning**: Supine, treatment leg toward physician, slight external rotation (frog-leg).

**Suggested Needle**: 2 ½ or 3 ½ inch, 22 or 25 gauge Quincke.

**Needle Approach**: Anterolateral to posteromedial, in-plane.

**LA Volume**: 10 mL.

With the patient supine in a frog-leg position, the adductor canal may be identified at or just proximal to mid-thigh, beneath the sartorius muscle (Fig. 31.5). The saphenous nerve courses just anterior to the femoral artery. After skin sterilization, the anterior to posterior, in-plane needle approach begins in short-axis to the canal, positioning the needle deep to the sartorius and just anterior to the femoral artery. Color doppler may aid in avoiding small vascular branches, such as the descending geniculate artery. Moving the transducer away from you while pressing the distal edge of the transducer into the skin will create a shallower angle and better needle visualization. This technique can be remembered as "move away and look back." Mild transient quadriceps weakness is possible as the NVM is responsible for 8% of the motor function of this muscle group [85].

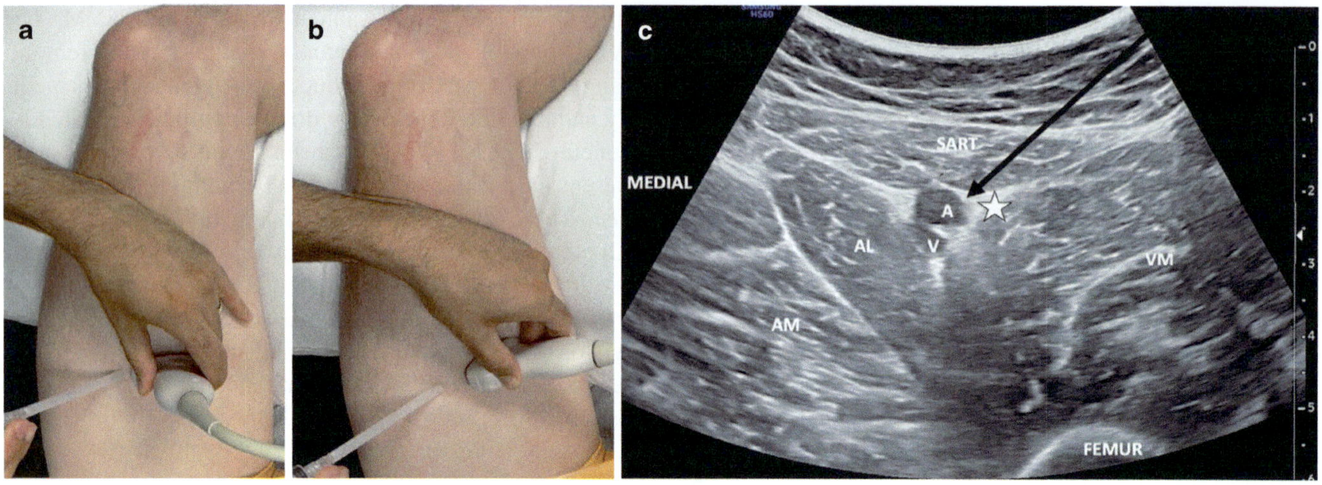


**Fig. 31.5** (**a**) Initial probe positioning at mid-thigh with subsequent "move away and look back" positioning (**b**) for an adductor canal block. (**c**) Corresponding US image of an in-plane injection of the adductor canal in short-axis. Note the sartorius overlying the adductor canal at this level. *A* femoral artery, *V* femoral vein, ☆ nerves, → needle path, *VM* vastus medialis, *Sart* sartorius, *AM* adductor magnus, *AL* adductor longus. (Photo courtesy of Michael Khadavi, MD)

**Technique**: Superior medial geniculate nerve (SMGN) and inferior medial geniculate nerve (IMGN) blocks.

**Indication**: Intraosseous injection to the medial femoral condyle and medial tibial plateau of the knee, medial meniscus and medial collateral ligament injection.

**Positioning**: Supine, treatment leg toward physician, external rotation (frog-leg).

**Suggested Needle**: 1 ½ or 2 ½ inch, 25-gauge.

**Needle Approach:** Proximal to distal (SMGN), distal to proximal (IMGN), both in-plane.

**LA Volume**: 2–5 mL for each nerve.

In the same supine, frog-leg position, the ultrasound probe is placed in a coronal plane along the distal medial thigh at the medial femoral epicondyle. Deep to the vastus medialis, the small SMGN and accompanying vasculature will be visualized as they course anteriorly along the boney surface at the flair of the epicondyle. Color doppler may be helpful to visualize the accompanying artery (Fig. 31.6). An in-plane approach is used, in short-axis to the nerve and from distal to proximal until periosteum is reached. The ultrasound probe is then moved to the medial tibia plateau and in a coronal plane at the base of the flair of the medial tibial plateau, the small IMGN is found next to its vasculature, hugging the periosteum. As this nerve also courses from posterior to anterior, the nerve should be blocked in its most posterior location. In short-axis to the nerve, a long-axis injection approach is taken from distal to proximal.

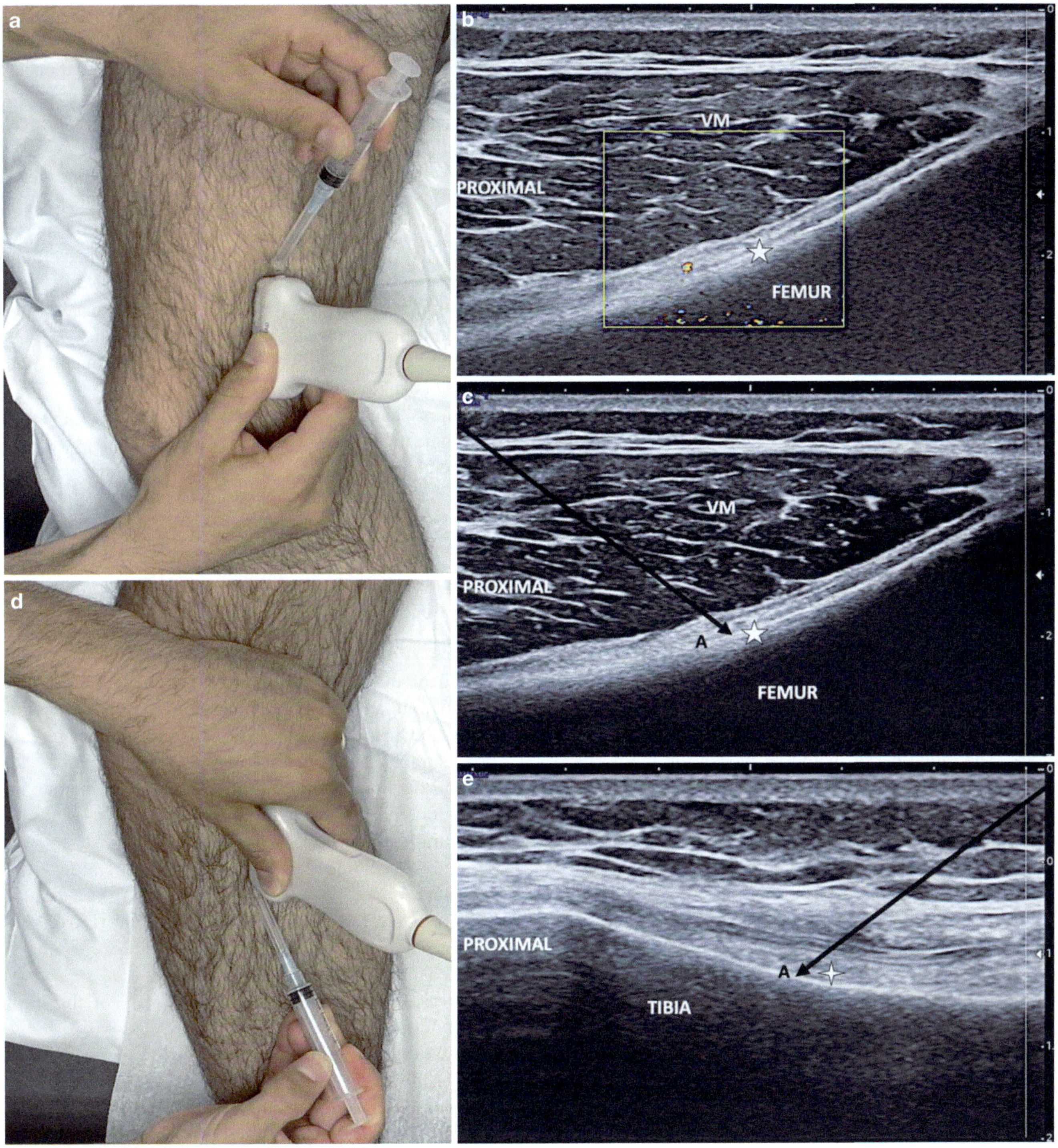


**Fig. 31.6** (**a**) Probe position for SMGN injection. Color doppler (**b**) may assistance in identifying the artery followed by an in-plane injection of the SMGN (**c**). Probe position (**d**) and in-plane injection of the IMGN. *VM* vastus medialis, *A* geniculate arteries, ☆ in (**b**) and (**c**) superior medial geniculate nerves, ✧ in (**e**) inferior medial geniculate nerves, ➔ needle path. (Photo courtesy of Michael Khadavi, MD)

## Conclusion

The goal of the orthobiologic physician is to provide safe, effective, and minimally painful procedures. In the right setting, PNBs may facilitate all of these goals. An understanding of local and systemic risks of PNBs and LAs such as LAST will improve their safety. LAST most often presents with CNS and cardiac signs and symptoms but atypical presentations are common. The risk of LAST increases with lower muscle mass, higher LA doses (especially bupivacaine), and comorbidities. We suggest performing PNBs with:

- An awake and communicative patient
- Ultrasound-guidance and a straight-line view of the screen
- Small volumes (2–3 mL) at multiple perineural locations around a nerve
- The lowest possible LA dose
- Constant monitoring for local and systemic symptoms during injection
- Lidocaine and not bupivacaine
- Brace and support for the peripherally blocked limb to help with resulting weakness

Identification of each of nerve on ultrasound is vital to a successful and safe block. For a TN block at the ankle, find the nerve in the tarsal tunnel and track it proximally. The posterior tibial artery is another useful landmark as they travel together in a neurovascular bundle. For a popliteal block, the CPN is easily identified at the fibular head and lies medial to the short head of the biceps femoris as you move proximally. Continue proximally on the CPN to find the TN as they bifurcate from the sciatic nerve. For a suprascapular block, the supraspinatus fossa is deep with the SSN at its lateral floor. SSN is easiest to find anteriorly as it exits the suprascapular notch, then travels posterolaterally. A 22-gauge is often easier to guide than a 25 gauge for this block. The radial nerve should be blocked in the posterior compartment before it passes the intermuscular septum for CET procedures. Find the nerve against the posterolateral mid-humerus and track it proximally until it is away from the shaft of the humerus for a good site for blockade. ACB will include the NVM if performed just proximal to mid-thigh, where the femoral artery is directly under the sartorius. "Move away and look back" for better needle visualization (Fig. 31.5). A well-performed PNB may provide excellent procedural comfort while eliminating complications from IV sedation and locally injected LA in the target tissue. The benefits and risks of a PNB must be considered and discussed with each patient. Further studies comparing PNBs to locally injected LA and IV sedation on orthobiologic outcomes, procedural pain levels, and safety are necessary. We also suggest further study on minimum necessary LA doses for PNBs when used for orthobiologic procedures and on PNB techniques for other common orthobiologic procedures.

## Key Points

1. Peripheral nerve blocks may improve the comfort and safety of many orthobiologic procedures.
2. Risks of peripheral nerve blocks, such as local anesthetic systemic toxicity (LAST) and nerve damage from local anesthetic, must be considered and discussed with each patient before these blocks are performed.
3. Lidocaine poses the lowest risk of LAST among local anesthetics.
4. Always use ultrasound guidance and a "straight line view" if possible when performing peripheral nerve blocks.
5. Multiple small-volume LA injections about the nerve with "halo" visualization may reduce risk of inadvertent intravascular injection of a large volume of LA.
6. A tibial nerve block at the ankle may be successfully used for plantar fascia orthobiologic injections.
7. A popliteal block provides anesthesia for a majority of foot and ankle orthobiologic procedures.
8. A suprascapular block may provide anesthesia for orthobiologic procedures to the supraspinatus and infraspinatus tendons.
9. Blockage of the radial nerve at the mid-arm provides anesthesia to the common extensor tendon and lateral epicondyle.
10. Further and continuing training on PNBs is suggested.

## Questions

1. Which of the following regenerative procedures is the least painful and most often does not require any analgesic premedication, local or regional anesthesia, or post-procedural analgesic medication?
   A. Intra-articular (joint) injection with PRP
   B. Intra-tendinous injection with PRP
   C. Intra-tendinous injection with BMC
   D. Intra-discal injection with PRP
   **Answer: A**
2. Which of the following is not a significant risk of performing a peripheral nerve block?
   A. Local anesthetic systemic toxicity
   B. Nerve damage from chemical toxicity of the local anesthetic
   C. Vasovagal reaction
   D. Discitis
   **Answer: D**
3. Which of the following suggestions may decrease risk during a peripheral nerve block?
   A. Perform the nerve block in a heavily sedated patient
   B. Performing the nerve block while the patient reports pain and paresthesias
   C. Performing the nerve block with ultrasound-guidance
   D. Performing a nerve block on a morbidly obese patient
   **Answer: C**

4. Blocking which of the following pair of nerves would provide regional anesthesia for a talonavicular intra-osseous orthobiologic procedure?
   A. Sural and deep peroneal nerves
   B. Common Peroneal and Tibial Nerves
   C. Femoral and Saphenous Nerves
   D. Tibial and Femoral Nerves
   **Answer: B**
5. Which local anesthetic carries the least risk of local anesthetic systemic toxicity (LAST)?
   A. Lidocaine
   B. Bupivacaine
   C. Ropivacaine
   D. Marcaine
   **Answer: A**

## References


1. Durant TJ, Dwyer CR, McCarthy MB, et al. Protective nature of platelet-rich plasma against chondrocyte death when combined with corticosteroids or local anesthetics. Am J Sports Med. 2017;45:218–25.
2. Dregalla RC, Uribe Y, Bodor M. Effect of local anesthetics on platelet physiology and function. J Orthop Res. 2021;39:2744–54.
3. Bedi A, Trinh TQ, Olszewski AM, et al. Nonbiologic injections in sports medicine. JBJS Rev. 2020;8:e0052.
4. Piper SL, Laron D, Manzano G, et al. A comparison of lidocaine, ropivacaine and dexamethasone toxicity on bovine tenocytes in culture. J Bone Joint Surg Br. 2012;94:856–62.
5. Zhang AZ, Ficklscherer A, Gülecyüz MF, et al. Cell toxicity in fibroblasts, tenocytes, and human mesenchymal stem cells-a comparison of necrosis and apoptosis-inducing ability in ropivacaine, bupivacaine, and triamcinolone [published correction appears in *arthroscopy* 2021; 37:1357]. Arthroscopy. 2017;33:840–8.
6. Jayaram P, Kennedy DJ, Yeh P, et al. Chondrotoxic effects of local anesthetics on human knee articular cartilage: a systematic review. PM R. 2019;11:379–400.
7. Schaufele MK, Marín DR, Tate JL, et al. Adverse events of conscious sedation in ambulatory spine procedures. Spine J. 2011;11:1093–100.
8. American Society of Anesthesiologists Task Force on sedation and analgesia by non-anesthesiologists. Practice guidelines for sedation and analgesia by non-anesthesiologists. Anesthesiology. 2002;96:1004–17.
9. Arepally A, Oechsle D, Kirkwood S, et al. Safety of conscious sedation in interventional radiology. Cardiovasc Intervent Radiol. 2001;24:185–90.
10. Liu SS, YaDeau JT, Shaw PM, et al. Incidence of unintentional intraneural injection and postoperative neurological complications with ultrasound-guided interscalene and supraclavicular nerve blocks. Anaesthesia. 2011;66:168–74.
11. Clarkson CW, Hondeghem LM. Mechanism for bupivacaine depression of cardiac conduction: fast block of sodium channels during the action potential with slow recovery from block during diastole. Anesthesiology. 1985;62:396–405.
12. Wiederhold BD, Garmon EH, Peterson E, et al. Nerve block anesthesia. In: StatPearls. Treasure Island (FL): StatPearls Publishing; 2022.
13. Bellolio MF, Gilani WI, Barrionuevo P, et al. Incidence of adverse events in adults undergoing procedural sedation in the emergency department: a systematic review and meta-analysis. Acad Emerg Med. 2016;23:119–34.
14. Mian A, Chaudhry I, Huang R, et al. Brachial plexus anesthesia: a review of the relevant anatomy, complications, and anatomical variations. Clin Anat. 2014;27:210–21.
15. Divella M, Vetrugno L, Orso D, et al. Which regional anesthesia technique is the best for arthroscopic shoulder surgery in terms of postoperative outcomes? A comprehensive literature review. Eur Rev Med Pharmacol Sci. 2021;25:985–98.
16. Ritchie ED, Tong D, Chung F, et al. Suprascapular nerve block for postoperative pain relief in arthroscopic shoulder surgery: a new modality? Anesth Analg. 1997;84:1306–12.
17. Cunningham DJ, LaRose MA, Zhang GX, et al. Regional anesthesia reduces inpatient and outpatient perioperative opioid demand in periarticular elbow surgery. J Shoulder Elb Surg. 2022;31:e48–57.
18. Lin E, Choi J, Hadzic A. Peripheral nerve blocks for outpatient surgery: evidence-based indications. Curr Opin Anaesthesiol. 2013;26:467–74.
19. Gadsden JC, Tsai T, Iwata T, et al. Low interscalene block provides reliable anesthesia for surgery at or about the elbow. J Clin Anesth. 2009;21:98–102.
20. Memtsoudis SG, Cozowicz C, Bekeris J, et al. Peripheral nerve block anesthesia/analgesia for patients undergoing primary hip and knee arthroplasty: recommendations from the International Consensus on Anesthesia-Related Outcomes After Surgery (ICAROS) group based on a systematic review and meta-analysis of current literature. Reg Anesth Pain Med. 2021;46:971–85.
21. Gibbons C, Bruce J, Carpenter J, et al. Identification of risk factors by systematic review and development of risk-adjusted models for surgical site infection. Health Technol Assess. 2011;15:1–iv.
22. Korol E, Johnston K, Waser N, et al. A systematic review of risk factors associated with surgical site infections among surgical patients. PLoS One. 2013;8:e83743.
23. Cheng H, Chen BP, Soleas IM, et al. Prolonged operative duration increases risk of surgical site infections: a systematic review. Surg Infect. 2017;18:722–35.
24. Schneider B, Kennedy DJ, Casey E, et al. Trainee involvement in transforaminal epidural steroid injections associated with increased incidence of vasovagal reactions. PM R. 2014;6:914–9.
25. Gitman M, Barrington MJ. Local anesthetic systemic toxicity: a review of recent case reports and registries. Reg Anesth Pain Med. 2018;43:124–30.
26. Macfarlane AJR, Gitman M, Bornstein KJ, et al. Updates in our understanding of local anaesthetic systemic toxicity: a narrative review. Anaesthesia. 2021;76:27–39.
27. Neal JM, Barrington MJ, Fettiplace MR, et al. The third American Society of Regional Anesthesia and Pain Medicine practice advisory on local anesthetic systemic toxicity: executive summary 2017. Reg Anesth Pain Med. 2018;43:113–23.
28. Helander EM, Kaye AJ, Eng MR, et al. Regional nerve blocks-best practice strategies for reduction in complications and comprehensive review. Curr Pain Headache Rep. 2019;23:43.
29. Brummett CM, Williams BA. Additives to local anesthetics for peripheral nerve blockade. Int Anesthesiol Clin. 2011;49:104–16.
30. Lirk P, Hollmann MW, Strichartz G. The science of local anesthesia: basic research, clinical application, and future directions. Anesth Analg. 2018;126:1381–92.
31. Barrington MJ, Uda Y. Did ultrasound fulfill the promise of safety in regional anesthesia? Curr Opin Anaesthesiol. 2018;31:649–55.
32. Neal JM, Bernards CM, Butterworth JF 4th, et al. ASRA practice advisory on local anesthetic systemic toxicity. Reg Anesth Pain Med. 2010;35:152–61.

33. Mazoit JX. Arrêt cardiaque et anesthésiques locaux [Cardiac arrest and local anaesthetics]. Presse Med. 2013;42:280–6.
34. Williams DJ, Walker JD. A nomogram for calculating the maximum dose of local anaesthetic. Anaesthesia. 2014;69:847–53.
35. El-Boghdadly K, Pawa A, Chin KJ. Local anesthetic systemic toxicity: current perspectives. Local Reg Anesth. 2018;11:35–44.
36. Wadlund DL. Local anesthetic systemic toxicity. AORN J. 2017;106:367–77.
37. Neal JM, Neal EJ, Weinberg GL. American Society of Regional Anesthesia and Pain Medicine local anesthetic systemic toxicity checklist: 2020 version. Reg Anesth Pain Med. 2021;46:81–2.
38. Kennedy DJ, Schneider B, Casey E, et al. Vasovagal rates in flouroscopically guided interventional procedures: a study of over 8,000 injections. Pain Med. 2013;14:1854–9.
39. Al-Assam H, Azzopardi C, McGarry S, et al. Vasovagal reactions in ultrasound guided musculoskeletal injections: a study of 2,462 procedures. J Clin Orthop Trauma. 2021;24:101706.
40. Rapp SE, Pavlin DJ, Nessly ML, et al. Effect of patient position on the incidence of vasovagal response to venous cannulation. Arch Intern Med. 1993;153:1698–704.
41. Akella K, Olshansky B, Lakkireddy D, Gopinathannair R. Pacing therapies for vasovagal syncope. J Atr Fibrillation. 2020;13:2406.
42. Vidri R, Emerick T, Alter B, et al. Managing vasovagal reactions in the outpatient pain clinic setting: a review for pain medicine physicians not trained in anesthesiology. Pain Med. 2022;23:1189–93.
43. Finn DM, Agarwal RR, Ilfeld BM, et al. Fall risk associated with continuous peripheral nerve blocks following knee and hip arthroplasty. Medsurg Nurs. 2016;25:25–49.
44. Muraskin SI, Conrad B, Zheng N, et al. Falls associated with lower-extremity-nerve blocks: a pilot investigation of mechanisms. Reg Anesth Pain Med. 2007;32:67–72.
45. Feigl GC, Schmid M, Zahn PK, et al. The posterior femoral cutaneous nerve contributes significantly to sensory innervation of the lower leg: an anatomical investigation. Br J Anaesth. 2020;124:308–13.
46. Chadha M, Si S, Bhatt D, et al. The comparison of two different volumes of 0.5% ropivacaine in ultrasound-guided supraclavicular brachial plexus block onset and duration of analgesia for upper limb surgery: a randomized controlled study. Anesth Essays Res. 2020;14:87–91.
47. Redborg KE, Antonakakis JG, Beach ML, et al. Ultrasound improves the success rate of a tibial nerve block at the ankle. Reg Anesth Pain Med. 2009;34:256–60.
48. Chin KJ, Wong NW, Macfarlane AJ, et al. Ultrasound-guided versus anatomic landmark-guided ankle blocks: a 6-year retrospective review. Reg Anesth Pain Med. 2011;36:611–8.
49. Liu SS, Ngeow J, John RS. Evidence basis for ultrasound-guided block characteristics: onset, quality, and duration. Reg Anesth Pain Med. 2010;35:S26–35.
50. Salinas FV. Evidence basis for ultrasound guidance for lower-extremity peripheral nerve block: update 2016. Reg Anesth Pain Med. 2016;41:261–74.
51. Munirama S, McLeod G. A systematic review and meta-analysis of ultrasound versus electrical stimulation for peripheral nerve location and blockade. Anaesthesia. 2015;70:1084–91.
52. Eichenberger U, Stöckli S, Marhofer P, et al. Minimal local anesthetic volume for peripheral nerve block: a new ultrasound-guided, nerve dimension-based method. Reg Anesth Pain Med. 2009;34:242–6.
53. Nakamura T, Popitz-Bergez F, Birknes J, et al. The critical role of concentration for lidocaine block of peripheral nerve in vivo: studies of function and drug uptake in the rat. Anesthesiology. 2003;99:1189–97.
54. Taboada Muñiz M, Rodríguez J, Bermúdez M, et al. Low volume and high concentration of local anesthetic is more efficacious than high volume and low concentration in Labat's sciatic nerve block: a prospective, randomized comparison. Anesth Analg. 2008;107:2085–8.
55. O'Donnell BD, Iohom G. An estimation of the minimum effective anesthetic volume of 2% lidocaine in ultrasound-guided axillary brachial plexus block. Anesthesiology. 2009;111:25–9.
56. Fredrickson MJ, Abeysekera A, White R. Randomized study of the effect of local anesthetic volume and concentration on the duration of peripheral nerve blockade. Reg Anesth Pain Med. 2012;37:495–501.
57. Herrick MD, Liu H, Davis M, et al. Regional anesthesia decreases complications and resource utilization in shoulder arthroplasty patients. Acta Anaesthesiol Scand. 2018;62:540–7.
58. Freygant M, Dziurzyńska-Białek E, Guz W, et al. Magnetic resonance imaging of rotator cuff tears in shoulder impingement syndrome. Pol J Radiol. 2014;79:391–7.
59. Dahan TH, Fortin L, Pelletier M, et al. Double blind randomized clinical trial examining the efficacy of bupivacaine suprascapular nerve blocks in frozen shoulder. J Rheumatol. 2000;27:1464–9.
60. Lewis RN. The use of combined suprascapular and circumflex (articular branches) nerve blocks in the management of chronic arthritis of the shoulder joint. Eur J Anaesthesiol. 1999;16:37–41.
61. Gleeson AP, Graham CA, Jones I, et al. Comparison of intra-articular lignocaine and asuprascapular nerve block for acute anterior shoulder dislocation. Injury. 1997;28:141–2.
62. Breen TW, Haigh JD. Continuous suprascapular nerve block for analgesia of scapular fracture. Can J Anaesth. 1990;37:786–8.
63. Schneider-Kolsky ME, Pike J, Connell DA. CT-guided suprascapular nerve blocks: a pilot study. Skelet Radiol. 2004;33:277–82.
64. Shah RV, Racz GB. Pulsed mode radiofrequency lesioning of the suprascapular nerve for the treatment of chronic shoulder pain. Pain Physician. 2003;6:503–6.
65. Basta M, Sanganeria T, Varacallo M. Anatomy, shoulder and upper limb, suprascapular nerve. In: StatPearls. Treasure Island, FL: StatPearls Publishing; 2021.
66. Parris WC. Suprascapular nerve block: a safer technique. Anesthesiology. 1990;72:580–1.
67. Dangoisse MJ, Wilson DJ, Glynn CJ. MRI and clinical study of an easy and safe technique of suprascapular nerve blockade. Acta Anaesthesiol Belg. 1994;45:49–54.
68. Gado K, Emery P. Modified suprascapular nerve block with bupivacaine alone effectively controls chronic shoulder pain in patients with rheumatoid arthritis. Ann Rheum Dis. 1993;52:215–8.
69. Laumonerie P, Blasco L, Tibbo ME, et al. Distal suprascapular nerve block-do it yourself: cadaveric feasibility study. J Shoulder Elb Surg. 2019;28:1291–7.
70. Schoenherr JW, Flynn DN, Doyal A. Suprascapular nerve block. In: StatPearls. Treasure Island, FL: StatPearls Publishing; 2022.
71. Hacıbeyoğlu G, Kocadağ G, Özkan A, et al. Brachial plexus block for elbow surgery in a patient with spinal muscular atrophy. J Clin Anesth. 2021;72:110275.
72. Zhai WW, Wang XD, Li M, et al. A study on diaphragm function after interscalene brachial plexus block using a fixed dose of ropivacaine with different concentrations. Zhonghua Yi Xue Za Zhi. 2016;96:2229–33.
73. Al-Qattan MM. The nerve supply to extensor carpi radialis brevis. J Anat. 1996;188:249–50.
74. Cho H, Lee HY, Gil YC, et al. Topographical anatomy of the radial nerve and its muscular branches related to surface landmarks. Clin Anat. 2013;26:862–9.
75. Jeon A, Kim YG, Kwon SO, et al. Relationship between the branching patterns of the radial nerve and supinator muscle. Biomed Res Int. 2021;2021:8691114–7.
76. Han JR, Tran J, Agur AMR. Overview of the innervation of ankle joint. Phys Med Rehabil Clin N Am. 2021;38:791–801.

77. Eldegwy MH, Ibrahim SM, Hanora S, et al. Ultrasound-guided sciatic politeal nerve block: a comparison of separate tibial and common peroneal nerve injections versus injecting proximal to the bifurcation. Middle East J Anaesthesiol. 2015;23:171–6.
78. Nakagawa A, Miyake R, Naito Y. Ultrasound-guided sciatic nerve block in the popliteal fossa for the postoperative pain control after Achilles tendon repair. Masui. 2012;61:546–8.
79. Standring S. Gray's anatomy. 41st ed. Edinburgh: Elsevier Churchill Livingstone; 2016.
80. Tran J, Peng PWH, Chan VWS, et al. Overview of innervation of knee joint. Phys Med Rehabil Clin N A m. 2021;32:767–78.
81. Roberts SL, Stout A, Dreyfuss P. Review of knee joint innervation: implications for diagnostic blocks and radiofrequency ablation. Pain Med. 2020;21:922–38.
82. Gardner E. The innervation of the knee joint. Anat Rec. 1948;101:109–30.
83. Jaeger P, Nielsen ZJ, Henningsen MH, et al. Adductor canal block versus femoral nerve block and quadriceps strength: a randomized, double-blind, placebo-controlled, crossover study in healthy volunteers. Anesthesiology. 2013;118:409–15.
84. Charous MT, Madison SJ, Suresh PJ, et al. Continuous femoral nerve blocks: varying local anesthetic delivery method (bolus versus basal) to minimize quadriceps motor block while maintaining sensory block. Anesthesiology. 2011;115:774–81.
85. Johnston DF, Sondekoppam RV, Uppal V, et al. Hybrid blocks for total knee arthroplasty: a technical description. Clin J Pain. 2018;34:222–30.
86. Sankineani SR, Reddy ARC, Eachempati KK, et al. Comparison of adductor canal block and IPACK block (interspace between the popliteal artery and the capsule of the posterior knee) with adductor canal block alone after total knee arthroplasty: a prospective control trial on pain and knee function in immediate postoperative period. Eur J Orthop Surg Traumatol. 2018;28:1391–5.
87. Tran J, Peng PWH, Lam K, et al. Anatomical study of the innervation of anterior knee joint capsule: implication for image-guided intervention. Reg Anesth Pain Med. 2018;43:407–14.

# 32 Rationale and Clinical Guidelines for Post-Orthobiologic Rehabilitation

Luga Podesta, Raymond Mattfeld, Michael Khadavi, and Eric S. Honbo

The study of orthobiologic and cell-based therapies has increasingly studied since platelet-rich plasma (PRP) was first used in 1987 in the United States to control wound healing after cardiac surgery [1]. Over the last decade, there has been a significant increase in the utilization of orthobiologic therapies and cell-based therapies by orthopedic, musculoskeletal, and sports medicine physicians as a treatment for a variety of musculoskeletal and orthopedic conditions Involving tendons, ligaments, muscles, the intervertebral disk, and synovial joints. Despite promising clinical results and the widespread use of PRP and other orthobiologic and cell-based therapies to treat musculoskeletal and orthopedic injuries, their use remains controversial due to the heterogeneity in study designs, lack of reporting standardization, and evidence of publication bias [2–4].

L. Podesta
Regenerative Sports Medicine, Bluetail Medical Group, Naples, FL, USA

Podesta Sports Medicine and Orthopedics, Naples, FL, USA

Florida Everglades, Homestead, FL, United States

R. Mattfeld
Sports Rehab Consultants and Bright Bay Physical Therapy, Adjunct Professor of Physical Therapy at Touro College School of Health Sciences and Post-Professional DPT Program, New York, NY, USA

M. Khadavi (✉)
Kansas City Orthopedic Alliance, Leawood, KS, USA

Kansas City Ballet, Sporting Kansas City, Kansas City, KS, USA

E. S. Honbo
Advanced Physical Therapy and Sports Medicine- Spine and Sport PT, Adjunct Clinical Faculty, USC Department of Biokinesiology and Physical Therapy, Consultant- Chinese Olympic Committee/Team, Los Angeles, CA, USA

## Tissue Healing

Healing is defined as a dynamic and complex biologic progression that results in the restoration of anatomic structure and function. When an injury occurs, it triggers a complex physiologic process characterized by a predictable cascade of biological tissue responses triggered by the injury itself. Physiologic healing progresses through four overlapping stages: stage 1, hemostasis; stage 2, the acute inflammatory phase; stage 3, the proliferative or repair phase; and stage 4, the remodeling phase, as shown in Fig. 32.1, the healing cascade.

Hemostasis phase, stage 1, is the first and shortest phase of the healing cascade, occurring within seconds to minutes, and this is the process of forming a blood clot to stop bleeding. Platelets are vital to hemostasis, also functioning as the physiological trigger to activate acute inflammation and program tissue repair [4, 5].

The inflammatory phase, stage 2, begins immediately following the injury and continues for 48 to 72 h. Platelets are stimulated to provide hemostasis by forming a clot. Platelets in the clot then degranulate and secrete several growth factors, hemostatic factors, and cytokines from alpha granules that are necessary in the early stages of the clotting cascade. Histamine and serotonin are released from the dense granules and function to increase capillary permeability, activate macrophages, and allow inflammatory cells greater access to the injury site [4, 6, 7]. The inflammatory phase can last up to 72 h and is characterized by localized pain, swelling, erythema, and increased local tissue temperature.

The proliferative phase (stage 3) begins with the activation of fibroblasts by growth factors and inflammatory mediators released during the acute inflammatory phase [8]. The proliferative phase is defined by cell proliferation, neovascularization and matrix synthesis in addition to other metabolic processes that aid and remodeling and organization of the healing tissue [9]. During the ensuing 48 h to 6 weeks, anatomic structures begin to be restored while tissue generation


A. Navani et al. (eds.), *Essentials of Regenerative Medicine in Interventional Pain Management*,
https://doi.org/10.1007/978-3-031-50357-3_32

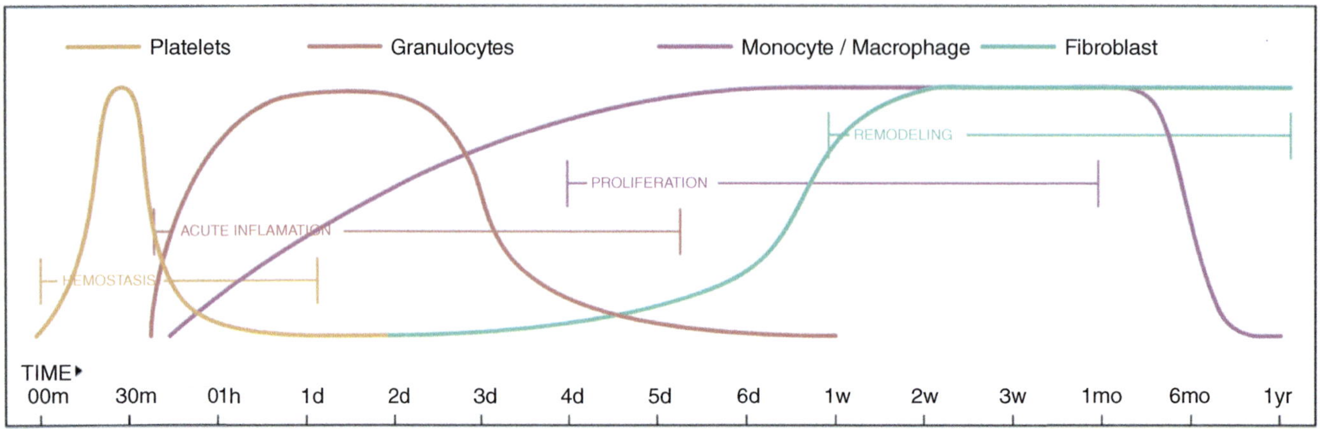


**Fig. 32.1** Healing cascade. The 4 stages of the healing cascade consisting of 4 partially overlapping stages, hemostasis, acute inflammation, proliferation, and remodeling. The particular cell type activity within each phase is crucial for the progression and successful execution of the healing cascade leading to tissue repair. (Modified from Honbo E, et al. Honbo ES, Mattfeld R, Khadavi M, Podesta L. Clinical rationale and rehabilitation guidelines for post biologic therapy. *Phys Med Rehabil Clin N Am* 2023; 34:239–263 [4])

occurs. Fibroblasts begin to synthesize scar tissue and capillary neoformation begins to reestablish nutrients to the injured tissue. Stage 3 ends with the beginning of wound contracture.

The remodeling phase, stage 4, is the longest phase of the healing cascade, beginning several weeks after the initial injury and may last months to over a year depending on the severity of the initial injury [4, 8]. Fibroblasts are responsible for remodeling, replacing the type 3 collagen matrix with the stronger type 1 collagen matrix. Fibroblasts either die through apoptosis or differentiate into myofibroblasts which align to the direction of force within the tissue. Failure to properly transition from the proliferation phase may lead to excessive or hypertrophic scarring [4, 8].

## Bioformulations

Platelet-rich plasma is the supinate fraction processed from autologous peripheral whole blood that has a platelet concentration above the normal baseline.

There are substantial differences in the content of platelet concentrations produced by the various automated and manual PRP recovery systems. PRP is produced by either single spin or double spin density gradient centrifugation producing platelet concentrations ranging from 1.7 to 14x baseline. There are substantial differences in the content of platelet concentrations produced by the various automated and manual PRP recovery systems. The most common methods of PRP production utilize either a plasma-based system or a buffy coat system, which produce either a leukocyte-poor PRP (LP-PRP) or a leukocyte-rich PRP (LR-PRP) respectively. Regardless of the system used, all systems must release growth factors from platelets. However, they all differ in platelet, WBC, RBC and plasma concentrations.

There has been a significant amount of literature published on the effects of PRP. We currently know that platelet-rich plasma has an analgesic effect by diminishing inflammation, augmenting the proliferation of tenocytes, osteoblasts, and myocytes and enhancing the secretion of hyaluronic acid by synovial joints. PRP is often used in combination with hyaluronic acid because of their synergistic effect. Finally, PRP through its paracrine effect can augment stem cell migration.

As with all orthobiologic therapies including PRP, there is great variability in reporting post orthobiologic rehabilitation rationales, exercise progressions and timelines for safe return to normal activities and sports. Several articles have proposed a standardized rehabilitation program following treatment with PRP [4, 10–12].

Townshend et al. [2], performed a systematic review of post-procedure protocols following platelet-rich plasma injections for tendinopathy. They discovered significant variability in reporting periods of relative rest, weight-bearing restrictions, rehabilitation exercises, full return to play, and the use of non-steroidal anti-inflammatory drugs (NSAIDs) medications pre and post-treatment.

## Therapeutic Guidelines for Post Orthobiologic Rehabilitation

The concept of prehabilitation or prehab is not foreign to physical therapy. Prehabilitation, first mentioned in the literature in 1946 [13], reported how a period of training for warfare with physical therapy, and whole-body strength training, improved the physical and mental well-being of army recruits during the Second World War. Recently, prehabilitation has been mentioned in various orthopedic conditions and settings i.e., anterior cruciate ligament (ACL) injury, joint replacements, regenerative cartilage surgery, and running [14–19].

The intent of prehabilitation is to prepare a patient for optimal recovery from a particular procedure. The focus is placed on the restoration of arthrokinematic and osteokinematic mobility, soft-tissue extensibility, muscle flexibility, balance, proprioception, and strength. Additionally, giving the clinician an opportunity to focus on improvement on other areas of the body that the patient would benefit from [20].

Emphasis is given to determine the etiology the injury; what, how, and why it has occurred. In the case of ligament injuries, before orthobiologic treatment, dynamic imaging is performed to establish baseline measurements of ligament morphology and joint stability. Baseline metrics include stressed and non-stressed ligament tensioning in comparison to the contralateral joint. These measurements are repeated at key time periods post-injection per Podesta and colleagues [4, 21]. These metrics are then used to confirm tissue healing and joint stability for the purpose of rehabilitation progression.

A key component to prehabilitation with respect to orthobiologics is patient education regarding the overall phases, time frames, and expectations for recovery from biologic treatment (Table 32.1) [4, 20].

Post orthobiologic rehabilitation progression is based on several individual factors including the combinations of the patient's age and health, the severity of injury, time since treatment, physiologic healing mechanism, tissue integrity, response to physical therapy treatment dosage, as well as adherence to appropriate home programs.

**Table 32.1** Pre- post- orthobiologic treatment instructions [4]

| |
|---|
| Before orthobiologic treatment |
| • Stop all NSAIDs 2 to 3 weeks before the procedure |
| • Prehab-gain general understanding of the healing process, time frames, use of heat to facilitate vascularization and local metabolism, and set patients up for best outcome |
| • Use of functional outcome tools such as the KJOCS, SPADI, lower extremity functional scale, patellofemoral index (SCOR), etc., are commonly used to establish the patient's baseline subjective functional status |
| • Establish baseline with imaging (dynamic musculoskeletal ultrasound, fluoroscopy, MRI) |
| Post- orthobiologic treatment |
| • Apply a thermal heat agent for 15 min four times per day when awake for 2 weeks |
| • Acetaminophen for pain as needed |
| • Increased pain and inflammation are expected in the treated tissue after the procedure |
| • Range-of-motion exercises as tolerated after the procedure |
| • Continue to use any braces, splints or crutches as recommended by your physician after the treatment |
| • Dynamic musculoskeletal ultrasound at 6 to 7 weeks to determine tissue continuity, integrity, and joint stability to safely progress later phase rehabilitation and higher-level functional activities |

*NSAID* nonsteroidal anti-inflammatory medications, *KJOCS* Kerlan–Jobe Orthopedic Score, *SPADI* shoulder pain and disability index, *MRI* magnetic resonance imaging

Optimal loading is both tissue and adaptation specific. Loading improves the mechanical properties of tissue and creates better fiber alignment and less disorganized matrix formation. The goal of rehabilitation following orthobiologic therapies such as PRP injections is to progressively and therapeutically place appropriate amounts of physical stress and optimally load through and along the injured tissue to help facilitate healing [21]. Characteristics of optimal loading include controlled and directed force to the selective tissue; early loading through functional ranges; an appropriate blend of compressive, tensile, and shear loading; variability in magnitude, direction, duration, and intensity; includes neural overload; needs to be dynamic and reflect demands of that tissue; and is functionally appropriate [4].

Loading is tissue-specific (Table 32.2), for example, soft tissues, ligaments, muscles, and tendons require linear loading due to fiber alignment. However, articular cartilage requires controlled and progressive compressive loading due to osteochondral matrix physiology. Articular cartilage responds optimally to progressive compression or shear, which creates a pumping effect' providing joint cell nutrition, molecule transport, and joint lubrication. Successful functional outcomes for articular cartilage repair necessitate the correct quantity of loading. For example: Insufficient loading (immobilization): leads to adverse effects, and delayed healing, thinning, and softening of articular cartilage, decreases matrix content, and chondrocytes, and yields less lubrication; Moderate loading (low impact exercise, bike, and swim): essential for cartilage health and aids repair; Excessive loading (high impact exercise, Plyometrics): is detrimental to cartilage.

Guidelines, expectations, and treatment interventions following orthobiologic treatments are based on loading and the natural tissue healing cascade and repair principles. A clinical framework that includes criteria-based progressions, goals within the healing phases, and clinical rationales for progression following orthobiologic treatments are suggested in the rehabilitation phase specific tables below (Tables 32.3, 32.4, and 32.6) [20].

Controlled physical stress to each tissue (muscle, tendon, ligament, and bone) is imparted throughout the duration of rehabilitation to facilitate repair and may include tension, torsion, compression, and shear [21]. The stress or loading is applied via manual therapy techniques, dosed medical exercise therapy progressions, in addition to functional strengthening and return to play phase exercises. There is limited evidence in the literature defining specific protocols following orthobiologic or cell-based therapies and limited documentation regarding tissue healing time frames following these unique therapies [4, 11, 12, 22, 23].

**Table 32.2** Tissue-specific loading and healing timeline [4]

| Tissue | Healing capacity, Vascularity | Force to facilitate healing | Tissue healing time | Loading projection |
|---|---|---|---|---|
| Muscle | Good/abundant | Contractile loading Isometric > concentric > eccentric muscle | 6–8 weeks | Range of motion parameters, SubMaxim isometrics > max isometrics > Concentrics > eccentrics > ballistic progressions |
| Tendon | Fair/less | Contractile loading Isometric > concentric > eccentric | 8–12 weeks | Range of motion parameters, SubMaxim isometrics > max isometrics > Concentrics > eccentrics > ballistic progressions |
| Ligament | Less/diminished | Tension | 7–14 weeks | Bracing with protected range of motion parameters |
| Articular cartilage, labrum, intraventricular disk | Limited/absent | Controlled fiber Tension in line of stress | 6–12 weeks | Bracing, unloading, progressing weight bearing, aquatics, stationary bike, weight bearing loading and controlled torsional stress |
| Subchondral bone/bone | | Controlled weight bearing | 6–12 weeks | Weight bearing status, unloading |

**Table 32.3** PHASE I-II hemostasis/inflammatory (0–7 days) criteria-based clinical rationale rehab progressions [4]

| Goal | Anticipated impairments and functional limitations | Intervention | Rationale | Criteria to progress to this phase |
|---|---|---|---|---|
| • Protect tissue<br>• Allow biologic to absorb<br>• Daily activity as tolerated within provided brace<br>• Avoid access loading or stress to treated area<br>• Improve tissue vascularity and joints synovialization via gentle movement of extremity<br>• Avoid tissue overload or exercise unless approved by doctor | • Day 1–2: Painful in treated tissue/ joint<br>• Day 3–6: Diminishing pain and improving significantly<br>• Day 7: Minimal pain, improved quality of range of motion | Restrictions:<br>• Avoid all varus, valgus, A/P and rotational loads, ligament stressing, or any loaded contractile activities or exercises**<br>• Maintain any limited weight-bearing precautions**<br>• Tissues/joint-specific protected bracing and weight bearing<br>• No exercise except for the rehab program<br>• Upper extremity treatments—No lifting > body weight<br>• Tylenol for pain<br>• Heath packs for 15 min, 4x/day for 1–2 weeks<br>• Avoid ice over the treatment site<br>• Shower okay 24 h after the procedure<br>• No submersion in water, bath, pool, hot tub, or ocean for 1 week<br>Physical therapy progression: (home-based)<br>• Progress passive range of motion to active range of motion, to point of initial resistance, within brace restraints, and only within physician range of motion restraints<br>• Gentle sub maximum isometrics (lower to mid-range submaximal holds) twice daily | • Minimize stress on the injection site<br>• Cross-link initiation and homeostasis occurring as biologic activating to prepare for cross bridging | • Post injection with no signs of infection |

There is no absolute progression or transition between phases and there can be variability between patients pending each individual case.

The goal following orthobiologic therapies is to promote adequate tissue healing such that the tissue is able to once again maximally withstand the physiologic stresses and forces placed upon it with daily functional demands or sporting activities [24]. Collagen fibers run in parallel alignment that affords the tissue to withstand tensile forces and unilateral stress placed upon it [24]. The following information is based on the authors' clinical experience treating patients who have undergone orthobiologic injections with PRP and bone marrow-derived progenitor cells to ligament, tendon, muscle, intra-articular joints, and subchondral bone since 2006.

## Phase I-II (Homeostasis/Inflammatory Phases)

Time: 0 to 7 days post-treatment.

Goals:

- Allow the PRP to absorb the injected tissue.
- Avoid cross-link disruption.
- Facilitate integrity of cross-link formation.

**Table 32.4** PHASE III proliferative (3–6 weeks) criteria-based clinical rationale rehab progressions [4]

| Goal | Anticipated impairments and functional limitations | Intervention | Rationale | Criteria to progress to this phase |
|---|---|---|---|---|
| • Protect tissue<br>• Facilitate collagen deposition<br>• Avoid disruption of collagen cross-link and facilitate parallel fiber alignment<br>• Minimize deconditioning<br>• Communication among physician, physical therapist and patient is essential during this key transitional phase | • Limited upper extremity/lower extremity strength and cardiovascular endurance<br>• Limited tissue tolerance to weight bearing, compression, shearing, tensile, and contractile loads, or functional activities<br>• Pain diminishing<br>• Limited tolerance with heavy lifting, pushing, pulling functional activities | Restrictions:<br>• Avoid all varus, valgus, loads, or ligament stressing activities and exercises for 7 weeks**<br>• Avoid all compressive torsional loads for articular cartilage tissue times 5–6 weeks**<br>• Continue use of assistive devices as instructed by a physician<br>• No overstressing of tissue through exercise or impact activity<br>• Initiate high repetition low load therapeutic concentric exercises for muscle/tendon<br>• Initiate unloaded cyclical high-volume compression–decompression exercises for cartilage/joints<br>• Improve tissue vascularity enjoyed synovialization by initiating upper extremity cardiovascular exercise for lower body procedures, or lower body exercise after upper body procedures<br>• No exercises except for supervised rehab program for 3–6 weeks<br>Physical therapy progression overview:<br>• Pain should not increase greater than 2 points on 10 points VAS scale<br>• Modalities for collagen synthesis in pain control:<br>• Moist heat, non-thermal ultrasound, cold laser, Russian stim, ES, shockwave therapy ESWT<br>Manual therapy techniques:<br>• Gentle soft tissue mobilization along line of tissue fibers<br>• Joint mobilization to maintain arthrokinematics<br>Therapeutic exercises:<br>• AROM to point of initial resistance submaximal to maximal isometrics<br>• Emphasize proper postural alignment, distal joint position.<br>Adjust exercise progression based on severity of injury<br>• Gradual progression of early to mid-phase weight-bearing activities (aquatics, unloaded treadmill, elliptical)<br>• Initiate low resistance, high repetition, concentric, open chain exercise<br>• Initiate BFR exercises<br>• Initiation and progression of eccentric exercises as concentric strength increases<br>Neuromuscular re-education:<br>• PNF and rhythmic stabilization exercises<br>• Proprioceptive training<br>• Use of taping techniques as indicated for facilitation/inhibition | • Pain threshold significantly reduced<br>• Collagen synthesis occurring, aligning in the longitudinal axis<br>• Cross bridging occurring and matrix integrity improving<br>• Tissue beginning to withstand tensile forces and loads<br>• Use modalities to facilitate collagen formulation and remodeling<br>• Soft tissue mobilization techniques have mechanical, physiologic, it's the logic, and neurological effects on the tissue which facilitate the healing mechanism and fiber alignment<br>• Progress toward light ligamentous loading by end of phase III<br>• Tissue tensile strength should be strong enough to initiate stress loading exercises<br>• BFR enables strengthening using a light load and a relatively low volume of work<br>• Cardiovascular training to improve endurance and tissue repair | • Full pain-free tolerate full<br>• Able to ROM weight bearing (dependent on treated tissue) |

*ESWT* extracorporeal shockwave therapy, *ROM* range of motion, *VAS* visual analog scale, *BFR* blood flow restriction, *AROM* active range of motion, *ROM* range of motion, *PNF* proprioceptive neuromuscular facilitation

Phase I consists of early mobilization, gentle self-stretching, and protected or unloaded weight-bearing functional activities to prevent the deleterious effects of immobilization and to promote tissue healing [21]. Due to the elevated inflammatory response, the patient commonly experiences an increase in pain for 1 to 3 days following the treatment. Following PRP injection the majority of the growth factors are released within the first hour of injection but continued release can occur for up to 7 days following injection [25, 26]. Thus, the home program for the first 7 days following PRP injection is aimed at avoiding disruption of this physiologic mechanism and includes: rest, gentle active motion, submaximal isometric holds in all pain-free planes and ranges to help fiber alignment, pain medications such as over-the- counter analgesics avoiding NSAIDs, to control symptoms. Use of a heating agent at home following biologics treatment is recommended. Thermotherapy increases tissue temperature, blood flow, and metabolism (providing a closer physiologic environment). Increased blood flow facilitates tissue healing by supplying protein, nutrients, and oxygen at the site of injury. A 1C increase in tissue temperature is associated with a 10% to 15% increase in local tissue metabolism [26]. This increase in metabolism aids the healing process by increasing both catabolic and anabolic reactions needed to degrade and remove metabolic by-products of tissue damage and provides the milieu for tissue repair [27].

## Phase II (Inflammatory/Early Proliferative Phase)

Time: 8 to 21 days post-treatment.
Goals:

- Avoid disruption of collagen cross-link bridging and formulation.
- Initiate early motion.
- Initiate high repetition no resistance loading exercises.
- Obtain 90% full range of motion.

Phase II consists of continued gentle active motion and increased activity at home. Patients should obtain 90% of full range of motion by the end of week 2 [21]. Light soft-tissue mobilizations of the treated tissue and surrounding fascia should commence toward the end of week 3. However, to avoid disruption of collagen cross-link bridging and formation, deeper soft-tissue techniques (transverse friction, etc.) are not implemented until the sixth week following treatment [21]. Gentle early motion of the local joints facilitate the physiologic tissue healing response following orthobiologic treatments as it proceeds through the inflammatory, reparative, and remodeling phases of the healing cascade. Furthermore, early motion and self-stretching prevent joint adhesions, and increase muscle contraction, muscle fiber size, and tension, in addition to increasing resting levels of glycogen and protein synthesis. Mechanical stimulation (ie: motion and controlled external loading) is a prerequisite for platelet effect in the early post-injection phase. Platelets influence the early stages of tissue repair, allowing mechanical stimulation to further drive the reparative phase healing [27].

Submaximal-maximal effort isometrics performed three times per day are also initiated in attempts to create light tension in the direction of the tendon fibers. However, when treating treated ligaments, any tensioning forces are avoided at this time (i.e., valgus elbow stress to ulnar collateral ligament [UCL]) until 7 weeks post-injection. Intermittent rest and care with resuming work and normal daily functional activities are also encouraged at home to help control post-injection symptoms and early inflammatory elevation.

Progressive full arc motion in the first two phases (0 to 21 days), prevents ligament atrophy, and increases ligament linear tissue stress and stiffness, particularly at a bone-ligament junction. Exercises that stress ligaments or functional activities that apply excessive muscle or tendon tension are avoided during this phase. There may be a 2- to 4- week delay with ligament healing due to decreased tissue vascularization [4, 21]. Exercises exerting tension on ligaments i.e., the elbow UCL (valgus stress) is not begun until later during phase 3, 6 to 7 weeks post-treatment.

Early motion restoration aids connective tissue lubrication between collagen cross-links, increases collagen mass, decreases abnormal collagen cross-links, and prevents the development of adhesion. Articular cartilage, following orthobiologic treatments, responds to early motion, intermittent compression, and decompression loading with improved metabolic activity and increased health of the cartilage matrix. Muscles, tendons, ligaments, and bone tissue all respond favorably to motion. Restoration of a full range of motion during the first 10 to 14 days following orthobiologic therapies is encouraged [4, 21].

Modalities used in the first two phases can include ultrasound, electrical stimulation, and laser. The use of modalities during this phase is aimed to further stimulate tissue healing and increase local perfusion and oxygen delivery to the site. Nonthermal ultrasound is commonly used to facilitate tissue repair and regeneration in damaged tissue. Research that supports the use of therapeutic ultrasound to increase boney and muscle tissue regeneration [28–30]. However, most studies that support the use of nonthermal ultrasound and laser to aid tissue healing are based on animal studies. It is still unclear if using a pulsed non-thermal ultrasound is more effective than a low-intensity continuous protocol in terms of proliferation and tissue healing. The use of laser treatment in patients with lateral epicondylosis was found to lower subjective overall pain levels with reports of 90% to 100% relief in more than 45% of the patients who were treated with laser [31]. In addition, studies have found the use of low-energy laser improves tensile strength and stiffness in repairing medial collateral ligament in rats at 3 and 6 weeks after injury [32]. We have found positive results using Russian electrical stimulation with the following parameters: 2500 Hz frequency, 50 pps, 10/10 s duty cycle, 2 s ramp time for 10 to 12 min after tendon PRP, to help increase endorphin release and minimize tissue response to loading and manual mobilizations.

In upper extremity patients, progressive loading with shoulder, elbow, and wrist exercises during phases 2 to 4 is a critical component of the post-biologic treatment plan. The first two phases include use of concentric low-load higher repetition exercise dosage (three sets of 20 to 25 repetitions) is recommended. Once three sets of 25 repetitions has been achieved, the weight is increased by 1 pound and progresses as tolerated. Proper postural alignment, proximal and distal joint positioning, and control throughout the range entire range of motion are emphasized. This submaximal intensity using higher repetition progression improves tissue vascularization, helps align collagen cross-links, promotes tissue healing, and enables the tissue to start adapting to controlled amounts of stress [21]. Endurance training versus strengthening has been used in the early phases with success following PRP injections. Submaximal loading exercises reduce homeostasis and tissue break-down and symptom exacerbation during the first 2 to 4 weeks. Studies have shown that resistance exercise is more effective in inducing acute muscle anabolism than high-load low volume or work-matched resistance exercise modes (isometrics) [33].

## Phase III (Proliferative/Reparative Phase)

Time: 3 to 6 weeks post-treatment.

Goals:

- Adjust exercise progression based on the type of tissue and severity of the injury.
- Use of modalities to aid tissue proliferation (recommend pulsed US, laser, electrical stimulation, extracorporeal shockwave therapy [ESWT], blood flow restriction [BFR] therapy).
- Begin high repetition loading and concentric exercises.
- Begin functional activities.
- *Avoid ligament stress for 7 weeks with activities of daily living and exercises. *
- **Progress to Eccentrics weeks 4 to 6. **

Pain levels have typically lessened by the end of the third week. By 3 weeks collagen synthesis is occurring and aligning in the longitudinal axis. Tissue is beginning to withstand tensile forces and loads. However, it is important to adjust exercise progression based on the type of tissue and the severity of injury (ligament healing and proliferation take longer) [4, 21]. Blood Flow restriction (BFR), has been shown to increase training and loading capacity for those who are not ready or are unable to tolerate heavier training loads and is often initiated in week 3 following orthobiologic injection. Specifically, BFR has been shown to produce increased muscle mass, and strength, as well as cross-sectional patellar and Achilles tendon stiffness at comparable levels to conventional high-load resistance training [4, 34]. The use of BFR in the early post-treatment phases of rehabilitation gives us the ability to provide skeletal muscle with a sufficient exercise stimulus using a light load and a relatively low volume of work. It is helpful for treating muscle, tendon, ligaments, and joints which is important in the earlier post-injection phase of healing [35].

Soft-tissue mobilizations and progressive loading via resisted exercise are key components of the post-injection proliferative and remodeling phases [4, 21]. During the proliferative and remodeling phase of muscle and tendon tissue healing, the use of soft-tissue mobilizations techniques (ASTYM®, deep transverse friction mobilization, Active Release, Graston Technique) in conjunction with appropriately timed exercise progressions is an important component of the healing process to help minimize this scar formation and promote anatomic tissue fiber healing in line of stress [4, 21]. These techniques, however, should be used with caution to avoid tissue alignment disruption and only once in the remodeling phase, after 6 to 8 weeks post-injection.

First described by Cyriax [36], deep transverse tissue mobilization (DTFM) is an aggressive form of soft-tissue mobilization to promote tissue healing. Deep Transverse Friction mobilizations and other soft-tissue manipulation techniques have mechanical, physiologic, histologic, and neurologic effects on the tissue which facilitate the healing mechanism of orthobiologic injections (Table 32.5). To promote normal resolution of the collagen tissue, the tissue to be treated should be in a moderate (not painful) stretch position. Tissue reactions to DTFM may include rapid desensitization, latent post-treatment soreness, and moderate tissue bruising covering the area of tissue contact is acceptable [36].

There has been increased interest and use of ESWT, over the past several years. Studies show it to be effective in reducing pain and improving function [4, 37–39]. The proposed mechanisms of ESWT involve inducing suppression of

**Table 32.5** Effects of deep transverse tissue mobilization [4]

| |
|---|
| Mechanical |
| • Distortion and elongation of collagen fibers |
| • Increased interstitial mobility |
| Histologic |
| • Prevents scar formation and haphazard collagen orientation |
| • Stimulate collagen orientation along lines of stress via "Piezzo-electric effect" |
| Physiologic |
| • Localized hyperemia |
| • Stimulate white blood cell invasion and healing production |
| • Destruction of substance P |
| Neurologic |
| • Initial nociceptor stimulation |
| • Mechanoreceptor stimulation |
| • Pain inhibition via "central biasing mechanism" |

inflammatory reactions and oxidative stress and eliciting neovascularization in degenerative tissues [40]. The use of ESWT combined with orthobiologic therapies such as PRP are felt to be beneficial on pain reduction and recovery of physical activity of patients with insertional Achilles tendinopathy and can relieve the pain synergistically for knee osteoarthritis [41, 42].

Eccentric loading is initiated early in the proliferative and remodeling phases at approximately weeks 5 to 6 for muscle belly and 6 to 8 for tendon pending individual patient status. Owing to its positive effect on improving tissue integrity, strength, and improved function, eccentric loading is the other important component of post-orthobiologic rehabilitation [4, 21]. Eccentric contractions function to decelerate a limb, provide shock absorption, and generate forces 14% to 50% greater than a maximal concentric contraction [43]. This increased force generation improves musculotendinous integrity by inducing muscle hypertrophy and increased tensile strength or by lengthening the musculotendinous unit [44].

Unlike concentric exercise in phases 1 and 2, eccentric exercises and tissue loading has been shown to aid in angiogenesis in early tendon injury [45]. Daily eccentric loading was found to not have any detrimental effect on tendon vascularity or microcirculation [45]. A systematic review of tendinopathy found that eccentric exercises had the most clinical efficacy in regenerating function [46]. Other studies have found that eccentric exercise progressions are an effective treatment for chronic tendinosis [47, 48]. Eccentric tendon loading exercise progressions are thus implemented into the post-orthobiologic rehabilitation by week 4 to 5 pending individual patient response [4, 21].

There is no clear consensus on the best non-operative treatment for muscle injuries beyond immediate rest and anti-inflammatory medications or modalities [29]. In chronic tendinosis injuries, rest has been found to be a less effective treatment efficacy [46, 49]. In turn, eccentric exercise and loading have been shown in many studies to be beneficial in treating patients with tendinosis [44, 45]. The optimal dosage and frequency of eccentric loading for treating chronic tendinosis have not yet been established [50]. However, the literature indicates that to optimize treatment, orthobiologics or cell-based therapies should be combined with the mechanical stimulus of the tissue in order to promote optimal post-injection tissue healing and repair [21].

## Phase IV (Remodeling Phase)

Time: 6 to 14 weeks post-treatment.

Goals:

- Eccentric loading, plyometric training return to sport/activity (pending individual sport and post-injection status variable from patient to patient)
- Continue tissue remodeling facilitation with deep transverse friction and soft-tissue mobilizations
- *Diagnostic ultrasound (approx. 6 to 7 weeks) may be repeated to determine the extent of healing
- Resume full functional or sporting activity for 10 to 12 weeks pending progress with the post-injection program.

Post-orthobiologically treated tissues commonly exhibit increased tensile strength during the remodeling phase [21]. Tissue remodeling facilitation is continued with the use of deep transverse friction and soft-tissue mobilizations. Depending on the response to the eccentric strengthening progression, the patient progresses to coordination and speed drills, plyometrics, ballistics, and more explosive sport-specific phase 4 exercises [4].

Throughout the remodeling phase, connective tissue fiber orientation is better aligned, it has improved tensile strength and suited to withstand more demanding tensile stress [51]. Functional strengthening exercises including, plyometrics, ballistics, neuromuscular power, and coordination exercises are performed at more intense levels enabling patients to meet the demands of work and or sport.

Studies regarding the efficacy of using scores on subjective functional tools or questionnaires to help determine readiness to safely resume a particular activity or sport have not yet been published. A grading system has been used to describe tendinopathy associated with jumper's knee [51].

Return to sport or work is recommended when selective tissue tension tests are non-provocative and follow-up functional tools to ascertain patient readiness to resume higher level exercises improve or normalize. Weight-bearing, ligament stress tests, resisted muscle-tendon tests in a lengthened position, and compression tests for bone are common examples.

There are not clearly defined or objective means of determining when an athlete is able to safely return to play or when a patient is able to return to a functional activity (job duty).

The use of a patient's subjective assertions, detailed clinical examination, functional index score as to whether the patient is ready to resume certain activities or sports, together with repeat diagnostic imaging such as dynamic ultrasound findings, are all utilized in determining when the patient is ready to resume the desired activity.

Interval running programs, interval throwing programs, and on-field agility progressions are initiated in phase 4. Although there are several notable documented cases in which an athlete has returned to play earlier following orthobiologic therapies, most patients are safely able to resume full functional or sporting activity by 10 to 14 weeks (Table 32.6) [4, 21].

**Table 32.6** PHASE IV remodeling (6–14 weeks) criteria-based clinical rationale rehab progressions [4]

| Goal | Anticipated impairments and functional limitations | Intervention | Rationale | Criteria to progress to this phase |
|---|---|---|---|---|
| • Restore normal tissue integrity and fiber alignment<br>• Maximize tissue vascularity and joint synovialization<br>• Increase tissue tensile strength<br>• Improve joint proprioception<br>• Improve force production, tissue elasticity and ability to withstand tensile stretching<br>Critical decision-making period—determine if tissue has sufficiently healed via dynamic imaging or if his second injection and or surgical intervention is warranted<br>• prepare for return to activity, sports | • Limited upper extremity/lower extremity strength<br>• Limited ligament in tendon tensile strength during early phase IV<br>• Diminished but improving weight-bearing or joint compressive exercise loading tolerance<br>• Limited joint proprioception<br>• Altered timing and mechanics with sports specific and functional activities | Diagnostic imaging:<br>• Diagnostic ultrasound (6–8 weeks) to determine extent of healing and exercise progression and return to activity or sport status<br>Physical therapy progression: (physical therapy)<br>Modalities:<br>• Continue as needed<br>Manual therapy:<br>• Can you deep transverse friction mobilization/massage to increase tissue vascularization and break up tissue adhesions<br>Therapeutic exercise:<br>• Progress exercise and functional mobility integrating upper extremity/lower extremity CKC exercises as appropriate<br>• Progress eccentric exercise<br>• Progressive plyometric loading from < body weight too full body weight—Bilateral to single<br>• Progress to ballistic, explosive training<br>• Sports specific training<br>≲ 50% effort up to week 8<br>≲below 75% effort up to week 10<br>≲below 90% effort up to week 12<br>• Initiate interval sport programs (throwing, running, on field drills) pending results of diagnostic ultrasound<br>• Return to sports 10–15 weeks depending on the sports/activity<br>Neuromuscular reeducation:<br>• Light eccentric resistance pulley or tubing patterns with control speed emphasis<br>• Light resistance proprioceptive neuromuscular facilitation exercises performed manually using distal hand placement and initiating joint specific motions and adding pulleys or tubing/bands<br>• Progress proprioception exercises to unstable surfaces<br>** Dynamic imaging (musculoskeletal ultrasound) is used to confirm ligament healing, joint stability, and load progression.<br>• Initiate ligament and joint loading when healing and joint stability are determined, exercise progression is initiated.<br>• Initiate eccentric tendon loading<br>If sufficient healing and stability has not occurred at 6–7 weeks, a second injection vs. surgical stabilization may be warranted<br>Week 8–10: Progress to fast twitch and dynamic exercises<br>• Increase speed, resistance, and functional strengthening<br>• Add kinetic chain functional and sports specific loading progressions<br>**Pending repeated ultrasound imaging findings progress to return to play phase 4<br>Week 10–12: Reassess objective exam results, functional testing, and subjective functional tool scores to determine return to higher level activity and/or sport specific play<br>• Begin interval return to sports program. Start interval throwing, batting, tennis serves, volleyball kidding programs pending repeat ultrasound imaging findings<br>Weeks 12–16: Progress from 75–90+ % in controlled setting. Gradual return to sport at 12–15 weeks | • Increase tissue integrity and strength of repair tissue<br>*Improved ability to produce force, withstand tensile stretching, absorb shock, and increased tissue elasticity<br>• Reassess functional index score and dynamic imaging to correlate with objective exam findings to determine soft tissue and subchondral bone healing, joint stability and pace of current exercise progression can continue, or if a second injection and/ or surgical intervention is warranted | • Pain-free joint stability with stress testing<br>• Show tissue integrity and joint stability with dynamic imaging<br>• Subjective functional index tool indicates patient is ready to progress through phase IV to return to play status<br>• Functional testing performed to determine return to activity |

## Summary

The use of orthobiologic therapies for musculoskeletal, orthopedic and sports medicine pathologies, to deliver high concentrations of naturally occurring biologically active growth factors, proteins and progenitor cells to the site of injury is extremely promising and ic continually evolving. Early protection and tissue-specific progressive loading are critical components to successful outcomes following orthobiologic intervention. Each tissue heals and responds differently. Ligament, tendon, muscle, and articular cartilage each have unique healing properties that require tissue-specific loading. The authors have found using a criteria-based loading and exercise progression guided by dynamic imaging when appropriate to further advance the goal-oriented rehabilitation program. Each patient and injured tissue are unique and require specific intervention and rehabilitation.

### Key Points

1. Healing cascade: consisting of four partially overlapping stages including hemostasis, acute inflammation, proliferation, and remodeling stages. Each phase is associated with a particular cell type activity which is crucial for the progression and successful execution of a human cascade eating to tissue repair.
2. Bioformulations and Deliverable platelets: Platelet-rich plasma can be prepared using various methods that will either maintain physiologic ratios of blood components such as leukocyte rich-platelet rich plasma (LR-PRP) or remove specific components such as leukocyte poor-platelet rich plasma (LP-PRP). In a normal wound repair reaction, all components of blood have important roles in driving the healing response, and the cooperative physiologic balance of cellular and biomechanical action results in the restoration of the injured tissue to a healthy state. Therefore a LR-PRP bioformulation with a higher concentration of deliverable platelets, 3–5 billion platelets per mL, is optimal for tissue repair.
3. Pre Orthobiologic instructions: Discontinue all nonsteroidal anti-inflammatory medications (NSAIDs), aspirin, and statins 2–3 weeks prior to the procedure. Prepare patients for optimal outcomes by educating and supplying them with a general understanding of the healing process anticipated for their specific injury, time frames necessary for optimal healing. Instruct patients to utilize heat to facilitate vascularization and local metabolism. Utilize functional outcome tools to establish a patient's baseline subjective functional status. Establish baseline objective data with imaging such as dynamic musculoskeletal ultrasound, fluoroscopy, or MRI.
4. Post Orthobiologic instructions: apply thermal heat agent for 15 min four times per day 1 week for 2 weeks. Instruct patients to utilize acetaminophen for post-injectional pain as needed. Educate patients that increase pain and inflammation are expected in the treated tissue after the procedure for approximately 24 to 72 h. Range of motion exercise is begun as tolerated after the procedure. Assistive devices such as braces, splints, or ambulatory aids are continued per physician recommendations after treatment. Utilize dynamic imaging 6 to 7 weeks post-treatment to determine tissue continuity, integrity, and joint stability to safely progress later phases of rehabilitation and higher-level functional activities.
5. Tissue specific loading and healing timelines: Loading is tissue-specific, for example, soft tissue, ligaments, muscles, and tendons require linear loading due to fiber alignment. Whereas articular cartilage requires controlled and progressive compressive loading due to osteochondral matrix physiology. Articular Cartlidge responds optimally to progressive compression for sheer, which would create a "pumping effect" providing joint cell nutrition molecular Transport and joint lubrication.
6. Articular Cartilage repair: Successful functional outcomes for cartilage repair necessitate the correct quantity of loading. Insufficient loading (immobilization): leads to adverse effects and delayed healing, thinning, and softening of articular cartilage, decreases matrix content and chondrocytes, yielding less lubrication. Moderate loading (low-impact exercise, bike, and swimming): is essential for cartilage health and aids repair. Excessive loading: (high impact exercise, and plyometric exercise) is detrimental to cartilage.
7. Phase I-II (Homeostasis/Inflammatory Phases) Goals: Time: 0 to 7 days post-treatment. Goals: Allow the orthobiologic to absorb the injected tissue; Avoid cross-link disruption; Facilitate integrity of cross-link formation.
8. Phase II (Inflammatory/Early Proliferative Phase) Goals: Time: 8 to 21 days post-treatment.

   Goals: Avoid disruption of collagen cross-link bridging and formulation; Initiate early motion; Initiate high repetition no resistance loading exercises; Obtain 90% full range of motion.
9. Phase III (Proliferative/Reparative Phase) Goals: Time: 3 to 6 Weeks Post-Treatment

   Goals: Adjust exercise progression based on the type of tissue and severity of the injury; Use therapeutic modalities to aid tissue proliferation, pulsed ultrasound, laser, electrical stimulation, extracorporeal shockwave therapy [ESWT], blood flow restriction [BFR] therapy; Begin high repetition loading and concentric exercises; Begin functional activities; *Avoid ligament stress for

7 weeks with activities of daily living and exercises; **Progress to Eccentric exercise in weeks 4 to 6.

10. Phase IV (Remodeling Phase) Goals: Time: 6 to 14 weeks post-treatment

    Goals: Eccentric loading, plyometric training, return to sport/activity (pending individual sport and post-injection status which is variable from patient to patient); Continue tissue remodeling facilitation with deep transverse friction and soft-tissue mobilizations; *Diagnostic ultrasound (approx. 6 to 7 weeks) may be repeated to determine the extent of healing; Resume full functional or sporting activity by 10 to 12 weeks, pending progress with the post-injection program.

## References


1. Ferrari M, Zia S, Valbonesi M, et al. A new technique for hemodilution, preparation of autologous platelet-rich plasma and inter-operative blood salvage in cardiac surgery. Int J Artif Organs. 1987;10:47–50.
2. Townsend C, Von Rickenbach KJ, Bailowitz Z, et al. Post-procedure protocols following platelet rich plasma injections for tendinopathy: a systematic review. PM R. 2020;12:904–15.
3. Moraes VY, Lenza M, Tamaoki M, et al. Platelet-rich therapies for musculoskeletal soft-tissue injuries. Cochrane Database Syst Rev. 2014;40:CD010071.
4. Honbo ES, Mattfeld R, Khadavi M, Podesta L. Clinical rationale and rehabilitation guidelines for post biologic therapy. Phys Med Rehabil Clin N Am. 2023;34:239–63.
5. Broughton G 2nd, Janis JE, Attinger CE. The basic science of wound healing. Plast Reconstr Surg. 2006;117:12S–34S.
6. Foster T, Puskas B, Mandelbaum B, et al. Platelet-rich plasma-from basic science to clinical applications. Am J Sports Med. 2009;37:2259–72.
7. Bennett NT, Schultz GS. Growth factors and wound healing: biochemical proper- ties of growth factors and their receptors. Am J Surg. 1993;165:728737.
8. Parrish WR, Roides B. Physiology of blood components in wound healing: an appreciation of cellular co-operativity in platelet rich plasma action. J Exec Sports Orthop. 2017;4:1–14.
9. Dean RS, DePhillipo NN, LaPrade RF. Ligament lesions: cell therapy. In: Filardo G, Mandelbaum BR, Muschler GF, et al., editors. Orthobiologics, injectable therapies for the musculoskeletal system. Cham: Springer Nature; 2021. p. 245–55.
10. Finnoff J, Fowler S, Lai J, et al. Treatment of chronic tendinopathy with ultrasound- guided needle tenotomy and platelet-rich plasma injection. PM R. 2011;3:900–11.
11. van Ark M, van den Akker-Scheek I, Meijer L, et al. An exercise-based physical therapy program for patients with patellar tendinopathy after platelet-rich plasma injection. Phys Ther Sport. 2013;14:124–30.
12. Kaux JF, Forthomme B, Namurois MH, et al. Description of a standardized rehabilitation program based on sub-maximal eccentric following a platelet-rich plasma infiltration for jumper's knee. Muscles Ligaments Tendons J. 2014;4:85–9.
13. Prehabilitation, rehabilitation, and revocation in the army. Br Med J 1946; 1:192197.
14. Shaarani S, O'Hare C, Quinn A, et al. Effect of prehabilitation on the outcome of anterior cruciate ligament reconstruction. Am J Sports Med. 2013;41:2117–27.
15. Carter H, Littlewood C, Webster K, et al. The effectiveness of preoperative rehabilitation programmes on postoperative outcomes following anterior cruciate ligament (ACL) reconstruction: a systematic review. BMC Musculoskelet Disord. 2020;21:647.
16. Potts G, Reid D, Larmer P. The effectiveness of preoperative exercise programmes on quadriceps strength before and following anterior cruciate ligament (ACL) reconstruction: a systematic review. Phys Ther Sport. 2022;54:16–28.
17. Linton L, Valentin S. Running coaches and running group leaders' engagement with, and beliefs and perceived barriers to prehabilitation and injury prevention strategies for runners. Phys Ther Sport. 2020;46:54–62.
18. Villers J, Burch J, Scheller M, et al. Physical therapy prehabilitation on a reverse total shoulder replacement candidate: a case study. J Phys Ther Sci. 2020;32:197–205.
19. Hirschmuller A, Schoch W, Baur H, et al. Rehabilitation before regenerative cartilage knee surgery: a new prehabilitation guideline based on the best available evidence. Arch Orthop Trauma Surg. 2019;139:217230.
20. Honbo E, Podesta L. Clinical applications of platelet rich plasma therapy. In: Maxey L, Magnusson J, editors. Rehabilitation for the postsurgical orthopaedic patient. 3rd ed. St. Louis, MI: Mosby; 2012. p. 171–92.
21. Podesta L, Crow S, Volkner D, et al. Treatment of elbow partial ulnar collateral ligament tears with platelet rich plasma. Am J Sports Med. 2013;41:1689–94.
22. Mautner K, Malanga G. Optimization of ingredients, procedures, and rehabilitation for platelet-rich plasma injections for chronic tendinopathy. Pain Manag. 2011;1:523–32.
23. Donatelli R, Owens-Burkhart A. Effects of immobilization on the extensibility of periarticular connective tissue. J Orthop Sports Phys Ther. 1981;3:67–72.
24. Hammond JW, Hinton RY, Curl LA, et al. Use of autologous platelet-rich plasma to treat muscle strain injuries. Am J Sports Med. 2009;37:1135–42.
25. Cameron MH. Thermal agents: physical principles, cold and superficial heat. In: Physical agents in rehabilitation: from research to practice. Philadelphia, PA: Saunders; 1999. p. 149–75.
26. Nadler S, Weingand K, Kruse R. The physiologic basis and clinical applications of cryotherapy and thermotherapy for the pain practitioner. Pain Physician. 2004;7:395–9.
27. Virchenko O, Aspenberg P. How can one platelet injection after tendon injury lead to a stronger tendon after 4 weeks? Interplay between early regeneration and mechanical stimulation. Acta Orthop. 2006;77:806–12.
28. Byl NN, Mckenzie AL, West JM, et al. Pulsed micro amperage stimulation: a controlled study of healing of surgically induced wounds in Yucatan pigs. Phys Ther. 1994;74:201–13; [discussion: 213-218].
29. Chan YS, Li Y, Foster W, et al. The use of suramin, an antifibrotic agent, to improve muscle recovery after strain injury. Am J Sports Med. 2005;33:43–51.
30. Gum S, Reddy G, Stehno-Bittel L, et al. Combined ultrasound, electrical stimulation, and laser promote collagen synthesis with moderate changes in tendon biomechanics. Am J Phys Med Rehab. 1997;76:288–96.
31. Zlato S, Tatjana T. Comparison between low level laser therapy, transcutanous electro-neural stimulation, visible incoherent polarized light, and placebo in the treatment of lateral epicondylitis: a pilot clinical study on 120 patients. Lasers Surg Med. 2002;Supplement:14.
32. Fung DT, Ng GY, Leung MC, et al. Therapeutic low energy laser improves the me- chanical strength of repairing medial collateral ligament. Lasers Surg Med. 2002;31:91–6.

33. Song JS, Spitz RW, Yamada Y, et al. Exercise-induced hypoalgesia and pain reduction following blood flow restriction: a brief review. Phys Ther Sport. 2021;50:89–96.
34. Centner C, Jerger S, Lauber B, et al. Low-load blood flow restriction and high- load resistance training induce comparable changes in patellar tendon properties. Med Sci Sports Exerc. 2022;54:582–9.
35. Courville XF, Coe MP, Hecht PJ. Current concepts review: non insertional Achilles tendinopathy. Foot Ankle Int. 2009;30:1132–42.
36. Cyriax JH. Textbook of orthopaedic medicine: volume 2 treatment by manipulation, massage, and injection. 10th ed. London: Bailliere Tindall; 1980.
37. Vulpani M, Trischitta D, Trovato P, et al. Extracorporeal shockwave therapy (ESWT) in Achilles tendinopathy: a long-term follow-up observational study. J Sports Med Phys Fitness. 2009;49:171–6.
38. Aldajah S, Alashram A, Annino G, et al. Analgesic effect of extracorporeal shock-wave therapy in individuals with lateral epicondylitis: a randomized controlled trial. J Funct Morphol Kinesiol. 2022;7:1–8.
39. Chang C, Chen L, Chou Y, et al. The effectiveness of platelet-rich plasma and radial extracorporeal shock wave compared with platelet-rich plasma in the treatment of moderate carpal tunnel syndrome. Pain Med. 2020;21:1668–75.
40. Erroi D, Sigona M, Suarez T, et al. Conservative treatment for insertional achilles tendinopathy: platelet-rich plasma and focused shock waves. A retrospective study. Muscles Ligaments Tendons J. 2017;7:98–106.
41. Wenzhen S, Yongjie L, Guowei W, et al. Prospective clinical study on extracorporeal shock wave therapy combined with platelet-rich plasma injection for knee osteoarthritis. Chin J Reparative Reconstr Surg. 2019;33:1527–8.
42. Dean E. Physiology and therapeutic implication of negative work. Phys Ther. 1988;68:233–7.
43. Alfredson H, Pietila T, Jonsson P, et al. Heavy-load eccentric calf muscle training for the treatment of chronic Achilles tendinosis. Am J Sports Med. 1998;26:360–6.
44. Nakamura K, Kitaoka K, Tomita K. Effect of eccentric exercise on the healing pro- cess of injured patellar tendon in rats. J Orthop Sci. 2008;13:371–8.
45. Magnussen RA, Dunn WR, Thompson AB. Nonoperative treatment of midportion Achilles tendinopathy: a systematic review. Clin J Sport Med. 2009;19:54–64.
46. Purdam CR, Jonsson P, Alfredson H, et al. A pilot study of the eccentric decline squat in the management of painful chronic patellar tendinopathy. Br J Sports Med. 2004;38:395–7.
47. Young MA, Cook JL, Purdam CR, et al. Eccentric decline squat protocol offers superior results at 12 months compared with traditional eccentric protocol for patellar tendinopathy in volleyball players. Br J Sports Med. 2005;39:102–5.
48. Rompe JD, Nafe B, Furia JP, et al. Eccentric loading versus eccentric loading plus shock-wave treatment for midportion Achilles tendinopathy: a randomized controlled trial. Am J Sports Med. 2009;37:463–70.
49. Andres BM, Murrell GA. Molecular and clinical developments in tendinopathy: editorial comment. Clin Orthop Relat Res. 2008;466:1519–20.
50. Kellet J. Acute soft-tissue injuries- a review of the literature. Med Sci Sports Exerc. 1986;18:489–500.
51. Blazina ME, Kerlan RK, Jobe FW, et al. Jumpers Knee. Orthop Clin North Am. 1973;4:665–78.

# Regulatory Aspects of Regenerative Therapy and Stem Cell Therapy: Current Perspectives

33

Rikin Patel and LoJ'ai Marks

## Introduction

While regenerative medicine is an exciting field with promising developments and a growing market demand, there are also concerns of unqualified providers, misleading marketing, and documented instances of unproven interventions performed without approval through proper channels, putting patients at risk and undermining the field [1–4]. Thus, agencies such as the United States Food and Drug Administration (FDA), the Federation of State Medical Boards (FSMB), the Federal Trade Commission (FTC), and others have worked to improve the regulation of these products, developing policy and enforcing relevant laws to promote compliance.

Providers have a duty to protect the integrity of the field, patients and society at large by navigating treatments, consent, marketing, and biologics processing with care and in accordance with government guidelines. Hence, this chapter will discuss guidelines set forth by the FDA, relevant legislation, and efforts of other governing bodies to promote safety, transparency, evidence-based practice, and operating in good faith, refraining from misleading or exploiting patients. Most importantly, the information covered in this chapter will outline ways to avoid being on the FDA's radar and disciplinary action, maintaining acceptable practices that support the integrity of the regenerative medicine field and promote the Hippocratic oath to do no harm.

## Background

The FDA is the government agency broadly responsible for regulating products such as food, cosmetics, medical devices, radiation-emitting devices, veterinary products, tobacco products, and biologics including regenerative medicine products to protect public health and safety in the United States [5]. While the FDA is responsible for oversight and regulation, they are not tasked with defining nomenclature and do rely on experts to help develop policy to aid in setting appropriate guidelines. A collaborative effort is crucial in this complex and ever-evolving field to develop appropriate policy and guidelines implementing updates accordingly as changes and advancements unfold. Moreover, while the FDA enforces laws relevant to their scope outlined above, the FTC enforces laws related to marketing and advertising protecting the public from deceptive, fraudulent, or unfair advertising [5]; this includes marketing and advertising in regenerative medicine practices.

In the early stages of regulatory development, the FDA published: *Regulatory Considerations for Human Cells, Tissues and Cellular and Tissue-Based Products: Minimal Manipulation and Homologous Use for HCT/Ps* as final guidance in November 2017, which they later updated in July 2020. The FDA gave a 3-year period of leniency in an effort to give businesses the opportunity to take appropriate action to ensure they were operating in accordance with FDA guidelines. This discretionary enforcement period was intended for certain products with respect to the FDA's investigational new drug (IND) application and premarket approval process. This period was initially planned to end in November of 2020, but was later extended to May of 2021. It is important to know that the FDA now considers that discretionary enforcement period to have ended and all practitioners remain responsible for full compliance under Title 21 of the Code of Federal Regulations (CFR) Part 1271, the Public Health Services (PHS) Act and the Federal Food Drug & Cosmetics (FD&C) Act. What a practitioner may have assumed was compliant in the discretionary period no longer applies and our current perspective must be based on adherence to this guidance.

While the FDA is the party responsible for regulating and enforcing the use of biologics including human cells, tissues, and cellular-based products (HCT/Ps), the FSMB has a

R. Patel (✉) · L. Marks
The Orthobiologic Institute, Los Angeles, CA, USA


A. Navani et al. (eds.), *Essentials of Regenerative Medicine in Interventional Pain Management*,
https://doi.org/10.1007/978-3-031-50357-3_33

mission that includes providing services and initiatives that promote patient safety, quality healthcare, and best regulatory practices for physicians (allopathic and osteopathic) and in most jurisdictions, other healthcare professionals as well [6]. In alignment with their mission and at the request of former U.S. Senator Lamar Alexander, in 2017 FSMB chair Gregory B. Snyder, MD developed a workgroup of diverse, qualified experts to create a report of best practices for state medical boards regulating promotion, communication, and treatments at stem cell clinics in the United States [1]. It is important for regenerative medicine providers to understand the FMSB recommendations from the *Report and Recommendations of the Workgroup to Study Regenerative and Stem Cell Therapy Practices*– which has been adopted as policy by the FSMB as of April 2018. The key fundamentals to these recommendations are to "focus on sensible and necessary principles of patient safety, autonomy, and non-exploitation" [1].

It maintains that new, experimental/unproven treatments should be reserved for when evidence-based treatments have been exhausted and must be based on clinically reasonable theory/peer-acknowledged practice [1].

The FSMB has encouraged state boards to:

- Make licensees aware of relevant legislation/guidelines concerning stem cell and regenerative therapies and keep licensees and the public informed about new developments and regulations [1].
- Consider expanding their rules/policies on informed consent to include a shared decision-making framework including:
  - explanation, discussion, and comparison of treatment options.
  - patient values/preferences.
  - partnered decision with the patient.
  - evaluation of partnered decision [1].
- Review professional marketing materials and claims as information sources in physician complaint investigations [1].
- Monitor publicly available licensee warning letters to consider investigating professional misconduct regarding stem cell and regenerative therapy [1].
- **Physicians must** be competent and appropriately trained to offer proposed intervention(s) and only offer treatment to those they have a bona-fide patient-doctor relationship with [1].

It further outlines that physicians should:

- Practice shared decision making when discussing treatments with patients; physicians must avoid claims that are deceptive, intentionally false or misleading, especially making promises about questionable outcomes [1].
- Not use gag orders or disclaimers to dodge liability [1].
- Be prepared to provide evidence to back any claims about treatment benefits that are documented [1].
- Not exploit patients financially (e.g., charge excessively for treatments, recommend, provide or charge for unnecessary services or intentionally deceive patients to increase payment) [1].
- Educate patients on stem cell interventions and update them on important community resources [1].

Now that we have introduced some relevant background information and general guiding principles, in the next section, we will give an overview of the current regulatory framework as it applies to regenerative medicine in spine intervention, pain management and orthopedics.

However, before we delve into the current regulatory guidelines, we would like to make the disclaimer that no regenerative medicine products are currently approved by the FDA for orthopedic use and generally, all stem cell products require FDA approval for use [7–9]. The only stem cell products currently FDA-approved for use in the United States are umbilical cord-derived hematopoietic progenitor cells and they are only approved for use in patients with hematopoietic disorders [9]. With the above in mind, please note that stem cell products, exosome products, and other promoted regenerative medicine products including adipose tissue-derived products (e.g., stromal vascular fraction), human umbilical cord blood, Wharton's Jelly, and amniotic fluid are not FDA-approved for any orthopedic conditions, are regulated by the FDA and hence, require the FDA's approval before use in humans [9].

## Current Regulatory Framework Overview

The guidelines outlined by the FDA and FSMB should be the basis for decision making by regenerative medicine providers. While many legal challenges continue to exist in the space and are aiming to overcome these hurdles, practitioners are reminded that the governing doctrine remains with the FDA. The optimal way for providers to protect themselves is to operate within the guidelines set forth. This current perspective is designed to provide a broader understanding of the regulatory landscape and is not to be considered a guide or expert legal advice.

The FDA regulates cell-based therapies including HCT/Ps through the Center for Biologic Evaluation and Research (CBER). CBER is the FDA division that regulates and oversees the development, approval, and post-marketing surveillance of biologic products, including vaccines, blood products, gene therapies, and allergenic extracts. CBER ensures the safety, efficacy, and quality of these products for public health. This section will detail regulatory pathways and relevant legislation enforced by the FDA. The FDA acknowledges that regenerative medicine is complex and continuously evolving.

For this reason, the FDA vows to continuously reassess their application of HCT/P regulatory framework, as scientific advances emerge [10]. With that in mind, although we have the current regulatory landscape summarized below, we recommend periodically checking the FDA website for updates.

The key to compliance in the United States for any practitioner of regenerative medicine is to have a fundamental understanding and acceptance of the definitions that guide our regenerative medicine products.

Autologous use in regenerative medicine refers to the use of a patient's own cells, tissues, or fluids for therapeutic purposes. The FDA defines autologous use as "the implantation, transplantation, infusion, or transfer of human cells or tissue back into the individual from whom the cells or tissue were recovered" [11]. Autologous regenerative medicine products can fit the classification of minimally manipulated and intended for homologous use, meaning they are used for the same purpose and function as the original tissue. Also, they may be regulated as biologics or medical devices, depending on their intended use and the level of manipulation involved in their production.

Allogeneic products in regenerative medicine refer to cells, tissues, or fluids obtained from a donor and used for therapeutic purposes in another individual. Unlike autologous products, allogeneic products are not derived from the recipient's own cells and are intended for transplantation or implantation. The FDA regulates allogeneic regenerative medicine products as biologics, and they must meet standards for safety, efficacy, and quality before they can be marketed for use in patients. Allogeneic products may be minimally manipulated or highly processed, depending on the intended use and the level of modification required to meet specific therapeutic goals.

Minimal manipulation is a term used by the FDA to describe cells, tissues, or fluid products that have undergone minimal processing and have not changed the original relevant characteristics of the cells or tissues [10]. According to the FDA, products must meet the following criteria to be considered minimally manipulated:

1. The product is not combined with another article (excluding water, crystalloids, or sterilizing, preserving, or storage agents).
2. The product does not undergo a procedure that alters the relevant biological characteristics of cells or tissues, or that changes their function or systemic distribution.
3. The product is intended for homologous use only (i.e., for the same function as the original tissue) [10].

Minimally manipulated products in regenerative medicine are generally considered to be regulated as medical devices and are subject to less stringent regulatory requirements compared to highly processed regenerative medicine products. According to 21 CFR Section 1271.3 (f), there are two different definitions for minimal manipulation; one is for structural tissue and the other is for cells or non-structural tissue. Each is defined with distinctions clarified below. For structural tissue, minimal manipulation means that HCT/P processing does not change the original relevant tissue characteristics relating to the tissue's utility for reconstruction, repair, or replacement [10].

For non-structural tissue or cells, minimal manipulation means that HCT/P processing does not change the relevant biological characteristics of the cells or tissues [10]. Relevant biological characteristics (of cells or non-structural tissues) are properties of the cells or non-structural tissues in the donor that contribute to the cells' or tissue's function(s) [10]. Examples include differentiation or activation state, proliferation potential, and metabolic activity [10]. Hence, processing that changes relevant biological characteristics of cells or tissues would generally be considered more than minimal manipulation [10].

Homologous use in regenerative medicine refers to the use of cells, tissues, or fluids for the same purpose and function as the original tissue. The FDA defines homologous use, per 21 CFR 1271.3(c) as, "the repair, reconstruction, replacement, or supplementation of a recipient's cells or tissues with an HCT/P that performs the same basic function or functions in the recipient as in the donor" [10]. Homologous use is an important consideration in the regulation of regenerative medicine products, as the intended use of the product can affect the regulatory classification and requirements for the product. Products intended for homologous use are generally considered to be minimally manipulated and regulated as medical devices, while highly processed products intended for non-homologous use are generally considered to be biologics and subject to more stringent regulatory requirements.

The FDA currently regulates all allogeneic products in regenerative medicine as biologics. These products must meet standards for safety, efficacy, and quality before they can be marketed for use in patients. The FDA requires allogeneic regenerative medicine products to undergo preclinical testing, clinical trials, and a pre-approval process, which includes review of the manufacturing process, labeling, and product testing, to ensure their safety and efficacy for use in patients.

The FDA also monitors allogeneic products after they are approved and marketed, to ensure they continue to meet regulatory standards and to identify and address any safety concerns that may arise. The FDA may take regulatory action, such as issuing warnings or imposing restrictions, to protect the public health if it determines that an allogeneic product poses a significant risk to patients.

In general, the FDA's current position on allogeneic products in regenerative medicine is one of careful oversight and regulation, with the goal of ensuring their safety, efficacy, and quality for use in patients. At the time of this perspective, there is currently no FDA approved allogeneic product on the market that can be used by a practitioner in any regenerative medicine therapy.

The FTC is responsible for protecting consumers from false or misleading advertising and promoting competition in the marketplace. In the context of regenerative medicine, the FTC provides guidelines for physicians to consider when marketing their services and products to consumers.

Some of the key marketing guidelines for physicians to consider in regenerative medicine per the FTC include:

1. Truthful and non-misleading advertising: Advertisements should not make false or misleading claims about the safety, efficacy, or effectiveness of regenerative medicine products or treatments.
2. Evidence-based claims: Advertisements should be supported by reliable scientific evidence and should not overstate the benefits or downplay the risks of regenerative medicine treatments.
3. Disclosing material information: Advertisements should include all material information that consumers need to make informed decisions, such as the risks and benefits of regenerative medicine treatments.
4. Endorsements: Endorsements and testimonials should be truthful and not misleading.
5. Pricing and refund policies: Advertisements should accurately represent the cost of regenerative medicine treatments and refund policies should be clearly disclosed.

The FTC enforces these guidelines through investigations, enforcement actions, and consumer education efforts. It is important for physicians to be aware of and comply with these guidelines to avoid potential legal and reputational consequences. In 2020, a study that reviewed 896 online websites found that 96% of them made at least one misinformation regarding stem cell therapy and the mean was five per site [12].

When marketing cell therapies, most providers of unproven stem cell interventions (lacking compelling evidence) have historically been in the "stem cell tourism" (warranting travel outside of the U.S.) category; However, in recent years, more U.S. physicians have started offering stem cell and regenerative treatments and many use direct-to-consumer (DTC) marketing to promote them on their websites [1].

- Data promoting unproven therapies tend to overemphasize benefits and undermine risks, misleading patients [1].
- Physicians who engage in deceptive practices or false advertising risk violating their state's ***Medical Practice Act*** [1].

Some clinics providing non-evidence-based treatments or treatment lacking appropriate clinical rationale use "co-opted tokens of scientific legitimacy" (e.g., celebrity testimonials, affiliations, etc.) to lend credibility that may be undue, misleading consumers [1].

Stem cell treatment advertising must be objective, and physicians should be prepared to support claims with documented evidence [1]. Physicians must be accurate and not intentionally deceptive about their qualifications and competency in treatments they can offer patients [1]. Physicians offering regenerative and stem cell interventions should be mindful that state medical boards can use information/marketing from their company website as data to investigate complaints [1]. The presence of misleading marketing/information on a company website invalidates the informed consent process [1]. Physicians must be accurate and transparent about their clinical trial processes and are responsible for ensuring any clinical trial data referenced in their marketing follows accepted research standards and guidelines [1].

Some general accepted recommendations for physicians to be compliant in practicing regenerative medicine include the following:

1. Stay informed of relevant regulations: The physician should be aware of and understand the regulatory requirements for regenerative medicine products and treatments, including FDA guidance, state laws, and professional standards.
2. Adhere to medical standards of care: The physician should provide treatment that is consistent with accepted medical standards of care, using evidence-based practices and following established protocols and within their clinical scope of practice or after having obtained appropriate competencies.
3. The physician should only use regenerative medicine products that have been approved by the FDA or that meet other relevant regulatory requirements regarding their use.
4. Obtain informed consent: The physician should provide patients with accurate and complete information about the risks and benefits of regenerative medicine treatments and obtain written informed consent before providing treatment.
5. Maintain accurate medical records: The physician should maintain complete and accurate medical records for each patient, including documentation of the patient's medical history, treatment plan, and response to treatment.
6. Comply with marketing regulations: The physician should comply with the marketing guidelines provided by

the FTC, including making truthful and non-misleading advertisements and disclosing all material information to patients.

7. Engage in ongoing education and training: The physician should stay up-to-date with the latest developments in regenerative medicine and participate in ongoing education and training to maintain and improve their knowledge and skills.

## Biologics License Applications

A biologic license, also known as a biologics license application (BLA), is a regulatory clearance required by the US FDA for the marketing of biological products, including allogeneic products used in regenerative medicine [10]. The BLA is a comprehensive submission that includes information about the product's manufacturing processes, quality controls, clinical data, and other relevant information to demonstrate the product's safety, efficacy, and purity [10].

Yes, a biologic license is needed to use an allogeneic product in regenerative medicine in the United States. The FDA requires that all biological products be licensed prior to marketing, and allogeneic products used in regenerative medicine are considered biological products under FDA regulations [10]. The BLA provides assurance to healthcare providers, patients, and the public that the product has been reviewed and found to be safe, effective, and pure for its intended use.

## 21 CFR 1271 and PHS Act Pathways

21 CFR 1271.3(d) is a regulation issued by the US FDA under the CFR Title 21, Part 1271. It governs the current Good Tissue Practices (cGTP) for HCT/P, and establishments that manufacture HCT/Ps for transplantation into humans [11].

The regulation outlines the requirements for donor screening and testing, product labeling, and donor eligibility determination, among other aspects, to ensure that HCT/Ps are safe and free from communicable diseases. In the context of regenerative medicine, 21 CFR 1271.3(d) is important because it sets the minimum standards for the manufacture of HCT/Ps, including allogeneic products used in regenerative medicine. These standards ensure that the products are safe and free from communicable diseases, and that they meet appropriate quality and purity standards [11].

Compliance with 21 CFR 1271.3(d) is mandatory for HCT/P establishments that manufacture products for transplantation into humans, including those used in regenerative medicine. Failure to comply with the regulation may result in enforcement action by the FDA, including product seizures and civil or criminal penalties. According to 21 CFR 1271.3 (d), HCT/Ps are defined as articles containing or consisting of human cells or tissues that are intended for implantation, transplantation, infusion, or transfer into a human recipient [10]. Examples of HCT/Ps include, but are not limited to the following: Bone, ligament, skin, dura mater, heart valve, cornea, hematopoietic stem/progenitor cells derived from peripheral and cord blood, manipulated autologous chondrocytes, epithelial cells on a synthetic matrix, and semen or other reproductive tissue [10].

Practitioners should also have a thorough understanding of sections 361 and 351 of the Public Health Service Act. This Act regulates HCT/Ps. If a product meets the definition of an HCT/P in 21 CFR 1271.3(d) but does not meet the same surgical procedure exception in 21 CFR 121.15(b), either section 351 or 361 of the PHS will apply. If it meets the four criteria outlined in 21 CFR 1271.10 (a), then the HCT/P is regulated solely under PHS section 361 and 21 CFR Part 1271. However, if an HCT/P does not meet the 21 CFR 1271.10(a) criteria, it is subject to regulation under PHS section 351 and other applicable regulations for premarket approval [10].

Section 351 is a stricter pathway requiring clinical trials under an IND application and an FDA license. Section 361 is only focused on preventing communicable disease [1]. When evaluating a product in the market it is important for the practitioner to have an understanding of these two pathways before making a decision on whether or not it is appropriate for use in the treatment of their patient. Practitioners should not rely on industry alone to make this determination. Instead, it is recommended to have a thorough understanding of these pathways prior to clinical decision making.

A cell therapy may follow a 361 HCT/P Pathway if cells are minimally manipulated, intended for homologous use, not combined with other materials, and only rely on metabolic activity of cells if autologous. In this pathway, a product only has to comply with manufacturing compliance to go to market but note that no claims of efficacy or potency can be made. A Cell Therapy should follow a 351 HCT/P pathway if it does not meet the criteria stipulated below for 361 products [13]. If an HCT/P does not meet the criteria for the 361 pathway, and is regulated as a biological product, it is subject to section 351 of the PHS Act along with the FD&C Act and hence, subject to premarket approval [10]. Also note, that in order to legally market a biologic product, a biologics license is required, which is issued by the FDA only after determining that the establishment(s) and biological products meet the applicable requirements to ensure the continued safety, purity, and potency of the products [10].

During the FDA discretionary period, which is now expired, it is the opinion of this author that many companies produce allogeneic products to market through the 361

HCT/P Pathway inappropriately. Practitioners unaware of these compliance benchmarks began using these products not fully understanding that the products should have undergone preclinical and clinical testing prior to commercial sale. This pathway permits potency claims to be made based on clinical testing outcomes. Many products that continue to make these claims and available on the market are doing so without approval following the 351 pathway. It is important to note that this pathway typically takes 10–15 years [13].

## Expedited Tracks

In an effort to facilitate access to innovative therapies, practitioners should be aware of FDA expedited tracks and implications of such provisions. The *Expanded Access to Investigational Drugs for Treatment Use* program allows treatment if tested in a clinical trial and is subject to FDA approval [1]. Other expedited tracks include **"Right to Try" state legislation** (and similar federal laws/proposed legislation) which allow terminally ill patients to try treatments that have passed phase-1 trials without seeking FDA approval) [1] and the ***twenty-first Century Cures Act*** (Federal legislative efforts to accelerate patient access to treatment and promotes innovation n medical products and treatments [1]; it also includes the **Regenerative Medicine Advanced Therapy (RMAT),** which is an FDA program and designation that facilitates and expedites review of regenerative medicine therapies intended to treat, modify, reverse or cure serious or life-threatening disease) [1, 14].

## Platelet-Rich Plasma

Platelet rich plasma (PRP), blood taken from an individual and given back to the same individual as (platelet rich plasma) is not an HCT/P under 21 CFR Part 1271 because it is a blood product. Accordingly, FDA does not apply the criteria in 21 CFR 1271.10(a) to PRP [10].

Moreover, the FDA further stated that because PRP is a blood product and not an HCT/P, that the same surgical procedure exception outlined in 21 CFR 1271.15(b) does not apply to PRP either [11]. Hence, the FDA states that PRP is beyond the scope of their published minimal manipulation/homologous use and same surgical exception guidance documents [10, 11].

## Bone Marrow

"Bone marrow is a source of hematopoietic stem/progenitor cells. Minimally manipulated bone marrow for homologous use and not combined with another article (with certain exceptions), is not considered an HCT/P (21 CFR 1271.3(d) [4]). However, bone marrow that is more than minimally manipulated, intended by the manufacturer for a non-homologous use, or combined with another article with limited exceptions, meets the definition of an HCT/P and is subject to the regulations in 21 CFR Part 1271" [10].

## Adipose Tissue

The FDA previously issued guidance on adipose tissue use in regenerative medicine in December of 2014. However, since releasing more current documents, the FDA has withdrawn the 2014 Adipose draft guidance and declared the adipose guidelines discussed in more current documents (*Same Surgical Procedure Exception Under 21 CFR 1271.15(b): Questions and Answers Regarding the Scope of the Exception* (November 2017) and the *Regulatory Considerations For Human Cells, Tissues and Cellular and Tissue-Based Products: Minimal Manipulation and Homologous Use for HCT/Ps)* (July 2020) to serve as official guidance on adipose tissue. There remains a great deal of controversy surrounding adipose tissue use in regenerative medicine. According to the guidance, adipose tissue-derived stem/stromal cells (ASCs) are considered "biological products" and are subject to regulatory oversight [5].

The guidance recommends that companies developing products using ASCs follow current good manufacturing practice (cGMP) requirements and submit a pre-market approval application (PMA) to the FDA if they are marketing the product for uses beyond minimally manipulated and homologous uses (i.e., treatment of the same condition or tissue from which the cells were derived).

Essentially, the FDA requires a BLA for stromal vascular fraction (SVF), a source of ASCs which is generally considered to be more than minimally manipulated [10]. This is necessary for the marketing and commercial distribution of ASC products for non-homologous uses.

However, the key debate and where there has been a gray area, is what constitutes minimal manipulation and homologous use in pain and orthopedic regenerative medicine.

The determination of whether a use is considered homologous can be complex and is dependent on a variety of factors, including the intended use, the level of manipulation, and the route of administration, among others [10].

However, the FDA has tried to clarify some of this confusion in recent years (and also since initially putting out adipose guidelines has published a Q&A-style same surgical exception document in addition to the 2020 homologous use guideline document that have effectively replaced the original adipose guideline publication and supersedes it) [10, 11]. For example, they defined minimal manipulation and homologous use for structural and non-structural tissues (in addition to defining structural and non-structural tissue) and also answered inquiries regarding homologous use and minimal manipulation in different situations.

For example, in response to the question, "I isolate cells from structural tissue to produce a cellular therapy product. What definition of minimal manipulation would apply?" [10] The FDA gave the following example pertaining to adipose-derived stromal/stem cells in example 14–1: "Original relevant characteristics of adipose tissue relating to its utility to provide cushioning and support generally include its bulk and lipid storage capacity. A manufacturer recovers adipose tissue by tumescent liposuction and processes (e.g., enzymatically digests, mechanically disrupts, etc.) the adipose tissue to isolate cellular components (with or without subsequent cell culture or expansion), commonly referred to as stromal vascular fraction, which is considered a potential source of adipose-derived stromal/stem cells. The definition of minimal manipulation for structural tissue applies. In this example, the HCT/P generally is considered more than minimally manipulated because the processing breaks down and eliminates the adipocytes and the surrounding structural components that provide cushioning and support, thereby altering the original relevant characteristics of the HCT/P relating to its utility for reconstruction, repair, or replacement" [10].

In the context of regenerative medicine, mechanical separation of adipose tissue is generally considered to be minimal manipulation, while enzymatic separation is considered a form of more than minimal manipulation [10].

Minimal manipulation refers to procedures that preserve the structural and functional integrity of the cells and tissues being used for regenerative medicine purposes. Mechanical separation involves physically separating the adipose tissue from other tissues, which can be done without affecting the cells within the tissue.

On the other hand, enzymatic separation involves using enzymes to dissolve the connective tissue and other structures within the adipose tissue, which can alter the cells and their functions [10].

This type of manipulation is considered more substantial and is often used to obtain a higher yield of adipose-derived stem cells, but it also means the cells are subjected to more processing before they can be used for therapeutic purposes and therefore require a BLA license.

When considering products for PRP and bone marrow, practitioners need to verify that the products they are selecting are at least 510(k) compliant. A 510(k) compliant device in regenerative medicine is a medical device that has been cleared by the US FDA under the 510(k) premarket notification process. This process requires that the manufacturer demonstrate that the device is substantially equivalent to a legally marketed device, or "predicate device," that is already on the market. The 510(k) clearance process allows manufacturers to bring new medical devices to market without conducting extensive clinical trials, provided that they can demonstrate that their device is substantially equivalent to a predicate device [15].

The IND pathway is a regulatory process that allows the testing of experimental drugs in human subjects. In the field of regenerative medicine, IND applications are typically required for development of new cellular therapies that are not yet approved by the FDA. This is frequently the case for products that are allogeneic, not minimally manipulated or intended for homologous use [16, 17]. The IND process typically involves the submission of preclinical data, including studies in animals, to the FDA to demonstrate the safety and efficacy of the experimental therapy. If the FDA determines that the therapy is safe for initial testing in human subjects, a clinical trial may be initiated. Another key component to an IND is the role of The Institutional Review Board (IRB). The IRB's role is to review and approve the proposed clinical trial plan for an experimental drug before it can be tested in human subjects. It is responsible for ensuring that the design is scientifically sound, and that the study participants are protected along with the study complying with all relevant regulations and ethical standards [16, 17].

An IRB approval is not the same as an IND approval. Often, biologics have an IRB approval however that does not mean it can be widely used in practice by practitioners. The IND process also includes ongoing reporting to the FDA of clinical trial results and any adverse events that occur during the trial. Once the clinical trial is complete, the sponsor of the IND application may submit a New Drug Application (NDA) to the FDA seeking approval to market the therapy. The NDA must include the results of the clinical trial, as well as information on the manufacture and labeling of the therapy [16, 17].

In regenerative medicine, for a practitioner to use an experimental drug on a patient, they should be named as a principal investigator of the clinical trial. A drug being in an IND does not allow it to be marketed to a practitioner for use if they are not named as a principal investigator. This is often an area where practitioners believe they are permitted to use a biological product because it is in an IND, when in fact it is actually only permitted to be used by actual investigators. This is not often reported to the practitioner when biologic products are marketed to them.

Figure 33.1 shows flow chart illustrating the application of the criteria in 21 CFR 1271.15(b) and 1271.10(a).

### Key Points

Now that we have covered the basics of current regenerative medicine regulation guidelines and legislation, in this section, we'll review some key takeaway points to remember to stay compliant, avoid disciplinary action and maintain good practice.

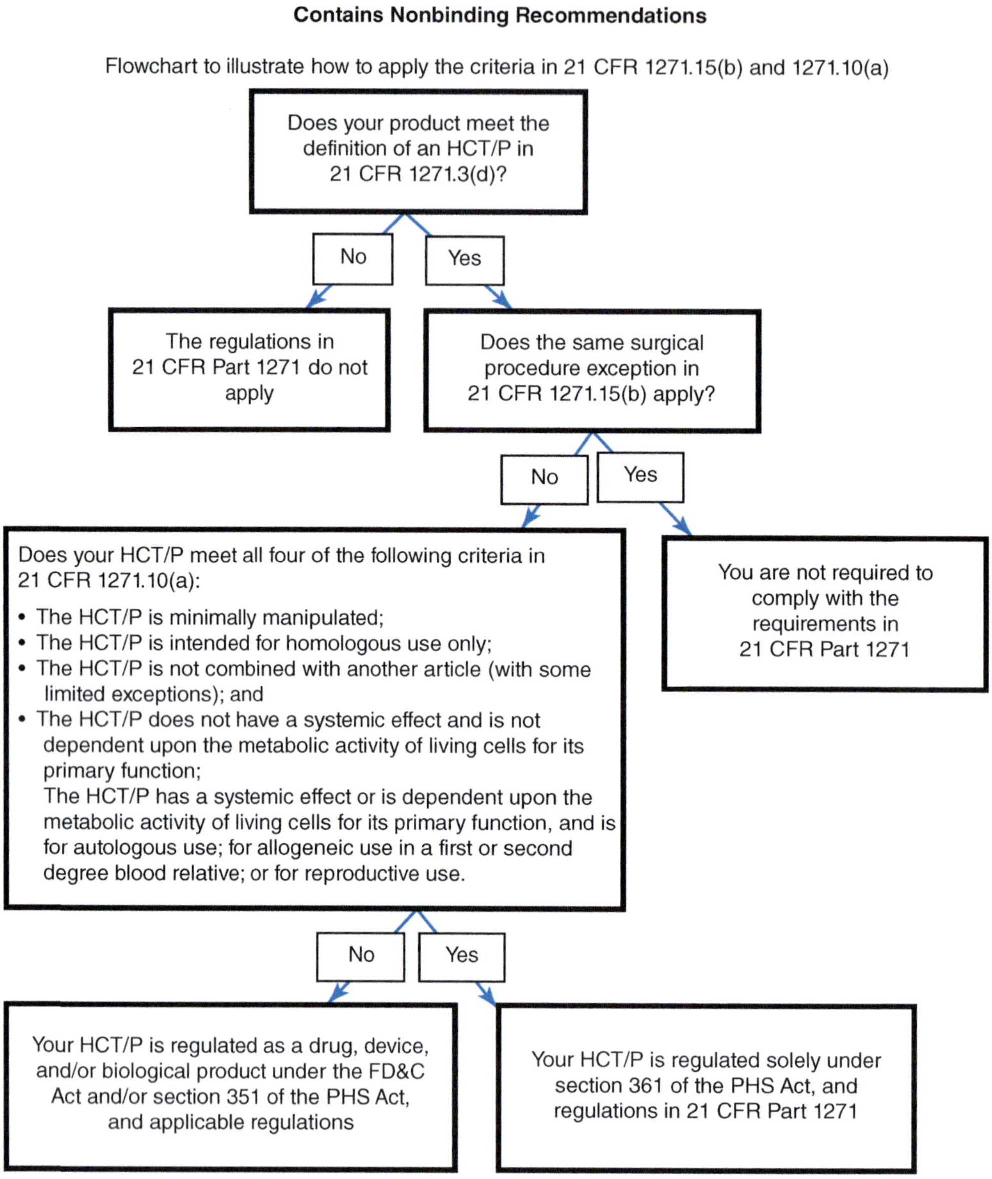


**Fig. 33.1** Flowchart to illustrate how to apply the criteria in 21 CFR 1271.15(b) and 1271.10(a) from the FDA

- Autologous SVF adipose is not exempt from HCT/P regulation based on the same surgical exception [11, 18] under 21 CFR 1271.15 (b) because the adipose tissue is processed beyond minor handling and stem cells are extracted [9].
- Amniotic fluid, Wharton's Jelly, Umbilical cord blood, exosomes, and Adipose SVF are not FDA approved (3,6,18,24). Hence, in order to use them in humans, prior to use, you MUST have an approved biologics license or be a listed investigator in an FDA approved study (e.g., IND or INE) [3, 14, 18]. Otherwise, do not use them.
- Remember to adhere to FTC regulation (i.e., the ***Federal Trade Commission Act***) as non-compliant advertising puts you at risk. This means avoiding any unfair or deceptive acts/practices [18, 19].
  – Anyone can report you for false advertising.
- Also keep in mind that the ***Lanham Act*** allows competitors to file a federal claim (under section 43 (a)) if their competitor has advertising that is false or misleading (this act only applies to competitors, not consumers) [18, 20].
  – Hence, it is important to be specific and have an abundance of supporting data if you make claims of any specific outcomes.
- Be careful using patient testimonials as advertisement and remember that you own the content in your patient testimonials [18]. You may even consider using a disclaimer in advertising/testimonials (i.e., "…results and claims are not typical or meant to represent your results…") [18].
- Be sure to use clear language/layman terminology and avoid jargon when explaining treatments to patients. This is an important part of the informed consent process. Also

note that any misleading language or advertising voids the informed consent.

- Additional things to avoid in regenerative medicine advertising:

  - Advertising treatments with unapproved drugs [18].
  - Advertising PRP, BMC, or microfat treatments since there are not any FDA-approved orthopedic regenerative therapies [18].
  - Advertising better outcomes than competitors or alternative treatments [18].
  - Advertising board certifications from unrecognized entities [18].
  - Using patient testimonials that violate any of the above [18].

In closing, it is important to note that the regulatory landscape is constantly evolving and new guidance and regulations may be issued by the FDA at any time. Therefore, we recommend following FDA updates and consulting with a qualified healthcare professional or regulatory expert for the most up-to-date information on the use of any biologic product and its use in the treatment of any condition in pain medicine.

**Acknowledgements** This book chapter is modified and updated from a previous book chapter, "Regulatory aspects of regenerative therapy and stem cell therapy: Current perspectives" by Richard M. Rosenthal, MD, and Elliot Lander, MD, in *Essentials of Regenerative Medicine in Interventional Pain Management* published by ASIPP Publishing.

## References

1. Federation of State Medical Boards. Report and recommendations of the workgroup to study regenerative and stem cell therapy practices. 2018; https://www.fsmb.org/siteassets/advocacy/policies/fsmb-stem-cell-workgroup-report.pdf. Accessed 20 Apr 2023.
2. Fang WH, Vangsness CT Jr. Governmental regulations and increasing Food and Drug Administration oversight of regenerative medicine products: what's new in 2020? Arthroscopy. 2020;36:2765–70.
3. U.S. Food and Drug Administration. Important patient and consumer information about regenerative medicine therapies. 2021; https://www.fda.gov/vaccines-blood-biologics/consumers-biologics/important-patient-and-consumer-information-about-regenerative-medicine-therapies. Accessed 20 Apr 2023.
4. Cook M, Richey A, Brafman DA, et al. Weighing up the evidence used by direct-to-consumer stem cell businesses. Stem Cell Rep. 2021;16:2852–60.
5. U.S. Food and Drug Administration. What does FDA regulate? 2022; https://www.fda.gov/about-fda/fda-basics/what-does-fda-regulate. Accessed 20 Apr 2023.
6. Federation of State Medical Boards. https://www.fsmb.org/about-fsmb/. Accessed 20 Apr 2023.
7. U.S. Food and Drug Administration. Public safety notification on exosome products. 2019; https://www.fda.gov/vaccines-blood-biologics/safety-availability-biologics/public-safety-notification-exosome-products. Accessed 20 Apr 2023.
8. U.S. Food and Drug Administration. FDA warns about stem cell therapies. 2019; https://www.fda.gov/consumers/consumer-updates/fda-warns-about-stem-cell-therapies. Accessed 20 Apr 2023.
9. U.S. Food and Drug Administration. Consumer alert on regenrative medicine products including stem cells and exosomes. 2020; https://www.fda.gov/vaccines-blood-biologics/consumers-biologics/consumer-alert-regenerative-medicine-products-including-stem-cells-and-exosomes. Accessed 20 Apr 2023.
10. U.S. Food and Drug Administration. Regulatory considerations for human cells, tissues, and cellular and tissue-based products: minimal manipulation and homologous use. Guidance for Industry and Food and Drug Administration Staff; 2020; https://www.fda.gov/regulatory-information/search-fda-guidance-documents/regulatory-considerations-human-cells-tissues-and-cellular-and-tissue-based-products-minimal. Accessed 20 Apr 2023.
11. U.S. Food and Drug Administration. Same surgical procedure exception under 21 CFR 1271.15(b): questions and answers regarding the scope of the exception. 2017; https://www.fda.gov/regulatory-information/search-fda-guidance-documents/same-surgical-procedure-exception-under-21-cfr-127115b-quetions-and-answers-regarding-scope.com. Accessed 20 Apr 2023.
12. Kingery MT, Schoof L, Strauss EJ, et al. Online direct-to-consumer advertising of stem cell therapy for musculoskeletal injury and disease: misinformation and violation of ethical and legal advertising parameters. J Bone Joint Surg Am. 2020;102:2–9.
13. Lverman LI, Flanagan F, Rodriguez-Granrose D, et al. Identifying and managing sources of variability in cell therapy manufacturing and clinical trials. Regen Eng Transl Med. 2019;5:354–61.
14. Vaggelas A, Seimetz D. Expediting drug development: FDA's new regenerative medicine advanced therapy designation. Ther Innov Regul Sci. 2019;53:364–73.
15. U.S. Food and Drug Administration. Premarket notification (510(k)). 2022; https://www.fda.gov/medical-devices/premarket-submissions/premarket-notification-510k. Accessed 20 Apr 2023.
16. U.S. Food and Drug Administration. How drugs are developed and approved. 2022; https://www.fda.gov/drugs/how-drugs-are-developed-and-approved/investigational-new-drug-ind-proces. Accessed 20 Apr 2023.
17. U.S. Food and Drug Administration. Regenerative medicine advanced therapy designation. 2023; https://www.fda.gov/medical-devices/regenerative-medicine-advanced-therapies/regenerative-medicine-advanced-therapy-rmat-designation. Accessed 20 Apr 2023.
18. Buford D. Regenerative medicine: practical pearls and pitfalls. Lecture given at TOBI Orthobiologics Symposium; 2022.
19. Federal Trade Commission Act. https://www.ftc.gov/legal-library/browse/statutes/federal-trade-commission-act. Accessed 20 Apr 2023.
20. U.S. Trademark Law; https://www.uspto.gov/sites/default/files/trademarks/law/Trademark_Statutes.pdf. Accessed 20 Apr 2023.

# Ethical and Policy Issues in Practices of Regenerative Medicine for the Treatment of Chronic Pain

34

John R. Shook and James Giordano

## Introduction

There is building evidence to support the use of select techniques of 'regenerative medicine' in the treatment of certain forms of chronic pain and musculoskeletal disorders. For example, musculoskeletal and spinal disorders can be mitigated through the use of injected biologics, in the form of cells, components of cells, extracellular fluids such as blood plasma, and other kinds of protein molecules such as collagen and growth factors. These materials can fortify physiological processes of repair and regeneration (i.e., to reduce inflammation, prevent further tissue deterioration, regrow tissues, and restore structural and functional integrity). Applied alone or in conjunction with other modes of treatment, such approaches offer promise for relieving both signs/symptoms as well as causes of chronic pain.

Biologics may be manipulated and modified before they are introduced into the patient. Technological modification is often required to (1) obtain biologics in quantities sufficient for effective treatment, (2) prepare the biologics for regenerative action/effect at the target site, and (3) ensure that these biologics do not incur deleterious side effects. The regenerative use of biologics is frequently described as a "natural" form of therapy, since it is unlike pharmacologics, surgical procedures, or transplants from nonhuman animals (i.e., xenotransplants). Such naturality should not obscure the fact that highly advanced technologies are involved in the development, manipulation and cultivation of regenerative biologics. This is true for stem cells, as well. Attaching the label of "natural" to a stem cell can distract attention from a number of relevant technical, medical and ethical issues and concerns, which must be balanced against the medical and social urgency to provide safe and effective relief from chronic pain. This urgency is all the more heightened as the search for alternatives and solutions to opioid abuse and illicit drug epidemic become an increasing scientific, clinical, social, and legal priority.

Most commonly, studies addressing applications of regenerative approaches to chronic pain focus on joint degeneration, intervertebral disc degeneration, and neuropathic conditions [1]. A majority of these studies target pain incurred by osteoarthroses of the major joints (knee, hip, and shoulder), and treatments for these disorders and resultant pain most frequently involves the use of platelet-rich plasma (PRP). In this technique, plasma of a patient's own blood is separated (from its red blood cells) to afford an injectate concentrated with leuokocytes, platelets, and platelet-secreted growth factors. PRP has been deemed as a safe treatment for particular forms of chronic pain and is approved as an "off label" application by the United States Food and Drug Administration (FDA), 21 CFR 1271 [2]. Also, there is growing evidence to support PRP's superiority over corticosteroids for mild to moderate osteoarthritic pain, and the effectiveness of PRP treatment in conjunction with physical rehabilitation. Although large studies of PRP alleviation of joint pain are lacking, surveying small randomized controlled trials demonstrates that significant pain relief can be achieved [3, 4].

## Stem Cells

Another common method of regenerative medicine involves injection of mesenchymal stem cells (MSC) that are taken from the patient's body. Like PRP, stem cells stimulate physiological mechanisms of recovery and repair, yet further contribute structural components (i.e., cells) as well. The use of

J. R. Shook
Department of Philosophy and Science Education, University of Buffalo, Buffalo, NY, USA
e-mail: jshook@pragmatism.org

J. Giordano (✉)
Departments of Neurology and Biochemistry, Chief of Neuroethics Studies Program, Pellegrino Center for Clinical Bioethics, Georgetown University Medical Center, Washington, DC, USA
e-mail: jgiordano@neurobioethics.org





A. Navani et al. (eds.), *Essentials of Regenerative Medicine in Interventional Pain Management*,
https://doi.org/10.1007/978-3-031-50357-3_34

MSC has been shown to produce clinically significant alleviation of pain of arthro- and/or chondro-degenerative conditions [4–7]. However, MSC therapy incurs issues as focal to the use of stem cells. MSCs are "multipotent", and therefore have a limited ability to differentiate into a constrained variety of connective tissues (such as cartilage or tendon). They are derived from bone marrow; similar types of stem cells are derived from adipose tissue. Stromal vascular fraction (SVF), which contains stem cells, and adult pluripotent stem cells (APSC) are among the more common methods of regenerative medicine used for treating chronic pain conditions. Studies employing these methods, while relatively few in number, have provided evidence of moderate control of arthro-degenerative pain [8]. Given that the source is the patient's own body, such cells are autologous in nature. Because autologous cells characteristically do not evoke an immune response (and rejection) by the host, such cells are generally preferred to allogeneic stem cells that have been obtained from another human source.

Autologous stem cells are also regarded as "ethically preferable" in that they are "adult" cells, as opposed to stem cells from an allogeneic embryonic source, which although could be derived from umbilical cord tissue and blood, and/or amniotic fluid, can also be taken directly from embryonic tissues that are obtained upon termination of pregnancy. Ethical issues relevant to, and reflective of differing cultural and religious perspectives regarding the moral status of the embryo have fostered tensions toward certain governmental policies and legal rulings in America and Europe [9–11]. Research involving embryonic stem cells (ESCs) is regulated with special restrictions in the United States, UK, and several other countries, due in large part to these moral sensitivities [12].

While there is no US federal law criminalizing research on stem cells or embryos, there are some state laws prohibiting such research due to anti-abortion sentiments. To meet National Institutes of Health (NIH) ethics, and thus to be considered for subsequent FDA approval, ESC research must use tissues that have already been legally created and destroyed. Most embryos used for federally- or private foundation-funded stem cell research are obtained from fertility facilities. Embryos are created and stored for in vitro fertilization (IVF) procedures, but many are never used, so surplus embryos that would ordinarily be destroyed can instead be donated for research purposes with the consent of the fertility patients. In the United States, much of Europe, and some Asian countries, governmental funds cannot be used to subsidize research involving embryos beyond the early stages of development [13]. This limitation does not affect the typical harvesting of stem cells from embryos during the 4–6-day period of the blastocyst stage but does inhibit investigations of cellular development and transformation during later stages of embryonic growth.

The use of ESCs stands as an ethical demarcation line. On one side, the source of these cells makes human ESCs morally tainted for those who believe that human life begins at conception. Nevertheless, since IVF will remain legal, and the disposal of surplus embryos will likewise remain legal and morally tolerable (at least to some; with emphatic note of opposing religious views), research on embryos destined for destruction appears to remain politically palatable. On the other side, only destroyable IVF-derived embryos seem to be morally legitimate sources in the contemporary political context. Such ethical debates—and perspectival dissonance—have both constrained the value and experimental use of ESCs, and at the same time, prompted research to explore methods of increasing the potency of APSCs. Attaining the pluripotent characteristics of ESCs by other means that avoid the destruction of embryos has proven to be difficult, either because of unpromising results, or due to impediments in governmental approval and support (in the United States and Europe). Yet, such challenges have also served to compel researchers to pursue alternative sources and procedures of stem cell acquisition. Perhaps most significant in this regard has been development of technologies and methods to genetically re-program autologous cells, so as to enable cultivation of directed cell lines of specific tissues (e.g., neurons, glia, etc.) in accord with Good Manufacturing Practice (GMP) standards. While reprogrammed autologous cell lines are not without current—and potential –problem(s), these methods are iteratively improving, with increasing specificity, selectivity and efficiency as demonstrable hallmarks of enhanced scientific, engineering and manufacturing capability (NB: a complete discussion of these methods and developments in the field is beyond the scope of the present chapter, for overview see [14]).

## Ethical Issues

While it is important to acknowledge and regard moral concerns about ESCs, simply avoiding their use does not fully abate both general and more specific ethical and legal issues evoked by the use of stem cells; a listing of which is provided in Table 34.1 [15–17].

These can be categorized as being: (1) inherent to scientific/technical aspects, and (2) generated by social factors, circumstances, and concerns [16], and are not mutually exclusive, but rather are interactive (Fig. 34.1). Issues inherent to science and technology are focal to safety, efficacy, clinical effectiveness, and cost. Safety is also a prime social concern and consideration as regards veridicality of claims (about relative benefits, burdens and risks) that are important to sustain the probity of informed consent. As well, safety, efficacy/effectiveness and costs contribute to the viability of any scientific, technological and/or medical approach in practice (as a social good), which in turn both reflects and contributes to the relative value of the approach(es) in question. Costs and value also contribute to and reflect need, demand, and provision (i.e., which patients will receive these state-of-the-art technologies, and the methods with which such questions about the allocation of resources can/should be addressed and resolved) [18]. And while justice characteristically refers to ethical integrity in the of provision of goods, the concept of "justice-as-lawfulness", and questions of the ways that legal claims could, and should be handled when dealing with cutting edge technologies (for which a casuistic approach might not be viable) must also be acknowledged [17, 19, 20].

**Table 34.1** Domains of neuroethico-legal issues (adapted from Giordano et al. [15–17])

| |
|---|
| *Science and/or technology inherent* |
| • Uncertainties of new, or novel applications of science and technology[a,b,c] |
| – Lack of prior application/durable use-in-practice/exemplars |
| • Unknown mechanisms of action and/or interaction(s)[a,b] |
| • Known burden on risk factors/effects[a,b,c] |
| • Unknown and/or unanticipated effects (inclusive of side and/or adverse effects)[a,b] |
| – Differing efficacy and effectiveness |
| • Costs of development, manufacture, and/or distribution |
| *Societally derived* |
| • Misperception/misinterpretation of capabilities, limits, and/or meaning/value[a,b,c] |
| • Inappropriate use/misuse in practice[a,b,c] |
| – Can lead to over- or under-use |
| • Economic inequalities and/or cost manipulation[b,c] |
| – Can lead to misuse |
| • Provision/distribution of resources[b,c] |
| – Based upon economics and/or political direction (and economic factors, as above) |
| • Cultural difference(s) in needs, values, use, and/or access[b,c] |
| – Inclusive of philosophical views and ethical perspectives and practices; and political influence |
| • Vulnerability of particular individuals and/or populations[a,b,c] |
| • Legal liabilities[a,b,c] |

[a] Influences/affects or affected by safety
[b] Influences/affects viability of use-in-practice
[c] Influences/affects or affected by costs

To be sure, there are persistent concerns about known risks, and possible unknown risks (intermediate and/or longer-term) effects of stem cells. Such risks include: (1) local tissue and/or systemic responses to implanted tissues; (2) mutagenicity; (3) aberrant alteration of the local geno- and phenotype—and resultant function—of the nervous system, which could incur induction of (perhaps previously unknown) mechanisms of non-local neural network activity [17]. And while autologous ASPCs may reduce if not negate risk of immunological reactions, these cells must also be monitored for consistent differentiation, stability and tumorigenicity [21, 22] for overview of intersecting technical and ethical concerns, see also [23]. Ongoing research is focused upon methods to assure full pluripotency, and to concomitantly avert such adverse effects remains a focus of ongoing research [24–26].

Given that stem cell-based therapeutics remain experimental, both the acquisition and use of stem cells of every sort should be governed by ethical standards that protect the rights and interests of research participants. Obtaining fully informed consent of autologous stem cell donors/recipients should detail the experimental plan, as well as the type of stem cells to be used, the source of those cells, the intended use of those cells, the probability (and/or likelihood) of benefits and risks, and mechanisms for, and assurance of continuity of clinical care [19, 20, 27]. Donors of tissues for use

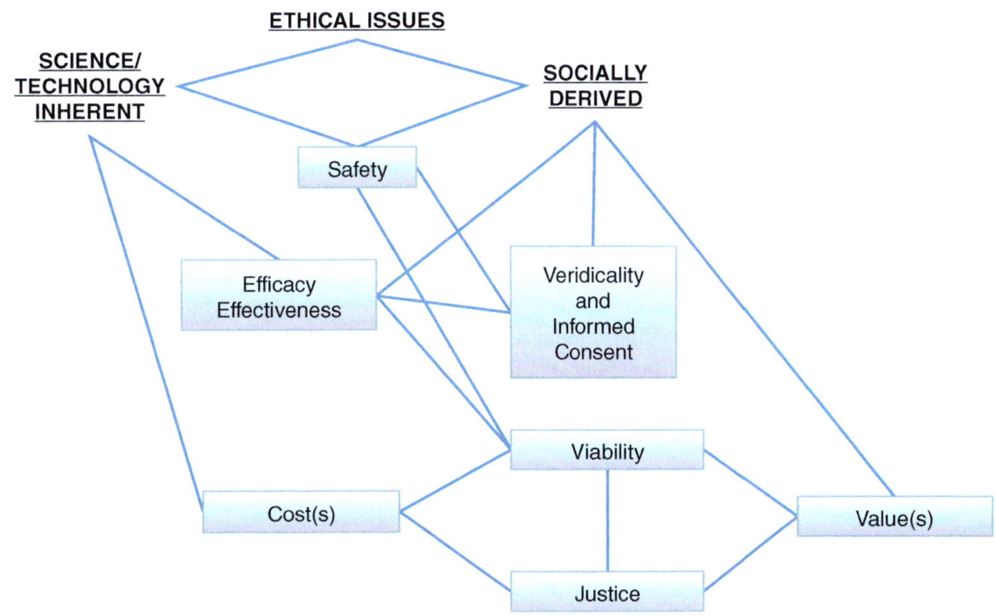


**Fig. 34.1** Interactive domains of ethical issues as inherent to scientific/technical aspects, and fostered by social factors, circumstances, and concerns, as presented in Table 34.1

in others (e.g., embryo donors) should be informed about whether cells are destined for a specific research or clinical project, or for indefinite use as a stem cell line. Some ethical questions remain unresolved, such as whether donors should be asked for unrestricted future use of cells in advance (or be subject to re-consent rules), and whether donors could receive some form of financial compensation [28, 29].

## Policy Issues

The transition from preclinical laboratory studies to "first-in-human" clinical trials must satisfy requisite assessments of preclinical benchmarks for the functional integrity, stability and purity of biologics, and reasonable expectations about effectiveness in vivo based on animal models. In addition, there are particular difficulties with testing the safety and efficacy of stem cells that are somewhat unlike the rigors typical for drug testing. For example, in the United States, navigating the three phases for FDA approval toward marketable stem cell treatments generates particular challenges where stem cells are concerned [30, 31]. The FDA has not approved SVF for marketing and consumer use, asserting that SVF and other stem cell procedures require premarket review because stem cell transfer involves "significant manufacturing", and stem cells are not performing "the same basic functions" (as the inherent biological cells/tissues) [32]. Moreover, while the FDA can expedite review of experimental stem cell therapies via the Regenerative Medicine Advanced Therapy (RMAT) designation, to date no stem cell therapy has yet received FDA approval through that route. At this writing, the FDA has only approved one widely used stem cell treatment, for bone marrow transplantation.

Any and all research and clinical trials of MSC, SVF, APSC, and ESC require advance approval to reach Phase I, II, or III study because the biological structure and/or function of those tissues is modified before application; thus such research falls under the same regulations as biologic drugs [32]. This "modification" is understood by the FDA to include the multiplication of stem cells cultured in vitro, even if the cells are from the patient and retain their structure and multipotency while multiplying. The FDA has therefore strictly enforced the regulation of autologous (and allogeneic) stem cell research in the same manner as manufactured drugs [32, 33]. Certain states, however, such as California, have liberalized and funded stem cell research [34], and although many European countries have roughly equivalent standards, less restrictive rules about the development and use of autologous and allogeneic stem cell therapies are typical in a number of countries, including Australia, India, and China [35, 36].

Phase III trials of stem cell therapies for pain have been rare, although a few have been completed or are currently underway. An appreciable number of randomized controlled studies have not accompanied even the most common use of autologous adipose SVF for orthopedic conditions; case reports and small cohort studies that primarily assess safety and feasibility comprise the preponderance of research. It is important to note that even early stages trials are susceptible to unintentionally or intentionally conveying unreasonable therapeutic expectations [37]. Despite lack of FDA approval or many phase III studies, over three hundred business enterprises are developing or offering stem cell interventions (almost all MSC/SVF) in the United States alone. Their advertisements typically suggest that authentic medical treatments for diseases, degenerative conditions, and pain relief are being provided [38, 39].

## Current and Future Vistas

Like many other domains of neurotechnological development, stem cell research, its translation toward clinical application(s), and markets of engagement are becoming ever more international enterprises [40, 41]. Indeed, a recent study of websites of over four hundred clinics located around the world reveals how typical marketing tends to pronounce far-reaching claims of successful treatments for a host of health problems, while providing fewer specifics or verifiable facts about the origin and quality of stem cells, academic legitimacy or certified legality [42]. There are legitimate concerns that fundamental expectations for safety and efficacy are not being sufficiently satisfied in this emerging arena of stem cell marketing. Scrutiny of, and consternation about the unethicality or illegality of stem cell products will only intensify when stem cells developed/produced in and by less-regulated nations enter the market with increasing regularity and numbers. The ethical failures accompanying this growing international business of "stem cell tourism" are not just mirages due to failures to appreciate cultural differences, although it is true that the economics of medicine is hardly the same around the world [43], and the values, needs and ethics that guide the economics, research and therapeutics of medicine in various countries and cultures may differ that well.

Appreciating such differences does not imply that medical experimentation and care should descend into scientific or moral relativism; nor should the ethics that guide such efforts assume an imperialist stance [44, 45]. In this light, we have proposed that both a cosmopolitan palette, and more particular communitarian constructs may be of value in establishing an internationally relevant neuroethics, at very least as a preliminary, tentative concept from which to advance both discourse and practice [46]. Important to any such effort are issues of (in)equality, (in)equities and economic and political power dynamics of allocation, provision and sustainability of the resources, goods and resources of biomedicine that each and all must be recognized, acknowledged and addressed [17, 47, 48]. Stem cell patients anywhere have the right to be well-informed about treatment options, benefits, and associated risks, and clinicians have a responsibility to protect and benefit patients by promulgating professional standards for stem cell therapeutics at both domestic and international levels [49]. Guidance provided by the International Society for Stem Cell Research (ISSCR) could be instrumental in this regard [50–52].

But such guidance cannot and should not be static, nor should it ascribe to a "wait and see" approach. Any efforts to shape brain research and its uses must apprehend and the accelerating pace and widening scope of international endeavors and (public, professional, commercial and political) interest in neuroscience and technology. The defined and intended "good" of such enterprises (i.e., to reduce the burden of disease and illness, inclusive of chronic pain) is evident; yet none the less obvious are the ethico-legal issues that such novel science and technological means generate within the social sphere. Furthermore, ethical deliberation and discourse must inform policy, which tends to be formulated somewhat more slowly, and in some cases may be more reactive than proactive [27, 41, 53]. As we have stated before [17, 19, 27, 44–46, 54–56], and reiterate here, addressing such issues need not impede neuroscientific and neurotechnological progress, but instead should serve more clearly elucidate: (1) those ways that applications of science and its tools could affect particular patients—and society; (2) the utility of current ethical systems, guidelines and policies; and (3) what types of new ethical and policy formulations might be needed to best direct research and use of neuroscience and technology.

**Key Points**

1. Regenerative medicine techniques, particularly the use of injected biologics like platelet-rich plasma (PRP), show promise in the treatment of chronic pain and musculoskeletal disorders. These techniques aim to reduce inflammation, prevent tissue deterioration, promote tissue repair and regeneration, and restore structural and functional integrity.
2. While regenerative biologics are often described as "natural," it is crucial to recognize that highly advanced technologies are involved in their development, manipulation, and cultivation. Attaching the label of "natural" to these therapies can overshadow important technical, medical, and ethical considerations that must be addressed.
3. Regenerative medicine techniques commonly involve the injection of mesenchymal stem cells (MSC) derived from the patient's own body. These stem cells contribute both physiological mechanisms of recovery and repair, as well as structural components in the form of cells.
4. The use of autologous stem cells, which are derived from the patient's own body, is generally preferred over allogeneic stem cells sourced from another individual. Autologous cells do not typically elicit an immune response or rejection by the host, making them more favorable.
5. autologous stem cells are considered preferable because they are "adult" cells, avoiding the ethical concerns associated with stem cells derived from allogeneic embryonic sources.
6. Ethical and legal issues related to stem cell research and therapy extend beyond the use of embryonic stem cells (ESCs). These concerns encompass both scientific/technical aspects (such as safety, efficacy, clinical effectiveness, and cost) and social factors, circumstances, and concerns.
7. Risks associated with stem cell therapies include local and systemic responses to implanted tissues, mutagenicity, and potential alterations in the geno- and phenotype of the nervous system.
8. Transition from preclinical studies to clinical trials for stem cell therapies requires meeting preclinical benchmarks and demonstrating functional integrity, stability, and purity of biologics. The FDA approval process for stem cell treatments presents unique challenges compared to traditional drug testing, and navigating the three phases of FDA approval for marketable stem cell treatments can be particularly challenging.
9. Research and clinical trials involving mesenchymal stem cells (MSC), SVF, adult pluripotent stem cells (APSC), and embryonic stem cells (ESC) require advance approval and are regulated by the FDA as biologic drugs.
10. The regulation and development of autologous and allogeneic stem cell therapies vary across countries. Some states, like California, have liberalized and funded stem cell research, while less restrictive rules exist in countries like Australia, India, and China.

**Acknowledgments** This work was supported in part by federal funds UL1TR001409 from the National Center for Advancing Translational Sciences (NCATS), National Institutes of Health (NIH), through the Clinical and Translational Science Awards Program (CTSA), a trademark of the Department of Health and Human Services, part of the Roadmap Initiative, "Re-Engineering the Clinical Research Enterprise" (JG); the European Union's Horizon 2020 Research and Innovation Programme under grant agreement 720270: HBP SGA1 (JG); the AEHS Foundation Project Neuro-HOPE (JG); an unrestricted research grant from Fortuna Fix Inc. (JG), and by the Austin and Ann O'Malley Distinguished Visiting Chair in Bioethics of Loyola Marymount University (JG).

## References


1. Chakravathy K, Chen Y, He C, Christo PJ. Stem cell therapy for chronic pain management: review of uses, advances, and adverse effects. Pain Physician. 2017;20:293–305.
2. Beitzel K, Allen D, Apostolakos J, et al. US definitions, current use, and FDA stance on use of platelet-rich plasma in sports medicine. J Knee Surg. 2015;28:29–34.
3. Attias M, Zhukova O, Azeem N. Sacroiliac joint injections for the treatment of pain in the rehabilitation patient. In: Carayannopoulos A, editor. Comprehensive pain management in the rehabilitation patient. Berlin: Springer; 2017. p. 469–91.
4. Shen L, Yuan T, Chen S, et al. The temporal effect of platelet-rich plasma on pain and physical function in the treatment of knee osteoarthritis: systematic review and meta-analysis of randomized controlled trials. J Orthop Surg Res. 2017;12:16.
5. Centeno C, Markle J, Didson E, et al. Treatment of lumbar degenerative disc disease-associated radicular pain with culture-expanded autologous mesenchymal stem cells: a pilot study on safety and efficacy. J Transl Med. 2017;15:197.
6. Guo W, Imai S, Yang J, et al. In vivo immune interactions of multipotent stromal cells underlie their long-lasting pain-relieving effect. Sci Rep. 2017;7:10107.
7. Huh Y, Ji RR, Chen G. Neuroinflammation, bone marrow stem cells, and chronic pain. Front Immunol. 2017;8:1014.
8. Fodor PB, Paulseth SG. Adipose derived stromal cell (ADSC) injections for pain management of osteoarthritis in the human knee joint. Aesthet Surg J. 2016;36:229–36.
9. Hyun I, Wilkerson A, Johnston J. Embryology policy: revisit the 14-day rule. Nature. 2016;533:169–71.
10. Schickl H, Braun M, Dabrock P. Ways out of the patenting prohibition: human parthenogenetic and induced pluripotent stem cells. Bioethics. 2017;31:409–17.
11. Taupitz J. Legal aspects of research with human embryonic stem cells. Eur Rev. 2017;25:121–30.
12. Ilic D, Ogilvie C. Human embryonic stem cells: what have we done? What are we doing? Where are we going? Stem Cells. 2016;35:17–25.
13. Lysaght T, Kerridge IH, Sipp D, et al. Ethical and regulatory challenges with autologous adult stem cells: a comparative review of international regulations. J Bioeth Inq. 2017;14:261–73.
14. Wu S, FitzGerald KT, Giordano J. On the viability and potential value of stem cells for repair and treatment of central neurotrauma: overview and speculations. Front Neurol. 2018;9:1–24.
15. Giordano J. The human prospect(s) of neuroscience and neurotechnology: domains of influence and the necessity—and questions—of neuroethics. Hum Prospect. 2014;4:1–18.
16. Giordano J. Neurotechnological progress: the need for neuroethics. In: Gonzalez F, editor. The next step: exponential life. Barcelona: BBVA Press; 2016.
17. Giordano J. Neurogenetic and neural tissue-implantation technology: neuroethical, legal and social issues. In: Giordano J, editor. Neurotechnology: premises, potential and problems. Boca Raton, FL: CRC Press; 2012. p. 59–67.
18. Giordano J, Boswell MV. Engaging philosophy, ethics and policy in, and for pain medicine. In: Giordano J, Boswell MV, editors. Pain medicine: philosophy, ethics and policy. Oxford: Linton Atlantic; 2009. p. 13–20.
19. Giordano J. A preparatory neuroethical approach to assessing developments in neurotechnology. AMA J Ethics. 2015;17:56–61.
20. Giordano J. Conditions for consent to the use of neurotechnology: a preparatory neuroethical approach to risk assessment and reduction. AJOB Neurosci. 2015;6:12–4.
21. de Miguel-Beriain I. The ethics of stem cells revisited. Adv Drug Deliv Rev. 2015;82-83:176–80.
22. Zheng YL. Some ethical concerns about human induced pluripotent stem cells. Sci Eng Ethics. 2016;22:1277–84.
23. Boer GJ. Transplantation and xenotransplantation. In: Giordano J, Gordijn B, editors. Scientific and philosophical perspectives in neuroethics. Cambridge: Cambridge University Press; 2010. p. 190–215.
24. Zheng D, Wang X, Xu RH. One stone for multiple birds: generating universally compatible human embryonic stem cells. Stem Cells. 2016;34:2269–75.
25. Kumar D, Anand T, Kues WA. Clinical potential of human-induced pluripotent stem cells: perspectives of induced pluripotent stem cells. Cell Biol Toxicol. 2017;33:99–112.
26. Miles DC, de Vries NA, Gisler S, et al. TRIM28 is an epigenetic barrier to induced pluripotent stem cell reprogramming. Stem Cells. 2017;35:147–57.
27. Giordano J. Toward an operational neuroethical risk analysis and mitigation paradigm for emerging neuroscience and technology (neuroS/T). Exp Neurol. 2017;287:492–5.
28. Ghosh D, Mehta N, Patil A. Ethical issues in biomedical use of human embryonic stem cells. J Reprod Health Med. 2016;2:S27–37.
29. Riezzo I, Pascale N, La Russa R, et al. Donor selection for allogenic hemopoietic stem cell transplantation: clinical and ethical considerations. Stem Cells Int. 2017;2017:5250790–11.
30. Barazzetti G, Hurst SA, Mauron A. Adapting preclinical benchmarks for first-in-human trials of human embryonic stem cell-based therapies. Stem Cells Transl Med. 2016;5:1058–66.
31. Habets MG, van Delden JJ, Niemansburg SL, et al. Ethical considerations for examining efficacy in first-in-human pluripotent stem cell studies. Mol Ther. 2016;24:2039–42.
32. Food and Drug Administration. Statement from FDA Commissioner Scott Gottlieb, M.D. on the FDA's new policy steps and enforcement efforts to ensure proper oversight of stem cell therapies and regenerative medicine, 2017; https://www.fda.gov/news-events/press-announcements/statement-fda-commissioner-scott-gottlieb-md-fdas-new-policy-steps-and-enforcement-efforts-ensure. Accessed 26 Jun 2023.
33. Centeno CJ, Bashir J. Safety and regulatory issues regarding stem cell therapies: one clinic's perspective. PM R. 2015;7:S4–7.
34. Acosta ND, Golub SH. The new federalism: state policies regarding embryonic stem cell research. J Law Med Ethics. 2016;44:419–36.
35. Denoon A, Hitchcock J, Lawford DJ. The regulation of stem cells in the UK and the EU. In: Vertès A, Qureshi N, Caplan A, et al., editors. Stem cells in regenerative medicine: science, regulation and business strategies. Chichester: Wiley; 2015. p. 125–46.
36. Lysaght T. Oversight and evidence in stem cell innovation: an examination of international guidelines. In: Pham VP, Rosemann A, editors. Stem cells in clinical applications: safety, ethics, and regulations. Berlin: Springer; 2015. p. 217–36.
37. Banja J. Ethical considerations in stem cell research on neurologic and orthopedic conditions. In: Pham VP, Rosemann A, editors. Stem cells in clinical applications: safety, ethics, and regulations. Berlin: Springer; 2015. p. 155–71.
38. Turner L, Knoepfler P. Selling stem cells in the USA: assessing the direct-to-consumer industry. Cell Stem Cell. 2016;19:154–7.
39. Petersen A, Munsie M, Tanner C, et al. Stem cell tourism and the political economy of Hope. London: Palgrave Macmillan; 2017.
40. Blank RH. Globalization: pluralist concerns and contexts. In: Giordano J, Gordijn B, editors. Scientific and philosophical perspectives in neuroethics. Cambridge: Cambridge University Press; 2010. p. 321–42.
41. Benedikter R, Giordano J. Neurotechnology: new frontiers for European policy. Pan Euro Network Sci Tech. 2012;3:204–7.
42. Berger I, Ahmad A, Bansal A, et al. Global distribution of businesses marketing stem cell-based interventions. Cell Stem Cell. 2016;19:158–62.

43. Sleeboom-Faulkner ME. The large grey area between 'bona fide' and 'rogue' stem cell interventions: ethical acceptability and the need to include local variability. Technol Forecast Soc Change. 2016;109:76–86.
44. Giordano J, Shook JR. Minding brain science in medicine: on the need for neuroethical engagement for guidance of neuroscience in clinical contexts. Ethics Biol Engineer Med. 2015;6:37–42.
45. Shook JR, Giordano J. Principled research ethics in practice? Reflections for neuroethics and bioethics. Cortex. 2015;5:1–4.
46. Lanzilao E, Shook J, Benedikter R, Giordano J. Advancing neuroscience on the 21st century world stage: the need for—and proposed structure of—an internationally relevant neuroethics. Ethics Biol Engineer Med. 2013;4:211–29.
47. Anderson MA, Fitz N, Howlader D. Neurotechnology research and the world stage: ethics, biopower and policy. In: Giordano J, editor. Neurotechnology: premises, potential and problems. Boca Raton, FL: CRC Press; 2012. p. 287–300.
48. Jesty J. Biotechnology: who benefits? Who is harmed? In: Giordano J, editor. Neurotechnology: premises, potential and problems. Boca Raton, FL: CRC Press; 2012. p. 301–20.
49. Cohen IG. Patients with passports: medical tourism, law and ethics. Oxford: Oxford University Press; 2015.
50. Tsou A. Ethical considerations when counseling patients about stem cell tourism. Continuum. 2015;21:201–5.
51. Frati P, Scopetti M, Santurro A, et al. Stem cell research and clinical translation: a roadmap about good clinical practice and patient care. Stem Cells Int. 2017;2017:5080259–8.
52. Marzotto T, Alt PM. Stem cell research: hope or hype? London: Routledge; 2017.
53. Swetnam M, McBride D, Herzfeld C, et al. Neurotechnology futures study. Arlington: Potomac Institute Press; 2013.
54. Giordano J, Benedikter R, Boswell MV. Pain medicine, biotechnology and market effects: tools, tekne and moral responsibility. Ethics Biol Engineer Med. 2010;1:135–42.
55. Giordano J. Neuroethics: coming of age and facing the future. In: Giordano J, Gordijn B, editors. Scientific and philosophical perspectives in neuroethics. Cambridge: Cambridge University Press; 2010. p. xxv–xxiv.
56. Shook JR, Giordano J. A principled, cosmopolitan neuroethics: considerations for international relevance. Philos Ethics Humanit Med. 2014;9:1.

# Index





A. Navani et al. (eds.), *Essentials of Regenerative Medicine in Interventional Pain Management*,
https://doi.org/10.1007/978-3-031-50357-3

**D**



**E**

## S

**T**